AAPG Treatise of Petroleum Geology

The American Association of Petroleum Geologists
gratefully acknowledges and appreciates the leadership and support
of the AAPG Foundation in the development of the
Treatise of Petroleum Geology

STRUCTURAL TRAPS VIII

COMPILED BY
NORMAN H. FOSTER
AND
EDWARD A. BEAUMONT

TREATISE OF PETROLEUM GEOLOGY
ATLAS OF OIL AND GAS FIELDS

PUBLISHED BY
THE AMERICAN ASSOCIATION OF PETROLEUM GEOLOGISTS
TULSA, OKLAHOMA 74101, U.S.A.

ISBN: 0-89181-590-2
ISSN: 1043-6103

Available from:
The AAPG Bookstore
P.O. Box 979
Tulsa, OK 74101-0979

Phone: (918) 584-2555
Telex: 49-9432
FAX: (918) 584-0469

Association Editor: Susan Longacre
Science Director: Gary D. Howell
Publications Manager: Cathleen P. Williams
Special Projects Editor: Anne H. Thomas
Science Staff: William G. Brownfield
Project Production: Custom Editorial Productions, Inc.

TABLE OF CONTENTS

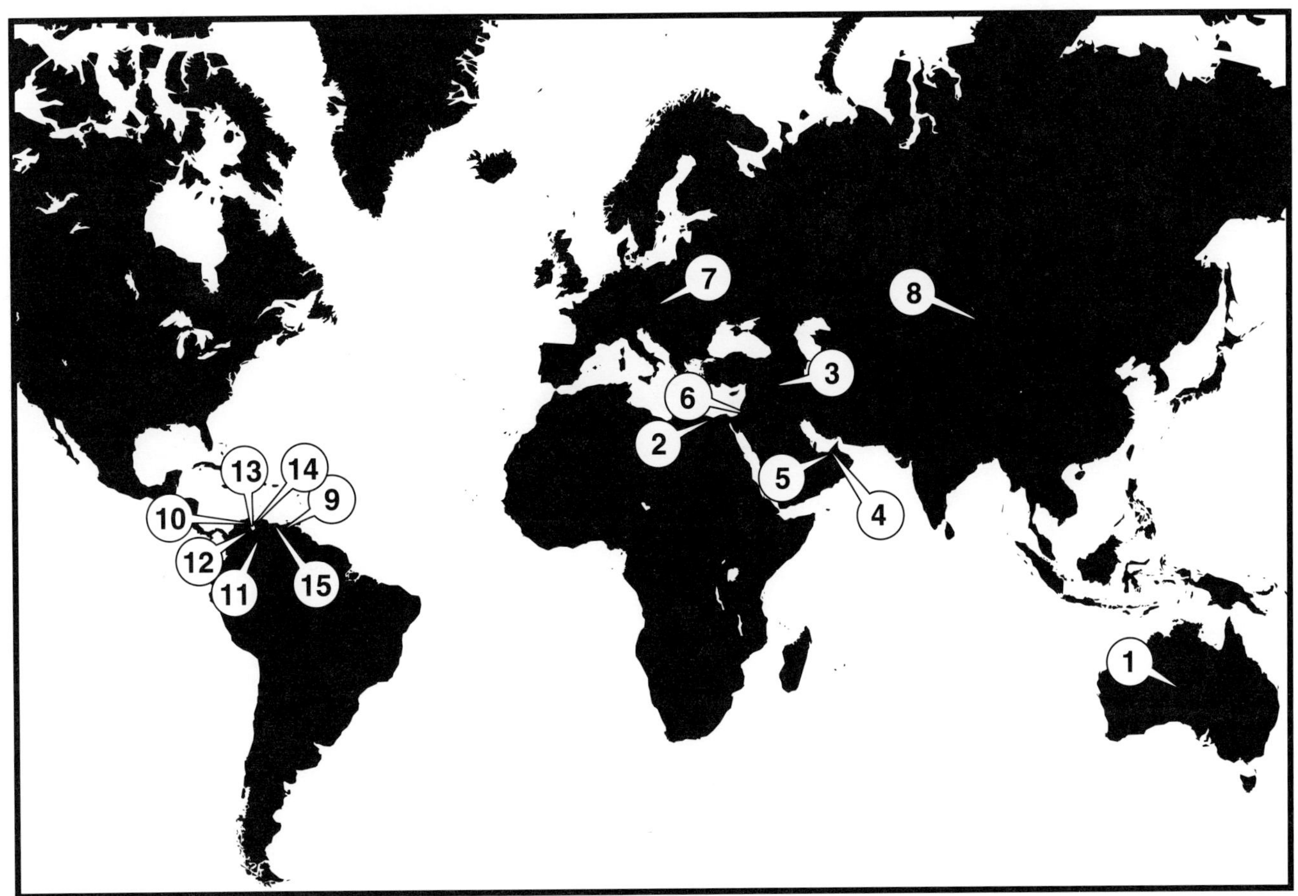

TREATISE OF PETROLEUM GEOLOGY
ADVISORY BOARD

John M. Parker
Stephen J. Patmore
Dallas L. Peck
William H. Pelton
Alain Perrodon
James A. Peterson
Edward B. Picou, Jr.
Max Grow Pitcher
David E. Powley
William F. Precht
A. Pulunggono
Bailey Rascoe, Jr.
Donald L. Rasmussen
Bradley S. Ray
R. Randy Ray
Dudley D. Rice
Edward P. Riker
Richard Douglas Robertson
Edward C. Roy, Jr.
Eric A. Rudd
Floyd F. Sabins, Jr.
Nahum Schneidermann
Robert T. Sellars, Jr.
Faroog A. Sharief
John W. Shelton
Synthia E. Smith
Robert M. Sneider
Frank P. Sonnenberg
Stephen A. Sonnenberg
William E. Speer
Bill St. John
William R. Stanton
Philip H. Stark
Richard Steinmetz
Per R. Stokke
Denise M. Stone
Donald S. Stone
Douglas K. Strickland
James V. Taranik
Harry Ter Best, Jr.
Bruce K. Thatcher, Jr.
M. Ray Thomasson
Jack C. Threet
Bernard Tissot
Don F. Tobin
Donald F. Todd
Harrison L. Townes
M. O. Turner
Peter R. Vail
B. van Hoorn
Arthur M. Van Tyne
Ian R. Vann
Harry K. Veal*
Steven L. Veal
Richard R. Vincelette
Cecil von Hagen
Fred J. Wagner, Jr.
William A. Walker, Jr.
Carol A. Walsh
Anthony Walton
Douglas W. Waples
Harry W. Wassall, III
W. Lynn Watney
N. L. Watts
Koenradd J. Weber
Robert J. Weimer
Dietrich H. Welte
James E. Wilson, Jr.
Martha O. Withjack
P. W. J. Wood
Homer O. Woodbury
Walter W. Wornardt
Marcelo R. Yrigoyen
Dr. Mehmet A. Yukler
Zhai Guangming
Robert Zinke

* Deceased

American Association of Petroleum Geologists Foundation
Treatise of Petroleum Geology Fund*

Major Corporate Contributors
($25,000 or more)

Amoco Production Company
BP Exploration Company Limited
Chevron Corporation
Exxon Company, U.S.A.
Mobil Oil Corporation
Oryx Energy Company
Pennzoil Exploration and Production Company
Shell Oil Company
Texaco Foundation
Union Pacific Foundation
Unocal Corporation

Other Corporate Contributors
($5,000 to $25,000)

ARCO Oil & Gas Company
Ashland Oil, Inc.
Cabot Oil & Gas Corporation
Canadian Hunter Exploration Ltd.
Conoco Inc.
Marathon Oil Company
The McGee Foundation, Inc.
Phillips Petroleum Company
Transco Energy Company
Union Texas Petroleum Corporation

Major Individual Contributors
($1,000 or more)

John J. Amoruso
Thornton E. Anderson
C. Hayden Atchison
Richard A. Baile
Richard R. Bloomer
A. S. Bonner, Jr.
David G. Campbell
Herbert G. Davis
George A. Donnelly, Jr.
Paul H. Dudley, Jr.
Lewis G. Fearing
Lawrence W. Funkhouser
James A. Gibbs
George R. Gibson
William E. Gipson
Mrs. Vito A. (Mary Jane) Gotautas
Robert D. Gunn
Merrill W. Haas
Cecil V. Hagen
Frank W. Harrison
William A. Heck
Roy M. Huffington
J. R. Jackson, Jr.
Harrison C. Jamison
Thomas N. Jordan, Jr.
Hugh M. Looney
Jack P. Martin
John W. Mason
George B. McBride
Dean A. McGee
John R. McMillan
Lee Wayne Moore
Grover E. Murray
Rudolf B. Siegert
Robert M. Sneider
Estate of Mrs. John (Elizabeth) Teagle
Jack C. Threet
Charles Weiner
Harry Westmoreland
James E. Wilson, Jr.
P. W. J. Wood

The Foundation also gratefully acknowledges the many who have supported this endeavor with additional contributions.

*Based on contributions received as of September 30, 1992.

PREFACE

The Atlas of Oil and Gas Fields and the Treatise of Petroleum Geology

The *Treatise of Petroleum Geology* was conceived during a discussion held at the annual AAPG meeting in 1984 in San Antonio, Texas. This discussion led to the conviction that AAPG should publish a state-of-the-art textbook in petroleum geology, aimed not at the student, but at the practicing petroleum geologist. The textbook gradually evolved into a series of three different publications: the Reprint Series, the Atlas of Oil and Gas Fields, and the Handbook of Petroleum Geology. Collectively these publications are known as the *Treatise of Petroleum Geology*, AAPG's Diamond Jubilee project commemorating the Association's 75th anniversary in 1991.

With input from the Advisory Board of the Treatise of Petroleum Geology, we designed this set of publications to represent, to the degree possible, the cutting edge in petroleum exploration knowledge and application: the Reprint Series to provide useful and important published literature; the Atlas to comprise a collection of detailed field studies that illustrate the many ways oil and gas are trapped and to serve as a guide to the petroleum geology of basins where these fields are found; and the Handbook as a professional explorationist's guide to the latest knowledge in the various areas of petroleum geology and related disciplines.

The Treatise Atlas is part of AAPG's long tradition of publishing field studies. Notable AAPG field study compilations include *Structure of Typical American Fields*, published in 1929 and edited by Sidney Powers; and Memoir 30, *Giant Fields of 1968–1978*, published in 1981 and edited by Michel T. Halbouty. The Treatise Atlas continues that tradition but introduces a format designed for easier access to data.

Hundreds of geologists participated in this first compilation of the Atlas. Authors are from all parts of the industry and numerous countries. We gratefully acknowledge the generous contribution of their knowledge, resources, and time.

Purpose of the Atlas

The purpose of the Atlas is twofold: (1) to help exploration and development geologists become more efficient by increasing their awareness of the ways oil and gas are trapped, and (2) to serve as a reference for both the petroleum geology of the fields described and the basins in which they occur.

Imagination is the primary tool of the explorationist. Wallace E. Pratt once said that the unfound field must first be sought in the mind. In part, what is imagined is based on what is remembered; memory is the direct link to what is created in the mind. To create ideas that lead to the discovery of new fields, the mind of the geologist builds from its knowledge of petroleum geology. To that end, the Atlas of field studies will be a primary source for locating much of the information necessary for creating prospects and will provide a connection to the phenomenon of oil and gas traps.

Next to the firsthand experience of having prospects tested with the drill bit, studying the many facets and concepts of developed fields is perhaps the best way for the geologist to develop the ability to create plays and prospects. Also, familiarity with the many ways oil and gas are trapped allows the geologist to see through the noise inherent to exploration data and to close gaps in that data.

Format of the Atlas

To facilitate data access, all field studies in the Atlas follow the same format. Once users become familiar with this format, they will know where to look for the information they seek. Different fields from different parts of the world can be easily compared and contrasted.

The following is a generalized format outline for field studies in the Atlas:

- Location
- History
  - Pre-Discovery
  - Discovery
  - Post-Discovery
- Discovery Method
- Structure
  - Tectonic History
  - Regional Structure
  - Local Structure
- Stratigraphy
- Trap
  - General Description
  - Reservoir(s)
  - Source(s)
- Exploration Concepts

Criteria for Inclusion of a Field

Fields described in the Atlas are selected using two main criteria: (1) trap type, and (2) geographic distribution. Our ultimate goal for the Atlas is to include a field study from each major petroleum-producing province and to include an example of each known trap type. Size or economic importance are not, of themselves, criteria. Many fields that are not giants are included because they are geologically unique, because they are significant examples of geological investigation and original thinking, or because they are historically important, having led to the discovery of many other fields.

Grouping of Fields into Separate Volumes

We considered several ways to group fields in these volumes. We chose trap type because the purpose of the Atlas is to make exploration geologists more effective oil and gas trap finders, regardless of where they search for traps.

Grouping oil and gas field studies into separate volumes by trap type is a difficult exercise. We decided to group the fields into volumes by designating them as structural or stratigraphic traps. Most traps are a combination of both structure and stratigraphy. Some traps are obviously more a consequence of one than the other, but many are not. The continuum that exists between purely stratigraphic and purely structural traps is what makes grouping difficult. A further complication is that many fields contain more than one trap type.

Papers selected for *Structural Traps VIII*

This volume in the *Atlas of Oil and Gas Fields* series is a collection of studies of fields with wide geographic distribution that have traps mainly related to anticlinal closure. Many of these fields have traps that also have strong components of stratigraphic control.

The first field described, Mereenie, is in the Amadeus basin of central Australia. The reservoir is an elongate anticlinal trap also controlled by diagenetic alteration. So far the Ordovician reservoir of Mereenie is the oldest found in Australia, where production is mainly from Tertiary rocks. Its existence is important because it suggests that other sub-Mesozoic reservoirs may await discovery in Australia.

The next five studies are of fields of the Middle East. The trap of Razzak field, of the Razzak/Alamein basin in Egypt, is a complex of three separate faulted anticlinal culminations associated with wrench faulting along a reverse fault. Ain Zalah, in the Zagros foldbelt of Iraq, is an elongate anticlinal trap. Asab and Bu Hasa are in Abu Dhabi. A dome over salt results in multiple-pay entrapments at Asab. The oil in the multiple pays of the Bu Hasa anticline is strongly controlled by stratigraphic variations. The traps of the Zohar-Kidod-Haqanaim fields in Israel are a series of three anticlines in an asymmetrically folded trend related to reverse faulting.

Ždánice-Krystalinikum, in Czechoslovakia, contains a trap that could be considered structural or stratigraphic, depending on your point of view. The principal reservoir is in fractured, crystalline basement rocks of a paleo hill sealed by the mudstones and claystones of an overthrust. There are also minor traps in sandstone pinch-outs.

The traps of Kelamayi field in the Zhungeer basin of China are a complex of four types: unconformity truncation, unconformity onlap, reverse fault, and updip tar seal.

The last six fields described in this volume are in Venezuela. Los Lanudos, Tarra, Lama, and Tiguaje are in the Maracaibo basin; Santa Rosa and Yucal-Placer are in the Eastern Venezuela basin; and Guafita is in the Barinas-Apure basin. Santa Rosa's trap is in an anticline along the leading edge of a thrust sheet. Los Lanudos is a faulted anticline trap with multiple sandstone reservoir bodies. Guafita's trap is a faulted anticline along a regional wrench fault system. The multiple pays of Tarra are in a faulted anticline within a thrust sheet. Tarra's deepest trap is an unconformity truncation. Lama is a complex of traps in an anticlinal wrench-fault structure with a fractured limestone reservoir bounded by faults, unconformity truncation, and lateral pinch-outs. Tiguaje's trap is in a wrench-fault-generated anticline with lateral fault seals.

The last field, Yucal-Placer, is a stratigraphic trap with a strong structural component. It is included in this volume of the atlas because it was drilled on a surface anticline as a structural prospect. The anticline was confirmed by seismic data. Even after a number of wells had been drilled, the field was thought to be structural. It is associated with complex, alternately extensive and compressive regimes that initially resulted in gravity faulting and later in thrust faulting along the same fault zones—and in parallel folding. The character of the updip pinch-outs of most of the reservoirs was not recognized until many wells had been drilled.

Drilling and discovery show how closely the original exploration concept matched reality. Knowing the history of discovery may help explorationists realize that others before them confronted seemingly insoluble problems that were eventually solved. It is instructive to learn about the sequences of thinking that solved these problems as well as the role of serendipity in discovery. If luck has a role in successful exploration, we still come away with the sense that creativity, logic, and knowledge guide the explorationist to those places where serendipity may play.

Careful study of these fields will enhance the prospect generator's knowledge base and, consequently, his or her ability to apply that knowledge toward future prospecting.

Edward A. Beaumont
Norman H. Foster, Editors

Mereenie Field—Australia
Amadeus Basin, Northern Territory

P. J. HAVORD
Winandu Pty Ltd
Sydney, Australia

FIELD CLASSIFICATION

BASIN: Amadeus
BASIN TYPE: Cratonic Sag
RESERVOIR ROCK TYPE: Sandstone
RESERVOIR ENVIRONMENT OF DEPOSITION: Marginal Marine
RESERVOIR AGE: Ordovician
PETROLEUM TYPE: Oil and Gas
TRAP TYPE: Anticline with Diagenetic Control of Reservoir Distribution

LOCATION

The Mereenie oil and gas field is located in the central Amadeus basin in Australia's arid heartland. It lies 152 mi (245 km) west-southwest of Alice Springs, Northern Territory, and 62 mi (100 km) west of the Palm Valley gas field—the only other currently producing field in the basin (Figure 1). The field takes its name from the Mereenie Range, 30 mi (48 km) to the northeast, a name originally given to the area by an aboriginal tribe. *Mereenie* literally means "yam totem center."

Total areal extent of the field exceeds 47 mi^2 (122 km^2). The field is operated by Australian Gas Light Petroleum (AGLP) Limited for the Mereenie Joint Venture.

Mereenie has the largest in-place oil reserves (186 million barrels, MMBO) of any Australian mainland oil field. Ultimate recovery has been estimated at up to 34 MMBO and 593 bcf of gas.

HISTORY

Pre-Discovery

After the Second World War the Australian government, through the Bureau of Mineral Resources (BMR), undertook a national geological mapping program to assist exploration for indigenous fuel and mineral resources. Comprehensive aerial photographic surveys were run including initial coverage of the Amadeus basin. BMR geologists, seconded to the Northern Territory Administration in Alice Springs, identified several large prospective structures within images obtained of the spectacular surface structural geology of the central and eastern areas of the basin. Among these was the vividly outcropping eastern portion of the Mereenie anticline where erosion has produced a typical valley anticline landform (Figure 2). Up until this time the basin's Paleozoic and Proterozoic sequences and its remoteness from markets had deterred petroleum exploration. Aroused by the BMR work, Frome Broken Hill Company Pty Ltd undertook the first investigation of the basin's petroleum potential. Frome conducted geological and gravity studies between 1958 and 1960 but relinquished their permit without exploration operations reaching the drilling stage, apparently concluding that source rock conditions were unfavorable.

About this time the Canadian geologist Duncan McNaughton was assessing the prospectivity of the Amadeus basin for Magellan Petroleum Corporation following a recommendation by Alan Condon of the BMR in Canberra that the basin had exploration merit. Terry Quinlan, a BMR geologist resident in Alice Springs at the time, brought to McNaughton's attention the structures previously identified by aerial photography, including the Mereenie anticline. McNaughton recommended Magellan take up a large exploration area in the north-central part of the basin. The Mereenie anticline was situated in the southwestern corner of this area, but because of an omission in the existing aerial photographic mosaics from which the acreage was selected, only the eastern half of the structure fell within the permit area initially awarded to Magellan in 1960. United Canso, a Canadian company associated with Magellan, took title to an adjacent exploration area in 1961, which included the western half of the Mereenie anticline. During 1961 Roy Hopkins, a Magellan geologist, carried out detailed field mapping of the structure. His work showed that the exposed part of the anticline plunged to the east. Continuation of the

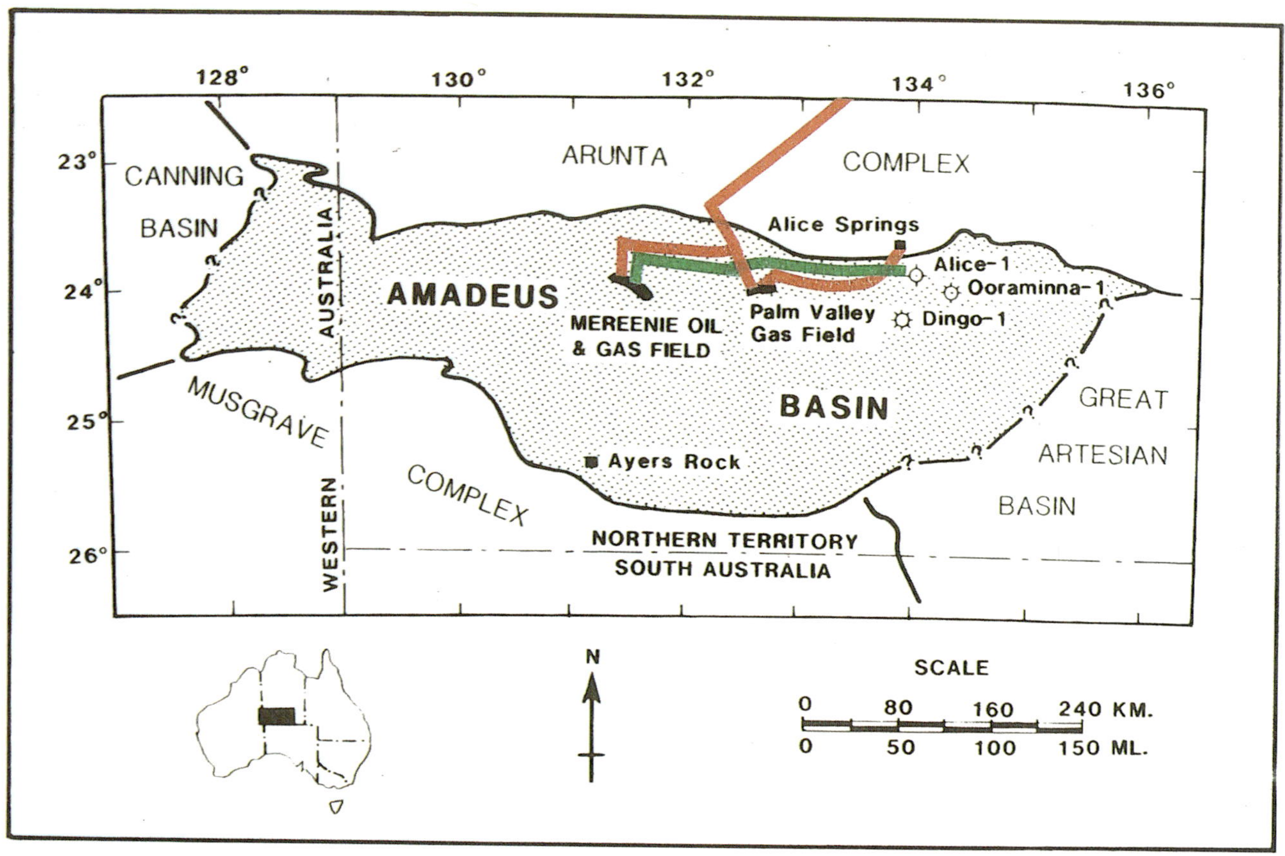

Figure 1. Regional setting of the Amadeus basin. Heavy lines are pipelines.

structure to the west beneath sand cover was suspected, but further exploration was needed to determine whether closure occurred in this area. Field investigations also revealed potential source and reservoir beds in the Ordovician rocks outcropping in the ranges to the north and south of the Mereenie anticline.

During the early 1960s, a young Australian exploration company known as Exoil N.L. (now AGLP) had explored various parts of Australia but had been unable to find large, well-defined drilling targets. A chance meeting between Exoil's Executive Chairman, C. W. (Bill) Siller—a geologist—and Duncan McNaughton led to Exoil's farming in to Magellan's Amadeus basin acreage where large untested structures were known. The original agreement was for Exoil to drill three exploration wells in the basin to earn a 50% undivided interest in blocks encompassing the prospects to be drilled. Two structures, Ooraminna and Alice, were ready for drilling at the time of the Exoil farm-in. The third well was tentatively planned to test the Mereenie anticline if Exoil could prove the structure was closed. During 1962, Exoil conducted a seismic reflection survey over part of the obscured western portion of the Mereenie anticline. Seismic data quality was poor but the survey indicated a culmination on the structure.

The two exploration wells drilled prior to the Mereenie discovery demonstrated the presence of hydrocarbons in the Amadeus basin. Exoil drilled the first exploration well in the basin, Ooraminna 1, in early 1963 to test Proterozoic section in the Ooraminna anticline (Figure 1). A small gas flow was obtained from the Late Proterozoic Areyonga Formation, but the well was plugged and abandoned. The second well, Alice 1 (Figure 1), unsuccessfully tested a seismically interpreted Cambrian carbonate reef and the overlying Cambrian and Ordovician drape section. Minor oil shows were recorded in tight Middle and Upper Cambrian intervals, but no evidence for a reef was found.

In mid-1963, while Alice 1 was being drilled, a BMR phosphate exploration core hole, AP1, situated 15 mi (24 km) southeast of the Mereenie anticline detected oil staining in core samples recovered from the Stairway sandstone, Horn Valley siltstone, and Pacoota sandstone (Figures 3 and 4). The proximity of this hole to the Mereenie structure raised hopes for the success of the third wildcat well in the basin—Mereenie 1.

Discovery

Mereenie 1 (Figure 5) was spudded in December 1963 by Exoil as a crestal test of the Mereenie

Figure 2. Landsat image of the Mereenie field and the area to the east. The circular feature known as Gosses Bluff, to the northeast of Mereenie, is believed to be an astrobleme.

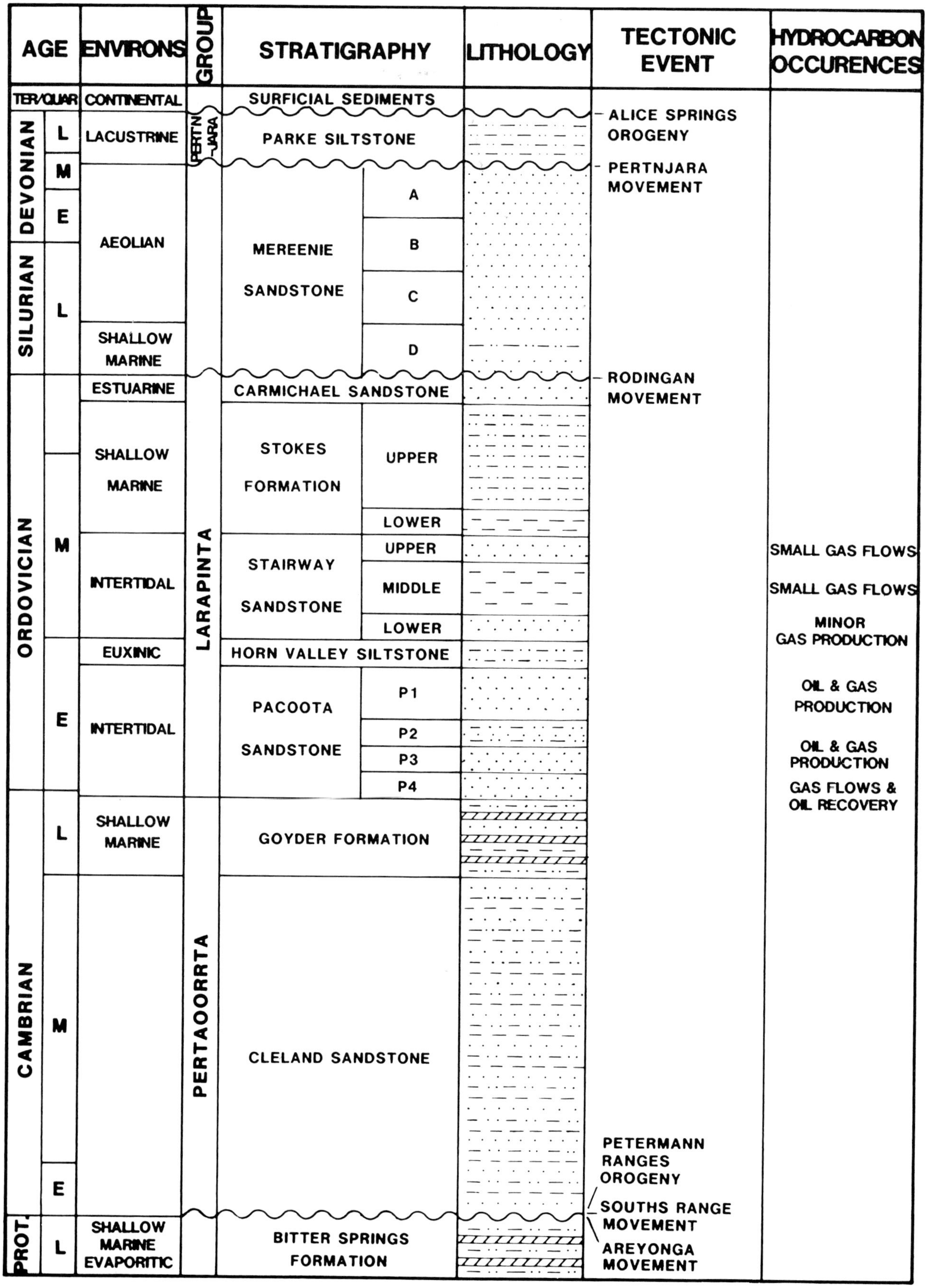

AGE		ENVIRONS	GROUP	STRATIGRAPHY		TECTONIC EVENT	HYDROCARBON OCCURENCES
TER/QUAR		CONTINENTAL		SURFICIAL SEDIMENTS			
DEVONIAN	L	LACUSTRINE	PERTN-JARA	PARKE SILTSTONE		ALICE SPRINGS OROGENY	
DEVONIAN	M	AEOLIAN		MEREENIE SANDSTONE	A	PERTNJARA MOVEMENT	
DEVONIAN	E	AEOLIAN		MEREENIE SANDSTONE	B		
SILURIAN	L	AEOLIAN		MEREENIE SANDSTONE	C		
SILURIAN	L	SHALLOW MARINE		MEREENIE SANDSTONE	D		
ORDOVICIAN		ESTUARINE	LARAPINTA	CARMICHAEL SANDSTONE		RODINGAN MOVEMENT	
ORDOVICIAN	M	SHALLOW MARINE	LARAPINTA	STOKES FORMATION	UPPER		
ORDOVICIAN	M	SHALLOW MARINE	LARAPINTA	STOKES FORMATION	LOWER		
ORDOVICIAN	M	INTERTIDAL	LARAPINTA	STAIRWAY SANDSTONE	UPPER		SMALL GAS FLOWS
ORDOVICIAN	M	INTERTIDAL	LARAPINTA	STAIRWAY SANDSTONE	MIDDLE		SMALL GAS FLOWS
ORDOVICIAN	M	INTERTIDAL	LARAPINTA	STAIRWAY SANDSTONE	LOWER		MINOR GAS PRODUCTION
ORDOVICIAN	E	EUXINIC	LARAPINTA	HORN VALLEY SILTSTONE			
ORDOVICIAN	E	INTERTIDAL	LARAPINTA	PACOOTA SANDSTONE	P1		OIL & GAS PRODUCTION
ORDOVICIAN	E	INTERTIDAL	LARAPINTA	PACOOTA SANDSTONE	P2		OIL & GAS PRODUCTION
ORDOVICIAN	E	INTERTIDAL	LARAPINTA	PACOOTA SANDSTONE	P3		
ORDOVICIAN	E	INTERTIDAL	LARAPINTA	PACOOTA SANDSTONE	P4		GAS FLOWS & OIL RECOVERY
CAMBRIAN	L	SHALLOW MARINE	PERTAOORRTA	GOYDER FORMATION			
CAMBRIAN	M		PERTAOORRTA	CLELAND SANDSTONE		PETERMANN RANGES OROGENY	
CAMBRIAN	E		PERTAOORRTA	CLELAND SANDSTONE		SOUTHS RANGE MOVEMENT	
PROT.	L	SHALLOW MARINE EVAPORITIC		BITTER SPRINGS FORMATION		AREYONGA MOVEMENT	

Figure 3. Stratigraphy of the Mereenie field.

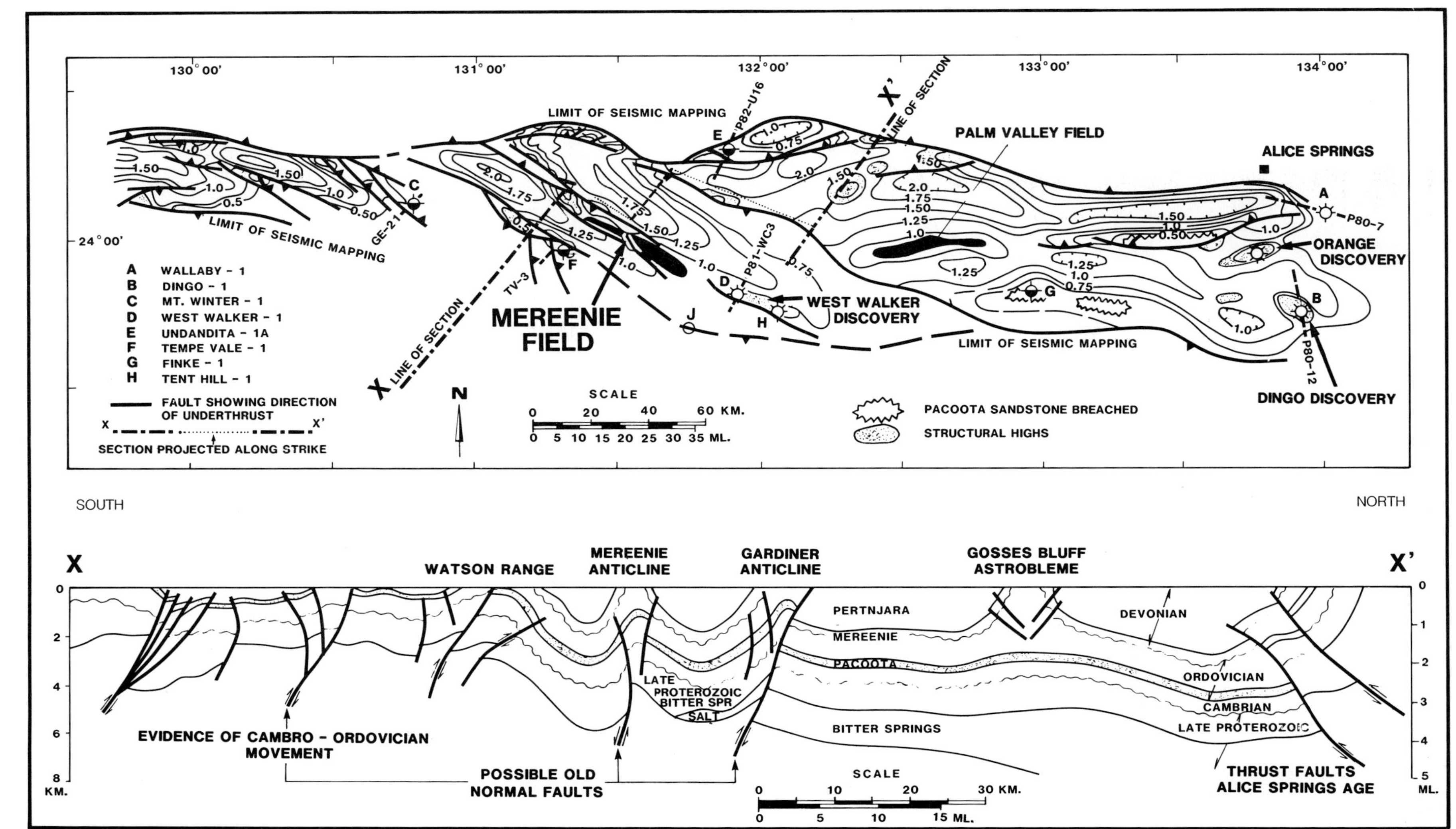

Figure 4. Regional time structure map for the top of the Pacoota sandstone showing a representative cross section XX′ of the western Amadeus basin. (After Schroder and Gorter, 1984.) J is approximate location of BMR AP1 corehole.

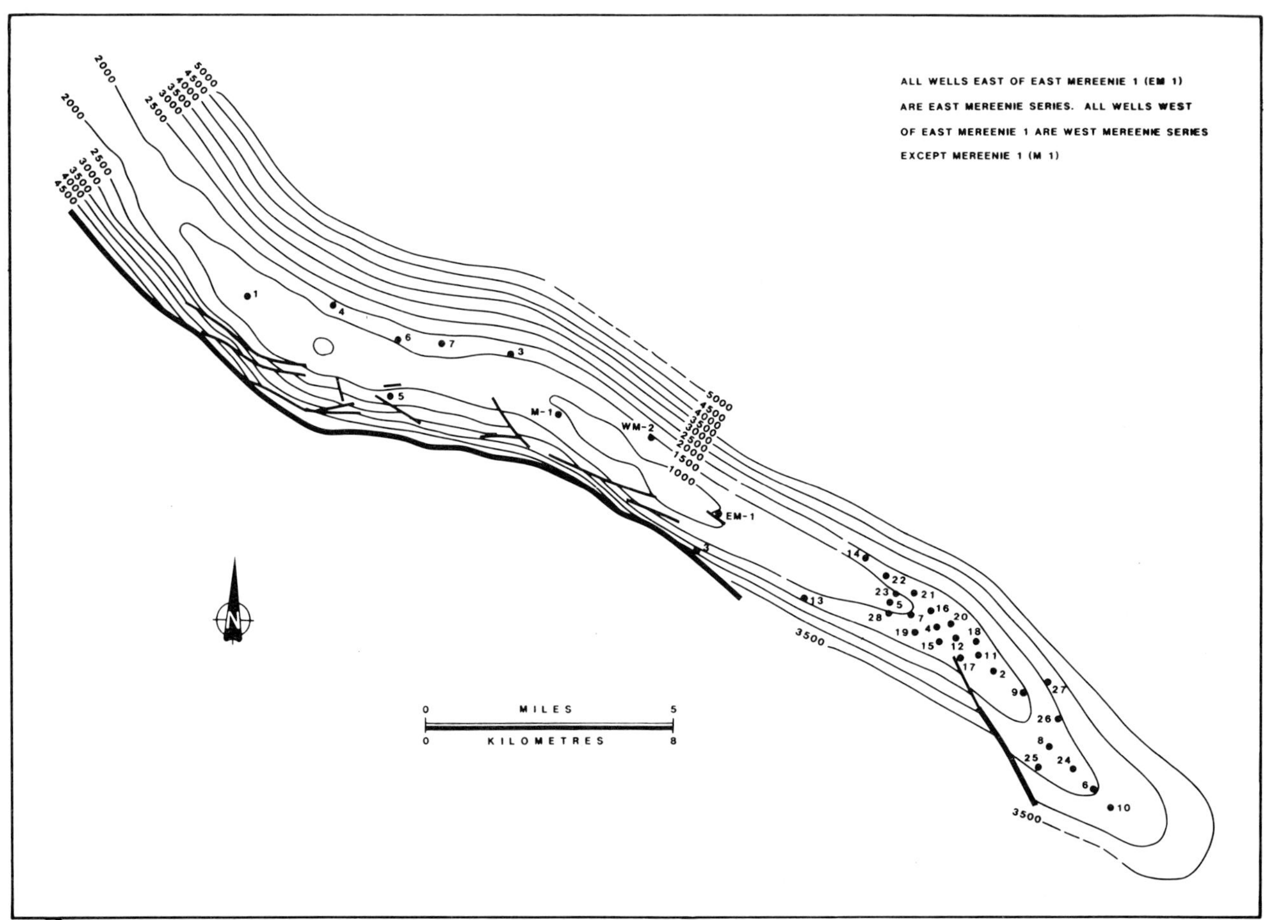

Figure 5. Simplified depth structure map of the top of the Pacoota sandstone for the Mereenie field. Depth contours are in feet below sea level.

anticline near the center of the structure to evaluate Ordovician and Cambrian sandstones. The first hydrocarbon shows in Mereenie 1 occurred in the Middle Ordovician Stairway sandstone. Gas flows of 350 MCFGD from the upper Stairway sandstone and 300 MCFGD from the lower Stairway sandstone were recorded. A drill-stem test in the P1 (uppermost) subunit of the Lower Ordovician Pacoota sandstone flowed gas at the rate of 4.8 MMCFGD. After this test the packer was pulled and gas, estimated at 11 MMCFGD, blew mud out of the hole. Mechanical difficulties then forced the well to be plugged and abandoned without reaching objectives within lower subunits of the Pacoota sandstone and the underlying Cambrian section. Despite these problems, Mereenie 1 was the first well to discover gas in potentially commercial quantities in the Northern Territory and proved the existence of reservoir quality in the sequence. Small quantities of condensate were associated with the gas, and traces of oil bleeding from core samples of the lower Stairway sandstone indicated oil legs could occur down-structure along the flanks of the anticline.

The first oil leg was discovered in the third well drilled on the field—East Mereenie 2 (Figure 5). This was the first deep well in Australia to be predominantly drilled by air. It followed the successful stepout gas appraisal well, East Mereenie 1 (EM1, Figure 5), which penetrated the entire Pacoota sandstone and produced significant gas flows from sandstones within the P1 and P3 subunits. East Mereenie 2 was also planned to prove additional gas but was located on the plunging eastern crest of the structure primarily to investigate the possibility of oil. Gas flows of up to 3.1 MMCFGD were obtained from sandstones within the P1 subunit and good oil shows were observed in cuttings and core samples from the P3 subunit. Oil was recovered but did not flow to the surface during drill-stem testing of these shows, but a flow rate equivalent to 150 BOPD was estimated for one of these tests. Other oil legs were discovered in later wells but usually in less permeable sandstones.

Post-Discovery

Following East Mereenie 2, four wells were drilled on the field between 1964 and 1967. This round of appraisal drilling confirmed the presence of several Pacoota sandstone reservoirs, each with a large gas cap overlying an oil rim around the flanks of the anticline. During 1969, Northwest Mereenie 1 was drilled by Magellan on trend with the Mereenie anticline but was plugged and abandoned as a dry hole. Later, improved resolution seismic data revealed that the well location was separated from the Mereenie anticline by a structural saddle.

There was no further drilling on the field until November 1981 when two production leases were awarded. The Mereenie Joint Venture then embarked on an oil appraisal drilling program in January 1982, which continued until March 1986 when the oil price collapse forced operations to be suspended. By this time, 36 wells had been drilled and six seismic surveys conducted over the field. Since 1986 another seismic survey has been carried out over the western half of the field, but as of December 1988, drilling had not resumed.

Several mud types have been used in drilling development wells at Mereenie. Owing to the presence of a strong aquifer in the Mereenie sandstone (used for field water supply), air and mist drilling were used from spud to surface casing point, which is generally 300 ft (92 m) below the top of the Stokes formation. Surface casing is 8⅝-in. in P1 oil development wells and 10¾-in. in P3 oil development wells. All wells are then air drilled to the top of the Pacoota sandstone or until gas flows from the Stairway sandstone forced changeover to mud drilling because of potential downhole fire or explosion. Any significant gas flow from the Stairway sandstone is evaluated by open-hole testing with a critical flow prover. Water-based muds were used during early drilling of the Pacoota sandstone at Mereenie but were replaced by invert oil emulsion muds in an attempt to reduce formation damage.

After reaching total depth, open-hole, wireline logs are usually run from the bottom of the hole to the top of the Stairway sandstone and consist of gamma-ray, induction, density, neutron, and caliper. Drill-stem testing usually precedes completion of development wells as a result of unpredictable reservoir performance. Cased-hole, wireline logs consist of gamma-ray, cement bond log, variable density log, and casing collar locator. Casing-gun or open-hole completions were originally used at Mereenie. Existing practice is for tubing conveyed perforation of 5½-in. production casing in P1 completions and 7⅝-in. open-hole below 8⅝-in. production casing in P3 completions. Tubing has been 2⅜-in. in all wells drilled since 1982. Attempts to stimulate some poorer wells at Mereenie by hydraulic fracturing have met with moderate success, while acidizing efforts have been hampered by mechanical problems.

The Mereenie field has only been partially developed with development drilling intended exclusively for oil production because a major gas market is lacking in the region. After initial appraisal drilling around the oil discovery made in East Mereenie 2, an oil production station and gathering system were established near the eastern end of the field in 1984. This led to more drilling on the eastern nose of the anticline and accounts for the present concentration of wells in this area where reservoir quality is generally better and where gentle dips cause the oil leg to be areally wider. Well spacing in this area averages 160 ac. Produced solution gas is re-injected into the gas caps of the oil-producing reservoirs for pressure maintenance.

The character of production trends and pressure histories are typical of reservoirs with solution gas and gas cap expansion drive mechanisms. Very little water has been produced from the field, and simulation studies show that negligible water influx has occurred. Pressure monitoring has indicated the presence of permeability barriers (faults?) on the eastern nose of the structure. These barriers have divided the eastern nose area into isolated blocks, causing reservoir pressure near some of the initially more productive P3 oil wells to drop below 1000 psi (6900 kPa). The average field reservoir pressure in the main oil-producing sands of the P3 subunit has, however, only dropped from 1780 to 1720 psi (12,270 to 11,860 kPa).

Oil production from the field commenced in September 1984 at approximately 1500 BOPD from seven wells. All oil produced from the western end of the field is trucked to the eastern production station. Transportation of oil from Mereenie was initially by truck to Alice Springs, but since October 1985, oil has been piped from the field to storage facilities at Brewer Estate near Alice Springs. From Alice Springs it is railed to the Port Stanvac refinery near Adelaide, South Australia. Peak field production of approximately 3900 BOPD occurred in March 1986. By the end of 1988, field production was approximately 1900 BOPD from 19 wells. Most oil production is from the P3 sandstones, particularly the P3-130 sandstone, with the remainder from the P1-80 sandstone in wells on the eastern flank. All oil and gas production has been by primary recovery methods. Apart from plunger lift and gas lift in a few wells, no other form of artificial lift has been used at Mereenie.

In May 1987, Mereenie gas was first transported through the pipeline that connects it and the Palm Valley field to Darwin, Northern Territory. Sales gas production from the field at the end of 1988 was nearly 6 MMCFGD. Another 6 MMCFGD was injected back into the field for pressure support of the eastern oil reservoirs. Most of the produced gas comes from oil wells completed in the P3 reservoirs. Sales gas production is to meet the 53 bcf of gas contracted by the Northern Territory Electricity Commission over a 25 year period.

Original oil-in-place at Mereenie is estimated to be 186 million barrels of which 34 million barrels (18%) are considered potentially recoverable given full field development under favorable economic conditions. Approximately 4 million barrels of oil have been produced. Original gas-in-place in the field is estimated to be 821 bcf of which 593 bcf (72%) are considered recoverable. Recoverable gas consists of 436 bcf of gas cap gas and 157 bcf of solution gas.

DISCOVERY METHOD

The exposed eastern portion of the Mereenie anticline was initially identified by aerial photography. This was eventually followed by surface and seismic structural mapping and the recognition of potential source, seal, and reservoir horizons outcropping in the area. Fortuitous oil shows in the nearby AP1 hole in the same horizons confirmed not only the source rock potential but also the presence of desirable maturity levels and that primary migration had occurred. The necessary elements for a hydrocarbon accumulation had therefore been demonstrated to exist in the Mereenie area prior to drilling.

STRUCTURE

The Amadeus basin is a 500 mi (804 km) long (Figure 1), east-west-trending intracratonic depression. It is a remnant of a much larger depositional province that covered a vast area of central Australia. The 150,000 mi^2 (388,500 km^2) basin area overlies an east-west zone of weakness that has undergone both tension and compression as the stable blocks to the north and south have moved. According to St. John et al. (1984), the basin is classified as IIA using the scheme of Klemme (1971) (i.e., Continental multicycle basin; Craton margin—composite) and is classified 41 under the modified scheme of Bally and Snelson (1980) (i.e., Folded belt; related to A-subduction).

Formation of the Amadeus basin commenced during a period of extension in the Late Proterozoic (ca. 900 Ma). Two separate eras of major diastrophism have subsequently influenced structural development and sedimentation within the basin.

Events of the first era during the Late Proterozoic set the basic structural framework of the basin. Initial activity, the Areyonga and Souths Range movements (Figure 3), were characterized by uplift and erosion in the southwest. Sediment was shed northward toward the east-west-trending depocenter of the subsiding basin and eventually was deposited in deltaic and shallow marine environments. The Petermann Ranges orogeny was the last and strongest event during this era. It caused severe folding and thrusting in the south, which sent thick sequences of upper Proterozoic strata sliding northward on a décollement surface within evaporites of the Bitter Springs Formation.

Basin subsidence continued during the early Paleozoic, bringing Cambrian marine carbonate sediments from an eastern sea. This process was followed by a major marine transgression from the north during the Ordovician as the seaway pushed further west across the Canning basin. It was during this latter period that reservoir, source, and seal clastic sediments derived from the west and southwest were deposited as a shelf facies in the Mereenie area.

The second era of diastrophism commenced in the Late Ordovician with movements originating in the northeastern area of the basin. Uplift and erosion of up to 10,000 ft (3050 m) of Ordovician and Cambrian sediments in the northeast caused by the Rodingan movement preceded folding and thrusting by the Pertnjara movement (Figure 3). The final event during this era, the Late Devonian Alice Springs orogeny (Figure 3), was the most dramatic. Originating in the northeast, thrusting and folding caused detachment and southward push of strata over the Bitter Springs décollement surface, as well as developing a secondary décollement surface in the evaporites of the Late Cambrian Chandler Formation in the eastern part of the basin. The Alice Springs orogeny heavily overprinted pre-existing structures and formed additional ones including the Mereenie anticline.

One of a series of trending en echelon structures in west-northwest to east-southeast trend (Figure 4), the Mereenie anticline is bounded to the south by a large thrust fault (Figure 5). The field is located on the upper plate of the thrust (Figure 6). Several smaller reverse and normal faults have been identified in seismic data across the structure and in the outcropping eastern portion of the anticline. However, at depth, seismic data quality is too poor to determine whether the main thrust is basement involved or is part of a Proterozic décollement feature.

The Mereenie structure is an easterly plunging, slightly asymmetrical anticline that is more than 25 mi (40 km) long and has an average width of 2.5 mi (4 km). The crest of the anticline is sinuous and has several culminations. The proximity of the southern limb to the main thrust causes it to dip more steeply than the northern limb. Seismically estimated dips of the flanks generally vary between 10° and 25°, with dips in excess of 25° at some locations on the southern flank.

STRATIGRAPHY

Almost all oil and gas production from the Mereenie field has been from the Pacoota sandstone, which ranges in age from Late Cambrian to Early Ordovician (Figure 3). The Middle Ordovician Stairway sandstone contains virtually undeveloped gas reserves in generally very low permeability sandstones. Both formations have been divided into subunits as follows:

Stairway sandstone	upper Stairway sandstone (USS)
	middle Stairway sandstone (MSS)
	lower Stairway sandstone (LSS)

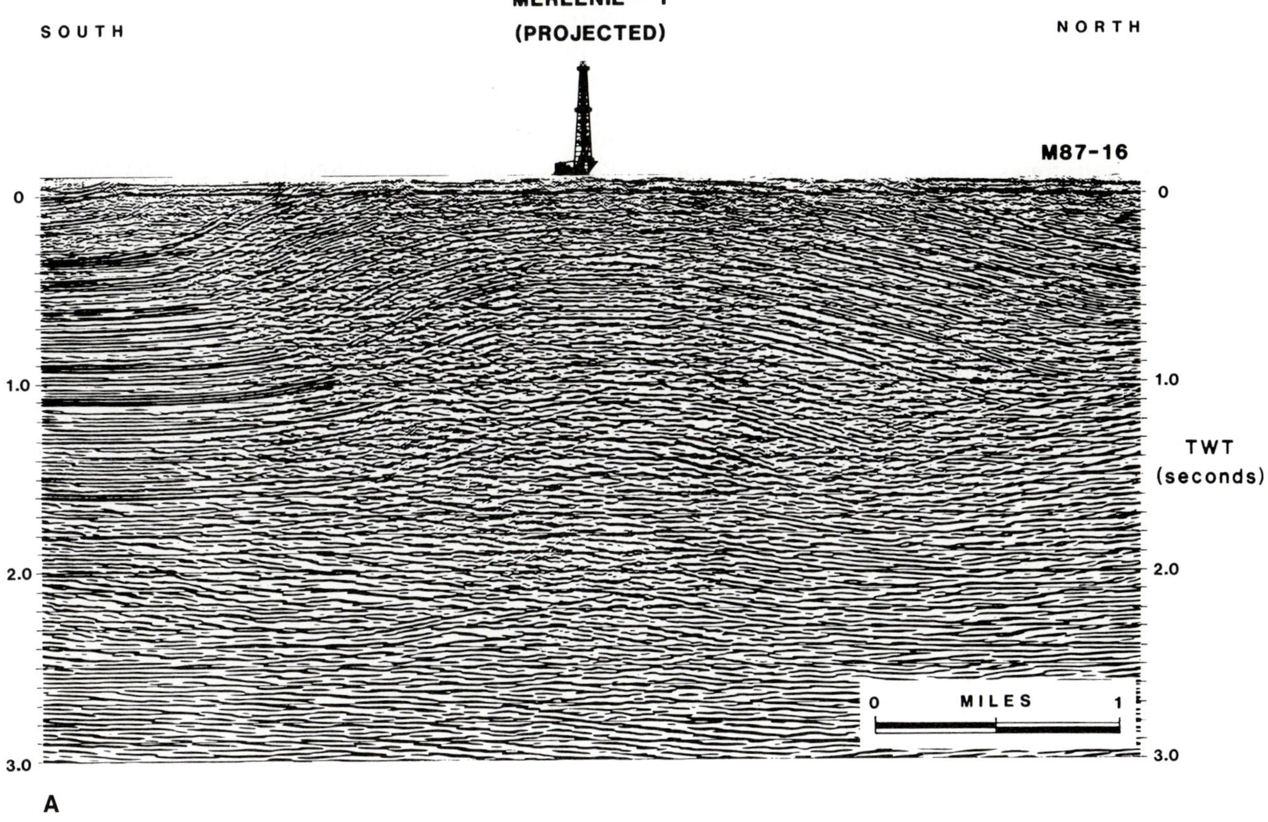

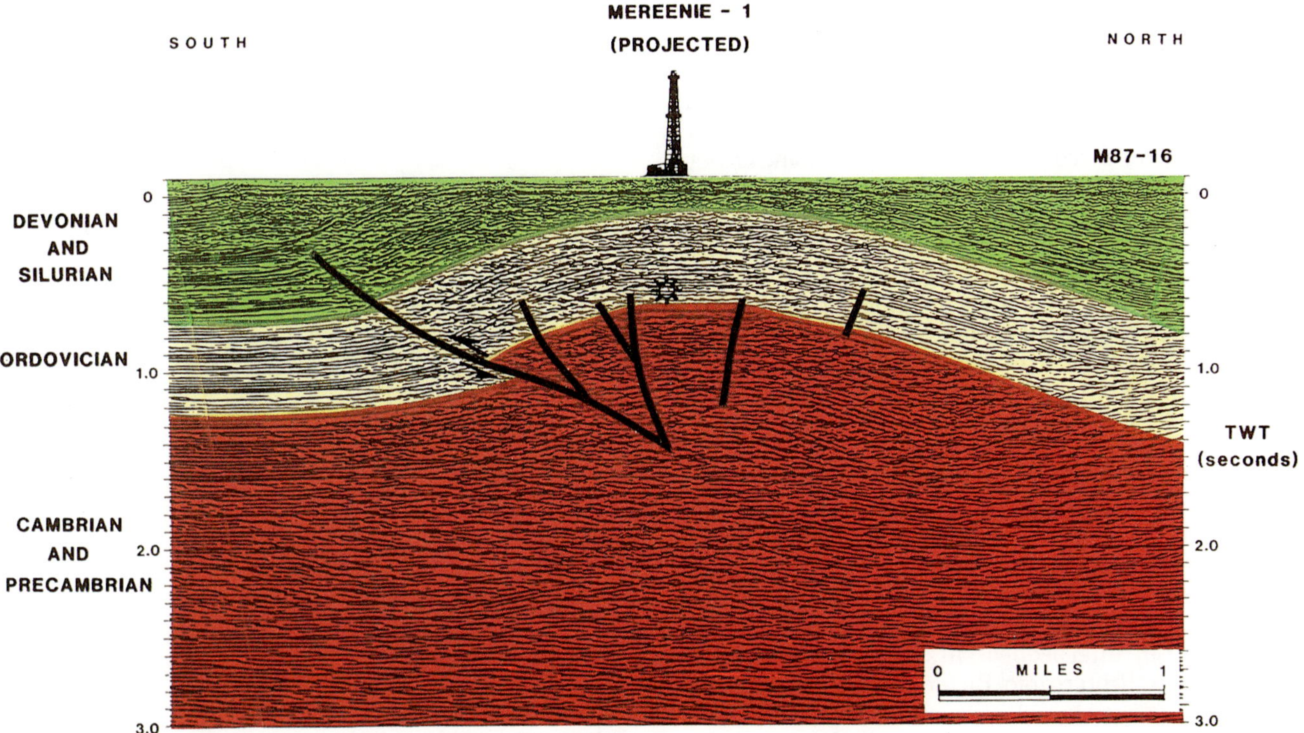

Figure 6. Migrated seismic dip line through the central portion of the Mereenie field. (A) Uninterpreted. (B) Interpreted. The apparent thickening of the Ordovician adjacent to the main thrust is not supported by drilling data and is believed to be related to ray path complexities caused by steep southern flank dips.

Pacoota sandstone	P1 subunit
	P2 subunit
	P3 subunit
	P4 subunit

As each subunit consists of a sequence of sandstones, siltstones, and shales, and as each sandstone horizon has different reservoir properties and distribution patterns, the Stairway and Pacoota subunits have been further subdivided into individual sandstones. At Mereenie, 24 reservoir sandstones have been identified in the Pacoota sandstone and 10 within the Stairway sandstone. Using mainly the gamma-ray response, sandstones in the Stairway and Pacoota formations have been named according to their stratigraphic position below the top of each subunit in the East Mereenie 1 (EM-1) type well, for example:

LS-160	160 ft (50 m) below the top of the lower Stairway sandstone in EM1
P1-80	80 ft (25 m) below the top of the P1 subunit in EM1
P3-130	130 ft (40 m) below the top of the P3 subunit in EM1

Pacoota and Stairway sandstones in other wells are correlated directly or indirectly with EM1 to ascertain their correct designation. Some oil-producing sandstones (e.g., P3-130) have been further subdivided for reservoir modeling purposes.

The Pacoota and Stairway formations hold all of the discovered oil and most of the discovered gas reserves in the Amadeus basin. All gas reserves in the Palm Valley field are contained in Stairway and Pacoota sandstones. Mereenie has been the only commercial oil discovery made in the basin. Other structures are either breached or have been found to possess very poor reservoir quality at the Pacoota level. The only other formation that is known to have commercial petroleum-bearing potential in the basin is the upper Proterozoic–Lower Cambrian Arumbera sandstone. A potentially economic gas discovery has been made in the Arumbera sandstone at Dingo, about 35 mi (55 km) south of Alice Springs (Figure 4).

Several Mereenie field wells have penetrated the entire Pacoota sandstone and reached total depth in clastics or carbonates of the underlying Upper Cambrian Goyder Formation. The deepest well drilled on the field, East Mereenie 4, is the only field well to to have penetrated below the Goyder Formation. This well reached a total depth of 8750 ft (2669 m) in interbedded siltstones and dolomites of the upper Proterozoic Bitter Springs Formation after passing through clastics and carbonates of the overlying Lower to Middle Cambrian Cleland sandstone. Like all other deep wells drilled on the field, East Mereenie 4 had very low porosity and no significant hydrocarbon shows below the base of the Pacoota sandstone.

A geohistory plot for the Horn Valley siltstone and Bitter Springs Formation in East Mereenie 4 (Figure 7) illustrates that the penetrated sequence in the field has had a fairly simple burial history. Slow subsidence of upper Proterozoic sediments was interrupted by Late Proterozoic diastrophism that caused uplift and erosion. Deposition on the Bitter Springs unconformity surface commenced in the Early Cambrian with relatively rapid sedimentation and burial continuing until the Late Devonian. The Ordovician reservoir sandstones and the Horn Valley siltstone were deposited during this phase, which was interrupted during the Silurian by uplift and erosion caused by the Rodingan movement. Since formation of the Mereenie anticline by the Alice Springs orogeny in the Late Devonian, the sequence has been steadily eroded with uplift continuing until the Cretaceous.

TRAP

Mereenie is a combination trap with multiple pay zones. While anticlinal trapping is the dominant component, other as yet unproven mechanisms are contributing to trapping in the field as indicated by the presence of hydrocarbons below simple structural closure.

The mapped vertical closure on the Mereenie anticline is approximately 1000 ft (305 m) for each reservoir sandstone. Vertical seals for the Pacoota and Stairway reservoir sandstones are intraformational siltstones and shales, with the Horn Valley siltstone and Stokes Formation providing, respectively, regional seals. The only defined spill point is at the western end of the field and occurs at the same location for each sandstone. Oil and gas have

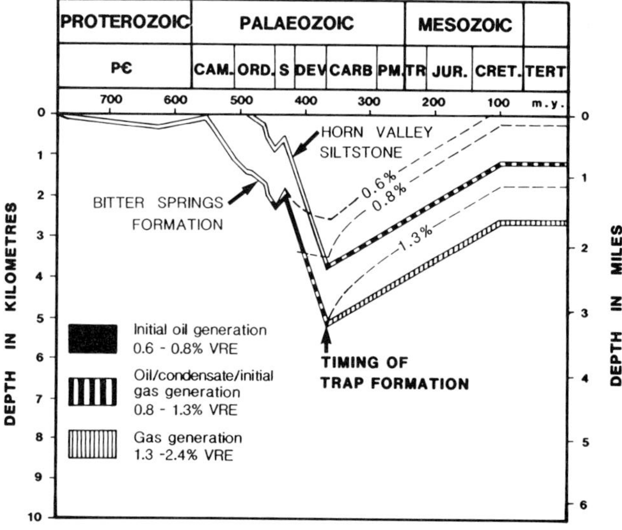

Figure 7. Burial and maturation histories for the Mereenie field based on data from East Mereenie 4. The assumed geothermal gradient is 84°F/mi (29°C/km). (Modified after Jackson et al., 1984.)

been detected below the respective western spill level in some Pacoota and Stairway sandstones in several wells in the eastern half of the field. In the extreme case, gas and condensate were recovered from the LS-160 sandstone in East Mereenie 10, the easternmost and structurally lowest well drilled on the field, approximately 750 ft (229 m) below the western structural spill level for this sandstone.

Fluid contact levels within the field have been determined for only a few sandstones. The gas-oil contact appears to be at the same level (-2130 ft [-650 m] subsea) for all Pacoota sandstones, indicating fluid communication between these reservoirs. However, some water-oil contacts (WOCs) occur at different levels in different sandstones. The main oil-producing sandstone, the P3-130, has a WOC at the western end of the field that is approximately 90 ft (27 m) higher than at the eastern end. While some aquifer pressure data indicate hydrodynamic trapping in this reservoir sandstone, the high aquifer water salinity (89,000 ppm) suggests otherwise. Also, the WOC levels in this sandstone have been determined only at the eastern and western ends of the field so that a tilted fluid contact in this sandstone is only apparent.

Likely trapping mechanisms controlling the distribution of hydrocarbons at Mereenie below simple structural closure are faulting and permeability pinch-outs. The improved vertical resolution of seismic data recently acquired over the western half of the field has revealed splinter faulting off the main thrust obliquely cutting the southern flank of the structure. Throws of up to 60 ft (18 m) on these faults juxtapose reservoir sandstones against nonreservoir units, thereby providing lateral seals, such as for the oil encountered nearly 300 ft (100 m) below simple structural closure level in the relatively thin P1-280 sandstone in well EM3, which is located on the central southern flank of the Mereenie anticline.

Known permeability barriers on the eastern nose of the structure account for the occurrence of oil and gas at the eastern end of the field below the western structural spill level. These barriers also explain the lower WOC in the P3-130 sandstone at the eastern end of the field. Seismic data resolution over the eastern half of the field is insufficient to determine whether these permeability barriers are cross-structure sealing faults. Some sandstones within the Pacoota and Stairway formations (e.g., P1-80 and LSS-160) exhibit significant permeability reduction in certain areas of the field. These permeability variations, which are primarily due to diagenetic effects, probably exert trapping control within particular reservoir sandstones.

RESERVOIRS

Depths to the top of the Pacoota and Stairway formations are variable as a result of the easterly plunge and asymmetry of the Mereenie anticline. The highest point of the Pacoota sandstone occurs on an undrilled culmination in the central-eastern portion of the field at a seismically interpreted subsurface depth of 3500 ft (1068 m). It deepens to about 4250 ft (1296 m) subsurface at the structure's interpreted western spill point and to 4766 ft (1454 m) in East Mereenie 6—the easternmost well of the field where hydrocarbons have been encountered in the Pacoota sandstone. Average thickness of the Pacoota sandstone is 1050 ft (320 m). Average thickness of the Stairway sandstone is 800 ft (244 m). Both formations thicken slightly from east to west across the field. Most Pacoota and Stairway sandstones show remarkably uniform gross thickness and log character along both strike and dip of the field (Figure 8). In contrast, net sandstone, porosity, and especially permeability are highly variable.

Lithology

The Pacoota sandstone reservoirs consist of fine- to medium-grained, moderately to well-sorted, quartz-rich sandstones with feldspars being the only other significant grain type. Clay content is low, consisting of authigenic illite and lesser chlorite. The sandstones are extensively cemented by quartz overgrowths with variable amounts of carbonate (mainly dolomite), anhydrite, and gypsum cements and with minor feldspar overgrowths. Martin (1986) interpreted sporadic zones of abnormally high radioactivity in the Pacoota sandstone at Mereenie to be due to anomalous concentrations of base-metal sulfide minerals and possible associated uranium.

The Stairway reservoirs are also quartz-rich sandstones but are generally more angular and finer grained with fewer and thinner coarser zones than the Pacoota reservoir sandstones. Clay content is low and consists of authigenic illite, chlorite, kaolinite, and illite/smectite. Cementation by quartz overgrowths is dominant, with sulfate and carbonate cements occasionally significant. Disseminated to nodular pyrite is sometimes present, as are phosphorites and collophane cement.

Depositional Environments

In the only detailed paleoenvironmental study of the Pacoota sandstone in the Mereenie field, Garside (1987) identified 16 depositional facies. No similarly detailed analysis has been carried out on the Stairway sandstone at Mereenie, but Catsoulis (1986) has included data from the field in developing a basinal depositional model of this formation. Both studies found that neither the entrapment of hydrocarbons nor existing reservoir quality trends depend primarily upon depositional facies.

The Pacoota sandstone in the Mereenie area was deposited in a transgressive series of marine to shoreline environments on the southwestern margin of a shallow epicontinental sea. Sediment source for

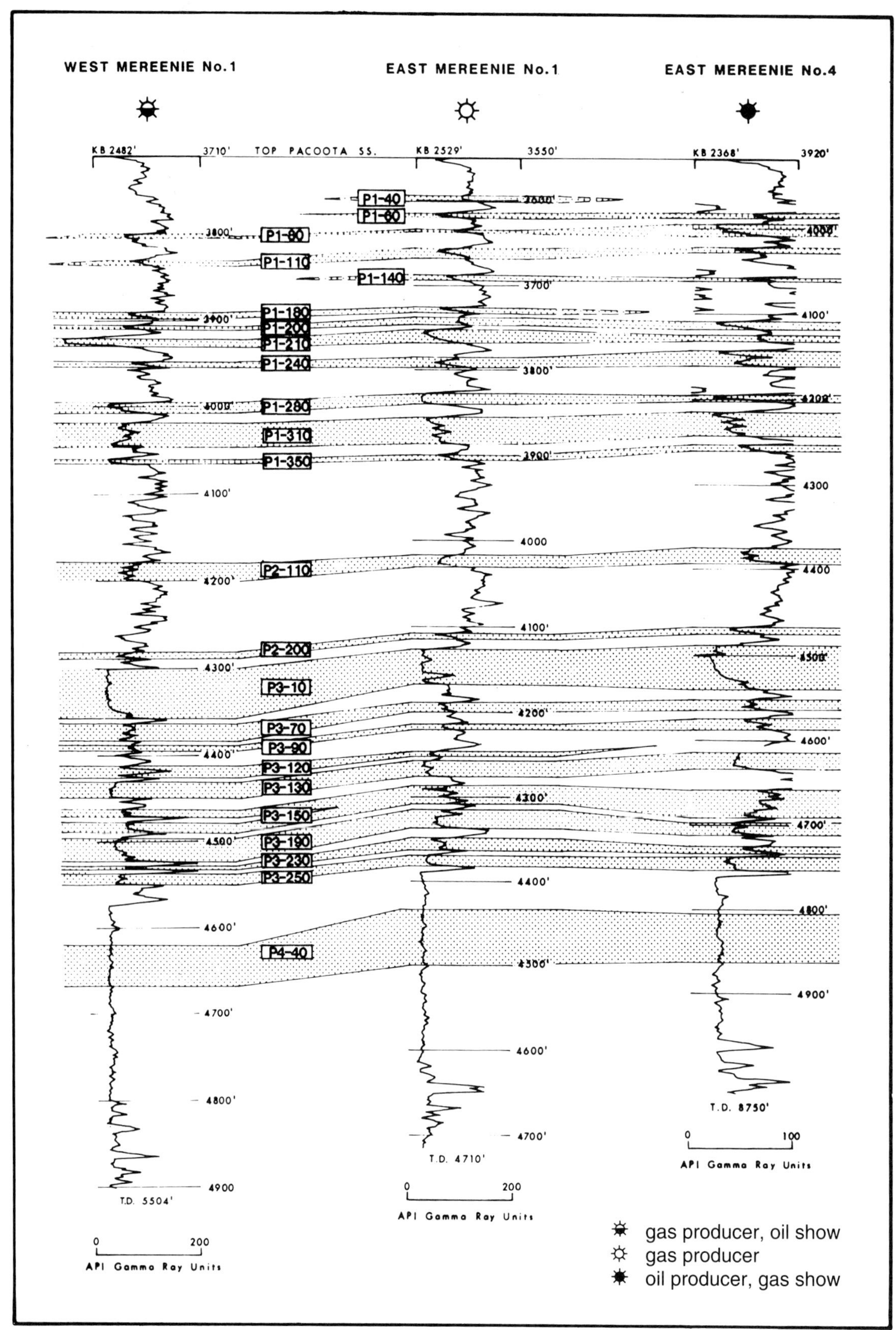

Figure 8. Stratigraphic section of the Pacoota sandstone along the crest of the Mereenie anticline. Note persistent log character and thickness of most Pacoota sands along this 15 mi strike section.

the area was from the west and southwest, transported by fluvial drainage systems and longshore drift. However, no direct evidence exists for fluvial deposition at Mereenie. No vegetation existed during the Ordovician, and the climate was hot and dry. At the time of deposition of the Pacoota sandstone, the Mereenie area was a shelf complex situated southwest of a basin hingeline. The basin's main depocenter was north of this hingeline.

At Mereenie the P4 and P3 subunits are shoreline deposits while the P2 and P1 subunits are moderately deep to shallow marine nearshore to shoreface deposits. The overall sequence is transgressive with regressive phases and is a precursor to the Horn Valley siltstone transgression (Figure 9). The following paleoenvironmental subdivisions of the Pacoota subunits at Mereenie have been made:

P4 Shoreface to intertidal. A homogenous sheet sand that extends throughout the western half of the basin (Figure 10).

P3 A generally transgressive shoreline deposit ranging from lagoon to beach to submergent/emergent barrier bar and thence to shoreface deposits. The main oil-producing reservoirs in the field, the P3-130 and P3-120 sandstones, are interpreted as subaqueous sand wave (reworked barrier bar) and shoreface deposits, respectively (Figures 10 and 11).

P2 Shallow to moderately deep epicontinental marine. A transgressive unit marked by a glauconite horizon occurs at its base (Figure 10).

P1 Shallow marine to shoreface epicontinental marine. Minor oil and gas production is obtained from the P1-80 reservoir, which consists of sand shoal and sand bar deposits (Figures 12 and 13).

Depositional strikes within the P3 (Figure 14) and P4 subunits appear to be subparallel to those of the present-day structural trend (Figures 4 and 5), but those of the P2 and P1 subunits have not been determined. Some of the P1 sand shoals appear to be off strike, with only the distal edges present, while other shoals are continuous across the field.

Unlike the Pacoota sandstone, which was deposited on the margin of an open sea, the Stairway sandstone was deposited as a regressive-transgressive shallow marine sequence in a restricted seaway. Elements of both epeiric sea and barrier island/lagoonal environments are recognized in the Stairway sandstone in the basin (Figure 9). Facies analysis of the Stairway sandstone subunits has led to the following environmental interpretations:

LSS Shallow marine to shoreface. Reworked shoreline deposits and channel lag gravels.

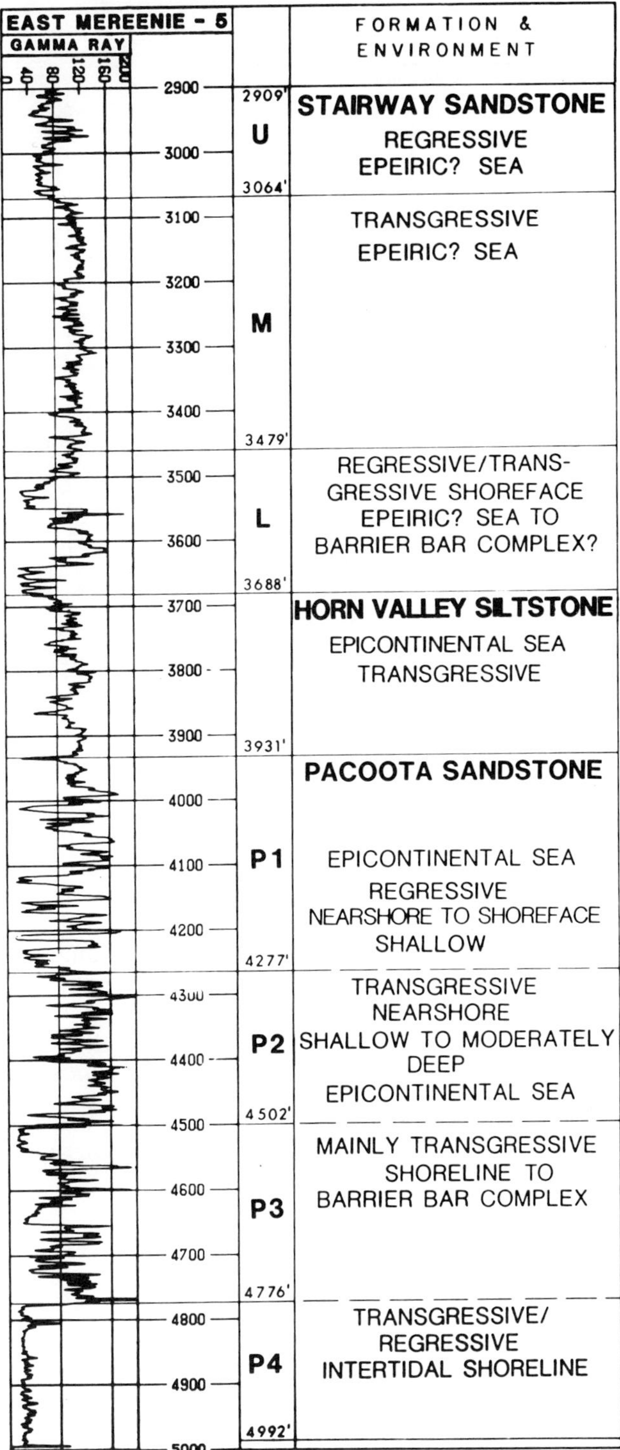

Figure 9. Interpretation of mega depositional environments within the Pacoota sandstone, Horn Valley siltstone, and Stairway sandstone in the Mereenie field. (After Garside, 1987.)

MSS Lagoonal with intensely bioturbated sediments.

USS Storm generated suspension clouds.

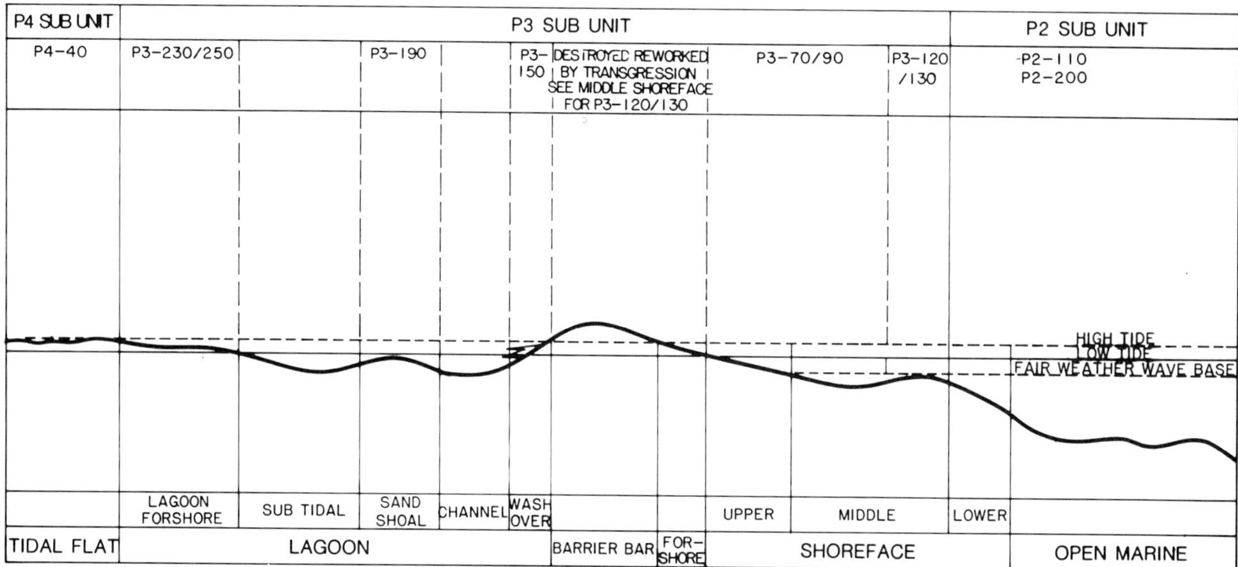

Figure 10. Paleosurface profile and environmental schematic of the Pacoota sandstone P4, P3, and P2 subunits in the Mereenie field. (After Garside, 1987.)

Reservoir sandstones within the Pacoota and Stairway formations are regressive units within overall transgressive sequences. These shoreline deposits and basal transgressive sands have traditionally been viewed as primary exploration targets, with good porosity and permeability development. The reservoir quality of certain sandstones within the P4, P3, P1, and lower Stairway sandstone subunits would have been excellent originally but has since been drastically diminished by diagenesis.

Diagenesis and Reservoir Characteristics

Petrological evidence indicates that as a result of diagenetic processes, particularly silicification, original textural characteristics such as grain size, shape, and sorting now have little or no influence on porosity and permeability in the sandstones (Martin, 1983). Porosity reduction has resulted from cementation by quartz overgrowths, carbonates, and sulfates, and, to a lesser extent, by feldspar overgrowths, microstylolite formation, and pore filling by authigenic clays. A sequence of diagenetic changes in the Pacoota sandstone at Mereenie has been deduced from textural relationships (Figure 15).

Four types of visible porosity have been recognized in the Pacoota sandstone. Types 1 and 2 are primary while types 3 and 4 are secondary. Type 1 occurs where clay or iron oxide coatings on sand grains have prevented infilling of pores by quartz overgrowths and type 2 is due to only partial filling of available pore space by quartz overgrowths. Type 3 results from the dissolution of individual unstable framework grains such as feldspars and makes only a small contribution to total porosity. Type 4 is due to the partial dissolution of carbonate or sulfate cement. Several of these porosity types are present in each reservoir sandstone at Mereenie. While no particular type is known to be predominant, evidence exists that a greater proportion of the porosity in the P1 sandstones is primary whereas the formation of secondary porosity by cement dissolution appears to be more widespread in the P3 sandstones. In both cases, interconnection between pores is limited because of the well-cemented nature of the sandstones and the often irregular pore distribution. Despite the low clay content of these sandstones, the potential for blockage of such pore systems by migration of clay fines has been known for some time (Figure 16). Recent studies have confirmed pore throat blockage by fines migration to be a primary cause of formation damage in the field (Figure 17).

Although variable, the average porosity of the producing Pacoota sandstones is 8% and the average permeability is 10 md (Figures 18 and 19). Analysis of Mereenie well test data by Towler (1986) reveals significant discrepancies between drawdown and build-up permeabilities for the oil-producing P3 sandstones in the eastern flank wells. Build-up permeabilities in this area are two to five times greater than the respective drawdown permeabilities. Such differences can be attributed to stress-dependent permeability (Vairogs and Rhodes, 1973), which is typically caused by a system of micro fractures. Rather than cause dual porosity behavior, these fractures exhibit permeability that is a function of the fluid pressure in the pores. At Mereenie, however, little evidence exists for open fractures at any scale. Core, log, production and pressure test,

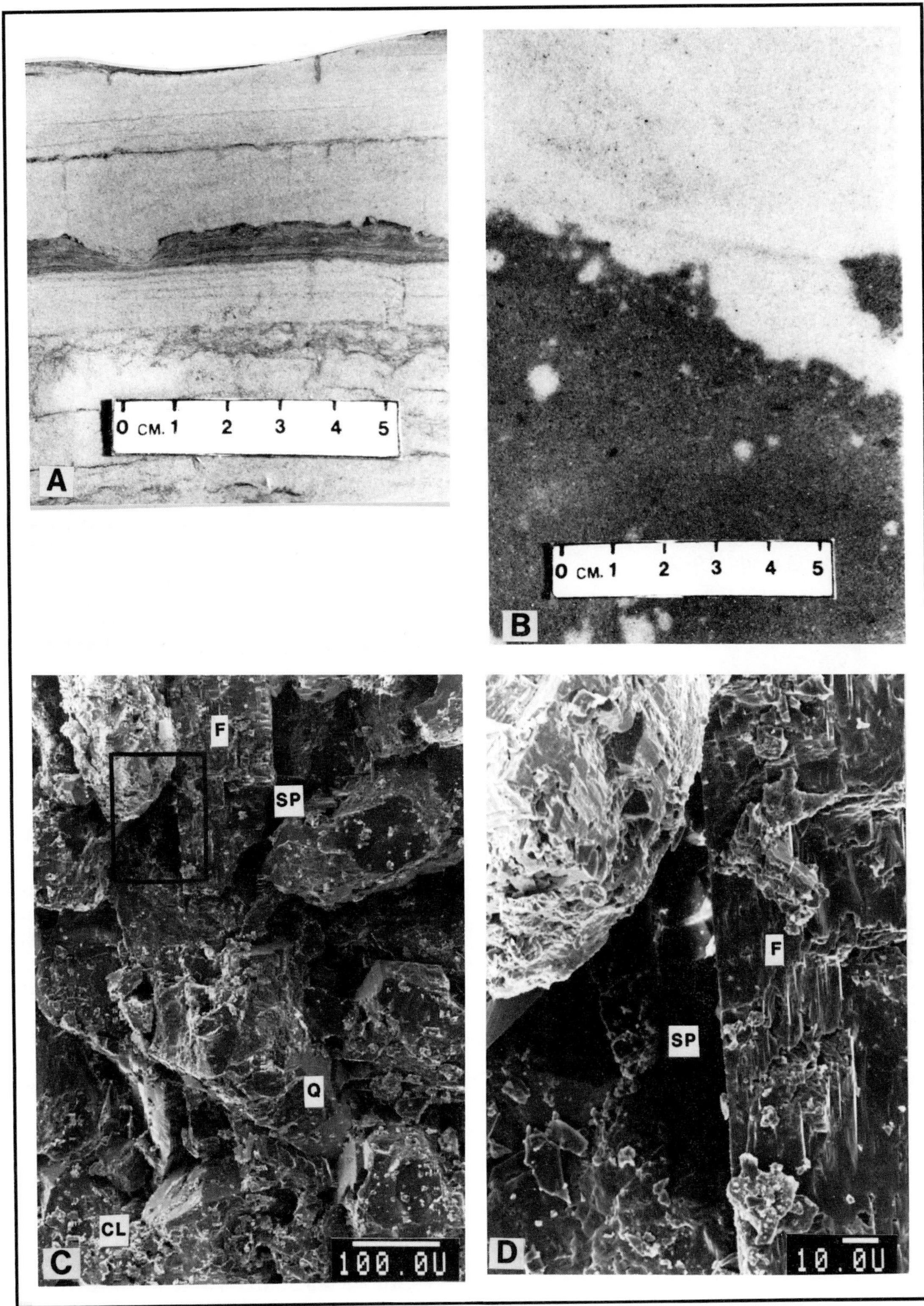

Figure 11. Shoreface facies. (A) Skolithis worm burrows are apparent in the upper area of the photograph and dewatering structures in the central and lower areas (P3-120 sandstone, East Mereenie 9). (B) Fine-grained quartzose sandstone with iron staining in central and lower areas (P3-120 sandstone, East Mereenie 7). (C) Low-magnification SEM photo of a relatively porous section of B. Many of the pores (SP) are large and irregular in shape, indicating they are secondary and probably result from the dissolution of carbonate or sulfate cement. Some patches of clay (CL) occur on grain surfaces. Overgrowths are visible on some quartz grains (Q). An elongated feldspar grain (F) is also marked. (D) Detail of the inset area in C showing a large pore (SP) of probable secondary origin bordered by a feldspar (F).

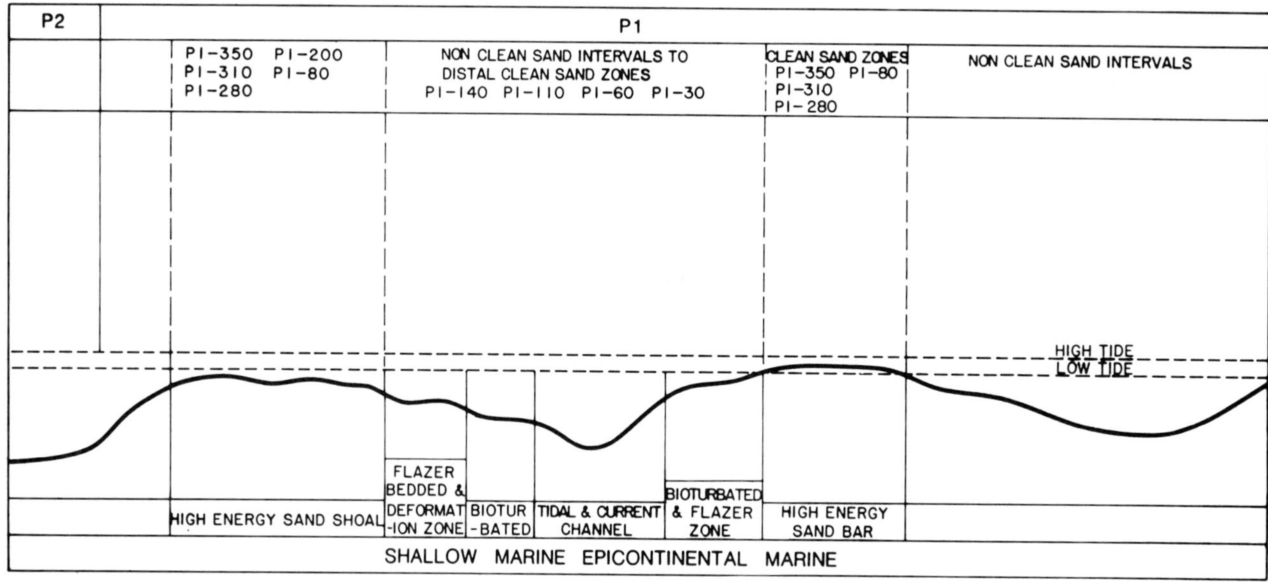

Figure 12. Paleosurface profile and environmental schematic of the Pacoota sandstone P1 subunit in the Mereenie field. (After Garside, 1987.)

and petrological evidence indicates that most faults and other fractures in the field have been naturally cemented and act as seals or barriers to flow. Results of recent core studies show that what was originally interpreted as stress-dependent permeability for a particular Mereenie well test is now believed to be due to a skin factor that was increasing during the drawdown period. Blocking of pore throats by fines migration is believed to occur at all formation fluid velocities, resulting in permeability progressively decreasing during flowing periods. As pressure declines in the near-wellbore region, oil flow is further impeded by increasing gas saturation (lower relative permeability to oil) and salt precipitation from formation brines.

SOURCE ROCK AND HYDROCARBON CHARACTERISTICS

Geochemical evidence presented by Kurylowicz et al. (1976), Gorter (1984), and Jackson et al. (1984) indicates the Horn Valley siltstone has sourced most of the oil at Mereenie, with some contribution from shales within the Pacoota sandstone. The isotopic composition of Mereenie oils and condensates ($\delta^{13}C$ = -30 to -31 per mil) falls well within the range of values displayed by extracts and kerogens isolated from oil-prone Horn Valley and Pacoota shales. An unusual odd-over-even predominance of C_{15} to C_{20} n-alkanes in Mereenie oils is likewise present in extracts of the Horn Valley siltstone (Figure 20). In conjunction with the low pristane/phytane ratios (pr/ph = 0.8 to 2.5) of the oils and source rock extracts, these data indicate an origin from sapropelic type II kerogen derived from fatty lipids of marine algae and bacteria (Figure 21). Gas chromatographic profiles of the saturated hydrocarbons isolated from alternative oil-prone source rocks within the underlying Cambrian and Proterozoic sequences do not match those of Mereenie oils.

Average total organic carbon (TOC) for the Horn Valley siltstone is 0.96% with a measured maximum of 2.30%. Shales within the Pacoota sandstone have an average TOC of 0.40% with a measured maximum of 0.89%. The average equivalent vitrinite reflectance is 1.05% for the Horn Valley siltstone and 1.1% for the Pacoota sandstone. Maturity modeling (Figure 7) shows that the Horn Valley siltstone commenced generating oil at Mereenie about 400 Ma (i.e., just prior to development of the Mereenie anticline) while pre-Ordovician source rocks in the area had become supermature by this time—further evidence for Ordovician shales having sourced the oil in the field.

Mereenie oil is unusual in that it is one of the few Australian oils of strictly marine origin (McKirdy, in Wells, 1976). According to Thompson's (1983) oil classification scheme, which is based on heptane and isoheptane values, the Mereenie crudes are all supermature. Such oils are the result of protracted thermal transformation and substantial

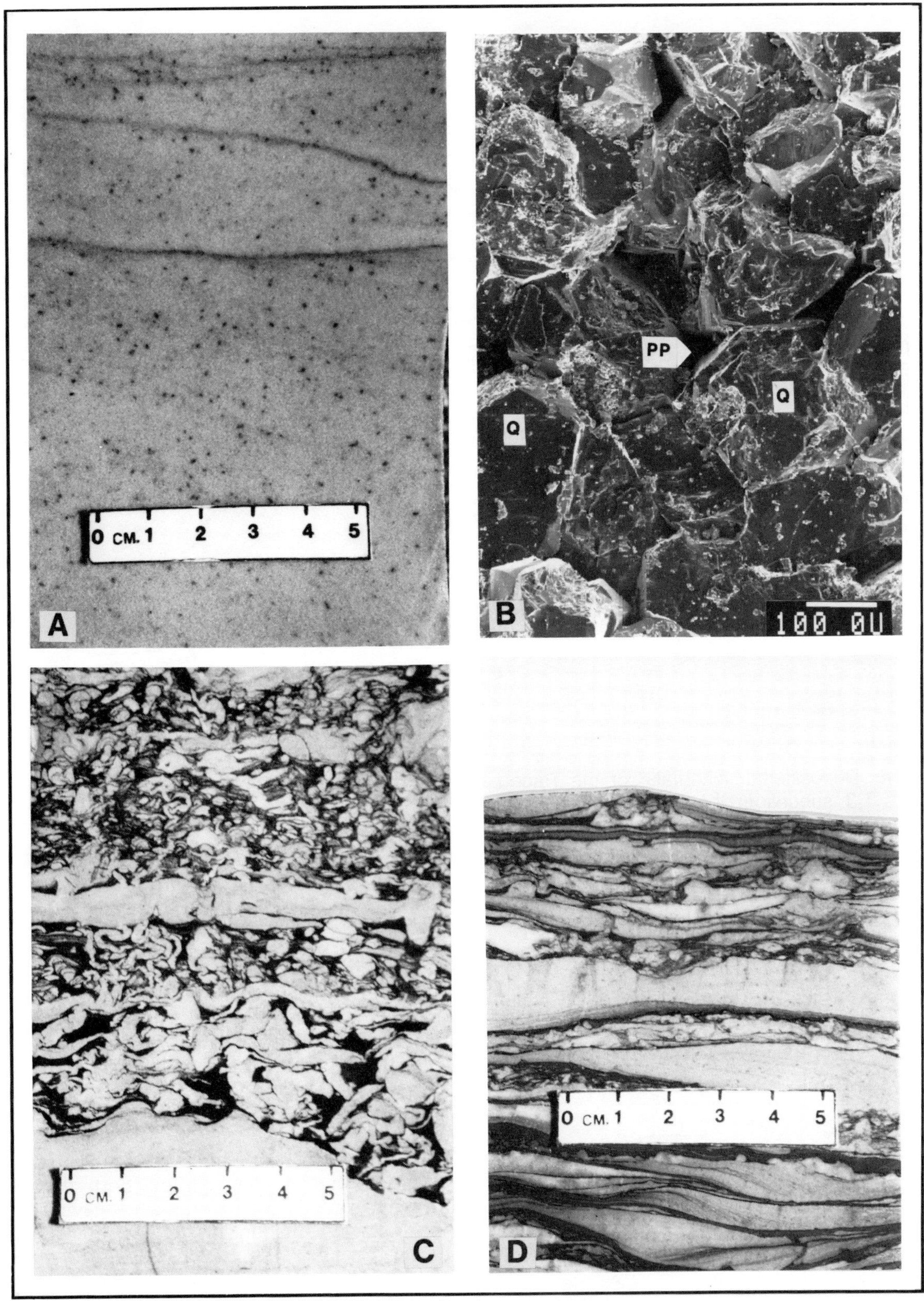

Figure 13. Sand shoal and bar facies. (A) Relatively porous, fine-grained quartzose sandstone with disseminated siderite nodules of probable diagenetic origin (P1-80 sandstone, East Mereenie 5). (B) Low-magnification SEM photo of A showing framework quartz grains (Q) with well-developed planar overgrowth faces giving the grains an angular appearance. Visible pores (PP) are of probable primary origin. (C) Hydroplastically deformed (dewatering) lithofacies occur immediately above the clean sand lithofacies of A. (D) Flaser bedding lithofacies also occur above the clean sand lithofacies of A.

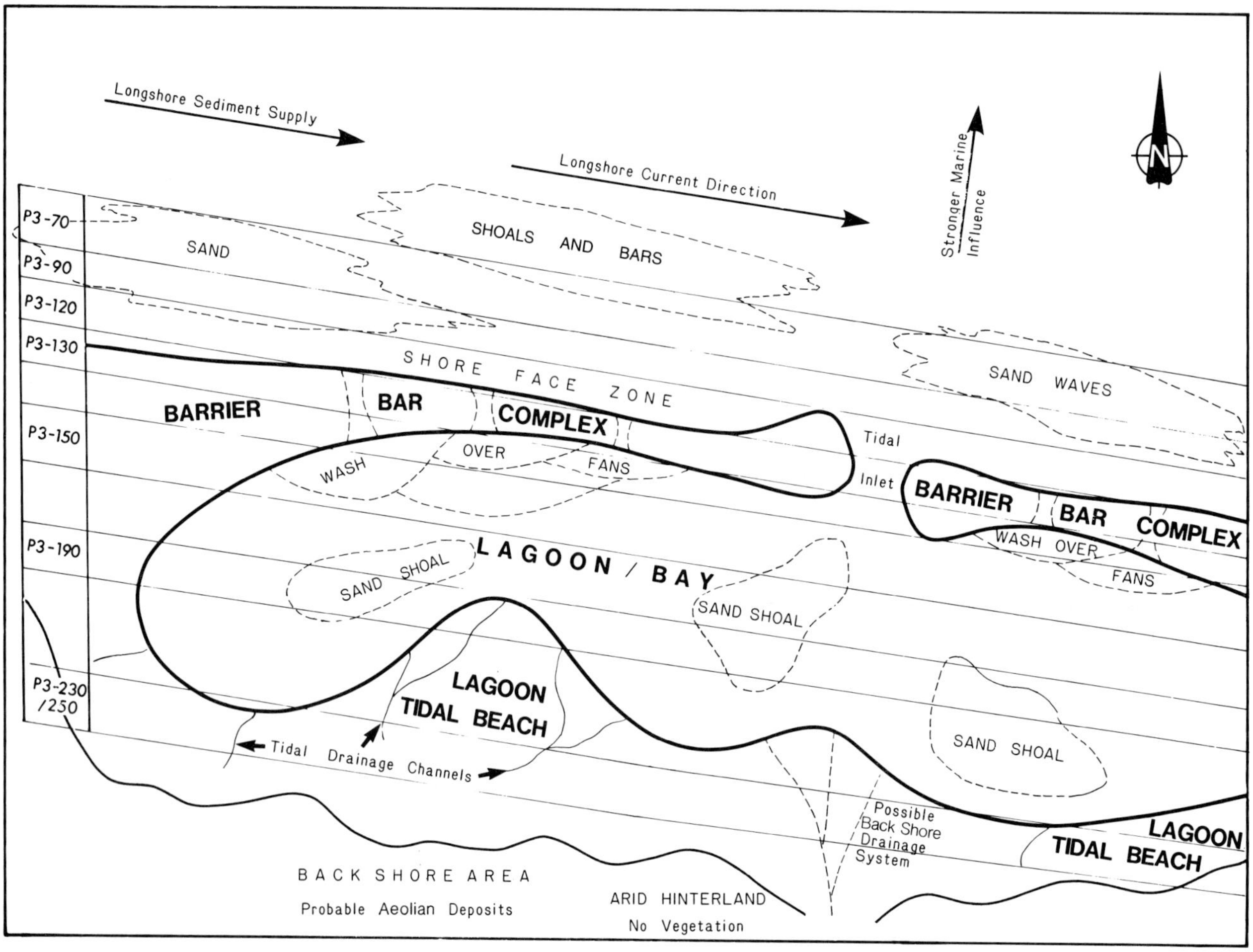

Figure 14. Depositional schematic of the Pacoota sandstone P3 subunit in the Mereenie field. (After Garside, 1987.)

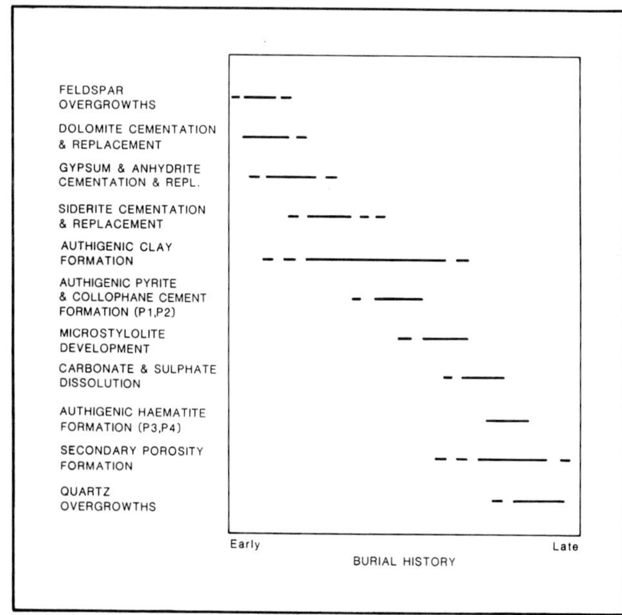

Figure 15. Diagenetic history of the Pacoota sandstone in the Mereenie field. (Modified after Martin, 1983.)

gasification in the reservoir. There are two distinct oil types in the field. Both are paraffinic and have a stock tank gravity of approximately 49° API. The oil from the P1-80 reservoir near the eastern end of the field is honey-colored and has a gas/oil ratio of 1000 SCF/STB and a formation volume factor of 1.59. The oil from the P3 reservoirs is dark green and has a gas/oil ratio of 840 SCF/STB and a formation volume factor of 1.50. Sulfur content is less than 0.1% in both cases.

Within the field there is significant stratigraphic variation in gas composition. Both dryness and nitrogen content increase steadily downward from the Stairway sandstone [$C_1/(C_1\text{-}C_4) = 0.68$, $N_2 = 2.6\%$] to the Pacoota P4 subunit [$C_1/(C_1\text{-}C_4) = 0.83$, $N_2 = 15.0\%$]. High nitrogen contents can occur where the gas has been generated from supermature organic matter and/or has migrated over large distances (Lutz et al., 1975; Stahl et al., 1977). Some of the gas in the Pacoota reservoirs therefore may have been sourced from the underlying upper Proterozoic strata or from the regionally supermature Ordovician sequence north of Mereenie.

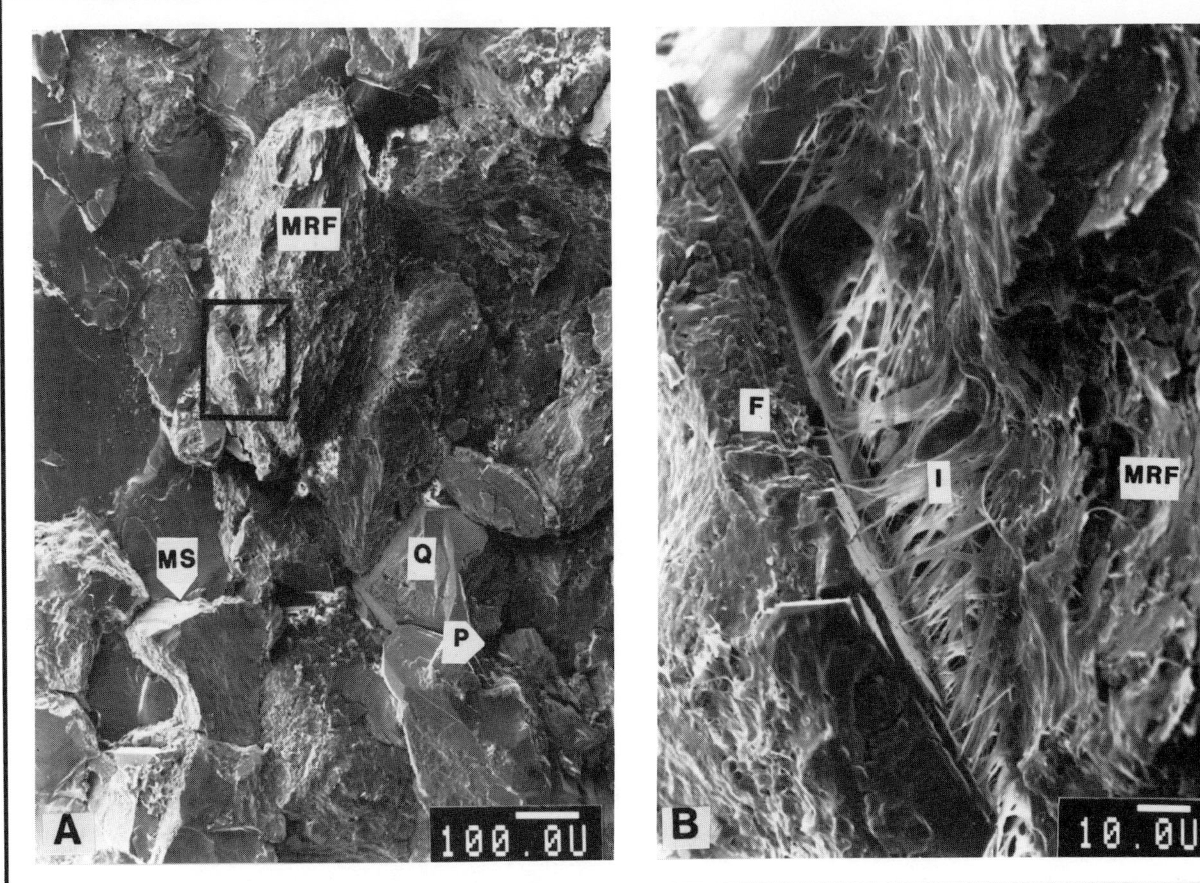

Figure 16 (A) Low-magnification SEM photo showing quartz grains (Q) with some overgrowths and sutured grain contacts where microstylolites (MS) have begun to form. Pores (P) vary considerably in size and shape and are erratically distributed with limited interconnection. A large micaceous metamorphic rock fragment (MRF) is shown (P3-190 sandstone, West Mereenie 3). (B) Detail of the inset area in A showing fibrous authigenic illite (I) bridging a pore between the micaceous metamorphic rock fragment (MRF) and a feldspar (F).

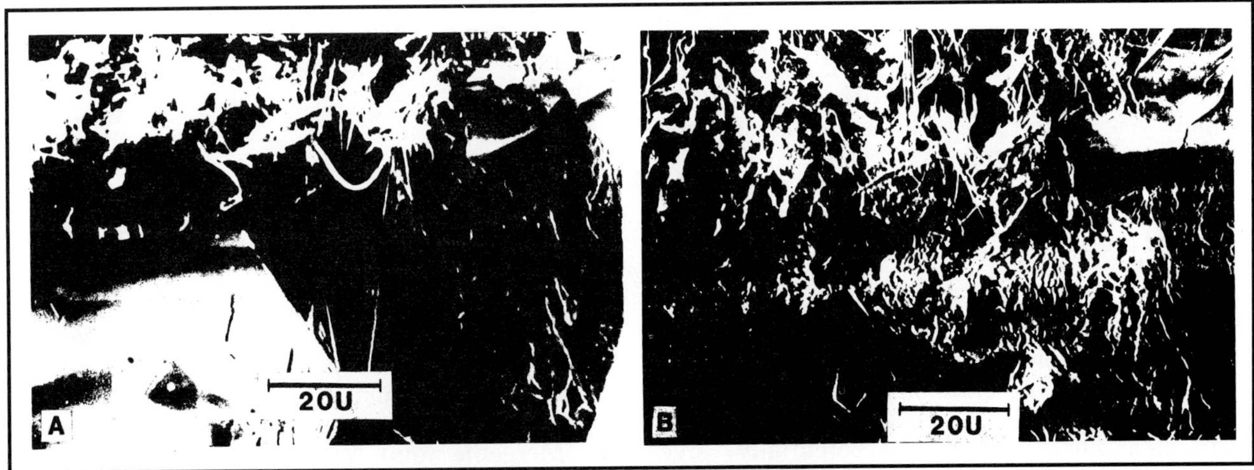

Figure 17. Laboratory study shows the effect of fines plugging by fluid movement. (A) SEM photo of fibrous illite adjacent to a pore before flow. (B) SEM photo of the same sample after toluene was passed through the core plug at 200 psi (1380 kPa) differential pressure. The pore in A has been extensively plugged by illite and clay-sized pieces of quartz and feldspar (P3-120 sandstone, East Mereenie 23). More recent studies indicate such formation damage will occur at any fluid velocity within certain Mereenie reservoir sandstones.

Figure 18. Capillary pressure curve (A) for the sandstone sample shown in the SEM photo (B). Planar overgrowth faces are apparent on the quartz grain (Q) and a secondary pore (SP) is developing in a dissolving feldspar grain (F) in this fine-grained, relatively clear, quartzose sandstone (P3-130 sandstone, East Mereenie 23).

EXPLORATION AND DEVELOPMENT CONCEPTS

Exploration and development of the Mereenie field have been technically and economically demanding. Erratic reservoir performance, low world oil prices, high transportation costs, and lack of a gas market delayed development and production from the field for many years. Protracted legal and compensation negotiations also hindered development.

Oil production at Mereenie from multiple, thin, generally poor quality reservoir sandstones is now marginally profitable within the prevailing economic climate of low world oil prices, unfavorable foreign exchange rates, and deregulation of the Australian oil industry. High gas/oil ratios are forcing some wells to be shut-in. Coupled with no development drilling since early 1986, the field oil production rate is presently declining at an average of 3% per month (Figure 22). The emergence of potentially large markets for Mereenie gas in mining centers in northern Australia offers new opportunities for the field but poses a considerable challenge for optimizing any co-development of the large remaining oil and gas reserves. Horizontal drilling for oil recovery is one strategy currently being investigated.

The presence of hydrocarbons below simple structural closure, spatial and temporal variations in permeability, and complex diagenetic alteration of reservoir sandstones are some of the distinguishing features of the Mereenie oil and gas field. Successful ongoing development will rely on an improved understanding of the special nature of this ancient and isolated field.

ACKNOWLEDGMENTS

The Mereenie Joint Venture permitted publication of the paper. Advice from C. W. Siller, D. McNaughton, and R. Hopkins formed the basis of the pre-discovery historical account of the field. I appreciate the assistance and contributions made by the Mereenie Joint Venture partners: Magellan Petroleum Australia Ltd and SANTOS Ltd; my former colleagues at AGL Petroleum Ltd, particularly C. S. Chen, I. E. Garside, W. Lawson, W. G. Mogg, W. Singley, and P. C. Sippe; and R. J. Korsch and V. Passmore of the Bureau of Mineral Resources. Diagrams were drafted by I. Lenneberg.

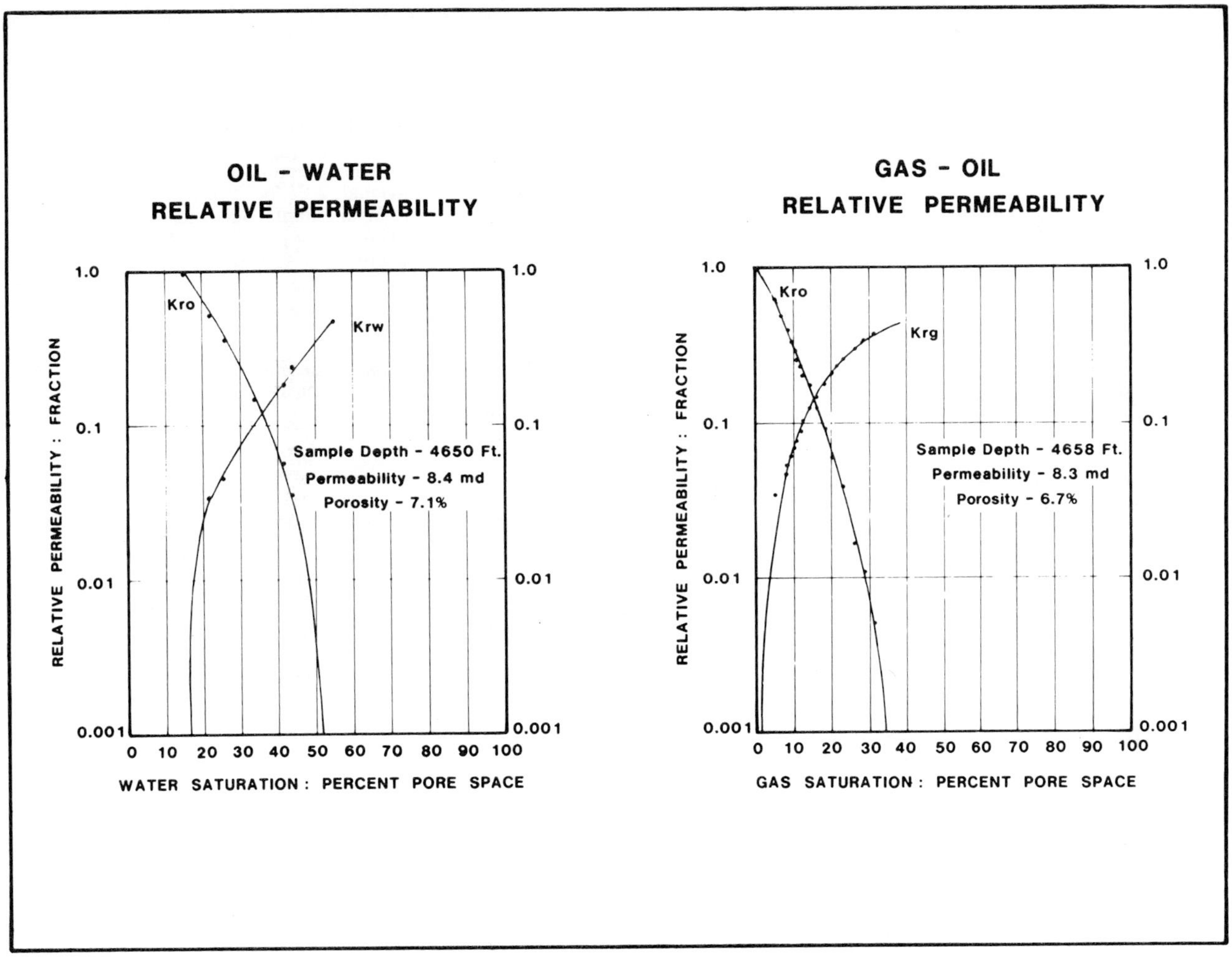

OIL - WATER
RELATIVE PERMEABILITY

RELATIVE PERMEABILITY : FRACTION

Kro
Krw

Sample Depth - 4650 Ft.
Permeability - 8.4 md
Porosity - 7.1%

WATER SATURATION : PERCENT PORE SPACE

GAS - OIL
RELATIVE PERMEABILITY

RELATIVE PERMEABILITY : FRACTION

Kro
Krg

Sample Depth - 4658 Ft.
Permeability - 8.3 md
Porosity - 6.7%

GAS SATURATION : PERCENT PORE SPACE

Figure 19. Oil-water and gas-oil relative permeability curves for core samples from the P3-130 sandstone in East Mereenie 23.

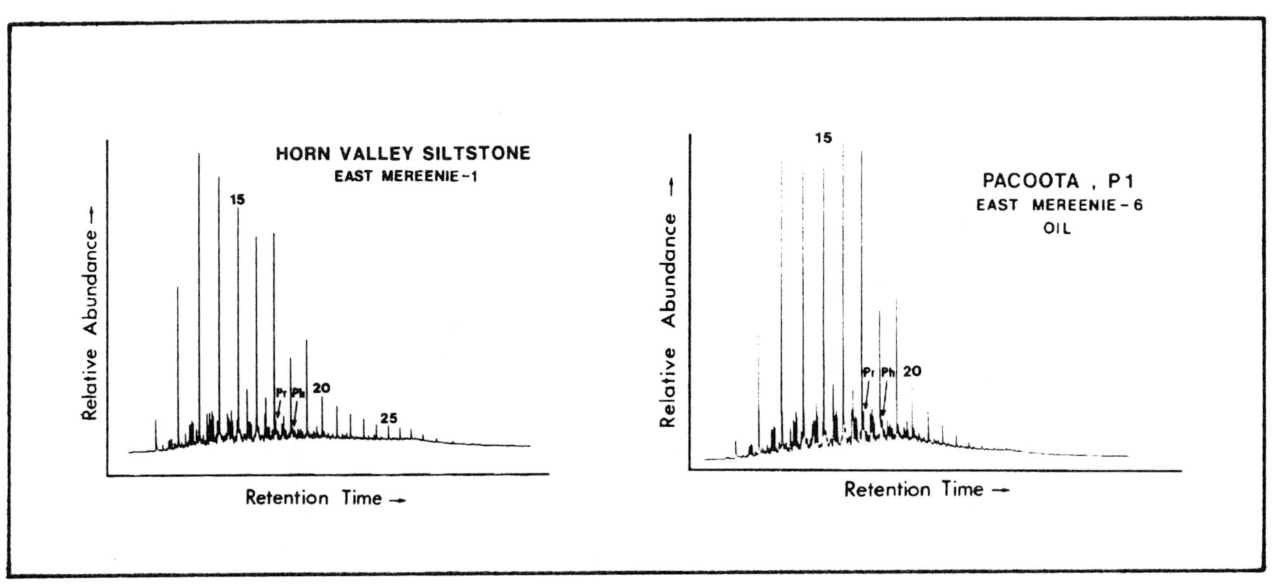

Figure 20. Saturated hydrocarbon (alkane) chromatograms for a Horn Valley source rock extract and a Pacoota sandstone P1 subunit oil sample from the Mereenie field. (After Jackson et al., 1984.)

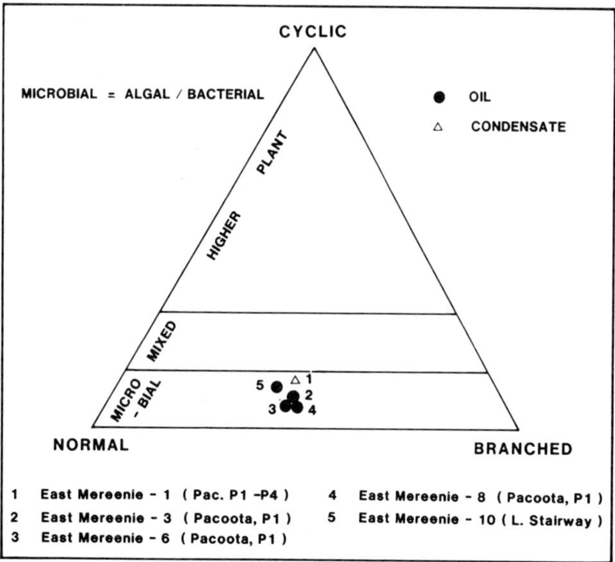

Figure 21. Typing of Mereenie oils and condensate by their gasoline range (C_5-C_7) composition. (Modified after Jackson et al., 1984.)

REFERENCES CITED

Bally, A. W., and S. Snelson, 1980, Realms of subsidence, *in* Facts and principles of world petroleum occurrence: Canadian Society of Petroleum Geologists Memoir 6, p. 9-94.

Catsoulis, D., 1986, Reservoir characteristics and sedimentology of the Stairway sandstone, central Amadeus basin, Northern Territory: Unpublished BSc honours thesis, Department of Geology and Mineralogy, University of Queensland, Australia.

Garside, I. E., 1987, Mereenie field facies study Pacoota sandstone: Unpublished company report, AGL Petroleum Ltd, Brisbane.

Gorter, J. D., 1984, Source potential of the Horn Valley siltstone, Amadeus basin: APEA Journal, v. 24, part 1, p. 66-90.

Jackson, K. S., D. M. McKirdy, and J. A. Deckelman, 1984, Hydrocarbon generation in the Amadeus basin, central Australia: APEA Journal, v. 24, pt. 1, p. 42-65.

Klemme, H. D., 1971, What giants and their basins have in common: Oil and Gas Journal, v. 69, n. 9, 10, 11; pt. 1, p. 85-90; pt. 2, p. 103-110; pt. 3, p. 96-100.

Kurylowicz, L. E., S. Ozimic, D. M. McKirdy, A. J. Kantsler, and A. C. Cook, 1976, Reservoir and source rock potential of the Larapinta Group, Amadeus basin, central Australia: APEA Journal, v. 16, pt. 1, p. 49-65.

Lutz, M., J. P. H. Kaasschieter, and D. H. Van Wijhe, 1975, Geological factors controlling Rotliegand gas accumulations in the Mid-European basin: Proceedings of the 9th World Petroleum Congress, Tokyo, 2, p. 93-103.

Martin, K. R., 1983, Petrology of the Pacoota sandstone in the Mereenie field, Amadeus basin, Northern Territory: Report to AGL Petroleum, Brisbane.

Martin, K. R., 1986, An investigation of the cause of abnormally high radioactivity within the Pacoota sandstone: Report to AGL Petroleum, Brisbane.

Schroder, R. J., and J. D. Gorter, 1984, A review of the recent exploration and hydrocarbon potential of the Amadeus basin, Northern Territory: APEA Journal, v. 24, pt. 1, p. 19-41.

Stahl, W., H. Boigk, and G. Wollanke, 1977, Carbon and nitrogen isotope data of Upper Carboniferous and Rotliegand gases from North Germany and their relationship to the maturity of the organic source material, *in* R. Campos and J. Goni, eds., Advances in organic geochemistry 1975: Madrid, Enadimsa, p. 539-559.

St. John, B., A. W. Bally, and H. D. Klemme, 1984, Sedimentary provinces of the world, hydrocarbon productive and nonproductive: AAPG Map Series, 35 p.

Thompson, K. F. M., 1983, Classification and thermal history of petroleum based on light hydrocarbons: Geochimica et Cosmochimica Acta, v. 47, p. 303-316.

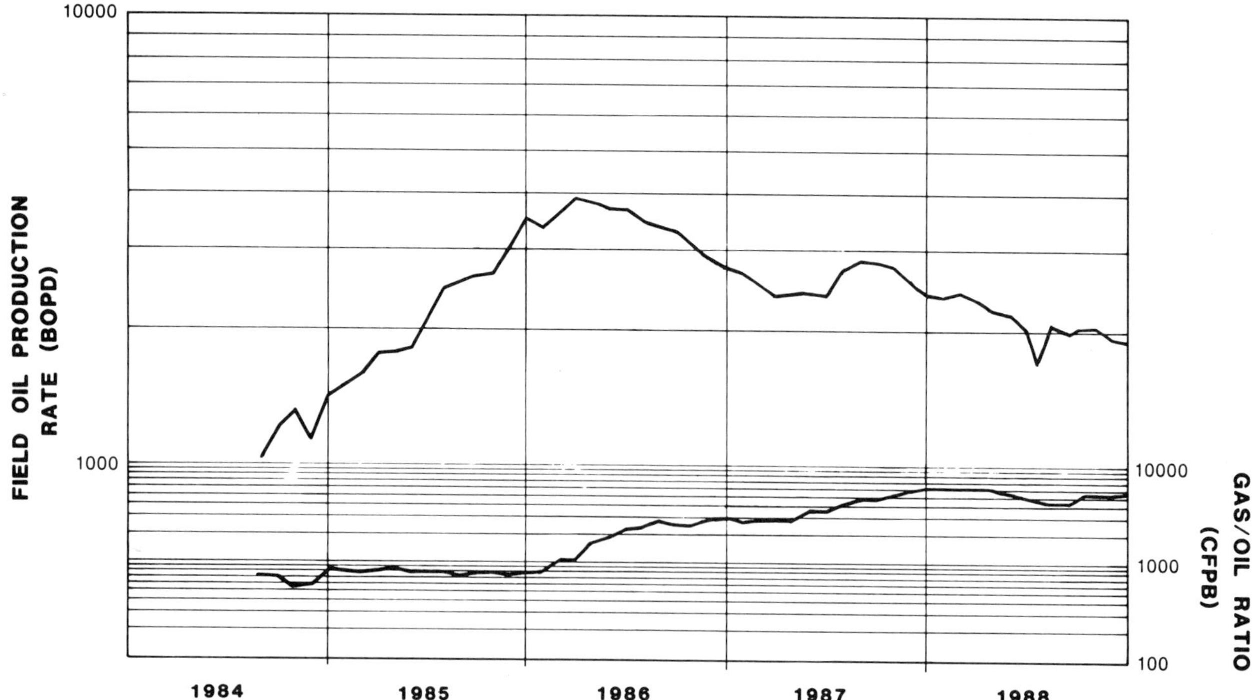

Figure 22. Oil production and gas/oil ratio history for the Mereenie field.

Towler, B. F., 1986, Well tests in a complex sandstone reservoir: SPE 15420 presented at the 61st SPE Annual Technical Conference and Exhibition, New Orleans.

Vairogs, J., and V. W. Rhodes, 1973, Pressure transient tests in formations having stress sensitive permeability: Journal of Petroleum Technology (August 1973); Trans AIME 255.

Wells, A. T., 1976, Geology of the Late Proterozoic-Palaeozoic Amadeus basin: Excursion Guide no. 48A, 25th International Geological Congress, Progress Press, Canberra.

SUGGESTED READING

Benbow, D. D., 1988, Gas from the NT's Amadeus basin: APEA Bulletin No. 25, p. 3-5.

Pearson, T. R., and D. D. Benbow, 1976, Amadeus basin, *in* Economic Geology of Australia and Papua New Guinea. 3. Petroleum: Monograph Series No. 7, p. 216-225.

Roe, L. E., 1988, The Amadeus basin, *in* Petroleum in Australia: the first century: APEA, p. 232-251.

Towler, B. F., 1986, Reservoir simulation in the Mereenie field: APEA Journal, v. 26, pt. 1, p. 428-446.

Wells, A. T., D. J. Forman, L. C. Ranford, and P. J. Cook, 1970, Geology of the Amadeus basin, Central Australia: Bureau of Mineral Resources Bulletin No. 100.

Wilkinson, R., 1983, A thirst for burning—the story of Australia's oil industry: Sydney, David Ell Press.

Field name .... *Mereenie field*

Ultimate recoverable reserves .... *34,000,000 bbl oil and 593 bcf gas*

Field location:

Country .... *Australia*

State .... *Northern Territory*

Basin/Province .... *Amadeus basin*

Field discovery:

Year first pay discovered .... *1964 (East Mereenie No. 1 gas discovery well in Lower Ordovician Pacoota sandstone also tested gas in Middle Ordovician sandstone)*

Year second pay discovered .... *1964 (East Mereenie No. 2 oil discovery well with gas in Pacoota oil leg)*

Discovery well name and general location:

First pay .... *Mereenie No. 1, 245 km west of Alice Springs and approx. 100 km west of Palm Valley gas field*

Second pay .... *East Mereenie No. 2, 230 km west of Alice Springs*

Discovery well operator .... *Exoil (NT) Pty Ltd (M1 and EM2)*

IP:

First pay .... *Gas discovery well (M1), 4.8 MMCFGD (DST 6, Mereenie 1)*

Second pay .... *Oil discovery well (EM2), est. 150 BOPD (DST 9, East Mereenie 2)*

All other zones with shows of oil and gas in the field:

Age	Formation	Type of Show
Upper Cambrian-Lower Ordovician	*Pacoota sandstone*	*All currently known oil and gas reserves*
Middle Ordovician	*Stairway sandstone*	*Oil and gas flows and shows*
Lower Ordovician	*Horn Valley siltstone*	*Minor oil shows*

Geologic concept leading to discovery and method or methods used to delineate prospect

The eastern half of the Mereenie structure is expressed at the surface. The recognition of this surface structure led to a seismic survey in 1962 indicating subsurface closure. In 1963 a government phosphate core hole drilled 15 mi (24 km) southeast of Mereenie intersected oil staining in the Stairway sandstone, Horn Valley siltstone, and Pacoota sandstone. These factors led to the drilling of Mereenie 1, which was the first significant petroleum discovery in the Amadeus basin.

Structure:

Province/basin type .... *Bally 41, Klemme II A*

Tectonic history

The structural evolution of the Amadeus basin has been influenced by at least six tectonic episodes of which the Petermann Ranges and Alice Springs orogenies appear to have had the greatest effect. The Late Proterozoic Petermann Ranges orogeny was characterized by northward thrusting on a salt décollement surface, recumbent folding, and deep crustal movements producing east-west-striking anticlines in the southwest of the basin. Structures initiated during the Late Proterozoic were subsequently modified and new ones created during the Late Devonian to Early Carboniferous Alice Springs orogeny. In the eastern part of the basin strike-slip faults, thrust sheets, and folding resulted, while in the central and western area numerous large-scale folds were generated (e.g., the Mereenie and Palm Valley anticlines).

Regional structure

The Mereenie structure is one of an en echelon series of approximately east-west-trending anticlines in the central western Amadeus basin.

Local structure

The Mereenie field occurs within a large elongate west-northwest to east-southeast anticlinal structure bounded to the south by a large thrust fault.

Trap:

Trap type(s)

One anticlinal trap with multiple pays. Some pays appear to have nonstructural trapping components.

Basin stratigraphy (major stratigraphic intervals from surface to deepest penetration in field):

Chronostratigraphy	Formation	Depth to Top in ft (m)
Middle-Upper Devonian	*Parke siltstone*	*Surface*
Upper Silurian-Middle Devonian	*Mereenie sandstone*	*250 (76)*
Upper Ordovician	*Carmichael sandstone*	*1850 (564)*
Middle Ordovician	*Stokes siltstone*	*2050 (625)*
	Stairway sandstone	*3050 (930)*
Lower Ordovician	*Horn Valley siltstone*	*3850 (1174)*
Upper Cambrian-Lower Ordovician	*Pacoota sandstone*	*4050 (1235)*
Upper Cambrian	*Goyder Formation*	*5100 (1552)*
Lower-Upper Cambrian	*Cleland sandstone and Tempe Formation*	*5700 (1738)*
Proterozoic	*Black Springs Formation*	*8400 (2562)*

Reservoir characteristics:

Number of reservoirs *2*

Formations .... *Pacoota sandstone (all production to date); Stairway sandstone (no production yet)*

Ages .... *Pacoota, Upper Cambrian-Lower Ordovician; Stairway, Middle Ordovician*

Depths to tops of reservoirs .... *Pacoota, 4050 ft (1235 m); Stairway, 3050 ft (930)*

Gross thickness (top to bottom of producing interval) .... *1930 ft (588 m)*

Net thickness—total thickness of producing zones

Average .... *80 ft (24 m)*

Maximum .... *126 ft (39 m)*

Lithology .... *Fine to medium quartzose sandstone*

Porosity type .... *Secondary solution porosity with minor remnant primary intergranular*

Average porosity *8%*

Average permeability .... *10 md*

Seals:

Upper

Formation, fault, or other feature .... *Horn Valley siltstone*

Lithology .... *Siltstone*

Lateral

Formation, fault, or other feature .... *Formation*

Lithology .... *Horn Valley siltstone*

Source:

Formation and age .... *Horn Valley siltstone (Lower Ordovician)*

Lithology .... *Gray-black siltstone*

Average total organic carbon (TOC) *0.96%*

Maximum TOC *2.3%*

Kerogen type (I, II, or III) .... *II*

Vitrinite reflectance (maturation) .... $R_o = 1.05$ *avg.*

Time of hydrocarbon expulsion .... *Oil generation commenced Late Devonian and continues today; gas generation commenced Early Carboniferous and continues today*

Present depth to top of source .... *3900 ft (1189 m)*

Thickness .... *200 ft (61 m)*

Potential yield .... *Unknown*

Appendix 2. Production Data

Field name .... *Mereenie field*

Field size:

- **Proved acres** .... *42,000 ac (17,000 ha)*
- **Number of wells all years** *36*
- **Current number of wells** *35*
- **Well spacing** .... *Variable but 160 ac at eastern end of field*
- **Ultimate recoverable** .... *593 bcf; 34 MMBO*
- **Cumulative production** .... *7.6 bcf; 3.2 MMBO (as of 31/12/87)*
- **Annual production** .... *2 bcf; 0.9 MMBO (for 1987)*
- **Present decline rate** .... *3% per month*
- **Initial decline rate** .... *3.5% per month*
- **Overall decline rate** .... *3% per month*
- **Annual water production** .... *4683 bbl*
- **In place, total reserves** .... *821 bcf; 186 MMBO*
- **In place, per acre foot** .... *387.9 bbl*
- **Primary recovery** .... *Dependent on economics*
- **Cumulative water production** .... *6320 bbl*

Drilling and casing practices:

- **Amount of surface casing set** .... *3000 ft (915 m)*
- **Casing program**

  8⅝-in. surface casing; 5½-in. production casing; 2⅜-in. tubing
- **Drilling mud** .... *Usually air/mist to top of pay then oil-based mud through pay*
- **High pressure zones** .... *None*

Completion practices:

- **Interval(s) perforated** *P3-70/90/120/130/150/190/230/250*
- **Well treatment** .... *Open-hole completions standard*

Formation evaluation:

- **Logging suites** .... *GR, neutron, density, resistivity*
- **Testing practices** .... *DST, then complete and production test*
- **Mud logging techniques**

  Samples collected every 30 ft down to the top of the Pacoota, then every 10 ft to T.D.; a hot wire gas detector is connected when drilling from the top of the Stairway sandstone to T.D.

Oil characteristics:

- **Type** .... *Paraffinic*
- **API gravity** .... *49°*
- **Initial GOR** .... *840 scf/bbl*
- **Sulfur, wt%** .... *750 ppm*
- **Viscosity, SUS** .... *0.36 cp*
- **Pour point** .... *-40°C (-40°F)*
- **Gas-oil distillate** .... *NA*

Field characteristics:

- **Average elevation** .... *2500 ft (762 m)*
- **Initial pressure** .... *1780 psi (12,270 kPa)*
- **Present pressure** .... *1730 psi (11,930 kPa)*
- **Pressure gradient** .... *0.28 psi/ft (6.33 kPa/m)*
- **Temperature** .... *146°F (63°C)*
- **Geothermal gradient** .... *0.019°F/ft (0.035°C/m)*

Drive .... *Gas cap expansion and solution gas drive*
Oil column thickness .... *320 ft (98 m)*
Oil-water contact .... *-2450 ft (-747 m)*
Connate water *25%*
Water salinity, TDS .... *89,000 ppm*
Resistivity of water .... *0.05 ohm-m*

Transportation method and market for oil and gas:
Oil is piped to Alice Springs, then railed to Adelaide; gas is pipelined to Darwin

Razzak Field—Egypt
Razzak-Alamein Basin, Northern Western Desert

A. SHAWKY ABDINE
WAFIK MESHREF
SAFI WASFI
A. NABIL SHAHIN
ARNE AADLAND
A. AAL
Gulf of Suez Petroleum Company
Cairo, Egypt

FIELD CLASSIFICATION

BASIN: Razzak-Alamein
BASIN TYPE: Intercratonic
RESERVOIR ROCK TYPE: Sandstones and Carbonates
RESERVOIR ENVIRONMENT OF DEPOSITION: Stable Shelf
RESERVOIR AGE: Cretaceous
PETROLEUM TYPE: Oil
TRAP TYPE: Structural/Complex Anticline

LOCATION

The Razzak field is located in the north-central portion of the Western Desert of Egypt (Figure 1). The Razzak field "complex," so named for its three separate culminations (West Razzak, Razzak Main, and East Razzak) (Figure 2), is developed along the northeast–southwest-trending Qattara-Alamein ridge; it is a prominent linear, structurally high trend along which the Razzak, Yidma, and Alamein fields are located (Figure 3). These fields all produce from Cretaceous Aptian through Cenomanian-age carbonates and clastics (Figure 4). The Qattara-Alamein axis is situated within the Alamein basin, which appears to have had its maximum basinal development during the Early Cretaceous. The basin area covers most of the north-central portion of the Western Desert north of the Sharib-Sheiba platform (Figure 3).

The entire Razzak field complex extends for 17 km (10.6 mi); East Razzak is 3.5 km by 1.5 km (2.2 by 0.9 mi); Razzak Main is 3.0 by 1.0 km (1.9 by 0.6 mi); and West Razzak is 4.5 by 2.0 km (2.8 by 1.2 mi). The complex is operated by the Gulf of Suez Petroleum Company (GUPCO), which is a joint venture company formed by Egyptian General Petroleum Corporation (EGPC) and AMOCO Egypt Oil Company (subsidiary of AMOCO Production Company). Razzak has produced above 50 million bbl of oil (MMBO). Reserve estimates are not available.

HISTORY

Pre-Discovery

The first integrated and systematic hydrocarbon exploration program in the Western Desert was carried out in the mid-1950s by Sahara Petroleum Company (SAPETCO), which was given exploration rights over the entire 200,000 km^2 (77,158 mi^2) "northern Western Desert" concession (Abdine, 1974). Before this period, only limited exploration was carried out by oil companies in the area. This included some semidetailed seismic acquisition, gravity surveys (Figure 3), and the drilling of four wells (Figure 3). The well Dabaa'-1 near Ras El Dabaa' on the Mediterranean coast, 130 km (81 mi) southwest of Alexandria, was drilled by Anglo Egyptian Oil Company (AEO) in 1939. Khatatba-1, 45 km (28 mi) northwest of Cairo, was drilled by South Mediterranean Oil Company (SOMED) in September 1944. Abu Roash-1 and -2, just west of Cairo (Figure 3), were drilled by Standard Oil Company (SOC) in 1946.

The subsequent SAPETCO activities in the "northern Western Desert" concession were carried out from 1954 to 1958 and included aerial photograph surveys and reconnaissance geologic and seismic surveys, together with Bouguer gravity data acquisition. In addition, SAPETCO drilled nine deep

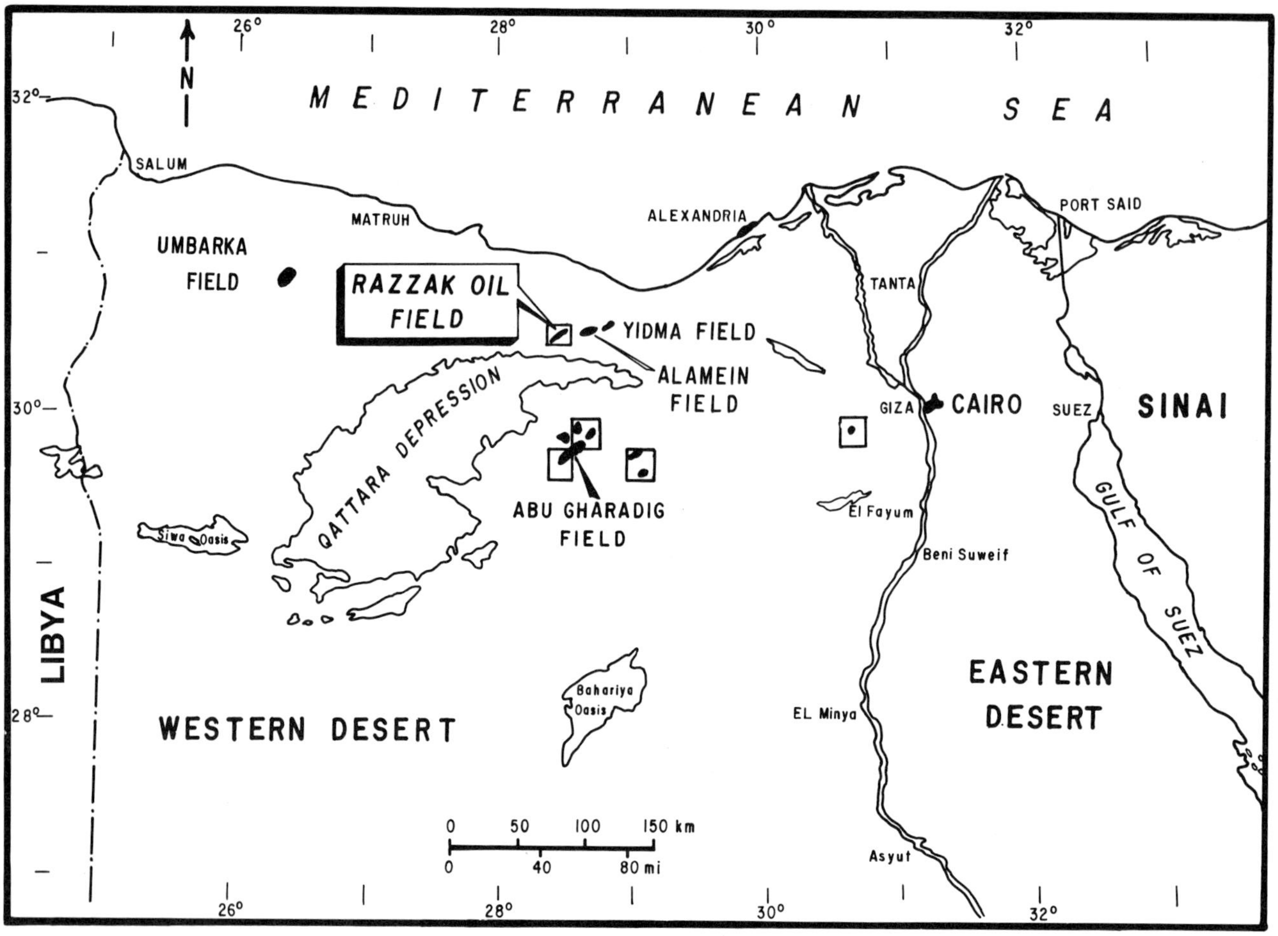

Figure 1. Location of Razzak field, Western Desert, Egypt.

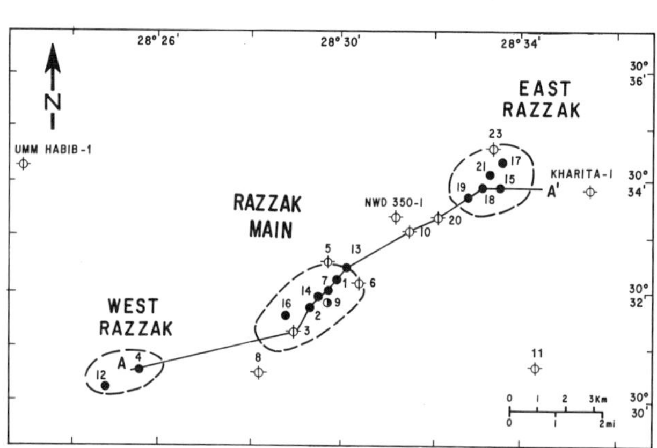

Figure 2. Razzak field "complex," so named for its three separate culminations (West Razzak, Razzak Main, and East Razzak). A-A' is the line of the correlation panel shown in Figure 8.

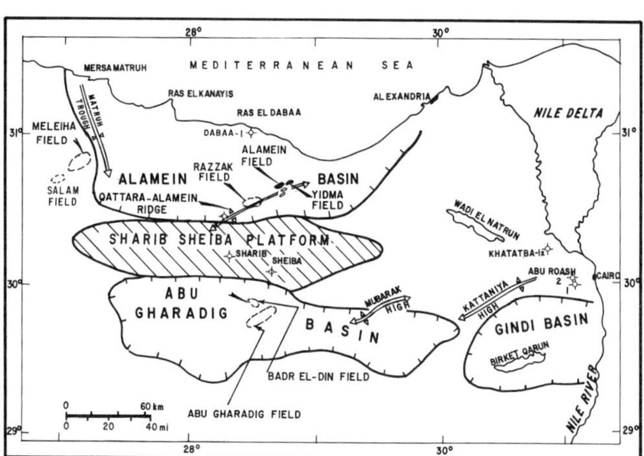

Figure 3. Simplified geologic provinces of the northern Western Desert. The Razzak field "complex" extends for 17 km (11 mi) and is developed along the northeast-southwest-trending Qattara-Alamein ridge along which the Alamein and Yidma fields are located. The dry hole at the southwest end of ridge is the Qattara Rim-1 well.

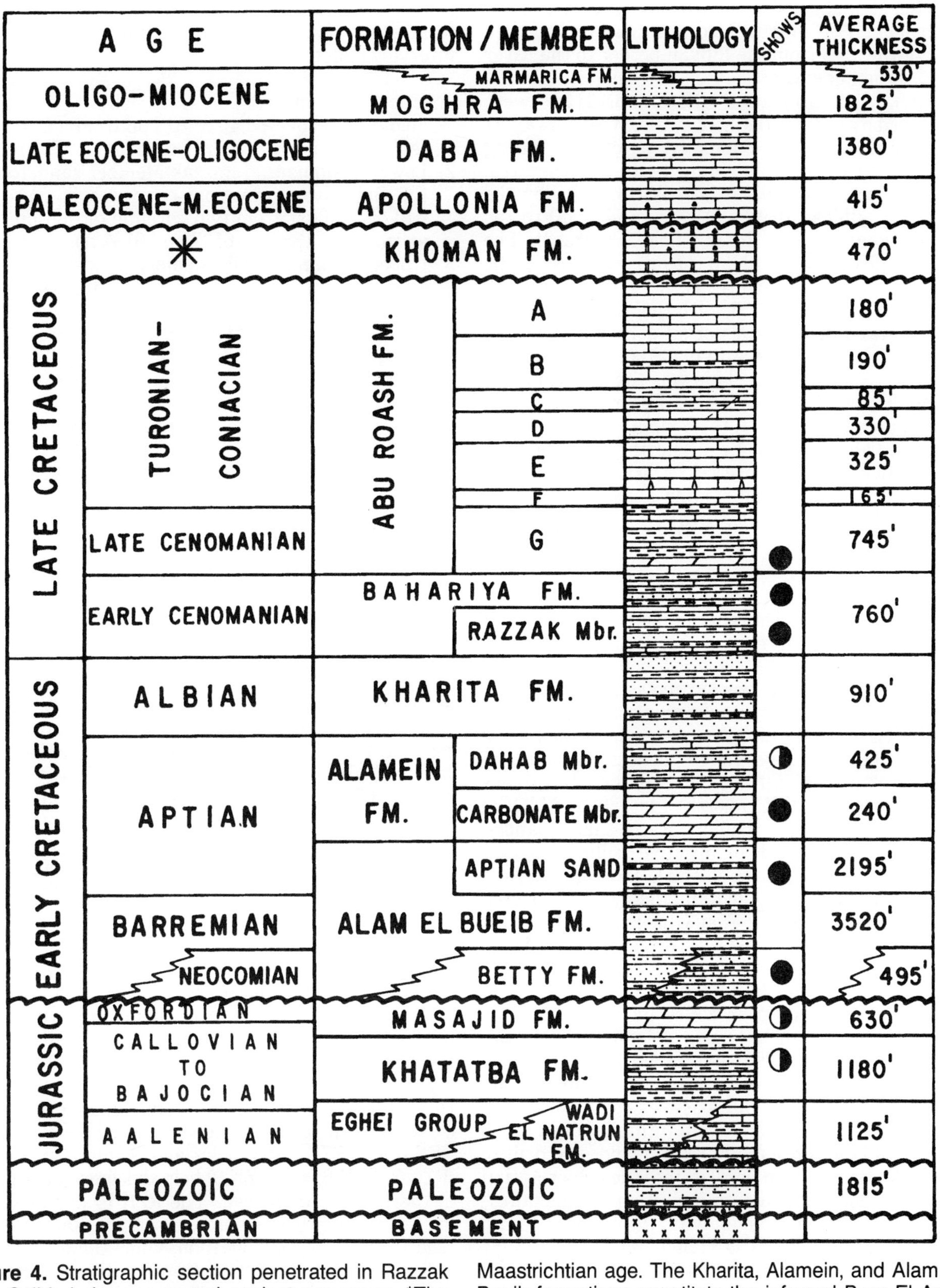

AGE		FORMATION / MEMBER		LITHOLOGY	SHOWS	AVERAGE THICKNESS
OLIGO-MIOCENE		MARMARICA FM.				530'
OLIGO-MIOCENE		MOGHRA FM.				1825'
LATE EOCENE-OLIGOCENE		DABA FM.				1380'
PALEOCENE-M.EOCENE		APOLLONIA FM.				415'
LATE CRETACEOUS	*	KHOMAN FM.				470'
LATE CRETACEOUS	TURONIAN-CONIACIAN	ABU ROASH FM.	A			180'
LATE CRETACEOUS	TURONIAN-CONIACIAN	ABU ROASH FM.	B			190'
LATE CRETACEOUS	TURONIAN-CONIACIAN	ABU ROASH FM.	C			85'
LATE CRETACEOUS	TURONIAN-CONIACIAN	ABU ROASH FM.	D			330'
LATE CRETACEOUS	TURONIAN-CONIACIAN	ABU ROASH FM.	E			325'
LATE CRETACEOUS	TURONIAN-CONIACIAN	ABU ROASH FM.	F			165'
LATE CRETACEOUS	LATE CENOMANIAN	ABU ROASH FM.	G		●	745'
LATE CRETACEOUS	EARLY CENOMANIAN	BAHARIYA FM.			●	760'
LATE CRETACEOUS	EARLY CENOMANIAN	BAHARIYA FM.	RAZZAK Mbr.		●	
EARLY CRETACEOUS	ALBIAN	KHARITA FM.				910'
EARLY CRETACEOUS	APTIAN	ALAMEIN FM.	DAHAB Mbr.		◑	425'
EARLY CRETACEOUS	APTIAN	ALAMEIN FM.	CARBONATE Mbr.		●	240'
EARLY CRETACEOUS	APTIAN		APTIAN SAND		●	2195'
EARLY CRETACEOUS	BARREMIAN	ALAM EL BUEIB FM.				3520'
EARLY CRETACEOUS	NEOCOMIAN	BETTY FM.			●	495'
JURASSIC	OXFORDIAN	MASAJID FM.			◑	630'
JURASSIC	CALLOVIAN TO BAJOCIAN	KHATATBA FM.			◑	1180'
JURASSIC	AALENIAN	EGHEI GROUP	WADI EL NATRUN FM.			1125'
PALEOZOIC		PALEOZOIC				1815'
PRECAMBRIAN		BASEMENT				

Figure 4. Stratigraphic section penetrated in Razzak field. Solid circles are posted against pay zones. *The Khoman Formation consists of beds of Campanian to Maastrichtian age. The Kharita, Alamein, and Alam El Bueib formations constitute the informal Burg El Arab formation.

exploratory wells, together with 13 shallow water wells, to obtain structural and stratigraphic information.

In 1963, Phillips Petroleum Company signed an exploration agreement with EGPC for an area covering about 96,000 km^2 (37,036 mi^2) that extended north of lat. 30°N into the Mediterranean and covered most of the area from the Libyan border to the River Nile. Aeromagnetic and extensive seismic surveys together with an active drilling program were conducted by Phillips (later WEPCO, Western Desert Operating Petroleum Company, a joint operating company between Phillips and EGPC) in the area. Prior to the Razzak discovery, WEPCO drilled the Kharita-1 wildcat (dry hole) 7.5 km (4.7 mi) to the east of what became Razzak field (Figure 2). The Alamein field, located 18 km (11 mi) northeast of the East Razzak field (Figure 1), was discovered by Phillips in December 1966, with the Alamein-1X well which flowed 8000 barrels of oil per day (BOPD) of 34.5° API oil from the Aptian "Alamein dolomite" reservoir (Abdine, 1974). This was the first commercial oil discovery in the Western Desert. In the same general area in July 1971, the Yidma oil field (Figure 1), located about 12 km (4.7 mi) northeast of the East Razzak field, was discovered by WEPCO with the completion of the well Yidma-1 as an Alamein dolomite producer (Deibis, 1978).

Discovery

In 1964, Pan American Oil Company (now AMOCO Production Company) signed an exploration agreement with EGPC that covered some 73,000 km^2 (28,162 mi^2). The area was located south of lat. 30°N and extended from Fayum Oasis west to long. 27°E (Figure 1). In 1969, AMOCO was granted the 28,000 km^2 (10,800 mi^2) "Nile Valley" concession south of lat. 28°N and a number of blocks that were relinquished by WEPCO north of lat. 30°N. AMOCO conducted extensive seismic work in this expanded concession and carried out an active drilling program that led to the discoveries of Abu Gharadig (AG) field and Razzak field. Abu Gharadig field was discovered in October 1969 in the Abu Gharadig basin south of lat. 30°N. Razzak field was discovered in March 1972, with the completion of the Razzak-1 (RZK-1) well, which tested 35° to 49° API oil from seven carbonate and sandstone zones in the Cretaceous section. The Aptian "Alamein dolomite" in Razzak field initially flowed 3974 BOPD. With the discovery of the Razzak field, the Cretaceous oil potential along the Qattara-Alamein ridge was well established.

The RZK-1 discovery well (initially designated NWD 349-1) was originally recommended to drill to a total depth of 2591 m (8500 ft) into the Lower Aptian clastic section below the Aptian "Alamein dolomite," which, as noted above, was previously found productive in the WEPCO Yidma and Alamein fields to the northeast. The well was to test an "asymmetric" anticlinal feature on the Aptian dolomite horizon with an estimated vertical closure of 76 m (250 ft) and an areal closure of approximately 8.4 km^2 (3.2 mi^2).

The well was spudded in January 1972 and drilled to a total depth of 3384 m (11,098 ft) into the Jurassic Khatatba Formation (Figure 4). The well tested oil from seven zones in an overall interval that extended from about 1738 (5700 ft) to 3320 m (10,890 ft). The intervals tested include:

1. Upper Cenomanian Abu Roash "G" member dolomite; recovered 40.6° API oil from a drill-stem test (DST) at 1738 to 1747 m (5701 to 5729 ft).
2. Lower Cenomanian Razzak Member sandstone of the Bahariya Formation; recovered 33.5° to 39° API oil and gas from two wireline formation tests at 1871 and 1888 m (6137 and 6194 ft).
3. Upper Aptian, Dahab Member sandstone; recovered 35.3° API oil from a wireline formation test at 2261 m (7415 ft).
4. Aptian, Carbonate Member (Alamein Formation) dolomite; recovered 70 barrels of 37° API oil from DST at 2321 to 2343 m (7612 to 7686 ft).
5. Lower Aptian sand member, upper zone (upper part of Alam El Bueib Formation); recovered 39 barrels of 49° API oil and oil and gas-cut mud from a DST at 2404 to 2413 m (7876 to 7917 ft).
6. Lower Aptian sand member, lower zone (upper part of Alam El Bueib Formation); recovered gas and 54° API condensate from wireline formation tests at 2579 and 2584 m (8460 and 8476 ft).
7. Neocomian–Barremian Betty Formation sandstone; recovered 10 bbl of 43° API oil from a DST at 3298 to 3320 m (10,818 to 10,890 ft).

Wireline formation analyses indicated that four separate hydrocarbon-water contacts (HWC) were present in the discovery well. An extended production test of the Alamein dolomite from perforations between 2333 and 2349 m (7652 and 7704 ft) flowed an average of 1116 BOPD of 39.2° API oil, with a low gas-oil ratio (GOR) of 52 ft^3 gas/barrel (cfg/bbl). The RZK-1 well was declared commercial and development/delineation drilling was initiated with RZK-2, which was spudded in April 1972 (Figure 2).

Post-Discovery

After the RZK-1 discovery, field development and drilling continued without interruption from April 1972 to April 1973 with the drilling of the wells RZK-2 through RZK-13. During this period, the Abu Roash "G" pay in the West Razzak field was discovered by the RZK-4 well, and Bahariya production was established in Razzak Main by the RZK-7 well. The RZK-6 and RZK-8 dry holes and marginal producers (RZK-3, RZK-5, and RZK-9) helped delineate the limits of the Razzak Main field. The RZK-10 dry hole was drilled 3 km (1.9 mi) northeast of the Razzak Main field to test a stepout location, while the RZK-11 dry hole was drilled about 8 km (5 mi) southeast of Razzak Main to test a separate prospect. Well RZK-12 was completed as an Abu Roash "G" producer in West

Razzak field, while well RZK-13 was drilled to test the Jurassic section (which recovered oil from a DST Masajid limestone).

Two more wells were drilled in Razzak Main field in 1974: RZK-16, which was completed in the Abu Roash "G," and RZK-14, which was drilled as a deep test in the field with a total depth of 3456 m (11,337 ft) in the lowermost Cretaceous, completed as a Bahariya producer.

From March 1978 to December 1978, RZK-15 through RZK-21 wells were drilled. The RZK-15 well established the Bahariya production in the East Razzak field. RZK-17, -18, -19, and -21 were all completed as Bahariya oil producers in East Razzak. The RZK-20 dry hole helped establish the southwestern limits of the East Razzak field.

In 1981, the RZK-23 dry hole established the northern limit of the East Razzak field and the RZK-22 (formerly NWD 350-1) was drilled to test a prospect between the Razzak Main and East Razzak fields. This well was also a dry hole. No drilling has taken place in the Razzak field complex since 1981.

A maximum of 11 wells were producing in West Razzak and Razzak Main (Figure 5) during the period 1973 to mid-1976. These wells were originally completed as Abu Roash "G" producers (RZK-4, -12, -13, and -16), Bahariya producers (RZK-7, -9, -13, and -14), and Alamein producers (RZK-1, -2, -3, and -5). A maximum of five wells were producing in the East Razzak field (Figure 6) in 1979 and 1980. All these produced from the Bahariya Formation and include RZK-15, -17, -18, -19, and -21.

At the present time throughout the entire field complex, five wells are producing from the Abu Roash "G," RZK-4, -9, -12, -14, and -15, while three wells produce from the Bahariya, RZK-17, -18, and -21 (comingled with the Abu Roash "G"); RZK-1 produces from the Dahab Member, while RZK-2 produces from the Alamein dolomite. The presently producing wells have in many cases been recompleted in zones other than the original producing zones.

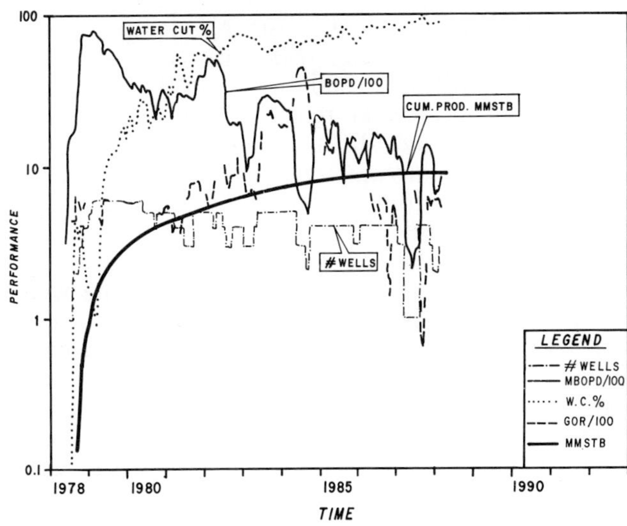

Figure 6. Production performance vs. time, East Razzak field.

The development program consisted of drilling vertical wells, usually through the Alamein Formation in Razzak Main and West Razzak fields, and within the Upper Kharita Formation in the East Razzak field. There was no regular well spacing pattern. In general, 20-in. surface casing was set at 91 m (300 ft) below ground level; 13⅜-in. casing set between 914 and 1220 m (3000 and 4000 ft), usually within the Apollonia limestone section; and 9⅝-in. casing set to total depth in producing wells. The wells were all drilled with water-based mud, and typically production was tested through perforations. For the Bahariya producers, multiple zones were perforated. These zones were variable in development among wells, while the entire Abu Roash "G" pay was perforated in the "G" producers. Only the upper Alamein dolomite pay was perforated in the Alamein producers. The Alamein and Abu Roash "G" dolomites were usually acidized.

Wireline log surveys in all the wells included standard spontaneous potential, gamma ray, neutron, density, sonic, dipmeter, dual induction/laterolog and proximity/micrologs. Wireline formation tests were also occasionally conducted in zones of interest.

In the West Razzak and Razzak Main field complex, production commenced in mid-1972 and reached a peak production rate of about 20,000 BOPD in mid-1975 (Figure 5). Production rates averaged between about 15,000 and 20,000 BOPD from mid-1973 to the end of 1975. By early 1976, cumulative production was about 20.8 MMBO. From 1976 to mid-1986, the average daily production rate steadily declined to about 1600 BOPD, with the cumulative production near the end of 1986 being about 40 MMBO (Figure 5). During the early production years, until about

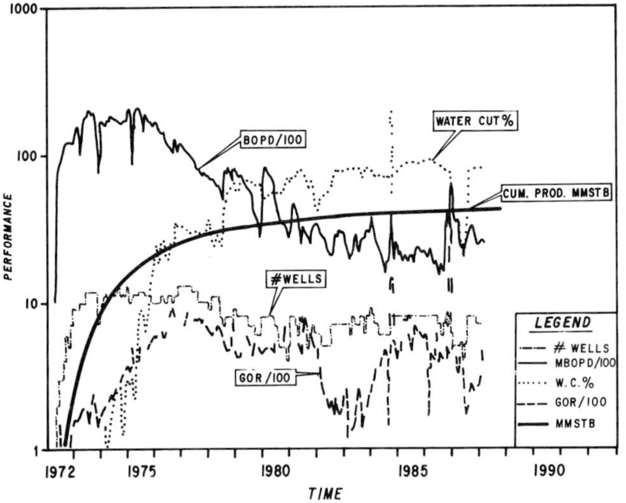

Figure 5. Production performance vs. time, West Razzak and Razzak Main fields. MBOPD, thousand barrels oil per day; MMSTB, million stock tank barrels; GOR, gas-oil ratio.

mid-1975, the water cut by volume generally remained below 3%. After this time, the water cut increased significantly (mainly within the Bahariya and Alamein reservoirs), reaching over 30% in 1976 and 1977. From mid-1978 to the present, the water cut in the West Razzak/Razzak Main field complex has risen steadily from 50 to over 90%. It should also be noted that in June 1982, a dump flood project was initiated using Bahariya water from the RZK-3 well to flood the Abu Roash "G" pay in the same well and to enhance Abu Roash "G" production in the RZK-16 well. This project has since been discontinued because of the flooding out of the RZK-16.

In the East Razzak field, Bahariya production started in May 1978 and reached a peak production rate of about 8000 BOPD early in 1979 (Figure 6). By mid-1979, cumulative production was about 2.0 MMBO. Since mid-1979, production has declined in a very discontinuous manner. The high and low production spikes (Figure 6) mainly depend on well workover schedules and other operational activities in the field. By mid-1986, the average daily production rate was about 1500 BOPD from four wells; cumulative production was 8.5 MMBO (Figure 6). During the peak production period through mid-1979, the water cut, though variable, remained below 10%. Thereafter, the percentage water cut rose steadily to the present level of 80 to 90%.

For the entire complex, the current production rate (June 1988) is about 3900 BOPD. The cumulative production to date is 48 MMBO. The present (and overall) production decline rate for Razzak Main and West Razzak is 23%, while East Razzak is 15%.

DISCOVERY METHOD

Stratigraphic studies and basin analysis were carried out in AMOCO blocks in the Western Desert beginning in 1969 and through the early 1970s. These studies were based primarily on subsurface well data and extensive geophysical seismic reflection surveys, initially using 10-fold analog common-depth-point stacked sections with "Vibroseis" as an energy source (Ezzat, 1972, 1975). Digital "Vibroseis" crews were mobilized in 1969 and 1970 in the area. Available gravity and aeromagnetic surveys complemented the seismic data.

It was clear from the early stages of exploration that the relatively deep-seated pre-Tertiary structures, near basinal deeps, were the most likely exploration targets. While some Tertiary rejuvenation of such structures is common, many are not reflected in the Tertiary section. Thus, it was important to map the pre-Tertiary events. In the Razzak-Alamein area, after the Razzak discovery, new data allowed mapping of seismic reflectors at the base of the Tertiary and the top of the Cenomanian (Abu Roash "G" member) and the Alamein Carbonate Member.

Structural mapping of the Alamein and top of the Cenomanian events in the Razzak area (designated as the "Q" block in the early 1970s) indicated the presence, at the Alamein horizon, of three separate anticlinal features or culminations oriented along a southwest to northeast trend, on line with the Yidma and Alamein oil fields (Figure 7). The RZK-1 well (formerly NWD 349-1) was recommended to test the central anticline, which was subsequently developed as the "Razzak Main" field. The southwestern culmination was later developed as the "West Razzak" field. The northeastern closure (NWD 350-1) was not found productive, but the East Razzak field was later mapped and developed some 3 km (1.9 mi) farther to the northeast of the early interpreted high.

The tested structure was originally described as an anticlinal seismic feature, bounded to the northwest and southeast by down-to-the southeast faults and to the south by a down-to-the south fault. Vertical closure was assumed to be about 76 m (250 ft) (50 milliseconds) at the Alamein horizon, with the area of closure to be 2.4 km^2 (2076 ac). It was recognized that the feature was on trend with established production in the Yidma and Alamein fields to the northeast and with the Qattara Rim-1 well, 23 km (14.2 mi) to the southwest, which tested 20° API oil from Bahariya sands (Figure 3). The Alamein carbonate section was considered the primary target. A stratigraphic-log cross section A–A′ (Figure 2) through Razzak field is shown in Figure 8.

It is evident that reflection seismic mapping of pre-Tertiary and Cretaceous (and, if possible, older) events was essential to the discovery of the field. There are no definitive magnetic or gravity anomalies in the field area, and the deep-seated Razzak field culminations are not reflected at the surface. Deep well data provided the main source of information for compiling the stratigraphic framework and sedimentary history of the Western Desert, as the known outcrops are scarce (except for the Tertiary, the upper Bahariya section that crops out in the Bahariya Oasis [Figure 1], and a small Abu Roash outcrop just west of Cairo).

STRUCTURE

Regional Structure

The Qattara-Alamein ridge or structural axis is located to the north of the Sharib-Sheiba structural platform (Figure 3) that separates the Abu Gharadig basin to the south from the Alamein basin to the north.

The Alamein basin is a somewhat broad term used to describe an extensive structural downwarp (or intracratonal basin) along the northern unstable shelf of the Arabian-Nubian shield developed primarily during the Early Cretaceous. Jurassic sediments, although present in the Alamein-Razzak area, appear to have their thickest and most well-

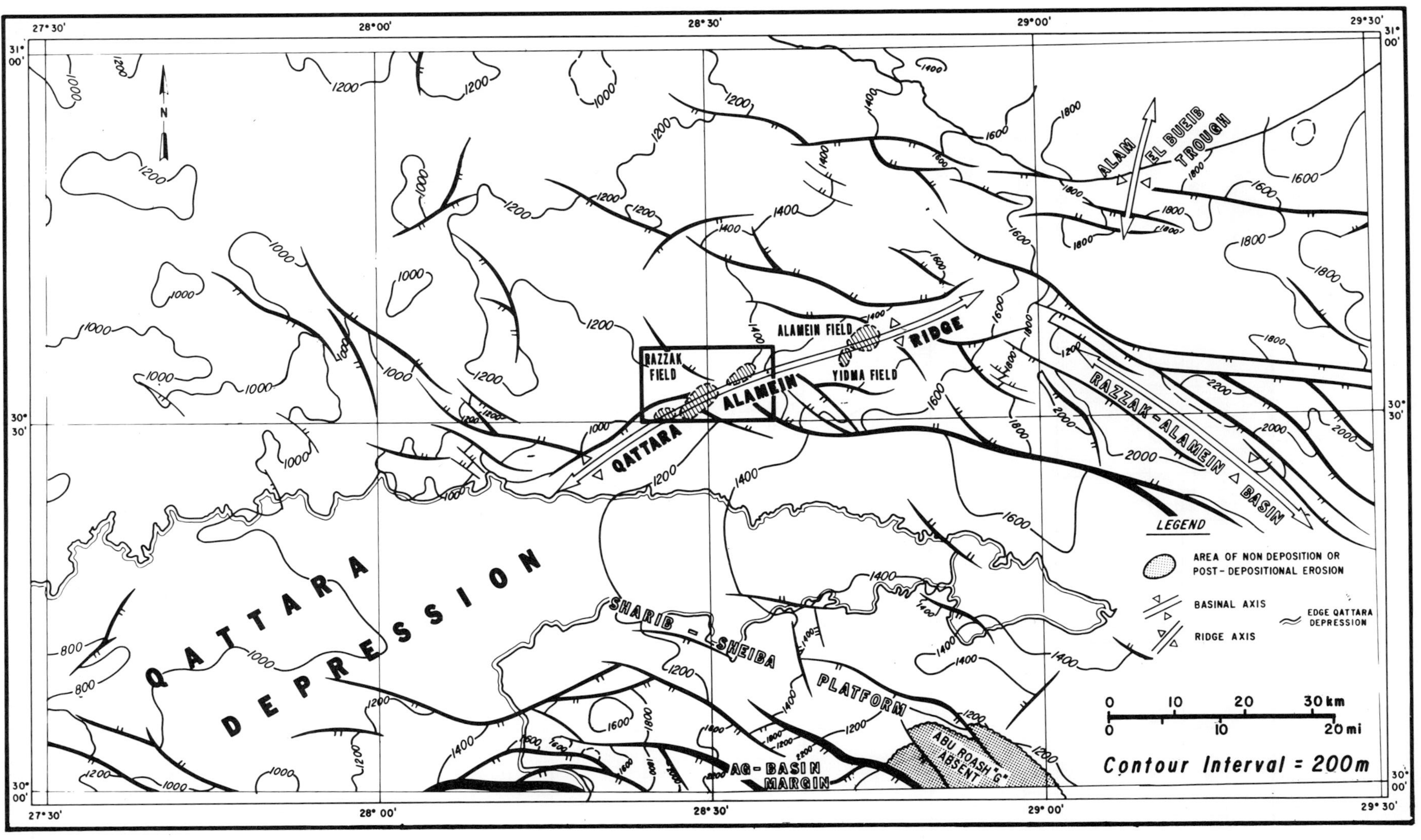

Figure 7. Top Abu Roash "G" structure map, showing Razzak-Alamein basin tectonic setting. The locations of three separate anticlinal features forming the Razzak field complex are shown oriented along a southwest to northeast trend, on line with the Yidma and Alamein oil fields. Fields shown by crosshatching. The map is derived by converting migrated seismic data to meters.

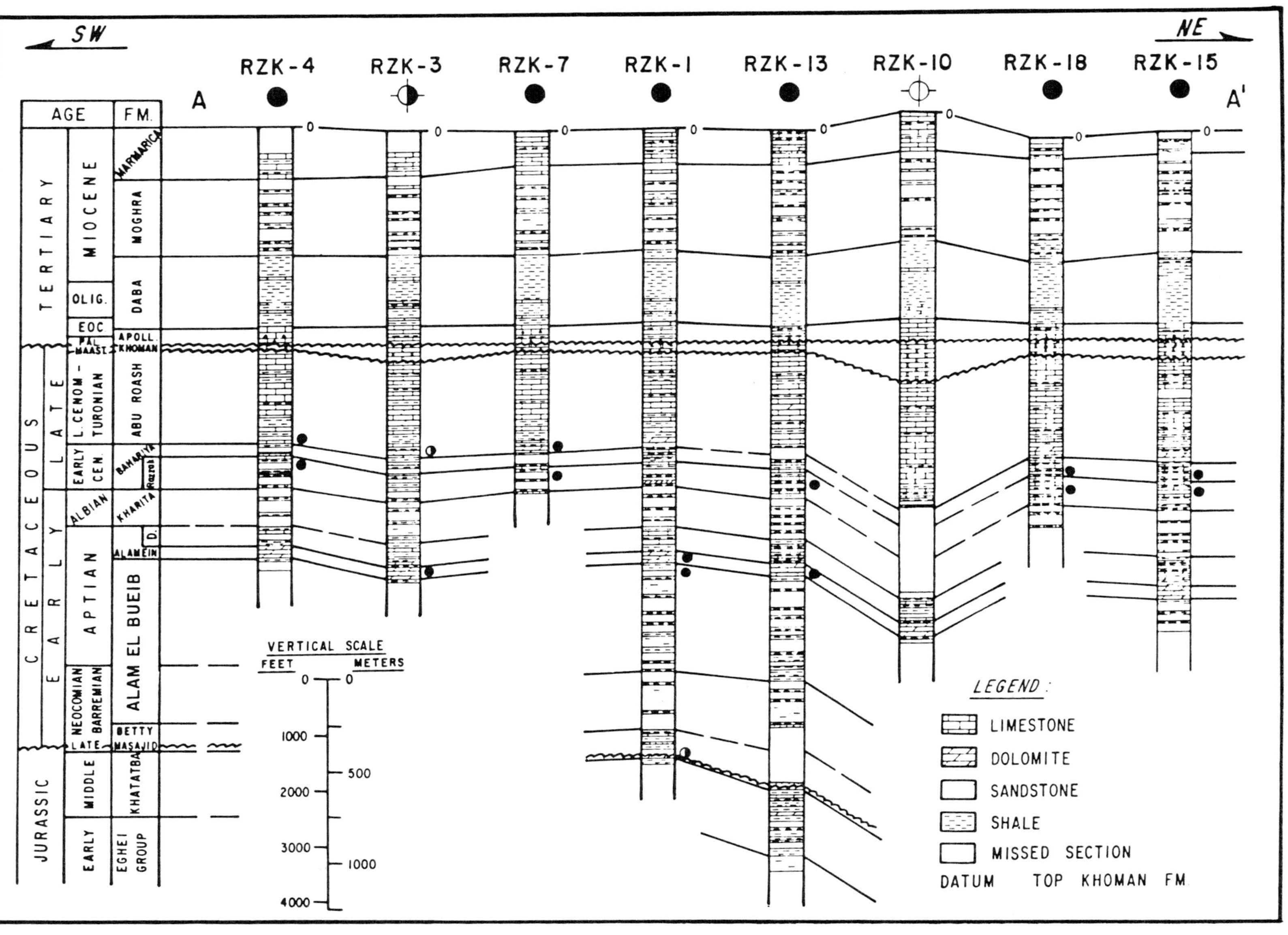

Figure 8. Stratigraphic cross section A–A′ (Figure 2) through Razzak field. Datum line is the pronounced unconformity on top of the Khoman Formation.

developed section in one basinal depocenter located west of Cairo and in another to the northwest along the Mediterranean coast east of Mersa Matruh (RRI, 1982). The Cretaceous-age Alamein basin extends from east to west along most of the northern tier of the Western Desert, with the thickest Neocomian through Albian section centered immediately to the north, or northeast, of the Razzak-Alamein area. The Qattara-Alamein ridge existed during the Cenomanian and Turonian, with the thickest upper Cretaceous section in the area centered relative to both to the east-northeast and west-northwest of the ridge. A Senonian downwarp developed to the north-northeast of the ridge which was cut by a series of northwest-southeast-oriented normal faults. Later, during the Tertiary, the area was part of a broad platform that was subject to only minor faulting and folding and on which relatively thin sediments were deposited compared to the Abu Gharadig basin to the south.

A simplified regional tectonic setting map (Figure 7) of the Razzak area (based on an available Abu Roash "G" time map) shows a dominant northwest-southeast structural grain in the area defined by most of the major fault trends and the Razzak-Alamein graben (basin). Conjugate to this trend is a northeast-southwest-oriented series of faults and structural axes, of which the Qattara-Alamein ridge is a prominent example. As noted above, the maximum growth along the ridge was during the Late Cretaceous.

Local Structure

A detailed time map of the Razzak field complex (Figure 9) at the Alamein carbonate level shows the three separate structural culminations of West Razzak, Razzak Main, and East Razzak. The maximum structural closure of all three culminations is to the southeast and on the downthrown side of the major northeast-southwest-trending "S" fault, which bounds all three field areas to the northwest. The "S" fault has both reverse and normal throw along its length, and the West and East Razzak culminations appear to be formed by southeast to northwest compression (Figures 10 to

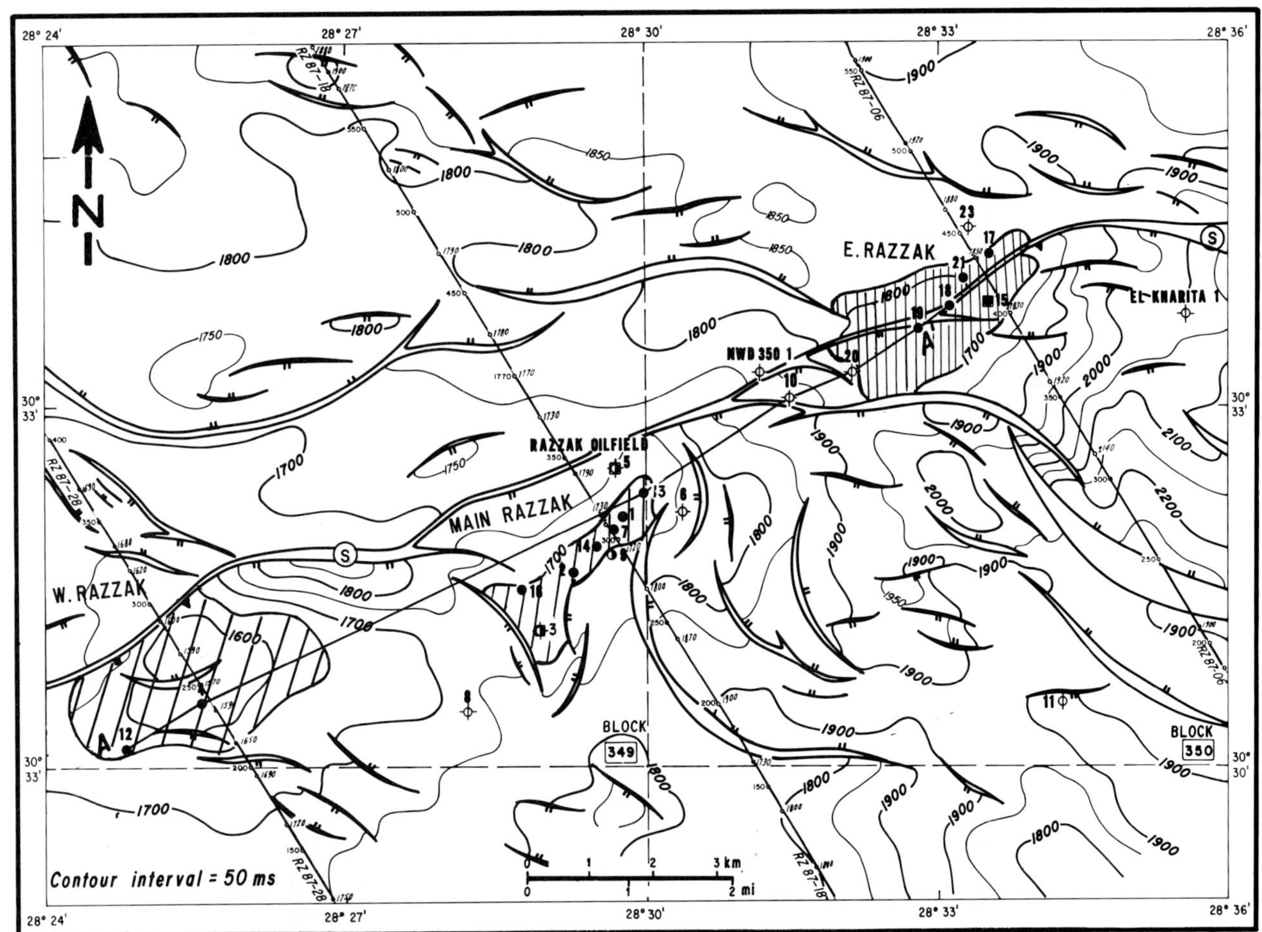

Figure 9. Time map of top Alamein Carbonate Member showing the three separate structural culminations of West Razzak, Razzak Main, and East Razzak fields. The maximum structural closure is to the southeast. Seismic lines RZ 87-28, RZ 87-18, and RZ 87-06 are profiled in Figures 10, 11, and 12. A–A′ is the line of cross section of Figure 13.

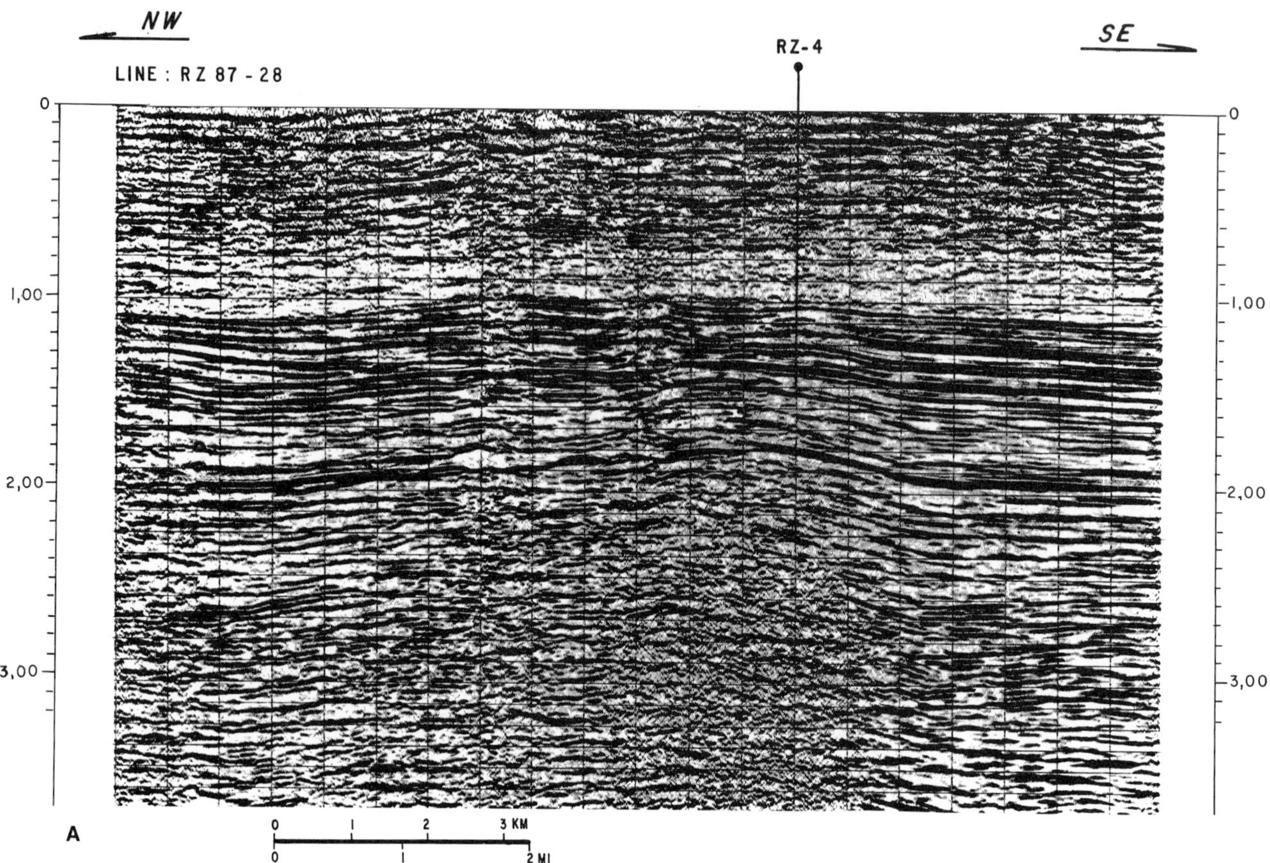

Figure 10. Seismic profile on line RZ 87-28 (Figure 9) showing that the West Razzak structure has culminated by a southeast to northwest compression. The Lower Cretaceous section is significantly thicker on the downthrown side of the "S" fault. (A) Uninterpreted. (B) Interpreted. Vertical scale is 2-way time in milliseconds.

12). A clear extensional axis oriented northeast to southwest is present in the area and is reflected by numerous northwest-oriented normal faults. Another point of interest is that the field complex is developed over a structural trough southeast of the "S" fault in which the Lower Cretaceous section is significantly thicker than that along the upthrown side of the fault (Figures 10 to 12). It is postulated that Cretaceous wrench fault movements along the "S" fault caused a structural "rebound" of this trough that gave rise to the Razzak field culminations. These become progressively higher structurally from northeast to southwest.

STRATIGRAPHY

The stratigraphic succession penetrated in the Razzak field ranges in age from Miocene to Early Jurassic and has a total thickness of more than 3963 m (13,000 ft) (Figure 4).

The Jurassic section has been divided by Robertson Research International (RRI, 1982) into three units:

1. A lowermost unit is the Bahrein Formation, which consists of a massive, predominantly continental sandstone section.
2. A middle section consists of interbedded siltstones, sandstones, shales, and some carbonates of the Khatatba Formation, which is about 358 m (1175 ft) thick. The sandstones predominate toward the bottom of the section. The depositional environment is shallow marine and the Khatatba Formation represents a marine transgression into the area. It is conformable with both the Bahrein Formation below and the overlying Masajid Formation.
3. An upper carbonate unit, the Masajid Formation, is about 24 m (80 ft) thick and consists of dense dolomites, limestones, and dolomitic limestones with shale intercalations. The unit has an open marine, neritic depositional environment and represents the maximum extent of Late Jurassic marine transgression in the area. A regional unconformity separates the Jurassic from the overlying Lower Cretaceous section.

The Cretaceous section consists, in a broad sense, of four alternating sedimentary cycles (RRI, 1982).

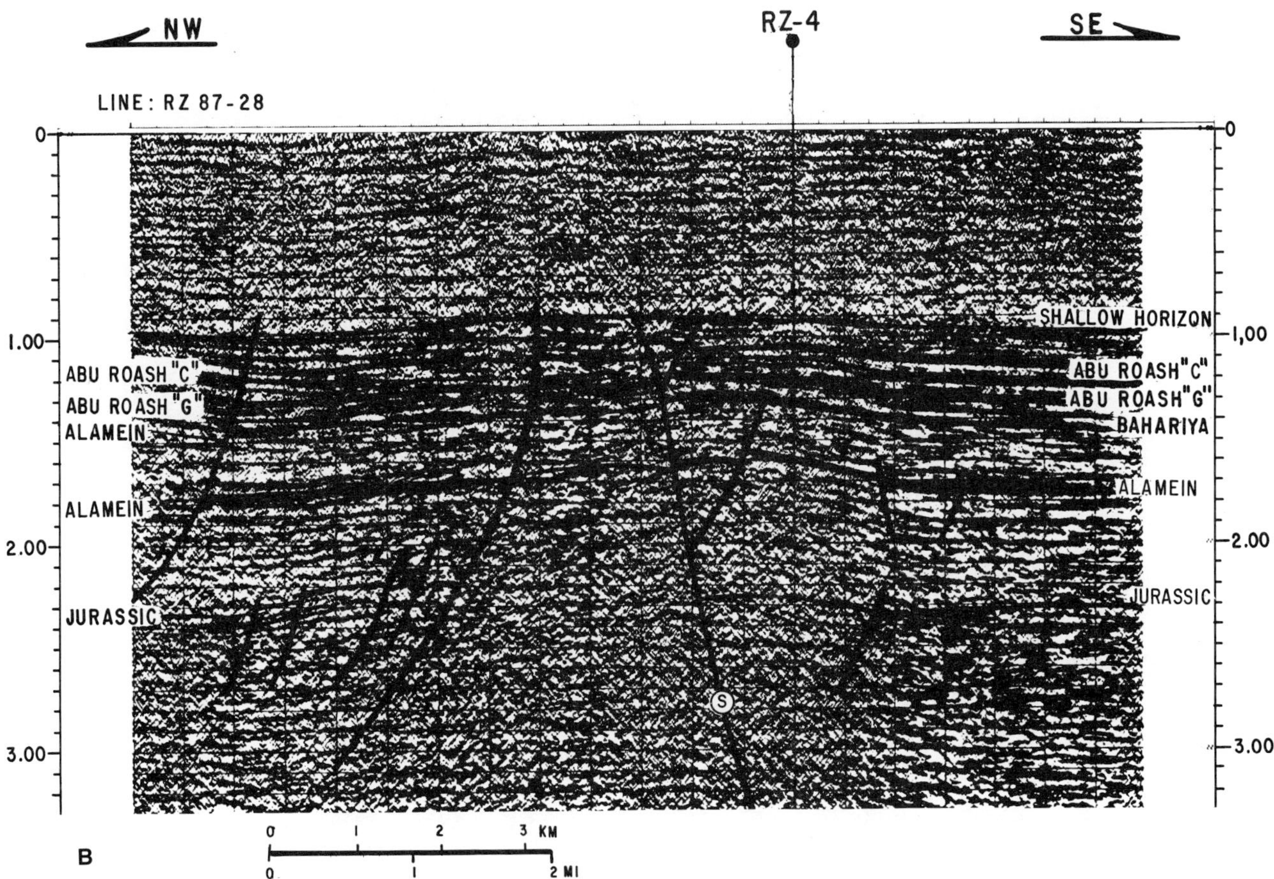

Figure 10. Continued

The first and third cycles, from the bottom, consist of predominantly massive sandstones with thick interbeds and intercalations of shales in some places. The depositional environments of these clastics range from open marine to sublittoral, continental, or estuarine, representing relatively rapid basinal downwarp, with clastics introduced into the sedimentary section from the exposed Afro-Arabian shield to the south.

The second and fourth cycles consist primarily of open to shallow (and possibly in part restricted) marine carbonates, which represent relatively quiet shelf conditions.

The first cycle has been divided into the following units in ascending order (Figure 4):

- The Neocomian Betty Formation, which is about 140 m (460 ft) thick and consists of interbedded varicolored shales, sandstones, and sandy shales. It was deposited under sublittoral to neritic marine conditions.
- Overlying the Betty unconformably is the Alam El Bueib Formation (Barremian), 320 m (1050 ft) thick and composed of marginal marine to deltaic sandstones with some carbonate and greenish-gray shale intercalations.
- The upper part of the Alam El Bueib Formation is the Aptian sand member (lower Aptian), which is some 350 m (1650 ft) thick and consists of a sequence of thick sandstones with some shale interbeds, especially toward the base. The depositional environment is considered to be mainly shallow marine to occasionally sublittoral.

The second cycle consists of the Alamein Formation that has been divided into two members:

- The Alamein dolomite of the Alamein Carbonate Member consists mainly of carbonates with subordinate shales. The unit ranges from 72 to 75 m (235 to 245 ft) thick and can be subdivided into a lower and an upper horizon (Metwalli et al., 1982). The lower horizon consists of a mixed limestone/dolomite sequence, while the upper horizon is mainly a vuggy, fractured, and finely crystalline dolomite. The depositional environment is shallow to possibly, in part, restricted marine.
- The Dahab Member, some 79 m (260 ft) thick, consists of shallow marine to sublittoral interbeds of shales, sandstones, siltstones, and dolomitic limestone. This member generally forms the seal for the underlying carbonate reservoir.

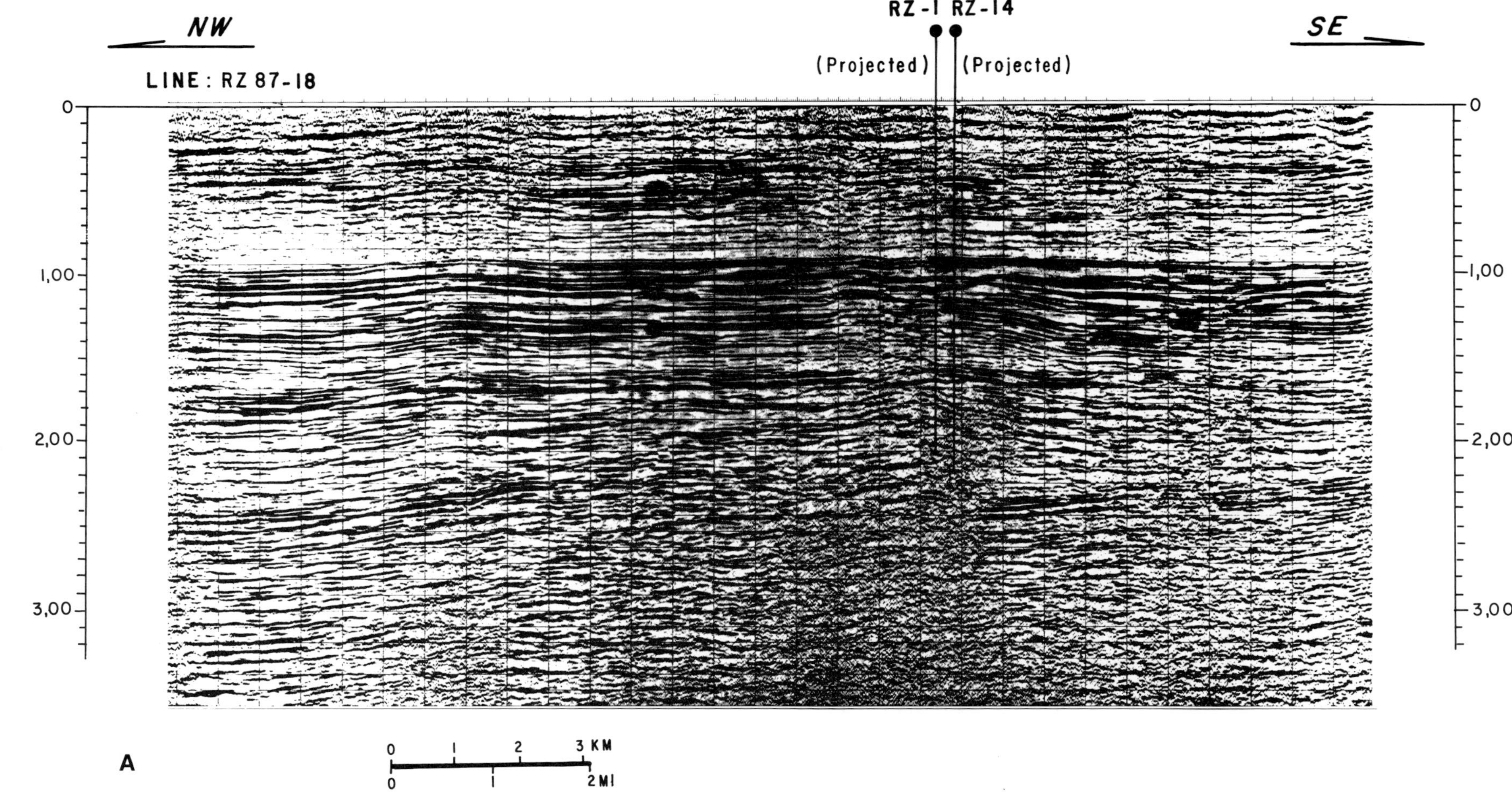

Figure 11. Seismic profile on line RZ 87-18 (Figure 9) illustrating a significantly thicker Lower Cretaceous section on the downthrown side of the "S" fault. (A) Uninterpreted. (B) Interpreted. Vertical scale is 2-way time in milliseconds.

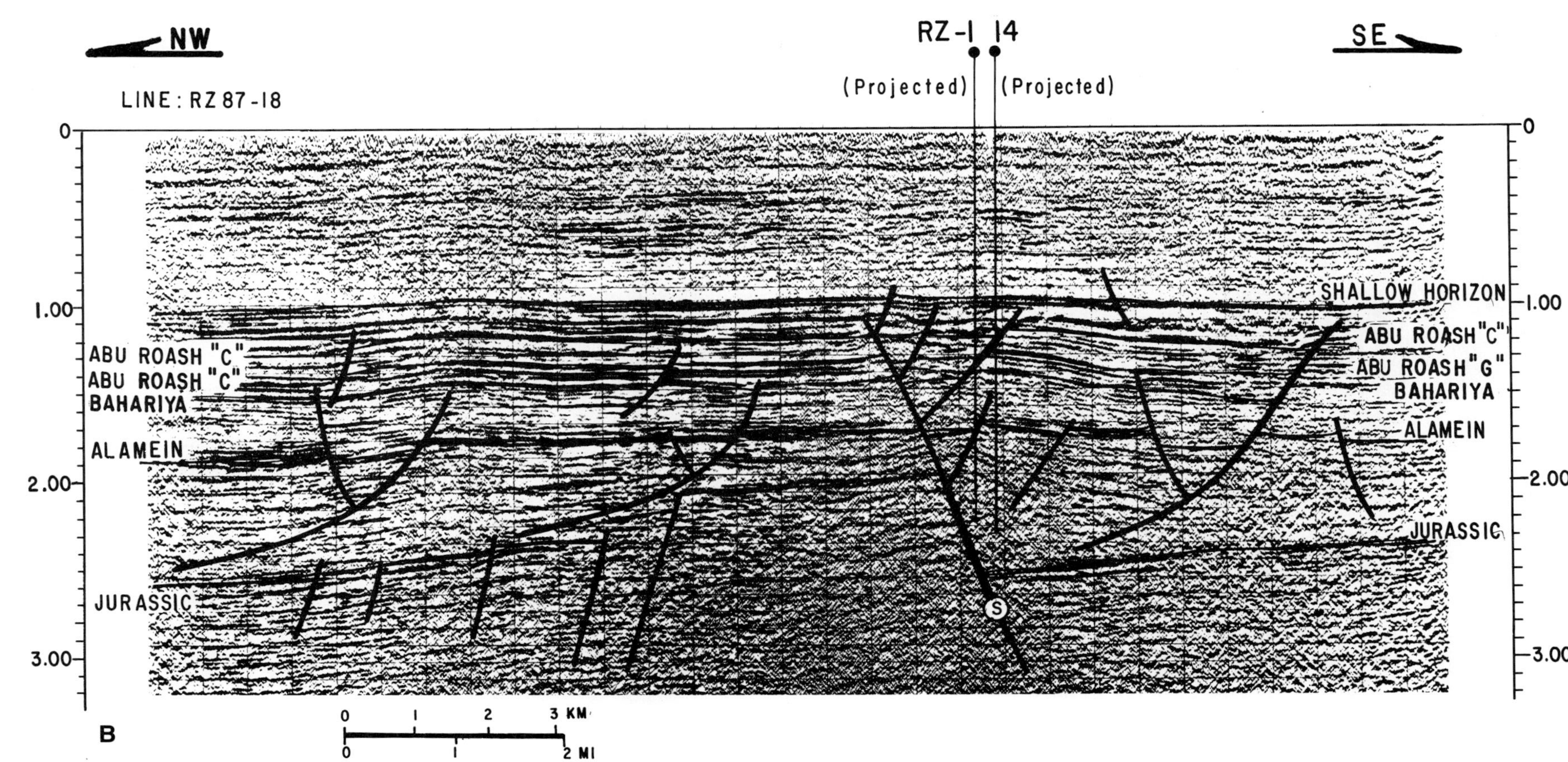

Figure 11. Continued

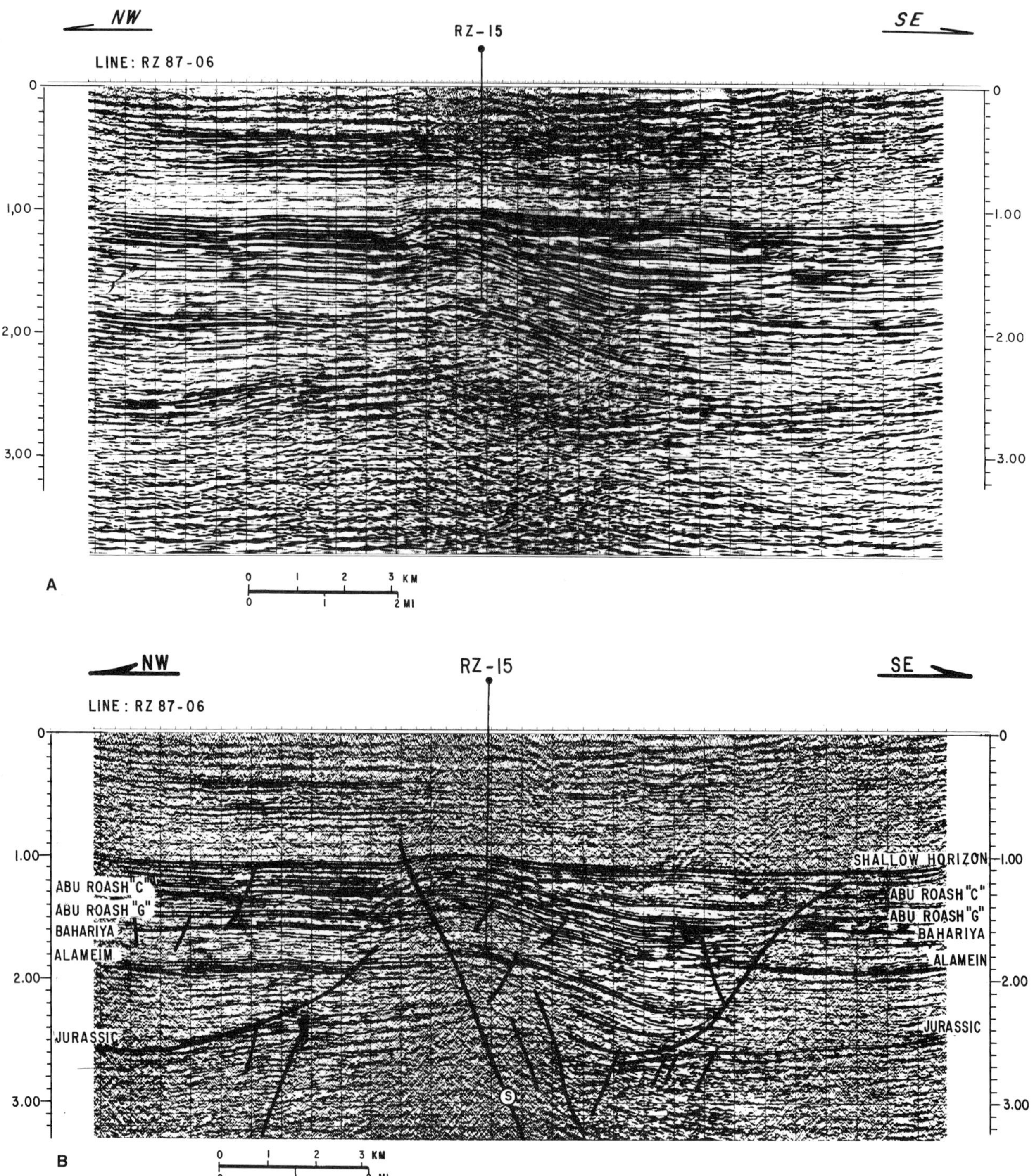

Figure 12. Seismic profile on line RZ 87-06 (Figure 9) showing that the East Razzak structure has culminated by a southeast to northwest compression. The Lower Cretaceous section is thickest on the downthrown side of the "S" fault. (A) Uninterpreted. (B) Interpreted. Vertical scale is 2-way time in milliseconds.

The third (clastic) cycle has been divided in ascending order into the following:

- Conformably overlying the Alamein Formation, the Albian Kharita Formation, some 274 m (900 ft) thick, is composed of marginal marine to deltaic sandstone with shale and rare carbonate interbeds.
- Unconformably overlying the Kharita Formation is the upper Albian to lower Cenomanian Bahariya Formation (including the Razzak Member), which is 213 m (700 ft) thick in the area. The formation is composed of interbedded sandstones, shales, and sandy shales, with occasional limestone stringers. The sandstones increase in thickness and number

toward the base of the section. The environment of deposition changes from shallow marine at the base to deep marine toward the top.

The fourth (carbonate) cycle consists of the Upper Cenomanian to Coniacian Abu Roash Formation, which is up to some 640 m (2100 ft) thick in the Razzak area and conformably overlies the Bahariya Formation. The Abu Roash is, in turn, unconformably overlain by the Campanian to Maastrichtian age, open marine, fine-grained chalky limestones of the Khoman Formation.

The Abu Roash has been subdivided into the members "A and B undifferentiated," "C," "D and E undifferentiated," "F," and "G." These represent cyclic shallow marine to open marine depositional environments. The Abu Roash "A and B undifferentiated" at the top of the formation is composed of limestones with shale and fine-grained lime silt and mud interbeds and was deposited under fluctuating high energy, shallow marine to relatively deep marine, transgressive conditions. The Abu Roash "C" member is composed of calcarenites with silty shale interbeds. The Abu Roash "D and E undifferentiated" consists of dense, glauconitic, chalky limestones with dolomitic crystalline limestone and shale intercalations. The Abu Roash "F" member consists of a thick pyritic, cream-colored calcarenite with abundant open marine fauna and forms a widespread marker in the Western Desert. The lower Abu Roash "G" member consists mainly of interbedded limestones and shales with a thin dolomite unit at the base that is oil bearing in the West Razzak and Razzak Main fields.

TRAP

The oil in the Razzak field is considered to be structurally trapped through a combination of structural dip and faulting (Figure 13). Separate oil columns have been noted in the Abu Roash "G" dolomite/Bahariya sandstones; in the Dahab Member sandstones; in the Alamein Carbonate Member/uppermost Aptian sandstone member; in the lower Aptian sandstone member; and in the thick sandstones of the Betty Formation. Seals to vertical migration for the Dahab Member and all the other underlying reservoir sections depend on relatively thin overlying shales or tight siltstones. Consequently, the traps for these reservoir sections, which are separated by thick porous sandstones, are very dependent, or sensitive to, the throw on the numerous faults that cut the Razzak field culminations. The Abu Roash "G" member and underlying Bahariya Formation are, perhaps, less sensitive to the fault throws since these units are capped by the thick impermeable carbonates and shales of the Abu Roash Formation.

The generalized northeast to southwest cross section of the field complex (Figure 13) indicates that each culmination in the field has a separate HWC. West Razzak has a HWC at a depth of 1607 m (5270 ft) for the Abu Roash "G"/upper Bahariya section. Razzak Main field has at least three HWCs at 1730 m (5673 ft) for the Bahariya, at 2217 m (7273 ft) for the Alamein/upper Aptian sand, and possibly at 3126 (10,253 ft) for the Betty Formation. East Razzak has a HWC at 1848 m (6061 ft) in the lower Bahariya section.

Reservoir

A number of stratigraphic units have proven to be hydrocarbon bearing in various wells in the field complex. These units include the lower Abu Roash "G" dolomite; the upper sandstones of the Bahariya Formation and the Razzak Member (the lower sandstones of the Bahariya Formation); the Aptian Alamein Carbonate Member, the lower Aptian sandstone, and the Neocomian Betty Formation sandstone. Depths to the producing reservoirs among the Razzak field structures vary with the stratigraphic position of the pay horizons as well as with the structural depth differences within the individual culminations (Figure 12).

For example, the depth to the Bahariya sandstone pay in Razzak Main field is about 1700 m (5575 ft), while that in the East Razzak field is about 1773 m (5815 ft). Also, the depth to the Alamein pay within the same Razzak Main varies from about 2155 m (7070 ft) to 2195 m (7200 ft). Petrographic characteristics of the oil-bearing rocks of Alamein, Bahariya, and Abu Roash formations in the Razzak field area are taken from Metwalli et al. (1982).

Reservoir Characteristics and Facies

Alamein Carbonate Member (Late Aptian)—The net thickness of the "clean" Alamein dolomite pay equivalent in the field complex ranges from about 38 to 69 m (125 to 225 ft) (Figure 14). The perforated producing intervals average 24 m (80 ft) with a maximum of 46 m (150 ft). The porosity is both intercrystalline and vuggy. Porosity in the "clean" section ranges from about 4 to 10% (Figures 15 and 16).

Permeability ranges from 33 to 222 md. Water saturation in the pay zones averages about 15%, and the oil recovery factor is about 33% through an active water drive.

Based on four cores from wells RZK-1 and RZK-3, the Alamein dolomite of Razzak field is a massive, hard to very hard dolostone and varies in color from creamy white and tan to relatively dark brown. Mostly, it is medium-grain sized crystalline, fractured, and vuggy. The vugs are lined and partially or completely filled with secondary calcite. X-ray analysis confirmed the presence of calcite.

The sparry calcite rhombs filling vugs within the finely crystalline Alamein dolomite may show bird's-

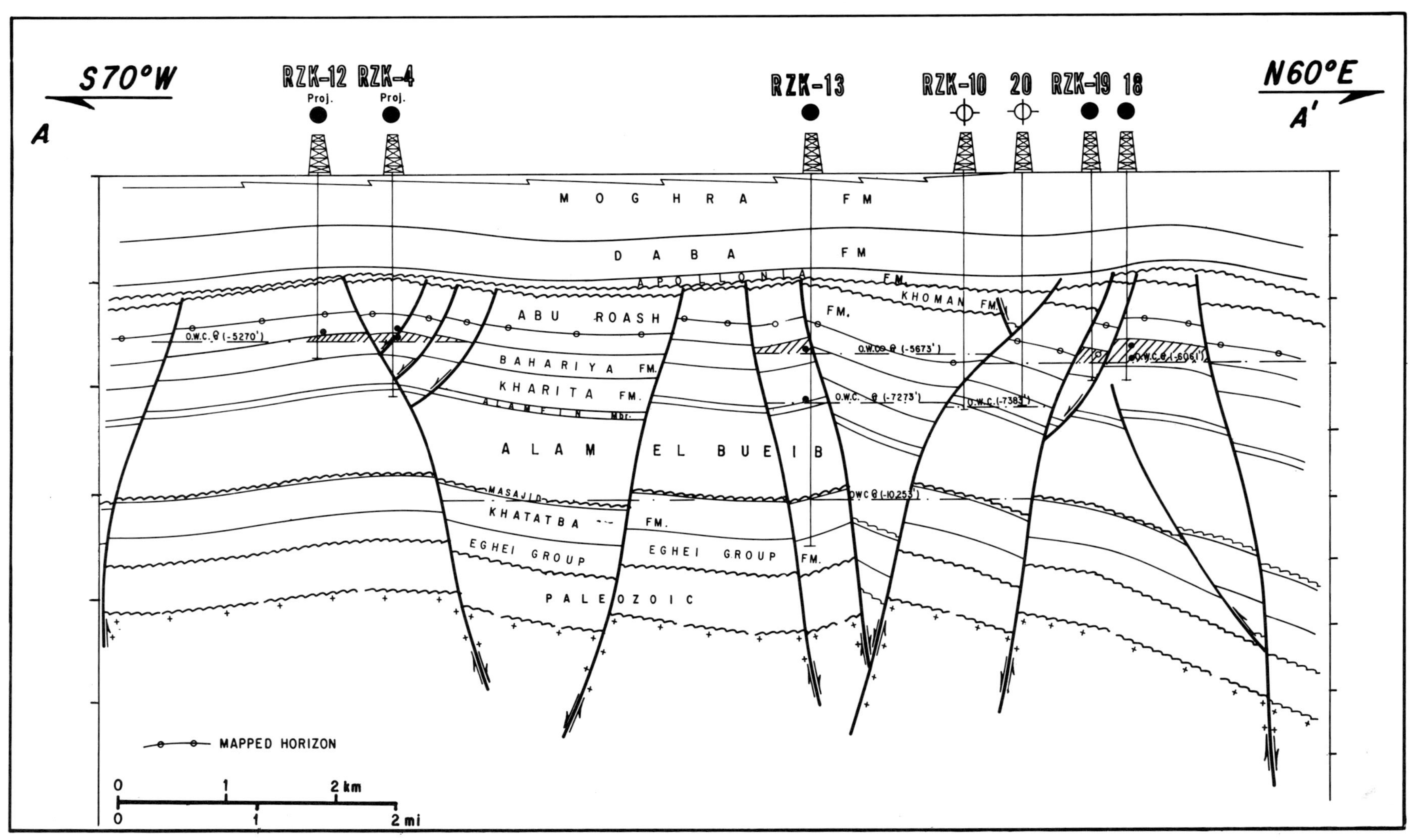

Figure 13. Regional structural cross section A–A′ (Figure 9) through Razzak field. The oil in Razzak field is considered to be structurally trapped through a combination of structural dip and faulting.

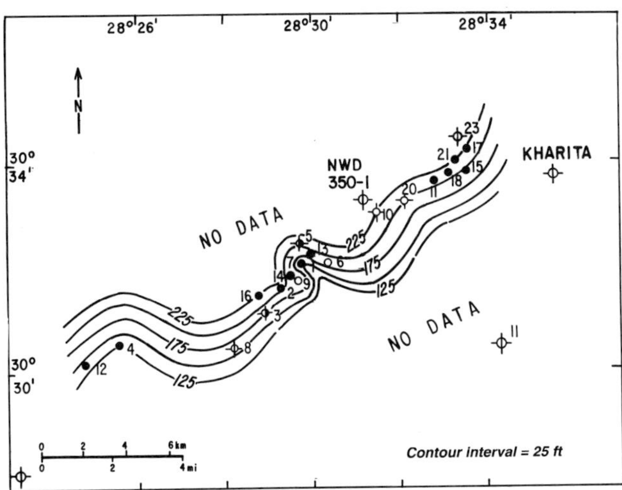

Figure 14. Net pay isopach map of "clean" Alamein dolomite. Porosity cutoff, 4%. Alamein dolomite varies in thickness from about 38 to 69 m (125 to 225 ft). The producing zone averages 25 m (80 ft), with a maximum of 46 m (150 ft).

eye structure (Hamed, 1972). Pyrite is also recorded as fissure or fracture fillings in the medium- to coarsely crystalline dolomite. In this case, it may replace the rhombs lining the fracture.

The Razzak Member of the Bahariya Formation (Late Albian)—The net thickness of reservoir quality sandstone in the Bahariya Formation in the field complex ranges from 46 to 91 m (150 to 300 ft) (Figure 17). The thickness of the producing zone is 14 m (46 ft), with the maximum being 30 m (100 ft). Porosity values in the pay section range from 21 to 25% in Razzak Main to about 26% in East Razzak (Figures 18 to 20). In Razzak Main, the water saturation is about 15%, and the permeabilities range to 400 md. In East Razzak, the initial water saturation is about 30%, and the permeabilities range up to 230 md. The recovery factor in Razzak Main field is about 30% through an active water drive; in East Razzak field it is about 20% through a partial water drive.

Cores from well RZK-16 show that the Bahariya Formation is represented by a moderately well-sorted, glauconitic sublitharenite (Folk, 1974; Pettijohn, 1975). The rock consists of fine-grained quartz, with an average grain size of 0.126 mm. The glauconite (average size 0.189 mm) is abundant, ranging between 45 and 57% of the framework of the studied sandstones (average 53%), which renders the rock a green color. Glauconite varies in color from pale to dark grass green in color, grading into yellowish brown, indicating oxidation to limonite. Most of the glauconite grains are produced by diagenetic alteration of lime mud, often present as rounded pellets (Figure 21). Micrite is also affected by aggrading recrystallization to reach as much as 15 microns crystal size. Glauconite grains of detrital origin are also recorded. These show the effect of transportation and are embraced by diagenetic glauconite, indicating the existence of more than one generation of the mineral.

Abu Roash "G" Member (Late Cenomanian)—The net thickness of the Abu Roash "G" dolomite pay zone ranges from about 3.6 to 5.5 m (12 to 18 ft) (Figure 22), and the porosities range from 26 to 36% (Figures 23 and 24). Porosity is intercrystalline and vuggy. In Razzak Main the water saturation is about 23%, while the permeabilities range up to 300 md. The recovery factor in this area is about 19% through a depletion drive. In West Razzak, the water saturation is also about 23%, and the permeabilities range up to 145 md. The recovery factor in West Razzak is 23.3% through a depletion drive.

Two cores were cut, representing the lowermost part of the Abu Roash "G" member, 18 m (60 ft) from well RZK-5 and 12.8 m (42 ft) from well RZK-12 in this member. The two cores consist of micrite and biomicrite at the top. The micrite is formed of calcite crystals 3–5 microns in diameter and is thus equivalent to type III of Folk (1974), mudstone of Dunham (1962), and the carbonate mudstone group of Pettijohn (1975).

The micrites were subjected to aggrading neomorphic recrystallization (Bathurst, 1971), resulting in a mosaic of sparry calcite with an average size of 44 microns, forming a bird's-eye structure. It is also partially dolomitized in well RZK-12, with perfect dolomite rhombs averaging 29 microns in diameter. Dolomitization increases with the relative increase in the pyrite content. These varieties were deposited in a tranquil water environment of low-energy conditions that were not capable of removing the clay-sized material.

Source

Potential Source Rocks

Total organic carbon (TOC) and pyrolysis results from available samples in the Razzak area (wells RZK-13 and RZK-14, Figure 2) suggest that the Jurassic Khatatba Formation (four samples) and Betty Formation (one sample) have marginal to excellent hydrocarbon source potential in terms of their organic content and hydrocarbon-generating capability.

The maximum TOC values of the Khatatba and Betty formations are 2.1 and 1.7 wt %, respectively.

Maturation

The maturity measurements available (vitrinite reflectance, R_0 %) were found not reliable for the limited number of samples and grains measured per sample. The thermal and burial histories of two sedimentary sections were modeled through time. One of these two sections represents the drilled Razzak structure and the other represents the Alamein basin where hydrocarbons are thought to have been generated, expelled, and migrated in to

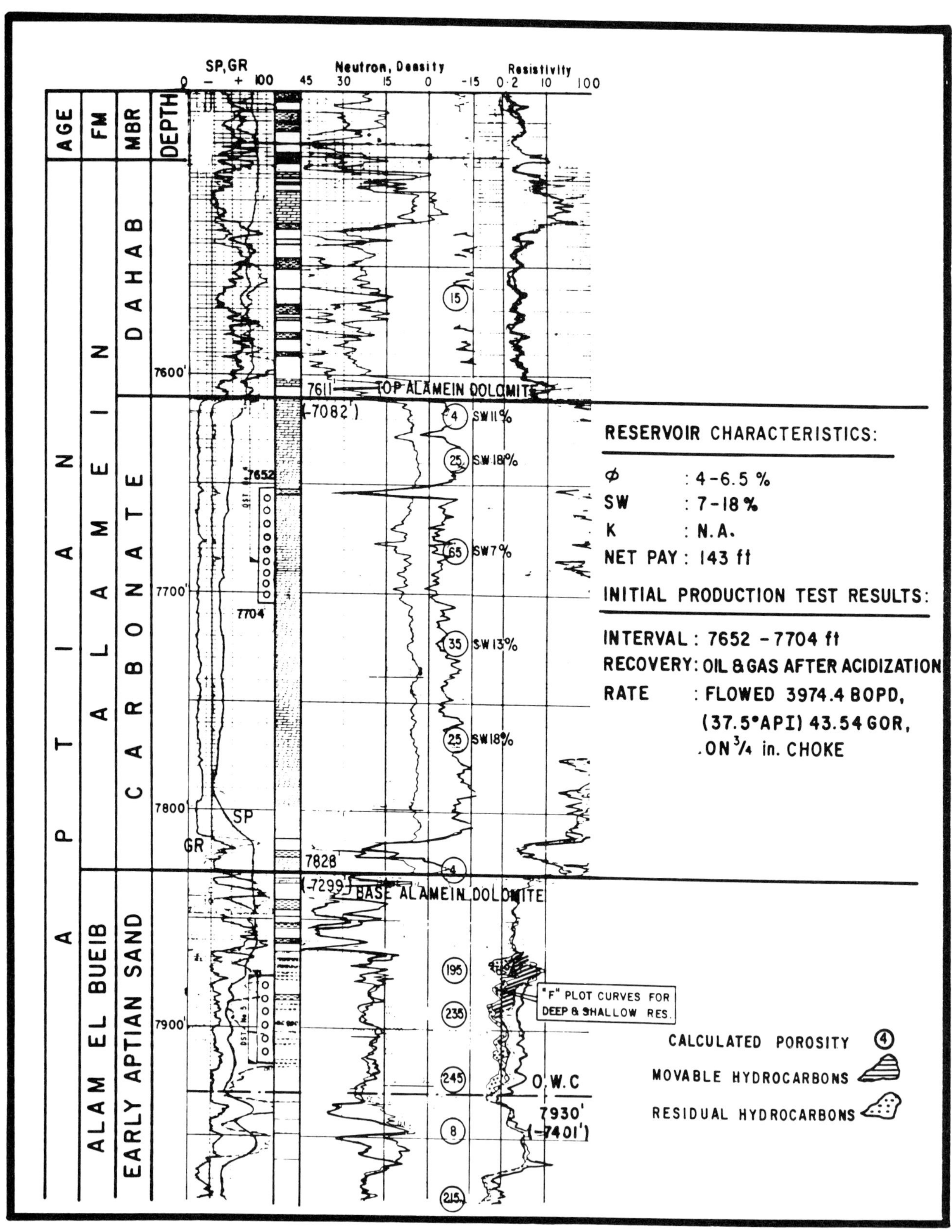

Figure 15. Reservoir characteristics of the Alamein dolomite, Razzak-1 well, Razzak Main field. The porosity is both intercrystalline and vuggy and ranges from about 6 to 10%.

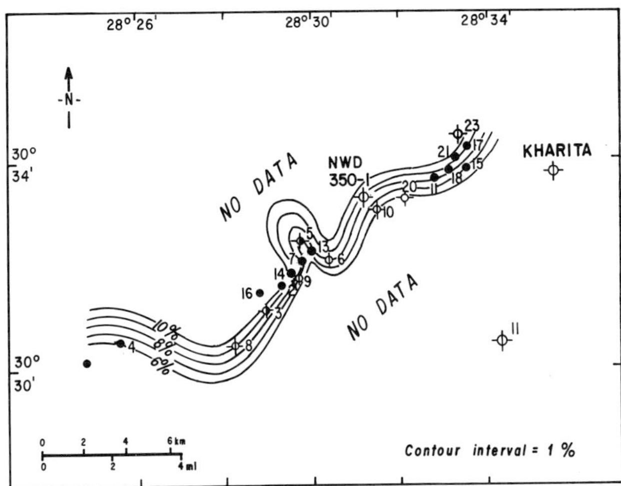

Figure 16. Average porosity in "clean" Alamein dolomite pay of the Alamein Carbonate Member, Razzak field. Porosity values in the clean section range from about 6 to 10%. Porosity is calculated from logs for the perforated interval in the Alamein Dolomite in the RZ-1 well (Figure 15). The value, 6.5%, is the net pay porosity used in the map. (RZ-1 is not shown but is next to RZ-13. See Figure 2).

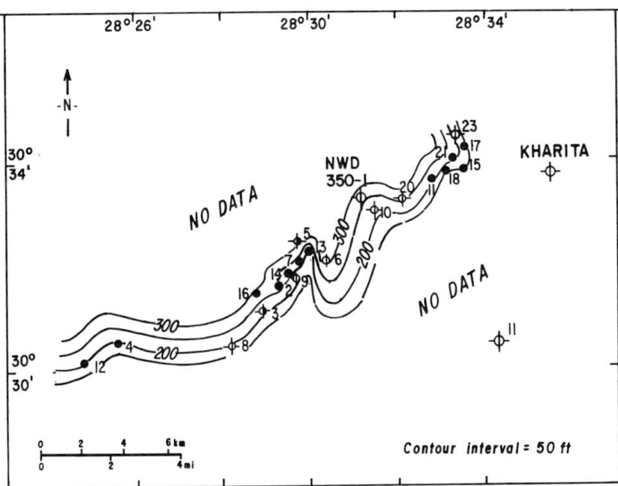

Figure 17. Net pay isopach map of the Bahariya Formation reservoir quality sand, Razzak field. Porosity cutoff, 10%. The producing zone is 14 m (46 ft) thick, with the maximum being 30 m (100 ft). Porosity calculated from logs.

the Razzak field. The cumulative time-temperature index (TTI) would then suggest a possible equivalent R_0 % threshold. The maximum recorded bottom-hole temperature from available well data was used to calculate the geothermal gradient in both investigated sections. Measured temperatures were corrected for loss of heat during mud circulation (Fertl and Wichman, 1977). An increase of 10% in temperature gradient was incorporated in the thermal history modeling during the post-Khoman early Tertiary event and during the post Apollonia early Oligocene event (Shahin et al., 1987).

The oil window is often defined as the depth interval between the onset of hydrocarbon generation (at $R_0 = 0.6\%$) and the oil floor (at $R_0 = 1.35\%$; Waples, 1980, 1985). An explorationist, however, would be more interested in the depth at which peak generation, expulsion, and migration took place. In this work, the oil window is defined as the interval between the depth at which the peak generation stage was reached (at TTI = 40, equivalent to $R_0 = 0.8\%$), that is the depth at which hydrocarbons are expelled and migrate, and the depth of the oil floor (at TTI = 180, equivalent to $R_0 = 1.35\%$), the depth below which liquid oil might not generate.

Thermal Burial History of the Razzak Structure (Field)

After burial of about 3780 m (12,400 ft) of sediments at a consistent, corrected temperature gradient of 1.45°F/100 ft (26°C/km), the burial history plot (Figure 25) indicates that the basal Khatatba Formation may have reached the peak generation/expulsion stage 20 Ma.

Thermal Burial History of the Alamein Basin

Deep burial by more than 4450 m (14,600 ft) of sedimentary section to the northeast of the Razzak field caused the potential source beds in this basin to reach peak generation earlier than in the shallower Razzak field area. The Jurassic Khatatba Formation is shown in Figure 26 to have reached the peak generation/expulsion stage at 65 Ma at about the time of maximum structural growth of the Razzak culminations during the Late Cretaceous. The basal 366 m (1200 ft) of the Khatatba Formation is presently below the oil floor (Figure 26). The present-day depth of peak generation is at 2896 m (9500 ft).

The maturation modeling of both the Razzak structure and the Alamein basin to the northeast of the field suggests that petroleum accumulations in the area are sourced only by the Jurassic Khatatba Formation.

From the analytical results, together with the thermal burial history data, it can be tentatively concluded that the hydrocarbons which charged the Razzak field must have migrated vertically from mature, probably Jurassic age source rocks buried in the structural trough that underlies the field. The entire Cretaceous section would be too immature in the vicinity of the field to have expelled oil. The analyses suggest that even at 3354 m (11,000 ft), the samples were at the pre-peak hydrocarbon generation stage. Long-range migration into the area from deeply buried possible Lower Cretaceous and Jurassic sources to the north or northwest is also possible.

Oil/Oil and Oil/Rock Correlation

All the oils from the Cretaceous age reservoirs in the Razzak field are interpreted to be genetically related. The Razzak oils are all waxy and have similar

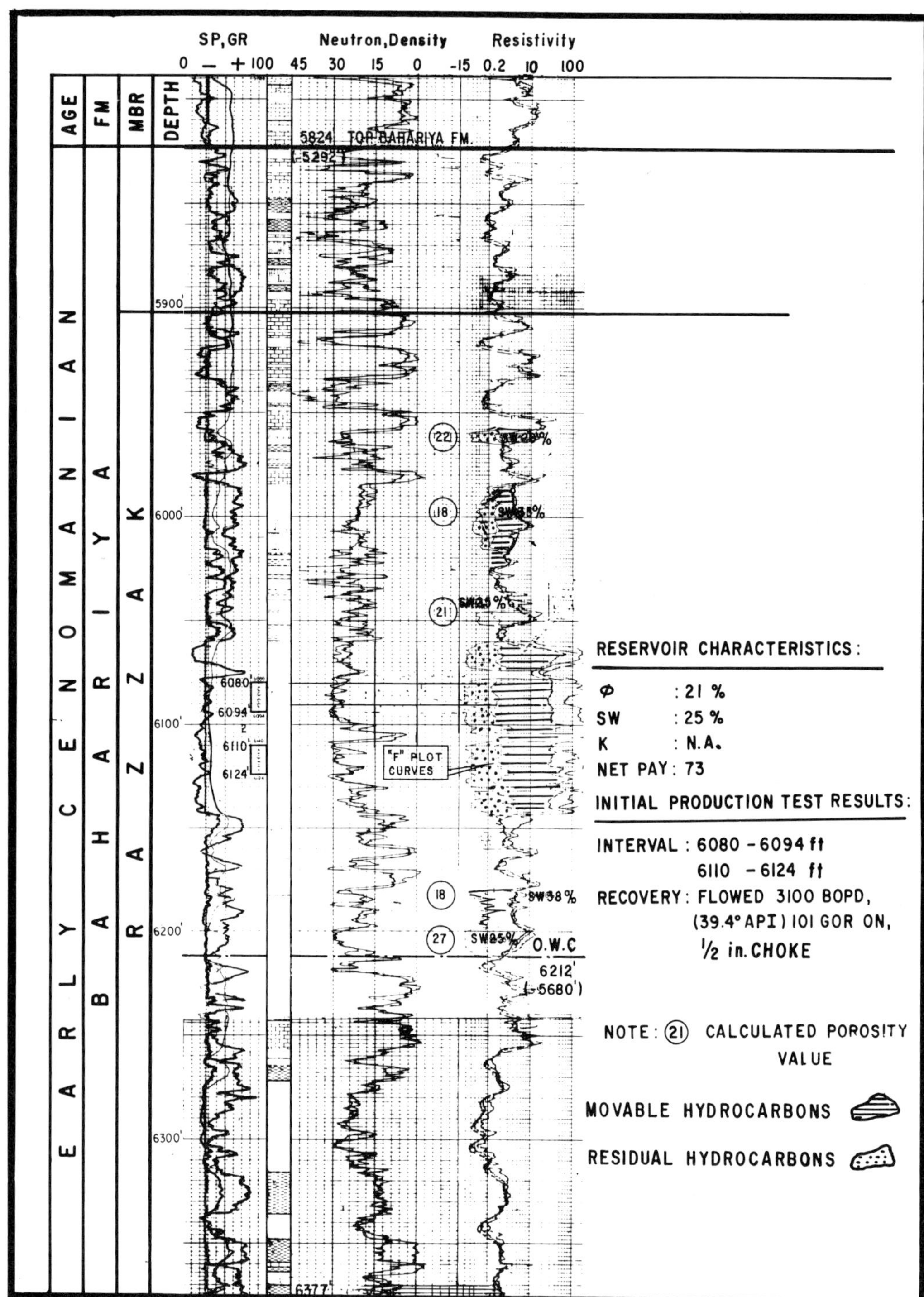

Figure 18. Reservoir characteristics of the Razzak Member (Bahariya Formation), Razzak Main field. The porosity values in the pay section range from 21 to 25% in Razzak Main to about 26% in East Razzak.

isoprenoid distributions (Figure 27). The sulfur content for all the oils is about 0.87%. The oil recovered from a Jurassic Masajid DST in well RZK-13 was, in part, interpreted to be genetically related to the Cretaceous oils.

Oil and Field Characteristics

The Abu Roash "G" oil from the Razzak Main field is about API 38.5°, and the viscosity is 0.456 cp. The initial solution GOR was 860 SCFG/STB and the

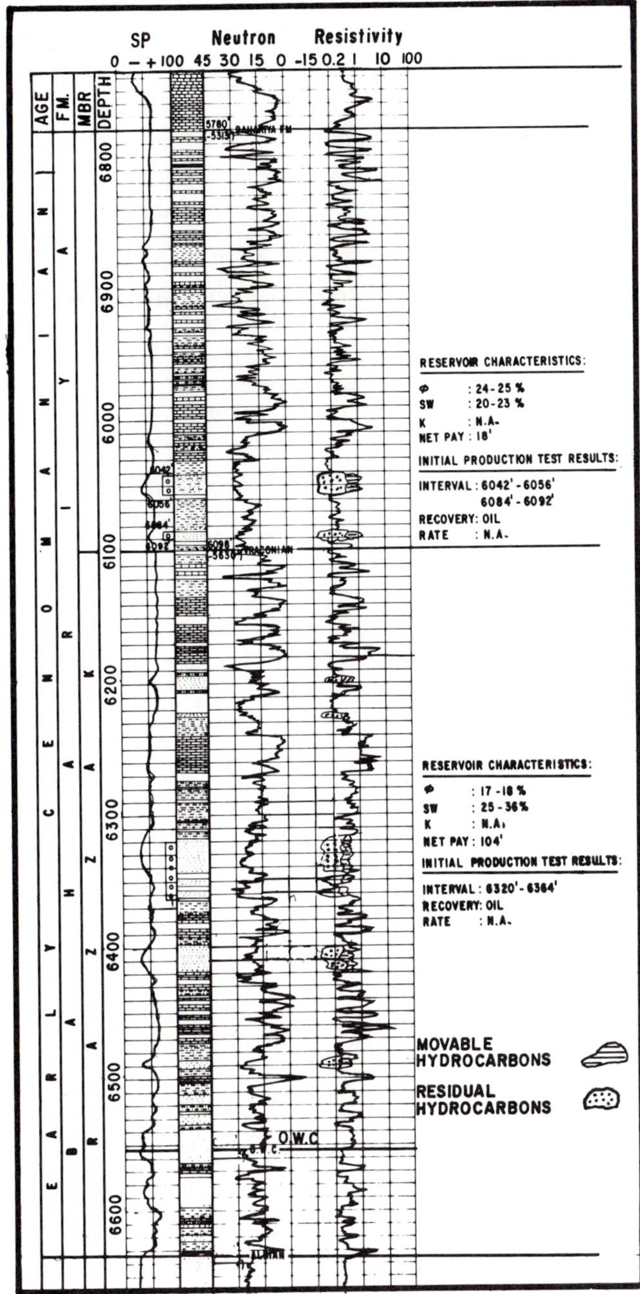

Figure 19. Reservoir characteristics of the Bahariya Formation and Razzak Member sandstones, East Razzak field. The water saturation in Razzak Main field is about 15% and the permeabilities reach 400 md.

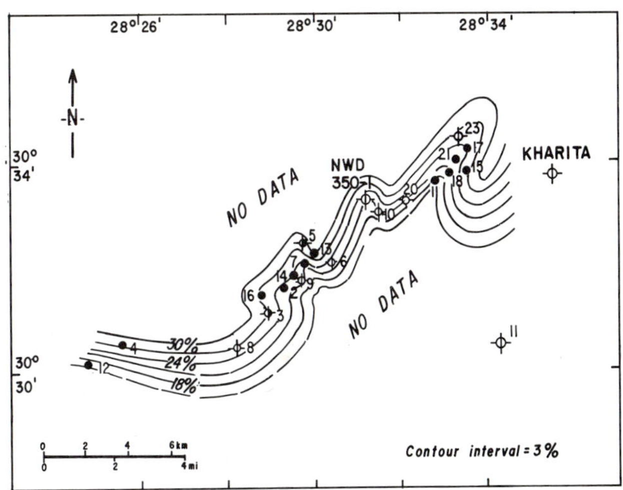

Figure 20. Average porosity in Bahariya sandstone pays, Razzak field. In East Razzak, the water saturation is about 30% and the permeability is 230 md. Porosity is calculated from logs for the perforated intervals (Figures 18 and 19). Porosity cutoff is 10%.

Figure 21. Thin section of large, rounded, pelletal, glauconitic sandstone. Some pellets suffered compaction prior to cementation with calcite. These pellets were produced by diagenetic alteration of lime mud.

formation volume factor is estimated to be 1.35 bbl/stb at the original saturation pressure. The initial saturation formation pressure was 2350 psi, decreasing to 875 psi by mid-1986. Abu Roash "G" oil from the West Razzak field is about API 34°. The initial GOR was 125 SCFG/STB, and the formation volume factor was estimated to be 1.12 bbl/stb at the original saturation pressure. The initial saturation formation pressure was 2400 psi. The current pressure is not available.

The Bahariya oil from the Razzak Main field is about API 37.5°, and the viscosity is 0.93 cp. The initial solution GOR was 80 SCFG/STB, and the formation volume factor was estimated to be 1.12 bbl/stb at the original saturation pressure. The initial saturation formation pressure was 2524 psi,

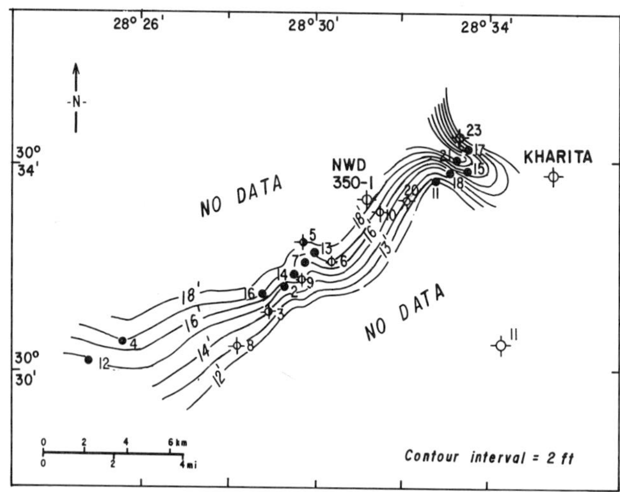

Figure 22. Net pay isopach map of reservoir quality Abu Roash "G" dolomite pay equivalent, Razzak field. Porosity cutoff, 8%. The thickness of Abu Roash "G" dolomite pay zone ranges from 3.6 to 5.5 m (12 to 18 ft). Porosity calculated from logs.

decreasing to 1600 psi by mid-1986. The Bahariya oil from the East Razzak field is about API 38.0°, and the viscosity is 0.545 cp. The initial solution GOR was 500 SCFG/STB, and the formation volume factor was estimated to be 1.343 bbl/stb at the original saturation pressure. The initial saturation formation pressure was 2528 psi. The current pressure is not available.

The Alamein dolomite oil from Razzak Main field is about API 37.0°, and the viscosity is 1.30 cp. The initial solution GOR was 127 SCFG/STB and the formation volume factor was estimated to be 1.115 bbl/stb at the original saturation pressure. The initial saturation pressure was 3225 psi, decreasing to 3190 psi by mid-1986.

It is of interest to note that in the Razzak Main/West Razzak field complex as a whole the GOR increased to as high as 900 SCFG/STB in mid-1976 as the daily production rate started to decline (Figure 5). Since then the ratio has remained generally between about 400 and 800 SCFG/STB except in 1982 and 1983 during the dump flood project period of the Abu Roash "G" in Razzak Main field, when the GOR dropped to around 200 SCFG/STB.

EXPLORATION AND DEVELOPMENT CONCEPTS

The Razzak field was discovered primarily as a result of improved reflection seismic data resolution of pre-Tertiary events that became available in the late 1960s and early 1970s. The ability to map in greater detail and accuracy the Cretaceous and earlier events with

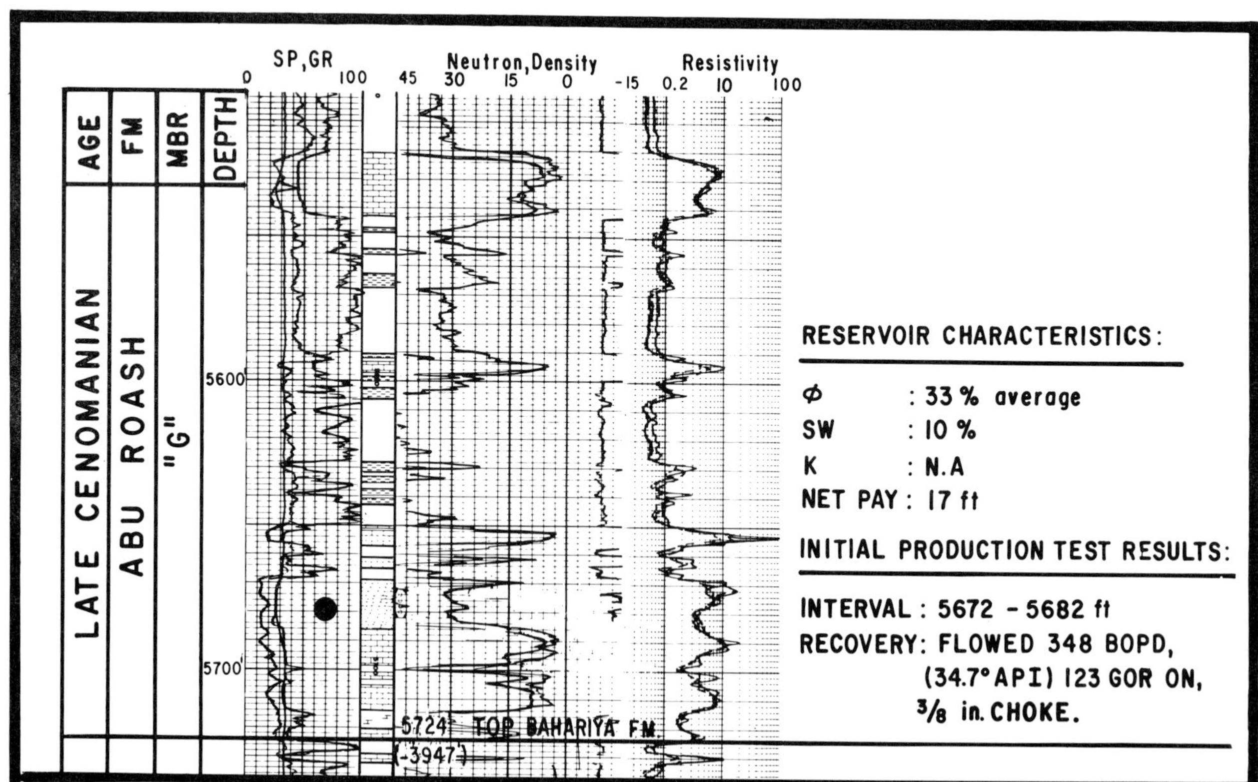

Figure 23. Reservoir characteristics of the Abu Roash "G" dolomite, West Razzak field. The porosity of Abu Roash "G" dolomite ranges from 26 to 36%. Black dot indicates pay zone.

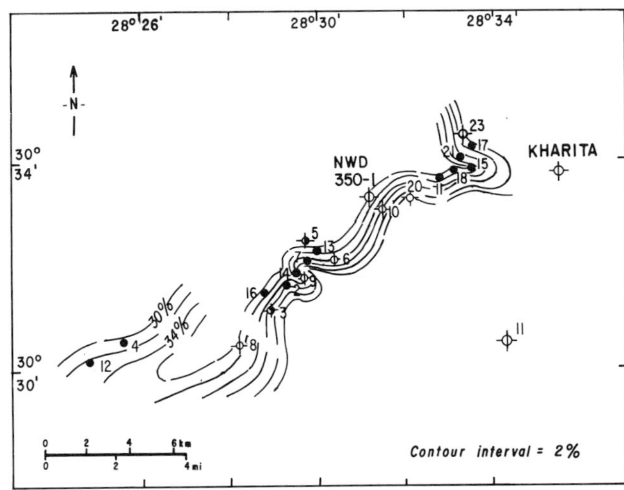

Figure 24. Average porosity of Abu Roash "G" dolomite pay, Razzak field. The porosity of Abu Roash "G" dolomite is intercrystalline and vuggy. Porosity is calculated from logs for the perforated interval (Figure 23).

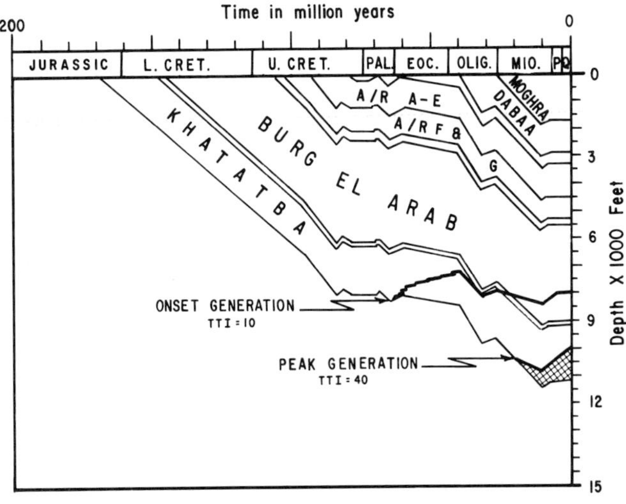

Figure 25. Razzak field thermal burial history. The Middle Jurassic basal Khatatba Formation may have reached the peak generation/expulsion stage 20 Ma (early Miocene). The Burg El Arab is the old informal name for the Kharita, Alamein, and Alam El Bueib formations (Figure 4).

up-to-date seismic techniques has led to a number of significant discoveries in recent years. Most notable among these are the Kharita Formation producing in the Badr El-Din (BED) field, northwest of the Abu Gharadig (AG) field, and the Bahariya Formation and deeper zones (including Jurassic) producing in the Salam and Meleha fields in the northwestern portion of the Western Desert (Figure 3).

Understanding the relationships between basin development history and structural growth of specific prospective areas with respect to timing of

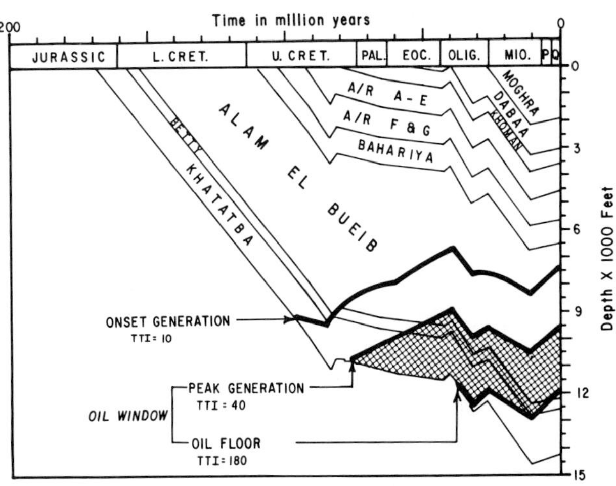

Figure 26. Alamein basin thermal burial history. The Middle Jurassic Khatatba Formation has reached the peak generation/expulsion stage at 65 Ma (Late Cretaceous), at about the time of maximum structural growth of the Razzak culminations. At present, the basal Khatatba Formation is below the oil floor.

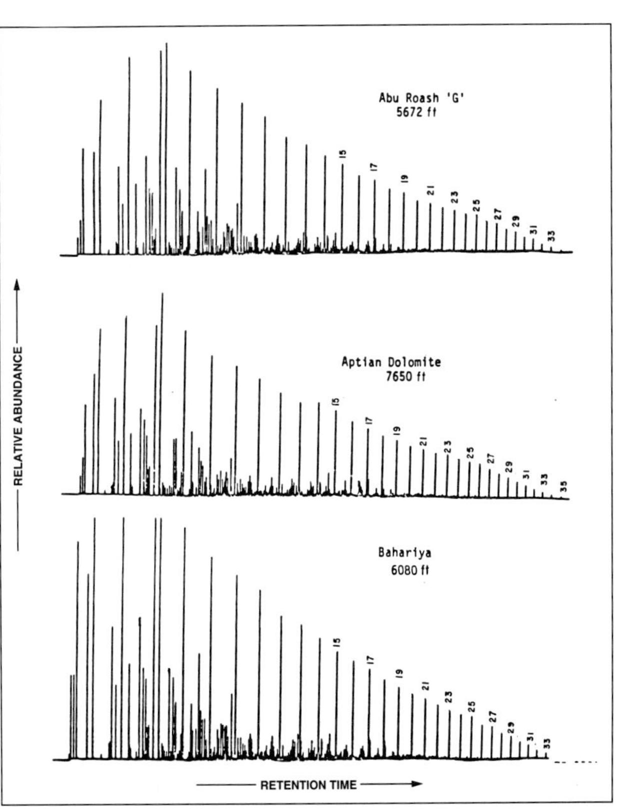

Figure 27. Representative gas chromatograms of Razzak field oils. All oils from the Cretaceous-age reservoirs in Razzak field are interpreted to be mature and genetically related. The Razzak oils are all waxy and have similar normal alkane distribution.

hydrocarbon generation and migration from the basinal areas is necessary for efficient exploration and development. This, with other tools used in basin analysis, such as geophoto, satellite imagery, magnetics, gravity, and geochemistry, would narrow the search for prospective structures. The well density distribution is presently too sparse to provide the number of subsurface control points needed for the detailed analysis required to locate future prospects. The well data, of course, provide the basic general stratigraphy, depositional history, and framework for the seismic interpretations. The wells also give essential information regarding the geographic and vertical stratigraphic distribution of seals, reservoirs, and source rocks in the section and the locations of the general areas of regional highs and basinal lows. Detailed facies maps and seismically controlled isopach maps, combined with sedimentation rate and maturation history profiles, should help isolate the likely mature source areas in the Western Desert and provide estimates of the timing of expulsion and direction of hydrocarbon migration.

ACKNOWLEDGMENTS

The Egyptian General Petroleum Corporation (EGPC) and AMOCO permitted the publication of this paper.

The study was prepared under the supervision of A. Shawky Abdine (Chairman of the Board), Wafik Meshref (Exploration General Manager), and Safi Wasfi (Chief Geologist), Gulf of Suez Petroleum Company (GUPCO).

The study was edited by A. Nabil Shahin and compiled by Arne Aadland and Ahmed Aal.

In addition to cited literature, several unpublished GUPCO reports were the basis for Razzak field study. These reports are authored by A. Abdine, A. Aal, H. Ali, I. Awad, S. Bishlawy, A. Darwish, M. Haddad, M. Hagras, R. Oslon, E. Osman, P. Papazis, O. Shaarawy, A. Shahin, and M. Shehab.

The manuscript was reviewed by H. Dalton, A. Gad, A. Hassouba (who also provided reservoir petrology and environment), J. Jenner, S. Moaty, R. Nelson, and T. Russell. Thanks are extended to AAPG reviewers B. Brownfield, T. Beaumont, and "anonymous" for their critical review and valuable comments. We are, however, responsible for any negative qualities that remain.

Technical assistance was provided by A. Salem, M. Aman, A. Badawi, M. Laboudy, H. Mostafa, and G. Mostafa.

REFERENCES CITED

Abdine, A. S., 1974, Oil and gas discoveries in the Northern Western desert of Egypt: Egyptian General Petroleum Corporation, Fourth International Exploration Conference, Cairo.

Bathurst, R. G. G., 1971, Carbonate sediments and their diagenesis; developments in sedimentology (12): Amsterdam, Elsevier, 620 p.

Deibis, S., 1978, Yidma discovery and its bearing on oil exploration, Western Desert, Egypt: Organization of Arab Petroleum Exporting Countries Seminar, Kuwait.

Dunham, R. J., 1962, Classification of carbonate rocks according to depositional texture, *in* W. E. Ham, ed., Classification of carbonate rocks: AAPG Memoir 1, p. 108-121.

Ezzat, M. R., 1972, Case history of seismic exploration in the Egyptian Western Desert: Eighth Arab Petroleum Congress, Algiers.

Ezzat, M. R., 1975, Factors affecting seismic exploration in the Razzak area: Ninth Arab Petroleum Congress, Dubai.

Fertl, H. W., and P. A. Wichman, 1977, How to determine static bottom hole temperature from well log data: World Oil, v. 184, n. 1, p. 105-106.

Folk, R. L., 1974, Petrology of sedimentary rocks: Austin, Texas University Press, 128 p.

Hamed, A. R. A., 1972, Environmental interpretations of the Aptian carbonates of the Western Egyptian Desert: Eighth Arab Petroleum Congress, Algiers.

Khalil, N. A., and S. El Mofty, 1972, Some geophysical anomalies in the Western Desert, Egypt, their geological significance and oil prospects: Eighth Arab Petroleum Congress, Algiers.

Metwalli, M. H., G. Philip, and A. M. Wali, 1982, Primary dolomites as oil reservoirs; their occurrence and significance in the Northern Western Desert, Egypt: Acta Geologica Academiae Scientiarum Hungoricae, v. 25, n. 3-4.

Pettijohn, F. J., 1975, Sedimentary rocks, 3rd edition: New York, Harper and Row, 614 p.

RRI (Robertson Research International), 1982, Petroleum potential evaluation, Western Desert, Egypt; Consultants Report: Egyptian General Petroleum Corporation, Cairo.

Shahin, A. N., M. M. Shehab, and H. F. Mansour, 1987, Quantitative evaluation and timing of petroleum generation in Abu Gharadig basin, Western Desert, Egypt: AAPG Treatise of Petroleum Geology Conference on Application of Geochemistry to Petroleum Exploration, Denver, Colorado.

Waples, D., 1980, Time and temperature in petroleum formation, application of Lopatin's methods to petroleum exploration: AAPG Bulletin, v. 64, n. 6, p. 916-926.

Waples, D. W., 1985, Geochemistry in petroleum exploration: Boston, International Human Resources Development Corporation, 232 p.

Appendix 1. Field Description

Field name .... *Razzak field (East, Main, and West)*

Ultimate recoverable reserves .... *NA*

Field location:

- **Country** .... *Egypt*
- **Basin/Province** .... *Razzak-Alamein basin*

Field discovery:

- **Year first pay discovered** .... *Lower Cretaceous Aptian dolomite January 1972*
- **Year second pay discovered** .... *Upper Cretaceous Abu Roash dolomite June 1972*
- **Year third pay discovered** .... *Upper Cretaceous Bahariya sandstone October 1972*

Discovery well name and general location:

- **First pay** .... *Razzak No. 1 (RZK-1), 261 km (162 mi) northwest of Cairo and 58 km (36 mi) south of Mediterranean Sea*
- **Second pay** .... *RZK-4, 7.1 km (4.4 mi) southwest of RZK-1 well*
- **Third pay** .... *RZK-7, 232 m (760 ft) southwest of RZK-1 well*

Discovery well operator

- **First pay** .... *AMOCO Egypt Oil Co.*
- **Second pay** .... *AMOCO Egypt Oil Co.*
- **Third pay** .... *Nipco Exploration Division*

IP

- **First pay** .... *Aptian dolomite, 3974 BOPD*
- **Second pay** .... *Abu Roash "G" dolomite, 190 BOPD*
- **Third pay** .... *Bahariya sandstone, 3100 BOPD*

All other zones with shows of oil and gas in the field:

Age	Formation	Type of Show
Early Aptian	*Burg El Arab**	*Oil and gas*
Neocomian	*Betty*	*Oil*
Jurassic	*Masajid*	*Oil*

**Comprises the Kharita, Alamein, and Alam El Bueib formations*

Geologic concept leading to discovery and method or methods used to delineate prospect

The field is on trend with the Alamein field to the northeast. The discovery well was located on reflection seismic data that indicated faulted anomalies with four-way closure on the Cretaceous mapped horizons.

Structure:

Province/basin type .... *Klemme II Ca/c; Bally 1141*

Tectonic history

The Razzak-Alamein basin apparently had its maximum development during the Jurassic-Early Cretaceous with subsidence continuing at a reduced rate during Late Cretaceous and Tertiary. Intermittent movements along one or more major northeast-southwest-oriented wrench faults continued since early Cretaceous or earlier to latest Cretaceous-early Tertiary and controlled the structure grain in the basin area.

Regional structure

The field lies along the northeast-southwest-trending Qattara-Alamein fault-controlled anticlinal ridge found in the northwestern portion of the Razzak-Alamein basin area.

Local structure

The field complex at the Cretaceous level appears to form three separate faulted anticlinal culminations associated with wrench faulting along major northeast-southwest-oriented reverse fault that bounds the field to the northwest.

Trap:

Trap type(s)

All hydrocarbon-bearing zones in the Razzak field are structurally (fault) trapped.

Basin stratigraphy (major stratigraphic intervals from surface to deepest penetration in field):

Chronostratigraphy	Formation	Depth to Top in m (ft)
Miocene	*Moghra*	*222 (730)*
Oligocene	*Daba*	*757 (2484)*
Paleocene/Eocene	*Appollonia*	*1216 (3990)*
Campanian/Maastrichtian	*Khoman*	*1327 (4352)*
Upper Cenomanian to Coniacian	*Abu Roash*	*1509 (4950)*
Lower Cenomanian	*Bahariya*	*2094 (6868)*
Barremian/Albian	*Burg El Arab*	*2307 (7566)*
Neocomian	*Betty*	*3226 (10,582)*
Middle Jurassic	*Masajid*	*3345 (10,971)*
	Khatatba	*3350 (10,989)*
Early Jurassic	*Eghei group*	*3626 (11,892)*

Reservoir characteristics:

Number of reservoirs *3*

Formations .... *Abu Roash Formation (Abu Roash "G" dolomite Member); Bahariya Formation; Alamein dolomite*

Ages .... *Late Cenomanian, Early Cenomanian, Aptian, respectively*

Depths to tops of reservoirs

Abu Roash "G," 1397 m (4581 ft) in Razzak Main and 1329 m (4358 ft) in West Razzak

Bahariya Formation, 1613 m (5292 ft) in Razzak Main and 1687 m (5533 ft) in East Razzak

Aptian dolomite, 2159 m (7082 ft) and 2174 m (7132 ft) in Razzak Main

Gross thickness (top to bottom of producing interval)

Abu Roash "G," 6.7 m (22 ft); Bahariya, 106.7 m (350 ft); Aptian dolomite, 64.6 m (212 ft)

Net thickness—total thickness of producing zones

Average .... *Abu Roash "G," 5 m (17 ft); Bahariya, 14 m (46 ft); Aptian dolomite, 24 m (80 ft)*

Maximum .... *Abu Roash "G," 6 m (20 ft); Bahariya, 30 m (100 ft); Aptian dolomite, 46 m (150 ft)*

Lithology

Abu Roash "G" dolomite: tan, light tan, sucrosic texture, medium-hard, fair porosity

Bahariya sandstone: light gray, tan, very fine grained, hard, tight, occasionally friable, partly silty to siltstone, glauconitic, partly pyritic, with fair to good porosity

Aptian dolomite: light tan, tan, partly white, fine crystalline, partly cryptocrystalline, with rare dark brown-black bituminous matter

Porosity type .... *Dolomite, vuggy and intercrystalline porosity; sandstones, intergranular connected porosity*

Average porosity .... *Abu Roash "G," 32%; Bahariya, 25%; Aptian dolomite, 8%*

Average permeability .... *Abu Roash "G," 300 md; Bahariya, 400 md; Aptian dolomite, NA*

Seals:

Upper

Formation, fault, or other feature .... *Abu Roash and Bahariya formations, and Burg El Arab formation (Dahab Member)*

Lithology .... *Shale, siltstone, and limestone*

Lateral

Formation, fault, or other feature .... *Fault contact with impermeable beds and/or structural dip*

Source:

Formation and age .... *Khatatba Formation (Jurassic)*

Lithology .... *Shales*

Average total organic carbon (TOC) .... *Khatatba Formation, 1.3 wt%; Betty Formation, 0.83 wt%*
Maximum TOC .... *Khatatba Formation, 2.1 wt%; Betty Formation, 1.7 wt%*
Kerogen type (I, II, or III) .... *II to III*
Vitrinite reflectance (maturation) .... *NA*
Time of hydrocarbon expulsion .... *Khatatba Formation, 50 Ma in West Razzak basin, 15 Ma in drilled section*
Present depth to top of source .... *Khatatba Formation, 3369–3628 m (11,050–11,900 ft); Betty Formation, 3456 m (11,337 ft)*
Thickness .... *Khatatba, 259 m (850 ft); Betty, 30 m (100 ft)*
Potential yield .... *NA*

Appendix 2. Production Data

Field name .... *Razzak field (East, Main, and West)*

Field size:

Proved acres .... *Abu Roash "G," 3325; Bahariya, 640 and 735; Aptian dolomite, 1060*
Number of wells all years .... *West Razzak, 2; Main Razzak, 11; East Razzak, 9*
Well spacing .... *Abu Roash "G," 3; Bahariya, 6; Aptian dolomite, 1*
Ultimate recoverable .... *NA*
Cumulative production .... *Abu Roash "G," 5.5 MMbbl; Bahariya, 22.3 MMbbl; Aptian dolomite, 19.7 MMbbl*
Annual production .... *Abu Roash "G," 0.4 MMbbl; Bahariya, 1.6 MMbbl; Aptian dolomite, 1.4 MMbbl*
Present decline rate .... *Main and West Razzak, 23%; East Razzak, 15%*
Annual water production .... *Abu Roash "G," 0.1 MMbbl; Bahariya, 5.0 MMbbl; Aptian dolomite, 1.1 MMbbl*
In place, total reserves .... *NA*
In place, per acre foot .... *NA*
Primary recovery .... *NA*
Secondary recovery .... *NA*
Cumulative water production .... *Abu Roash "G," 0.6 MMbbl; Bahariya, 32.7 MMbbl; Aptian dolomite, 11.4 Mmbbl*

Drilling and casing practices:

Amount of surface casing set .... *91 m (300 ft)*
Casing program .... *20-in. to 91 m (300 ft); 13⅜-in. to 914 m (300 ft); 9⅝-in. to T.D.*
Drilling mud .... *Water-based mud*
Bit program .... *Varies*
High pressure zones .... *None*

Completion practices:

Intervals perforated
Abu Roash "G" dolomite pay zone, multiple Bahariya sandstone intervals (variable among wells) and top part of Alamein pay
Well treatment .... *Abu Roash "G" and Aptian dolomite need acidization*

Formation evaluation:

Logging suites .... *SP, GR, neutron, density, sonic, resistivity*
Testing practices .... *Typically production tested*
Mud logging techniques .... *Not typically done on infill wells*

Oil characteristics:

Type .... *Naphthenic*

API gravity

Abu Roash "G": Razzak Main, 38.5°; West Razzak, 34°

Bahariya: Razzak Main, 37.5°; East Razzak, 38°

Aptian dolomite: 37°

Initial GOR

Abu Roash "G": Razzak Main, 860 SCF/STB; West Razzak, 125 SCF/STB

Bahariya: Razzak Main, 80 SCF/STB; East Razzak, 500 SCF/STB

Aptian dolomite: 127 SCF/STB

Viscosity, SUS

Abu Roash "G": Razzak Main, 0.456 cp

Bahariya: Razzak Main, 0.93 cp; East Razzak, 0.545 cp

Aptian dolomite: 1.30 cp

Pour point .... *NA*

Gas-oil distillate .... *NA*

Field characteristics:

Average elevation .... *West Razzak, 191 m (628 ft); Main Razzak, 165 m (541 ft); East Razzak, 193 m (456 ft)*

Initial pressure

Abu Roash "G": 2350 psi

Bahariya: Razzak Main, 2524 psi; East Razzak, 2528 psi

Aptian dolomite: 3225 psi

Present pressure

Abu Roash "G": 875 psi

Bahariya: Razzak Main, 1600 psi; East Razzak, NA

Aptian dolomite: 3190 psi

Temperature .... *Abu Roash "G," NA; Bahariya, 74 to 77°C (165-170°F); Aptian dolomite, 68°C (155°F)*

Geothermal gradient .... *26°C/km (1.45°F/100 ft)*

Drive .... *Water flood*

Oil column thickness .... *Abu Roash "G," 5.2 m (17 ft); Bahariya, 14 m (46 ft); Aptian dolomite, 24 m (80 ft)*

Oil-water contact .... *Abu Roash "G," 1607 m (5270 ft); Bahariya, 1729-1860 m (5670-6100 ft); Aptian dolomite, 2213 m (7260 ft)*

Connate water .... *Abu Roash "G," 22.5%; Bahariya and Aptian dolomite, 15%*

Water salinity, TDS .... *Abu Roash "G," 46,000 ppm; Bahariya, 45,000-95,000 ppm; Aptian dolomite, 95,000*

Resistivity of water .... *Abu Roash "G," 0.145 ohm; Bahariya, 0.148-0.079 ohm; Aptian dolomite, 0.079 ohm*

Bulk volume water (%) .... *NA*

Transportation method and market for oil and gas:

Pipelines

Ain Zalah Field—Iraq
Zagros Folded Zone, Northern Iraq

MOHAMED HOSSNY EL ZARKA*
Alexandria University
Alexandria, Egypt

FIELD CLASSIFICATION

BASIN: Zagros
BASIN TYPE: Foredeep
RESERVOIR ROCK TYPE: Limestone (Fractured)
RESERVOIR AGE: Lower and Upper Cretaceous
PETROLEUM TYPE: Oil
TRAP TYPE: Faulted Anticline
RESERVOIR ENVIRONMENT OF DEPOSITION: Subsiding Basin with Local Intertidal/ Supralittoral Conditions
TRAP DESCRIPTION: Steep flank anticline dissected by scissors fault, resulting in double culmination

LOCATION

The Ain Zalah oil field is located about 60 km northwest of the town of Mosul in northern Iraq (Figure 1A) between longitudes 42°30′ and 42°45′E, and latitudes 36°43′56″ and 36°48′13″N. It is to the northwest of a group of oil fields of the Kirkuk province. All are located in the Zagros folded zone (Figure 1B). The nearest field is Butma to the south (Figures 1A and 1B). Ain Zalah is a prominent surface anticline consisting of a series of elongated ridges—a low range of hills, similar to the Kettleman Hills of California. The outcropping rocks consist of limestone and gypsum beds belonging to the Euphrates and lower Fars formations (lower to middle Miocene). The topographic features are a reflection of structural folding. The producing structure is an east-west-trending symmetrical anticline 19 km in length and 5 km in width, with flank dips of 30°. The topographic feature runs concordant with the subsurface axis of the fold.

The field produces from two reservoirs. The upper one is the Shiranish Formation of Maastrichtian to late Campanian age, and the lower one is the Qamchuqa Formation of Albian age (Table 1). Initial reserves, estimated by material balance, were 37 and 158 MMBO for the upper and lower reservoirs, respectively. Annual production in 1974 was 1.7 and 1.5 MMBO for the upper and lower reservoirs, respectively. The cumulative production as of the end of 1976 was 20 MMBO and 146 MMBO for the upper and lower reservoirs, respectively. The proven reserves, as of 1 January 1975, were 20.5 MMBO and 15.4 MMBO, while the anticipated reserves, as of 1 January 1976, are 19 MMBO and 14.4 MMBO for the two reservoirs, respectively.

*A variant spelling of the author's name appears in the text and figure captions, citing a prior publication.

HISTORY

Pre-Discovery

The prominent surface anticline at Ain Zalah had attracted oil explorationists working in the adjoining prolific Kirkuk province, and thus it was a prospect based on surface geology. Gravitometer surveys in 1938 showed an anomaly interpreted to confirm the existence of the anticline in the subsurface.

Discovery

The discovery well (no. 1, Figure 1A) was drilled in 1939. It was completed with an initial production rate of approximately 1500 BOPD, but this production was lost in a later attempt to drill deeper. The well was finally completed with a production of less than 200 BOPD from fractured limestones of the Upper Cretaceous Shiranish Formation.

Post-Discovery

All drilling was shut down in 1941 because of the war and was not resumed until 1947. From 1947 until 1950, 13 wells were drilled to explore the extent of the subsurface structure of the field and to delineate

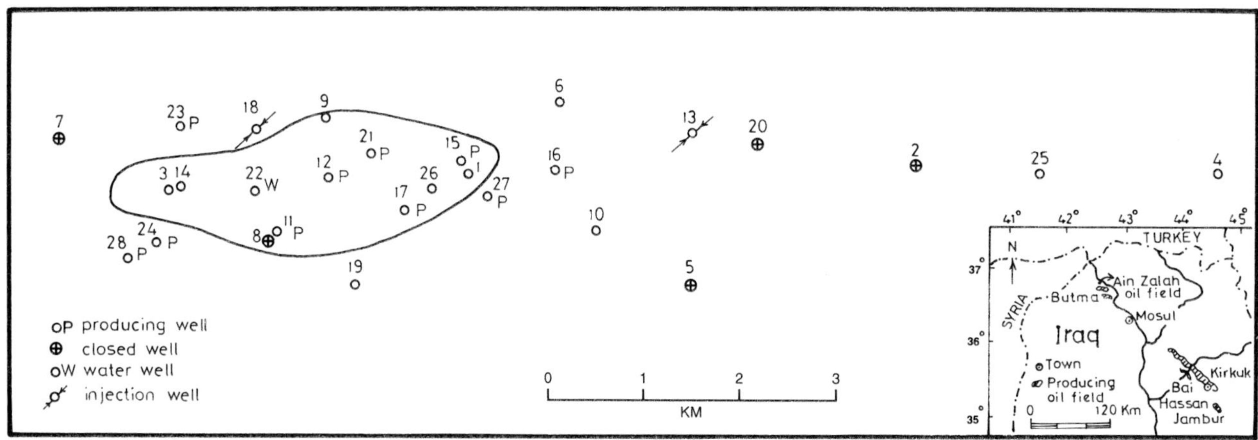

Figure 1A. Ain Zalah oil field showing location of wells, location of the field in northwestern Iraq, and location of some other fields. The best upper pay wells are in the pear-shaped outline in the western part of field. (From Elzarka and Ahmed, 1983. Used with permission of Journal of Petroleum Geology.)

the oil-water contact (wells no. 1-13, Figure 1A). In 1950, a deep exploration test, well no. 14, proved a valuable producing reservoir in a middle Cretaceous porous limestone (Qamchuqa). The depths, thickness, and the formations encountered at Ain Zalah are given in Table 1.

In 1957, the field was put on production from 28 wells. Until 1969, oil produced from the upper pay zone was replenished by oil from the lower pay zone, but at that time, it was believed that water invaded the reservoir along fault planes, fractures, and microstructures between the two pays cutting off the oil influx (Elzarka and Ahmed, 1983). The reservoir has an active water drive resulting in the volumetric replacement of oil by water. The field producing area is shown in Figure 1A from both the lower and upper pay zones.

DISCOVERY METHOD

Oil at the Ain Zalah field, as with many other known oil fields of northern Iraq, is trapped in an anticlinal feature. These structures can be mapped by surface geology (DeGolyer, 1946). Until 1950, the only known producing reservoir was the fractured Shiranish limestones of Upper Cretaceous age. Thirteen wells had been drilled to these upper beds by this time and the subsurface structure was defined. The fourteenth was drilled to 3333 m depth in the Lower Jurassic and was the discovery well of the second pay zone in the permeable, porous, and fractured Qamchuqa limestones of the middle Cretaceous. The discovery was the result of the surface geology and gravity interpretations that encouraged deeper drilling.

STRUCTURE

Ain Zalah is located in the northwestern portion of the extensive Zagros fold zone (Figure 1B). The Zagros fold belt conforms to an A-type subduction classification (41) of Bally and Snelson (1980) and is part of a continental multicycle basin, crustal collision zone of convergent plate margin (II Ca) of Klemme (St. John et al., 1984).

Hart and Hay (1974) described the Ain Zalah structure as a steeply folded foothills type anticline 12 mi (19 km) long and 3 mi (5 km) across. The folding occurred during the late Tertiary Taurus-Zagros orogeny. The structure below the Cretaceous is unknown, as only one well (well 14, Figure 1A) penetrates the deeper horizons.

The oldest faulting identified is related to an east-northeast, west-southwest graben formed during lower Shiranish (Maastrichtian) deposition. This is evidenced by Shiranish Formation thickness variations (Hart and Hay, 1974). There is little indication of thickness variations within the Mashurah and earlier formations. The next phase of faulting occurred during upper Shiranish deposition when, again, a graben-type structure developed with a northeast-southwest trend. The structure is thus interpreted to be the result of two phases of block fault movement upon which was superimposed late Tertiary folding. The faults are thought to have been initially nearly vertical but the fault planes are now folded (see Hart and Hay, 1974, p. 978).

The Arabian craton seems to have been subjected to progressive shearing and, in the resulting phase of major basement block adjustments (uplifting and downwarping), a fairly well defined belt of east-west-trending horst and graben structures developed. This can be traced through northeast Syria and northern

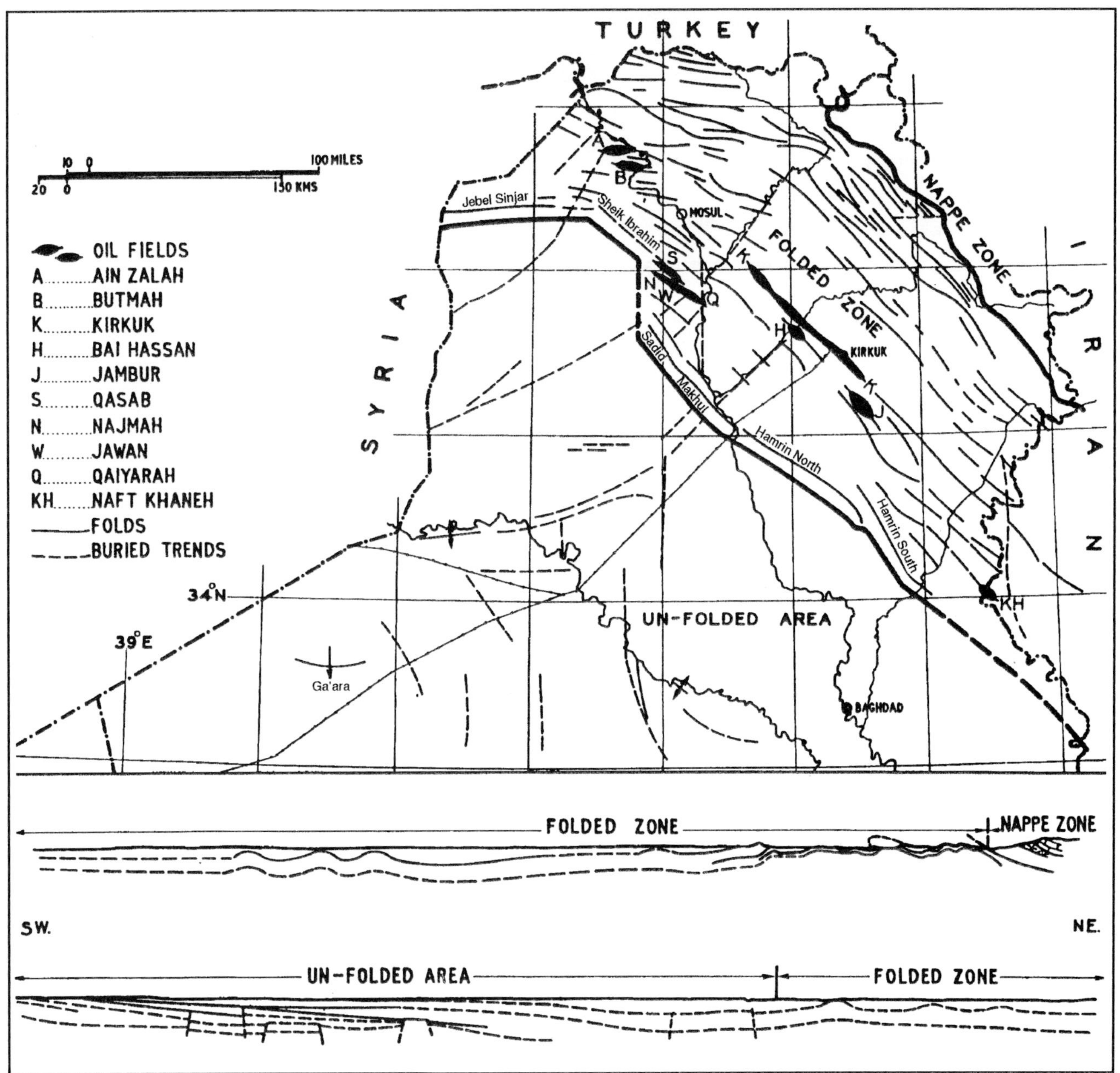

Figure 1B. Tectonic framework map of northern Iraq showing a portion of the Zagros folded zone, above, and schematic structural cross section, below. (After Dunnington, 1958.)

Iraq. The main graben associated with Ain Zalah follows this trend (Henson in Fairbridge and Badoux, 1960, p. 114–116, and Hart and Hay, 1974, p. 980–981).

Elzarka and Ahmed (1983) define the structure of the Cretaceous section as a bicrestal double-plunging anticline whose east-west axis is affected by a major east-west south-dipping fault (Figures 2 and 5). [Ed. note: Compare Figure 2A with Hart and Hay, 1974, figure 10, p. 981.] The western extension of the fault displays normal displacement, while the eastern extension of the same fault displays reverse displacement. The average throw on the fault is 200 m down to the south (normal) in the west and about 150 m down to the north (reverse) in the east. The faulting and folding seem to date from the Late Cretaceous and culminate with the late Tertiary Taurus-Zagros orogeny. Fault motion is rotational. Minor faults affect both Lower and middle Cretaceous rocks.

STRATIGRAPHY

In outcrop, the stratigraphic succession begins with the lower Fars Formation (lower Miocene) and ends with the Lower Jurassic section (Figure 3). Daniel (1954) describes the succession encountered,

Table 1. Summary and description of stratigraphic section for Ain Zalah oil field.

System	Series	Formation	Avg. thick. (m)	Lithology
Neogene	Miocene	Lower Fars Jeribe Dhiban Euphrates	335 41	Alternating gypsum, siltstone, and limestone beds in thicknesses varying from 0.3 to 15 m.
Paleogene	Oligocene	Anah/Azkand Bajwan/Baba Shurau/Sheikh Alas	167	Porous and crystalline limestone with some gypsum and anhydrite inclusions.
	Eocene + Paleocene	Avanah/Jadala	710	Argillaceous to crystalline limestone, and porous, shallow-water, and some current-bedded sandstone.
		Aaliji	517	Dark gray calcareous shale with minor thin limestone beds at the top. Fissile blue black to green shale with sand lenses.
Cretaceous	Upper	Shiranish	704	Marly, dense, gray to dark gray, shaly limestone in the upper part, less shaly at depth.
		Mashurah	85	Crystalline limestone with bedded chert and flint members. Dolomitized limestone and some thin beds of black bituminous shale at base.
		Wajnah	15	Dolomitic limestone with Ostracods and shale of local lithification.
	"Middle"	Gir Bir	45	Dolomitic, white, dense limestone.
		Qamchuqa	165	Dolomitic limestone and thin black bituminous shale. The limestone is made up of organic debris, pelleted and oil saturated. The shale is black gray and fissile.
	Lower	Sarmord	34	Dolomitic recrystallized limestone with partings of black shaly nature with green pyritic slickensided fossiliferous shale.
Jurassic	Middle	Sargelu	471	Recrystallized dolomitic limestone with thin shaly beds and thin breccias increasing in size and number with depth; also, some conglomerates.
	Lower		>56	Massive anhydrite with thin coarsely crystalline, dolomitic limestone and thin breccias and conglomerates.

Compiled and modified from Daniel (1954); Hart and Hay (1974); and Ahmed (1980).

and his succession is shown, with modifications, in Table 1. Possible source rocks occur in the middle Cretaceous Qamchuqa and Upper Cretaceous Mashurah formations where thin beds of black bituminous shale are encountered within the dolomitized limestone beds.

TRAP

Ain Zalah field is a faulted anticlinal trap (Figure 2) with two reservoirs (Figure 4). The oil was initially trapped at the crest of the downthrown, southern anticlinal block in the fractured limestones of the Qamchuqa Formation (Figure 5). Migration continued along and across the fault to be trapped at the crest of the northern, upthrown block. Further upward movement led to entrapment of oil at the crest of the Shiranish Formation structure. The unconformity surface, e.g., between the Upper and middle Cretaceous, as well as fractures, pores, and stylolites, provided an avenue for the movement of fluids. These shales and limestones have imperfectly sealed—maintaining, for the most part, the upper and lower pay zones—but oil has moved from the lower to the upper reservoir and is also found in Aaliji sands just above the Shiranish.

The paleotectonic profiles (Figure 5) show the postulated mode of migration and accumulation of

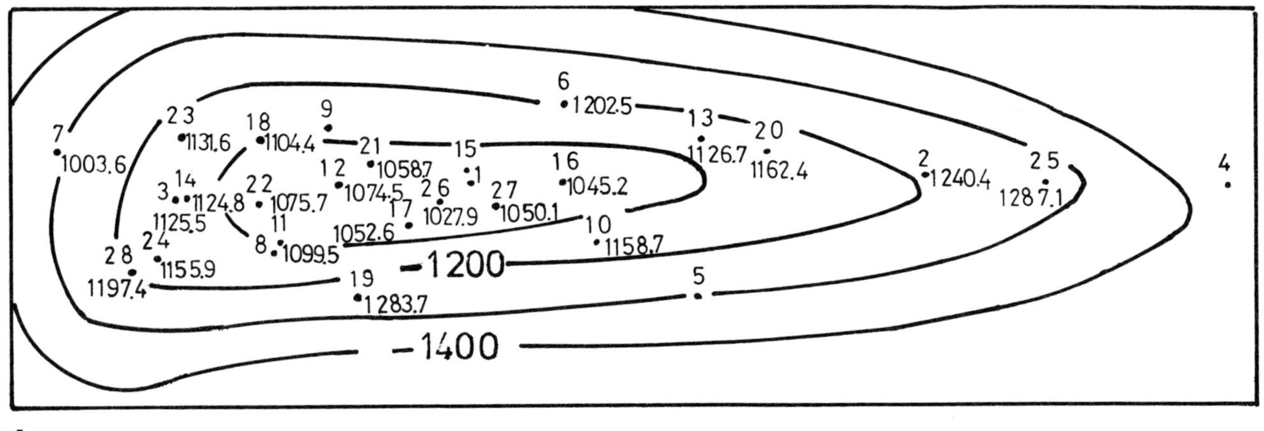

A

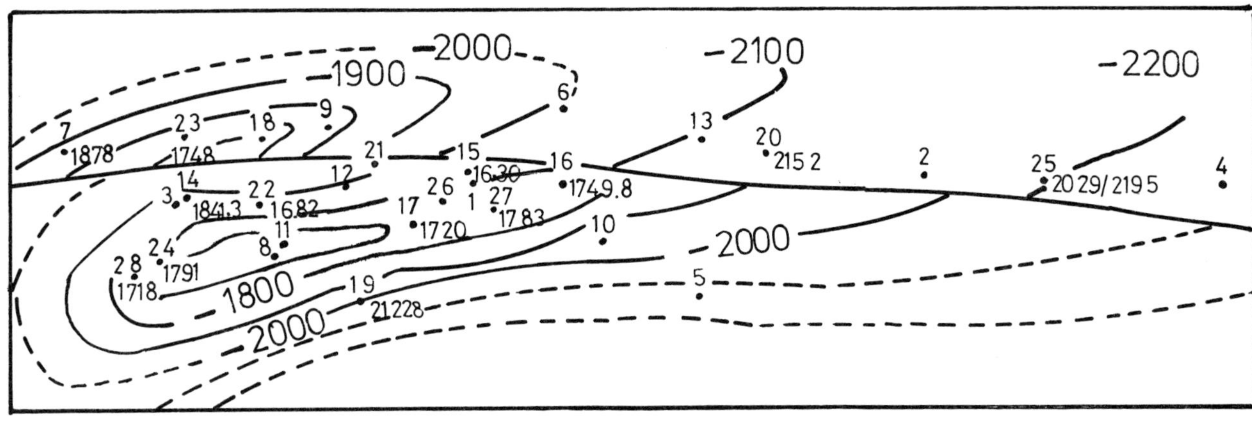

B

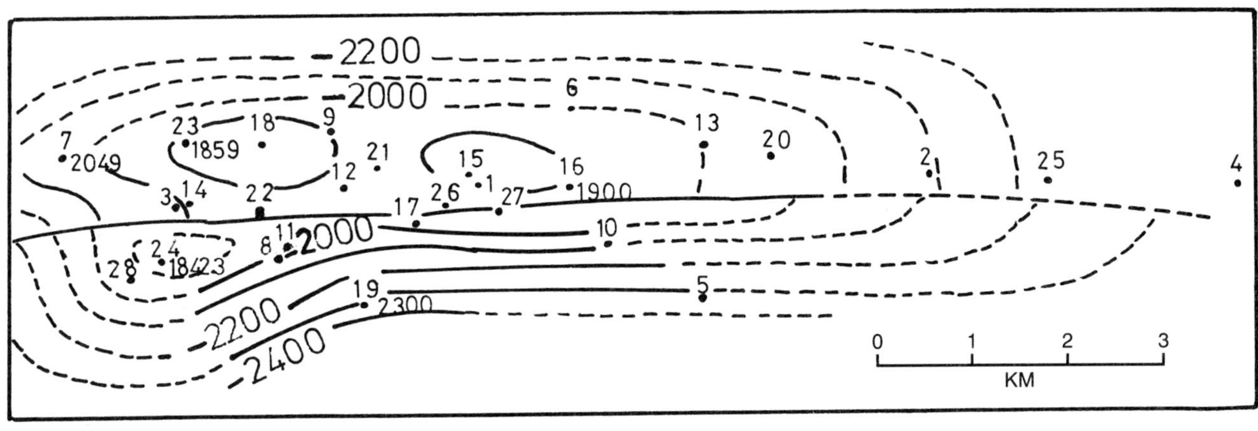

C

Figure 2. Structure contour maps, Ain Zalah oil field. (From Elzarka and Ahmed, 1983. Used with permission of Journal of Petroleum Geology.) (A) Structure on top of Shiranish Formation. (B) Structure on top of Mashurah Formation. (C) Structure on top of Qamchuqa Formation. Contour interval is 100 m and values are subsea. Contour lines are dashed where speculative.

oil in the field. This is summarized as follows. Oil was generated in possible source rocks of middle and Upper Cretaceous age (Qamchuqa and Mashurah formations) in the troughs of a basin to the south of the field (Figure 5F). Overburden load caused by continuous sedimentation and rock diagenesis squeezed the generating oil (Figure 5E). The process of oil and water migration started from the south of the field in the Late Cretaceous. Faulting may have opened up channels, causing inter-pay connection (Figure 5D). The oil, only partially trapped in the crest of the downthrown, southern anticlinal block in the fractured limestones of the Qamchuqa, continued migrating along and across the fault to

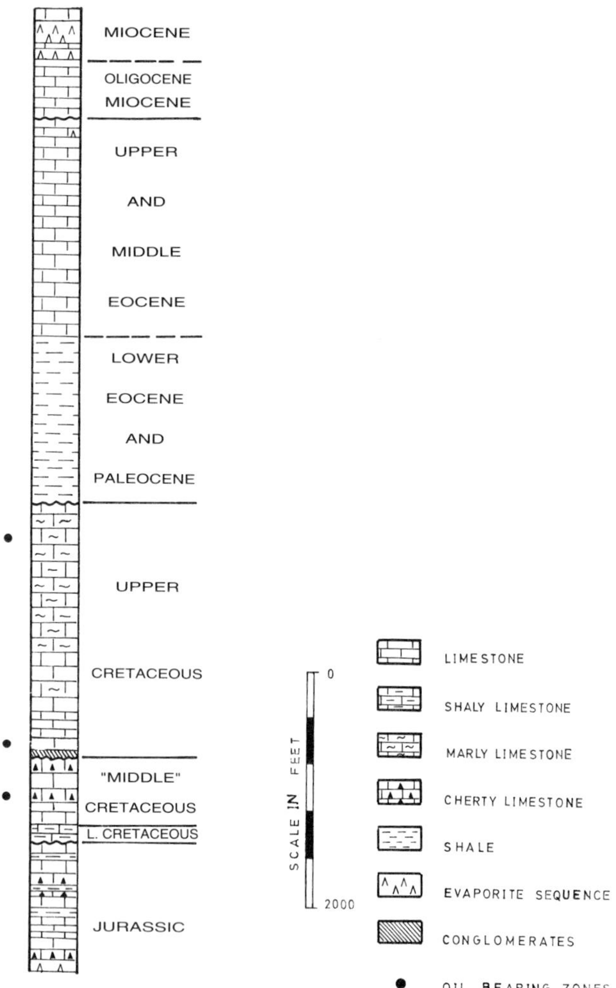

Figure 3. Stratigraphic section, Ain Zalah oil field. (After Daniel, 1954.) See Table 1 for formation names.

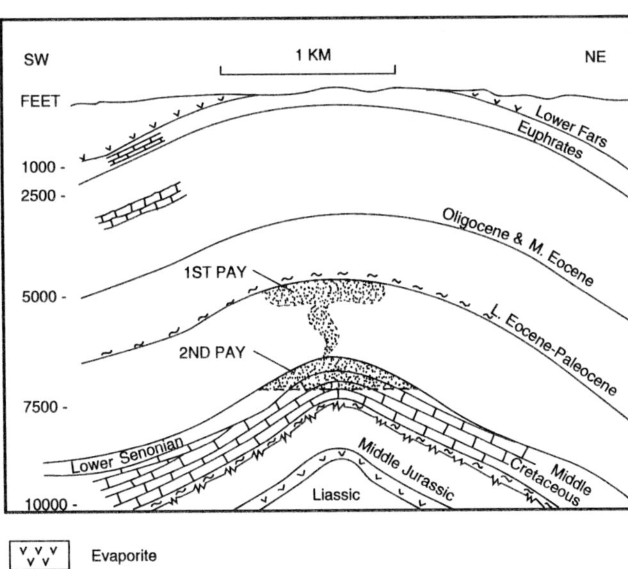

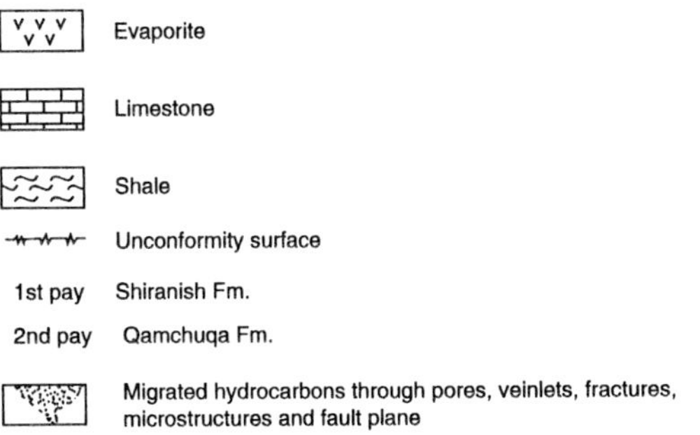

Figure 4. Cross section, Ain Zalah oil field, showing the generalized stratigraphy, the first and second pay (upper and lower reservoirs), and their connection through fractures. (After Dunnington, 1958.)

the crest of the northern upthrown block (Figure 5C). Further upward migration led to entrapment of oil in the crests of the Shiranish formation structure (Figure 5A and B). The surface of the unconformity, e.g., between the Upper and middle Cretaceous, provided an avenue for the movement of fluids, laterally and then vertically along fractures. The oil-water contact is tilted and may be evidence of hydrodynamics in the eastern section of the field.

Reservoirs

Stratigraphy, Lithology, and Depths

The upper reservoir consists of fractured, marly, globigerinal limestone of the upper Shiranish Formation (Maastrichtian to late Campanian age) at an average depth of 1814 m. The lower reservoir within the fractured limestones of the Qamchuqa Formation (Campanian age), which also contain appreciable amounts of oil, are assigned to the lower reservoir. This lower reservoir is comprised of dolomitic limestone and organic debris. It is pelleted and saturated with oil. The two reservoirs are separated by approximately 600 m of barren, tight, marly Shiranish limestone.

Depositional Setting

The Shiranish, Mashurah, and Wajnah sediments (Table 1) were deposited in a continuously subsiding basin (Ahmed et al., 1986). Some local parts of the basin oscillated between intertidal and supratidal conditions during deposition, especially in the middle Cretaceous. The basin was generally deeper during the Late Cretaceous.

Porosity and Permeability

The average porosity of samples from the upper part of the Shiranish Formation reservoir is 3.3% and decreases with depth. The porosity of the Mashurah and Qamchuqa formations averages 1.1% and increases with depth. The measured permeabilities for some rock samples of the producing

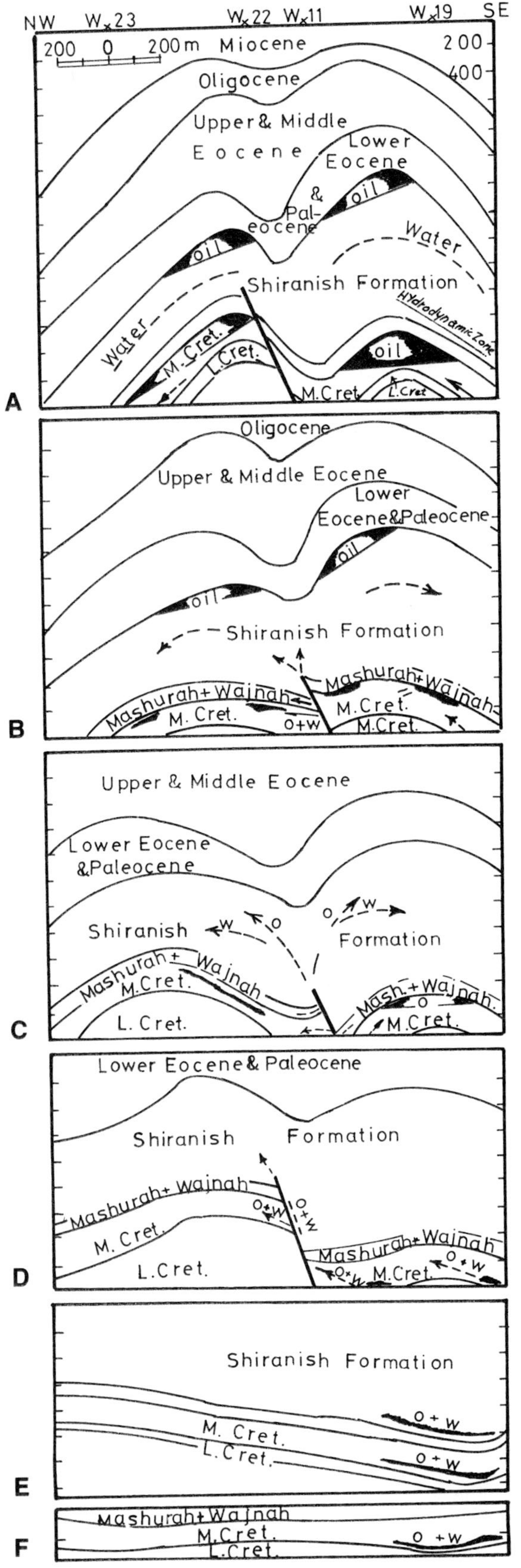

Figure 5. Northwest-southeast paleotectonic profiles across Ain Zalah oil field showing the structural development and mode of oil migration and accumulation. (From Elzarka and Ahmed, 1983. Used with permission of Journal of Petroleum Geology.)

formations reach a maximum of 1 md. Veinlets of mixed calcite-dolomite composition as well as stylolitic seams are common. The veinlets are characterized by a sharp boundary with matrix, indicating a tectonic origin. The calcite fillings that formed perpendicular to the wall of the veins suggest that the calcite cement was deposited in an open space. The stylolitic seams are always associated with dolomitization, where the euhedral dolomite rhombs are concentrated along these dark argillaceous seams. Only fracturing makes it possible for the two zones to be producing reservoirs.

Pay Zone Thicknesses

The gross reservoir thickness is 704 m for the Shiranish and 165 m for the Qamchuqa. The average net thicknesses are 91 and 100 m, respectively.

Hydrocarbon Characteristics

Hydrocarbon type is paraffinic and averages 31.5° API for Shiranish oil and 30.7° API for Qamchuqa oil. The initial gas/oil ratio (GOR) is 250 ft^3/bbl. The average wt% sulfur is 2.6 for Shiranish oil and 2.5 for Qamchuqa oil.

Fluid Flow Characteristics

The reservoir data indicate that the anticipated in-place oil reserves of the upper Shiranish pay zone are *more* than those of the Qamchuqa pay zone, while the recoverable reserves are *less*. A pressure drop, which took place during Shiranish pay zone production, was countered by water injection in the northern flank of the structure. This supplemented the natural water drive and maintained reservoir pressure. The oil-water contact is tilted, created by an active hydrodynamic regime along the eastern flank of the structure (Elzarka and Ahmed, 1983).

Development Consideration

Primary production is restricted mainly to the western part of the field, especially from the upper pay zone (Shiranish Formation), where the best wells are clustered in a pear-shaped area. The average well spacing is about 1.5 km. The oil wettability of the formation is responsible for a large quantity of oil that will be left behind after oil production from the upper pay zone. The production from this zone has a natural water drive system.

Faults

The faults were recognized by subsurface mapping techniques (isopach mapping, Hart and Hay, 1974). The earliest faulting identified is that forming east-northeast to west-southwest grabens during early Shiranish deposition. Hart and Hay (1974) delineated a graben in the western part of the field that is probably bounded by nearly vertical normal or gravity faults. They also delineated other cross faults of a more speculative position. In the east, a major reverse fault was deduced by well data that showed a repeated section (Hart and Hay, 1974, their figure 2). Elzarka and Ahmed (1983) interpreted a major east-west pivot fault, dipping due south. The eastern part of this rotational fault is a reverse fault, as evident from the repeated section, but is a normal fault in the western part.

Faulting created secondary openings, fractures, and veinlets. The veinlets may have opened new channels, causing inter-pay connection. Oil fractionation to lighter and heavier components is good evidence of the imperfect sealing and barriers to fluid movement by the faults (Al-Shahristani and Al-Atyia, 1972).

Oil Source Rocks

The Mashurah and Qamchuqa black bituminous shales (Upper and middle Cretaceous) are possible source rocks for the oil (Dunnington, 1958). There is no information concerning the TOC, kerogen type, and maturation level of these nonetheless organic-rich rocks. The process of oil formation and migration started in the south of the field during the Late Cretaceous.

EXPLORATION CONCEPTS

Regional Play

Ain Zalah field is located in the northwest portion of the Zagros fold zone, a highly folded area extending from Iran through northern Iraq and across Southern Turkey. Numerous oil fields have been found on anticlines within this folded zone (Figure 1B). Qayarah, Najmah, and Jawan fields are all associated with culminations on the same fold axis, while Qasab field is on a parallel structure. Kirkuk is a very large northwest-southeast-trending anticline consisting of three major domal culminations: Khurmala, Avanah, and Baba. Bai Hassan and Jambur fields near the Kirkuk field produce oil from similar structures. All of these fields were discovered by surface and gravity exploration methods a few years before World War II. Kirkuk, one of the giant oil fields of the Middle East, was discovered in 1925 near the gas seeps of the "Eternal Fires." The reservoirs in these fields are in Tertiary and Cretaceous age limestones.

General Application of Geologic Parameters

Ain Zalah and Butmah produce oil from fractured Cretaceous limestone reservoirs on anticlinal structures. These fields lie in a region where the depositional environments and structural events were quite favorable for oil generation, migration, and accumulation.

Lessons

Like most northern Iraq oil fields, the Ain Zalah field was discovered before World War II using surface geology and geophysical (gravity) investigations. The success ratio of wells drilled reached 60%. These same techniques would be applicable today, along with modern seismic mapping.

The drop in pressure in the upper pay zone was successfully countered by water injection into wells on the flanks of the structure.

ACKNOWLEDGMENTS

The management of Iraq Company of Oil Operations (I.C.O.O.) is acknowledged for providing data and materials needed to accomplish this work.

REFERENCES CITED

Ahmed, W. A., 1980, Subsurface studies of the Upper Cretaceous formation in Ain Zalah oil field, northern Iraq: M. Sc. Thesis, Baghdad University.

Ahmed, W. A., M. H. Elzarka, and M. El-Dabbas, 1986, Petrographical and geochemical studies of Cretaceous carbonate rocks, Ain Zalah oilfield, North Iraq: Journal of Petroleum Geology, v. 9, p. 429-438.

Al-Shahristani, H., and M. J. Al-Atyia, 1972, Vertical migration of oil in Iraqi oil fields, evidence based on vanadium and nickel concentrations: Geochemica and Cosmochemica Acta, v. 36, p. 929-935.

Bally, A. W., and S. Snelson, 1980, Realms of subsidence: Canadian Society of Petroleum Geologists, Memoir 6, p. 9-94.

Daniel, E. J., 1954, Fractured reservoir of Middle East: AAPG Bulletin, v. 38, p. 774-815.

DeGolyer, E., 1946, Oil exploration in the Middle East: The Mines Magazine, p. 493-494.

Dunnington, H. V., 1958, Generation, migration, accumulation and dissipation of oil in Northern Iraq, *in* L. G. Weeks, ed., Habitat of oil: AAPG, p. 1194-1251.

Dunnington, H. V., 1967, Stratigraphical distribution of oil fields in the Iraq, Iran, Arabia Basin: Journal of the Institute of Petroleum, London, v. 53, p. 129-161.

Elzarka, M. H., and W. A. Ahmed, 1983, Formational water characteristics as an indicator for the process of oil migration and accumulation at the Ain Zalah field, Northern Iraq: Journal of Petroleum Geology, v. 6, p. 165-178.

Fairbridge, R. W., and H. Badoux, 1960, Slump blocks in the Cretaceous of northern Syria (abs.): Geological Society of London Proceedings, no. 1581, p. 113-117.

Hart, E., and J. T. C. Hay, 1974, Structure of Ain Zalah field: AAPG Bulletin, v. 58, p. 973-981.

St. John, B., A .W. Bally, and H. D. Klemme, 1984, Sedimentary provinces of the world—hydrocarbon productive and nonproductive: AAPG, p. 2-35.

SUGGESTED READING

Aguilera, R., and H. K. van Poolen, 1979, Studies showing occurrence of fractured reservoirs: The Oil and Gas Journal,

p. 71-76. Brief description of the petrophysics of Ain Zalah reservoirs.

Al-Shahristani, H., and M. J. Al-Atyia, 1972, Trace elements in Iraqi oils and their relation to the origin and migration of these oils: Eighth Arab Petroleum Congress, Algeria, paper no. 98(B-3).

Al-Shahristani, H., and A. G. Hanna, 1974, Bromine as an indicator of oil migration within Iraqi oilfields: Geochemica et Cosmochemica Acta, v. 38, p. 1303-1306.

Avedisian, A. M., and A. H. Hammosh, 1970, Oil gravity and age variations in Middle East crudes: Journal of the Geological Society of Iraq, v. 3, p. 41-53.

Thode, H. G., and J. Monster, 1970, Sulphur isotope abundances and genetic relations of oil accumulations in Middle East basin: AAPG Bulletin, v. 54, p. 627-637.

Appendix 1. Field Description

Field name .... *Ain Zalah field*
Ultimate recoverable reserves .... *195,000,000 bbl*

Field location:

- **Country** .... *Iraq*
- **State** .... *Ninao, Northern Iraq*
- **Basin/Province** .... *Zagros Folded Zone*

Field discovery:

- **Year first pay discovered** .... *Upper Cretaceous Shiranish Formation 1939*
- **Year second pay discovered** .... *"Middle" Cretaceous Qamchuqa Formation 1950*

Discovery well name and general location:

- **First pay** .... *Well No. 1 about long. 42°35′E, lat. 36°47′N*
- **Second pay** .... *Well No. 14 about 3 km west of well no. 1*

Discovery well operator .... *IPAC (Iraqi Petroleum & Associated Companies)*

- **Second pay** .... *INOC (Iraqi National Oil Corporation)*

IP

- **First pay** .... *1500 (Shiranish Formation)*
- **Second pay** .... *8000 (Qamchuqa Formation)*

All other zones with shows of oil and gas in the field:

Age	Formation	Type of Show
Upper Cretaceous	*Mashurah*	*Oil stains*
"Middle" Cretaceous	*Gir Bir*	*Oil stains*
Lower Cretaceous	*Upper Sarmord*	*Traces of oil*

Geologic concept leading to discovery and method or methods used to delineate prospect
The prospect leading to discovery was delineated by surface geological mapping of a low range of anticlinal hills.

Structure:

Province/basin type .... *Bally 41; Klemme II Ca*

Tectonic history
Syndepositional normal faulting created horst and graben structures in upper Campanian rocks of northern Iraq. The first phase of faulting again created a horst and graben structure of northeast-southwest trend. The Taurus-Zagros orogenic movement of late Miocene-Pliocene time created the present structure of Ain Zalah field.

Regional structure
Steeply folded foothill type anticline the strike of which curves to the west and then to the southwest.

Local structure
Steeply folded double plunging anticline (19 km long, 5 km wide) affected by an east-west major fault, normal in its western trace and reversed in the eastern due to rotation affecting the eastern block, thus being of the pivot or scissor type.

Trap:

Trap type(s)

Faulted anticlinal trap with two fractured carbonate reservoirs.

Chronostratigraphy:

System	Series	Formation	Depth (m)
Neogene	*Miocene*	*Lower Fars, Jeribe, Dhiban, Euphrates*	*375*
Paleogene	*Oligocene*	*Anah Azkand, Bajwan, Baba Shurau, Sheik Alas*	*420*
	Eocene (m & u)	*Avanah, Jadala*	*585*
	Eocene (l) & Paleocene	*Aaliji*	*1295*
Cretaceous	*Upper*	*Shiranish*	*1815*
		Mashurah	*2520*
		Wajnah	*2605*
	"Middle"	*Gir Bir*	*2620*
		Qamchuqa	*2665*
	Lower	*Sarmord*	*2830*
Jurassic	*Middle*	*Sargelu*	*2860*
	Lower		*3335*

Reservoir characteristics:

Number of reservoirs *2*

Formations .... *Upper Cretaceous Maastrichtian-Campanian upper Shiranish Formation; "Middle" Cretaceous Albian Qamchuqa Formation*

Depths to tops of reservoirs .... *Shiranish, 1815 m; Qamchuqa, 2665 m*

Gross thickness (top to bottom of producing interval) .... *704 and 165 m for upper and lower, respectively*

Net thickness—total thickness of producing zones

Average .... *91 and 100 m for upper and lower*

Maximum .... *500 and 212 m for upper and lower*

Lithology

The upper Shiranish pay zone is fractured marly, gray to dark gray shaly limestone; the Qamchuqa pay zone is fractured dolomitic limestone containing much organic debris

Porosity type .... *Very low matrix porosity with fractures*

Average porosity .... *3.3% and 1.3% for upper and lower*

Average permeability .... *1 md*

Seals:

Upper

Formation, fault, or other feature .... *Paleocene/lower Eocene Aaliji Formation*

Lithology .... *Shale*

Lateral

Formation, fault, or other feature .... *Unconformity surface between middle and Upper Cretaceous*

Lithology .... *Shale*

Source:

Formation and age .... *Upper Cretaceous Mashurah and middle Cretaceous Qamchuqa*

Lithology .... *Shale*

Average total organic carbon (TOC) .... *Not determined*

Maximum TOC .... *Not determined*

Kerogen type (I, II, or III) .... *Not determined*

Vitrinite reflectance (maturation) .... *Not determined*
Time of hydrocarbon expulsion .... *Late to post-Cretaceous*
Present depth to top of source .... *Mashurah, 2520 m; Qamchuqa, 2665 m*
Thickness .... *Mashurah, 85 m; Qamchuqa, 165 m*
Potential yield .... *NA*

Appendix 2. Production Data

Field name .... *Ain Zalah field*

Field size:

Proved acres .... *9300 ha*
Number of wells all years *28*
Current number of wells *10*
Well spacing .... *0.5 to 2 km*
Ultimate recoverable .... *195 million bbl*
Cumulative production .... *170 million bbl (as of 1976)*
Annual production .... *Approx. 3 million bbl (as of 1976)*
Present decline rate .... *10% (after water injection commenced)*
Initial decline rate *5%*
Overall decline rate .... *70% (before water injection commenced)*
Annual water production .... *1 million bbl (as of 1976)*
In place, total reserves .... *226 million bbl*
In place, per acre foot .... *97 bbl*
Primary recovery .... *150 million bbl*

Oil characteristics:

Type .... *Paraffinic*
API gravity .... *31.5°–30.7°*
Base .... *Paraffinic*
Initial GOR .... *250 ft³/bbl*
Sulfur, wt% *2.6–2.5*

Field characteristics:

Average elevation .... *Approx. 700 m*
Initial pressure .... *16.70–21.92 MPa (upper and lower, respectively)*
Present pressure .... *13.26 MPA (after commencement of water injection)*
Pressure decline .... *476 kPa*
Temperature .... *64.4°–85.0°C (upper and lower)*
Geothermal gradient .... *2.8°C/100 m*
Drive .... *Gas-water*
Oil column thickness .... *Avg. 100 m*
Oil-water contact .... *Tilted*
Water salinity, TDS .... *11,822 and 39,542 ppm (upper and lower)*

Transportation method and market for oil and gas:
NA

Asab Field—United Arab Emirates
Rub Al Khali Basin, Abu Dhabi

A. S. ALSHARHAN
The Desert and Marine Environment Research Center
United Arab Emirates University
Al-Ain, United Arab Emirates

FIELD CLASSIFICATION

BASIN: Rub Al Khali
BASIN TYPE: Foredeep
RESERVOIR ROCK TYPE: Limestone
RESERVOIR ENVIRONMENT OF DEPOSITION: Platform
RESERVOIR AGE: Lower Cretaceous
PETROLEUM TYPE: Oil
TRAP TYPE: Anticline with Multiple Pays

LOCATION

Abu Dhabi is the largest emirate in the United Arab Emirates (U.A.E.), with a total area of about 25,500 mi^2 (66,000 km^2), and contains the largest oilfields in the U.A.E., such as Bab, Bu Hasa, Zakum, Umm Shaif, and Asab (Figure 1). Asab field is approximately 95 mi (150 km) southwest of Abu Dhabi Island, 55 mi (85 km) southeast of Bab field and 60 mi (100 km) east-southeast of Bu Hasa field (Figure 1). The field lies within the limits of the concession granted in 1939 to Petroleum Development Limited (now Abu Dhabi Petroleum Company or ADCO). The terrain is mostly sand dunes separated by flat sand corridors and occasional very small, inland sabkha flats.

Asab is one of the giant onshore fields producing from Lower Cretaceous carbonates, with estimated reserves of 10.5 billion barrels of oil (Appendix 1) (Mallinson and Sharp, 1975). It is located on the north side of the Rub Al Khali basin (Figure 2), on the eastern shelf of the Arabian platform. In this setting, thick sedimentary deposits—primarily carbonates—accumulated, and intermittent tectonic movement resulted in relatively gentle folding, faulting, and salt movements.

HISTORY

Exploration activities in Abu Dhabi Emirate were started in 1936. In that year, Petroleum Development Trucial Coast (PDTC), a subsidiary of the Iraq Petroleum Company (IPC), was established. In January 1939, the PDTC signed an agreement for oil exploration covering the area of the Trucial Coast (since 1971 called the United Arab Emirates).

The original concession, which covered all the onshore areas of Abu Dhabi, was made in 1939 with PDTC, which acquired a 75-year concession, covering the land, islands, and territorial waters. In 1950, test drilling was commenced, and the first well was drilled in Ras Sadr (RS-1 well) (Figure 1), but no oil was discovered. In 1953, drilling began on a well designated Murban-1 (renamed Bab-1), 65 mi (105 km) southwest of Abu Dhabi Island, but it was shut down because of technical problems. Oil was first discovered in commercial quantities in the Bab field in 1959; subsequently oil was discovered at Bu Hasa in 1962, at Asab in 1964, at Shah in 1966, and at Sahil in 1967 (Figure 1). In July 1962, PDTC was renamed Abu Dhabi Petroleum Company (ADPC), and in 1982, it was renamed Abu Dhabi Company (ADCO) for onshore oil operations. The original agreement was amended to include an annual relinquishment of a percentage of the concession. Part of the area relinquished by ADCO has been granted to and explored by other oil companies.

While development of Bu Hasa was in progress in the early 1960s, further seismic reconnaissance and semi-detailed regional studies were carried out by ADPC. These studies, with seismic surveys initiated in 1959 in eastern Abu Dhabi, identified the structural trend that includes the Shah, Asab, and Sahil fields (Figure 3). This led to the drilling of the

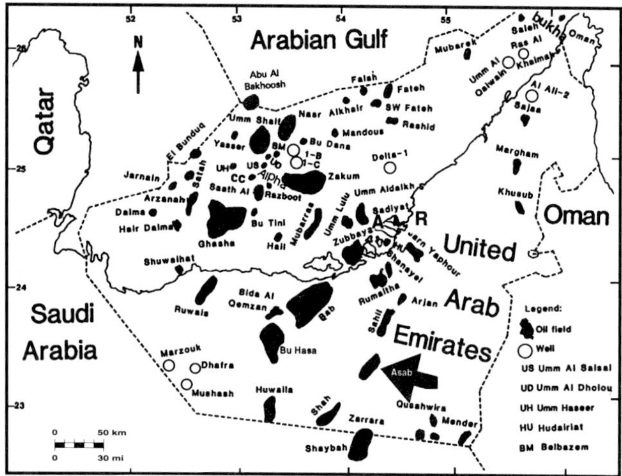

Figure 1. Location map of the United Arab Emirates (U.A.E.), showing Asab and other major oil fields. A is Abu Dhabi Island and R is Ras Sadr (RS-1).

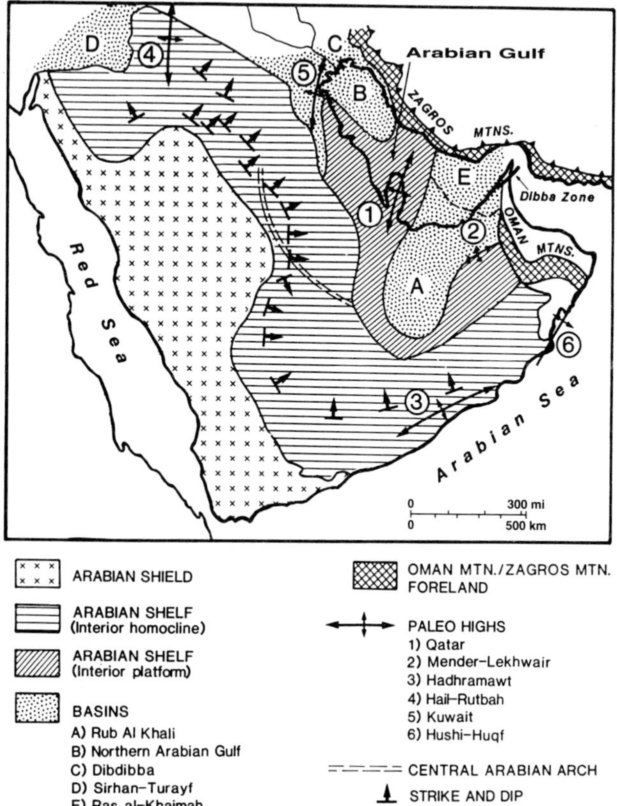

Figure 2. Geologic setting and structural provinces of the Arabian Peninsula. (Modified from Powers et al., 1966; Alsharhan and Nairn, 1986.)

discovery well, Abu Jidu-1 in May 1965 (Figure 5). The well, drilled to a total depth of 11,388 ft (3471 m) in the Upper Jurassic, discovered sweet oil of about 39° API in the Lower Cretaceous (Kharaib Formation, zones B and C). The field was originally named Abu Jidu, but subsequent drilling found it to be comprised of two separate structures that were designated as the Asab and Shah fields. As a result, the first six wells, which were originally named JD-1 through JD-6, were renamed Asab 1 to 6 with the SB prefix (e.g., Sb-1 through Sb-6).

After the successful exploratory well, four further appraisal wells were drilled in 1966 to (1) delineate the structure, (2) determine the fluid and rock properties, and (3) determine the production potential. About 460,000 BOPD were produced from 22 wells in January 1974 from the Lower Cretaceous carbonates of the Thamama Group (Kharaib Formation, zones B and C). Zone B produced about 400,000 BOPD and zone C about 60,000 BOPD.

Initial development plans for Kharaib B and C zones were based on a depletion rate of 5% per year for the initial reserves (Mallinson and Sharp, 1975). Reservoir engineering studies predicted that in order to maintain production at the same level, to maximize the ultimate recovery, and to achieve adequate sweep efficiency, full pressure maintenance by peripheral water injection should be adopted. Plans for pressure maintenance were completed two years after the field came on stream. Full implementation will be 740,000 barrels of water per day (bw/d) injected into the two reservoirs (zones B and C).

DISCOVERY METHOD

In the Emirate of Abu Dhabi, the first onshore gravity and magnetic surveys were begun in 1947 and completed in 1948. Gravity and magnetic surveys were again conducted during 1953 to 1955 and again in 1959 and 1960. These surveys indicated that the broad and gentle structural swells of Abu Dhabi were reflecting deep structural differentiation and some uplift at great depth in the vicinity of the Murban (Bab–Bu Hasa) area (Hajash, 1967). The results of this early survey provided a useful basis for later seismic surveys.

A seismic reflection survey first conducted in the area of Abu Dhabi in late 1949 and early 1950 led to the drilling of the first exploratory well in the region (Ras Sadr-1). In the vicinity of Murban, seismic surveys indicated a broad domal uplift generally coincident with the Murban gravity high, upon which Murban-1 was drilled in 1953 (Hajash, 1967). Seismic surveys were continued intermittently in the area and eventually led to the drilling of the discovery well Abu Jidu-1 (Asab-1) in 1965.

STRUCTURE

Abu Dhabi lies on the eastern part of the Arabian platform and is situated in the central part of the Rub Al Khali basin (Figure 2), a primarily Tertiary feature superimposed on the stable Arabian shield.

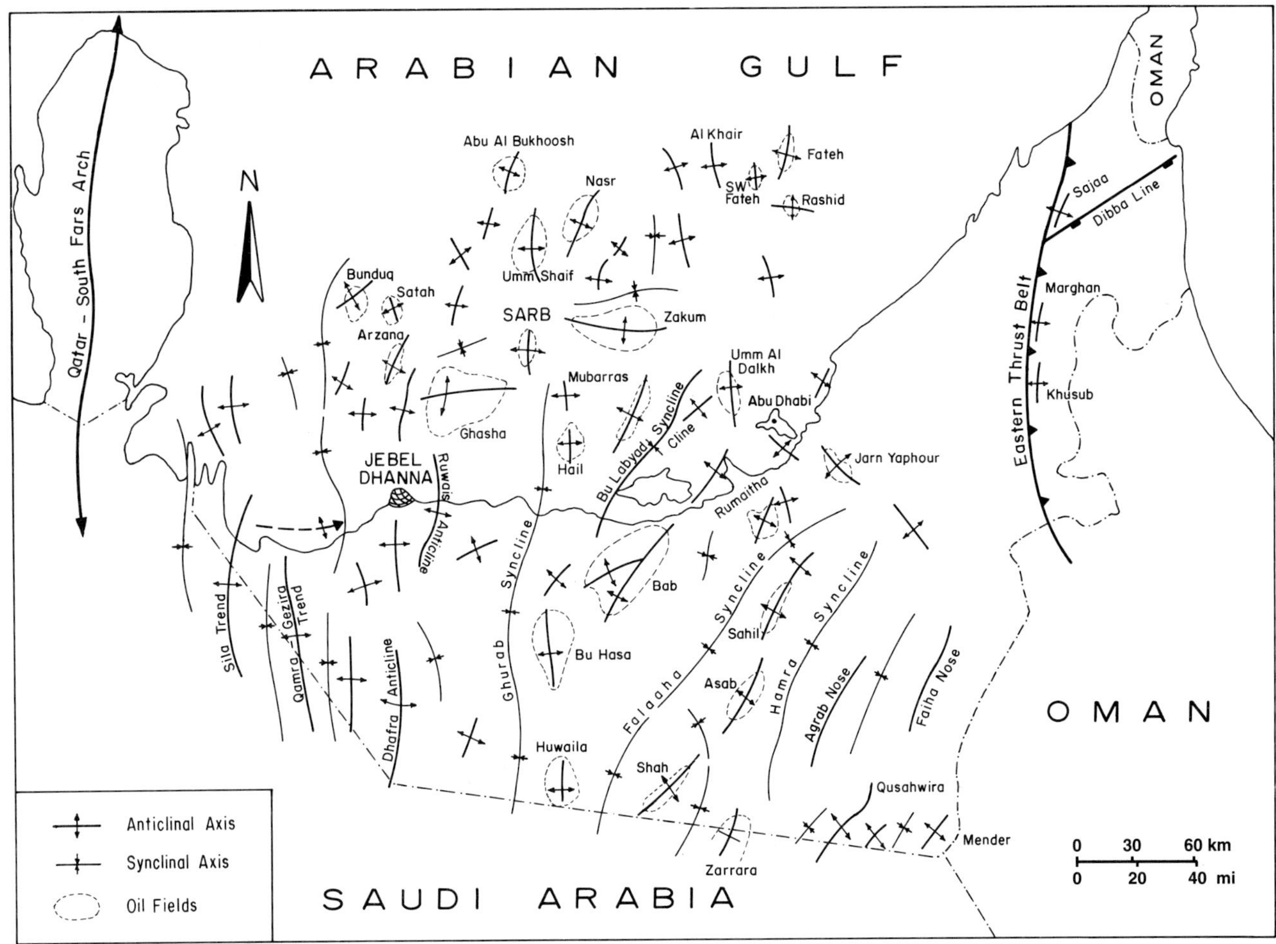

Figure 3. Main structural trends in Abu Dhabi, structural expression near top Thamama Group (Lower Cretaceous). (Modified from Schlumberger, 1981.)

This area has undergone subsidence relatively greater than that of the adjoining region, resulting in the accumulation of a thick sedimentary section. Relatively gentle folding, faulting, and salt movements were induced by periodic tectonic activity.

Obviously, structural features such as the central Arabian arch and the Rub Al Khali basin had a profound influence on the structural geology of Abu Dhabi (Figure 2). These two features dip gently northeast, i.e., toward the Arabian Gulf. General thickening of the sedimentary section in this direction has been observed, indicating an increasing rate of subsidence.

The structural character of the geology of Abu Dhabi is typical of the Arabian platform, generally characterized by gentle, simple folds, some of which are quite large. Between the anticlinal trends are corresponding synclinal trends, some regionally prominent, such as the Falaha and Ghurab synclines (Figure 3). Folds are mostly periclinal in style, with flank dips generally less than 5° (Schlumberger, 1981). The regional dip is eastwards. Figure 3 shows the trends of fold axes of Abu Dhabi as mapped near the top of the Thamama Group (Lower Cretaceous). Four main fold axes alignments can be recognized: north-south, east-west, northwest-southeast, and northeast-southwest (Alsharhan, 1989).

The north-south aligned folds (the Arabian folds) are dominant in the area and are believed to result from deep-seated basement faults (Schlumberger, 1981; Alsharhan, 1985a, b). This trend is clearly displayed in the western part of Abu Dhabi, formerly part of the southern Arabian Gulf Infracambrian salt basin. Western Abu Dhabi, especially the offshore, is characterized by small circular or subcircular salt-cored dome structures that possess flank dips in the 5 to 10° range (Figure 3). Salt piercement structures resulting in islands are characteristic of this area. It is probable that interference between the Zagros and Arabian trends has given rise to some of these salt dome islands.

The southeast of Abu Dhabi (where Asab is located), east of the Falaha syncline, is characterized by northeast-southwest alignment of fold axes (Shah, Asab, and Sahil trends). To the northeast, northwest-southeast fold trends are dominant (Jarn Yaphour trend). In the central part of offshore Abu Dhabi, east-west-aligned folds (Zakum and Ghasha trends) related to the Zagros folding are superimposed on the older north–south-trending growth structure of the Arabian folds. Interference between the Arabian and Zagros folds, together with Tertiary salt diapirism, has greatly modified these structures (Schlumberger, 1981; Alsharhan, 1985a, b). In the central offshore area, almost perpendicular folds swing around Zakum field, while onshore the north-south Bu Hasa trend swings north of Bu Hasa to a northeast-southwest alignment. Diapiric salt is believed to underlie most of the structures in Abu Dhabi.

The Abu Dhabi fields manifest the long period of structural growth typical of Arabian Gulf structures. It is generally accepted that deep-seated faulting and salt flow initiated and perpetuated structural growth in the region. There are indications of salt pillows and salt piercement domes in the southern Arabian Gulf area. Most of the structural growth took place during the period from early Late Cretaceous through middle Miocene (Figure 4). Regional tilting toward the northeast occurred during late Miocene. The Miocene–Pliocene sediments reflect little indication of the underlying structures, while Holocene deposits are essentially flat-lying.

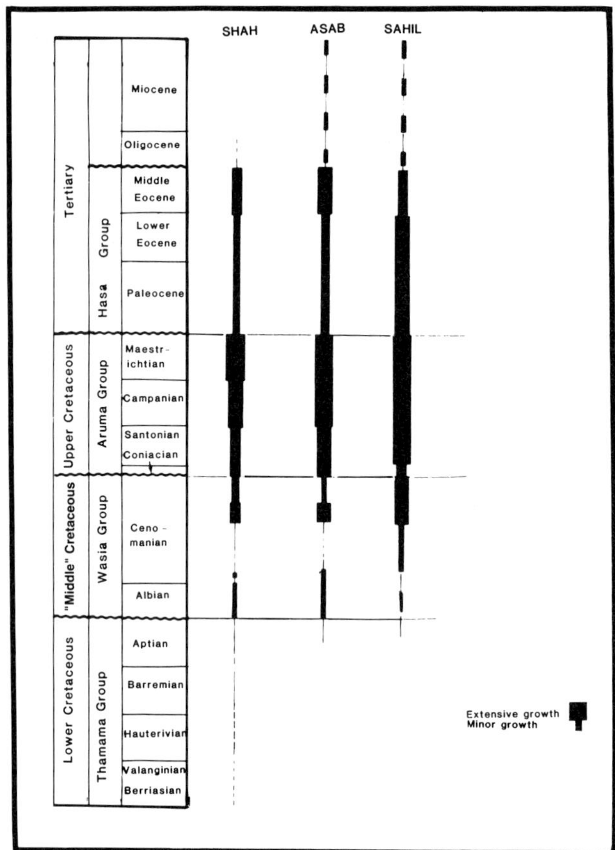

Figure 4. Timing of structural growth of Asab, Sahil, and Shah oil fields, Abu Dhabi, U.A.E. (After ADCO, 1986.)

Asab field is an elongated domal anticlinal structure, 16 mi (26 km) long and 6 mi (9 km) wide, trending northeast-southwest. The south flank is broader than the north flank (Figure 5). It has more than 600 ft (183 m) of structural relief at the top of the Lower Cretaceous (Thamama) level and an areal closure of about 90 mi^2 (230 km^2) at the top of the Thamama Group Kharaib Formation (ADCO staff, 1986). The field is thought to result from pillows formed from movement of deeply buried Hormuz salt of Infracambrian age. The structural history of the Shah-Asab-Sahil fields is shown in Figure 4. The Asab structure started to evolve during "Middle" Cretaceous times, particularly in the Cenomanian. However, a slight uplift of about 50 ft (15 m) started during Albian times. The main structural uplift took place during late Campanian–early Maastrichtian times. Only 25% of structural uplift occurred during the "Middle" Cretaceous, while about 65% took place during the Upper Cretaceous. The final structural uplift occurred during the Tertiary orogeny. Tilting toward the south is seen as sediment thickening southward and as an unconformity in the late "Middle" Cretaceous to the north.

The structural high is highly faulted, especially in the south, southeast, and central part of the field. The faults are believed to have developed in response to salt piercement or compressive forces because this field is located close to the orogenic zone that affected the Oman Mountains in the Late Cretaceous. Structural dip for the structural contour map shown on Figure 5 is derived, in part, from high-resolution dipmeter tests. Dips on the western flank are 3 to 4°, and on the eastern flank, about 2 to 3°.

Asab field is a normal faulted structure. Faults are concentrated in the crestal area and the southeast and the western flanks (see also Hassan et al., 1990) (Figures 5 and 6). Most of the faulting occurred in late "Middle" Cretaceous and Campanian times. During Maastrichtian time the Asab field was uplifted to its maximum amplitude when faulting was no longer active. All the faults encountered are normal, with a maximum throw of 90 ft (27 m) produced in Santonian time. Some growth faulting also occurred during Campanian time.

There is no evidence of effective vertical communication between any of the Thamama reservoir zones. However, lateral communication may occur within Kharaib zone B as a result of faulting. The evidences of faulting are abrupt changes in structural contour strikes, abrupt changes in dips between wells, anomalous porosities, and abrupt formation thinnings (fault-outs).

REGIONAL STRATIGRAPHY

The Permian to Recent section encountered in the deepest wells of Abu Dhabi has a maximum thickness of about 21,000 ft (6506 m). A feature of sedimentation since the beginning of the Late Permian has been

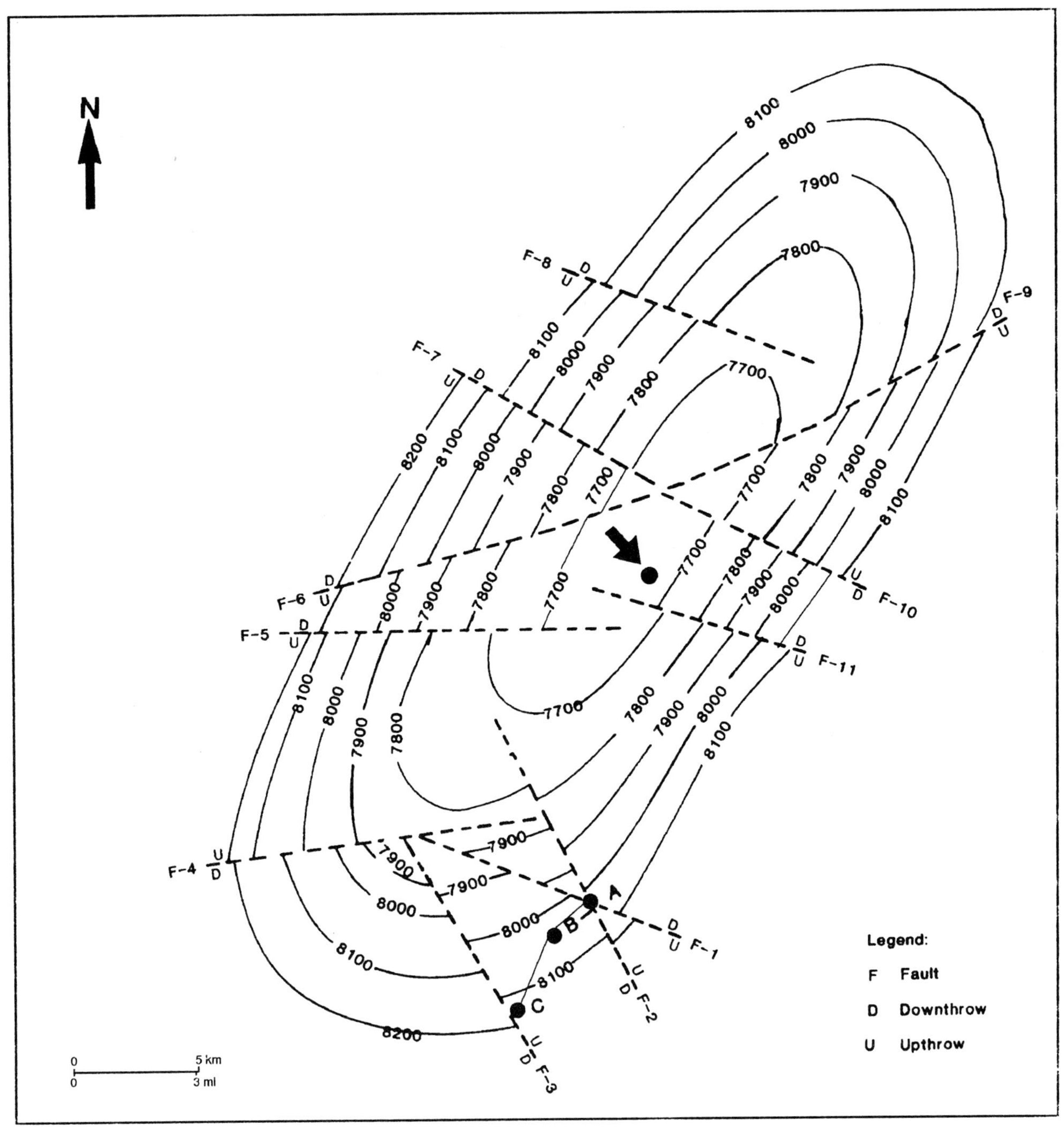

Figure 5. Structural contour map of top of zone B, showing 11 normal faults in Asab field. Discovery well is the well spot near center of structure. Correlation A, B, and C are shown on Figure 6.

the dominance of shelf carbonates (Figure 7). Evaporites are of secondary importance. There are minor influxes of argillaceous and arenaceous clastics, but these are rare (Alsharhan, 1989).

The first sediments laid down after the Hercynian orogeny were the continental clastic sediments of pre-Khuff time (Permian-Carboniferous?). In Late Permian time, a marine transgression occurred and a carbonate platform was established over the area. During this time, the limestone/dolomite and minor anhydrite beds of the Khuff Formation were developed. Recent exploratory drilling into this formation has resulted in significant gas discoveries at Zakum and Umm Shaif fields. The carbonate platform was maintained throughout Early Triassic times with the deposition of shales, limestones, and dolomites of the Sudair Formation. The Gulailah (Jilh) Formation above the Sudair is composed of alternating sequences of anhydrites, dolomites, limestones, and minor shales. Following these,

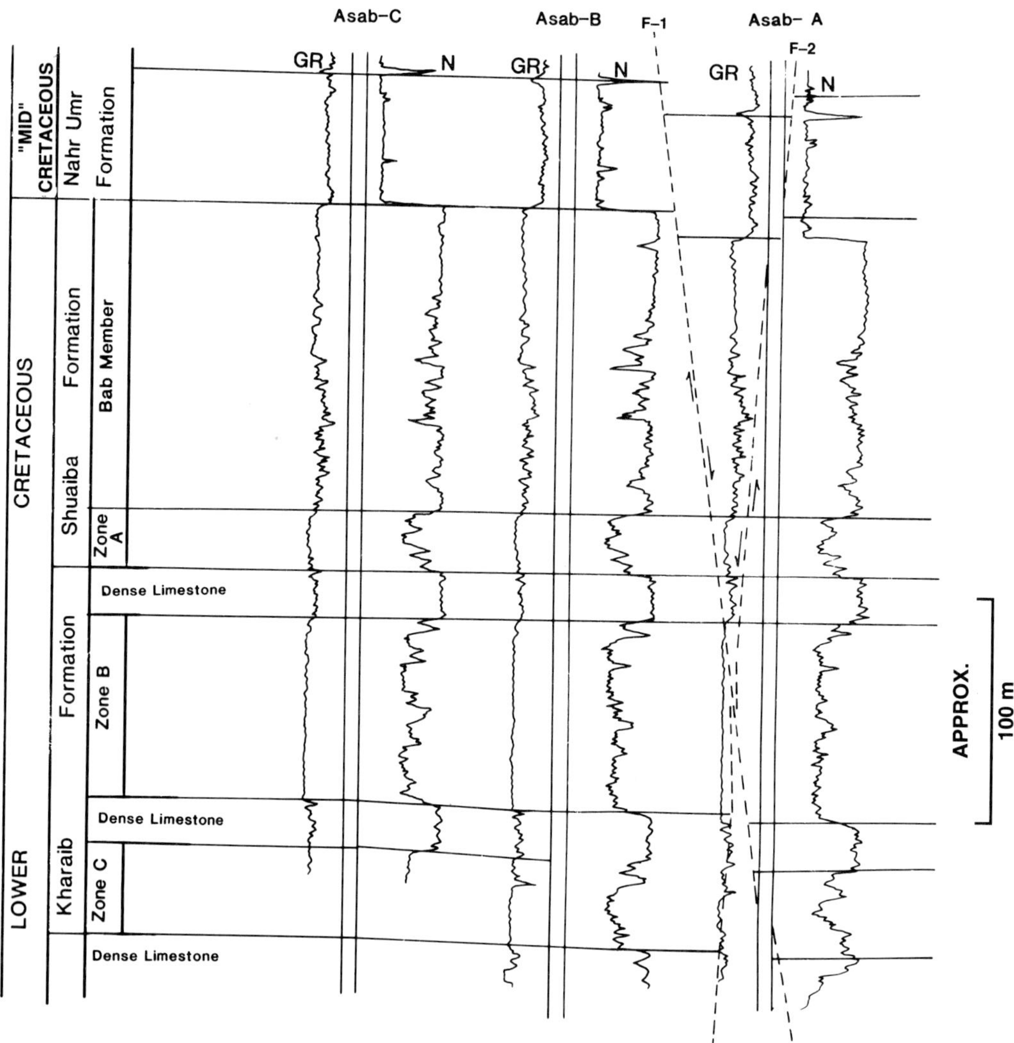

Figure 6. Well logs correlations and displacement of faults. The locations of wells Asab A, Asab B, and Asab C are shown on Figure 5.

during late Triassic time, the climate was less arid and a relative drop in sea level preceded the deposition of the continental sandstones and siltstones of the Minjur Formation in the onshore areas. The Triassic sequence has moderate source rock development, and gas shows occur in some offshore structures.

At the end of the Triassic, the Arabian block was subjected to a major period of uplift and erosion, resulting in the development of a regional unconformity. This was followed by regional subsidence, and a major marine transgression gave rise to a wide carbonate platform. During this marine megacycle spanning Early and Middle Jurassic, epeiric carbonates were deposited across the Arabian Gulf (Alsharhan and Kendall, 1986). Deposition commenced with a mixture of terrigenous clastics and carbonates of the Marrat Formation, followed by the deeper water argillaceous limestones and dolomites of the Hamlah and Izhara formations, and ended with the shallow water, moderate- to high-energy limestones of the Araej Formation. This formation contains considerable accumulations of hydrocarbon in the offshore areas (Alsharhan, 1989).

During Late Jurassic, there was a gradual transition from deep water (west) to shallow shelf (east) sedimentation that graded through shoal and lagoonal facies and culminated in supratidal conditions. In early Late Jurassic time, intrashelf basinal sediments, consisting largely of argillaceous limestones, were deposited to the west, while to the east a cleaner limestones facies, the Diyab/Dukhan Formation, was deposited. These rocks are an excellent hydrocarbon source for the Upper Jurassic–Lower Cretaceous reservoirs of the area. The Diyab/Dukhan was followed by the cyclic deposition of limestones, dolomites, and anhydrites of the Arab (Qatar/Fahahil) Formation, which forms the principal Upper Jurassic reservoirs in the western part of

Series and Stages		Group	Formation	Member	Generalized Lithological Description	Fields
Quaternary			Recent		Intercalation clastic and carbonates.	
Miocene	M.		L. Fars		Dolomitic limestone and Anhydrite.	
Miocene	E.		Clastic/Evaporite		Intercalation clastic, limestone and anhydritic dolomite.	Oil in Mandous
Oligocene			Asmari			
Eocene	Middle	Hasa	Damman		Nummulitic Pack/wackestones and dolomitic limestones. Gray shales.	
Eocene	Early	Hasa	Rus		Dolomitic limestones and anhydrite.	
Paleocene		Hasa	Umm Er Radhuma		Wackestones, dolomitic limestones. Gray shales.	
Late Cretaceous	Maestrichtian	Aruma	Simsima		Pack/wackestones with dolomite and dolomitic limestones.	Oil only in Shah from Simsima. Both Simsima and Fiqa pass to Gurpi in eastern offshore Abu Dhabi.
Late Cretaceous	Companian	Aruma	Fiqa		Marls with marly limestones.	
Late Cretaceous	Santonian	Aruma	Halul		Chalky bioclastic wackestones.	Oil in Mandous and Alkhair
Late Cretaceous	Coniacian	Aruma	Laffan		Shales with minor argillaceous limestones.	
"Middle" Cretaceous	Cenomanian	Wasia	Mishrif	Ruwayda; Tuwayil	Mishrif (M): rudistid limestones. Ruwayda (R): argillaceous limestones. Tuwayil (T): shales	Oil in Umm Al Dalkh and Hair Dalma from Mishrif
"Middle" Cretaceous	Cenomanian	Wasia	Shilaif		Argillaceous lime mudstones and wackestones.	
"Middle" Cretaceous	Albian	Wasia	Mauddud		Bioclastic wackestones and packstones.	
"Middle" Cretaceous	Albian	Wasia	Nahr Umr		Varigated shales with rare lenses of sandstone and limestone.	
Early Cretaceous	Aptian	Thamama	Shuaiba	Bab	Shuaiba: Shelf carbonates and/or rudist reefs. Bab Member: argillaceous limestones.	Oil in Bu Hasa, Zakum, Umm Shaif, Jarn Yaphour, and Mandous
Early Cretaceous	Barremian	Thamama	Kharaib		Mud supported sediments graded to grain supported sediments.	Oil in Bab, Asab, Sahil, Zakum, Umm Shaif
Early Cretaceous	Hauterivian	Thamama	Lekhwair	Zakum	Packstones, wackestones with some grainstones and mudstones.	Oil in Zakum
Early Cretaceous	Valanginian to Berriasian	Thamama	Habshan		Lime mudstones, wackestones, Oolitic grainstones, subordinate dolomites.	Oil in Bab, Asab and Zakum
Upper Jurassic	Tithonian	Si La	Hith	Asab	Hith: Anhydrite massive and nodular, subordinate limestones and dolomite. Asab: Oolitic limestones and dolomites	No anhydrite in eastern Abu Dhabi
Upper Jurassic	Kimmeridgian	Si La	Arab	Qatar	Grainstones, packstones, dolomite with subordinate anhydrite.	Oil in Arab Fm. in Umm Shaif, Bunduq, Nasr, Saath Al Razz Boot, Ghasha, Satah, Dalma, Jarnain. Gas in Hair Dalma and Bab
Upper Jurassic	L. Kimmeridgian to Oxfordian	Si La	Arab	Fahahil	Argillaceous lime mudstones, wackestones with subordinate dolomitic limestones.	
Upper Jurassic	L. Kimmeridgian to Oxfordian	Si La	Diyab / Dukhan	Dukhan; Tuwaiq		Very minor oil in Zakum from Diyab.
Middle Jurassic	Bathonian		Araej	U. Araej; Uwainat; L. Araej	Peloidal grainstones, packstones, wackestones and lime mudstones.	Oil and Gas in Umm Shaif and western offshore area. Gas in Zakum
Middle Jurassic	Bajocian		Izhara		Argillaceous limestones and subordinate shales.	
Early Jurassic	Liassic		Hamlah		Dolomite and dolomitic limestones.	
Early Jurassic	Liassic		Marrat		Dolomitic limestones, dolomite and subordinate shales. (Does not exist in offshore areas.)	
Triassic	Late		Minjur		Continental clastics with minor limestones. (Does not exist in offshore areas.)	Rates as a gas source in offshore area
Triassic	Middle		Jilh (Julailah)		Dolomite, anhydrite, dolomitic limestones and occasional shales.	
Triassic	Early		Sudair		Dolomite, shales and minor limestones.	
Permian	Late		Khuff		Dolomitic limestones, limestones, dolomite and some anhydrites. Anhydrite marker bed. Dolomite and dolomitic limestones. Minor limestones	Gas in Zakum, Umm Shaif, Nasr, Satah, Hair Dalma, Abu Al Bukhoosh and Hail
Permian	Early		Wajid		Quartzitic sandstones, shales, siltstones and thin beds of dolomites and anhydrites.	
Carboniferous			Wajid			

Figure 7. Generalized stratigraphic column in Abu Dhabi fields.

the offshore areas. The anhydrites of the Hith Formation were deposited in the west and provide an excellent seal over the Arab reservoirs. To the east, shallow shelf conditions prevailed and here the dolomites and limestones of the Asab Formation accumulated.

In the Lower Cretaceous a transgression over the marine shelf flooded most of the Arabian Gulf (Alsharhan and Nairn, 1986). The earliest Cretaceous sediments are the dominantly mixed oolitic, dolomitic limestones and lime mudstones (the Habshan Formation). These were followed by a long period of cyclic carbonate sedimentation with alternating shelf limestones and deeper water limestones of the Lekhwair and Kharaib formations. In central Abu Dhabi an intrashelf basin was formed in Aptian time where argillaceous limestones and shales (Bab Member) accumulated, and at the fringes or rim of this basin, rudistiid and algal buildups were deposited (Shuaiba Formation). Collectively, these formations make up the Thamama Group. This Lower Cretaceous sequence contains significant hydrocarbon accumulations in that area of the U.A.E. where the Kharaib Formation is the major reservoir in the Asab, Sahil, and Bab fields; the Shuaiba Formation is the main reservoir in the Bu Hasa field; and the Lekhwair Formation is the main reservoir in the Zakum field.

Toward the end of the Aptian, epeirogenic movement terminated the deposition of the Thamama Group and resulted in a period of subaerial exposure. This was followed by deposition of the Wasia Group ("Middle" Cretaceous), starting with the transgressive shales of the Nahr Umr Formation that forms the seal over the Shuaiba Formation reservoirs (Alsharhan, 1987a). Toward the end of Nahr Umr sedimentation, shale deposition diminished to the point that marine carbonate deposition commenced again across the area, beginning with the Mauddud Formation in which a transition occurred from shallow-marine sedimentation to somewhat deeper water conditions. A basin then developed in central Abu Dhabi in which the *Pithonella* limestones of the Shilaif Formation were deposited. These formed good source rocks for some "Middle" Cretaceous reservoirs. At the basin margins, shallow shelf sedimentation led to the development of the foraminiferal-algal-rudist wackestones/packstones/grainstones of the Mishrif Formation. Oil accumulations have been found in this formation in both structural and stratigraphic traps (e.g., Umm al Dalkh and Fateh fields). At the end of Shilaif deposition, a minor period of uplift affected the central basin area. As a result, the Tuwayil and Ruwaydha Members (Mishrif equivalents) accumulated in a gradually constricting basin. At the end of Cenomanian time, a major period of emergence and erosion terminated the deposition of the Wasia Group (Alsharhan and Nairn, 1988).

Deposition of the Upper Cretaceous Aruma Group began with the transgressive Laffan Shale, which unconformably overlies the Wasia Group in all parts of the basin and acts as a seal for the Mishrif reservoir (Alsharhan and Nairn, 1990). However, some areas, such as southeast and northeast Abu Dhabi, remained emergent throughout the whole of Late Cretaceous time. Above the Laffan, the Halul Formation is characterized by shallow shelf carbonates. Following their deposition, renewed subsidence and the associated transgression in Campanian times resulted in the deposition of the basinal shales and limestones of the Fiqa Formation. A shallowing of the basin led to the deposition of the shallow shelf carbonates of the Simsima Formation. In late Maastrichtian time, a regressive facies developed over a large part of the Arabian Gulf region, resulting in nondeposition at the end of the Cretaceous (Alsharhan and Nairn, 1990). Oil shows in the Aruma Group have been found in several offshore structures, and a commercial oil accumulation has been found in the Simsima Formation in the Shah field. The source for this latter oil is probably the underlying "Middle" Cretaceous Shilaif Formation.

A widespread transgression occurred during the Paleocene, which resulted in deposition of thin basal shales, followed by the shallow shelf limestones, both of the Umm Er Rhadhuma Formation. In the early Eocene, restricted shelf conditions prevailed and the carbonate evaporitic sequence of the Rus Formation was deposited, followed by the widespread nummulitic limestones of the Dammam Formation. During late middle Eocene time, widespread emergence of the Arabian platform occurred. However, to the east, the Asmari Formation (Oligocene) was deposited, consisting mainly of shelf limestones. It was overlain by a thick Miocene sequence of interbedded carbonates, salt, anhydrite, shales, and clastics. Minor oil accumulations have been discovered in the Asmari in the offshore areas of the U.A.E. (e.g., in Mandous field).

TRAP

The Asab structure is a simple elongate anticline (Figure 5). The trap is both structural, with closure provided by salt doming and growth faults, and stratigraphic, with a favorable combination of depositional and diagenetic trapping conditions. The field has more than 600 ft (183 m) of structural relief at the top of the Thamama and an areal closure of about 88 mi^2 (229 km^2) at the top of Thamama zone B.

Classification and Description of Asab Reservoir Rocks

The Barremian Kharaib Formation is the main producing formation in the field. It is divided into two zones (B and C). The zones are lithologically heterogeneous and their variations in composition and texture (Figure 8), reflecting variable conditions of deposition, affect reservoir behavior. An understanding of the characteristics, relationships, and

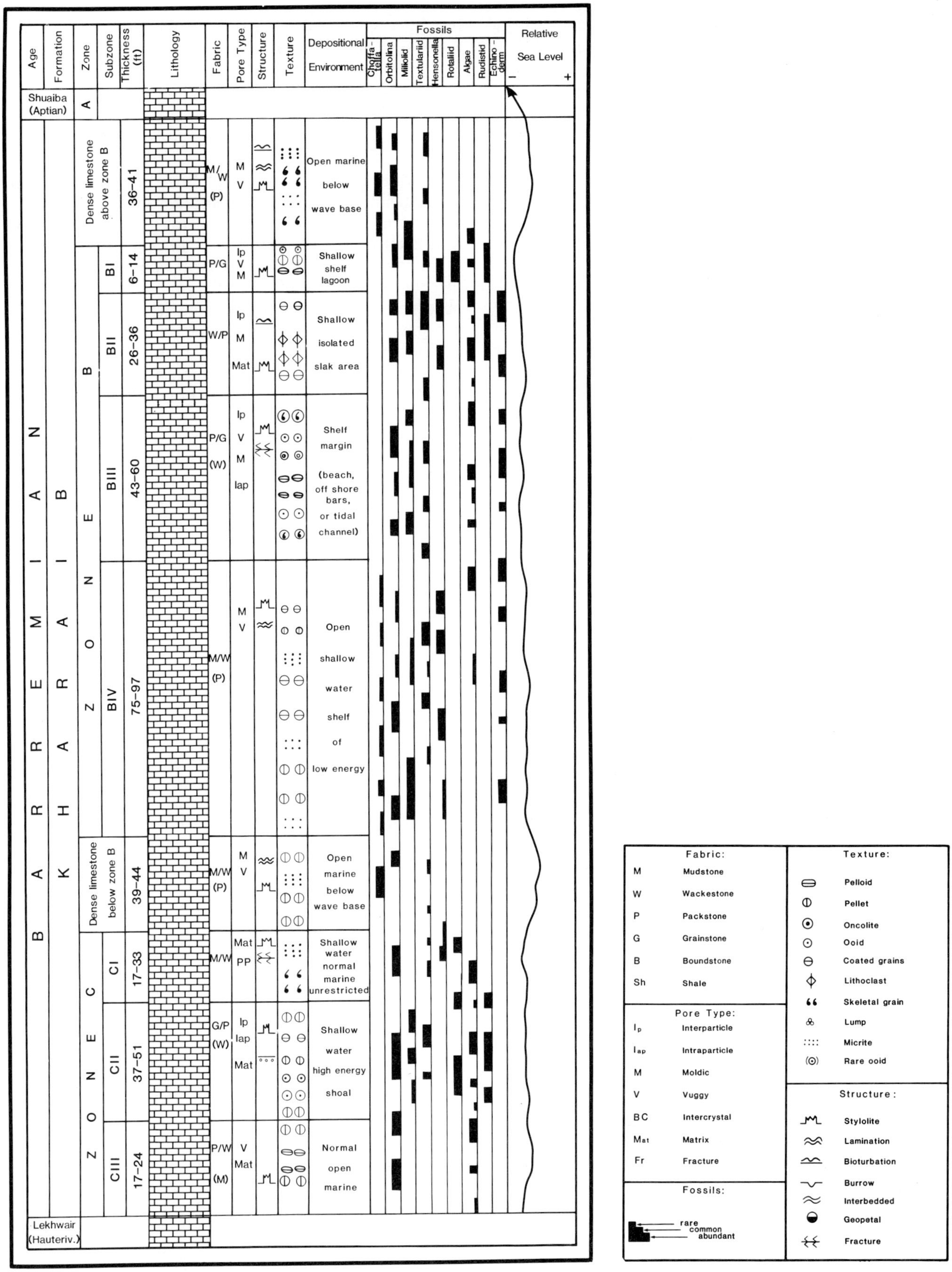

Age	Formation	Zone	Subzone	Thickness (ft)	Lithology	Fabric	Pore Type	Structure	Texture	Depositional Environment	Fossils: Choffatella, Orbitolina, Miliolid, Textulariid, Hensonella, Rotaliid, Algae, Rudistid, Echinoderm	Relative Sea Level − +
Shuaiba (Aptian)		A										
BARREMIAN	KHARAIB	Dense limestone above zone B		36–41		M/W (P)	M V			Open marine below wave base		
		B ZONE	BI	6–14		P/G	Ip V M			Shallow shelf lagoon		
			BII	26–36		W/P	Ip M Mat			Shallow isolated slak area		
			BIII	43–60		P/G (W)	Ip V M Iap			Shelf margin (beach, off shore bars, or tidal channel)		
			BIV	75–97		M/W (P)	M V			Open shallow water shelf of low energy		
		Dense limestone below zone B		39–44		M/W (P)	M V			Open marine below wave base		
		C ZONE	CI	17–33		M/W	Mat PP			Shallow water normal marine unrestricted		
			CII	37–51		G/P (W)	Ip Iap Mat			Shallow water high energy shoal		
			CIII	17–24		P/W (M)	V Mat			Normal open marine		
Lekhwair (Hauteriv.)												

Fabric:	
M	Mudstone
W	Wackestone
P	Packstone
G	Grainstone
B	Boundstone
Sh	Shale

Pore Type:	
Ip	Interparticle
Iap	Intraparticle
M	Moldic
V	Vuggy
BC	Intercrystal
Mat	Matrix
Fr	Fracture

Fossils: rare, common, abundant

Texture: Pelloid, Pellet, Oncolite, Ooid, Coated grains, Lithoclast, Skeletal grain, Lump, Micrite, Rare ooid

Structure: Stylolite, Lamination, Bioturbation, Burrow, Interbedded, Geopetal, Fracture

Figure 8. Composite log shows lithology, pore types, environment of deposition, fossil occurrences, and relative sea level in the Kharaib Formation.

distribution of these facies types is important to the evaluation of the potential and performance of the subzones. Zone B is about 178 ft (52 m) thick and is the main reservoir in the field, with the best reservoir characteristics in the upper part where there is good permeability. Zone C is about 90 ft (28 m) thick and is separated from zone B by about 45 ft (14 m) of dense limestone; it is another producer in the field, but deteriorated permeability of the reservoir reduces its potential.

Earlier studies of zone B reservoir of the Bab field by Harris et al. (1968) have emphasized the marked change between the lower, dominantly lime mud-supported section, and the upper, predominantly grain-supported section. This division holds true in the Asab field as well, and here, reservoir behavior is controlled by lithological variations within the grain-supported subdivision. To facilitate this understanding of Asab subzones, Johnson and Budd (1975) and Alsharhan (1985b) classified zone B, based on lithology, into six facies units. These are the lime mud facies (LM); the Miliolid-algal lump facies (M3A), the Rudist facies (R1 and R2); and the Miliolid-pellet facies (M1 and M2) (Figure 9). Each is laterally continuous with only minor variations in thickness, analogous to facies of the Kharaib Formation of Bab field (Harris et al., 1968; Alsharhan, 1985b). Zone B can therefore be divided into four subzones (BI to BIV), which are labeled and described in ascending order. Subzones may be defined as porous intervals separated by correlatable stylolites. The relationships between these subzones, thicknesses, stylolites, and petrophysical characteristics are shown in Figure 10.

Zone C also can be subdivided into five lithological facies units (Figure 9). These are the wackestone/packstone facies (WP); the oolitic/intraclast grainstone/packstone facies (G1 and G2); the peloidal/intraclast packstone facies (P); and lime mudstone/wackestone facies (MW). The relationships between the reservoir subzonation and the facies units are shown in Figure 9. The initial study of zone C by Johnson and Budd (1975) introduced alphanumeric units as in zone B. In the Asab field, two prominent stylolitized zones, with varying degrees of associated stylolithification, are present in all flank wells in zone C. This latter zone therefore can be further subdivided into three subzones (Johnson and Budd, 1975) that are labeled and described in ascending order—CI, CII, and CIII (Figure 11).

Dense Limestones

The dense limestones that underlie and overlie zone B (Figures 6 and 8) range in thickness from 30 to 46 ft (9.4 to 14.4 m). They are characterized by very fine grained, slightly argillaceous lime mudstones, peloidal wackestones/packstones, and dolomitic mudstones (Figure 12A, B). Foraminiferal skeletal debris bands also occur, containing abundant *Choffatella* sp., orbitolinids, and textulariids. Abundant seams of stylolites and streaks of shales have been observed. Porosity and permeability are typically very poor in these limestones, and consequently they are effective seals. The sediments probably accumulated under slightly deeper, normal open marine conditions (outer shelf margin).

Zone B Reservoir

Subzone BI—B1 consists of fine- to coarse-grained, gray, peloidal packstone to grainstone (Figure 12C), occurring in well-bedded units and occasionally displaying graded bedding. This rock contains intraclasts, peloids, and micrite envelopes; a few microstylolites and grain-to-grain contacts are observed. Echinoderm fragments with syntaxial calcite overgrowth and some baroque dolomite also occur.

The thickness of BI ranges from 6 to 14 ft (2 to 4 m). In both the southern and northern crestal areas, thinner sections are recorded. Thicknesses generally decrease down the flanks.

Porosity is mainly interparticle, but leached vugs and moldic porosity are present. The average porosities are relatively low when compared to those of the other subzones. This is due to the development of stylolite layer D1 over the whole of Asab field. Porosity values range from 27 to 12% (Figure 9). The southern crestal portion of the field has slightly better porosities than does the north. Maximum vertical permeability is 71.4 md (Figure 9), while very low permeabilities occur on the southern flank of Asab as a result of the strong development of stylolite layer D1. The higher permeabilities are developed in the peloidal grainstone facies.

The main biota are miliolids, orbitolinids, textulariids, *Hensonella* sp., *Pseudochrysallidina* sp., *Bacinella irregularis* algae, rotaliids, a few rudistid fragments, and sponge spicules.

BI was deposited in a shallow shelf lagoon. The grainstone fabric and lack of carbonate mud indicate deposition took place in a high-energy environment, probably around islands and tidal channels.

Subzone BII—BII consists of buff, porous wackestone to packstone (Figure 12D), with an abundance of rudistids randomly stacked together in the section and frequently fragmented because of compaction. Denser and slightly argillaceous peloidal packstones and grainstones are also present. Some stylolites and an abundance of wispy seams are observed in the section. Calcite, baroque dolomite, some rhombic dolomite, and authigenic kaolinite are present.

The thickness of BII ranges from 26 to 36 ft (8 to 11 m) and gradually thins down the flanks. This subzone exhibits extremely leached, vuggy, and moldic porosity related to accumulations of rudists. Some interparticle porosity is present. The porosity is rather high and the average values range from 22 to 35% (Figure 9). Anomalously high values observed in some wells (porosity up to 30%) might be caused by fracturing related to faulting. The maximum permeability is 42 md and the minimum is less than 1 md (Figure 9). The best permeability occurs in the crestal wells. The high values are

Formation	Zone	Subzone	Unit	Lithology	Stylolite	Thickness ft (m) max	Thickness ft (m) min	Porosity (%) max	Porosity (%) min	V. Permeability (md) max	V. Permeability (md) min	H. Permeability (md) max	H. Permeability (md) min
KHARAIB FORMATION (BARREMIAN)	ZONE B	I	M1	Miliolid, Pelloidal Pack/grainstones		14 (4.4)	6 (1.8)	26.9	11.9	71.4	0.1	790	0.34
					D1								
		II	R1	Rudistid debris Wacke/packstones		36	26	34.5	22.3	41.5	0.9	504	14.4
			M2	Miliolid, Pelloidal Pack/grainstones									
			R2	Rudistid debris Wacke/packstones		(11)	(8)						
					D2								
		III	M3A	Miliolid, Pelloidal Algal Pack/grainstones		60 (10.3)	43 (13.1)	37.2	21.6	21.3	0.7	83.5	5.8
					D3								
		IV	LM	Lime Mudstones locally Wackestones		97 (29.6)	75 (23)	34	19.7	10.8	0.2	14.3	0.9
	ZONE C	I	MW	Lime Mudstones / wackestones		23 (7.0)	17 (5.2)	20.1	12.2	0.3	0.13	4.6	0.6
					S1								
		II	G1	Oolitic, intraclast Pack/grainstones		51	37	31.6	22.8	15.2	0.73	797	39.8
			P1	Packstones		(15.6)	(11.3)						
			G2	Oolitic, intraclast Pack/grainstones									
					S2								
		III	WP	Well bedded Wacke/packstones		24 (7.3)	17 (5.2)	29.8	18	3.5	0.43	18.5	2.1

Figure 9. Lithofacies units and average porosity and permeability in the Kharaib Formation.

usually related to the presence of well-sorted grainstone facies. Deterioration of the porosity takes place gradually down structure in all directions.

There is an abundance of rudistids (mainly monopleurids). Some textulariids, *Hensonella* sp., miliolids, orbitolinids, echinoderm plates and spines, *Pseudochrysallidina* sp., and *Bacinella irregularis* algae are also present.

The nature of these deposits suggests probable accumulations in small, isolated slack-water areas that periodically developed in shallow water; such areas may have been created by local topographic irregularities. Laminations were produced by the periodic influx of skeletal debris from adjacent areas. The depositional setting ranges from a protected shallow subtidal and intertidal to a high-energy shoal.

Subzone BIII—BIII is a medium- to coarse-grained peloidal packstone to grainstone (Figure 12E) with a few oncolites and micrite envelopes of algal origin. Some wackestone and mudstone textures (Figure 12F) are observed in some parts of the section. Stylolites, insoluble residues, and extensive fractures occasionally developed immediately adjacent to the stylolites. Scattered dolomite with cloudy centers replaces the lime mudstone matrix in the lime mud part of the unit.

The thickness distribution of the BIII is more regular than in subzones BI and BII. The subzone thins from the crest down the flanks. The maximum thickness is 60 ft (18 m) and the lowest value so far is 43 ft (13 m).

Interparticle porosity occurs within the grainstone intervals, and some leached vug and moldic porosity is also present. This subzone has the best porosity of all the subzones. Porosity varies from 21.6 to 37.2%. Porosities decrease on the eastern flank, and the porous interval seems to change rapidly and to thin on the western flank. The average permeability is considerably lower than in subzone BII. The distribution varies eccentrically from the crest, with a maximum 21.3 md on the east flank. The upper part of subzone BIII has, in general, a higher permeability than the lower part. This is due to the

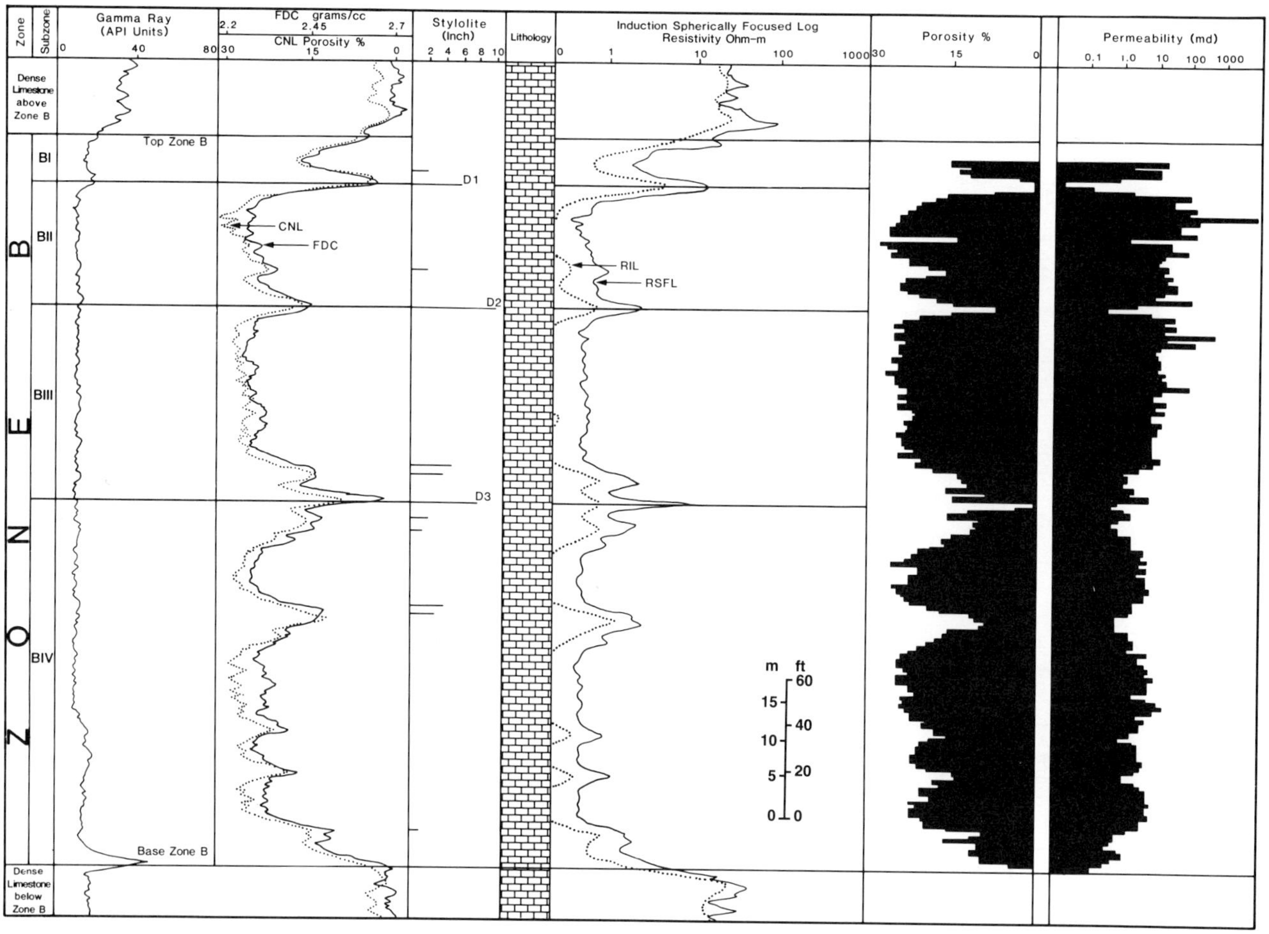

Zone
Subzone
Gamma Ray (API Units)
0 40 80
FDC grams/cc
2.2 2.45 2.7
CNL Porosity %
30 15 0
Stylolite (Inch)
2 4 6 8 10
Lithology
Induction Spherically Focused Log Resistivity Ohm-m
0 1 10 100 1000
Porosity %
30 15 0
Permeability (md)
0.1 1.0 10 100 1000
Dense Limestone above Zone B
Top Zone B
BI
D1
BII
CNL
FDC
RIL
RSFL
D2
BIII
D3
BIV
ZONE B
m ft
60
15
40
10
5 20
0 0
Base Zone B
Dense Limestone below Zone B

Figure 10. Log characteristics and porosity/permeability and stylolites in zone B.

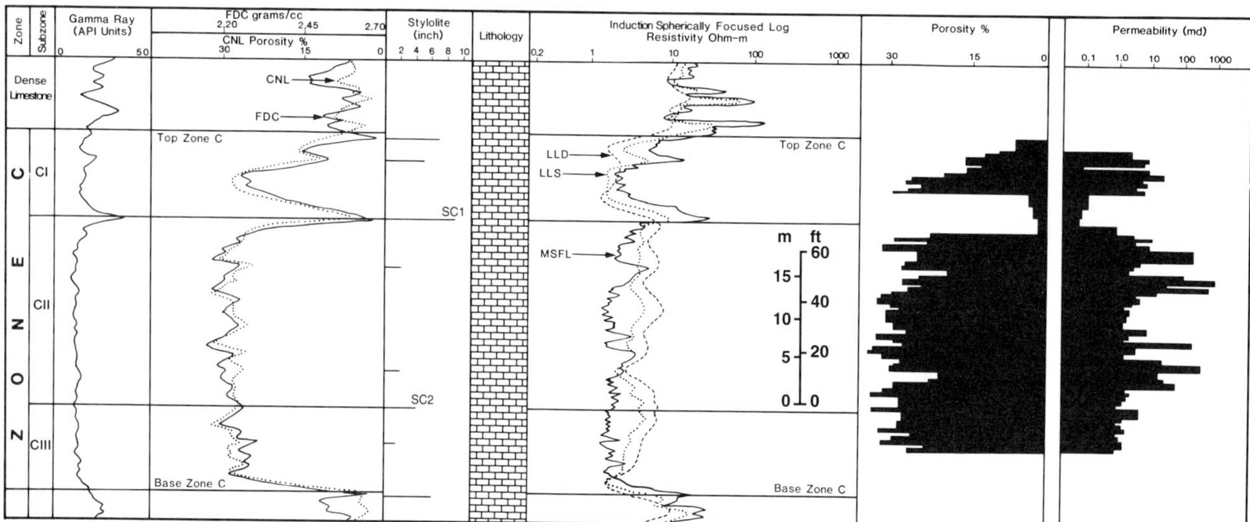

Figure 11. Log characteristics and porosity/permeability and stylolites in zone C.

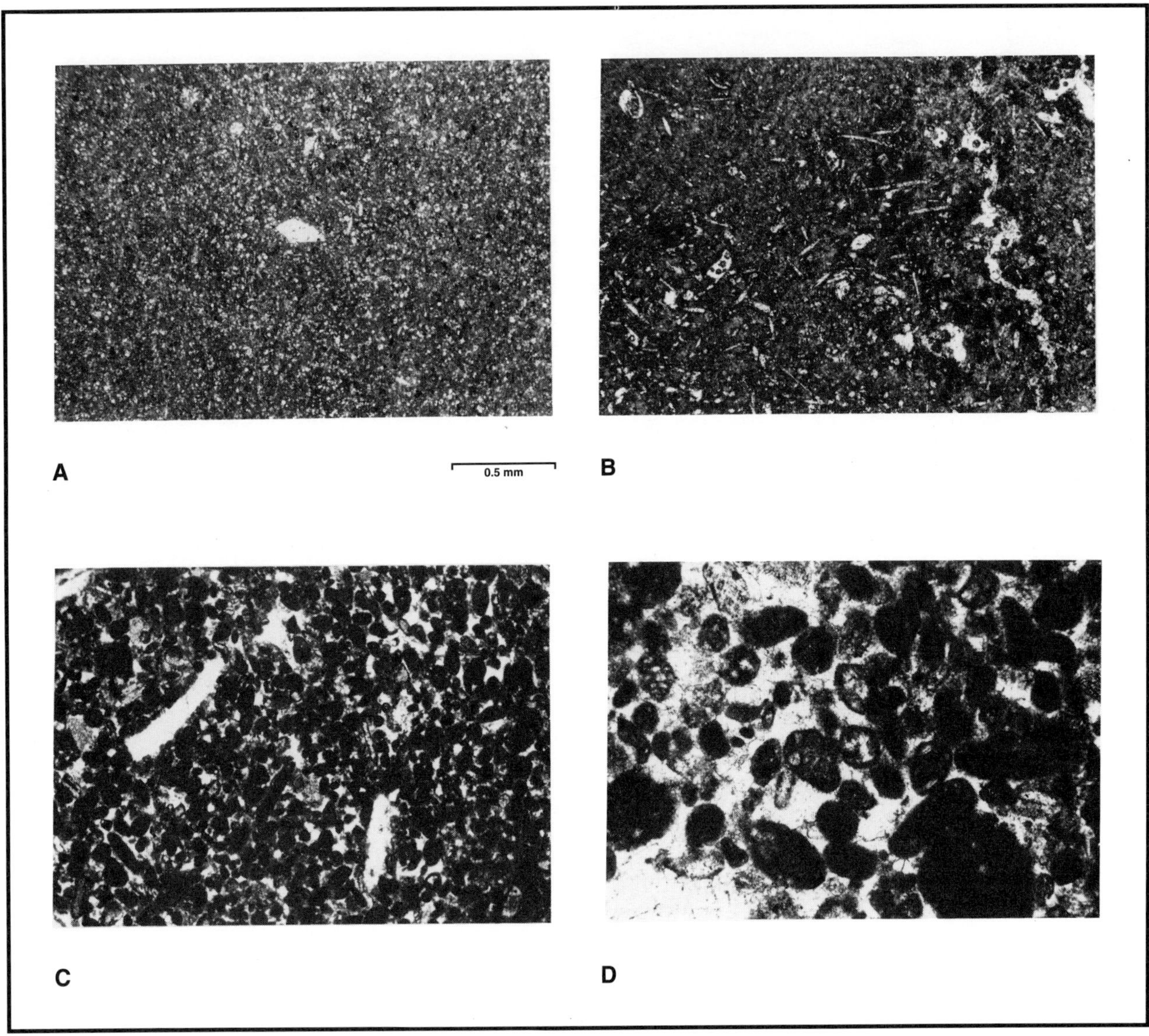

Figure 12. (A) Photomicrograph of dolomitized mudstone with silt-sized grains of detrital quartz. Medium size of authigenic quartz in the center of the photo. The bar scale is 0.5 mm and is the same for all photomicrographs. **(B)** Lime mudstone/wackestone with abun-dant sponge spicules and some textulariids, echinoids, and gastropods. The photo shows fracture filled by calcite. **(C)** Peloidal packstone/grainstones with reworked orbitolinids, miliolids, *Choffatella*, debris of echinoids, and pelecypods. The grains are partially cemented and most of them are blackened. **(D)** Peloidal packstone with abundant *Orbitolina*, miliolids, and textulariids. Most of the grains are blackened with micrited envelope.

textural change from grainstone/packstone to mudstone/wackestone.

The biota are mainly miliolids and orbitolinids with an abundance of codiacean green algae (*Bacinella irregularis* and *Lithocodium aggregatum*), echinoderms, textulariids, and *Pseudochrysallidina* sp.

BIII was deposited in a shelf margin. Variations in lithology suggest that sediments accumulated as beaches, offshore bars, and tidal channels with moderate to high energy.

Subzone BIV—Subzone BIV is the thickest subzone and is composed mainly of microporous, dense mudstone to wackestone (Figure 12G) that grades to intraclastic packstone toward the top of the subzone. Peloids and intraclasts are the major limestone particles of this unit. Dolomite takes the form of cloudy-centered, euhedral rhombohedra scattered throughout the section, replacing lime mud and as local patches of dolomite associated with burrows (Figure 12H). Stylolites occur and are

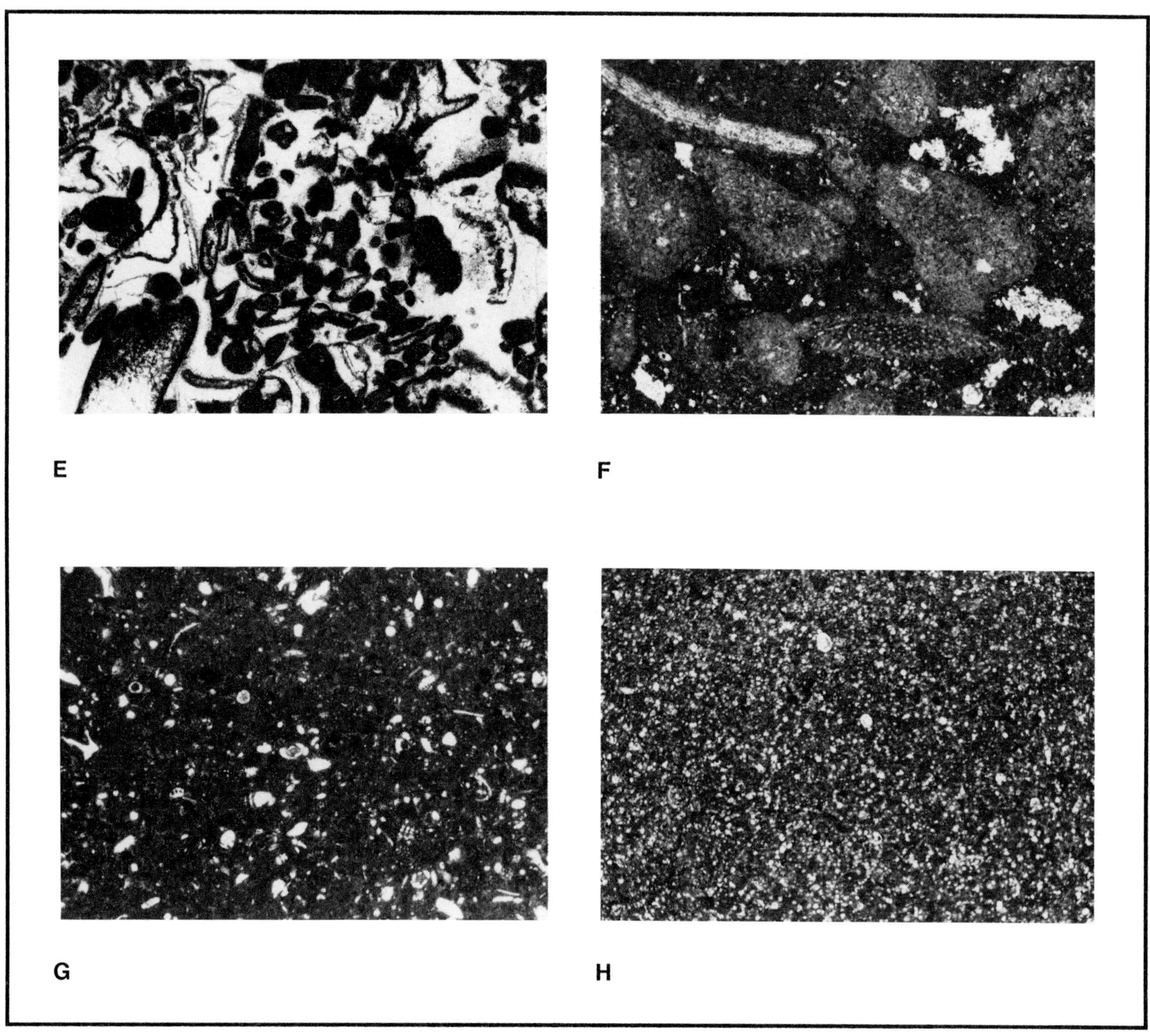

Figure 12. Continued. (E). Cemented ooid-skeletal grainstone with coated micrite envelope. An early fringe of cement crystals around the grains is overlain by coarse, blocky spar. Grains include pelecypod shells, orbitolinids, algae, and echinoids. Most of the grains are blackened. **(F)** Wackestones with abundant orbitolinids. **(G)** Lime mudstones/wackestones with some sponge spicules, miliolids, orbitolinids, and fine-grained quartz. **(H)** Dolomitized lime mudstone.

hydrocarbon saturated; they have high amplitudes and some thin discontinuous insoluble residues that exhibit minimal associated matrix cementation.

This subzone has the greatest thickness in the northern area (97 ft; 29 m). Down flank it thins to 77 ft (24 m) in the south of the field and 75 ft (23 m) in the north. Thickening occurs locally in some wells on the west flank.

The dominant pore type throughout the section is an extensively leached matrix porosity with minor moldic porosity. The highest porosity is 34% and the lowest is 19.7% (Figure 9). Due to a lithology of fine and microporous lime mudstones, the permeability is the lowest of all the subzones. The porosity deteriorates rather rapidly down flank in all directions. The permeability is slightly better in the northern part of the field than in the southern area. The maximum permeability value measured is about 10 md, and the lowest is about 0.2 md.

Orbitolinids and echinoderm fragments are common; miliolids, textulariids, and *Pseudochrysallidina* sp. are rare; a few *Choffatella* sp. and *Hensonella* sp. are present, and *Dictyoconus* sp. is locally abundant.

Asab was part of an open shallow-water shelf on which sediments accumulated in a low-energy zone below wave base during a possible stillstand after

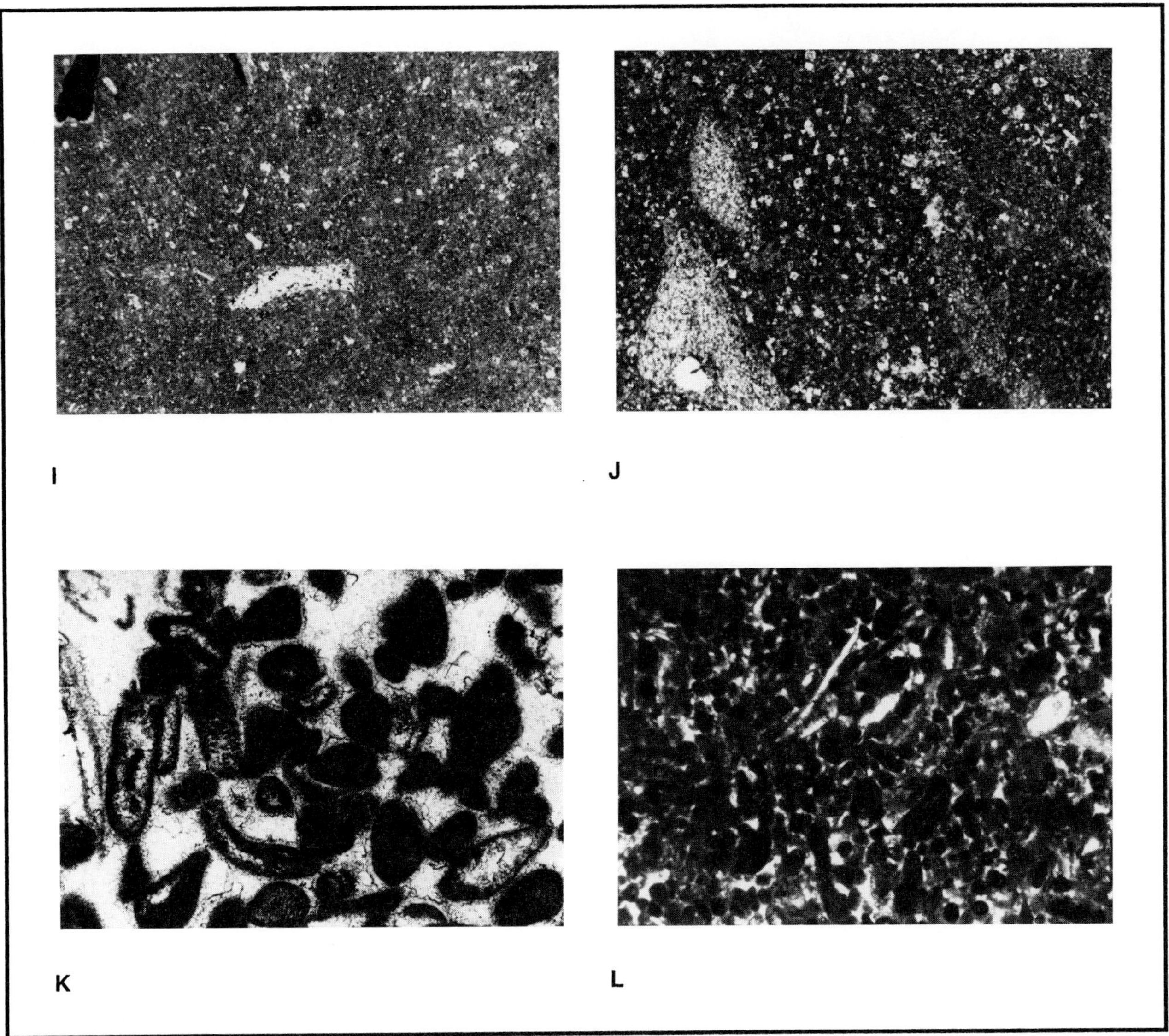

Figure 12. Continued. (I) Lime mudstone with scattered bioclasts and fine-grained quartz. **(J)** Dolomitized wackestones with reworked orbitolinids and algae. **(K)** Grainstone with lithoclasts and bioclasts. These consist of *Choffatella, Orbitolina*, fragments of echinoids, and pelecypods. Most of the grains are blackened. The grainstone shows two generations of cement; i.e., fibrous fringe of calcite followed by blocky sparite. **(L)** Peloidal packstone with thin, fibrous cement. Few forams, echinoid debris, and pelecypod shells. Most of the grains are blackened.

an initial transgression period that is suggested by the underlying dense limestone section.

Zone C Reservoir

Subzone CI—CI is a gray brown, fine-grained, dense lime mudstone to wackestone (Figure 12I), which, near stylolite zones, grades to buff, microporous and porous limestones downward. Stylolites and fractures tend to be patchy and are not extensive. There is scattered dolomite throughout the section (Figure 12J). This subzone corresponds to lime mudstone/wackestone facies (WM).

The thickness of CI varies from 23 ft (7.0 m) on the crest to 17 ft (5.0 m) on the west flank. In general the thickness variation is small but a gradual downflank thinning is observed.

Mainly solution vuggy and moldic porosity is observed in this subzone. Porosity values range from 12 to 30% (Figure 11). The lowest values are mainly related to the occurrence of lime mudstone facies and the well-developed stylolite zone (SC-1) in the lower part of this subzone. Due to its lithological content (lime mudstones/wackestones), this subzone has the lowest permeability of all the subzones. The highest

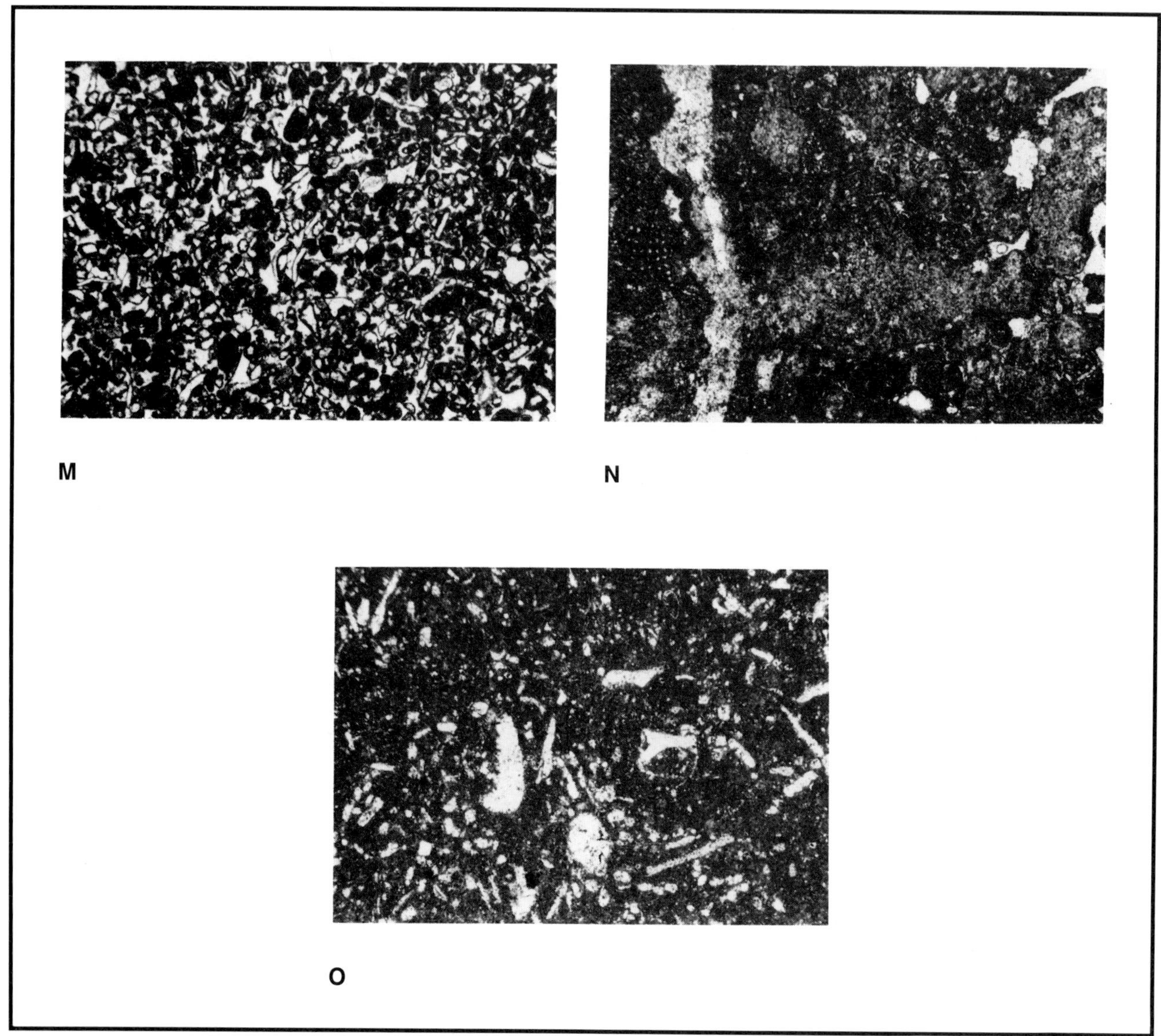

Figure 12. Continued. (M) Bioturbated ooid-skeletal grainstone. Grains include pelecypod shells, miliolids, algae, and peloids. Most of the grains are blackened. **(N)** Lime mudstones/wackestones with orbitolinids, miliolids, gymnocodiacians, and echinoid fragments. Most of the bioclasts are blackened. **(O)** Wackestone with abundant debris of organisms: dasycladaceans, echinoids, and pelecy-pods, with rare quartz.

values occur at the structural crest, with local high values in some wells to the south.

The main biota are orbitolinids, textulariids, rotaliids, *Pseudochrysallidina* sp., and *Hensonella* sp.

CI was deposited in shallow water typical of normal marine low-energy conditions.

Subzone CII—Subzone CII contains four units that are characterized by the following lithology (from top to bottom): (1) well-developed bedding of grainstone to packstone (facies G1) (Figure 12K); (2) thick-bedded coarse- to fine-grained packstone, peloidal and with intraclasts; occasional thin beds of grainstone and also thin beds of microporous lime mudstone (facies P) (Figure 12L); (3) grainstone to packstone, oolitic and intraclastic, sometimes finely laminated and bioturbated (facies G2) (Figure 12M); and (4) microporous micritic lime mudstone and packstone with intraclasts and algal lumps, corresponding to the upper part of wackestone/packstone facies (facies WP) (Figure 12N).

All of this succession grades upward from coarse to fine grained. In this subzone there are stylolites, pressure solution feature at grain-to-grain contacts, some baroque dolomite, thin rim calcite cementation, and isolated large calcite crystals as overgrowths on echinoderm plates.

Subzone CII is the thickest of all the subzones and represents approximately half of zone C. The maximum thickness of 51 ft (16 m) is on the crest, and a minimum thickness of 37 ft (11 m) is in the southernmost wells, resulting from the cutting out of the base of the subzone by a fault.

CII porosity is mainly interparticle and moldic. The porosity range of this subzone is the highest of all of the subzones. The maximum of 30% was found in some wells in the crestal area, but local high percentages occur in some wells on the west flank (26.8%) and on the east flank (26.5%). Faulting may have played a role in the formation of these localized high porosity anomalies on the flanks. This subzone has the highest permeability values of zone C, the high permeability values mainly due to the presence of well-sorted medium- to coarse-grained grainstone. The absolute maximum permeability is about 800 md and is most likely due to fracturing created by faulting.

The dominant organisms in the subzone are the algae *Bacinella irregularis*, rudistiid fragments, echinoderms, orbitolinids, and miliolids; textulariids and rotaliids are rare.

CII was formed in a setting that ranged from shallow water with sediments deposited in high-energy shoals to bank-type environments and to shallow marine conditions with moderate to high energy, semi-restricted environments behind ooid shoals.

Subzone CIII—CIII consists of fine- to medium-grained, microporous wackestone to packstone, and some lime mudstone (Figure 12O). Peloidal and micritic intraclasts are the dominant particles. There are some microstylolites. The subzone corresponds to the middle and lower part of wackestone/packstone facies (WP).

Subzone CIII varies in thickness from 24 ft (7 m) in the southeast to 17 ft (5 m) in the south. The subzone generally thins down the structural flanks although there are many anomalies fieldwide, probably because of variations in paleotopography.

Porosity is mainly leached vugs and matrix and is slightly lower than subzone CII because more local porosity anomalies are present on the west and north flanks of the structure. This phenomenon is thought to be related to secondary porosity development. Porosity ranges from 18 to 30%. The distribution trend is uniform, with the better values at the crestal area. Local maximum permeability present in wells on the eastern flank cannot be explained by faulting but could be due to lithologic variations.

There is an abundance of *Bacinella irregularis* algae and orbitolinids in subzone CIII.

The sediments were deposited under semi-restricted shelf conditions with low to moderate energy.

Stylolite Distribution

Three main types of stylolites identified in the Lower Cretaceous reservoirs of Asab are subdivided on the basis of amplitude, morphology, lateral continuity, and thickness of accumulated insoluble residue. These types are rectangular stylolites, solution seams or wave-like stylolites, and wispy seams or horsetail stylolites (Alsharhan, 1987b).

The amplitude of the stylolites is measured in cores from the apex to the lowest point of the stylolite (Figure 13). Measured amplitude varies with the type of stylolite (e.g., the rectangular type has the largest amplitudes; the wispy seams have the smallest ranges). The high amplitude stylolites (more than 35 cm) are observed in the most lithified facies and the stylolitic parting (less than 1 cm) occurs in the less lithified facies.

Petrographic features attributable to pressure dissolution are very common in the core material. Low amplitude and wispy solution seams occur both above and below the oil-water contact. Porosity and permeability are significantly reduced adjacent to stylolites.

Stylolite intensity (SI) is expressed by the empirical formula (Figure 13) proposed by Johnson and Budd (1975) to provide quantified mappable data for reservoir analysis: $SI = H/\Phi \times 100$, where:

SI = stylolite intensity
H = thickness of the stylolite zone (e.g., D1, D2, and D3 on Figure 10) on logs (feet)
Φ = minimum porosity of the stylolite interval (%)

Stylolite intensity is inversely proportional to the porosity and permeability of the zone; therefore, the high-intensity stylolite zones are more likely to act as barriers to fluid flow, especially to vertical flow.

The percentage of thinning (which also gives the minimum thinning) attributable to the stylolitization

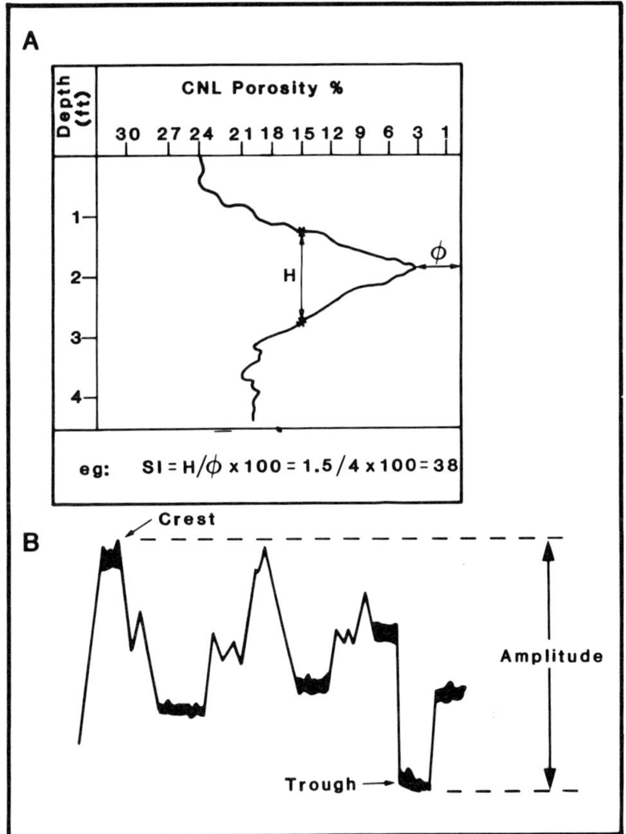

Figure 13. Measured amplitude of stylolites. CNL, calculated neutron log.

of a limestone unit may also be related to the average porosity before and after stylolitization. Dunnington (1967) provided an empirical formula for calculating the percentage of thinning in a section caused by stylolitization. The total loss of porosity within the growing stylolite zone is interpreted as mainly a result of chemical compaction. Some of the dissolved material is redistributed within this zone, but a great deal of the carbonate must have passed out to the system to precipitate elsewhere.

The thicker and the more laterally continuous the insoluble residue portion of the stylolites, the more effective it is as a barrier to fluid flow. Johnson and Budd (1975) and Koepnick (1984) believe that, in zone B, swarms of wispy seams may form a significant barrier to fluid cross flow and that these are commonly associated with impermeable intervals in the reservoir. They also add that it is the lateral continuity and thick buildup of insoluble residue that makes these solution seams important barriers to fluid flow.

Stylolites are common throughout zone B. Three major stylolite zones (D1, D2, and D3) have been identified (Johnson and Budd, 1975; Alsharhan, 1985b), and these are used to subdivide the reservoir into four subzones (Figure 10). The D1 stylolite zone, which separates subzone I and II, is made up of a number of stylolites that reduce the permeability and act as a barrier to fluid flow. The D2 and D3 stylolite zones separate subzone II and III and III and IV, respectively, and are less intense than D1.

DI stylolitic interval (Figure 14) is the best-developed stylolite swarm of the three and is present in all wells. The intensity varies from moderate to strong. The higher intensity values occur in nearly all the flank wells, and strong stylolite development is noticeable locally in some wells to the south and along the crest of the structure.

The D2 stylolite swarm is not present over the whole field (Figure 14); consequently, localized cross flow is expected between subzones II and III. The amplitude of D2 stylolites is highest on the flanks and is, in general, much higher than those of zone D1. This seems to confirm the observation that high amplitude only results in moderate stylolite intensity.

The stylolite intensity of the D3 stylolite swarm varies more widely than do those of D1 and D2. The higher values are present on the southern north-western flanks of the structure. The D3 stylolitic interval is not present along the crest area and is moderately developed in the rest of the field (Figure 14). The amplitude of D3 stylolites is highest for the entire Asab field; in the south the amplitude reaches 12 in. (30 cm) in some wells, again supporting the remarkable association of high amplitudes with the only moderate intensities just mentioned.

Koepnick (1984) and Johnson and Budd (1975) concluded that the D1 stylolite horizon in zone B constitutes a significant barrier to vertical fluid cross flow by virtue of its lateral continuity and its associated zone of matrix cementation; the D2 and D3 stylolite swarms appear to be less significant flow barriers owing to their lateral discontinuity, weak cementation, and level of hydrocarbon saturation. The D1 stylolite swarm was largely developed before hydrocarbon entrapment, whereas the D2 and D3 grew mostly during and after hydrocarbon entrapment (Koepnick, 1984). Johnson and Budd (1975) reported that the D1 stylolite swarm in zone B, found over much of the Asab field, particularly down flank, is a complete barrier to the movement of fluids, while toward the crest it is less of a barrier. The D2 and D3 stylolite swarms present no real barrier over the crest of the field, and they probably restrict vertical fluid movement but little.

Koepnick (1984) and Alsharhan (1987b) found that stylolite distribution in zone B is controlled by stress concentrations developed during growth of the anticlinal closure and by differential inhibition of stylolite growth by hydrocarbon saturation.

The Thamama zone C in Asab field is subdivided into three subzones (CI to CIII) by two stylolite swarms: SCI at the base of subzone CI and SC2 at the base of subzone CII (Figure 11). Stylolite swarm SCI is present over all the field (Figure 15), separating subzones CI from CII, and is very strongly developed in both the northern sector of the field and in some wells in the south. It is considered to have a significant effect upon interreservoir communication between the subzones. Stylolite swarm SC2 is the boundary between subzone CII and CIII. It is less well developed than SC1—moderately developed along the south flank and weakly developed on the north and east flanks. In parts of the field where stylolite swarm SC2 is poorly developed or absent, interreservoir fluid flow between subzone CII and CIII during production is very likely.

Reservoir Data

The reservoirs of the Kharaib Formation are located in zones B and C and consist of about 180 and 80 ft (56 and 25 m), respectively, of lime mudstones/wackestones and peloidal packstones/grainstones. These are sealed by overlying dense lime mudstones.

The isopach of zone B (Figure 16) shows a maximum thickness of 195 ft (61 m) in the northern part of the field. The minimum thickness recorded is on the southern plunge of the structure. In general the map shows a northeast to southwest trend with a thick reservoir in the crestal area, thinning toward the flanks.

The thickest Thamama zone, zone B, is estimated to contain 8.8 MMBO, while zone C contains 1.4 MMBO (Johnson and Budd, 1975). Sulfur content is 0.94 wt %. In its initial conditions, the reservoir crude was undersaturated by about 1800 psi. Differential degassing at the reservoir temperature of 250°F (120°C) gave a solution GOR of 997 scf/bbl and a crude of 38.4° API at standard conditions (Mallinson and Sharp, 1975). In 1975 the recoverable reserves in the Thamama zone B were estimated at 3 billion stock tank barrels (B STB), with an initial rate of production of 400,000 b/d, while in Thamama zone C, reserves were estimated at 0.6 B STB, with an

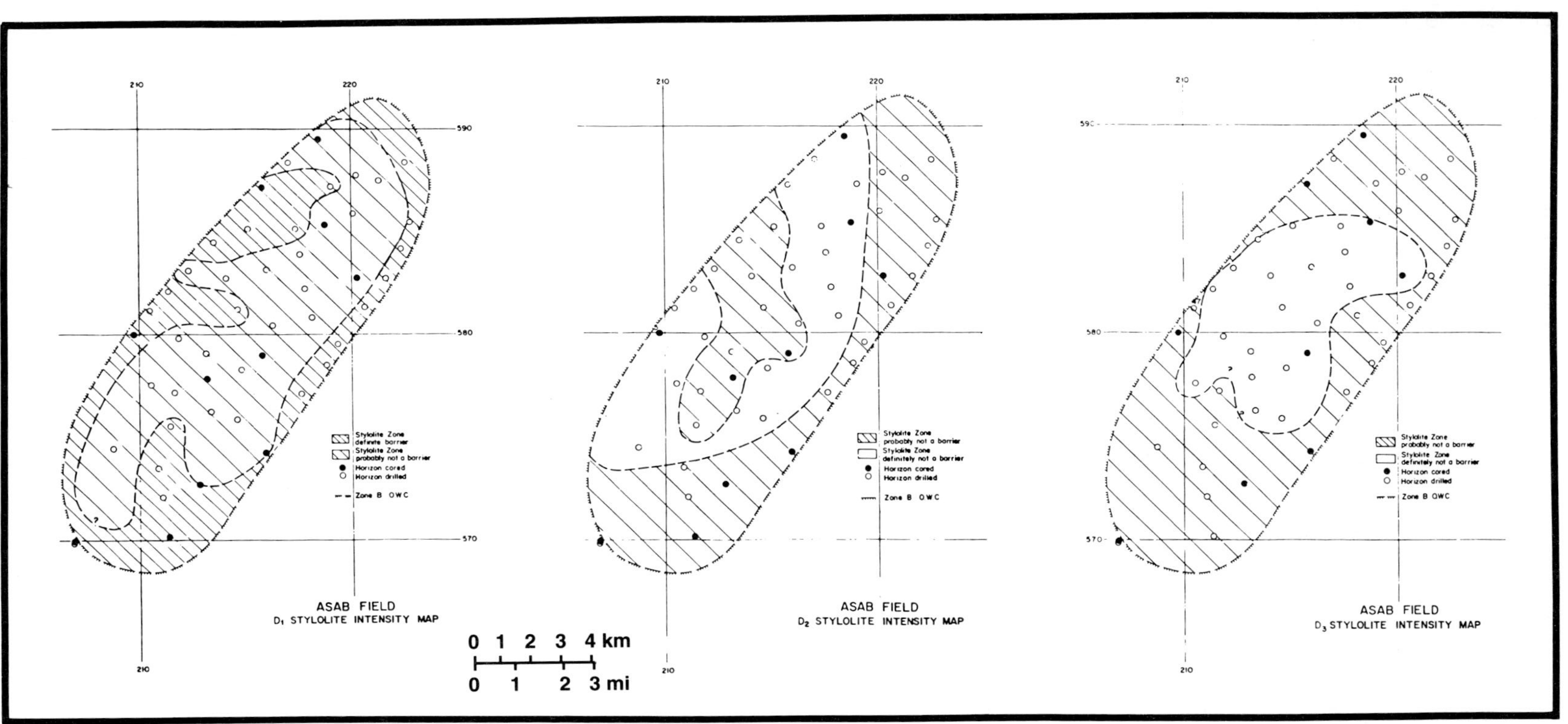

Figure 14. Stylolite intensity map of zone B. (After Johnson and Budd, 1975.)

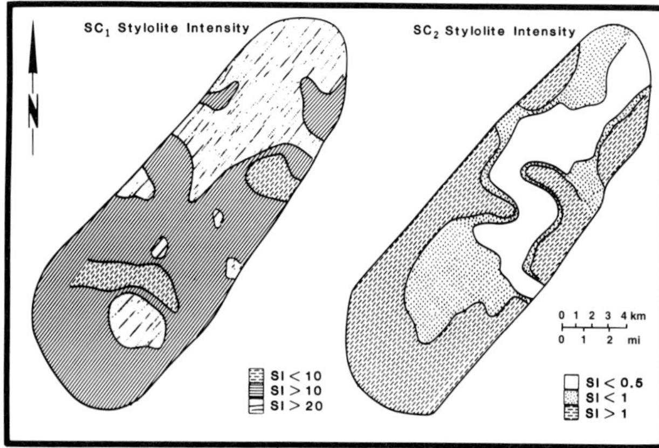

Figure 15. Stylolite intensity map of zone C.

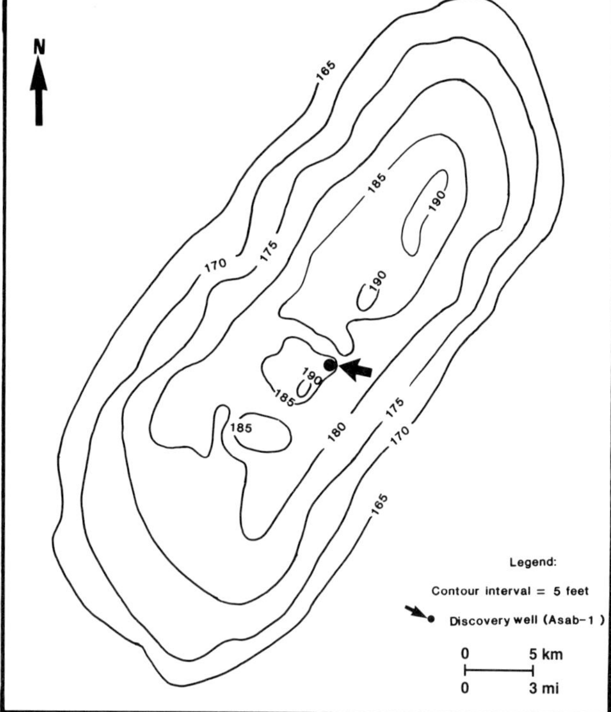

Figure 16. Isopach map of zone B of Kharaib Formation.

initial production rate of 60,000 b/d (Mallinson and Sharp, 1975).

From detailed descriptions of cores and thin sections and from calculations of porosity and permeability values from core plugs and logs, the following can be concluded:

1. The mud-supported rocks have good porosity and permeability because of abundant interparticle pores associated with an important phase of dissolution.
2. The grainstones show higher peaks of permeabilities. This is due to the association of the interparticle pores and solution vugs. The slight decrease of the porosity values is probably the result of compaction.
3. Some porosities measured in the grainstones are high (20–30%), while the permeabilities are rather low (1–10 md). These low permeability values are a close match to those of the mudstones and are probably due to more complete calcite cementation.
4. The lowest porosity and permeability values are a result of stylolitic partings, laminations, and bioturbation; low values also result from the influence of cementation coupled with the lack of solution in the grainstones and from action of the cementation partly plugging the interparticle pores in the mud-supported textures.
5. A large percentage of higher porosity and lower permeability observed in some parts of the section is the result of dissolution of carbonate grains, accompanied at the same time by reprecipitation of the carbonates in adjacent primary pores, as described by Harris et al. (1985). The end product is secondary porosity (moldic and vugs) and variable permeability.

The main factors controlling *porosity distribution* are those of depositional facies change and dissolution, which are greatest on the top of the structures, decreasing down the flanks. Another phenomenon that gives lower porosity values on the flanks is a greater degree of cementation as compared to that on the top. The better *permeability distribution* is *also* a response to dissolution that is greatest at the top of the structure.

Hydrocarbon Habitat

In the western areas of Abu Dhabi, structural traps started to develop in the Late Jurassic, while in the eastern areas, where Asab is located, structural trap development seems to have been later than Jurassic. The most prolific reservoirs in Abu Dhabi occur in Upper Jurassic and Lower Cretaceous formations, which were sourced (as to hydrocarbons) from the same rock unit in the Upper Jurassic, the Diyab/Dukhan Formation.

The Diyab (offshore terminology) and the Dukhan (onshore terminology), usually referred to jointly as the Diyab/Dukhan Formation, is Oxfordian to mid-Kimmeridgian in age. It is about 1600 ft (500 m) thick in the western areas, thinning gradually toward the east and northeast. It consists of finely laminated, dark gray, argillaceous lime mudstones/wackestones with calcareous shales; the lower part (100 to 150 ft [31 to 47 m] thick) is more carbonaceous, and the laminated argillaceous limestones are rich in organic matter.

In the eastern area of Abu Dhabi, in which Asab is located, the unit gradually changes facies to domination by dolomites and dolomitic limestones, with some peloidal packstones/grainstones. This change reflects the different depositional settings of the formation. In western Abu Dhabi, an intrashelf basin was developed in the Late Jurassic where local euxinic conditions produced the starved-basin sequences of laminated, bituminous lime muds, shales, and marls, locally forming good source rock (Alsharhan, 1989). Similar intrashelf basins were found in eastern Saudi Arabia and the northern Arabian Gulf (Alsharhan and Kendall, 1986). In eastern Abu Dhabi, shallow water with subtidal lagoonal deposits prevailed.

The present-day maturity of the Diyab/Dukhan Formation is shown in Figure 17A. The formation in most of the onshore areas (where Asab is located) lies below the oil window (R_o = 1.3%), except in southeast Abu Dhabi, where the lower maturity is associated with the Mender/Lekhwair high. The Falaha syncline, west of Asab field, the area eastward and northward from Jarn Yaphour, and the area westward from Bu Hasa have the highest maturity levels, with an R_o = 1.7% (Alsharhan, 1989).

Figure 17B shows the paleo-maturity map of the Diyab/Dukhan Formation for Eocene time (49 Ma). The maturity pattern is very similar to the present one and parallels present structural trends. In most of the offshore area, hydrocarbon generation did not commence until the early Eocene, and by this time, the Diyab/Dukhan Formation in onshore areas had reached the phase of maximum oil generation (ADNOC, 1984).

Figure 17C shows the paleo-maturity map of the Diyab/Dukhan Formation in Campanian time (73 Ma). A general increase in maturity to the south and a low maturity in the southeast, and with these, the first signs of movement of the Falaha syncline, suggest that significant salt movement was starting; the earliest hydrocarbon generation in the onshore areas probably was initiated at this time (ADNOC, 1984).

Total organic carbon (TOC) of the Diyab/Dukhan Formation ranges between 0.3 and 5.5% by weight in the offshore areas, while in the onshore areas it ranges between 0.5 and 0.82%; in the southeast (at Mender and Qusawirah), it ranges from 0.5 to 1.2% by weight.

Pyrolysis data show that the residual source rock potential (P_2 yield) ranges between 0.5 and 5.0 kg/ton. The kerogen content is 60 to 80% and is of a sapropelic oil-prone type in some wells, while in other wells there is a mixed sapropelic and humic (oil and gas prone) kerogen grading to highly humic type (Loutfi and El-bishlawy, 1986).

Gas chromatography-mass spectrophotometry (GC-MS) of the sterane and triterpane fingerprints of the Diyab/Dukhan extracts closely resemble those of the Thamama and Arab formations (Figure

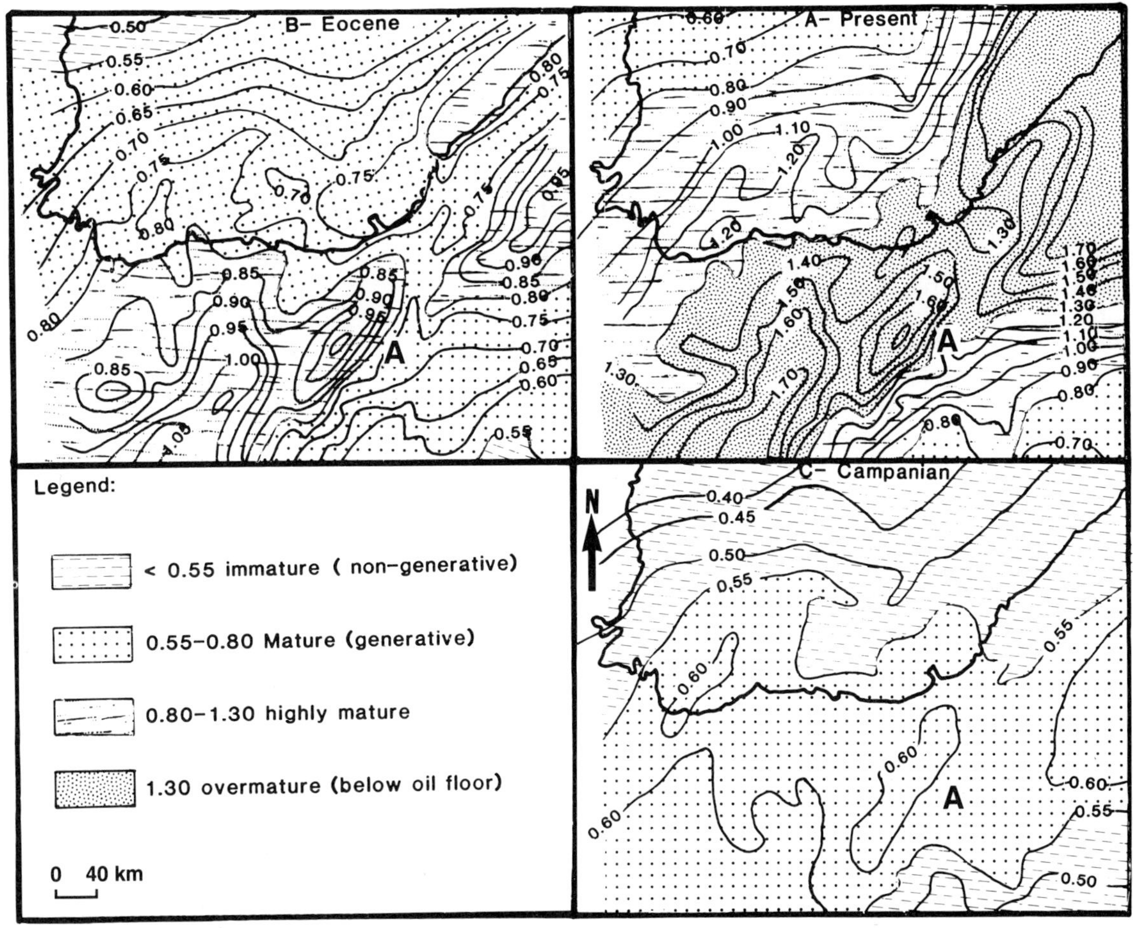

Figure 17. Paleo-maturity, calculated vitrinite reflectance (R_o%), of the Diyab/Dukhan Formation, Upper Jurassic source rock, Abu Dhabi. (After ADNOC, 1984; Loutfi and El-bishlawy, 1986.) A is location of Asab field.

18A, B). Also, carbon isotope values for the Diyab/Dukhan Formation source-rock extracts (-24.7‰ to -27.0‰) closely match the Arab and Thamama oils (-25.2‰ to -26.9‰) (Loutfi and El-bishlawy, 1986).

These results indicate that oil was generated in the Diyab/Dukhan Formation in western Abu Dhabi, where it has moderate to good source potential, and then migrated eastward. All the Arab Formation reservoirs to the west of the Hith edge should be progressively filled if we assume that they were on this migration path. Oil migrating eastward of the Hith edge would then migrate vertically into the Thamama reservoirs.

Figure 19 shows the hydrocarbon source potential of the Diyab/Dukhan Formation in Abu Dhabi. The maps indicate three other possible source areas to be considered in migration modeling: the Falaha syncline (3 in Figure 19), Sir Abu Nuair syncline (1 in Figure 19), and northeastern part of Abu Dhabi (2 in Figure 19). Assuming that the Diyab/Dukhan Formation in the Falaha syncline had reached maturity in Campanian time, the major part of the generated oil would have migrated vertically into the Thamama reservoirs to charge most of the existing structures at Asab, Sahil, Bab, and Bu Hasa (Loutfi and El-bishlawy, 1986).

In Eocene time (49 Ma) (Figure 19), the western Diyab/Dukhan source area was still charging the Jurassic reservoirs in some of the western offshore fields. Also the three possible source areas would still have been charging the Thamama reservoirs to the east (Loutfi and El-bishlawy, 1986). The southeast source area would have been charging the Qusahwira and Mender fields as the result of the emplacement of the basal Umm Er Rhadhuma shale as a seal.

Oil migrating to the southwest along the eastern margin of the Falaha syncline progressively filled the Sahil and Asab structures. The absence of Thamama oil in Shah can be explained by diversion of oil migrating from the northeast into Zarrara by a saddle existing between the Asab and Shah structures since the early Late Cretaceous (Figure 20), predating the onset of oil migration, causing diversion of the migration pathway toward the Zarrara structure (ADNOC, 1984; Loutfi and El-bishlawy, 1986). If oil had migrated from western Abu Dhabi around the southern margin of the Falaha syncline, then the Shah structure would be charged with Thamama oil.

The distribution of Thamama oils across the area appears to be related to the occurrence of the Hith Anhydrite; Thamama oils are mainly reservoired to the east of the "Hith edge," which runs approximately north-south through central Abu Dhabi (Figures 19 and 21). The Hith Anhydrite forms the regional seal for the Upper Jurassic Arab oils in the offshore areas. Because of the absence of this seal in the eastern part of Abu Dhabi, oil generated from the Diyab/Dukhan Formation is able to migrate vertically, directly into the Thamama reservoirs. Thamama reservoirs only contain oil where the Hith seals of the Arab Formation were not initially anhydrite (probably anhydrite and dolomite to porous

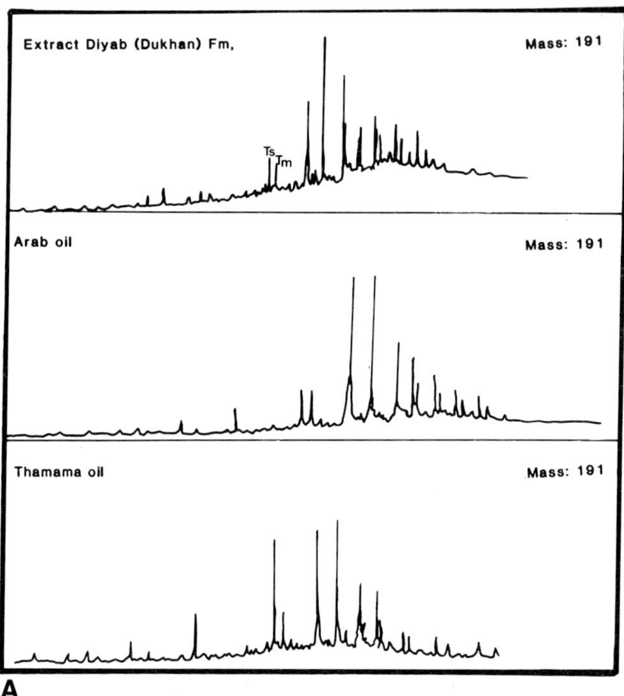

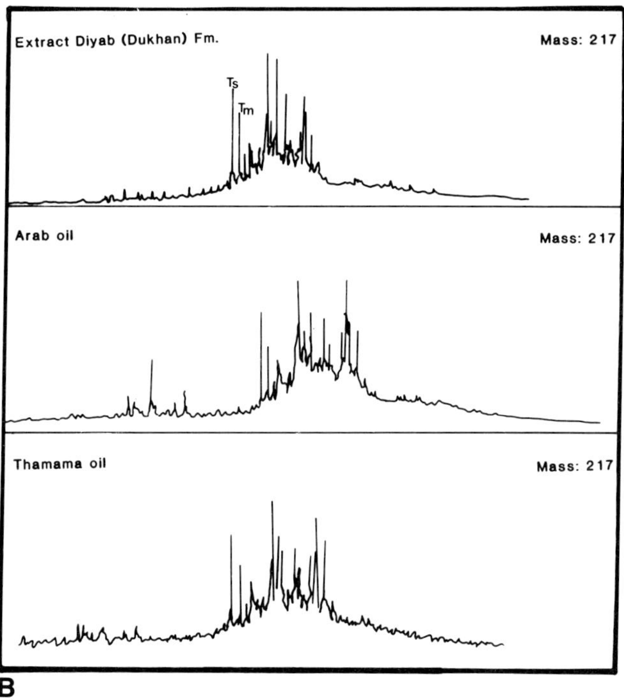

Figure 18. (A) Gas chromatography-mass spectrophotometry (GC-MS) triterpane fingerprints for extract from Diyab/Dukhan Formation and of Arab and Thamama formations oils, Abu Dhabi oil fields. (After Lutfi and El-bishlawy, 1986.) (B) Gas chromatography-mass spectrophotometry (GC-MS) sterane fingerprints for extract from Diyab/Dukhan Formation and of Arab and Thamama formations oils, Abu Dhabi oil fields. (After Loutfi and El-bishlawy, 1986.)

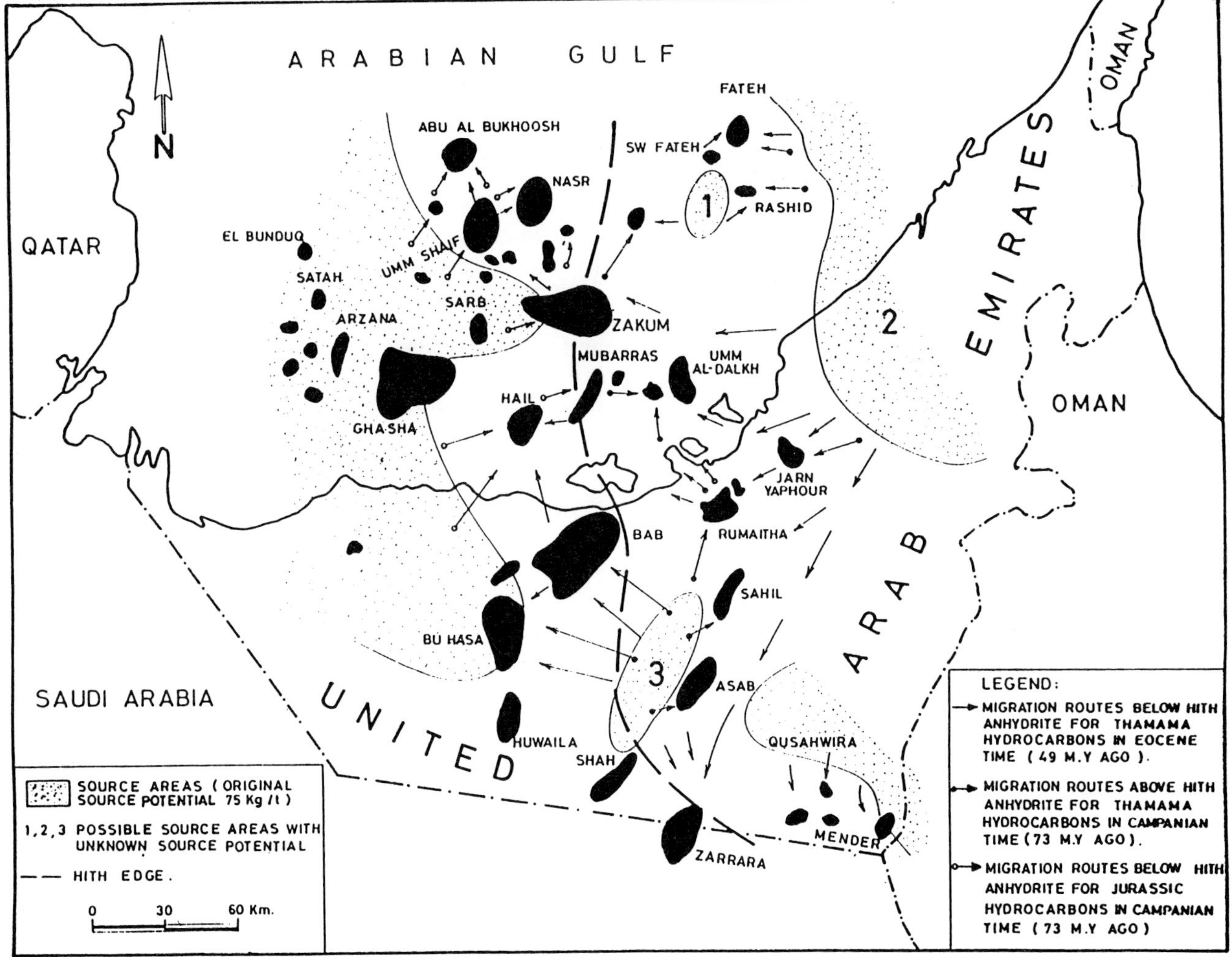

Figure 19. Possible migration pathways from Diyab/Dukhan Formation source and other possible source areas in Campanian and Eocene times. (Compiled from Loutfi and El-bishlawy, 1986.) The Hith Formation, west of the line (Hith edge), is an important barrier to vertical migration and is the seal of the Arab Formation.

limestones) or where it has been breached by faulting (Murris, 1981; Alsharhan, 1985a).

The vertical migration from the Diyab/Dukhan source to the Thamama reservoirs through the dense limestone intervals is possible, as pointed out by Schowalter (1979), who showed that when the buoyant pressure of a hydrocarbon column in a trap exceeds the resisting capillary pressure force of the cap rock, then hydrocarbons can migrate vertically through the seal.

EXPLORATION CONCEPTS

Salt halokinesis up to the end of the Mesozoic created many structural traps in the U.A.E. Exploration strategy of the Cretaceous carbonate platform necessarily takes into consideration the porosity trends. Primary porosity is found in carbonate mounds that have been subsequently subjected to subaerial exposure that resulted in secondary porosity. In the reservoirs of the Kharaib Formation, good porosity/permeability characteristics in zones B and C are associated with grainstones and packstones. Poor permeabilities are associated with the microporous lime mudstones and wackestones that usually occur in the lower part of zone B and the upper part of zone C.

At the Bab, Asab, and Sahil fields, the main oil accumulation is in the Kharaib Formation. Oil is found in the porous intervals of zones B and C at Umm Shaif and Zakum fields, while at Zubbaya, Rumaitha, Jarn Yaphour, Zibara, Bu Hasa, Fateh, Southwest Fateh, Delta, and Mubarras fields, oil is in zone B only.

The Kharaib Formation reservoir sequences are initiated by the decrease in argillaceous sediments and subsequent leaching of the lime mudstones and wackestones, which is the essential difference between the porous parts adjacent to nonpermeable parts of the producing reservoirs and the dense argillaceous limestone seals.

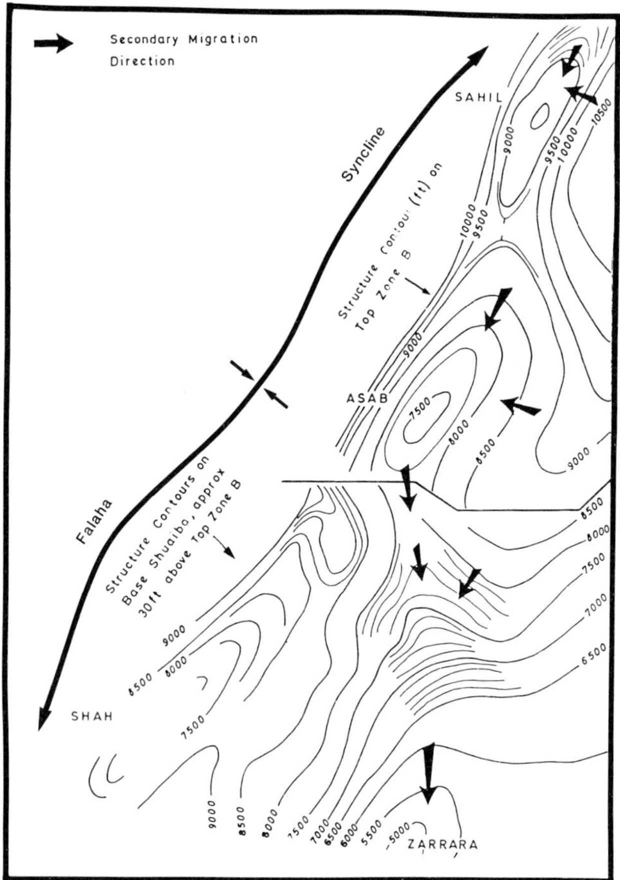

Figure 20. Secondary migration directions of Thamama oil in the Sahil, Asab, Shah region. (After ADNOC, 1984.)

The best reservoir permeability is found in the grain-rich sediments that appear to have had porosity enhanced by freshwater leaching. Johnson and Budd (1975) show porosity values for the producing cycles in a range from 30 to 35% extending through the whole sequence in the Asab field, while the reservoir-quality permeability is confined usually to the grain-rich upper third of each cycle of deposition.

The porous zones of the Kharaib Formation extend to northeast Abu Dhabi as far north as the western Dubai, and in southeast Abu Dhabi, to the Lekhwair high. This formation has proved to be a major exploration target in the Gulf area, producing from peloidal packstone/grainstone sediments in which the primary porosity has been augmented by subaerial leaching by fresh water. Dolomitization has been a relatively minor factor in creating reservoir-quality porosity, as in Arab field, where dolomitization destroyed reservoir porosity rather than enhanced it (Johnson and Budd, 1975; Alsharhan, 1985b).

Future exploration efforts will concentrate on smaller structures and subtle traps. Known hydrocarbon-bearing structures need detailed geological and geophysical studies and reinterpretation of well and seismic data.

ACKNOWLEDGMENTS

I thank the geologists past and present of ADCO for their assistance and fruitful discussions of different aspects of the Asab reservoirs. Part of this paper is extracted from my doctoral dissertation written at the University of South Carolina. Professor Christopher G. St.C. Kendall supervised the research of this dissertation, and his advice and encouragement is greatly appreciated. I express my deep gratitude for the editors and reviewers of AAPG who read the manuscript and offered valuable advice and recommendations. Figures 1, 2, 3, 17, and 19 are reproduced with permission from the Journal of Petroleum Geology, U.K.

REFERENCES

Abu Dhabi Company for Onshore Oil Operations (ADCO), 1986, Tectonics of the oil fields in Abu Dhabi onshore areas: Paper presented at the symposium on the hydrocarbon potential of intense thrust zones, held in Abu Dhabi 14-18 December 1986, 12 p.

Abu Dhabi National Oil Company (ADNOC), 1984, Geochemical studies, source rock determination and migration of oil in the U.A.E.: Paper presented at the conference on the origin and migration of Arab oils, held in Kuwait and sponsored by OAPEC, 36 p.

Alsharhan, A. S., 1985a, Depositional environments, reservoir units evolution and hydrocarbon habitat of Shuaiba Formation (Lower Cretaceous) of Abu Dhabi, United Arab Emirates: AAPG Bulletin, v. 69, p. 899-912.

Alsharhan, A. S., 1985b, Petrography, sedimentology, diagenesis and reservoir characteristics of the Shuaiba and Kharaib Formations (Barremian-mid Aptian carbonate sediments of Abu Dhabi, United Arab Emirates: Unpublished Ph.D. dissertation, University of South Carolina, Columbia, South Carolina, 175 p.

Alsharhan, A. S., 1987a, Geology and reservoir characteristics of carbonate buildup in giant Bu Hasa oil field, Abu Dhabi, United Arab Emirates: AAPG Bulletin, v. 71, p. 1304-1318.

Alsharhan, A. S., 1987b, Distribution of stylolites in the carbonate reservoirs of the United Arab Emirates (Abstr.): Paper presented at 4th SEPM mid-year meeting, Austin, Texas.

Alsharhan, A. S., 1989, Petroleum geology of the United Arab Emirates: Journal of Petroleum Geology, v. 12, p. 253-288.

Alsharhan, A. S., and C. G. St.C. Kendall, 1986, Precambrian to Jurassic rocks of Arabian Gulf and adjacent areas: their facies, depositional setting and hydrocarbon habitat: AAPG Bulletin, v. 70, p. 977-1002.

Alsharhan, A. S., and A. E. M. Nairn, 1986, A review of the Cretaceous formations in the Arabian Peninsula and the Gulf: Part I. Lower Cretaceous (Thamama Group) stratigraphy and paleogeography: Journal of Petroleum Geology, v. 9, p. 365-392.

Alsharhan, A. S., and A. E. M. Nairn, 1988, A review of the Cretaceous formations in the Arabian Peninsula and the Gulf: Part II. Mid Cretaceous (Wasia Group) stratigraphy and paleogeography: Journal of Petroleum Geology, v. 11, p. 89-112.

Alsharhan, A. S., and A. E. M. Nairn, 1990, A review of the Cretaceous formations in the Arabian Peninsula and the Gulf: Part III. Upper Cretaceous (Aruma Group) stratigraphy and paleogeography: Journal of Petroleum Geology, v. 13, p. 247-266.

Dunnington, H. V., 1967, Aspects of diagenesis and shape change in stylolitic limestone reservoirs: 7th World Petroleum Congress, v. 2, p. 339-352.

Frost, S. H., D. M. Bliefnick, and P. M. Harris, 1983, Deposition and porosity evolution of a Lower Cretaceous rudist buildup, Shuaiba Formation of eastern Arabian Peninsula, *in* Carbonate buildups: SEPM Core Workshop 4, Dallas, p. 381-410.

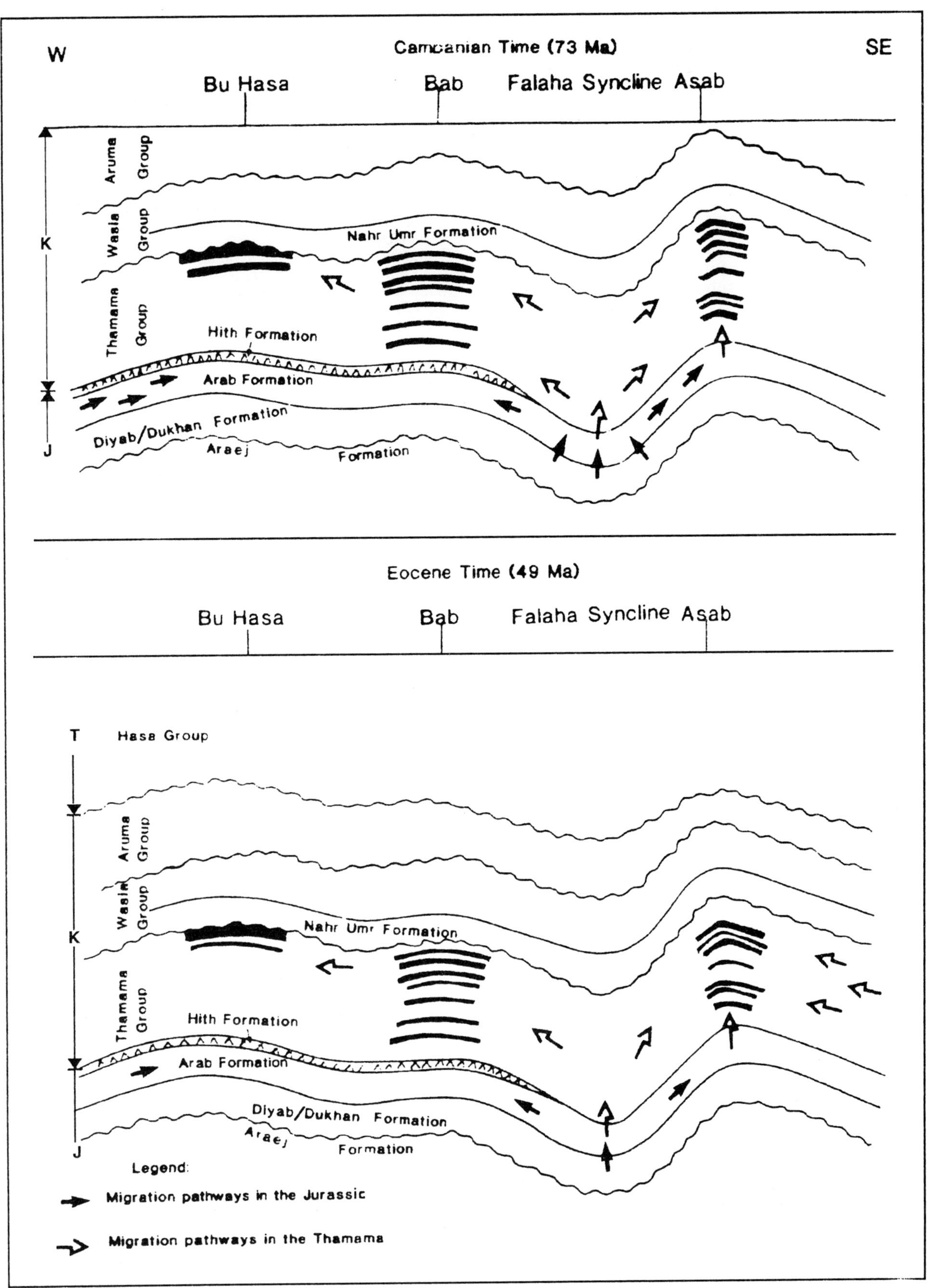

Figure 21. Possible migration pathways along the Hith edge from Diyab/Dukhan Formation source in Campanian and Eocene times. (Modified from Loutfi and El-bishlawy, 1986.)

Hajash, G. M., 1967, The Abu Dhabi Sheikdom—the onshore oilfields history of exploration and development: 7th World Petroleum Congress, v. 2, p. 129-139.

Harris, P. M., C. G. St.C. Kendall, and I. Lerche, 1985, Carbonate cementation—a brief review, *in* Carbonate cements: eds., N. Schneidermann and P. M. Harris, SEPM Special Publication 36, p. 79-96.

Harris, T. J., J. T. Hay, and B. N. Twombley, 1968, Contrasting limestone reservoirs in the Murban field, Abu Dhabi: Proc. 2nd AIME Reg. Techn. Symp., Dhahran, Saudi Arabia, p. 149-187.

Hassan, A., F. Youssef, and M. Ayoub, 1990, Controlling water flood advance in Asab Thamama B reservoir: 4th Abu Dhabi Petroleum Conference, 14-16 May 1990, organized by ADNOC and SPE, p. 292-306.

Johnson, J. A. D., and S. R. Budd, 1975, The geology of the Zone B and Zone C Lower Cretaceous limestone reservoirs of Asab field, Abu Dhabi: 9th Arab Petroleum Congress, Dubai, 24 p.

Koepnick, R. B., 1984, Distribution and vertical permeability of stylolites within a Lower Cretaceous carbonate reservoir, Abu Dhabi, United Arab Emirates: Paper presented at seminar on stylolites held in Abu Dhabi, 6-8 November 1983, sponsored by Abu Dhabi National Reservoir Research Foundation, 19 p.

Loutfi, G., and S. El-bishlawy, 1986, Habitat of hydrocarbon in Abu Dhabi, U.A.E.: Paper presented at the symposium on the hydrocarbon potential of intense thrust zones, held in Abu Dhabi, 14-18 December 1986, 60 p.

Mallinson, J. E., and J. W. G. Sharp, 1975, Reservoir engineering aspects of the development of the Asab field: 9th Arab Petroleum Congress, Dubai, Paper 137 (BO1), 22 p.

Murris, R. J., 1981, Middle East: stratigraphic evolution and oil habitat: Geologie en Mijnbouw, p. 467-486.

Powers, R. W., L. F. Ramirez, C. D. Redmond, and E. L. Elberg, 1966, Geology of the Arabian Peninsula: sedimentary geology of Saudi Arabia: U.S. Geological Survey Professional paper 560D, 147 p.

Schlumberger, 1981, Well Evaluation Conference, United Arab Emirates and Qatar, 271 p.

Schowalter, T. T., 1979, Mechanics of secondary hydrocarbon migration and entrapment: AAPG Bulletin, v. 63, p. 723-760.

Appendix 1. Field Description

Field name .... *Asab field*

Ultimate recoverable reserves .... *3.6 billion bbl*

Field location:

Country .... *Abu Dhabi*

State .... *United Arab Emirates*

Basin/Province .... *Rub Al Khali basin*

Field discovery:

Year first pay discovered .... *Lower Cretaceous Kharaib Formation zone B 1965*

Year second pay discovered .... *Kharaib Formation zone C 1965*

Discovery well name and general location:

First pay .... *Abu Jidu-1 (Asab-1) 54°16′E, 23°24′N (zone B Kharaib Formation)*

Second pay .... *Abu-Jidu-1 (Asab-1) (zone C Kharaib Formation)*

Discovery well operator .... *Abu Dhabi Petroleum Co. (ADPC)*

IP

First pay .... *400,000 BOPD (zone B Kharaib Formation)*

Second pay .... *60,000 BOPD (zone C Kharaib Formation)*

All other zones with shows of oil and gas in the field:

Age	Formation	Type of Show
Aptian	*Zone A (lower Shuaiba Formation)*	*Oil*
Valanginian	*Zone F (lower Lekhwair Formation)*	*Gas*

Geologic concept leading to discovery and method or methods used to delineate prospect

Gravimeter and seismic surveys.

Structure:

Province/basin type .... *Bally 221; Klemme II Ca; Rub Al Khali basin*

Tectonic history

Geological growth of the anticline with which this field is associated is probably controlled by deep-seated faults and the movement of Infracambrian (Hormuz) Salt in depth. This salt is thought to have formed salt pillow structures under large low-dip anticlines and domes.

Regional structure

U.A.E. is part of the eastern margin of the Arabian Peninsula, situated between two major structural highs (Qatar arch and Mender-Lekhwair high). The structural pattern is relatively simple, with gentle folding and warping. Periodic epeirogenic movements of faulted basement blocks or salt domes appear to have been the main tectonic factors controlling the structural development in the area.

Local structure

Asab structure is an elongated domal anticline, its long axis trending northeast-southwest, with maximum dip of 6° on the western flank. The field is a normal faulted structure, and most of the faulting occurred in the "Middle" Cretaceous.

Trap:

Trap type(s) .... *Anticlinal trap with multiple pays*

Basin stratigraphy (major stratigraphic intervals from surface to deepest penetration in field):

Chronostratigraphy	Formation	Depth to Top in ft (m)
Recent-upper Eocene	*Recent-upper Eocene*	*0-1710 (0-520)*
Paleocene-lower Eocene	*Hasa Group*	*1710-5150 (520-1570)*
Upper Cretaceous	*Aruma Group*	*5150-6135 (1570-1870)*

Chronostratigraphy	Formation	Depth to Top in ft (m)
"Middle" Cretaceous	*Wasia Group*	*6135–7755 (1870–2365)*
Lower Cretaceous	*Thamama Group*	*7755–10,575 (2365–3225)*
Upper Jurassic	*Sila Group*	*10,575–11,200 (2365–3415)*

Reservoir characteristics:

Number of reservoirs *2*
Formations .... *Kharaib Formation (zones B and C)*
Ages .... *Barremian–early Aptian*
Depths to tops of reservoirs .... *8460 ft (2580 m) and 8700 ft (2560 m), respectively*
Gross thickness (top to bottom of producing interval) .... *Zone B, 190 ft (58 m); zone C, 90 ft (27 m)*
Net thickness—total thickness of producing zones
Average .... *Zone B, 185 ft (56 m); zone C, 70 ft (21 m)*
Maximum .... *Zone B, 190 ft (58 m); zone C, 80 ft (24 m)*
Lithology .... *Peloidal-bioclastic packstones and grainstones grading to lime mudstone and wackestones*
Porosity type .... *Moldic, vuggy, and interparticle porosities; matrix porosity in lime mudstones*
Average porosity *20–30%*
Average permeability .... *100–400 md*

Seals:

Upper
Formation, fault, or other feature .... *Intraformational seal*
Lithology .... *Dense, argillaceous lime mudstone*
Lateral
Formation, fault, or other feature .... *Downdip OWC*
Lithology .... *Shales*

Source:

Formation and age .... *Diyab/Dukhan Formation, Upper Jurassic*
Lithology .... *Finely laminated, argillaceous limestone and shales*
Average total organic carbon (TOC) *0.5–1.2%*
Maximum TOC *5.5%*
Kerogen type (I, II, or III) .... *Sapropelic and humic (oil and gas prone)*
Vitrinite reflectance (maturation) .... $R_o = 1.3\text{–}1.7\%$
Time of hydrocarbon expulsion .... *Eocene*
Present depth to top of source .... *13,000 ft (4063 m)*
Thickness .... *1500 ft (469 m)*
Potential yield .... $P_2 = 5.2\text{–}0.2$ *kg/MT*

ASAB

Appendix 2. Production Data

Field name .... *Asab field*

Field size:

Proved acres .... *229 km^2 (88 m^2)*
Number of wells all years *226*
Current number of wells *226*
Well spacing .... *1 km (3280 ft)*
Ultimate recoverable .... *3.6 billion STB oil*
Cumulative production .... *NA*
Annual production .... *300,000 BOPD*
Decline rate .... *NA*

Annual water production .... *NA*
In place, total reserves .... *10.5 billion bbl*
In place, per acre-foot .... *NA*
Primary recovery .... *3 billion STB oil*
Secondary recovery .... *NA*
Enhanced recovery .... *NA*
Cumulative water production .... *NA*

Drilling and casing practices:

Amount of surface casing set .... *300–400 ft (94–125 m)*
Casing program
At 300–400 ft (94–125 m), 27-in.; at 4000 ft (1220 m), 13⅜-in.; at 8000 ft (2440 m), 9⅝-in.; at 8500 ft (2590 m), 7-in.
Drilling Mud .... *Bentonite*
Bit program .... *26-in., 12¼-in., 8½-in.*
High pressure zones .... *None*

Completion practices:

Intervals perforated .... *14 ft (4.3. m) in zone B and 12 ft (3.7 m) in zone C*
Well treatment .... *Acidize with HCl*

Formation evaluation:

Logging suites .... *FDC/CNL, GR, and ISF*
Testing practices .... *Communication test dual producer*
Mud logging techniques .... *Used only in exploration wells*

Oil characteristics:

Type .... *Sweet*
API gravity .... *40.7°*
Base .... *NA*
Initial GOR .... *850 SCF/bbl*
Sulfur, wt% *0.94%*
Viscosity, SUS .... *0.3 cp*
Pour point .... *10°F (-12°C)*
Gas-oil distillate *1.6%*

Field characteristics:

Average elevation .... *300–400 ft (90–120 m)*
Initial pressure .... *Zone B, 3930 psi at 7700 ft (27.096 MPa at 2350 m); zone C, 3960 psi at 7800 ft (27.303 MPa at 2380 m)*
Present pressure .... *NA*
Pressure gradient .... *NA*
Temperature .... *250°F (121°C)*
Geothermal gradient .... *1.5°F/100 ft (2.7°C/100 m)*
Drive .... *Water*
Oil column thickness .... *>200 ft (>60 m)*
Oil-water contact .... *Zone B, -7915 ft (-2412 m)*
Connate water *4–10%*
Water salinity, TDS .... *NA*
Resistivity of water .... *0.017 ohm-m*
Bulk volume water (%) .... *NA*
Formation volume factor .... *1.6 RB/STB*

Transportation method and market for oil and gas:
Pipeline to Jebel Dhanna terminal; market, western Europe, U.S.A., and Japan

Bu Hasa Field—United Arab Emirates
Rub al Khali Basin, Abu Dhabi

A. S. ALSHARHAN
The Desert and Marine Environment Research Center
United Arab Emirates University
Al Ain, United Arab Emirates

FIELD CLASSIFICATION

BASIN: Rub al Khali
BASIN TYPE: Foredeep
RESERVOIR ROCK TYPE: Limestone
RESERVOIR ENVIRONMENT OF DEPOSITION: Platform with Reefs
RESERVOIR AGE: Lower Cretaceous
PETROLEUM TYPE: Oil
TRAP TYPE: Anticline with Stratigraphic Controls
TRAP DESCRIPTION: Elongate anticline containing multiple pays; oil distribution strongly controlled by stratigraphic variations

LOCATION

Bu Hasa field is located in Abu Dhabi, United Arab Emirates, approximately 30 mi (50 km) southwest of the Bab field (Figure 1). The field is an ovate anticline with a north-south major axis about 22 mi (35 km) long. The field lies within the limits of the concession granted in 1939 to Petroleum Development Limited, now known as Abu Dhabi Company for Onshore Oil Operations (ADCO). The terrain is mostly sand dunes separated by flat sand corridors and occasional very small, inland sabkha flats.

Bu Hasa is a giant onshore field producing from Lower Cretaceous rudist buildups, with estimated reserves of 20 billion bbl of oil. It is located on the northern side of the Rub al Khali basin, on the eastern shelf of the Arabian platform (Figure 2). In this setting, thick sedimentary deposits, primarily carbonates, accumulated and were deformed by intermittent tectonic movements that resulted in relatively gentle folding, faulting, and salt movements.

HISTORY

Exploration activities in Abu Dhabi were started in 1936. In that year, Petroleum Development Trucial Coast (PDTC), a subsidiary of the Iraq Petroleum Company (IPC), was established, and in the same year a few geological reconnaissance trips were initiated. In January 1939, PDTC signed an agreement for oil exploration covering the area of the Trucial Coast (now called the United Arab Emirates).

The original concession, which covered all of the onshore areas of Abu Dhabi, was given to PDTC in 1939. The concession was for 75 years and covered the land, islands, and territorial water. In 1950, exploratory drilling began. The first well was drilled in Ras Sadr (RS-1) and terminated at a depth of 13,000 ft (4062.5 m) without any significant oil indication (Figure 1). In 1953, drilling of a well designated Murban-1 (later renamed Bab-1) commenced 65 mi (105 km) southwest of Abu Dhabi Island but was shut down owing to technical problems. Oil was first discovered in commercial quantities in the Bab field in 1959. Subsequently, oil was discovered at Bu Hasa in 1962, at Asab in 1964, at Shah in 1966, and at Sahil in 1967.

In July 1962, PDTC was renamed Abu Dhabi Petroleum Company (ADPC), and in 1982, it was renamed Abu Dhabi Company for Onshore Oil Operations (ADCO). The agreement was amended to include an annual relinquishment of a percentage of the concession, and part of the areas relinquished by ADCO have been granted to and explored by other oil companies.

The discovery well of Bu Hasa was the Bu Hasa-12. It was drilled in 1962 on the crestal area of the Bu Hasa anticline, considered at that time to be a southwestern extension of the Bab field (Figures 4 and 5). The common structural feature was known as the Murban structure. The Bu Hasa-12 encoun-

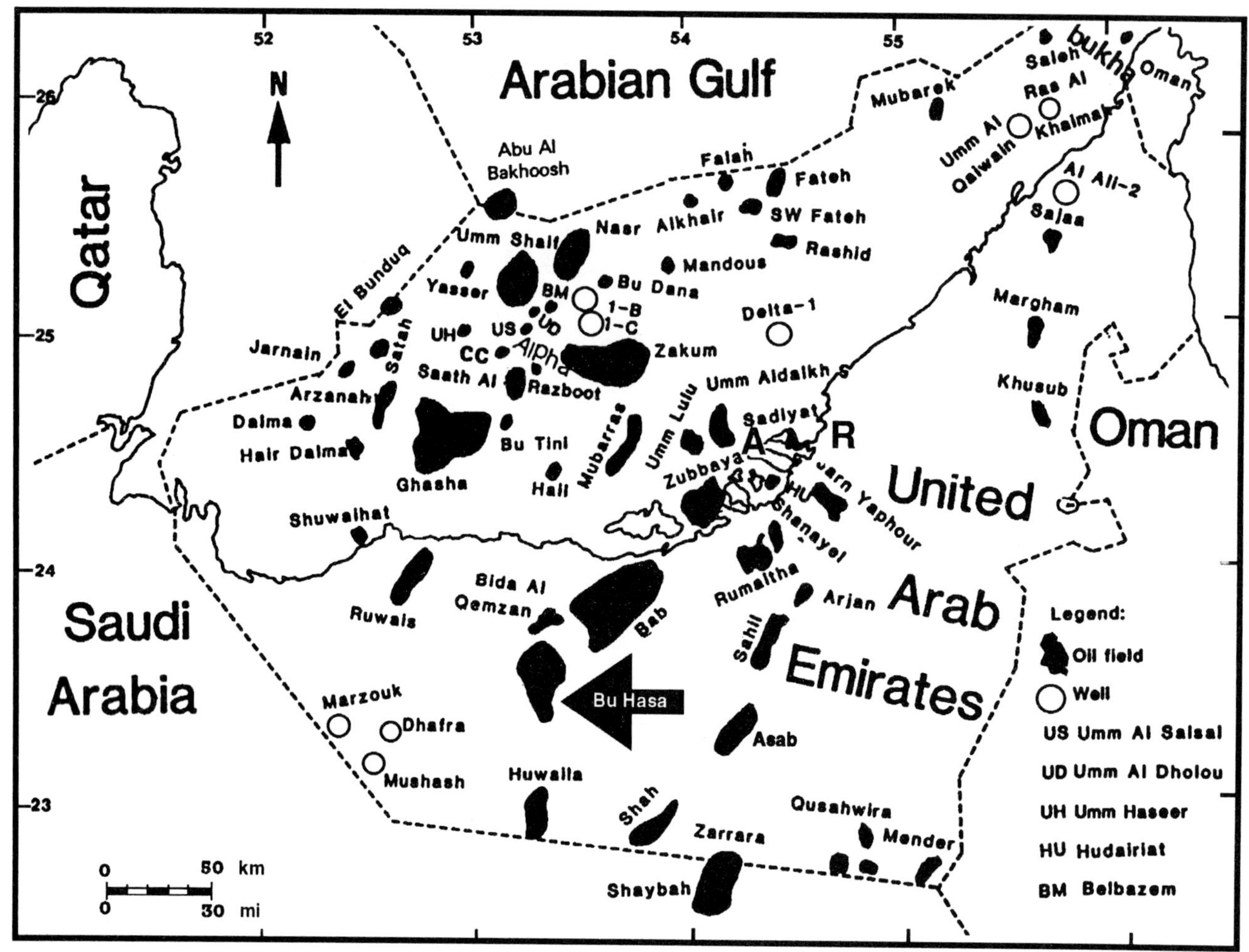

Figure 1. Location map of United Arab Emirates showing Bu Hasa and other major oil fields. A is Abu Dhabi Island and R is Ras Sadr (RS-1).

tered an unexpected development of reefal sediments at the top of the Lower Cretaceous (Thamama Group, Figure 3). The development of 485 ft (151.5 m) thick rudistid and algal facies was hydrocarbon-bearing.

DISCOVERY METHOD

In Abu Dhabi, the first onshore gravity and magnetic surveys were conducted in 1947 and completed in 1948. Gravity and magnetic surveys were again conducted in 1953–1955 and in 1959–1960. Interpretation of these gravity and magnetic surveys indicated broad, gentle structural features that appeared to reflect deep structural differentiation and some uplift at great depth in the Murban (Bab/Bu Hasa) area (Hajash, 1967). The results of these early surveys provided a useful basis for later seismic surveys. A seismic reflection survey was first conducted in the area of Abu Dhabi in late 1949 and early 1950. This survey indicated a broad domal uplift, in general coinciding with the Tarif (Murban) gravity high upon which the Murban-1 well was later drilled in 1951 (Hajash, 1967).

While the development of Bab (Murban) continued, a well designated the Murban-12 (later renamed Bu Hasa-12) was located in March 1962 near the crest of the structure that had been defined by seismic reflection, centered some 30 mi (50 km) southwest of the Bab field. The well bottomed at 8667 ft (2709 m) in the dense unit underlying the Thamama zone B. The well encountered undersaturated 39° API oil in the Lower Cretaceous, both in the Shuaiba Formation and in zone B of the Kharaib Formation (Figure 3). The top of the Lower Cretaceous Thamama Group was encountered higher than expected, and several hundred feet of this unit had been drilled before the operator stopped to log (Hajash, 1967). An oil-water contact (OWC) in zone B, some 425 ft (133 m) structurally higher than in the Bab field, indicated a separate pool to which the name Bu Hasa was applied (Hajash, 1967). The field went on stream in May 1965 at an initial rate of 116,000 BOPD. In 1965, the drilling of appraisal and delineation wells began in the southern part of the

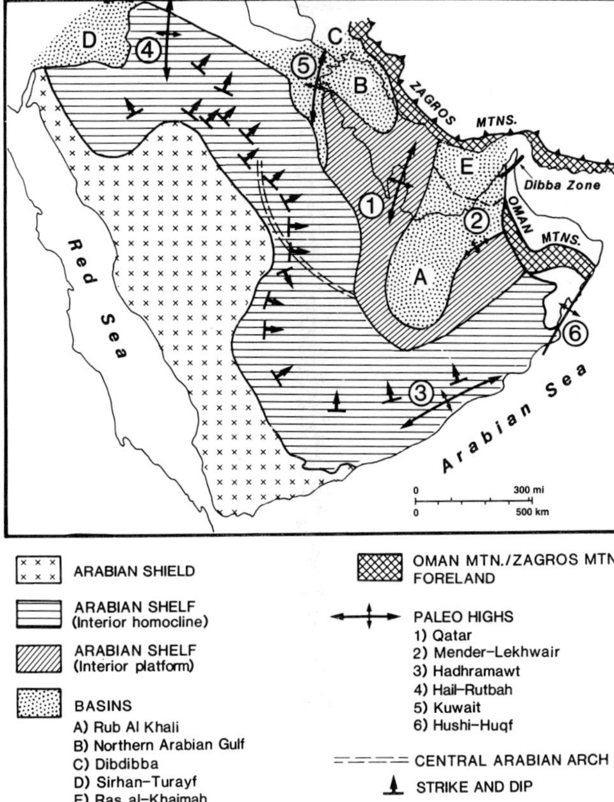

Figure 2. Geologic setting and structural provinces of the Arabian Peninsula. (Modified from Powers et al., 1966, and Alsharhan and Nairn, 1986.)

field, and in 1969 it was proved that the Bu Hasa structure was full to the spillpoint at the saddle between Bu Hasa and the Huwaila structure to the south (Twombley and Scott, 1975).

STRUCTURE

Abu Dhabi lies on the eastern part of the Arabian platform and is situated in the central part of the Rub al Khali basin (Figure 2). Tectonic features such as the central Arabian arch, Qatar–South Fars arch, and the Rub al Khali basin had a pronounced influence on structural developments in the Abu Dhabi area.

The Abu Dhabi area appears as a wide trough at the Thamama level (Figure 4). This trough is oriented roughly northeast-southwest and deepens toward the northeast in the onshore area. In the northwest and southeast areas, the trough is shallower. The burial depth of the Thamama is variable from place to place. In southeast Abu Dhabi (Mender area), the Thamama is considerably shallower, at about 4000 ft (1220 m) subsea; in western Abu Dhabi, it is about 7000 ft (2135 m) subsea; while in central Abu Dhabi the Thamama lies at about 11,000 ft (3355 m) subsea and deepens further toward the northeast to reach about 15,000 ft (4575 m) subsea (ADCO Staff, 1986). Two principal tectonic trends are evident: In the eastern areas, the northeast-southwest trend is expressed by the Shah-Asab axis, while in the western areas, the north-south trend is expressed by the Huwaila–Bu Hasa axis (Schlumberger, 1981; Alsharhan, 1989). These anticlinal ridges are separated by major synclines. The Falaha syncline parallels the Shah-Asab-Sahil trend on the west (Figure 4), while the Ghurab syncline lies to the west of the Bu Hasa-Bab trend.

The Bu Hasa field is oval shaped (Figure 5), with a 22 mi (35 km) north-south major axis and a 12.5 mi (20 km) east-west minor axis. The structure is a north-south-trending anticline with a broad nose at the northern end and a narrower plunging ridge southward. The highest well at the crest of the structure was the discovery well. The northern flank dip is on the order of 1°. The western flank, with a structural dip of 2°, is approximately twice as steep as the eastern flank. The dip of the southern ridge toward Huwaila is about ¾°. The structural contour map of the top of the Shuaiba has minor irregularities that possibly were caused by local erosion at the unconformity surface or by differential compaction. The structural relief of the field is about 800 ft (250 m), and the areal closure is about 207 mi^2 (538 km^2) at the top of the Shuaiba. No faults have been identified in Cretaceous formations.

Structural growth at Bu Hasa was initiated during Albian time (Figure 6) and was moderate until the late Cenomanian, at which time a tilting of the newly born structure occurred toward the south (Huwaila structure) (ADCO Staff, 1986). The majority of the structural uplift (about 85%) occurred during the Late Cretaceous; i.e., growth was re-initiated in the earliest part of the Late Cretaceous and continued until reaching maximum paroxysm during the late Campanian-Maastrichtian. Then there is evidence of further moderate structural growth until the end of the Eocene (ADCO Staff, 1986).

CRETACEOUS STRATIGRAPHY

Following the deposition of the regressive Upper Jurassic Hith anhydrite, a marine transgression occurred that resulted in flooding of most of the Arabian Gulf area. The Qatar arch (Figure 2), which remained as a positive feature, appears to have affected sedimentation throughout the Early Cretaceous (Thamama Group deposition), with sediments generally increasing in thickness in all directions away from it. In Abu Dhabi, the Thamama Group (Figure 3) began with the deposition of the Habshan Formation, which is dominated by lime mudstones and wackestones of lagoonal deposits to the west and by oolitic packstone/grainstones interbedded with dolomite toward the east. This was followed by the

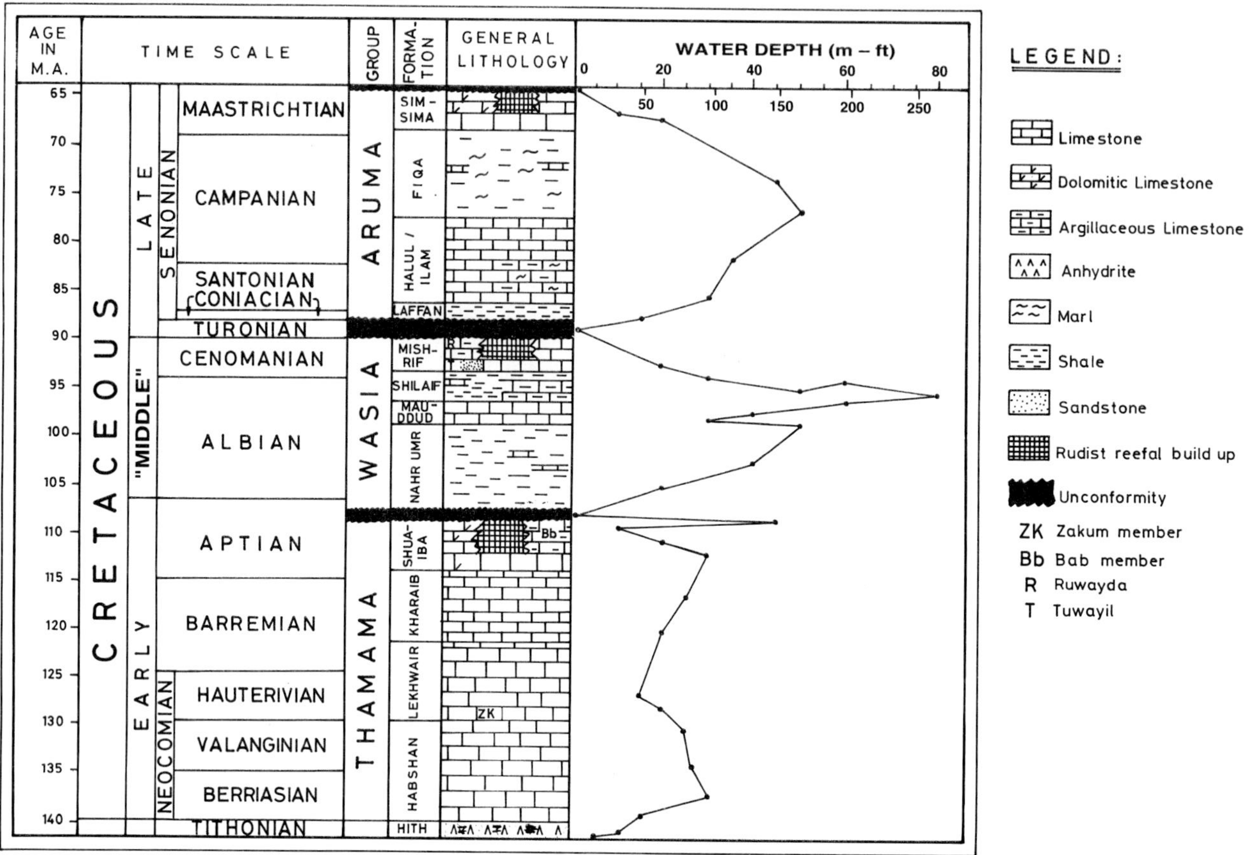

Figure 3. General stratigraphic chart of the Cretaceous in Abu Dhabi, U.A.E.

deposition of the Lekhwair Formation, which is characterized by normal marine cyclical shelf carbonates. The Kharaib Formation overlays the Lekhwair and is dominated by mud-supported sediments in the lower part, grading upward to packstones and grainstones; it exhibits the cyclic pattern of carbonate sedimentation already seen in the Lekhwair (Alsharhan, 1989). The Shuaiba Formation is the uppermost rock unit in the Thamama Group and the main oil-bearing horizon in the Bu Hasa field; it represents thick, porous shelf carbonates characterized by rudistid and algal platform sediments in western areas (Alsharhan, 1985a). The Bab Member, equivalent to the Shuaiba buildup, developed in an intrashelf basin in central Abu Dhabi. It is characterized by mud-supported limestones with shaley and argillaceous partings and a deep marine fauna (Alsharhan and Nairn, 1986). Toward the end of the Aptian, epeirogenic movements and/or a sea level fall terminated the deposition of the Thamama Group and caused a period of emergence and erosion.

Deposition of the "Middle" Cretaceous Wasia Group (Figure 3) began with the sheet-like neritic deposits of the transgressive Nahr Umr Formation, which consists of a sequence of variegated shales with some lenses of limestone and sandstone. This shale formation acts as a regional seal for the Shuaiba reservoir in the Bu Hasa field. At the end of Nahr Umr deposition, shallow marine carbonate sediments were deposited as bioclastic wackestone/packstones of the Mauddud Formation. A basin developed in central Abu Dhabi in which bituminous shale and dark gray bituminous mudstones of the Shilaif Formation, with abundant *Pinthonella*, were deposited. At the end of Shilaif time, a minor period of uplift affected the central basin area. On the edges of this intrashelf basin, the Mishrif Formation formed as a carbonate shelf deposit characterized by mollusc-fragment grainstones and packstones, rudist and algae grainstones and boundstones, and muddy bioclastic miliolid limestones (Alsharhan, 1989). At the end of the Cenomanian, a major period of emergence and erosion terminated the deposition of the Wasia Group.

Deposition of the Upper Cretaceous Aruma Group (Figure 3) began with the transgressive shales of the Laffan Formation, which lie unconformably on the Wasia Group. In Santonian time, a neritic carbonate shelf developed over much of the Arabian Gulf. The Halul Formation represents a shallowing of the epeiric sea, with the occurrence of high-energy shelf limestones at the top of the formation. Renewed subsidence and the associated transgression in the

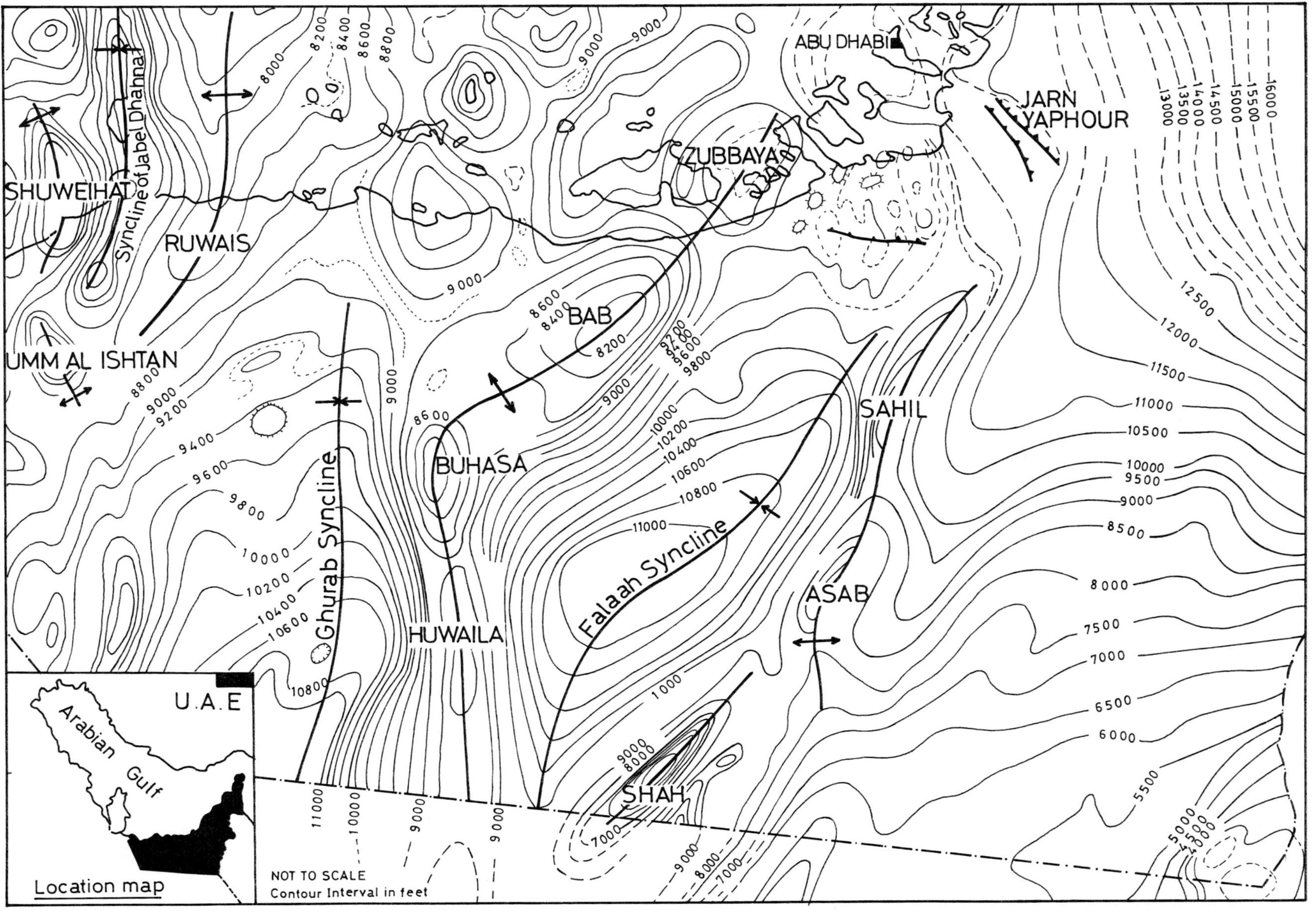

Figure 4. Structural configuration of Abu Dhabi onshore area at the top of Zone B of the Kharaib Formation (Thamama Group). (Modified from ADCO, 1986.) Contour interval, 200, 500, and 1000 ft. Note contour interval change at Sahil-Asab axis.

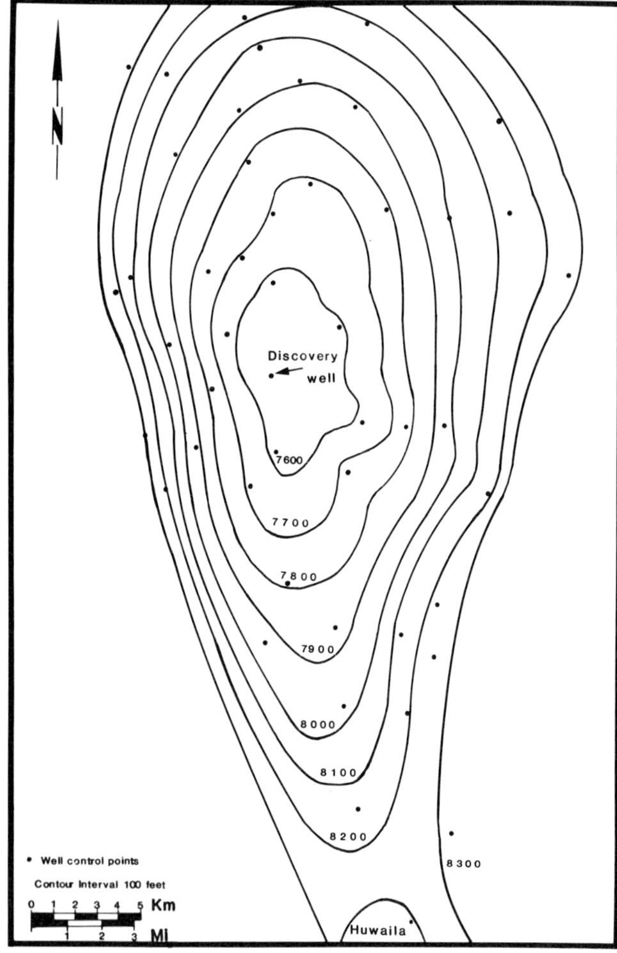

Figure 5. Structural contour map of the top of the Shuaiba Formation in the Bu Hasa field, Abu Dhabi. Contour interval, 100 ft.

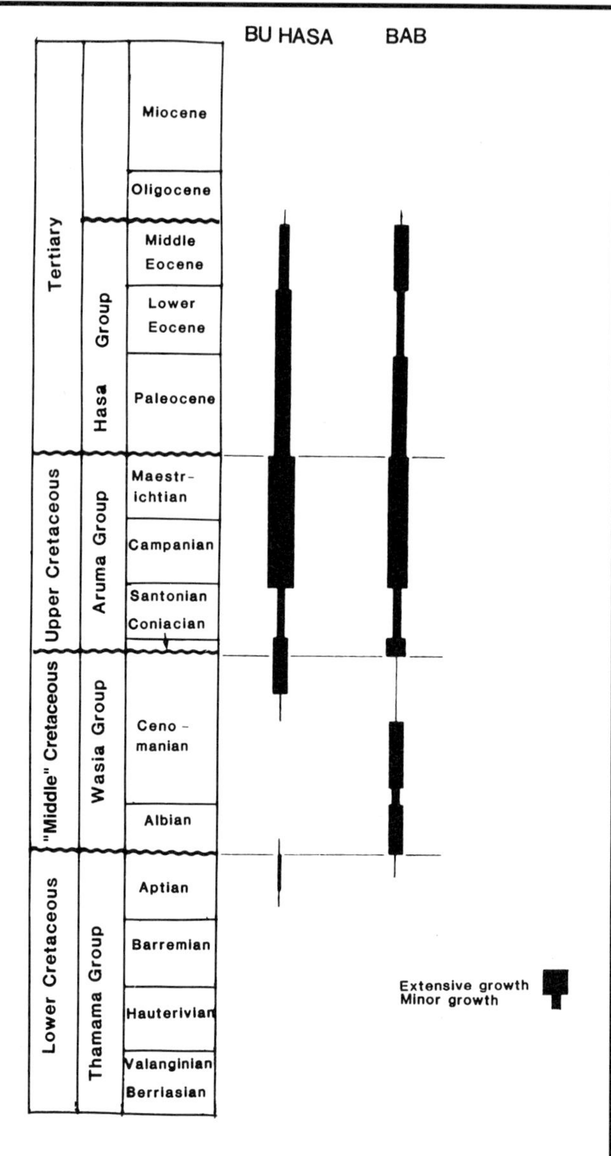

Figure 6. Timing of structural growth of Bab and Bu Hasa fields, Abu Dhabi. (Modified from ADCO, 1986.)

Campanian resulted in the deposition of the basinal shales and limestones of the Fiqa Formation. A shallowing of the basin led to the deposition of the neritic shoal and shelf limestones and dolomites of the Simsima Formation. A late Maastrichtian regional unconformity marks the boundary between the Cretaceous and Tertiary formations. For more information regarding overall stratigraphy, see the Asab field study in this volume.

TRAP

Thickness of Hydrocarbon Trap

The isopach map of the Shuaiba Formation (Figure 7) shows a zone of maximum thickness within and across the structural crest of the field. This zone is about 2 to 3 mi (3 to 5 km) wide and extends northwest-southeast for at least 14 mi (23 km).

The south-central portion of the field has irregular sediment thickness variations caused by a combination of post-depositional erosion and varying depositional rates of different environmental settings (e.g., lagoon versus shoal and bank). A general thinning trend occurs to the south-southwest away from the crest of the buildup, with a minor thickening trend in the southernmost area. Approximately 6 to 8 mi (10 to 12 km) northeast of and parallel to the axis of the maximum buildup, the stratigraphically highest Shuaiba buildups occur in a basinward, downdip setting. These buildups are characterized by *Orbitolina* limestones and rudistid accumulations. Some stromatoporoid boundstones have been noted in wells along this trend. The Shuaiba grades into a basinal facies and thins gradually north-northeast toward the Bab field.

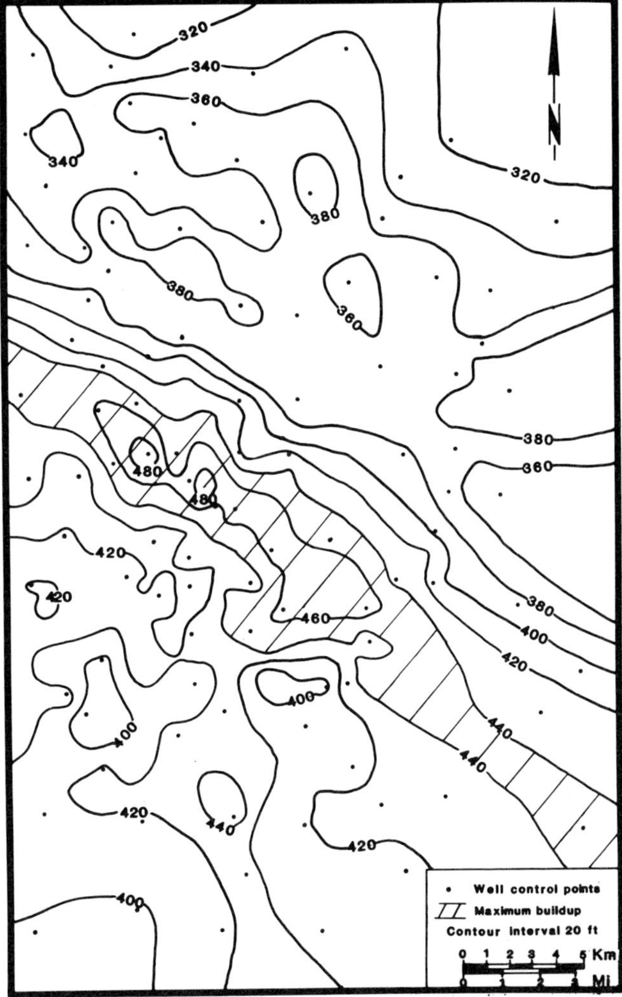

Figure 7. Isopach map of Shuaiba Formation in the Bu Hasa field.

Reservoir Units

Classification and Description

The Shuaiba Formation displays pronounced vertical and lateral lithologic variations generated by the development of a rudist reef complex over an algal platform. Nine distinct petrophysical/reservoir subzonation units (A to I, Figure 8) have been identified (Harris et al., 1968; Twombley and Scott, 1975; Alsharhan, 1985a, b, 1987; Hulstrand et al., 1985) that are significant to reservoir mechanics. The nine distinctive units (Figure 8) were defined by their porosity, permeability, and lithology. The properties of these units as to porosity, permeability, and log character can be seen in Figure 9. The logs suggest that the Shuaiba reservoir contains little terrigenous mud and is essentially carbonate. The stylolites common to the section are characteristically associated with a reduction in porosity and sometimes by an increase in gamma-ray intensity (Figure 9). Permeability variations between the different units are reflected mainly by changes in irreducible water saturation. The general lithologies and depositional settings of these units are plotted on the composite log of Figure 10.

Reservoir Unit I—Unit I contains pale grayish-brown packstones that grade into poorly sorted grainstones containing skeletal grains. The matrix appears to consist in part of lime mud, although locally it contains sufficient grains that packstones are developed. The skeletal grains include rounded, sand-size shell fragments. Some dense, bioturbated, serpulid wackestones occur in some intervals.

Unit I developed in the southern part of the field and the zero isopach has a northwest-southeast trend (Figure 11C). The unit becomes thinner toward the south. It ranges in thickness between 10 and 130 ft (3 and 40 m), averaging 80 ft (24.4 m).

Secondary porosity (solution vugs and molds) dominates and occurs within rudists, corals, and algal colonies. The reef debris has a high percentage of primary interparticle, intraparticle, and framework porosity. Local leaching of gastropod shells forms molds and vugs. Porosities are partially to completely occluded by small amounts of sparry calcite (drusy and blocky crystal) and baroque dolomite.

Porosity and permeability are distributed erratically, corresponding to heterogenous facies. Porosities average 18% and permeabilities 12 md. However, where porous sections of the mainly rudistid limestone facies alternate with dense argillaceous and stylolitic limestones, permeabilities show sharp variations from over 100 md in the rudist facies down to less than 1 md in the dense limestones.

The base of Unit I is picked where there is a clear contrast between the dense character of this unit and the porous character of the underlying unit (Figure 12). The top is easily picked on logs where this unit is always capped by the shales of the Nahr Umr Formation. The unit is characterized on logs by a gamma-ray intensity decrease downward and a neutron increase where the dense and argillaceous limestone occurs.

Fossils consist predominantly of rudists with subordinate gastropods, corals, *Bacinella irregularis* algae, orbitolinids, miliolids, sponge spicules, and stromatoporoids. In slabbed cores, the rudistid bivalves appear unbroken, although this usually cannot be determined in thin sections.

Small amounts of pyrite and nodules of anhydrite have been observed as well as well-developed stylolites and microstylolites. The lime-mud matrix contains appreciable amounts of fine-grained peloids, some of which possess thin oolitic coatings.

Medium-grained skeletal sands (locally oolitic), a coarse-grained skeletal packstone, rudist fragments associated with stromatoporoids, encrusting algae, and miliolids suggest accumulation in a very shallow marine lagoonal setting.

Reservoir Unit H—Unit H consists of buff to light brown, medium- to coarse-grained packstone and grainstones. The unit also contains abundant

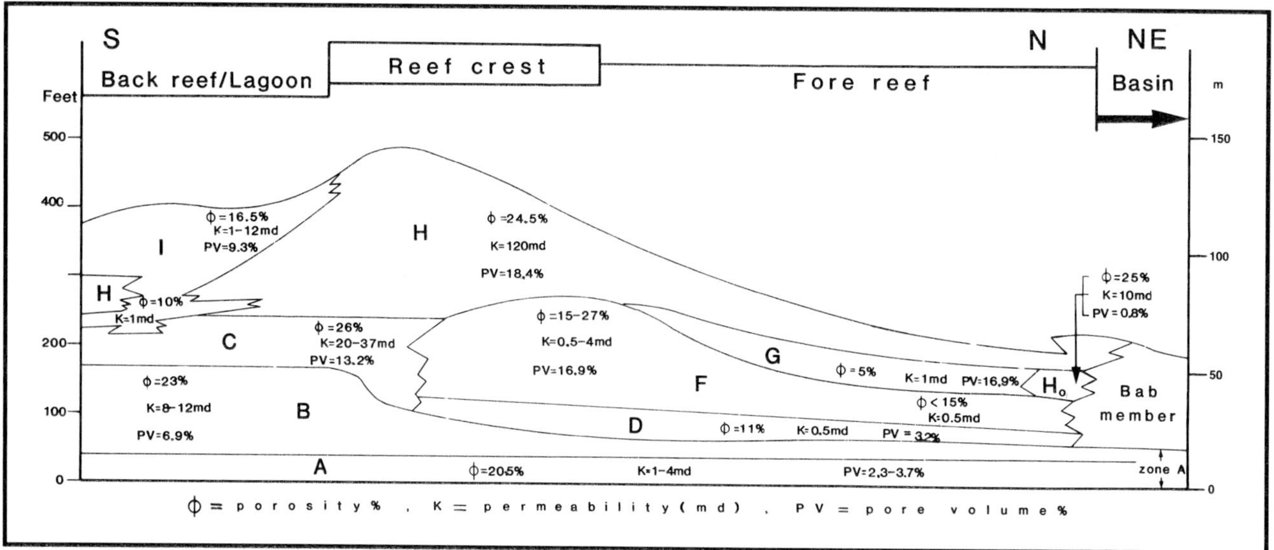

Figure 8. Schematic cross section showing distribution of reservoir units, average porosity (∅), permeability (K), and pore volume (PV) in the Shuaiba Formation of the Bu Hasa field. (Modified from Harris et al., 1968, and Schlumberger, 1981.) Letters (A–I) denote the different reservoir units.

caprinid rudistid fragments, the dominant grains of these buildups, which sometimes occur in growth position. The accumulation of these organisms is somewhat comparable to present-day oyster banks. Monopleurid, requienid, and caprotinid rudists are also present but are less abundant than caprinids. Some layers of lime mudstone and wackestone occur in parts of the section.

Unit H thickness varies between 9 and 258 ft (3 and 78 m) but increases toward the crest of the field (Figure 11A). In the center of the field, this unit covers an area about 1 to 2 mi (2 to 3 km) wide and extending northwest-southeast for 14 mi (23 km) along the crest of the structure, reaching a maximum thickness of 257 ft (78 m) of rudistid and *Orbitolina* sediments. Some isolated areas of this unit to the south are 150 ft (46 m) thick. Scattered patches of rudist-rich beds are also found in the south. These beds were probably patch reefs in early lagoons. Dense limestones of Unit I were formed later in these beds.

Unit H pore types include interparticle, matrix, vuggy, and moldic leached porosity. Large crystals of calcite and baroque dolomite line the pores and locally occlude them.

This unit has good to very good porosity and permeability. The porosity averages 25% while permeability is erratic, averaging 100–120 md. In some beds, permeability is over 500 md.

The base of Unit H is taken to be at the depth of slightly increased gamma ray signal and decreased porosity or, where present, at a dense streak at the top of the underlying unit (Figure 13). The top is recognized by the first limestone or shale that appears where Unit H is capped by the Nahr Umr Shale or at the point of slightly decreased porosity where Unit H is overlain by Unit I. The unit is characterized by low gamma-ray excursion and neutron curve, which are related to a porous to very porous section.

Besides the rudists, abundant stromatoporoids, corals, peloids, and lithoclasts occur in Unit H. Similarly, abundant orbitolinids, dasycladacean and codiacean green algae, miliolids, vulvulinids, echinoderm plates and spines, brachiopods, and ostracod fragments are present.

Locally, the rudist shells are packed so closely that they resemble oyster banks, suggesting growth in a shallow marine setting. The wave energy and currents during the deposition of these sediments were low since most of the rudistid shells are unbroken and some are filled with lime mud. These rudistid limestones may have been deposited in a shoal-bank, with low to moderate wave and current energy or on the forebank below wave base.

Reservoir Unit H_0—Unit H_0 consists of gray, medium-grained, *Orbitolina* packstone to grainstones and rare wackestone.

This unit is developed only in the northern part of Bu Hasa field, and its thickness increases rapidly to over 120 ft (37 m) along a wide northwest-southeast linear trend (Figure 11A). In the extreme north, the unit decreases to about 30 ft (9.1 m).

Secondary porosity (leached, vuggy, and moldic) dominates Unit H_0. Calcite crystals fill the vugs locally, and some scattered baroque dolomite is present.

The porosity of Unit H_0 averages 20% and permeability around 10 md.

The top of the unit is taken to be at the point where there is marked increase in porosity upward, though where it is overlain by Unit I, it may be marked by a stylolitized zone. The base is taken at the dolomitic streak developed at the top of the

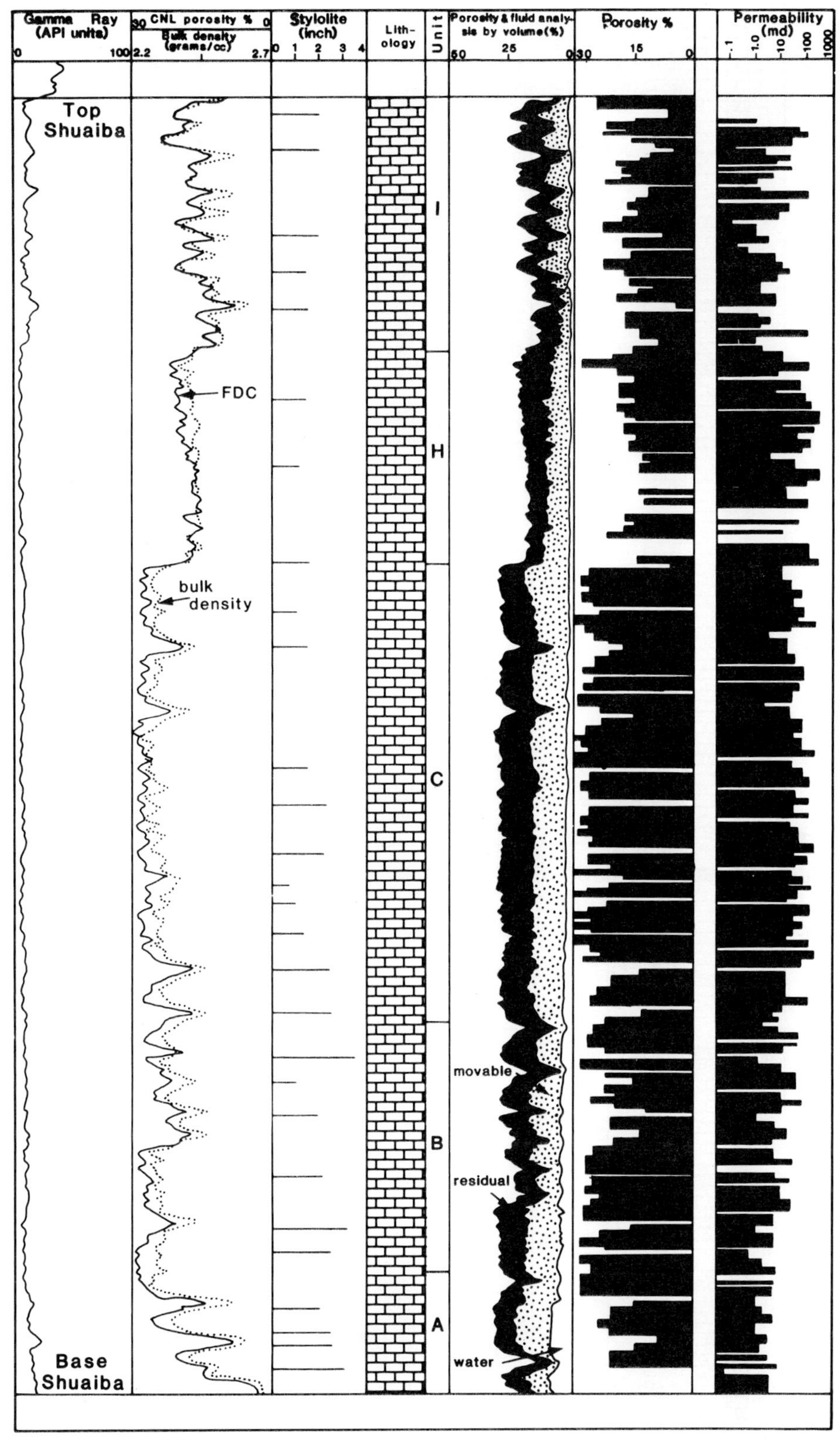

Figure 9. Relationship between FDC/CNL logs, porosities, permeabilities, and stylolites in Shuaiba Formation of the Bu Hasa field. Total thickness is 420 ft.

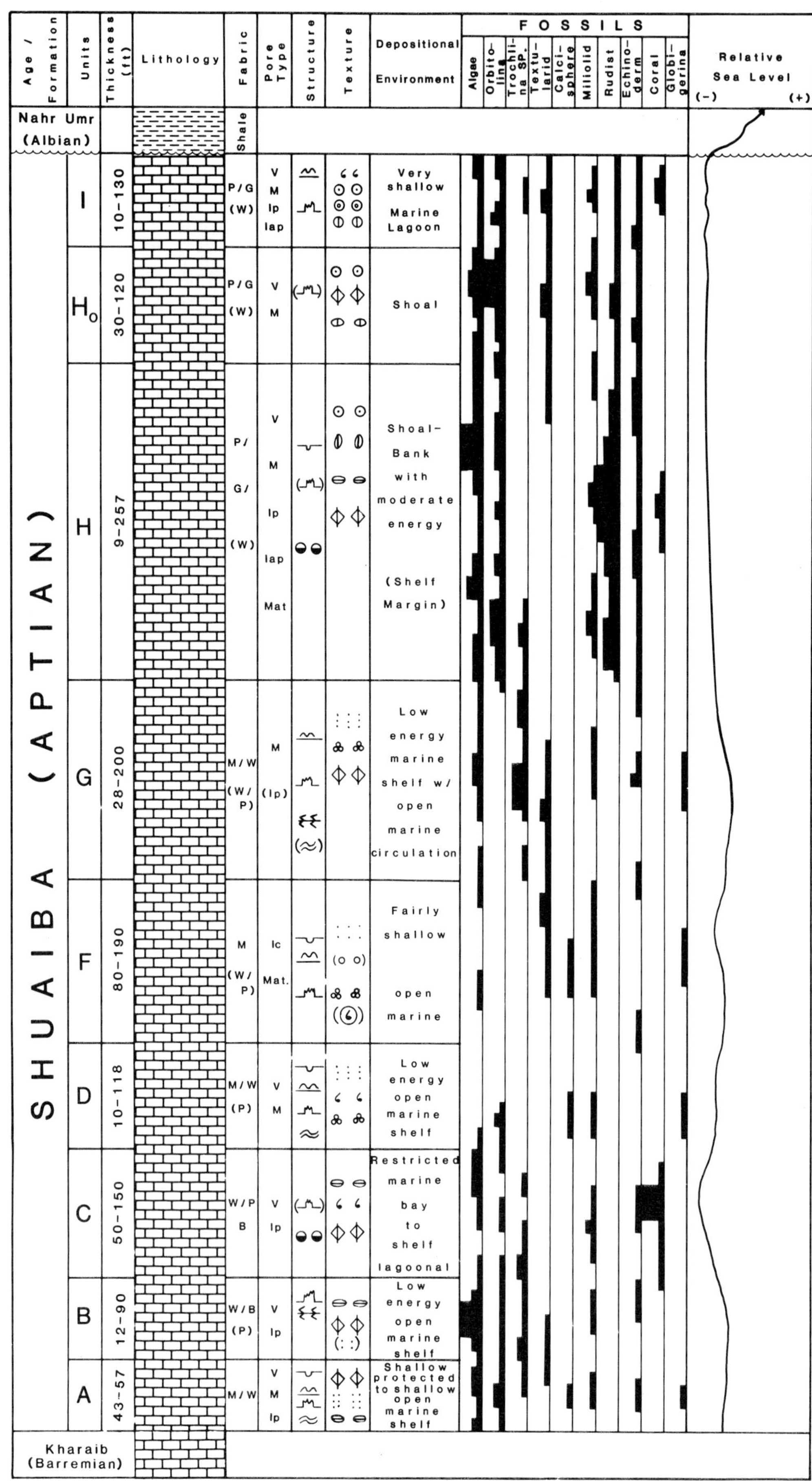
Age / Formation
Units
Thickness (ft)
Lithology
Fabric
Pore Type
Structure
Texture
Depositional Environment
FOSSILS
Algae
Orbitolina
Trochilina SP.
Textularid
Calcisphere
Miliolid
Rudist
Echinoderm
Coral
Globigerina
Relative Sea Level
(-)
(+)
Nahr Umr (Albian)
Shale
SHUAIBA (APTIAN)
I
10-130
P/G (W)
V M Ip lap
Very shallow Marine Lagoon
H0
30-120
P/G (W)
V M
Shoal
H
9-257
P/ G/ (W)
V M Ip lap Mat
Shoal-Bank with moderate energy (Shelf Margin)
G
28-200
M/W (W/P)
M (Ip)
Low energy marine shelf w/ open marine circulation
F
80-190
M (W/P)
Ic Mat.
Fairly shallow open marine
D
10-118
M/W (P)
V M
Low energy open marine shelf
C
50-150
W/P B
V Ip
Restricted marine bay to shelf lagoonal
B
12-90
W/B (P)
V Ip
Low energy open marine shelf
A
43-57
M/W
V M Ip
Shallow protected to shallow open marine shelf
Kharaib (Barremian)

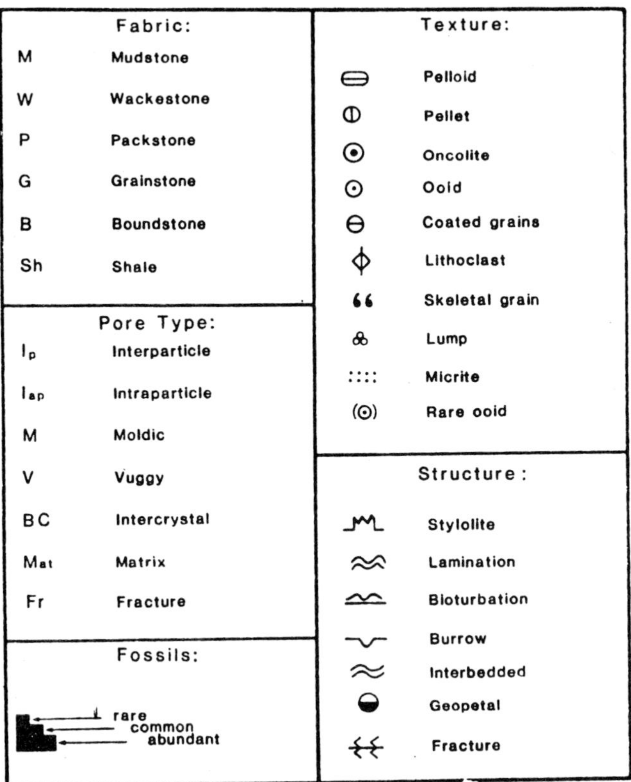

Figure 10. Composite log (page 108) showing lithology, pore types, environments of deposition, fossil occurrences, and relative sea levels in the Shuaiba Formation. Legend for log appears above.

underlying unit. Unit H_0 is characterized by a low gamma-ray response and a highly porous section that can be recognized on the neutron log curve.

Fossils in Unit H_0 are abundant orbitolinids, common codiacean algae, echinoderms, textulariids, some broken rudists, and a few miliolids.

The presence of abundant *Orbitolina* in these sediments is indicative of moderate-energy shoal to forebank deposits.

Reservoir Unit G—Unit G is a fine- to medium-grained, dark gray, dense, lime mudstone with scattered, fine, bioclastic debris, grading to wackestones. It is slightly fossiliferous, containing small bivalve shells and gastropods. Packstones occur particularly where an influx of rounded, sand-size micritic allochems and a few ooliths occur.

Shaley streaks and the dark limestones may be due to either the presence of finely disseminated clay within the matrix or the effects of lithification. Scattered crystals of pyrite occur in conjunction with small vugs, and fractures are filled with calcite. Well-developed stylolites associated with microstylolites are common.

Unit G is over 200 ft (61 m) thick at its maximum, but thins northward to 28 ft (8.5 m) (Figure 11B).

The matrix is very hard and is thought to be extensively cemented to neomorphic microspar. The cement may have been derived from the effects of pressure solution, evidenced by stylolites in parts of the section. Most of the primary interparticle porosity that had been present throughout the sequence has been filled by drusy calcite and microcrystalline cements. Localized areas of moldic porosity were formed by the leaching of fossil shells that have also been filled by these cements.

Unit G porosity averages 3% and permeability is extremely low (less than 1 md). However, the porosity of some interbeds of packstones and grainstones attains 20%, with permeabilities of 10 md.

The top of Unit G is easily recognizable as a point of inflection of the gamma-ray and neutron log curves (Figure 14). The base is picked at the top of dolomitic streaks associated with gamma-ray peaks at the top of the underlying unit. The dense unit is characterized by excursions representing a low gamma-ray and high neutron curve section.

Fossils in Unit G are mainly dominated by skeletal debris, including calcareous benthonic faunas, miliolids, globigerinid foraminifera, echinoderm and bivalve fragments, sponge spicules, and dasycladacean algae. Textulariids are common and gastropods are rare.

The sequence shows a continuation of low energy marine shelf to ramp deposition with open-marine circulation. The limestones contain coarse, reworked, abraded carbonate grains (oolitic coatings) that most likely were transported into a protected environment from adjacent mobile shoals of carbonate sands.

Reservoir Unit F—Unit F consists mainly of lime mudstones that are dark gray to pale gray, fine-grained, microporous, and dense. It is faintly bedded and burrowed, containing abraded and/or micritized shell debris and some oncolites. Occasional shale partings are present and stylolites are well developed. In some intervals, fossiliferous intraclastic packstones occur, and in some samples, grainstone textures are developed.

Unit F developed only in the northern part of the field. The thickness ranges from 70 to 110 ft (21 to 33 m) near its southern limit and then increases to 170 ft (52 m) (Figure 11C) within a short distance. This belt of thicker sediment is 0.6 to 1.2 mi (1 to 2 km) wide and 13.7 mi (22 km) long. The unit then thins gradually toward the north and wedges to a thickness of about 15 ft (4.5 m) in northern wells (Figure 11C).

Unit F pores are fine to medium, formed by rhombic and mosaic dolomite with relic leached fossils. The pores also contain small amounts of anhydrite and pyrite. Dolomite replacement is so widespread that it has produced small amounts of intercrystalline porosity that, in part, has been reduced by later blocky calcite cementation. The dolomitization has formed a fine to medium crystalline mosaic. The small amounts of primary interparticle porosity that were present have been completely filled by drusy calcite and microcrystalline silica cements; some

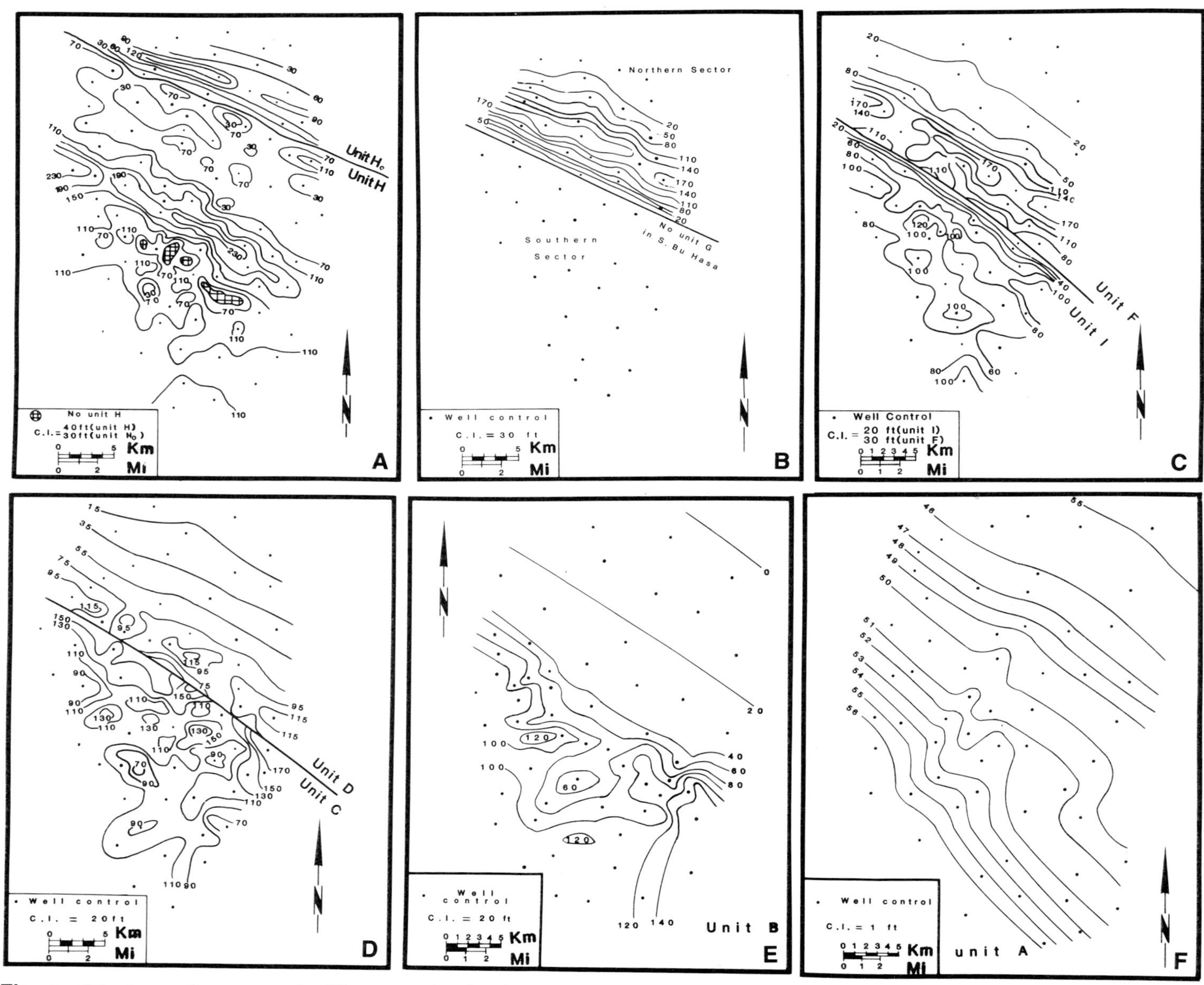

Figure 11. Isopach maps of different units in the Shuaiba Formation, Bu Hasa field. (Modified from Hulstrand et al., 1985.)

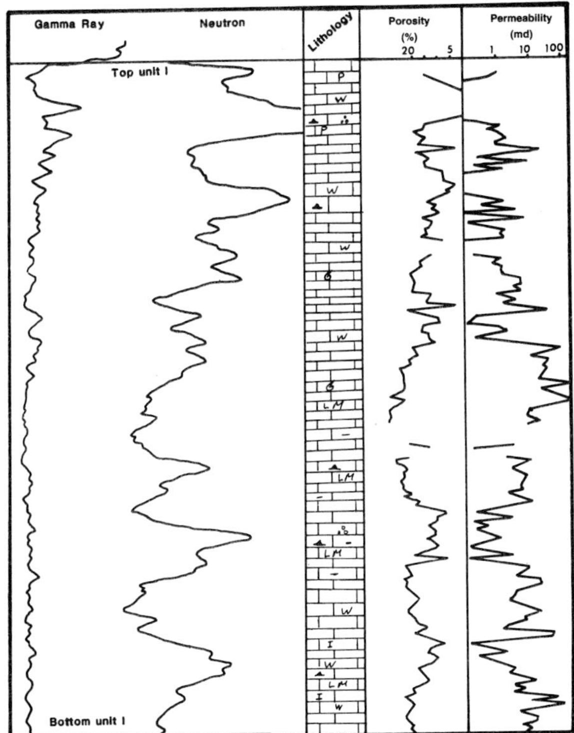

Figure 12. General lithology, log character, and porosity-permeability measurement of reservoir Unit I. Thickness of unit shown is 100 ft.

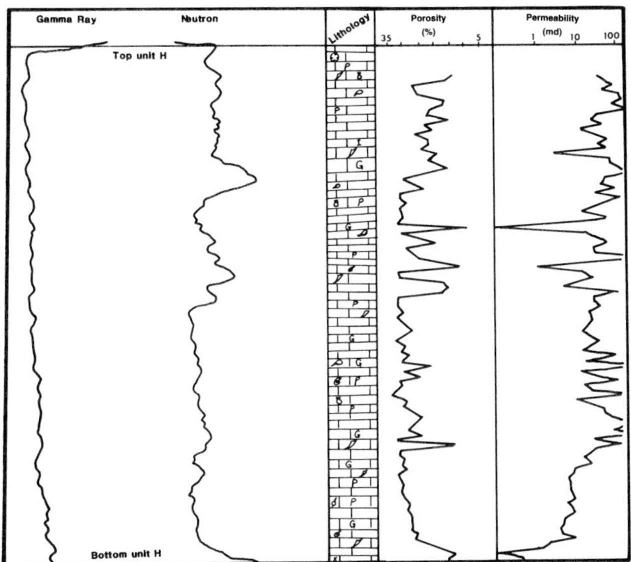

Figure 13. General lithology, log character, and porosity-permeability measurement of reservoir Unit H. Thickness of unit shown is 180 ft.

matrix porosity has been observed throughout the section.

Porosity averages 27% and permeability 5 md. However, in northern wells, porosity and permeability deteriorate as the lithology grades into

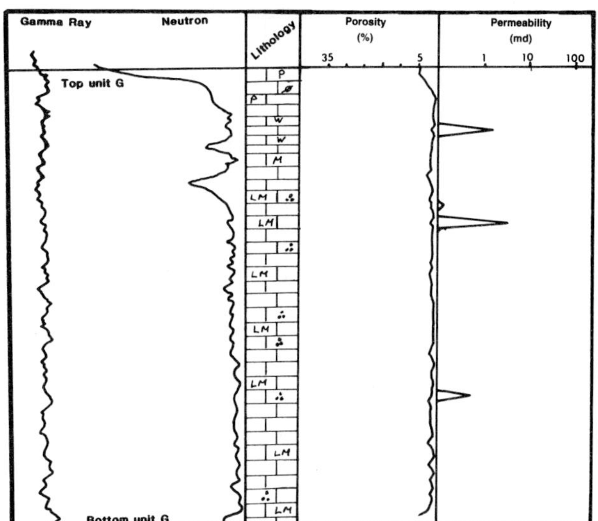

Figure 14. General lithology, log character, and porosity-permeability measurement of reservoir Unit G. Thickness of unit shown is 160 ft.

progressively denser limestones where the porosity averages 15% and the permeability is less than 1 md.

The top of Unit F is defined by a higher gamma-ray reading and an upward increase in porosity; it is commonly characterized by the presence of a dense and/or dolomitic streak shown by the neutron log and a downward diminution in porosity (Figure 15). The gamma-ray log has low response in this unit, though it tends to be higher in the basal part. The neutron log shows porous sections interspersed with some dense beds.

The section contains small, calcareous, benthonic foraminifers, globigerinids, miliolids, echinoderms, skeletal debris, sponge spicules, and algal-encrusted grains.

In the northern part of the field, dense lime-mud sediments are thought to represent continued mud-supported limestone deposited below wave base on an open-marine shelf. The presence of planktonic foraminifera (open sea) combined with the thinness of the packstone to grainstone interval leads to the belief that fairly shallow shelf, open-marine deposition continued (with influx of very shallow water carbonate detritus). Deposition became progressively more restricted in response to a shallowing of water depths toward the central part of the field, as indicated by extensive dolomitization processes close to the reef crest area.

Reservoir Unit D—Unit D consists of pale grayish-brown, medium-grained, microporous lime mudstones that are bioturbated, slightly argillaceous, and burrowed. Occasional horizons of wackestone grading to packstone contain peloids and intraclasts. These fabrics may have resulted either from localized increased activity of burrowing organisms or the periodic influx of these allochems during storms. Laminated shales and microstylolites have been observed and occasional pyrite nodules also occur.

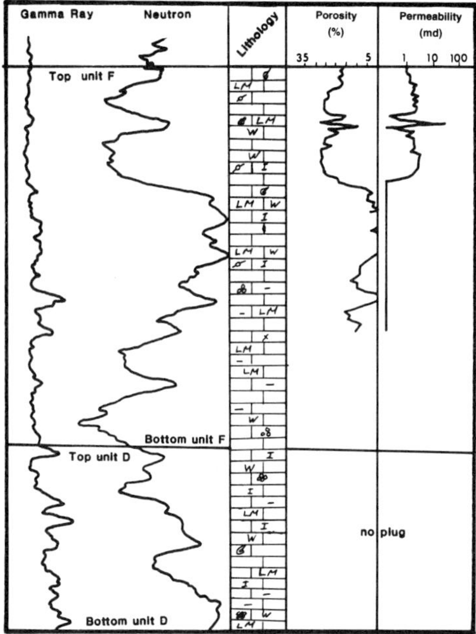

Figure 15. General lithology, log character, and porosity-permeability measurement of reservoir Units D and F. Thicknesses of units shown are 90 and 130 ft, respectively.

Thickness of the unit in the northern part of the field ranges from a maximum of 114 ft (35 m) to a minimum of about 10 ft (3 m) in the northern wells (Figure 11D).

Pore types are mainly vuggy and moldic pores formed by leaching of fossils. Both types were greater and more widespread prior to fill by drusy calcite cement.

The average porosity is 11% and the average permeability is generally less than 1 md (around 0.5 md).

The top of Unit D is picked where the gamma ray decreases and where porosity in the overlying unit increases (Figure 15). The base of Unit D or the top of Unit B is picked where the gamma-ray and neutron curves show a very high response accompanied by a decrease in porosity. Unit D has a high to very high gamma-ray response and the neutron log indicates poor porosity.

Fossils in Unit D consist mainly of bioclastic debris, gastropods, scattered orbitolinids, common miliolids, and globigerinids.

The common planktonic foraminifers suggest an unrestricted open-marine circulation and a low-energy shelf setting, with deposition occurring below wave base, although no evidence exists to indicate that the shelf was particularly deep.

Reservoir Unit C—Unit C consists of buff to brown, fine- to medium-grained wackestones, containing abundant and frequently large algal colonies (boundstones). Peloidal packstones occur, and in these there is an abundance of stylolites. A thick biostrome of corals occurs in this unit, which is also characterized by: (1) lower gamma-ray response than in any other part of the unit, (2) lower average porosity because the corals are calcitized and dolomitized, and (3) a virtual absence of stylolites.

Unit C developed only in the south. The thickest part coincides with the thickest coral buildups, ranging from 66 to 181 ft (22–55 m) (Figure 11D).

The widespread coral buildup contains large vugs that are partially filled with calcite. Porosities include the presence of large cavities partly filled with lime mud, coarse crystalline calcite, and some dolomite. Secondary leached porosity occurs, but interparticle porosity is dominant.

The average porosity is about 25% and the average permeability is 37 md. Better permeability coincides with greater quantities of algae and corals.

The top of this unit, where it is in contact with Unit H, is characterized by a sharp, local decrease of porosity related to a slightly higher gamma-ray activity (Figure 16). In this unit, the gamma-ray and neutron curves are fairly constant. The base of Unit C, where it can be seen, is in contact with Unit B.

Fossils found in Unit C are *Bacinella irregularis* algae, corals, stromatoporoids, miliolids, orbitolinids, and *Trocholina* sp.

The depositional setting ranges from low to high turbulence. On the one hand, sedimentation under quiet conditions probably occurred in a restricted shallow marine bay to a shelf-lagoonal complex in which lithoclast peloidal grains and algae occur. Coral and stromatoporoids indicate other areas where energy levels were moderate to high.

Reservoir Unit B—Unit B consists of light gray to buff, medium-grained wackestones and boundstones with abundant codiacean algae. Occasional fine-grained microporous lime mudstones with abundant stylolites also occur, along with scattered pyrite and peloidal grains.

This unit is encountered in all wells and has the overall shape of a wedge thinning toward the north (Figure 11E). The wedge block has a maximum thickness of 75 ft (23 m) in the western flank and a minimum of 12 ft (4 m) in the northern wells. In the southern portion of the field, the maximum thickness is 120 ft (37 m) and the minimum is 46 ft (14 m). Local patterns of thinning or thickening are not prominent to the south.

Pore types are dominantly vuggy and moldic from leached foraminifera in conjunction with interparticle porosity. Some calcite filling of the vugs and fractures occurs, as well as pyrite and baroque dolomite filling.

In northern Bu Hasa, porosity ranges from 10 to 25% and permeability averages 1 md or less. In southern Bu Hasa, the average porosity is 23% and the average permeability is 12 md. The permeability decrease northward is mainly due to the presence of numerous stylolites.

The base of Unit B is taken at the point of inflection on the neutron log at the top of the stylolite interval S4 of Unit A (Figures 17 and 18). The top is defined

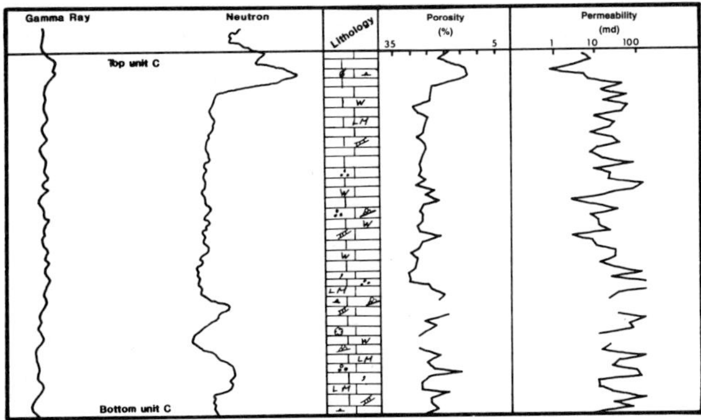

Figure 16. General lithology, log character, and porosity-permeability measurement of reservoir Unit C. Thickness of unit shown is 140 ft.

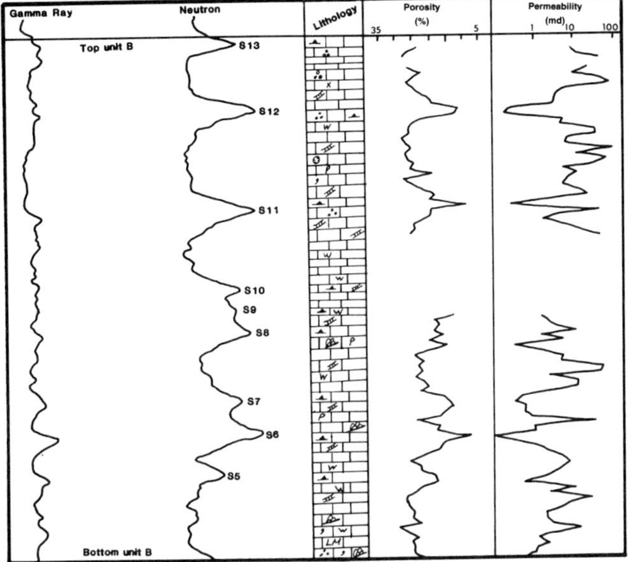

Figure 17. General lithology, log character, and porosity-permeability measurement of reservoir Unit B. Thickness of unit shown is 75 ft.

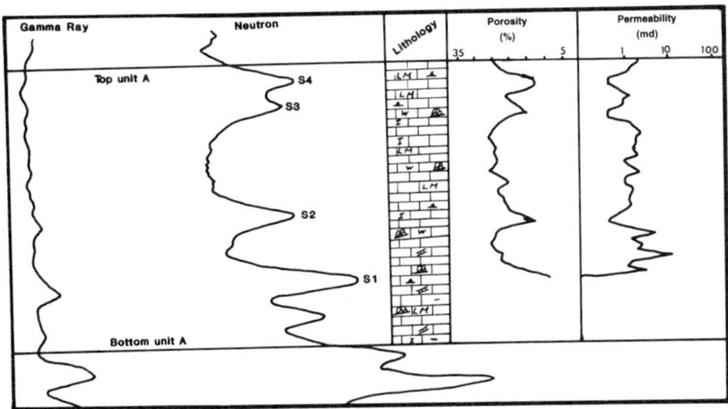

Figure 18. General lithology, log character, and porosity-permeability measurement of reservoir Unit A. Thickness of unit shown is 45 ft.

at the point where the gamma-ray and the neutron curves show a higher response, and the characteristic porosity of the overlying unit decreases. Unit B exhibits a low gamma-ray response and is mainly a porous section as seen on the neutron log. Numerous stylolites are developed in this unit and are associated with an increase of gamma-ray response.

Bacinella irregularis and *Lithocodium aggregatum* are abundant along with orbitolinids, echinoderm fragments, dasycladacean algae, and subordinate *Trocholina* sp., with a few corals, stromatoporoids, and crinoids.

The depositional setting is thought to be a low-energy open-marine shelf. The planktonic foraminifera are small and not very abundant, suggesting that the shelf was not very deep.

Reservoir Unit A—The lower part of Unit A is composed of wackestone and packstones with local grainstones and colonies of codiacean algae (*Bacinella* and *Lithocodium*), followed by light gray to buff, microporous lime mudstones and wackestones that are slightly argillaceous, burrowed, and bioturbated and that contain some peloids and lithoclasts. Stylolites and bands of argillaceous muds are observed.

This unit is encountered in all wells and ranges from 45 to 56 ft (15 to 17 m) across the field, thinnest along the far northern plunge and thickest along the southeastern flank (Figure 11F).

Pore types include vuggy, moldic, and interparticle porosity. Stylolites with amplitudes of 3 to 6 in. (8 to 15 cm) reduce porosity, as do some baroque dolomite and calcite crystals.

The average porosity is 19% and the average permeability is 4 md, with the highest values in the algal layers.

The base of Unit A lies at the boundary between the porous section of this unit and the underlying dense limestone of the Kharaib Formation (Figure 18). The top of Unit A is marked by the point of inflection on the neutron curve above a stylolite interval (S4, Figure 18). The neutron log shows Unit A as a porous sedimentary sequence subdivided by four stylolite zones. Stylolite intervals S1 and S2 are very distinct. Stylolite interval S3 below stylolite S4 always shows the lowest log porosity.

Fossils include common to abundant echinoderms, a few *Lenticulina* sp., calcispheres, *Orbitolina*, and colonies of *Bacinella irregularis* and *Lithocodium aggregatum* algae, miliolids and globigerinids, *Choffatella* sp., *Salpingoporella* dinarica, and valvulinids.

The abundance of algal colonies suggests that deposition occurred within shallow water in the photic zone. The occurrence of calcispheres, miliolids, and other arenaceous benthonic foraminifers indicates a lack of open-marine circulation and suggests a slight restriction (inner shelf or deep lagoonal setting).

Lithofacies and Depositional History

Eleven main geological lithofacies have been defined in the Shuaiba from cores, based on clear biolithological characteristics (Alsharhan, 1987). The predominant lithology and the main fossils were used to support assignment of rock intervals to the various facies. The main features of the lithofacies are described in detail below. Their spatial relationships within three main phases of deposition are shown schematically in Figure 19.

In order to fully demonstrate the nature of the Shuaiba buildup, the lithology, biogenic constituents, and phases of deposition are described in detail.

Early Phase—Shuaiba sedimentation was initiated as a microporous lime mudstone and wackestone with some colonies of *Lithocodium aggregatum* and *Bacinella irregularis* and with a few dasycladacean algae, miliolids, and arenaceous benthonic foraminifera (textulariids and valvulinids), suggesting a lack of open-marine circulation and a slight restriction (lithofacies I), probably in a deep lagoon or inner shelf setting. The fine-grained nature of the lime-mud matrix surrounding the algae indicates that the energy levels were low. Thus a shallow, restricted marine environment can be attributed to the basal Shuaiba Formation.

The overlying sediments (lithofacies II) consist of lime mudstones and wackestones with common *Orbitolina* (e.g., *O. discoidia*, Figure 20A), echinoderms, scarce *Lenticulina* sp., calcispheres, and some planktonic foraminifera, indicating a deepening sea, which resulted in good open-marine circulation and a low-energy, shallow, open-marine shelf setting.

Middle Phase—A return to the shallow water conditions similar to that of lithofacies I was marked by extensive colonization of codiacean algae to form boundstone sediments (Figure 20B, C). Growth was more rapid in the southern and central areas, and an algal platform formed on which stromatoporoids (Figure 20D) were common (lithofacies III). The energy conditions remained low, and sedimentation probably occurred in a restricted marine bay to shelf lagoon. Widespread algae and coral (Figure 20E) developed. These corals form secondary vuggy porosity partially filled by calcite. The encrusting and binding organisms—stromatoporoids, corals, and algal colonies—correspond to a bank-to-reef buildup. Energy levels were moderate to high as indicated by the presence of broken corals, stromatoporoids, and grain-supported sediments. The sediments of lithofacies I, II, and III are fairly uniform over the field, but are thicker in the south (300 ft or 92 m) than in the north (70 ft or 21 m).

The open-shelf marine planktonic facies (lithofacies IV), composed mainly of dense and argillaceous lime mudstones and wackestones interbedded with seams of shales and stylolites, presumably was deposited during an increase in water depth. This increase in water depth was probably caused by quick subsidence of a portion of the shelf near the central and northern part of the field or by a gradually rising global sea level. Simultaneously, to the south, the algal platform continued to accumulate (lithofacies III). The sediments of lithofacies IV were deposited on a very shallow ramp or rim setting adjacent to the reef. These sediments thicken abruptly toward the north because of aggradation and progradation and then gradually thin to the northeast. The thickness of lithofacies IV ranges from 80 to 200 ft (24 to 61 m) and can be correlated with a condensed section underlying the Bab Member (Figure 19, well A).

Late Phase—All studies of the Shuaiba sedimentation during the later phases of deposition are very complicated. The rudistid buildup became more rapid and was characterized by Caprinidae, Caprotinidae, Monopleuridae, and Requieniidae. These occurred in different associations with each other and varied markedly across the field.

This late phase is also characterized by shallowing in the Shuaiba sedimentation so that the rudistid colonies accumulated in the south under predominantly low-energy conditions, whereas to the north the rudistids are broken and more sparse, suggesting deposition under higher-energy conditions.

Lithofacies V is composed of porous caprotinids and caprinids in a packstone and grainstone matrix (Figure 20F and G). The matrix with the rudistids was accumulated over most of the field as shoal bank deposits. Encrusting and binding organisms are rarely seen, and locally, the rudistid shells are packed, resembling oyster banks that probably accumulated in a very shallow marine environment.

Rudistid colonization continued throughout the remainder of the Shuaiba and was characterized by caprinids with common corals, stromatoporoids, and algae (lithofacies VI). Sparse rudistid colonization occurred over an extensive area to the north and continued over the central part of the field (Figure 20H) (lithofacies VI).

Grain-supported sediments sparsely colonized by rudists (Figure 20I) (lithofacies VII) accumulated in an area of high-energy shoals that prograded toward the north. At the same time, in some of the northern areas, the rudistid debris was extensively reworked (Figure 20J) and is characterized by miliolid grainstone beach deposits (lithofacies VIII). Elsewhere, these grainstones are very coarse and are characterized by a sequence of cyclic sedimentation suggesting products of periodic storms.

During the later phases of Shuaiba deposition, evidence exists of a marked change in water depth across the field. Generally, shallower water conditions prevailed in the south, paralleling northward progradation of the facies across the field. The southern facies is characterized by lagoonal sediments (Figure 20K), sparsely colonized by rudists (lithofacies IX). In the north shelf, lagoonal deposits with characteristic miliolids and dasycladacean algae in lime-mud-supported sediments were deposited, probably during a relative sea-level fall (lithofacies X). Even farther to the north, a markedly different depositional character is suggested by the presence of *Orbitolina* shoal deposits (Figure 20L) (lithofacies XI).

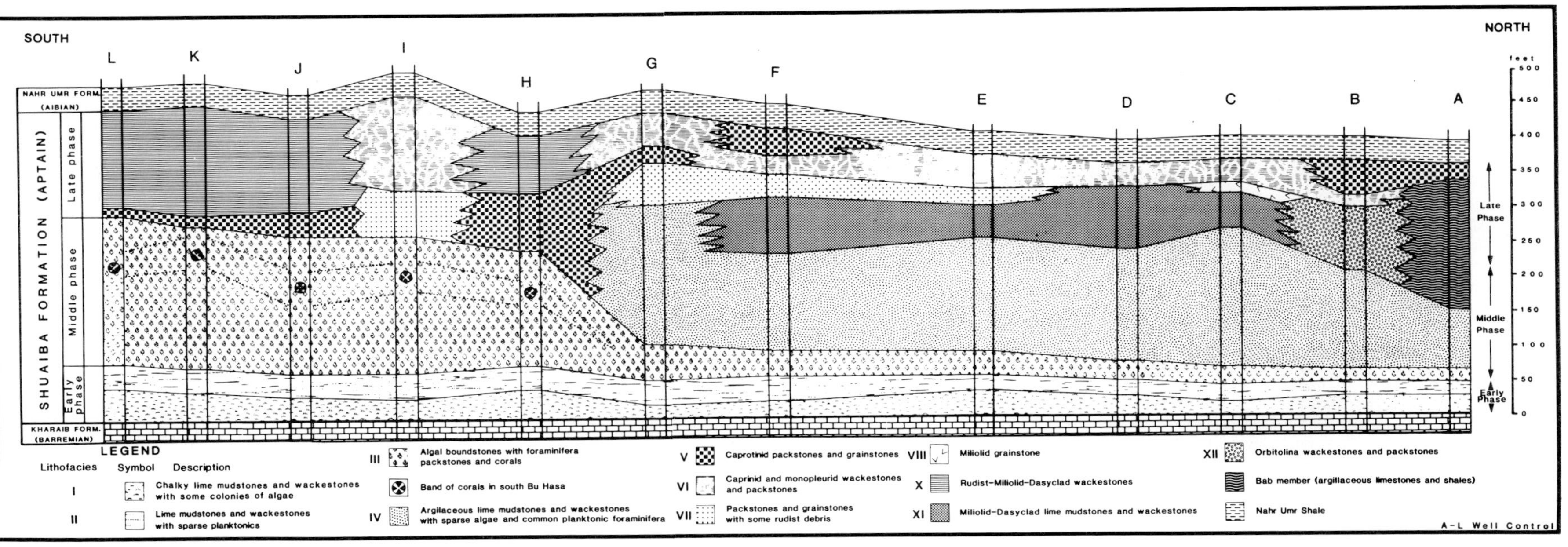

Figure 19. South-to-north schematic cross section across the Bu Hasa field, showing the lithofacies distribution in the Shuaiba Formation.

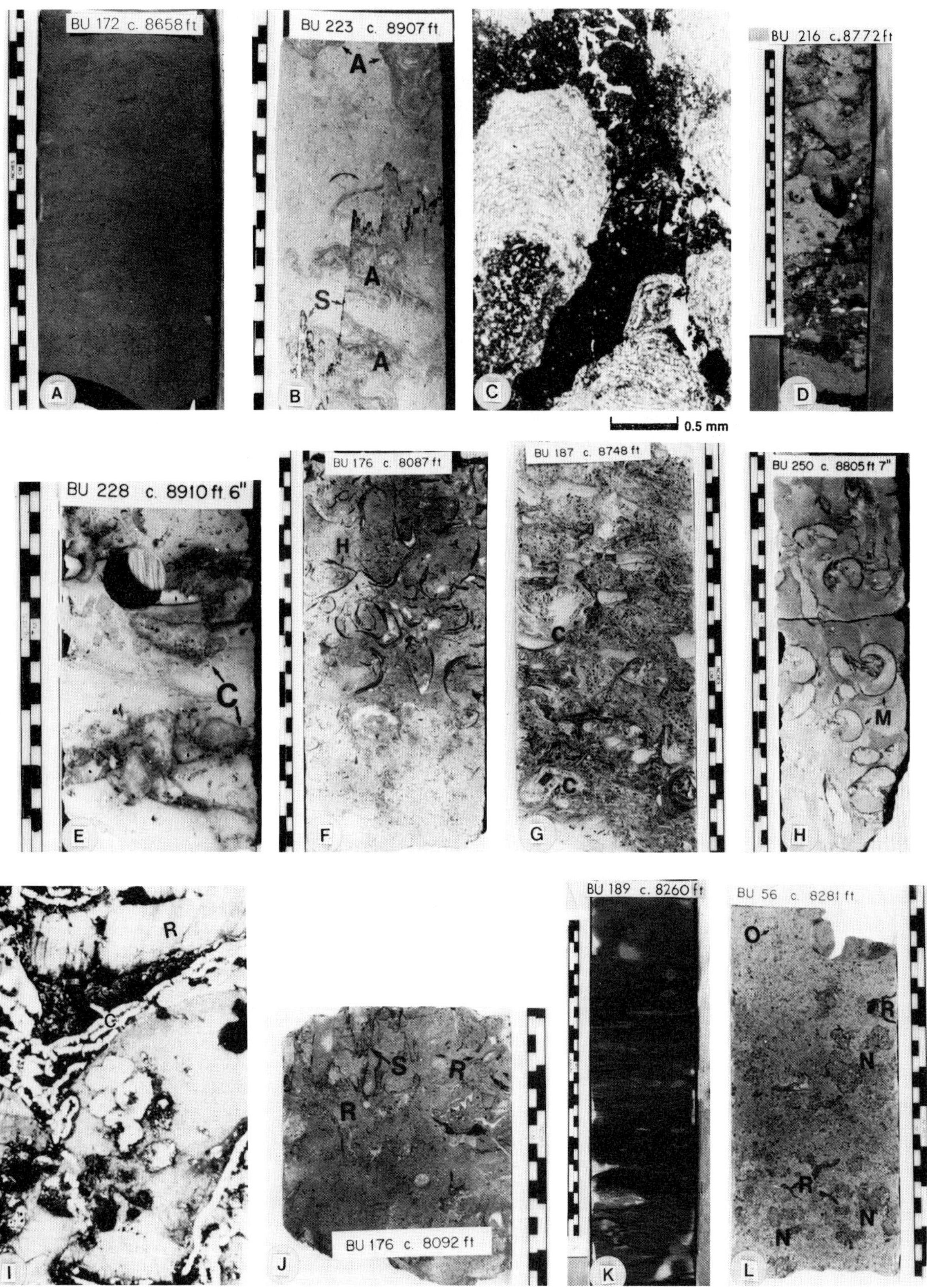
BU 172 c. 8658 ft
A
BU 223 c. 8907 ft
A
A
S
A
B
C
BU 216 c. 8772 ft
D
0.5 mm
BU 228 c. 8910 ft 6"
C
E
BU 176 c. 8087 ft
H
F
BU 187 c. 8748 ft
C
C
G
BU 250 c. 8805 ft 7"
M
H
R
C
I
0.5 mm
S
R
R
BU 176 c. 8092 ft
J
BU 189 c. 8260 ft
K
BU 56 c. 8281 ft
O
R
N
R
N
N
L

The final emergence of the top of the Shuaiba and erosion of its surface (before deposition of the terrigenous muds of the capping Nahr Umr shale) were caused by regional uplift and exposure during lowstands of eustatic sea level (Murris, 1981; Harris et al., 1984; Alsharhan, 1985a). This emergence and erosion were evidenced by the irregular and bored surface at the Shuaiba/Nahr Umr contact as found in cores (Harris et al., 1968; Twombley and Scott, 1975; Alsharhan, 1985a, b).

Rudist Distribution

Paleoecological studies of the late phase of Shuaiba sedimentation indicated that four families of rudists are present. They occur in different associations with each other and vary markedly across the Bu Hasa. These are Caprinidae (*Caprina*), Caprotinidae (*Horiopleura*), Monopleuridae (*Monopleura*), and Requieniidae (*Toucasia*). *Caprina* shells are the most abundant rudists in the Bu Hasa. Over large areas of northern Bu Hasa *Caprina* dominate, with fewer *Horiopleura*, *Monopleura*, and *Toucasia*, while in the central part of the field, *Caprina* are associated with *Horiopleura*. In the southern part of the field, caprinids associated with monopleurids formed the initial extensive colonization by rudists. *Horiopleura* and *Toucasia* occur in the south sparsely or in thin beds; these organisms dwelled on the surface of the carbonate muds in the lagoonal area and became widespread over the entire south part of Bu Hasa. The presence of requienids in a lagoon is analogous to the rudists distribution in the El Abra Limestone of Mexico described by Coogan et al. (1972).

Figure 20. Photographs of core slabs and photomicrographs of more representative lithofacies. (A) Light brown fine-grained lithology with abundant *Orbitolina* sp. (B) Wackestone and boundstone with *Lithocodium aggregatum* algae, A, and a large stylolite, S, cutting the specimen. (C) Photomicrograph showing algal boundstone with coral development. (D) Light brown packstone and boundstone containing abundant Stromatoporoidea. Stylolite visible in lower half. (E) Common corals, C, are visible. Blocky calcite cement fills pores almost entirely. (F) Wackestone and packstone containing unfragmented common caprotinid rudists, some of which are filling with geopetal lime mud. (G) Packstone with common *Caprina* rudists fragments, C. (H) *Monopleura* (M) rudist wackestone. (I) Photomicrograph showing rudistid grainstone, R, with good interparticle porosity (black) and thin linings of drusy calcite, C, around rudist shells. (J) Packstone with abundant rudistid fragments, R, and stylolite, S, in left upper corner. (K) Dolomitic peloidal packstone and wackestone that have been extensively burrowed. Argillaceous partings occur throughout the specimen. Top of specimen consists of dense lime mudstone. (L) Wackestone with abundant algae, N; *Orbitolina*, O; and few rudist shells, R.

The initial phase of rudist colonization in the southern part of the field was dominated by caprinids, while in the central part of the field, the assemblage comprised mainly caprotinids with subordinate caprinids. The lagoonal environments of south Bu Hasa were characterized by a caprotinid-requienid assemblage. In the northwestern areas, rudist debris was extensively reworked in a shoal environment, and, toward the north, shoal facies (mainly orbitoids) gave way to extensive caprinids and caprotinids rudist colonization, continuing throughout the remainder of the Shuaiba. By the end of the Shuaiba sedimentation, rudist colonization was very widespread throughout the whole of the north and central areas. A more sparse rudist colonization had by this time covered extensive areas of the north Bu Hasa and this colonization continued over the central area.

Hydrocarbon Habitat, Maturation, and Migration

In the western areas of Abu Dhabi, where Bu Hasa field is located, the structural traps started developing in the Late Jurassic, while in eastern areas the initiation of the structural development of the traps seems to be younger than Jurassic. The most prolific reservoirs in Abu Dhabi occur in the Upper Jurassic and Lower Cretaceous formations, which were hydrocarbon sourced from the same rock unit in the Upper Jurassic (Diyab/Dukhan Formation).

At Bu Hasa, the Diyab (offshore terminology) and the Dukhan (onshore terminology), usually called the Diyab/Dukhan Formation, is Oxfordian/mid-Kimmeridgian in age. It is about 1600 ft (500 m) thick in western areas, thinning gradually toward the east and northeast. It consists of finely laminated, dark gray, argillaceous lime mudstones/wackestones with calcareous shales, the lower part of which (100 to 150 ft [31 to 46 m] thick) is more carbonaceous. The laminated argillaceous limestones are rich in organic matter.

In the eastern areas of Abu Dhabi the facies gradually changes, and the formation is dominated by dolomite, dolomitic limestones, and some peloidal packstone/grainstones. This change reflects the depositional setting of the formation. In western Abu Dhabi, the intrashelf basin was developed in the Late Jurassic where local euxinic conditions produced starved-basin sequences of laminated bituminous lime muds, shales, and marls (Diyab/Dukhan), locally forming a good source rock. Similar intrashelf basins were found in eastern Saudi Arabia and the northern gulf (Alsharhan and Kendall, 1986). In eastern Abu Dhabi shallow water of subtidal-lagoonal deposits occurred.

The present-day maturity of the Diyab/Dukhan Formation is shown in Figure 21A. Most of the onshore areas lie below the oil window ($R_o = 1.3\%$). The Falaha syncline, east of Jarn Yaphour, and the Ghurab syncline, west of the Bu Hasa, have the highest maturity levels, with $R_o = 1.7\%$.

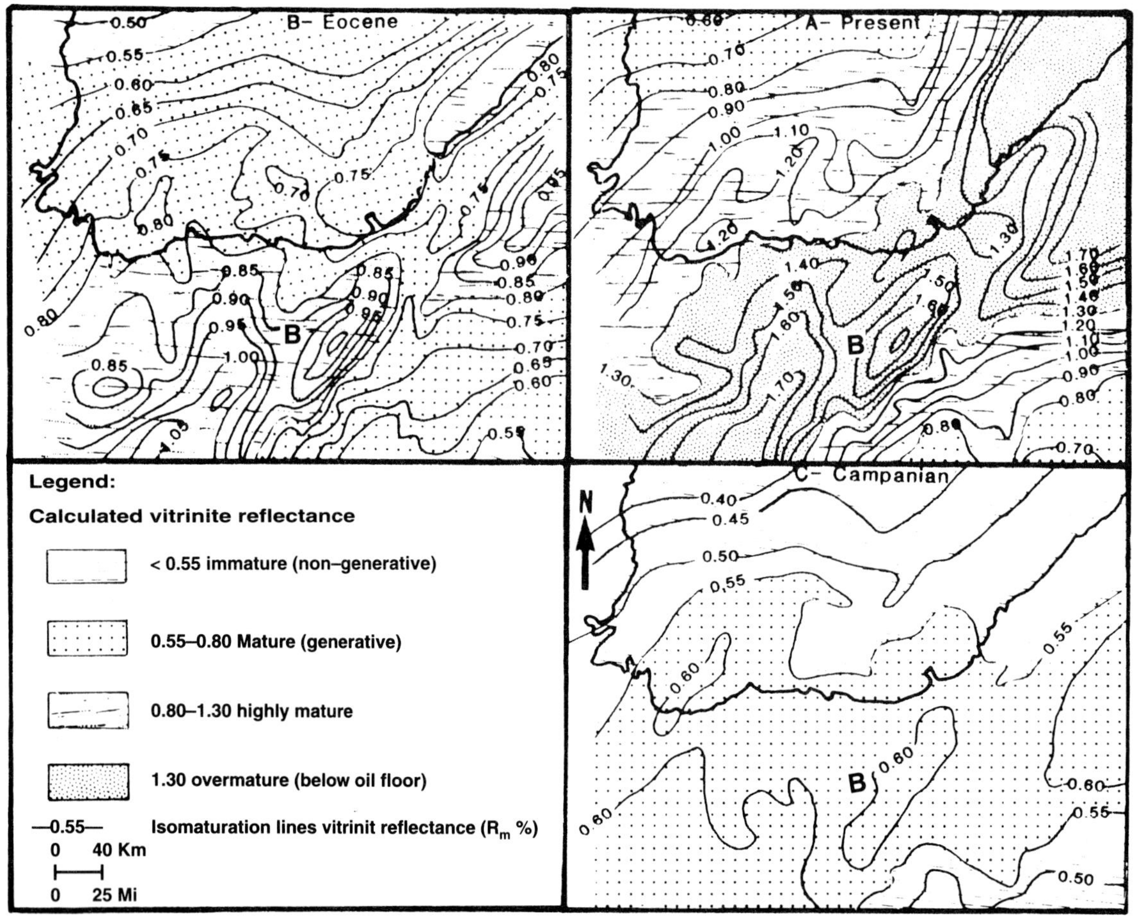

Figure 21. Paleo-maturity of the Diyab/Dukhan Formation, Upper Jurassic source rock, Abu Dhabi. (Compiled by Alsharhan, 1989, from ADNOC, 1984; Lutfi and El-bishlawy, 1986.)

Figure 21B shows the paleo-maturity map of the Diyab/Dukhan Formation for Eocene time (49 Ma). The maturity pattern is very similar to the present one and parallels present structural trends. In most of the offshore area, hydrocarbon generation did not commence until the early Eocene, and by this time, the Diyab/Dukhan Formation in the onshore area had reached the phase of maximum oil generation (ADNOC, 1984; Alsharhan, 1989).

Figure 21C shows the paleo-maturity map of the Diyab/Dukhan Formation in Campanian times (73 Ma). A general increase in maturity to the south, a low maturity in the southeast, and the first signs of the Falaha syncline moving suggest that significant salt movement was starting and also that the earliest hydrocarbon generation in the onshore areas probably was initiated at this time (ADNOC, 1984; Alsharhan, 1989).

Total organic carbon (TOC) determined from the Diyab/Dukhan Formation samples ranges between 0.3 and 5.5% by weight in the offshore areas, while in the onshore areas it ranges between 0.5 and 0.82%; in the southeast (at Mender and Qusahwira, Figure 1), it ranges from 0.5 to 1.2% by weight.

Pyrolysis data show that the residual source rock potential (P_2 yield) ranges between 0.5 and 5.0 kg/MT. The kerogen content is 60 to 80% and is of a sapropelic oil-prone type in some wells, while in other wells, there is a mixed sapropelic and humic (oil and gas prone) kerogen grading to highly humic type (Lutfi and El-bishlawy, 1986; Alsharhan, 1989). Using the pyrolysis results, a source potential map for the Diyab/Dukhan Formation has been constructed (Figure 22). The area of highest residual source-rock potential in Abu Dhabi is in the western offshore area and part of the onshore area. As revealed from analysis, the rich source interval only occurs at the base of the formation and is characterized by a high gamma-ray kick (Figure 23), whereas the rest of the formation, which has uniformly low source potential, gives a low gamma-ray response.

Gas chromatography-mass spectrophotometry (GC-MS) of the sterane and triterpane fingerprints of the Diyab/Dukhan extracts closely resemble those of the Thamama and Arab formations (Figure 24A, B). Also, carbon isotope values for the Diyab/Dukhan Formation source-rock extracts (-24.7‰ to -27.0‰) closely match the Arab and Thamama oils (-25.2‰ to -26.9‰) (Lutfi and El-bishlawy, 1986).

These results indicate that oil was generated from the Diyab/Dukhan Formation in western Abu Dhabi, where it has moderate-to-good source potential, and

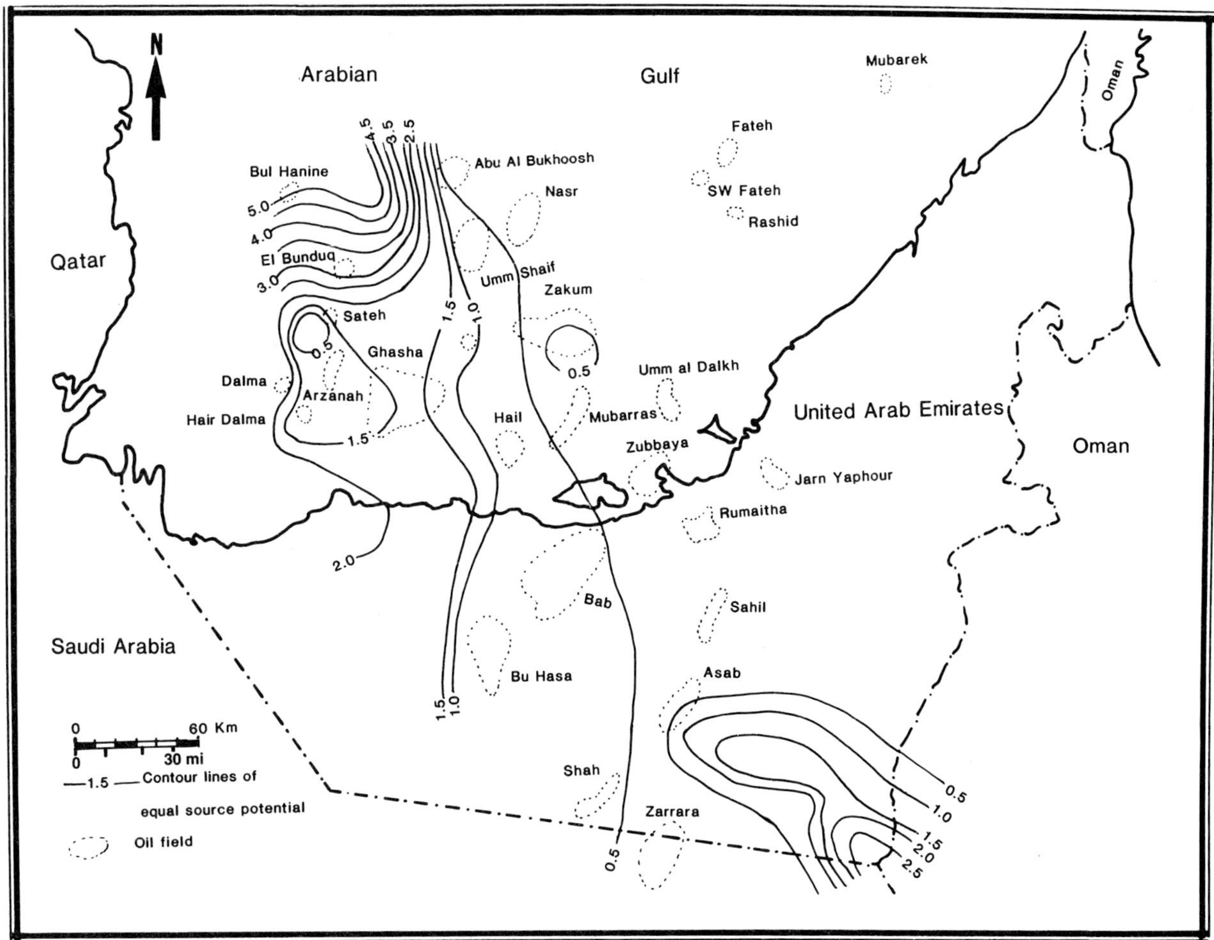

Figure 22. Present source-rock potential map of the Diyab/Dukhan Formation, Abu Dhabi. (Modified from ADNOC, 1984.)

then migrated eastward. All the Arab reservoirs to the west of the Hith edge (Figures 25 and 26) have been progressively filled if they were on this migration path; oil migrating eastward to the Hith edge would then have migrated vertically into the Thamama reservoirs.

Figure 25 shows the hydrocarbon source potential of the Diyab/Dukhan Formation in Abu Dhabi. The maps indicate three other possible source areas to be considered in the migration modeling. These are the Falaha syncline, Sir Abu Nuair syncline, and northeastern part of Abu Dhabi. Assuming that the Diyab/Dukhan Formation in the Falaha syncline had reached maturity in Campanian time, then the major part of the generated oil would have migrated vertically into the Thamama reservoirs to charge most of the existing structures at Asab, Sahil, Bab, and Bu Hasa (Lutfi and El-bishlawy, 1986; Alsharhan, 1989).

In Eocene time (49 Ma), the western Diyab/Dukhan source area was still charging the Jurassic reservoirs in some of the western offshore fields (Figure 25). Also the three possible source areas still would have been charging the Thamama reservoirs to the east (Lutfi and El-bishlawy, 1986; Alsharhan, 1989).

The source-rock potential of the pre-Shuaiba part of the Thamama Group (Lower Cretaceous) is generally poor. Intervals with moderately richer material occur in the Kharaib and Lekhwair formations (TOC < 0.5 to 0.46%) and the Habshan Formation (TOC 0.81 to 0.61%). These dense intervals have been postulated (Clarke, 1975) to be the source for the lower Thamama reservoired oils, but the results of gas chromatography indicate that this hypothesis is not tenable. The pre-Shuaiba Thamama lacks source potential and is not capable of generating large quantities of hydrocarbon. However, the richer intervals noted above could have made a minor contribution to some of the Thamama reservoired oil (Alsharhan, 1985a).

The sediments deposited in the Shuaiba intrashelf basin (Bab Member) have a fair source rock character (TOC 0.5 to 0.8%). Some samples from south and eastern onshore Abu Dhabi showed moderate potential, while at Umm Shaif-88, the ADNOC staff (1984) reported the presence of some rich source rock

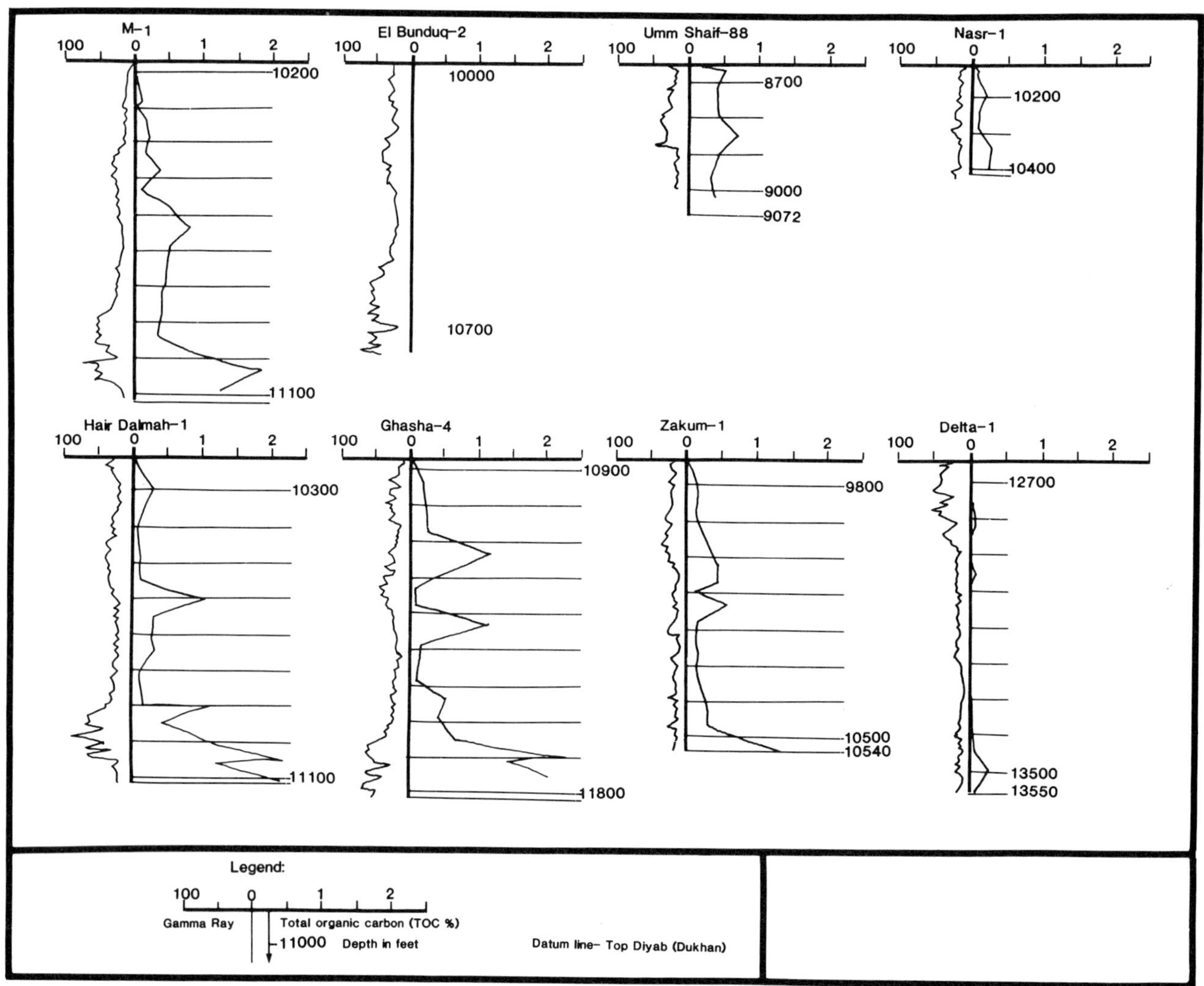

Figure 23. Correlation diagram of gamma-ray and total organic carbon (TOC) of the Diyab/Dukhan Formation, Abu Dhabi. (Modified from ADNOC, 1984.)

in Bab Member where pyrolysis yields reached 10 kg/MT. This member is composed of microporous argillaceous limestones and shales. It is widely believed that these sediments contain hydrocarbon source rocks which have contributed to the charging of the main reservoir at the Bu Hasa field (Murris, 1981; Forst et al., 1983; Alsharhan, 1985a).

The distribution of Thamama oils across the area appears to be related to the occurrence of the Hith Anhydrite; Thamama oils are mainly reservoired to the east of the Hith "Edge," which runs approximately north-south through central Abu Dhabi (east of the Zakum to west of the Shah fields). The Hith Anhydrite forms the regional seal for the Upper Jurassic Arab oils' offshore areas. Because of the absence of this seal in the eastern part of Abu Dhabi, oil generated from the Dukhan is able to migrate vertically, directly into the Thamama reservoirs (Figure 26). Thamama reservoirs only contain oil where the Hith seals of the Arab Formation were not initially anhydrite (probably anhydrite and dolomite), where facies changes occur (as in eastern Abu Dhabi, passing to porous limestones), or where it is breached by faulting (Murris, 1981; Alsharhan, 1985a).

The vertical migration from the Diyab/Dukhan source to the Thamama reservoirs through the dense limestone intervals is possible, as pointed out by Schowalter (1979), who indicated that when the buoyant pressure of a hydrocarbon column in a trap exceeds the resisting force of the capillary pressure of the cap rock, then hydrocarbons can migrate vertically through the seal.

EXPLORATION PLAY CONCEPTS

Significant stratigraphic traps could be developed in the Abu Dhabi intrashelf basin in an environment

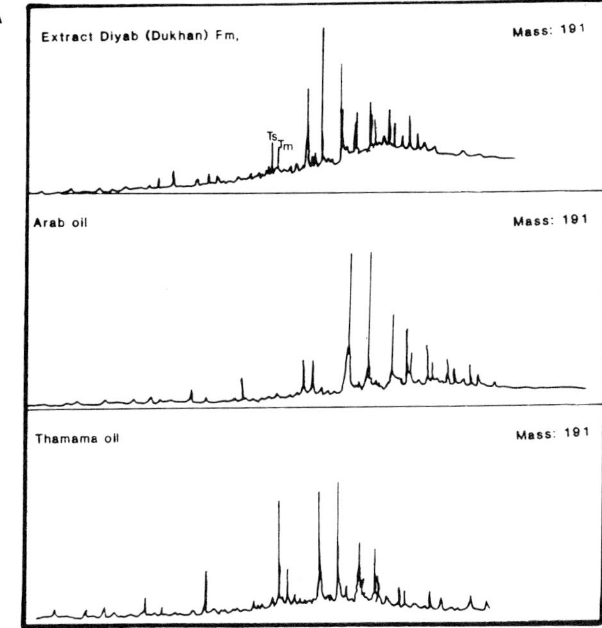

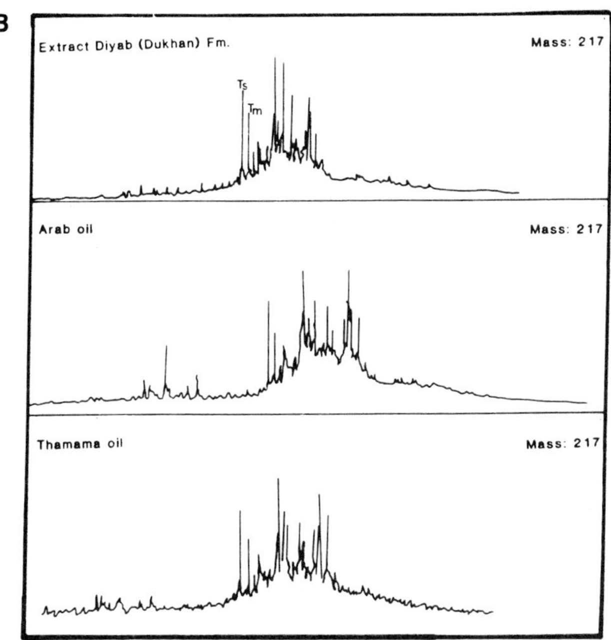

Figure 24. (A) Gas chromatography-mass spectrophotometry (GC-MS) triterpane fingerprints for extracts from Diyab/Dukhan Formation and from Arab and Thamama oils of Abu Dhabi oil fields. (Modified by Alsharhan, 1989, from Lutfi and El-bishlawy, 1986.) (B) Gas chromatography-mass spectrophotometry (GC-MS) sterane fingerprints for extracts from Diyab/Dukhan Formation and from Arab and Thamama oils of Abu Dhabi oil fields. (Modified from Lutfi and El-bishlawy, 1986.)

of isolated open-marine mounds and shoal deposits. Such basinal mounds and shoal facies encountered in some areas in Abu Dhabi could have very good primary porosity and would have excellent reservoir potentials. This concept resembles some areas in North America, such as the Rainbow-Zama trend in Alberta, Canada, and the Niagaran reef complex in the Michigan basin in the United States.

On a worldwide basis, rudist buildups occur as discrete mounds (Frost et al., 1983), rather than in the continuous or semi-continuous fringing of barrier reefs, thus the buildup-prone belt interpreted to be present in this area, as in Bu Hasa, Ruwais, Huwaila, Zarrarah, Shaybah, and Shah fields, consists of a band of scattered mounds with a wide range of sizes.

To provide a better understanding of the facies relationships and exploration potential for the future, we should integrate the seismic stratigraphic framework with core descriptions and wireline log data and then construct regional scale cross sections across Abu Dhabi to illustrate age relationships, correlations of lithologic units, and facies relationship within the Shuaiba Formation.

Prospective conditions occur in the Shuaiba Formation in different areas of Abu Dhabi. The Shuaiba buildup is oil-producing at the Bu Hasa field, while the Bab Member porous intervals were found to be oil- and gas-bearing in the southern part of Asab, as well as in some wells in Ruwais, Jarn Yaphour, and Mender. The basinward pinch-out of the Bab Member porosity development on the southern flank of the Asab field creates a stratigraphic-structural trap (one well tested 1000 BOPD). At Zibara, low volumes of oil and gas were tested from the Bab Member sequence; therefore, a potential hydrocarbon stratigraphic trap could exist within the Bab Member basinward porosity pinch-out. Minor amounts of hydrocarbons were tested in the Bab Member sequence in wells at Mash Hur and Qusahwira. The Shuaiba Formation is hydrocarbon-bearing at Zarrarah-Shaybah and Mandous fields.

The Shuaiba buildup can be traced as discrete mounds in a narrow belt from northwestern to southeastern Abu Dhabi (Figure 27). The Shah, Bu Hasa, and Huwaila fields are in this belt, as well as some producing wells in western Abu Dhabi that are thought to penetrate prolific rudist mounds. This narrow belt can be extended farther to encompass Zarrarah-Shaybah (Abu Dhabi and Saudi Arabia), Yibal, and Huwaisah fields (Oman), and then continuing as a loop into the southern gulf area to include fields such as Reshadat (formerly Rakash) of Iran and Idd El Shargi of Qatar (Alsharhan, 1985a). Most of these fields produce from the Shuaiba Formation.

Frost et al. (1983) gave some analogous examples from Lower Cretaceous carbonate buildups in the United States and Mexico, in which rudists coexisted with other reef-building organisms, similar to those in the Shuaiba of Arabia. Some of these are the Sligo buildups in the subsurface of south Texas; in the prolific Black Lake field of central Louisiana; in the Cupido Formation exposed in southern Mexico; in outcrops of the early Albian Mural limestone in south Arizona and Mexico; and in middle and late Albian outcrops of the Glen Rose Formation in central Texas.

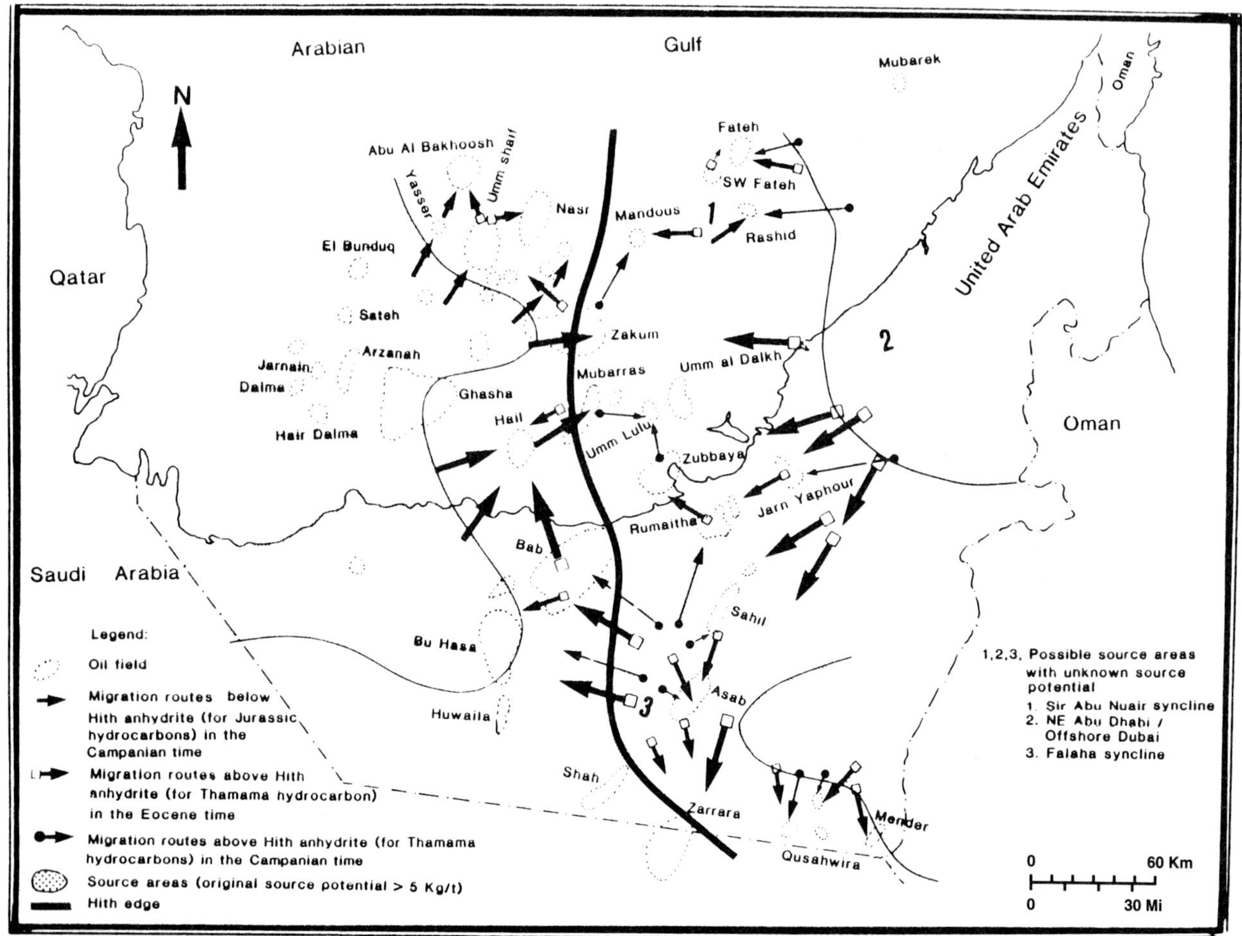

Figure 25. Possible migration pathways from Diyab/ Dukhan Formation source and other possible source areas in Campanian and Eocene times. (Compiled and modified from Alsharhan, 1989, Lutfi and El-bishlawy, 1986.)

ACKNOWLEDGMENTS

I thank the geologists (past and present) of ADCO for their assistance and fruitful discussions of different aspects of the Bu Hasa reservoir.

I express my deep gratitude for the assistance of Christopher G. St. C. Kendall of the University of South Carolina who read the manuscript and offered valuable advice and recommendations. Figures 1, 2, and 21 were reproduced with permission from the Journal of Petroleum Geology, U.K.

REFERENCES CITED

Abu Dhabi Company for Onshore Oil Operations (ADCO), 1986, Tectonics of the oil fields in Abu Dhabi onshore areas: Paper presented at the symposium on the hydrocarbon potential of intense thrust zones, held in Abu Dhabi Dec. 14-18, 1986, 12 p.

Abu Dhabi National Oil Company (ADNOC), 1984, Geochemical studies, source rock determination and migration of oil in the U.A.E.: Paper presented at the conference on the origin and migration of Arab oils, held in Kuwait and sponsored by OAPEC, 36 p.

Alsharhan, A. S., 1985a, Depositional environments, reservoir units evolution and hydrocarbon habitat of Shuaiba Formation (Lower Cretaceous) of Abu Dhabi, United Arab Emirates: AAPG Bulletin, v. 69, p. 899-912.

Alsharhan, A. S., 1985b, Petrography, sedimentology, diagenesis and reservoir characteristics of the Shuaiba and Kharaib formations (Barremian-mid-Aptian) carbonate sediments of Abu Dhabi, United Arab Emirates: Unpublished dissertation, Columbia, University of South Carolina, 175 p.

Alsharhan, A. S. 1987, Geology and reservoir characteristics of carbonate buildup in giant Bu Hasa oil field, Abu Dhabi, United Arab Emirates: AAPG Bulletin, v. 71, p. 1304-1318.

Alsharhan, A. S., 1989, Petroleum geology of the United Arab Emirates: Journal of Petroleum Geology, v. 12, p. 253-288.

Alsharhan, A. S., and C. G. St. C. Kendall, 1986, Precambrian to Jurassic rocks of Arabian Gulf and adjacent areas: their facies, depositional setting and hydrocarbon habitat: AAPG Bulletin, v. 70, p. 977-1002.

Alsharhan, A. S., and A. E. M. Nairn, 1986, A review of the Cretaceous formations in the Arabian Peninsula and the Gulf: Part 1. Lower Cretaceous (Thamama Group) stratigraphy and paleogeography: Journal of Petroleum Geology, v. 9, p. 365-392.

Clarke, R. H., 1975, Petroleum formation and accumulation in Abu Dhabi: 9th Arab Petroleum Congress, Dubai, 20 p.

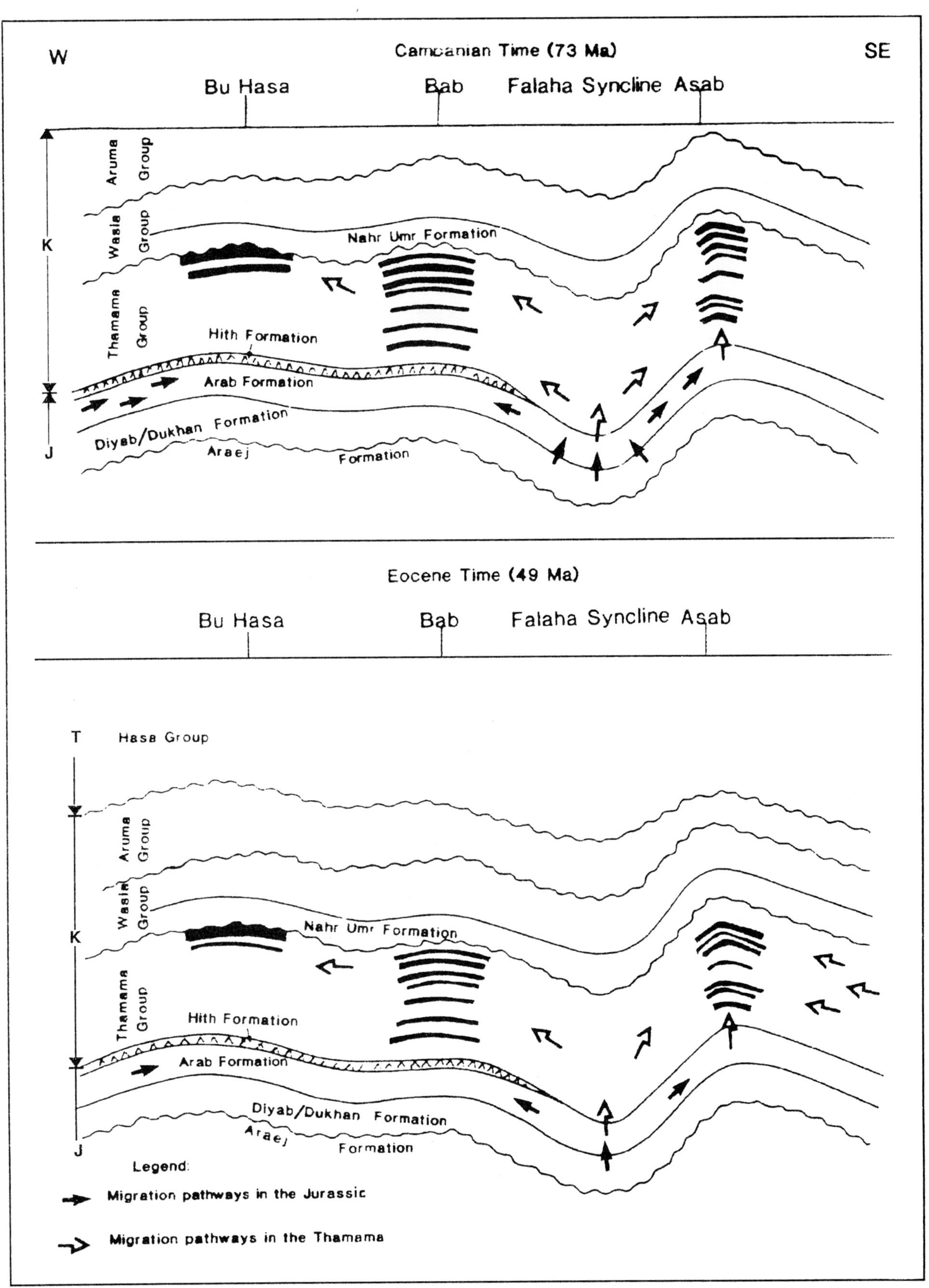

Figure 26. Possible migration pathways along the Hith "Edge" from Diyab/Dukhan Formation source in Campanian and Eocene times. (Modified from Lutfi and El-bishlawy, 1986.)

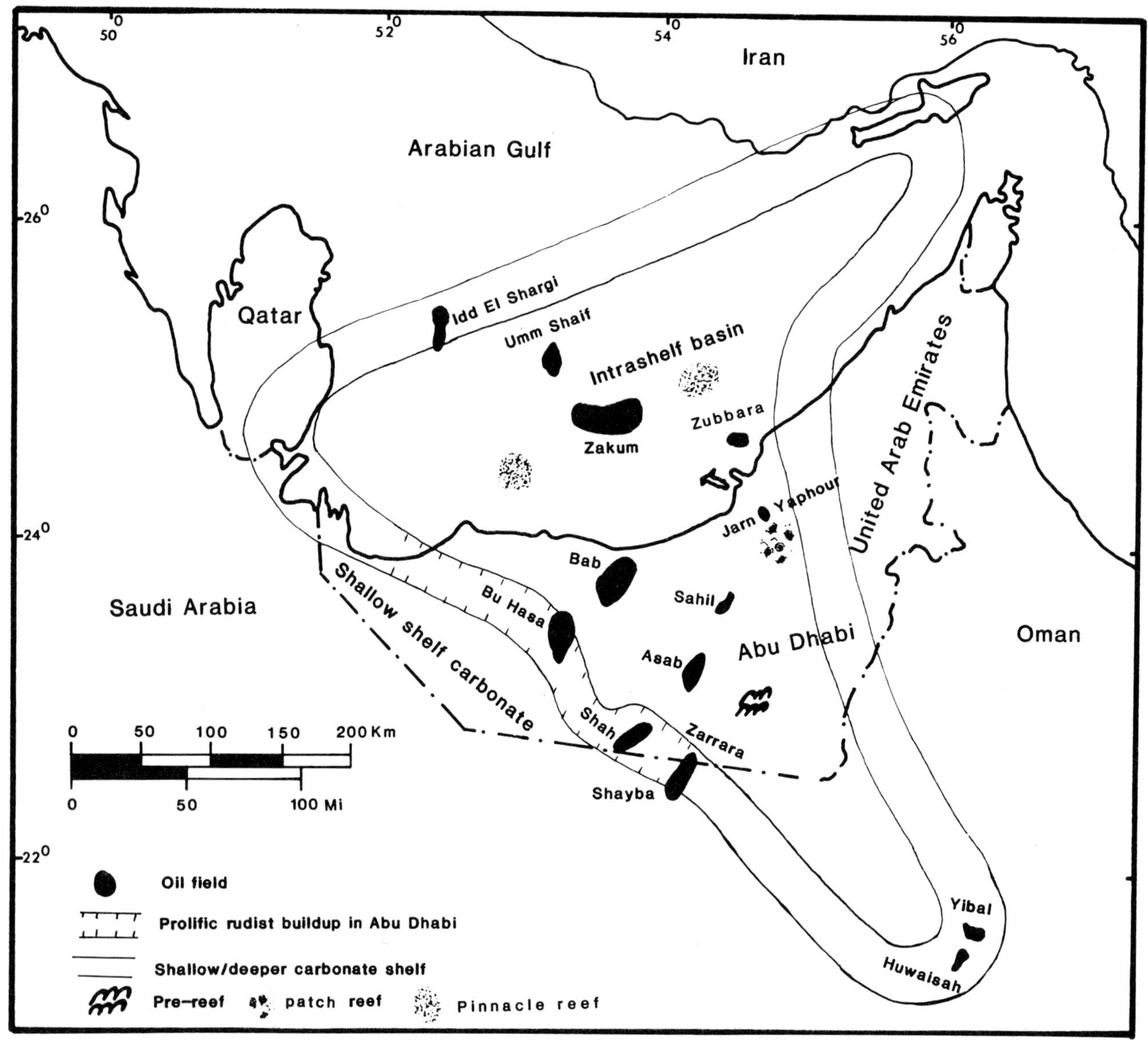

Figure 27. Approximate trend of rudist buildup in southern Arabian Gulf.

Coogan, A. H., D. G. Bebout, and C. Maggio, 1972, Depositional environments and geologic history of Golden Lane and Poza Rica trend, Mexico, an alternative view: AAPG Bulletin, v. 56, p. 1419-1447.

Frost, S. H., D. M. Bliefnick, and P. M. Harris, 1983, Deposition and porosity evolution of a Lower Cretaceous rudist buildup, Shuaiba Formation of eastern Arabian Peninsula: Carbonate Buildups, SEPM Core Workshop 4, Dallas, p. 381-410.

Hajash, G. M., 1967, The Abu Dhabi Sheikhdom—the onshore oilfields history of exploration and development: 7th World Petroleum Congress, v. 2, p. 129-139.

Harris, P. M., S. H. Frost, G. A. Seiglie, and N. Schneidermann, 1984, Regional unconformities and depositional cycles, Cretaceous of the Arabian Peninsula, *in* J. S. Schlee, ed., Interregional Unconformities and Hydrocarbon Accumulation, AAPG Memoir 36, p. 67-80.

Harris, T. J., J. T. Hay, and B. N. Twombley, 1968, Contrasting limestone reservoirs in the Murban field, Abu Dhabi, *in* Proceedings 2nd AIME Reg. Techn. Symposium, Dhahran, Saudi Arabia, p. 149-187.

Hulstrand, R. F., M. K. A. Abou Choucha, and S. M. Albaker, 1985, Geological model of the Bu Hasa Shuaiba reservoir, Abu Dhabi: 4th Middle East Oil Technical Conference, Bahrain, March, Society of Petroleum Engineers paper 13699, 8 p.

Lutfi, G., and S. El-bishlawy, 1986, Habitat of hydrocarbon in Abu Dhabi, U.A.E.: Paper presented in Abu Dhabi at the symposium on the hydrocarbon potential of intense thrust zones, Dec. 14-18, 60 p.

Murris, R. J., 1981, Middle East: stratigraphic evolution and oil habitat: Geologie en Mijnbouw, v. 601, p. 467-486.

Powers, R. W., L. F. Ramirez, C. D. Redmond, and E. L. Elberg, 1966, Geology of the Arabian Peninsula: Sedimentary geology of Saudi Arabia, USGS professional paper 560D, 147 p.

Schlumberger, 1981, Well Evaluation Conference, United Arab Emirates and Qatar, 271 p.

Schowalter, T. T., 1979, Mechanics of secondary hydrocarbon migration and entrapment: AAPG Bulletin, v. 63, p. 723-760.

Twombley, B. N., and J. Scott, 1975, Application of geological studies in the development of the Bu Hasa field, Abu Dhabi: 9th Arab Petroleum Congress, Dubai, 27 p.

Appendix 1. Field Description

Field name .... *Bu Hasa field*

Ultimate recoverable reserves .... *7×10^9 bbl*

Field location:

Country .... *Abu Dhabi*
State .... *United Arab Emirates*
Basin/Province .... *Rub al Khali basin*

Field discovery:

Year first pay discovered .... *Aptian Shuaiba Formation June 1962*
Year second pay discovered .... *None*

Discovery well name and general location .... *Murban-12 53°10′ E, 23°30′ N*

Discovery well operator .... *Abu Dhabi Petroleum Co.*

IP

First pay .... *120,000 STB/day*

All other zones with shows of oil and gas in the field:

Age	Formation	Type of Show
Barremian	*Kharaib*	*Oil*

Geologic concept leading to discovery and method or methods used to delineate prospect

Gravimeter and seismic surveys.

Structure:

Province/basin type .... *Bally 221; Klemme II Ca*

Tectonic history

Geologic growth of the structure with which this field is associated is probably controlled by deep-seated faults and the movement of the Infracambrian (Hormuz) salt in depth. This salt is thought to have formed pillow structures under large, low-dip anticlines and domes.

Regional structure

The United Arab Emirates is part of the eastern margin of the Arabian peninsula situated between two major structural highs (the Qatar arch and Mender-Lekhwair high). The structural pattern is relatively simple, with gentle folding and warping. Periodic epeirogenic movements of faulted basement blocks or salt domes appear to have been the main tectonic factors controlling the structural developments of the area.

Local structure

Bu Hasa structure is an oval-shaped anticline, with a 35 km north-south major axis and a 20 km east-west minor axis. The field has a broad nose at the northern end and a narrower plunging ridge southward. The northern flank dip is of the order of 1°. The western flank has a structural dip of 3°, while the eastern flank dip is of the order of 2°.

Trap:

Trap type(s) .... *Combination structural and stratigraphic trap*

Basin stratigraphy (major stratigraphic intervals from surface to deepest penetration in field):

Chronostratigraphy	Formation	Depth to Top in ft (m)
Recent–Miocene	*Recent–Miocene*	*0–1250 (380)*
Lower Tertiary	*Hasa Group*	*1250–4500 (380–1370)*
Upper Cretaceous	*Aruma Group*	*4500–6400 (1370–1950)*
"Middle" Cretaceous	*Wasia Group*	*6400–7990 (1950–2440)*

Chronostratigraphy	Formation	Depth to Top in ft (m)
Lower Cretaceous	*Thamama Group*	*7990–10,900 (2440–3320)*
Upper Jurassic	*Sila Group*	*10,900–13,100 (3320–4000)*
Middle Jurassic	*Areaj/Izhara Fms.*	*13,100–14,250 (4000–4350)*
Lower Jurassic	*Hamlah/Marrat Fms.*	*14,250–14,950 (4350–4560)*
Triassic	*Minjur/Jilh/Sudair Fms.*	*14,950–17,750 (4560–5410)*
Upper Permian	*Khuff Fm.*	*17,750–20,360 (5410–6210)*

Reservoir characteristics:

Number of reservoirs *1*
Formations .... *Shuaiba Formation*
Ages .... *Early Mid-Aptian*
Depths to tops of reservoirs .... *8000 ft (2440 m)*
Gross thickness (top to bottom of producing interval) .... *450 ft (137 m)*
Net thickness—total thickness of producing zones
Average .... *290 ft (88 m)*
Maximum .... *400 ft (120 m)*
Lithology .... *Reefal limestone*
Porosity type

Due to subareal exposure, good secondary porosities were developed as large pores and vugs, with some inter- and intraparticle porosities

Average porosity *5–27%*
Average permeability .... *0.1–120 md*

Seals:

Upper
Formation, fault, or other feature .... *Nahr Umr Formation*
Lithology .... *Shale*
Lateral
Formation, fault, or other feature .... *Permeability barrier*
Lithology .... *Limestone*

Source:

Formation and age .... *Dukhan Formation, Upper Jurassic*
Lithology .... *Finely laminated argillaceous limestones and shales*
Average total organic carbon (TOC) *0.5–1.2%*
Maximum TOC *5.5%*
Kerogen type (I, II, or III) .... *Sapropelic*
Vitrinite reflectance (maturation) .... $R_0 = 1.3$
Time of hydrocarbon expulsion .... *Eocene*
Present depth to top of source .... *12,000 ft (3660 m)*
Thickness .... *1500 ft (460 m)*
Potential yield .... $P_2 = 5.2\text{-}0.2$ *kg/ton*

BU HASA

Appendix 2. Production Data

Field name .... *Bu Hasa field*

Field size:

Proved acres .... *155,673 ac (63,000 ha)*
Number of wells all years *320*
Current number of wells *320*

Well spacing .... 1 km (3280 ft)
Ultimate recoverable .... 7×10^9 bbl
Cumulative production .... NA
Annual production .... 300,000 bbl/day
Decline rates .... NA
Annual water production .... NA
In place, total reserves .... 20×10^9 bbl
In place, per acre foot .... NA
Primary recovery .... 4×10^9 bbl
Secondary and enhanced recovery .... 3×10^9 bbl
Cumulative water production .... NA

Drilling and casing practices:

Amount of surface casing set .... 300–400 ft (94–125 m)
Casing program

27-in. at 300 ft; 13⅜-in. at 4000 ft; 9⅝-in. at top Shuaiba; 7-in. in the Shuaiba

Drilling mud .... Bentonite
Bit program .... 26-in., 12¼-in., 8½-in.
High pressure zones .... NA

Completion practices:

Interval(s) perforated .... Shuaiba porous unit H
Well treatment .... HCl

Formation evaluation:

Logging suites .... FDC/CNL, GR, ISF
Testing practices .... Communication test dual producer and injectivity test
Mud logging techniques .... Used only in exploratory wells

Oil characteristics:

Type .... Sweet
API gravity .... 39° API
Base .... NA
Initial GOR .... 790 SCF/STB
Sulfur, wt% 0.95%
Viscosity, SUS .... 0.28 cp
Pour point .... 10–20° F (−12 to −28.9° C)
Gas-oil distillate .... NA

Field characteristics:

Average elevation .... 100–200 ft (30-61 m)
Initial pressure .... 3956 psi (27.28 MPa)
Present pressure .... NA
Pressure gradient .... NA
Temperature .... 250° F (121° C)
Geothermal gradient .... 1.5° F/100 ft (2.7° C/100 m)
Drive .... Water
Oil column thickness .... 280 ft (88 m)
Oil-water contact .... Varies
Connate water 5–10%
Water salinity, TDS .... 181,886 mg/L
Resistivity of water .... 0.0452 ohm-m at 76° F (24° C)
Bulk volume water (%) .... NA

Transportation method and market for oil and gas:

Pipeline to Jebel Dhanna terminal on the Arabian Gulf; market—western Europe, U.S.A., and Japan

Zohar-Kidod-Haqanaim Fields—Israel
Eastern Mediterranean Basin

Y. GILBOA
Naphtha - Israel Petroleum Corporation Ltd.
Tel Aviv, Israel

H. FLIGELMAN
Oil Exploration (Investments) Ltd.
Tel Aviv, Israel

B. DERIN
Israel National Oil Company
Tel Aviv, Israel

FIELD CLASSIFICATION

BASIN: Eastern Mediterranean
BASIN TYPE: Passive Margin
RESERVOIR ROCK TYPE: Carbonates and Sandstones
RESERVOIR AGE: Jurassic
PETROLEUM TYPE: Gas
TRAP TYPE: Anticline with Multiple Pays
RESERVOIR ENVIRONMENT OF DEPOSITION: Platform
TRAP DESCRIPTION: Series of three anticlines in an asymmetrically folded trend related to reverse faulting

LOCATION

The Zohar-Kidod-Haqanaim gas fields are located in the Eastern Mediterranean basin in the southern Judean Desert of Israel, 30 km southeast of Beer Sheva and close to the small town of Arad (Figure 1). The fields are located along a trend of Upper Cretaceous surface anticlines. They are situated to the west of the Dead Sea graben and southeast of the Judean hills (Figure 2).

The Zohar field was the first gas field discovered (1958) in the Eastern Mediterranean and the largest of the gas fields in Israel. Production is from the Jurassic Zohar Formation.

The productive areas are approximately 10 km^2 at Zohar, 5 km^2 at Kidod, and 8 km^2 at Haqanaim. The fields are operated in the Arad lease by Naphtha - Israel Petroleum Corporation Ltd. The gas produced is supplied through a pipeline system to several industrial plants situated in the southern part of Israel (Negev). Ultimate recovery from the three fields is estimated to be 69 bcf.

HISTORY

Pre-Discovery

Geological surveys have been carried out since the early fifties by the Israel Geological Survey and various oil companies in the Judean Desert and the northeastern Negev (Ball and Ball, 1953). The studies were followed by the drilling of fault block structures near the Dead Sea (Mazal 1 in 1953–1954, Mazal 2 in 1954, and Massada 1 in 1954–1955 by Lapidoth; and 'Ein Geddi 1 in 1955 by Israel American Co.) and surface structures (like Rekhme 1 in 1954–1955 by Israel American and Kurnub 1 in 1955–1956 by Pan Israel, Figure 1). These early activities were described by Grader (1957).

Naphtha - Israel Petroleum Corporation Ltd. (Naphtha), founded in 1956, completed a series of geological studies in that area using detailed plane table mapping of the Upper Cretaceous surface structures. The rough topography and difficult

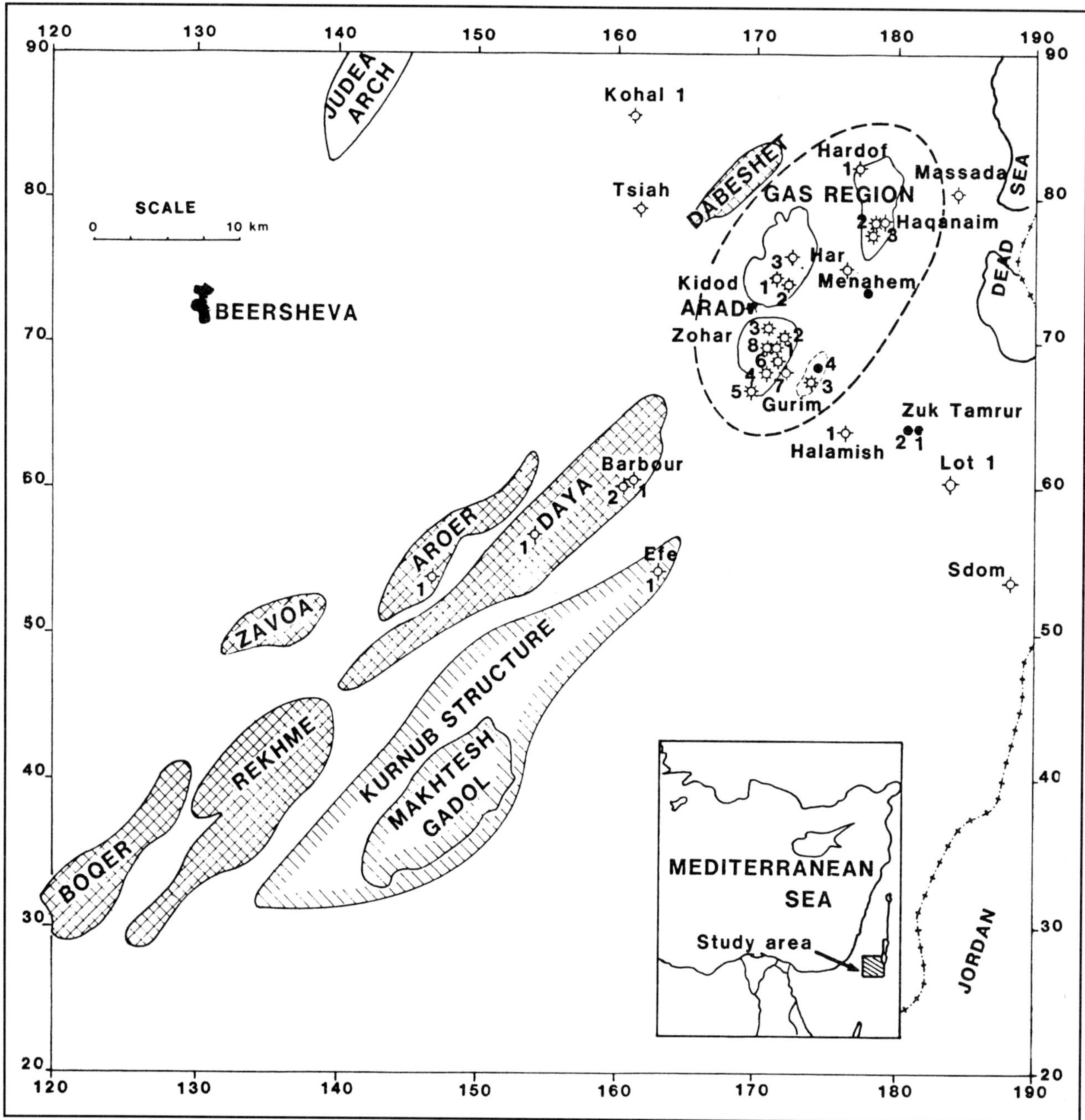

Figure 1. Location map of the Zohar gas region. The grid system of this and following maps is U.T.M. Zone 36.

access, as well as very poor seismic data results of that period, discouraged the further use of seismic surveys in the area. After one dry hole in a fault block near the Dead Sea (Ein Geddi 2, 1957), Naphtha started to test the well-defined surface structures of the southern Judean Desert, beginning with the largest, the Zohar uplift (Figure 2).

Discovery

The discovery well, Zohar 1, was spudded on 17 June 1957 and completed as a gas producer on 3 July 1958. The well was drilled to a total depth (T.D.) of 1999.5 m in Upper Triassic beds. The initial gas production of the discovery well was approximately 440 MCFGPD from the Jurassic Zohar Formation (Coates et a1., 1963).

Post-Discovery

Post-discovery activity was concentrated mainly in development of the Zohar field; however, the drilling of stepout exploratory wells led to the

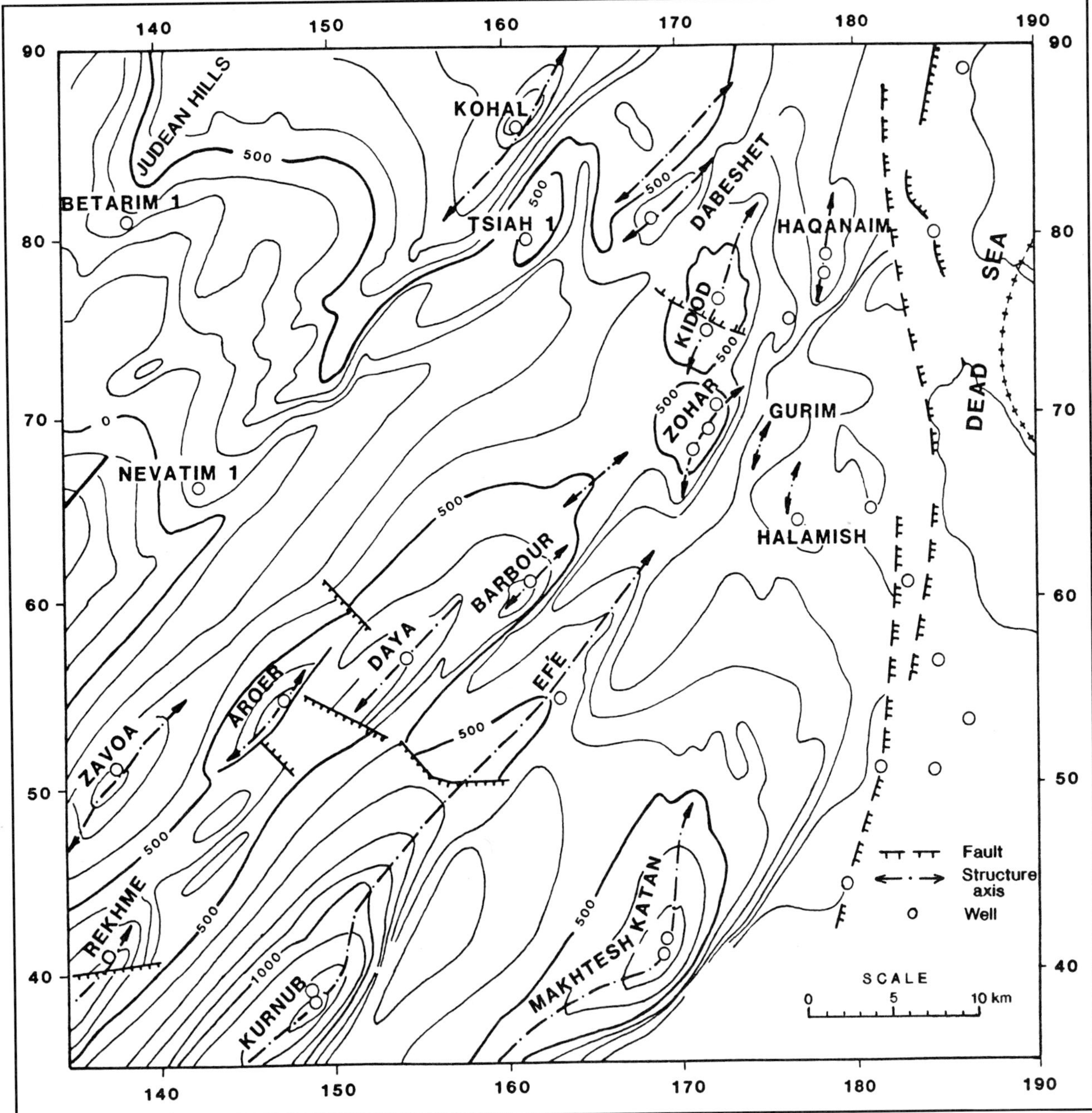

Figure 2. Structural map of the top of the Turonian. Contours in meters above sea level. Contour interval, 100 m.

discovery of two new gas fields, Kidod and Haqanaim (Picard and Eliezri, 1964). Since 1959, the development of these three fields has proceeded in the following stages, assisted by continuous mapping update, using new data as available (Aharoni, 1964, 1969, 1976) (Figure 3):

1. January 1959–June 1960: Drilling of five development wells took place in the Zohar field. Gas production was limited to various tight carbonate zones and a small prolific sandstone lens of the Zohar Formation. These wells did not penetrate deeper than the Upper Triassic.
2. July 1960–January 1961: Discovery of the Kidod field by the Kidod 1 well occurred in August 1960, followed by the drilling of Kidod 2, a development well. Both wells reached the Middle Jurassic. Commercial production was obtained only from Kidod 1. In the Zohar field one stepout well, Zohar 7, reached the Lower Jurassic but was dry as the result of the productive interval being on the downthrown side of the eastern bounding reverse fault (Figure 4).

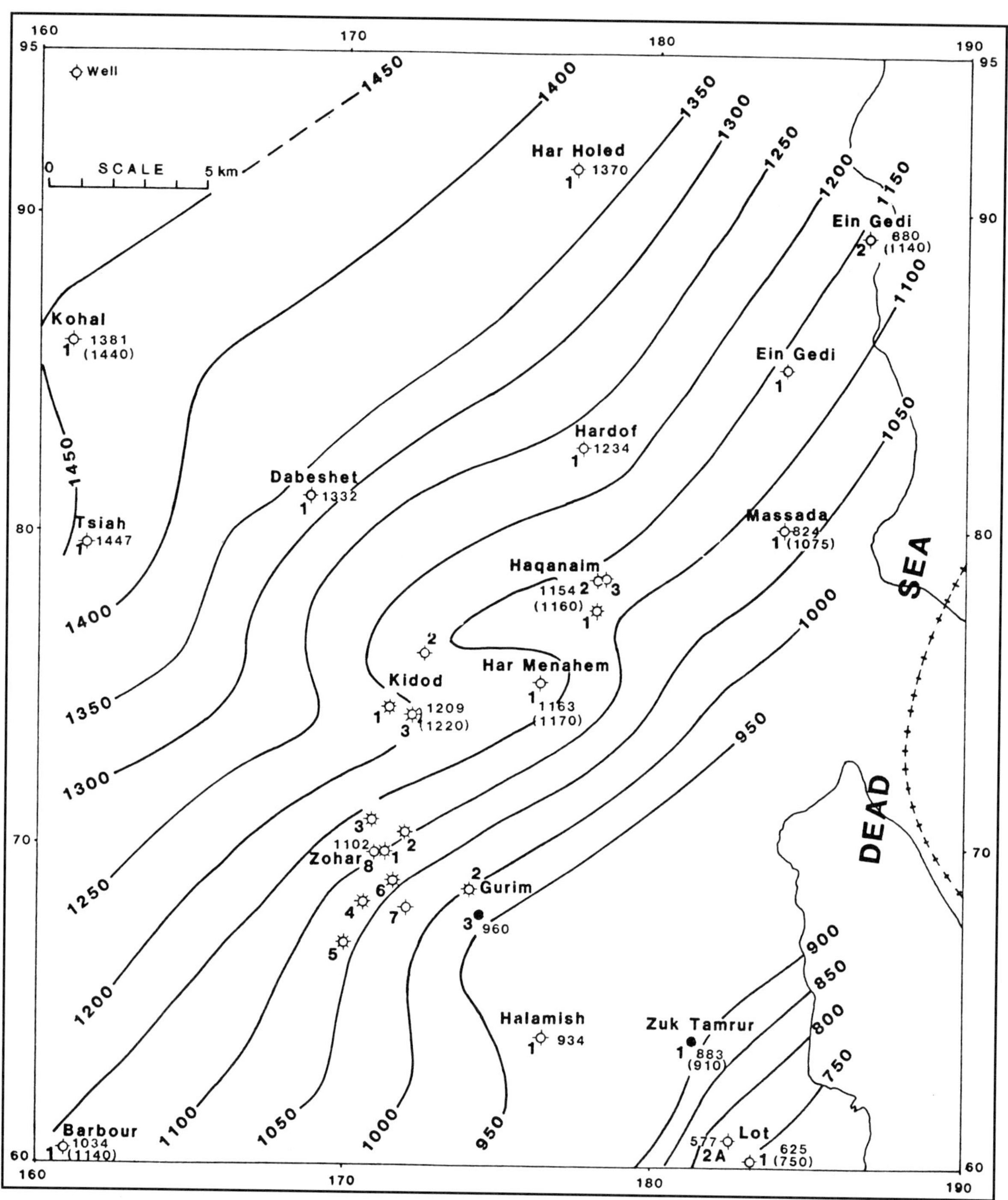

Figure 3. Isopach map of the interval from the top of the Judea Group to the top of the Zohar Formation. Drilled thicknesses in meters are shown by the wells, with calculated thicknesses in parentheses. Isopach interval, 50 m. (From Marcus, 1982.)

3. July 1961–May 1962: Discovery of the Haqanaim field by the Haqanaim 1 well occurred in November 1961, followed by the drilling of one development well, Haqanaim 2, reaching T.D. in Middle Jurassic beds but abandoned because of low porosity and permeability in the reservoir rock.
4. July 1964–October 1968: Two deep test wells, Zohar 8 and Haqanaim 3, were drilled to Paleozoic beds, as another development well in the Kidod field, Kidod 3 (May 1967), and two wildcats on the same structural trend (at Har Menahem and Hardof, Figure 1) were drilled.

No drilling activity has occurred since the end of 1968 in the Zohar trend. Sixteen wells were drilled within the Arad lease: eight wells at Zohar, three

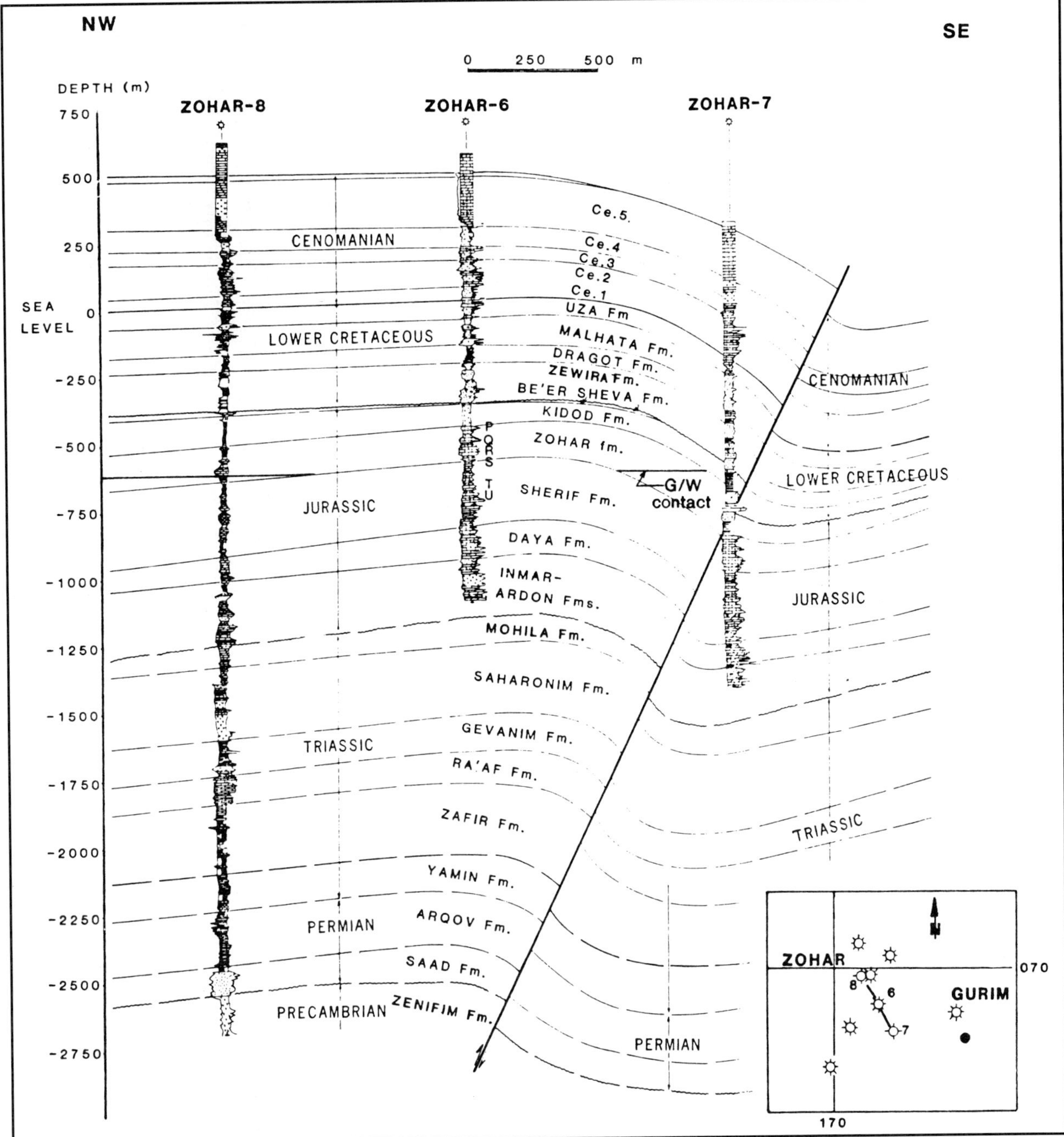

Figure 4. Zohar gas field geological cross section. Legend for lithologies shown on Figure 9. (From Schlein et al., 1980.)

at Kidod, and three at Haqanaim. Only recently, improved seismic data were acquired that revealed newly defined structures deep below the producing zones (Figure 5).

The Zohar gas field produced from June 1962 to August 1979. The field was then used for injection and storage of gas imported from the Sadot field until April 1982, when Sadot was handed over to Egypt. Then, production was resumed and injected gas was produced until November 1984. Since then gas has again been produced from the reservoir's original content.

Gas sales from the Kidod field began in March 1962 and continued until August 1979. Most of the gas was produced from Kidod 3. Because of poor performance, Kidod 1 was shut in as an observation

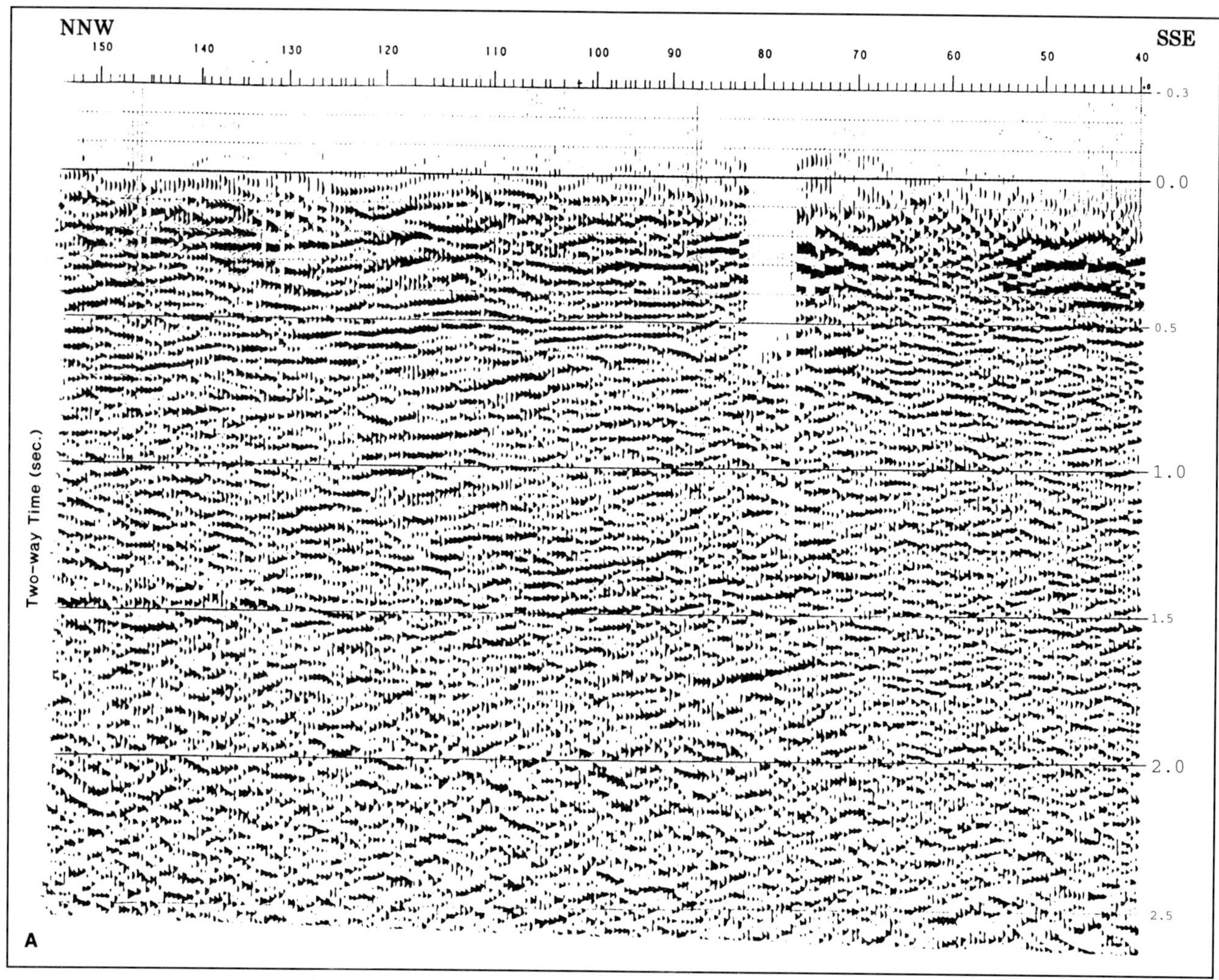

Figure 5. Transverse seismic section of the Zohar gas field. See Figure 4 for location. (A) Uninterpreted. (B) Interpreted.

well after the Kidod field produced about 1.2 bcf of gas. Because demand was satisfied by the Sadot gas and because of the delay in rehabilitation of the Kidod 1 field, it remained shut in until June 1988 when production was resumed from the Kidod 3 well.

Gas sales from the Haqanaim field began in 1969 and continued until 1978 when the producing well (Haqanaim 1) was shut in because of water encroachment. Haqanaim 2 was used as an observation well.

Various wildcats were drilled in the vicinity of the Arad lease, and most had hydrocarbon shows. In the area between the Dead Sea graben and the Zohar trend, two small oil fields, Gurim and Zuk Tamrur, were discovered; however, their production was insignificant (Figure 1).

All of the wells were drilled with rotary equipment using water-base mud. Lost circulation zones above the regional water table were drilled with air. A typical completion consisted of 300–500 m of surface casing (13⅜-in., 12¼-in., or 9⅝-in.), a 1050–1150 m intermediate casing (9⅝-in. or 7-in.), and a producing string to T.D. (7-in., 5½-in., 5-in., or 2⅞-in. tubing).

The electric logs run in these gas fields were of the old type, including electric/SP, microlog, induction, gamma ray, neutron, and acoustic. The early logs were used mainly for correlation since they were inadequate for reliable porosity and fluid saturation calculations, especially from carbonate formations.

DISCOVERY METHOD

The fields were discovered by surface mapping of the exposed structures in the highlands of the northeastern Negev, Israel. Exploration was directed at structural traps in Jurassic carbonates and sandstones covered by shales. Upper Cretaceous surface anticlines were found to be in structural

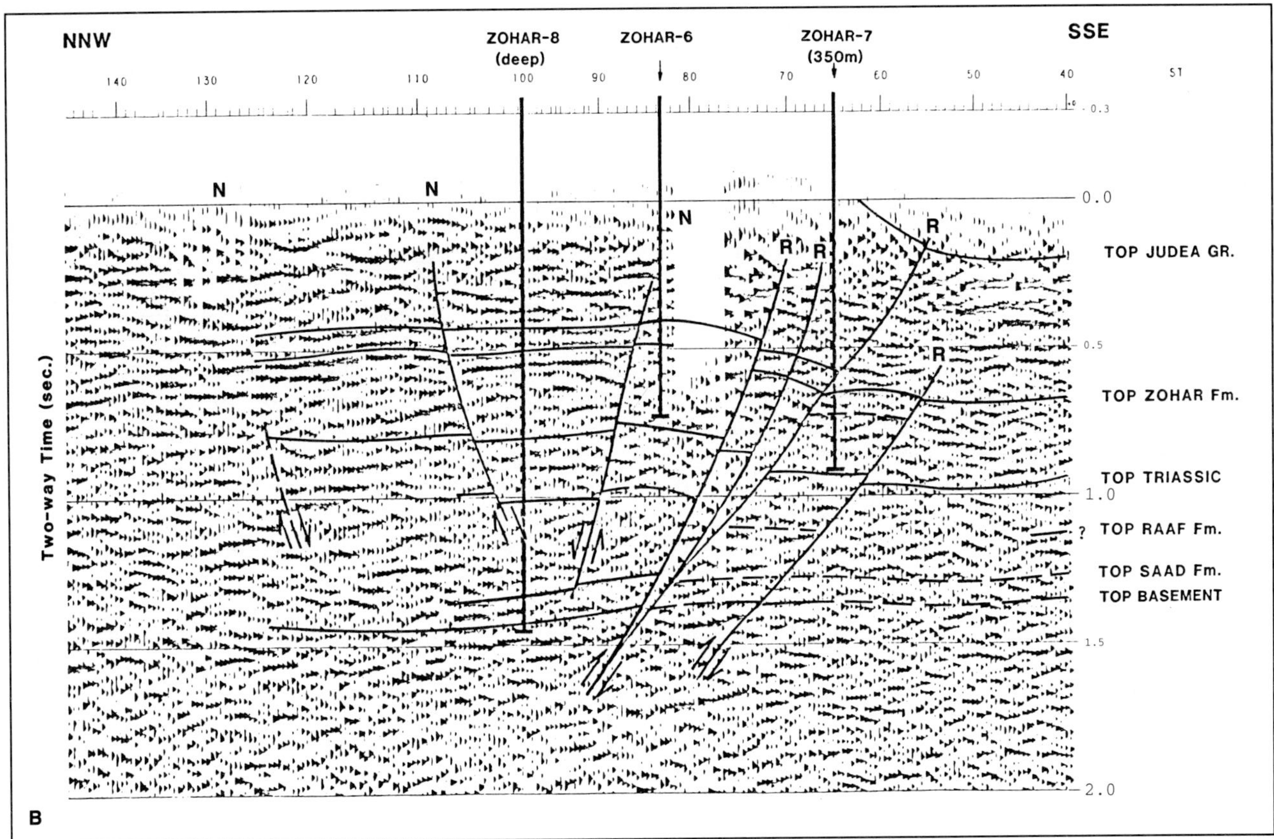

Figure 5. Continued

harmony with the Middle Jurassic pay zones, and drilling near anticlinal crests mapped at the surface led to the discovery of the gas reservoirs quite above the water level. Production on the eastern flank of the Zohar structure failed because of reverse faulting that had been undetected by surface mapping.

Prevailing rough terrain conditions impeded seismic support in the subsequent discovery of the three gas fields and in additional exploration below the producing reservoirs or in other nearby surface structures.

Only recently (1988) a mixed seismic sources technique using a combination of vibroseis and explosives, accomplished with the aid of a helicopter, was successful in providing useful data. Seismic subsurface mapping of the area has enabled the identification of relative structural positions, particularly of the structural disharmony between the Cretaceous and Lower Jurassic–Triassic formations. The success of these techniques will encourage future exploration for the deeper targets.

STRUCTURE

The Eastern Mediterranean basin, originating in early Mesozoic rifting (Monod et al., 1974; Bein and Gvirtzman, 1977; Garfunkel and Derin, 1983), is located on the northwestern margin of the Arabo-Nubian crystalline basement platform (Picard, 1959; Bentor, 1960), which was stabilized during the Pan African orogeny of the late Precambrian (DeSitter, 1962; Engel et al., 1980; Bielski, 1982). The geological history of the basin, of which Israel occupies the southeastern corner, is closely related to the interplay of the huge, rigid, cratonic mass and the sea that lay to the north and northwest.

The structural evolution of the area can be divided into three periods of tectonic activity:

1. Paleozoic (pre-Permian). Late Paleozoic movements manifested by a regional unconformity deeply truncating lower Paleozoic and Precambrian rocks.
2. Permian to Early Cretaceous. The main event was the formation of the passive continental margin in early Mesozoic times (Garfunkel, 1988). Platform sediments were accumulated over the area in several depositional cycles separated by periods of uplift and erosion, with a short period of uplift at the end of the Triassic and an important regional uplift, tilting northwestward, at the end of the Jurassic. The uplift at the end of the Jurassic broadly coincided with the highly alkaline intrusive magmatism related to interplate, hot

spot activity (Garfunkel and Derin, 1988). Associated volcanism occurred after the erosion phase.

3. Late Cretaceous to Recent. Rifting and the continental breakup stage of the Arabo-African continent was manifested in Israel by faulting, vertical movements, and volcanism (Garfunkel, 1988). The Late Cretaceous-Tertiary Alpine orogenic belt (Syrian Arc type) in Israel is delineated to the east by the Neogene Dead Sea transform.

The Zohar-Kidod-Haqanaim gas fields are located in the northeastern Negev highland, Israel, that consists of several chains of northeast-southwest anticlinal structures forming part of the Syrian Arc. The Zohar-Kidod-Haqanaim gas fields are located about 10 km west of the Dead Sea graben on a northeast-southwest and north-northeast-south-southwest-oriented, asymmetrically folded structural trend (Makhtesh Gadol to Zohar, Figure 1). These anticlinal features have broad crestal areas with gentle northwest flanks, narrow and steep southeast flanks, and gentle dips within the adjacent syncline (Aharoni, 1964, 1976; Figures 2 and 6). The elevation difference between the crest and nearby (less than 2 km) synclinal axis surpasses 500 m. The folds are broken by normal faults that are perpendicular to the axial plain but seldom cross their asymmetric flanks (Aharoni, 1976).

In the subsurface, the asymmetric anticlinal structure of the Zohar gas field is somewhat more complex than at the surface. The steep southeastern flank at the surface is underlain at depth by a northeast-southwest-striking reverse fault (Figure 4). As shown by the only well that crossed this fault (Zohar 7), the dip of the reverse fault is 65°, and at the level of the Jurassic Zohar Formation, the throw is 160 m. In the northwestern upthrown block where the gas field is located, dips of 60° were measured close to the fault at Zohar level. The steep flank of the other folded structures also appears to have buried reverse faults, and at depth, these structures resemble tilted fault blocks. The Kidod anticline has a gentle southeast flank. In the Haqanaim structure, horst-type faulting is an important element. The features of the folding indicate that pre-existing basement faults controlled the formation of many folds (DeSitter, 1962; Mimran, 1976; Eran, 1982; Salamon, 1987; Figure 4).

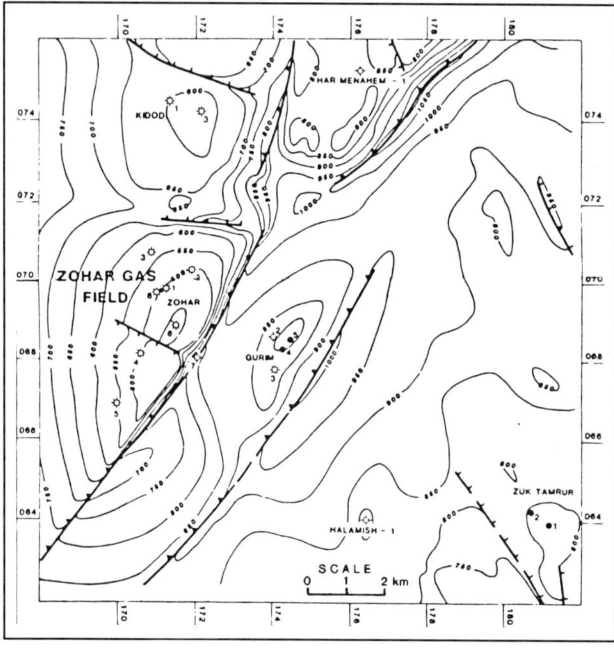

Figure 6. Zohar gas field structure—top of Zohar Formation (Callovian). Contour values ssl. Contour interval, 50 m. (From O.E.(I.)L., 1988.)

A reprocessed transverse seismic line yields a clear picture of the asymmetrical Zohar structure (Figure 5). The crest of the Jurassic Zohar Formation is located about 500 m west of the bounding reverse fault, whereas the deeper Lower Triassic–Paleozoic formations are close to the fault and generally dip westward.

Toward the southeast, the Lower Cretaceous Zeweira Formation overlies deeper and deeper Jurassic beds, leading to the inference of pre-Cretaceous northwest tilting and subsequent peneplanation (Garfunkel and Derin, 1988). Consequently, in the northwest flank of the Zohar structure, the Beer Sheva limestone is the first Jurassic formation below the Lower Cretaceous, whereas in the eastern flank, it is directly underlain by the Kidod shales (Figures 4 and 8). The effect of the eastward angular truncation of the top of the Jurassic accentuates the westward dip of the pre-Cretaceous section compared to that of the overlying Cretaceous beds.

The folding phase started during the late Cenomanian and continued intermittently until the Miocene. The perpendicular or transverse faults developed between post-Campanian and late Pliocene time (Aharoni, 1976).

The Dead Sea graben collapse occurred along major wrench faults having throws of a thousand to several thousands of meters associated with intensive faulting and shattering of the nearby folded areas. The resultant difference in elevation between the top of the Turonian at the Zohar field and that speculated in the Dead Sea graben is about 6500 m (Figure 7).

STRATIGRAPHY

Vast, shallow epeiric seas characterized the Eastern Mediterranean basin area during long geological periods (Derin, 1974; Druckman, 1974; Goldberg and Friedman, 1974). Most of the sediments were deposited on a platform under varying continental and epicontinental environments. The stratigraphic column in the platform area of Israel is considered to be not more than 6.5 km thick, mostly

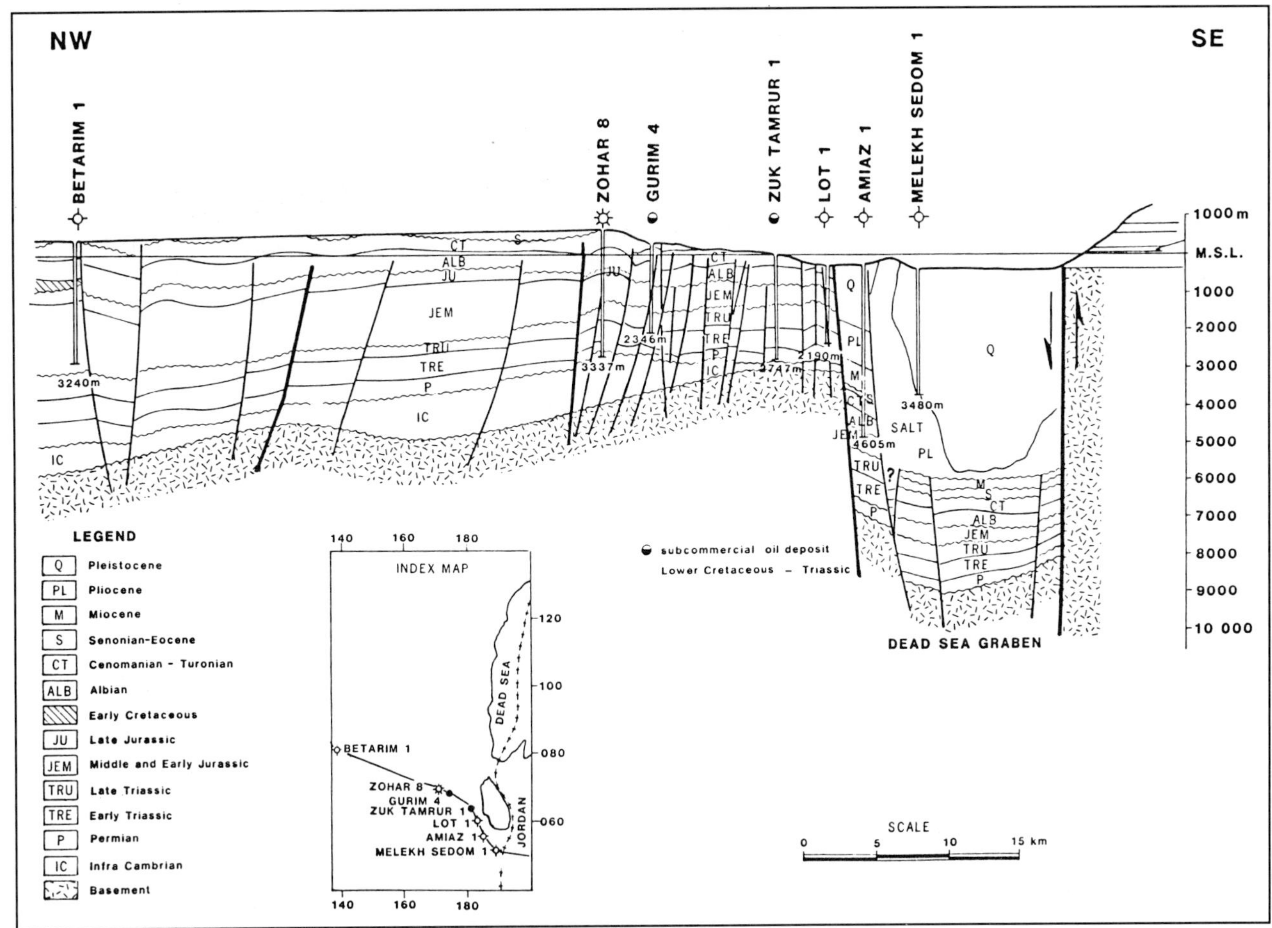

Figure 7. Geological cross section from Dead Sea northwest to Betarim 1. (From O.E.(I.)L., 1988.)

of Mesozoic age. In the Paleozoic, shallow sea transgressions were rare, whereas erosion periods were very common. Only Permian strata have been identified. In the early Tertiary, on the other hand, the relief was high and the sea began to retreat. West of this platform, a thicker sedimentary column was deposited on the shelf edge and continental slope during the Cretaceous and Neogene periods, underlying the present-day Coastal Plain and shelf (Bein and Gvirtzman, 1977; Garfunkel and Derin, 1983).

The Arabo-Nubian massif contributed enormous quantities of clastic matter to the shallow, adjacent basin. The clastics intermingled with shallow sea carbonates of biogenic origin. Evaporites were deposited in periods of regression in lagoonal and supertidal environments (sabkhas), and black shales accumulated in euxinic inner basins of the epeiric sea (Derin, 1974; Druckman, 1974, 1976).

The Zohar-Kidod-Haqanaim fields are located on the platform area about 70 km east of the Middle Jurassic shelf break. Most of the wells penetrated the Upper Cretaceous to Lower Jurassic section. One well, Haqanaim 3, penetrated the Permian sandstone (Sa'ad Formation), while another, Zohar 8, reached Precambrian igneous rocks (Zenifim Formation). The stratigraphy of the area is shown on Table 1 and Figures 8 and 9.

Paleozoic

The Precambrian Zenifim Formation is overlain by approximately 300 m of Late Permian rocks (the Sa'ad and part of the Arqov formations). The lignitic sandstones of the Sa'ad Formation and the shales, sands, and carbonates of the Arqov Formation were deposited in shallow marine and fluviatile environments during the Permian transgressive phase. Both formations are rich in organic matter.

Triassic

The Triassic sequence in the Zohar area is nearly 900 m thick. The lower marine formations are composed mainly of limestone (Yamin Formation), thick impervious shale (Zafir Formation), and carbonates (Ra'af Formation). The regressive phases in the Scythian and Anisian stages are characterized by deposition of sands in the Arqov and Yamin formations and sand beds within shales in the

Table 1. Generalized stratigraphic column for the Zohar area.

System		Series	Group	Formation	Thick. (m)	Lithology
CRETACEOUS	Upper	Senonian	Mount Scopus	Mishash Menuha		Chalk and chert Chalk or bitum. chalk
		Turonian-Cenomanian	Judea		500	Limestone Dolomite & marlstone
	Lower	Albian-Hauterivian?	Kurnub		400	Limestone and shales sandstone
JURASSIC		Oxfordian	Arad	Beer Sheva	0-80	Limestone
				Kidod	80-100	Marl and shales
		Callovian		Zohar	120	Limestone, shales and some sandstone
		Bathonian		Sherif	250	Limestone, sandstone and shales
		Bajocian		Daya	110	Limestone, dolomite, sandstone and shales
		Aalenian		Inmar	120	Sandstone
		Lias		Ardon	80	Dolomite
TRIASSIC		Carnian	Ramon	Mohilla	90	Dolomite, anhydrite?
		Ladinian		Saharonim	260	Limestone and shale
		Anisian		Gevanim	120	Sandstone and shale
				Ra'af	110	Dolomite, limestone
		Scythian	Negev	Zafir	270	Shales, limestone and sandstone
				Yamin	100	Limestone, argillaceous, some sand
PERMIAN				Arqov	200	Sand, shale and limestone
				Sa'ad	180	Sandstone, lignitic
Pre-Cambrian				Zenifim		Igneous and arkose

Gevanim Formation. Shales and limestones of Ladinian-Carnian age (Saharonim Formation) are replaced near the top of the Triassic by limestone interbedded with anhydrite originating in a tidal lagoon environment (Mohilla Formation) related to a third regressive phase (Druckman, 1974, 1976; Freund et al., 1975; Eshet, 1987).

Jurassic

During the Jurassic time the Zohar area was dominated by two environments of deposition, sea marginal flats and shallow marine (epeiric) (Figure 9 and Goldberg and Friedman, 1974). The 800 m thick Jurassic sequence is subdivided into three distinct units:

1. The Lower Jurassic unit consists of shallow, shelf carbonates (Ardon Formation).
2. The Middle Jurassic unit is characterized by an interplay between braided streams that deposited terrigenous sandy sediments, and carbonates originating in tidal flats (Inmar, Daya, Sherif, and Zohar formations).
3. The Upper Jurassic is characterized by varying low and high energy shelf, inner basin, and patch reef sediments (shales of the Kidod Formation and limestones of the Beer Sheva Formation).

The regional uplift and tilting at the end of the Jurassic exposed the Beer Sheva limestone to erosion in the Zohar area, forming the Late Jurassic-Early Cretaceous unconformity phase (Figures 7 and 8). This erosional phase was more intense in the eastern uplifted reaches of the area (Goldberg, 1970; Goldberg and Friedman, 1974; Derin, 1974).

Lower Cretaceous

During Early Cretaceous time, the Zohar area was exposed and covered by eolian and fluviatile sands, silts, and shales interbedded with some carbonates of marine origin (400 m of the Kurnub Group). North and westward, the essentially continental environ-

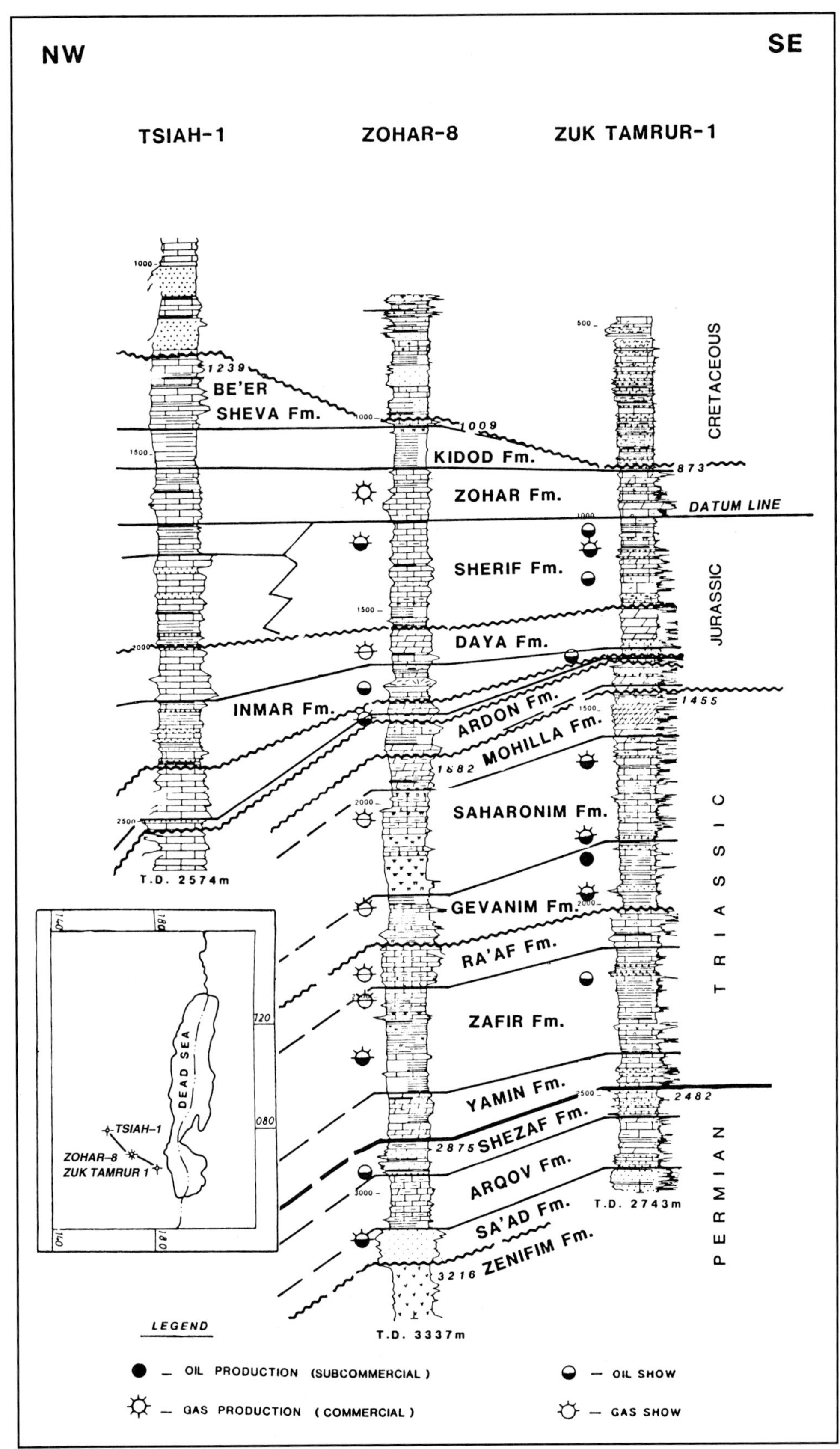

Figure 8. Electric and lithologic log correlation of Zohar 8 with Tsiah 1 and Zuk Tamrur 1. (From OE(I)L, 1988.)

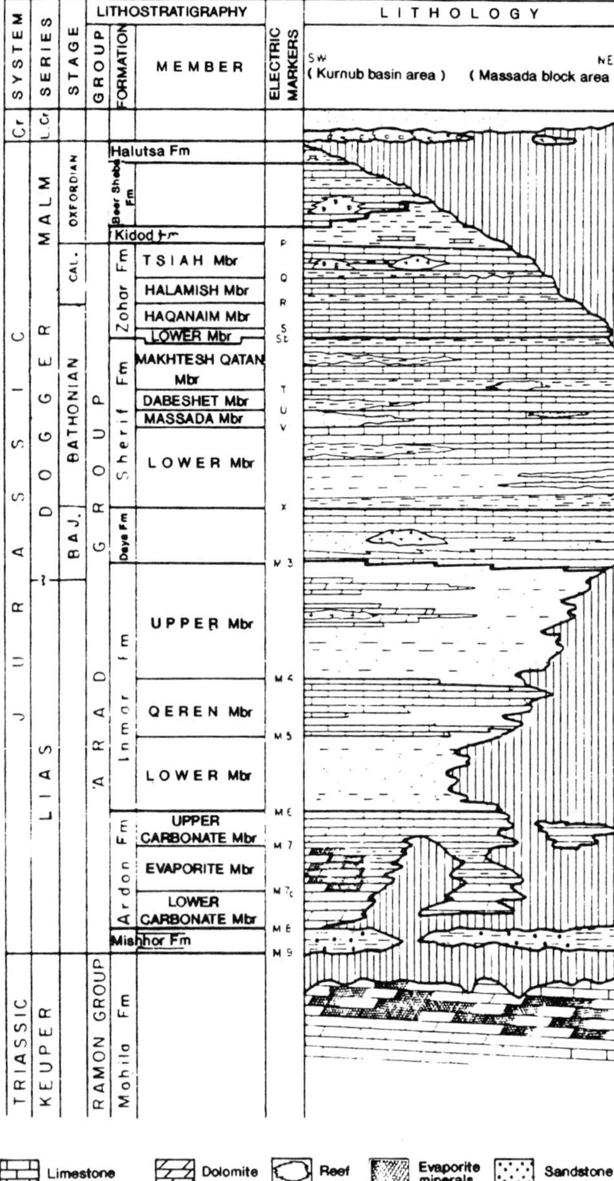

Figure 9. Schematic composite columnar section for the Jurassic in southern Israel. (From Goldberg and Friedman, 1974.)

ment of deposition grades into one of a shallow shelf, which persisted until the Turonian (Aharoni, 1966).

Upper Cretaceous

The Late Cretaceous sequence consists of shelf carbonates, dolomite, and limestone of Cenomanian-Turonian age (Judea Group). A lost circulation zone characterizes this porous and vuggy carbonate (mainly dolomitic) complex, which is the most prominent aquifer of Israel. Senonian beds present in the bordering synclines of the Zohar area consist of chalks, marls, phosphates, and chert beds (Menuha and Mishash formations). In some parts, bituminous chalks and marls are common.

TRAP

The Zohar-Kidod-Haqanaim gas fields are structural traps located on a northeast-southwest-trending faulted anticline, characterized by a gently dipping northwest flank and a narrow steeply dipping southeast flank (Figures 4 and 6). This structural configuration probably developed during the Late Cretaceous-Eocene folding phase. The Middle Jurassic Zohar limestone and sandstone reservoir rocks are overlain and sealed by the impermeable early Oxfordian Kidod shale.

Reservoirs

General Description

In the three gas fields of Zohar, Kidod, and Haqanaim, the primary gas reservoir, the Zohar Formation, consists mainly of limestone and is subdivided by shale interlayers into three units: the upper P, the middle Q, and the lower R (see Figures 4 and 9). Part of unit S also produces in the Zohar field. These units are easily correlated by electric logs within the fields and for tens of kilometers beyond. At the base of the Q unit, a 5 m thick sandstone lens is productive in the southeastern part of the Zohar gas field (wells Zohar 4, 5, and 6). This lens is in apparent communication with the surrounding limestone.

The top of the primary pay zone in Zohar field is encountered at 1000–1100 m depth (370–470 m ssl). The gross thickness of the producing interval is 110 m; of these 40 m are above the Q sandstone and 65 m are below it. The total thickness of the limestone in the Zohar Formation is 70 m.

A lower sequence of zones that belong to the Sherif Formation (remainder of S, T, and U) consists of limestone and sandstone beds that produce only in Zohar 6 (lower gas range). The gas-water level of Zohar field is uncertain. West of the reverse fault, none of the wells penetrated the gas-water interface in the Q sandstone. The Zohar 3 and 8 wells penetrated the R limestone, probably through the water level. However, because the limestone is very tight in Zohar 5 and fractured in Zohar 3 and because the Zohar log saturation calculations are unreliable, the depth of the gas-water contact remains questionable. Based on pressure data from measurements in the Zohar 1 and 7 wells and assuming that the Zohar reservoir fluids communicate across the eastern reverse fault, the water level would be at 632 m ssl.

In the two other gas fields, the top of the pay zone was encountered deeper. At Kidod field it is 1200–1240 m and at Haqanaim, 1150 m. The gross thickness of the Zohar Formation is 130 m at Kidod and 95 m at Haqanaim.

Limestone of peritidal facies composes most of the Zohar Formation reservoirs (Figure 10), with sandstones, siltstones, and shales being minor components (Goldberg and Friedman, 1974). The limestones are predominantly biopelmicrites with minor intrabiopelsparite. Faunal remnants are mostly mollusk fragments, foraminifera, solitary corals, and echinoids, with sponge spicules and calcareous algae. Quartz grains are abundant in thin interbeds. Argillaceous pellets and disseminated pyrite are common (Barzel and Friedman, 1970).

Most of the porosity is of the intercrystalline-interskeletal types in interstices of the secondary calcite. Less porosity is developed in the empty interiors of leached fossil fragments.

The Q sandstone is a medium-grained, rounded to subangular calcareous sandstone that grades from a poorly consolidated sandstone to dolomitic sandstone and sandy dolomite.

Rock and Fluid Characteristics

The different compositions of the gas samples taken from the three reservoirs are summarized in Table 2. The gas compositions of Zohar and Kidod fields are similar and consist of more than 95% methane with more than 2-3% ethane, propane, and butane, and 0.3-3.5% CO_2. The gas gravity is 0.58 (air = 1). The Haqanaim gas is heavier, characterized by a gravity of 0.62 and, in addition, contains H_2S (0.08% mole). Such variations in gas composition are expected. As discussed later in the *Source* section concerning the genetic relationship between the gases and oil, the variations in the gas composition reflect the different products of separation in stages of the original oil.

The composition and properties of the Jurassic water in the gas fields have been affected by the intrusion of meteoric water and by the degree of the invasion (Starinsky, 1974). In some rock formations

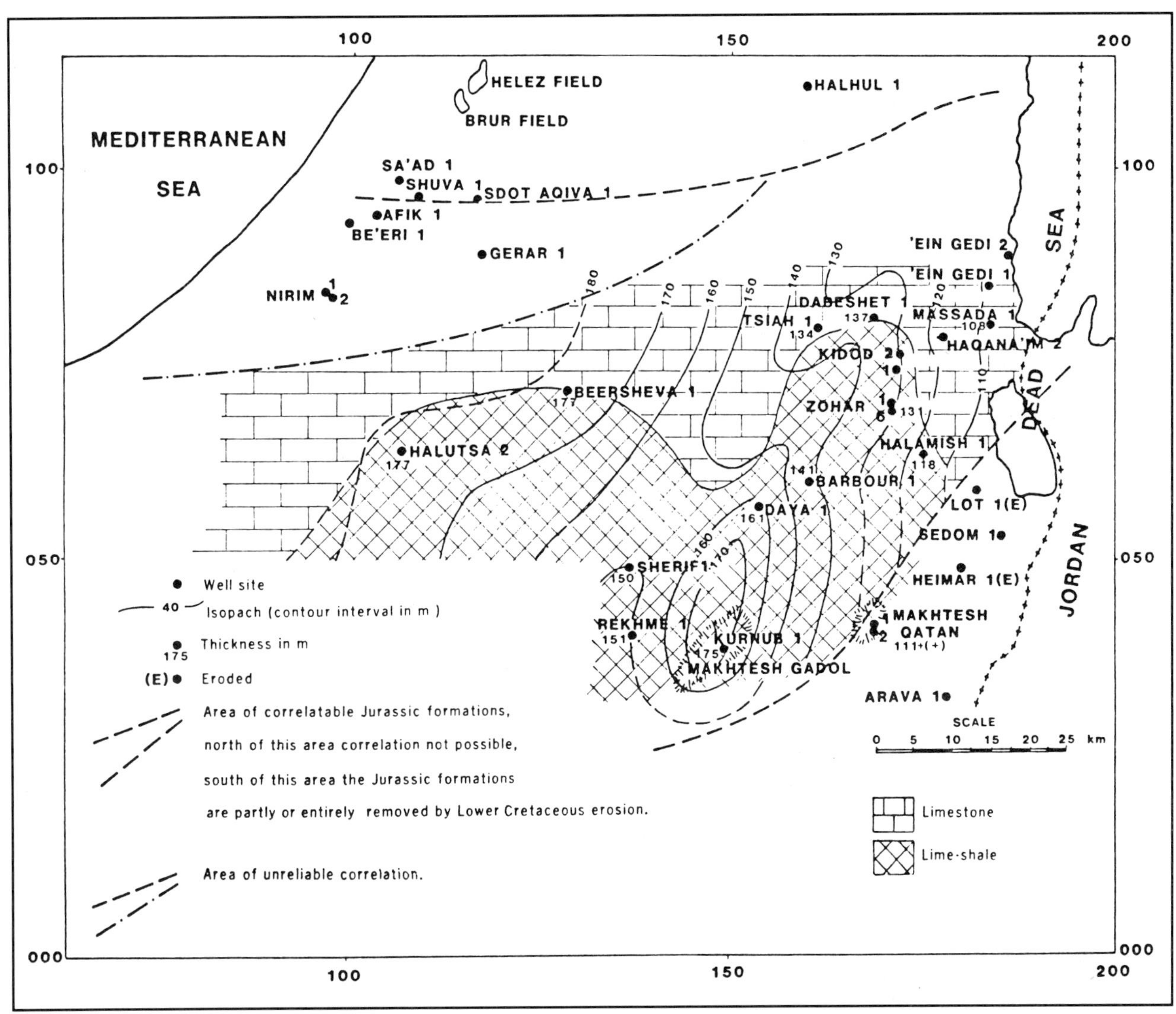

Figure 10. Lithofacies and isopach map of Zohar Formation, southern Israel. (From Goldberg, 1970.)

Table 2. C_1-C_4 Concentration in the hydrocarbon fraction and the isotope carbon composition of the CH_4 and CO_2.

Example (well name)	C_1 (%)	C_2 (%)	C_3 (%)	iC_4 (%)	nC_4 (%)	$\frac{C_1}{C_{1-4}}$	CH_4 (‰)	CO_2 (‰)
Zohar 3	97.90	1.94	0.08	0.02	0	0.98	-51.78	—
Zohar 6	96.86	2.70	0.12	0.04	0.17	0.97	—	—
Haqanaim 1	95.55	3.99	0.34	0.12	0.01	0.96	-52.00	+7.33
Zohar 8	98.37	1.58	0.05	0	0	0.98	-51.30	—
Kidod 3	96.72	2.93	0.26	0.8	0	0.97	-51.54	+7.09

From Amit and Bein, 1979b.

where the water could flow easily, the original saline water has been flushed to the east toward the Dead Sea graben by fresh water. The fresh water flowed in from the southwestern erosional cirques (Makhtesh Gadol and Makhtesh Katan, Figures 1 and 2). Nevertheless, formation water that kept the characteristics of the Group III water [ratio Na/Cl 0.75, high Ca/Mg ratio and (Ca+Mg)/Na 0.27; Fleisher, 1987], but with a relatively low ion concentration (25,000–30,000 ppm TDS) (Table 3), accompanies the hydrocarbon accumulations (Figure 11).

Analyses of the numerous recovered cores give a relatively reliable picture of the gas fields' porosities and permeabilities. Porosities in the Zohar limestone are low, varying significantly from one bed to another. The limestone permeability values are extremely low, both absolutely and in relation to the corresponding porosities. The permeabilities vary from less than 0.01 md in the case of 6% porosity to 0.15 md in the case of 12% porosity. In the Zohar field, the overall average porosity is 6.0% whereas the average useful porosity value of the total Zohar limestone is 6.9%, reaching 8.4% in the Q limestone zone; 5% is regarded as minimum effective porosity. The average thickness of the productive limestone is 50.4 m. Of the three main productive zones, the R zone has the greatest volume.

The Q sandstone has excellent reservoir properties. Its porosity averages 25% and its permeability is 1400 md. Unfortunately, the low recovery of the cores taken in the Q sandstone possibly resulted in the loss of the best reservoir sections needed to reliably measure the hydraulic parameters. The wells completed in this porous sandstone have significantly greater productivities than those completed in the limestone.

The Zohar Formation limestone of the Kidod field has porosities and permeabilities similar to those of Zohar field. The thickness of the productive limestone in Kidod field, including the P, Q, and R zones, is 27.4 m and has an overall average porosity of 9.3%. The Haqanaim field has similar porosities but higher permeability because of a greater amount of natural fracturing.

Capillary pressure data indicate that the limestone has a very long two-phase transition section. At its top, the irreducible water saturation is 12–35% in the relatively porous limestone sections and 30–45% in the tight sections. Water capillary rise is very high, 24–67 m above full saturation to the first point of measurable gas and over 160 m to the irreducible water saturation level (Figure 12). By way of contrast the Q sandstone has very little capillary water rise and produces gas at levels where the limestone is almost fully saturated with water.

Connate water saturations obtained from old electric logs are in many instances inconclusive. Nevertheless, for the most part, minimum water saturations are in the range of 40 to 55%, with the exception of the Q sandstone in which water saturations are less than 35%.

Reserves, Production, and Performance

The gas reservoirs of the three fields are nearly depleted. The largest and most important reservoir is in the Zohar field. Its original reservoir pressure at the average depth of the Q zone was 879 psia (6060 kPa) at 1174 m (543 m ssl, Figure 13). The lithology of the producing formation varies from north to south. In the northern part of the field, Zohar 2 and 3 wells were completed in a tight limestone, whereas the Zohar 1, 4, 5, 6, and 8 wells were completed in the prolific Q sandstone section. In spite of the intervening shales, it seems that the P, Q, and R zones form a common reservoir. The material balance plot of the field, in terms of p/z and cumulative gas production, exhibits a linear trend (Figure 14), where p is the average reservoir pressure and z is the gas compressibility factor. From the (p/z) plot, the Zohar field has behaved as a classic, volumetrically controlled gas reservoir.

Although the field is underlain by an aquifer, the water influx effect is not noticeable because of the extremely tight nature of the rock in which macroscopic fissures are not appreciable. The original gas in place is estimated to be 59 bcf (Clarke, 1963; Coates, 1964). The reservoir pressure is not uniform over the field. In the limestone of the northern part of the field, the pressure measured in Zohar 2 declines at a lower rate than that in the sandstone measured in the observation well, Zohar 5. It is believed that the more productive reservoir rock depletes faster but is recharged by significant gas migration from the comparatively impermeable limestone section. The current cumulative gas

Table 3. Chemical composition of middle aquifer formation waters from borehole Zohar 6.

Date	Depth (m)	TDS (gr/L)	millequivalents/liter Ca	Mg	Na	K	Cl	SO_4	HCO_3	Cl/Br	Na/Cl	Ca/Na	Ca+Mg/Na
7/60	1100	5.7	7	7	75	2.3	75	15	4	536	1.0	0.09	0.18
8/60	1172	32.0*	194	32	322	1.1	496	18	28	627	0.65	0.60	0.70
12/61	1172	8.2	2	7	123	1.3	116	11	3	228	1.06	0.02	0.07
10/60	1195	20.6	57	23	258	3.3	340	12	6	—	0.76	0.22	0.31
8/60	1226-1229	31.1*	217	39	288	3.9	498	17	36	—	0.58	0.75	0.89
12/61	1220-1223 1226-1229	24.6*	169	20	245	2.4	415	12	—	—	0.59	0.69	0.77
7/60	1300	11.3	12	8	168	3.8	167	14	6	349	1.00	0.07	0.12
8/60	1311-1314	6.9	16	14	76	1.5	79	30	8	—	0.96	0.21	0.39
7/60	1349-1352	6.2	7	8	93	1.7	86	1	24	—	1.08	0.07	0.16
7/60	1520-1539	3.3	1	2	56	—	28	4	27	—	2.00	0.02	0.05
7/60	1550-1553	3.2	3	6	51	2.1	31	1	27	—	1.64	0.06	0.18
7/60	1600	4.6	2	5	66	—	52	15	7	459	1.27	0.03	0.11
7/60	1628-1633	3.7	1	5	64	2.4	48	6	22	—	1.33	0.01	0.09

Note: asterisks represent calcium chloride water.
From Fleisher, 1987.

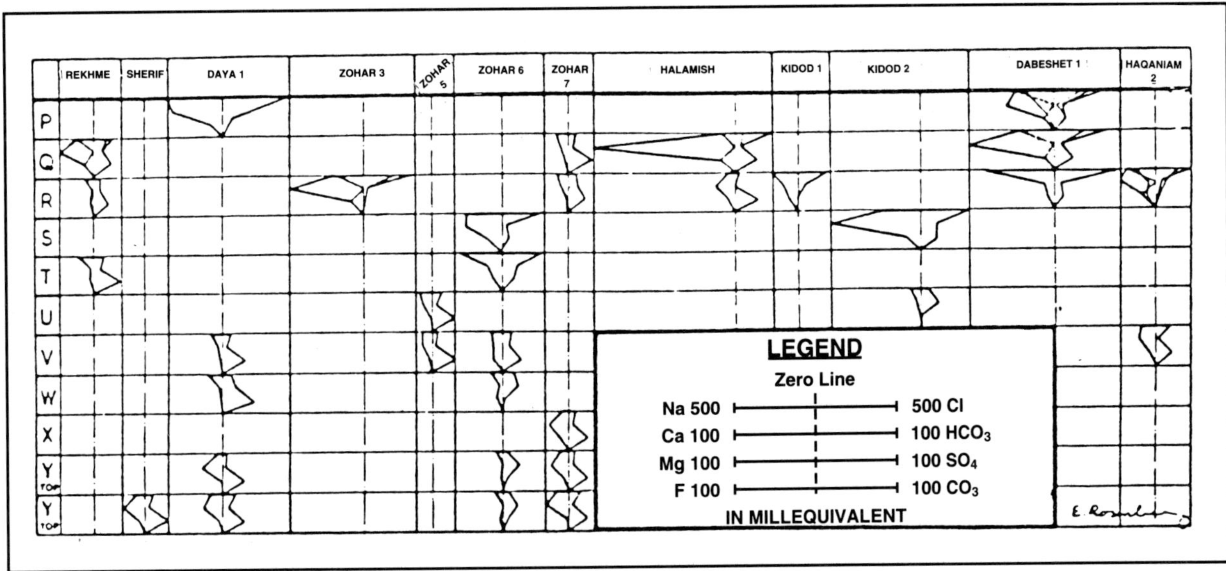

Figure 11. Jurassic water composition patterns in the Zohar area. (From Coates et al., 1963.) Water analysis results in the form of Schiff diagrams obtained by stretching lines between values in the legend. Ions Cl^- and Na^- meq/L are multiplied by 500 and other ions by 100. E.g., in Kidod 2 the Ca^{++} contents are very high in the S-horizon. The letters in the left column are limestone horizons in the Zohar and older Jurassic formations (Figure 9). P, Q, and R are the major gas producers in the Zohar field.

production from the field at the end of 1988 is about 44 bcf.

The present pressure of the sandstone reservoir is 225 psi (1550 kPa). The remaining reserves as of 1 January 1989, assuming that the Zohar 5 wellhead pressure is 75 psi (515 kPa) at abandonment, are about 10 bcf of gas.

The Kidod field is second in reservoir size to the Zohar field. Its original pressure was 879 psi (6060 kPa) at 1294 m and original gas in place was estimated at about 10.1 bcf. The cumulative gas produced is about 6.0 bcf. During most of its life the Kidod field behaved as though it were under volumetric control. When the reservoir pressure declined below 525 psi (3620 kPa), water influx became noticeable. The amount of the remaining recoverable reserves will be dependent upon the rate of water invasion. If the gas reserves can be depleted before water encroach-

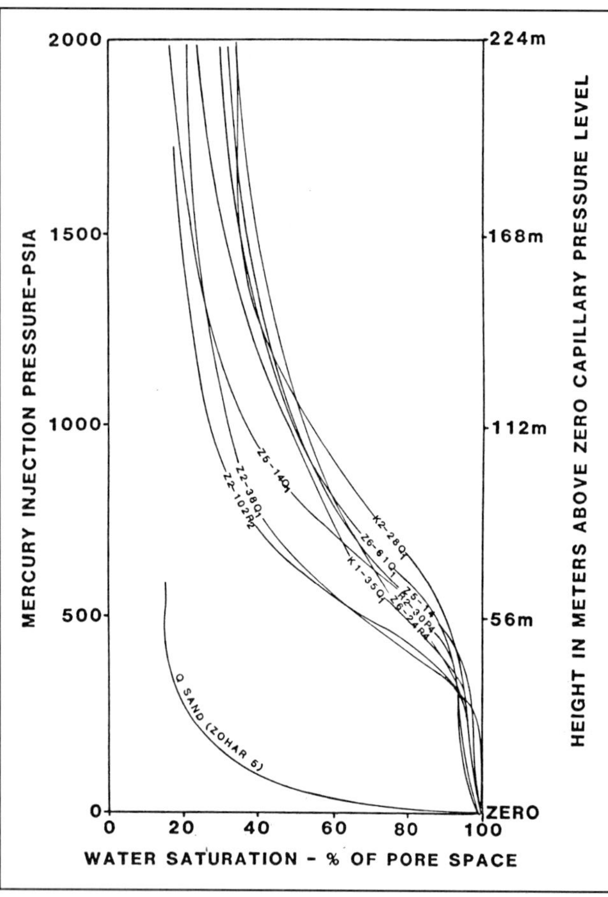

Figure 12. Capillary pressure curves of the Zohar limestones in the Zohar area. (From Coates et al., 1963.)

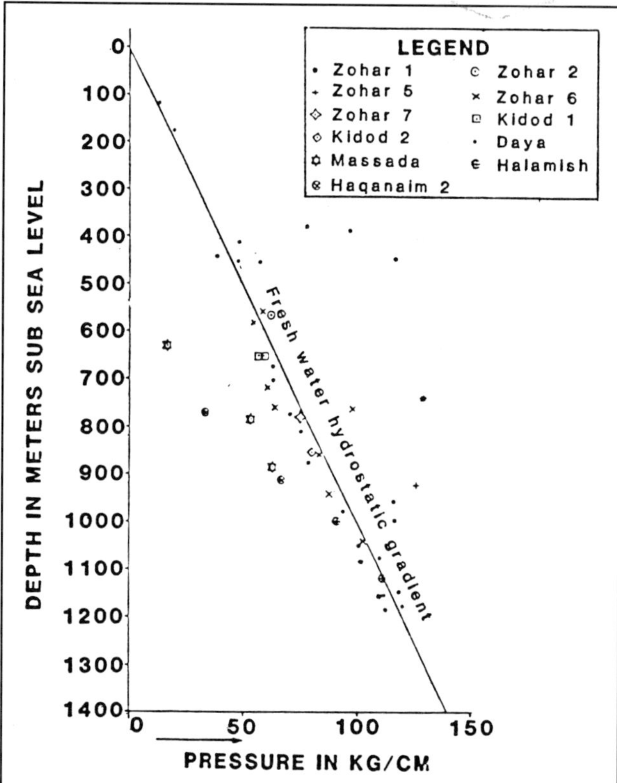

Figure 13. Formation pressures versus depth in the Zohar area. (From Coates et al., 1963.)

ment, the amount to be produced for a wellhead abandonment pressure of 100 psi (690 kPa) is estimated at about 8.9 bcf of gas.

The smallest in reserves of the three reservoirs is that of Haqanaim field, which had an original pressure of 978 psia (6740 kPa) at 1140 m. Its original gas in place is estimated at about 8.0 bcf. The cumulative gas produced is 5.1 bcf. The ultimate gas recovery at an assumed wellhead abandonment pressure of 100 psig (690 kPa) at the Haqanaim 2 well is about 7.1 bcf. The field is not producing because of water invasion.

Source

The source rock of the Zohar gas and the hydrocarbon migration paths are speculative (Ettinger and Langotzki, 1969). The overlying Kidod shales, although in contact with the gas-containing Zohar limestone, are too immature to generate hydrocarbons, and their organochemical characteristics differ from the gas and oil discovered in the region (Figure 15).

The hydrocarbon composition ($C_1/\Sigma C_1$-$C_4 = 0.96$–0.90) and the carbon isotope ratio of the methane (-51.4 to -52.1%, Table 2) from the Zohar gas field indicate that the gas was generated in association with an oil generation process. Vitrinite reflectance values of 0.5% in the Zohar Formation suggest that the gas was trapped far from its generation site (Table 4). Perhaps the hydrocarbons are related to a source in Triassic or older formations located in the west (Glattstein, 1976; Kisch, 1978; Amit and Bein, 1979b; Bein and Amit, 1980; Bein et al., 1984).

The origin of the various asphalts, heavy and light oil, and gas observed in the southwest rim of the Dead Sea graben, where the Zohar gas field is located, was recently studied by Spiro et al. (1983) and Tennenbaum (1983). Their studies concluded that Senonian-aged bituminous beds buried in the Dead Sea graben form the source rock of all the hydrocarbon occurrences in the area (Figure 7). As a consequence, and based on the depositional history of the graben fill, the major phase of oil generation had to occur in the Pliocene. The differences between the various types of hydrocarbon occurrences were thought to be from a combination of several effects such as variations in source material, thermal history of the different downwarped blocks, water-washing caused by regional hydrodynamic conditions, and biodegradation.

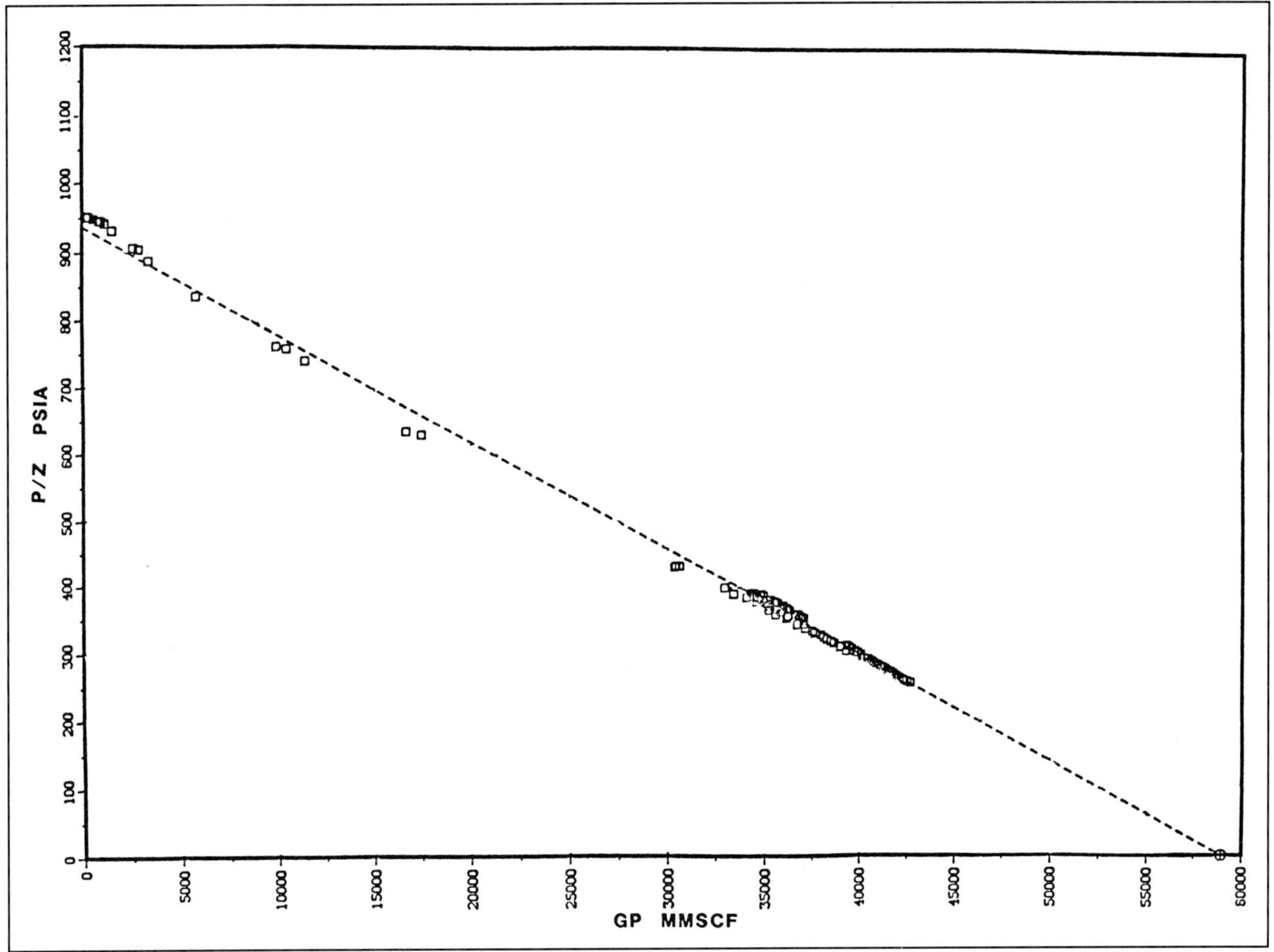

Figure 14. Zohar gas field, *p*/*z* vs. cumulative gas production, March 1987.

A recent modeling study (Braester et al., 1987) provides a better understanding of the combined effects of the regional geology, hydrogeological conditions, and physical properties of rock and fluids on the different types of hydrocarbon accumulations noted in the west side of the Dead Sea graben. It was shown that most of the hydrocarbon occurrences in the region could have been derived from a single type of light oil, like that related to Senonian source, generated in the graben. Upward migration associated with gas liberation could have resulted in natural fractionating and development of various types of oil and gas that spread laterally and vertically over the region. The best oil migration conduits from the graben to its surroundings are the Triassic formations with highly saline waters which increased the buoyancy driving force. Inasmuch as the Jurassic beds containing relatively fresh water flowed eastward to the graben in reverse direction to the possible hydrocarbon migration path, it is most likely that the Jurassic rocks were impregnated by oil and gas that came from the Triassic beds through faults. Under these conditions, structural traps in the Jurassic formations, like those at Zohar, Kidod, and Haqanaim fields, were filled with gas. Oil was flushed toward the graben and may have accumulated only in down-faulted blocks as a result of restrictions in flow passage, such as low permeability zones (Figure 16).

EXPLORATION AND DEVELOPMENT CONCEPTS

Based on the production history of the three gas fields, which gives considerable information on the nature of the reservoirs, no further drilling is needed to deplete the fields. The primary gas field, Zohar, is currently being drained by three active producers, Zohar 4, 6, and 8. Because the reservoir pressure has fallen below the minimum pressure required to supply all the nearby industrial consumers, the gas transmission system will probably be redesigned and compressors installed. Conversion of the Zohar gas field to underground gas storage is also under study.

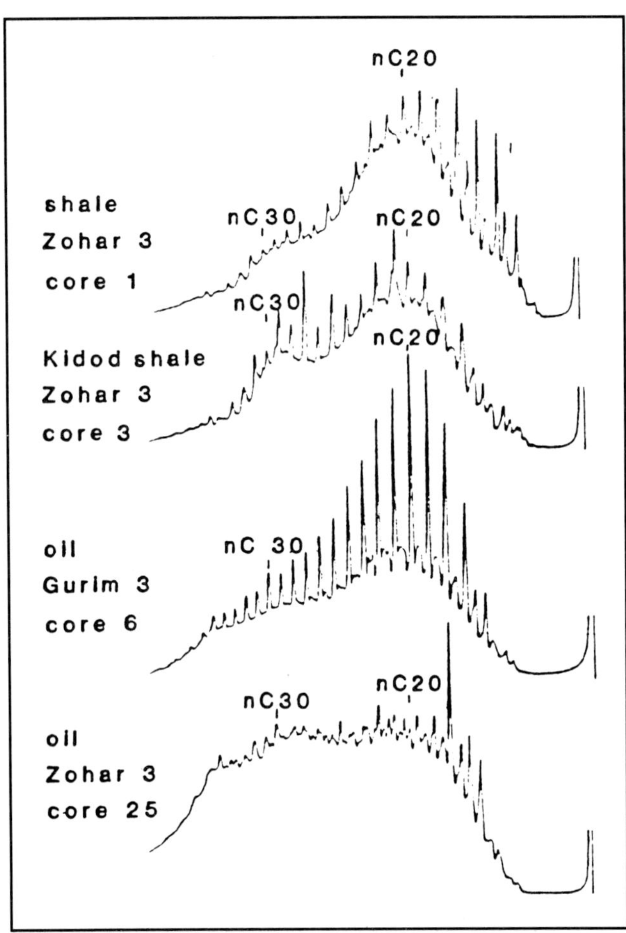

Figure 15. Gas chromatographs of saturated hydrocarbons in oils and the Kidod shales from the Zohar area. (From Amit and Bein, 1979b.)

The field can be converted into an excellent energy storage area with the additions of new wells and a gas-gathering system. For this purpose, a detailed computer model study of the Zohar field utilizing all available geologic and engineering data will be undertaken.

Findings in the boreholes on and close to the Zohar structural trend are encouraging to exploratory drilling for deeper targets (Triassic and Paleozoic); such drilling will be preceded by complementary seismic shooting. The supporting data for such exploration are:

1. Oil shows below the gas-producing zones in the tight Triassic Ra'af limestone and the Permian Sa'ad Formation (Zohar 8 well).
2. Recovery of 27° API gravity oil from the Ra'af limestone in the Massada 1 well (located just northeast of Haqanaim field).
3. Large-scale fractures developed in the lost circulation zones of the Ra'af Formation (Haqanaim 3 well).
4. The production of 9000 bbl oil from a sandstone of the Middle Triassic Gevanim Formation in the Zuk Tamrur 1 well.
5. The production of 9000 bbl of heavy oil (17° API gravity) from Lower Cretaceous sandstones in the Gurim 4 well before early water breakthrough occurred.

Small nearby structures should be tested for shallow gas targets to find additional reserves and increase the supply to the existing local industrial and domestic demand.

Table 4. Source rock and oil analysis.

Sample (well name)	Depth (m)	Resource	Formation	Organic Carbon (%)	Total Extractables (ppm)	Saturated hydrocarbons (%)	Aromatic hydrocarbons (%)	Resins + Asphaltenes (%)	Aromatic saturated hydrocarbons (%)
Zohar 1 Core 10	1036	Shales	Kidod	0.67	417	26.43	9.35	64.23	0.35
Zohar 3 Core 1	1065	Shales	Kidod	0.64	300	26.67	11.60	61.73	0.44
Zohar 3 Core 3	1140	Shales	Kidod	1.0	831	23.23	17.08	69.92	0.74
Zohar 6 Core 18	1350	Oil	Sherif	—	—	25.48	43.33	31.19	1.70
Zohar 6 Core 25	1350	Oil	Daya	—	—	27.79	38.01	34.20	1.37

From Amit and Bein, 1979b.

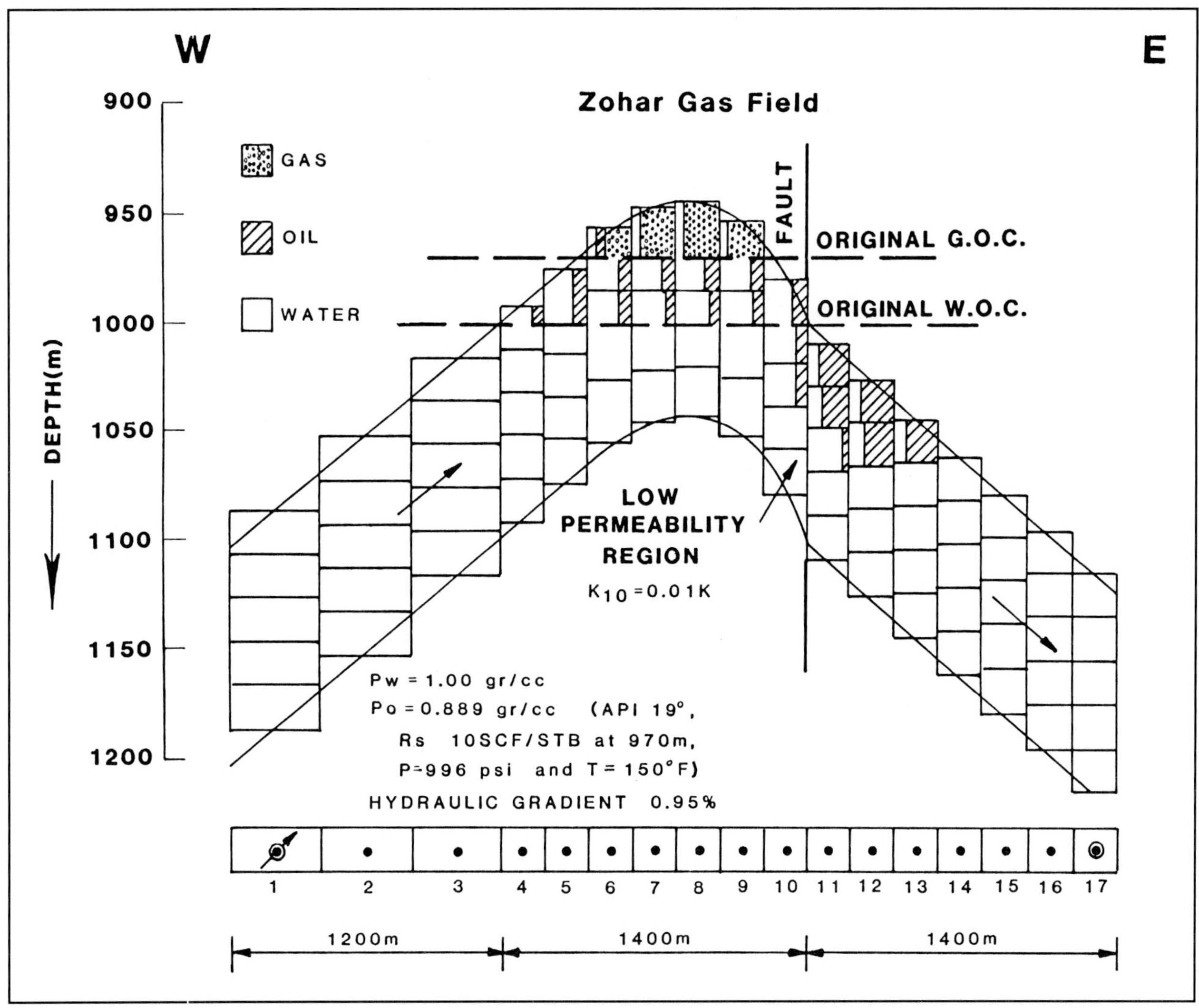

Figure 16. Model results of flushing of oil and gas in the Jurassic of the Zohar structure. (From Braester et al., 1987.)

ACKNOWLEDGMENTS

Mr. J. Fisher, Managing Director of the Israel National Oil Company Ltd. and Mr. E. Roih, Managing Director of Naphtha - Israel Petroleum Corporation Ltd. permitted the publication of the paper.

Unpublished studies and reports of Naphtha - Israel Petroleum Corporation Ltd. are the basis of the Zohar-Kidod-Haqanaim fields' presentation.

REFERENCES

Aharoni, E., 1964, The geology of the Rosh-Zohar anticline and west Hatrurim: Unpub. M.Sc. thesis, The Hebrew University, Jerusalem.

Aharoni, E., 1966, Oil and gas prospects of Kurnub Group (Lower Cretaceous) in southern Israel: AAPG, v. 50 (11).

Aharoni, E., 1969, Hydrocarbon prospects of the west side of the Dead Sea Graben, Israel: Israel Journal of Earth Sciences, v. 18, p. 137-141.

Aharoni, E., 1976, The surface and subsurface structure of southern Judean Desert, Israel: in Hebrew (unpublished).

Amit, O., and A. Bein, 1979a, The origin of the asphalt in the Dead Sea area: Journal of Geochemical Exploration.

Amit, O., and A. Bein, 1979b, The genesis of the Zohar gas as deduced from its chemical and carbon isotope composition: Journal of Petroleum Geology, v. 2, n. 1, p. 95-100.

Ball, M., and D. Ball, 1953, Oil prospects of Israel: AAPG Bulletin, v. 37, n. 1.

Barzel, A., and G. M. Friedman, 1970, The Zohar Formation (Jurassic) in southern Israel: a model of shallow water marine carbonate sedimentation: Israel Journal of Earth Sciences, v. 19, p. 183-207.

Bein, A., and O. Amit, 1980, The evolution of the Dead Sea floating asphalt blocks—simulation by pyrolysis: Journal of Petroleum Geology, v. 2, p. 429-447.

Bein, A., and G. Gvirtzman, 1977, A Mesozoic fossil edge of the Arabian plate along the Levant coastline and its bearing on the evolution of the eastern Mediterranean, *in* B. Biju-Duval and L. Montadert, eds., Structural history of the Mediterranean basins: Paris, Edition Technip, p. 95-110.

Bein A., S. Feinstein, Z. Aizenshtat, and Y. Weiler, 1984, Potential source rocks in Israel: a geochemical evolution: Geological Survey of Israel, Report GSI/17/84.

Bentor, Y. K., ed., 1960, Israel, *in* Lexique Stratigraphique International, v. III (Asie), Fasc. 10 C 2, Paris, CNRS, 151 p.

Bielski, M., 1982, Stages in the evolution of the Arabian-Nubian massif in Sinai: Ph.D. Thesis, Hebrew University of Jerusalem, Israel. 155 p. (In Hebrew, English abs.)

Braester, C., H. Fligelman, and E. Kashai, 1987, Hydrodynamic hydrocarbon accumulation in the southwestern flank of the Dead Sea Graben: Technion, Israel Institute of Technology.

Clarke, G. J., 1963, Gas reserves—Zohar field: Geopetrol study No. 12.

Coates, J., 1964, Geological aspects of gas reserve estimations in the Zohar region: Israel Journal of Earth Sciences, v. 13.

Coates, J., E. Gottesman, M. Jacobs, and E. Rosenberg, 1963, Gas discoveries in the Dead Sea region: Proc. 6th World Petroleum Congress, Frankfurt, Sec. 1, No. 26.

Derin, B., 1974, The Jurassic of central and northern Israel: Ph.D. thesis, Hebrew University, Jerusalem, 152 p. (In Hebrew, English abs.)

DeSitter, U., 1962, Structural development of the Arabian Shield in Palestine: Geologie en Mijnbouw, 41e Jaarg, p. 116-124.

Druckman, Y., 1974, The stratigraphy of the Triassic sequence in southern Israel: Geological Survey of Israel Bulletin No. 64, 92 p.

Druckman, Y., 1976, The Triassic in southern Israel and Sinai. A sedimentological model of marginal epicontinental marine environments: Ph.D. thesis, Hebrew University, Jerusalem, 188 p. (In Hebrew, English abs.)

Engel, A. E. J., T. H. Dixon, and R. J. Stern, 1980, Late Precambrian evolution of Afro-Nubian crust from ocean arc to craton: GSA Bulletin, v. 91, p. 699-706.

Eran, G., 1982, The geometry of the monoclines in the Negev: Unpublished M.Sc. thesis, Hebrew University, Jerusalem. (In Hebrew, English abs.)

Eshet, Y., 1987, Palynological aspects of the Permo-Triassic sequence in the subsurface of Israel: Unpublished Ph.D thesis, New York City University.

Ettinger, M., and Y. Langotzki, 1969, Hydrodynamics of the Mesozoic formations in the northern Negev, Israel: Geological Survey of Israel Bulletin No. 46.

Fleisher, E., 1987, The geochemistry of formation water in Israel: OEIL (Israel) Report 87/20.

Freund, R., M. Goldberg, T. Weissbrod, Y. Druckman, and B. Derin, 1975, The Triassic-Jurassic structure of Israel and its relation to the origin of the eastern Mediterranean: Geological Survey of Israel Bulletin No. 65, 26 p.

Garfunkel, Z., 1988, The pre-Quaternary geology of Israel, *in* Y. Yom Tov and E. Tchernov, eds., The zoogeography of Israel: Dr. W. Junk publ., Dordrecht, p. 7-34.

Garfunkel, Z., and B. Derin, 1983, Permian-early Mesozoic tectonism and continental margin formation in Israel and its implication for the history of the Eastern Mediterranean, *in* J. E. Dixon and R. H. F. Robertson, eds., The geological evolution of the Eastern Mediterranean: Geol. Society Special Publication No. 12, Blackwell, p. 187-202.

Garfunkel, Z., and B. Derin, 1988, Reevaluation of latest Jurassic-Early Cretaceous history of the Negev and the role of magmatic activity: Israel Journal of Earth Sciences, v. 37, p. 43-52.

Glattstein, B., 1976, Organo-chemical determination of maturation processes of the Gevar'am, Kidod and Inmar formations as potential source rocks: M.Sc. thesis, Hebrew University, Jerusalem. (In Hebrew.)

Goldberg, M., 1970, The lithostratigraphy of the Arad Group (Jurassic) in the Northern Negev: Ph.D. thesis, Hebrew University, Jerusalem, 137 p. (In Hebrew, English abs.)

Goldberg, M., and G. M. Friedman, 1974, Paleoenvironments and paleogeographic evolution of the Jurassic system in southern Israel: Geological Survey of Israel Bulletin 61.

Grader, P., 1957, New geological data on the Rekhme anticlinal crest: Research Council of Israel Bulletin, Biol. and Geol., v. 6B, p. 245-250.

Hubbert, M. K., 1953, Entrapment of petroleum under hydrodynamic conditions: AAPG Bulletin, v. 37, n. 8, p. 1954-2027.

Kisch, H. J., 1978, Coal ranks in Jurassic and Lower Cretaceous clastic rocks of the northern Negev and the southern Coastal Plain of Israel: Interpretation and implications of the maturity of potential petroleum source rocks: Israel Journal of Earth Sciences, v. 27, p. 23-35.

Marcus, E., 1982: Har Holed 1—Geological Completion Report: OE(I)L (Israel) Report 82/32.

Mimran, Y., 1976, Deep-seated faulting and the structures of the northern Negev (Israel). An analytic and experimental study: Israel Journal of Earth Sciences, v. 25, p. 111-126.

Monod, O., J. Marcoux, A. Poisson, and Y. F. Dumont, 1974, Le domaine d'Antalya, temoin de la fracturation de la plateforme africaine au cours du Trias: Bull. Soc. Geol. Fr. (7), 17, p. 116-125.

0E(I)L (Oil Exploration (Investments) Ltd.), 1988, Hydrocarbon potential of Israel—prepared for the Ministry of Energy & Infrastructure, Israel.

Picard, L., 1959, Geology and oil exploration of Israel: Bull. Res. Council, Israel, v. 6-8, p. 1-30.

Picard, L., and I. Z. Eliezri, 1964, Oil exploration of Israel: B'Olam Hadelek, n. 9-10, p. 77-104.

Salamon, A., 1987, Structural analysis of the reverse fault in the Zohar structure with respect to the "Zohar 9" drilling: Naphtha report.

Schlein, N., S. Salhov, and D. Spivak, 1980, Zohar 9—recommendation for drilling: OE(I)L (Israel) Report 80/25.

Spiro, B., D. H. Welte, J. Rullkotter, and R. G. Shaefer, 1983, Asphalts, oil and bituminous rocks from the Dead Sea area—a geochemical correlation study: AAPG Bulletin, v. 67, p. 1163-1175.

Starinsky, A., 1974, Relationship between Ca-Chloride brines and sedimentary rocks in Israel: Ph.D. thesis, Hebrew University, Jerusalem, 176 p. (In Hebrew, English abs.)

Tennenbaum, E., 1983, Researches in the geochemistry of oils and asphalts in the Dead Sea area: Ph.D. thesis, Hebrew University, Jerusalem, 135 p. (In Hebrew, English abs.)

Appendix 1. Field Description

Field name .... *Zohar-Kidod-Haqanaim fields*
Ultimate recoverable reserves .... *69 bcf*

Field location:

- **Country** .... *Israel*
- **Basin/Province** .... *Eastern Mediterranean*

Field discovery:

- **Year first pay discovered** .... *Zohar Formation (Zohar gas field) 1958*
- **Year second pay discovered** .... *Kidod gas field 1960*
- **Year third pay discovered** .... *Haqanaim gas field 1961*

Discovery well name and general location:

- **First pay** .... *Zohar 1, 41 km east of Beer Sheva*
- **Second pay** .... *Kidod 1, 4.5 km north of Zohar 1*
- **Third pay** .... *Haqanaim 1, 10 km northeast of Zohar 1*

Discovery well operator .... *Naphtha - Israel Petroleum Corporation Ltd.*

IP

- **First pay** .... *Zohar, 440–4300 Mcf/d*
- **Second pay** .... *Kidod, 2160 Mcf/d*
- **Third pay** .... *Haqanaim, 4270 Mcf/d*

All other zones with shows of oil and gas in the field:

Age	Formation	Type of Show
Triassic	*Ra'af*	*Oil, 21° API recovered in test*
Permian	*Sa'ad*	*Oil*

Geologic concept leading to discovery and method or methods used to delineate prospect

Surface geology and confirmation wells

Structure:

Province/basin type .... *Bally 1141; Klemme IIC a/b*

Tectonic history

Folding started in upper Cenomanian and continued to Miocene. Transverse faulting developed after Campanian, until late Pliocene. The folded area was affected also by the nearby graben collapse in the Dead Sea rift valley.

Regional structure

Field lies on a folded structural trend (Makhtesh Gadol-Zohar) west of the Dead Sea graben.

Local structure

Northeast-southwest and north-northeast-south-southwest asymmetric anticlines crossed by normal faults. Crests are broad with gentle northwest flank and narrow steep southeast flank. Steep flank is underlain by a reverse fault.

Trap:

Trap type(s)

Anticlinal trap with multiple pays (P, Q, and R limestone zones or units) and one sandstone pay of the Jurassic Zohar Formation.

Basin stratigraphy (major stratigraphic intervals from surface to deepest penetration in field):

Chronostratigraphy	Formation	Depth to Top in m
Senonian	*Mishash and Menuha*	*0*
Cenomanian-Turonian	*Judea Group*	*0-200*
Hauterivian-Albian	*Kurnub Group*	*500-700*
Oxfordian	*Beer Sheva*	*900-1100*
	Kidod	*900-1180*
Bathonian-Callovian	*Zohar*	*980-1280*
Bathonian	*Sherif*	*1100-1400*
Bajocian	*Daya*	*1350-1650*
Aalenian	*Inmar*	*1460-1760*
Lias	*Ardon*	*1580-1880*
Carnian	*Mohilla*	*1660-1960*
Ladinian	*Saharonim*	*1750-2050*
Anisian	*Gevanim*	*2090-2300*
Scythian-Anisian	*Ra'af*	*2130-2430*
Scythian	*Zafir*	*2230-2530*
	Yamin	*2500-2800*
Permian-Scythian	*Arqov*	*2600-2900*
Permian	*Sa'ad*	*2800-3100*
Infra Cambrian	*Zenifim*	*2980-3400*

Reservoir characteristics:

Number of reservoirs .... *3 limestone units (zones P, Q, R), 1 sandstone (Q)*
Formations .... *Zohar*
Ages .... *Jurassic*
Depths to tops of reservoirs .... *Zohar, 1000-1100 m; Kidod, 1200-1240 m; Haqanaim, 1150 m*
Gross thickness (top to bottom of producing interval) .... *Zohar, 110 m; Kidod, 130 m; Haqanaim, 95 m*

Net thickness—total thickness of producing zones

Average .... *Limestone, 70 m; sandstone, 0-5 m*

Lithology

Limestone: biopelmicrites with minor intrabiopelsparite, abundant quartz grains, interbeds, common argillaceous pellets and disseminated pyrites
Sandstone: medium-sized, rounded, subangular, little consolidated to dolomitic sandstone

Porosity type .... *Intercrystalline, porosity—in secondary calcite*
Average porosity .... *Limestone, 6.9%; in developed parts, 8.4%; Q sandstone, 25%*
Average permeability in limestone .... *0.01 to 0.15*

Seals:

Upper and lateral

Formation, fault, or other feature .... *Kidod Formation*
Lithology .... *Shales*

Source:

Formation and age

Unconfirmed, thus far can only suggest a possible source; Senonian beds buried in the Dead Sea graben

Lithology .... *Bituminous chalk and marl*
Average total organic carbon (TOC) *0.5%*
Maximum TOC *2.6%*
Kerogen type (I, II, or III) .... *II*
Vitrinite reflectance (maturation) .... *NA*

Time of hydrocarbon expulsion .... *Pliocene*
Present depth to top of source .... *6000-6500 m*
Thickness .... *150–250 m*
Potential yield .... *10^9 bbl of oil-generating capacity*

Appendix 2. Production Data

Field name .... *Zohar-Kidod-Haqanaim fields*

Field size:

Proved acres .... *Zohar, 1000 ha; Kidod, 500 ha; Haqanaim, 800 ha*
Number of wells all years .... *Zohar, 8 (7 prod.); Kidod, 3 (2 prod.); Haqanaim, 3 (1 prod.)*
Current number of wells .... *Zohar, 3 producers; Kidod, 1 producer*
Well spacing .... *Zohar, over 160 ac, in one case 40 ac; Kidod, 160 ac*
Ultimate recoverable .... *Zohar, 53 bcf; Kidod, 8.9 bcf; Haqanaim, 7.1 bcf*
Cumulative production .... *Zohar, 44 bcf; Kidod, 6.1 bcf; Haqanaim, 5.1 bcf (end 1988)*
Annual production .... *Zohar, 1.07 bcf; Kidod, shut in temporarily; Haqanaim, shut in*
Decline rates .... *NA*
Annual water production .... *20 m^3/producing well*
In place, total reserves .... *Zohar, 59 bcf; Kidod, 10.9 bcf; Haqanaim, 8 bcf*
Primary recovery .... *Zohar, 53 bcf; Kidod, 8.9 bcf; Haqanaim, 7.1 bcf*

Drilling and casing practices:

Amount of surface casing set .... *300–500 m*
Casing program

13⅜-in. or 9⅝-in. to 300–500 m; 9⅝-in. or 7-in. to 1050–1250 m; 7-in., 5½-in., or 2$^{2}/_{8}$-in. to T.D.; tubing 2⅜-in. or 2⅞-in.

Drilling mud .... *Fresh-water base*
Bit program .... *Varies*
High pressure zones .... *None*

Completion practices

Intervals perforated .... *Varies; each pay zone separately*
Well treatment .... *Acidized in limestone*

Formation evaluation:

Logging suites .... *Electric/SP, induction, microlog, GR, acoustic, neutron*
Testing practices .... *Typically production tested (mainly swabbing)*

Oil characteristics:

Type .... *Gas*
API gravity .... *Zohar and Kidod, 0.58; Haqanaim, 0.6 (air = 1)*
Viscosity, SUS .... *2.0 cp*

Field characteristics:

Average elevation .... *NA*
Initial pressure .... *Zohar, 879 psia (6060 kPa) at 1174 m; Kidod, 879 psia (6060 kPa) at 1294 m; Haqanaim, 998 psia (6880 kPa) at 1140 m*
Present pressure .... *NA*
Pressure gradient .... *0.0157 psi/ft (0.355 kPa/m)*
Temperature .... *125° F (51.6° C) ± 5%*
Geothermal gradient .... *NA*

Drive .... *Gas expansion*
Gas-water contact .... *Zohar, 632 m ssl*
Connate water .... *NA*
Water salinity, TDS *20,000–30,000*
Resistivity of water .... *0.15 ohm-m at 125° F (51.6° C)*
Bulk volume water (%) .... *NA*

Transportation method and market for oil and gas:

Pipeline system to industrial consumers

Ždánice-Krystalinikum Field—Czechoslovakia
Carpathian Foredeep, Moravia

J. J. KREJČÍ
Moravské naftové doly, a.s.
Hodonín, Czechoslovakia

FIELD CLASSIFICATION

BASIN: Carpathian Foredeep
BASIN TYPE: Cratonic Foreland
RESERVOIR ROCK TYPE: Crystalline Basement
RESERVOIR ENVIRONMENT OF DEPOSITION: Plutonic
RESERVOIR AGE: Precambrian
PETROLEUM TYPE: Oil and Gas
TRAP TYPE: Topographic-Tectonic

LOCATION

The Ždánice-Krystalinikum field is the largest oil and gas field producing from basement crystalline rocks in Czechoslovakia (Figure 1). It is located in the Ždánický les Highlands of Moravia at the junction of two geological provinces—the Bohemian Massif and the Western Carpathian nappe system (Figure 2). The reservoir is fractured and weathered crystalline igneous rocks of the Bohemian Massif. The seal was formed during late Miocene by tectonically emplaced upper Eocene through Oligocene pelites of the Ždánice-Subsilesian nappe. The Ždánice-Subsilesian nappe is in the outer part of the Western Carpathian nappes (Figures 3 and 4). The traps are paleotopographic highs of basement rocks that are segregated laterally by impermeable fault zones.

Geologically, the field is located in the outer flysch zone on the southeastern slope of the Bohemian Massif. Exploration and development are in progress and may extend the area of production beyond the current 12 km^2. The Ždánice-Krystalinikum field is operated by the corporation Moravské naftové doly (Moravian Petroleum Company), Hodonín. In-place total reserves for the Ždánice-Krystalinikum field at this time are 3.7 million MT of oil (27 MMBO) and 230 million m^3 (8.2 bcf) of gas. Other fields in the crystalline basement beneath the Carpathian nappes include East and West Kostelany, Koryčany, and West Zdánice fields (Figure 4).

HISTORY

Pre-Discovery

Early exploration in Moravia included gravity, magnetic, electric, and seismic surveys. The first gravity surveys that included the Ždánice-Krystalinikum area were conducted in 1941 by the German Association for Practical Research Investigation of Deposits (Berlin, Germany) and in 1944 by the Bohemian Office for Soil Research (Prague, Czechoslovakia). Běhounek (1957) compiled a Bouguer gravity map using measurements from both the 1941 and 1944 surveys. This map showed no anomaly in the Ždánický les Highlands area where the field was subsequently discovered (Figures 2 and 4).

A detailed gravity survey was conducted in the Ždánický les Highlands and surrounding area in 1964 by Odstrčil (1965). The resulting Bouguer residual map identified the Ždánice anomaly southeast of the village of Nížkovice. Additionally, an extensive west-east-trending gravity minimum south of the town of Bučovice was mapped and named the Rašovice depression.

Běhounek (1957) and Odstrčil (1965) acknowledge that unambiguous interpretation of gravity data is impossible. They state that although the gravity field is strongly influenced by the structure of the pre-Tertiary basement there is the difficulty that gravity

Figure 1. Location of Ždánice-Krystalinikum field in Czechoslovakia.

maxima and minima are also often associated with density variations in the basement rather than with the configuration of the basement surface. Differentiating those that result from structures (topographic) from those that result from density variations is very problematic. Additional geological and geophysical studies in the Moravian Province include Ibermajer and Doležal (1962), Dlabač (1962), Dlabač and Menčik (1964), Doležal (1962, 1964, 1965), and Adam et al. (1964).

Magnetic surveys in Moravia were conducted from 1942 to 1944 by Běhounek (1957). The entire area was resurveyed in 1962 by Mann et al. (1963), and again by Shánělec et al. (1964). The resulting vertical magnetic intensity mapping showed a large positive anomaly with the maximum near the village of Lovčice. This anomaly was interpreted as pre-Tertiary relief; however, subsequent drilling confirmed only the absence of high relief in the crystalline rocks.

Electric surveys were conducted using vertical electric sounding (measures resistivity and depths to different conductive layers). The main electrical horizon is the contact between the overlying low-resistivity rocks of the Ždánice-Subsilesian nappe and the underlying high-resistivity Precambrian crystalline rocks and Devonian carbonates. The contact was interpreted to be between 500 and 1000 m deep and did not appear to indicate a structural high in the Ždánický les Highlands area (Krejčí, 1974).

The first seismic survey in the Ždánický les Highlands area was conducted in 1964 when two refraction profiles, 66(R)/64 and 67(R)/164 (Figure 5), were shot using the refraction method (Fejfar et al., 1965). Line 67 (R) 64 was shot across the Ždánický les Highlands from southwest to northeast and indicated a large paleotopographic high under the Ždánice-Subsilesian nappe (Figures 6 and 7).

The first well drilled in the Ždánický les Highlands, 1965–1966, was the Žarošice #1 well (Figure 5), which reached a total depth of 2867 m and had a slight oil show between 1900 and 1980 m in Paleozoic sedimentary rocks. The second well, Žarošice #2 (Figure 5), drilled in 1966–1967, reached total depth at 1620 m and recovered gas with water in three zones in the Ždánice-Subsilesian nappe (Chmelík, 1969).

Two reflection profiles, 100/70 and 101/70, were shot in 1970 (Ciprys et al., 1971). Line 100/70 crosses profile 67(R)/164 near the southern edge of the Ždánice feature (Figure 5). The presence of Paleozoic rocks in the Bučovice #1 well, drilled in 1965, led to an interpretation that included Paleozoic rocks under the Ždánice-Subsilesian thrust along the whole profile (Figures 8A and 8B). The surface of the crystalline rocks was interpreted to be at a subsea depth of -1400 m where line 100/70 crossed 67(R)/64. The prospect on which the discovery well was drilled was delineated by seismic profiles 66(R)/64, 67(R)/164, 100/70, and 101/70 (Krejčí, 1971).

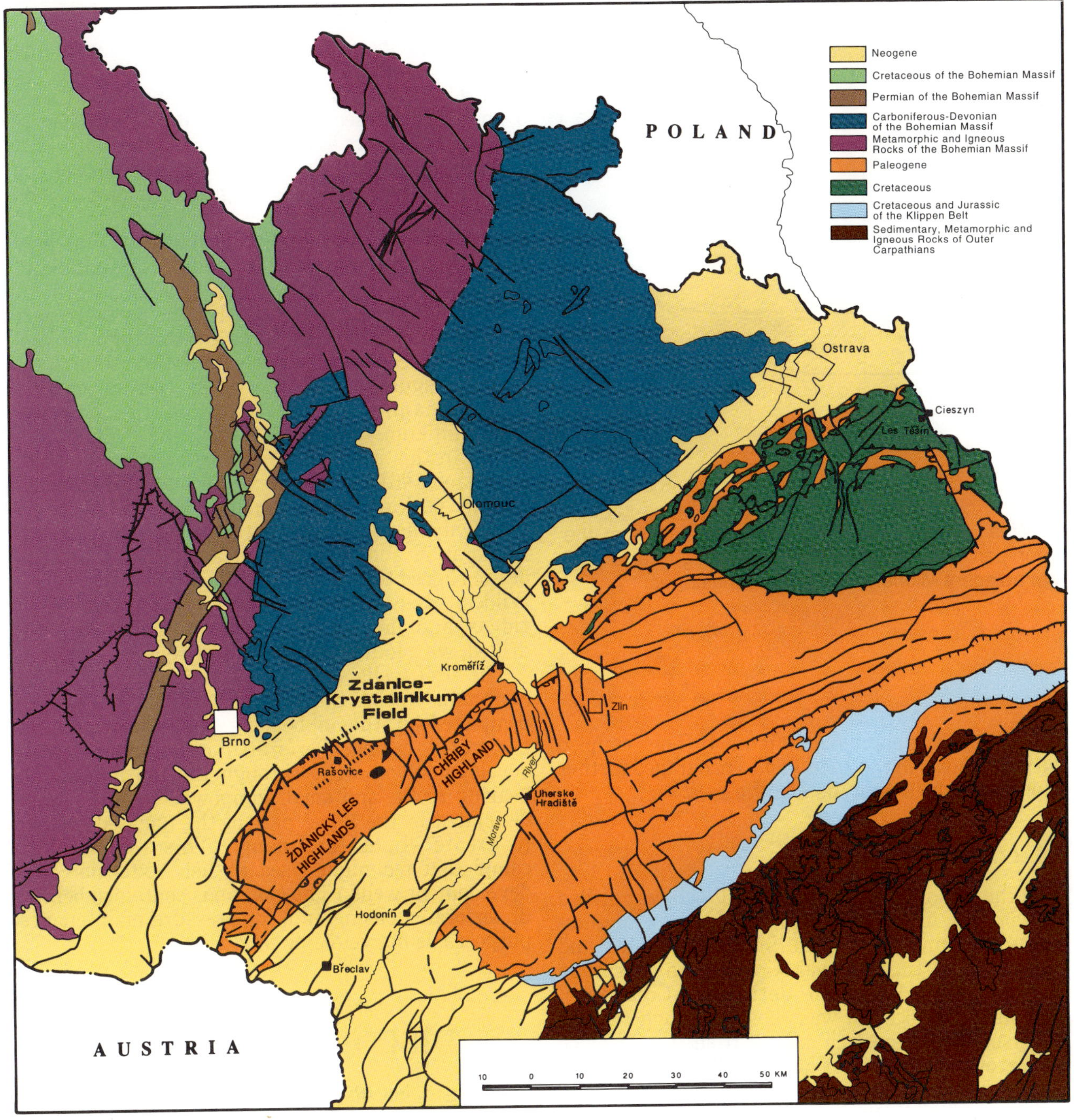

Figure 2 Geologic map of the central part of Czechoslovakia with the location of the Ždánice-Krystalinikum field in the Ždánický les Highlands area. The regional geologic provinces are distinguished on Figure 4.

Another seismic survey was conducted in 1971 (Cahelová et al., 1972), and these data were combined with the previous seismic data to delineate the prospects tested by the Ždánice #2 and #3 wells (Krejčí, 1972). Škárová (1971) postulated the presence of 2000 m of Paleozoic rocks, but this, too, was later disproved by drilling.

After the 1970-1971 seismic surveys, new wells were proposed for this area. The drilling of Ždánice #2 commenced at the end of 1972 but was a dry hole (Kostelníček, 1973). The Ždánice #3 well was drilled in 1973 and was also a dry hole. Oil and gas shows, however, in both the #2 and #3 encouraged drilling the Ždánice #1 in late 1973 at the location interpreted to be structurally high to the show wells. Production from sub-thrust Precambrian had been established in the foredeep in 1968 with the discovery of the East Kostelany field and in 1971 with the discovery

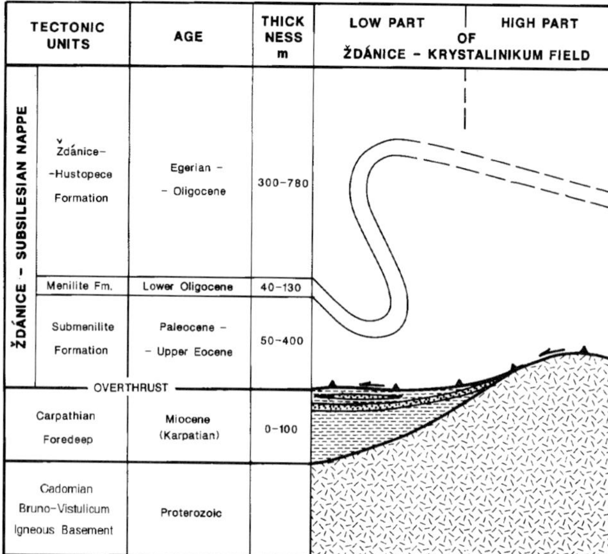

Figure 3. Generalized stratigraphic column for the Ždánice-Krystalinikum field in the outer flysch zone of the West Carpathians. The field produces from the upper part of crystalline topographic highs and also from adjacent Miocene Karpatian sandstones.

of the West Kostelany field (Figure 4). This, coupled with the shows in the Ždánice #3, encouraged further exploration drilling in the Ždánice area.

Discovery

At the end of 1973, the Ždánice #1 well was drilled and proved to be the discovery for the Ždánice-Krystalinikum field (Figure 5). It was drilled and completed by Czechoslovak Corporation Moravské naftové doly (Moravian Petroleum Company), Hodonín. The location was chosen through the use of seismic data and well control from the Bučovice #1, Žarošice #1 and Žarošice #2 wells. The prospect on which the discovery well was drilled was originally mapped by Krejčí (1971).

Paleozoic rocks were expected in the Ždánice #1 because of their presence in the Bučovice #1, and Žarosiče #1 and #2 wells. Prior to drilling, the base of the Ždánice-Subsilesian thrust was predicted at 830 m and the actual penetration was at 958 m. Underlying the thrust, 620 m of lower Carboniferous rocks and 300 m of Devonian rocks were expected but were absent. The discovery well was drilled in an area of the highest elevation of present-day topography in the Ždánický les Highlands (Figure 2) and is in the south-central part of the currently defined field. It had an initial production of approximately 17 m^3 (107 bbl) of oil per day pumping from open hole in igneous rocks from a depth of 964 to 990 m.

Post-Discovery

Initial development of the field was stepout drilling around the discovery well (Figure 9) beginning in 1974. The first development well, Ždánice #5 (Figures 9 and 12), intersected 87 m of oil-bearing igneous rocks. The Ždánice #11 well (Figures 9 and 12), drilled in 1976, discovered the gas cap for this part of the field. The Ždánice #76 well (Figure 9), drilled in 1986 to 950 m, located the highest buried topography of the crystalline rocks at -390 m (subsea). This part of the field was named the Ždánice-Krystalinikum #1 pool.

The Ždánice #30 well, drilled in the northern part of the field in 1980, discovered gas-bearing reservoir at the same elevation as the oil in the Krystalinikum #1 pool (Figure 9). This new pool, with its separate hydrodynamic system, is called the Ždánice-Krystalinikum #2 pool. Development drilling continues in this pool. In the eastern part of the field, the Ždánice #61 well (Figures 9, 10, and 13) discovered another pool in 1989, the Ždánice-Krystalinikum #3 pool. Only the central part of the field is well explored. Exploration and development are difficult because of the rugged topography and thickly forested areas. Wildcat wells are drilled mostly along mountain ridges and development wells are directionally drilled, with four or five wells from one location.

The first wells were completed open hole in the crystalline rocks. Subsequently drilled wells were completed in the oil-bearing zones using 5½-in. perforated casing. Perforated casing was set in development oil wells through the top half of oil-bearing rocks, and small gunpowder charges were detonated in the lower part of the uncased section to create slight fracturing. (This remains the completion technique used in field extension and development wells.) The gas cap is contained behind unperforated casing and, for pressure maintenance purposes, is not yet being produced.

DISCOVERY METHOD

The field was discovered through the construction of a structure map of the top of pre-Tertiary rocks using refraction and reflection seismic data (Figure 5). The contact between the Devonian carbonates and the Precambrian igneous rocks is not a good reflector. Reflection seismic surveys proved to be a very effective method for mapping the topography of the top of concealed crystalline rocks only where they are directly beneath Tertiary rocks (Figure 11). Overlying the crystalline rocks are thrusted Tertiary pelitic strata. Gravity surveys were not utilized extensively because mineralogical variations (density changes) in the crystalline rocks made interpretation very difficult. Magnetic surveys also did not work because of the wide range of variability in magnetic mineral composition of the Precambrian crystalline rocks.

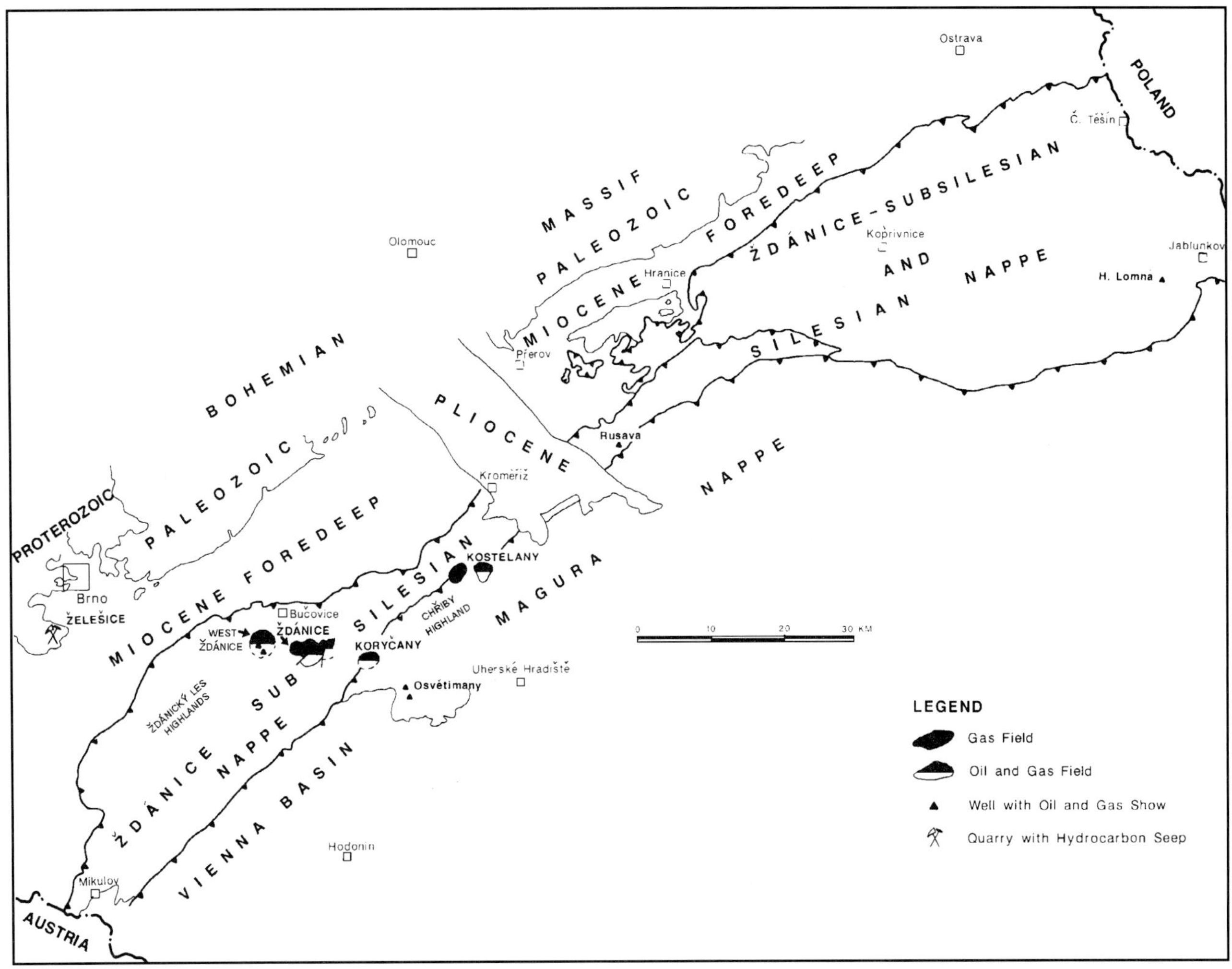

Figure 4. Fields that currently produce from crystalline basement beneath the Ždánice-Subsilesian and Magura nappes. Also shown are a quarry and wells with oil and gas shows.

STRUCTURE

Tectonic History and Regional Structure

The buried Precambrian hills of Ždánický les Highlands are located on the southeast flank of the Bohemian Massif, partly covered by the late early Miocene deposits of the Carpathian foredeep and overridden by the Ždánice-Subsilesian nappe of the Carpathian thrust belt. These hills are composed of crystalline rocks of the Bruno-Vistulicum terrane that were deformed and metamorphosed during the Cadomian orogeny (560 to 660 Ma, between the Proterozoic and Paleozoic). The sedimentary cover, consisting of Middle Devonian to Paleogene strata, was partly or completely removed from the Ždánický les Highlands by erosion prior to inundation by the Miocene sea and subsequent to Karpatian overthrusting of Carpathians.

The Ždánice-Subsilesian nappe conceals Eocene through Oligocene sediments, Paleozoic rocks, and Precambrian rocks. Topographic highs of Precambrian rocks in the Ždánický les Highlands were not inundated by the Karpatian sea and are consequently in direct contact with the base of the Ždánice-Subsilesian nappe. Thrust movement ceased between Karpatian and Badenian (early and late Miocene).

Regional Structure

The Precambrian surface dips to the southeast toward the Carpathian thrust-fold belt and includes thrust-covered paleotopography of knolls, ridges, and hills. One of the ridges with several paleotopographic highs is in the area of the Ždánický les Highlands and the Chřiby Highland (Figures 2 and 4). These basement highs are concealed by pelites of the Ždánice-Subsilesian nappe. On the west and northwest slopes of this thrust-covered ridge, Miocene

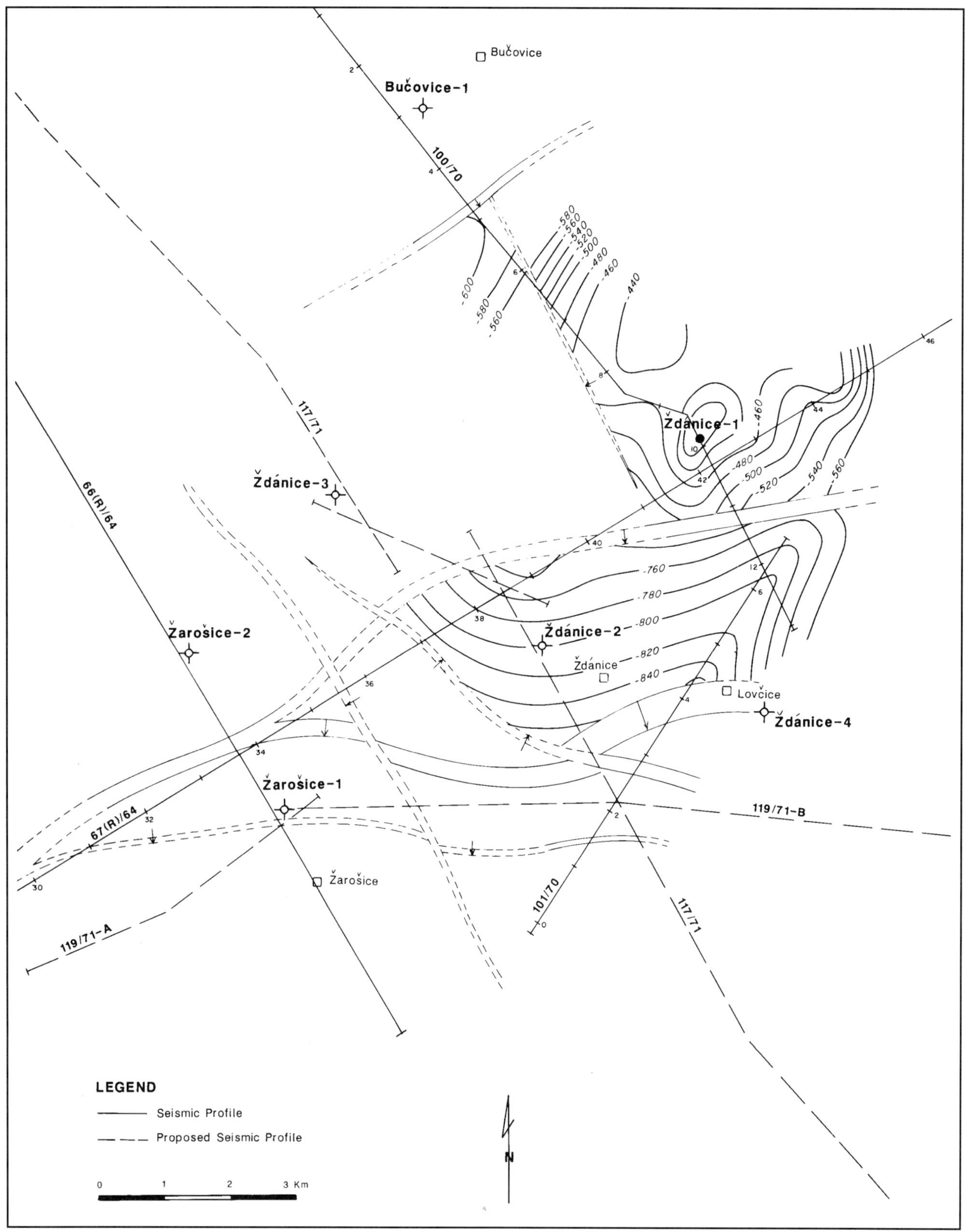

Figure 5. Structural interpretation (1971) of the pre-Tertiary relief in the Ždánický les Highlands area prior to the drilling of the Ždánice #1. Contour interval, 20 m. Depths are meters subsea. Existing seismic profiles and proposed seismic program are both shown. Three wells drilled in the 1960s are the Žarošice #1 and #2 and the Bučovice #1. The Ždánice #2 was drilled in 1972; the #3 was drilled in 1973; and the discovery well, Ždánice 1,was drilled near the end of 1973 and early 1974.

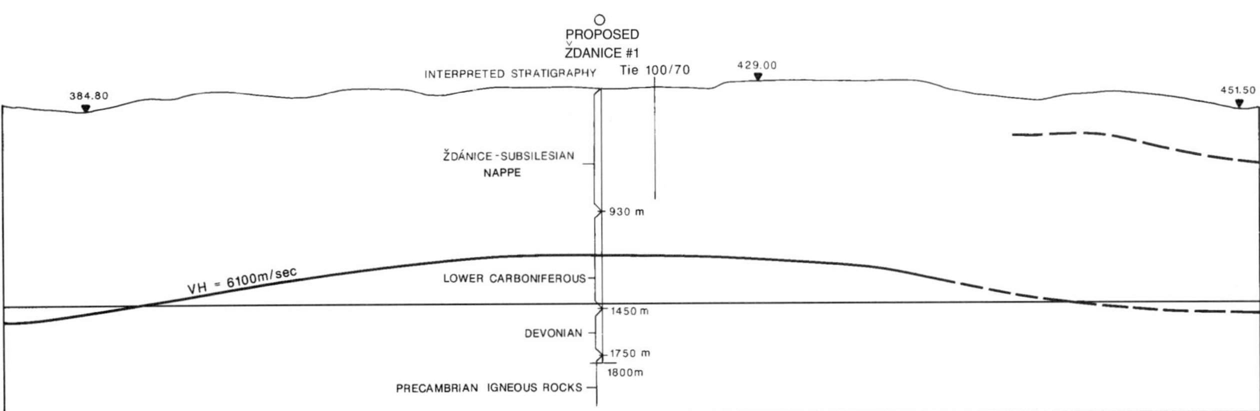

Figure 6. Interpretation of a 1965 refraction seismic line (67(R)/64) shows a high between 40 and 44 km. The Ždánice #1 well was drilled at 41.85 km. Depths of interpreted stratigraphy in meters below ground level. The refraction boundary (VH = 6100 m/sec) was interpreted to be the top of the carbonates. (See Figure 5 for location of profile.) Elevations of surface in meters; depths are below surface; horizontal line is -1000 m datum.

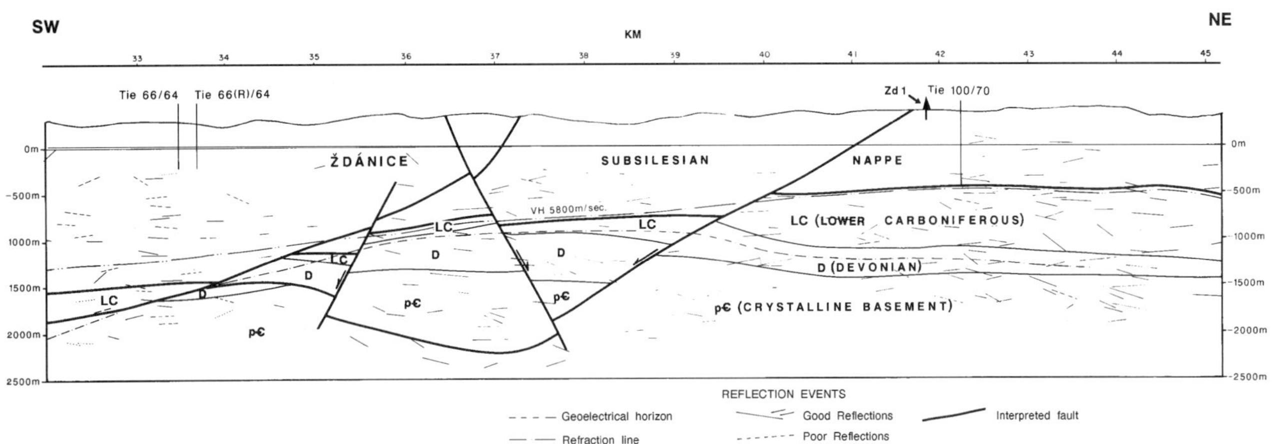

Figure 7. In 1971, refraction line 67(R)/64 was reinterpreted (Krejčí, 1971) showing more structural detail. A refraction velocity of VH = 5800 m/sec replaced the older version shown on Figure 6. A geoelectric horizon (refer to text) is shown along the profile as well as the proposed location of the Ždánice #1 (Zd 1) well at 41.85 km. (See Figure 5 for location of profile.) Paleozoic rocks were interpreted on this section because of their presence in the Bučovice #1 well drilled in 1965.

sediments have been preserved between the Precambrian rocks and the thrust. In places some Devonian carbonates and basal clastics have been preserved.

Local Structure

In Czechoslovakia, four fields produce from buried Precambrian hills along the northwestern fringe of the Carpathian fold belt. These range in size from 2.5 to 12 km^2. The Ždánice-Krystalinikum field is the largest of these Precambrian reservoir fields. It is an irregular, elongate, east-west-trending buried ridge of Precambrian rocks with two paleotopographic highs. About 6 km long and a maximum of 3.5 km wide, it has 60 to 180 m of topographic relief.

STRATIGRAPHY

There are three main stratigraphic units in the Ždánice Krystalinikum field: the Precambrian basement, the Miocene sediments overlying basement on the west and northwest part of the field,

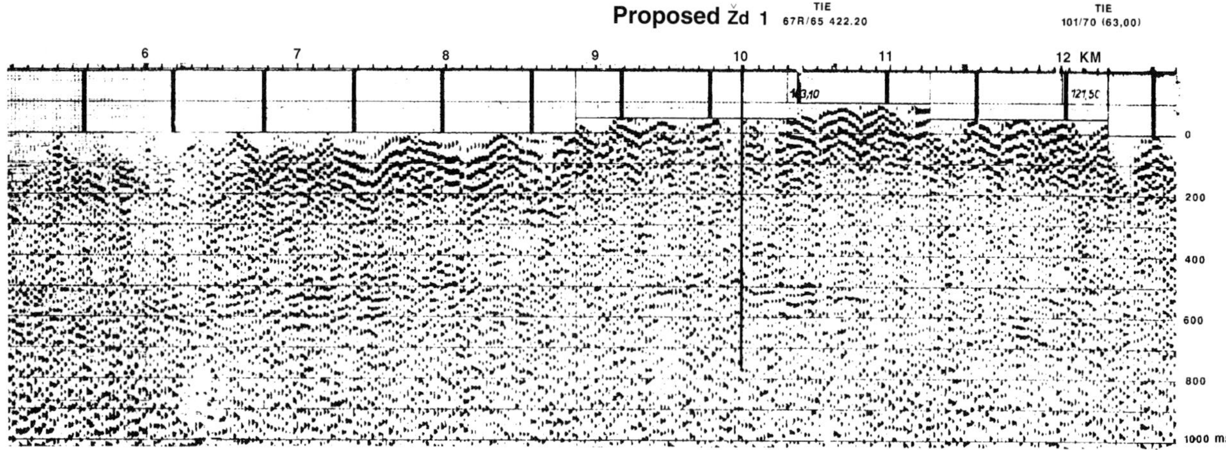

Figure 8A. Seismic line 100/70 shot in 1970 shows very poor reflections. At shot point 10 where the Ždánice #1 well was drilled in 1973, there are slight reflections at a depth of 500 ms. (See Figure 5 for location of profile.)

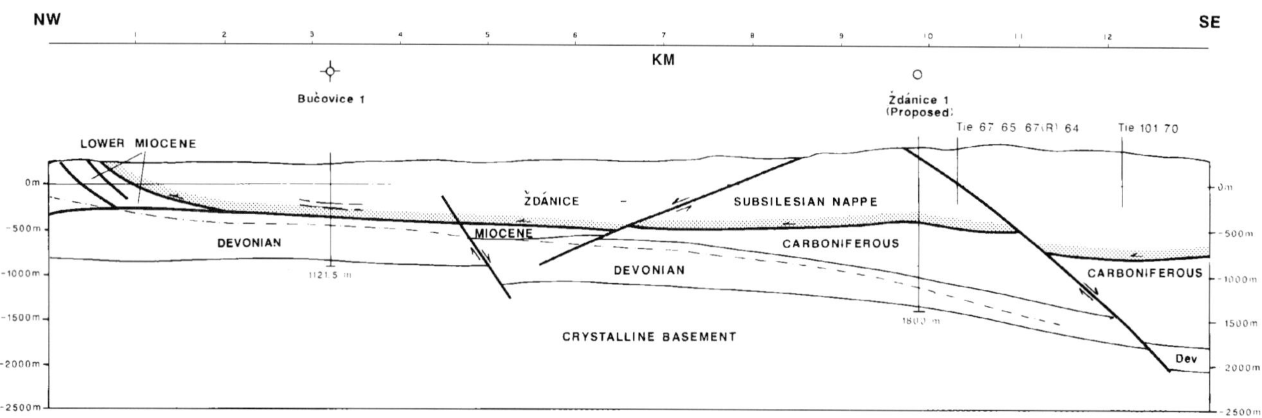

Figure 8B. Pre-drilling interpretation (Krejčí, 1971) of line 100/70 projected Carboniferous and Devonian rocks beneath the Ždánice-Subsilesian nappe, but these older rocks were absent. Devonian rocks had been found in the Bučovice #1 well, drilled in 1965. (See Figure 5 for location of line of profile.) Arrows indicate fault movement directions and stippled zone is the base of the interpreted décollement.

and the upper Eocene through Oligocene sediments within the Ždánice-Subsilesian nappe.

The basement is Precambrian (560 to 660 Ma; Dudek, 1980) and highly variable lithologically. These rocks belong to the Bruno-Vistulicum tectonic event and were formed at the close of Cadomian orogeny (Figure 3). Rock types include amphibole-biotite, granodiorite, diorite, and quartz diorite. Granite, granitic aplite, and meta-andesite are locally present. These rocks crop out near Brno, 25 km northwest of the Ždánice-Krystalinikum field.

Paleozoic rocks are absent in the Ždánice-Krystalinikum field but are preserved to the northwest. At the beginning of the Middle Devonian, part of the Ždánický les Highlands area was probably a depression, and basal clastic sediments of Middle Devonian (Givetian) age were deposited.

During the Miocene, the Rašovice depression formed on the Carpathian foredeep north and northwest of the Ždánice-Krystalinikum field. Sedimentation in the central part of the Rašovice depression began with the deposition of freshwater mudstones. Unsorted basal clastic sediments, including clasts of the underlying igneous rocks and occasional thin coal beds, are interbedded with mudstone. In the uppermost part of this section and on the slopes of the crystalline highs, the strata are dominantly sandstone and commonly directly overlie the crystalline basement. The sandstones are typical strand sands (Zádrapa and Krejčí, 1983). Claystone conglomerates underlie and are interbedded with the sandstones. The conglomerates are composed mainly of Oligocene claystone boulders with Karpatian (Miocene) claystone and sandstone matrix. The Oligocene claystone boulders are oval and do not appear to have been transported very far. These sediments are not autochthonous to the area of the Ždánice high. Cores from the base of the nappe show that these lower strata are tectonically very strained and served as a "lubricant" during thrusting.

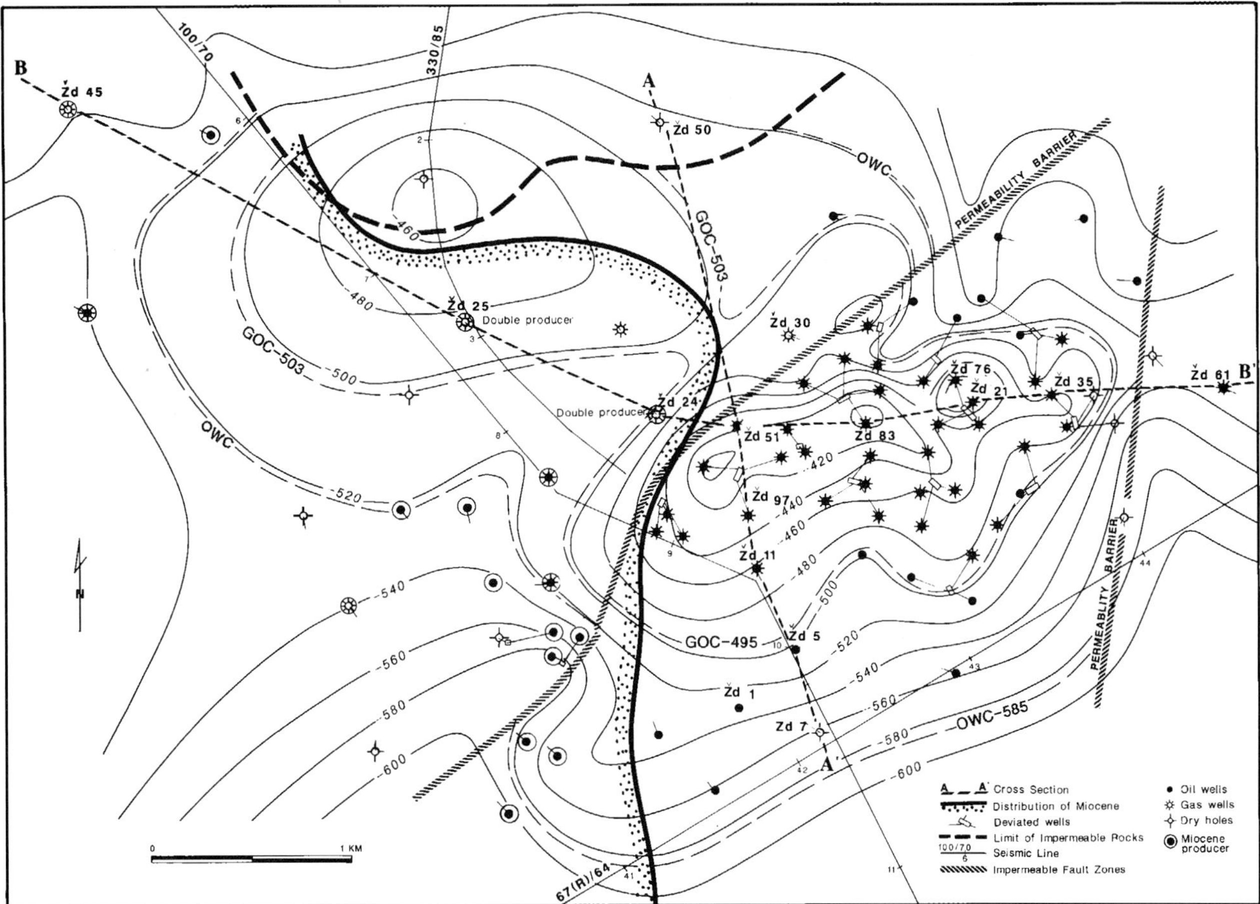

Figure 9. Structure map of the top of the basement in the northwestern part of the Ždánický les Highlands. The map was made after the discovery well and many development wells had been drilled. The eastern limit of Miocene sediments between the thrust and the crystalline basement is shown by the stippled pattern. The dashed line shows the area of impermeable crystalline rocks on the north flank of the field. Production from Miocene sandstones on the west side of the field is indicated by circled well spots. The Ždánice #61 recovered gas in the upper part of the crystalline basement and oil in the lower part. The oil-gas contact is at -600 m and the oil-water contact is unknown.

The Ždánice-Subsilesian nappe contains three formations in the Ždánický les Highlands: the Submenilite, Menilite, and the Ždánice-Hustopeče (Figure 3).

The Upper Eocene-Paleocene Submenilite Formation, found in the lower part of the Ždánice-Subsilesian nappe, contains mainly pelitic rocks interbedded with thin beds of calcareous sandstones. At the base of the nappe, it sometimes includes Oligocene and Miocene sediments of autochthonous origin that have been tectonically incorporated into the nappe.

The lower Oligocene Menilite Formation is also a pelitic sequence interbedded with chert layers that resulted from temporary coolings of sea water during deposition.

The Oligocene Egerian (Chattian) Ždánice-Hustopeče Formation, in the upper part of the nappe, is characterized by alternation of nonflysch and flysch, medium to coarse, rhythmic sequences of calcareous, gray claystones, and calcareous sandstones. The latter range from thin- to thick-bedded, are micaceous, and have irregular and convolute lamination and curvilaminar jointing.

The hydrocarbons trapped at Ždánice-Krystalinikum are believed to have migrated from Upper Jurassic marls that lie to the southeast beneath the Carpathian nappes.

TRAP

Trap Description

The Ždánice-Krystalinikum field is a unique trap—not because the principal reservoirs are in crystalline rocks but because they are sealed by the thrust sheet that overlies them, the Ždánice-Subsilesian nappe of Czechoslovakia. Thus the field is a combined topographic and tectonic trap. Buried topographic hills are not unique as reservoirs; however, a

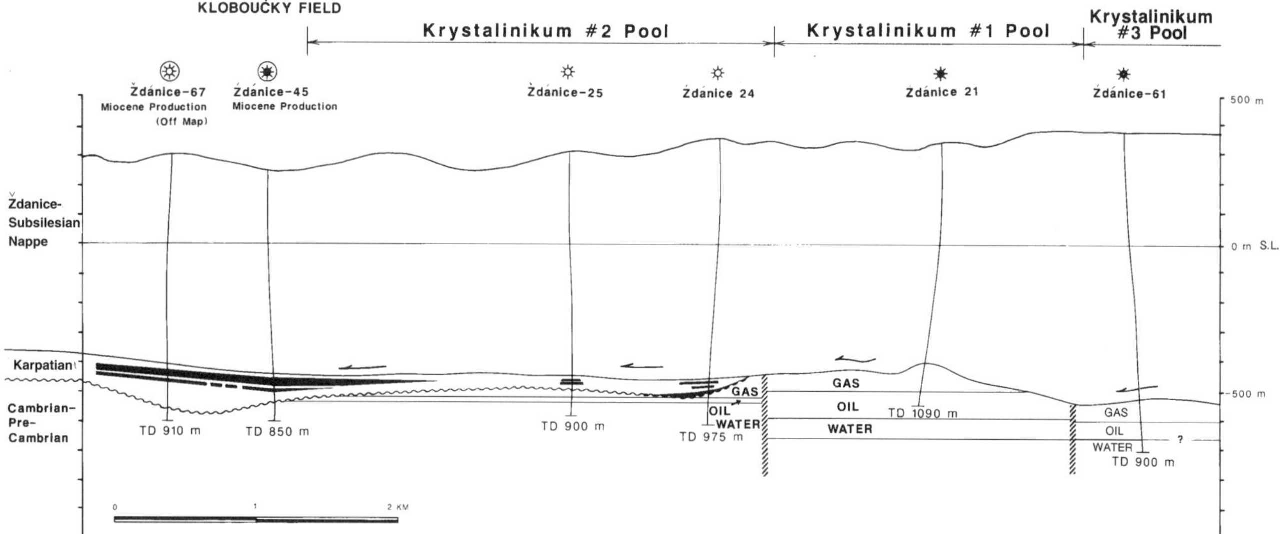

Figure 10. Cross section B-B′ across the Ždánice-Krystalinikum field showing the three separate pools with different oil-water contacts. Black zones in the Karpatian are productive sandstones. (See Figure 9 for location of cross section.) The cross-hatched vertical bars represent permeability barriers that are the lateral seals segregating the three pools. These are thought to be mylonized fault zones. Arrows at the base of the nappe indicate direction of thrusting.

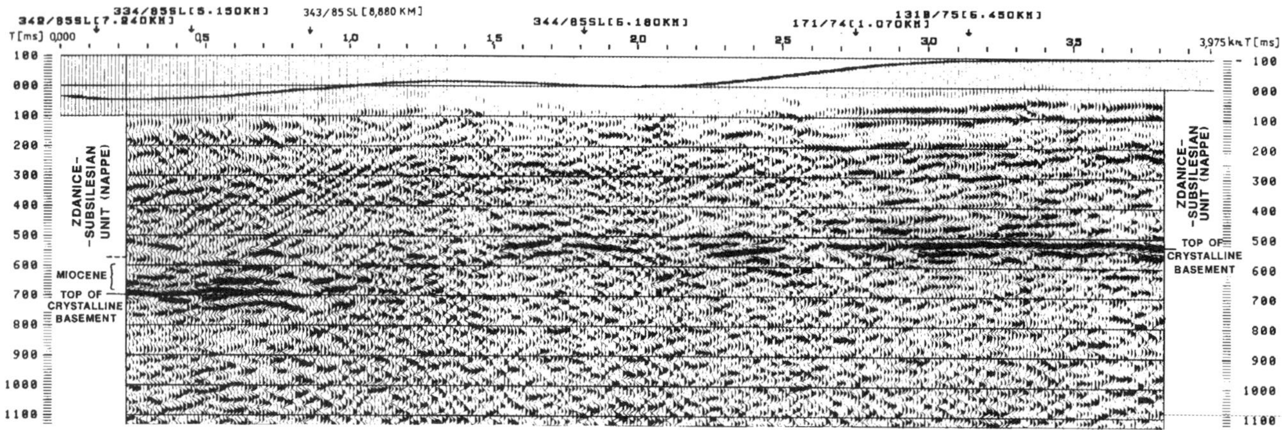

Figure 11. Seismic reflection line 330/85 (1985) shows the top of the crystalline basement on the right, but the contact between the Miocene sediments and the Ždánice-Submenilite nappe is not evident on the left. Between 1.0 and 1.5 km, the top of the basement also is not evident; here the basement may be very weathered or disaggregated. (See Figure 9 for location of profile.) The 1985 seismic data are much better quality data than the 1960–1970 data.

tectonically emplaced seal for these unusual reservoirs *is*. The main trap is a buried topographic hill in which the reservoir was developed through a combination of fracturing and weathering. The primary seal of overthrust sediments was tectonically emplaced, and secondary seals were formed by both tectonic and diagenetic mechanisms. The overlying and sealing pelites of the Submenilite Formation were emplaced by thrust action. Some peripheral seals were formed both from cataclasis along fault zones and from subsequent altering of the fault gouge material to clay. These zones of altered fault gouge cross the field and divide it into three segregated (independent) pools, the Krystalinikum #1, #2, and #3 pools (Figures 9, 10, 12, and 13).

The Krystalinikum #1 pool is the site of the initial discovery in the Ždánice-Krystalinikum field. This pool is in the central part of the field and has the

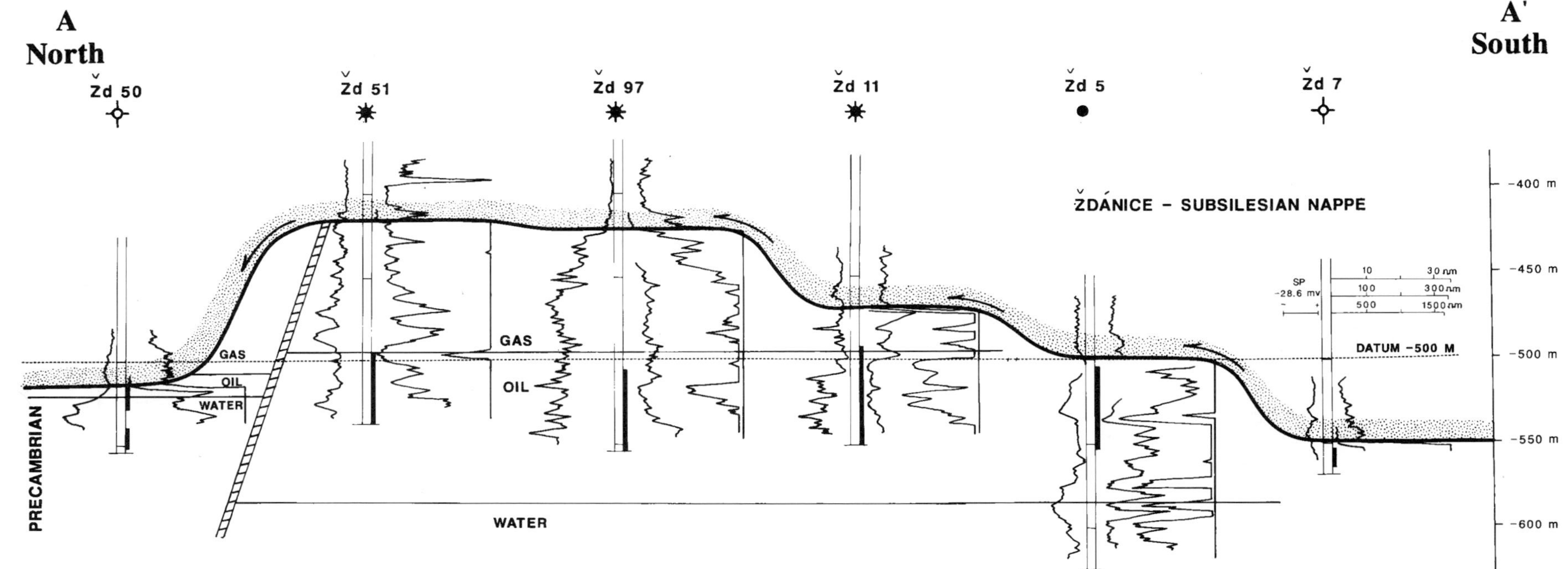

Figure 12. North-south electric log structural cross section showing Pools 1 and 2 of the Ždánice-Krystalinikum field (see Figure 9 for line of section). Depths in meters subsea. As in Figure 10, the cross-hatched bar near the northern end is a permeability barrier, probably a mylonized fault zone; pool #2 is north of pool #1 and this barrier. Arrows indicate direction of thrusting. The stippling represents the contact of the sealing nappe with the underlying reservoirs.

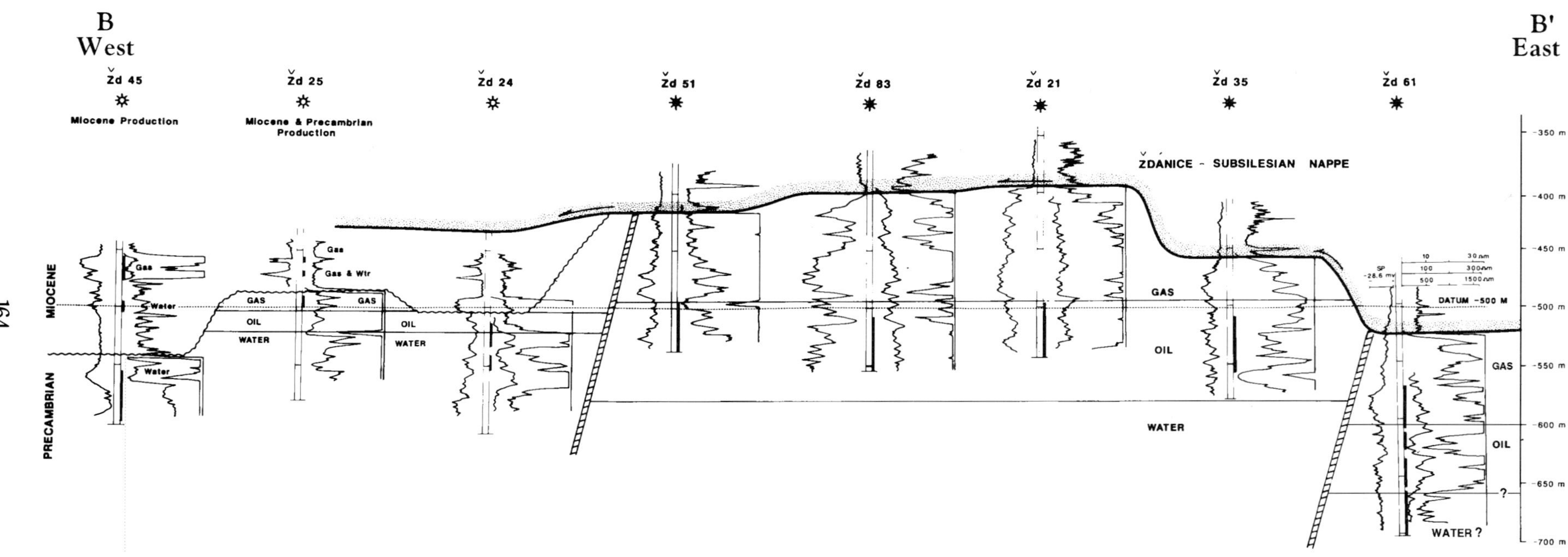

Figure 13. West-east electric log structural cross section showing pools 1, 2, and 3 of the Ždánice-Krystalinikum field. (See Figure 9 for line of section.) Depths in meters subsea. Symbols as in Figure 12. Pool #2 is west of the permeability barrier on the west side. Pool #3 is east of the eastern barrier.

thickest reservoir rocks. The igneous pay zone ranges from -390 m to -585 m (subsea) (Figure 9). The top of the pay zone occurs at drill depths that range from 720 m to 980 m. The Krystalinikum #1 pool reservoir ranges up to 205 m thick, with a gas cap comprising the upper 115 m. The gas-oil contact is -495 m and oil-water contact is -585 m (subsea). Water is always found below the oil in this reservoir.

The Krystalinikum #1 pool is bounded by impermeable zones on the northwest and east that are believed to be altered fault zones. The elongate, northeast-southwest-trending pool is 4.5 km long and 2 km wide. To the northwest, on the other side of the bounding fault zone, is the Krystalinikum #2 pool. To the east, on the other side of another fault zone, is the newly discovered Krystalinikum #3 pool.

The Krystalinikum #2 pool is the northwestern part of the Ždánice-Krystalinikum field (Figures 10 and 13). Drill depth to the top of the reservoir is 750 to 870 m, depending on the topographic relief of the crystalline rocks and surface relief. Subsea depths of this pool are from -460 m to -525 m.

The Krystalinikum #2 pool has a very thin oil column (about 18 m) and produces primarily gas. The maximum thickness of the gas cap is about 45 m. The gas-oil contact is -503 m. The northern limit of this pool has not yet been defined, but two wells to the north penetrated dense, unfractured, and impermeable rocks on the highest elevations of the basement topography. As illustrated by Figures 9, 10, and 13, unnamed Miocene sandstones are minor reservoirs in the lower part of Ždánice-Krystalinikum field on the west flank. Exploration to the northwest is currently in progress.

The Krystalinikum #3 pool is the easternmost part of the Ždánice-Krystalinikum field and was discovered in 1989 with the drilling of the Ždánice #61 well, which flowed gas and oil (Figures 9, 10, and 13). The oil-water contact in the Ždánice #61 well has not yet been determined. The gas-oil contact is approximately -660 m (subsea). Seismic data appear to show that the crystalline basement "high" continues to the northwest for about 3 km, and exploration and development are expected to continue in this direction.

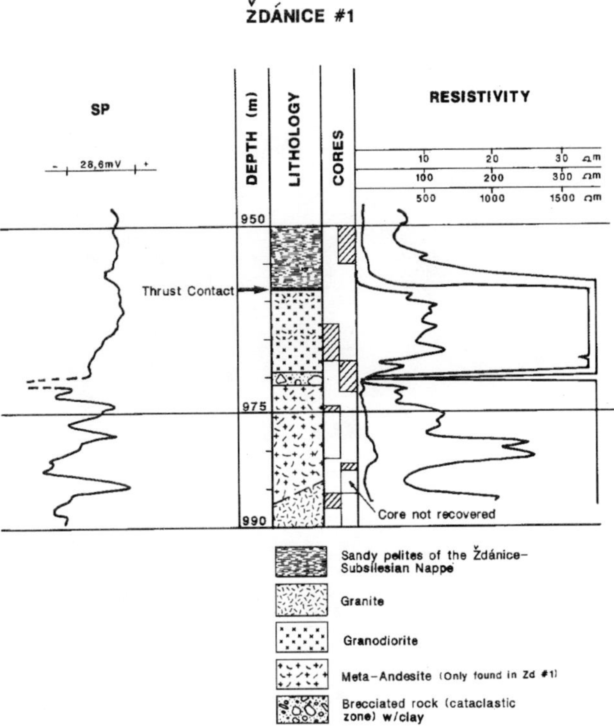

Figure 14. Resistivity/SP log from the Ždánice #1 discovery well, with core descriptions showing the variation of igneous rocks from the well bore. The density variation and complexity of the basement rocks make interpretation of gravity data very difficult.

Reservoir

The main reservoir in the Ždánice-Krystalinikum field, the fractured Precambrian rocks, varies in composition and includes amphibole-biotite, granodiorite, diorite, and quartz diorite (Figures 14, 15, and 16). Granite, granitic aplite, and meta-andesites are also locally present.

The reservoir rock is brecciated, with abundant cataclastic and mylonitic texture. It is weathered at the top. The intensity of deformation of the reservoir varies from well to well and strongly affects the reservoir quality in each well. Reservoir quality is also vertically variable in each well bore. Approximately 25% of the reservoir zone is devoid of any porosity and permeability. Three-fourths of the reservoir has porosity that ranges up to 14.7% and permeability up to as much as 22.25 md. The permeability and porosity of the rocks are dependent on the extent of fracturing and weathering, but the extremely weathered rocks are usually impermeable. These rocks contain a lot of clay that plugs the vugular porosity. The best reservoir rock has a combination of joint, vuggy, and intercrystalline porosity that was created by weathering and tectonic episodes.

The initial reservoir pressure is lower than normal hydrostatic pressure at that depth. In the middle of the oil zone (-540 m), the initial reservoir pressure in the Krystalinikum #1 pool is 7.43 MPa, and the temperature is 27.4°C. In the Krystalinikum #2 pool, the pressure at a depth of -487 m was measured at 6.69 MPa, and the temperature is 28°C. In the Krystalinikum #3 pool the pressure and temperature as measured at -606 m below sea level were 7.84 MPa and 29.5°C.

Oil and Gas Characteristics

The composition of the oil in the Krystalinikum #1 pool varies from well to well and varies even from different zones in the same well bore. It ranges from a light crude oil of 0.8579 g/cm^3 to a very heavy crude

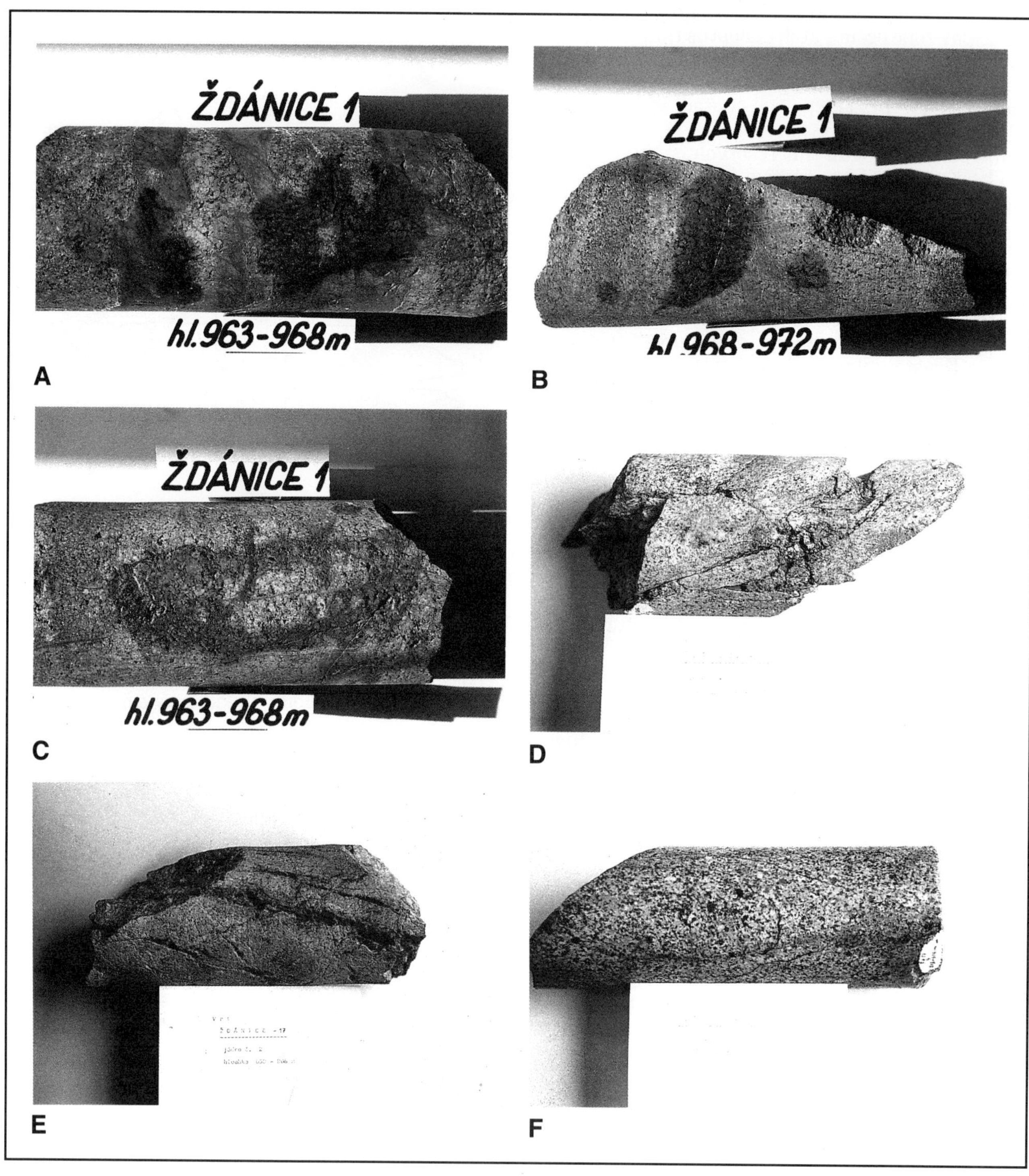

Figure 15. Photographs of cores. (A) and (B) Ždánice #1 well, core 7, depth 963-968 m. Medium-grained biotite granodiorite to leucocratic granite. There are small veins and irregularly oriented crystals of aplite in the granodiorite. The rock also contains irregular, strongly mylonitized red veins: Hydrothermal metamorphism decomposed biotites and feldspars. The fractures contain dark oil stains. (C) Ždánice #1 well, core 8, depth 968-972 m. Medium-grained biotite granodiorite with dark oil stains along the fractures. (D) Ždánice #15 well, core 5, depth 910-915 m. Strongly fissured and disaggregated biotite granodiorite. The rock is partly very weathered and has oil saturation in the fractures. (E) Ždánice #17 well, core 2, depth 858-868 m. Strongly weathered and tectonically brecciated, fine- to medium-grained biotite granodiorite saturated with oil. Note oil saturation along the fracture. (F) Ždánice #17 well, core 4, depth 913-918 m. Tight, unweathered, medium-grained biotite granodiorite with slight oil staining.

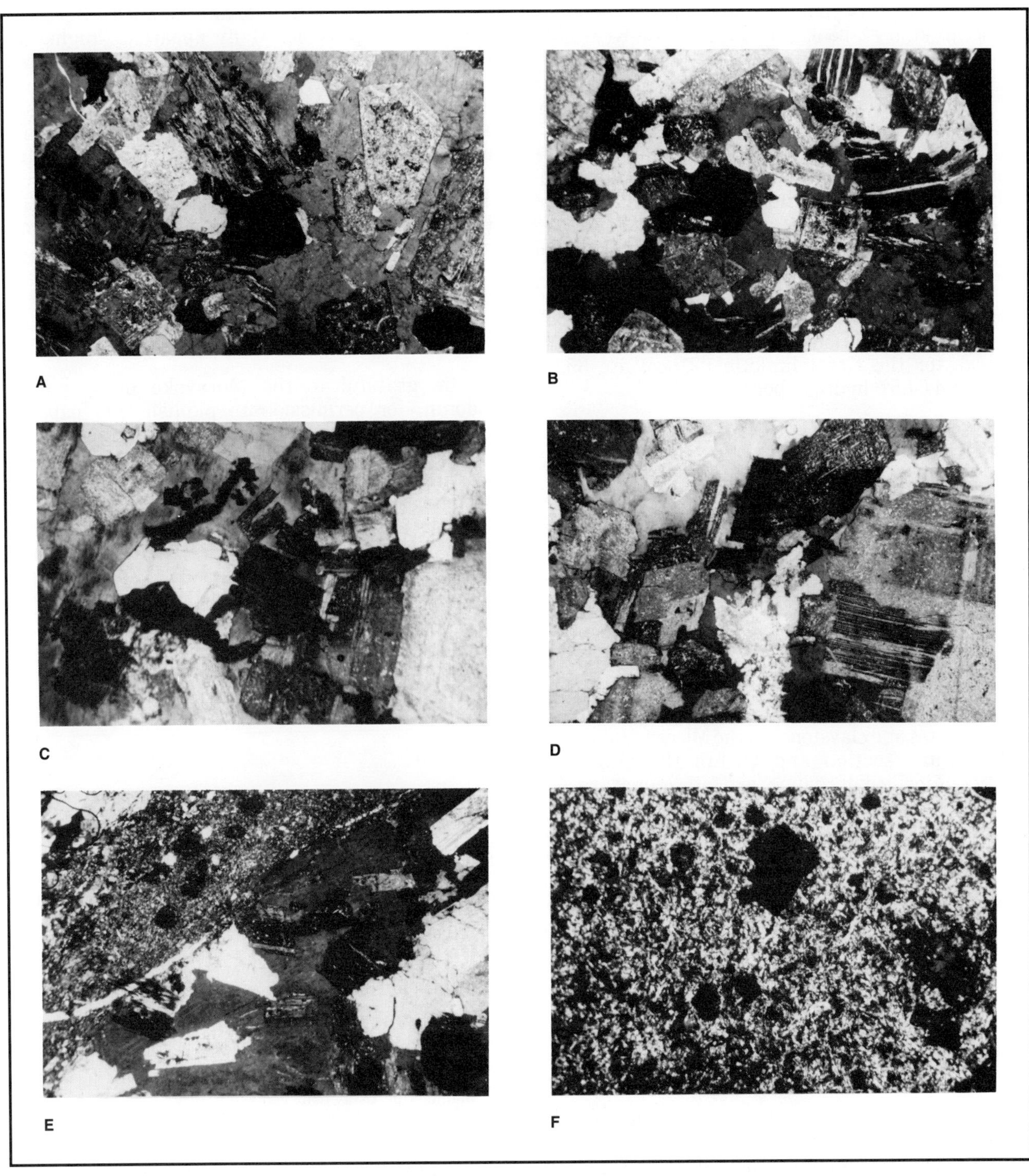

Figure 16. Photomicrographs of crystalline rocks from the Ždánice-Krystalinikum field. (A) and (B) Ždánice #61 core, 934-936.3 m, crossed nicols, width of view 4.1 mm. Biotite granodiorite with subautomorphic poikilitic texture. (C) and (D) Ždánice #18 core, 960-965 m, crossed nicols, width of view, 4.1 mm. Leucocratic granodiorite with poikilitic texture. (E) Ždánice #92 core, 905-910 m, crossed nicols, width of view 4.1 mm. Mylonitized microzone with fine mortar texture in biotite granite. (F) Ždánice #12 core, 919-924 m, crossed nicols, width of view 4.1 mm. Strong hydrothermal alteration of dioritic porphyry with chloritic pseudomorphs after mafic minerals.

of 0.9436 g/cm³ (33° to 19° API). The gas is 90% methane and the other lighter hydrocarbons. The oil from the Krystalinikum #2 pool and overlying sandstones of Karpatian age is heavy, its gravity ranging from 0.9112 to 0.9594 g/cm³ (24° to 16° API). The gas from this pool is similar to that from the Krystalinikum #2 pool. The oil from the Krystalinikum #3 pool also is heavy, 0.9401 g/cm³ (19° API). The gas is 91.5% methane, 4.2% ethane and the other lighter hydrocarbons (up to pentane), 1.5% nitrogen, and 1.6% carbon dioxide.

Oil produced from the Ždánice-Krystalinikum #1 pool is paraffinic-naphthenic and has no sulfur. Oil viscosity at 20°C is estimated at 40 MPa. Initial solution gas-oil ratio was approximately 24.48 m³/m³ (137 ft³/bbl). Initial fluid saturations calculated from logs are 32.5% water and 67.5% hydrocarbons for the Krystalinikum #1 pool and 48% water and 52% hydrocarbons for the Krystalinikum #2 pool. Initial fluid saturations, determined by the Messer method, for the Krystalinikum #2 pool are 35.5% water and 64.5% hydrocarbons.

Reservoir Water

Water underlies all of the oil-bearing crystalline rocks except in impermeable intervals. Chemical analysis of the water in the Krystalinikum #1 and #2 pools shows it is very similar. It is strongly mineralized (12 g/L of sodium chloride type and calcium chloride subtype). The water contains 10 to 12 mg/L iodine and 32 to 37 mg/L bromine.

Source Rocks

The pelites and claystones in the Miocene subthrust sedimentary section and within the Oligocene Ždánice-Subsilesian nappe are all immature for hydrocarbon generation. As stated in the *Stratigraphy* section, hydrocarbons are believed to have migrated from Upper Jurassic marls that lie to the southeast beneath the Carpathian nappes. These marls are restricted marine basin facies and have been identified as the main source rocks for oil, gas, and condensate in the Vienna basin (Ladwein, 1988).

DEVELOPMENTAL AND EXPLORATION CONCEPTS

Oil and gas prospecting in the buried crystalline rocks in Moravia has many problems including difficult surface terrain, narrow ridges, and inaccessible locations. Drilling in the overlying folded nappes is not a problem, but problems arise during pumping tests after casing the well. Sometimes the reservoir is damaged while the casing is being cemented. Marginal wells with scant oil flow might have better flow potentials if hydro-perforations and hydraulic fracturing procedures were used. These methods have not yet been used in the Ždánice-Krystalinikum field but possibly could improve production.

Seismic data acquisition is difficult because of the topographic relief and poor access in the Ždánický les Highlands. Geological interpretation of seismic lines is also very difficult, and detailed interpretations are not possible. Many apparent "highs" on seismic lines can be depressions. Complications are introduced by the propagation of seismic waves through folded nappes. Faults in the crystalline basement that form lateral barriers (impermeable zones) are also not visible on seismic profiles.

Development drilling of the Ždánice-Krystalinikum #1 pool was near completion in 1990. Approximately 66,000 MT (480,000 bbl) of oil has been produced from the field to 1989.

ACKNOWLEDGMENTS

I am grateful to the Moravské naftové doly, Hodonín, for permission to publish this paper. I appreciate the assistance from Robbie Gries for the major reviewing and editing of this paper, including assisting with the English translation and revising the figures. I appreciate the original drafting of V. Levayova and subsequent preparation of the figures for publication by Ann Priestman, Geo Graphics Drafting. I also thank Miroslav Zádrapa, Jan Řehánek, and L. Osvald, who provided and photographed the slabs of igneous rocks. Greg Wessel assisted in editing and Lynn Gries typed the manuscript. Bill Brownfield reviewed the study for the AAPG.

REFERENCES

Adam, Z., B. Beránek, V. Čekan, and M. Zounková, 1964, Interpretace lokálních anomállí, v prostoru západokarpatského flyše (Interpretation of local anomalies in West Carpathian Flysch area): Unpublished manuscript, Československé naftové doly—Geofysika, Brno.

Běhounek, R., 1957, Magnetický a gravimetrický průzkum na Moravě (Magnetic and gravimetric survey in Moravia): Sbornik Ústředního ústavu geologického, oddil geologický, pt. 2, v. 23, p. 299–319.

Cahelová, J., V. Ciprys, and M. Škárová, 1972, Zpráva o reflexním průzkumu jv. svahů Českého masívu, oblast Střed, v roce 1971 (Report on reflection seismic survey of southeastern slope of the Bohemian Massif, middle area, in 1971): Unpublished manuscript, Ústav užite geofysiky, Brno.

Chmelík, F., 1969, Kompexní geololgické zhodnocení vrtu Žarošice 1 a Žarošice 2 (Complex geological evaluation of Žarošice 1 and Žarošice 2 wells): Unpublished manuscript, Ústřední ústav geologický, Praha.

Ciprys, V., K. Soukeník, and Z. Adam, 1971, Zprava o seismickem pruzkumu svahu Ceskeho masivu (Report on seismic survey of the Bohemian Massif): Unpublished manuscript, Ustav uzite geofyziky, Brno.

Dlabač, M., 1962, Výzkum vzniku naftových ložisek oblasti vněkarpatského neogénu, geologické zpracování (Investigation of the origin of oil fields in the outer-Carpathian Neogene, geologically working): Unpublished manuscript, Výzkumný ústav—Československé naftové doly, Brno.

Dlabač, M., and E. Menčík, 1964, Geologická stavba autochtonního podkladu západni části vnějších Karpat na území ČSSR (Geological building of the autochthonous basement of western part of the Outer Carthpathian region of USSR): Rozpravy ČSAV, řada matematických a přírodnich věd, v. 74, issue 1.

Doležal, J., 1962, Kvatitativni interpretace tíhových profilů v oblasti západokarpatského flyše (Quantitative interpretation of gravimetric lines in the West Carpathian Flysch area): Unpublished manuscript, Československé naftové doly—Geofyzika, Brno.

Doležal, J., 1964, Kvantitativní interpretace tíhových profilů v oblasti západokarpatského flyše, 2. etapa (Quantitative interpretation of gravimetric lines in the West Carpathian Flysch area, 2nd stage): Unpublished manuscript, Československé naftové doly—Geofyzika, Brno.

Doležal, J., 1965, Kvantitativni interpretace tíhových profilů v západokarpatském flyši (Quantitative interpretation of gravimetric lines in the West Carpathian Flysch area): Sbornik geologických věd—Užitá geofyzika, v. 4, p. 59-83.

Dudek, A., 1980, The crystalline basement block of the Outer Carpathians in Moravia: Bruno-Vistulicum: Rozpravy Československé akademie věd, řada matematických a přírodnich věd, v. 90, issue 8.

Fejfar, M., V. Blaha, J. Cahelová, and J. Jarý, 1965, Zpráva o seismickém průzkumu v oblasti jihovýchodnich svahů Českého masívu (Report on seismic survey in southeastern slopes of Bohemian Massif area): Československé naftové doly—Geofyzika, Brno.

Ibermajer, J., and J. Doležal, 1962, Souborné zpracování a interpretace gravimetrických měření ve flyšové oblasti ČSSR (Collective working and interpretation of gravimetric measurements in the Flysch area of USSR): Užitá geofyzika, Sbornik prací Ústavu užité geofyziky, Praha.

Kostelníček, P., 1973, Závěrečná zpráva o strukturně stratigrafickém vrtu Ždánice 2 (Final report on structural stratigraphic Ždánice 2 well): Unpublished manuscript, Moravské naftové doly, Hodonin.

Krejčí, J., 1971, Projekt geologických praci. Hluboký strukturni průzkum oblasti Ždánického lesa, 1. etapa (Project of geological works. Deep structural research in Ždánický les Highlands area, 1st stage): Unpublished manuscript, Moravské naftové doly, Hodonin.

Krejčí, J., 1972, 1. Dodatek k projektu geologických praci. Hluboký strukturní průzkum oblasti Ždánického lesa, 1. etapa (First addition to the project of geological works. Deep structural research in Ždánický les Highlands area, 1st stage): Unpublished manuscript, Moravské naftové doly, Hodonin.

Krejčí, J., 1974, Závěrečná zpráva o strukturně stratigrafickém vrtu Ždánice 1 (Final report on structural stratigraphic Ždánice 1 well): Unpublished manuscript, Moravské naftové doly, Hodonin.

Ladwein, H. W., 1988, Organic geochemistry of Vienna basin: model for hydrocarbon generation in overthrust belts: AAPG Bulletin, v. 72, p. 586-599.

Mann, O., J. Jarý, and A. Možný, 1963, Magnetický průzkum v oblasti Vídeňské pánve a západokarpatského flyše (Magnetic survey in Vienna basin and West Carpathian Flysch areas): Unpublished manuscript, Československé naftové doly—Geofyzika, Brno.

Odstrčil, J., 1965, Detailní tíhový průzkum Ždánického lesa v roce 1964 (Detailed gravimetric survey in the Ždánicky les Highlands in 1964): Unpublished manuscript, Československé naftové doly—Geofyzika, Brno.

Shánělec, V., O. Mann, and Z. Adam, 1964, Magnetický průzkum na jihozápadní Moravě v roce 1963 (Magnetic survey in southwestern Moravia in 1963): Unpublished manuscript, Československé naftové doly—Geofyzika, Brno.

Zádrapa, M., and J. Krejčí, 1983, Geologicko-petrografické (granulometrické) zhodnoceni sedimentů z vrtu Ždánice 29 (Geological and petrographic [granulometric] evaluation of lower Miocene clastic sediments from Ždánice 29 well): Zemni plyn a nafta, v. 28, p. 241-265.

Appendix 1. Field Description

ŽDÁNICE

Field name .... *Ždánice-Krystalinikum field*
In-place reserves .... *3.71 × 10^6 MT (26.0 × 10^6 bbl) oil; 233 × 10^6 m^3 (8.2 bcf) gas*

Field location:

Country .... *Czechoslovakia*
Basin/Province .... *West Carpathian foreland, outer flysch zone*

Field discovery:

Year first pay discovered .... *Pool #1 1973**
Year second pay discovered .... *Pool #2 1980*
Year third pay discovered .... *Pool #3 1989*

**Miocene Karpatian sandstone production was discovered in 1977 on the southwestern slope of the crystalline basement topographic high.*

Discovery well name and general location:

First pay .... *Ždánice #1, 400 m southwest of highest peak ("U Slepice") of Ždánicky les Highlands and 6.5 km southeast of town of Bučovice*
Second pay .... *Ždánice #30, 1.8 km north of Ždánice #1 (gas pay only)*
Third pay .... *Ždánice #61, 2.9 km northeast of Ždánice #1*

Discovery well operator .... *Moravské naftové doly, Hodonín, Czechoslovakia*

IP

First pay .... *Approx. 17 m^3 (107 bbl) oil per day*
Second pay .... *NA*
Third pay .... *NA*

All other zones with shows of oil and gas in the field:

Age	Formation	Type of Show
Lower Oligocene	*Menilite*	*Gas*
Paleocene-upper Eocene	*Submenilite*	*Gas*

Geologic concept leading to discovery and method or methods used to delineate prospect
Seismic delineation of highs beneath the Ždánice-Subsilesian nappe. The prediscovery objective was Carboniferous and Devonian reservoirs. These rocks were absent, but buried topographic basement highs were found to be good reservoir rocks.

Structure:

Province/basin type .... *Bally 41; Klemme II Cb; cratonic foreland basin*

Tectonic history
Devonian to Tertiary crystalline basement rocks of the Bruno-Vistulicum on the Moravian craton were exposed and weathered. Periodic tectonic activity caused fracturing and brecciation of these rocks. During the Miocene the area was overthrust by Ždánice-Subsilesian nappe, predominantly pelites (mudstones) in their lower part.

Regional structure
This field is in the Carpathian foreland southeast of the Bohemian Massif and northwest of the main Carpathian foreland. The field underlies the Submenilite formation of the Ždánice-Subsilesian nappe.

Local structure
Northeast-southwest-trending buried topographic highs.

Trap:

Trap type(s)

Combined topographic/tectonic with minor stratigraphic (a) topographic (buried crystalline basement hills); (b) tectonic (top seal is tectonically emplaced mudstones of the Ždánice-Subsilesian thrust; lateral seals are fault gouges that have been altered to clay); (c) stratigraphic (discontinuous Karpatian sandstones overlying basement rocks in the western part of the field)

Basin stratigraphy (major stratigraphic intervals from surface to deepest penetration in field):

Chronostratigraphy	Formation	Depth to Top in m
Egerian-Oligocene	*Ždánice-Hustopece Formation*	*0*
Lower Oligocene	*Menilite Formation*	*300-780*
Paleocene-upper Eocene	*Submenilite Formation*	*300-910*
Miocene (Karpatian)	*Unnamed Carpathian foredeep*	*760-970*
Precambrian	*Bruno-Vistulicum (Tectonic Unit)*	*760-1010*

Reservoir characteristics:

Number of reservoirs *2*

Formations

Unnamed Karpatian (Miocene) sandstones; Cadomian Bruno-Vistulicum granodiorite, diorite, quartz diorite, granite, and granite aplite

Ages .... *Miocene and Precambrian (Bruno-Vistulicum)*

Depths to tops of reservoirs .... *Miocene, 760-970 m; Bruno-Vistulicum, 760-1010 m*

Gross thickness (top to bottom of producing interval) .... *205 m maximum*

Net thickness—total thickness of producing zones

Average .... *Approx. 100 m*

Maximum .... *205 m*

Lithology

Tectonically brecciated, weathered, and faulted amphibole-biotite, granodiorite, diorite, quartz diorite, granite, and granite aplite

Porosity type .... *Combination of fracture, vuggy, and intercrystalline porosity*

Average porosity .... *2-4% (ranges up to 0 to 14.7%)*

Average permeability .... *Ranges from 0 to 22.25 md and greater*

Seals:

Upper

Formation, fault, or other feature .... *Ždánice-Subsilesian nappe*

Lithology .... *Claystone/mudstone*

Lateral

Formation, fault, or other feature .... *Altered fault gouge*

Lithology .... *Altered reservoir rock from faulting*

Source:

Formation and age .... *Upper Jurassic Mikulov Marlite Formation*

Lithology .... *Marine marls*

Average total organic carbon (TOC) .... *NA*

Maximum TOC .... *NA*

Kerogen type (I, II, or III) .... *NA*

Vitrinite reflectance (maturation) .... *NA*

Time of hydrocarbon expulsion .... *After early Miocene*

Present depth to top of source .... *About 3000 m*

Thickness .... *More than 1000 m*

Potential yield .... *NA*

Appendix 2. Production Data

Field name ... *Ždánice-Krystalinikum field*

Field size:

- **Proved area** ... *12 km^2*
- **Number of wells all years** ... *82 wells drilled; 59 wells producing*
- **Current number of wells** ... *NA*
- **Well spacing** ... *About 4–8 ha (40,000–80,000 m^2)*
- **Ultimate recoverable (primary)** ... *370,000 to 550,000 MT (2.6 $\times 10^6$ to 3.9 $\times 10^6$ bbl) oil*
- **Cumulative production (through 1991)** ... *69,840 MT (488,740 bbl) oil; 6.221 $\times 10^6$ m^3 (220 $\times 10^6$ ft^3) gas**
- **Annual production (1991)** ... *5884 MT (41,180 bbl) oil; 372 $\times 10^3$ m^3 (13.14 $\times 10^6$ ft^3) gas**
- **Decline rates** ... *NA*
- **Annual water production** ... *NA*
- **In place, total reserves (original)** ... *3.71 $\times 10^6$ MT (26.0 $\times 10^6$ bbl) oil; 270 $\times 10^6$ m^3 (9.5 bcf) gas*
- **In place, per acre foot** ... *NA*
- **Primary recovery** ... *370,000 to 550,000 MT (2.6 to 3.9 $\times 10^6$ bbl) oil; 26 $\times 10^6$ m^3 (918 $\times 10^6$ ft^3) gas**
- **Secondary recovery** ... *NA*
- **Cumulative water production (through 1991)** ... *176,396 bbl; 28,028 m^3*

**Solution gas produced with the oil; gas cap gas not produced.*

Drilling and casing practices:

- **Amount of surface casing set** ... *Usually 200 m*
- **Casing program** ... *9⅝-in. to 200 m; 6⅝-in. to T.D.; 2⅞-in. tubing*
- **Drilling mud** ... *Water-based mud*
- **Bit program** ... *Roller bit, diamond core bit (in igneous rock only)*
- **High pressure zones** ... *None*

Completion practices:

- **Interval(s) perforated** ... *Variable; 20 to 50 m of perforated casing*
- **Well treatment** ... *Small gunpowder charges are sometimes detonated*

Formation evaluation:

- **Logging suites** ... *Resistivity, SP, and gamma ray (also cores)*
- **Testing practices** ... *Run pipe and swab*
  - **Mud logging techniques** ... *NA*

Oil characteristics:

- **Type** ... *Paraffinic-naphthenic*
- **API gravity** ... *19° to 33°*
- **Base** ... *NA*
- **Initial GOR** ... *NA*
- **Sulfur, wt%** ... *None*
- **Viscosity, SUS** ... *14–58 cp at 20°C*
- **Pour point** ... *Below –40°C*
- **Gas-oil distillate** ... *NA*

Field characteristics:

- **Average elevation** ... *350 m*
- **Initial pressure** ... *6900–7400 kPa*
- **Present pressure** ... *NA*
- **Pressure gradient** ... *NA*

Temperature .... *27°–30°C*
Geothermal gradient .... *NA*
Drive .... *Water*
Oil column thickness .... *90 m, 18 m, 60 m (pools 1, 2, 3)*
Oil-water contact .... *–585, –521, –660 m (pools 1, 2, 3)*
Connate water .... *NA*
Water salinity, TDS .... *Total mineralization 11–13 g/L*
Resistivity of water .... *NA*
Bulk volume water (%) .... *NA*

Transportation method and market for oil and gas:
Pipeline

Kelamayi Field—People's Republic of China

Zhungeer Basin, Xinjiang Province

ZHAI GUANG-MING
ZHAO WEN-ZHI
Scientific Research Institute of Petroleum Exploration and Development
Beijing, People's Republic of China

FIELD CLASSIFICATION

BASIN: Zhungeer
BASIN TYPE: Chinese
RESERVOIR ROCK TYPE: Conglomerate, Sandstone, and Dolomite
RESERVOIR AGE: Permian, Triassic, Jurassic
PETROLEUM TYPE: Oil
TRAP TYPE: Unconformity Truncation and Onlap, Fault, and Updip Tar Seal
RESERVOIR ENVIRONMENT OF DEPOSITION: Alluvial Fan, Braided Stream, and Carbonate Platform
TRAP DESCRIPTION: Complex of four trap types: unconformity truncation, unconformity onlap, reverse fault, and updip tar seal

LOCATION

Kelamayi oil field is located in the far northwest of China within the administration of Xinjiang Province and is 300 km from Wulumuqi, the capital of the province. Along the northeast-trending Kelamayi-Wuerhe thrust fault belt (abbreviated as Ke-Wu fault belt below) are the Hongshanzui, Kelamayi, Baijiantan, Baikouquan, Wuerhe, Fengchengcheng, and Xiazijie oil fields. They comprise the Kelamayi-Xiazijie oil-bearing trend, which is 250 km long. Kelamayi field is the largest in reserves and production (Figure 1).

Located on the northwest edge of Zhungeer basin (Figure 1), the field is close to the Zhayier mountains, the northwest boundary of the basin. The climate of this area is arid to semiarid continental. Along the mountain front are a proluvial-alluvial conglomerate zone and alluvial plains formed by intermittent streams; the area is gradually transformed into desert toward the heart of the basin. The Kelamayi-Xiazijie trend including the Kelamayi field is on the alluvial conglomerate-plain zone and partly on the foothills along the mountain front, with an elevation of approximately 300 m.

The Kelamayi field was discovered in October 1955 with production commencing in 1958. Several other oil fields and oil reservoirs have been discovered since 1955. Presently, the trend has six oil fields (Baijiantan is now part of Kelamayi), with an annual production of 5 million tons (76.5 million barrels) of oil. Kelamayi has produced about 27.8 million tons (426 million bbl) of oil and annual production (as of 1986) is about 2 million tons (30 million bbl).

HISTORY

In the early twentieth century, *Xinjiang Map Chronicles* and *Geography of Northwest China* were published, giving some description of oil seeps in the Kelamayi area. After the founding of the People's Republic, Chinese and Soviet geologists, while conducting geological investigations of the Zhungeer basin, discovered the "asphalt dune" in Heiyoushan, located near Kelamayi city, and the bituminous veins at Wuerhe, in the northern part of what would become the Kelamayi-Xiazijie trend.

After the Sino-Soviet Petroleum Company was set up in 1951, a geological field party constructed a large-scale (1:2500) geologic map of the outcrop area. This map depicted many "asphalt dunes" near Kelamayi city. In the southeast-dipping Mesozoic outcrop area, they discovered some small anticlines and noses with an area of not more than several square kilometers; they also described the asphalt and oil seeps located on the axis of the Heiyoushan anticline (Figure 2). In 1952, the field party recommended four shallow wells (1, 2, 1/3, and 4, Figure 2) on the Heiyoushan

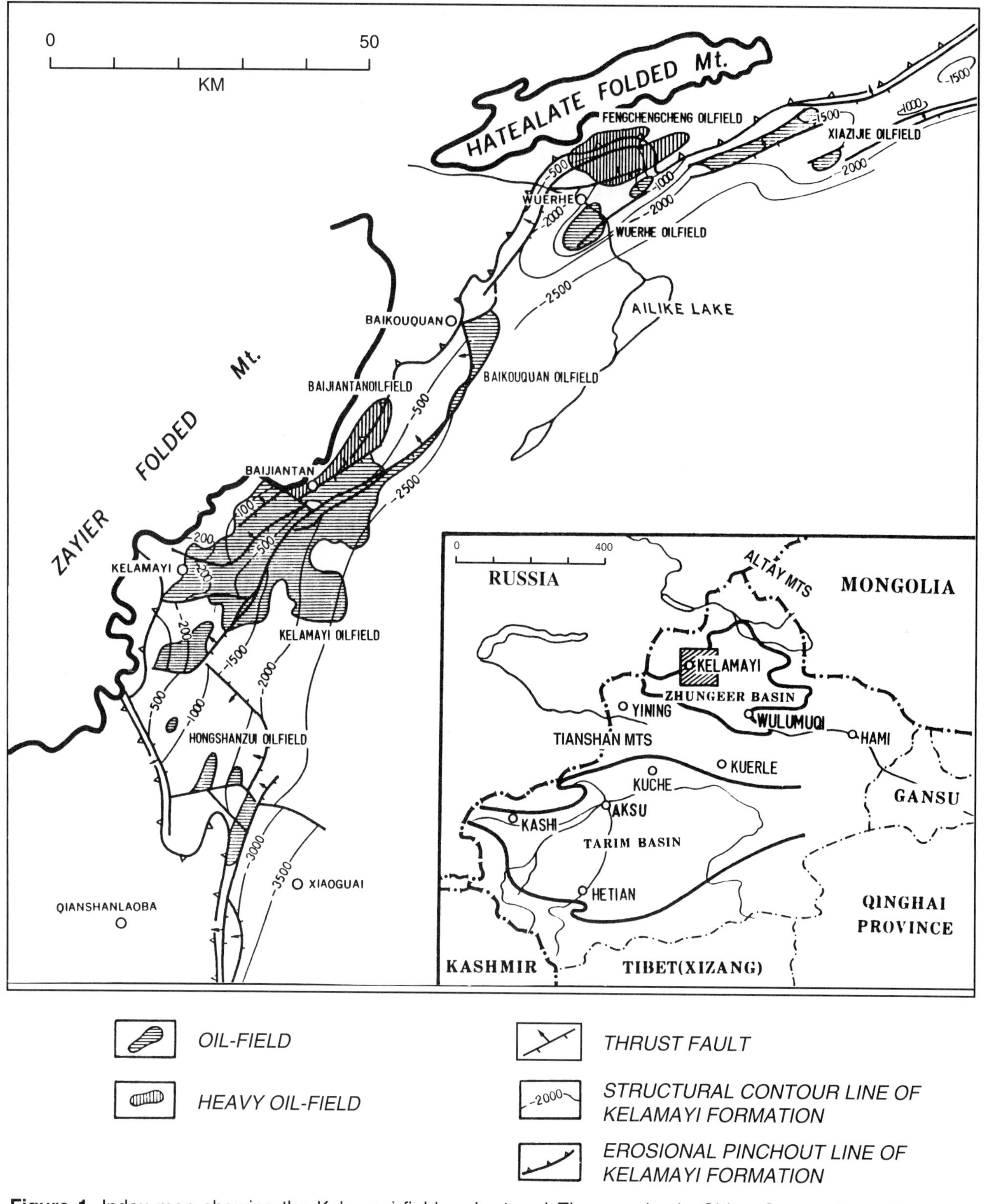

Figure 1. Index map showing the Kelamayi field and the distribution of oil fields in the Kelamayi oil-bearing trend, Zhungeer basin, China. Contour interval, 100 and 500 m.

anticline. The No. 2 well, on the top of the structure, obtained a little oil and gas but no commercial flow (Figure 2). Thus the drilling was suspended. In the same year, a survey in this area proved that the substrata were gently monoclinal, with some structural terraces and small anticlines.

In 1954, the Sino-Soviet Petroleum Company sent a second party to draw a 1:100,000 scale geologic map

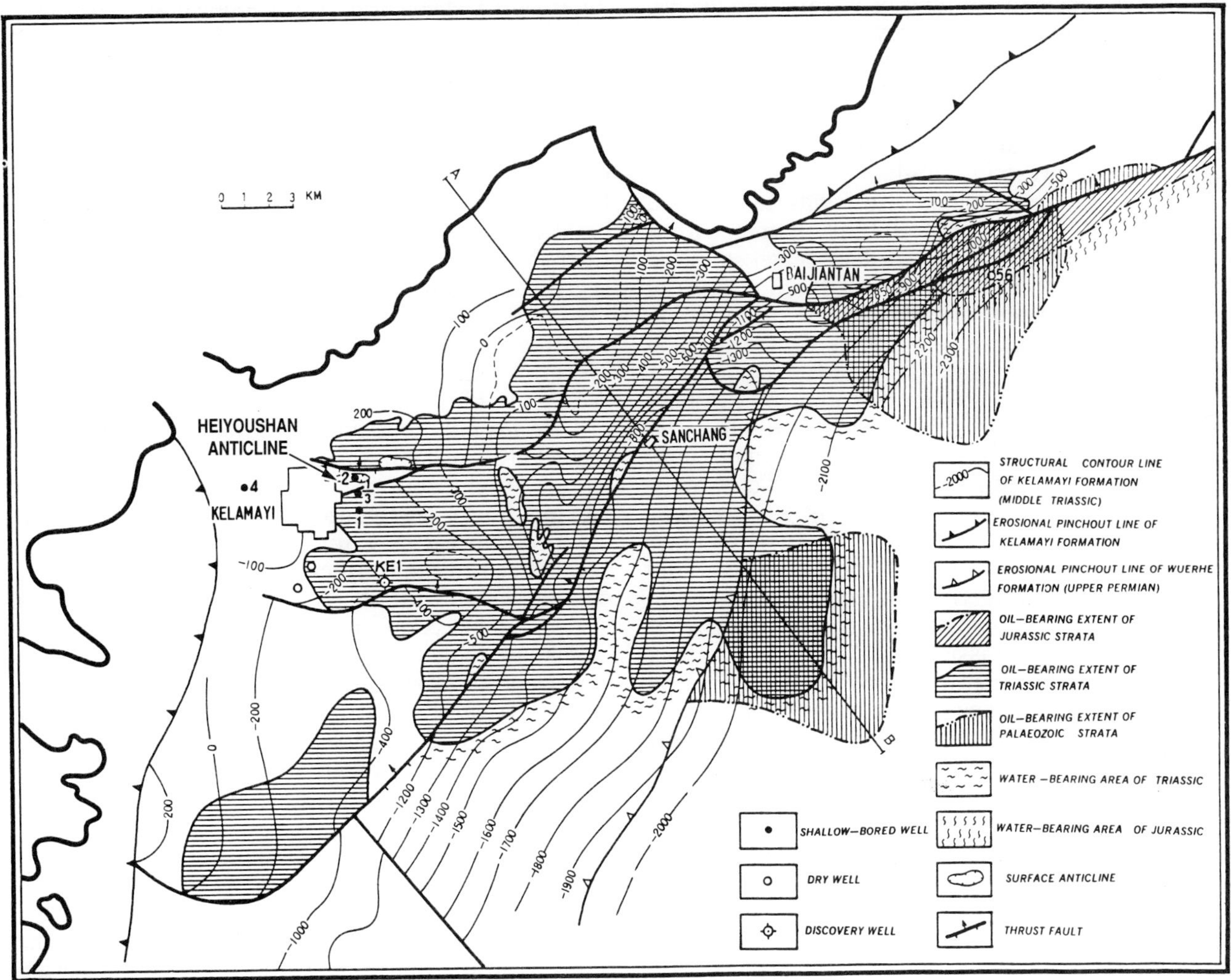

Figure 2. Structure map of Kelamayi Formation. Contour interval, 100 m. Structural cross section A-B is shown by Figure 3.

of the Kelamayi-Wuerhe area. Apart from the small regional structure discovered in 1951, it confirmed that the area was monoclinal, dipping from the northwestern edge of the basin to the center. Triassic, Jurassic, Cretaceous, and Neogene strata transgressed bed by bed toward the edge of the basin with a dip angle less than 4°. In the outcrop area on the basin flank, from Hongshanzui in the southwest to Wuerhe in the northeast, there were many types of oil and gas indications such as "asphalt dunes," bituminous veins, oil sands, oil and gas seeps, etc. These were mainly located near the overlapping unconformity interfaces between different geologic systems.

The party considered that the Zhungeer basin was deep and steeply dipping in the south and shallow and gently dipping in the north, so a large amount of hydrocarbons that originated in the sedimentary depression would have migrated to the north edge of the basin. The large number of oil and gas seeps discovered on the northwest edge is the final result of this northward migration of oil and gas. The party predicted that a large oil and gas field could be found on the northwest flank of the basin and that hydrocarbons would not necessarily be trapped by the small, local structures but would be widely distributed on the regional slope.

Under the guidance of this idea, a drilling program of three wells was designed in the downdip direction in the Kelamayi area. Drilling in 1955, the Ke-1 well was located 5 km southeast of Kelamayi city. The well penetrated the Cretaceous, Jurassic, and Triassic and reached the basement rock at a depth of 517 m, with total depth (TD) at 620 m. A test of Ke-1 in Triassic strata from 487.5 to 507.5 m obtained the first oil flow of 52 bbl/day. Later, three oil flows of 36, 59, and 75 bbl/day were obtained from three other pay zones of the Middle Triassic in the interval from 484 to 402 m depth.

At the same time as the drilling of the Ke-1 well, detailed seismic and geological surveys were implemented in this area. Apart from the small structures already mapped at the surface, several small subsurface uplifts were delineated with areas ranging from several to tens of square kilometers. At first, these small structures were used as the basis for the location of wildcats, but it was soon confirmed by drilling that the oil-bearing area was primarily stratigraphic and greatly exceeded the area of the structures. Hydrocarbon accumulations were not limited by local structural highs.

After the discovery of oil, Ke-1 was taken as a starting-point to extend the oil-bearing area with a spacing of 1 to 2 km. In order to get a clear understanding of the oil-bearing feature on the northwest margin, ten "drilling profiles" (each profile containing two to three wells) were drilled in 1956 along the northeast-trending Kelamayi-Wuerhe fault zone. This fault zone was found to be 130 km long and 20 to 30 km wide. This greatly extended the proven oil-bearing area, and after 1958, Baikouquan, Wuerhe, Hongshanzui, and Xiazijie oil fields were discovered.

In the early stages of exploration, the Kelamayi Formation of the Middle Triassic was considered the main exploratory target. With continued development, a secondary faulted pool in the Badaowan Formation (Lower Jurassic) was discovered in 1960. In 1964, an Upper Permian pool in the Wuerhe Formation in the Kelamayi field was found. In the 1980s, pools were discovered in fractured basement rocks. In 1981, an oil pool in the Lower Permian Fengcheng (Fengchengcheng) Formation (upper part of Jaimuhe Formation, facies change) was discovered. Meanwhile, drilling and production testing proved that the heavy oil in shallow reservoirs was also commercial and verified the two oil-bearing areas of Baijiantan (Figure 1) and Fengcheng. The Kelamayi area contains multiple oil pools, forming a complex oil-bearing play.

At present, the exploration of the Kelamayi oil-bearing trend is in the development stage. The reserves in the major pools are in the Kelamayi Formation. However, exploration for deeper hydrocarbons and shallow heavy oil is still underway, and the reserve potential of the trend is great. In the entire oil-bearing trend, 77% of the reserves are in Kelamayi field where the experience and knowledge of the geological parameters and exploration and development procedures can be used to develop a model for exploration and exploitation.

STRUCTURE

Tectonic History

Surrounded by mountain ranges, the Zhungeer basin (Figure 1), in which the field is located is an intermontane basin in the mid-Asiatic Mongolian folded mountain system, with the Tianshan Mountains to the south, the Altay Mountain Range to the northeast, and the Zhayier-Hatealate Mountains to the northwest. The basin, of triangular outline, has an area of 135,000 km^2. The framework of the basin is Hercynian folded basement, but there is no direct evidence indicating that the basin does not have Precambrian crystalline basement in the deepest, central part.

The Zhungeer basin was formed in the late stage of the Hercynian movement. The oceans around the basin were restricted at the time of the collision between the surrounding plates during the Middle Carboniferous to the Early Permian. As sea level regressed, there was still a remnant sea in the basin area in the Early Permian. The basin was filled with continental sediments during Late Permian and Mesozoic–Cenozoic times. Total thickness of the sediments reached 10 to 14 km. Lacustrine deposits of Late Permian age and shallow marine sequences from Late Carboniferous to Early Permian age are the dominant source beds of the Zhungeer basin. During Mesozoic–Cenozoic time, the Indian plate collided with the China plate. Strong pressure transferred to the southern margin of the Zhungeer basin through the Tibet-Tarim subplate and formed a deep depression in the northern part of the Tianshan folded mountains. Therefore, the Mesozoic–Cenozoic deposits of the Zhungeer basin are asymmetrical, thick in the south and thinner in the north. The northern part of the basin is a broad monocline with a sedimentary sequence of Mesozoic–Cenozoic sediments overlapping and thinning northward.

Hercynian, Indo-China, Yanshan, and Himalayan movements all caused strong deformation on the periphery of the basin but much weaker deformation within the basin. Folding, accompanied by some faulting, is the dominant form of rock deformation in the south margin where Himalayan movement played a leading role and formed several rows of anticlinal structures. On the northwest margin where faulting is dominant, with some slight folding, the early stage of thrust movement was very strong, but turned weaker later (Yanshanian and Himalayan movements). This margin was slightly uplifted during the Late Jurassic and Early Cretaceous.

Regional Structure

The regional structure of the northwest margin, in which the Kelamayi field is located, is a monocline, gently dipping southeastward. The dip angle of Permian–Triassic strata is 3 to 5°. The dip angle of Lower to Middle Jurassic is 2 to 4°. The Upper Jurassic and its overlying strata are more gently dipping, generally less than 3°, but in the area near the Ke-Wu fault trend, the strata of the Lower to Middle Jurassic and older are nearly vertical. Unconformities between strata in the area exist only in certain places. The Ke-Wu fault trend strikes northeastward, cutting through the oil-bearing beds, and is made up of a series of thrust faults. Dipping northwestward, the fault planes normally have a dip angle of 40 to 60°, becoming less with increasing depth. The vertical displacement ranges from tens to thousands of meters. These faults cut the slope into three terraces descending into the basin (Figures 3 and 4) and a few faults, striking northwestward, cut the terraces into several fault blocks (Figures 1 and 2).

The Ke-Wu fault trend began forming at the end of the Carboniferous, the activity continuing during the Triassic, weakening during Middle Jurassic, and gradually ceasing by Late Jurassic. Therefore the faults only cut the strata of pre-Late Jurassic age. Based on this structural background, the oil and gas migration and entrapment were controlled by both the regional monocline and by the Kelamayi-Wuerhe thrust faulting. Accumulations were sealed laterally and vertically by lithologic changes. The oil and gas accumulated in the top of the tilted blocks and local highs, each accumulation having its own oil-water contact along the flanks of the structures.

It should be noted that the structural features of Wuerhe-Xiazijie oil field in the northern part of the oil-bearing trend (Figure 1) are slightly different from those in the southern and central part of the trend. The east-west-trending basement faults caused the folding of overlying strata and formed the Wuerhe and Xiazijie anticlines. The hydrocarbon accumulations in this area are mainly anticlinally controlled, while the accumulations of hydrocarbons in the southern and central part of the trend are mainly controlled by stratigraphy and thrust- and block-faulting.

STRATIGRAPHY

The rocks of the oil-bearing trend, underlain by the folded Paleozoic basement, are of Permian, Mesozoic, and Cenozoic ages (Table 1). Their general lithological features are described as follows.

Basement (Paleozoic)

The outcrops of basement older than Late Carboniferous are found in the Zhayier Mountains to the northwest of the trend. Most of the drilling in the Kelamayi field has gone to the basement, and wells penetrating these rocks have drilled into them by hundreds or even thousands of meters. The basement complexes, which have experienced some slight metamorphism, consist of pyroclasts, clastics, neutral and basic volcanic rocks, and intrusive rocks. They have low porosity but well-developed fractures. The porosity and fractures in the volcanic rocks are the major reservoirs for oil and gas accumulation if they have good conduits to the source areas, such as in thrust belts where they directly contact source rocks and encounter trapping conditions.

Permian

The *Jiamuhe Formation (P1j)* is an interbedded greenish-gray conglomerate and sandstone with bands of mudstone in the lower part; light grayish-green, scarlet, and brown tuff and sandy conglomerate occur in the upper part. Locally dark mudstone is banded with limestone and dolomite. Fractures, solution voids, and cavities form oil-bearing beds. Hydrocarbons are derived from the dark mudstone within the formation.

The *Xiazijie Formation (P2x)* is a grayish-brown, grayish-green, and dark-grayish conglomerate and brown sandstone. It is the major pay in the Wuerhe area.

The *Wuerhe Formation (P2w)* is a thick-bedded grayish-green conglomerate with a few sandstone and mudstone layers on the margin of the basin, but it becomes finer grained basinward. The dark shale and oil shale interbedded with siltstone and dolomite in the southern part of the basin are considered to be the major source beds in the basin as well as in the

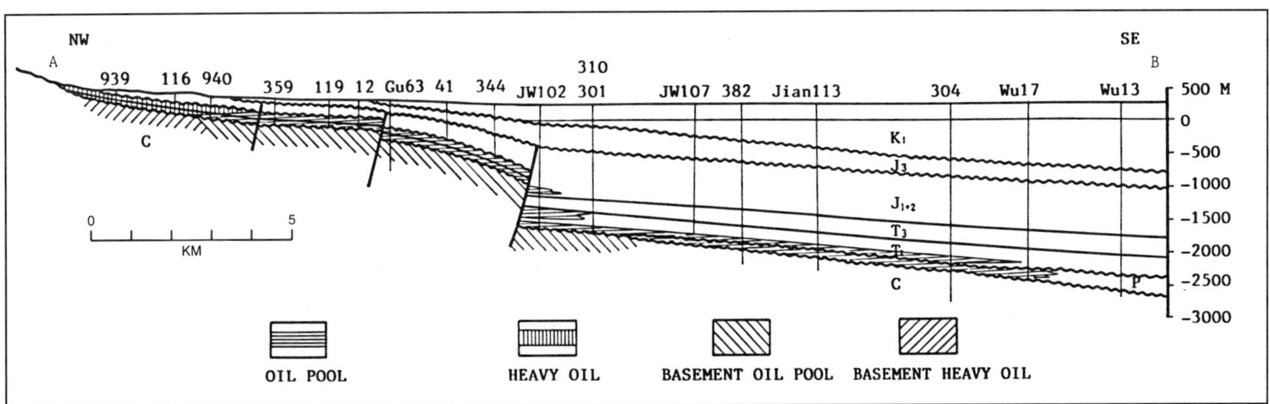

Figure 3. Northwest-southeast structure cross section through Kelamayi field. The location of the section (A-B) is shown on Figure 2.

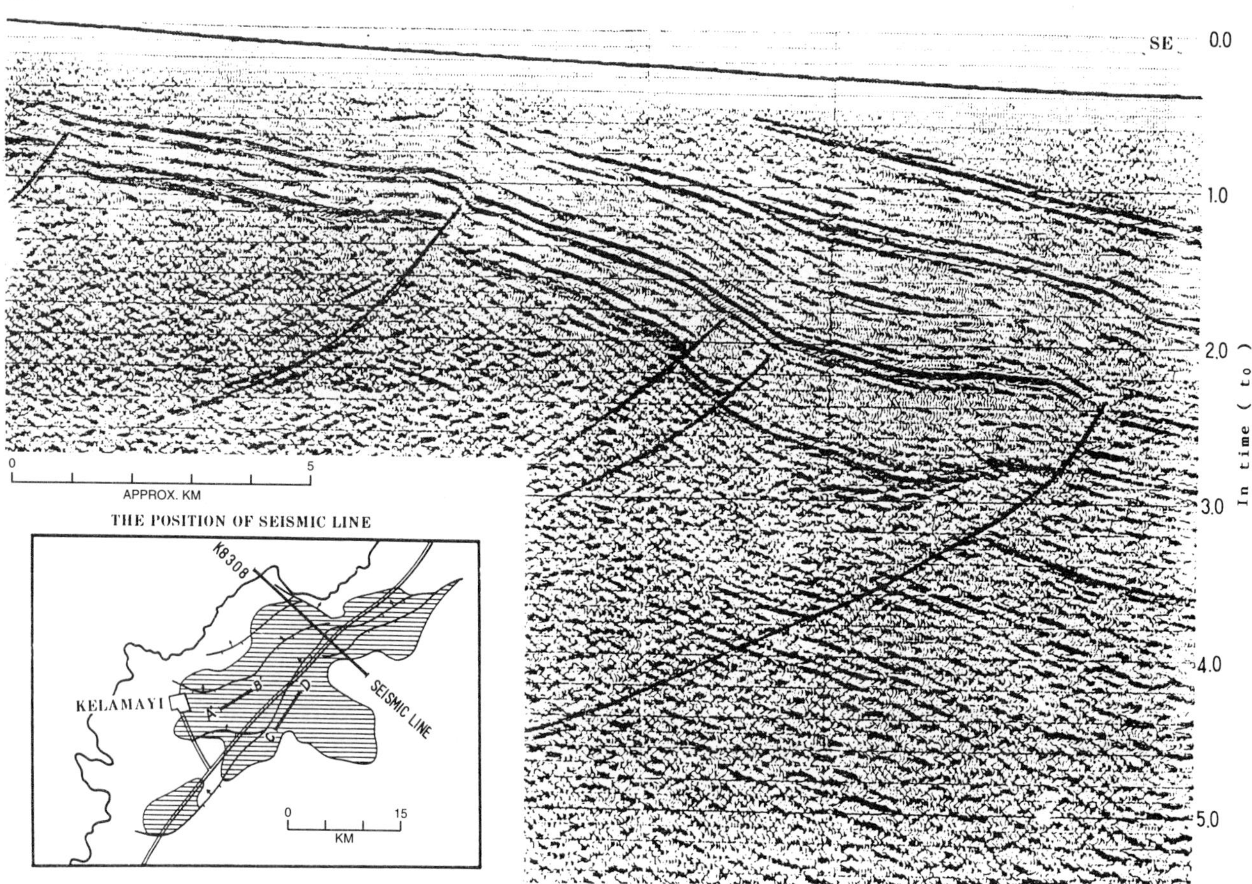

Figure 4. Seismic reflection profile. Inset also shows locations of profiles A-B and C-D, Figure 5. Length of profile is about 18 km.

Kelamayi oil-bearing trend. This formation is mainly made up of conglomerate in the oil-bearing area and is a secondary oil-producing zone in the Kelamayi field.

Triassic

The *Baikouquan Formation (T1b)* (Lower Triassic) is a suite of poorly sorted, subrounded, red, coarse conglomerate, missing in the Kelamayi field but distributed over the Baikouquan area. The thickest deposit of the Baikouquan Formation is located in the Baikouquan field and is its major producing zone.

The *Kelamayi Formation (T2k)* (Middle Triassic) is grayish-green, brown, proluvial and alluvial conglomerate interbedded with mudstone and sandstone. The conglomerate is poorly sorted and stratified and has poor lateral continuity. In the Kelamayi field, this bed of conglomerate unconformably overlies the basement and is the major reservoir of the field.

The *Baijiantan Formation (T3b)* (Upper Triassic) serves as the cap rock of the oil-bearing zone and is a gray to dark-gray mudstone with thin-bedded, calcareous siltstone, argillaceous limestone, brown hematite, and coal streaks. Sandstone is present in the upper part and hydrocarbons may accumulate near the fault. The Baijiantan Formation is a secondary producing bed in the field.

Jurassic

The *Coal Series* (Lower to Middle Jurassic) are fining-upward, cyclic sediments of light gray sandstone, conglomerate, dark mudstone, and coal beds. Locally, the contact with the Cretaceous is an angular unconformity. The coal series is divided into four formations as follows:

1. The *Badaowan Formation (J1b)* is a sandy conglomerate interbedded with mudstone and coal beds and forms a good reservoir zone on both sides of the Kelamayi-Wuerhe fault.
2. The *Sangonghe Formation (J1s)* is dominantly mudstone without coal.
3. The *Xishanyao Formation (J2x)* is a sandy conglomerate with coal beds.
4. The *Toudunhe Formation (J2t)* is a colored mudstone and sandstone with coal.

Table 1. Stratigraphic table of the northwest Zhungeer basin. Possible source rocks indicated by vertical lining, sandstone, dolomite, and conglomerate reservoirs by stippling, and basement reservoir rocks by "stars." Roman numerals I through VI are the major pay horizons from oldest to youngest.

System	Series	Formation	Symbol	Brief Lithological Features	Source Beds	Reservoir Beds
Tertiary			N	Light brown mudstone and sandstone		
Cretaceous	Upper	Legou	K21	Brown sandy mudstone and conglomerate		
	Lower	Tugulu	K1t	Gray-green sandstone interbedded with brown mudstone; basal sandstones contain heavy oil		
Jurassic	Upper	Qigu	J3q	Quartz sandstone and colored mudstone; heavy oil in sandstone		
	Middle	Toudunhe	J2t	Sandstone and mudstone, without coal		
		Xishanyao	J2x	Sandstone and mudstone with coal beds		
	Lower	Sangonghe	J1s	Lacustrine mudstone and sandstone, without coal beds		
		Badaowan	J1b	Sandstone and mudstone with coal beds; oil sand occurred in lower part		VI
Triassic	Upper	Baijiantan	T3b	Dark mudstone with band of sandstone in upper part which contains oil		
	Middle	Kelamayi	T2k	Fluvial and alluvial conglomerate, sandy mudstone; conglomerate bears oil		V
	Lower	Baikouquan	T1b	Conglomerate interbedded with mudstone; conglomerate is pay bed		IV
Permian	Upper	Wuerhe	P2w	Shales are source beds in southern margin; conglomerate is pay bed in northwest margin		III
		Xiazijie	P2x	Conglomerate bears oil		
	Lower	Fengcheng-cheng Jiamuhe/ Hongzhongse	P1j	Dark mudstone, tuffs and dolomite; mudstone is source bed; dolomite is reservoir		II
Pre-Permian		Xibeikulas Basement	C1+2	Volcanic rocks, metamorphic clastics as fractured reservoirs		I

The *Qigu Formation (J3q)* (Upper Jurassic) is a conglomerate and sandstone at the base with colored mudstone and sandstone in the upper part. The formation overlaps the basement at the basin margin, and hydrocarbons have migrated into the basal conglomerate through the unconformity interface to form a heavy oil pool.

Cretaceous

The *Tugulu Formation (K1t)* (Lower Cretaceous) is a grayish-green, fine sandstone interbedded with thin-bedded brown mudstone. A basal conglomerate is unconformably in contact with the underlying strata. The formation overlaps the basement at the basin margin where there are oil sands and oil seeps.

Neogene

Light-brown, sandy mudstone of Tertiary age is interbedded with sandstone. A layer of quartz conglomerate at the bottom unconformably overlaps all the older strata, and oil traces or seepages can be found occasionally in these sandstones at outcrops on the margin of the basin.

TRAPS

Types of Traps and Pools

Kelamayi field consists of four different trap types.

Stratigraphic-Lithologic

The main reservoir of the stratigraphic-lithologic type is in the Triassic Kelamayi Formation. The structural framework related to this trap is a southeast-dipping monocline with a dip angle of 2 to 5° overlapping unconformably onto the basement. Hydrocarbons generated in the basin migrated updip along the unconformity interface and accumulated in the conglomerate of the Kelamayi Formation. The conglomerate bed, deposited in an alluvial environment, has poor stratification and continuity. It is reduced in thickness in the updip direction and also abruptly pinches out laterally into the fluvial-swampy and fluvial-lacustrine argillaceous deposits. Thus the reservoir is actually several imbricated and irregular lens-shaped lithosomes. This type of oil pool has a weak water drive owing to its low permeability and pronounced heterogeneities; thus, it is regarded as a stratigraphic-lithologic oil pool with a monoclinal background, but also partly as an unconformable stratigraphic overlap pool.

Unconformable Stratigraphic Overlap

Upper Jurassic Qigu Formation reservoirs and part of the reservoirs in the Lower Cretaceous Tugulu Formation are of the unconformable stratigraphic overlap type. At the edge of the basin, these formations lie directly on the basement unconformity interface. Hydrocarbons migrated along the unconformity interface to the edge of the basin and entered the lower sandstone of the Qigu or Tugulu formations, thus forming an "unconformable stratigraphic overlap oil pool." In some uplifted areas where the Kelamayi Formation oil pool was partly eroded, the hydrocarbons possibly migrated again and entered the overlying Qigu and Tugulu formations, thus forming a stratigraphic unconformity pool. The strata mentioned above normally have a dip angle of 2 to 3° and are sealed by heavy oil or bitumen in the updip direction and are controlled by facies changes in the downdip and lateral directions. Oxidation as the result of surface water and biodegradation has often destroyed these oil pools. This type of pool is generally found at the updip edge of the formation at a shallow depth and contains heavy oil that results from biodegradation.

Fault-Sealed

This type of pool commonly occurs on the downthrown side of a fault or faulted block. Its reservoirs include the Wuerhe Formation (P2w), Kelamayi Formation (T2k), Baijiantan Formation (T3b), and Badaowan Formation (J1b). The hydrocarbons generated in the basin migrated in the updip direction after entering the Wuerhe and Kelamayi formations. When hydrocarbon migration was blocked by the Kelamayi-Wuerhe fault and sealed by the lateral lithology and structure, it forms a "fault-sealed pool." The hydrocarbons in the Triassic pool migrated updip into the Baijiantan (T3b) and Badaowan (J1b) formations along fault planes. Hydrocarbons became sealed in the T3b and J1b formations, forming a secondary fault-sealed pool. Fault-sealed pools occur in the Badaowan Formation on the downthrown side of the Kelamayi-Wuerhe fault and in District 7 of the Kelamayi field, as well as in the Baikouquan Formation on the downthrown side of the Baijiantan fault.

Basement

Migration of hydrocarbons along the unconformity interface of the basement and the overlying Kelamayi Formation was blocked by the Kelamayi-Wuerhe fault belt in an updip direction and formed the pool on the downthrown side of the fault as for the above-mentioned fault-sealed pools. Some of the trapped hydrocarbons could penetrate the fault plane and enter the upthrown side of the fault block of the basement rocks, forming a "basement oil pool." The basement was capped by argillite of the Kelamayi and Baijiantan formations and was wholly or partly sealed by bitumen and because of lithology in the updip direction and loss of fracture permeability laterally. Fractures are the main reservoir properties of the basement, and the oil output was related to the physical properties and the amount of fractures of the basement. Individual well productivities are quite variable.

It should be noted that the oil pool types in the northeastern part of the Kelamayi oil-bearing trend, which are dominantly structural, are somewhat different from those in the Kelamayi field. As mentioned above, basement faulting caused the overlying Permian and Triassic strata to be slightly folded and to form the anticlinal oil pools. Among these pools, the top of some higher anticlines were eroded in the late stage of the Indo-China movement, thus the oil and gas accumulated in the top was eliminated, but there were still some pools on the flanks or downdip edge, e.g., the main pool in Wuerhe field and the Xiazijie field. Accumulations such as Fengcheng field are generally located at depth of 2000 to 3000 m, and the hydrocarbons accumulated at the top of the anticline. Pools such as North Xiazijie field are located on a gentle anticline or fault block; hydrocarbons accumulated on the axis of such anticlines.

Reservoirs

The commercial reservoir rocks in the Kelamayi field include ten formations that belong to five systems (Carboniferous, Permian, Triassic, Jurassic, and Cretaceous). The main reservoir rocks include six formations. Good reservoirs developed in the top and middle parts of alluvial fans (Figures 5 and 6). Highly variable porosities and permeabilities range from 8.8% to 35% and 0.37 to 3000 md in the main reservoirs of the Kelamayi-Wuerhe oil-bearing trend. There are five important oil-bearing formations, discussed below. Refer to Figure 7 for their stratigraphic relationships. (See also Table 1.)

Basement (Lower to Middle Carboniferous) is composed of metamorphic and volcanic rocks of the Carboniferous period. Oil has accumulated in fractures and pores. The oil pool is tens to hundreds of meters thick and is buried to depths of from 400 to 2500 m. Single well production ranges from 75 to 1170 BOPD.

The Fengcheng Formation (Lower Permian) consists of a dolomitic volcanic tuff, muddy tuff, and dolomite, 20 to 30 m thick and 3000 to 3500 m in depth. Oil has accumulated in fractures and solution pores. Single well production is 440 BOPD.

The *Wuerhe Formation* (Upper Permian) is a thick, massive conglomerate with fracture-pores containing oil at depths of 2500 to 3000 m. Single well production ranges from 75 to 220 BOPD.

The *Kelamayi Formation* (Middle Triassic) is the primary oil pay. It is a series of proluvial-alluvial conglomerates at depths of 300 to 3000 m. Single well production is 7 to 440 BOPD.

The *Badaowan Formation* (Lower Jurassic) is a river-channel sandstone and conglomerate; oil pay is 20 to 30 m thick at depths from 1500 to 2500 m. Single well production ranges from 60 to 75 BOPD.

Faults

Since the faults generally control oil migration and accumulation in the Kelamayi field, it is of great importance to understand and map the faults in the Kelamayi trend. For this reason, very detailed seismic work was implemented. The seismic interpretations were proved or amended by data from later wildcats and production wells.

The Kelamayi fault strikes northeastward and cuts across the Kelamayi field. It is a major fault that formed during the time from Late Carboniferous to Late Jurassic. Continuing fault movement caused a great variation in the thicknesses of deposits of the upthrown and downthrown blocks. The fault thrusts from the edge to the interior of the basin (Figure 3) have displacements ranging from hundreds of meters to over 1000 m.

Apart from the major fault, there are also subordinate faults, some parallel or subparallel to the major fault in the Kelamayi field (Figures 1 and 2). All these faults have the same characteristics as the major fault except as to their smaller displacements. The intersection of the major fault with the

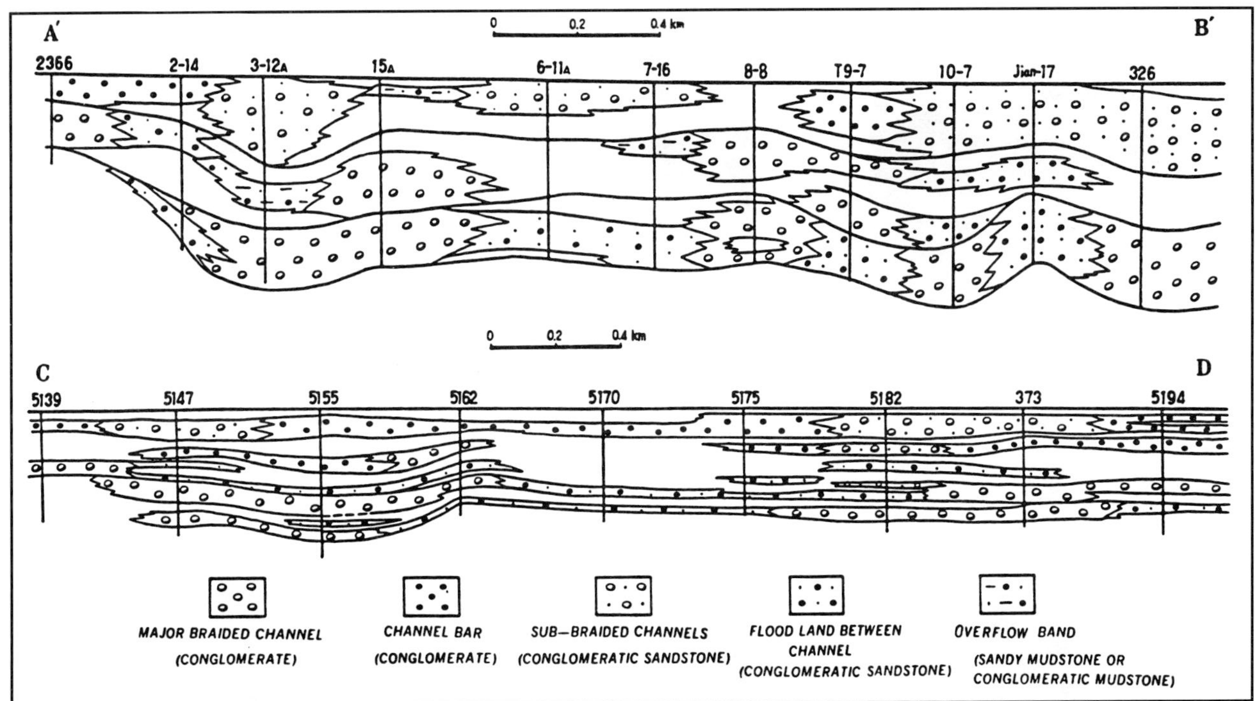

Figure 5. The lithological change profiles, A-B and C-D, showing I-type and II-type reservoirs of the Middle Triassic Kelamayi Formation. Figure 4 shows the locations of the profiles. I-type reservoirs are mainly alluvial fans: porosities > 20% and permeabilities > 100 md. II-type reservoirs are mainly alluvial-plain, braided-stream deposits: porosities 15 to 20% and permeabilities 50 to 100 md.

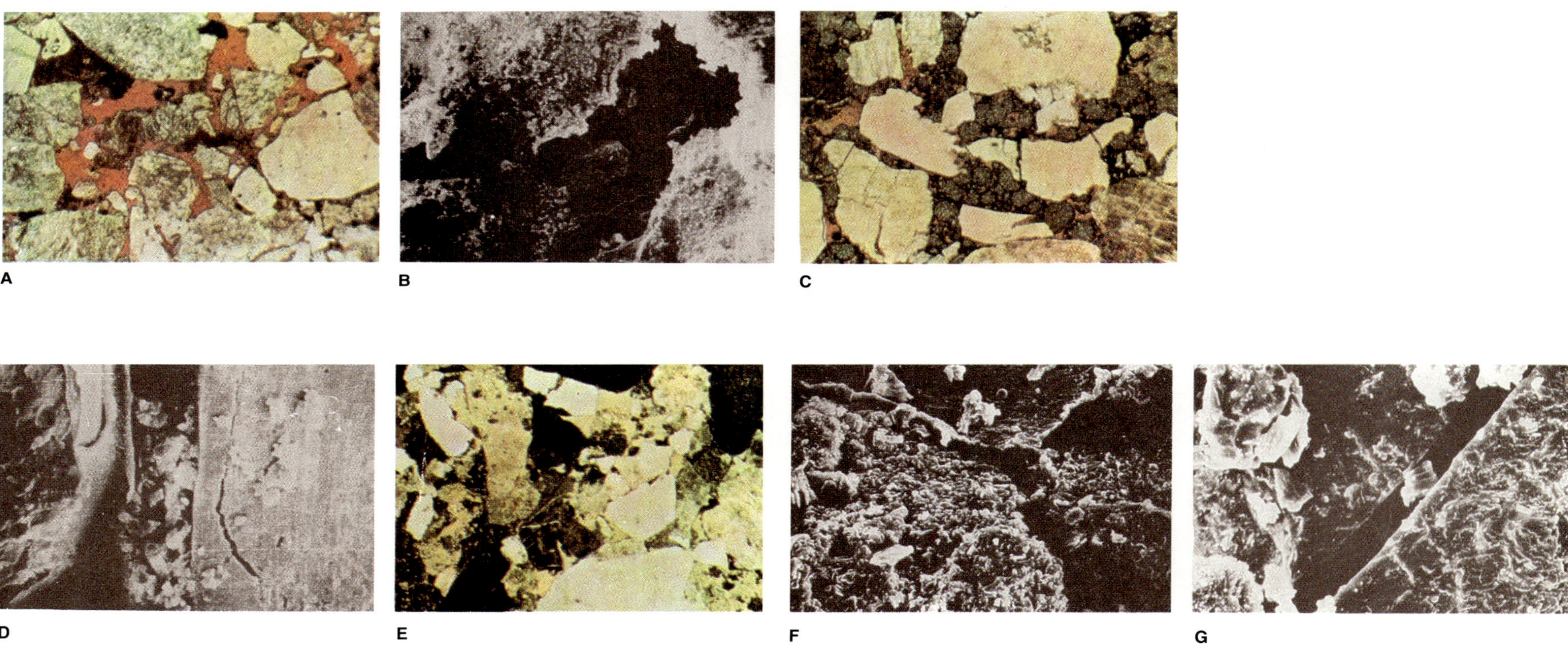

Figure 6. (A) Resin-impregnated photomicrograph of a thin section of sandy conglomerate, I-type reservoir, in a member of the Middle Triassic Kelamayi Formation, showing the development of intergranular pores with connecting network. (B) SEM micrograph of intergranular pore (dark area) in center part of (A). The pores are 100μ in diameter and are quartz lined. (C) Resin-impregnated photomicrograph of poorly sorted conglomerate, I-type reservoir, showing the well-connected dissolution pores in analcime. (D) SEM micrograph of (C), showing the intergranular fissures partly filled with poorly formed crystalline kaolinite. (E) Resin-impregnated photomicrograph of I-type reservoir, showing the conglomeratic sandy texture. The main porosity (gray areas) is intergranular, well connected, with little infilling. Intragranular pores are rare. (F) SEM micrograph of sandy conglomerate, II-type reservoir, showing the well-connected minor fissures. Kaolinite and illite are intergrown. Scale bar, 100μ. (G) Larger magnification SEM micrograph of (F), giving a better view of minor fissure. Scale bar, 100μ.

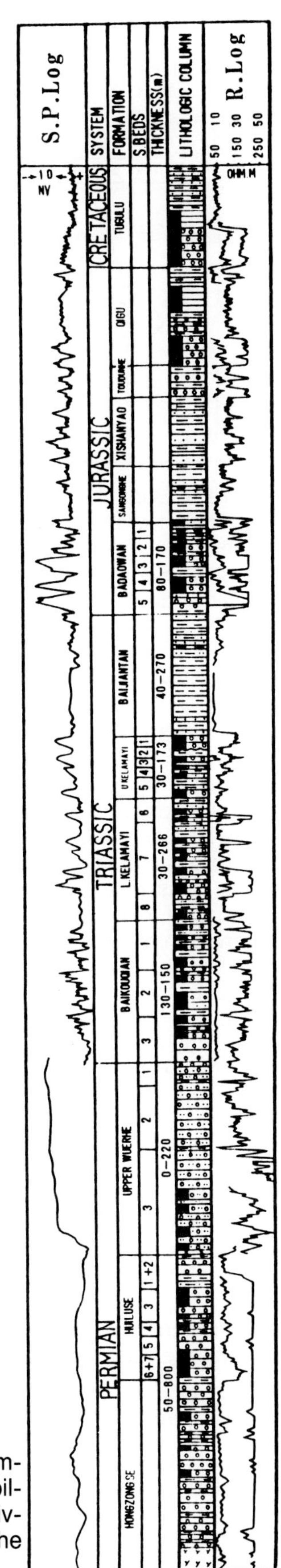

Figure 7. Generalized composite column of Kelamayi oil-bearing trend. R. Log, resistivity curves. Basement is the Xibeikulas Formation.

subordinate faults divides the field into many blocks. Faulting plays an important role in forming pools and their distribution, as indicated below:

1. The faults divided the field into several blocks that have their own oil-gas-water system.
2. They act as a seal in the updip direction of a pool.
3. Some faults are the channels through which the oil and gas migrate vertically and form secondary pools in the Upper Triassic and Lower to Middle Jurassic reservoirs.
4. The imbricated fault zone results in thickening of the reservoirs along the fault belt.
5. Fracturing along the fault zone increases the permeability within the reservoir.

As a result of faulting, traps have been created along the Kelamayi-Wuerhe fault belt that contain many thick reservoirs with abundant hydrocarbons, large reserves, and high-capacity wells.

Source Rocks

Based on geochemical data from correlation studies of different source rocks, the Wuerhe Formation of Late Permian age is considered to be the major source rock (Figure 8). This formation is a thick series of conglomerates in the Kelamayi field, however, and it presents no conditions for hydrocarbon generation. Sediments in the field become more fine grained from the edge to the center of the basin. The Manas Lake area, which is southeast of the Kelamayi field, is the depositional center for the Wuerhe Formation where argillite contains an average 1.01% of organic carbon and an average of 283 ppm of chloroform bitumen "A"; it is therefore considered a poor source rock. The argillite in the deeper part of Manas depression has not been drilled so far. These same layers have numerous outcrops along the southern edge of the basin (Wulumuqi and its eastern suburbs, Figure 1) where it is a series of dark lacustrine argillite and oil shale with some bitumen and heavy oil. The organic carbon may be as high as 10%, and chloroform bitumen "A" is up to 1000 to 3000 ppm; kerogen type II is dominant, while types I and III are rare. According to the geochemical data from the northwest edge of the basin, the Permian source rock of the Manas Lake depression passed the generation threshold during Late Triassic (Figure 9). The Jurassic might be the major hydrocarbon generation and drainage period. The structural framework on the northwest edge of the basin, such as monocline, faults, and folds, had been formed by the end of the Triassic, so there was an excellent concert among generation, drainage, and migration of hydrocarbons.

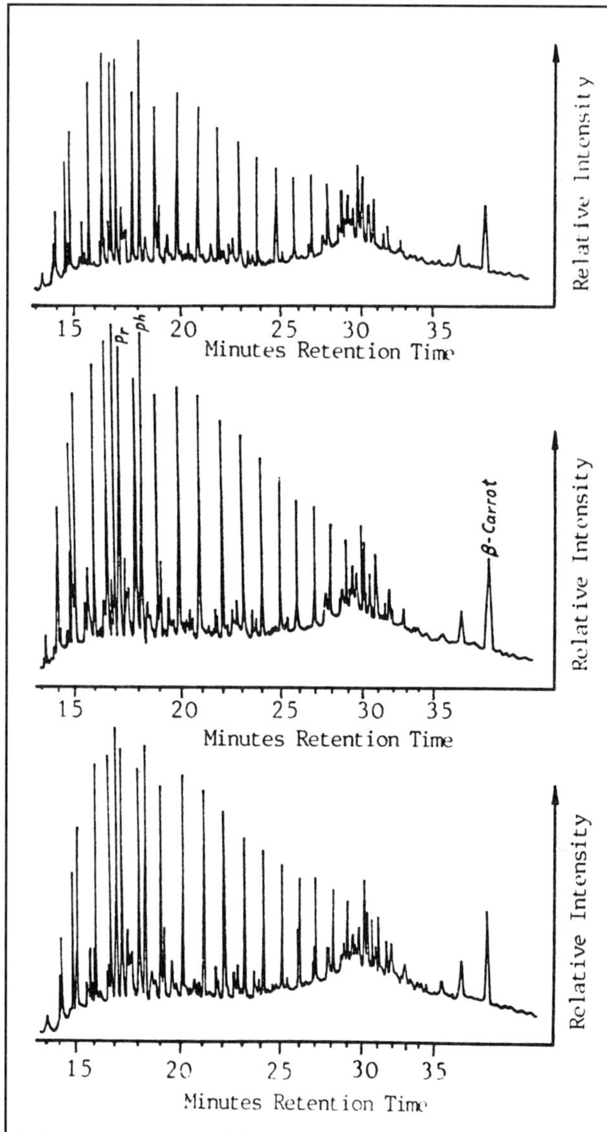

Figure 8. Gas chromatography charts of crude oil from Permian (top), Triassic, and Jurassic reservoirs in northwest edge of Zhungeer basin.

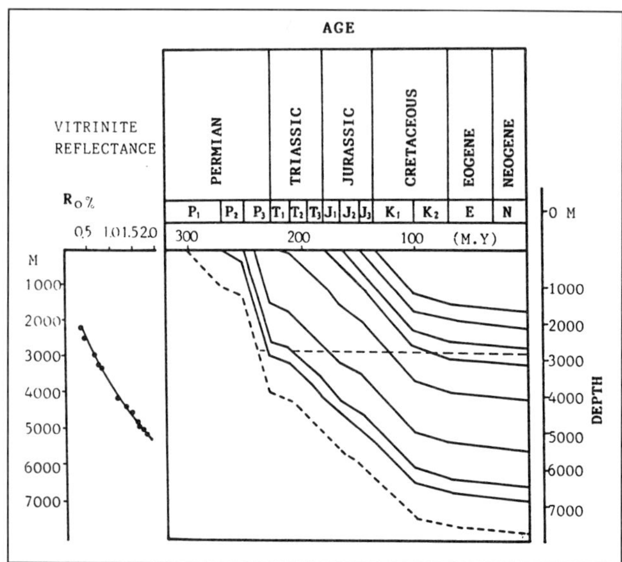

Figure 9. Burial history curve of Permian source bed in Aichan-1 well near Manas source depression. To the left vitrinite reflectance, R_o, is plotted against depth in meters. Subscripts indicate "lower" (as P_1), "middle" (P_2), and "upper" (P_3).

EXPLORATION CONCEPTS

Regional Play

The Kelamayi oil area is a regional oil-bearing trend controlled by both the Kelamayi-Wuerhe fault belt and the regional monocline. As mentioned above, on the downthrown side of the fault the hydrocarbons are trapped mainly by the composite action of fault sealing, the regional monocline, and impermeable lithologies. Therefore, a series of "fault-lithologic oil traps" were formed on the downthrown side of the fault. In the Wuerhe-Xiazijie area, in the northern part of the oil trend, most of the oil pools are in anticlinal traps, and the main oil-bearing formations are Permian and Triassic in age. In the area adjacent to the main fault, the hydrocarbons are trapped primarily in the fault blocks and confined by the faults; thus a series of fault-block oil entrapments have been formed. The major oil-bearing formations are Jurassic in age. In basement-faulted blocks other than Triassic and Permian, the oil-bearing beds are missing. On the upthrown side of the faults, the migration of hydrocarbons was controlled by the overlapping unconformity of the Upper Jurassic against Lower Cretaceous sediments, and also self-sealing by bitumen formation. Thus, basement and "unconformable overlap entrapments" and heavy oil pools were formed on the upthrown block. Therefore, along the fault belt a number of oil accumulations occur in a variety of traps, all of them related to the fault system.

General Application of Geological Parameters

As one of China's numerous fields producing oil of continental origin, the Kelamayi oil field possesses distinctive features as to geological conditions:

1. The Permian lacustrine oil-generating depression, which is situated on the downdip side of the regional slope on the northwest margin of the basin, provided a sufficient source of oil for the Kelamayi oil pools; the hydrocarbons originated from the downwarp formations and migrated updip along the slope. Kelamayi, located along the

migration path, has good associations of "generation-reservoir-trapping" conditions for oil accumulations.

2. The Kelamayi-Wuerhe fault belt, which began forming at the end of the Paleozoic, is made up of structural terraces descending toward the center of the basin. This made it possible for the foreland fan deposits of Mesozoic age to develop along the downthrown side of the fault belt. The series of conglomerates were deposited on the downthrown side before the hydrocarbons migrated out of the area and thus became the main reservoirs.
3. The unconformities between the basement and the overlying sediments are the pathways for oil migration.
4. The Kelamayi-Wuerhe thrust fault belt, which cuts across the regional monocline on the northwest margin of the basin, prevented the hydrocarbons from migrating further updip and caused their accumulation along the fault belt.
5. After the sedimentation in Early and Middle Triassic, a large lake developed during the Late Triassic in which were deposited a series of thick-bedded mudstones over a wide area, including the oil area. The Baijiantan Formation is the cap rock of the Kelamayi field area. From Late Jurassic to Early Cretaceous time, the sedimentation extended continuously toward the peripheral mountain region, and in combination with faulting, produced some basement traps by sealing the fractures and pores of the basement which had been eroded in an earlier stage.
6. The reservoirs are heterogeneous formations with medium to low permeabilities (Table 2 and Figure 6). The center parts of alluvial fans are the best oil-bearing reservoirs.
7. The oil pools lack active water drive but are highly saturated with gas.

Lessons

After more than 30 years' experience in exploration and development of the Kelamayi oil field, the following lessons may be used for reference:

1. After discovering oil and gas, we focused our attention on regional tectonics and explored further along the entire northwest margin of the basin with the use of seismic evaluations and ten drilling profiles. As a new oil field was discovered, regional geologic parameters were made clear in a short time, and a large oil-bearing area was confirmed on the northwest margin of the basin.
2. In the early stage of exploration, we studied the facies and distribution peculiarities of the oil pools of the Kelamayi Formation. We concluded that the hydrocarbon accumulations were controlled by a paleostructure and the fan-shaped alluvial deposits. The Kelamayi-Wuerhe fault belt had a positive effect upon oil accumulation, resulting in

Table 2. Kelamayi field reservoir characteristics.

Age	Reservoir Lithology	Depth (m)	Lithofacies	Pore Types	Porosity (%)	Permeability (md)	Pressure Coefficient	Thickness of Pay Beds (m)	Properties of Crude Oil (density, viscosity, and gas-oil ratio [GOR])
Tugulu Fm. (K1t)	Sandstone	50 to 300	fluvial lacustrine	Intergranular	35	1700 to 3000	1	5 to 10	r = 0.98
Qigu Fm. (J3q)	Sandstone	150 to 500	fluvial	Intergranular	34	310 to 6700	1	14 to 31	r = 0.98
Badaowan Fm. (J1b)	Sandstone conglomerate	800 to 1500	fluvial	Intergranular	20	137	1.1 to 1.3	20	r = 0.865 μ = 8.5 cp
Kelamayi Fm. (K2k)	Conglomerate, conglomerate sandstone	300 to 3500	proluvial to alluvial	Intergranular	18	279	1.3 to 1.5	15	r = 0.85-0.87 μ = 4-12 cp GOR = 50-150
Wuerhe Fm. (P2w)	Conglomerate	2000 to 3500	fluvial alluvial	Secondary pores	9	0.8	1.3	83	r = 0.84 μ = 0.7 cp GOR = 158
Basement (C1+2)	Andesite-basalt, volcanic detrital rock	500 to 3000	marine	Gas pore dissolved pore fracture	8.8	0.37		20 to 200	

high production capacities. This knowledge was helpful for later exploration.

3. When there are oil-forming conditions in a deep sedimentary depression of an asymmetric basin, then the gentle slope is an important "direction index" for hydrocarbon migration. If there are fault, stratigraphic, lithological, and other types of traps on the slope in an updip direction, then hydrocarbons can accumulate. If there are numerous oil and gas shows in outcrop areas, the exploration along the slope in a downdip direction will encounter oil fields similar to the Kelamayi field.
4. If the basement unconformity interface serves as the passageway for oil migration, basement oil pools may be formed if there are fractures and pores in the basement rocks. Trapping conditions would be best developed where faulting and upfolding caused the basement to be locally higher than the unconformity interface.
5. In the Kelamayi field, the recovery factor is very low because of the reservoir drive mechanism, a solution gas drive. It is very difficult to form a homogeneous frontal advance of the water level by means of linear flooding patterns, thus areal flooding patterns work more efficiently.
6. The Kelamayi oil field is a field with multiple oil pay zones and multiple types of reservoirs. During the early period of field development, exploration was aimed only at the oil-bearing beds of the Kelamayi Formation. The oil-bearing beds in the Wuerhe Formation were discovered in the 1960s, and the basement oil pool was not found until the 1980s. It took more than 30 years to develop the field's full potential. If the factors governing oil distribution of the Kelamayi field had been discovered earlier, the duration of exploration might have been shortened and costs would have been reduced.

REFERENCES

Chang Chi-yi, 1981, Alluvial-fan coarse clastic reservoirs in Karamay, *in* J. F. Mason, ed., Petroleum geology in China: Pennwell Books.

St. John, B., A. W. Bally, and H. D. Klemme, 1984, Sedimentary provinces of the world—hydrocarbon productive and nonproductive: AAPG map + 35 p.

Tissot, B. P., and D. H. Welte, 1984, Petroleum formation and occurrence: New York, Springer-Verlag, 669 p.

Appendix 1. Field Description

Field name .... *Kelamayi field*

Ultimate recoverable reserves .... *Approximately 2240 million bbl*

Field location:

- **Country** .... *People's Republic of China*
- **State** .... *Xinjiang Uygar Autonomous Region*
- **Basin/Province** .... *Zhungeer basin*

Field discovery:

- **Year first pay discovered** .... *Middle Triassic Kelamayi Formation 1955*
- **Year second pay discovered** .... *Lower Jurassic Badaowan Formation 1960*
- **Year third pay discovered** .... *Upper Permian Wuerhe Formation 1965*
- **Year fourth pay discovered** .... *1980s (in Paleozoic basement which obtained commercial flow in 1957, but not brought to geologists' attention at that time)*

Discovery well name and general location:

- **First pay** .... *Ke-1 well located 7 km northwest of Kelamayi city*
- **Second pay** .... *Ke-56 well, situated 29 km northeast of Kelamayi city*
- **Third pay** .... *Jianwu-1 well, located 25 km southeast of Kelamayi city*
- **Fourth pay** .... *Ke-222 well, located 41 km southeast of Kelamayi city*

Discovery well operator ...... *A team of explorationists jointly organized by Sino-Soviet Petroleum Company*

- **Second pay** .... *Same as first pay discovery*
- **Third pay** .... *Same as first pay discovery*

IP:

- **First pay** .... *50 bbl/day*
- **Second pay** .... *67 bbl/day*
- **Third pay** .... *73 bbl/day*
- **Fourth pay** .... *50 bbl/day*

All other zones with shows of oil and gas in the field:

Age	Formation	Type of Show
Upper Jurassic	*Qigu*	*Heavy oil*
Lower Cretaceous	*Tugulu*	*Heavy oil, bituminous sands*

Geologic concept leading to discovery and method or methods used to delineate prospect

Kelamayi oil-bearing trend is located on the northwest flank of Zhungeer basin, and the sedimentary strata covering this area dip gently toward the center of the basin. The surface geological survey made in the 1950s discovered many seeps such as asphalt dunes, veins of bitumen, and oil shows. It is recognized that various shows distributed on the northwest flank are the result of oil and gas generated in the basin migrating updip to the northwest margin. Under the guidance of this knowledge, a drilling program along the downdip direction was proposed in 1955. Ke-1 well, drilled in this exploration effort, was the discovery well of the Kelamayi field.

Structure:

Province/basin type .... *Bally 23; Klemme IIIA*

Tectonic history

Zhungeer basin, formed during the late Paleozoic, has a basement of Middle to Lower Carboniferous. Sediments of Upper Carboniferous to Permian are marine-continental intertonguing facies, forming lacustrine source beds of Permian age. Mesozoic–Cenozoic sediments are entirely continental. The Kelamayi-Wuerhe thrust fault formed during Triassic to Jurassic time. The hydrocarbons migrated from the source depression to the northwest margin and accumulated along the fault belt.

Regional structure

Field lies on the northwest monoclinal flank of the basin, and the hydrocarbons accumulated along both sides of the Kelamayi-Wuerhe thrust fault. The pay beds dip southeastward with dip angles of 2 to 5°.

Trap:

Trap type(s)

Four major types of oil pools have so far been determined in the field: (1) fault block pool with multi-oil reservoirs; (2) unconformity pool; (3) secondary fault sealed pools; and (4) basement reservoir

Basin stratigraphy (major stratigraphic intervals from surface to deepest penetration in field):

Chronostratigraphy	Formation	Depth to Top in m
Lower Cretaceous	*Tugulu*	*0-500*
Upper Jurassic	*Qigu*	*500-600 m*
Lower and Middle Jurassic	*Coal Measures*	*600-800*
Lower and Middle Triassic	*Baijiantan, Kelamayi*	*700-1000*
Upper Permian	*Wuerhe and Xiazijie*	*500-1200*
Lower and Middle Carboniferous	*Xibeikulas (basement)*	*1500-3200*

KELAMAYI

Reservoir characteristics:

Number of reservoirs *6*

Formations .... *Kelamayi (major reservoir), Wuerhe (Permian), Badaowan and Qigu (Jurassic), Tugulu (Cretaceous), and Basement (Xibeikulas)*

Ages .... *From Carboniferous to Cretaceous*

Depths to tops of reservoirs .... *300 to 3000 m*

Gross thickness (top to bottom of producing interval) .... *230 m (avg.)*

Net thickness—total thickness of producing zones

Average .... *13 m (8 to 18 m in general)*

Maximum .... *36 m*

Lithology .... *Poorly sorted, fine-grained sandstone to conglomerate*

Porosity type .... *Intergranular, highly variable porosity in conglomerate reservoirs of alluvial fans with interstitial sand, silt, and mud*

Average porosity .... *18.09% (14 to 21% range)*

Average permeability .... *279 md (50 to 150 range)*

Seals:

Upper

Formation, fault, or other feature .... *Baijiantan Formation, Ke-Wu fault*

Lithology .... *Mudstone*

Lateral

Formation, fault, or other feature .... *Fault, formation, and lithological change*

Lithology .... *Mudstone, bitumen, and unconformity*

Source:

Formation and age .... *Wuerhe Formation of Permian age*

Lithology .... *Shale and oil shale*

Average total organic carbon (TOC) *10.2%*

Maximum TOC *20.2%*

Kerogen type (I, II, or III) .... *II*

Vitrinite reflectance (maturation) .... R_o *= 0.51 to 1.34*

Time of hydrocarbon expulsion .... *Jurassic*

Present depth to top of source .... *1500 to 6000*

Thickness .... *500 m*

Potential yield .... *NA*

Appendix 2. Production Data

Field name .... *Kelamayi field*

Field size:

- **Proved acres** .... *407 km²*
- **Number of wells all years** *4893*
- **Current number of wells** .... *3467 (only 2724 wells are producing at present; the rest are shut-in)*
- **Well spacing** .... *Average 310 m (200 to 500 m in general)*
- **Ultimate recoverable** .... *Approx. 2240 million bbl; approx. 1400 million bbl developed*
- **Cumulative production** .... *Approximately 656.4 million bbl (1990)*
- **Annual production** .... *Approximately 47.6 million bbl in 1990*
- **Decline rate** .... *NA*
- **Annual water production** .... *32 million bbl*
- **In place, total reserves** .... *Approximately 8260 million bbl*
- **In place, per acre foot** .... *NA*

Drilling and casing practices:

- **Amount of surface casing set** .... *NA*
- **Casing program**

  Deep well: 20-in., 14-in., 10-in., 6-in./5-in.
  Shallow well: 16-in., 12-in./10-in., 6-in./5-in.

- **Drilling mud** .... *Water-based mud*
- **Bit program** .... *Deep well: 19 1/3-in., 13 3/4-in., 9 3/4-in., 7 3/4-in.*
- **High pressure zones** .... *None*

Completion practices:

- **Interval(s) perforated** .... *NA*
- **Well treatment** .... *Hydraulic fracturing*

Formation evaluation:

- **Logging suites** .... *Resistivity, spontaneous potential, lateral, laterolog, GR, sonic, isotope tracer, well-temperature, dip log, density, induction log, microspherically focused log, microlog, caliper, etc.*
- **Testing practices** .... *Drill-stem test, wireline formation tester*
- **Mud logging techniques** .... *Gas instrument (includes testing light hydrocarbon [C1 to C4], heavy hydrocarbon [>C4], and total hydrocarbon)*

Oil characteristics:

- **Type** .... *NA*
- **API gravity** .... *35–24°*
- **Base** .... *Cycloalkyl*
- **Initial GOR** .... *22–147.6 m³/ton*
- **Sulfur, wt%** .... *NA*
- **Viscosity, SUS** .... *3.2 to 210 cp*
- **Pour point** .... *+18° to –16° C*

Field characteristics:

- **Average elevation** .... *260 m*
- **Initial pressure** .... *62.86 to 314.36 bar (911 to 4556 psi)*
- **Present pressure** .... *51.75 to 267.72 bar (750 to 3880 psi)*
- **Pressure gradient** .... *0.131 to 0.138 bar/m (0.58 to 0.61 psi/ft)*
- **Temperature** .... *18.1 to 54.4° C*
- **Geothermal gradient** .... *0.0411 to 0.0241° C/m*

Drive .... *Dissolved gas drive, elastic-dissolved gas drive*
Oil column thickness .... *8 to 18 m*
Oil-water contact .... *–1200 m to 2300 m*
Connate water .... *30 to 40%*
Water salinity, TDS .... *2.8 to 55.4 g/L*
Resistivity of water .... *0.4 to 1.0 ohm*
Bulk volume water (%) .... *NA*

Transportation method and market for oil and gas:

Pipeline carries crude oil directly from Kelamayi field to the Dushanzi refinery in the southern margin of the basin and Wulumuqi refinery, and then part of crude oil is transported by train from Wulumuqi to Lanzhou refinery, the capital of Gansu province

Santa Rosa Field—Venezuela
Eastern Venezuela Basin

VIRGILIO VILLAROEL
Corpoven S.A.
Puerto La Cruz, Venezuela

FIELD CLASSIFICATION

BASIN: Eastern Venezuela
BASIN TYPE: Foredeep
RESERVOIR ROCK TYPE: Sandstone
RESERVOIR ENVIRONMENT
OF DEPOSITION: Distributary Channel and Mouth Bar Sandstones
RESERVOIR AGE: Oligocene to Miocene
PETROLEUM TYPE: Oil and Gas
TRAP TYPE: Anticline
TRAP DESCRIPTION: Anticline along leading edge of thrust sheet; reservoir pinch-outs control distribution of production on structure

LOCATION

The Santa Rosa field is in the Eastern Venezuela Basin, which underlies the states of Guárico, Anzoátegui, Monagas, and Delta Amacuro, for the most part (Figure 1A). The southern boundary is the Guayana shield; the northern boundaries are the Venezuelan Andes, Cordillera de la Costa, and the Eastern Serrania del Interior. The basin is separated from the Barinas-Apure basin to the west by the El Baul arch (Figure 1B). The eastern end extends to the sea. The basin is subdivided into a western subbasin, the Guárico, and an eastern subbasin, the Maturín, by the Anaco-Altamira fault trend. The area is further separated by the Anaco thrust fault into the Greater Oficina and Greater Anaco areas (Figure 1C). The Santa Rosa field lies in the Greater Anaco area, about 6.2 mi (10 km) northeast of the town of Anaco, Anzoátegui.

In addition to the Santa Rosa field, the Santa Ana, San Joaquín, and Guario fields were originally included in the Anaco trend by Funkhouser et al. (1948); at present, the El Toco, La Ceiba, La Vieja, and El Roble fields (Figure 1C) also are included since all were formed by the same tectonic forces. The Santa Rosa field is the largest and most important of the Anaco trend fields because of its large reserves of condensate, light oil, and gas.

The Santa Rosa field has a proven area of 51,362 ac (20,800 ha), an oil column of 9870 ft (3010 m), and an estimated ultimate recovery of 693.7 million bbl (MMBO); and it is thus classified as a giant field.

Of the nearly 300 oil fields discovered in Venezuela to the end of 1976, 32 could be classified as giant, according to Martinez (1976). Santa Rosa field was recognized as a giant in 1948, seven years after its discovery in 1941.

The topography is low and shrub covered, with elevations ranging from 400 to 600 ft (120–180 m) above sea level in the El Toco, Santa Ana, San Joaquín, El Roble, and Guario fields to more than 700 ft (214 m) in the Santa Rosa field and 1400 ft (427 m) in La Ceiba field. Locally maximum topographic relief of the area is some 300 ft (92 m), as reported by Funkhouser et al. (1948).

The principal small streams (*quebradas*) crossing the Anaco trend area from south to north are the Misa Cantada, Guere, Meria, Guaro, Chiguapo, Orocopiche, Aragua, Guario, and Anaco. Most of their water drains from the Mesa Formation sands and gravels that crop out in the area. They are intermittent on the surface but flow continuously below the stream beds through sand deposits, thereby helping to maintain a good hydrodynamic gradient in the sandstones of the upper Oficina formation. They also provide water for drilling operations.

Only the crest of the structural trend is exposed, and here beds dip from 11° to 20°, thus outlining reliably the surface geologic and topographic features, as mentioned by González de Juana et al. (1980) and shown by Banks and Driver (1957) (Figure 2).

HISTORY

Figure 3 shows the regional structure at the top of the Colorado-E sandstone and the chronology of the discovery of the producing fields along the Anaco trend.

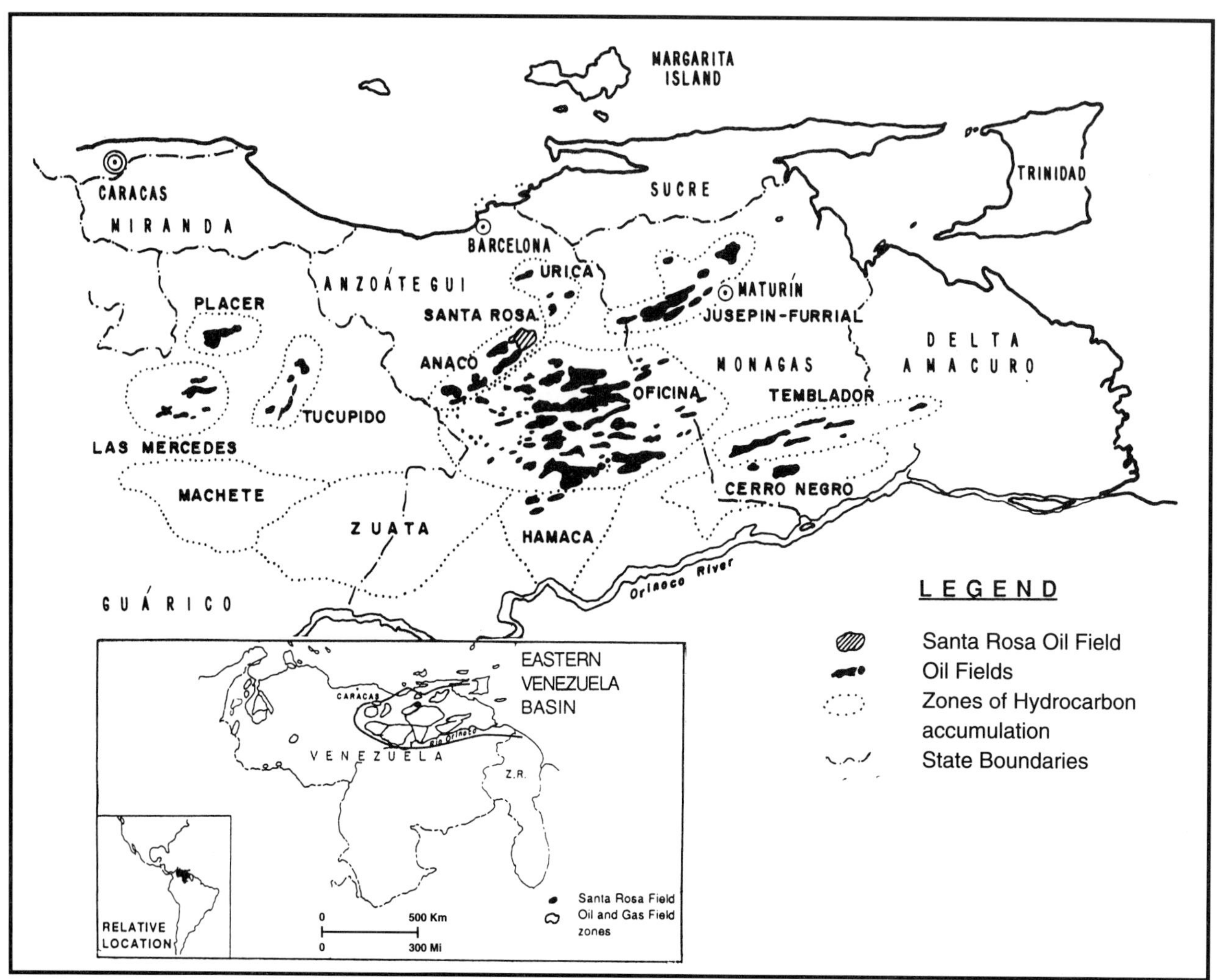

Figure 1A. Geographic location of the Santa Rosa oil field in relation to the other oil fields in the Eastern Venezuela Basin.

Pre-Discovery

The early exploration in the area was described by Funkhouser et al. (1948). The first well drilled on the Anaco trend was Santa Rosa No. 1 (RG-1, Figure 3) at a location based on a single seismic refraction seismograph line run through the area in 1931. RG-1 was spudded by Mene Grande Oil Company on 14 February 1934 and was plugged and abandoned on 8 May 1936 after reaching a final depth of 7214 ft (2200 m). A test showed gas at 5800 ft (1769 m), and an excellent cored section was of great value in defining the general stratigraphy of the region. Additional seismic lines and new drilling showed the well to be downdip on the east flank of the Santa Rosa structure.

The actual discovery of the Santa Ana and San Joaquín anticlines was made in 1934, following aerial photographic studies and surface geological reconnaissance (Figure 2). During 1935 and 1936, geologists of the Creole Petroleum Corporation and the Mene Grande Oil Company (subsidiary of the Gulf Oil Corporation) carried out detailed surface geological work over the San Joaquín and Santa Ana domes as well as extensive reflection seismograph surveys.

In 1937, AM-1 (Figure 3) in Santa Ana field was drilled to a total depth of 7628 ft (2326 m) and tested at rates of 3 MMCFG and 50 BPD of condensate. By that time, geologists had a good idea of the structure of the Anaco trend.

In 1938, Mene Grande spudded Santa Ana-2 (AM-2, Figure 3) and obtained a wet gas test, and later drilled AM-3 on the north flank of the west dome. This latter well was completed in April 1941 as the first oil well of the Santa Ana field, producing 666 bbl/day of 34° API oil from the Merecure Formation. Also in 1938, Creole Petroleum Corporation spudded JMN-1 and then JM-2 (Figure 3) in San Joaquín field. The JMN-1 well initially flowed 850 bbl/day of 39° API oil through a ½-in. choke, and JM-2 tested 1338 bbl/day of 37° API oil through a 19/32-in. choke.

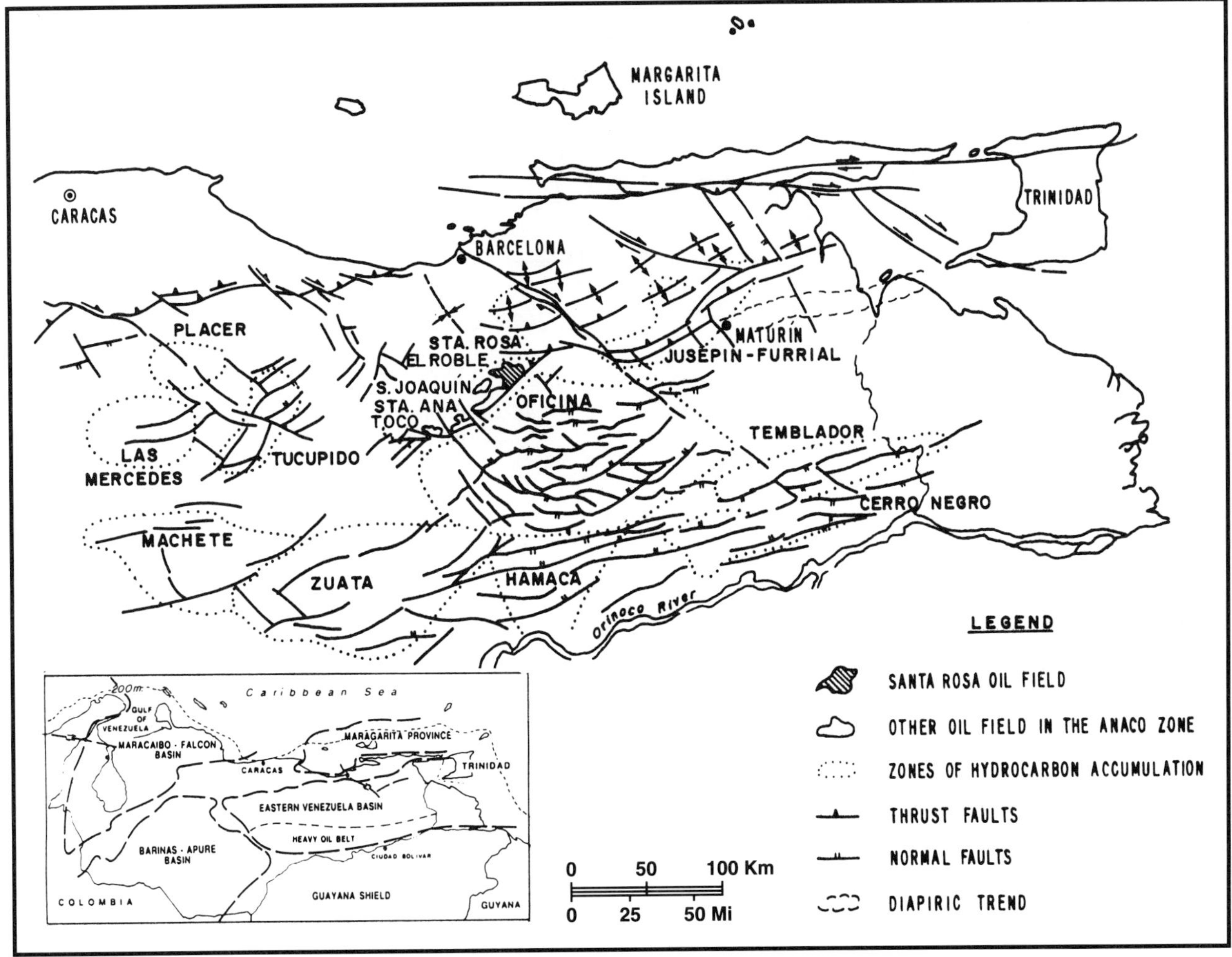

Figure 1B. The structural setting of the Eastern Venezuela Basin and the relative location of the Santa Rosa field and the other fields of the Anaco anticlinal trend.

Discovery

Mene Grande Oil Company drilled well RG-2 (Figure 3) on the Santa Rosa dome in May 1940, 5 mi (8 km) southwest of RG-1 and 6.2 mi (10 km) northeast of the town of Anaco. The well location was based on results of drilling in Santa Ana and San Joaquín fields and seismic surveys over the Santa Rosa dome that indicated RG-2 would be structurally higher than RG-1. Well RG-2 was the discovery well for the Santa Rosa field, completed on 24 January 1941, flowing 715 bbl/day of 42° API oil from the Verde-J sand unit of the Oficina Formation of Miocene age (Figure 4).

Post-Discovery

The second pay, discovered in well RG-9 (Figure 3) by Mene Grande in July 1948, was in the Merecure Formation, where the Merecure-O sand unit (Figure 4) was completed for 581 bbl/day of 37.7° API oil.

The third pay was discovered by Mene Grande's well RG-134 (Figure 3), drilled in 1966 and completed after redrilling in October 1972, at a rate of 107 bbl/day of 45.6° API oil from the San Juan-A sand unit (San Juan Formation of Cretaceous age, Figure 4). Other shows of oil occur in the Cretaceous San Antonio Formation.

The field was developed and operated by Mene Grande Oil Company until January 1976, and thereafter by the national oil company, S.A. Meneven, which became part of Corpoven, S.A. in June 1986. At present, the field has 259 wells that have penetrated 125 oil sand units and have been completed in 440 reservoirs in 122 of those sand units.

DISCOVERY METHOD

Discovery of the field was made possible by exploration seismic surveys, photogeological studies,

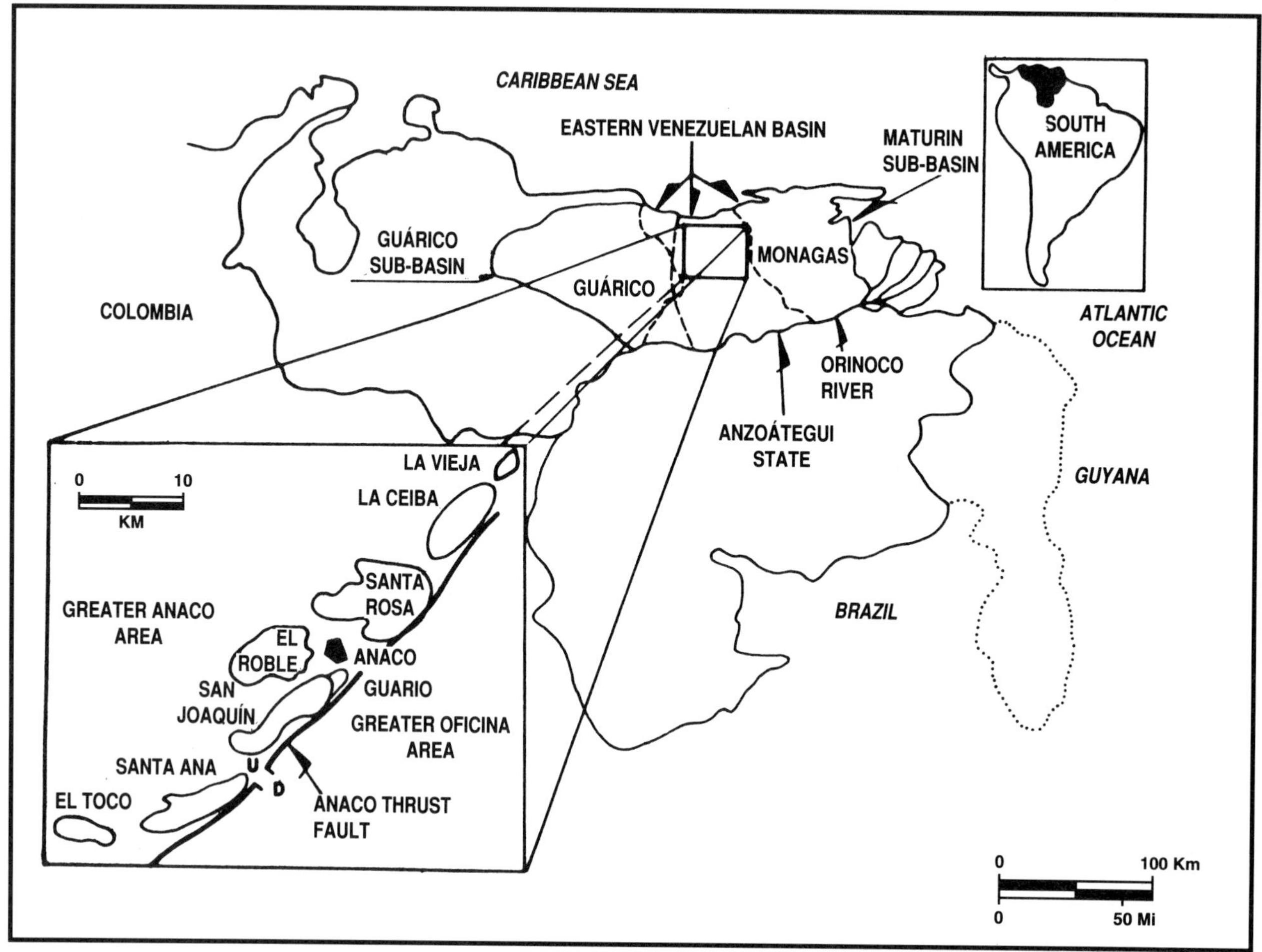

Figure 1C. Greater Anaco area showing the location of the Santa Rosa field and the other major producing fields.

and surface geologic mapping. These were followed by: (1) drilling and completion of RG-1 and analyses of its cores; (2) drilling and completion of AM-1 and subsequent stratigraphic correlation of strata penetrated with those of RG-1, followed by successful completion of other wells in the Santa Ana and San Joaquín fields; and (3) drilling and completion of the discovery well Santa Rosa-2 (RG-2).

Using today's exploration techniques, a similar discovery would be accomplished by (1) gravity surveys and photogeology, (2) reflection seismic surveys, (3) drilling, logging, and coring of first wells, and (4) high resolution 3-D seismic surveys for the development of the field.

STRUCTURE

Tectonic History

The structure of the Anaco trend has been known for more than 50 years because it forms the southernmost thrust-faulted anticlinal structure of the compressional forces acting from the north against the South America plate and causing the deformation and uplift of the Serranía del Interior (Figure 5). The structural history was interpreted by Banks and Driver (1957), who considered the origin of the trend to be a northwest-dipping normal fault with doming in its downthrown segment. These structures began to form at least as early as Oligocene time, contemporaneously with sedimentation.

While folding has continued uninterruptedly, possibly until the Recent, movement of growth faults stopped toward the end of middle Miocene time. Reversal of fault movement produced the present thrust as a result of compressional stresses in the basin that were active at different times along the trend. These deformative stresses were focused along a pre-existing line of weakness forming the Anaco trend in its present tectonic setting; the absence of pre-existing structures north of this trend might explain the lack of similar parallel, large features in that part of the basin.

The uplift and thrusting continued throughout Miocene time, causing erosion of the Oficina

Figure 2. Aerial photograph of the Santa Ana structure, which is similar to the Santa Rosa anticline, located farther to the north along the Anaco thrust fault. The discovery well for the trend was located in the Santa Ana structure. (From Funkhouser et al., 1948.)

Formation over the Santa Rosa dome and truncation of the overlying Freites Formation. One seismic section (Figure 6A) shows the Santa Rosa dome and the Anaco thrust fault. Another seismic section (Figure 6B) across the San Joaquín field provides a good example of the tectonic features of the Anaco trend: the anticlinal structure, the Anaco thrust fault, and related faults.

Regional Structure

The Anaco thrust consists of a northwest-dipping fault whose upthrown northwest flank is folded into a series of elongated domes oriented to the northeast, all asymmetrical toward the southeast, indicating pressure from the northwest. Most of the domes are structurally simple with only a few normal faults and (or) reverse faults. The Santa Rosa field occupies the northeastern, overthrust segment of the fault trend (Figure 3). The Guario, San Joaquin, Santa Ana, and El Toco fields are the other domes along this trend. The similarity between these individual structures requires a common origin.

Funkhouser et al. (1948) and Murany (1972) have suggested that the dip of the fault plane diminishes with the depth. Measured displacements decrease from northeast to southwest with a maximum of 7000 ft (2135 m) in Santa Rosa and a minimum of 700 ft (214 m) in Santa Ana.

Local Structure

The Santa Rosa dome is elongated in a N45°E direction. It is asymmetrical with a gentle northwest flank of 8° to 11° and a steep southeast flank dipping 20° into the Anaco thrust fault (Figure 7).

The dome is cut by two southeast-dipping faults that parallel the axis of the Santa Rosa dome, passing to the west of and near the crest of the structure (Figure 7). One is a normal fault in the northeastern and central part of the structure (first recognized in RG-15). The second fault (discovered in RG-2 and RG-5) is normal in the central part of the area but becomes an overturned fault in the southwest part of the dome. The northeast and southwest plunges

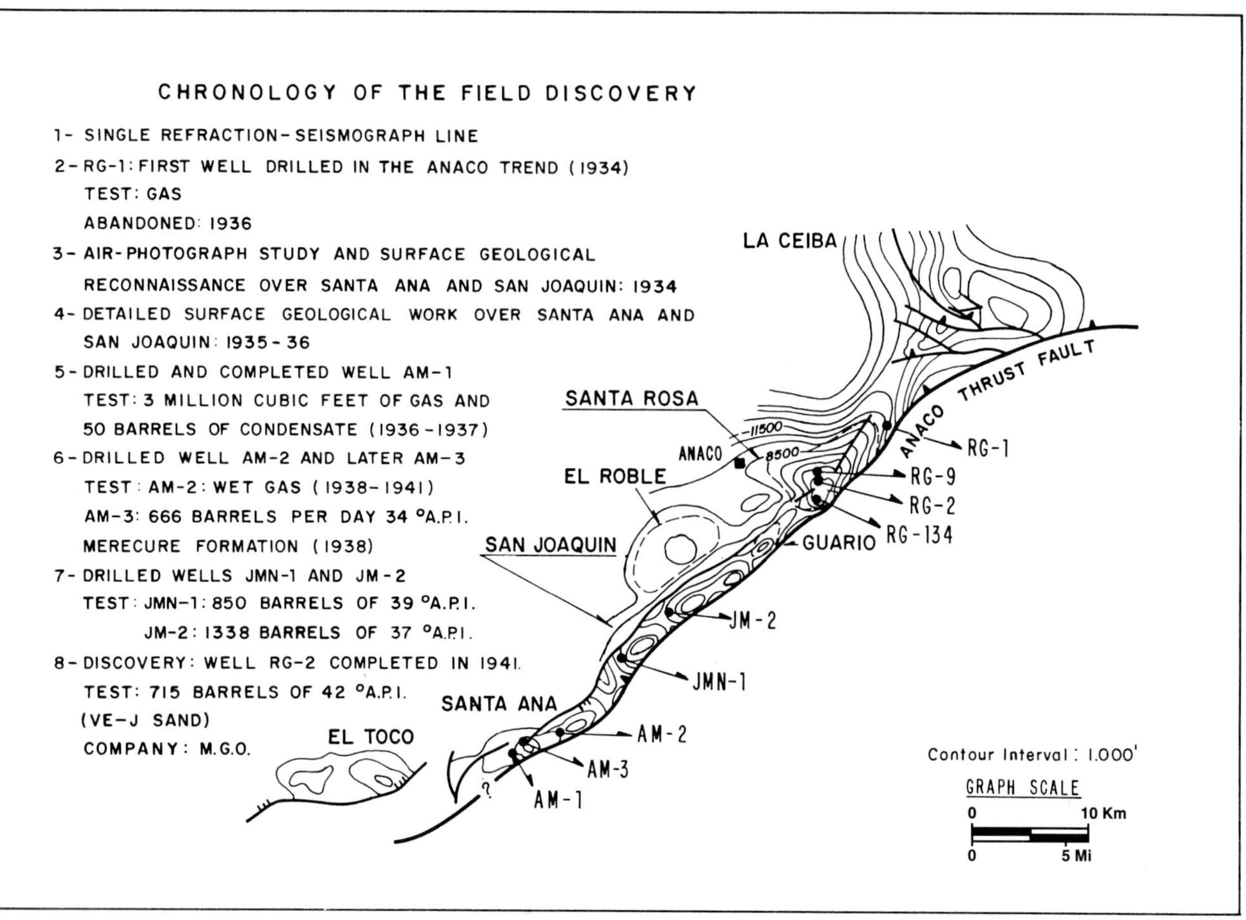

Figure 3. Regional structure of the Colorado-E sand unit along the northwest side of the Anaco thrust fault, the location of the wells mentioned in this study, and the chronology of the discoveries of the fields along the Anaco trend. Contour interval, 1000 ft.

of the anticline are illustrated by the well log cross section along the axis of the anticline (Figure 8).

The best data concerning the Anaco thrust fault have come from wells in the overthrust segment. However, Driver (1949), in attempting to establish stratigraphic equivalents between the Greater Oficina and Anaco areas by a detailed study of seismic lines, indicated that (1) there is an abrupt and abnormally high rate of basinward stratigraphic thickening of lower Oficina members northwestward across the Anaco fault and that (2) seismograph profiles EIR and ESR (Driver, figures 3 and 4, 1949), crossing the fault at Santa Ana, show an extraordinary variation of thrust throw and structural relief with depth.

STRATIGRAPHY

The general basin stratigraphy is shown by the columnar section of the Greater Anaco area (see Appendix 1). The oldest formation penetrated in the Santa Rosa field is the Cretaceous San Antonio Formation, whose base has not been reached. For convenience in stratigraphic work, the Oficina Formation has been divided, rather arbitrarily, into seven members based on a combination of lithologic, paleontologic, and electric log characteristics. The sandstones within each member have been grouped into sand units, which are designated by letter names. Individual sandstone beds within each sand unit have been numbered (Figure 4). The sandstones of the Merecure Formation have also been grouped into sand units, but are not shown in Figure 4. These are, in descending order:

Oficina Formation	
Blanco Member	eroded in field area
Azul Member	sand units A-A to A-R (A-A to A-C eroded in field area)
Moreno Member	sand units M-A to M-U
Naranja Member	sand units N-A to N-L1
Verde Member	sand units V-A to V-J
Amarillo Member	sand units AM-A to AM-L (A-A to A-L on Figure 4)
Colorado Member	sand units C-A to C-S

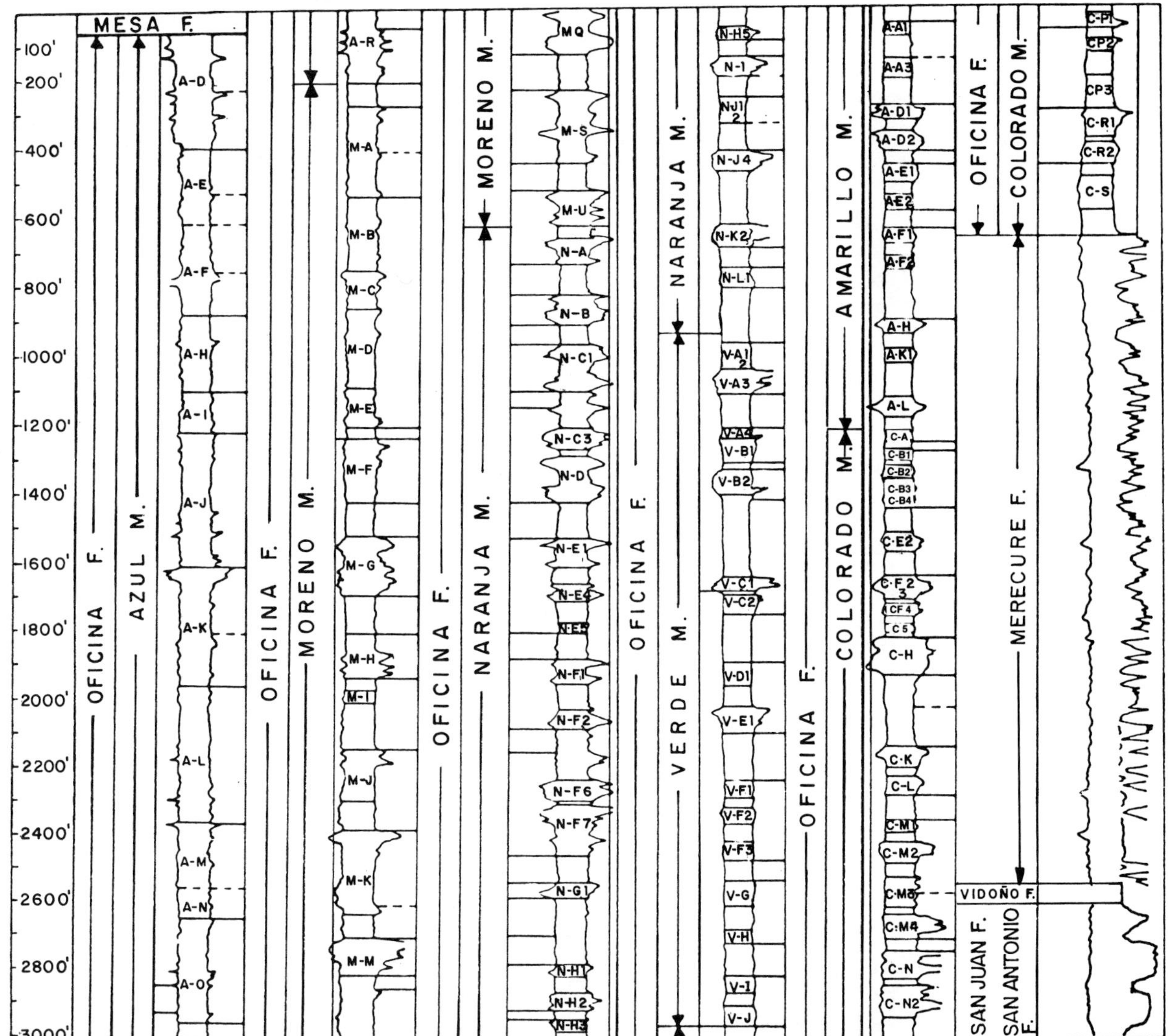

Figure 4. Stratigraphy of the Santa Rosa field, illustrated by a type log section. Formations and sand members or groups are shown in descending order from left to right. Sand units are distinguished by letters and numbers, such as Naranja sands N-C1 and N-F6.

Merecure Formation (No members)	sand units MER-A to MER-T (not indicated on Figure 4)

Lithologic characteristics of the formations encountered in Santa Rosa field are:

San Antonio	gray sandstones and dark shales with intercalated limestones
San Juan	massive gray well-sorted sandstones
Merecure	abundance of light gray to dark gray bedded sandstones interbedded with thin shales; also secondary growth of quartz grains
Oficina	gray and light beige-colored sandstones interbedded with gray shales; lignites, thin limestones, and green claystones are minor constituents

Typically, the sandstones are thought to represent bar or channel deposits in a deltaic environment. Thus the geometrical relationships among these deposits are complex and the reservoirs are characterized by several sedimentary environments (Total, 1985).

Lockwood and Aymard (1982) have divided the lower Oficina Formation into sedimentary packages. These reflect depositional environments and serve as a guide to facies interpretation (Table 1). Illustrating these criteria, a core from the Colorado-G sand of RG-16

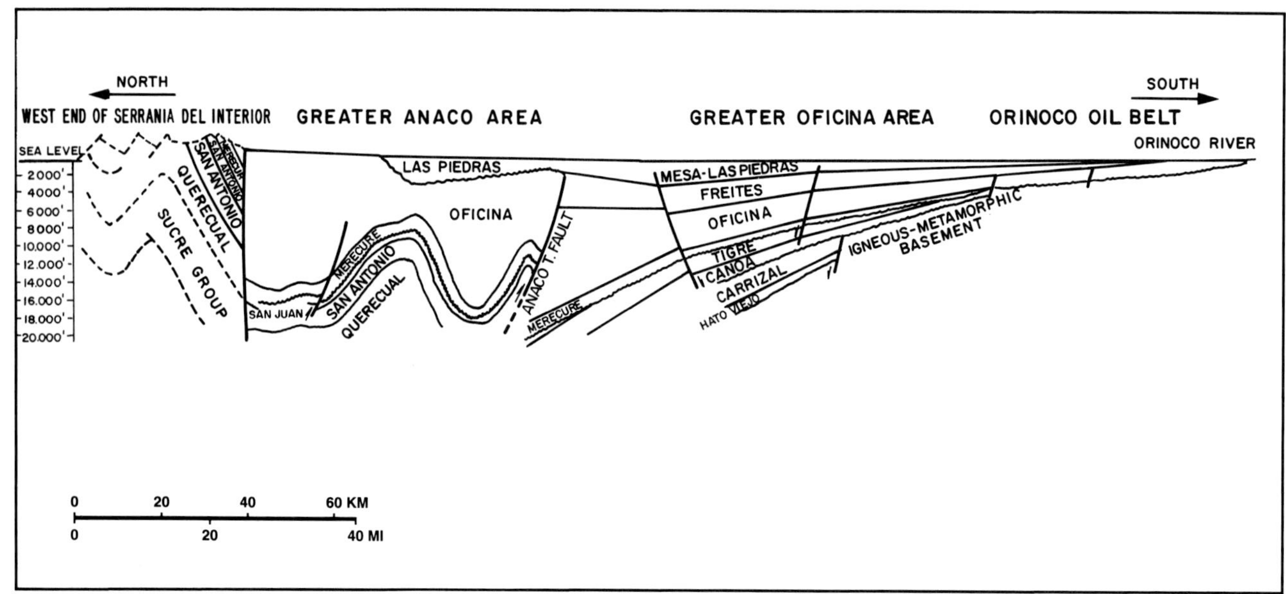

Figure 5. North-south cross section of the Eastern Venezuela Basin, from the Caribbean Sea (near Barcelona, see Figure 1A) to the Orinoco River. The Anaco trend was formed as the southernmost structure resulting from the compressional forces acting from the north against the Venezuelan plate.

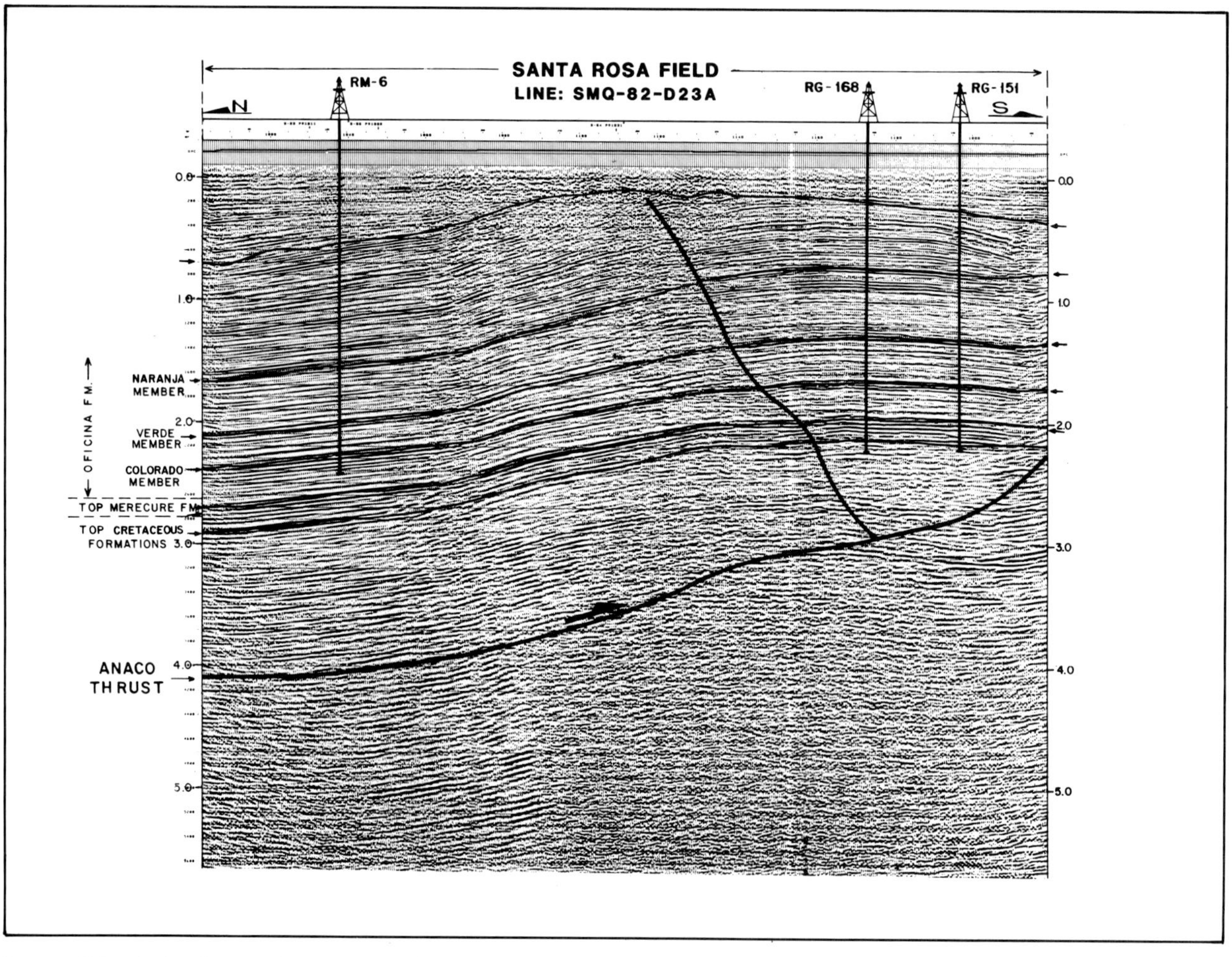

Figure 6A. Seismic section across the Santa Rosa field showing the anticlinal structure and the Anaco fault. See Figure 7 for location.

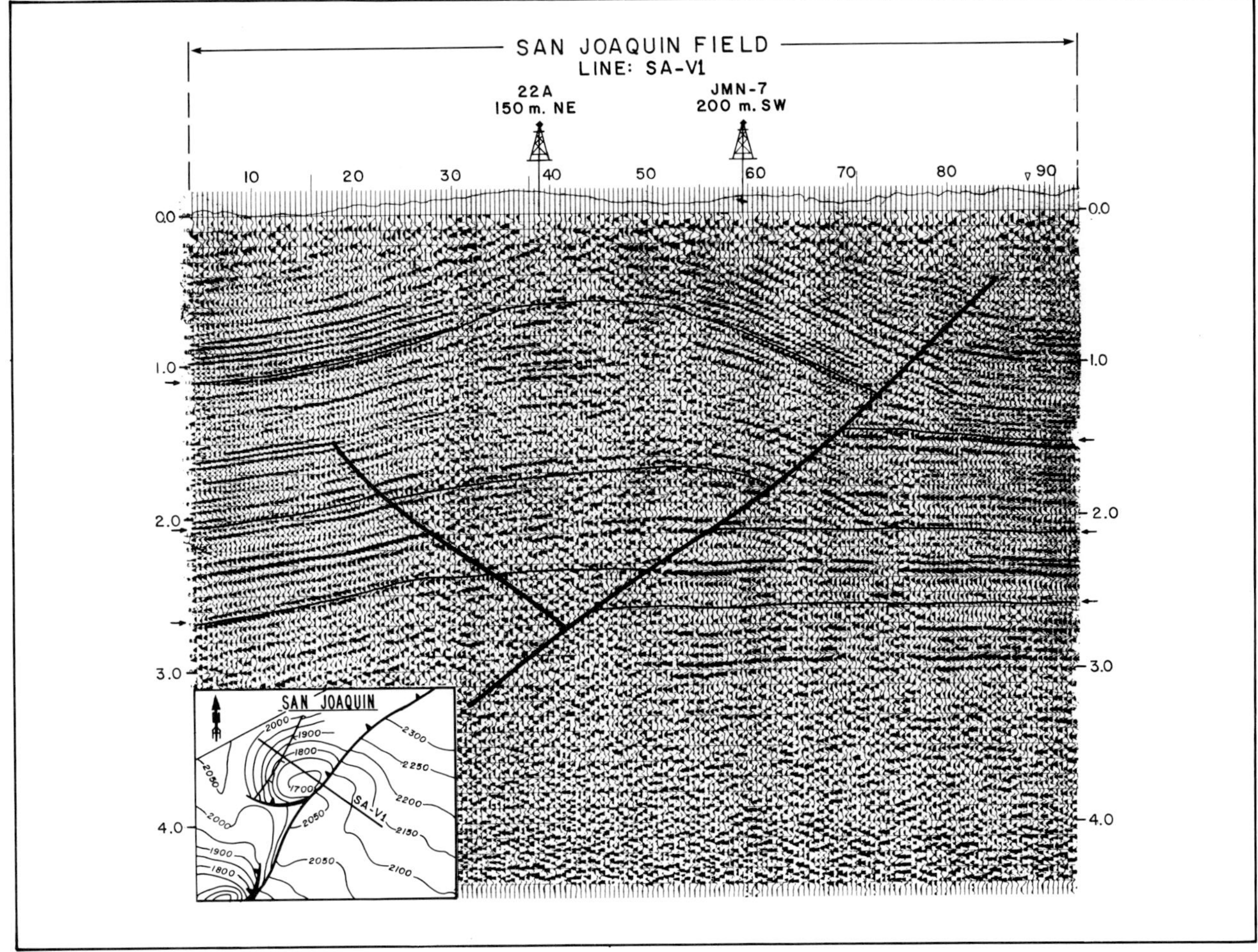

Figure 6B. Seismic section across the San Joaquín field as an example of the tectonic feature of the Anaco trend.

(Figure 9) shows the sedimentation corresponding to a coarsening-upward bar deposit in a coastal marine environment. The presence of calcite fragments, pyrite, and numerous siderite nodules in the associated shales, as well as polymorphic bioturbation, suggests a series of superimposed marine bars.

TRAPS

The trapping mechanism of the Santa Rosa accumulations, as in other fields of the Anaco trend, is an overthrust anticline, with minor normal and reverse faults, with multiple pays formed by numerous pinch-outs acting as lateral seals and shales as vertical seals interbedded with the producing sandstones (Figure 9). However, faults are not seals where the sandstone thickness is greater than the fault displacement. Also, in some areas different sandstones are juxtaposed across the main fault, thus allowing migration across the fault. The majority of the producing sandstones of the Anaco trend are massive to lenticular, providing different lateral extensions of each reservoir.

The isopach map of the Colorado-E sand unit (Figure 10) is typical of the thick channel-like sandstones that extend across the structure. The lateral pinch-outs of the sandstone bodies may occur anywhere over the structure, thereby forming traps against the updip pinch-out, as shown on Figure 10.

Reservoirs

At present, the 259 wells drilled have discovered 440 reservoirs of light oil and condensate, most of which are being produced or are awaiting workovers or enhanced recovery projects. The reservoirs are distributed throughout the San Antonio, San Juan, Merecure, and Oficina formations, the last two giving the best yield in the field. The principal reservoir

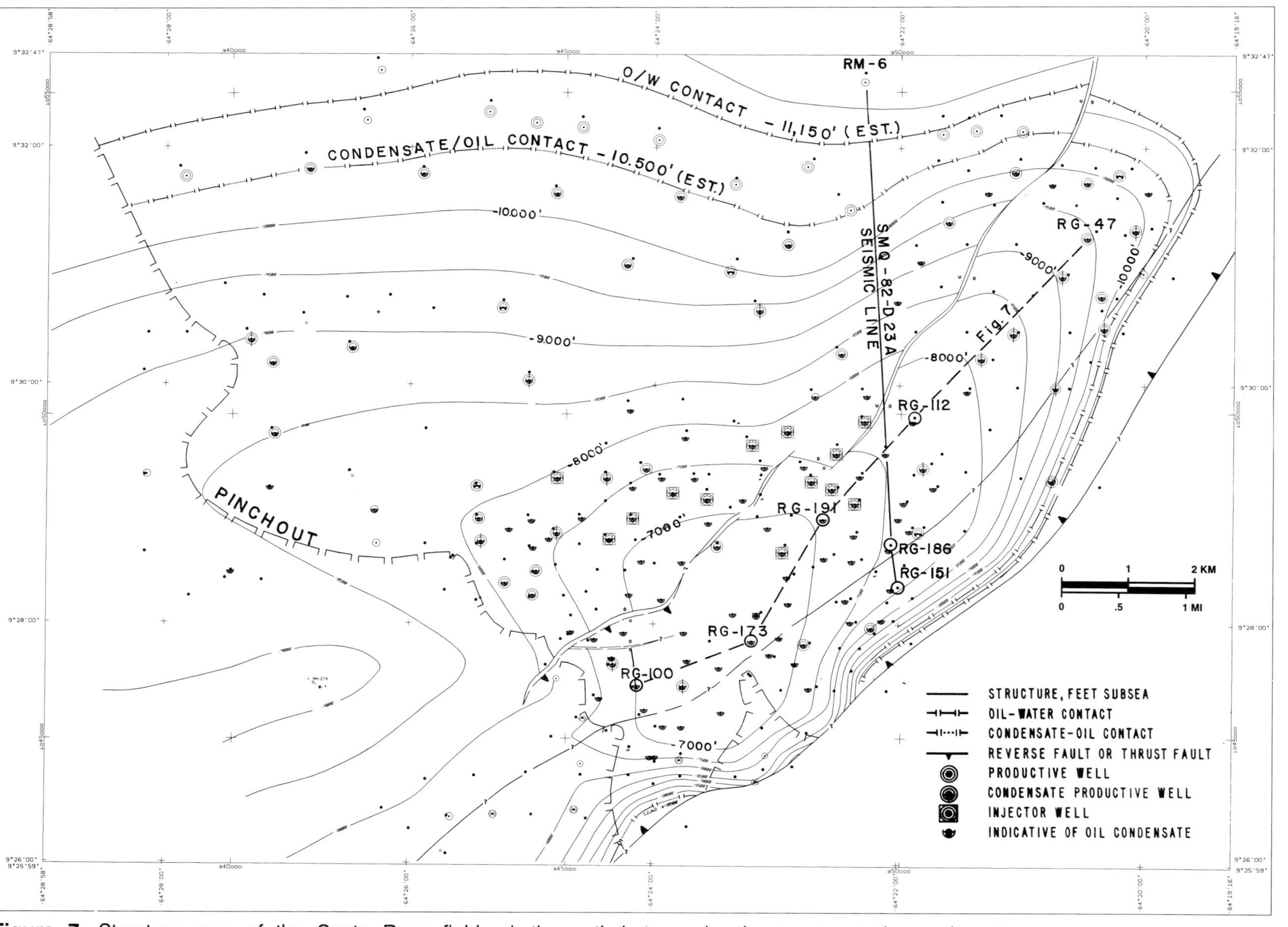

Figure 7. Structure map of the Santa Rosa field. Contoured on top of the Colorado-E sand unit, Oficina Formation. Contour interval, 500 ft. The reservoir is roughly fan-shaped with oil-water contacts running parallel to the major Anaco thrust fault and the structure in the north, but crossing the structure to the southwest along the sand pinch-out. Location of seismic section, Figure 6A, and well-log cross section, Figure 8, are shown.

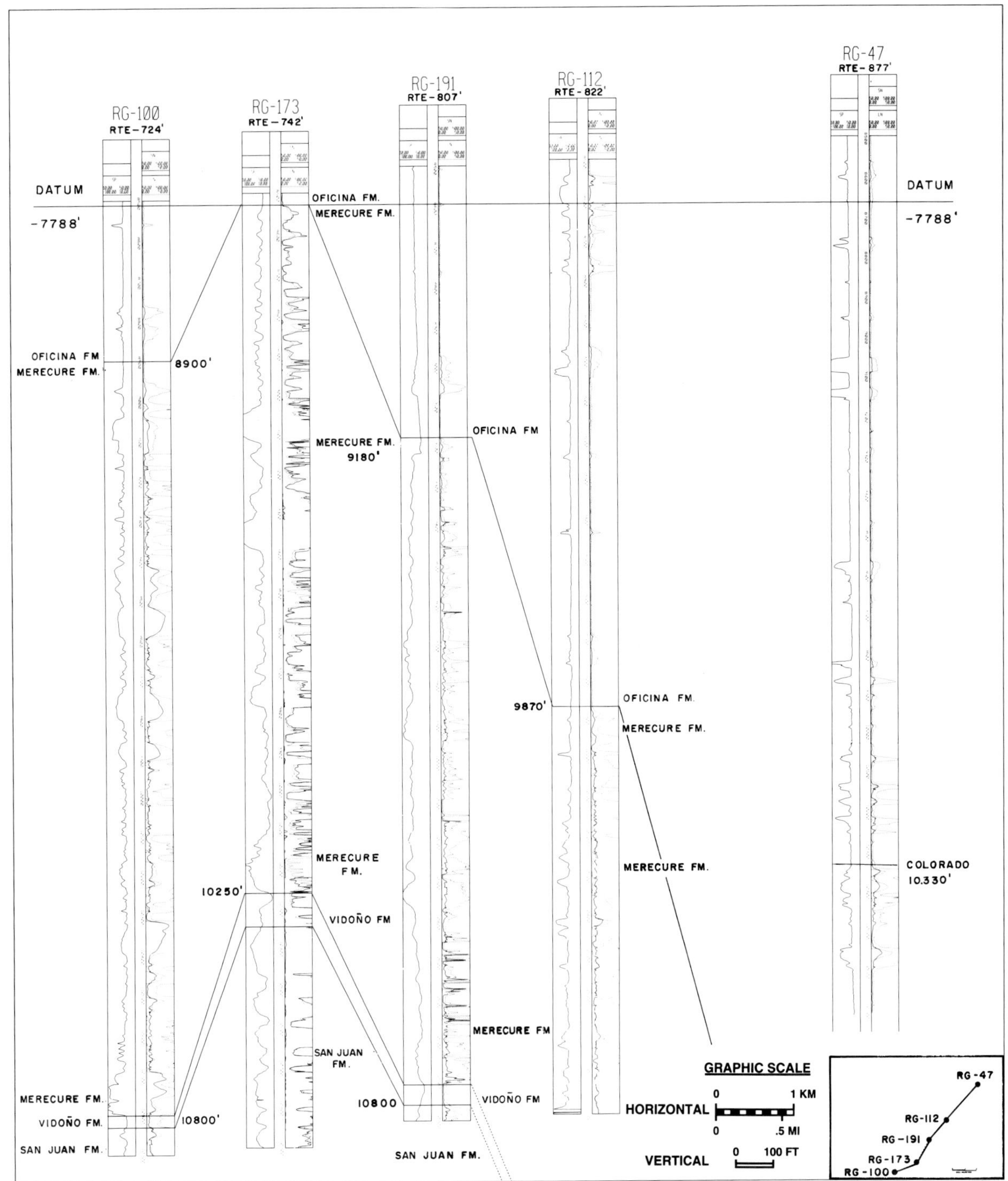

Figure 8. Well-log cross-section, parallel to the anticlinal axis of the Santa Rosa field, showing the facies changes in the Merecure Formation along the plunge of elongated dome.

characteristics of these formations are shown in Table 2.

Petrophysical Data from Cores and Logs

Cores and log data have been used to define the petrophysical characteristics of the reservoirs; Table 3 shows average values for the field.

The Archie equation, $(S_w)^n = \frac{a\,R_w}{\phi^m R_t}$, as well as other similar equations were used to obtain the water saturation for each reservoir, where:

R_w = water resistivity
S_w = water saturation
ϕ = porosity
R_t = resistivity

Table 1. Characteristics of the depositional environments in each sedimentary package, Verde to Colorado interval, Oficina formation.

Character	Associations	Position	Rock Type	Environment	Log Shape
Calcareous fossiliferous usually multiple thin beds	Shale, lignite	Upper	Transgressive beach sandstone	Beach	Variable
Thin-bedded	Transgressive and crevasse splay sandstones	Upper, middle	Lignite	Swamp	Resistivity spike
Massive, medium or coarse-grained upward-fining, always single	Crevasse splay sandstones, lignites, interdistributary bay shale	Middle	Distributary channel sandstone	Distributary channel	Cylinder or bell
Lignitic, silty, fine-grained thin-bedded often stacked	Interdistributary bay shale, distributary channel sandstone, lignite	Middle	Crevasse splay sandstones	Interdistributary bay	Spike or serrate cylinder
Thin-bedded	Distributary channel and crevasse splay sandstones, lignite	Middle	Shale	Interdistributary bay	Subdued serrate to smooth funnel
Silty, lignitic, fine-grained medium-bedded	Prodelta shale	Lower	Distributary mouth bar sandstone	Delta fringe	Spike, thin cylinder
Thin-bedded	Lenses of distributary mouth bar sandstone	Lower	Shale	Prodelta	Serrate spike buildup

n = saturation exponent (+2)
m = cementation factor exponent (+2)
a = rock constant (tortuosity factor) (+81)

The S_w values for the Oficina Formation were taken from core statistics prepared by the French company Total, in its "Synthesis of the Greater Anaco Area" (1985).

Examples of the sandstone characteristics in Santa Rosa field area are shown by Figures 11 and 12, which are thin sections of cores from well RG-16. The average porosity of the sandstones of the Oficina Formation is 19%, ranging from 9 to 25%. The Colorado E-F sand unit (Oficina Formation) has an average porosity of 15%, but the porosity may be up to 20% in some areas (Figure 11). The Colorado-G sand unit, which consists of six productive reservoirs, averages 16.5% porosity, with permeabilities ranging from 150 to 350 md (Figure 12).

Permeability and Pressures

In the reservoirs of this field the relative permeability is utilized in addition to the absolute permeability because of the effect of the presence of many fluid phases on the effective permeability of each phase. This concept is important for enhanced and numerical simulation studies. Figure 13 depicts the relationship between the amounts of the saturations of gas (S_g) and of connate water (S_{wc}) in the Colorado-E sandstones and their effect on the relative permeability to oil/gas (K_{rog}). The curve, based on data from the three wells indicated on the figure, can be used to estimate the K_{rog} in other wells, given the water and gas saturations, and to apply that estimated percentage to the effective permeability.

In the reservoirs, the pressure gradient is 0.57 (Miocene) and 0.35 (Oligocene–Cretaceous) psi/ft (12.9 and 7.9 kPa, respectively). The average initial pressure was 4300 psi/ft (12.9 and 7.9 kPa, respec-

Figure 9. Cored section of the Colorado-G sand unit from well RG-16, showing the sedimentation corresponding to a coarsening-up marine bar environment. The presence of polymorphic bioturbation and parallic foraminifera suggest a marine environment.

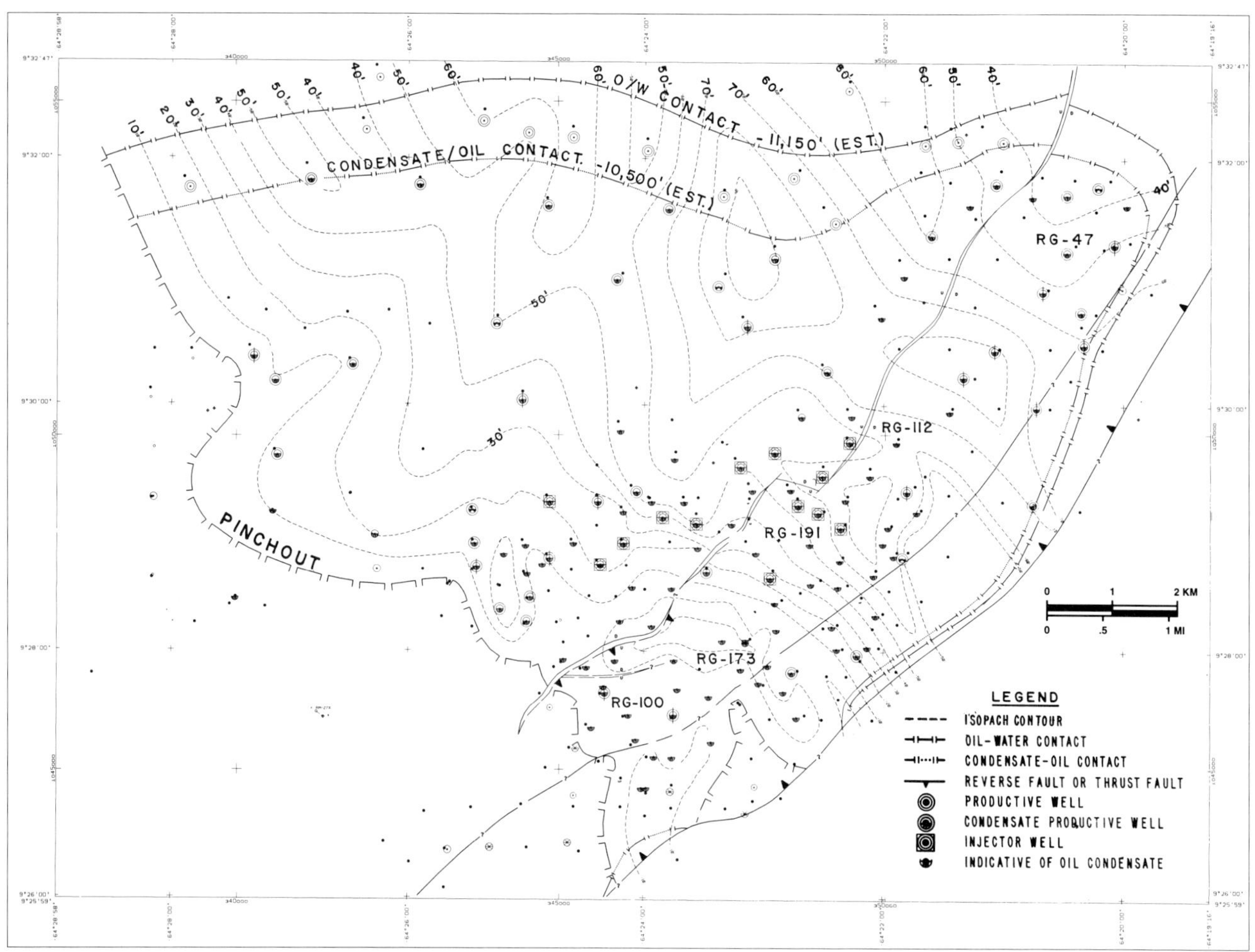

Figure 10. Isopach map (gross sand) of the Colorado-E sand unit, Oficina Formation, showing a thick channel-like sand body crossing the domal structure in a northwest direction, with the lateral pinch-out near the crest of the structure. Contour interval, 10 ft.

Table 2. Depths and thicknesses.

Formation	Age	Depth to Tops of Reservoir in ft (m)	Gross Thickness of Producing Interval in ft (m)	Total sandstone Thickness of Producing Zones in ft (m) Average	Maximum
Oficina	Miocene	1325 (405)	5750 (1755)	250 (65)	350 (105)
Merecure	Oligocene	8530 (2600)	1650 (505)	300 (90)	670 (205)
San Juan	Cretaceous	10,210 (3115)	2400 (730)	230 (70)	350 (105)
San Antonio	Cretaceous	11,250 (3430)			

Table 3. Average core and log data values.

Formation	Age	Porosity %	Permeability (Md)	S_w (%)
Oficina	Miocene	19	286	15-40
Merecure	Oligocene	13	150	15-30
San Juan	Cretaceous	8	100	25
San Antonio	Cretaceous	8	100	25

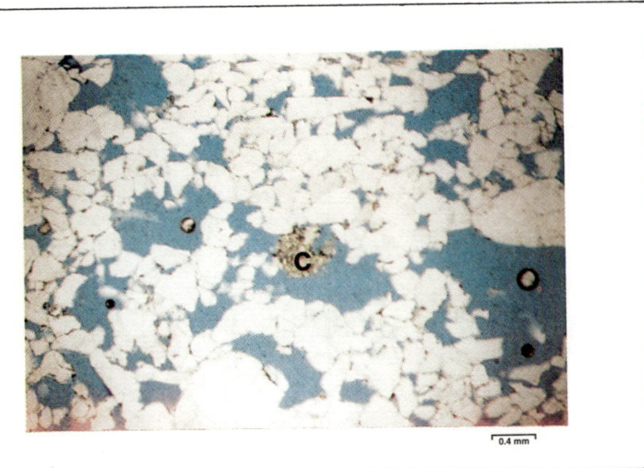

Figure 11. Thin section, under natural light, of the Colorado-E sand unit showing typical sandstone character of channel sands. Optical porosity is 18%, with clean and well-connected pore space. Original permeability was calculated at 390 md.

Figure 12. Thin sections of the Colorado-G sand unit, under natural light (above) and polarized light (below). The optical porosity is 16% and the calculated permeability is 150 md. The lower photo (polarized light) shows the remains of carbonates, marked C, and silica overgrowths, indicated by an arrow.

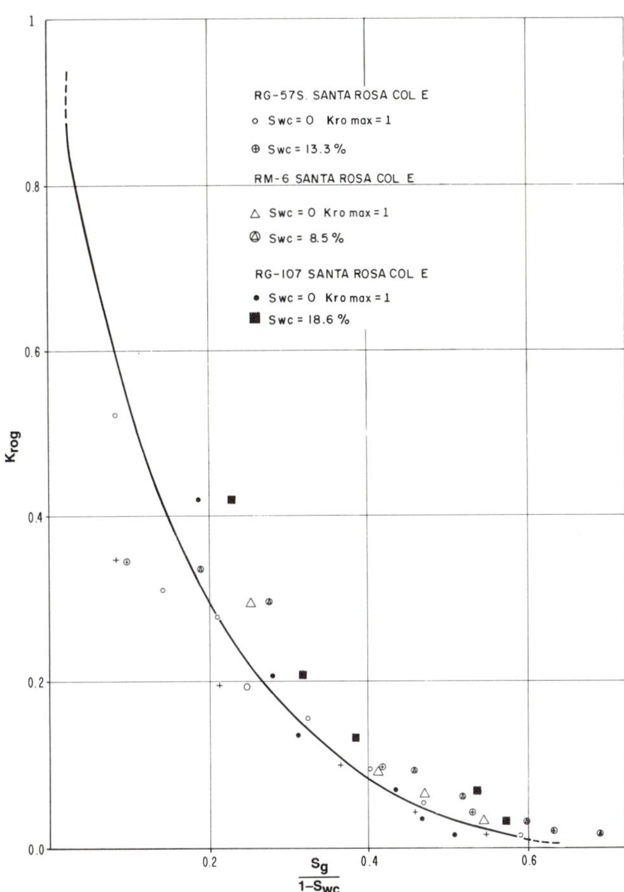

Figure 13. Relative permeability to oil/gas (K_{rog}) in the Colorado-E sand in three wells (RG-57s, RM-6, and RG-107). S_{wc} is the connate water saturation (fraction) and S_g is gas saturation. K_{rog} max = 1, varies from a minimum of 0 to a maximum of 1. S_{wc} values shown are actual saturations (Colorado-E sandstone) from the three wells indicated.

tively). The average initial pressure was 4300 psi (97.3 MPa) at an average subsea depth of 9100 ft (2776 m) (Figure 14). Present reservoir pressures are around 2900 psi (65.6 MPa).

Formation Temperature

The bottom hole temperature is 280°F (138°C) at an average subsea depth of 9100 ft (2776 m) and the geothermal gradient is 2.00°F/100 ft (3.6°C/100 m) (Figure 14).

Hydrocarbon Characteristics

The oils in the Santa Rosa field are paraffin condensates, averaging 41° API, with a GOR of 9900 ft^3/bbl. Characteristic components of the condensates, according to the gas chromatograph analyses, are shown in Table 4.

Characteristic components of oils in the Greater Anaco area are shown by Figure 15.

The hydrocarbon physical characteristics are shown in Table 5.

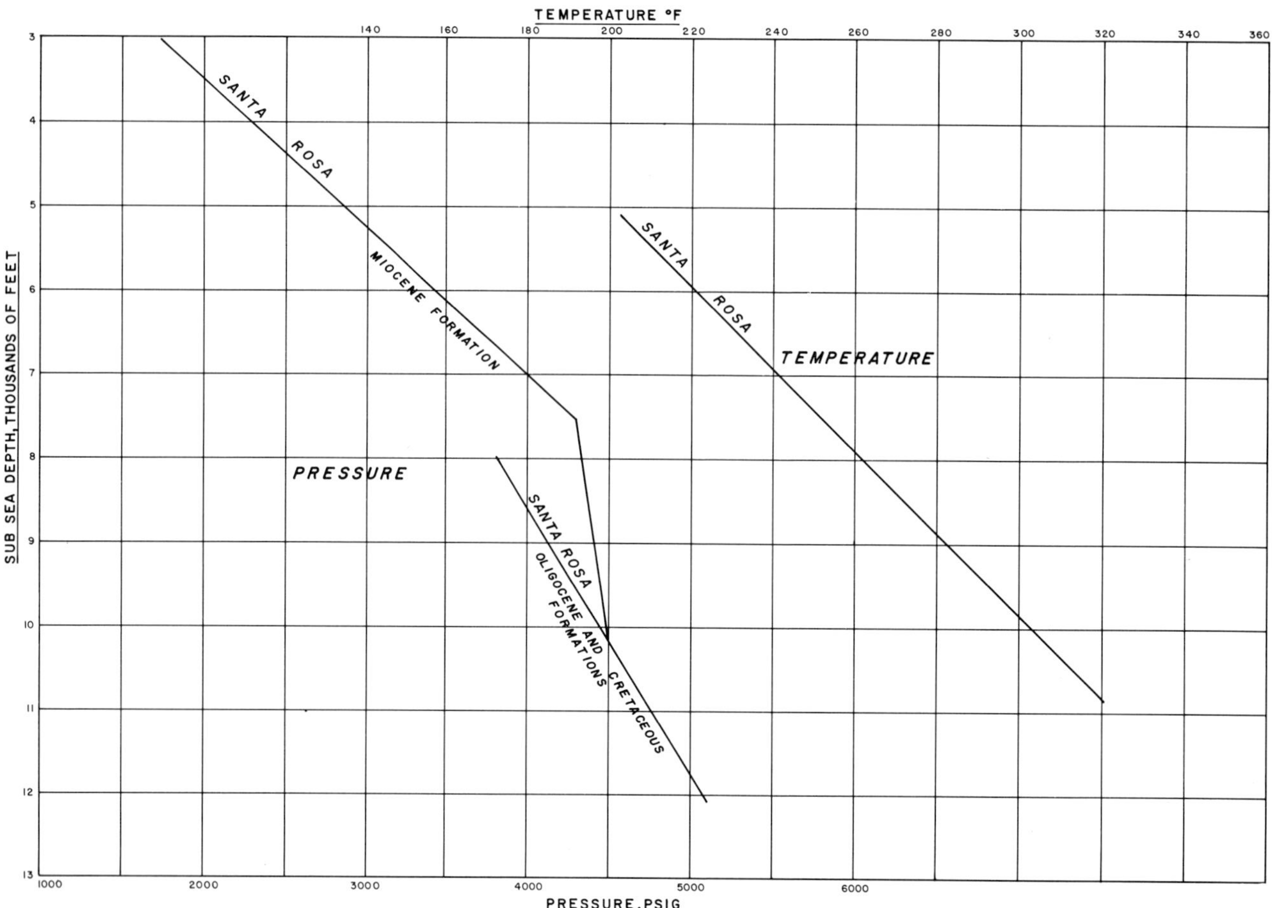

Figure 14. Graph of pressure and temperature vs. subsea depth.

Table 4. Gas chromatograph of oils.

Well	Sa	Ar	Re	As	P/F	P/C_{17}	F/C_{18}
RG-85	71	27	2	0	9.70	1.58	0.19
RG-168	71	23	5	1	4.80	0.68	0.16
RG-134	81	18	1	0	5.00	0.76	0.24

Sa, saturated; Ar, aromatic hydrocarbon; Re, resins; As, asphaltenes; P/F, pristane/phytane; C, carbon.

Water analyses combined with petrophysical interpretations give an average connate water content of 14.0%, a salinity of 14,540 ppm Cl, and a resistivity of 0.15 ohm-m.

Hydrocarbon Recovery

The field covers a proven area of 51,362 ac (20,802 ha) and presently has 259 wells, of which 154 are producing.

Total oil-in-place is 2310 MMBO (Table 6), with ultimate recoverable reserves of 708.4 MMBO (38% of the recoverable reserves in all of the fields in the Anaco trend), of which 532.4 MMBO will be by primary recovery and 176 MMBO by secondary recovery.

The cumulative production is 397 MMBO and 29.7 million bbl of water; annual production was 3.7 MMBO in 1989.

It is worthwile to note that the RG-14 reservoir in the Colorado E-F sand, subjected to gas injection, has an ultimate recovery of 132 MMBO (23% of the recoverable reserves of the field). Cumulative production is 115.8 MMBO at the end of 1989.

Gas and NGL Reservoirs

The Santa Rosa field is also an important source of gas and natural gas liquids, containing proved reserves of over 6.49 trillion cubic feet (tcf) of associated and solution gas as of 1989. Production has amounted to 4.7 tcf, with 2.6 tcf reinjected in

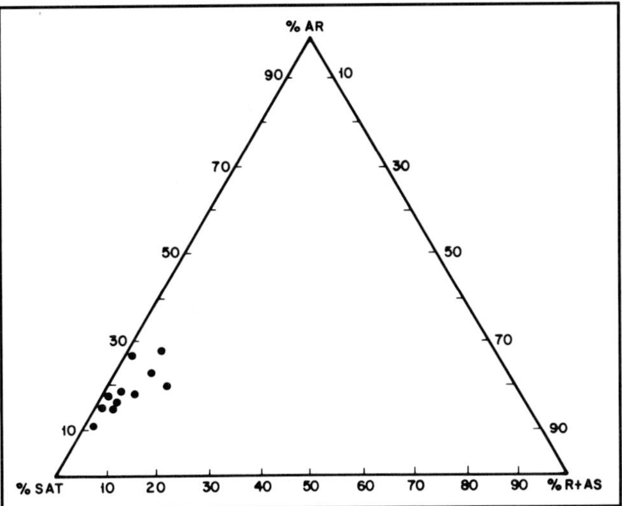

Figure 15. Triangular diagram of three characteristic components of oil (% saturated hydrocarbons, % aromatics, and % asphaltenes plus resins) for paraffinic oils in the Greater Anaco area.

projects. Remaining reserves are estimated to be 4.98 tcf, distributed in 440 reservoirs.

The Colorado E and F sand units contain the largest proved gas reserve of any reservoir in eastern Venezuela. The large gas cap on this reservoir originally contained about 110 bbl of condensate per mmcfg. The original estimates of the gas cap size indicated 1.248 tcf of gas, containing 137.9 MM bbl of condensate. The original estimates of the oil leg volume indicated 63.9 MMBO with formation volume factor of 2.198, and 1881 ft^3 gas/STBO at the surface. Estimates based on reservoir studies and 40 years of production indicate that the gas cap contained 1.23 tcf of gas—close to the original estimate of 1.248 tcf.

Production from the reservoir started in 1950 and was limited until the start of gas injection in 1955. Production since 1955 has included production from wells in the oil leg as well as from wells in the gas cap. Nine wells have produced from the oil leg, 28 wells from the gas cap, and 13 wells have served as gas reinjection wells. Some of the gas injection wells produced small volumes of liquid. As of 1989, the reported cumulative gas production has been 1.54 tcf, while gas reinjection was reported as 1.98 tcf (including solution gas stripped from the crude oil).

The condensate of Santa Rosa field is processed in plants in Santa Rosa and San Joaquín along with gas and condensate from other nearby fields, so it is difficult to obtain liquid production for any particular reservoir or field. The two refrigeration plants and one cryogenic plant have processed 126 MMB of liquids through 1988, principally butane-propane, and smaller amounts of pentane, methane-ethane, and gasoline.

Fluid-Flow Characteristics

Recovery mechanisms include solution gas drive, gas cap drive, water drive, and gas and water injection. The main obstacle to flow through the sands was the low pressure in some reservoirs; for that reason, gas or water injection was used to slow the decline of reservoir pressures and thus to increase ultimate recovery.

Including enhanced recovery efforts, the overall average decline rate for the last 20 years is 8% per

Table 5. Hydrocarbon characteristics.

	Miocene	Oligocene	Cretaceous
Sulfur, wt%	0.14	0.29	1.25
Sayboldt viscosity (100°F or 37.8°C)	35.2	43.8	—
Pour point	75.5°F (24.0°C)	95.0°F (35.0°C)	75.0°F (24.0°C)
Gas-oil distillate (%)	11.7	9.2	—

Table 6. Field reserves of the Anaco trend, showing the Santa Rosa field to be the most important in the trend.

	Total Oil in Place (MMbbl)	Recovery Factor (%)	Ultimate Potential Recovery (MMbbl)	Cumulative Production (MMbbl)	Remaining Reserves (MMbbl)
Santa Rosa	2312.8	30.6	708.4	397.0	311.4
(% of total reserves in place)	(30%)		(37.5%)	(49.2%)	(28.8%)
San Joaquín	1618.0	22.1	358.4	112.9	245.5
La Ceiba	404.5	24.4	98.7	49.5	49.2
Santa Ana	865.7	28.5	246.3	130.3	116.0
Guario	664.3	20.4	135.9	45.0	90.9
El Roble	1440.7	18.5	266.8	40.4	226.4
El Toco	319.9	23.6	75.4	31.5	43.9
TOTAL	7625.9	24.8	1889.9	806.6	1083.3

annum, and the present decline rate (based on data for the last 6 years) is 8.7% per annum, the low value due primarily to pressure maintenance of reservoirs under enhanced recovery (Figure 16). The present average decline rate per reservoir, excluding secondary recovery, is 30% per annum.

Source

The Cretaceous Querecual Formation has been recognized by many authors as the obvious source bed in the Eastern Venezuela Basin. A recent study on source beds by Audemard et al. (1986), from INTEVEP (Venezuelan Research Company), proposed the Oligocene Merecure Formation, in addition to the Querecual Formation, as an important source rock for hydrocarbons now trapped in the Santa Rosa field. They evaluated 325 samples from cores and ditch cuttings from 45 wells distributed throughout the basin, and arrived at these conclusions. (The following abreviations are used in this discussion: TOC, total organic carbon; H/C, hydrogen/carbon; P/F, pristane/phytane; P/C-17, pristane/carbon-17; F/C-18, phytane/carbon-18; HC, hydrocarbon.)

The Cretaceous Querecual Formation of well VZ-1 in the La Vieja field, a few kilometers northeast of Santa Rosa field, consists predominantly of limestones and calcareous shales. The kerogen of these rocks is composed of amorphous material with a high content of herbal and woody material. The average total organic material (TOC) is 3.24% with a maximum of 5.00%. The H/C ratio of 1.10, determined from the elemental kerogen analysis, characterizes the rock as type II (Waples, 1981).

The bitumen extracted from the rocks is predominantly naphthenic with low molecular weight. The relations P/F (1.1), P/C-17 (1.08), and F/C-18 obtained from chromatographic analysis are characteristic of material in a marine environment. The Querecual Formation is mature, having a vitrinite reflectance (R_0) of 0.90. The time of hydrocarbon expulsion is estimated at 15 Ma based on the burial chart of the La Vieja field, to the north of Santa Rosa (Figure 17).

The depth to top of source in Santa Rosa is estimated to be 17,000 ft (5185 m), and thickness is 3300 ft (1006 m).

The Oligocene Merecure Formation also contains source material; the shales of this formation were studied in wells RG-11 of Santa Rosa field and G-5 and G-66 of Guario field (Figures 1 and 3), where they contain an average TOC of 2.2% and a maximum of 3.4%. The Merecure kerogen is herbaceous (type II–III), has a H/C ratio of 0.98, and has a hydrogen index of 280 mg HC/g of paraffinic crude oil and gas. The organic material comes from high order plants, and the oil has a low F/C-18, P/F > 3, and a predominance of n-alkanes of high molecular weight with odd-number carbon isotopes. Vitrinite reflectance oscillates from 0.55 to 0.88%. Time of hydrocarbon expulsion is estimated at 8 Ma. Present depth to the top of the Merecure Formation source rock in Santa Rosa field is 9000 ft (2745 m).

Production, Transportation, and Market

As shown by Figure 18, the oil is transported by pipelines to the Puerto La Cruz Refinery where it is processed and held in tank farms for shipment. Part of the wet gas goes by pipeline to a cryogenic plant where the liquid (light oil) is separated, and the dry gas is transmitted to the national gas distribution system for downstream sales.

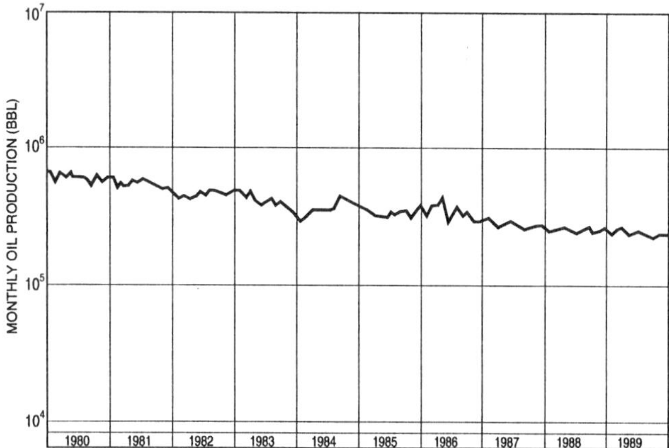

Figure 16. Production curve for the Santa Rosa field during the past ten years, 1980 to 1989. The average annual decline increased slightly from 8% to 8.7% as a result of other uses for the gas normally utilized in some of the gas injection projects. Decline is slight since the production rate was maintained by other secondary recovery projects.

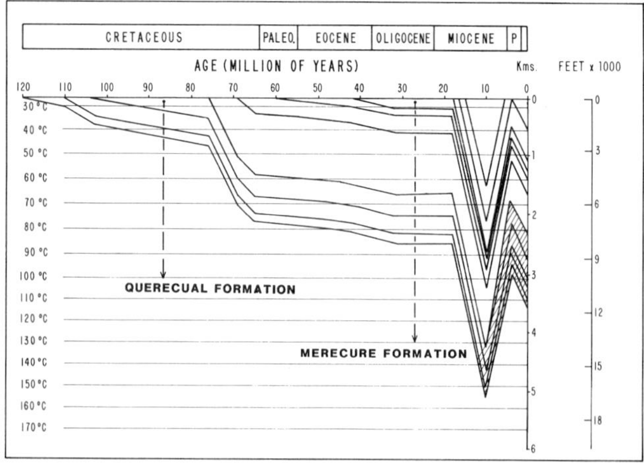

Figure 17. Burial history curve and generation of hydrocarbons for the well La Vieja-1, based on the Lopatin method. La Vieja is located ±30 km (18.6 mi) to the north of Santa Rosa field, but it is representative of the Anaco trend.

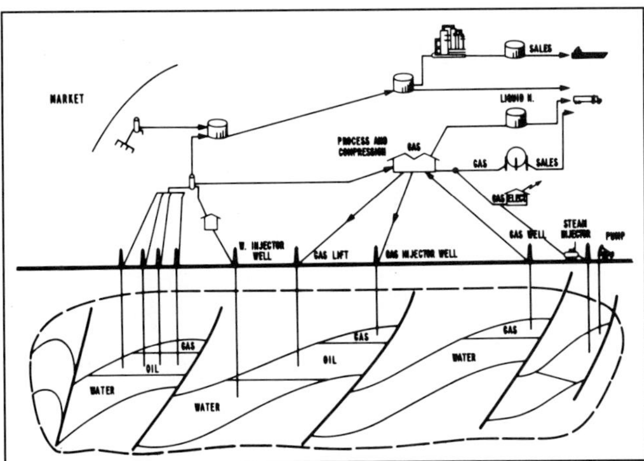

Figure 18. Schematic sketch showing the system for the production, transportation, and marketing of oil and gas from the Santa Rosa and other fields in the Greater Anaco and Oficina areas. Oil is produced and transported to the refinery for processing and shipment overseas and to local markets. Gas is used for projects and sales; rich gas is separated in cryogenic and separator plants and used for local sales and overseas shipments.

EXPLORATION CONCEPTS

Using today's exploration techniques, a discovery similar to Santa Rosa could be accomplished by using photogeology to locate the surface structures and a gravity survey to identify the subsurface trend of the highs. A reflection seismic survey then could reveal a detailed picture of the type of anticline and the position of the faults. Drilling, logging, and coring could confirm the structure, provide the stratigraphy, and test for the presence of hydrocarbons. Since many of the pay zones are stratigraphic and controlled by pinch-outs, a high-resolution 3-D seismic survey could be useful in field development.

ACKNOWLEDGMENTS

The author wishes to thank Petróleos de Venezuela S.A. for permission to publish this paper, and Corpoven S.A. for allowing adjustment of my schedule so as to complete it, and for permission to use unpublished company reports. I also wish to thank the many Corpoven geologists, geophysicists, and engineers who gave me advice and assistance. The author wishes particularly to acknowledge the help provided by Gordon Young and G. Don Kiser, advisors in the realization of this study.

REFERENCES CITED

Audemard, N. M., F. Audemard, E. Rodriguez, and N. Jordan, 1986, Génesis y distribución de los crudos en la Cuenca Oriental y F.P.O.: Unpublished report of INTEVEP, p. 28-35, 45-48, 58-73, 87-105.

Banks, L. M., and E. S. Driver, 1957, Geologic history of Santa Ana structure, Anaco Structural Trend, Anzoátegui, Venezuela: AAPG Bulletin, v. 41, p. 308-325.

Driver, E. S., 1949, Progress Reports Nos. 6 and 8, Seismic Project No. 63, Chrono 1091: 5: Unpublished report of Mene Grande Oil Co.

Funkhouser, L. C., L. C. Sass, and H. D. Hedberg, 1948, Santa Ana, San Joaquín, Guario, and Santa Rosa oil fields (Anaco fields), Central Anzoátegui, Venezuela: AAPG Bulletin, v. 32, p. 1851-1908.

González de Juana, C., J. M. Iturralde, and X. Picard Cadillat, 1980, Geologia de Venezuela y de sus Cuencas Petrolíferas: Fonninves Tomo I y II, p. 286-292, 909-939, 955-974.

Lockwood, R. P., and R. L. Aymard, 1982, Integrated geological study, Eastern Venezuela Basin: Unpublished report of Meneven, Part 5, p. 10-26.

Martinez, A. R., 1976, Geology of giant petroleum fields: AAPG Memoir 40, p. 326-336.

Murany, E. E., 1972, Tectonic basis for Anaco fault, Eastern Venezuela: AAPG Bulletin, v. 56, p. 860-870.

Total Compagnie Francaise des Petroles, 1985, Proyectos de inyección de fluido (area mayor de Anaco. Síntesis de los estudios de yacimiento): Unpublished report, p. 1-8, 52.

Waples, D., 1981, Organic geochemistry for exploration geologists: Burgess Publishing Company, CEPCD Division, p. 95-106.

SUGGESTED READING

Mendez Orlando, Historia geológica graficada de la Cuenca Oriental de Venezuela: VI Congreso Geológico Venezolano, Tomo II, p. 1000-1040.

Shaffer B., C. Cigler, and L. F. Vidllegas, 1982, Historia Geológica termal ("Geohistory") en la Cuenca Oriental: VI Congreso Geológico Venezolano, Tomo VI, p. 3585-3605.

Total Compagnie Francaise des Petroles, 1983, Determinación de las permeabilidades relativas, presión capilar y de saturaciones residuales en un flujo bifásico, p. 14-15.

Appendix 1. Field Description

Field name .... *Santa Rosa field*

Ultimate recoverable reserves .... *708.4 MMBO*

Field location:

- **Country** .... *Venezuela*
- **State** .... *Anzoátegui*
- **Basin/Province** .... *Eastern Venezuela Basin/Maturin subbasin*

Field discovery:

- **Year first pay discovered** .... *Miocene sandstones 1941*
- **Year second pay discovered** .... *Oligocene sandstones 1948*
- **Year third pay discovered** .... *Cretaceous sandstones 1966*

Discovery well name and general location:

- **First pay** .... *RG-2, 10 km northeast of town of Anaco*
- **Second pay** *RG-9*
- **Third pay** *RG-134*

Discovery well operator

- **First pay** .... *Mene Grande Oil Co.*
- **Second pay** .... *Mene Grande Oil Co.*
- **Third pay** .... *Mene Grande Oil Co.*

IP

- **First pay** .... *715 BOPD, 42° API*
- **Second pay** .... *581 BOPD, 37.7° API*
- **Third pay** .... *107 BOPD, 45.6° API*

Geologic concept leading to discovery and method or methods used to delineate prospect

Aerial photographs and torsion balance followed by structural drilling, plus reflection seismograph and subsurface geological work.

Structure:

Province/basin type .... *Bally 221, Klemme II Ca; compressional*

Tectonic history

The structural history applied to the entire Anaco trend is outlined as follows:

A northwest-dipping normal fault with doming in its downthrown block started at least as early as Oligocene time and was active during sedimentation. While folding continued uninterruptedly possibly until the Recent, growth along normal faults stopped toward the close of middle Miocene time. Reversal of fault movement produced the present thrust as a result of stresses set up in the basin and this thrusting may have started at different times along the trend. The deformative stresses occurred at the pre-existing line of weakness to localize the Anaco trend in the tectonic setting where we now find it; the lack of pre-existing structures north of the area of deformation might explain the absence of similar parallel features in that part of the basin. The uplift and thrust development continued throughout Miocene time causing the Freites Formation to thin over the Anaco area.

Regional structure

The Anaco thrust structure consists of a series of northeast-trending elongate domes that are without exception asymmetrical toward the southeast, indicating pressure from the northwest. The Santa Rosa field is situated in the northwestern overthrust segment of the fault.

Local structure

The Santa Rosa dome trends N45°E. It is an asymmetrical feature with a gentle northwest flank and a steep southeast flank dipping into the Anaco thrust fault.

Trap:

Trap type(s)

Thrust-faulted anticline with multiple pays and lateral pinch-outs

Basin stratigraphy (major stratigraphic intervals from surface to deepest penetration in field):

Chronostratigraphy	Formation	Depth to Top in ft (m)
Pleistocene	Mesa	150 (46)
Pliocene	La Piedras	523-2435 (159-743)
Miocene	Freites	Eroded over field area
	Oficina	673-2585 (205-788)
Oligocene	Merecure	8532-10,460 (2602-3190)
Eocene	Caratas	Eroded over field area
Paleocene	Vidoño	10,110-12,120 (3084-3697)
Cretaceous	San Juan	10,208-12,190 (3113-3718)
	San Antonio	11,250-11,880 (3431-3623)
	Querecual	Not reached by drill
	Chimana	
	El Cantil	
	Barranquin	

Reservoir characteristics:

Number of reservoirs *440*

Formations .... *San Antonio; San Juan; Merecure; Oficina*

Ages .... *Cretaceous; Cretaceous; Oligocene; Miocene*

Depths to tops of reservoirs .... *Cretaceous, 10,208 ft (3113 m); Oligocene, 8532 ft (2602 m); Miocene (Moreno Member), 1325 ft (404 m)*

Gross thickness (top to bottom of producing interval) .... *Cretaceous, 2400 ft (732 m); Oligocene, 1650 ft (503 m); Miocene, 5750 ft (1754 m)*

Net thickness—total thickness of producing zones

Average .... *Cretaceous, 230 ft (70 m); Oligocene, 300 ft (92 m); Miocene, 250 ft (76 m)*

Maximum .... *Cretaceous, 350 ft (107 m); Oligocene, 670 ft (204 m); Miocene, 350 ft (107 m)*

Lithology

San Antonio: gray sandstones and dark shales intercalated with limestones

San Juan: massive gray well-sorted sandstones

Merecure: abundance of light gray to dark gray bedded sandstones and thin shales; also secondary growth of quartz grains

Oficina: gray and light tan sandstones; lignites, thin limestones and green claystones are minor constituents

Porosity type .... *Intergranular*

Average porosity .... *Miocene, 19%; Oligocene, 13%; Cretaceous, 8%*

Average permeability .... *Miocene, 286 md; Oligocene, 150 md; Cretaceous, 100 md*

Seals:

Upper

Formation, fault, or other feature .... *Shales interbedded with producing sandstones*

Lithology .... *Shales*

Lateral

Formation, fault, or other feature .... *Faults and pinch-outs*

Lithology .... *Shales*

Source:

Formation and age .... *Querecual, Cretaceous; Merecure, Oligocene*

Lithology .... *Querecual, limestones, calcareous shales; Merecure, shale*

Average total organic carbon (TOC) .... *Querecual, 3.24%; Merecure, 2.2%*

Maximum TOC .... *Querecual, 5.0%; Merecure, 3.4%*

Kerogen type (I, II, or III) .... *Querecual, II; Merecure, II–III*
Vitrinite reflectance (maturation) .... *$R_o = 0.90$ (Querecual); 0.55–0.88 (Merecure)*
Time of hydrocarbon expulsion .... *Querecual, 15 Ma; Merecure, 8 Ma*
Present depth to top of source .... *Querecual, 17,000 ft (5185 m); Merecure, 9000 ft (2745 m)*
Thickness .... *Querecual, 3300 ft (1006 m); Merecure, 170 ft (52 m)*
Potential yield .... *Querecual, 7.5 to 12 bbl/MT rock; Merecure, 2.3 to 8.5 bbl/MT rock*

Appendix 2. Production Data

Field name .... *Santa Rosa field*

Field size:

Proved acres .... *51,362 ac (20,802 ha)*
Number of wells all years *259*
Well spacing .... *1312 ft (400 m)*
Ultimate recoverable .... *708.4 MMBO*
Cumulative production .... *397 MMBO*
Annual production (1989) .... *3.7 MMBO*
Present decline rate *8.7%/annum*
Initial decline rate .... *NA*
Overall decline rate *8.0%/annum*
Annual water production (1989) .... *2.2 MMBW*
In place, total reserves .... *2310 MMBO*
In place, per acre foot .... *300 bbl/ac ft*
Primary recovery .... *532.4 MMBO*
Secondary recovery .... *176 MMBO*
Enhanced recovery .... *NA*
Cumulative water production .... *29.7 MMBW*

Drilling and casing practices:

Amount of surface casing set .... *1500 ft (460 m)*
Casing program
20-in. at 500 ft (150 m); 13⅜-in. or 10¾-in. or 9⅝-in. at 1500 ft (460 m) or 3000 ft (900 m); 7-in. or 5½-in. at +10,000 ft (3000 m)
Drilling mud .... *Lignosulfonate (water-based mud)*
Bit program
17½-in. for casing 13⅜-in.; 14¾-in. for casing 10¾-in.; 12¼-in. for casing 9⅝-in.; 8⅜-in. or 8½-in. for casing 7-in.; 8⅜-in. for casing 5½-in.
High pressure zones .... *Colorado Member, Oficina Formation (lower and middle Miocene)*

Completion practices:

Intervals perforated .... *Different intervals in Cretaceous, Oligocene, and Miocene*
Well treatment .... *None*

Formation evaluation:

Logging suites .... *ISF-SFL-SP, FDC-CNL-CAL-GR, microlog, HDT, dipmeter, CBL-VDL-CCL-GR*
Testing practices .... *RFT, DST*
Mud logging techniques .... *Standard ditch samples (San Antonio)*

Oil characteristics:

Type .... *Oficina and Merecure, paraffin (intermediate paraffin)*
API gravity .... *44.1°*
Base .... *Intermediate paraffinic*
Initial GOR *9900:1*
Sulfur, wt% .... *Miocene, 0.14; Oligocene, 0.29; Cretaceous, 1.25*
Viscosity, SUS .... *Miocene, 35.2 (100°F); Oligocene, 43.8*
Pour point .... *Miocene, 75.5°F (24.2°C); Oligocene, 95°F (35.0°C); Cretaceous, 75°F (23.9°C)*
Gas-oil distillate .... *Miocene, 11.7%; Oligocene, 9.2%*

Field characteristics:

Average elevation .... *Surface, 400 to 600 ft (120–185 m); avg. depth producing fm -9100 ft (2775 m)*
Initial pressure .... *4300 psi (30 MPa) at -9100 ft (2775 m)*
Present pressure .... *2900 psi (20 MPa) at -9100 ft (2775 m)*
Pressure gradient ... *Miocene, 0.57 psi/ft (13.08 kPa/m); Oligocene-Cretaceous, 0.35 psi/ft (8.03 kPa/m)*
Temperature .... *280°F (138°C)*
Geothermal gradient .... *2.00°F/100 ft (3.7°C/100 m)*
Drive .... *Solution gas drive, associated gas drive, water drive, gas injection, water injection*
Oil column thickness .... *9870 ft (3010 m)*
Oil-water contact .... *Various depths in different segments (-1600 to -10,600 ft; -488 to -3233 m)*
Connate water *14.0%*
Water salinity, TDS .... *14.540 ppm Cl*
Resistivity of water .... *0.15 ohm-m*
Bulk volume water (%) .... *NA*

Transportation method and market for oil and gas:

Oil: by pipeline to the Puerto La Cruz refinery or for shipment
Gas: one part by gas pipeline to a cryogenic plant where gas is separated from the liquid (light oil); the other part to national gas distribution system

Los Lanudos Field—Venezuela
Maracaibo Basin, Zulia State

FREDDY A. SANCHEZ NOGUERA
Maraven, S.A.
Caracas, Venezuela

FIELD CLASSIFICATION

BASIN: Maracaibo
BASIN TYPE: Foredeep
RESERVOIR ROCK TYPE: Sandstone
RESERVOIR AGE: Eocene
PETROLEUM TYPE: Gas
TRAP TYPE: Faulted Anticline
RESERVOIR ENVIRONMENT OF DEPOSITION: Tidal Flat
TRAP DESCRIPTION: Faulted anticline containing multiple individual sandstone reservoir bodies

LOCATION

Los Lanudos gas field is located in northwestern Venezuela in the state of Zulia. It is west of Lake Maracaibo, about 22 km (14 mi) northwest of Maracaibo City and 6.5 km (4 mi) northwest of La Concepción field (Figure 1A). Other fields in the area are Mara, La Paz, and Boscan. It is estimated that the ultimate recovery will be 140 bcf of gas.

HISTORY

Los Lanudos was first considered to be part of La Concepción field but later drilling in the area proved it was a separate gas field.

The first well drilled in this field, by the Venezuelan Oil Company Ltd., was the C-147 in 1947 (Figures 2A and 3). In this well in the upper sandstone unit (Figure 4), the existence of gas was indicated, but the well was abandoned without further testing. In 1952, the Compañía Shell de Venezuela, Ltd., drilled the C-152 (Figures 2A and 3) in Los Lanudos, which was still considered at that time to be part of La Concepción field. This well is now considered the discovery well, but because gas was not needed, it was abandoned in 1962. Later, in 1957, Shell drilled the C-222 and C-223 wells, the former being abandoned during the same year after the electrical log interpretation showed no oil or gas potential. The C-223 well tested gas, but it was shut in owing to lack of gas-collecting facilities and markets. In 1981, Maraven, S.A. tested the C-223 well again and

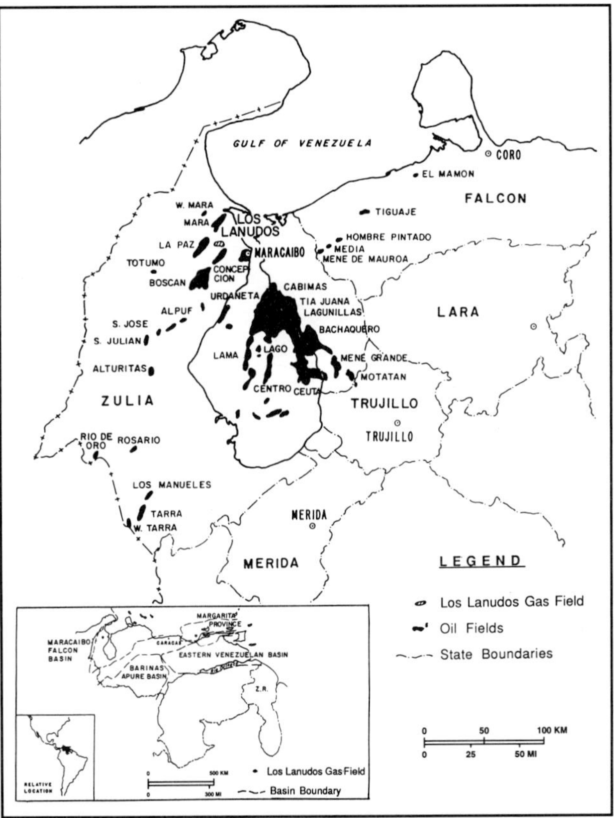

Figure 1A. Geographic location of the Los Lanudos gas field in western Venezuela and other fields of the Maracaibo-Falcón Basin. (Ed. note: Figures 1A and 1B, almost identical to Figures 1A and 1B in the Lama field paper in this volume, are included here for convenience.)

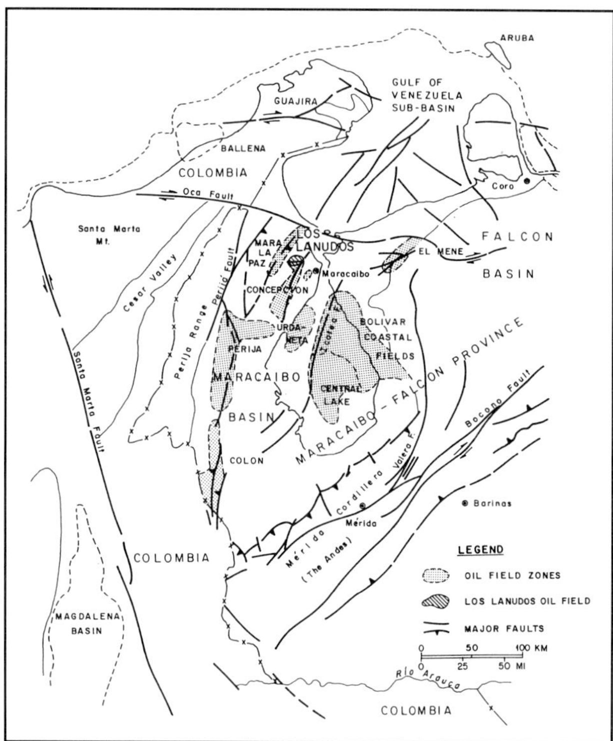

Figure 1B. Tectonic map of the Maracaibo-Falcón province showing the Los Lanudos field, other oil fields, and their relationship to the major fault trends.

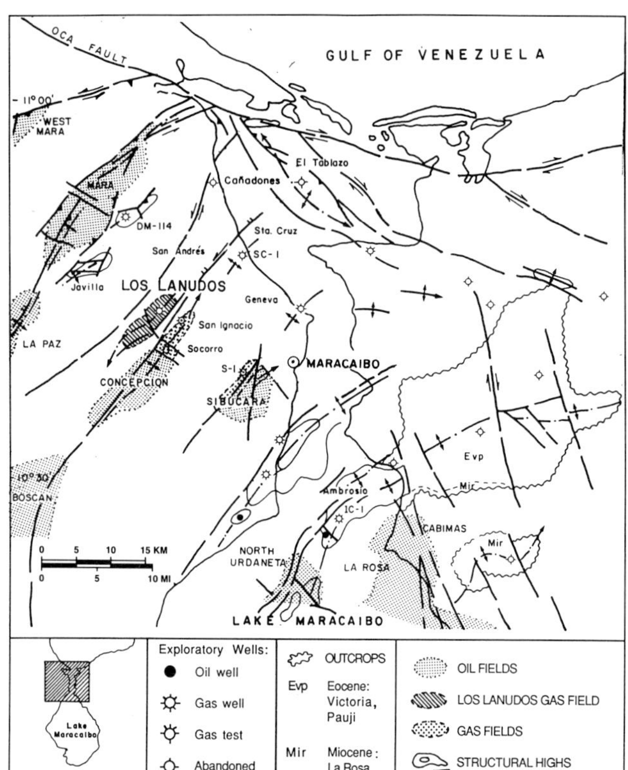

Figure 1C. Structural setting of the Mara-Maracaibo area showing the Los Lanudos gas field, other oil fields, and their relationship to structural trends.

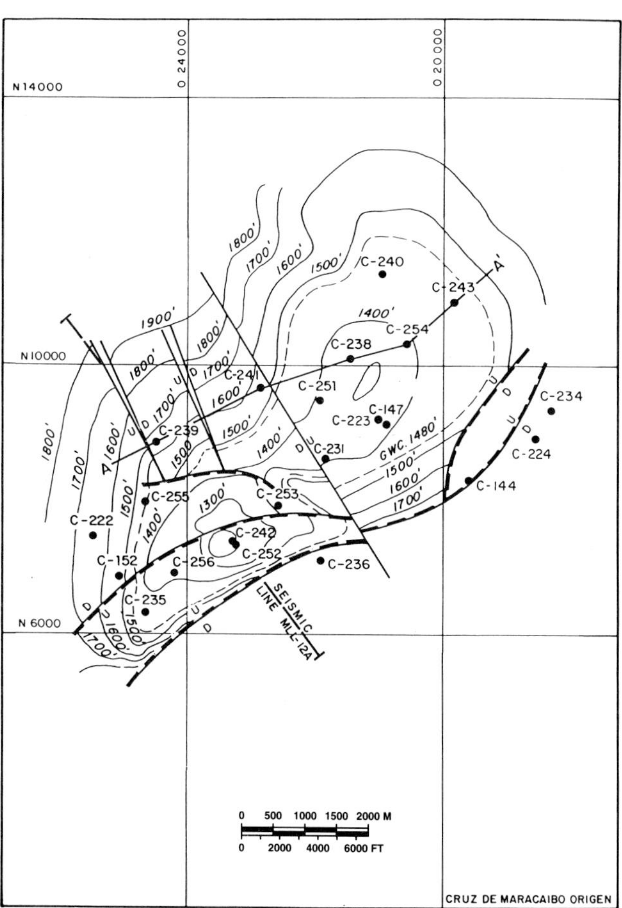

Figure 2A. Structure contour map of top of the A sandstone of the Misoa Formation, Los Lanudos gas field. Contour interval, 100 ft (30.5 m). Locations of seismic line MLL-12A, Figure 2C, and structure cross section AA′, Figure 6, are indicated. C-147 was the first well drilled, but C-152 was the discovery well.

reconditioned it. During the same year, Maraven, S.A. drilled the C-231, C-235, C-238, and C-239 wells, obtaining satisfactory results, and increasing gas reserves in the upper sandstone unit (Figure 5). It was then determined that Los Lanudos was a separate field from La Concepción inasmuch as Los Lanudos tested gas and La Concepción was an oil producer.

Production started in 1982. The gas-bearing sedimentary sequence of Los Lanudos field is contained within the upper sandstone unit of the middle Eocene Misoa Formation (Figure 4). The average thickness of the gas-bearing interval is 1470 ft (448 m), of which only 215 ft (66 m) can be considered as net gas sand. The main gas-producing member consists of six intervals, the A, B, C, D, E_1, and E_2 sandstones (Figures 4 and 5). These sandstone zones are interbedded with shaley intervals. The upper sandstone unit is shallow, with depths ranging between 1200 and 3100 ft (366 and 946 m). Initial production of the field from five wells was 22 million ft^3/day (MMCFG) during 1982.

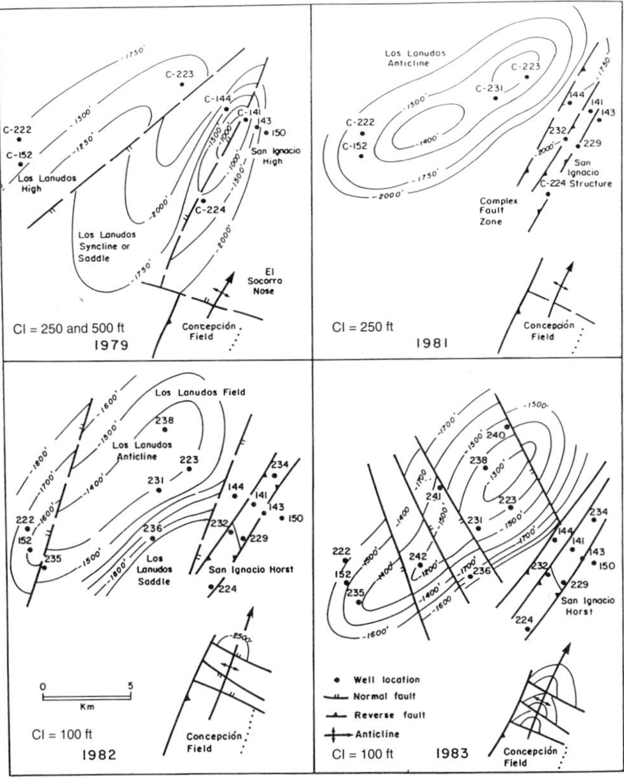

Figure 2B. Evolution of the structural interpretation of the Los Lanudos anticline during the years 1979 to 1983. All maps are of the top of the A sandstone.

DISCOVERY METHOD

The initial method used for discovery of the Los Lanudos field was surface geology. Afterwards, logs from adjacent fields—La Concepción and La Paz—were used, together with seismic data from regional studies. The interpretation of recent seismic data accounted for the definition of the structural features upon which drilling during 1981–1983 was based. In 1987, Maraven, S.A. and Intevep applied the SEISLOG™ technique (psuedo-sonic logs from seismic records) to available seismic data with good results, which aided in the interpretation of the field's structure and productive sandstone distribution.

STRUCTURE

The Maracaibo Basin is located within the unstable area bounded by the Oca, Perijá and Boconó transcurrent fault systems in northwestern Venezuela (Figure 1B). These systems, which have been active since the Jurassic, are related to right-lateral strike-slip movement along the Caribbean and South American plate boundaries. Changes in the relative rate and direction of displacement of these plates account for the orientation and location of depositional basins, uplifted land masses, and, on a regional scale, the character and rate of sedimentation. During the Tertiary, these movements reactivated preexisting lines of weakness in the basement underlying the Maracaibo Basin, causing compressive stresses along antithetic strike-slip faults aligned in a north-northeast-south-southwest direction (Figures 1B and 1C). These compressive stresses resulted in extremely deformed flower structures in the fault zones along the crests of the anticlinal trends.

Van Andel (1954) describes the Barinas-Mérida uplift, which took place at the end of the Paleocene, and its prolongation northwestward along the present Bolívar Coast and across the Mara-Maracaibo area. As a result of this uplift, partial erosion of the

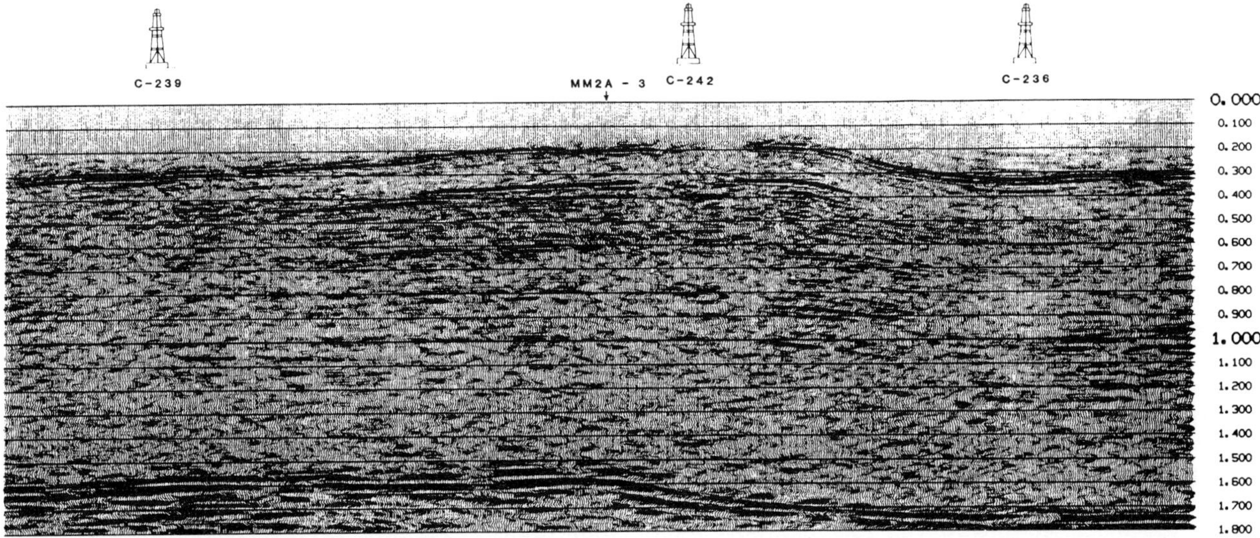

Figure 2C. Uninterpreted seismic line MLL-12A across the field. See Figure 2A for location. Length of section shown about 2.6 mi (4.2 km).

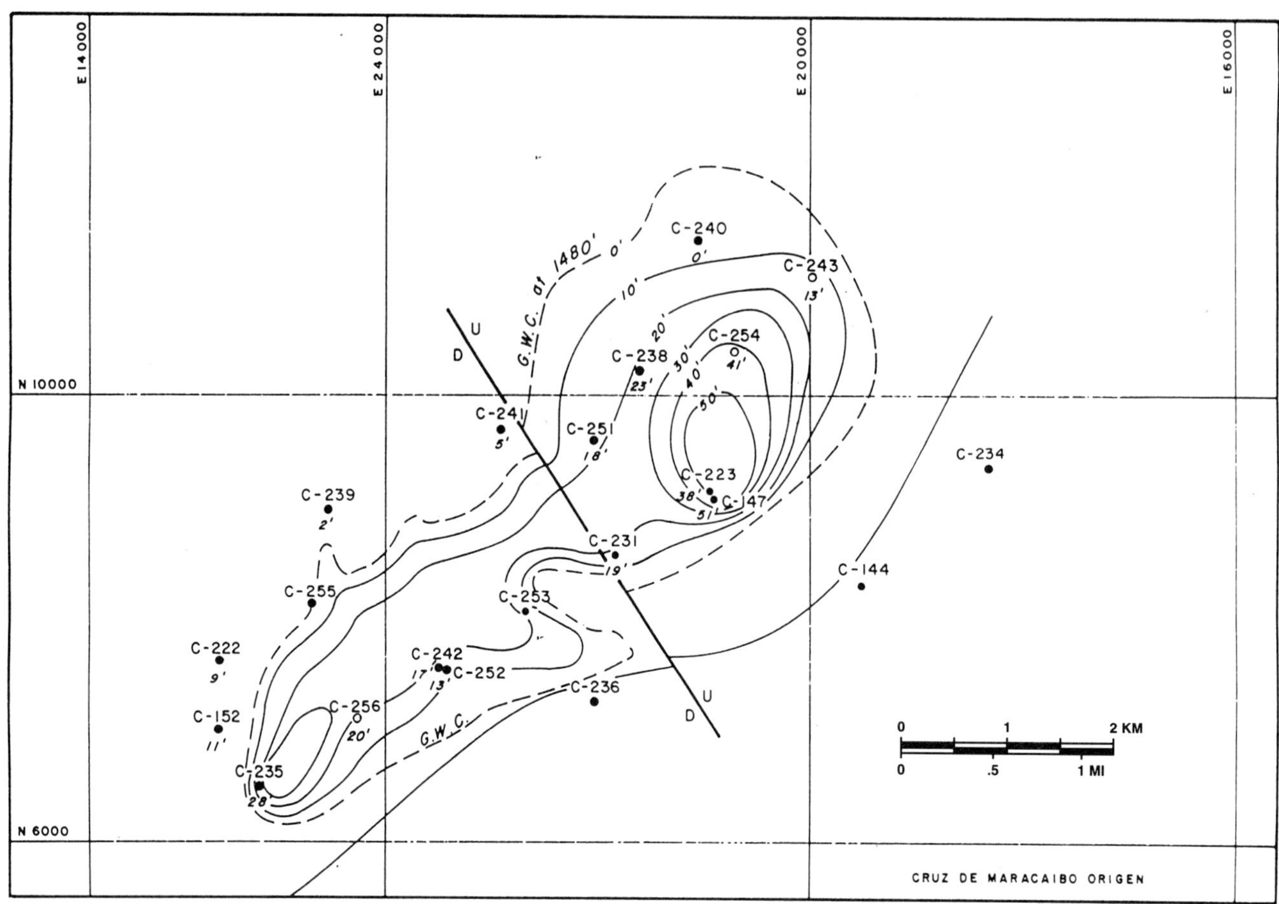

Figure 3. Isopach map of net gas sand, A sandstone. Contour interval, 10 ft (3 m).

Paleocene rocks took place prior to Eocene sedimentation. During this time, the Mara-Maracaibo area was located on the seaward unstable border of the stable Maracaibo platform area (principally granitic). Van Andel (1954) portrays this platform as an eccentric basin, fluvio-deltaic toward the east and south, showing transgressive coastal areas toward the Mérida high, with beach ridges extending northwest-southeastward across the platform's northern edge. Habicht (1958), not believing in lateral fill, stated that the major basin fill was longitudinal rather than transgressive. At the end of the middle Eocene, tectonic activity began again along the eastern margin of the Maracaibo platform. Intense thrusting and folding took place in the Guajira area (Stalder, 1981). The Mara-Maracaibo areas were almost certainly uplifted during this time, and reactivation of the main antithetic fault systems passing through La Paz-Mara and La Concepción-San Ignacio-Cañadores took place (Figure 1C). Toward the east, subsidence of the Bolívar Coast continued together with the deposition of sediments.

The Eocene structure of the Los Lanudos field consists of an asymmetric anticline, with dips of 0° to 5° except for the southeastern flank, where dips of 10° to 12° have been measured. This anticline extends in a southwest-northeast direction at a 25-30° angle to the overall orientation of the structure of the Mara-Maracaibo area. Up to 1982, structural interpretations of the field were based on well data, at which time a seismic survey was carried out to better define both the Eocene and Cretaceous structures. The evolution of the structural interpretations during the period 1979 to 1983, occasioned by the drilling program in the search for gas in the area, is shown by Figure 2B. The present interpretation, as shown on Figure 2A, dates from 1987, following analysis of the seismic data. The seismic interpretation indicated the existence of reverse faults, at both Cretaceous and Eocene levels, which formed the shallow Eocene anticline in which the gas was trapped, as well as a flattish syncline or saddle between the Los Lanudos faults and the adjacent San Ignacio horst or flower structure (Sánchez, 1987).

Reinterpretation of the structural framework was carried out at Eocene level, based upon data pertaining to the last six wells drilled within the field and a seismic interpretation utilizing the SIESLOG™ technique, which is a process of inversion of seismic profiles into sections of pseudo-sonic logs, where the frequencies (and other factors) are

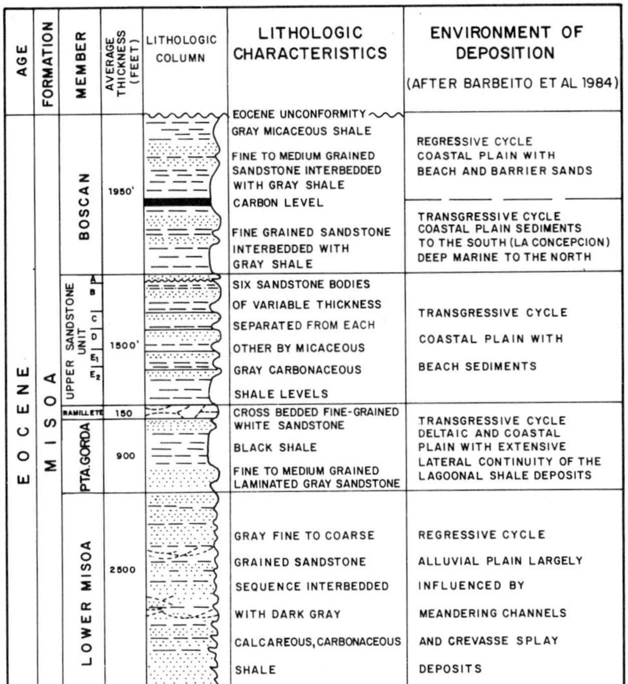

Figure 4. Stratigraphic column of the Eocene Misoa Formation.

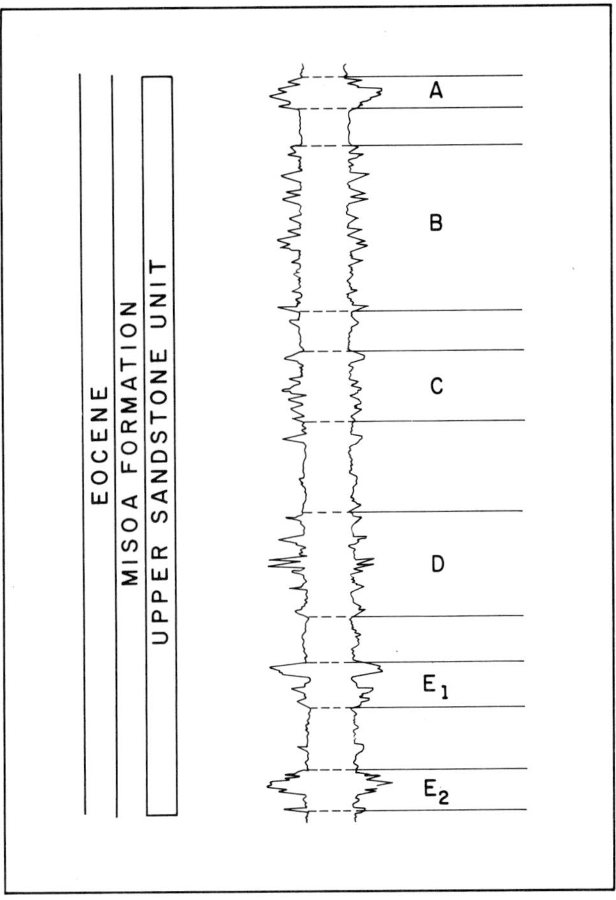

Figure 5. Type log of the upper sandstone unit of the Misoa Formation. Log represents about 1500 ft (460 m). Unquantified log curves indicate relative position of alphabetically designated sandstone intervals in the unit and typical sandstone/shale ratios.

accentuated by a range of colors (Reguiero et al., 1985). The inversion technique also permitted the delineation of the areal extent of the sandstone bodies and the saturation of gas in the reservoirs (Figure 3).

The structural model suggested for the Los Lanudos field is based on the compressive stresses that occurred throughout the Mara-Maracaibo area and which, in Los Lanudos, resulted in reverse, antithetic, and wrench faulting associated with the flower structure of the San Ignacio horst, which is the northern continuation of the central fault zone of the La Concepción field. Although it was thought that a major thrust fault may exist under the Los Lanudos Eocene anticline, the seismic sections revealed a simple reverse fault with little folding and the existence of a flattish saddle between the fault and the San Ignacio horst (Figure 2C, seismic line MLL-12A).

Seismic sections show a main sinuous reverse fault, extending northeast-southwest and forming the eastern limits of the field (Figure 2C). This fault shows left-lateral movement with a throw of about 300 ft (92 m) and dipping between 50° to 60° toward the northwest. The dip has been estimated and increases progressively with increased depth. Secondary normal faults associated with the main one can be observed in the seismic data. One of these secondary faults intersects the main one and divides the field in two fault blocks (Figure 2A). It dips toward the southwest and has a throw of about 20 ft (6 m).

All of the faults, excluding one encountered during drilling of the C-240 and C-254 wells, which extends through the deeper layers (namely C, D, E_1, and E_2 sandstones) have been identified on the seismic lines (Sánchez, 1987).

STRATIGRAPHY

In the Los Lanudos field, the middle Eocene Misoa Formation lies unconformably over the Paleocene sediments. The lowermost section consists of a sequence of gray sandstones and interbedded dark gray shales belonging to the lower Misoa Member, which is overlain by the Punta Gorda Member, a section of laminated and cross-bedded sandstones (Figure 4). There is a black shale sequence, about 600 ft (183 m) thick, in the upper part of this member. Sandstones belonging to the Ramillete Member, about 150 ft (46 m) thick, overlie the Punta Gorda. The upper sandstone unit overlies the Ramillete Member and consists of the A, B, C, D, E_1 and E_2 sandstones. These are separated by gray-colored fossiliferous, occasionally micaceous and carbonaceous shaley zones, with thicknesses varying between 30 and 500 ft (9 and 153 m) (Figure 5). The upper sandstone unit has an average thickness of 1470 ft (448 m). The sand-shale ratio is 0.42. The sandstones are quartzose, fine to medium grained, subangular to subrounded, moderately sorted, and well consolidated. Average thickness of each sandstone interval is 250 ft (76 m). Thinner bodies with no lateral continuity can be observed, however, in the major sandstone bodies.

The upper sandstone member is overlain by the Boscan Member, which consists of whitish fine-grained sandstones interbedded with gray shales.

The Cretaceous La Luna Formation (not shown in Figure 4; see Figure 9) is the main source rock in the Maracaibo Basin, and consists of limestones, calcareous black shales, and cherts.

TRAP

The major trap in the field is a faulted anticline containing multiple individual sandstone reservoirs. The trapping mechanism of the upper sandstone unit is a combination of folded and faulted structures and of stratigraphy. A few deep-penetrating faults favor vertical communication with the main Cretaceous source rocks (La Luna Formation). It has been determined from repeat formation test (RFT) measurements that gas-water contacts exist at subsea 1480, 1740, 2000, 2420, 2690, and 2900 ft (450, 530, 610, 740, 820, and 885 m) for A, B, C, D, E_1, and E_2 reservoirs, respectively (Broeckers et al., 1983) (Figure 6). Figure 2 shows a contour map of the faulted A sandstone unit, and Figure 3 indicates the variability in the thickness of the net gas sand in the A sandstone unit over the top of the anticline, with the gas-water contact at -1480 ft (-450 m). The average gross pay-zone thickness of the upper sandstone unit is 1470 ft (448 m), with a maximum of 1667 ft (508 m).

Individual sandstones are almost certainly not continuous, and varying amounts of depositional clay content should favor microstratigraphic traps within the middle Eocene interval. Individual reservoirs, sealed above and below by interbedded shales, are between 0.5 and 1 mi (1 and 2 km) wide and between 6 and 9 mi (10 and 15 km) long. Thicknesses range from a few feet to tens of feet. Since no migration path exists between them and the source rock, such sand bodies are nonprospective (Twombley, 1984).

Reservoir

The Los Lanudos field reservoirs are represented by the upper sandstone unit A, B, C, D, E_1, and E_2 intervals (Figures 4 and 5). The C-239 well cores of the upper sandstone unit show 16% sandstones (including beds as thin as 2 to 4 in. [50 to 100 mm]), 60% shales, and 24% of interlaminated facies of sandstones, siltstones, and mudstones. It is evident that the prevailing lithology consists of shales interbedded with minor quantities of sandstones (Gosh, 1983).

The complexity of the sedimentary environments within the Misoa Formation can be attributed to Eocene transgressive and regressive cycles. In the area of Los Lanudos, a shallow marine environment existed, characterized by a vast tidal plain that extended over the entire area (Figure 7).

Faunal associations pertaining to the Ramillete and Punta Gorda members suggest that the Mara-Maracaibo area underwent a gradual change from a coastal environment in the area of the Los Lanudos field to an upper neritic marine environment toward the east (Barbeito et al., 1984). Toward the south, however, cores from the C-152 well suggest the existence of a fluvio-deltaic environment (Gosh, 1983). Core evidence from the C-239 well indicates that sedimentation took place in a low energy environment (Twombley, 1984). An environment of alternately high and low energy periods could also have prevailed.

Porosity in the upper sandstone unit, both for Los Lanudos and La Concepción fields, shows moderately high values, with an average of 21%. Permeability ranges between 1.5 and 1000 md. Pressures in the reservoirs are the lowest in this area (700 psig [4800 kPa]).

The best producing sandstone interval is the E_2, with permeabilities ranging only between 20 and 170 md but with relatively high pressures (1280 psig [8800 kPa]). Unlike the A, E_1, and E_2 sandstones, the other intervals (B, C, and D) consist of thin interbedded shale layers with low permeability values (Sánchez, 1987). The B and C sandstones consist of individual producing reservoirs with varying pressures up to 50 psig (345 kPa). Initial pressures of the A, B, and C sandstones have dropped very little, possibly because of the small number of producing wells. In those reservoirs, however, which have a higher number of producing wells, pressure drops are greater.

The porosity-permeability plot for the C-239 well (Figure 8) shows a large scatter between points. For constant porosity values, permeability may vary between 0 and 500 md, whereas for a constant almost zero permeability level, porosity may range between 2% and 30% (Gosh, 1983).

Hydrocarbon characteristics that have been determined are based upon geochemical analyses of samples from the C-235, C-238, C-239, and C-242 wells. Samples analyzed were dry gases, probably bacterial. However, samples from the C-235 and C-239 wells seem to be mixtures of bacterial gas and gas related to oil generation.

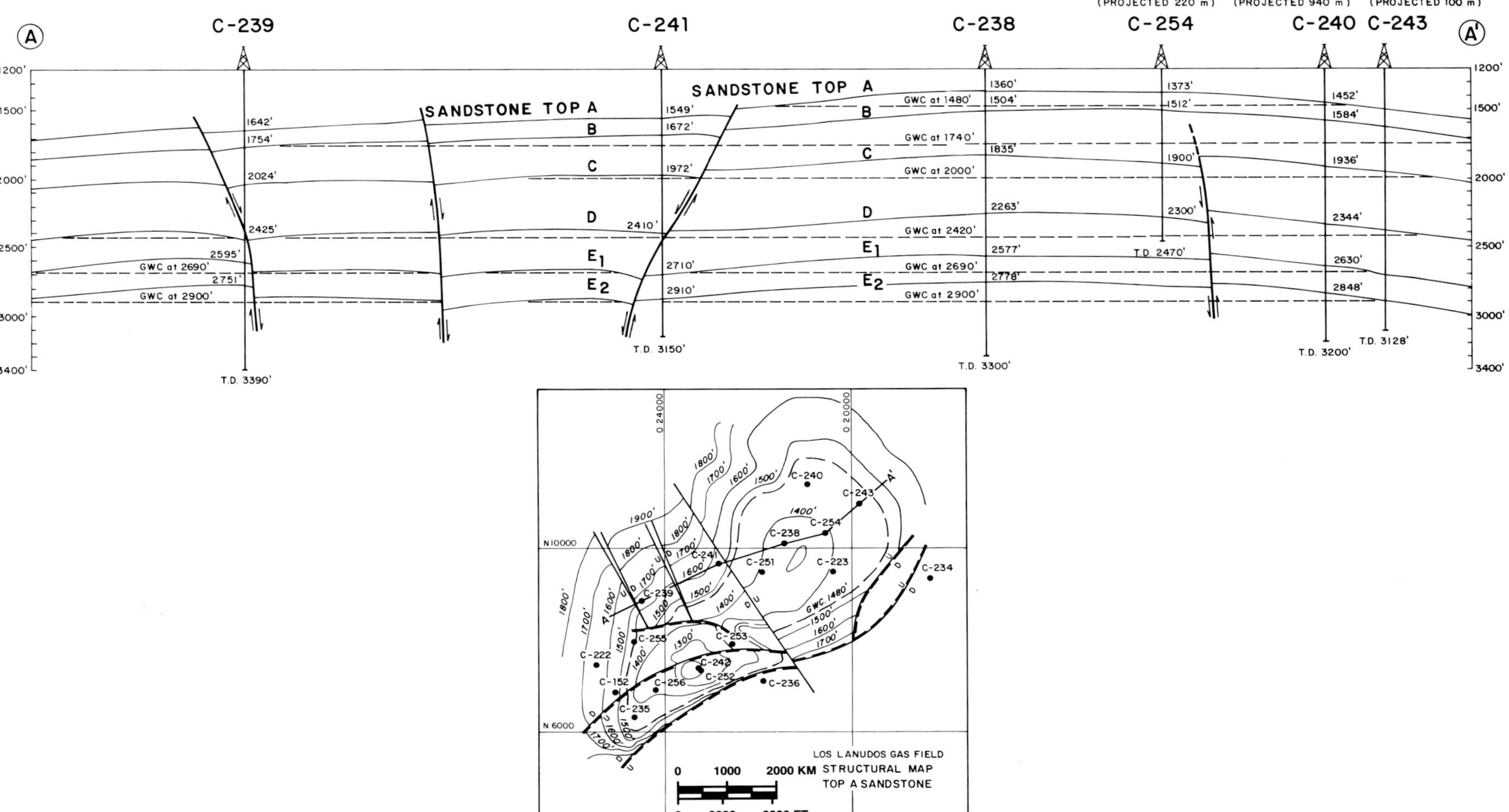

Figure 6. Structure cross section AA′ across the field showing the gas-water contacts (GWC) for the various sands of the upper sandstone unit. Length of cross section about 3.7 mi (6.0 km). (Location on Figure 2A.)

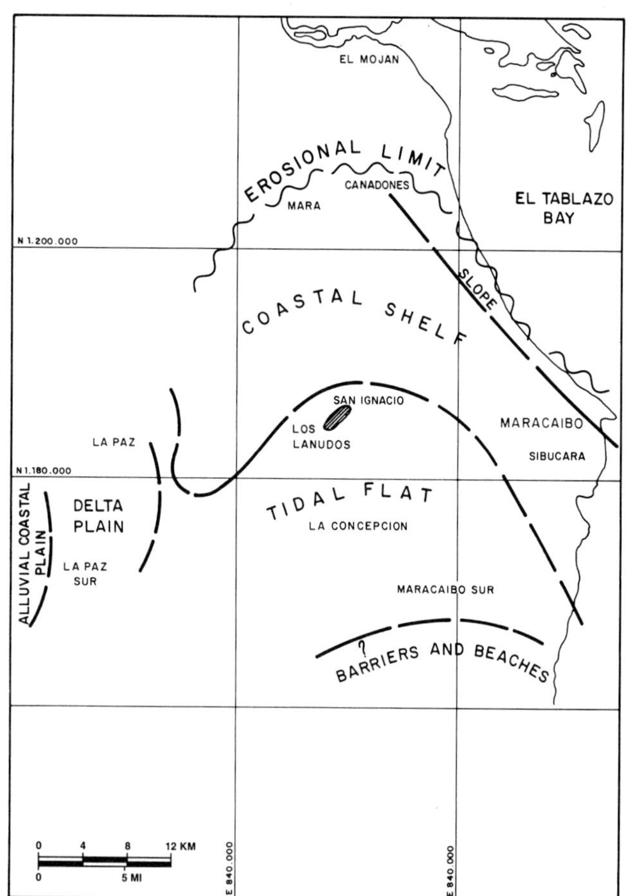

Figure 7. Paleogeographic map for the upper sandstone unit, Misoa Formation, showing Los Lanudos gas field and its location relative to the various depositional environments.

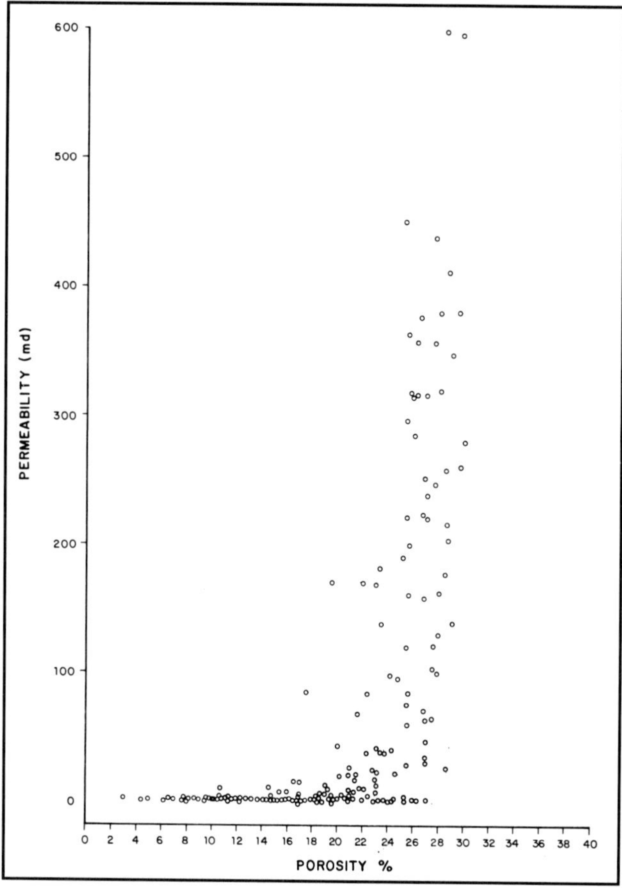

Figure 8. Porosity-permeability cross-plot of samples from well C-239.

Faults

Most faults were interpreted from seismic sections. Only one fault has been identified in boreholes; it was encountered during drilling of the C-240 and C-254 wells and is not observed on seismic sections. The fault, evident from well correlations, originated in deeper strata and shows no expression at shallower depths. It appears, however, to be associated with folds. Seismic sections show the main, sinuous reverse fault extending northeast-southwest along the southeast side of the field. This fault forms the eastern boundary and acts as a trapping mechanism for the gas (Figure 2A).

Source

The Cretaceous La Luna Formation consists of gray to black, calcareous, laminated shales, limestones, and cherts. It is considered to be the main source rock in the entire area of the Maracaibo Basin, with an average TOC of 3.8% and a maximum of 9.6%. Kerogen is type II. As shown by Figure 9, the outset of oil generation probably occurred in the Cogollo Group, also a source rock, and the La Luna in early middle Eocene and ended in late Eocene time, with the gas-forming phase continuing through the time of the Eocene uplift. With respect to the gases in Los Lanudos, Lew (1984) and Young (1984) classified the gases in a genetic manner using carbon isotopes $^{13}C_1$ and hydrogen (D_1) in methane and the concentration of C_{2+}, according to the classification of Schoell (1983). Lew distinguished three types of gases in Los Lanudos field and in the surrounding areas of San Ignacio, Socorro, and DM-114 (Figure 1C):

1. Thermogenic gas associated with oil generated in the Miocene by the La Luna Formation near the Mara area.
2. Dry gas (metagenic), nonassociated, marine type, generated at the end of the Eocene by the La Luna, near the San Ignacio trend.
3. Mixed biogenic and thermogenic gas, generated by the La Luna Formation at the end of the Eocene and mixed with biogenic gas present in the Eocene sediments in the Los Lanudos area.

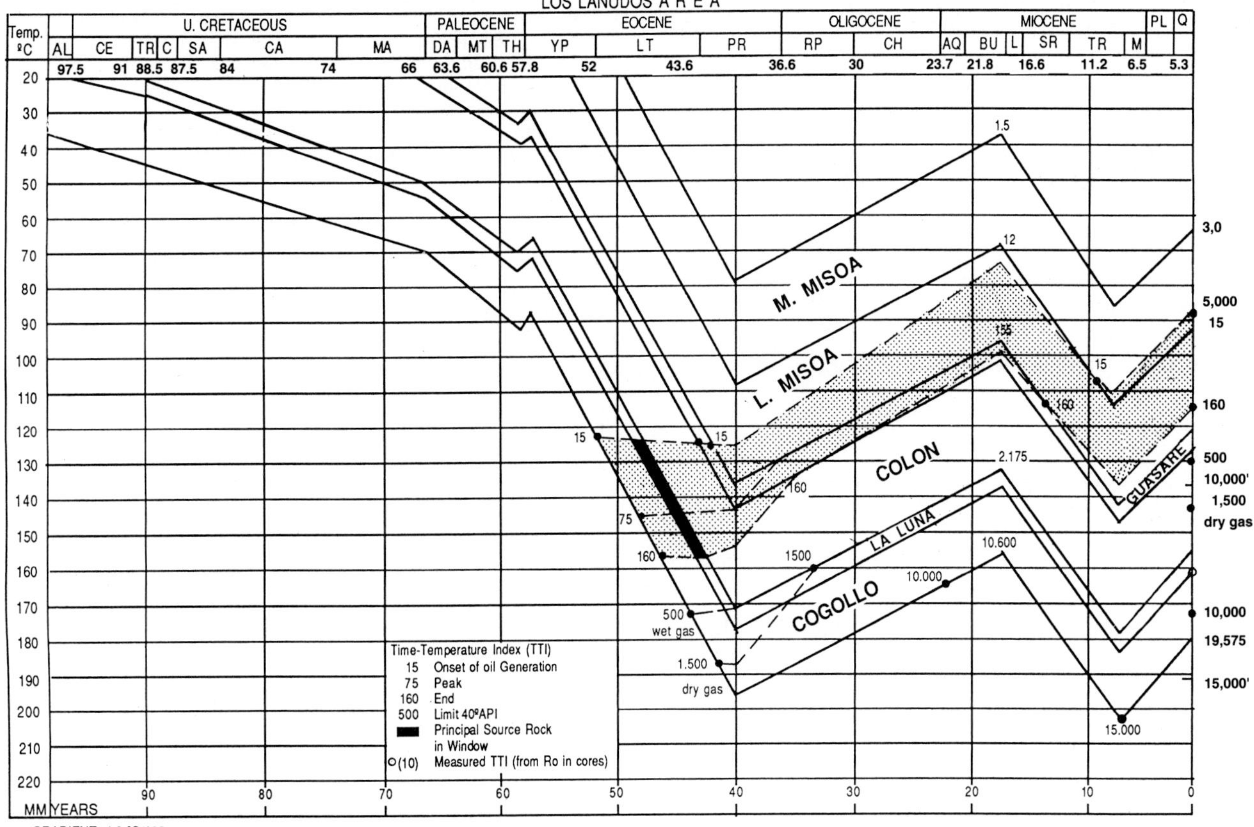

Figure 9. A Lopatin TTI reconstruction showing the iso-maturity lines of the oil generative window for the Los Lanudos area.

Young pointed out that the low $^{13}C_1$ and C_{2+} relationship in the mixed gas could possibly be due to loss of C_{2+} through migration instead of mixing with biogenic gas, especially in some of the E sandstones, where the $^{13}C_1$-deuterium relationship indicates thermal gas.

The Eocene lower Misoa sediments, with type III kerogen, have a TOC greater than 1.0% and are mature (TTI 15–160, Figure 9), but none of the analyses have detected gas generated from humic material.

EXPLORATION CONCEPTS

The regional play consists of a series of anticlinal structures trending toward the northeast. These are associated with wrench fault systems. The Los Lanudos structure consists of one of these asymmetrical anticlines, with flanks dipping between 2° and 4°. The anticline has been divided into blocks by a northwest-southeast fault and is bounded along its southeast flank by a transcurrent fault system more or less parallel to the anticlinal axis.

An anticlinal trap exists within the Los Lanudos field, consisting of multiple individual sandstone reservoirs. The reservoirs consist of coastal bars with some fluvio-deltaic channels and flood-plain splay deposits.

The Los Lanudos field shares many features with other fields in Venezuela, mainly those located southwest of Lake Maracaibo (Bonito, Las Cruces, Tarra) and some located in the Mara-Maracaibo area (San Ignacio, Alcaraván and El Socorro). All of these are gas-producing fields, consisting of anticlines bounded by a major fault on one side.

During the development of the field, the SEISLOG™ technique (pseudo-sonic logs from seismic records) aided in gas well locations, all of which were successful. A three-dimensional seismic survey has been scheduled to aid in the definition of the complex structural framework of the field. These modern techniques proved invaluable in structurally complex areas containing thin gas-bearing sands of limited extent. Because of structural complexity, standard seismic methods were of little help and subsurface data on faulting were scarce. Nevertheless, by using only the available well information and keeping the structural maps of the Misoa Formation simple, without involving the structural complexities of the underlying Cretaceous and Paleocene section and the many small faults, the exploratory and development program has been successful.

ACKNOWLEDGMENTS

The author wishes to thank Maraven, S.A. and Petróleos de Venezuela, S.A. for permission to publish this paper. Also I would like to thank Donald Goddard for his suggestions in writing the paper and the reviewers for their comments and suggestions.

REFERENCES CITED

Barbeito, P. J., A. M. Evans, and R. Pittelli, 1984, Estudio del Eoceno del área Mara-Maracaibo, Estado Zulia, Venezuela: Maraven Report EPC-7649.1.

Broeckers, M. P., W. D. Hartung, and P. L. Lingen, 1983, Review of the Los Lanudos, San Ignacio and El Socorro gas fields: Maraven Report EPC-7258.

Gosh, S. K., and B. Aguado, 1983. Diagénesis de las Areniscas del Eoceno en el área de Mara-Maracaibo: Intevep, INT-0092183.

Habicht, K., 1958, C-sands, Lake Maracaibo: a regional study of Eocene sand development: Maraven Report EPC-1562.

Lew, M., 1984, Estudio del eoceno del área de Mara-Maracaibo, Estado Zulia; Parte 3, Origen del gas libre y gas asociado: Maraven Report EPC-7647.3.

Reguiero, J., J. De Mena, and M. Rampazzo, 1985, Establecimiento de una metodologia para la delimitacion de yacimientos de gas en el area de Los Lanudos: Intevep Report INT-01351.85.

Sánchez, F. A., 1987, Campo Los Lanudos: Revisión geológica: Maraven Report EPC-10904.

Schoell, M., 1983, Genetic characterization of natural gases: AAPG Bulletin, v. 67, n. 12, p. 2225–2238.

Stalder, P., 1981, Geology and hydrocarbon potential of the Maracaibo, Northern Perijá and Sinamaica areas (Western Venezuela): Maraven Report EPC-6368.

Twombley, B. N., 1984, Estudio del Eocene del área Mara-Maracaibo: Parte II: Un estudio integrado de la sedimentación del Eoceno y el potencial de los yacimientos: Maraven Report EPC-7647.

Uzcátegui, E., 1985, Evaluación de la campaña de Perforación de Pozos de avanzada en el campo Los Lanudos, Maracaibo, Venezuela: Maraven, I.T. N. IPDT/1-IT65-02.

Van Andel, T. H., 1954, The Eocene in Western Venezuela: Part II: Facies distribution and regional geological history: Maracaibo, Venezuela, C.S.V., EPC-1327, EPC-1327 Report.

Van der Veen, F. M. and J. Posthuma, 1983, Geochemical analysis of 22 gas samples from Los Lanudos, Falcón, Mara-La Paz and Zuata wells, Venezuela: Maraven/Shell, EPC-7610.

Young, G. A., 1982, Evalución de las reservas de gas y su potencial de producción en Venezuela: Petróleos de Venezuela, S.A., Coordination of Exploration report.

Young, G. A., 1984, Sintesis de estudios sobre la geoquímica y diagénesis de sedimentos en la cuenca de Maracaibo: Petróleos de Venezuela S.A., Coordination of Exploration report.

Appendix 1. Field Description

Field name .... *Los Lanudos field*

Ultimate recoverable reserves .... *140 bcf*

Field location:

Country .... *Venezuela*

State .... *Zulia*

Basin/Province .... *Maracaibo Basin*

Field discovery:

Year field discovered .... *Eocene Misoa Formation 1943*

Discovery well name and general location: .... *Well C-147, 22 km northeast of Maracaibo*

Discovery well operator .... *Venezuelan Oil Concessions, Ltd.*

IP .... *NA*

All other zones with shows of oil and gas in the field:

Age	Formation	Type of Show
Eocene	*Misoa (upper sands)*	*Gas*

Geologic concept leading to discovery and method or methods used to delineate prospect

Surface geology, subsurface geology from electric logs of wells located in adjacent fields, and seismic interpretations. Interpretation of more recent seismic data was responsible for defining the structural framework which led to the 1981–1983 drilling activity.

Structure:

Province/basin type .... *Bally 222; Klemme III Bc*

Tectonic history

The Maracaibo Basin is located within the unstable zone bounded by the Oca, Perijá, and Boconó transcurrent fault systems in northwestern Venezuela. These systems are related to right-lateral strike-slip movement along the Caribbean and South American plates boundary, active since the Jurassic. Major periods of tectonism occurred in the Tertiary, including during the late Miocene Andean uplift.

Regional structure

Regional structure consists of a series of northeasterly trending anticlines associated with wrench fault systems.

Local structure

The field is associated with a northeast-trending asymmetrical anticline whose flanks have 2° to 5° dips except southeast dips of 10° to 12°. The anticline is divided into blocks by a reverse fault and bounded along the southeast by a transcurrent fault system more or less parallel to the anticlinal axis.

Trap:

Trap type(s)

Anticlinal trap with multiple, individual sandstone bodies. Stratigraphic traps: Coastal barrier bar and some fluvio-deltaic channel and crevasse splay deposits.

Basin stratigraphy (major stratigraphic intervals from surface to deepest penetration in field):

Chronostratigraphy	Formation	Depth to Top in ft (m)
Oligocene	*Peroc Formation*	*Outcrop*
Eocene	*Misoa Formation:*	*360 (110)*
	Boscan Member	*360 (110)*
	upper sandstone member	*1363 (416)*
	Ramillete Member	*2863 (873)*
	Punta Gorda Member	*3013 (919)*
	Lower Misoa Member	*3913 (1193)*

Reservoir characteristics:

Number of reservoirs *6*

Formations .... *Misoa-upper sandstone member with producing intervals A, B, C, D, E_1, and E_2*

Ages .... *Eocene*

Depths to tops of reservoirs .... *-1190 ft (-363 m) to top of A interval of the upper sandstone member*

Gross thickness (top to bottom of producing interval) .... *2350 ft (717 m)*

Net thickness—total thickness of producing zones

Average .... *1470 ft (448 m)*

Maximum .... *1667 ft (508 m)*

Lithology

Fine- to medium-grained, moderately sorted sandstone (5 to 40 ft [1.5 to 12 m] thick beds) interbedded with thinly laminated shales with abundant plant remains

Porosity type

Primary intergranular porosity, intercrystalline porosity, secondary porosity, abundant calcite cement, and minor moldic porosity

Average porosity *21%*

Permeability range .... *1.5 to 1000 md*

Seals:

Upper

Formation, fault, or other feature .... *Misoa Formation Boscan Member bottom shale*

Lithology .... *Gray shale*

Lateral

Formation, fault, or other feature .... *Fault/lithology change*

Lithology .... *Gray shale*

Source:

Formation and age .... *Cretaceous La Luna*

Lithology .... *Limestone, calcareous black shales, and cherts*

Average total organic carbon (TOC) *3.8%*

Maximum TOC *9.6%*

Kerogen type (I, II, or III) .... *II*

Vitrinite reflectance (maturation) .... $R_o = 0.6\text{-}1.2$

Time of hydrocarbon expulsion .... *Middle Eocene-Miocene*

Present depth to top of source .... *-9734 ft (-2969 m)*

Thickness .... *300 ft (92 m)*

Potential yield .... *10 kg HC/MT*

Appendix 2. Production Data

Field name .... *Los Lanudos*

Field size:

Proved acres .... *344,699 ac (139,498 ha)*

Number of wells all years *19*

Current number of wells *12*

Well spacing .... *1970 ft (600 m)*

Ultimate recoverable .... *140 bcf*

Cumulative production .... *50 bcf*

Annual production .... *9.1 bcf*

Present annual decline rate *12%*
Initial decline rate *15%*
Overall decline rate *5%*
Annual water production .... *None*
In place, total reserves .... *140 bcf*

Drilling and casing practices:

Amount of surface casing set .... *900 ft (275 m)*

Casing program

16-in. casing at ±150 ft (46 m) (J-55, STC, 84 lbs/ft); 10¾-in. casing at ±900 ft (275 m) (J-55, STC, 40.5 lbs/ft); 7-in. casing at casing point (through reservoir)(J-55, LTC, 23 lbs/ft)

Drilling mud

From 0–300 ft (90 m), CBM 80–85 lbs/ft³; from 300 ft (90 m) to T.D., Benex 85–88 lbs/ft³

Bit program .... *20-in., OSC 3AJ; 13¾-in., 111; 9⅞-in., X 3A, X 3AJ or 12*
High pressure zones .... *No overpressured zones*

Completion practices:

Interval(s) perforated .... *Misoa upper sandstones (A, B, C, D, E_1, and E_2 intervals)*
Well treatment .... *NA*

Formation evaluation:

Logging suites

ISF/SP/GR 1/200, 1/500, 1/1000 (±900-Sup) ISFL/MSFL ML/MC/GR 1/200; ISFL/GR 1/500, 1/1000; FDC (ϕ) FDC (pb) C 1/200, CNL ϕ/c 1/500, from T.D. to casing shoe

Testing practices .... *Perforate and test sandstone intervals from bottom to top individually*
Mud logging techniques .... *Lithologic description of cuttings, fluorescence analysis, and gas chromatography*

Field characteristics:

Average elevation .... *150 ft (46 m)*
Initial pressure .... *(average) 1280 psi (8.825 MPa)*
Present pressure .... *(average) 650 psi (4.481 MPa)*
Temperature .... *130° F (54° C)*
Drive .... *Gas expansion*
Oil column thickness .... *NA*
Oil-water contact .... *A interval, –1480 ft (–451 m)*
B interval, –1740 ft (–531 m)
C interval, –2000 ft (–610 m)
D interval, –2420 ft (–738 m)
E_1 interval, –2690 ft (–820 m)
E_2 interval, –2900 ft (–884 m)
Connate water *52%*
Water salinity, TDS .... *2000 ppm*
Resistivity of water .... *0.55 ohm-m*
Bulk volume water (%) .... *NA*

Transportation method and market for oil and gas:

Gas pipeline from Los Lanudos field to the city of Maracaibo

Guafita Field—Venezuela
Barinas-Apure Basin, Apure State

NESTOR CHIGNE
LEROY HERNANDEZ
Corpoven, S.A.
Caracas, Venezuela

FIELD CLASSIFICATION

BASIN: Barinas-Apure
BASIN TYPE: Foredeep
RESERVOIR ROCK TYPE: Sandstone
RESERVOIR AGE: Cretaceous and Oligocene
PETROLEUM TYPE: Oil
TRAP TYPE: Faulted Anticline
RESERVOIR ENVIRONMENT OF DEPOSITION: Fluvio-deltaic
TRAP DESCRIPTION: Faulted anticline associated with regional wrench fault system

LOCATION

The Guafita field is located in southwestern Venezuela in the state of Apure, adjacent to the Colombian border (Figure 1). Geologically, it lies over the crest of the Arauca arch, which separates the Barinas-Apure basin from the Llanos basin of Colombia; both basins are located along the eastern flank of the Andean Cordilleras (Figure 2). The Arauca River is the southern limit of the Guafita field as well as the border between Venezuela and Colombia. Current estimates of reserves are given in *History, Post-Discovery,* below.

HISTORY

Pre-Discovery

Exploration activities began in Apure in 1924. Until 1951, the effort was concentrated in northwestern Apure, where gravity and analog seismic surveys of limited extent were run. These surveys provided the first evidence of structure in the Burgua–La Ceiba area, where oil seeps had been known for years, and led to the drilling of the dry wells Burgua-1, -2, -3, and -4 between 1953 and 1954 (Figure 2). In the rest of Apure, geophysical and geological activities continued until 1958, extending southward and southeastward in Apure. At the same time, wells Yaure-1 and -2, Agua Linda-1, and Rosalia-1 were drilled in southwestern Barinas state, and Apure 2 farther south (Figure 2). All five wells were dry and exploration was discontinued for 20 years.

Regional maps prepared by Corpoven in 1978 (Figure 3) showed that all of the Barinas oil fields are concentrated on the northern flank of the northwest-southeast-trending paleoarch called the Mérida arch. These maps also showed, farther to the southwest, a similar paleoarch that was unexplored and thought to have a very similar geological history. As a consequence, it was believed that this paleoarch, called the Arauca arch, was the most prospective area left to be explored in the Barinas-Apure basin. All this information led to the planning of seismic acquisition to confirm the presence of the Arauca arch and associated structures.

The geophysical operations were reactivated in 1979, and by 1982, most of the region was covered by approximately 2400 km of seismic lines. The regional seismic intepretation defined three major plays, with the following order of priority: Burgua–La Ceiba, La Victoria, and Guafita (Figure 2). The exploration drilling began in 1983 in the Burgua–La Ceiba area of northwestern Apure with well LCB-1X, followed by Jordan-1X and MS-1X (Figure 2). All these wells found minor shows of oil and were abandoned.

The successful Caño Limón-1 well, drilled by Occidental Petroleum across the Arauca river from Guafita in Colombia in 1983, caused a change in the order of priorities in the exploration activity, and the Guafita structure became the next exploratory target.

Discovery

The discovery well, GF-1X, was drilled in February and March of 1984 by Corpoven, S.A. (Figures 2 and 6). The well reached a final depth of 9008 ft (2747

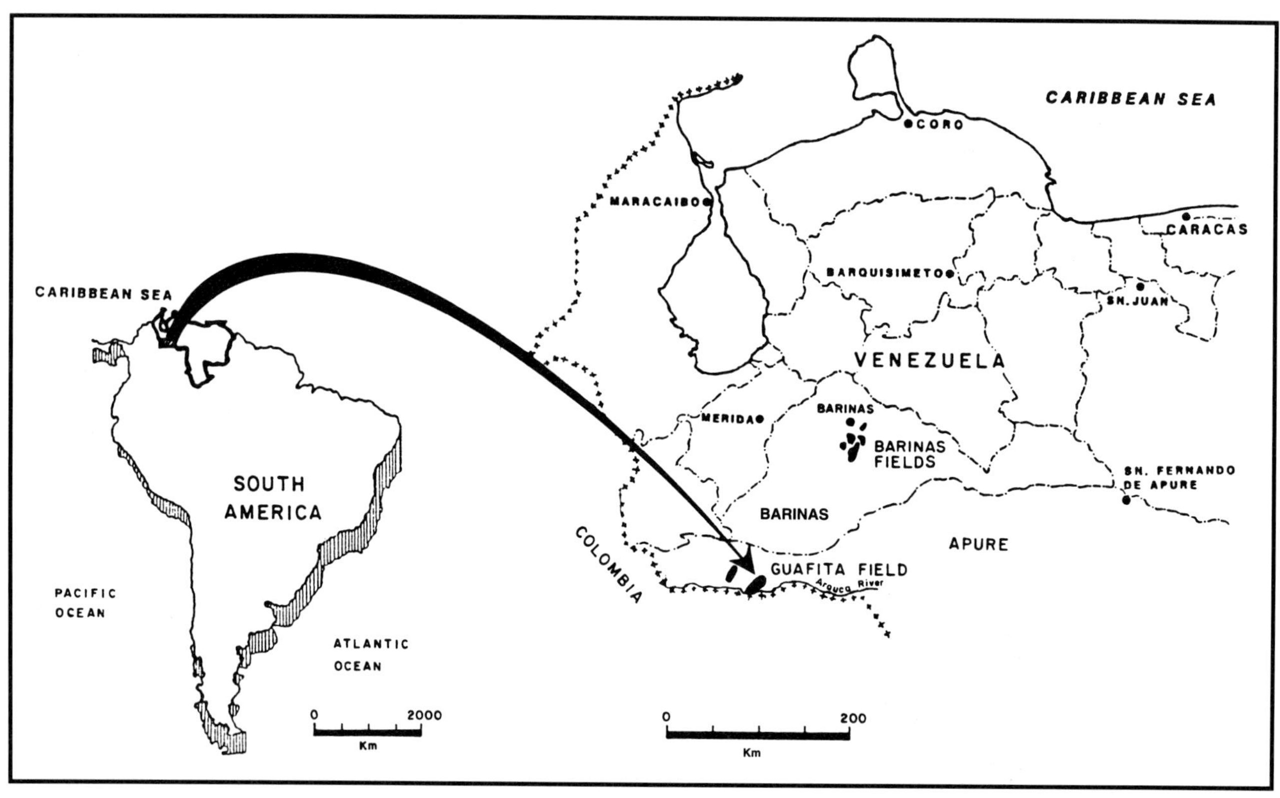

Figure 1. Location map of Venezuela showing Guafita field, the Barinas fields, and the states of Apure and Barinas.

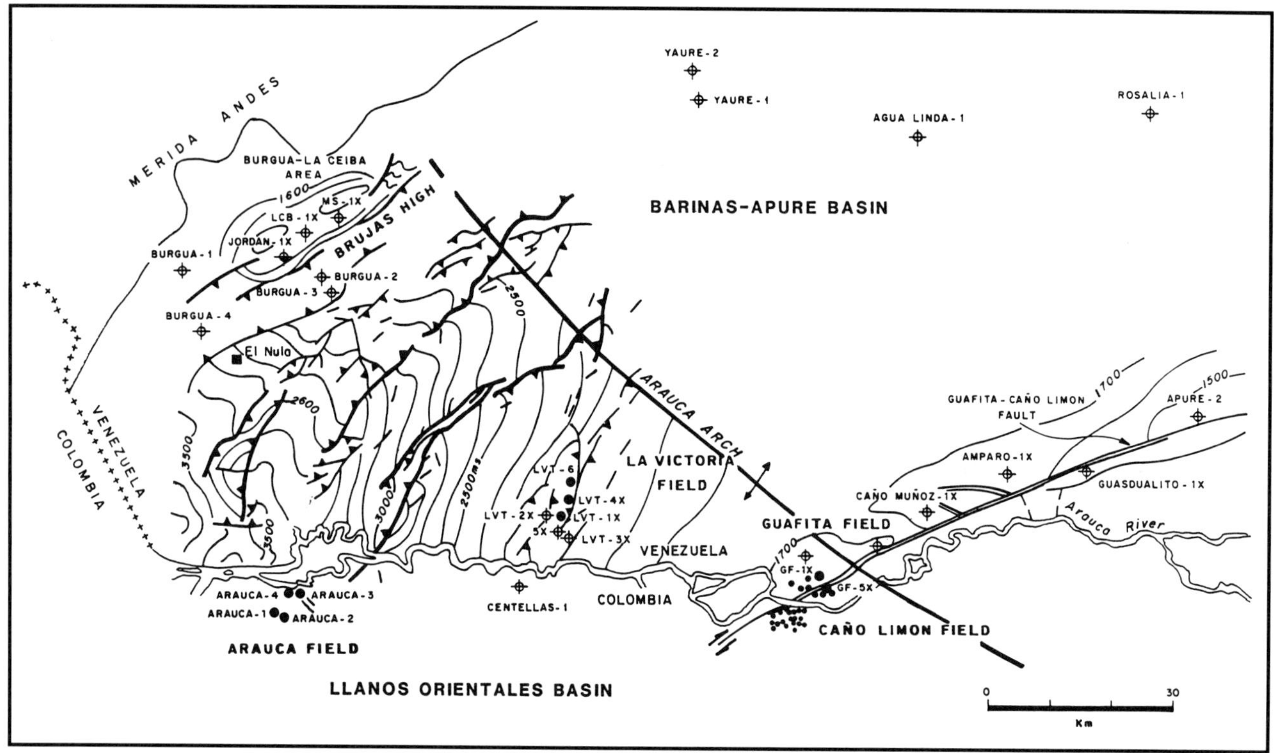

Figure 2. Southwestern Apure regional time structural map of the top of the Cretaceous. Contour interval, 100 ms.

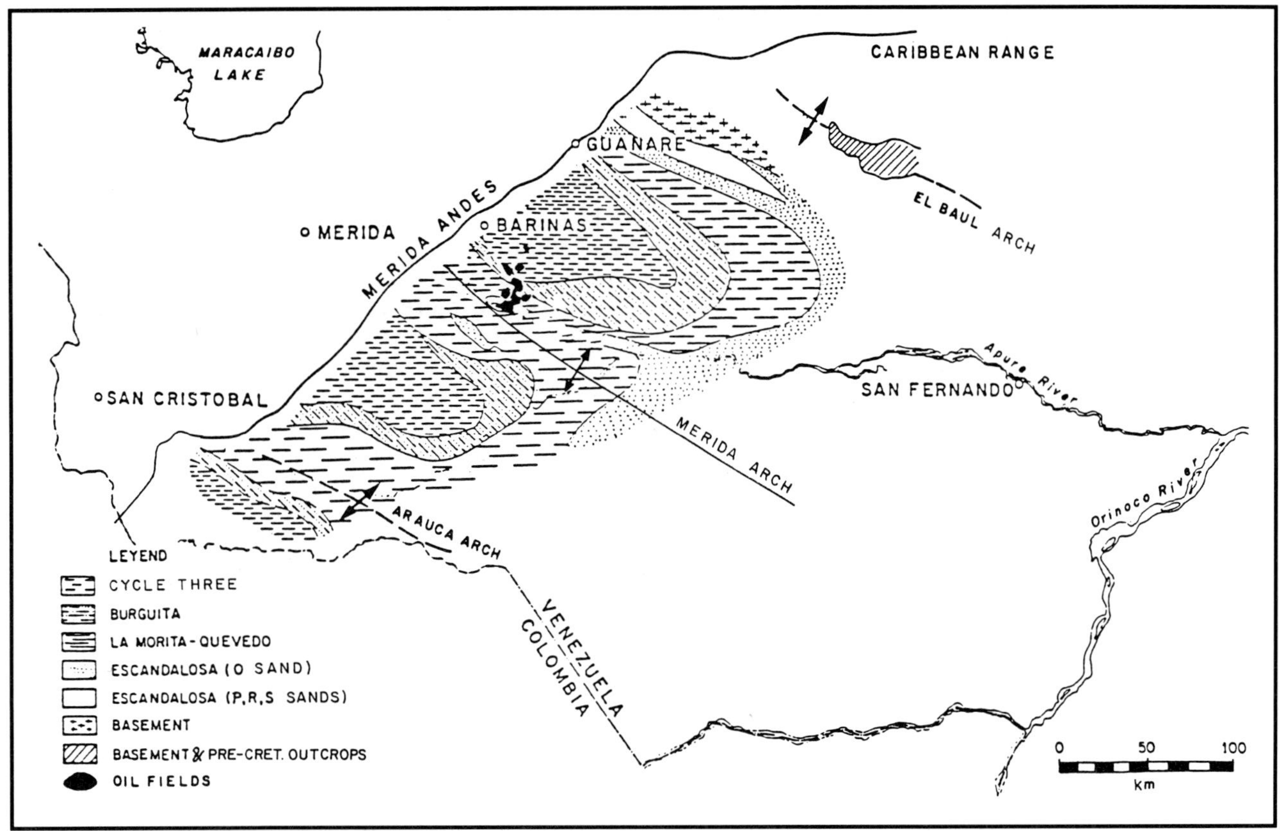

Figure 3. Subcrop map below the Eocene unconformity. (Modified from Russomanno and Velarde, 1982.)

m), the last 100 ft in red beds of pre-Cretaceous age. The drill site had been spotted near the top of a broad, gentle anticlinal feature located in the northern block of the Guafita–Caño Limón fault, previously delineated by seismic mapping.

Well GF-1X produced 1260 BOPD of 30° API from the lower part of the Oligocene basal sandstones of the Guafita Formation (Figure 13). At the same time, a significant oil seep named San Benito was detected about 7 km southeast of the Guafita anticline, close to the Arauca River. Oil gravities from this seep range from 36 to 52° API. This additional information reinforced the prospectivity of the area.

Following the results of the Caño Limón field discoveries in the southern block of the Guafita–Caño Limón fault, where the entire section of Oligocene basal sands is oil bearing, Corpoven drilled the GF-5X well south of this fault, finding the basal sandstones to be light oil bearing as well as the upper part of the Cretaceous (Figures 2 and 6). This well discovered the South Block of the Guafita field.

Post-Discovery

Following the discovery of North Block production by well GF-1X, wells GF-2X, -3X, and -4X were drilled to better define the geological and reservoir characteristics of the producing sands. Production limits in the North Block were delineated by dry holes GF-9X and GF-15. The dry holes GF-6X and GF-12X defined the northeastern limit of production in the South Block (Figure 6).

To explore the oil possibilities of other structural closures along the northeastern extension of the Guafita trend, wells Guasdualito-1X, Amparo-1X, and Caño Muñoz-1X were drilled, 78 km, 57 km, and 34 km northeast of GF-1X, respectively. All three wells were dry (Figure 2).

Guafita field has been on production since mid-1986. To date, 22 wells have been drilled, of which six are dry. The production rate of the field for September 1988 was 46.8 thousand bbl per day (46.8 MBOPD). Cumulative oil production as of 30 September was 25.2 million bbl (25.2 MMBO). Original oil-in-place for the field is estimated to be 1000 MMBO, of which approximately 250 MMBO are recoverable. However, according to the latest reservoir studies and production behavior, the recovery factor could be considerably higher, even up to 40%.

The produced oil is transported through a 16-in. pipeline, built in 1987, to El Toreño refinery, 220 km to the northeast; from there, it goes to the El Palito refinery, located 380 km farther northeast on the Caribbean coast, through a 20-in. pipeline.

DISCOVERY METHOD

The seismic interpretation in the Guafita area showed the presence of a broad, gentle anticlinal feature associated with a fault system. The main seismic reflector was almost at the same depth as the prospective Tertiary sandy section discovered in the Caño Limón field. Drilling of well GF-1X, near the crest of this anticline, confirmed the prospectivity of the structure.

STRUCTURE

Regional Structure

The sub-Andean basins belong to a series of foreland basins aligned along the eastern side of the Andean Cordilleras (Figure 4). These basins are separated by tectonic ridges or arches whose axes trend sub-perpendicular to the Cordillera (Figures 4 and 5). From this point of view, the Barinas-Apure basin is limited to the northeast by the El Baul arch and to the southwest by the Arauca arch. With the Arauca arch as a geologic boundary between the Barinas-Apure and Llanos basins, the part of Apure to the south of the Arauca arch belongs to the Llanos basin. The Guafita field is located over the crest of the Arauca arch (Figures 2 and 4). This arch is believed to be a broad, subtle early Tertiary feature striking approximately N45°W and crossing the Colombia-Venezuela border near Guafita.

The structural contours on the Cretaceous in southwestern Apure show that the basin deepens to the west-northwest, varying from 7500 ft (2300 m) at Guafita to more than 30,000 ft (9100 m) in the sub-Andean trough of the Mérida Andes (Figure 2). The fault pattern shown here is very consistent. It shows a high degree of alignment of long faults oriented in a southwest-northeast direction subparallel to the Mérida Andes (Figure 2). This is the most notable feature of the structural style in southwestern Apure. These structural trends continue for 40 to 80 km. East of the La Victoria field, structural dip is quite gentle, with only one relevant structure, the Guafita–Caño Limón fault, which is more than 100 km long and which extends into Colombia. The Guafita and Caño Limón oil fields are located along this fault, close to the international border between Colombia and Venezuela. The regional fault system, which includes the Guafita–Caño Limón fault and other faults close to the Mérida Andes, could be

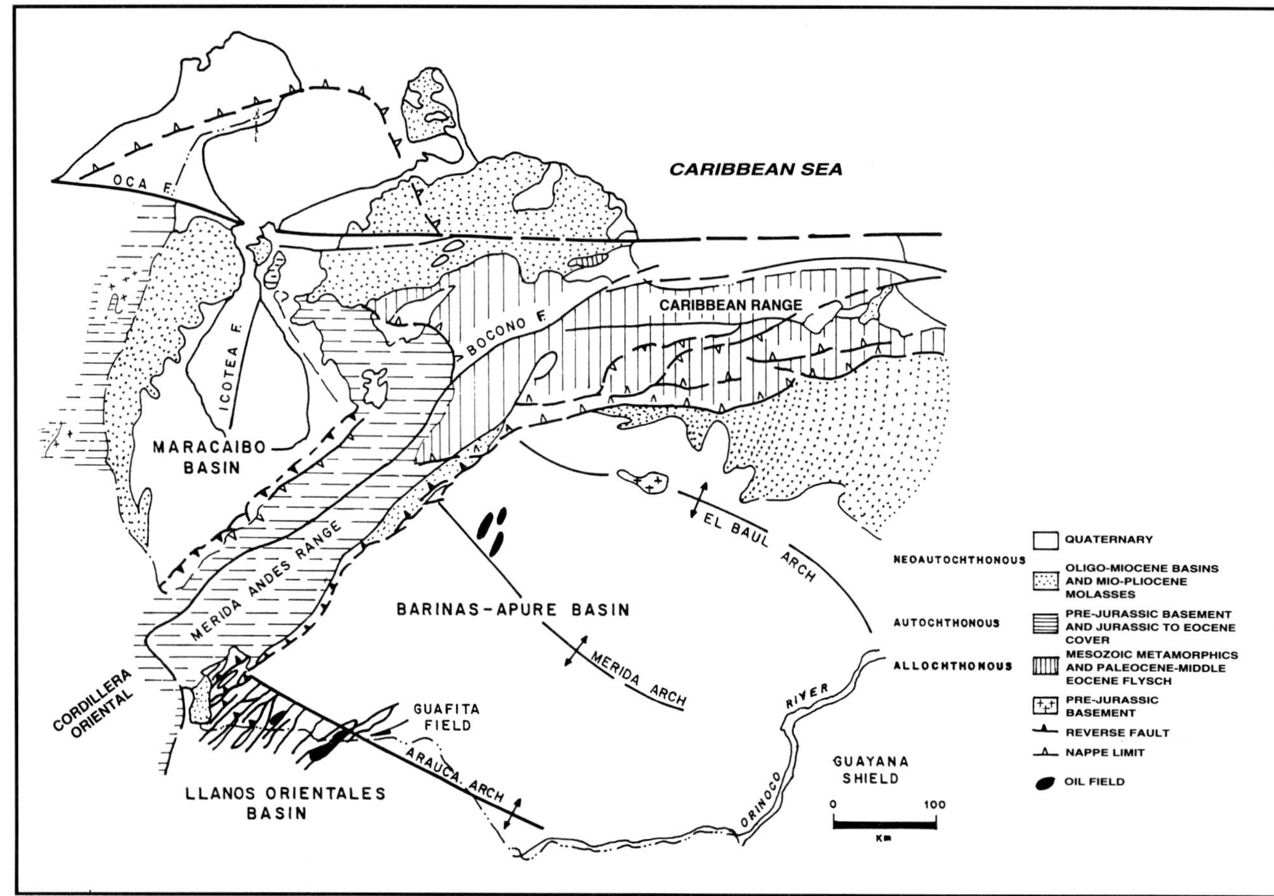

Figure 4. Simplified geological map of western Venezuela.

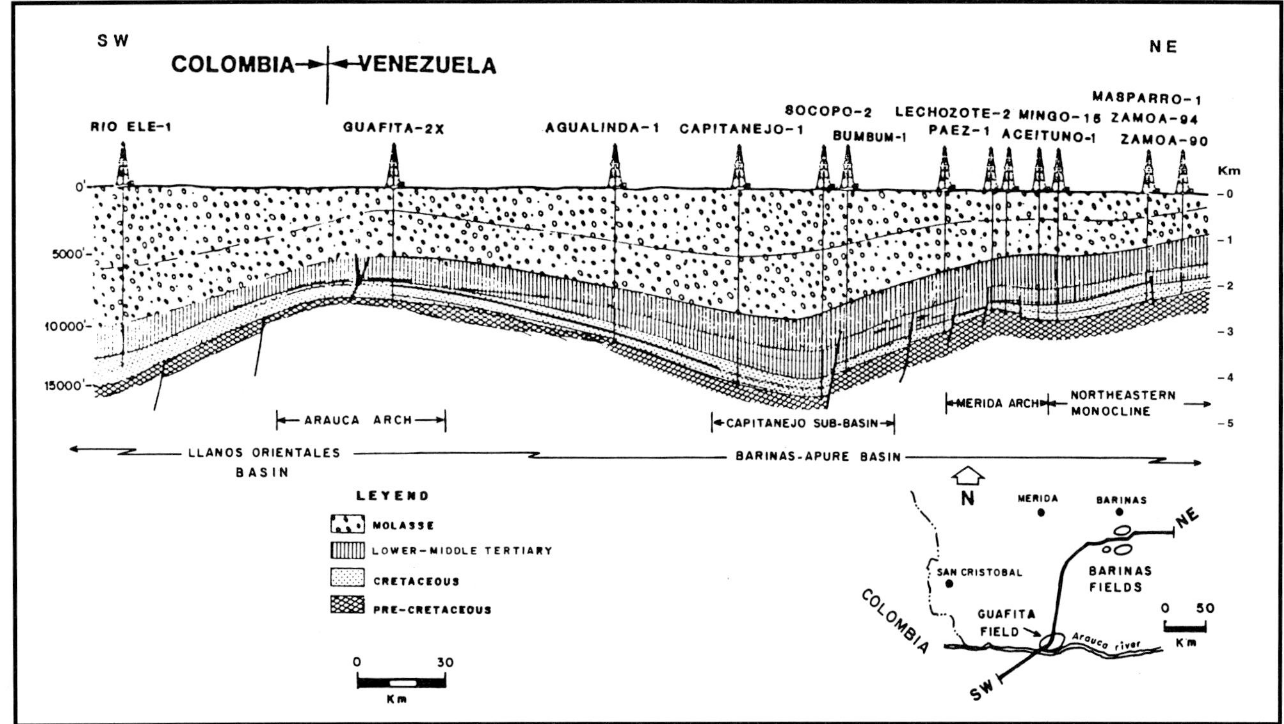

Figure 5. Northern Llanos and Barinas-Apure basins, regional geological section. (From Chigne, 1985.)

interpreted from seismic data to be transcurrent fault systems. The main evidence to support this interpretation is the notable variation of the fault throw and the relative position of associated fault blocks.

According to the hypothesis of the regional tectonic evolution, the post-Jurassic subduction of the Pacific plate below the South American plate produced a volcanic arc along the west coast of South America, generating a large back-arc basin that corresponds to the present Magdalena, Maracaibo, and Barinas-Apure basins. The Guayana shield was always the eastern limit, and the Cretaceous basins deepened toward the northwest, giving rise to the formation of normal faults trending southwest-northeast, with the fault surfaces dipping toward the depocenter of the basins. This Cretaceous normal faulting was reactivated by tectonic forces that gave rise to the post-early Miocene uplift of the Mérida Andes, thus producing the present-day structure of the Barinas-Apure foreland basin.

Local Structure

The Guafita field is on a northeast-trending anticline, split into two blocks by a strike-slip fault zone with variable vertical displacements of up to 500 ft (150 m) at the Cretaceous level (Figures 2 and 6). The southern block extends into Colombian territory and includes the Caño Limón field.

In the North Block of Guafita field, this anticline is about 14 km long and 7 km wide. Its north flank has a very gentle northwest dip of no more than 2° (Figures 6, 7, and 8). The block has two secondary normal faults that divide it into three reservoir segments. These faults of less than 50 ft (15 m) of throw cannot be seen in the seismic sections at the reservoir level. They have been interpreted from pressure build-up tests in the wells and from displacements of three oil-water contacts. The areal closure of this block is approximately 11,000 ac (4450 ha); structural closure is about 270 ft (82 m).

The structure of the South Block is a monocline dipping about 8° to the southeast. The La Yuca reverse fault, coming from Colombia, divides the block into two segments and terminates against the Guafita–Caño Limón fault (Figure 6). The areal closure is about 1400 ac (570 ha) and vertical closure is approximately 500 ft (150 m), in Venezuela.

The Tertiary basal sandstones thin from northeast to southwest in both blocks as can be seen in Figures 9, 10, 11, and 12. This suggests that the Guafita–Caño Limón structure was actively building during Oligocene sedimentation.

STRATIGRAPHY

The geologic column of Apure includes Paleozoic to Recent deposits. This sequence increases in thickness from east to west, reaching more than

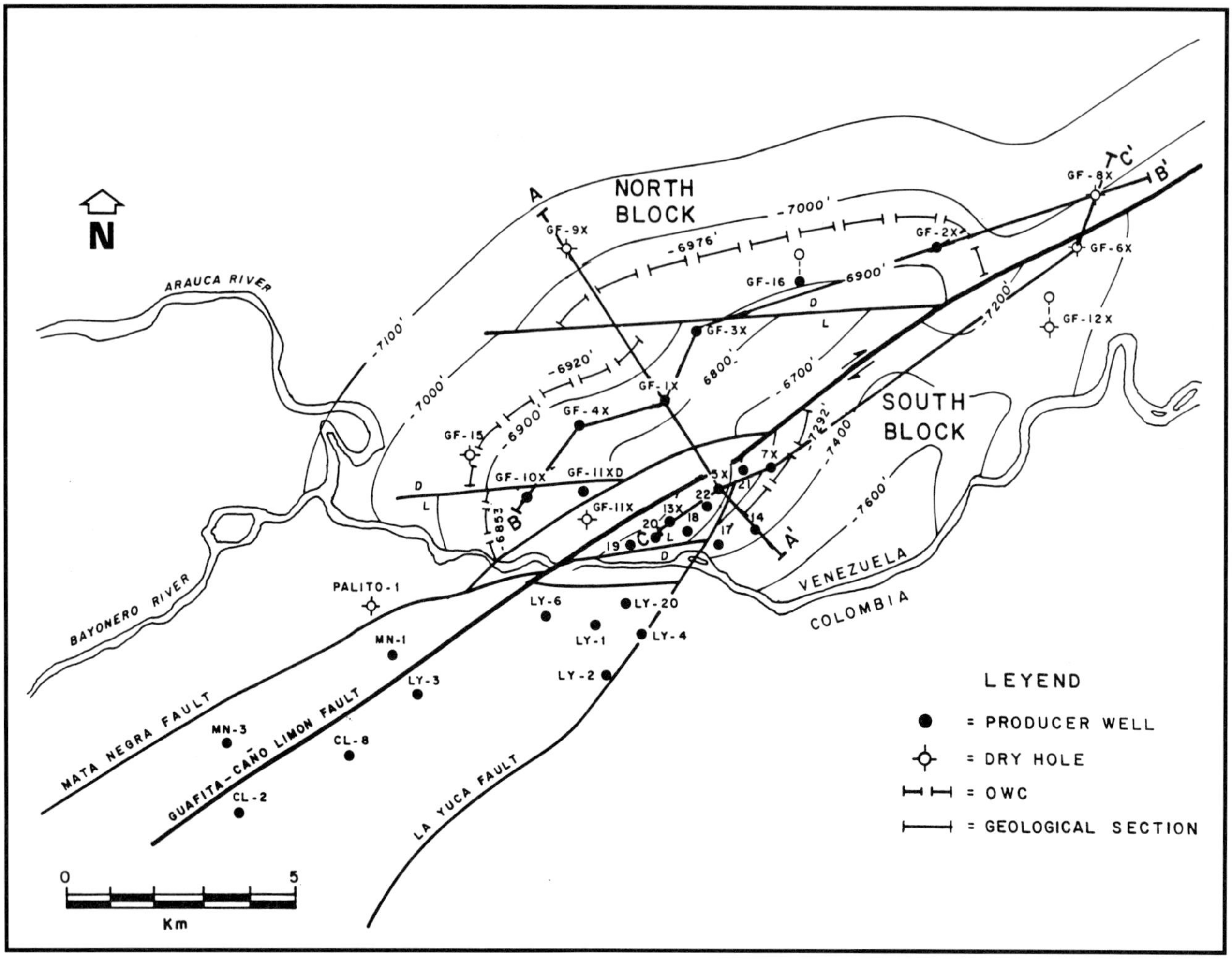

Figure 6. Structural map of the top of the Guafita shale. Contour interval, 100 ft.

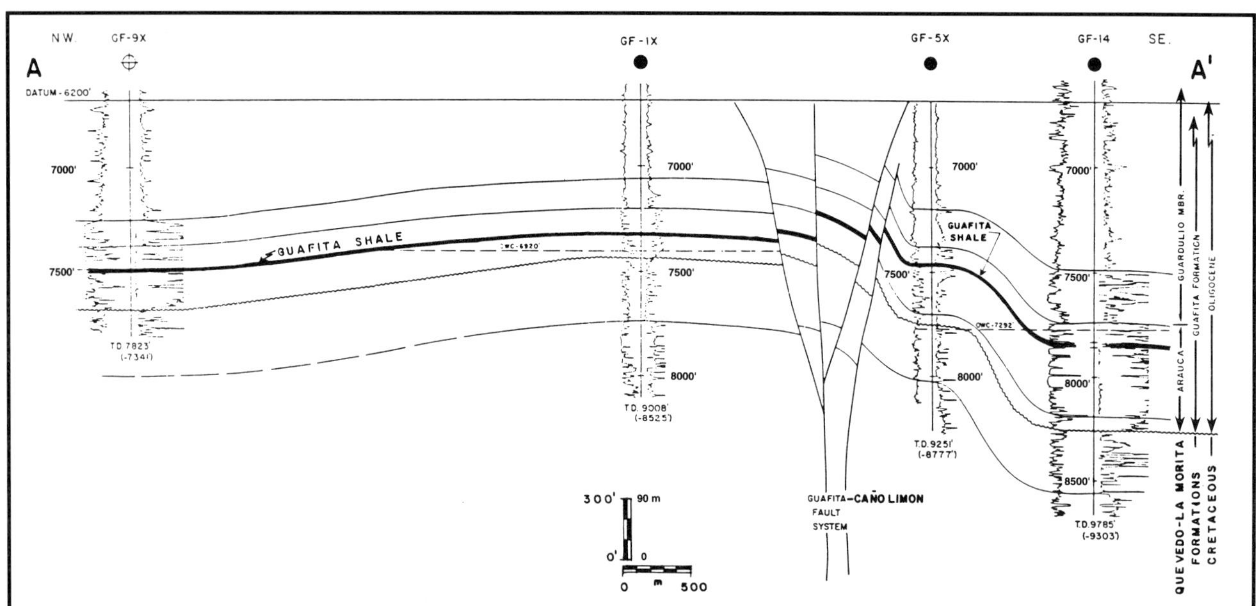

Figure 7. Guafita field, structural cross section A–A′. See Figure 6 for location.

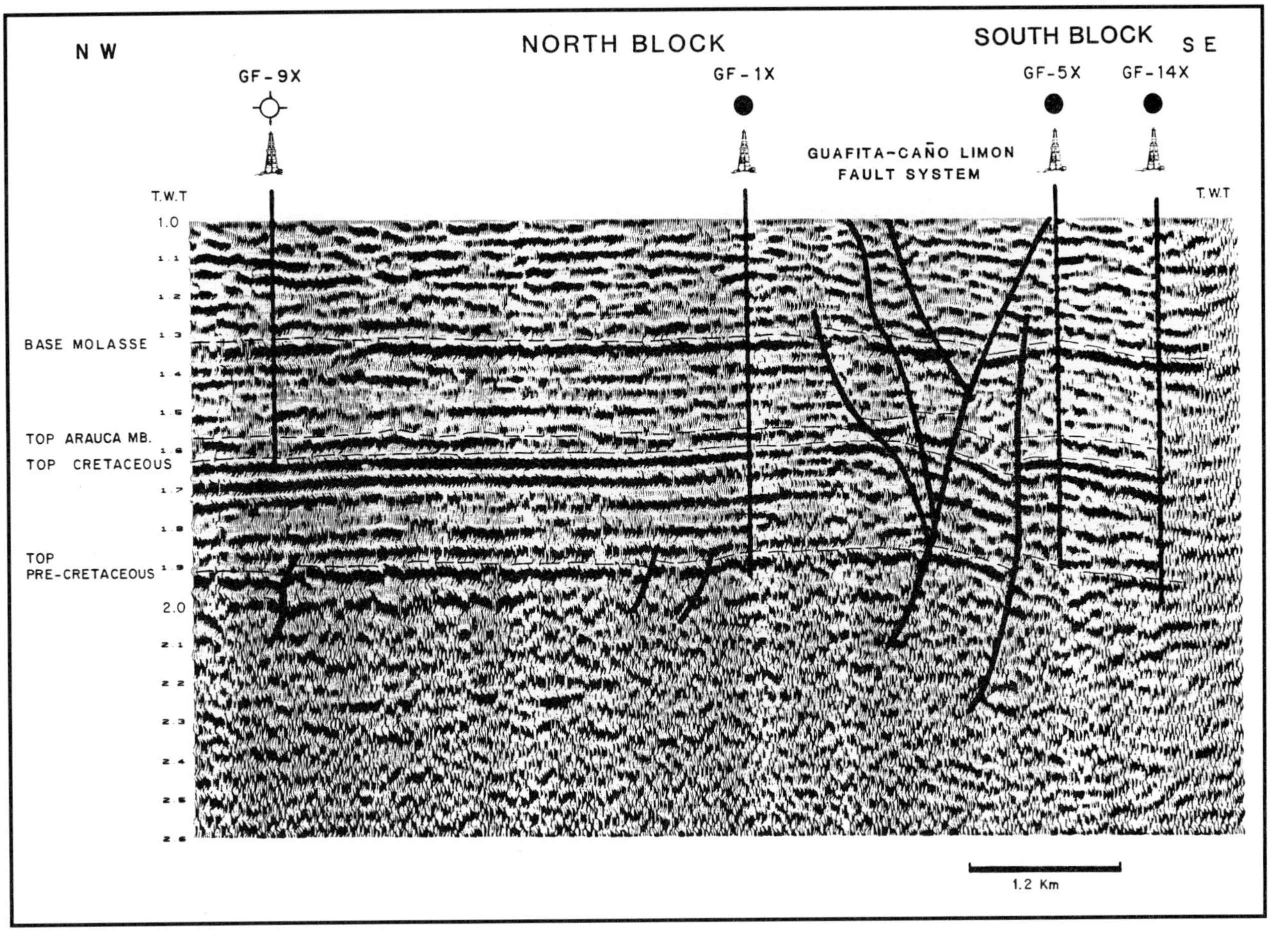

N W
NORTH BLOCK
SOUTH BLOCK
S E
GF-9X
GF-1X
GF-5X
GF-14X
GUAFITA-CAÑO LIMON
FAULT SYSTEM
T.W.T
T.W.T
1.0
BASE MOLASSE
TOP ARAUCA MB.
TOP CRETACEOUS
TOP
PRE-CRETACEOUS
2.0
1.2 Km

Figure 8. Guafita field, seismic interpretation, line 1. Location is same as A-A′ (Figure 6).

SW
NE
B
B'
GF-10X
GF-4X
GF-1X
GF-3X
GF-2X
GF-8X
DATUM-6200'
7000'
7500'
8000'
GUAFITA SHALE
GUAFITA SHALE
T.D.7902' (-7420')
T.D.8202' (-7716')
T.D.9008' (-8525')
T.D.7858' (-7382')
T.D.9106' (-8637')
T.D.8103' (-7632')
GUARDULIO MBR.
ARAUCA
GUAFITA FORMATION
OLIGOCENE
QUEVEDO-LA MORITA FORMATIONS
CRETACEOUS
300'
200
100
0
100 m
0 1 2
Km

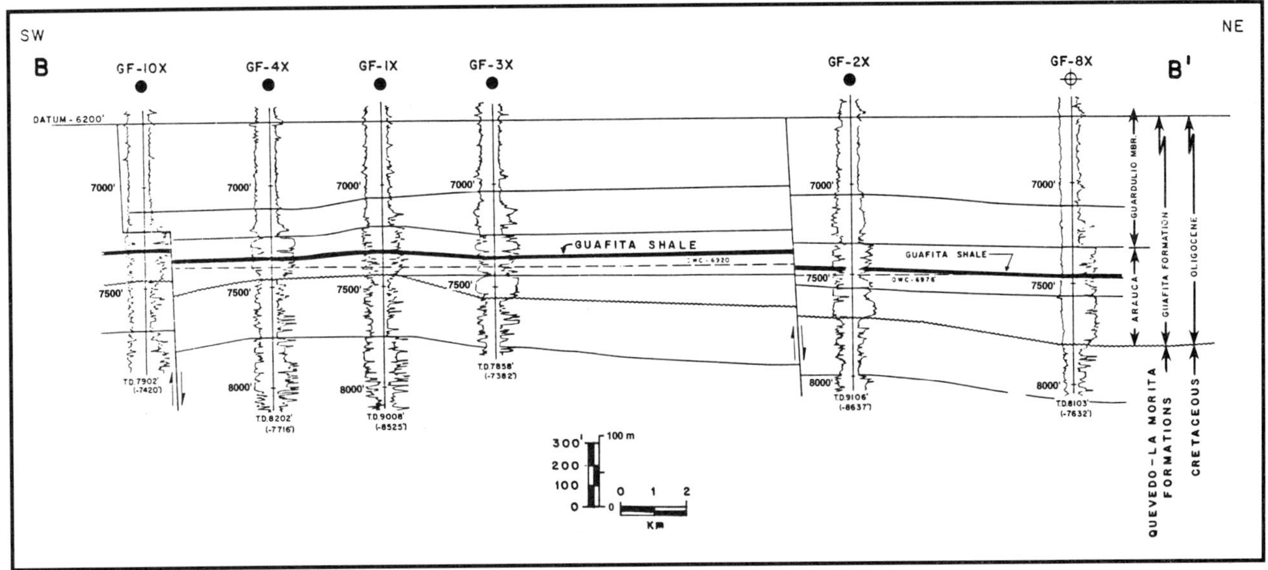

Figure 9. Guafita field, structural cross section B-B′. See Figure 6 for location.

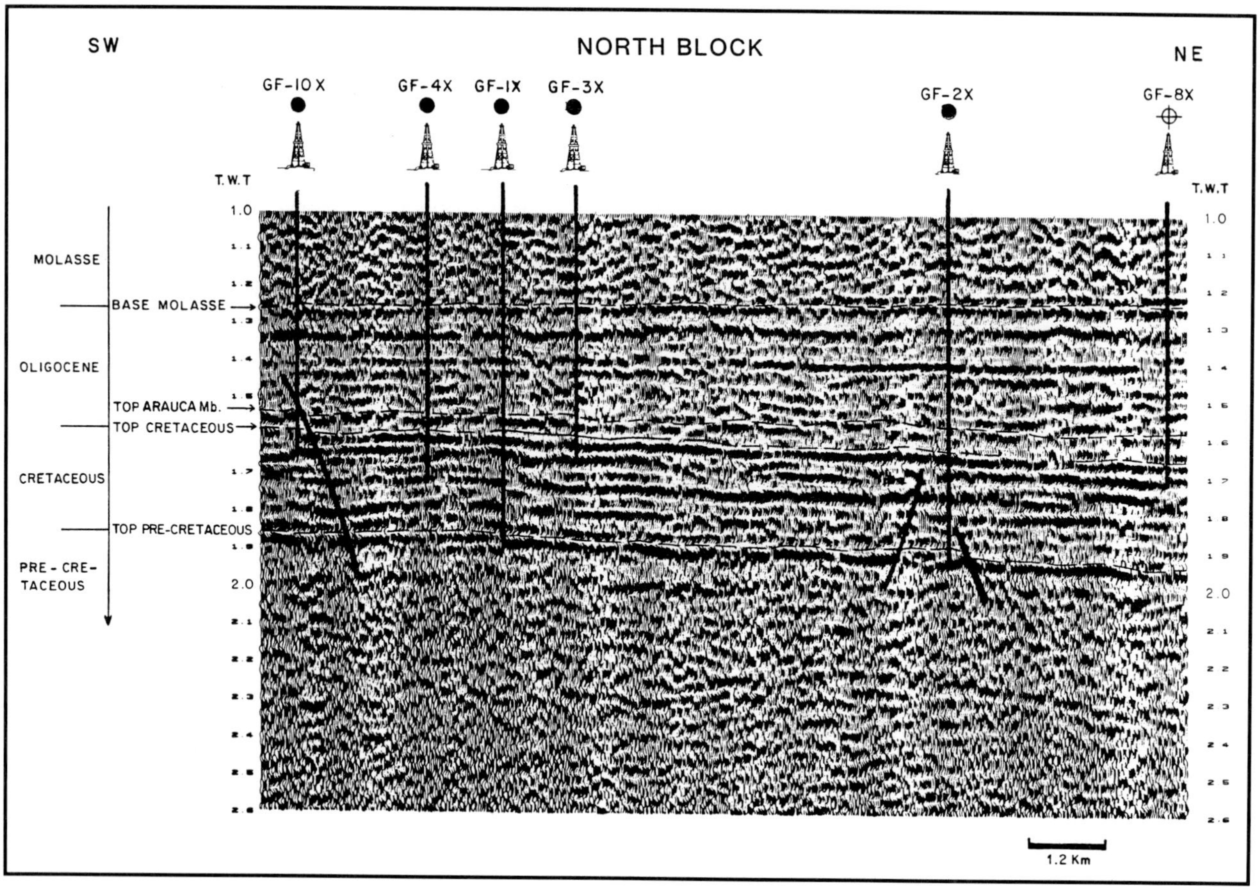

Figure 10. Guafita field, seismic interpretation, line 2. Location is same as B-B′ (Figure 6).

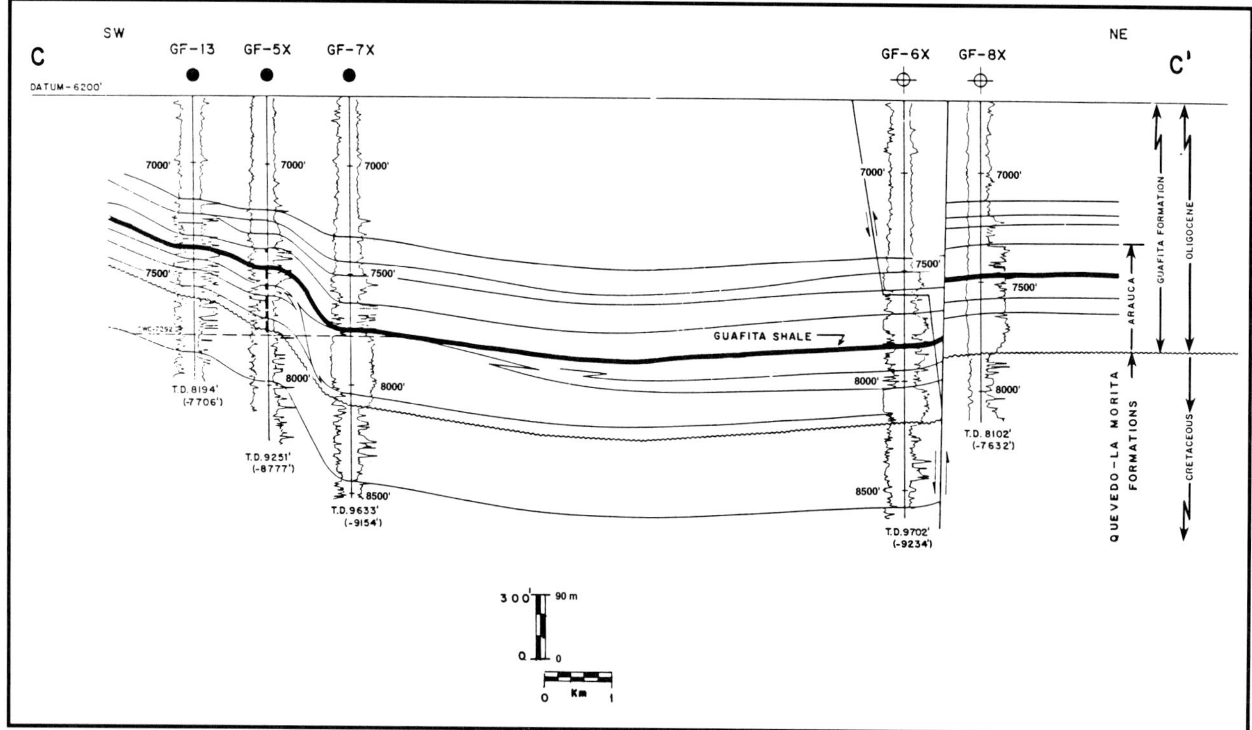

Figure 11. Guafita field, structural cross section C-C′. See Figure 6 for location. Vertical exaggeration × 9.

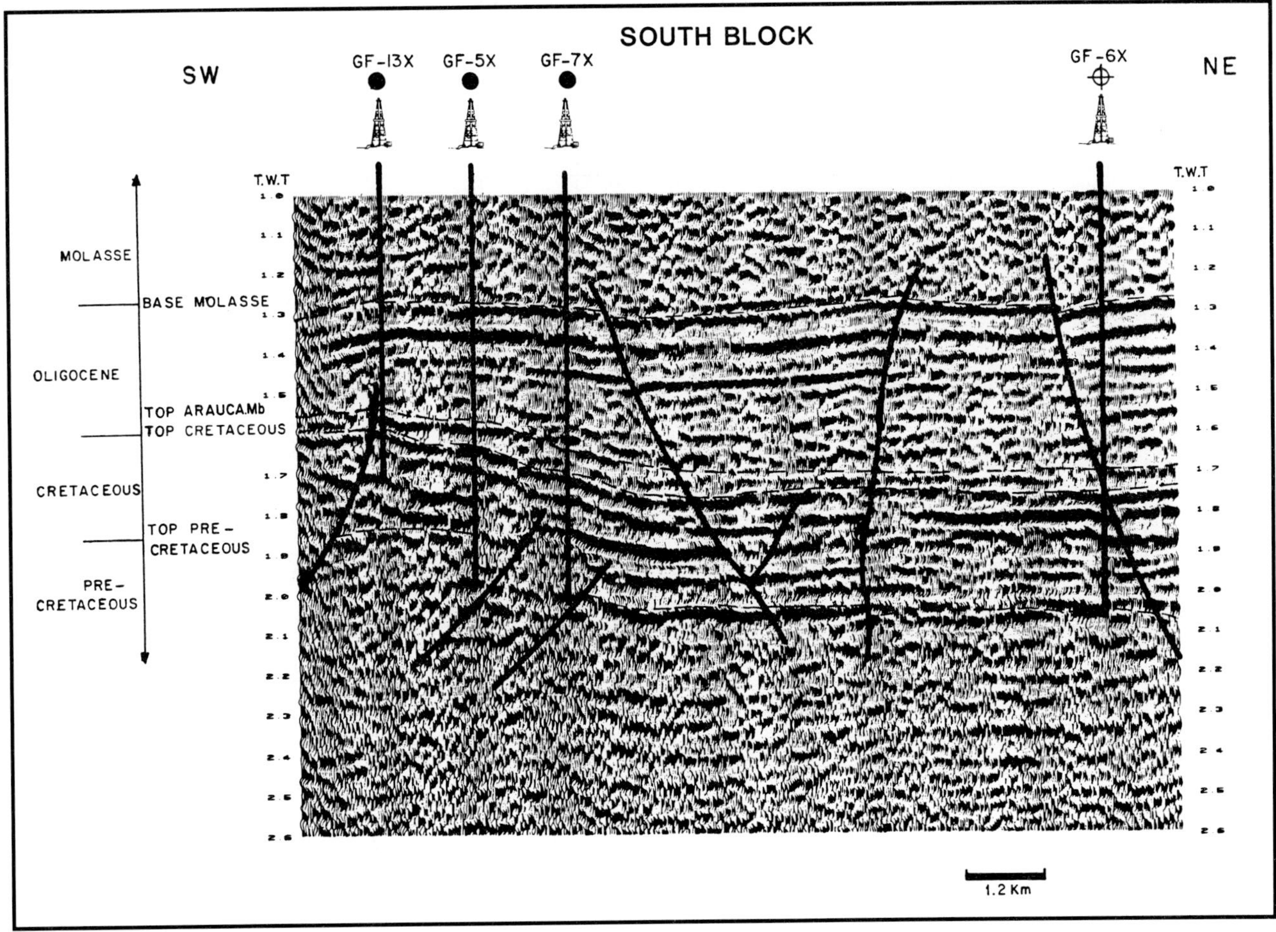

Figure 12. Guafita field, seismic interpretation, line 3. Location is same as C-C′ (Figure 6.)

30,000 ft (9150 m) in the deepest part of the Sub-Andean trough, adjacent to the Cordillera Oriental and Mérida Andes (Figure 4). The sediments contain few microfauna and relatively few microflora, which makes age determinations difficult and precludes reliable regional geological correlations. Nevertheless, recent publications (Ortega et al., 1987; de Monroy, 1988; and Portilla, 1988) indicate it is possible to extend correlations throughout the Barinas and Apure areas. Figure 13 shows the sedimentary sequence of Barinas and Apure. All the formations and their equivalents defined in the Barinas-Apure and Llanos basins crop out in the adjacent Mérida Andes and the Cordillera Oriental to the southwest.

The presence of oxidized continental sediments found in some wells and attributed to the La Quinta Formation could be the result of deposition within rift valleys that formed during the Jurassic.

The basal clastics of the Rio Negro Formation, of Barremian–Aptian age, mark the beginning of the Cretaceous transgression in this region. They are composed of medium- to coarse-grained fluvial sandstones. This formation has been found only in the subsurface of the Burgua–La Ceiba area where well Jordan-1X penetrated more than 1200 ft (370 m) of Rio Negro.

The Aguardiente Formation of Albian age usually unconformably overlies pre-Cretaceous rocks in the subsurface of the Barinas-Apure and Llanos basins. This is a transgressive unit, about 200 ft (60 m) thick, of clean sandstones interbedded with thin shales. The upper part of the formation is composed of glauconitic sandstones and sandy limestones that indicate the initiation of the regressive phase.

The Escandalosa Formation, of Cenomanian age, conformably overlies the Aguardiente Formation (BEICIP-CVP, 1978). The Escandalosa is predominantly a sandy section interbedded with thin beds of shale. The "P" sandstones are very good reservoir rocks in the Barinas fields, and, in La Victoria field, the Guayacan Limestone defines the upper limit of this formation.

The La Morita Formation, of Turonian–Coniacian age, represents the maximum transgression in the region (Gonzalez de Juana et al., 1980) and is partially time equivalent to the La Luna Formation, the prolific source rock of the Maracaibo basin. La Morita is composed of laminated dark shales that also constitute the main source rock in the Barinas-

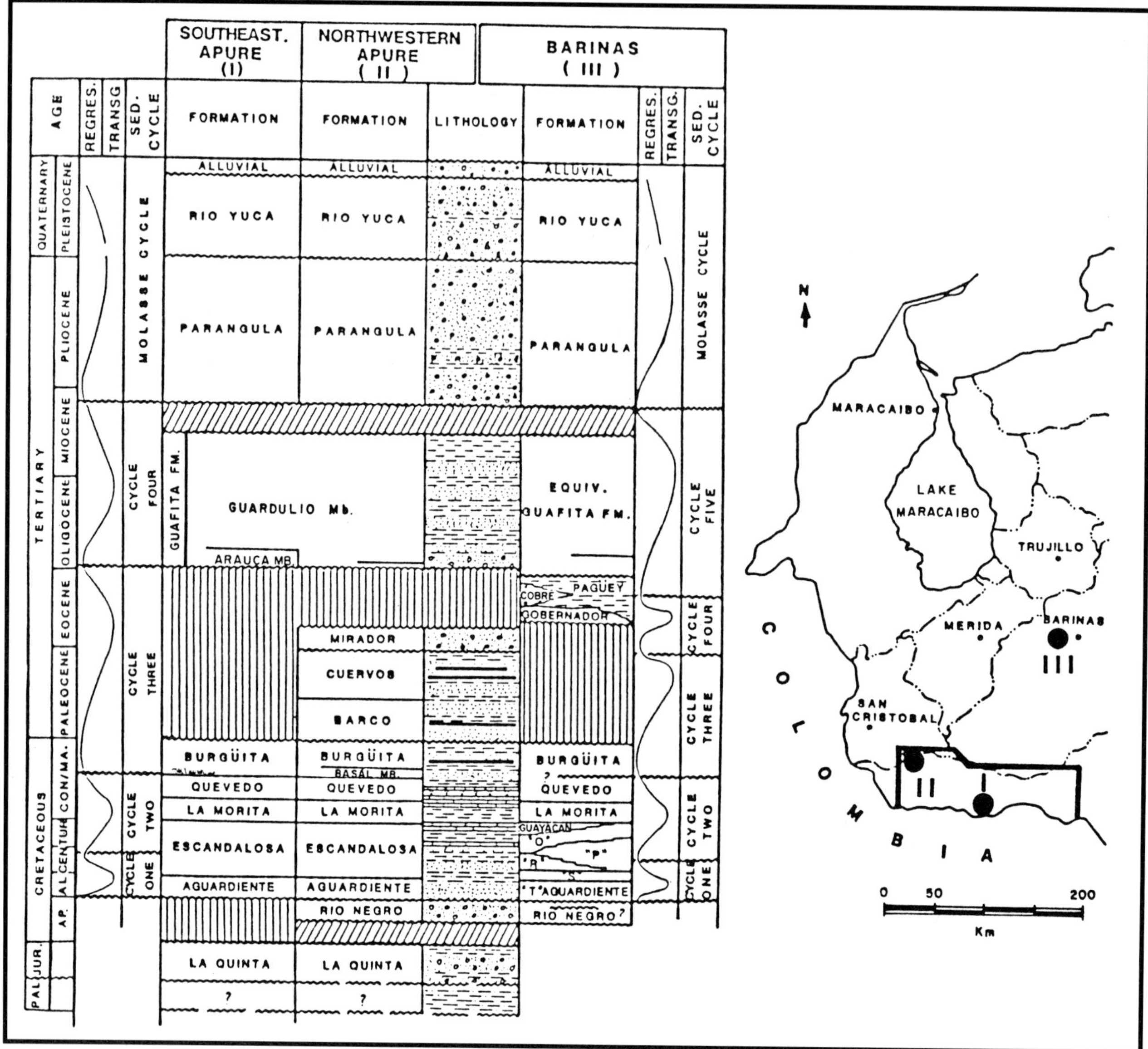

Figure 13. Geological columns for the Barinas-Apure area.

Apure basin. Sedimentation continued with deposition of the Quevedo Formation of Coniacian-Santonian age. The Quevedo rocks are composed of calcareous sandstones with thin phosphatic and shaley layers. The thickness of the combined La Morita and Quevedo formations averages 400 ft (120 m).

The Burguita Formation, of Campanian-Maastrichtian age, completes the Upper Cretaceous sequence. It is composed of clean sandstones interbedded with gray shales and averages 300 ft (90 m) in thickness. This formation unconformably underlies the Tertiary rocks and, except to the west, has been deeply eroded over the area. In the Guafita field, the Tertiary rests directly on the Quevedo Formation.

The Oligocene marks the beginning of a new sedimentary cycle in southeastern Apure during which medium to coarse clastics of the Arauca Member of the Guafita Formation were deposited. The Arauca Member is divided into two sandstone portions, the upper and lower Arauca sandstones separated by a thin but persistent bed of laminated shale called the Guafita shale. The Arauca Member passes into more marine facies of the Guardulio Member of early Miocene age. Using the palynological zonation of Muller et al. (1985), the presence of *Cicatricosisporites dorogenesis* and *Magnastriatites grandiosus* suggests that the age of the Arauca Member is Oligocene (Ortega et al., 1987). On the other hand, the main sandy section productive in Caño Limón field has been reported to be of late Eocene age (Gavela, 1985).

The Guardulio Member, in turn, underlies continental sediments of the molassic cycle, indicating that the regressive phase of the Oligocene-early

Miocene cycle had been eroded. This contact is an unconformity that presents clear evidence of the rise of the Mérida Andes which resulted in a molasse (early Miocene to Recent) that attains more than 16,000 ft (4900 m) thickness in the Mérida Andes foothills. The molasse is also thicker on the flanks of the arches. The younger section of the molasse, probably of Pliocene–Pleistocene age (Stephan, 1977), brings sedimentation to an end in the Barinas–Apure basin.

The integration of sedimentological and biostratigraphical studies indicated the presence of five sedimentary cycles in Apure and six cycles in the oil fields of Barinas to the north (Figure 13).

In Figure 13, the column to the left shows the stratigraphic sequence in southeast Apure and the Guafita field, the only difference being that the Burguita Formation is deeply eroded in the field. The sedimentary section here is 9500 ft (2900 m) thick, of which 8000 ft (2450 m) correspond to the Tertiary section. The main Tertiary reservoirs occur in the Arauca Member of the Guafita Formation with the regional sealing rocks corresponding to the Guardulio Member of the same formation. The Cretaceous section is 1500 ft (460 m) thick, with reservoirs in the Quevedo Formation. Figure 14 shows the lithosedimentary column of well GF-2X, holostratotype of the Guafita Formation in the Guafita field.

The pre-Cretaceous history in western Venezuela is poorly known. The post-Cretaceous geologic history indicates that during Albian to Maastrichtian times, the Guayana craton was bordered by neritic and coastal provinces with platform deposits, passing progressively into deeper water deposits toward the west-northwest. At the end of the Cretaceous, the uplift of the Cordillera Central of Colombia modified the orientation of the depositional systems, causing them to be rearranged perpendicularly to the sedimentary strike of the Cretaceous in such a way that the Tertiary depositional environments became progressively deeper toward the northeast.

The orogeny at the end of the Cretaceous caused the seas to retreat to the north (Gonzales de Juana et al., 1980) and the reactivation of the Arauca, Mérida, and El Baúl arches uplifts. The reactivation of the Arauca arch in Late Cretaceous time limited Paleocene sedimentation to its northern flank. Clastic Paleocene deposits (Barco and Los Cuervos formations) are present in the La Ceiba area, northwestern Apure. These clastics are also present southwest of the Arauca arch (Chigne, 1985) and north of the Arauca oil field in Colombia, where Paleocene sediments have been found in wells.

At the end of the middle Eocene, epeirogenic movements caused the retreat of the sea to the north; uplift of the arches was reactivated, permitting peneplanation and deep erosion of their crests. During this period, incipient uplift of the Mérida Andes began (Stephan, 1977, 1985). The Eocene formations in Barinas pinch out on the northern flank of the Arauca arch and against the Guayana shield, as they are not present in southern Apure. Evidence from regional seismic studies supports this configuration (Portilla, 1988). In the Llanos basin of Colombia, late Eocene formations have been found in several wells (Bogotá-Ruiz, 1988), demonstrating the importance of the Arauca arch as an active element influencing the distribution of the Paleogene sediments in the Llanos and Barinas-Apure basins (Figure 14).

At the end of Miocene, significant uplift of the Mérida Andes was renewed with consequent deposition of the postorogenic molasse in depocenters of the sub-Andean trough and the flanks of the Mérida and Arauca arches. The uprising Andes, at the same time, generated the foreland Barinas-Apure basin.

TRAP

The Guafita field is basically a structural trap, but because of its relatively complex fluvio-deltaic depositional environment, the possibility exists of having updip stratigraphic pinch-outs in sandy reservoirs of the Arauca Member. Stratigraphic thinning of various intervals over the crest of the anticline suggests upward movements contemporaneous with Oligocene deposition.

In the North Block, the Guafita shale is an effective seal dividing the Arauca Member into two intervals of which only the lower one is oil bearing; however, in the South Block, the entire Arauca Member is oil bearing. This situation can be explained by considering that the oil migrated from the southwest and filled the Caño Limón trap and the South Block of Guafita first. The Guafita–Caño Limón fault placed oil sandstones of the South Block in juxtaposition with the lower Arauca sandstones of the North Block (Figure 15), thus explaining why the upper Arauca sandstones are wet in that block; this also confirms the sealing efficiency of the Guafita shale (Chigne, 1985). Later tectonic movements sealed reservoirs along the Guafita–Caño Limón fault, which became an effective barrier between oil accumulations of the North and South blocks of the field. On the other hand, there is no seal between the lower Tertiary sandstones and the underlying Cretaceous sandstones, so these sections are in communication and act as a single reservoir.

RESERVOIR

As discussed previously, the main reservoir corresponds to the Arauca Member of the Guafita Formation. In the North Block, the gross thickness is about 240 ft (73 m), of which an average of 20 ft (6 m) is net pay sandstone. In the South Block, the gross thickness of the Arauca can reach 650 ft (200 m), with an average thickness of 150 ft (46 m) of net pay sandstone. Additionally, the Cretaceous reservoir present in the South Block has a gross

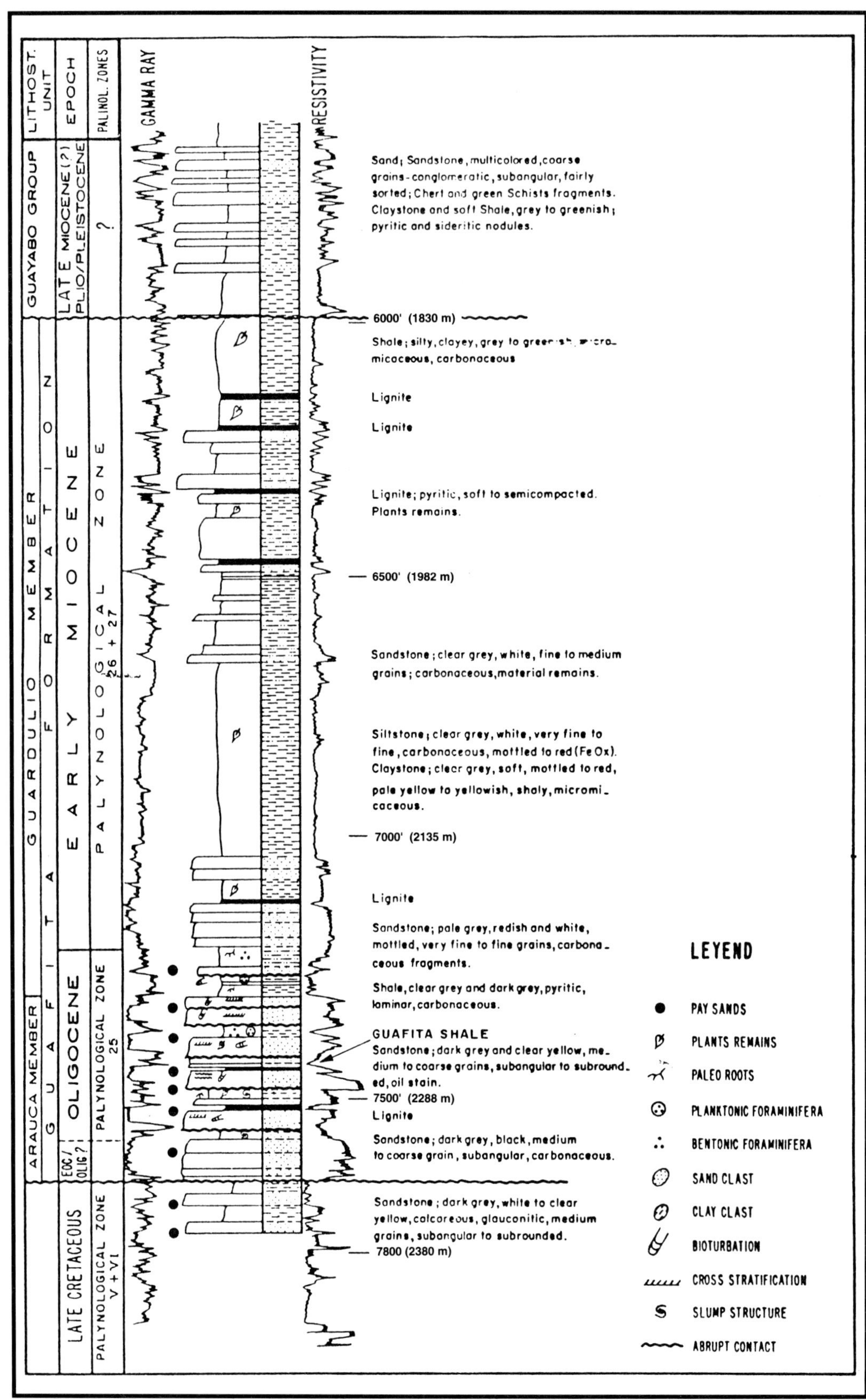

Figure 14. Lithosedimentologic column and electric log of well GF-2X (holostratotype) (Chigne, 1985). (From Ortega et al., 1987.)

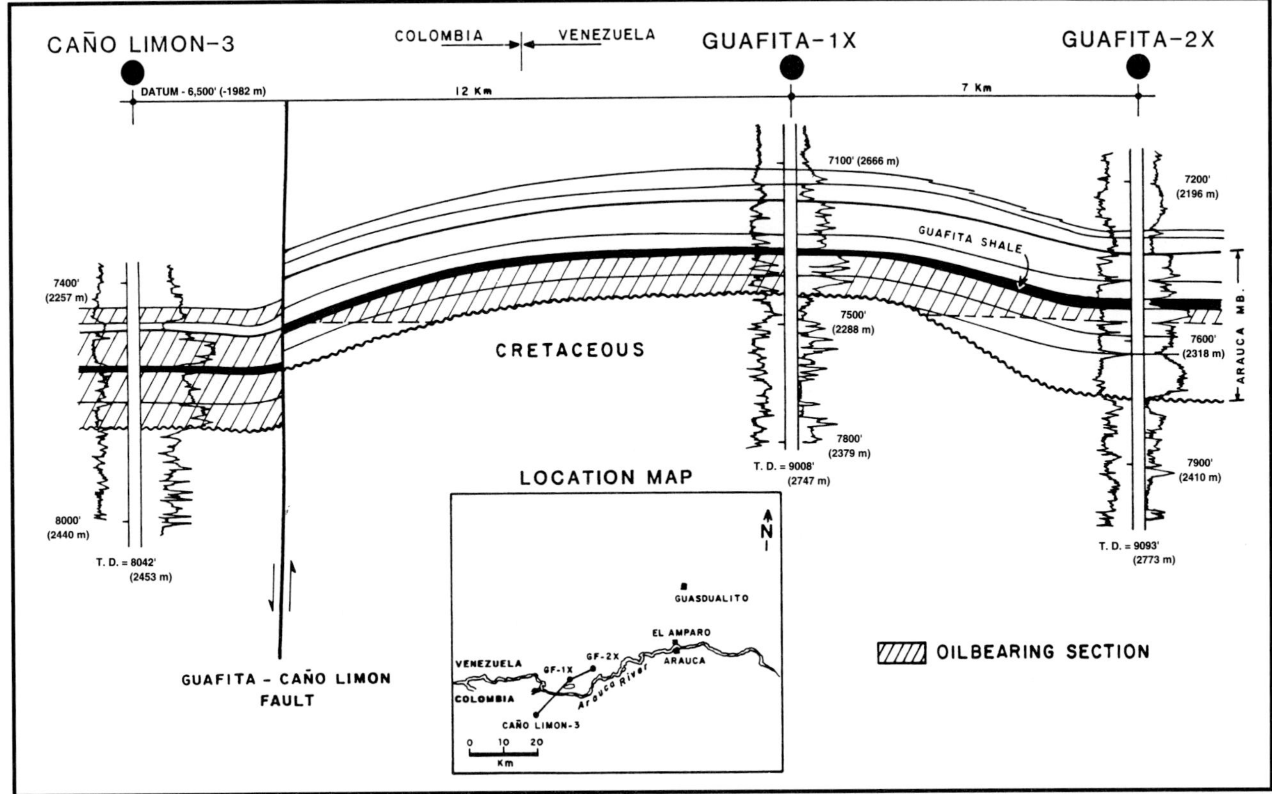

Figure 15. Cross section showing distribution of oil-bearing strata resulting from combined effects of faulting and sealing shale. Well depths indicated are subsea.

thickness of 270 ft (82 m), with the net pay sandstone thickness varying from 15 to 120 ft (4.5 to 37 m).

The Oligocene basal sandstones of the Guafita Formation were deposited in a lower deltaic plain close to the coast, with occasional marine episodes, all part of a constructive deltaic system (Maguregui et al., 1985; Ortega et al., 1987). Sedimentological studies of conventional cores, combined with electrofacies interpretation (Isea et al., 1986, and de Salazar et al., 1988), have led to the identification of several sedimentary deposits.

Active distributary channels (Figure 16) represent the main reservoir sandstone bodies broadly distributed in the field. In general, the sandstones are yellow to light green, poorly consolidated, medium to coarse grained, subangular to subrounded, and poorly to moderately sorted. Tabular cross-stratification is the main sedimentary structure. In lesser proportion, there are some abandoned distributary channels, usually with very fine sandstones interbedded with clay and carbonaceous detritic clasts. The interbedded shales are thin and correspond to tidal flat environments. They are dark gray, compact, with moderate parallel lamination. These shales include small sandy lenses, pyritic nodules, and some carbonaceous material. Sand bars of a deltaic margin showing gradational contacts with the underlying shales have also been identified (Figure 17).

The characteristics of the Tertiary reservoir rocks are shown in a series of photographs, Figures 18–21.

The rocks of the Arauca Member are considered to have originated from igneous-felsic rocks located in the Guayana shield southeast of Guafita field (Maguregui et al., 1985).

The reservoir characteristics and production tests for North and South blocks are presented in Tables 1 and 2. The reservoirs produce with very low GOR (7 scf/stb). The formation water is fresh, with a chloride content that varies from 500 to 1300 ppm.

SOURCE

Geochemical analyses carried out on shale samples of the Cretaceous from wells of the Barinas-Apure region (BEICIP-CVP, 1978; De Toni et al., 1986) indicate a type II kerogen content with predominantly marine organic matter to be responsible for the hydrocarbon generation in the area. Oil generation in the deepest part of the sub-Andean kitchen of Apure and northern Llanos basins began 11 Ma during the middle Miocene (Chigne, 1985). In the easternmost part of this hydrocarbon kitchen, 50 km west of the Guafita field, oil generation began even later, about 3 million years ago during the Pliocene.

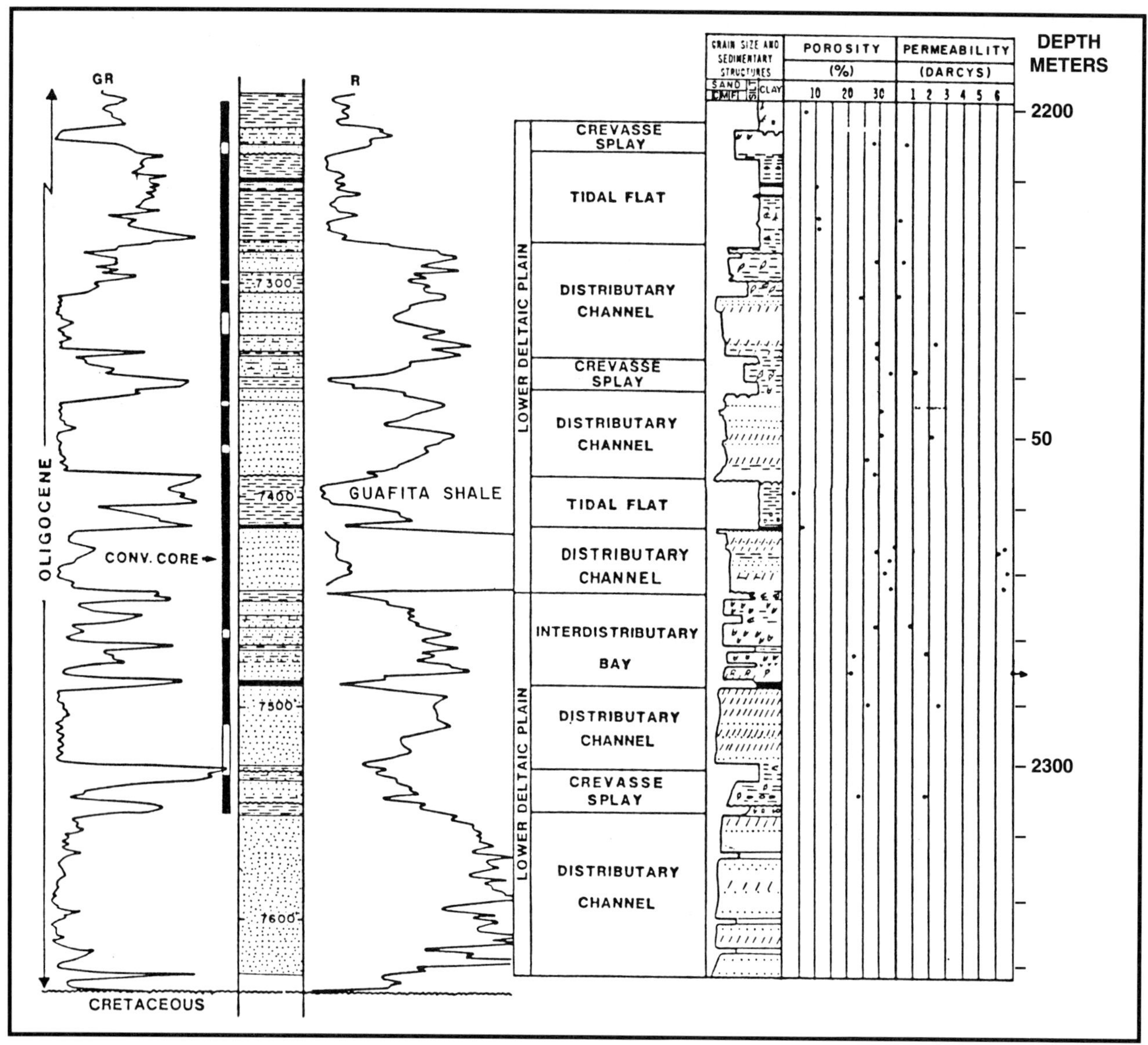

Figure 16. Well GF-2X. Sedimentology and petrophysics of basal sands of the Guafita Formation. (Modified from Maguregui et al., 1985.)

Figure 22 shows the thermal evolution map, where two main sub-Andean kitchens are delineated. Notice that the Guafita-Caño Limón area falls outside of the mature area; therefore, oil accumulated in these fields has migrated from the southwest about 60 km. It is also interesting to note that the shape of the hydrocarbon generation area is partially controlled by the Arauca and Mérida arches.

From mathematical simulations, it has been calculated that the amount of hydrocarbons generated is 65,000 MMBO. It is assumed that from this amount, 10% was expulsed. This means that from the middle Miocene to the present, some 6500 MMBO have migrated to structurally higher areas. Considering that 3200 MMBO in place have been discovered in the Caño Limón-Guafita trend and La Victoria and Arauca fields, and that 300 MMBO have been lost during the migration process, there are still 3000 MMBO to be discovered, of which 1000 MMBO may be recovered in structures on the southern flank of the Arauca Arch (Chigne and Hernández, 1990).

Crude oil in Guafita field is light gravity, subsaturated, with very low GOR and low metal content (vanadium, 14 ppm; nickel, 44 ppm; and sulfur, 0.39 ppm). Chromatographic analyses from one GF-5X oil sample (De Toni et al., 1986) indicate the sample to be composed of 53% saturated hydrocarbon, 63% aromatic hydrocarbon, 11% resins, and 5% asphaltenes.

The saturated hydrocarbon fraction C_{15} has abundant n-alkanes in the interval C_{16}–C_{25}, together with the values of pristane/phytane (1.2), pristane/

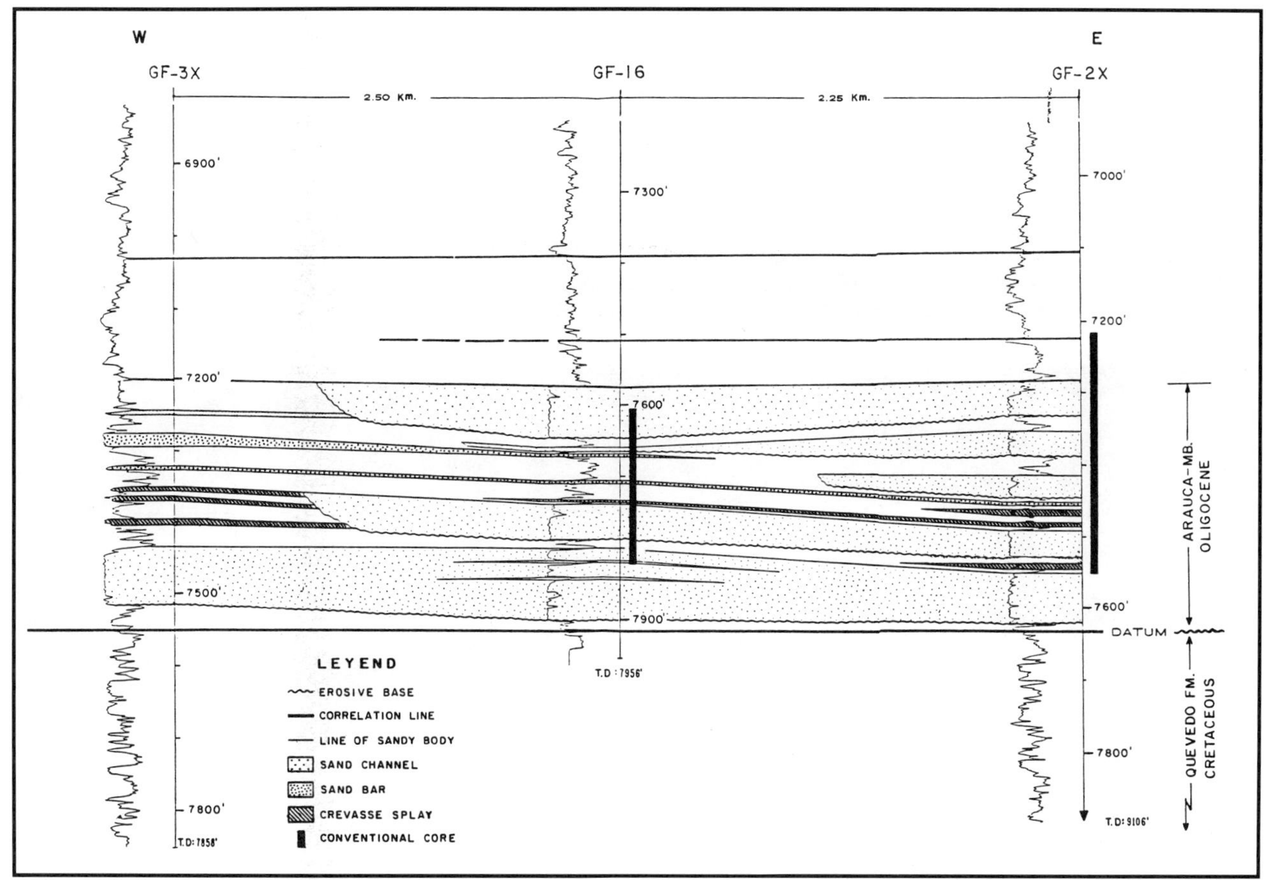

Figure 17. Guafita field, sedimentologic correlation. (From de Salazar et al., 1988.)

C_{17} (0.7), phytane/C_{18} (0.6), and Carbon Preferential Index (1.1), suggest that this crude oil came from predominantly marine organic matter. The low content of nickel, vanadium, and sulfur indicates that the crude oil could have originated from marine organic matter contained in shales.

EXPLORATION AND DEVELOPMENT CONCEPTS

The frequency content of the seismic sections recorded in Guafita is well under the resolution needed to show the typical 20 to 40 ft (6 to 12 m) thickness of the reservoir sands; also the minor faulting in the area, which has proven to play an important role in the oil accumulation, cannot be seen with this seismic data. It is therefore clear that for a better description of the Guafita reservoirs, high-resolution and 3D seismic data have to be acquired.

The high porosity and permeability values measured at reservoir level correspond to a high transmissibility isotropic medium, allowing wells to be produced at high rates. The low solution GOR, low compressibility fluid system in the reservoir, and the continuously increasing water cuts make necessary the use of electrocentrifugal pumping. However, in order to avoid early water coning and sand production during artificial lift, very careful production planning is necessary.

According to Bristow (1986), low permeability oil sands look like water sands on the electric logs and were not included in the net oil sand counts. A test of these low permeability oil sands must be made to determine if they are effective reservoir sands. If they are, these sands could contribute to reservoir performance and constitute a significant increase in the estimate of reservoir volume and reserves.

ACKNOWLEDGMENTS

We gratefully acknowledge the permission of Corpoven, S.A. (affiliate of Petróleos de Venezuela, S.A.) to publish this paper, and thank G. D. Kiser and F. Russomanno of Corpoven, S.A. and Dr. Amos Salvador of the University of Texas at Austin, for their suggestions and review.

Figure 18. Well GF-2X. Conventional core photos from 7437 to 7443 ft. Shows sandstones of an active distributary channel impregnated with hydrocarbons. Cross stratification is clearly observable in the medium-grained sandstone.

Figure 19. Well GF-5X. Conventional core slab photo from 7627 to 7633 ft. From left to right are a brown, coarse-grained, moderately to well-sorted sandstone impregnated with oil, followed by incipient cross-stratified and parallel laminated sandstones, as well as shales, which are light to dark gray, well laminated, with abundant load structures, mud cracks, and occasional flaser structures.

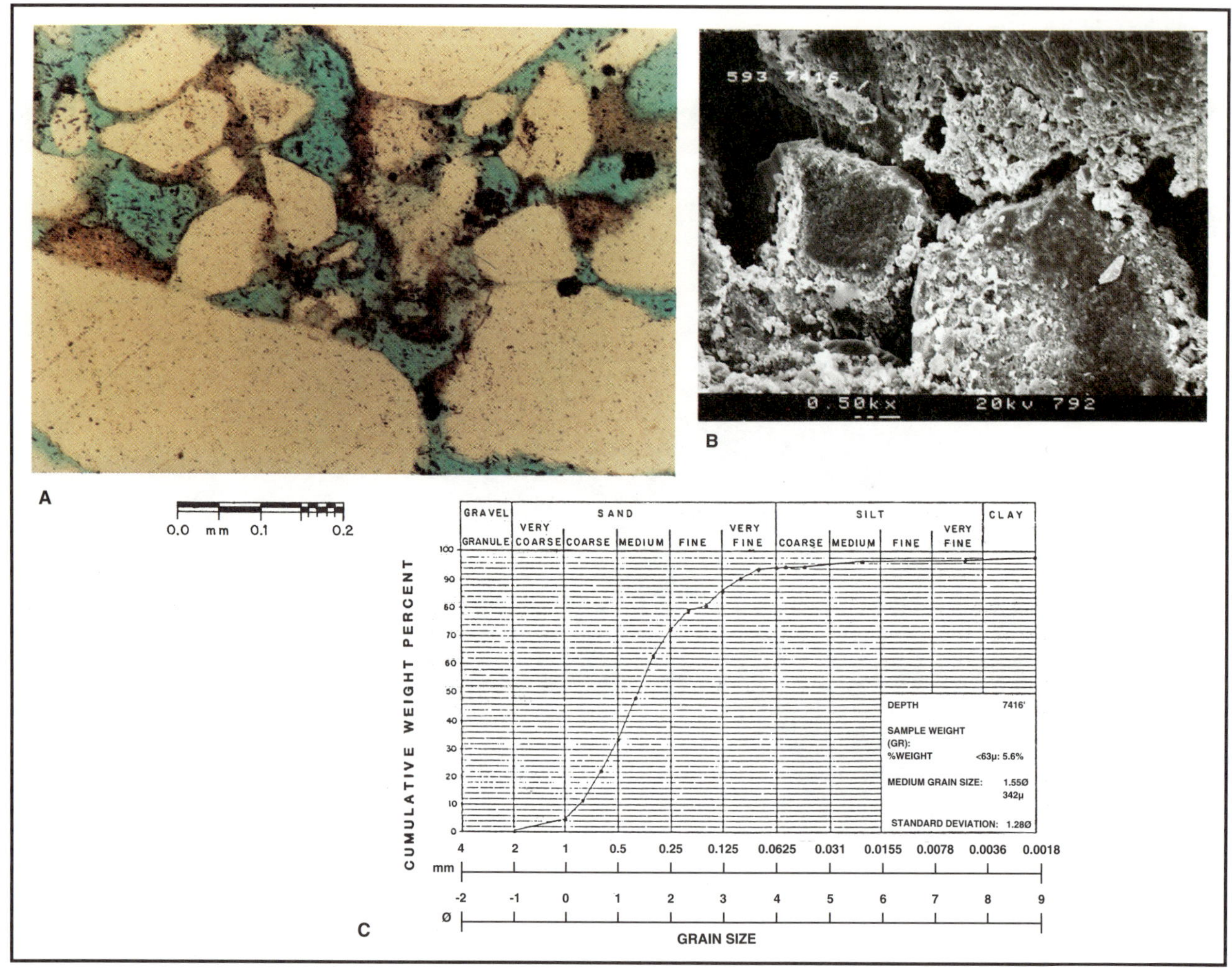

Figure 20. Well GF-2X. Representative sample, petrographic thin sections and grain size distribution at 7416 ft. (A) A clear contrast in the grain size of the sands, as well as the clay matrix and framboidal pyrite. (B) Clay minerals adhere to grain surfaces without plugging the pores. (C) A gradation from medium to very fine grain size with a very low percentage of silt and clay.

REFERENCES CITED

BEICIP-CVP., 1978, Evaluación del potencial petrolifero de la Cuenca Barinas-Apure: Corpoven, S.A., Internal Report.

Bogotá-Ruiz, J., 1988, Contribución al conocimiento estratigráfico de la cuenca de los Llanos, Colombia: III Simposio Bolivariano, Exploración Petrolera en las Cuencas Subandinas, Caracas, Venezuela, Marzo 13-16.

Bristow, J. D., 1986, The geology of the Guafita field. Apure state, Venezuela: Corpoven, S.A., Internal Report.

Chigne, N. A., 1985, Aspectos relevantes en la exploración de Apure: III Simposio Bolivariano, Exploración petrolera en las Cuencas Subandinas, Bogotá, Colombia, Agosto 13-16.

Chigne, N.A., and L. Hernandez, 1990, Main aspects of petroleum exploration in the Apure area of southwestern Venezuela 1985-1987: The Geological Society of London Special Publication 50, p. 55-75.

De Toni, B., S. Talukdar, and C. Toro, 1986, Estudio geoquimico orgánico del pozo GF-5X: Intevep, S.A., Internal Report.

Gavela, V. H., 1985, Campo Caño Limón, llanos orientales de Colombia: II Simposio Bolivariano, Exploración Petrolera en las Cuencas Subandinas, Bogotá, Colombia, Agosto 13-16.

Gonzalez de Juana, C., J. Iturralde, and X. Picard, 1980, Geologia de Venezuela y de sus cuencas petroliferas: Foninves, v. 1, p. 181-191 and v. 2, p. 420-43, Caracas, Venezuela.

Isea, A., A. Euribe, and G. Guiffuni, 1986, Evaluación geológica de los núcleos del pozo GF-5X: Intevep, S.A., Internal Report.

Maguregui, J., P. de Ramos, G. Guiffuni, and B. de Toni, 1985, Estudio geológico de los núcleos del pozo GF-2X: Intevep, S.A., Internal Report.

Monroy, Z., de, 1988, Revisión palinoestratigráfica preliminar del Cretáceo y Terciario de Apure, Venezuela Suroccidental: III Simposio Bolivariano, Exploración Petrolera de las Cuencas Subandinas, Caracas, Venezuela, Marzo, 13-16.

Muller, J., E. di Giacomo, and A. Van Erve, 1985, A palynological zonation for the Cretaceous, Tertiary and Quaternary of Northern South America: Mem., VI Cong. Geol. Venezolano, Caracas, T. II; p. 1041-1079.

PETROGRAPHIC THIN SECTIONS
WELL GF-5X

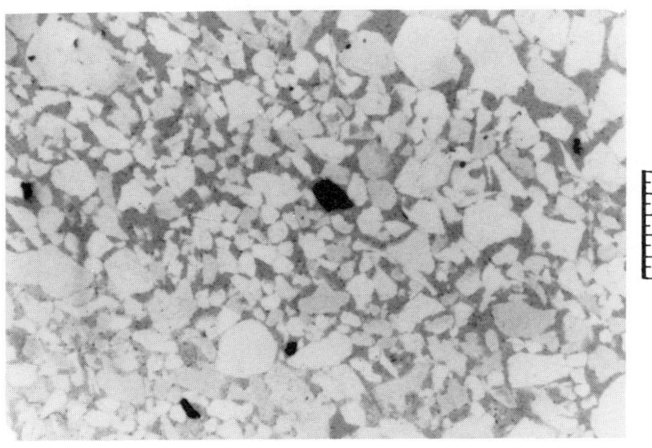

DEPTH: 7628' 10"

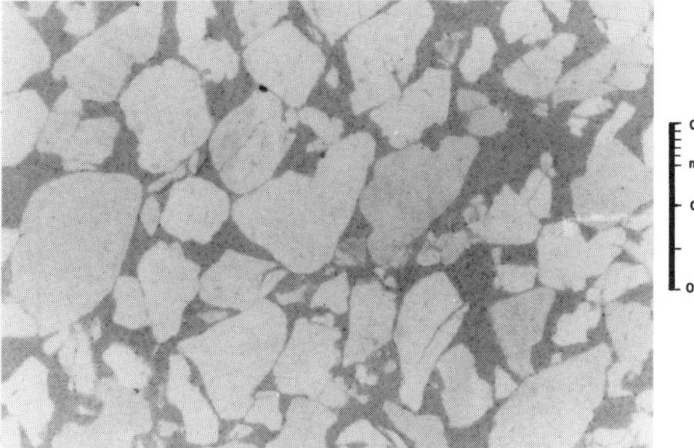

DEPTH: 7651' 10"

Figure 21. Well GF-5X. Petrographic thin sections of samples at 7628 ft 10 in. and 7651 ft 10 in. Upper photomicrograph, a quartzose sandstone, medium to coarse grained and poorly sorted. The cleanness and interconnection of the pores can he observed in most of the sample. The lower photomicrograph shows a quartzose sandstone, coarse grained and moderately sorted. The high porosity of the sample and the cleanness of the pores allow nearly complete interconnection.

Table 1. Reservoir characteristics.

	North Block	South Block
Net Pay Sand, ft (m)	20-45 (6-14)	20-150 (6-46)
Porosity, %	26	25-30
Permeability, md	1500-3400	7000
Water saturation, %	22	20-25
Oil gravity, °API	30	29
Oil viscosity, cp, @ 200°F, 3000 psia*	4.5	6.0
Bubble point pressure, psia	45	50
Solution GOR, scf/stb	7	7
Initial reservoir pressure, psi (kg/cm^2)	3130 (220)	3000-3500 (211-246)
Average reservoir depth, ft s.s. (ms.s.)	6850 (2090)	7100 (2165)
Production mechanism	Water drive	Water drive

*Oil viscosity, cp @ 93.3°C and 210.9 kg/cm^2 absolute.

Ortega, J., A. Van Erve, and Z. de Monroy, 1987, Formación Guafita, Nueva Unidad Litoestratigráfica del Terciario en el subsuelo de la Cuenca Barinas-Apure, Venezuela Suroccidental: Sociedad Venezolana de Geólogos Bull., v. 31, p. 9-35.
Portilla, A., 1988, Relaciones estructurales y estratigráficas de la Cuenca Barinas-Apure: Memorias IV Congreso Venezolano de Geofisica, Septiembre, 1988.
Russomanno, F., and H. Velarde, 1982, Geologia petrolera de la Cuenca Barinas-Apure: I Simposio Bolivariano, Exploración Petrolera en las Cuencas Subandinas, Bogotá, Colombia, Agosto, 18-20.
Salazar, A., de, S. de Cabrera, and Z. de Monroy, 1988, Estudio sedimentológico y bioestratigráfico de los núcleos del pozo GF-16: Corpoven, S.A., Internal Report.
Stephan, J., 1977, El contacto Cadena Caribe-Andes Merideños entre Carora y El Tocuyo, Estado Lara: V Congreso Geológico Venezolano, Caracas, Venezuela, Noviembre 19-23.
Stephan, J., 1985, Andes et Chaine Caraibe sur la transversale de Barquisimeto, Venezuela, Evolution Geodynamique: Symposium Caribbean Geodynamics, Paris.

Table 2. Production tests.

Well	Formation	Porosity	Net Oil Sand ft (m)	BOPD	BS&W %	Gravity °API	TP psi (kg/cm^2)	Choke inches (mm)
North Block								
GF-1X	Guafita	25	42 (12.8)	1264	1.5	29.7	360 (25.3)	¼ (6.4)
GF-2X	Guafita	27	30 (9.1)	1046	0.6	28.8	350 (24.6)	¼ (6.4)
GF-4X	Guafita	26	36 (10.0)	1122	1.2	29.0	394 (27.7)	¼ (6.4)
South Block								
GF-5X	Guafita	29	36 (10.0)	4985	0.1	30.4	110 (7.7)	¾ (19.0)
	Guafita	30	20 (6.1)	3322	0.2	30.4	80 (5.6)	¾ (19.0)
GF-7X	Guafita	28	12 (3.7)	3381	0.1	30.0	55 (3.9)	¾ (19.0)
GF-13X	Navay	27	22 (6.7)	3388	0.2	29.4	40 (2.8)	¾ (19.0)
	Navay	28	11 (3.4)	1113	0.1	29.4	60 (4.2)	⅜ (9.5)

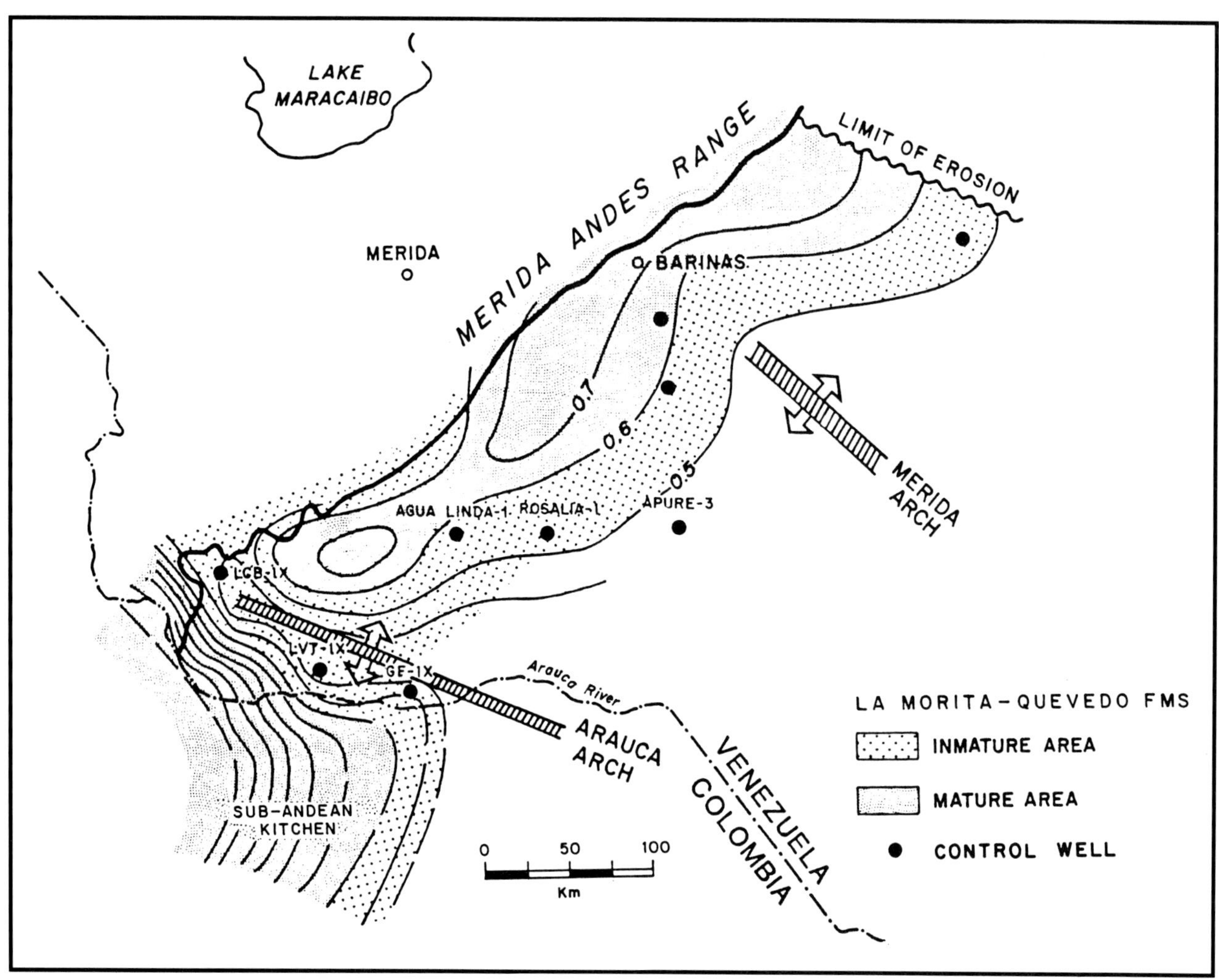

Figure 22. Northern Llanos and Barinas-Apure basins: thermal evolution based on vitrinite reflectance.

Appendix 1. Field Description

Field name .... *Guafita field*

Ultimate recoverable reserves .... *250 MM bbl*

Field location:

Country .... *Venezuela*
State .... *Apure*
Basin/Province .... *Sub-Andean basin*

Field discovery:

Year first pay discovered .... *Oligocene sands, Guafita Formation/Arauca Member 1984*
Year second pay discovered .... *Cretaceous sands, Quevedo Formation 1985*

Discovery well name and general location:

First pay .... *Guafita-1X, 45 km southwest of town of Guasdualito*
Second pay .... *Guafita-5X, 43 km southwest of town of Guasdualito*

Discovery well operator .... *Corpoven, S.A.*

IP

First pay .... *3500 BOPD on ½-in. choke*
Second pay .... *2500 BOPD on ¾-in. choke*
Third pay .... *350 BOPD oin ¼-in. choke*

Geologic concept leading to discovery and method or methods used to delineate prospect

Modern seismic delineated a large, gentle anticline bounded by a fault zone. The main seismic reflector was almost at the same depth as the discovery zone in the sandy section of Caño Limón field near the Colombian border.

Structure:

Province/basin type .... *Bally 221; Klemme II A*

Tectonic history

The Arauca arch was uplifted at the end of the Cretaceous, separating the Barinas basin in Venezuela from the Llanos Orientales of Colombia. The uplift of Mérida Andes in Pliocene time produced thick molassic section and the ultimate configuration of the area.

Regional structure

The regional structure consists of an anticlinal trend development along a strike-slip fault, genetically related to the regional fault system parallel to the Mérida Andes. The structure occurs on the southwestern flank near the crest of the Arauca arch.

Local structure

Northeast-trending anticline split into two blocks by a strike-slip fault zone of approximately 400 ft vertical displacement at the Oligocene level.

Trap:

Trap type(s)

A combination of structural and stratigraphic traps: one anticlinal trap with multiple reservoirs and one pinch-out trap

Basin stratigraphy (major stratigraphic intervals from surface to deepest penetration in field):

Chronostratigraphy	Formation	Depth to Top in ft (m)
Miocene-Pliocene	*Rio Yuca/Parángula*	*5000 (1500)*
Lower Miocene	*Leon*	*6000 (1800)*
Oligocene	*Guafita*	*6800 (2000)*
Cretaceous		*7300 (2200)*

GUAFITA

Reservoir characteristics:

Number of reservoirs *5*

Formations .... *Guafita: G-7; G-8; G-9; G-10; Navay Formation; Quevedo Formation*

Ages .... *Oligocene and Cretaceous*

Depths to tops of reservoirs

G-7, 6800 ft (2075 m); G-8, 6950 ft (2120 m); G-9, 7050 ft (2150 m); G-10, 7150 ft (2180 m); Quevedo, 7300 ft (2225 m)

Gross thickness (top to bottom of producing interval)

Guafita Formation, 600 ft (183 m); Navay Formation, 340 ft (104 m)

Net thickness—total thickness of producing zones

Average .... *Guafita Formation, 60 ft (18.3 m); Navay Formation, 12 ft (3.6 m)*

Maximum .... *Guafita Formation, 150 ft (45.8 m); Navay Formation, 35 ft (10.7 m)*

Lithology

Tertiary Arauca Member: massive quartz sand, yellow to light green, medium to coarse grained, friable, subrounded to subangular and fairly well sorted

Cretaceous Quevedo Member: Glauconitic sandstones, fairly well cemented, medium grained, interbedded with shales and medium grained, friable sands and thin limestones

Porosity type .... *Primary intergranular porosity*

Average porosity .... *Oligocene, 27%; Cretaceous, 17%*

Average permeability .... *Oligocene, 1700 md; Cretaceous, 200 md*

Seals:

Upper

Formation, fault, or other feature .... *Shales and clays interbedded with producing sands*

Lithology .... *Shales*

Lateral

Formation, fault, or other feature .... *Faults and pinch-outs*

Lithology .... *Shales*

Source:

Formation and age .... *La Morita-Quevedo formations*

Lithology .... *Shales*

Average total organic carbon (TOC) *2%*

Maximum TOC *3.5%*

Kerogen type (I, II, or III) .... *II*

Vitrinite reflectance (maturation) .... $R_o = 0.6\text{–}1.2$

Time of hydrocarbon expulsion .... *12 m.y.*

Present depth to top of source .... *16,000 ft (4900 m)*

Thickness .... *400 ft (122 m)*

Potential yield .... *65 billion bbl*

Appendix 2. Production Data

Field name .... *Guafita field*

Field size:

Proved acres .... *8590 ha*

Number of wells all years *22*

Current number of wells *22*

Well spacing .... *1200 and 600 m (3940 and 1670 m)*

Ultimate recoverable .... *250 (possibly up to 400) million bbl*

Cumulative production .... *25.2 million bbl*

Annual production .... *NA*

GUAFITA

Present decline rate .... *NA*
Annual water production .... *NA*
In place, total reserves .... *950 to 1000 million bbl*
In place, per acre foot .... *NA*
Primary recovery .... *NA*
Secondary recovery .... *NA*
Enhanced recovery .... *NA*
Cumulative water production .... *653 bbl*

Drilling and casing practices:

Amount of surface casing set .... *1500 ft (460 m)*

Casing program

13⅜-in. to 1500 ft (460 m); 9⅝-in. to 1500–8300 ft (460–2560 m)

Drilling Mud

Cal/Bentonite base water mud, 0–1500 ft (460 m); polymeric base mud, 1500–8300 ft (460–2560 m)

Bit program

17½-in. R-1, 0–1500 ft (460 m); 12¼-in. × 3A, 1500–7500 ft (460–2300 m); 12¼-in. J-1, 7500–8300 ft (2300–2560 m)

High pressure zones .... *None*

Completion practices:

Interval(s) perforated .... *North Block, G-9 sand; South Block, G-7 to G-10 sands, Cretaceous sands*
Well treatment .... *NA*

Formation evaluation:

Logging suites .... *DDL-MSFL-GR-SP-CAL; LDT-CNL-GR-CAL; LSS-ITT-GR; SHDT-DV; CST*
Testing practices .... *Repeat fluid tester*
Mud logging techniques .... *Standard ditch samples*

Oil characteristics:

Type .... *Intermediate paraffinic-naphthenic*
API gravity .... *29°*
Base .... *Naphthenic*
Initial GOR .... *7 scf/stb*
Sulfur, wt% *0.62*
Viscosity, SUS .... *44.5 kinematic viscosity at 100°F (37.8°C) centistokes*
Pour point .... *10°F (−12.2°C)*
Gas-oil distillate .... *NA*

Field characteristics:

Average elevation .... *7000 ft (2135 m) ss*
Initial pressure .... *3120 psi (21.5 MPa)*
Present pressure .... *3065 psi (21.1 MPa)*
Pressure gradient .... *0.37 psi/ft (8.37 kPa/m)*
Temperature .... *194°F (90°C)*
Geothermal gradient .... *1.5°F/100 ft (0.027°C/m)*
Drive .... *Water drive and fluid expansion*
Oil column thickness .... *20–150 ft (6–46 m)*
Oil-water contact .... *Various depths in different blocks; generally 7300 ft (2225 m) ss*
Connate water *20–25%*
Water salinity, TDS .... *500 to 1300 ppm Cl⁻*
Resistivity of water *1.2*
Bulk volume water (%) .... *NA*

Transportation method and market for oil and gas:

Pipeline transportation from Guafita field to Barinas Refinery

Tarra Field—Venezuela
Maracaibo Basin, Zulia State

ANGEL MOLINA
Maraven, S.A.
Caracas, Venezuela

FIELD CLASSIFICATION

BASIN: Maracaibo
BASIN TYPE: Foredeep
RESERVOIR ROCK TYPE: Sandstone and Limestone
RESERVOIR AGE: Eocene and Cretaceous
PETROLEUM TYPE: Oil and Gas
TRAP TYPE: Faulted Anticline and Unconformity Truncation
RESERVOIR ENVIRONMENT OF DEPOSITION: Fluvio-Deltaic and Shelf Platform
TRAP DESCRIPTION: Faulted anticline with multiple pays located in the hanging wall of thrust sheet; deeper trap is an unconformity truncation

LOCATION

The Tarra field is located in the Colón District in Zulia State, Venezuela (Figures 1A and 1B) near the West Tarra, Bonito, Concordia, and Los Manueles fields, about 124 mi (200 km) southwest of Maracaibo City. The area is somewhat hilly, with elevations up to 900 ft (275 m). It is bounded on the west by the Perija Range and on the east by a vast plain extending to Lake Maracaibo.

Original estimated ultimate recovery for the Tertiary reservoirs was 117 million bbl (MMBO) (Petroleum Information Corporation, 1980), but the most recent reserves estimate is 136.6 MMBO and 124 bcf gas.

HISTORY

Exploration began with surface geology around 1913, at which time the existence of an anticlinal structure with oil seeps was detected. The field was discovered by Compãnia Shell de Venezuela, Ltd., in 1916 with the drilling of the T-1 well (Figure 2). It was located on the crest of the surface structure in an area with many oil seeps, about 59 mi (95 km) west-northwest of the town of El Vigià. Once the well reached 830 ft (253 m), oil began to flow intermittently every two hours. At a depth of 875 ft (267 m), oil began to flow at a rate of 800 bbl/day. The well was then completed between 800 and 836 ft (244 and 255 m) in the upper Eocene Arenas del Cubo Member (Figure 3). The well produced up to 1230 BOPD, 22° API.

In 1939, the T-99 well (Figure 2) was drilled to a total depth of 8795 ft (2682 m). It penetrated 186 ft (57 m) into the Cretaceous reservoir comprised of the Aguardiente Formation and the Apon Formation Mercedes Member (Figure 4), producing small quantities of gas and condensate with a high water content; the well was therefore abandoned. The T-230 well (Figure 2) was drilled during 1982 and was completed between 9500 ft and 10,000 ft (2900 and 3050 m) in the Aguardiente-Mercedes reservoir with an initial production of 130 bbl/day of 42° API condensate and 0.34 million ft^3/day of gas (MMCFG). Completion of this well was based on density neutron, gamma ray, and resistivity logs and on analysis of oil-impregnated ditch samples and gas chromatography (Molina, 1987).

The field has been developed by the drilling and completion of 200 wells prior to 1949 and 19 wells between 1949 and 1960. All were completed in Eocene and Paleocene reservoirs (Petroleum Information Corporation, 1980). Three seismic surveys were carried out between 1979 and 1985 in order to define both the deeper Cretaceous and the shallower Tertiary structures. Such definition was necessary to evaluate and develop the Cretaceous reservoirs and to continue the development of the Tertiary reservoirs (Molina, 1987).

DISCOVERY METHOD

The T-1 well (Figure 2) was drilled on the crest of the structure close to the seeps. During drilling, oil flowed naturally from an upper Eocene sandy sequence currently known as the Arenas del Cubo Member (Figure 3). The methodology used was surface geology mapping in the proximity of oil seeps. Today, however, the latest seismic sections (Figures

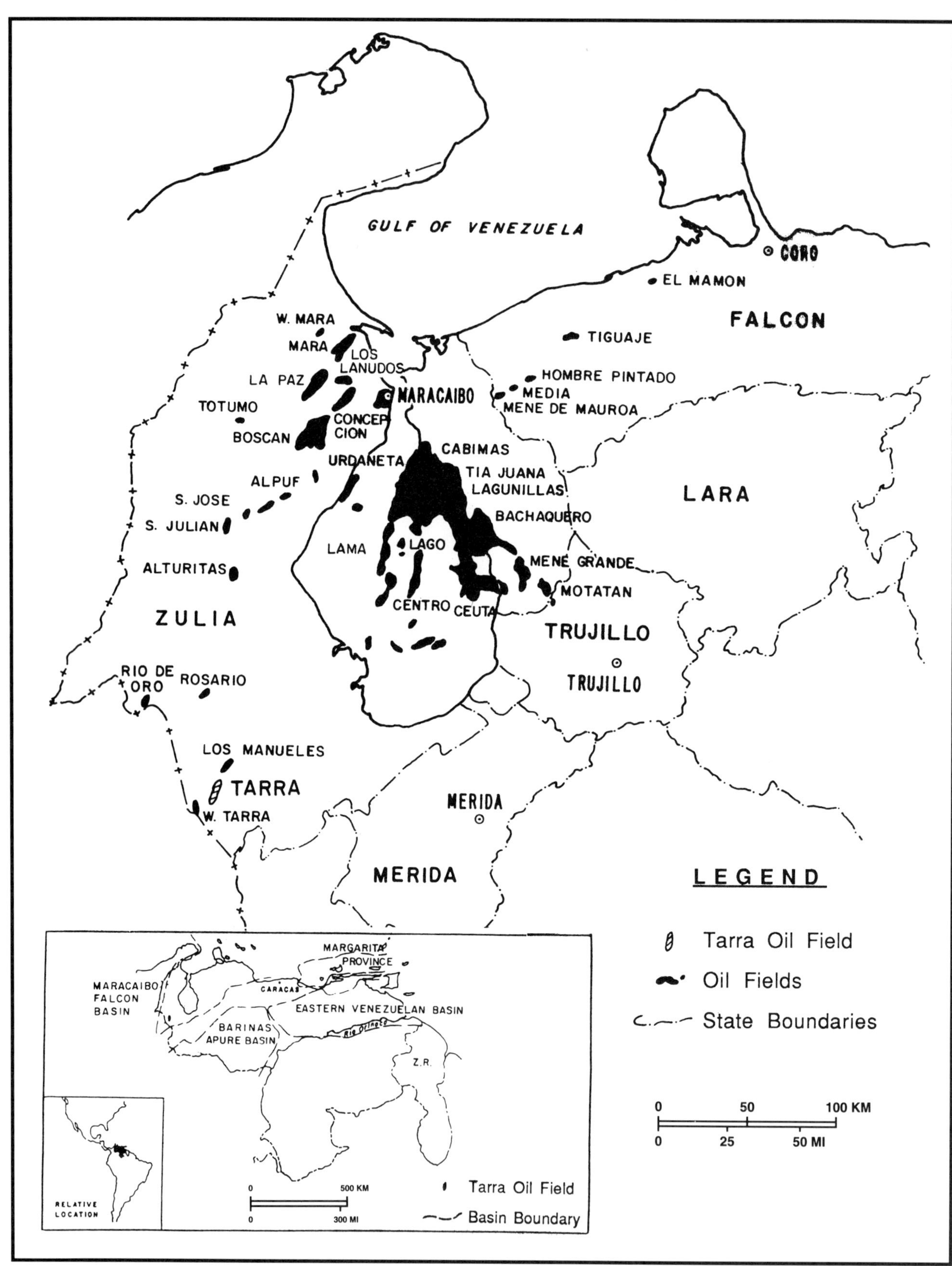

Figure 1A. Geographic location of Tarra and other oil fields in the Maracaibo-Falcón Basin of western Venezuela.

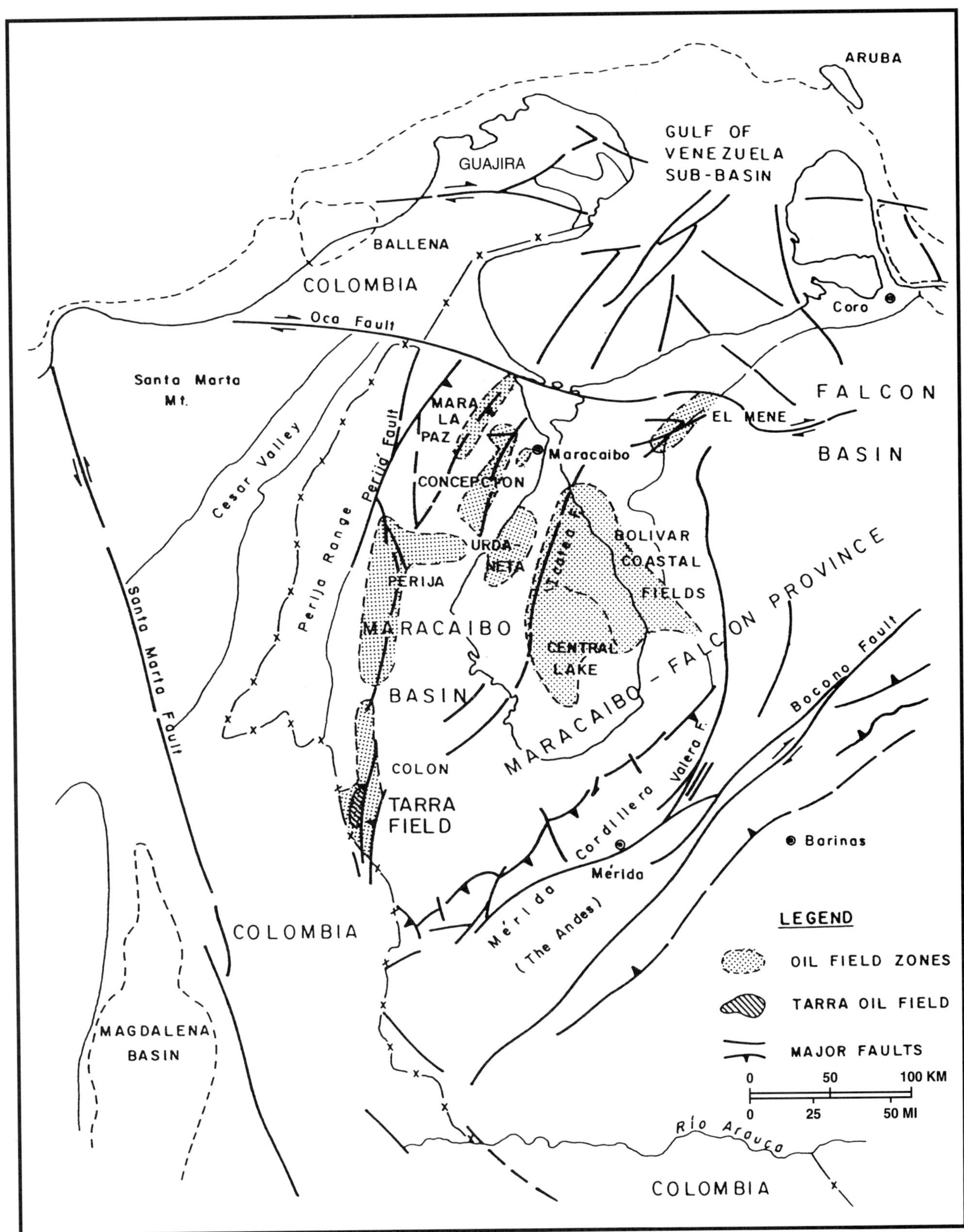

Figure 1B. Tectonic map of the Maracaibo-Falcón province showing the Tarra oil field and other oil field zones in relation to major fault trends.

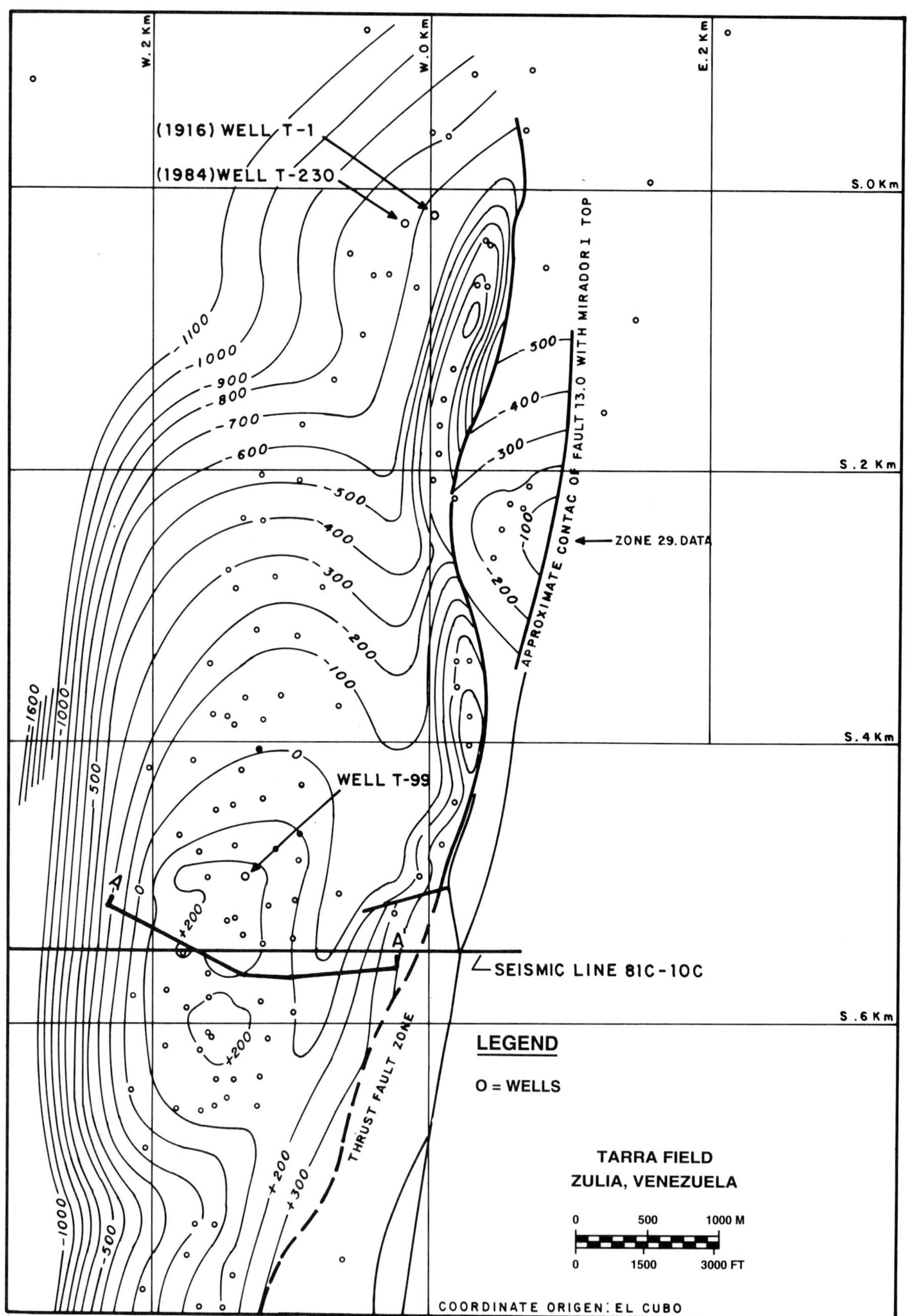

Figure 2. Structure contour map of the top of the Eocene Mirador Formation I Member. (After Petroleum Information Corp., 1979.) Contour interval, 100 ft (30.5 m). Locations for structural section AA′ (Figure 7) and seismic line 8K-10C (Figures 5 and 6) are indicated.

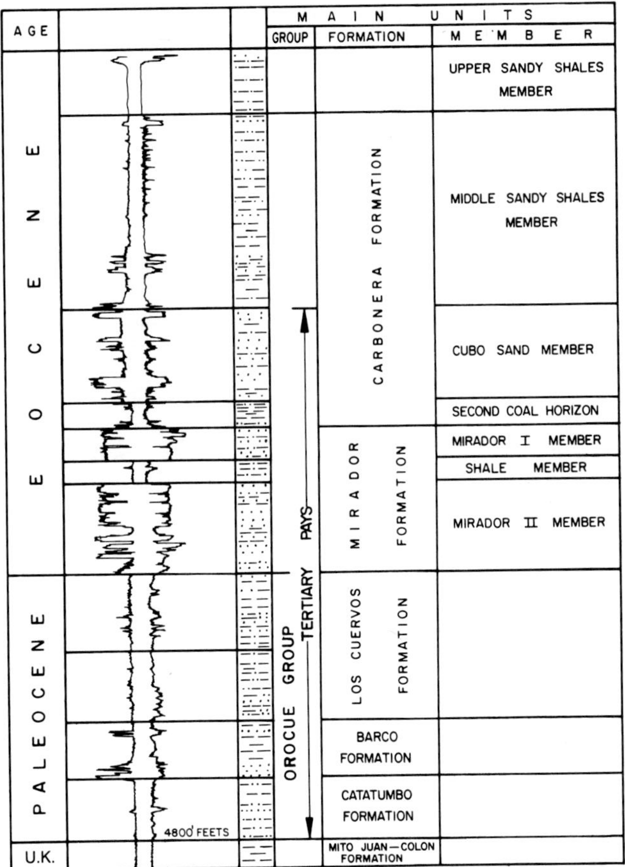

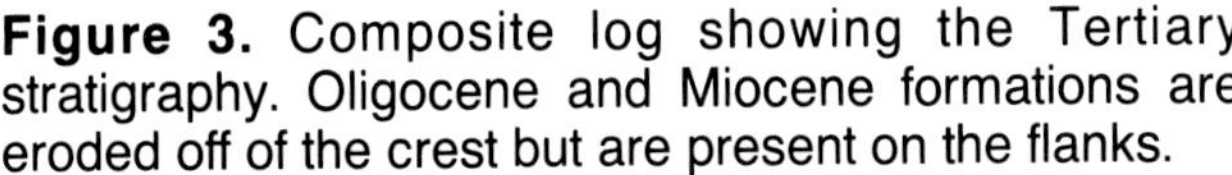

Figure 3. Composite log showing the Tertiary stratigraphy. Oligocene and Miocene formations are eroded off of the crest but are present on the flanks.

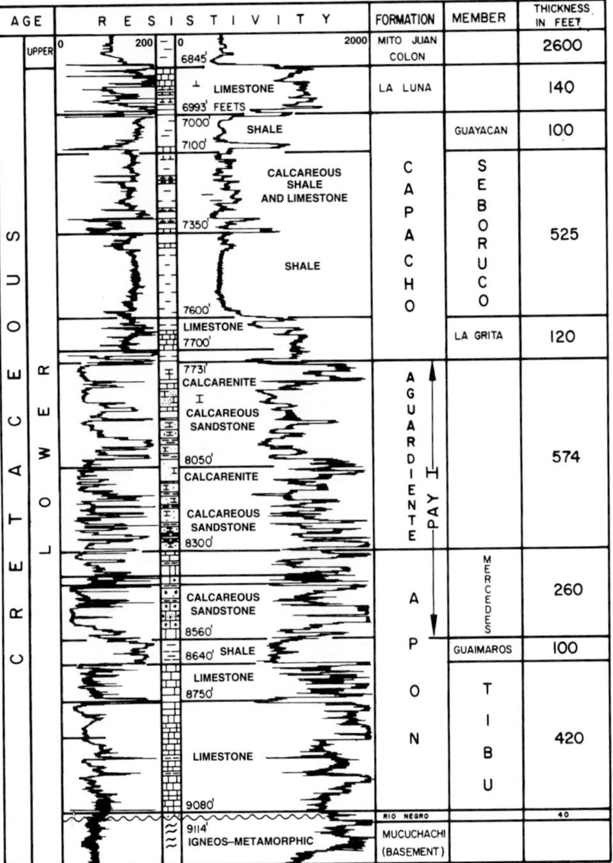

Figure 4. Composite log showing the Cretaceous stratigraphy of Tarra oil field.

5 and 6) show disharmonic structures between the Cretaceous and Tertiary levels. This information has to be analyzed prior to further development since a somewhat different model is needed to explain the structural incongruities between the two levels.

STRUCTURE

The Tarra field is located within the unstable Maracaibo Basin, which is bounded by the Oca, Perijá, and Boconó transcurrent fault systems in northwestern Venezuela (Figure 1B). These systems are related to right-lateral strike-slip movement along the Caribbean and South American plate boundaries.

During the Cretaceous, the Maracaibo Basin was a vast platform situated within the South American plate (González de Juana et al., 1980; Stobie, 1982). During this time, 4000 ft (1220 m) of transgressive marine sediments were deposited. They consist mainly of carbonates, sandstones, and shales. Later, during the Paleocene, compressional tectonic movements began to take place, and a regressive sequence was deposited. During the upper Eocene, strong tectonic movements related to the Oca, Boconó, and Perijá faults (Figure 1B) resulted in a sinistral north-northeast transcurrent fault system that affected the entire basin. This system was further developed during the Miocene and Pliocene, resulting in a series of anticlinal folds along these fault trends in western Venezuela (Figure 1B). Sedimentary environments during this time were mainly fluvial and coastal.

Tarra field is located on one such anticlinal fold (Figure 2), an asymmetric fold with its western flank thrust over the eastern flank. Using seismic line 81C-10C (Figure 5), an interpreted structural section was produced (Figure 6) with the help of geological information obtained from several wells drilled in the area. On the structural section, the Tertiary anticlinal structures can be observed to be in disharmony with the deeper Cretaceous structures. These structures are related to the plastic condition of the thick Colón–Mito Juan shales that absorbed the deformational stresses in the Cretaceous and also provided conditions for faulting and overthrusting in the overlying Tertiary. Figure 7 presents the structural relationship between the main Tarra thrust fault and the overlying and underlying Tertiary beds. As much as 2000 ft (610 m) of vertical displacement can be measured along the fault.

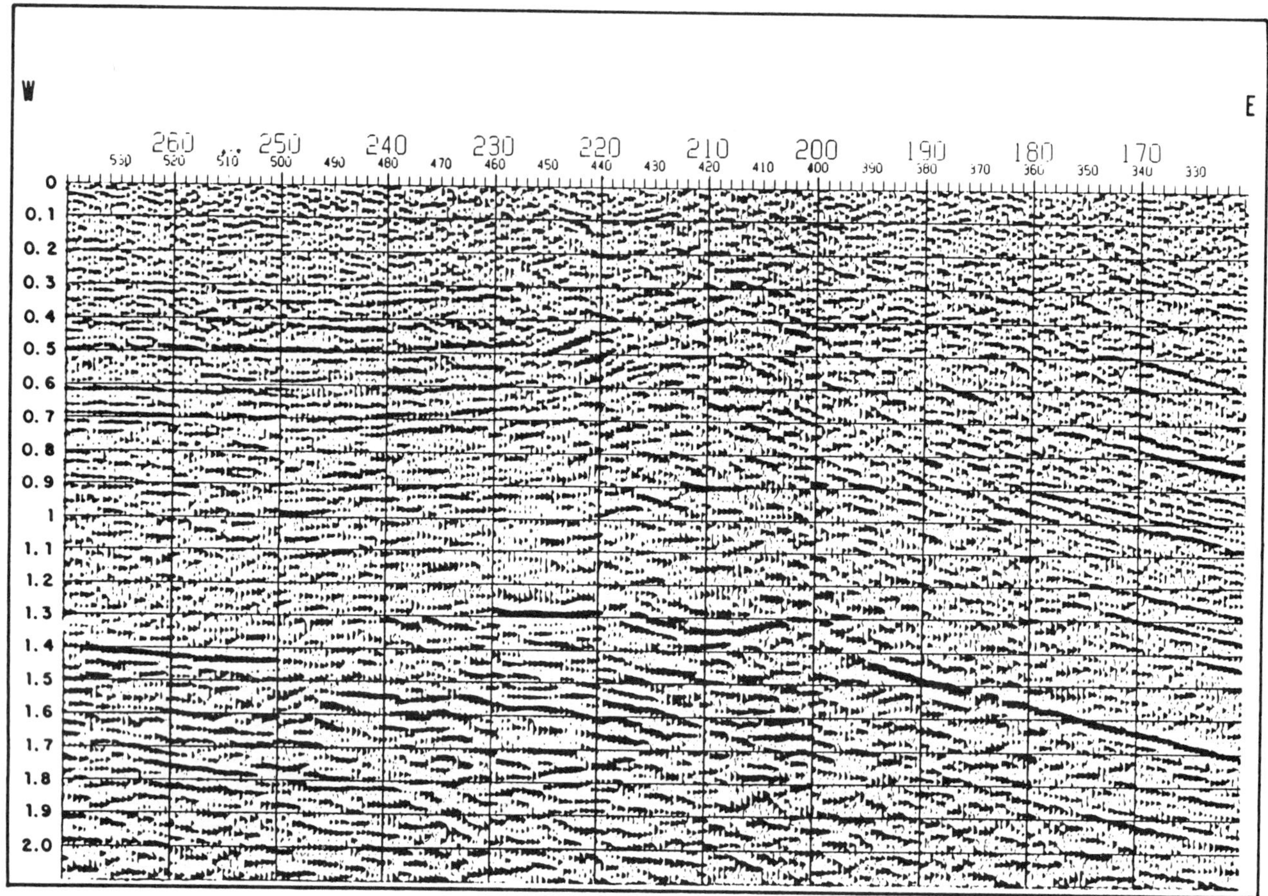

Figure 5A. Uninterpreted seismic line 81C-10C. (After Molina, 1987.) See Figure 2 for location. Length of section about 3.4 mi (5.5 km).

Multiple Tertiary reservoirs are located both above and below the thrust zones.

STRATIGRAPHY

The stratigraphy of Tarra field is illustrated in Figures 3 and 4, composite type logs of the field showing the log characteristics of the Tertiary and Cretaceous reservoirs. The following is a brief description of each formation.

Carboniferous

The Mucuchachi Formation is the oldest zone drilled and consists of slates, phyllites, and sericitic graphitic schists. The Mucuchachi is considered to be basement in this area.

Cretaceous

The Rio Negro Formation lies unconformably over the Mucuchachi Formation and consists of medium- and coarse-grained sandstones, partly feldspathic and agglutinated with silica.

The Apon Formation is divided into three parts or members. Its lower part (Tibu Member) is characterized by recrystallized, dense, and impermeable limestones (originally lime mudstones), containing minor quantities of shales throughout the sequence. Its middle part (Guaimaros Member) consists of a sequence of shale, sometimes dolomitized and 100 ft (30 m) thick. Its upper part (Las Mercedes Member) consists mainly of limestones, which are very sandy toward the surface. They are interbedded with shales and sandstones.

The Aguardiente Formation consists mainly of fine- to coarse-grained sandstones cemented with silica and calcite. Toward its lower part this formation shows interlayering of sandy limestones and shales, but consists only of shales toward the top. The formation and the Mercedes Member of the Apon Formation make up the Cretaceous reservoir.

The Capacho Formation is mainly shale. The lower, middle, and upper parts of this sequence contain some limestone layers. The La Grita, Seboruco, and Guayacán are members of this formation.

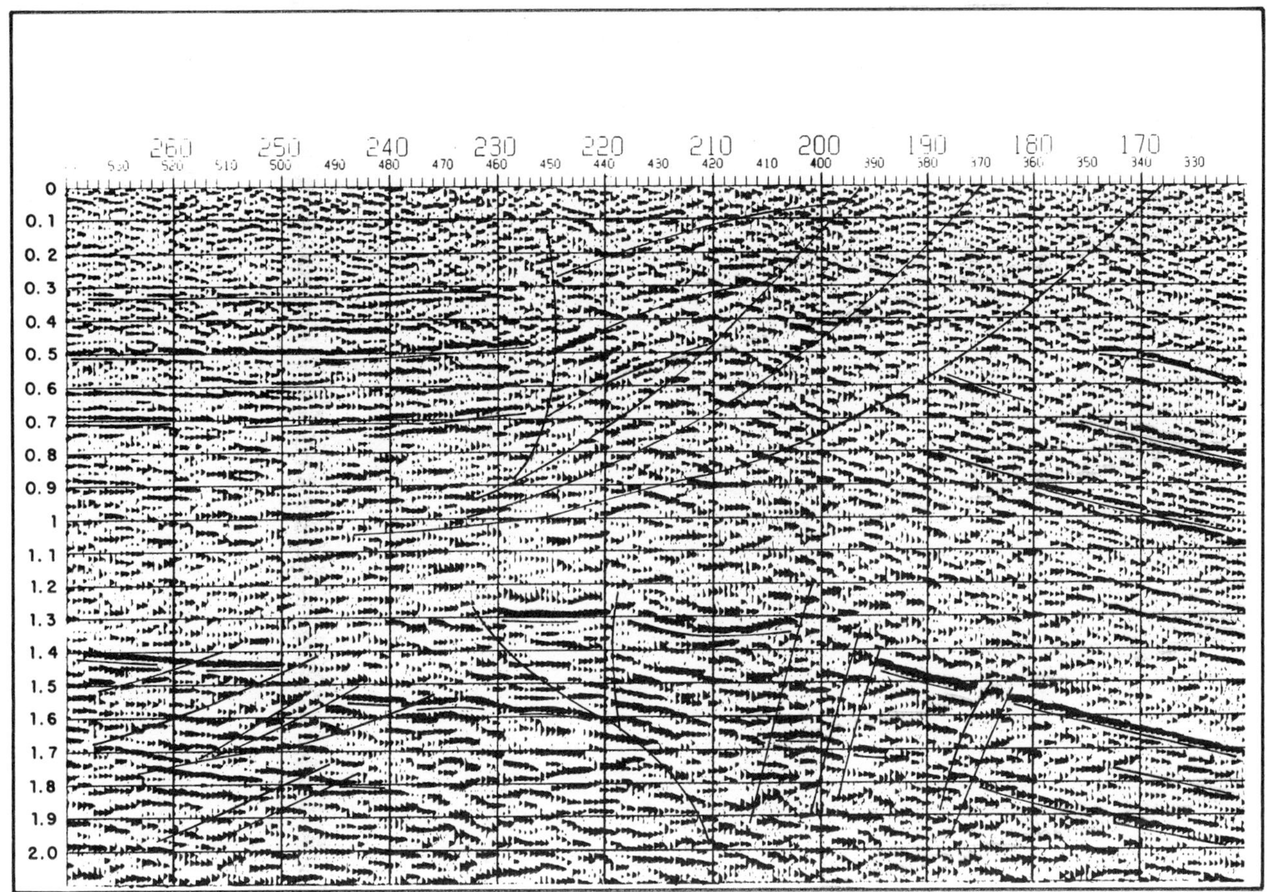

Figure 5B. Interpreted seismic line 81C-10C.

The La Luna Formation is considered to be the most important source rock in the Maracaibo Basin because of its high organic matter content and the degree of maturity it has reached (Blaser, 1979; Lew, 1984; Intevep, 1984; Talukdar et al., 1986). It consists of mudstones interbedded with minor quantities of calcareous shales.

The Colón–Mito Juan Formation consists of a sequence of dark gray-colored shales that become progressively more silty toward the top.

Paleocene

The Orocue Group is made up of three formations. The Catatumbo Formation is the lowest and consists of a sequence of silty shales interbedded with minor quantities of silty sandstones, siltstones, and coals. The Barco Formation is the middle unit and consists mainly of sandstones interbedded with siltstones, shales, and, occasionally, coal layers. The Los Cuervos Formation, the upper unit, is mainly a silty one. Its lower part consists of siltstones with a high detrital coal content, and it becomes more sandy toward the top.

Eocene

The Mirador Formation consists of three parts. The lower part (Mirador II member) consists mainly of medium-grained sandstones interbedded with siltstones and shales. The middle part (Shale member) consists of shales. The upper part (Mirador I member) consists of sandstones interbedded with shales. The Mirador Formation contains the second reservoir section of the field.

The Carbonera Formation in its lower part, which includes the Cubo Sand Member, consists of sandstones, silty sandstones, and minor quantities of siltstones and shales and its upper part, of siltstones and shales interbedded with some sandstones. It shows a thick regional layer of coal at 100 ft (30 m) above its base (Stobie, 1982).

Oligocene

The León Formation in its lower part consists of siltstones and shales. Its upper part is predominantly shaly. This formation is eroded away over the field area but occurs downdip on the structure.

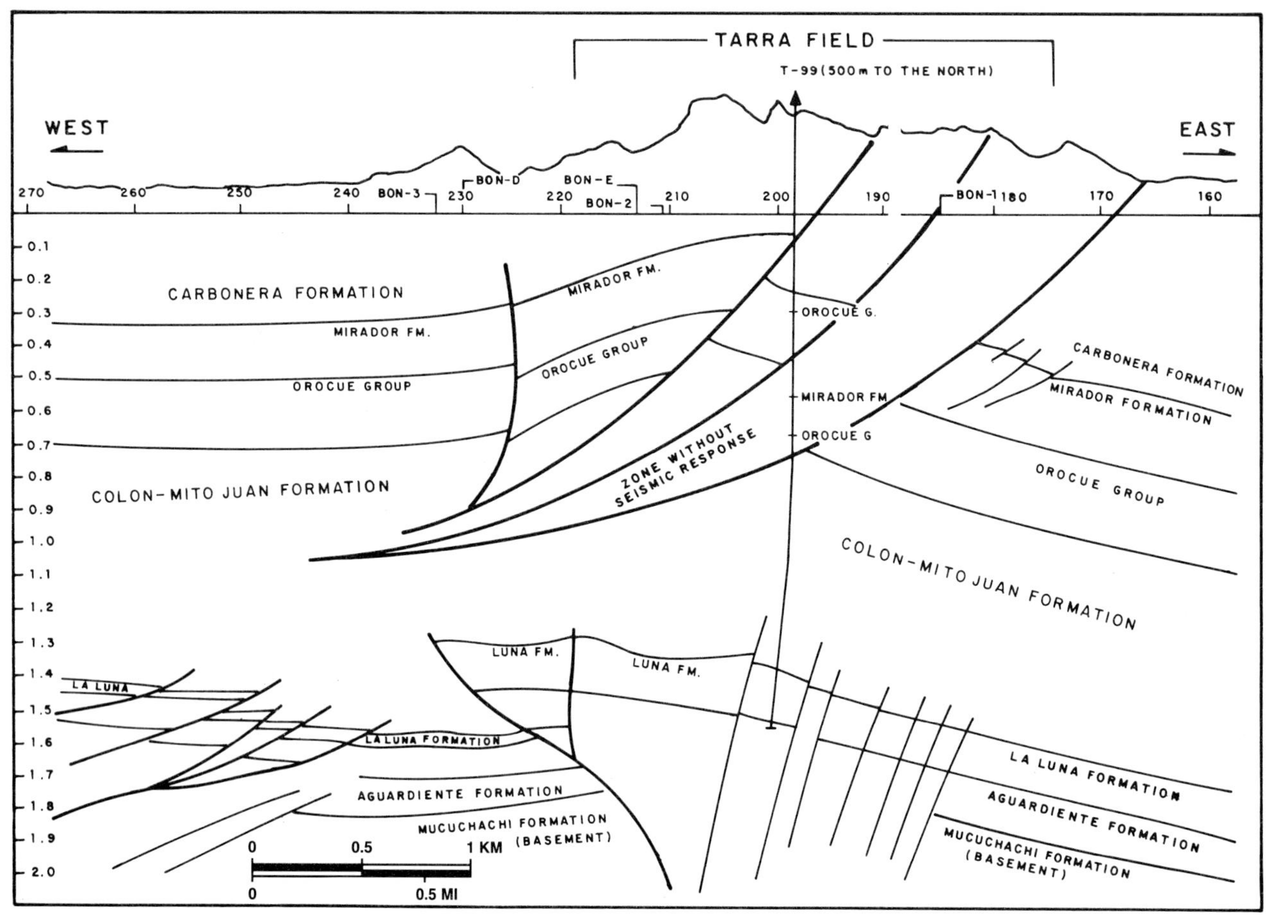

Figure 6. Interpretation of seismic line 81C-10C. (After Molina, 1987.)

Miocene

The Guayabo Formation consists of siltstones with a conglomeratic sequence toward the surface. Miocene deposits are missing from the field area as a result of erosion but occur downdip on the structure

TRAPS

The Paleocene and Eocene traps are the result of structural and sedimentological factors. Vertical shaly seals cover sequences of permeable sandstones. Laterally these sandstones change to shalier facies, thus losing their permeability. Some of the thrust faults that affect the entire Paleocene and Eocene sequence are believed to be sealing (Petroleum Information Corporation, 1980). However, other younger faults may have served as migration paths to the higher reservoirs or even to the surface seeps on the crest of the structure.

The updip traps in this field result from vertical seals in the thrust blocks in the form of facies changes and, in many, of the faults themselves.

Reservoirs

The reservoirs date from Paleocene to upper Eocene. The Paleocene reservoirs consist of sequences of sandstones and silty sandstones of the Orocue Group (Figure 3) at depths between 2500 and 6000 ft (760 and 1800 m) below the thrust zone, and between 1200 and 3000 ft (360 and 920 m) above it (Figures 6 and 7).

The Eocene reservoirs belong to the lower and upper members of the Mirador Formation and the Arenas del Cubo Member of the Carbonera Formation. The former consist of medium- to coarse-grained sandstones and occasional lenses of conglomerates. The Arenas del Cubo Member consists of many sequences of fine-grained silty sandstones. Above the thrust zone these reservoirs are at depths between 500 and 2000 ft (150 and 600 m) and below it, between 1800 and 3500 ft (550 and 1060 m).

The sedimentary environment that existed in this area during the Paleocene was fluvio-deltaic and consisted of low relief zones with vast flood plains. Lateral extent of the sand bodies generally ranges between 0.6 and 1.2 mi (0.9 and 1.9 km), but may be less.

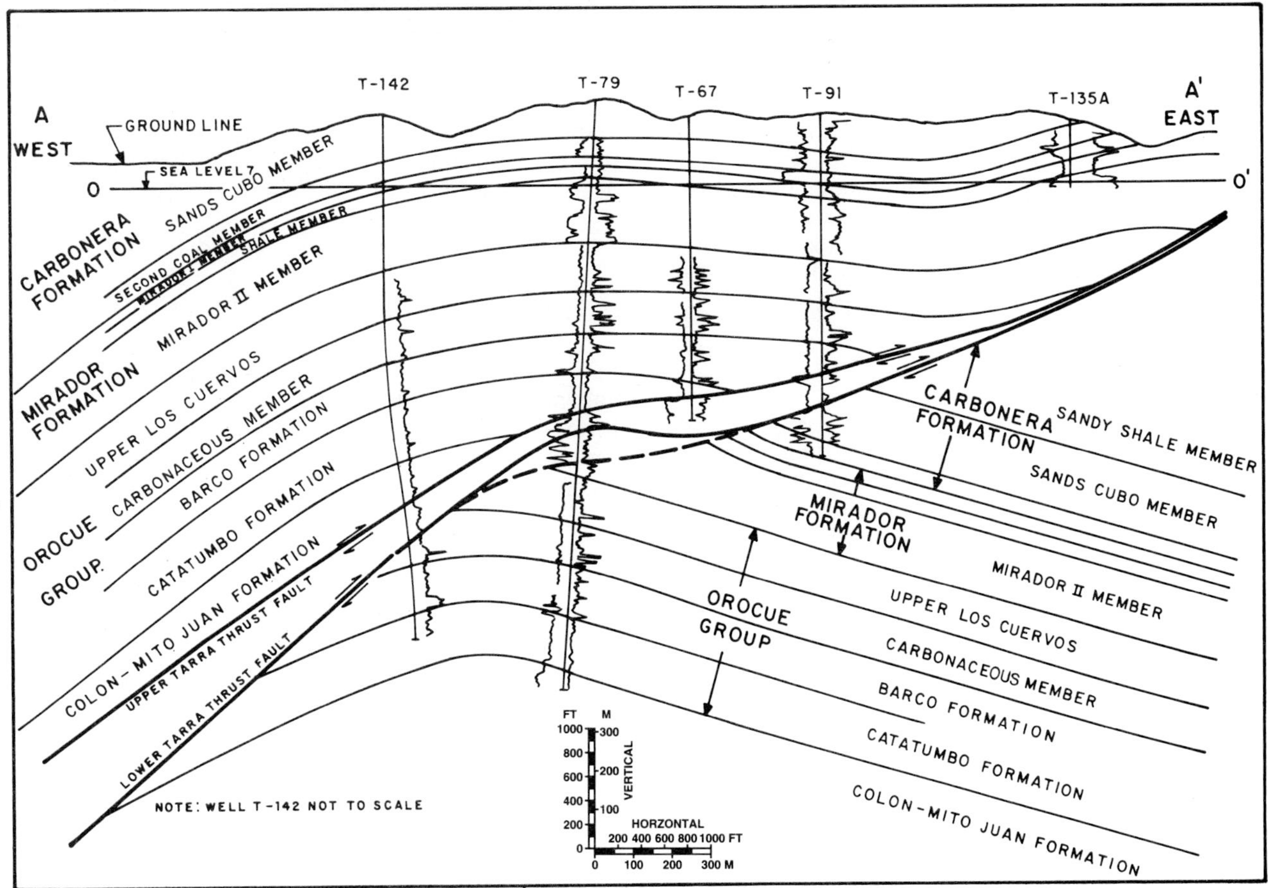

Figure 7. Structure cross section AA′, pre-seismic interpretation. (After Petroleum Information Corp., 1979.) See Figure 2 for location.

The Mirador Formation was deposited in a medium-energy fluvial environment with interbraided channels. Lateral extent of these sandstones is more continuous than those of the Paleocene sequences. The Paleocene sediments, together with the Mirador Formation, make up the end of a regressive sequence that began during the Upper Cretaceous.

The Arenas del Cubo Member represents the beginning of a change from a fluvial environment to one related to a low coastal deltaic plain. These sandstones have considerable lateral extension and may be interpreted as a coastal area between distributary channels. The sequences above them represent a swampy environment with laterally continuous coal beds.

These Orocue Group reservoirs tend to lose their permeability laterally because of facies changes to finer sediments. Average porosity for reservoirs of the group is 14%. Permeability is 50 md. Eocene sandstones show porosity values ranging between 14 and 20%. Permeability values are 600 md for the lower Mirador Formation (Mirador II), 150 md for the upper Mirador Formation (Mirador I), and 150 md for the Arenas del Cubo Member. Total Paleocene and Eocene reservoir thickness of the overthrusted and underthrusted blocks is 5000 ft (1500 m). The oil is 31° API, and initial GOR was 478 ft^3/bbl (85.1 m^3/m^3).

Well spacing has been based upon reservoir engineering considerations, such as the lateral drainage coefficient (estimated according to oil gravity), pressure, temperature, and lateral continuity of the reservoirs. No secondary oil-recovery operations have been considered to be necessary for this field because of the lenticular character of the sandstones and the fact that the production mechanism in most areas is a primary, very limited water drive (Petroleum Information Corporation, 1980).

Faults

The main thrust fault zone of the Tarra field was initially recognized by its surface expression and by the abrupt contact existing between the lower Eocene and Oligocene sediments. The drilling of wells through this thrust zone confirmed its existence and helped to better define its dip and orientation. Later, during 1979, 1981, and 1985, seismic surveys were carried out, which aided in a better interpretation and definition of the different structures in the Cretaceous and Tertiary strata (Figures 5 and 6).

It is believed that most faults associated with this thrust zone are sealing. The zone, however, is a complex one, not fully understood. Sandstone superposition and communication along some of the fault planes between different blocks could result in a single interconnected reservoir complex. Structural interpretation at the present stage is still considered to be a simplification of actual conditions.

Source

The source rock that accounts for the generation of most of the hydrocarbons in the Maracaibo Basin is the La Luna Formation (Blaser, 1979; Lew, 1984; Intevep, 1984; Talukdar et al., 1986). Other less important source rocks exist in the Cretaceous Capacho Formation and the Paleocene Orocue Group. The La Luna Formation shows an average TOC of 3.8%, maximum value being 9.6%. Kerogen is type II. Potential yield is about 10 kg of hydrocarbons per ton of rock. The rocks began to generate oil during the Eocene, and the process is still taking place today in different areas of the basin (Blaser, 1979; Talukdar et al., 1986).

It has been determined that the migration of hydrocarbons originated in the eastern part of the Maracaibo Basin because greater depths were reached by the La Luna Formation in the east than in the west and because regional dip during migration time had been toward the east.

A Lopatin/Waples type TTI chart for the Tarra-Rosario area (Figure 8) indicates the timing of oil generation and the degree of maturity of the source rocks. It also shows the different tectonic and diagenetic events that occurred during burial history. As shown by the figure, the outset of oil generation probably occurred in the principal source rocks, the black shaly limestones of the La Luna Formation, during the Miocene and in the underlying massive gray limestones of the Apon during the Oligocene, before the Andean uplifts. The base of the Apon apparently passed through the end of the generation period to the limit of 40° API oil, whereas the La Luna reached the end of oil generation (TTI = 160). Analysis of vitrinite reflectance in cores from the upper Capacho and lower La Luna provide an R_0 of 1.1 to 1.2% (approx. 100 to 150 TTI), which corresponds to the estimates of TTI on the chart. The Orocue Group and the Mirador have vitrinite reflectance values of 0.6 and 0.5%, respectively, approximately equivalent to 10 and 5 TTI (Young, 1984).

In the Maracaibo Basin there are two major problems in making estimates of TTI: One is the estimation of the thicknesses of Eocene and Oligocene-Miocene sediments originally deposited and later eroded away; and the other is the estimation of the temperature gradient, which varies considerably throughout the basin (1.7 to 2.5° F/100 ft or 3.1 to 4.6° C/100 m). Thus more Eocene and Miocene sediments could have been deposited in the area, especially toward the east in the Andean downwarp, than is indicated on Figure 8, and if so, the Orocue and Mirador formations also could have generated oil in areas adjacent to the Tarra structures.

EXPLORATION CONCEPTS

According to vitrinite reflectance and burial diagrams pertaining to the La Luna Formation (Intevep, 1984; Lew, 1984; Waples, 1980), hydrocarbon generation and migration started during the Eocene in a zone parallel to the eastern shoreline of Lake Maracaibo. Furthermore, the structures currently found in the Maracaibo Basin date from this time. Oil generation and the development of these structures continued during the Miocene and Pliocene. Possibly the first reservoirs to become saturated with hydrocarbons in this field were those in the Cretaceous, such as the Aguardiente Formation. At that time, an important vertical seal existed, consisting of 2000 ft (600 m) of shales of the Colón Formation. Fracturing of this seal as a result of regional tectonism was related to the development of overlying structures; this fracturing favored migration of hydrocarbons from the Cretaceous to the Paleocene and Eocene reservoirs that occurred when pressure limits across the fault zones were exceeded.

The Tarra field is similar to the Rosario, Los Manueles, Concordia, West Tarra, Rio de Oro, Mara, and La Paz fields. All consist of faulted anticlinal structures with reservoirs dating from the Cretaceous and Eocene.

Exploration methods used during initial discovery of the field were surface geology as related to oil seeps. Seismic surveys of the area prior to drilling did not exist. They have been recently carried out in order to define the structure of the deeper Cretaceous, where future development is being considered. An attempt will be made during future drilling to intersect the fault planes at the level of the Aguardiente Formation and the Mercedes Member and thus to optimize the drainage of this reservoir. A coring program will be required for detailed carbonate sedimentological and diagenetic studies. These will be carried out in the future. Correlation of electric logs and cores will aid in understanding the production potential of the fractured carbonate reservoirs.

ACKNOWLEDGMENTS

I thank Maraven, S.A. for permission to publish the results of work carried out in Tarra field through the many years of its production history. Reviews

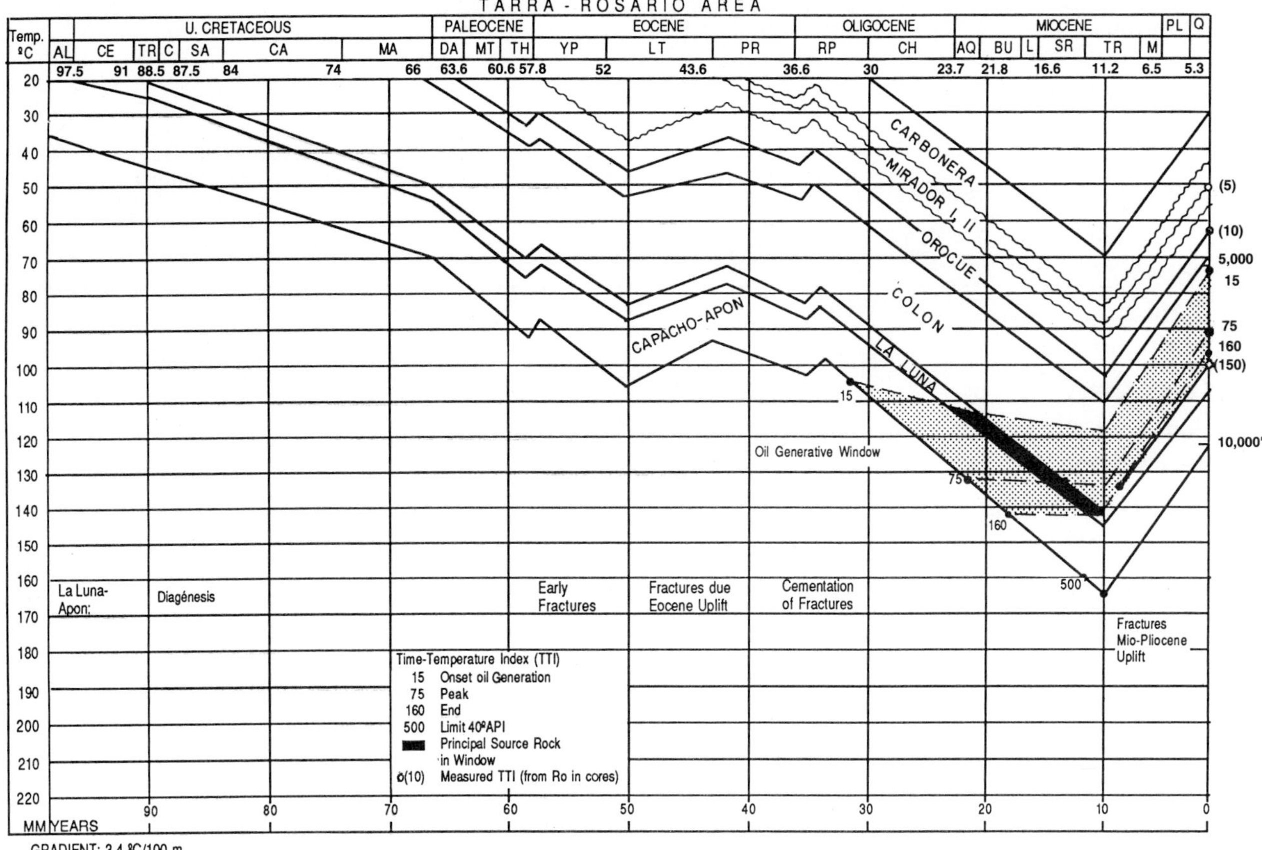

Figure 8. Lopatin TTI reconstruction showing isomaturity lines of the oil generative window in the Tarra area. Wavy lines are hiatal intervals.

by Gordon Young, Donald Goddard, and Ted Beaumont helped considerably in the improvement of the manuscript.

REFERENCES CITED

Blaser, R., 1979, Source rock and hydrocarbon generation in the Maracaibo Basin, Western Venezuela: Maraven Report EPC-5841.69.

González de Juana, C., J. Iturralde de Arozena, and X. Picard, 1980, Geologí a de Venezuela y sus cuencas petrolíferas: Ediciones Foninves, Caracas, 1031P, v. I y II.

Intevep, 1984, Estudio geoquímico regional de la cuenca de Maracaibo: Maraven Report EPC-7643.3.

Lew, M., 1984, Estudio del Eoceno del area de Mara-Maracaibo, Estado Zulia; Parte 3, Origen del gas libre y gas asociado: Maraven Report EPC. 7647.3.

Molina, A., 1987, Campos Bonito-Las Cruces: revision geológica del Cretaceo: Maraven Report EPC-11.636.

Petroleum Information Corporation, 1979, Tarra field: geological and engineering study: Maraven Report EPC. 6251. 9.

Petroleum Information Corporation, 1980, Las Cruces field: geological and engineering study: Maraven Report EPC-6539.9.

Stobie, R., 1982, Geological evaluation and summary of Cretaceous exploration within the area of the Tarra anticline system, South Colón, Zulia State: Maraven Report EPC-7172.

Talukdar, S., O. Gallango, and M. Chin-A-Lien, 1986, Generation and migration of hydrocarbons in the Maracaibo Basin, Venezuela: an integrated basin study, *in* Advances in organic geochemistry: Organic Geochemistry, v. 10, p. 261–279.

Waples, D.W., 1980, Time and temperature in petroleum formation: application of Lopatin's method to petroleum exploration: AAPG Bulletin, v. 64, n. 6, p. 916.

Young, G. A., 1984, Sintesis de estudios sobre la geoquímica y diagenesis de sedimentos en la Cuenca de Maracaibo: Petróleos de Venezuela Report, Coordination of Exploration.

TARRA

Appendix 1. Field Description

Field name .... *Tarra field*

Ultimate recoverable reserves .... *136.6 million bbl oil and 124 bcf gas*

Field location:

Country .... *Venezuela*

State .... *Zulia*

Basin/Province .... *Maracaibo Basin*

Field discovery:

Year field discovered .... *Lower Eocene and Paleocene sandstone 1916*

Year second pay discovered .... *Lower Cretaceous calcareous sandstone 1941*

Discovery well name and general location:

First pay .... *T-1, 95 km west-northwest of the town of El Vigia*

Second Pay .... *T-1-6, 99 km west-northwest of the town of El Vigia*

Discovery well operator .... *Cia Shell de Venezuela, Ltd.*

Second pay .... *Cia Shell de Venezuela, Ltd.*

IP .... *NA*

All other zones with shows of oil and gas in the field:

Age	Formation	Type of Show
Upper Cretaceous	*Colón*	*Gas*

Geologic concept leading to discovery and method or methods used to delineate prospect

Surface geology recognized a thrust-faulted anticlinal system associated with oil seeps. The discovery well drilled on the crest of the anticline encountered multiple Tertiary reservoir units ranging in age from Paleocene to Eocene.

Structure:

Province/basin type .... *Bally 222; Klemme III Bc*

Tectonic history

The Tarra field is located within the unstable zone bounded by the Oca, Perijá, and Boconó transcurrent fault systems in northwestern Venezuela. These systems are related to right-lateral strike-slip movement along the Caribbean and South American plates boundary. Major period of tectonic activity occurred in the Tertiary including the late Miocene Andean uplift.

Regional structure

Consists of a north-northeast-trending wrench fault system associated with several thrust-faulted anticlines.

Local structure

A low relief, north-northeast-trending anticline with a number of overthrust faults on the east flank parallel to the main axis.

Trap:

Trap type(s)

The Tarra field has (1) anticlinal trap combined with various overthrust blocks and multiple pays; (2) unconformity truncated traps; and (3) lateral pinch-outs.

Basin stratigraphy (major stratigraphic intervals from surface to deepest penetration in field):

Chronostratigraphy	Formation	Depth to Top in ft (m)
Eocene-Oligocene	*Carbonera*	*0*
Eocene	*Mirador*	*500 (152)*
Paleocene	*Orocene Group*	*1200 (366)*

Chronostratigraphy	Formation	Depth to Top in ft (m)
Cretaceous	*Colón*	*5000 (1525)*
	La Luna	*7000 (2135)*
	Capacho	*7100 (2166)*
	Aguardiente	*7700 (2348)*
	Apon	*8400 (2562)*
	Rio Negro	*9300 (2836)*

Reservoir characteristics:

Number of reservoirs *8*
Formations .... *Mirador Formation; Orocue Group; and Aguardiente Formation*
Ages .... *Eocene, Paleocene, and Cretaceous*
Depths to tops of reservoirs .... *500 ft (152 m); 1200 ft (366 m); 7700 ft (2348 m)*
Gross thickness (top to bottom of producing interval) .... *2600 ft (793 m)*
Net thickness—total thickness of producing zones
Average .... *1648 ft (5026 m)*
Maximum .... *1900 ft (580 m)*
Lithology
Mirador Formation; light gray, fine- to medium-grained sandstone interbedded siltstones and shales Orocue Group: dark gray, silty shale sequence, interbedded with siltstone, and carbonaceous intervals Aguardiente Formation: white to light gray, fine- to coarse-grained subrounded to subangular, variably but well cemented with silica and calcite; moderately well sorted and relatively tight, interbedded limestone and shales.
Porosity type .... *Intergranular primary porosity with minor secondary fracture porosity*
Average porosity *18%*
Average permeability .... *300 md*

Seals:

Upper
Formation, fault, or other feature .... *Carbonera Formation*
Lithology .... *Shales*
Lateral
Formation, fault, or other feature .... *Faults and juxtaposition of shales*
Lithology .... *Shales*

Source:

Formation and age .... *Cretaceous La Luna*
Lithology .... *Carbonaceous limestone, calcareous shales, and chert*
Average total organic carbon (TOC) *3.8%*
Maximum TOC *9.6%*
Kerogen type (I, II, or III) .... *II*
Vitrinite reflectance (maturation) .... $R_o = 1.2$
Time of hydrocarbon expulsion .... *Eocene–Miocene*
Present depth to top of source .... *7000 to 8000 ft (2135 to 2440 m)*
Thickness .... *100 ft (30 m)*
Potential yield .... *10 kg of hydrocarbon per ton of rock*

Appendix 2. Production Data

Field name .... *Tarra*

Field size:

Proved acres .... *1640 ac (664 ha)*

Number of wells all years *221*
Current number of wells *38*
Well spacing .... *300 m (984 ft)*
Ultimate recoverable .... *136.6 million bbl oil (1989 est. 90 million bbl) and 124 bcf gas*
Cumulative production .... *117 million bbl oil and 61 bcf gas*
Annual production .... *0.434 million bbl oil and 0.685 bcf gas*
Present decline rate .... *NA*
  Initial decline rate .... *8.0% (in first year)*
  Overall decline rate .... *3.9% (in 61 years)*
Annual water production .... *1 million bbl*
In place, total reserves .... *532 million bbl oil and 239 bcf gas*
In place, per acre foot .... *720–980 bbl*
Primary recovery .... *136.6 million bbl oil and 124 bcf gas*
Secondary recovery .... *NA*
Cumulative water production .... *104 million bbl*

TARRA

Drilling and casing practices:

Amount of surface casing set .... *15½-in. to 353 ft (108 m)*
Casing program
15½-in. from surface to 353 ft (108 m); 12-in. to 760 ft (232 m); and 10-in. to 800 ft (244 m)
Drilling mud .... *Lignosulfonate-clay base mud*
Bit program .... *NA*
High pressure zones .... *At 2000 ft (610 m) sandy shales of Colón Formation*

Completion practices:

Intervals perforated .... *Mirador Formation and Orocue Group, 782 to 854 ft (238 to 260 m)*
Well treatment .... *NA*

Formation evaluation:

Logging suites .... *Electrical and deviation logs*
Testing practices .... *Standard production test and some drill-stem tests*
Mud logging techniques .... *Analysis of cuttings; gas chromatograph*

Oil characteristics:

API gravity .... *Eocene, 22–25°; Cretaceous, 41–50°*
Base .... *Naphthenic*
Initial GOR .... *Tertiary, 150 to 370 ft³/bbl*; Cretaceous, 1750 ft³/bbl*
Sulfur, wt% .... *Not determined*
Viscosity, SUS .... *Eocene, 44 to 52; Cretaceous, 46 to 56*
Pour point .... *Below 32° F*
Gas-oil distillate .... *NA*
**Some productive upper reservoirs gas TSTM.*

Field characteristics:

Average elevation .... *500 ft (152 m)*
Initial pressure .... *3000 psi (20.7 MPa)*
Present pressure .... *950 psi (6.6 MPa)*
Pressure gradient .... *0.46 psi/ft (10.4 kPa/m)*
Temperature .... *170° F (76.7° C)*
Geothermal gradient .... *0.05° F/ft (0.09° C/m)*
Drive .... *Water*
Oil column thickness .... *2500 ft (762 m)*
Oil-water contact .... *Unknown*
Connate water .... *NA*
Water salinity, TDS .... *NA*

Resistivity of water .... *0.164 ohm-m² m*
Bulk volume water (%) .... *NA*

Transportation method and market for oil and gas:
From the field via pipeline, gas is sent to La Fria for industrial use; oil is sent via pipeline to oil dock on Lake Maracaibo for export.

Lama Field—Venezuela
Maracaibo Basin, Zulia State

ISMAEL DELGADO
Maraven, S.A.
Caracas, Venezuela

FIELD CLASSIFICATION

BASIN: Maracaibo
BASIN TYPE: Foredeep
PETROLEUM TYPE: Oil
RESERVOIR ROCK TYPE: Sandstones, Limestones
RESERVOIR AGE: Cretaceous, Paleocene, Eocene, Oligocene, Miocene
TRAP TYPE: Anticline, Fault Block, Unconformity Truncation
RESERVOIR ENVIRONMENT OF DEPOSITION: Fluvial-Deltaic and Marine Platform
TRAP DESCRIPTION: Anticlinal wrench-fault structure with fractured limestones, fault-bounded segments, unconformity truncations, and pinch-outs

LOCATION

The Lama field is located in the central part of Lake Maracaibo, Zulia State, in northwestern Venezuela, which is the most prolific oil-producing area of the country (Figure 1A). The field is on the western side of the Merida arch (Venezuelan Andes), a broad, north-plunging, regional feature formed in pre-Cretaceous time (Figure 1B).

The Lama field covers an area of 89 mi^2 (230 km^2), of which 53% contains proven reserves. The field was originally owned by the Superior Oil Co. of Venezuela but was acquired by Texaco in January 1969. Subsequently the field was acquired by Maraven, S.A., affiliate of Petroleos de Venezuela, S.A.

Other important oil-producing fields are located near the Lama field and have correlative reservoirs (Figure 1A). To the north are the Eocene and Cretaceous fields of Lagunillas, formerly operated by Exxon and Shell; to the east are the Miocene, Eocene, and Cretaceous reservoirs associated with the Centro field, originally owned by Exxon and Mene Grande Oil Co.; and to the south is the Miocene and Eocene Lamar field, initially operated by Shell, Venezuela Oil Co. (CVP), and Mene Grande Oil Co.

Estimated ultimate recovery for the Lama field is 3.5 billion bbl of oil (BBO) (Padron, 1961). Original oil reserves in place were about 9.7 BBO. The estimated potential yield for the field is about 150,000 bbl per day of 33° API oil from the various reservoirs (Miocene, Eocene, Paleocene, and Cretaceous).

HISTORY

Completion of the Lama-1 (Figure 2) well in 1957 was very important for the development of a large prolific area within Lake Maracaibo. Information from this well demonstrated the prolongation of the outstanding anticlinal Icotea structure, which extends from Cabimas, located on the northeast shore of the lake, south-southwestward in the subsurface into the Lama area (Figures 1A and 1B). The presence of the Icotea fault under Lake Maracaibo north of the Lama field was known from well information and seismic and magnetic surveys before the discovery of the Lama field. When concessions in the Lama area were offered in 1956, interested companies used the technology of the time to assess the possibility of the extension of the fault into the area to evaluate them. Mene Grande Oil Company, for one, used state-of-the-art magnetic and seismic surveys, which were at least sufficient to show the fault. Following this, additional drilling by the Superior Oil Company of Venezuela delineated the structure and "proved up" the Lama field. Its prolongation toward the south in the concessions of the Venezuelan Sun Oil Company was later confirmed.

Drilling and completion of the Lama-1 and SVS-1 wells (Figure 2) in 1957 led to the discovery of pay zones within the Cretaceous (SVS-1), Eocene (Lama-1), and Miocene (SVS-1). These wells also proved the existence of oil within the Paleocene, with economic

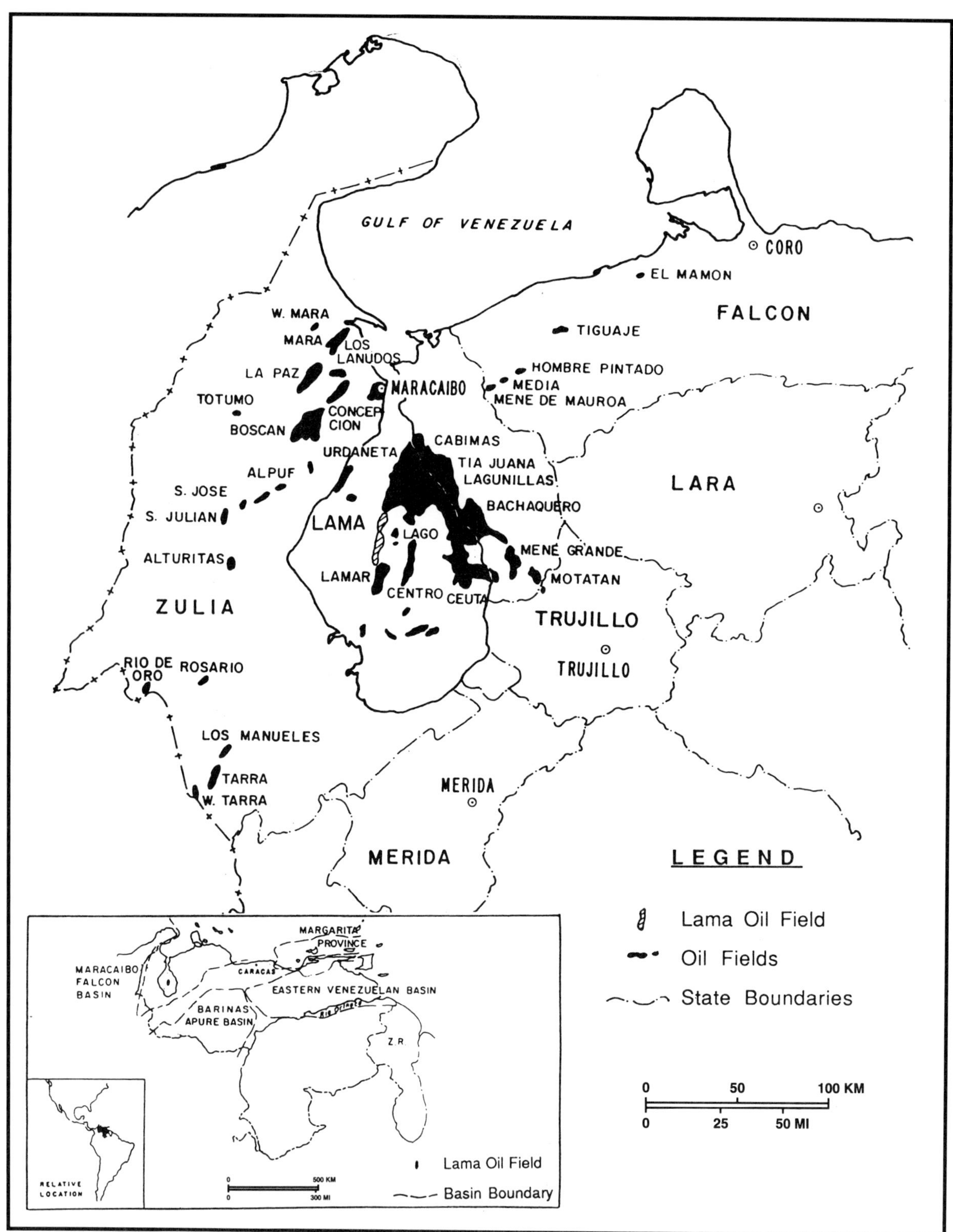

Figure 1A. Geographic location of the Lama oil field in Western Venezuela in relation to the other fields in the Maracaibo-Falcón Basin. (Ed. note: Figures 1A and 1B, almost identical to Figures 1A and 1B of the Los Lanudos field in this volume, are included here for convenience.)

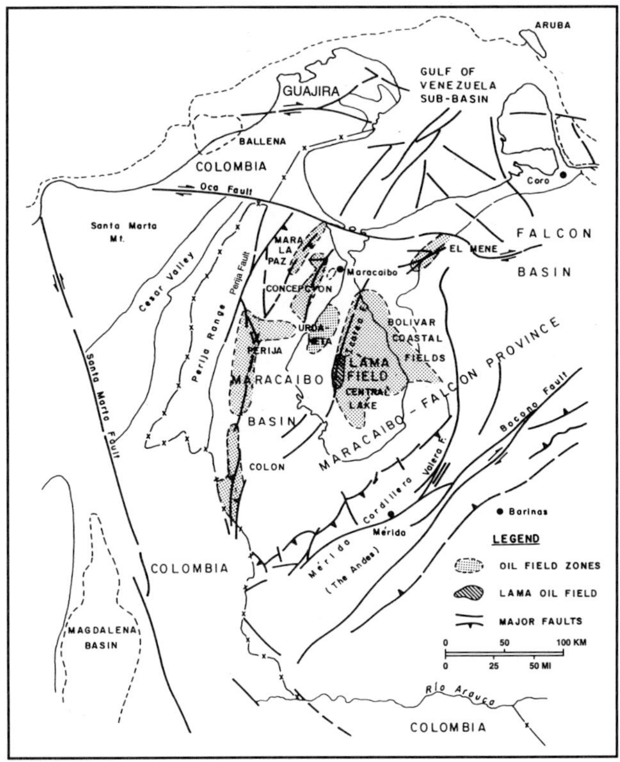

Figure 1B. Structural setting of the Maracaibo-Falcón Province and the relative location of the Lama field.

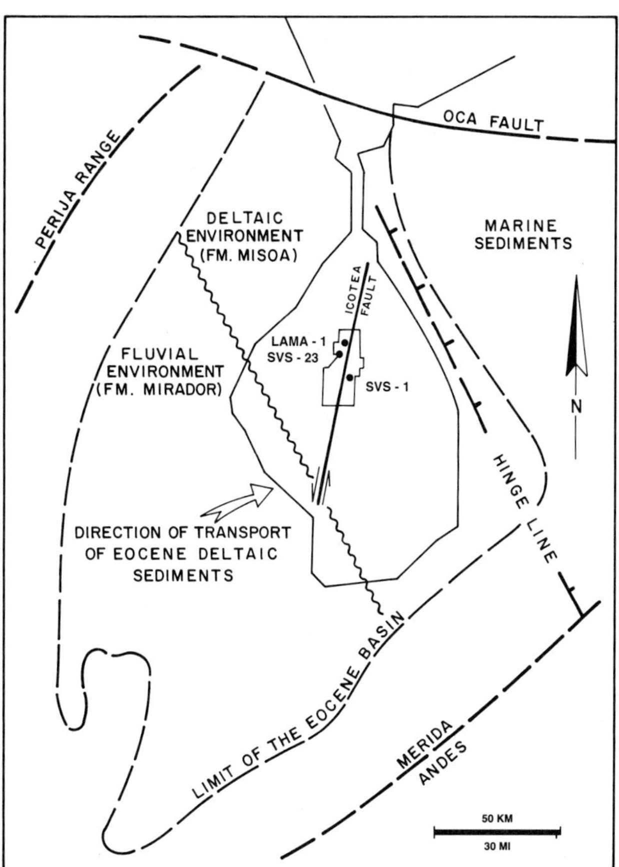

Figure 2. Sedimentary environments during the Eocene in the Maracaibo Basin.

quantities obtained two years later from the completed SVS-23 well in 1959 (Figure 2).

The Lama-1 well was drilled by the Superior Oil Company in the concession known as Grupo 75, 44 mi (71 km) south of Maracaibo City. It was completed in 1957 in the C sandstones of the lower Eocene Misoa Formation (Figure 3). Initial production was 2675 bbl per day of 29° to 34° API oil.

The SVS-1 well was drilled by the Venezuelan Sun Oil Company in the concession known as Bloque 1, 51 mi (82 km) south of Maracaibo City. It was initially completed in 1957 in the arenaceous sediments of the Cretaceous Colón–Mito Juan Formation (Figure 3) for 548 BOPD, 32° to 42° API. Ditch samples indicated the existence of deeper limestone reservoirs. This later led to the discovery of the most important hydrocarbon-bearing strata within the Cretaceous: a thick calcareous sequence, consisting of fractured limestones, corresponding to the La Luna, Maraca, Lisure, and Apon formations. The latter three make up the Lower Cretaceous Cogollo Group. The well was later recompleted (1958) in the basal sandstones of the Miocene Santa Barbara Member of the Santa Rosa Formation because of problems in producing the Cretaceous reservoir. Initial production was 729 BOPD, 28° to 32° API.

Although the Lama-1 and SVS-1 wells proved the existence of economic quantities of oil in the Paleocene, this was not confirmed until 1959 when drilling of the SVS-23 well (Figure 2) through the limestones of the Guasare Formation (Figure 3) led to the discovery of interesting prospects within this reservoir. This well was drilled by the Venezuelan Sun Oil Company, 50 mi (80 km) south of the city of Maracaibo, in the concession known as Bloque 1. Initial production was 1345 BOPD, 35° to 38° API.

Electric logs of subsequent wells drilled in this area and to the north and south indicated the existence of local oil reservoirs within the sandy Miocene interval belonging to the Lagunillas Formation (Zamora, 1977).

The depths of these reservoirs and the thicknesses of the oil-producing sequences vary notably. The average depth to the Cretaceous is about 12,800 ft (3900 m), with a maximum of 20,000 ft (6100 m), and the pay zone average thickness is 320 ft (100 m). The depth to the Paleocene is about 10,700 ft (3260 m), the maximum being 17,500 ft (5340 m), and the average thickness of the oil-producing sequence is 70 ft (20 m). The depth to the top of the Eocene structures is about 8500 ft (2590 m), with a maximum of 13,500 ft (4120 m), and the pay-zone average thickness is 2800 ft (855 m). The Miocene Santa Barbara Member lies at depths of about 8400 ft (2560 m), the maximum being 11,000 ft (3350 m), but the average thickness for oil prospects is only 60 ft (20 m).

AGE				FORMATION	
CENOZOIC	TERTIARY	RECENT			
		PLEISTOCENE		EL MILAGRO	
		PLIOCENE			
		MIOCENE	UPPER	LA PUERTA	
			MIDDLE	LAGUNILLAS	
			LOWER	LA ROSA	Sta Barbara Mbr ICOTEA
		OLIGOCENE			
		EOCENE	UPPER		
			MIDDLE	PAUJI	
				MISOA "B" SANDS	
			LOWER	MISOA "C" SANDS	
		PALAEOCENE		GUASARE	
MESOZOIC	CRETACEOUS	MAESTRICHTIAN		MITO JUAN	
		CAMPANIAN		COLON	
		SANTONIAN			
				SOCUY MEMBER	
		CONACIAN		LA LUNA	
		TURONIAN			
		CENOMANIAN			
		ALBIAN		COGOLLO GROUP	MARACA LISURE
		APTIAN			APON
		BARREMIAN		RIO NEGRO	
		NEOCOMIAN			
	JURASSIC/TRIASSIC			LA QUINTA	

Figure 3. Generalized stratigraphic column of central Lake Maracaibo. Icotea Formation is not present at Lama. The Santa Barbara Member is the basal unit of the La Rosa Formation.

Subsequent drilling operations during field development have extended the limits of these four reservoirs to as much as 50% of the entire Lama area for the Santa Barbara reservoir and to a minimum of 15% for the reservoir contained within the Paleocene Guasare Formation. This drilling activity, which included a total of 472 wells, has led to the definition of proven areas including 27,700 ac (11,218 ha) for the Miocene (472 wells), 63,675 ac (25,788 ha) for the Eocene (472 wells), 682 ac (276 ha) for the Paleocene (105 wells), and 10,926 ac (4425 ha) for the Cretaceous (30 wells). Some completions are comingled and others are multiple (dual and triple) completions.

Well spacing varies within a grid ranging between 1970 ft (600 m) for the Eocene wells and 3940 ft (1200 m) for the Cretaceous wells. Well spacing of less than 1970 ft (600 m) is used for secondary recovery. Main production operations used to develop and maintain the potential production of the field include directional drilling, using secondary recovery methods (water, gas, and vapor injection), producing selectively, fracturing, acidizing, and reservoir stimulating (Venezuelan Sun Oil Co., 1968).

The Lama field is structurally related to the east, north, and south to other fields developed within the same reservoirs by various companies (Exxon, Shell, Texaco, etc.).

DISCOVERY METHOD

The methods used to delineate and define the Lama area included the extrapolation of surface geology and oil seep information to the lake as well as analyses of seismic data related to the land west of Lake Maracaibo. These data were also extrapolated to the downthrown side of the Icotea fault, the main structural feature in the area (Figure 1B). Subsurface geological information pertaining to areas drilled by the Shell Company of Venezuela within and outside the lake was also useful (Staff of Caribbean Petroleum Company, 1948).

STRUCTURE

The Lama field is located in the Maracaibo Basin, bounded by the Santa Marta, Oca, and Boconó transcurrent fault systems (Figure 1B), in northwestern Venezuela (Sutton, 1946). In addition to these, another group of large, transcurrent, left-lateral, north–northeast-trending faults exist within this basin. The Lama-Icotea fault (Figures 1B and 2), which crosses the Lama field diagonally and is the main structure in the area, belongs to the latter group. These systems are related to right-lateral strike-slip movement along the Caribbean and South American plate boundaries.

Major periods of tectonic activity were during the Tertiary, and included the upper Miocene Andean uplift. The tectonic history of this basin suggests that it is separated from the Gulf of Venezuela by the Oca fault system. During the Paleozoic, the basin underwent many orogenic episodes only slightly affecting its configuration, in contrast to the major changes caused by tectono-thermal activity in the Permian–Triassic. The latter included the subsequent formation of the Merida arch and the deposition of the Cretaceous limestone sequence in a shallow marine shelf environment.

Tectonic and metamorphic activities in the Perijá area, west of the basin, resulted in regional deformation in Late Cretaceous, giving rise to geomorphic structures such as the north-Andean boundary uplift and the Perijá Eastern Plain, which bounded the Maracaibo Basin to the southeast and northwest, respectively. It is currently accepted that the Late Cretaceous orogeny produced the major structures of the Maracaibo Basin. These structures consist of low-relief anticlines and normal faults that continued to develop during the Paleocene and lower Eocene.

During the lower Tertiary, the basin received huge quantities of clastic sediments from the adjacent uplifted areas, mainly from the southwest, which were transported toward the Misoa delta, accounting for the sedimentation in the basin during the Eocene (Figure 2). The major deformation of the basin, together with the withdrawal of the sea toward the north, occurred during the upper Eocene (Figure 4). The structural features developed during the Late Cretaceous were reactivated, completing the north-south-trending anticlines and faults, and some new east-west-trending structures were developed. Left-lateral strike-slip movement along the north-south-trending faults resulted in displacement of the east-west-trending structures (Borger and Lenert, 1959). During the upper Eocene and lower Oligocene, the uplifted areas were intensely eroded. During the Miocene, inversion of the basin took place (subsidence of its southern part, previously the shallow, fluvial part of the basin, uplift of its northern part, formerly the platform slope, and the reactivation and further extension of the east-west fault system) (Zambrano, 1971).

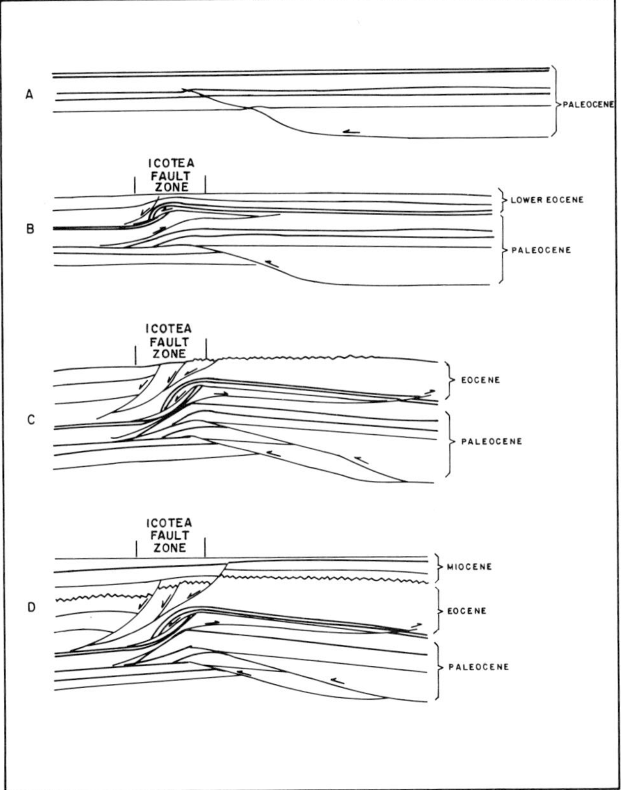

Figure 4. Sequence of structural development of the Icotea fault through time (based on a conceptual interpretation of seismic data across the fault zone). (A) At the end of the Paleocene. (B) At the end of the lower Eocene. (C) Formation of the intra-lower Eocene unconformity. (D) After upper Eocene deposition and erosion (Oligocene-Miocene) and later Miocene deposition.

During the Miocene, tectonic activity was almost nonexistent (Figure 4). Some longitudinal faults penetrated into the younger sediments of the Miocene, which fold slightly over the eroded Eocene beds. During the middle Miocene, folds and faults formed during the Eocene continued their structural development or were rejuvenated by a later orogeny that occurred during the upper Miocene and Pliocene. The final uplift of the Venezuelan Andes took place during this time, resulting in the current configuration of the Maracaibo Basin.

The initial, major periods of sedimentation were during the Cretaceous. A thick sequence of bioclastic limestones was deposited in a marine environment. The sand and shale deposition increased during Late Cretaceous, giving rise to an impermeable seal, the Colón shale, which trapped the hydrocarbons that accumulated in the bioclastic limestones below. Opinions concur that the La Luna Formation, the main source rock in the basin, was deposited in a euxinic environment during Cenomanian to Santonian time.

The oldest unconformity occurs between the Lower Cretaceous and the Triassic-Jurassic La Quinta Formation (Figure 3) or the igneous/metamorphic basement (Figure 8). Late Cretaceous and lower Paleocene orogeny caused uplifting and subsequent erosion of the Paleocene formations, resulting in an unconformity identified within the basin sediments. A vast and complex deltaic system was formed toward the end of the Paleocene erosion period. The combination of fluvial, fluvio-deltaic, and deltaic sedimentation resulted in the thick, sandy/shaley Eocene sequence that covers the entire basin (Figure 2). The most important oil reservoirs of the Maracaibo Basin accumulated during this period.

At the end of the upper Eocene, uplifting, folding, faulting, and intense erosion took place (Figures 3 and 4), giving rise to an unconformity between the Eocene and the overlying continental valley-fill sandstones of the Oligocene Icotea Formation. This was followed by a transgressive period of sedimentation, beginning with a regional basal conglomerate, the Santa Barbara Member of the La Rosa Formation, in which were later accumulated huge quantities of hydrocarbons. A major impermeable unit, the upper part of the La Rosa Formation, seals the Santa Barbara conglomerate. Later deposition of fluvial-deltaic and very sandy sediments forms the Lagunillas Formation, which produces oil in several areas of the basin.

The regional structure consists of a major northeast-trending wrench-fault system (the Lama-Icotea fault), along which many anticlines and associated antithetic normal faults were formed (Figure 6). The major geological feature is a north-northeast-trending anticlinal dome, with the Lama-Icotea fault extending along its crest. A series of en

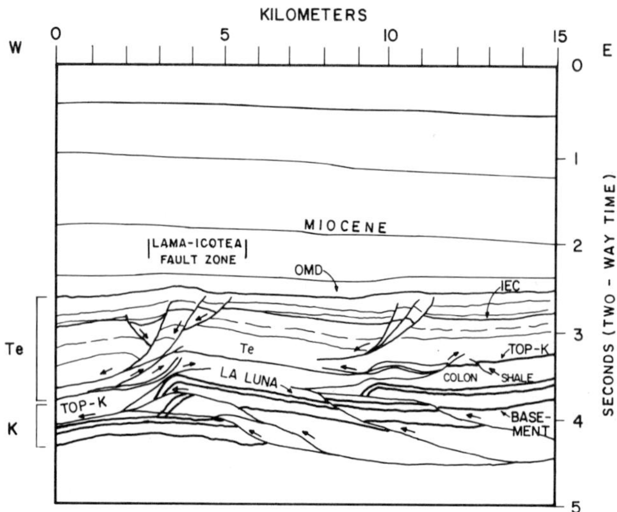

Figure 5. Interpretation of seismic line 73-48, central Lake Maracaibo. Note different structural styles in the Cretaceous (K) and Eocene (Te), the lack of structure above the Oligocene-Miocene unconformity (OMD), and the inter-Eocene unconformity (IEC).

echelon faulted anticlinal folds located on the flanks of the structure are closely related to the Lama-Icotea fault zone (Venezuelan Study Team, 1986).

In addition to the above-mentioned structural features, the latest seismic interpretations (Figure 5), based upon abundant, good quality seismic data together with geological and reservoir-related information, show that a wedge-shaped block arises along and parallel to the Lama-Icotea fault, bounded by two convergent reverse faults (Figure 6 and cross section B–B′ of Figure 7). Figure 6 shows the structure of the top of the Socuy. Current production from this structure is 150,000 BOPD. Seismic data (Stenson, 1969) also indicate the existence of two secondary fault systems, one of them stretching east-west and dipping toward the south, and the other paralleling the main northeast structural orientation. The faults and folds resulted from orogenic movements during the upper Eocene and lower Oligocene.

The major geological features pertaining to the Lama field are the Lama-Icotea wrench-fault system, the normal and reverse secondary fault systems, the folds associated with the Lama-Icotea system (Figures 6 and 7), and the synclinal structure parallel to it (Venezuela Study Team, 1986). Locally, the Lama-Icotea fault is a left-lateral transcurrent fault, showing dips ranging between 75° and 85°. This fault is crossed at the center of the Lama field by a large northwest-trending normal fault (N20W), dipping toward the southwest, with a throw of 2000 ft (610 m). An inversion of the fault throw takes place where these faults cross, and here, throws range between 500 and 1500 ft (150 and 460 m). The north-northeast-trending Lama-Icotea fault originated during the Triassic–Jurassic in the basement of the Maracaibo

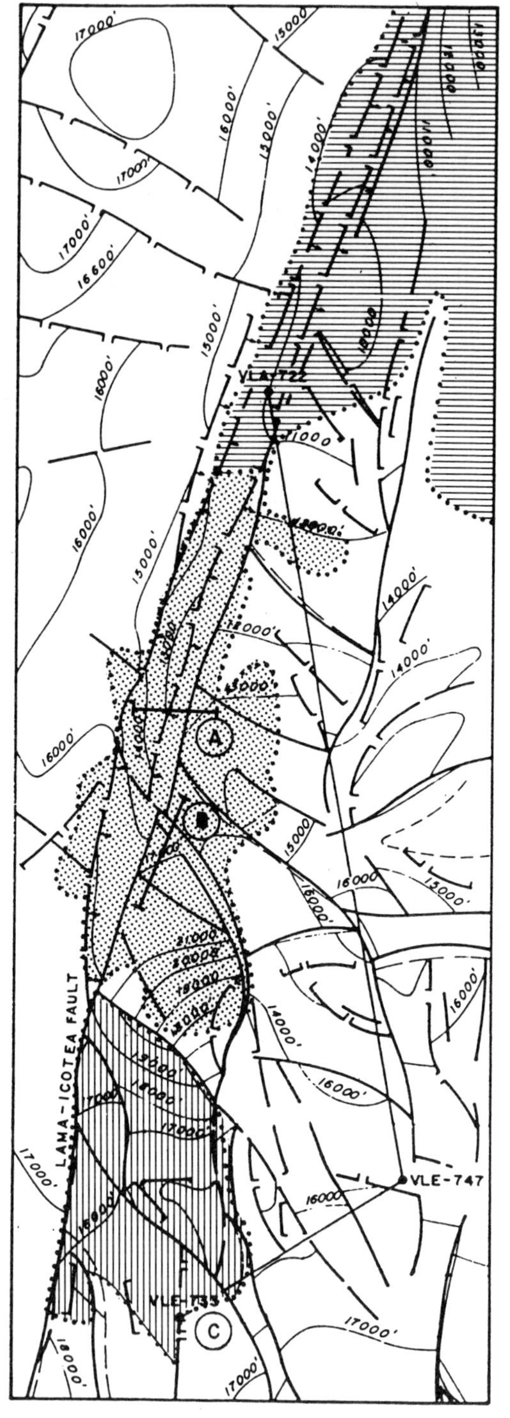

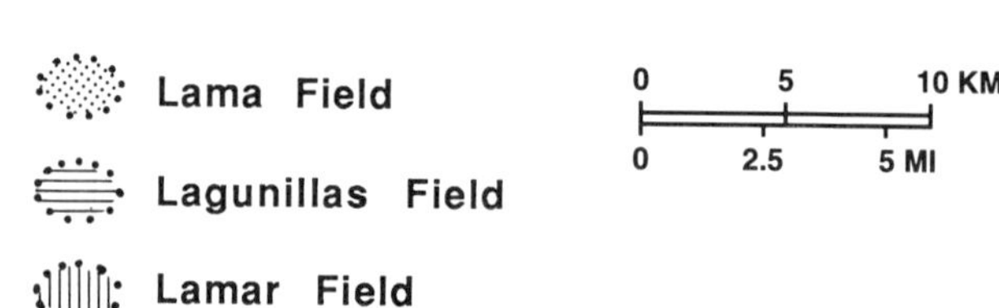

Figure 6. Structure map of the Lama field and adjacent fields along the Icotea fault complex. Contour map on top of the Socuy Member. Contour interval, 1000 ft. Values are subsea. Cross section at A shown on Figure 7 (B–B′). Cross section at B is shown by Figure 10 and at C by Figure 11.

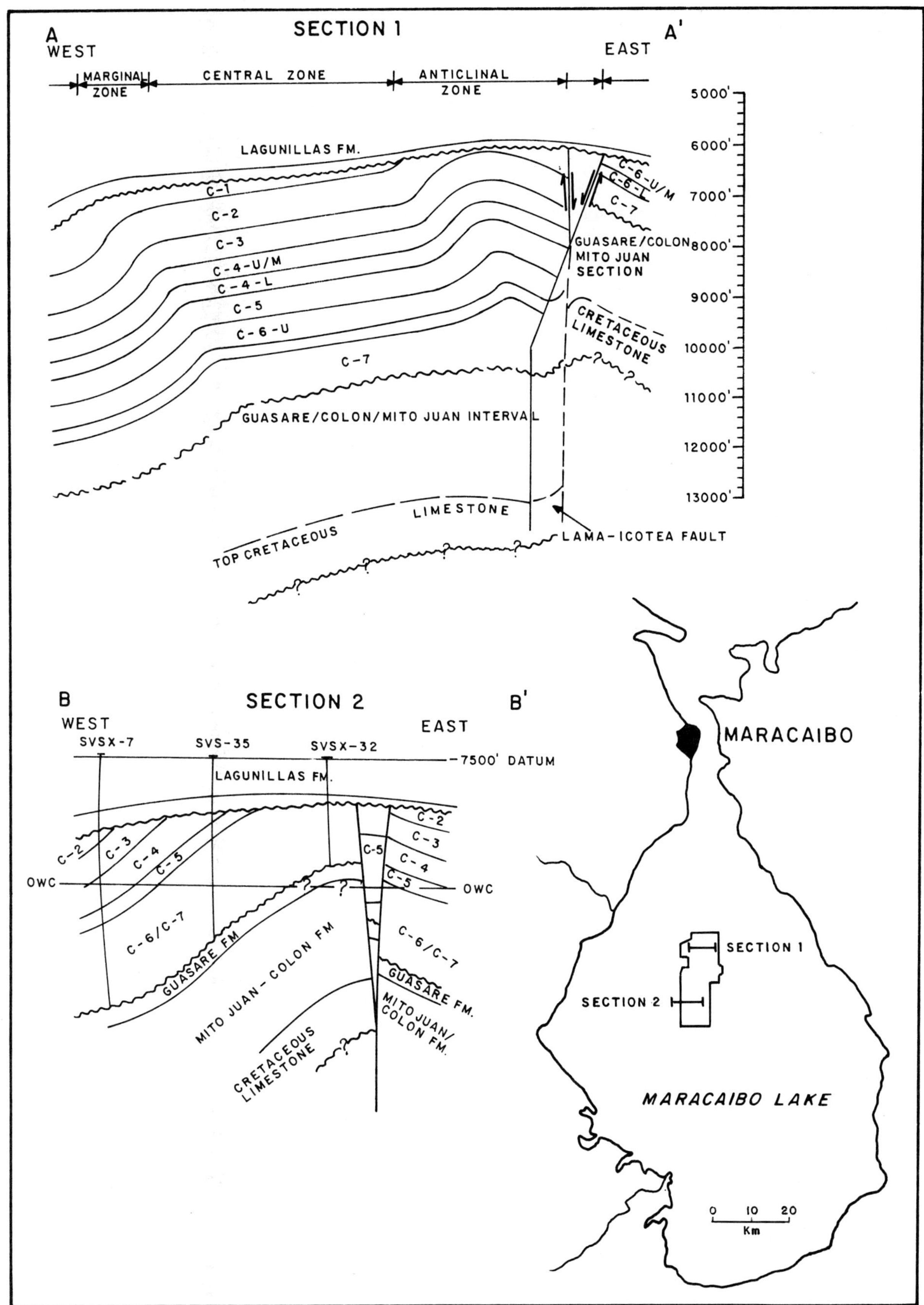

Figure 7. East-west structural sections across the Lama field, based on subsurface data. See Figure 6 for location.

Basin, and it extends upward through the entire geologic column to the upper Miocene, as evidenced by the latest seismic surveys carried out in the area. Folding originated during this faulting forms an angle of about 15° to the Lama-Icotea fault.

The trap of the field consists of an asymmetric anticline, deeply eroded and intensely deformed by the faults and folds, which particularly affect its crest. The main domal structure trends northeast-southwest. It is about 4 mi (7 km) wide and 43 mi (70 km) long. As with other anticlines along the trend, it is faulted on its crest. Upper Eocene orogenic movements, which account for normal faults parallel to the Lama-Icotea system, gave rise to a depression west of it (Figure 6 and cross section A–A′ of Figure 7). This is a downthrown block, and it contains approximately 3 BBO trapped by the Lama-Icotea and graben faults (Figures 6 and 7; also see maps of Figure 9). The flanks of the anticlinal structure dip toward the northwest and southeast at angles between 15° and 25°. The eastern flank, more highly uplifted, underwent greater erosion prior to the Oligocene, in some places removing sediments as deep as the lower Eocene.

It should be emphasized that the fault pattern varies considerably along the length of the Icotea fault zone so that a cross section in one area may be quite different from one in another area. The cross sections shown by Figures 4 and 5 are based on a conceptual interpretation of seismic line 73-48 and differ from the more conventional interpretations derived from subsurface information, as indicated by sections A–A′ and B–B′ on Figure 7. The structure of the Lama field shown on Figure 6 (top of the Socuy limestone member) is the current interpretation.

STRATIGRAPHY

The stratigraphic sequence drilled within the Lama field includes Recent through Jurassic–Triassic sediments and igneous/metamorphic basement (Figure 3) (Dikkers, 1964; Salvador and Hotz, 1963; Van Veen, 1972; Zambrano, 1971).

The conglomerates of the Cretaceous Rio Negro Formation, the limestones of the Cogollo Group (Apon, Lisure, and Maraca formations) and the La Luna Formation, and the shales of the Colón-Mito Juan Formation lie unconformably on top of the red beds belonging to the Triassic-Jurassic La Quinta Formation or upon the igneous/metamorphic basement (Figure 8). This entire sedimentary sequence consists mainly of 1200 to 1300 ft (366 to 396 m) of limestones covered by about 1700 ft (518 m) of marine shales. A thin (150 ft [46 m]) Paleocene shale sequence, interbedded with limestones and calcareous sandstones (Guasare Formation), overlies the Cretaceous sediments and unconformably underlies Eocene age rocks (Figure 8) (Venezuela Direccion de Geologia, 1956).

The preserved Eocene section (Figure 2) contains 80% coarse sand (Mirador Formation) at the southern part of the Lama area and 30% coarse sand (Misoa Formation) at its northern end (Asociacion Venezolana de Geologia, Mineria, y Petroleo, 1968). The lithology reflects both upward coarsening and the loss of the upper shalier section of the Eocene as a result of pre-Oligocene–Miocene erosion that took place southward.

The Eocene sediments in the Lama area consist of the informal B and C sandstones of the Misoa Formation, which underlie the partially eroded shales of the Pauji Formation (Figure 3) in some areas (Brondijk, 1967). This formation represents a deltaic distributary channel fill and marginal deltaic environment. The Misoa Formation consists of a transgressive sequence, interrupted by several regressive phases of deltaic buildup that contain less important reservoirs upward (Figure 8) (Edmonson, 1961). The thicknesses of the Eocene deltaic and marginal marine sediments range from less than 1000 ft (305 m) to over 7000 ft (2135 m) along the Lama-Icotea trend.

The eroded Eocene surface underlies the thick Miocene section, which consists of a thin layer (between 0 and 100 ft [0–31 m]) of basal sands, the Santa Barbara Member of the La Rosa Formation, and a thick sequence of marine shales (between 150 and 350 ft [46 and 107 m]) of the La Rosa Formation, which is locally interbedded with some sand bodies. The middle Miocene Lagunillas Formation, which consists of the Bachaquero, Laguna, Ojeda, and lower Lagunillas members, overlies the La Rosa Formation. The clays of the La Puerta Formation (upper Miocene), the Pliocene and Pleistocene beds, and the Recent surface deposits (Onia Formation) overlie the Lagunillas Formation.

TRAPS

Oil accumulation has been mainly affected by the Lama anticline together with faults and sandstone truncations (Bristow, 1974). Traps consist mainly of lateral fault-bound segments of the anticline. Some of the sandstones of the traps are truncated at the top, the Eocene unconformity. Oil accumulation has been locally related to thin sandstone layers and to small, lateral sandstone extensions. Faulting does not appear to be the most important element affecting oil accumulation, but it accounts for the individual production characteristics of some reservoirs.

The main oil reservoirs exist within the Eocene B and C sandstones. The second most important reservoir is in the Miocene Santa Barbara Member, and the third reservoir of importance is in Cretaceous rocks, which were affected by fractures and both lateral and vertical diagenesis. Other less important oil reservoirs in the Lama field are in the Guasare and Lagunillas formations, where the traps are partly

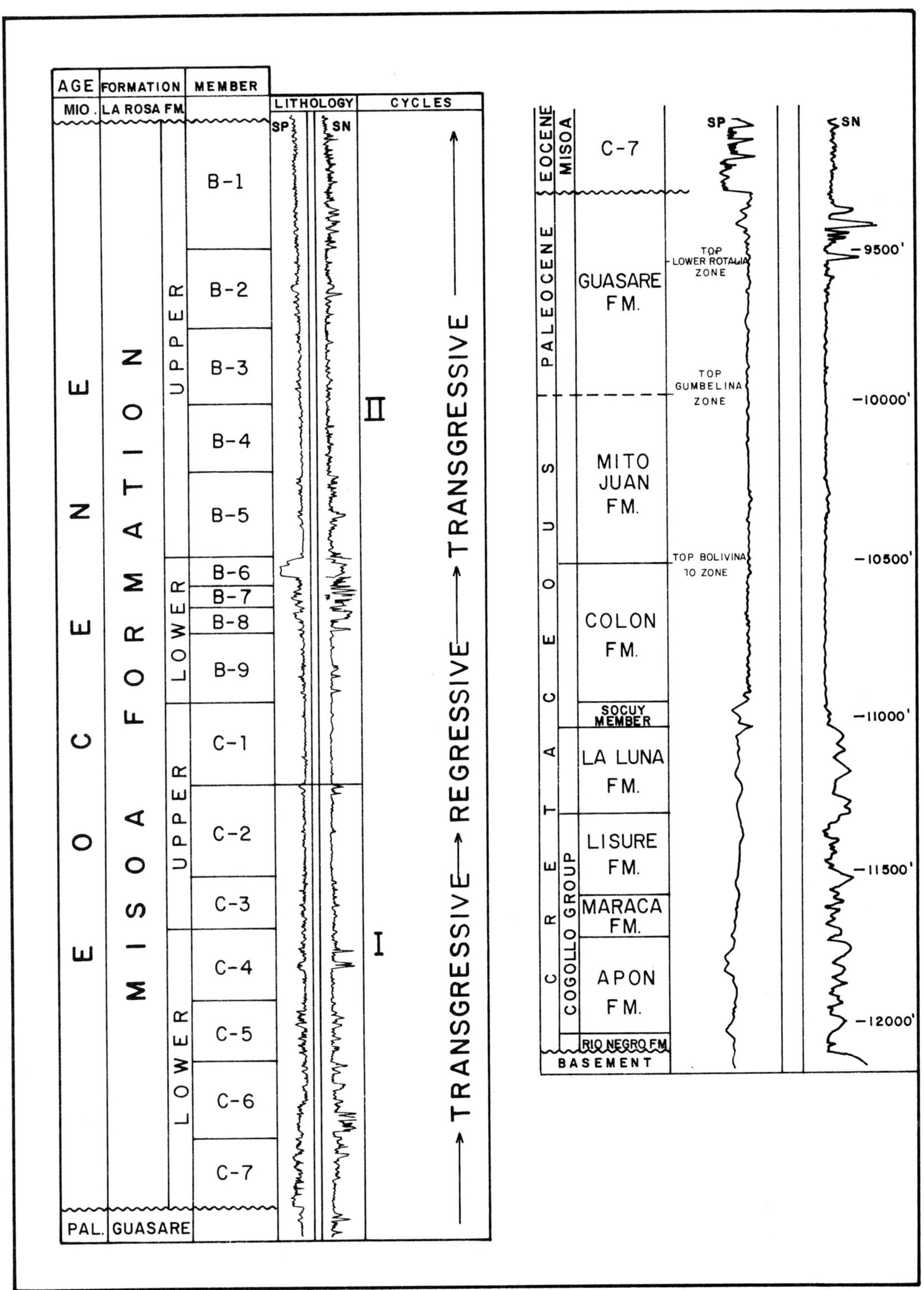

Figure 8. Type logs of the Eocene and Cretaceous intervals, Lama field.

stratigraphic and partly structural (Gonzalez de Juana et al., 1980).

Shales interbedded in the upper C sandstones (C-1, C-2, and C-3) and those belonging to the higher part of the La Rosa Formation, overlying the Santa Barbara Member, and the Eocene sandstones provide the seal over the Lama structure. Eocene shales are also effective seals. On the crest, however, where the upper C sands were truncated by the post-Eocene unconformity, the shales belonging to the La Rosa Formation are essential for sealing the Eocene oil reservoirs. Within the Cretaceous limestones, where fractures favored oil accumulation, the Colón shales are a regional natural seal acting as a barrier to vertical hydrocarbon migration.

Reservoirs

Lagunillas Formation (Middle Miocene)

The Lagunillas Formation is comprised of quartzose sandstones intercalated with sandy clays. Lateral changes to sandier lithofacies also occur. Occasional thin layers of shales, clays, and lignite occur. Oil reserves in the lenticular sandstones have been estimated at over 200 MBO.

Santa Barbara (Middle-Lower Miocene Santa Rosa Formation)

The lithology of the Santa Barbara Member consists of fine-grained, slightly compacted sands and a deeper sequence of coarse-grained, loose sand with high porosity and permeability. Porosity is intergranular and averages 25%. Average permeability is approximately 400 md. Maximum and minimum depths vary between 7000 and 11,000 ft (3135 and 3355 m). Total reserves have been estimated at about 750 MMBO in place and 224 MMBO recoverable.

The area of the reservoir is 27,702 ac (11,219 ha) (Figure 9A), with a total of 472 wells, spaced at 1970 ft (600 m). Of these, 121 wells are currently producing. Cumulative production is 142 MMBO, or 18.9% of original oil volume in place. Annual production (1988) is 5.6 MMBO, and the current decline rate is 7.5%. Annual water production is 1 million bbl, and the cumulative water production is 14.2 million bbl.

The production mechanisms are water and dissolved-gas drives, and the oil-water contacts are at depths ranging between 10,400 and 10,800 ft (3170 and 3300 m). Initial pressure in the reservoir was approximately 2500 psi at 7500 ft (17.0 MPa at 2290 m) to 4600 psi at 10,200 ft (32.0 MPa at 3110 m). Current pressure is about 3700 psi (25 MPa). Pressure gradient is 0.43 psi/ft (9.73 kPa/m) at formation temperatures between 190° and 220° F (88° and 104° C).

Misoa Formation (B and C Eocene Members)

The main oil reservoirs of the Lama field are in the sandstones of the Eocene Misoa Formation. This formation consists of two sandy or sandy/shaly sequences, the B (upper) and C (lower) sandstones (Figure 8). The former consists of a transgressive sequence divided into nine sandy or sandy/shaly bodies (B-1 to B-9). The lower part (B-6 to B-9) is mainly sandy, consisting of thick sandstones interbedded with thin shaly layers. This interval contains the most important oil reserves. Its shale content increases upward where the B-1 to B-5 intervals consist almost entirely of shales.

The C sandstone consists also of a transgressive sequence, divided into seven sandy or sandy/shaly bodies (C-1 to C-7). Its deeper C-6 and C-7 intervals consist mainly of very productive sandstones interbedded with thin shaly layers. The shale content increases upward, where the C-1 and C-2 intervals consist almost exclusively of shales. The B and C sandstone-shale members are correlatable across the field (Figure 10).

The depth to the top of the Misoa Formation reservoir varies between 7100 and 13,500 ft (2170 and 4120 m). It has an average intergranular porosity of 22%. Permeability is 300 md. Original oil in place was estimated at 7600 MMBO in 1987. Reservoir area covers about 63,675 ac (25,788 ha) (Figure 9B), with a total of 472 wells, initially spaced at 1970 ft (600 m). Of these, 397 wells are currently producing. Cumulative production is 2337 MMBO (25%). Annual production is 27.8 MMBO, and the current decline rate is 7%. The average annual water production is 2.2 million bbl, and the cumulative water production is 76 million bbl (Aristigueta, 1967; Maraven, 1988).

The production mechanisms are water, dissolved gas, and gas-cap drives. Oil-water contacts are at depths between 14,500 and 14,600 ft (4420 and 4450 m). Initial pressure in the Misoa B reservoir was 4200–4600 psi (29–31 MPa), and in the Misoa C, 3400–7100 psi (23–49 MPa). The current reservoir pressure is 4200 psi (29 MPa). Pressure gradient is 0.44 psi/ft (9.9 kPa/m) at reservoir temperatures between 190° and 240° F (88° and 116° C).

Guasare Formation (Paleocene)

The Guasare Formation consists of a thick sequence of fossiliferous, glauconitic limestones intercalated with thin layers of shales and sandstones (Van Andel, 1958). Porosities are of the intercrystalline, dolomitic, and secondary fracture-related type, with an average value of 10%. Permeability is 50 md.

The formation lies at depths between 10,000 and 17,500 ft (3050 and 5340 m). Pay zone area is 682 ac (276 ha) (Figure 9C), with reserves of original oil in place (1988) of about 10.2 MMBO and a cumulative production of 170 MBO. A total of 105 wells have been drilled to the reservoir, spaced at 1970 ft (600 m). Only four wells are currently producing. Recoverable reserves have been estimated at 1.7 MMBO (16.6%) through primary recovery. Annual production is 25 MBO, and cumulative water production is 10,000 bbl.

The production mechanisms are dissolved gas and gas-cap drives. Initial pressure in the reservoir was 1280 psi (8.82 MPa), and the current pressure has been measured at 1009 psi (6.95 MPa).

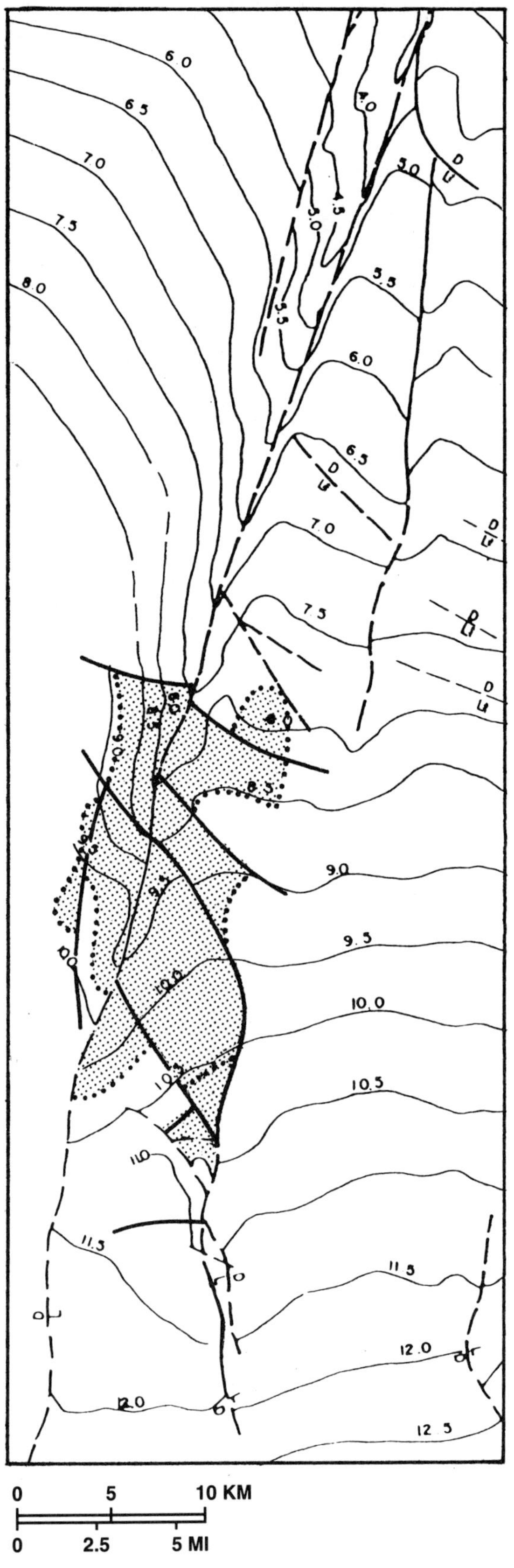

0 5 10 KM

0 2.5 5 MI

Figure 9A. Composite map of the Oligocene-Miocene reservoirs (Santa Barbara Member, La Rosa Formation). Structure map of base of supra-Eocene sediments (Eocene unconformity). Contour interval, 500 ft. Area of Oligocene-Miocene reservoirs stippled.

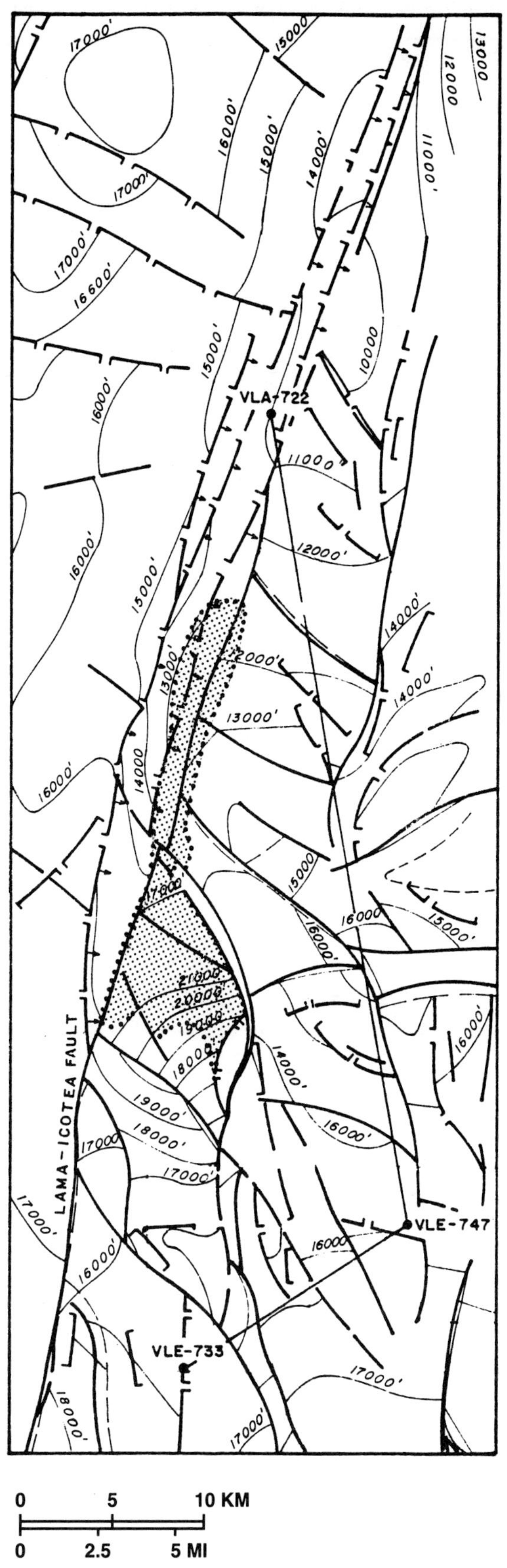

0 5 10 KM

0 2.5 5 MI

Figure 9B. Composite map of the Eocene reservoirs (Misoa Formation, B and C sand units) (stippled). Structure map of top of Socuy Member. Contour interval, 1000 ft.

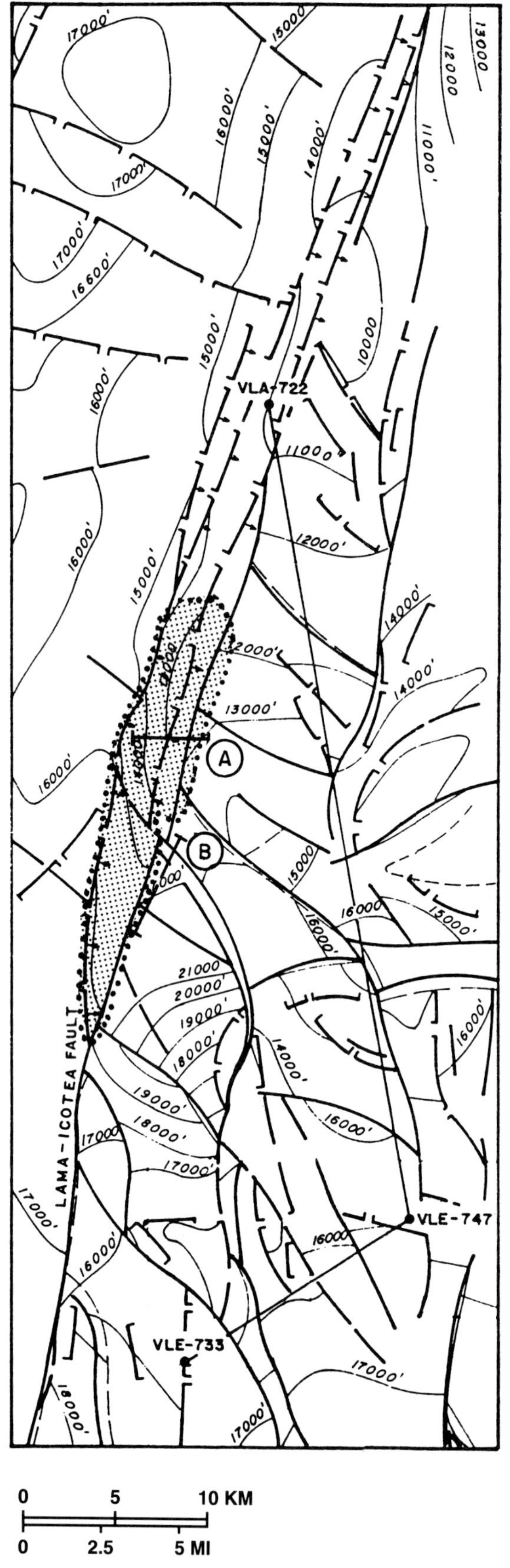

Figure 9C. Composite map of the Paleocene reservoirs (Guasare Formation) (stippled). Structure map of top of Socuy Member. Contour interval, 1000 ft.

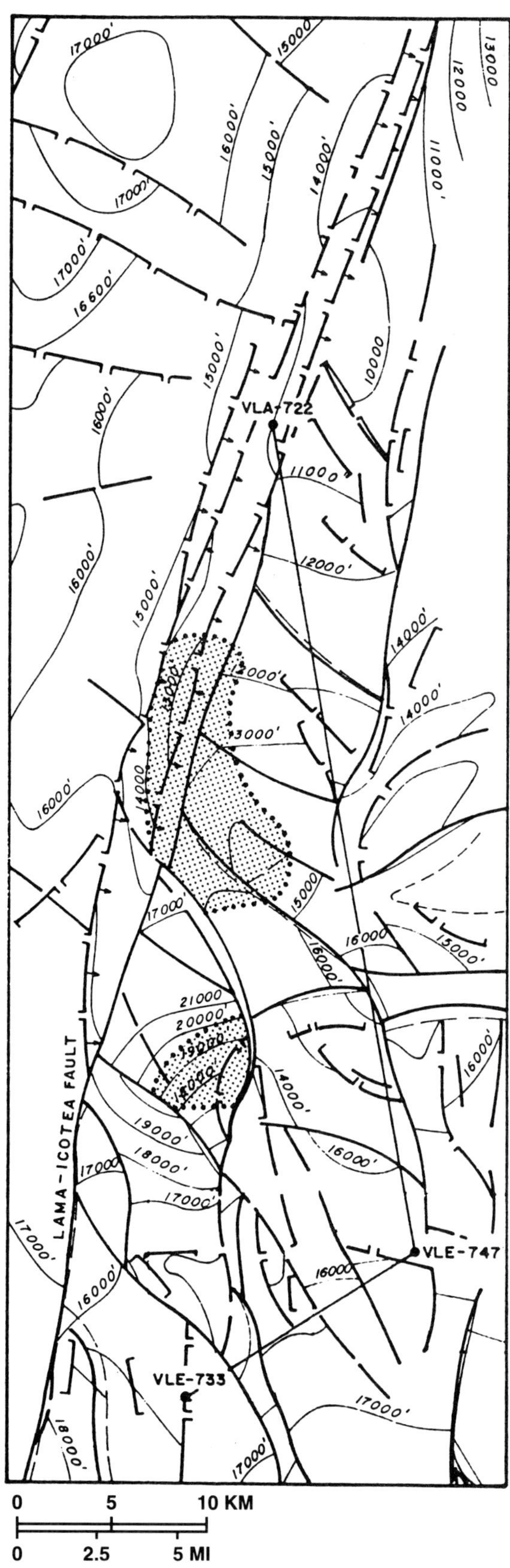

Figure 9D. Composite map of the Cretaceous reservoirs (La Luna-Apon formations) (stippled). Contour map of top of Socuy Member. Contour interval, 1000 ft.

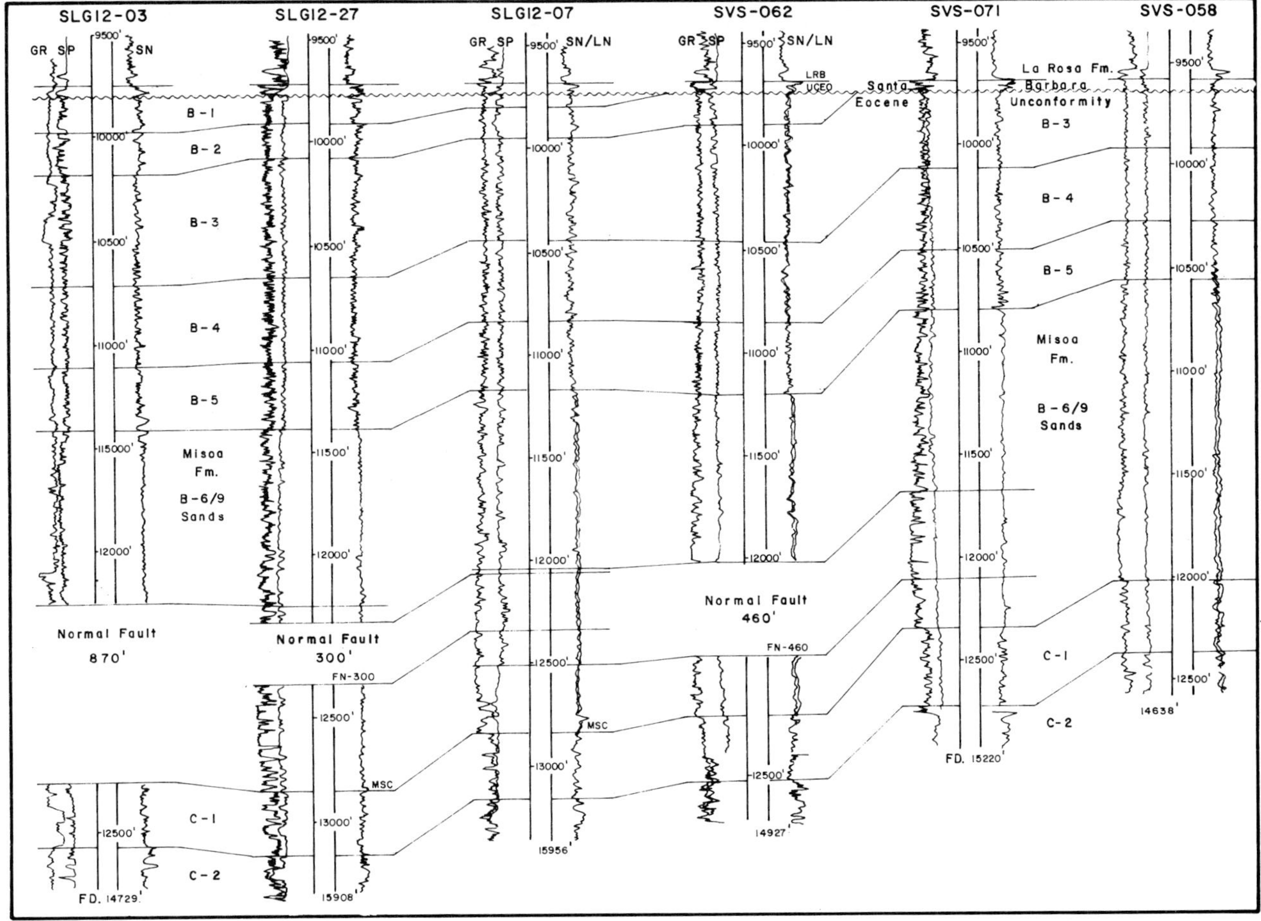

Figure 10. Eocene correlation section across the Lama field. See Figure 6 for location.

Cogollo Group (Lower Cretaceous)

The Cogollo Group contains the most important Cretaceous oil reservoirs in the Maracaibo Basin. It lies conformably under the La Luna Formation and is made up of three formations: Apon, Lisure, and Maraca (Figure 11) (Renz, 1977). The group consists of deposits of hard, fossiliferous, light to gray-colored limestones, interbedded with shales and with glauconitic intervals (Bartok and Reijers, 1977). Cores taken in different areas of Lake Maracaibo indicate the existence of intensely fractured intervals within the formations comprising the group, which suggest that production is due to fracture porosity. Cores have allowed characterization of the Cogollo Group as pelecypod biostromes and as grainstone and packstone bars.

The depth to the top of the Cretaceous reservoirs in the Lama field ranges between 12,400 and 20,000 ft (3780 and 6100 m). Porosities are of the fracture, vuggy, dolomitic, and biomoldic type, with fractures being the most important for production. Fracture porosity has been estimated at 3%, and permeability has been estimated at 0.1 md (León, 1975). The developed reservoir area is 10,926 ac (4425 ha) (Figure 9D), in which 30 wells have been drilled, initially spaced at 3940 ft (1200 m). Of these, 18 wells are currently producing.

Cumulative production, estimated in 1988, is 86 MMBO, 11.3% of original oil in place of 758 MMBO. Recoverable proven reserves are 155 MMBO (24.5%). Annual production is 1.3 MMBO, and annual water production is 645,000 bbl. Cumulative water production is 12.5 million bbl. The decline rate is 10.5% (Aristigueta, 1967).

The production mechanisms are water, dissolved gas, and gas-cap drives. Owing to the irregularity of the fracture patterns in the reservoir, oil-water contacts exist at various levels within the Cogollo Group. Initial pressure within the reservoir was estimated between 10,000 and 12,000 psi (69 and 83 MPa). Subsequent measurements suggest that current pressure ranges between 10,000 and 4000 psi (69 and 28 MPa), with a hydrostatic pressure gradient of 0.7 psi/ft (15.8 kPa/m) at formation temperatures ranging between 300° and 320° F (149° and 160° C). Production enhancement techniques include the use of solvents, acidizing, and hydraulic fracturing (Hernández, 1979).

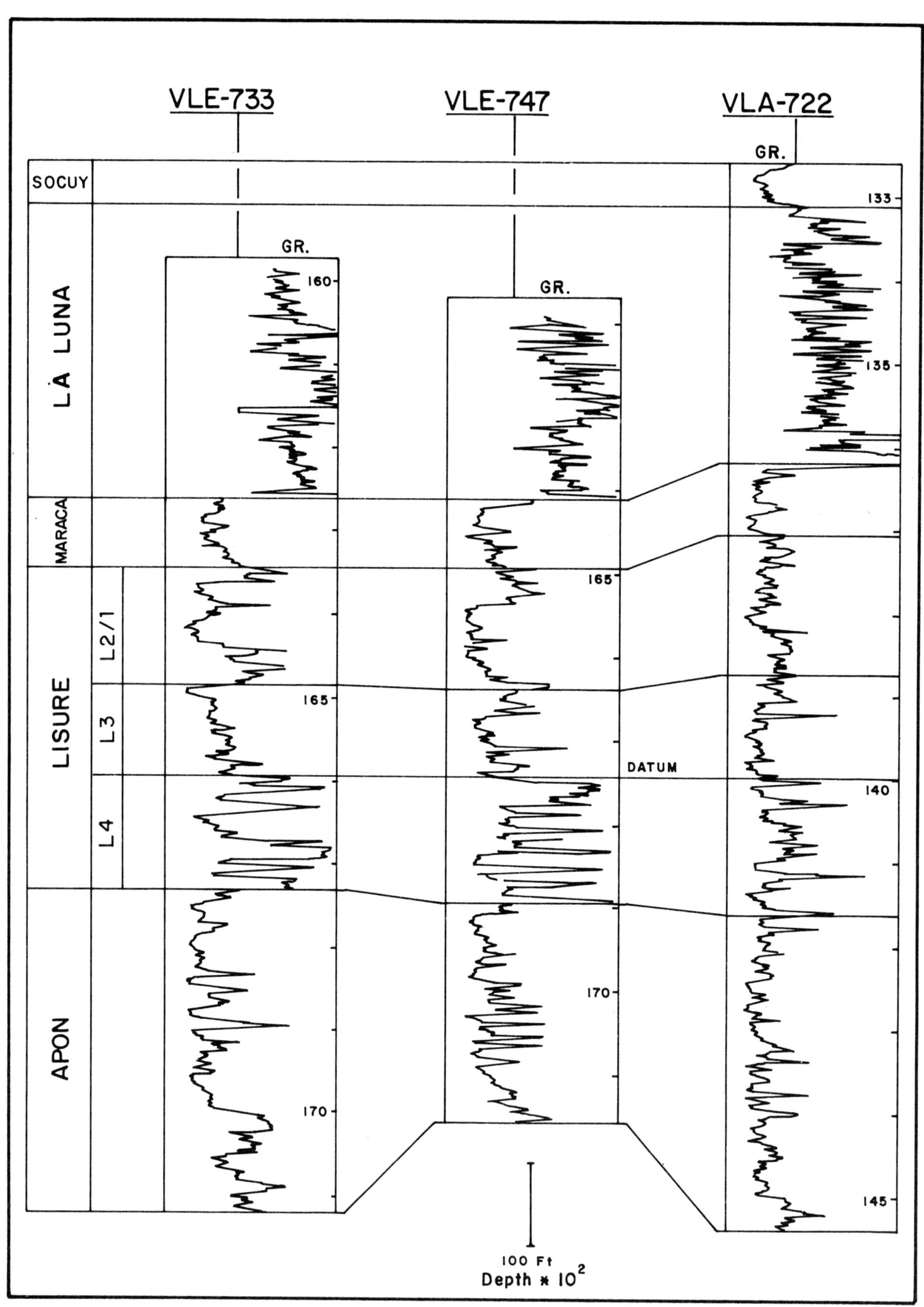

Figure 11. Correlation section of the Cretaceous in the area of the Lama field. See Figure 6 for location.

Sources

Oil migration from the source rocks to the traps in different areas, stratigraphic levels, and rock types was a very complex process (Young, 1956). It is assumed that oil contained in the Miocene reservoirs migrated into these sands during the late Miocene. The oil is believed to be of Eocene origin, and thus was not generated in the reservoir rocks. The Eocene oil migrated across the supra-Eocene unconformity, saturating the Santa Barbara porous beds, where it was sealed above by the thick, impermeable shales of the La Rosa Formation and, laterally, by minor faults.

These assumptions are based upon the facts that gas-oil ratios measured from Santa Barbara reservoirs and the oil gravities are always higher than those found within the underlying Eocene sand reservoirs, thus indicating gravity segregation of oil during migration. During a subsequent migration period, oil generated in Eocene rocks passed directly to the adjacent sandstones and migrated laterally toward the upper strata of the basin (Zambrano, 1971).

Lama field oils range between 28° and 42° API. Initial gas-oil ratios varied between 800 and 1260 ft^3/bbl (142 and 224 m^3/m^3). Sulfur content is 1.17%. Viscosity is 8.72 cp at 100° F (38° C). Pour point is -15° F (-26° C), and gas-oil distillate is 16.76%.

The source rocks belong to two different sequences: the Cretaceous La Luna Formation and the lower and middle Eocene shales, with a time gap of 35 million years between them (Breneman, 1960). The degree of metamorphism reached by the La Luna Formation indicates that the main hydrocarbon generation phase took place during the middle Eocene (Boesi, 1978). The oil generation phases corresponding to the Eocene shales occurred, however, during the lower and middle Miocene. This suggests that hydrocarbon migration within the basin did not begin prior to the Eocene (Hedberg, 1931).

The La Luna Formation characteristics identify it as the main source rock in the Maracaibo Basin. Its TOC is above 9%. Hydrocarbon yield is above 50 L/m^3 of rock (Maraven, S.A.). Organic matter contained in the Cretaceous sediments is of kerogen type II, as opposed to that in the Eocene and supra-Eocene formations, which is of humic type III terrestrial sediments.

Maturation level (DOM) for the Eocene rocks is 60, which suggests that they are mainly gas source rocks. Maturation level for the La Luna Formation is 56. It has been estimated that for an area of 11,580 mi^2 (30,000 km^2) in the Maracaibo Basin, an average thickness of 50 m and a yield of 50 L/m^2 of hydrocarbons (420 bbl per km^3), the La Luna Formation could have generated a theoretical total of 480 BBO, against an estimated 350 BBO in place discovered as of 1989. Its average TOC is 3.8%, with a maximum value of 9.6%; kerogen is type II; vitrinite reflectance varies from 0.4 to 4.8% from west to east across the basin.

In the Lama area, the minimum depth to the top of source is at 12,800 ft (3904 m); it has a maximum thickness of 24 ft (78 m) and a potential yield of 15.6 bbl/ton (2730 L/MT) of rock (Miller, 1958).

EXPLORATION CONCEPTS

Regional play associated with the field consists of an anticlinal structure faulted and eroded on its crest, with oil accumulations on both sides. Traps are mainly structural. The geological methods employed to discover the field were the standard ones used during any large oil exploration endeavor; that is, taking known geologic conditions in one area and extending them into an undrilled area. For example, as described in *Discovery Method*, the existence of the Icotea fault in the area north of Lama was known from wells, seismic, and magnetic data prior to discovery. The fault interpretation was extended by seismic and magnetic surveys southward into the Lama area to evaluate the various concession blocks that were offered in 1956.

Knowing what we know today, we would probably explore this basin in the same way, drilling along trends while gradually extending the field. It is also important to point out that many pools were discovered by accident in downthrown blocks while drilling for shallower stratigraphic traps. Examples are the B-1, B-2, B-3, and B-6/9 reservoirs, which were discovered while drilling for the Santa Barbara Member on the southern parts of blocks 1 and 12. We should have considered the possibility of finding oil in structural lows during the early exploration phases.

We did not place much emphasis on the Cretaceous play during early seismic surveys because the seismic profiles did not adequately define the continuity of this reflector. Modern seismic techniques have helped define this structure, leading to the discovery of new traps down-flank on the nearby Mara structure. What makes this field unique is its narrowness, together with the existence of a very thick oil column, measuring over 1000 ft (305 m).

ACKNOWLEDGMENTS

The author wishes to thank Maraven, S.A. and Petróleos de Venezuela, S.A. for permission to publish this paper and, in addition, to thank Donald Goddard for his suggestions in writing the paper and the reviewers for their comments and suggestions.

REFERENCES CITED

Aristigueta, G., 1967, Evaluación de las posibles reservas de petróleo en las reservas nacionales del Grupo 74: Editores, MEM, Informe privado, Caracas, p. 17.

Asociación Venezolana de Geologia, Mineria y Petróleo, 1968, La estratigrafia del Eoceno en la Cuenca de Maracaibo: Asoc. Venez. Geolo, Min. y Petrol., Publ. esp. 1, p. 99.

Bartok P., and T. Reijers, 1977, Sedimentoliogia, diagénesis y potencial petrolifero del Grupo Cogollo, Cretáceo Inferior, Cuenca de Maracaibo: V Congreso Geol. Venez. Caracas, Noviembre 1977, Memoria A. Espejo et al., Editores, MEM-Soc. Venez. Geol., IV, p. 529-557.

Boesi T., 1978, Resumen sobre el origen de las acumulaciones de hidrocarburos en la Formación Misoa, área del Lago de Maracaibo: Informe privado, Maraven, S.A., Caracas.

Borger, H. D., and E .F. Lenert, 1959, The geology and development of the Bolivar coastal field at Maracaibo, Venezuela: Asoc. Venez. Geolo. Min. y Petrol., Bol Inform., v. 2(9), p. 236-256.

Breneman, M. D., 1960, Estudio quimico de los petróleos crudos de la Cuenca de Maracaibo: III Cong. Geol. Venez., Caracas, Nov. 1959, Memoria, Bol., Geol. Caracas, Publ. Esp. 3, p. 1025-1069.

Bristow, J. D., 1974, Producción cretácica de la Creole y evaluación de registros en el Lago de Maracaibo: Asoc. Venez. Geol. Min y Petrol., Bol. Inform., v. 17 (10, 11, 12), p. 167-178.

Brondijk, J. F., 1967, The Misoa and Trujillo Formations: Asoc. Venez. Geol Min. y Petrol., Bol. Inform., v. 10 (1), p. 3-19.

Dikkers, A. M., 1964, Development of the La Paz field, Venezuela: Jour. Inst. Petrol., v. 50 (492), p 330-333.

Edmonson, C. L., 1961, Core analysis, North Lama area: C-6-X and C-6-X.32 reservoirs, well UD-60: Private report for Creole Petroleum Co., Caracas.

González de Juana, C., J. Iturralde de Arozena, and X. Picard, 1980, Geologia de Venezuela y de sus cuencas petroliferas: Ediciones Foninves, Caracas, 1031P, v. I y II.

Hedberg, H. D., 1931, Cretaceous limestones as petroleum source rocks in Northwestern Venezuela: AAPG Bulletin, v. 15(3), p. 229-244.

Hernández, L., 1979, Evaluación de las calizas cretácicas del Campo Lama, Lago de Maracaibo: Informe inédito, Corpoven, S.A., Caracas, p. 22.

Lehr, S., 1973, Lama field drilling efficiency study: Confidential report for Texaco Oil Co., Maracaibo, p. 19.

León, H., 1975, Intervalos productores del cretácico, campo Urdaneta Este, Lago de Maracaibo, (Preprint): I Jor. Venez. Geol. Min. y Petrol. Maracaibo, Septiembre, 1978, p. 31.

Miller, J. B., 1958, Habitat of oil in the Maracaibo Basin, Venezuela: Editor, AAPG, p. 1384.

Padrón, J. L., 1961, Geologia del campo Lama, Lago de Maracaibo, Informe iné dito para la Corporación Venezolana del Petróleo, Caracas, p. 72.

Renz, O., 1977, The lithologic units of the cretaceous in western Venezuela: V Cong. Geolg. Venez., Caracas, Noviembre, 1977, Memoria A. Espejo et al., Editores, MEM-Soc. Venez. Geol., v. I, p. 45-58.

Salvador, A., and E. F. Hotz, 1963, Petroleum occurrence in the Cretaceous of Venezuela: VI World Petroleum Congress, Frankfurt/Main, June, 1963, Proceedings, v. I, p. 115-140: Resumen en Asoc. Venez. Geol. Min. y Petrol., Bol. Inform., v. 6 (9), p. 278.

Staff of Caribbean Petroleum Co., 1948, Oil fields of Royal-Dutch Shell Group in Western Venezuela: AAPG Bulletin, v. 32 (4), p. 517-628.

Stenson, H., 1969, Exchange of seismic data Shell-Texaco, Block 1, Lama area, Lake Maracaibo: Confidential report by Shell Texaco, Caracas, p. 58.

Sutton, F. A., 1946, Geology of Maracaibo Basin, Venezuela: AAPG Bulletin, v. 30 (10), p. 1621-1741.

Van Andel, T. H., 1958, Origin and classification of Cretaceous, Paleocene and Eocene sandstones of western Venezuela: AAPG Bulletin, v. 42(4), p. 734-763.

Van Veen, F. R., 1972, Ambientes sedimentarios de las formaciones Mirador y Misoa del Eoceno Inferior y Medio en la Cuenca del Lago de Maracaibo: IV Cong. Venez., Noviembre, 1969, Caracas, Memoriam, Bol. Geol. Caracas, Publ. Esp. 5 v. II, p. 1074-1104.

Venezuela Direccion de Geologia, 1956, Léxico estratigráfico de Venezuela: Bol. Geol., Caracas, Publ. Esp. 1, p. 728.

Venezuelan Study Team, 1986, Lama South Block, Lama field, east flank, south area study: Progress report No. 1 for Maraven S.A. by Shell, The Hague, Holland.

Venezuelan Sun Oil Company, 1968, Reservoir performance: review of block 12, Lama field: Maraven private report, p. 31.

White, C., 1978, Cuenca de Maracaibo, habitat de hidrocarburos: Informe de Pregrso inédito, Maraven, S.A., Caracas, p. 14.

Young, G. A., 1956, Geologia de las cuencas sedimentarias de Venezuela y de sus campos petroliferos: Bol. Geol., Caracas Publ. Esp. 2, p. 140.

Zanbrano, E., 1971, Sintesis petrográfica y petrolera del Occidente de Venezuela: IV Cong. Geol. Venez., Caracas, Noviembre, 1969, Memoria, Bol. Geol. Caracas, Publ. Esp. 5, v. I, p. 483-545.

Zamora, L., 1977, Uso de perfiles en la identificacion de ambientes sedimentarios del Eoceno del Lago de Maracaibo: V Cong. Geol. Venez., Noviembre, 1977, Memoria, A. Espejo et al. Editores, MEM-Soc. Venez. Geolo., v. IV, p. 1359-1376.

Appendix 1. Field Description

Field name .... *Lama field*
Ultimate recoverable reserves .... *3464 million bbl oil*

Field location:

- **Country** .... *Venezuela*
- **State** .... *Zulia*
- **Basin/Province** .... *Maracaibo Basin*

Field discovery:

- **Year first pay discovered** .... *Oligocene-Miocene sandstones 1957*
- **Year second pay discovered** .... *Eocene sandstones 1957*
- **Year third pay discovered** .... *Cretaceous limestones 1957*
- **Year fourth pay discovered** .... *Paleocene limestones 1959*

Discovery well name and general location:

- **First pay** .... *SVS-1X: in lake 51 mi (82 km) south of Maracaibo (Oligocene-Miocene)*
- **Second pay** .... *Lama-1: in lake 44 mi (71 km) south of Maracaibo (Eocene)*
- **Third pay** .... *SVS-1X: 51 mi (82 km) south of Maracaibo (Cretaceous)*
- **Fourth pay** .... *SVS-23; 50 mi (80 km) south of Maracaibo (Paleocene)*

Discovery well operator .... *Sun Oil Co. (SVS wells)*

- **Second pay** .... *Superior Oil Co. (Lama wells)*

IP

- **First pay** .... *729 BOPD*
- **Second pay** .... *2675 BOPD*
- **Third pay** .... *548 BOPD*
- **Fourth pay** .... *1345 BOPD*

All other zones with shows of oil and gas in the field:

Age	Formation	Type of Show
Miocene	*Lagunillas*	*Electric log*

Geologic concept leading to discovery and method or methods used to delineate prospect

The geology along the Icotea fault system north of the Lama field was known to some extent. With additional seismic lines and extrapolating the known geology along the fault to the south, the companies had an idea that the area would be prospective.

Structure:

Province/basin type .... *Bally 222; Klemme III Bc*

Tectonic history

Lama field is located within the unstable zone bounded by the Oca, Perijá, and Boconó transcurrent fault systems in northwestern Venezuela. These systems are related to right-lateral strike-slip movement along the Caribbean and South American plates boundary. Major periods of tectonic activity occurred in the Tertiary including during the late Miocene Andean uplift.

Regional structure

A major northeast-trending wrench-fault system (Icotea fault) along which a number of anticlinal structures and associated antithetic normal faults have formed.

Local structure

Faulted monoclinal flanks on both sides of the Icotea fault with 10° to 15° dips and a southwest-trending anticline parallel to the Icotea fault to the southeast.

Trap:

Trap type(s)

Traps are mainly structural, being anticlinal traps laterally sealed by faults. Also the producing horizons are truncated by the overlying Eocene unconformity. Diagenesis controls accumulations in the Cretaceous.

Basin stratigraphy (major stratigraphic intervals from surface to deepest penetration in field):

Chronostratigraphy	Formation or Member	Depth to Top in ft (m)
Miocene	*Santa Barbara Member*	*7000-11,000 (2135-3355)*
Eocene	*Misoa Formation*	*7100-13,500 (2166-4118)*
Paleocene	*Guasare Formation*	*10,000-17,500 (3050-5338)*
Cretaceous	*Socuy Member/La Luna Formation/ Cogollo Group*	*12,400-20,000 (3782-6100)*

Reservoir characteristics:

Number of reservoirs *4*

Formations

Socuy Member/La Luna Formation/Cogollo Group (SLC); Guasare (G) Formation; Misoa (M) Formation; Santa Barbara (SB) Member

Ages .... *SLC, Cretaceous; G, Paleocene; M, Eocene; SB, Miocene*

Depths to tops of reservoirs .... *See Basin Stratigraphy, above*

Gross thickness (top to bottom of producing interval) .... *Cretaceous, 1200 ft (366 m); Paleocene, 200 ft (61 m); Eocene, 2800 ft (854 m); Miocene, 60 ft (15 m)*

Net thickness—total thickness of producing zones

Average .... *Cretaceous, 200 ft (61 m); Paleocene, 70 ft (21 m); Eocene, 3150 ft (961 m); Miocene, 33 ft (10 m)*

Maximum .... *Cretaceous, 320 ft (98 m); Paleocene, 190 ft (58 m); Eocene, 3565 ft (1087 m); Miocene, 120 ft (37 m)*

Lithology

Cretaceous: limestone, grainstone/bars, pelecypod biostromes
Paleocene: glauconitic limestones and shales
Eocene: interbedded fine- to medium-grain sandstones and shales
Miocene: conglomeratic sandstones

Porosity type

Cretaceous: fractures, vugs, dolomites, biomoldic (secondary)
Paleocene dolomites: fractures, intercrystalline
Eocene: intercrystalline
Miocene: intercrystalline

Average porosity .... *Cretaceous, 3%; Paleocene, 10%; Eocene, 22%; Miocene, 25%*

Average permeability ... *Cretaceous, 0.1 md; Paleocene, 50 md; Eocene, 300 md; Miocene, 450 md*

Seals:

Upper

Formation, fault, or other feature ..... *Cretaceous: Colón Formation; Miocene: La Rosa Formation*

Lithology .... *Colón Shale and La Rosa Shale*

Lateral

Formation, fault, or other feature .... *Cretaceous, diagenesis and faults; Eocene, faults and porosity reduction; Miocene, faults*

Source:

Formation and age .... *La Luna, Upper Cretaceous*

Lithology .... *Carbonaceous limestone, calcareous shales, and cherts*

Average total organic carbon (TOC) 3.8%
Maximum TOC 9.6%
Kerogen type (I, II, or III) II
Vitrinite reflectance (maturation) .... $R_o = 0.9\%$
Time of hydrocarbon expulsion .... Early expulsion, Eocene; late expulsion, Miocene-Pliocene
Present depth to top of source .... 12,800 ft (3904 m)
Thickness .... 256 ft (78 m)
Potential yield .... 10.0 kg hydrocarbons/MT rock

Appendix 2. Production Data

Note: Production data for reservoirs are given separately for each one in this appendix. Combined data, all reservoirs, is as follows (1991):

Original oil in place .... *9680 million bbl*
Ultimate recoverable .... *3464 million bbl*
Average recovery factor *35.7%*
Cumulative production .... *2612 million bbl*
Remaining proven reserves .... *852 million bbl*
Productive area .... *104,480 ac*
Productive volume .... *14,242,600 ac-ft*
Probable and possible .... *554 million bbl*
Ultimate potential recovery .... *4.0 billion bbl*
Average bbl/ac-ft in-place *674*

Reservoir name .... *Miocene Santa Barbara*

Reservoir size:

- **Proved acres** .... *27,702 ac (11,219 ha)*
- **Number of wells all years** *472*
- **Current number of wells** .... *121 (producing)*
- **Well spacing** .... *1970 ft (600 m)*
- **Ultimate recoverable** .... *224 million bbl*
- **Cumulative production** .... *142 million bbl*
- **Annual production** .... *5.6 million bbl*
- **Present decline rate** *7.5%*
  - **Initial decline rate** *1.5%*
  - **Overall decline rate** .... *NA*
- **Annual water production** .... *1.0×10^6 bbl*
- **In place, total reserves** .... *750 million bbl*
- **In place, per acre foot** .... *820 (605 to 930 range) bbl*
- **Primary recovery** .... *190 million bbl*
- **Secondary recovery** .... *34 million bbl*
- **Enhanced recovery** .... *NA*
- **Cumulative water production** .... *14.2×10^6 bbl*

Drilling and casing practices:

- **Amount of surface casing set** .... *1200 ft (370 m)*
- **Casing program** .... *10¾-in. and 9⅝-in.; tubing, 2⅞-in.*
- **Drilling mud** .... *Lignosulfonate, oil emulsion*
- **High pressure zones** .... *None*

Completion practices:

Interval(s) perforated .... *Santa Barbara Member*
Well treatment .... *None*

Formation evaluation:

Logging suites .... *Electrical logs, radioactive logs, production logs*
Testing practices .... *Standard production test through pre-perforated liner*
Mud logging techniques .... *Standard cutting analyses, gas chromatograph*

Oil characteristics:

Type .... *NA*
API gravity .... *29–32°*
Base .... *Paraffinic*
Initial GOR .... *800 ft^3/bbl (142 m^3/m^3)*
Sulfur, wt% *1.17**
Viscosity, SUS .... *8.72 cst at 100° F (37.8° C)**
Pour point .... *–15° F (–26° C)**
Gas-oil distillate *16.76%**

**Note: All reservoir crudes are mixed; values shown are combined crudes. (These items are therefore the same for this reservoir and all to follow.)*

Reservoir characteristics:

Average elevation .... *In lake (S.L.)*
Initial pressure .... *2500 psi at 7500 ft (17.0 MPa at 2290 m) to 4600 psi at 10,200 ft (32.0 MPa at 3110 m)*
Present pressure .... *3700 psi (26 MPa) (avg.)*
Pressure gradient .... *0.43 psi/ft (9.7 kPa/m)*
Temperature .... *190–220° F (88–104° C)*
Geothermal gradient .... *1.1° F/ft (2.0° C/m)*
Drive .... *Gas solution and water drive*
Oil column thickness .... *1500 to 1700 ft (460 to 520 m)*
Oil-water contact .... *–10,400 to –10,800 ft (–3170 to –3290 m)*
Connate water .... *10 to 35%*
Water salinity, TDS .... *5000 to 6000 ppm*
Resistivity of water .... *0.30 to 0.35 ohm-m^2 m*
Bulk volume water (%) .... *25 to 30%*

Transportation method and market for oil and gas:

Pipeline to Puerto Miranda; tankers to foreign markets or pipeline from Puerto Miranda to Cardon refinery

Reservoir name .... *Eocene Misoa*

Reservoir size:

Proved acres .... *63,675 ac (25,788 ha) (total B and C sequences)*
Number of wells all years *472*
Current number of wells .... *397 (producing)*
Well spacing .... *1970 ft (600 m)*
Ultimate recoverable, proven .... *3048 million bbl*
Ultimate recoverable, possible .... *3250 million bbl*
Cumulative production .... *2337 million bbl*
Annual production .... *27.8 million bbl*
Present decline rate *7%*
Initial decline rate *1.6%*
Overall decline rate .... *NA*
Annual water production .... *2.2×10^6 bbl*

LAMA

In place, total reserves .... *7600 million bbl*
In place, per acre foot .... *Misoa B, 725 to 1180 bbl; Misoa C, 540 to 1160 bbl*
Primary recovery .... *3048 million bbl*
Secondary recovery .... *8 million bbl*
Enhanced recovery .... *NA*
Probable and possible reserves .... *195 million bbl*
Cumulative water production .... *76×10^6 bbl*

Drilling and casing practices:

Amount of surface casing set .... *2200 ft (670 m)*
Casing program .... *Casing, 10¾-in., 9⅝-in., 7-in.; tubing, 3½-in. to 2⅞-in.*
Drilling mud .... *Lignosulfonate, Benex, oil emulsion*
High pressure zones .... *None*

Completion practices

Intervals perforated .... *Misoa formation B and C sand groups*
Well treatment .... *Acidization (wash and frac)*

Formation evaluation:

Logging suites .... *Electric logs, radioactive logs, production logs, acoustic logs*
Testing practices .... *Standard production tests through pre-perforated liner*
Mud logging techniques .... *Standard cutting analyses*

Oil characteristics:

Type .... *NA*
API gravity .... *29-34°*
Base .... *Paraffinic*
Initial GOR .... *800-1100 ft³/bbl (142-194 m³/m³)*
Sulfur, wt% *1.17*

Reservoir characteristics:

Average elevation .... *In lake (S.L.)*
Initial pressure .... *Misoa B, 4200 psi at 9500 ft (29.0 MPa) to 4600 psi at 10,200 ft (32.0 MPa); Misoa C, 3400 psi at 7300 ft (23.0 MPa) to 7100 psi at 15,900 ft (49.0 MPa)*
Present pressure .... *4200 psi (29 MPa) (avg.)*
Pressure gradient .... *0.43 psi/ft (9.7 kPa/m)*
Temperature .... *190-240° F (88-116° C)*
Geothermal gradient .... *1.3° F/ft (2.4° C/m)*
Drive .... *Gas solution, water drive, liquid expansion*
Oil column thickness .... *500 to 3000 ft (150 to 915 m)*
Oil-water contact .... *-14,500 to -10,600 ft (-4420 to -3230 m)*
Connate water .... *10 to 35%*
Water salinity, TDS .... *10,000 to 20,000 ppm Cl^-*
Resistivity of water .... *0.20 ohm-m² m*
Bulk volume water (%) .... *25 to 40%*

Transportation method and market for oil and gas:

Pipeline to Puerto Miranda; tankers to foreign markets or pipeline from Puerto Miranda to Cardon refinery

Reservoir name .... *Cretaceous Socuy/La Luna/Cogollo*

Reservoir size:

Proved acres .... *10,926 ac (4,425 ha)*
Number of wells all years *30*
Current number of wells *7*
Well spacing .... *3940 ft (1200 m)*
Ultimate recoverable, proven .... *155 million bbl*

Ultimate recoverable, possible ... *349 million bbl*
Cumulative production ... *86 million bbl*
Annual production ... *1.3 million bbl*
Present decline rate ... *10.5%*
  Initial decline rate ... *2%*
  Overall decline rate ... *NA*
Annual water production ... *645,000 bbl*
In place, total reserves ... *721 million bbl*
In place, per acre foot ... *180 to 270 bbl*
Primary recovery ... *134 million bbl*
Secondary recovery ... *NA*
Enhanced recovery ... *NA*
Cumulative water production ... *12.5 $\times$ 10^6 bbl*

Drilling and casing practices:

Amount of surface casing set ... *3200–3500 ft (975–1065 m)*
Casing program

Casing, 9⅝-in. or 7-in.; liner, 4½-in.; tubing, 2⅜-in. to 2⅞-in.; usually set casing at top of zone, then drill to T.D. and complete open hole

Drilling mud ... *KCl oil-base mud, lignosulfonate*
High pressure zones ... *None*

Completion practices:

Interval(s) perforated ... *La Luna/Cogollo (but usually open hole below 7-in. casing)*
Well treatment ... *Acidization (wash and frac), solvents*

Formation evaluation:

Logging suites ... *Electrical logs, radioactive logs, production logs, acoustic logs*
Testing practices ... *Open hole or through pre-perforated liner*
Mud logging techniques ... *Standard ditch sample analyses*

Oil characteristics:

Type ... *NA*
API gravity ... *32–42°*
Base ... *Paraffinic*
Initial GOR ... *1200 ft^3/bbl (211 m^3/m^3)*

Reservoir characteristics:

Average elevation ... *In lake (S.L.)*
Initial pressure ... *7700 psi at 12,600 ft (53.0 MPa) to 10,800 psi at 15,000 ft (74.0 MPa)**
Present pressure ... *12,000–4000 psi (83 to 28 MPa) (avg.)*
Pressure gradient ... *0.7 psi/ft (16 kPa/m)*
Temperature ... *300–320° F (150–160° C)*
Geothermal gradient ... *1.7° F/100 ft (3.1° C/100 m)*
Drive ... *Gas solution, water drive, liquid expansion*
Oil column thickness ... *±1500 ft (±460 m)*
Oil-water contact ... *Various depths in different zones*
Connate water ... *30 to 50% in matrix porosity*
Water salinity, TDS ... *10,000 to 20,000 ppm Cl^-*
Resistivity of water ... *0.1–0.04 ohm-m^2 m*
Bulk volume water (%) ... *2 to 25%*

**Limestones are overpressured; e.g., one is at 11,600 psi at 12,000 ft (80.0 MPa at 3660 m).*

Transportation method and market for oil and gas:

Pipeline to Puerto Miranda; tankers to foreign markets or pipeline from Puerto Miranda to Cardon refinery

Reservoir name .... *Paleocene Guasare*

Reservoir size:

- **Proved acres** .... *682 ac (276 ha)*
- **Number of wells all years** *105*
- **Current number of wells** .... *0 (producing)*
- **Well spacing** .... *1970 ft (600 m)*
- **Ultimate recoverable, proven** .... *1.7 million bbl*
- **Cumulative production** .... *170 thousand bbl*
- **Annual production** .... *25 thousand bbl*
- **Present decline rate** .... *NA*
  - **Initial decline rate** .... *NA*
  - **Overall decline rate** .... *NA*
- **Annual water production** .... *NA*
- **In place, total reserves** .... *10.2 million bbl*
- **In place, per acre foot** .... *466 to 760 bbl*
- **Primary recovery** .... *1.7 million bbl*
- **Secondary recovery** .... *NA*
- **Enhanced recovery** .... *NA*
- **Cumulative water production** .... *10,000 bbl*

Drilling and casing practices:

- **Amount of surface casing set** .... *2100 ft (640 m)*
- **Casing program** .... *Casing, 9⅝-in.; liner, 7-in.; tubing, 2⅞-in. to 3½-in.*
- **Drilling mud** .... *Oil emulsion*
- **High pressure zones** .... *None*

Completion practices

- **Intervals perforated** .... *Guasare Formation*
- **Well treatment** .... *None*

Formation evaluation:

- **Logging suites** .... *Electric logs, production logs*
- **Testing practices** .... *Standard production tests*
- **Mud logging techniques** .... *Standard ditch sample analyses*

Oil characteristics:

- **Type** .... *NA*
- **API gravity** .... *35–38°*
- **Base** .... *Paraffinic*
- **Initial GOR** .... *1260 ft^3/bbl (224 m^3/m^3)*

Reservoir characteristics:

- **Average elevation** .... *In lake (S.L.)*
- **Initial pressure** .... *4300 psi at 8500 ft (39.0 MPa) to 6700 psi at 9650 ft (46.0 MPa)*
- **Present pressure** .... *NA*
- **Pressure gradient** .... *NA*
- **Temperature** .... *NA*
- **Geothermal gradient** .... *NA*
- **Drive** .... *Gas solution, liquid expansion*
- **Oil column thickness** .... *2000 ft (610 m)*

Oil-water contact .... *NA*
Connate water .... *NA*
Water salinity, TDS .... *NA*
Resistivity of water .... *NA*
Bulk volume water (%) *5%*

Transportation method and market for oil and gas:

Pipeline to Puerto Miranda; tankers to foreign markets or pipeline from Puerto Miranda to Cardon refinery

Tiguaje Field—Venezuela
Maracaibo/Falcón Basin, Falcón State

ANGEL MOLINA
Maraven, S.A.
Caracas, Venezuela

FIELD CLASSIFICATION

BASIN: Maracaibo/Falcón
BASIN TYPE: Foredeep
RESERVOIR ROCK TYPE: Sandstone
RESERVOIR AGE: Middle Miocene
PETROLEUM TYPE: Oil
TRAP TYPE: Anticline
RESERVOIR ENVIRONMENT OF DEPOSITION: Coastal Plain
TRAP DESCRIPTION: Anticline with lateral fault seals, structure related to regional wrench fault system

LOCATION

The Tiguaje field is located in Venezuela, Falcón State, Buchivacoa District (Figure 1), near the Hombre Pintado, Media, and Meno de Mauroa fields in a vast plain bounded on the north by the Venezuelan Gulf coast and on the south by a mountainous area with elevations up to 2600 ft (800 m).

Estimated ultimate recovery of the Tiguaje field is 10.18 MMbbl of oil.

HISTORY

Exploration of Western Falcón began in 1921 with the drilling of the El Mene-1X well and the discovery of the El Mene de Mauroa field by British Controlled Oilfields, Ltd. (Halse, 1947). This well and the subsequent exploratory wells were drilled on geological surface structures, many of which had oil seeps.

In 1922, the well Berjadin-1 was drilled near the present site of the Tiguaje field by Standard Oil Company of Venezuela (SOV) on an anticline formed as part of the flower structure associated with the lateral movement of the Oca fault (Figure 2). The well was abandoned "wet" at 2110 ft (644 m), with a strong show of gas in the La Puerta Group at 1560 ft (476 m).

The following year, SOV drilled well Ojo de Agua-1, located to the north of the Oca fault, on the flank of the Dabajuro uplift. The well was abandoned at 2563 ft (782 m) with gas shows in two sandstones of the La Puerta Group.

During 1925, the American-British Oil Company drilled seven wells on the anticlinal flower structures associated with or adjacent to the Oca fault zone (Figure 2). The well El Alambre-1 had many gas shows from 713 to 2350 ft (217 to 717 m) in the La Puerta Group, and La Danta-1, located in the Oca fault zone, 400 ft (122 m) north of the Tiguaje reservoirs, tested 10 million ft^3/day from 2650 ft (808 m). The other wells, Agua Viva-1, Vicentote-1, Agua Dulce-1, and Macunare-1, all tested gas or had gas shows from several intervals in the La Puerta Group.

The Tiguaje field was discovered in 1931 when the American-British Oil Company deepened the well Berjadin-3, an old well, which had been temporarily abandoned at 4389 ft (1339 m), to a final depth of 5426 ft (1655 m). The well tested 100 bbl/day of 28° API oil at 3740 ft (1141 m). The well produced 3000 barrels of oil in 1935 all of which was used as fuel in field operations. Well Berjadin-4 was drilled too far downdip on the Tiguaje structure and was abandoned in 1937 at a depth of 5503 ft (1678 m). Operations in this new field were suspended for economic reasons by American-British Oil Company after the drilling of Berjadin-4.

Titles to the concessions in this area were later transferred to the Texas Petroleum Company, which carried out an extensive geological survey resulting in the drilling and completion of well Tiguaje-1 in 1953 at a depth of 5282 ft (1611 m). Tiguaje-1, located 2.7 km (1.7 mi) west of Berjadin-3 on the same structure, is now generally considered to be the discovery well for the field.

Many drill-stem tests were made in Tiguaje-1 at depths between 3200 and 3916 ft (976 and 1194 m) because of evidence of hydrocarbons in the form of high resistivities on electric logs and shows in ditch samples. Completions were made in the intervals 3242–3246 ft (989–990 m), 3291–3307 ft (1004–1009 m), and 3328–3348 ft (1015–1021 m) through the basal sandstones of the La Puerta Group, which are middle

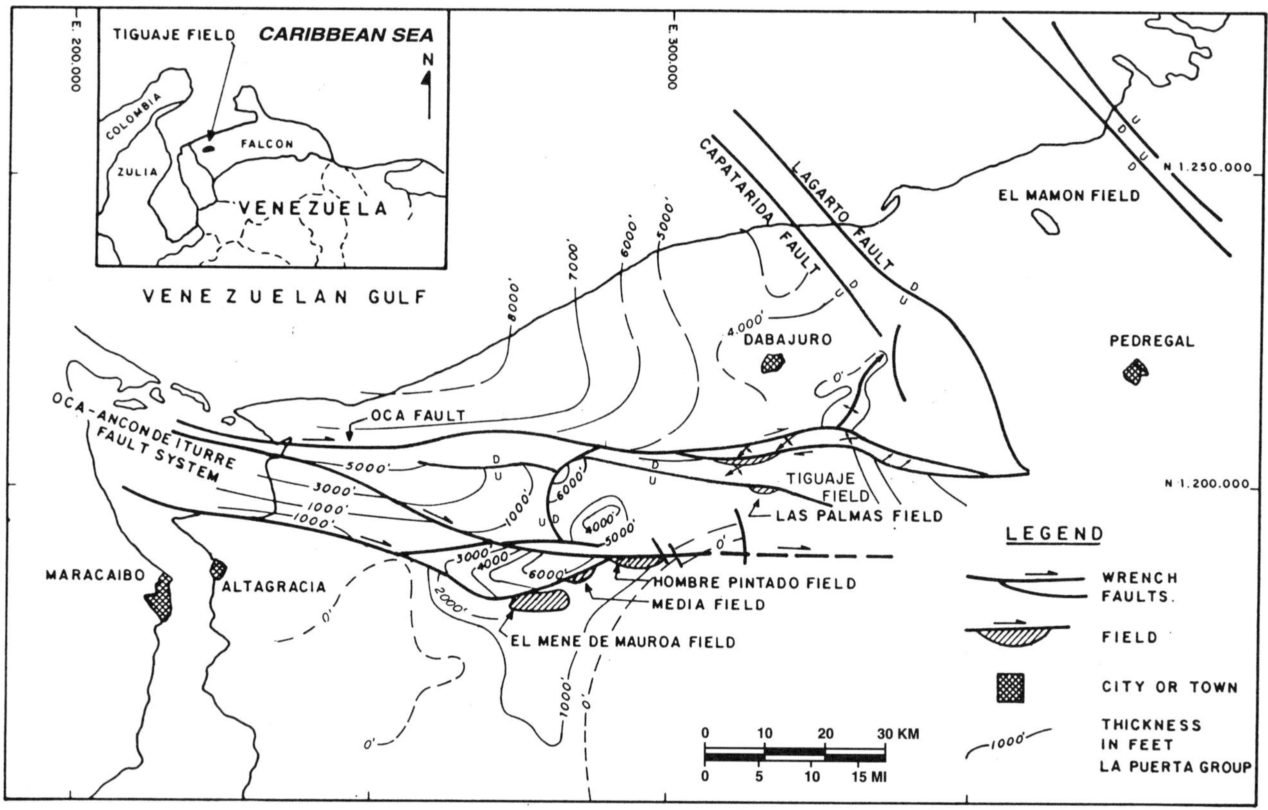

Figure 1. Relative situation of Tiguaje, Hombre Pintado, and El Mene de Mauroa fields, Venezuela, Falcón State.

Miocene, and in the interval 3906-3916 ft (1091-1194 m) through sandstones that are part of the upper Eocene La Victoria Formation. Formerly, this formation was mistaken for the middle Miocene Cerro Pelado Formation or for the Agua Clara and Castillo formations, which are middle and lower Miocene.

The field developed rapidly after 1955, because of the shallow depth of the reservoirs, and by the end of 1958, the field was producing over 800,000 bbl/year of 26.4° API oil. A total of 40 wells were drilled, of which 20 were producers, 11 were abandoned dry, 7 suspended dry, and 2 suspended for evaluation. The principal productive pay consists of the basal sandstones of the upper Miocene La Puerta Group. The sandstones of the Eocene La Victoria Formation are gas-bearing and have not been produced except to supply gas for special purposes.

During 1979 and 1980, Maraven, S.A., the affiliate of the National Oil Company, Petroleos de Venezuela, S.A., which had the exploratory responsibility for the region, carried out a detailed seismic survey of the area surrounding the Tiguaje field to obtain a better definition of the field and the major structures related to the Oca fault.

During 1982-1983, Maraven carried out an exploratory drilling program for gas, which was required by the nearby refineries, and drilled several wells on the flower structures of the Oca fault and in the Tiguaje field. One, Tiguaje-41X, was drilled as a deeper-pool test for gas and oil to a depth of 7506 ft (2289 m) and was suspended awaiting evaluation (Figures 2 and 6). Another well, Ti-59X, drilled in 1983 to a depth of 4513 ft (1376 m), produced oil on test but remains suspended.

DISCOVERY METHOD

The first phase of the field exploration consisted of surface geological surveys. These studies helped in understanding part of the stratigraphy and the structural and sedimentological characteristics of the mountainous region where many oil seeps had been observed (Halse, 1947). The second phase consisted of the drilling of wells in those areas that showed more favorable characteristics, such as structurally high areas bounded by faults and anticlinal structures, especially those close to oil seeps.

Seismic methods were used during 1979 and 1980 to aid in the definition of already known structures and to obtain a better understanding of the tectonic history of the area—nature of the faults, effects of unconformities, truncations, etc.

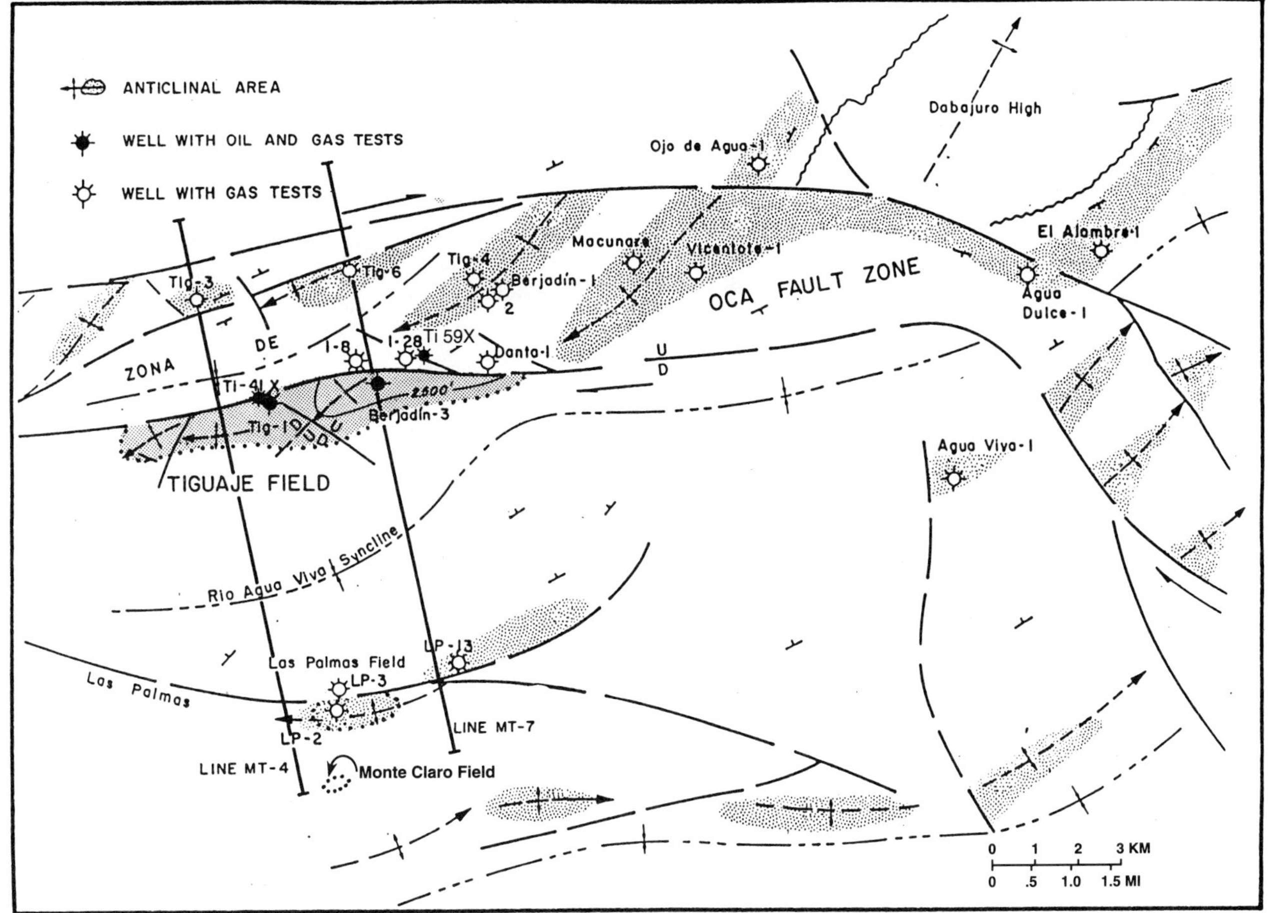

Figure 2. Generalized structural map of the Oca fault zone and the area adjacent to the Tiguaje field. Seismic sections along lines MT-4 and MT-7 are shown by Figures 7 and 8, respectively.

STRUCTURE

The Tiguaje field, located along the northern edge of the Falcón basin, is in an area of right-lateral transcurrent fault movements between the Caribbean and South American tectonic plates (Vásques and Dickey, 1972; Muessing, 1984; K. James, personal communication, 1984). The movements along the Oca and Ancon de Iturre system of faults have produced various pull-apart grabens and horst blocks (Wheeler, 1960; Habicht, personal communication, 1979; Muessing, 1984). The sediments comprising the La Victoria Formation, which unconformably overlie the lower and middle Eocene strata, were deposited principally in the grabens. Erosion of lower Eocene sediments continued in some positive areas.

Between the end of the late Eocene and middle Miocene, the positive areas continued to be eroded. The grabens were involved in the deformation process during this time. Between the end of the middle Miocene and the Pliocene, an inversion of the Falcón basin took place, during which the existing grabens were activated again and filled with sediments of the La Puerta Group. These were deposited unconformably in some areas over upper and lower Eocene strata.

At the end the Pliocene, the entire sequence was deformed by extreme folding and faulting. Uplifted and faulted Quaternary terraces, which can be seen in outcrop in some places, are evidence of the final phase of this tectonic activity.

To summarize, the regional structure consists of various pull-apart basins and tectonic highs formed by right-lateral displacement along the east-west Oca–Ancon de Iturre strike-slip fault system (Boesi and Goddard, 1991). The field is located on the northern edge of one of these basins, and its major features can be observed on the structure map (Figure 2), the seismic section MT-6 (Figure 3; see also seismic sections shown on Figures 7 and 8), and the structural cross section A–A′ (Figure 4).

STRATIGRAPHY

The stratigraphy of the Tiguaje field is illustrated in Figure 5. This type log of the field shows both

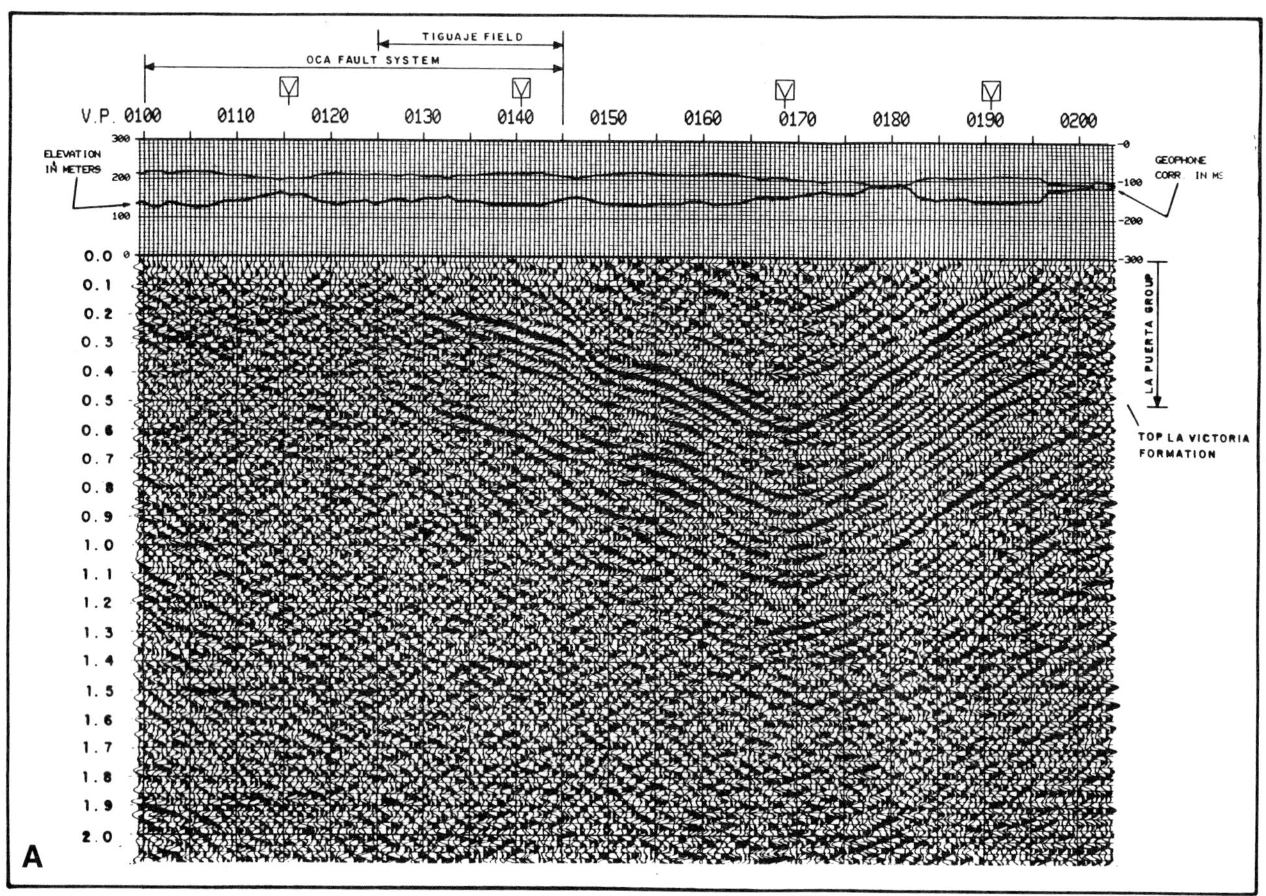

Figure 3. Seismic section 80-MT-6 across Tiguaje field. The location of the line is shown in Figure 6. The section shown is about 70 km (44 mi) long. (A) Uninterpreted. (B) Interpreted.

the existing reservoirs and the primary lithology of each unit.

The La Victoria Formation is 90% shale, interbedded with calcareous sandstones and sandy limestones, which vary in thickness from 1 to 50 ft (0.3 to 15 m). Some white, quartzose, coarse-grained limestones occasionally may be found, with thicknesses varying between 1 and 7 ft (0.3 and 2.1 m).

The La Puerta Group consists of three informal units (Halse, 1947; Molina, internal report, 1985). The lower La Puerta, a subsurface unit, consists of shales interbedded with sandstones, siltstones, and abundant coal seams. The middle La Puerta consists of intercalations of sandstones, argillaceous ferruginous siltstones, and in lesser proportions, very silty shales. Some very thin coal strata occasionally may be found. The upper La Puerta consists of light gray to reddish-colored claystones. It contains intercalations of friable sandstones with thicknesses up to 15 ft (4.6 m).

In this area, the likely source rocks belong to the shales from lower to middle Eocene. This inference is based on the content of organic matter and the degree of thermal maturity reached by these shales (according to Lopatin's method, Waples, 1980).

TRAPS

The local structure of the Tiguaje field is related to the right-lateral movement along the Oca fault and the resulting flower structure with parallel anticlinal folds both within and adjacent to the 3 to 4 km (2 to 2.5 mi) wide fault zone (Figure 2). The Tiguaje field is located on a plunging anticline that was formed against the southern boundary fault outside of the flower structure. The structural position permitted the accumulation of hydrocarbons migrating updip from the Rio Agua Viva syncline. This area provides an excellent sample of the sharp folds and complexly faulted structures that are formed by the movement along a lateral fault zone.

The folding of the basal sandstones of the La Puerta Group, which are truncated against the Oca fault, forms the major trap for oil and gas (Figures 4 and 6). Minor traps result from the truncation of the thin sandstones of the La Victoria Formation against the Eocene–Miocene unconformity and small closures against small faults. For the La Puerta sandstone reservoir, the oil column amounts to over 800 ft (244 m); the closure for the gas-bearing sandstone reservoirs of the La Victoria Formation is not known,

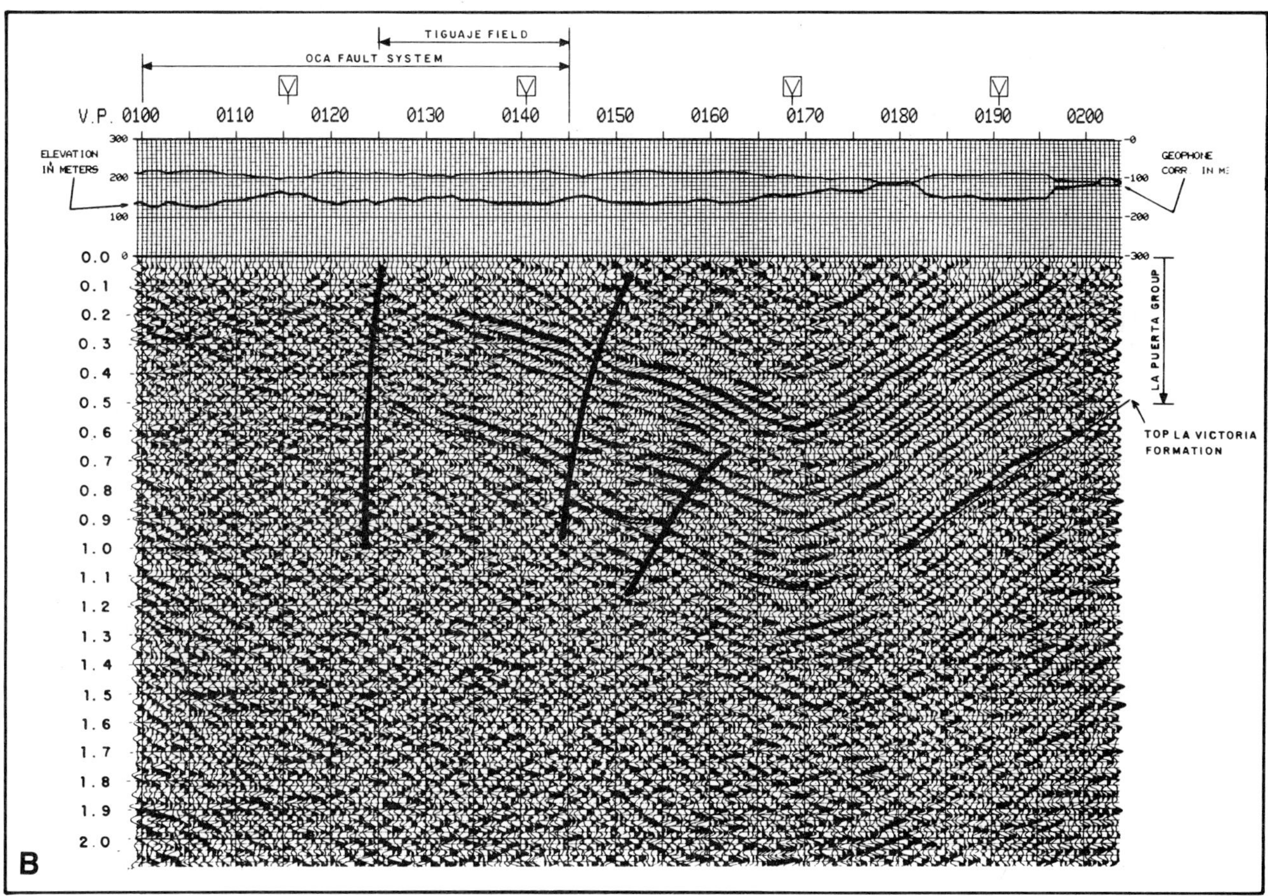

Figure 3. Continued

as only a few wells have penetrated sands and there has been little interest in producing gas reservoirs.

The La Victoria Formation contains highly calcareous sandstones that represent stream mouth bars of a prograding deltaic coastal plain. The extent of these sandstone reservoirs, at depths around 4400 ft (1342 m), is limited by facies changes to finer sediments, both vertically and laterally. Porosity is intergranular.

The primary reservoir of this field belongs to the lower unit of the La Puerta Group (Figures 5 and 6). This unit consists of slightly compacted, coarse, medium, and fine-grained sandstones at depths of 2000 to 3400 ft (610 to 1037 m). These sandstones were deposited in a deltaic coastal plain and swamp environment by distributary channels. The lateral and vertical extent of the lower La Puerta Group reservoirs is limited by the facies grading into finer sediments. The reservoir facies are oriented north-northeasterly and coalescence of some of them may occur. Like the La Victoria reservoirs, the porosity of the La Puerta reservoirs is intergranular. Porosity of 18–20% and permeability of 50 md are approximately the same for both the La Puerta and La Victoria reservoirs. The entire interval is 2500 ft (762 m) thick, and the net oil-bearing sandstones have an average thickness of 32 ft (9.8 m). Oil ranges between 25° and 27° API. Initial gas-oil ratio was 265 ft^3/bbl (47 m^3 gas/m^3oil. Pressures vary from 3000 to 5000 psi (21–34 MPa).

FAULTS

The Oca fault, which strikes east-west, limits both of the existing reservoirs to the north. This fault has been determined mainly by means of the interpretation of seismic data (Figures 3, 7, and 8 showing seismic lines 80-MT-6, 80-MT-4 EXT, and 80-MT-7/E).

These seismic sections illustrate the flower-like structure of the Oca fault, bringing Eocene and Cretaceous sediments near the surface and the complex minor faulting related to the lateral movement. It is believed that movement was initiated along the Oca fault at the end of the Paleocene with small displacements in anticlines far to the west, such as Mara, and with continued, greater displacements at the end of the Eocene and the end of Pliocene time. The amount of displacement is unknown but has been estimated by different authors to be between 10 and 300 km (6 and 185 mi).

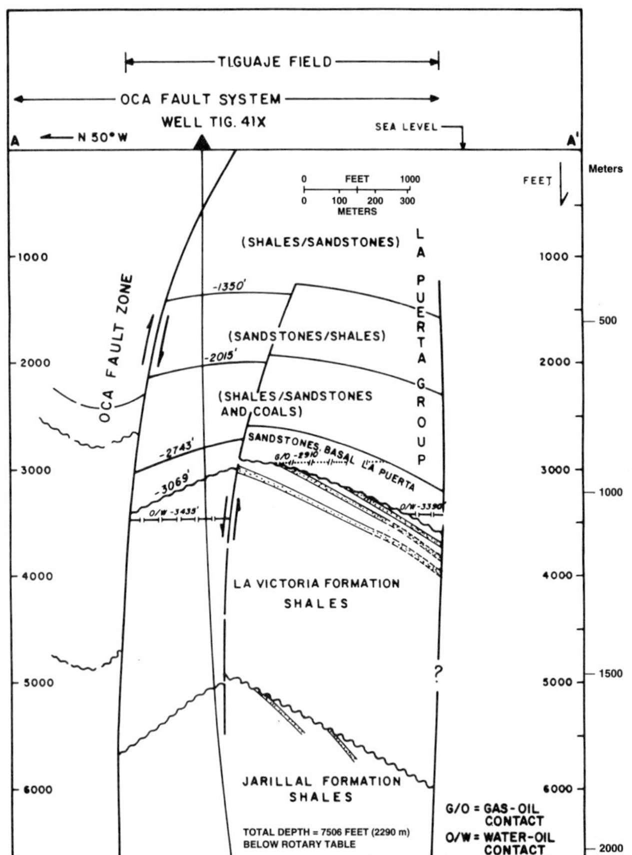

Figure 4. Northwest-southeast structural section across Tiguaje field showing La Victoria and basal La Puerta sandstones (pay zones I and II). Line of cross section is shown on Figure 6. See Figure 5 for stratigraphy and type log.

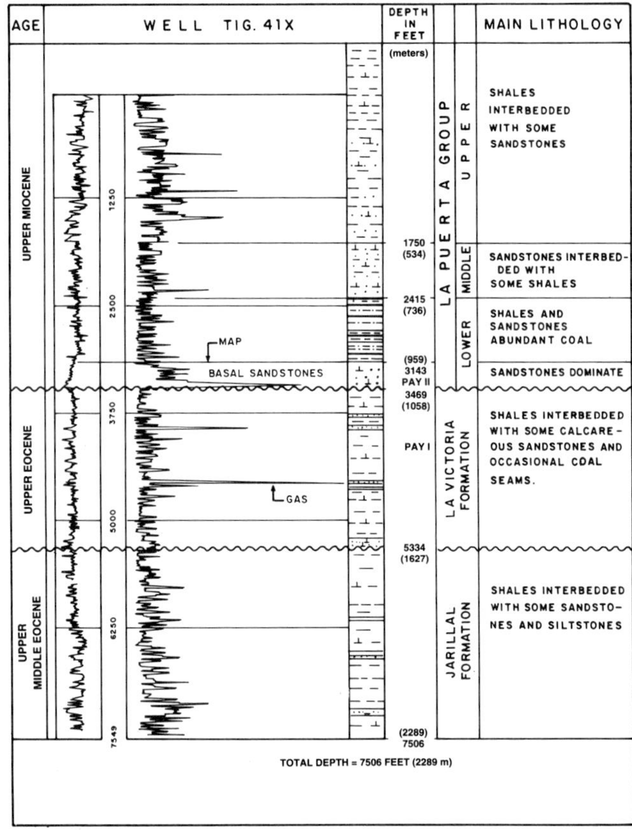

Figure 5. Stratigraphy and type log of Tiguaje field.

SOURCE

The likely source rocks belong to the upper middle Eocene Jarillal and lower Eocene Misoa formations. Toward the southwest of the field, these rocks began to generate oil 38 million years ago during the upper Eocene. Based on burial history studies, in the manner of Waples (1980), it is believed that they are currently generating oil, gas, and condensate, which suggests the existence of some migration pathways between these areas and the field (Figure 9). The samples of organic matter analyzed show TOC values ranging between 7 and 12% and are mainly of the humic type (M. Lew, personal communication, 1985; Boesi and Goddard, 1988).

EXPLORATION CONCEPTS

The tectonic history of the area indicates that the maturing process of the source rocks took place simultaneously and even after the forming of the structures, which are located updip of the generating areas. The fluvial sandstones corresponding to the basal unit of the La Puerta reservoir were deposited between the end of the middle and upper Miocene. This suggests that the oil generated between the upper Eocene and middle Miocene time was lost to the surface, though some oil remained in the upper Eocene sandstones.

The Tiguaje field resembles the Media, El Mene, and Hombre Pintado fields, which show similar characteristics regarding location, structure, depositional timing of reservoir rocks, and oil generation of the source rocks. This is due to the fact that they are included in an area that evolved in a similar tectonic environment between the Caribbean and South American plates.

The discovery of these fields can be attibuted to the easily recognizable surface expressions of the subsurface structures and to the discovery of associated oil seeps. These were enough to draw the attention of the first geologists who worked in this

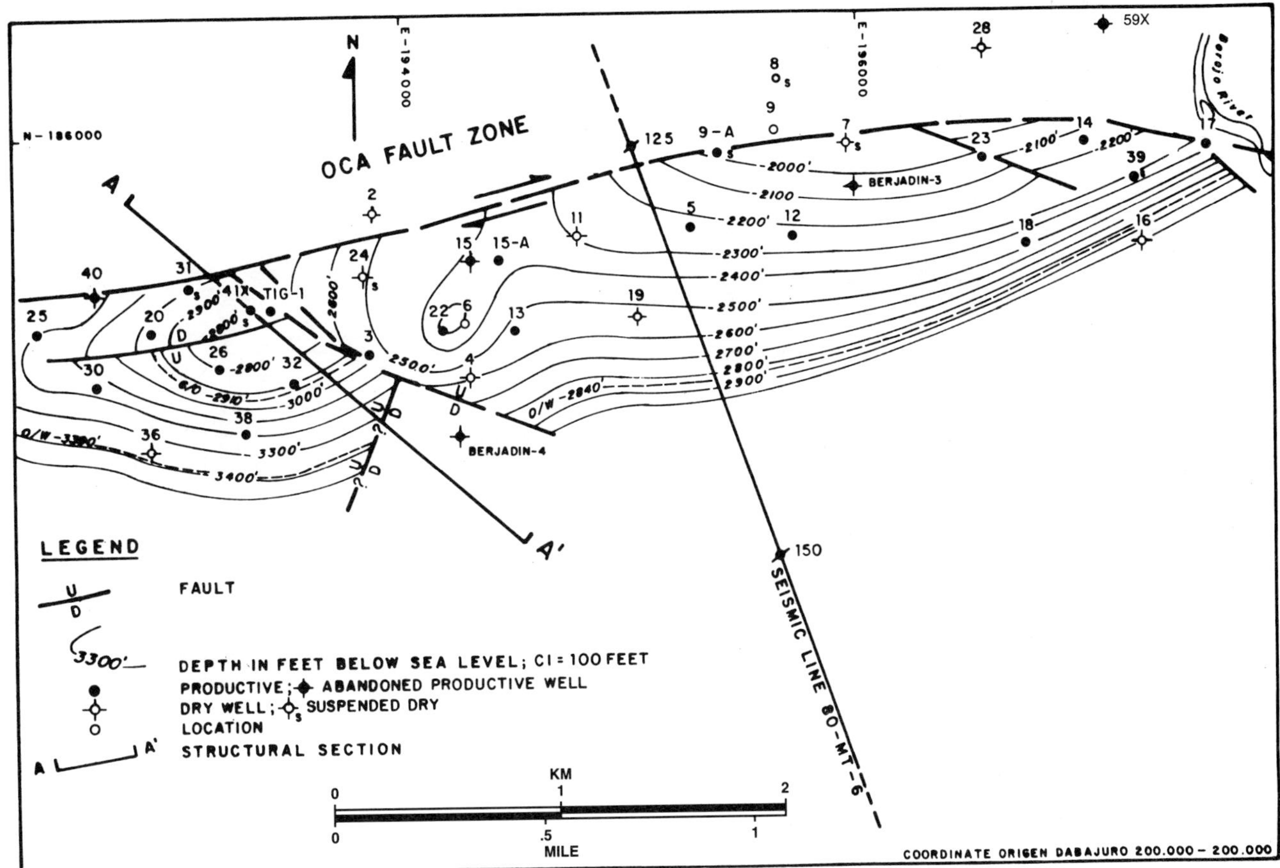

Figure 6. Structural map of Tiguaje field, contours on basal La Puerta Group (Sandstone 152).

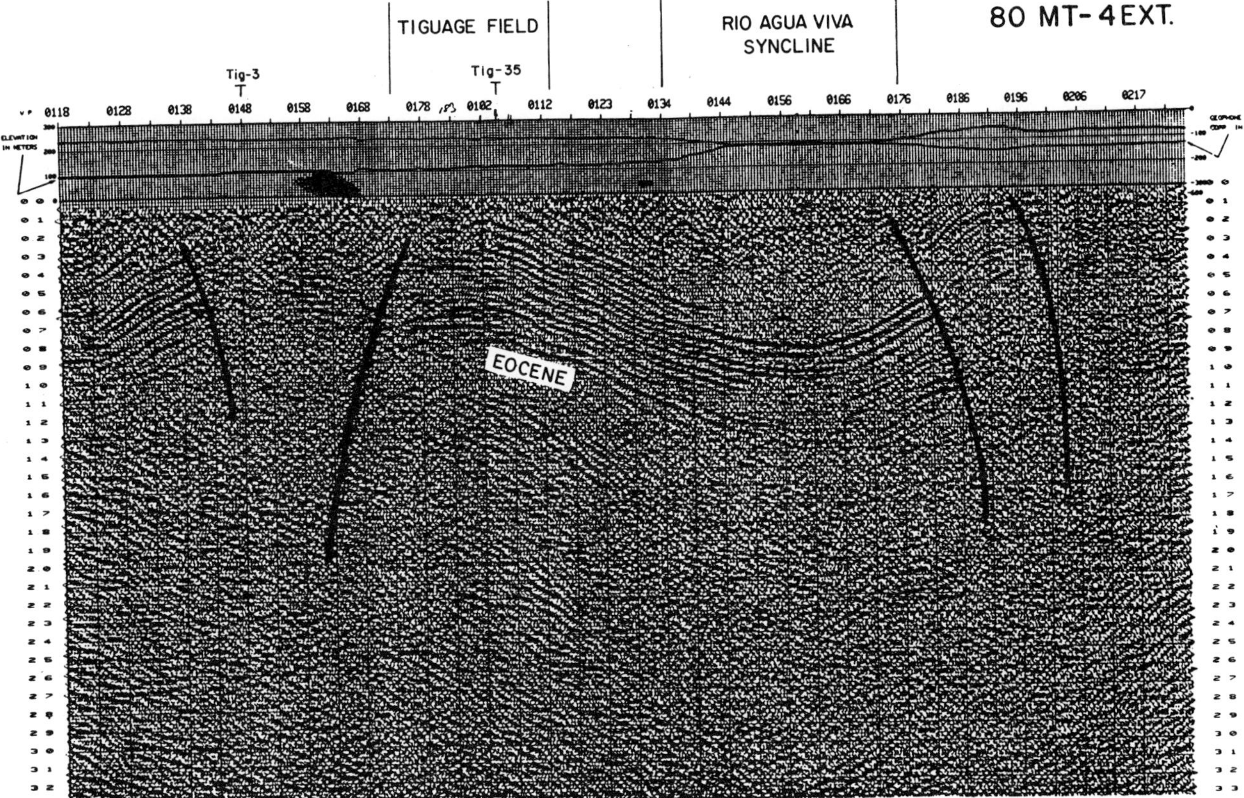

Figure 7. Seismic section 80-MT-4 EXT across Tiguaje field. Length of the section is about 130 km (81 mi).

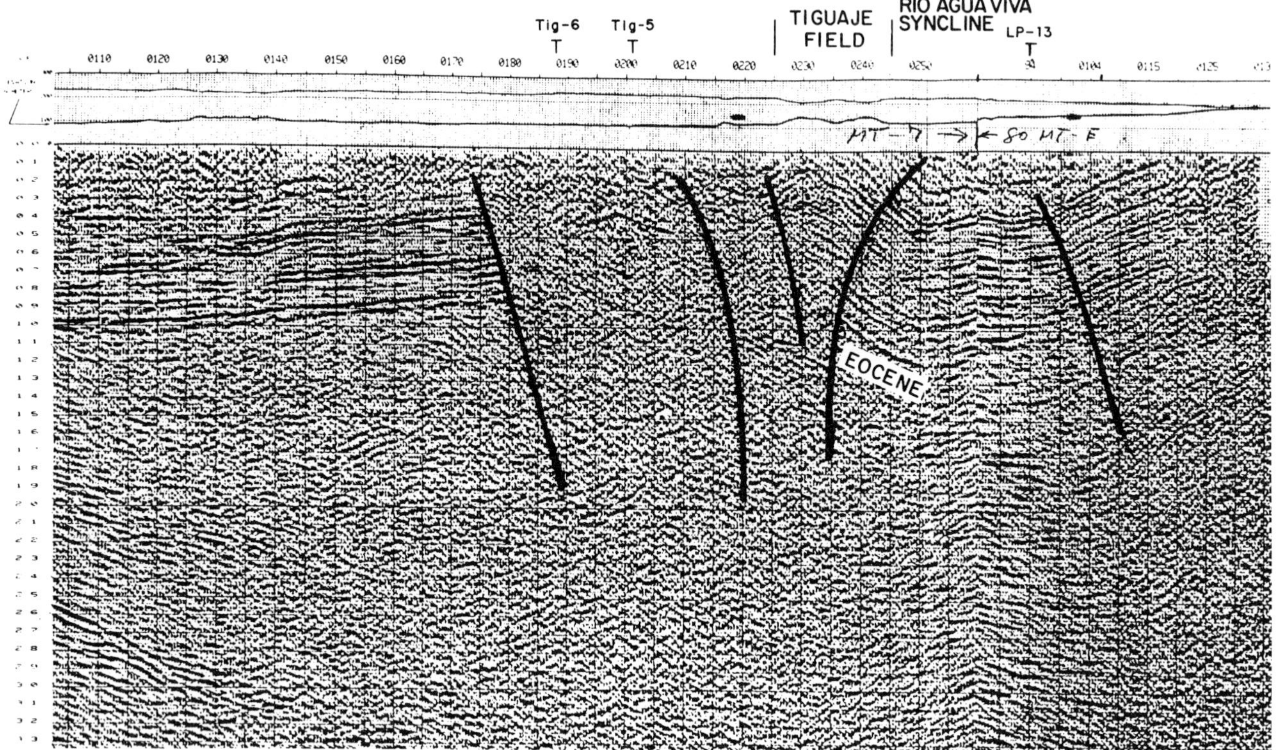

Figure 8. Seismic section 80-MT-7 across Tiguaje field. The location of the line is shown in Figure 2. Length is about 135 km (84 mi).

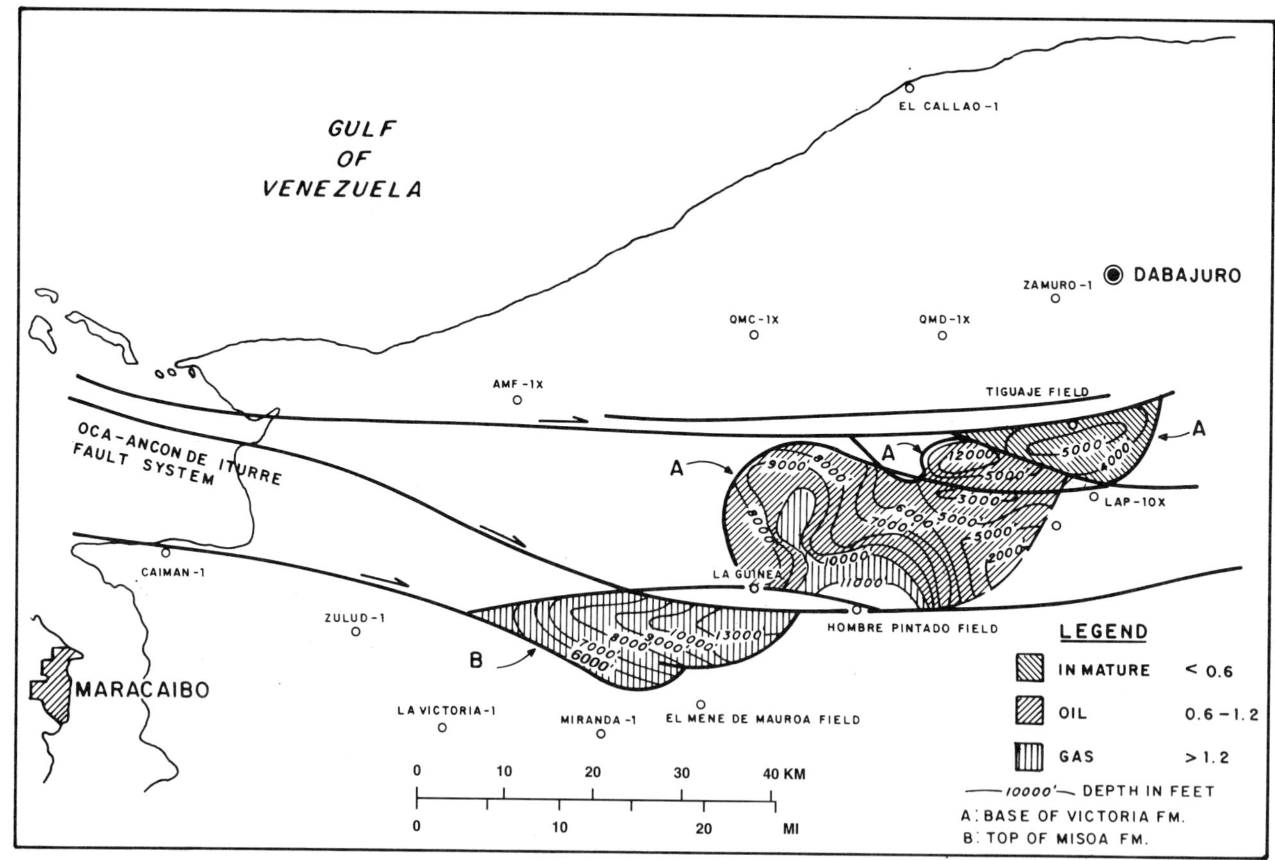

Figure 9. Actual maturity: (A) Base of Victoria Formation. (B) Top of Misoa Formation.

area. Future discoveries should be a consequence of interdisciplinary studies in the fields of seismic stratigraphy, palynology, paleontology, sedimentology, geochemistry, and surface geology that could lead to the discovery of new hydrocarbon reservoirs.

ACKNOWLEDGMENTS

The author thanks the management of Maraven, S.A. and Petróleos de Venezuela, S.A. for permission to publish this paper. He also wishes to thank Gordon Young and Donald Goddard for their help and suggestions in writing the paper and the AAPG staff and reviewers for their constructive comments.

REFERENCES CITED

Boesi, T., and D. Goddard, 1991, A new geologic model related to the distribution of hydrocarbon source rocks in the Falcón Basin, Northwestern Venezuela, *in* K. T. Biddle, ed., Active margin basins: AAPG Memoir 52.

Halse, G. W., 1947, Oil fields of West Buchivacoa, Venezuela: AAPG Bulletin, v. 31, p. 2170-2192.

Muessing, K. W., 1984, Structure and Cenozoic tectonics of the Falcón Basin, Venezuela and adjacent areas, *in* W. E. Bonini, R. B. Hargraves, and R. Shagam, eds., The Caribbean-South American plate boundary and regional tectonics: Geological Society of America Memoir 162, p. 217-230.

Vásquez, E., and P. A. Dickey, 1972, Major faulting in northwestern Venezuela and its relation to global tectonics: Transactions, VI Caribbean Geological Conference, Margarita, Venezuela, 1971, p. 191-202.

Waples, D. U., 1980, Time and temperature in petroleum formation: application of Lopatin's method to petroleum exploration: AAPG Bulletin, v. 64, n. 6, p. 916.

Wheeler, C. B., 1960, Estratigrafía del Oligoceno y Mioceno Inferior de Falcón Occidental y Nororiental: Boletín de Geología, Publicación Especial No. 3, Memoria Tercer Congreso Geológico Venezolano, Tomo I, p. 407-465.

Appendix 1. Field Description

Field name .... *Tiguaje field*

Ultimate recoverable reserves .... *10.18 million bbl oil*

Field location:

Country .... *Venezuela*
State .... *Falcón*
Basin/Province .... *Falcón basin*

Field discovery:

Year field discovered .... *Upper Eocene La Victoria Fm. and upper Miocene La Puerta sandstones 1953*

Discovery well name and general location:

First pay .... *Tiguaje-1, 16 km (9.5 mi) southwest of the town of Dabajuro in Falcón State*

Discovery well operator .... *Texaco Petroleum Co.*

IP .... *164 bpd*

All other zones with shows of oil and gas in the field: .... *None*

Geologic concept leading to discovery and method or methods used to delineate prospect

Oil seeps discovered close to the villages of Dabajuro, Mene Mauroa, and Mene de Acosta encouraged field surface geology in the Tiguaje area. The first wells were on surface anticlinal evidence. Recent (1980–1981) seismic surveys encouraged the drilling of 10 additional wells in 1982.

Structure:

Province/basin type .... *Bally 332; Klemme III Bb*

Tectonic history

The Falcón basin originated as a pull-apart basin in the early Tertiary as a consequence of right-lateral, east-west strike-slip movement along the Caribbean-South American plate boundary. Major tectonic movements during the Miocene were responsible for folding and horst and graben structures.

Regional structure

Regionally, the area consists of a system of east-trending, right-lateral strike-slip faults collectively known as the Oca fault system, along which were formed a series of more or less parallel folds.

Local structure

A small anticlinal structure bounded on the north by the Oca fault system and associated with the flower structures of the Oca wrench fault system.

Trap:

Trap type(s)

Tiguaje structure is an anticlinal trap associated with fault traps and truncations against the Eocene unconformity.

Basin stratigraphy (major stratigraphic intervals from surface to deepest penetration in field):

Chronostratigraphy	Formation	Depth to Top in ft (m)
Upper Miocene	*La Puerta*	*Outcrop*
	La Puerta basal	*3270 (1246)*
Upper Eocene	*La Victoria*	*4400 (1676)*

Reservoir characteristics:

Number of reservoirs *2*
Formations .... *La Puerta, La Victoria*
Ages .... *Upper Miocene and upper Eocene*
Depths to tops of reservoirs .... *3300 ft (1257 m); 4400 ft (1676 m)*
Gross thickness (top to bottom of producing interval) .... *±2500 ft (952 m)*

TIGUAJE

Net thickness—total thickness of producing zones

Average .... *32 ft (12 m)*

Maximum .... *NA*

Lithology

Both reservoirs: shales interbedded with sandstones, siltstones, and lignites

Porosity type .... *Primary, intergranular*

Average porosity *18–20%*

Average permeability .... *50 md*

Seals:

Upper

Formation, fault, or other feature .... *La Puerta Formation*

Lithology .... *Shales*

Lateral

Formation, fault, or other feature .... *Transcurrent and normal sealing faults*

Source:

Formation and age .... *Jarillal and Misoa (Eocene)*

Lithology .... *Shales interbedded with sandstones, siltstones, and lignites*

Average total organic carbon (TOC) .... *NA*

Maximum TOC .... *NA*

Kerogen type (I, II, or III) .... *NA*

Vitrinite reflectance (maturation) .... *NA*

Time of hydrocarbon expulsion .... *Miocene–Pliocene*

Present depth to top of source .... *±6000 ft (2286 m)*

Thickness .... *7000 ft (2667 m)*

Potential yield .... *14 kg of hydrocarbon/ton of rock*

Appendix 2. Production Data

Field name .... *Tiguaje field*

Field size:

Proved acres .... *1464 ac (593 ha)*

Number of wells all years *44*

Current number of wells .... *NA*

Well spacing .... *400 m (1312 ft)*

Ultimate recoverable .... *10.18 million bbl oil*

Cumulative production .... *10.15 million bbl oil*

Annual production .... *0.28 million bbl oil*

Decline rates .... *NA*

Annual water production .... *NA*

In place, total reserves .... *NA*

In place, per acre foot .... *899 bbl*

Primary recovery .... *10.18 million bbl*

Secondary recovery .... *NA*

Cumulative water production .... *NA*

Drilling and casing practices:

Amount of surface casing set .... *9⅝-in. to 957 ft (292 m)*

Drilling mud .... *Mud treated with barite and bentonite*

Bit program .... *NA*
High pressure zones .... *None*

Completion practices:

Intervals perforated .... *3442-3246 ft (1050-990 m)*
Well treatment .... *HCl acid wash on few wells*

Formation evaluation:

Logging suites .... *Electric log, microlog, and dipmeter*
Testing practices .... *Drill-stem tests on a few wells and standard production tests*
Mud logging techniques .... *Cutting analyses; gas chromatograph on the later wells*

Oil characteristics:

API gravity .... *25-27°*
Base .... *Paraffinic*
Initial GOR .... *265 ft^3 gas/bbl oil (47 m^3 gas/m^3 oil)*
Sulfur, wt% .... *±0.30*
Viscosity, SUS .... *34 SUS*
Pour point .... *32° F (0° C)*
Gas-oil distillate *30.5*

Field characteristics:

Average elevation .... *400 ft (122 m)*
Initial pressure .... *1400 psi (9.6 MPa)*
Present pressure .... *400 psi (2.8 MPa)*
Pressure gradient .... *0.55 psi/ft (12.4 kPa/m)*
Temperature .... *86-106° F (30-41° C)*
Geothermal gradient .... *0.03° F/ft (0.05° C/m)*
Drive .... *NA*
Oil column thickness .... *24 ft (7.3 m)*
Oil-water contact .... *3450± ft (1052± m)*
Connate water .... *NA*
Water salinity, TDS .... *1700-1800 ppm*
Resistivity of water .... *NA*

Transportation method and market for oil and gas:
Pipeline from the field ties into main line that goes to Cardon refinery.

Yucal-Placer Field—Venezuela
Eastern Venezuela Basin, Guárico Subbasin

J. Q. DAAL
R. LANDER
Corpoven, S.A.
Puerto La Cruz, Venezuela

FIELD CLASSIFICATION

BASIN: Eastern Venezuela
BASIN TYPE: Foreland
RESERVOIR ROCK TYPE: Sandstones
RESERVOIR ENVIRONMENT OF DEPOSITION: Marine Bar Sandstones and Platform Edge and Slope Sandstones
RESERVOIR AGE: Oligocene, with Cretaceous and lower Miocene
PETROLEUM TYPE: Gas
TRAP TYPE: Updip Pinch-Outs and Lenticular Facies
TRAP DESCRIPTION: Updip pinch-out of regional sandstones and lateral pinch-out of local sandstones, in part associated with faults and folds

LOCATION

The Yucal-Placer field is located in the State of Guárico on the northern edge of the Guárico subbasin of the Eastern Venezuelan Basin. It is limited to the north by the overthrust belt bounding the southern limits of the Caribbean Mountain System, and to the east, west, and south by sandstone pinch-outs (Figure 1).

The areal extent of the field is approximately 235,300 ac (368 mi^2 or 952 km^2). Estimated ultimate recovery of nonassociated gas from the field is 4.5 trillion ft^3 (tcf) based on a possible recovery factor of 70%, which makes the Yucal-Placer field one of the largest gas fields in South America.

HISTORY

Pre-Discovery

Drilling began in 1939 in the Guárico subbasin and was concentrated in the south where depths to pre-Cretaceous basement are generally less than 5000 ft (1525 m). Most of these wells encountered heavy to extra-heavy asphalt and were abandoned.

Exploration during the following years proceeded northward with minor discoveries of medium to light oil and associated gas. On 25 November 1941, the Mercedes-2 well discovered the giant Greater Las Mercedes area, followed in 1946 by discovery of the major Piragua-Guavinita-Palacios and Tucupido-Saban-Ruiz trends and other smaller producing areas, such as Grico (1946), Punzón (1947), and the Valle 17-7 and Dakoa-16 (1955) areas.

The first dry gas discoveries were Lechozo (1947), Placer (1948), Barbacoas-1 (1950), Copa-1 (1951, Cocomon field), and Yucal-1 (1957). Final depths of these wells vary from 5800 to 9000 ft (1770 to 2750 m).

Discovery

The Yucal-Placer field was discovered by the Placer-1 well, drilled in late 1947 by Sociedad Anónima Petrolera Las Mercedes (SAPLM) in partnership with Venezuelan Atlantic Refining Company (VARCO) and Creole Petroleum Corporation (Figure 2). The well was spudded on 21 April 1947 and reached a final depth in the Cretaceous Infante Limestone of 9041 ft (2758 m) on 12 September 1947. During the drilling, the well blew out for 17 days with gas coming from Tertiary Roblecito sandstone (6055 ft, 1847 m, depth) at an estimated rate of 32 million ft^3/per day (MMCFD), with minor amounts of 48° API condensate. Once brought under control, the well was completed in the Roblecito R-38 sandstone in 1950 and was placed on production to supply the city of Caracas. It was abandoned in 1959 after producing 7 billion ft^3 (bcf) of gas.

Placer-1 was followed by Yucal-1, 8 mi (13 km) to the northwest, drilled by the Mene Grande Oil

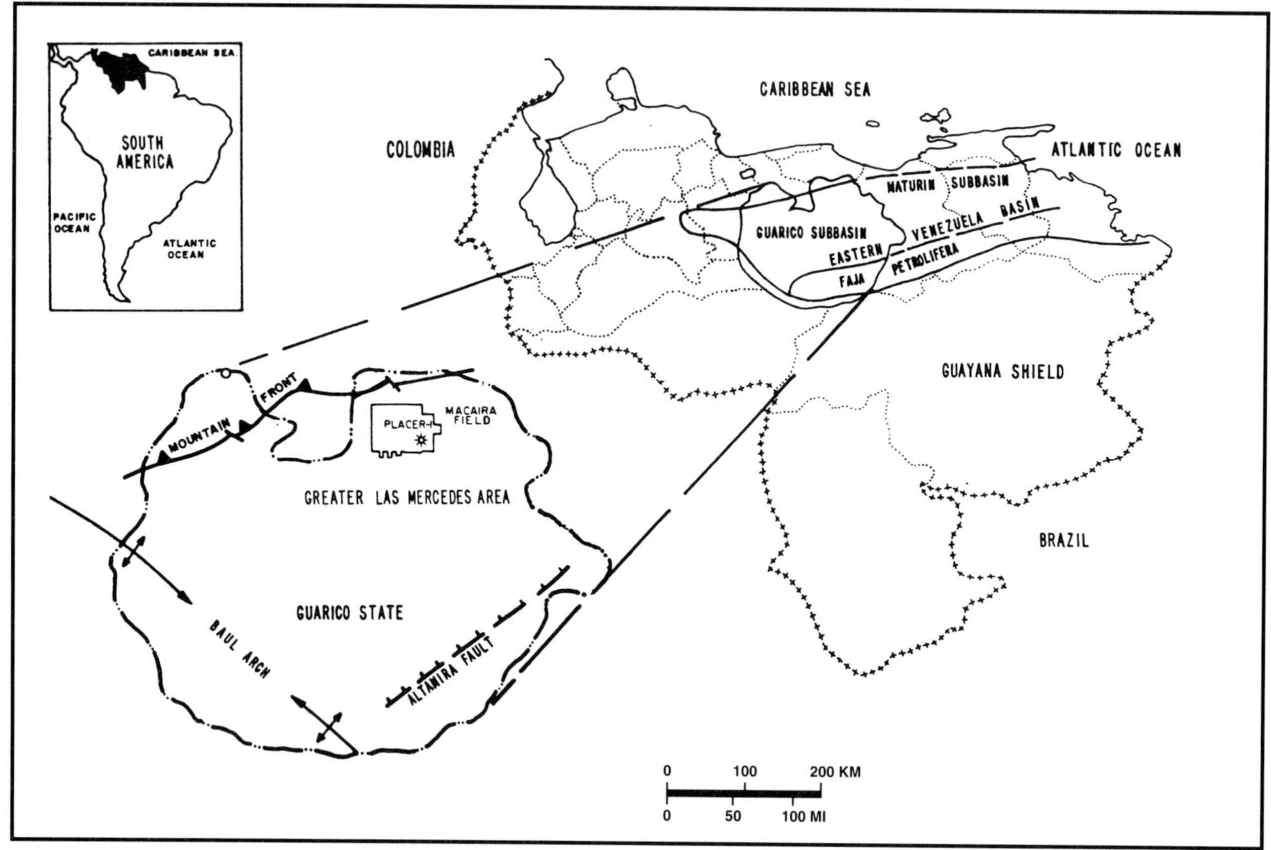

Figure 1. Geographic location showing the Yucal-Placer field in the Eastern Venezuela Basin, Guárico subbasin. The Greater Las Mercedes area includes the Mercedes, Piragua, Guavinita, and Palacios fields and the Tucupido, Saban, and Ruiz fields to the east.

Company in 1957 on concessions obtained from VARCO (Figure 2). Drilled on the crest of the Yucal anticline, the well bottomed at 8200 ft (2500 m) and tested 16 million ft^3 per day of gas from two sandstones in the lower Roblecito and La Pascua formations, both of Tertiary age. It was abandoned as noncommercial. In 1959, Varco drilled Placer-2 very close to Placer-1. Placer-2 was also completed in the Roblecito R-38 sandstone and produced 80 bcf of gas before being abandoned in 1977.

The Corporación Venezolana de Petróleo (CVP) initiated a new exploration campaign in the 1970s consisting of 354 mi (570 km) of seismic lines and four wells (29-PLA-1 through 29-PLA-4). Two wells were completed as gas producers from the Roblecito Formation, one from the La Pascua Formation, and one well was abandoned for mechanical reasons. During 1981–1982, Corpoven, S.A. (successor of CVP) completed 284 mi (457 km) of additional seismic lines and drilled another nine wells (29-PLA-5 through 29-PLA-13). Of these, five were completed in the La Pascua P-4 sandstone, two in Roblecito sandstones, and one in the Cretaceous Infante limestones; the ninth well was abandoned as noncommercial (Figure 2).

Post-Discovery

On 1 January 1983, the Corpoven acreage of northern Guárico was assigned by PDVSA to S.A. Meneven, which drilled eight more wells within and to the north and northeast of the field. The next drilling campaign in 1985–1986 consisted of eight wells designed to define the field limits.

On 1 June 1986, S.A. Meneven was merged into Corpoven, S. A., which was currently reevaluating the gas potential of northern Guárico and preparing an integrated development program for the field in order to supply gas to central and western Venezuela through the NURGAS distribution system whose construction was under way.

Planning studies utilized the productive potential of the 33 wells drilled to date and considered possible gas throughputs in the range between 100 and 150 MMCFGD. Seventeen existing wells were selected to be produced from the main reservoirs (R-26, R-38, R-51, R-54, R-56, P-2/3, P-4, and P-5 sandstones). The production/operational scheme adopted will maintain production at a rate of 125 MMCFGD over a 21-year period (1990–2011), requiring the drilling

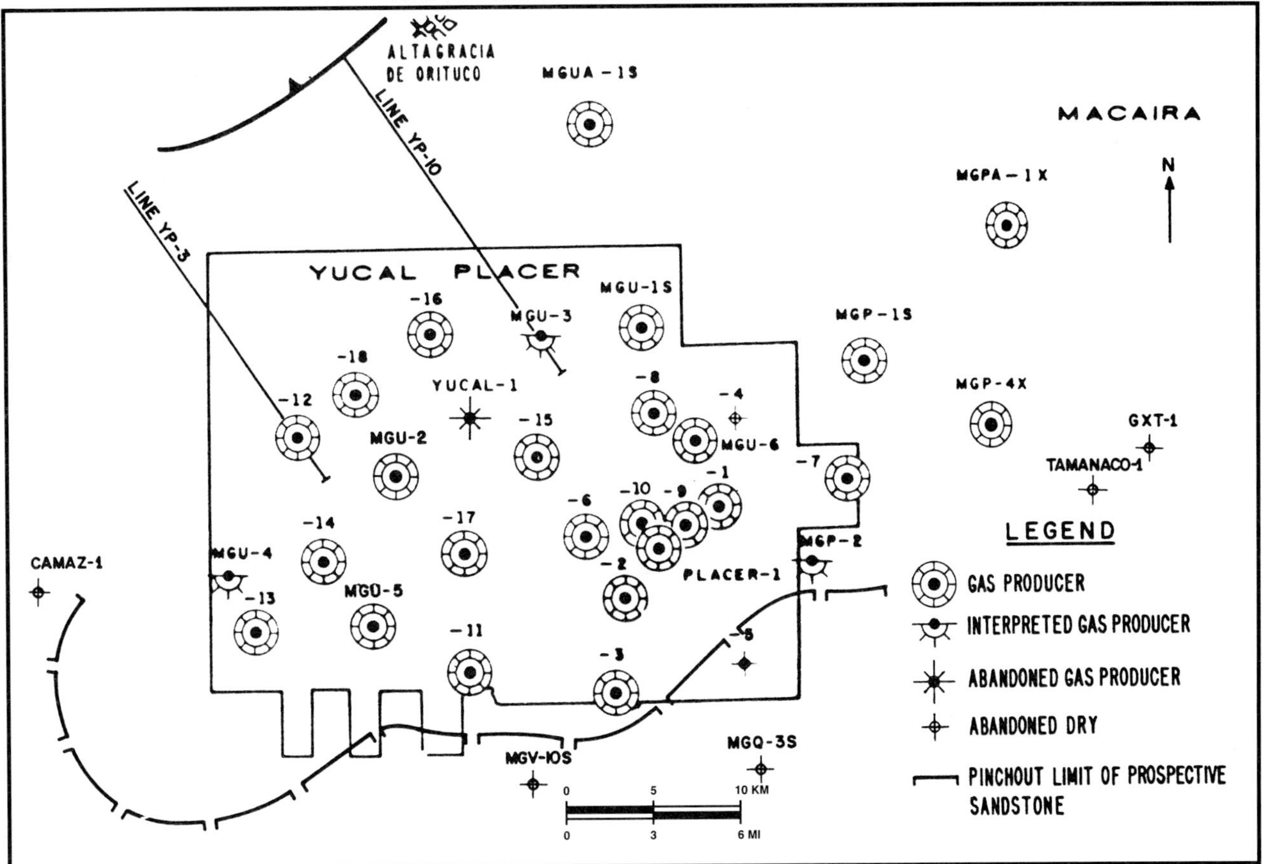

Figure 2. Wells drilled in and around the Yucal-Placer field. The location of the nearby town of Altagracia de Orituco is shown at the top of the map. Seismic lines YP-3 and YP-10 are shown by Figures 5 and 6, respectively.

of eight additional wells in the period 1998–2009. The field is presently (1987–1988) producing 35 to 40 MMCFGD, and regular restoration tests are being made to study reservoir behavior. After the year 2011, production from the field is expected to decline to economic limit in 2020.

DISCOVERY METHOD

Little information concerning the discovery made by the Placer-1 well is available. Surface geology, followed by gravity and torsion balance surveys, delineated the Yucal anticline and led to the location of Yucal-1; later analog reflection seismic lines confirmed the anticlinal structure as well as thrust faulting identified in the well. The updip stratigraphic pinch-out was not known until 1981 when Corpoven carried out geological studies based on the numerous wells drilled within the area.

The quality of present-day seismic data and seismic-stratigraphy techniques, including seismic inversion studies, allows reliable mapping of the structure and stratigraphy of the reservoir sandstones, thus guiding new development and exploratory wells.

STRUCTURE

The Yucal-Placer field shows evidence of several tectonic regimes that are intimately related to the sedimentary evolution of the Eastern Venezuelan Basin (Figures 3 and 4). During the Cretaceous, the tectonic regime was extensional and was accompanied by synsedimentary normal faults. Toward the end of the Cretaceous, a compressional regime predominated, resulting in shortening of the sedimentary section and in uplift and erosion. Reverse faulting occurred along rejuvenated normal fault lines of weakness; the geometry of some of these reverse faults is not typical, inversed throw being less than the original normal displacement.

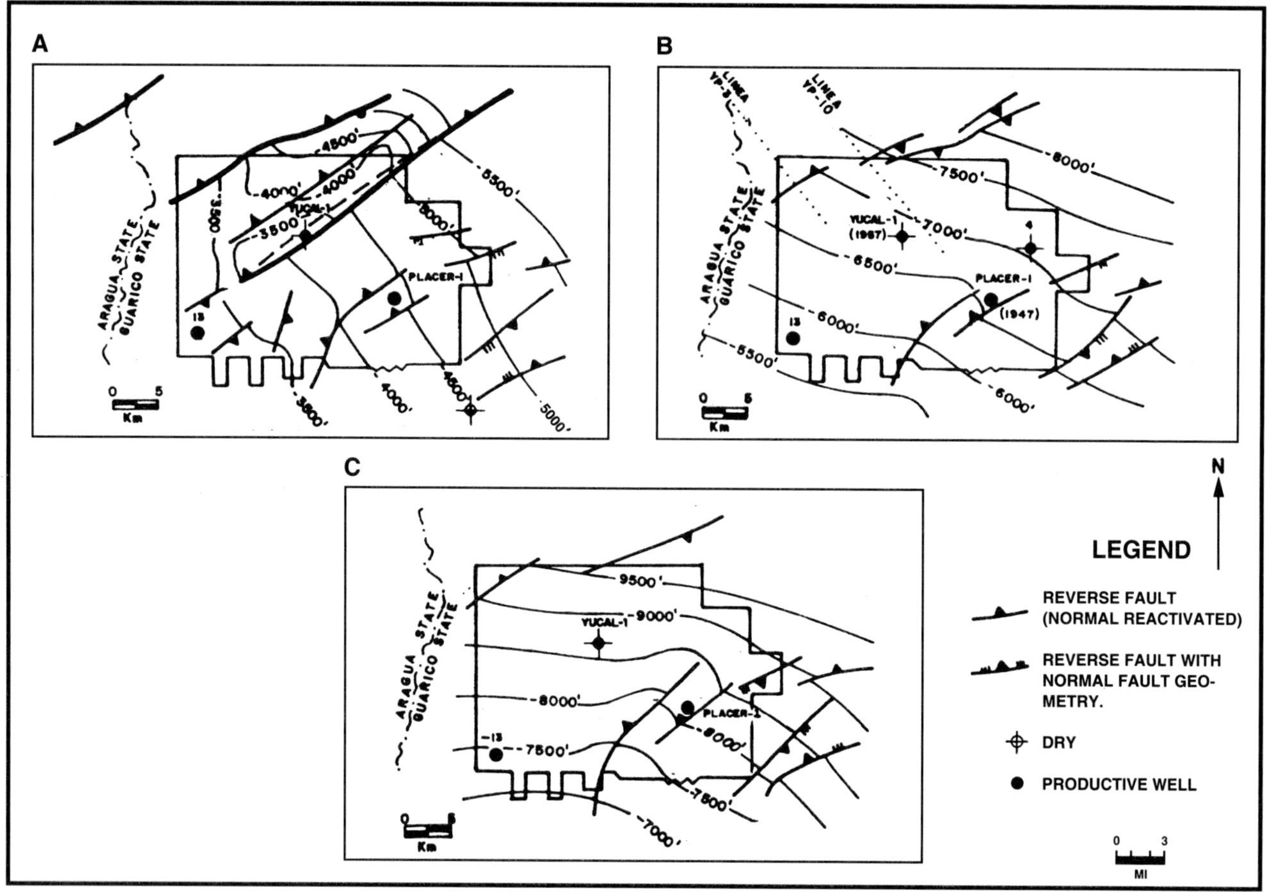

Figure 3. Structure maps, elevations in feet subsea. Contour intervals, 500 ft. (A) Middle and upper Oligocene Roblecito Formation top. (B) Lower Oligocene La Pascua Formation top. (C) Cretaceous Tigre Formation, Infante limestone top. (From Aymard et al., 1985.)

The entire area was affected by regional differential subsidence from south to north during the Oligocene, resulting in the formation of a deep depositional trough to the northwest bounded to the southeast by a broad hinge zone. This differential subsidence allowed the deposition of the transgressive La Pascua and the lower Roblecito sandstones whose sediment source was in the south.

A second compressive regime again caused shortening and uplift in the Coast Ranges, which originated a new sediment source in the north during the middle and upper Miocene. This sediment transport from the north is evident on seismic sections as southward progradation and other seismic anomalies at the base of the continental shelf.

Finally, the upper Miocene–Pleistocene orogeny caused reactivation of all previous structural elements and the formation of new ones: folds, high-angle reverse faults, and low-angle thrust faults.

The foregoing description demonstrates the importance (and the difficulty) of structural analysis in separating structural elements into time and space and in reconstructing their evolution through geologic time. It should be noted that structure is not a determining factor in the trapping in this area.

Faults

In the area, four tectonic regimes occurred (Figures 5 and 6): (1) an extensional regime during the Cretaceous; (2) a weak Late Cretaceous compressional phase giving rise to partial inversion of preexisting normal faults of a northeast-southwest trend; (3) an extensional regime associated with the differential subsidence that provoked Tertiary sedimentation; and (4) an active compressional regime culminating in the upper Miocene, which originated the reverse fault systems and cylindrical folding with northeast-southwest orientation. These structures involve principally the Roblecito and Chaguaramas formations and are evidenced by the Yucal anticline and its associated southward-thrust faults, which extend for 35 km (22 mi), with vertical displacements of up to 1500 ft (460 m) and which serve as southern barriers to the R-24 and R-26 reservoirs.

Regional Structure

The Guárico subbasin, as the western half of the Eastern Venezuela Basin, is limited to the east and

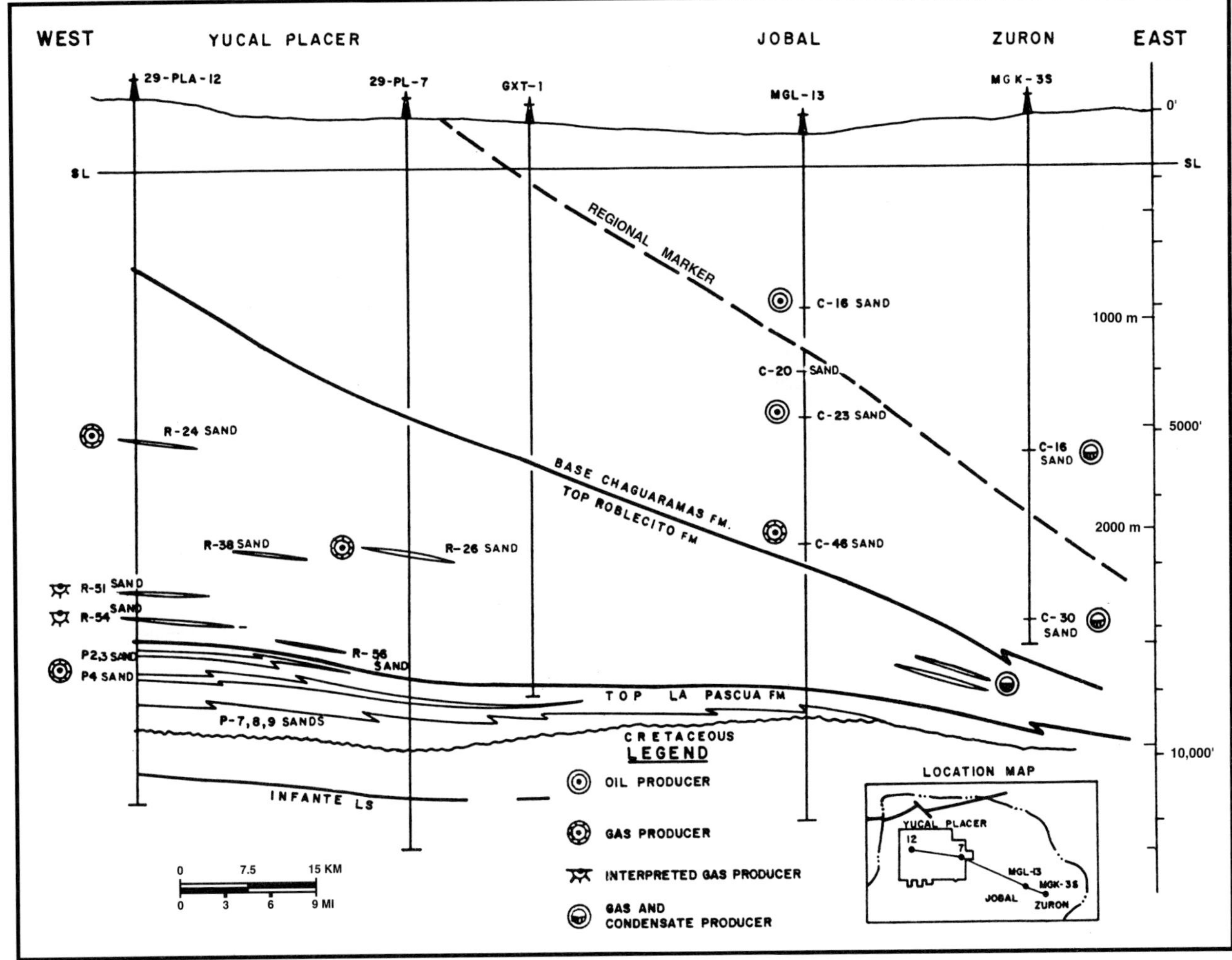

Figure 4. Regional geologic cross section showing the stratigraphic positions of alphanumerically designated productive sandstones.

southeast by the Anaco-Altamira fault trend, to the west by the El Baúl arch, and to the south by the Guayana shield. Structurally, the Yucal-Placer field is on the southern flank of the sedimentary basin, near the axis of the depocenter, whose northern flank is overthrust by the Caribbean Mountain System (Figure 7).

Local Structure

In the area of the field, two structural provinces are recognized: to the north, upper Miocene to Pleistocene thrust faulting and folding parallel to the mountain front and extending to the surface, involving principally the Chaguaramas-Roblecito strata; to the south, a monocline dipping gently (2–3°) northeast to east. The latter monocline is only slightly affected by tectonism and involves the entire Cretaceous–Tertiary sedimentary sequence.

STRATIGRAPHY

The stratigraphic sequence consists of two sedimentary cycles, Cretaceous and Tertiary, separated by a pre-Oligocene unconformity (Figure 8).

Cretaceous Cycle

The deepest sediments encountered belong to the Tigre and Canoa formations of the Temblador Group. The Canoa Formation, less known in the field, is estimated to be 900 ft (275 m) thick in well 29-PLA-6E and is composed of essentially lenticular continental sandstones with minor shale intervals. The overlying better-known Tigre sediments consist predominantly of Maastrichtian massive glauconitic sandstones, carbonaceous shales, fossiliferous limestones, silts, and phosphates, all of marine origin, and average 900 ft (275 m) in thickness.

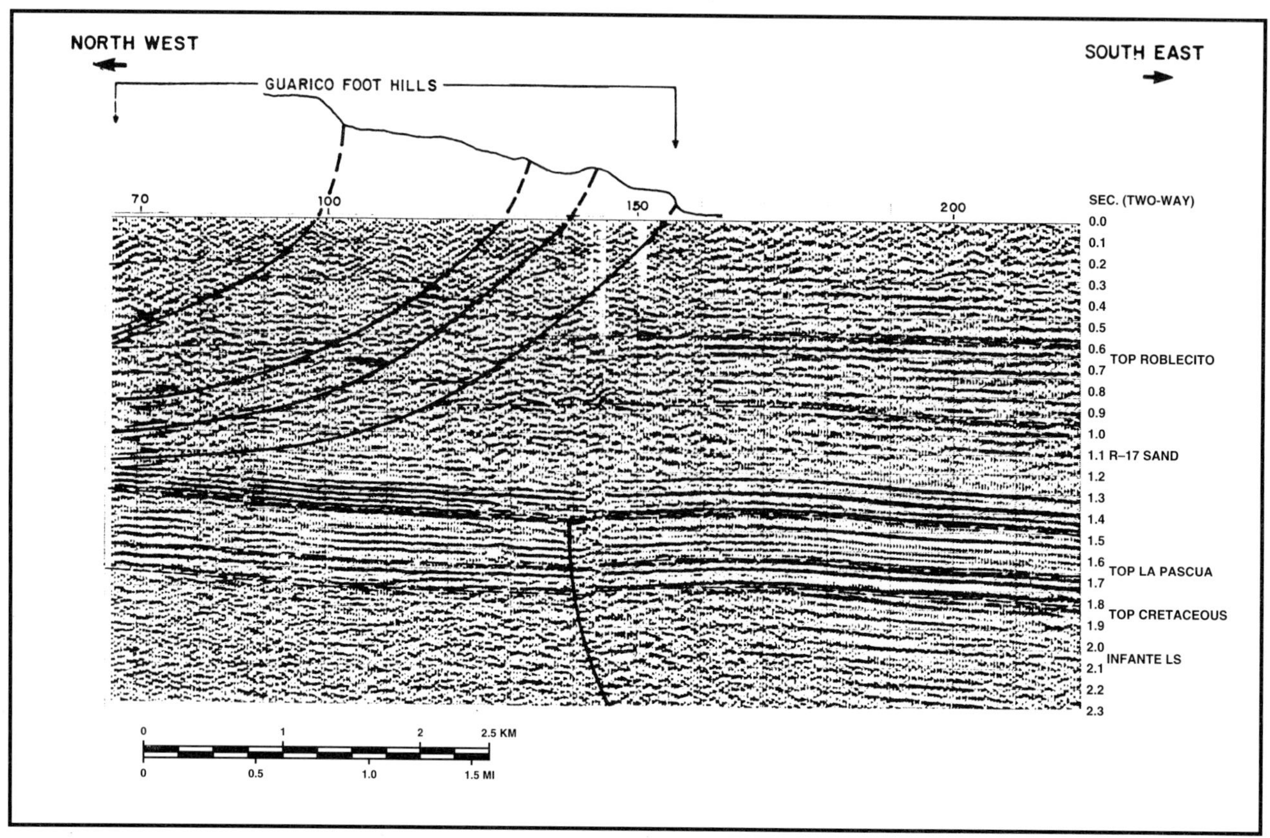

Figure 5. Interpreted seismic section YP-3 showing extensional and compressional structural styles of the Yucal-Placer area (see text, *Structure, Faults*). The location of the section is shown on Figure 2. (From Aymard et al., 1985.)

The Tigre Formation is divided into three members; in chronostratigraphic order, these are the La Cruz, Infante ("N" limestone), and the Guavinita members.

The La Cruz member is composed basically of paralic to shallow marine sandstones with minor amounts of shale, showing lateral and vertical transition to the underlying, more continental Canoa clastic sediments.

The Infante member, as described from cores in well 29-PLA-14E, is composed of fossiliferous glauconitic limestones (biomicrites and biospartites) deposited in a marginal marine/platform environment.

The overlying Guavinita member consists of some 300 ft (90 m) of glauconitic, calcareous, and kaolinitic sandstones, black shales, silty limestones, and dolomitic claystones, overlain unconformably by the Oligocene La Pascua sandstones.

Tertiary Cycle

The Tertiary cycle is initiated by transgressive littoral sandstones, followed by open-marine shales, and terminated by alternations of regressive sands, shales, and silts.

The La Pascua formation, of lower to middle Oligocene age, begins at the base with massive, very compacted sandstones, interrupted by occasional thin shale streaks, deposited in longshore bars of a coastal marine environment. The middle and upper parts of the formation are comprised of similar sandstones, but with thicker alternating shale intervals, considered to represent migrating marine bars. The thickness of La Pascua varies gradually from 750 ft (230 m) in the southwest to approximately 1700 ft (520 m) in the northwest (Figure 9). The sandstones rest unconformably upon the eroded Guavinita strata and are laterally and vertically transitional with the overlying Roblecito shales. Individual sandstone lenses pinch out updip to the southeast.

The Roblecito formation, of middle to upper Oligocene age, is composed almost entirely of open-marine, dark gray to black, glauconitic, calcareous shales with ocasional fine- to medium-grained bedded sandstones. Platform deposits occur in the lower Roblecito, turbidite deposits occur in the middle Roblecito, and shallow marine to littoral deposits in the upper Roblecito. The Roblecito sandstones pinch out individually toward the southeast. The thickness of the formation varies from 3000 ft (915 m) in the southwest to 5600 ft (1710 m) in the northwest (Figure 9).

The lower Miocene Chaguaramas Formation forms the regressive phase of the Tertiary sedimentary

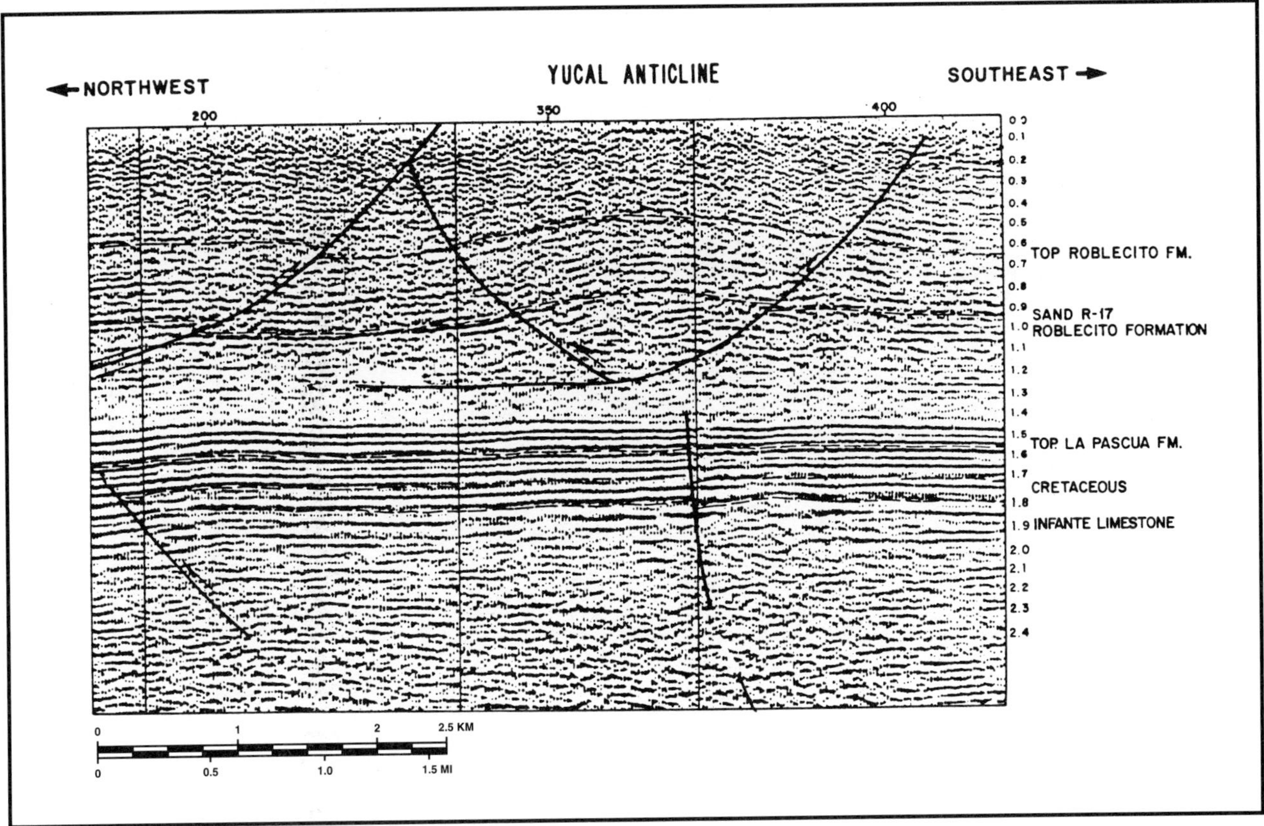

Figure 6. Interpreted seismic section YP-10, structural styles. The location of the section is shown on Figure 2. (From Aymard et al., 1985.)

cycle and is characterized by alternating lenticular sandstones, shales, and numerous thin lignites deposited in a marginal marine to tidal swamp environment. The thickness of the Chaguaramas Formation varies from 4000 ft (1220 m) in the east to 750 ft (230 m) in the west (Figure 9). The formation is exposed at the surface over most of the subbasin, being eroded progressively deeper westward toward the El Baúl arch.

TRAP

Trap Types

Trapping mechanisms for the gas accumulations are almost entirely related to the updip stratigraphic pinch-outs of regional sandstone bodies and the lateral pinch-out of lenticular sandstones. Excellent updip, lateral, and vertical seals are assured by the impermeable Roblecito and upper La Pascua shales. Tertiary reservoirs are restricted exclusively to the sandstones, principally in the La Pascua and Roblecito formations (Figure 10). Fault barriers and areas of low porosity-permeability are of secondary importance. Cretaceous sandstone reservoirs are volumetrically less important, mainly as a result of the loss of porosity through diagenetic processes. Fracture porosity seems to be locally important in both Cretaceous limestones and sandstones of the Infante and Guavinita members (Tigre Formation).

Lithofacies and Sedimentary Environment Studies

The interpretation of lithofacies and sedimentary environment was accomplished in several stages.

1. Detailed electric log correlations and identification of lithofacies interval packages, using chronostratigraphic markers and spontaneous potential curve shapes.
2. Description and environment interpretation of cores from six wells.
3. Integration of the above data with paleobathymetric data.
4. Formulation of a sedimentary model that conforms to regional patterns and preparation of facies and sandstone distribution maps for the main reservoir sands: R-24, R-26, R-38, R-54, and R-56 (lower to middle Roblecito); P-2/3, P-4, and P-5 (upper La Pascua).

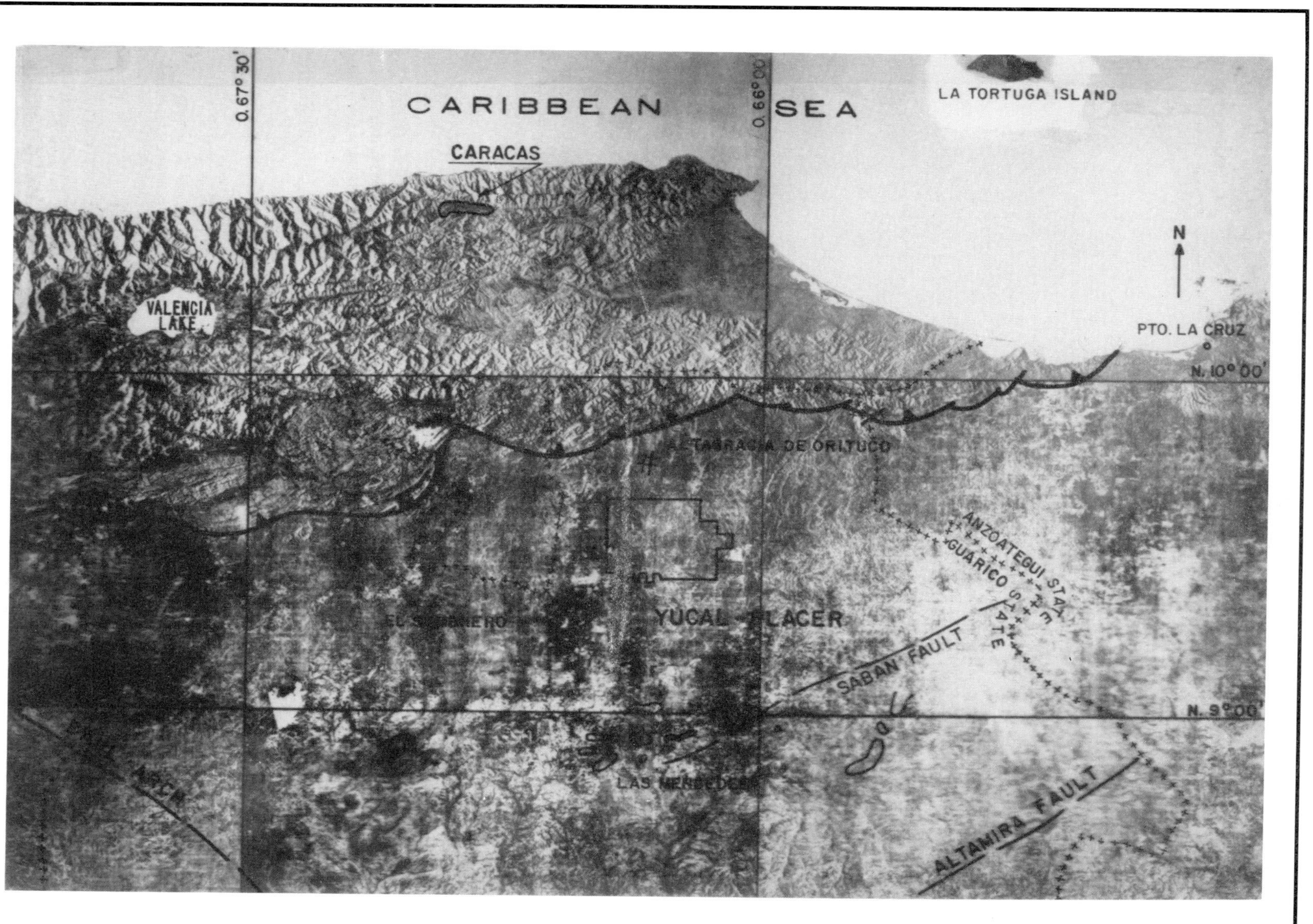

Figure 7. Position of Yucal-Placer field relative to major tectonic and physiographic features.

PERIOD	AGE		FORMATION	LITHOLOGY	DESCRIPTION
	PLEISTOCENE		ALLUVION		FLUVIAL SANDSTONES
	PLIOCENE				
TERTIARY	MIOCENE		CHAGUARAMAS 600-1400 m		SAND-SHALE-LIGNITE SERIES
	OLIGOCENE	UPPER	ROBLECITO 100-1500 m		PREDOMINANTLY MARINE SHALE, DARK GRAY TO BLACK WITH FEW SANDSTONES
		MIDDLE			
		LOWER	LA PASCUA 300-430 m		BRACKISH-WATER TO MARINE SANDSTONES AND DARK SHALES WITH A FEW LIGNITES.
	EOCENE				
	PALEOCENE				
CRETACEOUS	MAESTRICHTIAN to TURONIAN		GRUPO TEMBLADOR — TIGRE 250 m — GUAVINITA		SANDS AND THIN LIMESTONES ASSOCIATED WITH FISH REMAINS AND CHERT BEDS
			INFANTE		LOCALLY GLAUCONITIC, COMPACT, FOSSILIFEROUS LIMESTONE.
			LA CRUZ		KAOLINITIC SANDS WITH MINOR INTERCALATIONS OF BLACK CARBONACEOUS FOSSILIFEROUS SHALE.
	CENOMANIAN to APTIAN		CANOA		LENTICULAR CONTINENTAL SANDSTONES WITH MINOR SHALE INTERVALS

Figure 8. Stratigraphic column for the Yucal-Placer field. Tertiary cycle rocks rest on the Cretaceous cycle above the pre-Oligocene unconformity. Productive gas zones are indicated by the symbols in the lithology column. The half circle indicates an interpreted but untested productive zone and the vertically cross-hatched symbol indicates condensate with the gas.

Facies and Sand Distribution Maps

A facies map (Figure 11) of the P-4 sandstone (largest reservoir of the field) shows the gas well producers and the sandstone limits to the south and west. The interpretation indicates a maximum sand thickness of 70 ft (21 m) that pinches out to the south, west, and northwest. The east-northeast to west-southwest orientation of the sandstone body is parallel to the La Pascua sedimentary distribution.

Conventional cores from wells MGP-1S, MGU-2, MGU-3, and MGU-5 permit interpretation of the paleogeography as a system of marine platform bars with lateral coalescence, locally crossed by distributary tidal channels (Figure 12). These channel sandstones show very low porosity (3–8%), which probably explains why MGU-2 tested dry at this level.

Composite sandstone distribution maps were also prepared for the R-24, R-26, and R-38 reservoir sandstone of the middle Roblecito formation and the R-51, R-54, and R-56 sandstone of the lower Roblecito (Figure 13). The middle Roblecito sandstone, based on biostratigraphy, is interpreted as slope deposits, especially the R-38 sandstone. The basal Roblecito sandstones, as well as the upper La Pascua sands, appear to have been deposited on the continental slope or the edge of the platform. All these sandstones

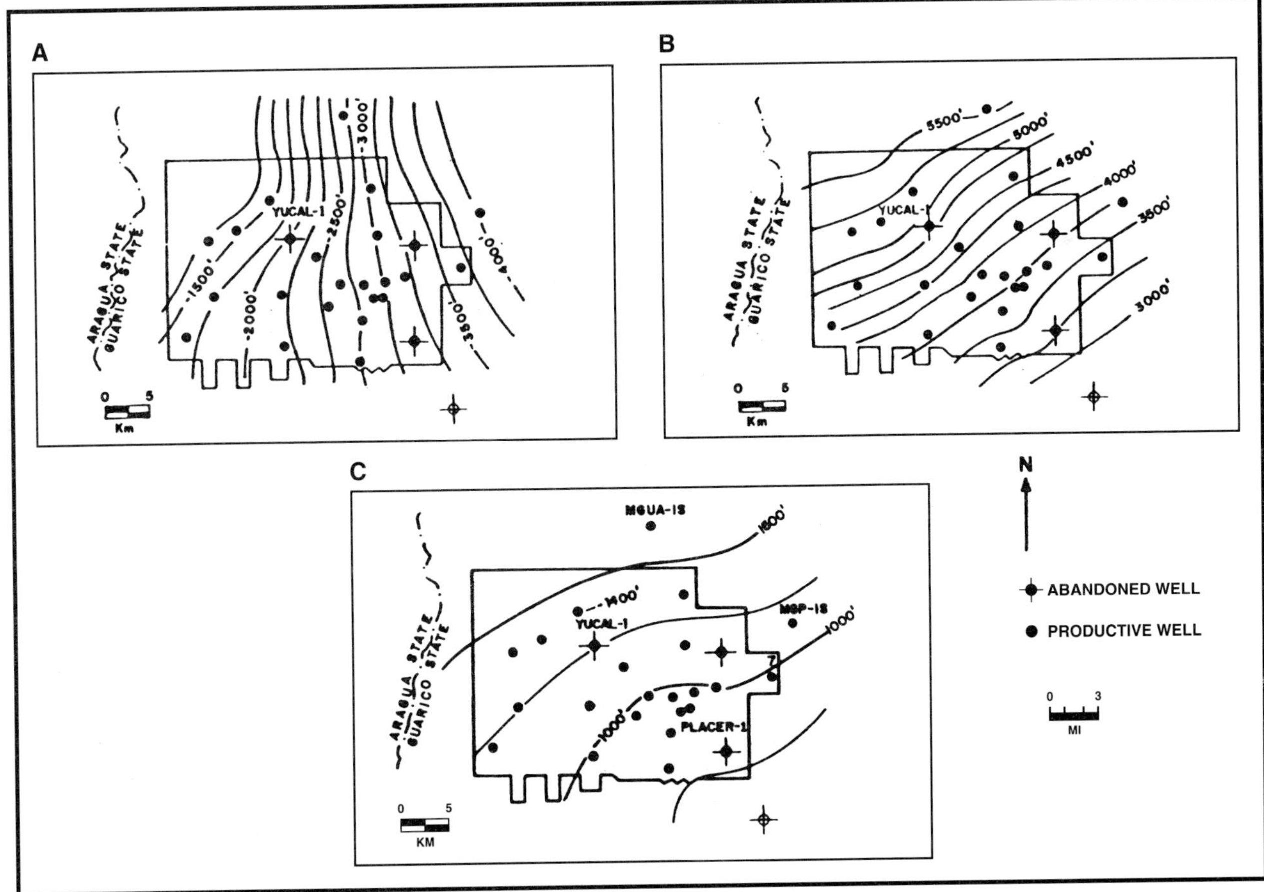

Figure 9. Isopachs of Tertiary formations. Contours in feet with 250 ft intervals. (A) Miocene Chaguaramas Formation. (B) Middle and upper Oligocene Roblecito Formation. (C) Lower Oligocene La Pascua Formation. Compare with structure maps (Figure 3) and regional cross section (Figure 4). (From Aymard et al., 1985.)

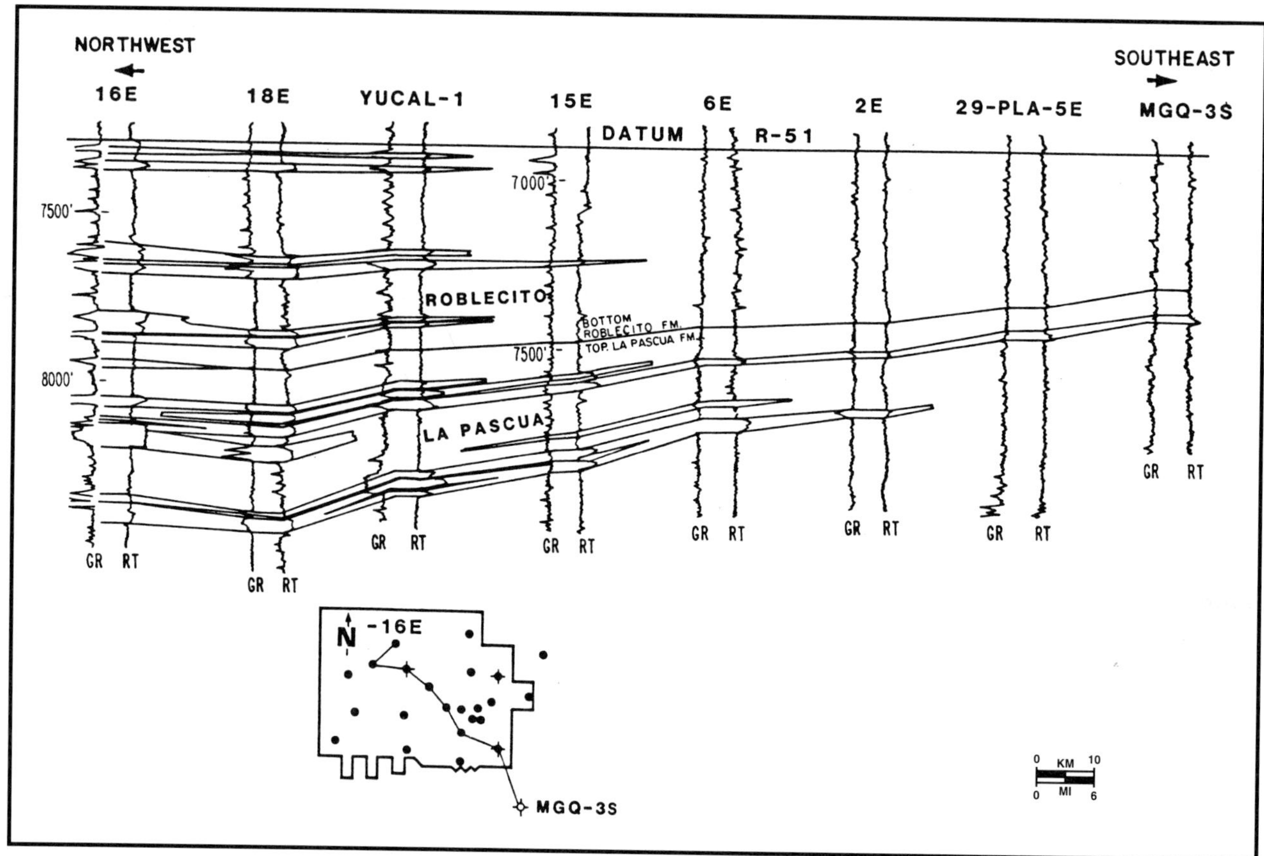

Figure 10. Northwest-southeast stratigraphic cross section of Roblecito/La Pascua formations showing the southward pinch-out of reservoir sandstones. Datum is the top of the Roblecito R-51 sandstone. (The electric logs show characteristic GR and RT curve signatures to demonstrate correlations of sands and their pinch-outs. No quantification intended.) (Modified from Aymard et al., 1985.)

present common characteristics: All thicken northward or northeastward, all pinch out to the south, and all are persistently developed throughout their area of extension. All sandstone bodies are oriented with their long axes approximately normal to the regional sedimentary trends.

Reservoirs

The more important gas reservoirs are Oligocene marine sandstones at depths between 3900 and 9300 ft (1190 and 2835 m). These are identified as the P-5, P-4, and P-2/3 sands of the La Pascua Formation and the R-56, R-54, R-38, R-26, and R-24 sands of the Roblecito formation, all in ascending order. Locally, the uppermost sandstones of the Roblecito formation and the lowermost sandstones of the Chaguaramas formation contain gas associated with liquid hydrocarbons and condensate. Also, one well produces gas from the Cretaceous Infante Limestone.

Stratigraphy and Facies

The La Pascua sandstones are related to a marginal marine environment. The P-4 sandstone is described from cores in well MGU-2 (Figure 12). The P-4 is the major gas reservoir of the field because of its permeability and has received the most attention in studies and investigations. The core shows two coastal bar sandstones, 7732 to 7708 ft (2358 to 2351 m), with well-developed "coarsening-upward" sequences separated by black shales, and in the upper part (7706 to 7677 ft; 2350 to 2341 m) a tidal channel sand with a "fining-upward" grain-size gradation and a basal erosional contact. Above this channel sandstone there are silty sandstones and shales with abundant fossil debris and other organic material.

Roblecito and Chaguaramas are less well documented. The Roblecito Formation, predominantly shaly and showing no carbonaceous strata, represents the period of maximum basin deepening and marine transgression.

Some sandstones are developed in bar systems up to 100 ft (30 m) thick (R-56, R-S4, R-38, R-26, and R-24). They are gray, fine-grained wackes and quartz

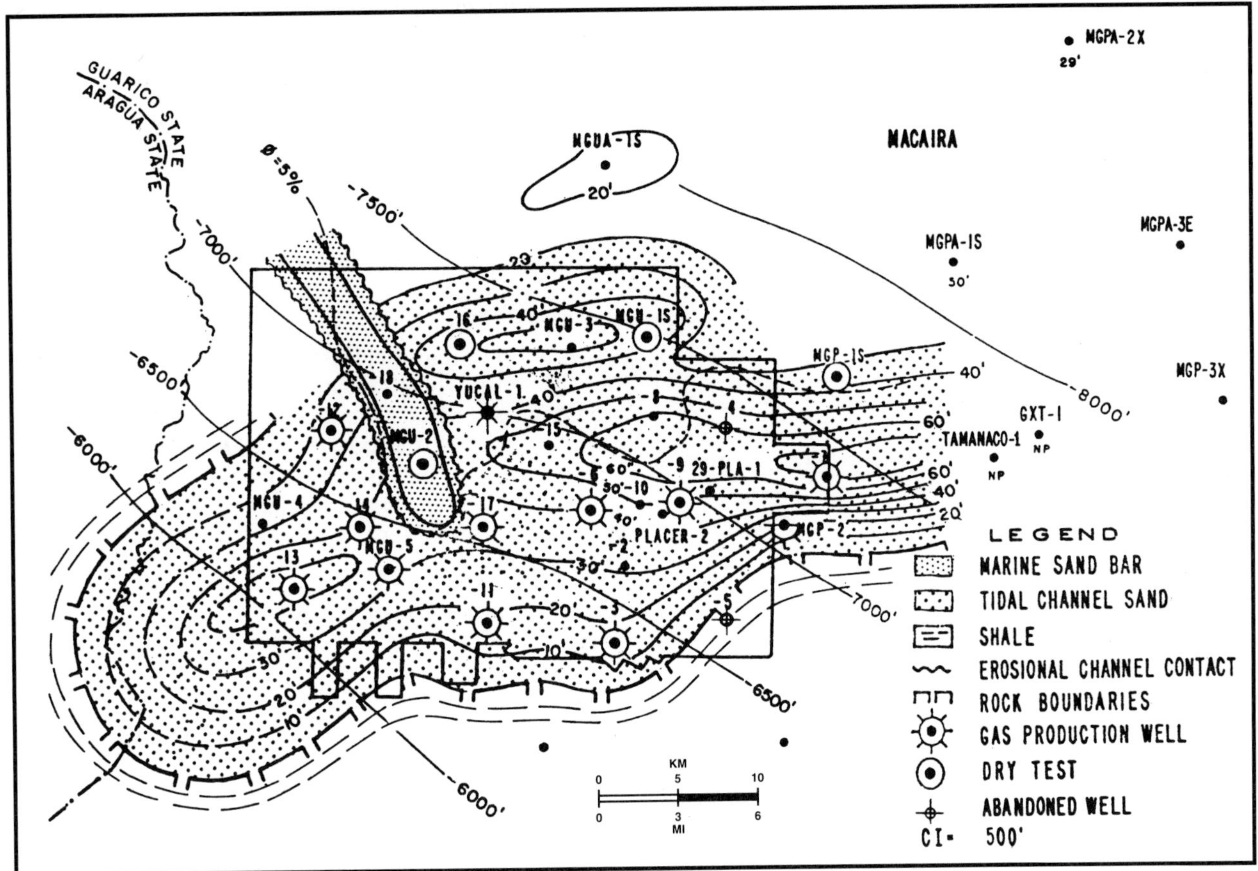

Figure 11. Structure/isopach/facies map of La Pascua P-4 sandstone, the largest reservoir of the Yucal-Placer field. (Modified from Laurier et al., 1987.)

arenites with wavy laminations and current ripple-marks with 1 to 4 cm of relief. Tabular cross-stratification, undulations, and ripple-marks are common. The shales are dark gray, laminated, and horizontally bioturbated at a millimetric scale. Sandstone-shale contacts are gradational, with shales grading progressively to shaly bioturbated siltstones and then to sands.

Porosity Types and Diagenesis

An important porosity phenomenon occurs in the field. As Roblecito and La Pascua sandstones thicken increasingly to the northeast and become progressively deeper, porosity and permeability decrease (Figure 14). Petrophysical and other laboratory investigations performed by Reistroffer (1987a) and Laurier and Gendrot (1987) have led to conclusions that could explain the porosity-permeability variations. Petrographic studies were made of the P-4 sandstone in two wells. In MGU-5, at a high structural position near the sand pinch-out in the southern part of the field, the P-4 has good porosities of 8 to 17%; in MGU-2, at an intermediate structural position in the middle of the field, low porosities of 3 to 7% have led to dry tests of the sandstone (Figures 15 and 16). In this well, diagenetic minerals were recognized where secondary silicification (pressure dissolution), strong compaction, and carbonate cementation played decisive roles in porosity reduction throughout the central and northern parts of the field (Figures 17 and 18). High secondary porosity in MGU-5, resulting in very good reservoir conditions, is possibly the result of dissolution of the carbonate cements.

Black amorphous products occurring as interstitial filling between sand grains occupy a volume of 3%. This material may be the altered remains of petroleum trapped during an intermediate maturation phase and is possibly responsible for protecting the structurally high parts of the reservoir from the silicification phenomenon characteristic of structurally lower sandstones below the paleo oil-water contact.

These observations have led to the postulation of two evolutionary schemes, one related to chemical composition and mineral diagenesis (Figure 19), the other to an early phase of oil entrapment and the existence of a paleo oil-water contact (Figure 20).

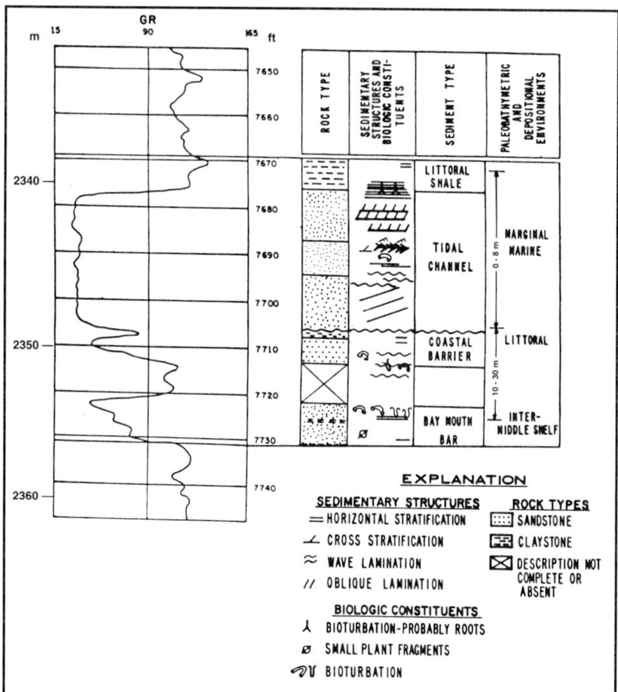

Figure 12. Lithologic and sedimentologic features of La Pascua P-4 sandstone cores from well MGU-2 (location shown on Figure 2). Arrowed depths in meters are water depths of deposition. (Interpretation by D. Laurier.)

Gas Characteristics

A summary of the characteristics of the proven reservoirs can be seen in Table 1. Yucal-Placer gas is predominantly thermogenic and is derived from marine organic material found in Cretaceous source rocks. This type of gas is accumulated in middle to upper sandstones of La Pascua and lower to middle Roblecito sandstones.

A second type of gas tested in upper Roblecito (R-1) and lower Chaguaramas (C-49) sands in wells MGP-1S and MGU-1S has a high C_2^+ content, also indicating thermogenic origin, but the Cl_1 to C_7 and CO_2 contents suggest an origin in terrestrial source rocks of possible Tertiary age. The large amounts of CO_2 (up to 20%) could have derived from carbonates, either from carbonate rock or from carbonate cement of detrital rocks.

Methane content increases from north to southeast and southwest at the expense of CO_2 content. Gas analyses from several wells of the field are shown in Figure 21.

Field Development Plans

Studies of future gas requirements for national markets show the need for production up to 125 MMCFD of gas from this field (dehydrated and untreated for removal of H_2S and CO_2). Seventeen wells are available for production from Roblecito and La Pascua sandstones; Cretaceous sandstones (with

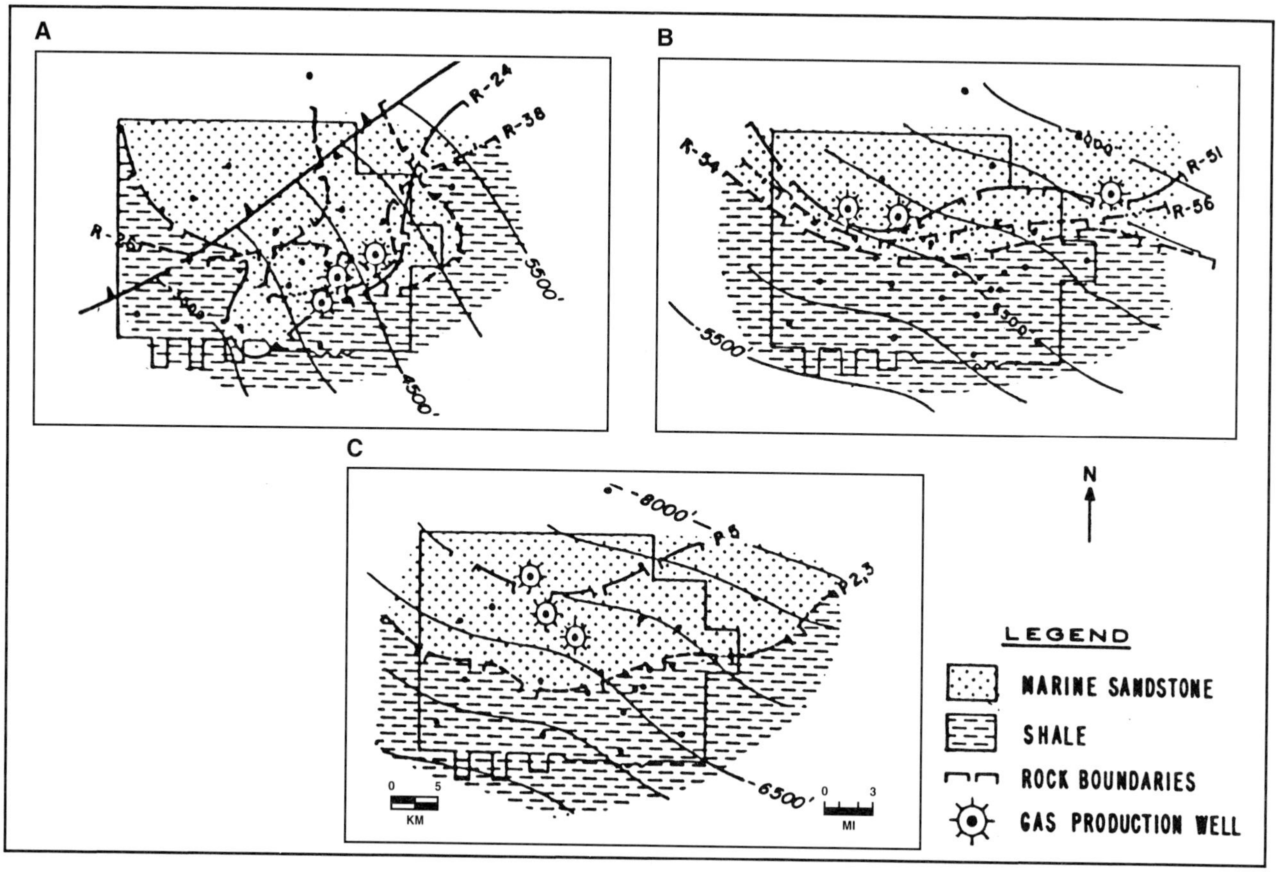

Figure 13. Marine sandstone pinch-out traps. Depths are in feet subsea with 500 ft contour intervals. (A) Middle Roblecito slope-deposit sandstones R-24, R-26, and R-38. R-24 structure. (B) Lower Roblecito slope and shelf sandstones R-51, R-54, and R-56. La Pascua structure. (C) La Pascua shelf marine bar sandstones P-2/3 and P-5 sandstones. La Pascua structure. (From Aymard et al., 1985.)

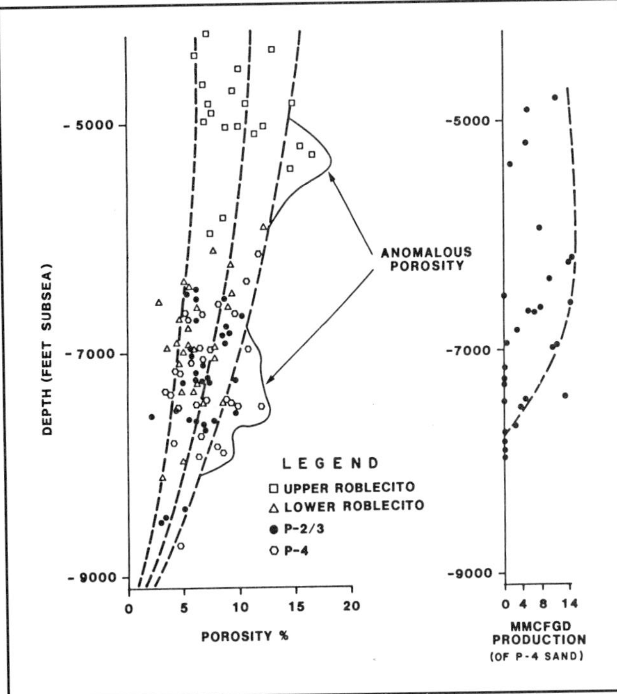

Figure 14. Porosity versus depth for upper and lower Roblecito sandstones and for La Pascua P-2/3 and P-4 sandstones. At the right, productivity versus depth of the P-4 sandstone. (Modified from Reistroffer, 1987a.)

high H_2S content) and the R-1 and C-49 sandstones are not considered at present for this need. Eight wells are presently connected to the Placer-Lechozo gas line, of which five are producing.

In order to maintain production at 125 MMCFD during the next 20 years, some additional wells are needed. The total wells planned will reach 25 at the end of the period. The production profile shown in Figure 22 follows the above development plan. It shows constant production levels during 21 years as well as the distribution among the various reservoirs: R-SUP (R-24/26 and R-38), P-4, and others (Roblecito Inferior, P-2/3 and P-5). No new well will be needed before 1998.

Hydrocarbon Source, Maturation, and Migration

Source Rock

Geochemical studies carried out by the industry laboratory (INTEVEP in 1984, 1985, and 1986) determined that the Cretaceous Tigre Formation (La Cruz, Infante, and Guavinita members) was the principal source rock for the gas found in the Guárico subbasin. This regional study involved organic, petrographic, and geochemical analysis of 350 rock samples from 45 wells of which three are located in the Yucal-Placer field; 15 gas samples from

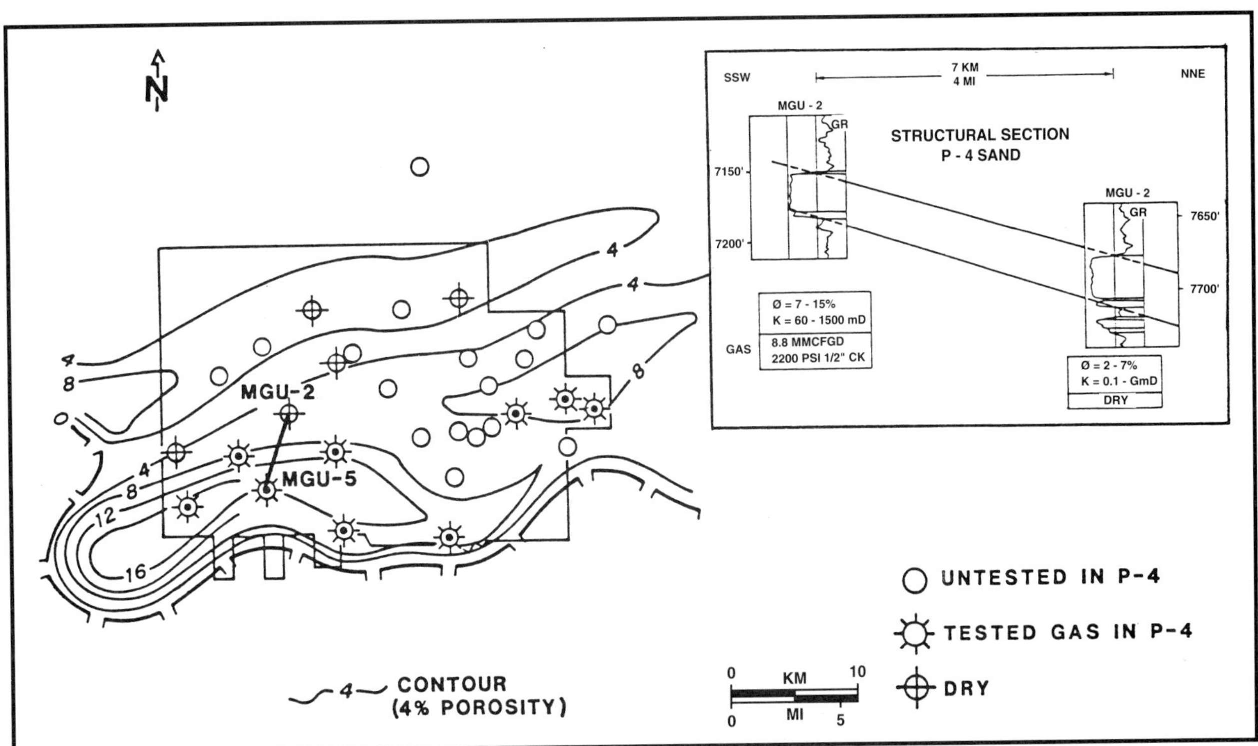

Figure 15. Porosity map of the La Pascua P-4 sandstone. Contour interval, 2%. The structural section at the right shows the relative structural positions of wells MGU-5 and MGU-2 and their similar P-4 sandstone gamma ray signatures. But the productive MGU-5 well has excellent porosity and permeability in the P-4 sandstone, while MGU-2 in the middle of the field has poor porosity and permeability and is dry in the P-4 sand. Depths shown in the structural section are in feet. (Modified from Reistroffer, 1987a.)

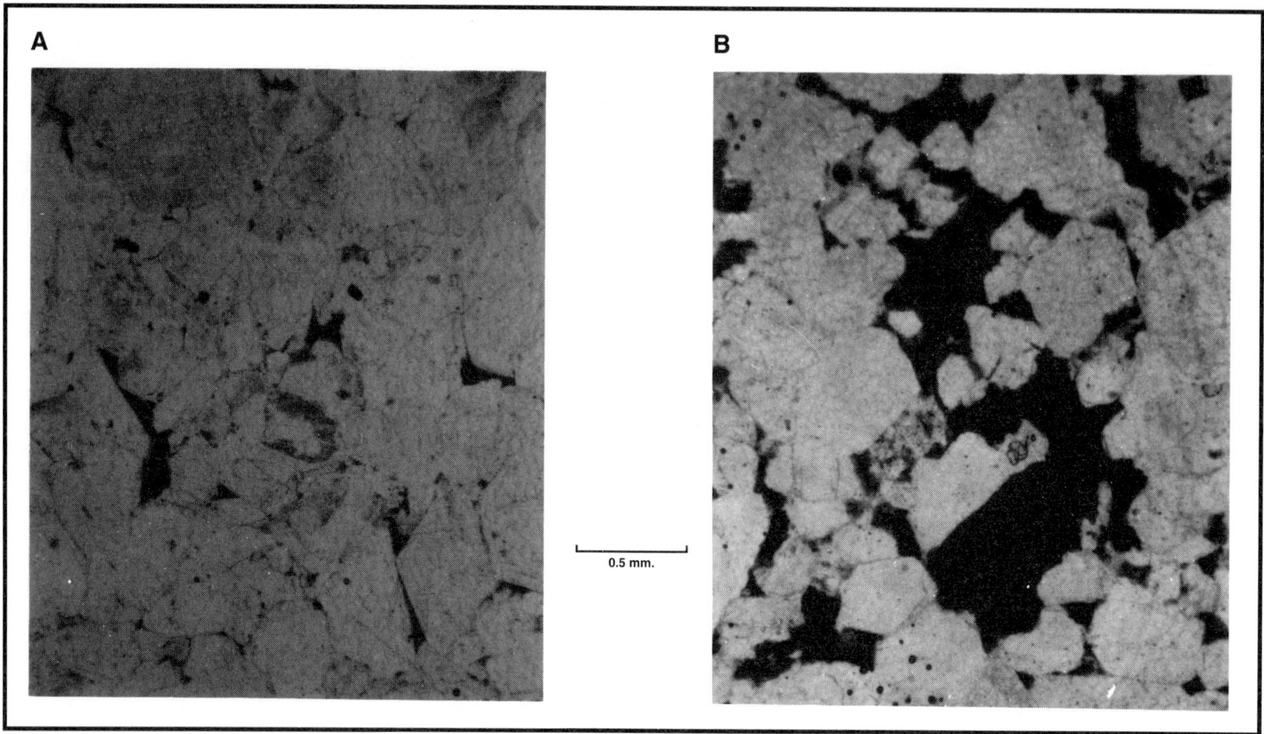

Figure 16. Thin sections, natural light, showing petrographic facies, P-4 sandstone, well MGU-5. (From Laurier and Gendrot, 1987.) (A) Compacted quartzites at 7179 ft (2190 m) depth, with 3.1% porosity and 0.34 md permeability. (B) Secondary porosity at 7154 ft (2182 m) depth. Porosity 15%. Permeability 1170 md.

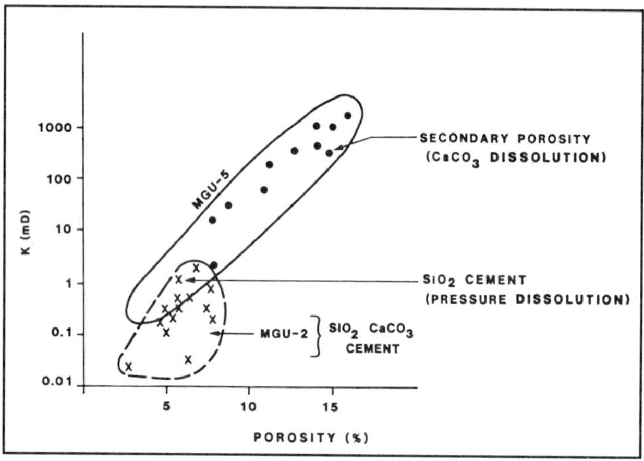

Figure 17. Porosity versus permeability of sandstones from wells MGU-2 and MGU-5. (From Laurier and Gendrot, 1987.)

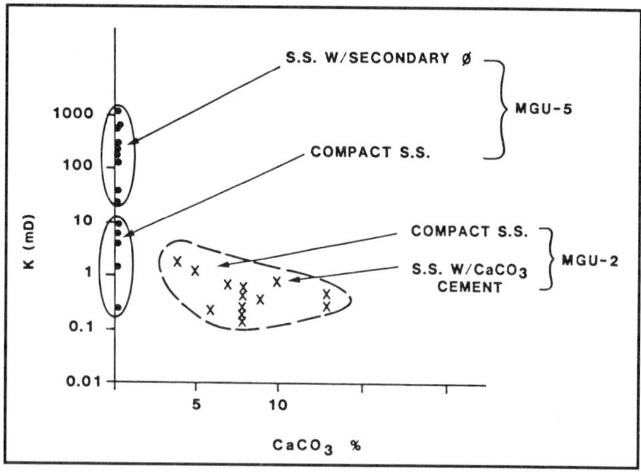

Figure 18. Percentage calcium carbonate cement versus permeability in sandstones from wells MGU-2 and MGU-5. (From Laurier and Gendrot, 1987.)

Roblecito and La Pascua reservoirs of the field were also analyzed.

(The following abbreviation are used in this discussion: TOC, total organic carbon; HC, hydrocarbon; MOE/TOC, mobil oil extract/total organic carbon; P/F, pristane/phytane; P/C_{17}, pristane/carbon 17; F/C_{18}, phytane/carbon 18; H/C, hydrogen/carbon; S/TT, steranes/tricyclic terpanes; HC/T, hydrocarbon weight/ton of rock.)

The La Cruz member, in 29-PLA-14E, has a total organic carbon (TOC) of 0.23 to 2.25%, with predominantly amorphous marine kerogen (80%) and a hydrocarbon index from 150 to 260 mgHC/g that could be diminished by maturation effects and influence of matrix minerals. The oil potential reaches 6.5 g HC/T of rock with an extractability of 9% (MOE/TOC). The source rock is thermally overmature and may be the source of Yucal-Placer gas.

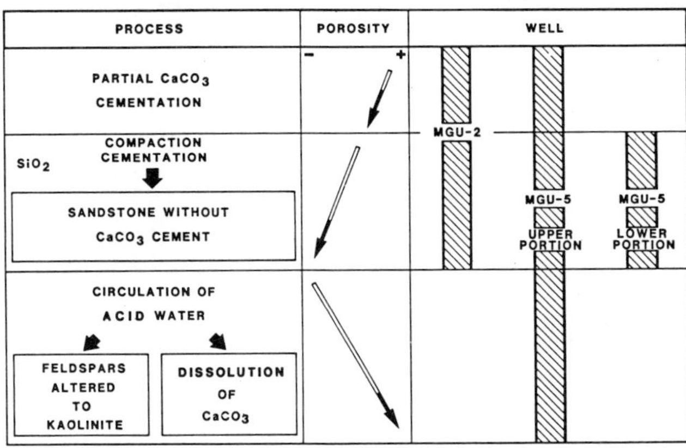

Figure 19. Evolutionary scheme of mineral diagenesis of P-4 sandstone resulting in variations of rock properties. An alternative postulated scheme is shown by Figure 20. (From Laurier and Gendrot, 1987.)

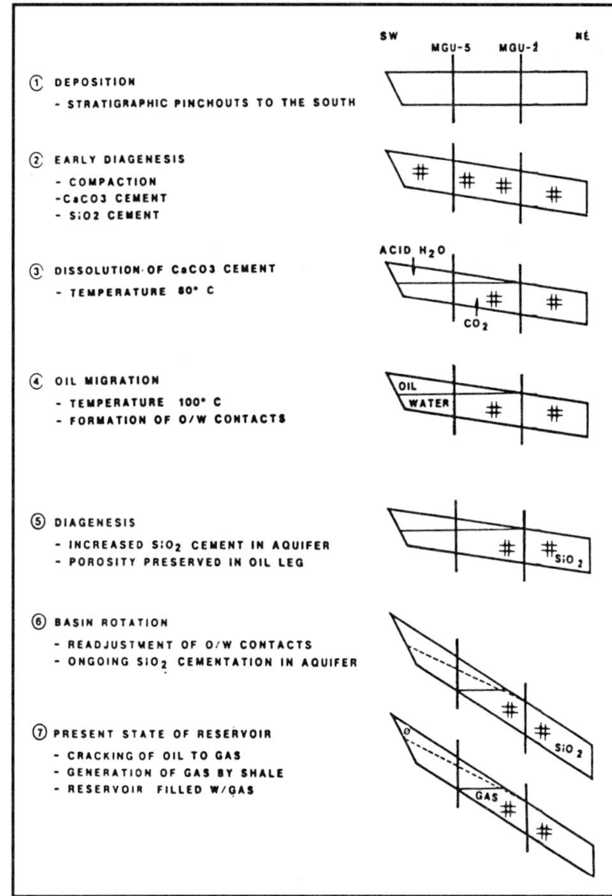

Figure 20. Alternative scheme of diagenetic evolution of P-4 sandstone resulting from an early oil entrapment phase, a paleo oil-water contact, and later tilting. This scheme is based on petrographic facies as seen in wells MGU-2 and MGU-5. The symbol # indicates water-filled reservoir and Φ is porosity. (Modified from Reistroffer, 1987b.)

The Infante limestone contains a medium to high organic matter content of 0.5 to 2.71% TOC, with 80% to 90% of amorphous kerogen deposited in a marine platform environment in water depths from zero to 165 ft (50 m). It is thermally overmature. But this rock could have generated liquid hydrocarbons before reaching the overmature stage when it would have produced gas.

Shales of the Guavinita member proved to have abundant organic material with 1 to 3% TOC and a tendency toward overmaturity. The kerogen is amorphous, type II, of marine origin, with a H/C ratio of 1.17. The material was deposited in bathyal marine (985–1970 ft; 300–600 m) depth. The low hydrocarbon index obtained from Rock-Eval pyrolysis may be attributed in part to thermal maturity and the effects of matrix minerals. Extractability attains a value of 14.5% (MOE/TOC) with the extracted bitumen showing an alkane distribution of C_{15}^{+}, which is characteristically marine and is reflected in the ratios of P/F (1.1), P/C_{17} (0.65), F/C_{18} (0.65), and S/TT (0.34) and in the abundance of tricyclic terpanes and the predominance of stereoisomers from the sterane C_{27} over those of C_{29}.

Thus, the Tigre Formation, whose maturity is seen in a refractive index of 1.2 to 2.0% R_o, is the best source rock for the gas produced in the field.

The La Pascua was analyzed for source rock characteristics in wells 29-PLA-14E and MGP-1S. Total organic content is medium to poor (0.5 to 1.2%) and is dominated by 60% to 80% of amorphous marine kerogen deposited in shallow platform waters (0–165 ft; 0–50 m depth). The kerogen is overmature. These

Table 1. Characteristics of Yucal-Placer field proven reservoirs.

Formation	% Porosity	MMCFGD I.P.	Original Reservoir Pressure psi (kPa)	Temperature °F (°C)
Roblecito	8-18	4-10	1500-3900 (10,342-26,890)	238-276 (114-136)
La Pascua	5-18	3-16	4000-4100 (27,580-28,270)	305-346 (152-174)

Parameters related to definition of reservoir limits are:

1. Gradual lateral loss of porosity down to limits of 3 to 4%.
2. Cementation and other diagenetic processes that limit and locally eliminate acceptable permeabilities.
3. Gas-water contacts have yet to be determined.

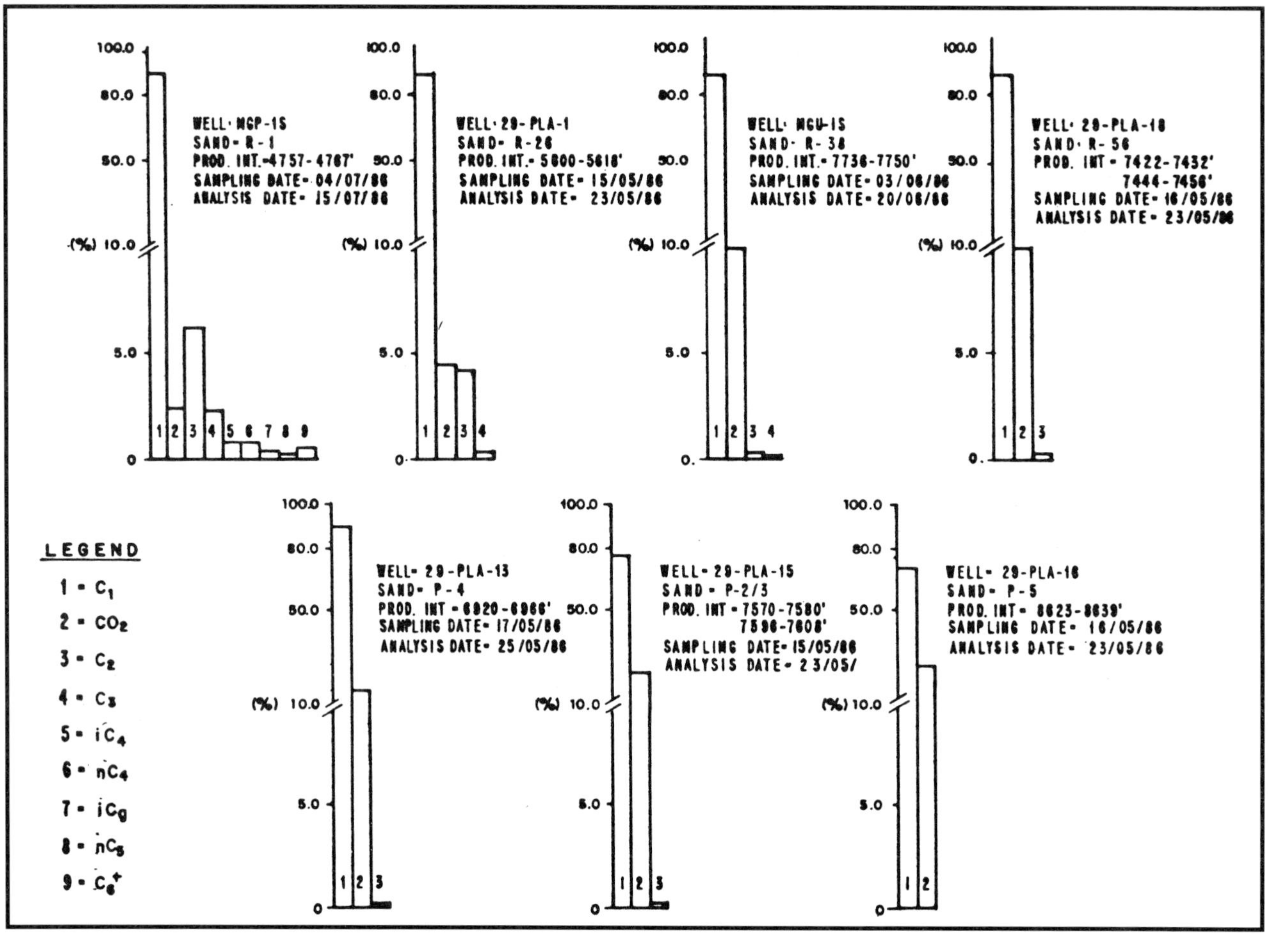

Figure 21. Gas analyses from several wells of the field showing percentage chemical compositions.

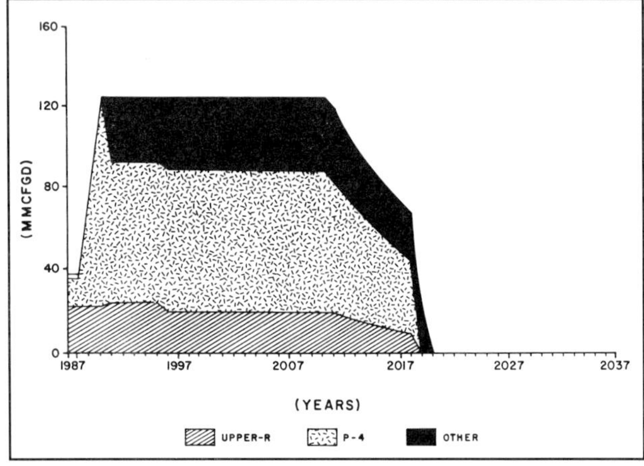

Figure 22. Future production rate profile resulting from development plan described in the text. The area patterns under the curves show the proportions of gas produced from different sandstones. The scheme is intended to provide a level production rate for domestic use for 21 years. Eight additional wells will be drilled between 1998 and 2009 to sustain production levels.

values, plus the high content of herbaceous kerogen (resins, waxes, etc.), indicate a possible source for paraffinic crude oils and gas.

Roblecito shales were analyzed in 29-PLA-15E. Organic material content is low (0.5 to 1.0%), with a high percentage of woody and only 15% of amorphous kerogen, low H/C ratios (0.8 in immature samples), and a low hydrogen index of 25–100. The shales are interpreted to be poor sources for liquid hydrocarbons. Because of overpressured zones, the well samples are overmature at depths greater than 5000 ft (1525 m) and marginally mature at the top of the formation.

Subsidence History

The subsidence diagram (Figure 23) is based on age and temperatures for the Canoa, Tigre, La Pascua, Roblecito, and Chaguaramas tops in well 29-PLA-15. The greatest source of uncertainty is the amount of Miocene sediments eroded and their absolute age at present ground level, for which reason compaction effects were ignored. For other wells (MGP-1S, 29-PLA-13, and 29-PLA-14), sonic logs and velocity data

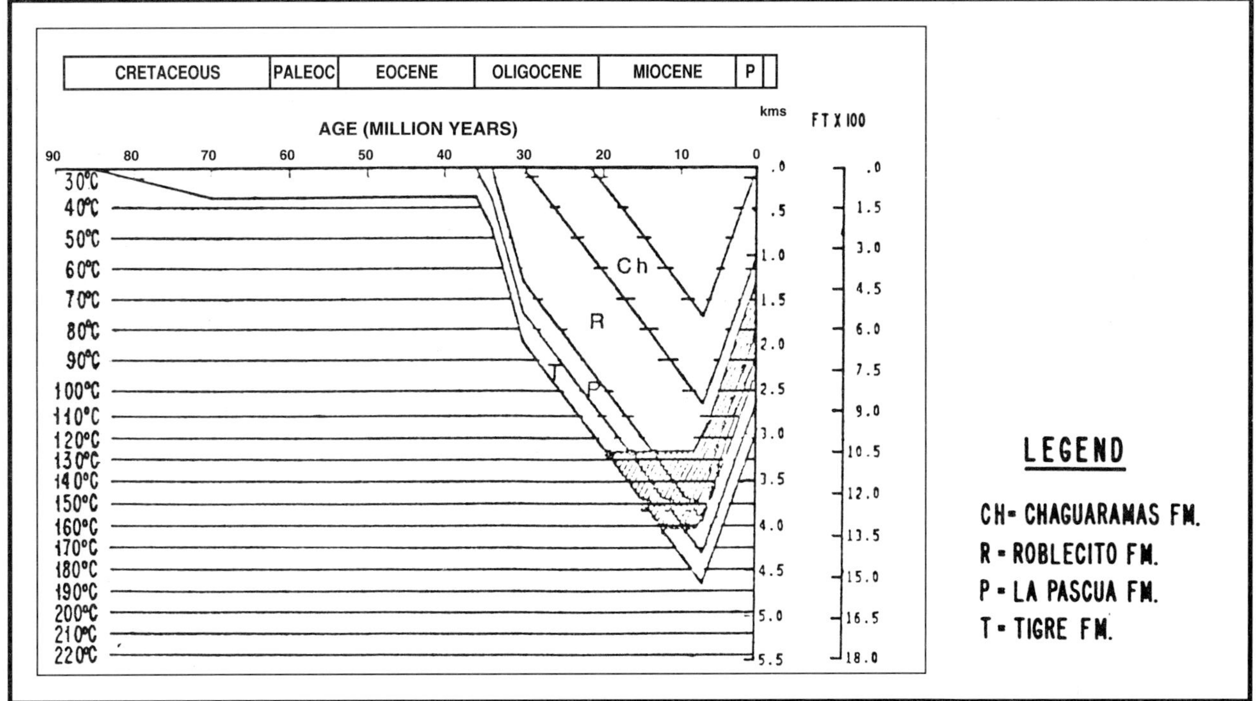

Figure 23. Subsidence history diagram based on age and temperature of the formations listed in the legend (well 29-PLA-15).

were used to detect overpressured intervals through the thick Oligocene shales.

Thermal History

The maturity level of organic material was calculated using bottom-hole temperatures corrected according to the Fertl and Wichmann method (1976). Present-day temperature gradients vary from 1.6° to 2.3°F/100 ft (0.27° to 0.39°C/m), the latter being the highest gradient known in the Eastern Venezuela Basin. Paleo temperature gradients were ignored owing to lack of data.

Estimation of Hydrocarbons Generated, Migrated, and Accumulated

Waples's (1981) equation was used to estimate generated gas volumes:

$$\text{volume generated} = Q_1Q_2M$$

where: Q_1 = volume of original organic material (TOC).
Q_2 = quality of hydrocarbon-generating kerogen.
M = maturity level of source rock.

Cretaceous Cycle—Gas volumes generated from the Tigre Formation were estimated for the volume of source rock with a maturity above 160 TTI (R_o = 1.35). These gas volumes were generated, first, from the carbon left in the kerogen after reaching the end of the oil-generation window and, second, from the bitumen retained in the source and converted to gas by thermal cracking.

The volume of gas generated from residual kerogen during the last 20 million years is estimated to be 1705×10^{12} ft^3. Of this, 75% or 1279×10^{12} ft^3 were expulsed from the source rock. Of the amount of expulsed gas, 5% or 71.7×10^{12} ft^3 migrated and was preserved in present-day traps during the last 5 million years, gas migrated previously not having been retained in the structures.

Unexpulsed bitumen is considered to represent 80% to 60% of the total generated crudes; of this, 4.0% was converted to gas at maturity levels equal to or greater than 10,000 TTI. If 75% of the bitumen-generated gas was expulsed from the source rock, and 5% of that volume was trapped and preserved, the bitumen-derived volume of gas would measure between 47 and 62.6×10^{12} ft^3.

Oligocene Cycle—The greater part of Oligocene-sourced gas is attributed to the La Pascua source rocks, estimated to have generated 84.1×10^{12} ft^3 of gas from overmature kerogen, of which 2% or 1.683×10^{12} ft^3 reached traps. Gas derived from thermal cracking of residual bitumen in La Pascua source rocks and accumulated in traps is estimated at 1.334×10^{12} ft^3. Thus, the total amount of La Pascua-sourced gas is estimated to be about 3.02×10^{12} ft^3 (see Table 2 for a summary of the gas volumes).

Table 2. Summary of gas volumes generated and entrapped.

Time (Ma)	Volume Generated ($\times 10^{12}$ ft³)	Volume Expelled ($\times 10^{12}$ ft³)	(%)	Volume Trapped ($\times 10^{12}$ ft³)
20-0	1706.5	1279.5	75	72.57
8-5	84.1	63.1	75	3.02
Total	1790.6	1342.6	75	75.59 (±8)

Combined Cretaceous and Tertiary gas generated in this area is, therefore, on the order of 68×10^{12} to 83×10^{12} ft³.

EXPLORATION CONCEPTS

The first exploratory wells in the Yucal-Placer area were drilled on surface structures and it was originally thought that the reservoirs were fault controlled. However, during the early 1980s, with the drilling of several wells that provided good subsurface data, it was found that the updip pinch-outs of the sandstones were the controlling factors in the entrapment of gas. By utilizing new seismic techniques, such as seismic inversion and sonic logs, it was possible to pinpoint the stratigraphic traps and thus better locate new exploratory wells and delineate the sand bodies.

For this type of stratigraphic play, it is believed that sophisticated seismic techniques are the best tools to use for pinpointing the traps. With a few regional seismic lines to first depict the major trends of sand development, localized patterns of seismic grids can then be placed so as to trace pinch-out areas and possible traps. A series of stratigraphic test holes to determine sandstone thickness, porosity, and diagenetic characteristics combined with the seismic interpretation should provide a good delineation and evaluation of the prospective traps.

ACKNOWLEDGMENTS

The authors would like to express their thanks to Corpoven, S.A. (affiliate of Petróleos de Venezuela, S. A.), which permitted publication of this paper, Lisys Pérez who supplied production data, D. Kiser for his contributions and English translation, and J. Reistroffer for his suggestions.

REFERENCES CITED

Aymard, R., J. Q. Daal, and M. Coriat, 1985, Campo Yucal-Placer. Trampa Estratigráfica Gigante de Gas en la Cuenca Oriental de Venezuela. VI Congreso Geológico de Venezuela—Caracas, p. 2779-2803.

Bolivar, E., and L. Vierma, 1987, Geoquímica del gas natural en Yucal Placer: Unpublished technical report, Intevep, S. A., Los Teques.

Fertl, W. H., and P. A. Wichmann, 1976, Static formation temperature from well logs: Dresser-Atlas Technical Memorandum, v. 7, n. 3, Dresser-Atlas Corporation.

Laurier, D., and C. Genrot, 1987, Efectos de la diagénesis mineral en las areniscas del Campo Yucal-Placer (Estado Guárico): IV Jornadas Técnicas Corpoven, Puerto La Cruz.

Laurier, D., J. Q. Daal, and P. Alfonsi, 1987, Litofacies y ambientes de sedimentación del intervalo P-4, Formación La Pascua, Area Yucal Placer-Macaira: IV Jornadas Técnicas Corpoven, Puerto La Cruz.

Reistroffer, J., 1987a, Modelo exploratorio basado en un análisis petrofísico de los campos Yucal-Placer, Macaira, Cardonal y Uveral, al noroeste de Guárico: IV Jornadas Técnicas Corpoven, Puerto La Cruz.

Reistroffer, J., 1987b, Suministro de 125 MMPCND de gas al Mercado Interno desde el campo Yucal-Placer. Vol. 1 and II: Unpublished technical report, Corpoven, S.A., Puerto La Cruz.

Waples, D. W., 1981, Organic geochemistry for exploration geologists: Burgess Publishing Company, CEPCD Division, 151 p.

Appendix 1. Field Description

Field name .... *Yucal-Placer field*

Ultimate recoverable reserves .... *4500 bcf of nonassociated gas*

Field location:

- **Country** .... *Venezuela*
- **State** .... *Guárico*
- **Basin/Province** .... *Eastern Venezuela Basin/Guárico subbasin*

Field discovery:

- **Year first pay discovered** .... *Middle Oligocene sandstones 1947*
- **Year second pay discovered** .... *Lower Oligocene sandstones 1957*
- **Year third pay discovered** .... *Cretaceous limestones 1981*
- **Year fourth pay discovered** .... *Lower Miocene sandstones 1983*

Discovery well name and general location:

- **First pay** .... *Placer-1, 150 km southeast of Caracas, Distrito Federal*
- **Second pay** .... *Yucal-1, 12.6 km northwest of Placer-1*
- **Third pay** .... *29-PLA-8E, 6 km north of Placer-1*
- **Fourth pay** .... *MGU-1S, 11 km north of Placer-1*

Discovery well operator .... *S.A.P. Las Mercedes in conjunction with the Venezuelan Atlantic Refining Co. and Creole Corp.*

- **Second pay** .... *Mene Grande Oil Co.*
- **Third pay** .... *Corporación Venezolana del Petróleo*
- **Fourth pay** .... *S.A. Meneven*

IP

- **First pay** .... *9.8 MMCFD (½-in. choke)*
- **Second pay** .... *4.3 MMCFD (½-in. choke)*
- **Third pay** .... *14.7 MMCFD (½-in. choke)*
- **Fourth pay** .... *1.1 MMCFD + 16 bbl/day condensate (½-in. choke)*

All other zones with shows of oil and gas in the field:

Age	Formation	Type of Show
Cretaceous	*Tigre*	*Dry gas*
Oligocene	*La Pascua and Roblecito*	*Dry gas*
Miocene	*Chaguaramas*	*Associated gas/condensate*

Geologic concept leading to discovery and method or methods used to delineate prospect

Surface geology followed by gravity and torsion balance methods, with better apparent prospective areas covered by reflection and refraction seismic methods, and last, structural drilling.

Structure:

Province/basin type .... *Bally 221; Klemme II Ca; foreland basin*

Tectonic history

The Cretaceous sediments were deposited in a great platform area. During Oligocene and earlier Miocene deposition, the section dipped to the north-northwest, but in the upper Miocene, tectonic compression from the north caused folding, thrust faulting, reactivation of older faults, and a change in the structural dip to the northeast.

Regional structure

The Yucal-Placer field is located on the northwestern flank of the Eastern Venezuela Basin at the foot of the Overthrust belt of the Caribbean Mountain System.

Local structure

Two important structural aspects are present in the Yucal-Placer field: (1) a general 2° to 3° homoclinal dip to the northeast in the Cretaceous, La Pascua, and lower Roblecito formations; (2) the upper Roblecito and Chaguaramas formations are overthrusted in the northwest part of the field, with the middle Roblecito, a shaly section, serving as décollement surface. The thrusts form northeast-southwest-trending anticlines.

Trap:

Trap type(s)

The Yucal-Placer field is composed of multiple updip stratigraphic pinch-outs and one anticlinal trap with three pay zones.

Basin stratigraphy (major stratigraphic intervals from surface to deepest penetration in field):

Chronostratigraphy	Formation	Depth to Top in ft (m)
Lower-middle Miocene	*Chaguaramas*	*Outcrops at surface*
Middle-upper Oligocene	*Roblecito*	*2350-6500 (720-1980)*
Lower Oligocene	*La Pascua*	*6500-9300 (1980-2840)*
Cretaceous	*El Tigre*	*7750-11,000 (2360-3360)*

Reservoir characteristics:

Number of reservoirs *4*

Formations .... *Tigre Formation, La Pascua Formation, Roblecito Formation, and Chaguaramas Formation*

Ages .... *Tigre, Cretaceous; La Pascua, lower Oligocene; Roblecito, middle and upper Oligocene; Chaguaramas, lower Miocene*

Depths to tops of reservoirs .... *Tigre, Cretaceous 10,200 ft (3110 m); La Pascua, 6500 ft (1980 m); Roblecito, 2350 ft (720 m); Chaguaramas, 2280 ft (700 m)*

Gross thickness (top to bottom of producing interval) .... *Tigre, 120 ft (37 m); La Pascua, 1250 ft (380 m); Roblecito, 4200 ft (1280 m); Chaguaramas, 3000 ft (920 m)*

Net thickness--total thickness of producing zones

Average .... *Tigre, 20 ft (6.1 m); La Pascua, 30 ft (9.2 m); Roblecito, 15 ft (4.6 m); Chaguaramas, 10 ft (3.0 m)*

Maximum .... *Tigre, 24 ft (7.3 m); La Pascua, 200 ft (61.0 m); Roblecito, 100 ft (30.5 m); Chaguaramas, 70 ft (21.3 m)*

Lithology

Tigre: subdivided into La Cruz, Infante, and Guavinita members

La Cruz Member: composed of lenticular, coarse, kaolinitic sandstones with minor intercalations of black, carbonaceous, fossiliferous shale

Infante Member: gray to brown, locally glauconitic, compact, fossiliferous limestone

Guavinita Member: sandstones and thin limestones associated with fish remains and chert beds

La Pascua: brackish water to marine sandstones and dark shales with a few lignites

Roblecito: predominantly marine shale, dark gray to black, with few sandstones

Chaguaramas: sandstone-shale-lignite series

Porosity type .... *Intergranular secondary porosity by dissolution of unstable minerals*

Average porosity .... *Cretaceous, 5%; Oligocene, 12%; Miocene, 15%*

Average permeability .... *Cretaceous and Oligocene, 850 md; Miocene, 1500 md*

Seals:

Upper

Formation, fault, or other feature .... *Shales interbedded with producing sandstones*

Lithology .... *Shales*

Lateral

Formation, fault, or other feature .... *Sandstone pinch-outs*

Lithology .... *Shales*

Source:

Formation and age .... *Tigre, Cretaceous; La Pascua, Oligocene*
Lithology .... *Shales*
Average total organic carbon (TOC) .... *Tigre, 1.6%; La Pascua, 0.75%*
Maximum TOC .... *Tigre, 2.71%; La Pascua, 10%*
Kerogen type (I, II, or III) .... *Tigre, II; La Pascua, III*
Vitrinite reflectance (maturation) .... R_o = *Tigre, 2.0%; La Pascua, 1.3%*
Time of hydrocarbon expulsion .... *Tigre, 20–0 m.y.; La Pascua, 8–5 m.y.*
Present depth to top of source .... *Tigre, 10,000 ft (3050 m); La Pascua, 4500 ft (1370 m)*
Thickness .... *Tigre, 400 ft (120 m); La Pascua, 4000 ft (1200 m)*
Potential yield .... *Tigre, 1706* $\times 10^{12}$ *ft*^{3}*; La Pascua 84.1* $\times 10^{12}$ *ft*^{3}

Appendix 2. Production Data

Field name .... *Yucal-Placer field*

Field size:

Proved acres .... *235,300 ac (95,300 ha)*
Number of wells all years (including 8 future wells) *41*
Current number of wells *33*
Well spacing .... *2500 m (8200 ft)*
Ultimate recoverable .... *4500 bcf*
Cumulative production .... *1800 mmcf on 9/30/87*
Annual production (1987) .... *6000 mmcf*
Decline rate .... *Not enough production history to calculate*
Annual water production .... *36,000 bbl*
In place, total reserves .... *6400 tcf*
In place, per acre foot .... *442 mcf/ac-ft*
Primary recovery .... *4500 bcf*
Secondary and enhanced recovery .... *NA*
Cumulative water production .... *110,000 bbl*

Drilling and casing practices:

Amount of surface casing set .... *10,000 ft (3020 m)*
Casing program .... *13⅜-in. at 1500 ft (455 m); 9⅝-in. at 7500 ft (2260 m); 7-in. at 10,000 ft (3020 m)*
Drilling mud .... *Inverted emulsion and lignosulfonate*
Bit program .... *17½-in. to 1500 ft (455 m); 12¼-in. to 7500 ft (2260 m); 8⅜-in. to 10,000 ft (3020 m)*
High pressure zones .... *Middle Oligocene (Roblecito Formation)*

Completion practices:

Perforated intervals
Tigre, Guavinita and Infante members; La Pascua, P-8, 9; P-4 and P-2 sandstones; Roblecito, R-56, R-54, and R-50 sandstones; Chaguramas, C-46 sandstone
Well treatment .... *None*

Formation evaluation:

Logging suites .... *DIL/GR, BHC, LDL/CNL/GR and WST*
Testing practices .... *RFT and DST*
Mud logging techniques .... *Standard ditch samples*

Gas characteristics:

Type .... *Dry gas (methane 90%) with 48° API condensate*
Viscosity, SUS .... *0.022227 cp (at 4050 psi or 27,925 kPa)*

Field characteristics:

Average elevation .... *776 ft (235 m)*
Initial pressure .... *4050 psi (27,925 kPa)*
Present pressure .... *3950 psi (27,235 kPa)*
Pressure gradient .... *0.068 psi/ft (1.538 kPa/m)*
Temperature .... *330°F (166°C)*
Geothermal gradient .... *3.1°F/100 ft (0.056°C/m)*
Drive .... *Gas expansion*

Transportation method and market for oil and gas:
Pipeline of 20-in. diameter, internal market

AAPG TREATISE OF PETROLEUM GEOLOGY

The American Association of Petroleum Geologists
gratefully acknowledges and appreciates the leadership and support
of the AAPG Foundation in the development of the
Treatise of Petroleum Geology

STRUCTURAL TRAPS IV
TECTONIC AND NONTECTONIC FOLD TRAPS

COMPILED BY
EDWARD A. BEAUMONT
AND
NORMAN H. FOSTER

TREATISE OF PETROLEUM GEOLOGY
ATLAS OF OIL AND GAS FIELDS

PUBLISHED BY
THE AMERICAN ASSOCIATION OF PETROLEUM GEOLOGISTS
TULSA, OKLAHOMA 74101, U.S.A.

ISBN: 0-89181-584-8
ISSN: 1043-6103

Available from:
The AAPG Bookstore
P.O. Box 979
Tulsa, OK 74101-0979

Phone: (918) 584-2555
Telex: 49-9432
FAX: (918) 584-0469

Association Editor: Susan Longacre
Science Director: Gary D. Howell
Publications Manager: Cathleen P. Williams
Special Projects Editor: Anne H. Thomas
Science Staff: William G. Brownfield
Project Production: Custom Editorial Productions

TABLE OF CONTENTS

TREATISE OF PETROLEUM GEOLOGY
ADVISORY BOARD

Eric A. Rudd
Floyd F. Sabins, Jr.
Nahum Schneidermann
Peter A. Scholle
George L. Scott, Jr.
Robert T. Sellars, Jr.
Faroog A. Sharief
John W. Shelton
Phillip W. Shoemaker
Synthia E. Smith
Robert M. Sneider
Stephen A. Sonnenberg
William E. Speer
Ernest J. Spradlin
Bill St. John
Philip H. Stark
Richard Steinmetz
Per R. Stokke
Denise M. Stone
Donald S. Stone
Doug Strickland
James V. Taranik
Harry Ter Best, Jr.
Bruce K. Thatcher, Jr.
M. Ray Thomasson
Jack C. Threet
Bernard Tissot
Donald F. Todd
M. O. Turner
Peter R. Vail
B. van Hoorn
Arthur M. Van Tyne
Ian R. Vann
Harry K. Veal*
Steven L. Veal
Richard R. Vincelette
Cecil von Hagen
Fred J. Wagner, Jr.
William A. Walker, Jr.
Anthony Walton
Douglas W. Waples
Harry W. Wassall, III
W. Lynn Watney
N. L. Watts
Koenradd J. Weber
Robert J. Weimer
Dietrich H. Welte
Alun H. Whittaker
James E. Wilson, Jr.
John R. Wingert
Martha O. Withjack
P. W. J. Wood
Homer O. Woodbury
Walter W. Wornardt
Marcelo R. Yrigoyen
Mehmet A. Yukler
Zhai Guangming
Robert Zinke

* Deceased

American Association of Petroleum Geologists Foundation
Treatise of Petroleum Geology Fund*

Major Corporate Contributors
($25,000 or more)

BP Exploration Company Limited
Chevron Corporation
Exxon Company, U.S.A.
Mobil Oil Corporation
Oryx Energy Company
Pennzoil Exploration and Production Company
Shell Oil Company
Union Pacific Foundation
Unocal Corporation

Other Corporate Contributors
($5,000 to $25,000)

Ashland Oil, Inc.
Cabot Oil & Gas Corporation
Canadian Hunter Exploration Ltd.
Conoco Inc.
Marathon Oil Company
The McGee Foundation, Inc.
Phillips Petroleum Company
Texaco Philanthropic Foundation Inc.
Transco Energy Company

Major Individual Contributors
($1,000 or more)

John J. Amoruso
C. Hayden Atchison
Richard A. Baile
Richard R. Bloomer
A. S. Bonner, Jr.
David G. Campbell
Herbert G. Davis
Paul H. Dudley, Jr.
Lewis G. Fearing
James A. Gibbs
George R. Gibson
William E. Gipson
Robert D. Gunn
Merrill W. Haas
Cecil V. Hagen
Frank W. Harrison
William A. Heck
Roy M. Huffington
Harrison C. Jamison
Thomas N. Jordan, Jr.
Hugh M. Looney
Jack P. Martin
John W. Mason
George B. McBride
Dean A. McGee
John R. McMillan
Grover E. Murray
Rudolf B. Siegert
Robert M. Sneider
Jack C. Threet
Charles Weiner
Harry Westmoreland
James E. Wilson, Jr.
P. W. J. Wood

The Foundation also gratefully acknowledges the many who have supported this endeavor with additional contributions.

*Based on contributions received as of October 12, 1990.

PREFACE

The Atlas of Oil and Gas Fields and the Treatise of Petroleum Geology

The *Treatise of Petroleum Geology* was conceived during a discussion held at the annual AAPG meeting in 1984 in San Antonio, Texas. This discussion led to the conviction that AAPG should publish a state-of-the-art textbook in petroleum geology, aimed not at the student, but at the practicing petroleum geologist. The textbook gradually evolved into a series of three different publications: the Reprint Series, the Atlas of Oil and Gas Fields, and the Handbook of Petroleum Geology. Collectively these publications are known as the *Treatise of Petroleum Geology*, AAPG's Diamond Jubilee project commemorating the Association's 75th anniversary in 1991.

With input from the Advisory Board of the Treatise of Petroleum Geology, we designed this set of publications to represent, to the degree possible, the cutting edge in petroleum exploration knowledge and application: the Reprint Series to provide useful and important published literature; the Atlas to comprise a collection of detailed field studies that illustrate the many ways oil and gas are trapped and to serve as a guide to the petroleum geology of basins where these fields are found; and the Handbook as a professional explorationist's guide to the latest knowledge in the various areas of petroleum geology and related disciplines.

The Treatise Atlas is part of AAPG's long tradition of publishing field studies. Notable AAPG field study compilations include *Structure of Typical American Fields*, published in 1929 and edited by Sidney Powers; and Memoir 30, *Giant Fields of 1968-1978*, published in 1981 and edited by Michel T. Halbouty. The Treatise Atlas continues that tradition but introduces a format designed for easier access to data.

Hundreds of geologists participated in this first compilation of the Atlas. Authors are from all parts of the industry and numerous countries. We gratefully acknowledge the generous contribution of their knowledge, resources, and time.

Purpose of the Atlas

The purpose of the Atlas is twofold: (1) to help exploration and development geologists become more efficient by increasing their awareness of the ways oil and gas are trapped, and (2) to serve as a reference for both the petroleum geology of the fields described and the basins in which they occur.

Imagination is the primary tool of the explorationist. Wallace E. Pratt once said that the unfound field must first be sought in the mind. In part, what is imagined is based on what is remembered; memory is the direct link to what is created in the mind. To create ideas that lead to the discovery of new fields, the mind of the geologist builds from its knowledge of petroleum geology. To that end, the Atlas of field studies will be a primary source for locating much of the information necessary for creating prospects and will provide a connection to the phenomenon of oil and gas traps.

Next to the firsthand experience of having prospects tested with the drill bit, studying the many facets and concepts of developed fields is perhaps the best way for the geologist to develop the ability to create plays and prospects. Also, familiarity with the many ways oil and gas are trapped allows the geologist to see through the noise inherent to exploration data and to close gaps in that data.

Format of the Atlas

To facilitate data access, all field studies in the Atlas follow the same format. Once users become familiar with this format, they will know where to look for the information they seek. Different fields from different parts of the world can be easily compared and contrasted.

The following is a generalized format outline for field studies in the Atlas:

- Location
- History
  - Pre-Discovery
  - Discovery
  - Post-Discovery
- Discovery Method
- Structure
  - Tectonic History
  - Regional Structure
  - Local Structure
- Stratigraphy
- Trap
  - General Description
  - Reservoir(s)
  - Source(s)
- Exploration Concepts

Criteria for Inclusion of a Field

Fields described in the Atlas are selected using two main criteria: (1) trap type, and (2) geographic distribution. Our ultimate goal for the Atlas is to include a field study from each major petroleum-producing province and to include an example of each known trap type. Size or economic importance are not, of themselves, criteria. Many fields that are not giants are included because they are geologically unique, because they are significant examples of geological investigation and original thinking, or because they are historically important, having led to the discovery of many other fields.

Grouping of Fields into Separate Volumes

We considered several ways to group fields in these volumes. We chose trap type because the purpose of the Atlas is to make exploration geologists more effective oil and gas trap finders, regardless of where they search for traps.

Grouping oil and gas field studies into separate volumes by trap type is a difficult exercise. We decided to group the fields into volumes by designating them as structural or stratigraphic traps. Most traps are a combination of both structure and stratigraphy. Some traps are obviously more a consequence of one than the other, but many are not. The continuum that exists between purely stratigraphic and purely structural traps is what makes grouping difficult. A further complication is that many fields contain more than one trap type.

Papers Selected for *Structural Traps IV*

The volume in hand is a collection of field studies that describe traps resulting from folding, whether the trap was caused by tectonic or nontectonic forces.

The first four studies are of Paleocene–Cretaceous chalk reservoirs domed over salt diapirs in the North Sea: historic Ekofisk, discovered in 1969; Eldfisk and West Ekofisk, discovered within the following year; and Tommeliten Gamma, discovered in 1977, after a great deal was known about the chalk reservoirs. Despite common origins, reservoirs, lithologies, and proximity, each of these fields has unique characteristics. Collectively, their discovery and development comprise the history of important seismic and field development techniques. Baffling gas effects that distorted seismic records and led to costly misinterpretations at the time have since become signposts of significant reservoirs.

The volume also includes studies of fields located elsewhere in the North Sea that produce from Jurassic and Paleocene sandstone reservoirs associated with folds—the Fulmar, Midgard, Sleipner Øst, and Sleipner Vest fields. Some of these are profoundly affected by faults and some are modified by facies changes.

Two Australian fields are from fold traps with stratigraphic modifications: the Jackson oil field, with dip closure and multiple pays; and the Palm Valley gas field, from a fractured reservoir.

Other equally fascinating studies include the giant Uzen anticlinal trap with multiple pays in the Caspian basin; the Helez-Brur-Kokhav, Israel's most significant field; the structural stratigraphic Appleton oil field in Alabama; and the Lindsborg field of the Kansas Salina basin.

Another aspect described in each field study is the history of its exploration and development. What seems obvious today usually was not obvious when the fields were discovered. Sometimes what was expected was not what was encountered. Geologists found these fields by creating concepts based on a limited amount of information. The information at their disposal was limited by the technology available at the time. Drilling and discovery show how often and how closely concept matches reality. Knowing the history of discovery may help explorationists realize that problems, seemingly insoluble at one time, eventually were solved. It also is instructive to learn about the sequence of thinking that solved these problems.

Study these fields carefully to become a better creator of prospect concepts and enjoy learning about the fascinating science of petroleum geology. Good hunting!

Edward A. Beaumont
Norman H. Foster, Editors

Ekofisk Field—Norway
Central Graben, North Sea

CHARLES T. FEAZEL
IAN A. KNIGHT
LAWRENCE J. PEKOT
Phillips Petroleum Company
Bartlesville, Oklahoma

FIELD CLASSIFICATION

BASIN: North Sea
BASIN TYPE: Rift
RESERVOIR ROCK TYPE: Chalk (Fractured)
RESERVOIR ENVIRONMENT OF DEPOSITION: Pelagic Coccolith Ooze Resedimented Mainly by Debris Flow
RESERVOIR AGE: Cretaceous–Paleocene
PETROLEUM TYPE: Oil
TRAP TYPE: Elongate Anticline

LOCATION

Ekofisk field and its satellite fields (West Ekofisk, Albuskjell, Tor, Eldfisk, and Edda), collectively known as the Greater Ekofisk area, lie within the Central graben of the North Sea (Norwegian Sector) (Figure 1). Water depth at Ekofisk is 230 ft (70 m).

First of the North Sea commercial oil discoveries, Ekofisk is a giant field, with an estimated 6.70 billion bbl of stock tank oil and 10.330 tcf of gas in place at the time of discovery. Ultimate recoverable petroleum includes approximately 1.5 billion bbl of oil and 5 tcf of gas. The field produces from 39 wells, and a waterflood project is underway.

HISTORY

Pre-Discovery

Offshore exploration activity began in the shallow waters of the Southern Gas basin of the North Sea, primarily encouraged by the discovery of the giant Groningen field on the Netherlands coast. From there, exploration drilling progressed northward into the deeper waters of the northern North Sea with limited success. By 1969, over 200 offshore exploratory wells had been drilled, of which 32 were within the Norwegian Sector. Increasing pessimism toward the prospectivity of the North Sea was finally removed at the end of 1969, when the Phillips Norway Group (Phillips, AGIP, Elf, Norsk Hydro, Total, and Petrofina) drilled the 2/4-1X well on the upturned edge of what was believed to be a block-faulted dome.

Discovery

Common-reflection-point seismic data (Figure 2) showed a reflection immediately above the top of the Danian section that bowed upward to form a dome with 800 ft (244 m) of closure. The initial interpretation of this structure included a graben at its crest, so a drilling program was designed to test the upturned rims of the strata surrounding the central faulted region.

The first well, the 2/4-1X, was drilled as high as possible on the structure without penetrating the apparent collapse zone in the center. The well was abandoned after repeated well flow and lost circulation problems occurred while drilling through Miocene carbonates, where an oil show was noted. In December 1969, a second well, the Phillips 2/4 A-1X, was drilled through the Miocene without difficulty and encountered 600 ft of oil shows in the Danian and Maastrichtian chalk. Earliest tests of the interval from 10,364 to 10,464 ft (3159–3190 m) flowed 1071 BOPD, with API gravity of 37.2°. Later tests sustained flow rates exceeding 10,000 BOPD (Van den Bark and Thomas, 1981).

It was not until other wells were drilled on the periphery of the "collapse zone," initial production was underway, and permanent platforms were installed on the seabed that a well was drilled in the center of the structure. What had been interpreted as a graben proved to be the crest of a dome

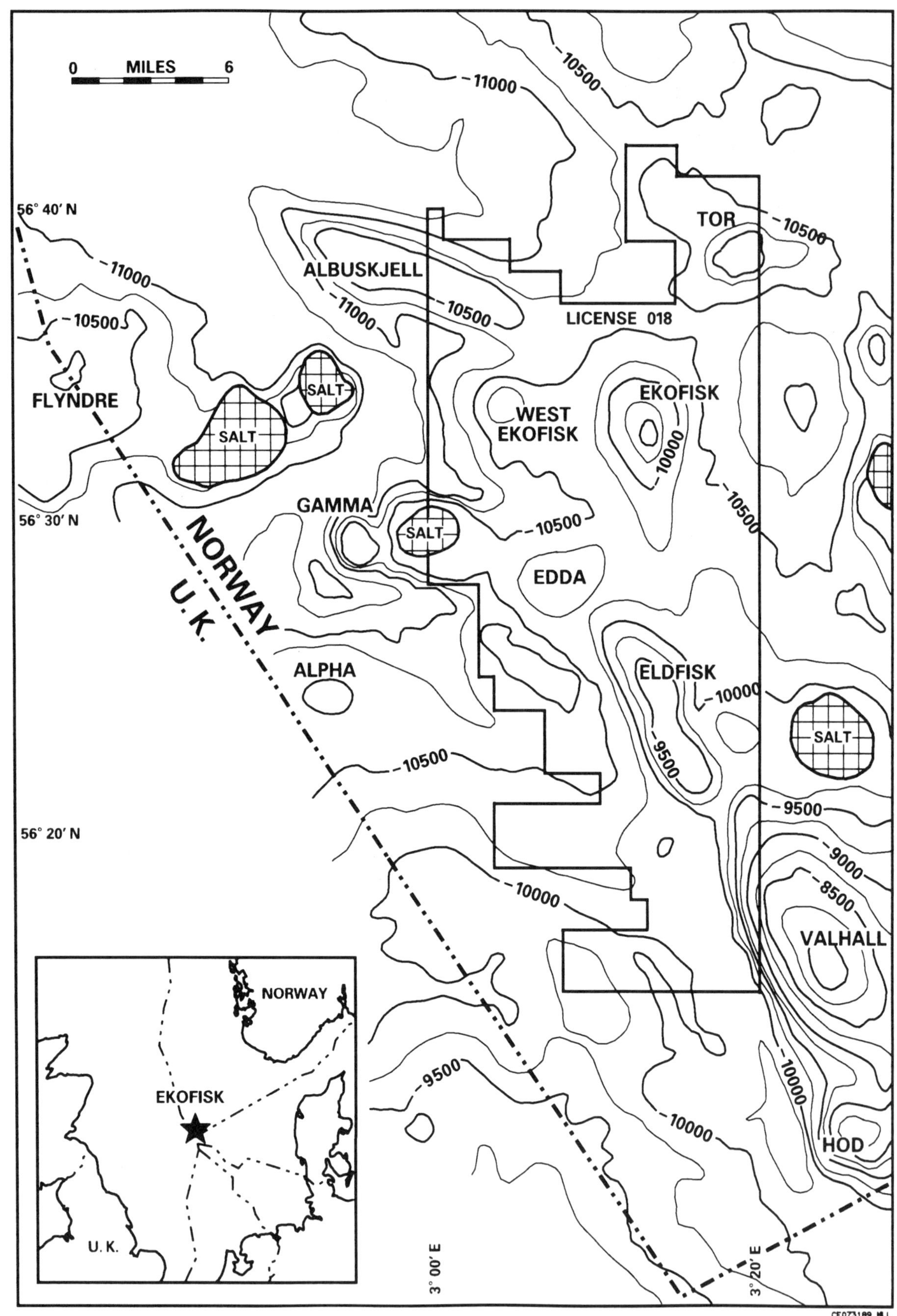

Figure 1. Location map showing structure at the top of the chalk (Ekofisk Formation) in the Central trough of the North Sea. Subsea contour interval 250 ft. Cross-hatched areas indicate salt piercement through the chalk.

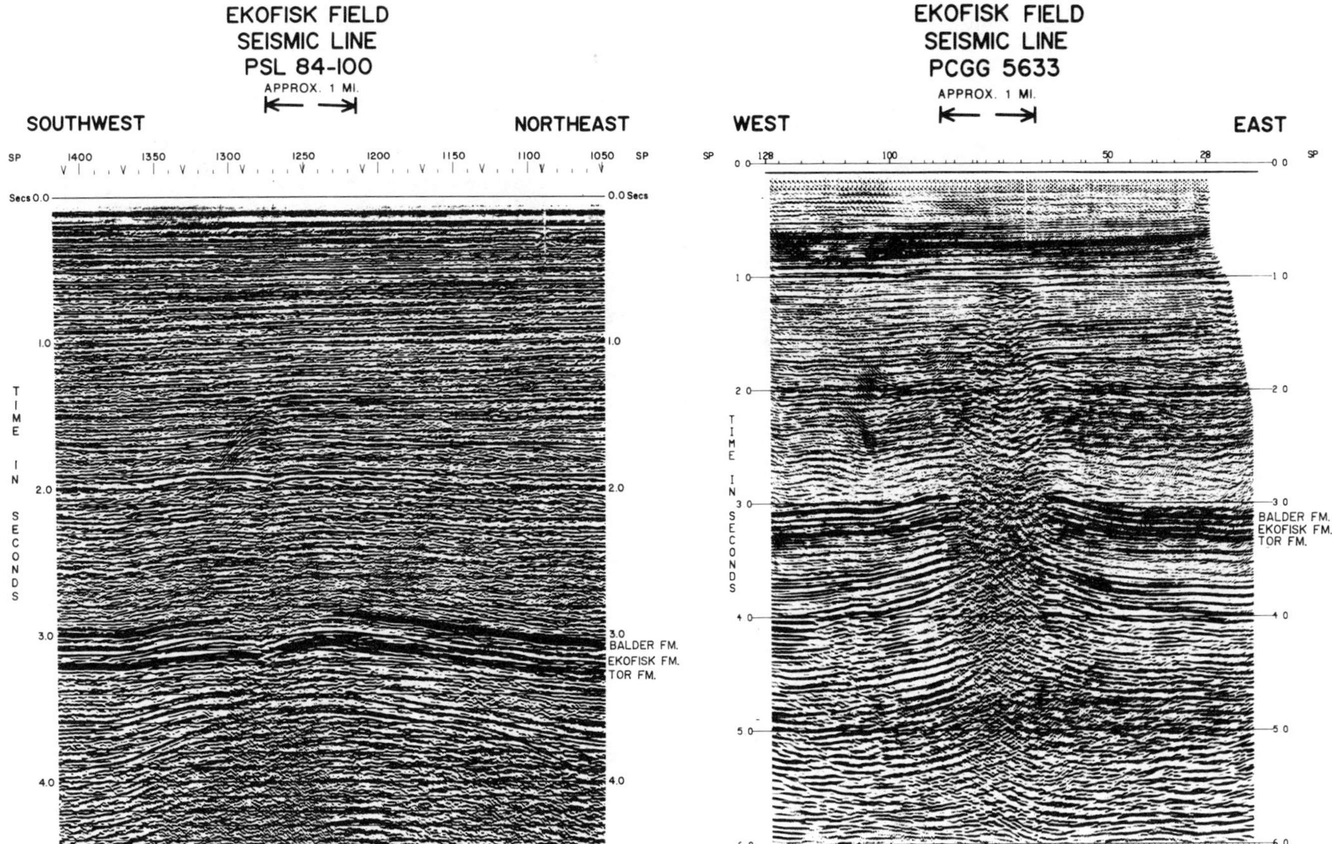

Figure 2. Seismic lines across Ekofisk field (line locations shown on Figure 3). Poor seismic returns from the crestal area, originally mapped as a collapse zone, are now known to be due to a low-velocity gas-charged column above the reservoir. Vertical scale is two-way travel time.

containing 1033 ft (315 m) of pay. The apparent collapse zone was the result of a gas-charged low-velocity section penetrated in the 2/4 C-8 crestal well between 5800 and 7190 ft (1768–2192 m). Sonic velocity in this interval is estimated to be 4950 ft (1509 m) per second—lower than that of any other North Sea sediments buried to similar depth. This velocity anomaly created the pronounced pull-down of reflections visible in Figure 2.

Post-Discovery

Development began in 1971 with the placement of two single-point mooring buoys for loading tankers with oil from the first four exploration wells at Ekofisk. These original four wells were brought on production via subsea completions within 18 months of the Ekofisk discovery. Production to tankers continued until 1975, when a crude oil pipeline was completed to Teeside, England (220 mi, 34-in. diameter). Platforms were installed at Ekofisk for drilling, production, and housing, and included a million-barrel-capacity concrete oil storage tank. Forty development wells and eight gas-injection wells were drilled at Ekofisk (a total of 163 wells have been drilled in the Greater Ekofisk area). Gas is processed at Ekofisk Center and began flowing in 1977 through a 275 mi, 36-in. pipeline to Emden, Germany.

DISCOVERY METHOD

Then

As described above, the initial interpretation of CRP seismic data suggested that the Ekofisk structure was a faulted dome (Figure 3).

Now

From today's perspective, two factors are regarded as essential to the discovery of Ekofisk field. First is the ongoing use of seismic data to identify the structure. The gas-charged low-velocity zone that

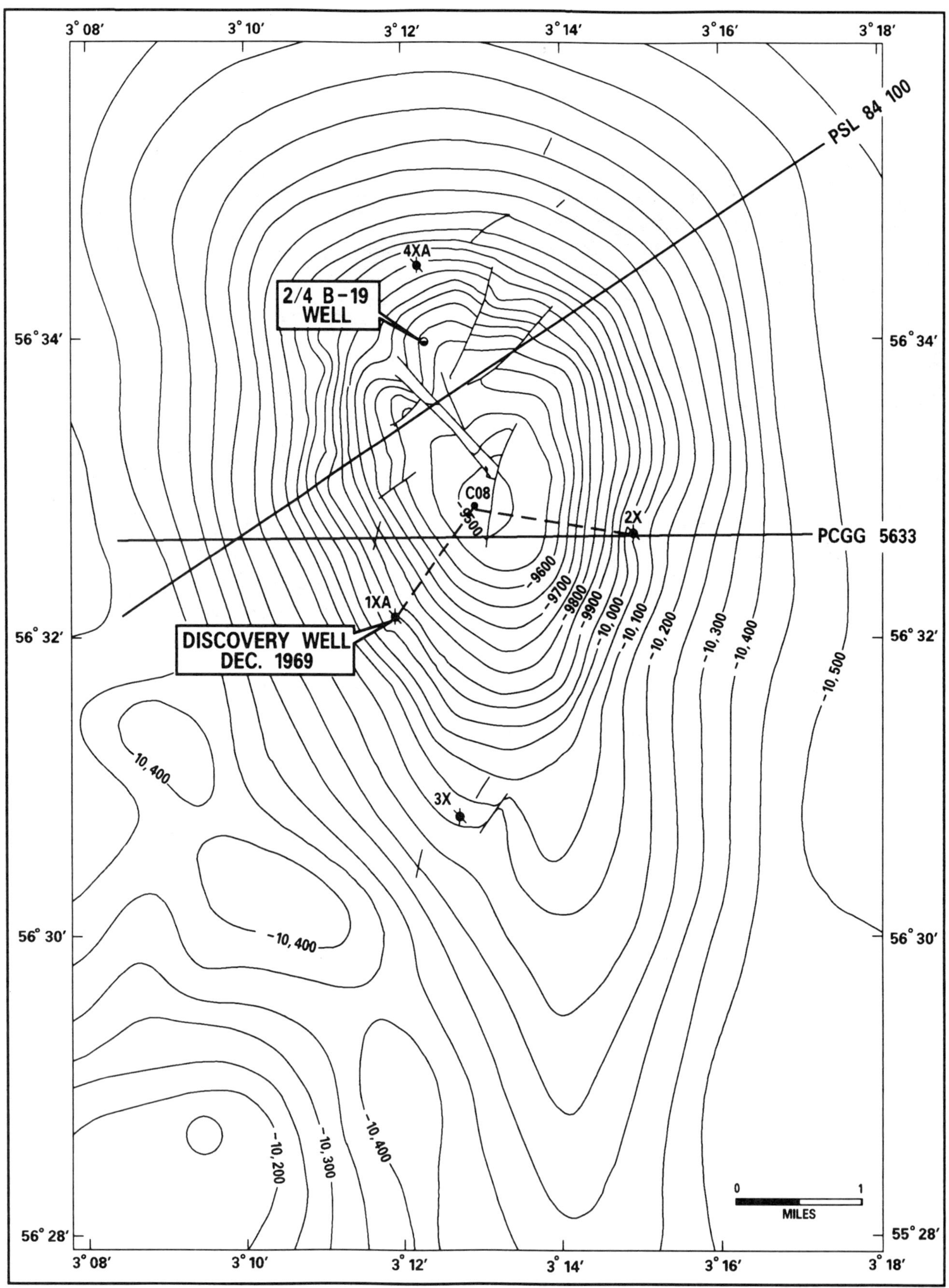

Figure 3. Structure map of Ekofisk field at the top of the Ekofisk Formation. Subsea contour interval 50 ft. Seismic lines PCGG 5633 and PSL 84-100 are shown in Figure 2. Dashed line is path of cross section shown in Figure 6.

gives a velocity pull-down effect is now recognized as responsible for the early interpretation of the Ekofisk structure. Developments in seismic processing and application of vertical seismic profiling have now minimized this difficulty. A second major factor was a commitment to an aggressive drilling program despite increasing pessimism during the early exploration of the North Sea.

STRUCTURE

Tectonic History

The North Sea Central graben is a failed rift and consists of a series of en echelon normal and strike-slip faults initiated in the Triassic Period and active throughout the Mesozoic Era. The dominant northwest-southeast orientation of these faults reflects a complex interaction of structural lineaments originating both before the rifting phase (pre-Permian) and during the main episode of fault movement in the Late Jurassic Period (Hatton, 1986). Continued subsidence within the graben during the Jurassic and Early Cretaceous Periods was associated with further strike-slip motion and fault inversion to create smaller highs and basins in which locally thick and laterally complex sedimentary sequences were deposited. During this period, both shallow marine and deep water clastic sedimentation prevailed. During the Late Jurassic Period, the Central graben served as a restricted basin in which thick organic-rich shales (Mandal Formation) accumulated. These shales form the major petroleum source rock for the area.

Tectonic activity peaked in the Late Jurassic Period, after which erosion of structural highs resulted in a prominent unconformity and subsequent redistribution of clastic sediments into Early Cretaceous basins. Halokinesis of thick evaporites (Permian) further complicated the structural and sedimentological evolution of the graben.

The onset of the Late Cretaceous Period saw a significant change in both structural and depositional style. Widespread basinal subsidence beyond the confines of the Central graben was accompanied by eustatic and climatic changes that resulted in the deposition of chalk sequences in excess of 4000 ft thick. Continued tectonic activity both at the graben margins and at the fault blocks within the graben contributed to extensive remobilization and reworking of chalks into the graben, particularly toward the end of Cretaceous and Early Danian times (see stratigraphic column, Figure 4). Ongoing halokinesis also promoted sediment redistribution, although at a more local scale.

Tectonic activity throughout the Tertiary Period was of limited duration and magnitude. Uplift of the surrounding Tertiary forelands, and influx of cold waters from the north as the Norwegian Sea opened,

	CHALK FORMATIONS	STAGE	
CHALK GROUP	EKOFISK FORMATION	DANIAN	PALEOCENE
	TOR FORMATION	MAASTRICHTIAN	UPPER CRETACEOUS
	HOD FORMATION	CAMPANIAN	
		SANTONIAN	
		CONIACIAN	
		TURONIAN	
	PLENUS MARL FORMATION	CENOMANIAN	
	HIDRA FORMATION		

Figure 4. Generalized stratigraphy of the North Sea chalk.

prompted a return to clastic deposition during the late Danian Age. This resulted in a thick accumulation of rapidly deposited and overpressured claystones subjected only to ongoing halokinesis until these sediments were folded and faulted by minor compressional stresses during Miocene–Oligocene time.

Regional Structure

The regional structure of the area at the top of the Chalk Group is shown in Figure 1. This map suggests three prominent structural controls:

1. The inverted Late Jurassic Lindesnes ridge, upon which Valhall and Eldfisk fields are located.
2. The anticlines of Albuskjell and Tor fields, whose structural orientation appears to reflect the influence of deeper wrench movements on salt diapirism.
3. Salt diapirism, confirmed beneath Albuskjell, Tor, and Eldfisk fields and identified in piercement structures throughout the area.

Local Structure

The structure map for Ekofisk field at the top of the chalk reservoir (i.e., top of the Ekofisk Formation) is shown in Figure 3. The map demonstrates four-way structural closure over an assumed salt diapir. The field is elliptical, elongate north-south. Faults at the reservoir level have been identified on the structure, with an apparent northwest-southeast component dominant in the northern part of the field.

A crestal graben with approximately 150 ft of throw at the Ekofisk Formation level has been identified by vertical seismic profiling and recent drilling. Recent mapping has revealed faults trending north-northeast.

Core, wireline log, and production test data clearly demonstrate that the reservoir is highly fractured. Fractures are related to tectonic processes (halokinesis and/or regional stresses) and diagenetic processes including stylolite formation.

STRATIGRAPHY

A generalized summary of the Chalk Group stratigraphy in the Greater Ekofisk area is seen in Figure 4. Sandstone reservoirs of Triassic, Upper Jurassic, and Paleocene age have all tested hydrocarbons in the central North Sea. However, the main reservoirs within the Greater Ekofisk area are chalks of Campanian, Maastrichtian, and Danian age. The main source rock for the area is organic-rich shale of the Upper Jurassic Mandal Formation.

The stratigraphy of the 2/4 B-19 well on Ekofisk field is shown in Figure 5. A thick Upper Jurassic shale sequence includes the source interval in the Mandal Formation. Above the basal Cretaceous unconformity, siltstones and claystones of Early Cretaceous age pass conformably upward into the Chalk Group (Upper Cretaceous and lower Tertiary). The Ekofisk field reservoirs lie within allochthonous chalk sections of the Tor and Ekofisk formations (D'Heur, 1984; Feazel and Farrell, 1988).

A thin transitional marl interval equivalent to the Maureen Formation forms an imperfect seal above the chalk reservoirs (Figure 5). This unit grades upward into a sequence of overpressured claystones with scattered thin siltstones and volcanic tuffs of the Paleocene–lower Eocene Rogaland Group. Above the thin volcanics that identify the Balder Formation, a thick sequence of highly overpressured claystones is interrupted only by siltstones and thin carbonates of Miocene age and periglacial sediments of Pliocene–Pleistocene age near the surface.

Trap

Ekofisk field is a domal trap with two pay zones and a seal resulting from overpressured shales above with a downward pressure decrease into the reservoir. The dome, elongated in a north-south direction, has areal closure of 12,070 ac (49 km^2) and vertical closure of 800 ft (244 m) at the top of the Ekofisk Formation. The oil column extends below the spill point of the reservoir and is approximately 1000 ft (305 m) thick. Chalk at the crest of the Ekofisk structure is generally more porous than chalk on its flanks, a consequence of fracturing, overpressure, early introduction of hydrocarbons into the crestal chalk pores, and compaction of surrounding areas (D'Heur, 1984; Feazel et al., 1985; Feazel and Schatzinger, 1985).

Between the two productive reservoirs lies a low-porosity interval known as the Tight Zone (Figure 6). This brittle, siliceous layer is fractured, permitting limited fluid communication and pressure continuity between the two chalk reservoirs of the Ekofisk and Tor formations.

The base of hydrocarbons is found at a different elevation in each well, and the oil-water contact is diffuse. This appears to be a consequence of the gradual downward decrease in porosity in the chalk. As seen in Figure 7, oil-filled porosity in the Tor Formation tapers off to near-zero below the main oil accumulation.

In summary, trap development at Ekofisk field was a consequence of the following processes:

1. Deposition of chalk with high primary porosity (originally as high as 75–80% in coccolith ooze), overlain by low-permeability claystones and shales, which form an effective seal.
2. Early development of anticlinal closure.
3. Retention of high reservoir porosity (>45%) at abnormally great depth as a result of abnormally high formation pressure in the chalks and early hydrocarbon entry.
4. Development of a pervasive natural fracture system that facilitated hydrocarbon migration into the reservoir.

Reservoir

The Ekofisk and Tor formations that comprise the petroleum reservoir for Ekofisk and its satellite fields consist of chalk: the weakly cemented skeletal debris of unicellular algae that live in the upper levels of the ocean known as the photic zone (the near-surface zone of sunlight penetration). In the Ekofisk field, this material has been reworked from a pelagic seafloor ooze: Its reservoir characteristics result from resedimentation processes (Schatzinger et al., 1985; Feazel and Farrell, 1988). Chalk is almost pure calcium carbonate—in the Central graben of the North Sea, where chalk deposition was far-removed from the influence of terrigenous sediments, the noncarbonate fraction (insoluble residue) of the chalk is commonly less than 10%. This noncarbonate fraction increases in abundance upward in the Ekofisk Formation: Changing environmental conditions after the Cretaceous–Tertiary transition brought more terrigenous clays into the Central graben.

The Ekofisk field reservoir chalk is poorly to well-cemented biomicrite consisting mainly of coccolithophorid skeletal debris, chemically stable as low-magnesium calcite. Silica is the most important accessory mineral, occurring in varying concentrations as disseminated microcrystalline quartz or nodular chert. Clays, macrofauna bioclasts, and other components are present in low concentrations.

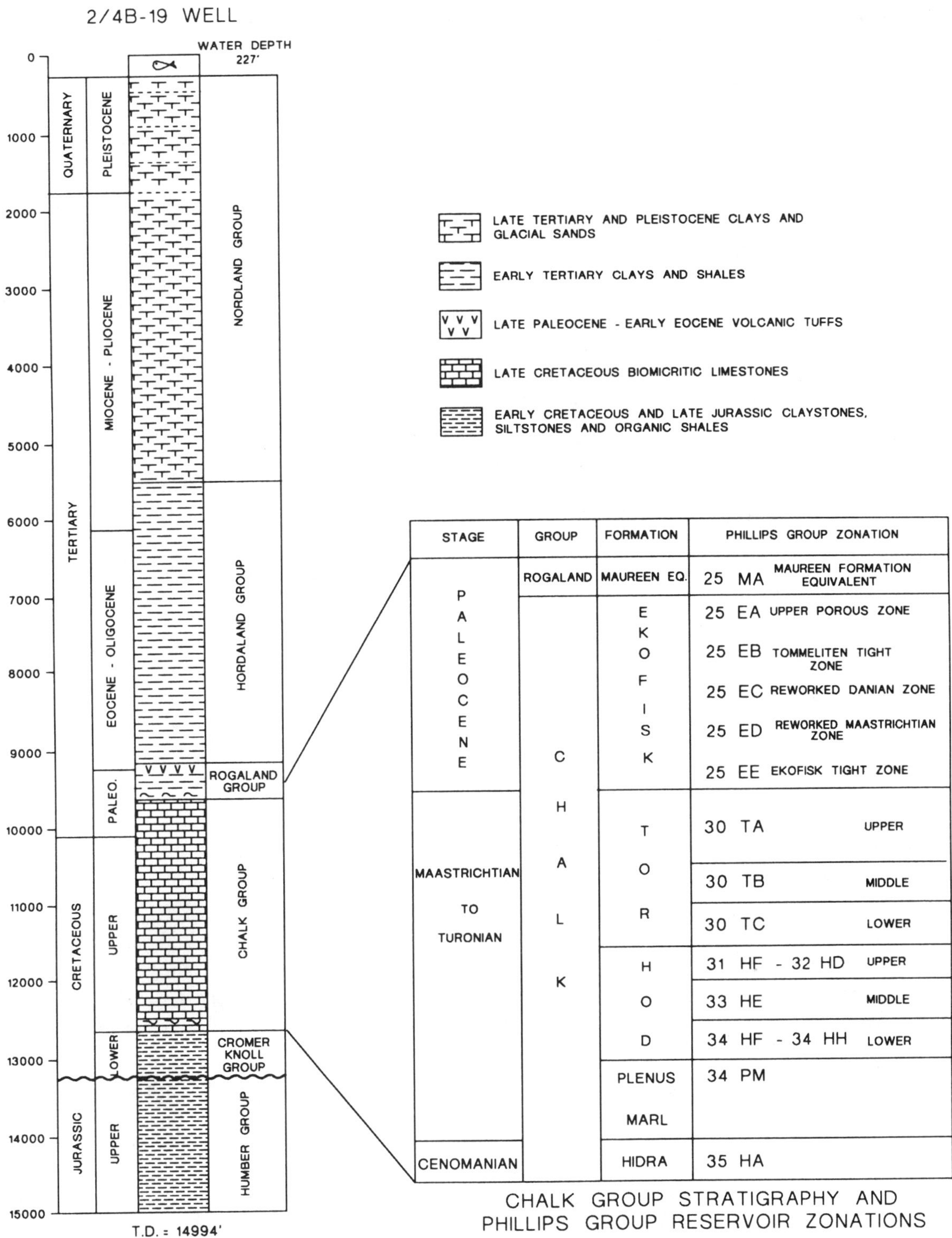

Figure 5. Stratigraphy of the Ekofisk 2/4 B-19 well.

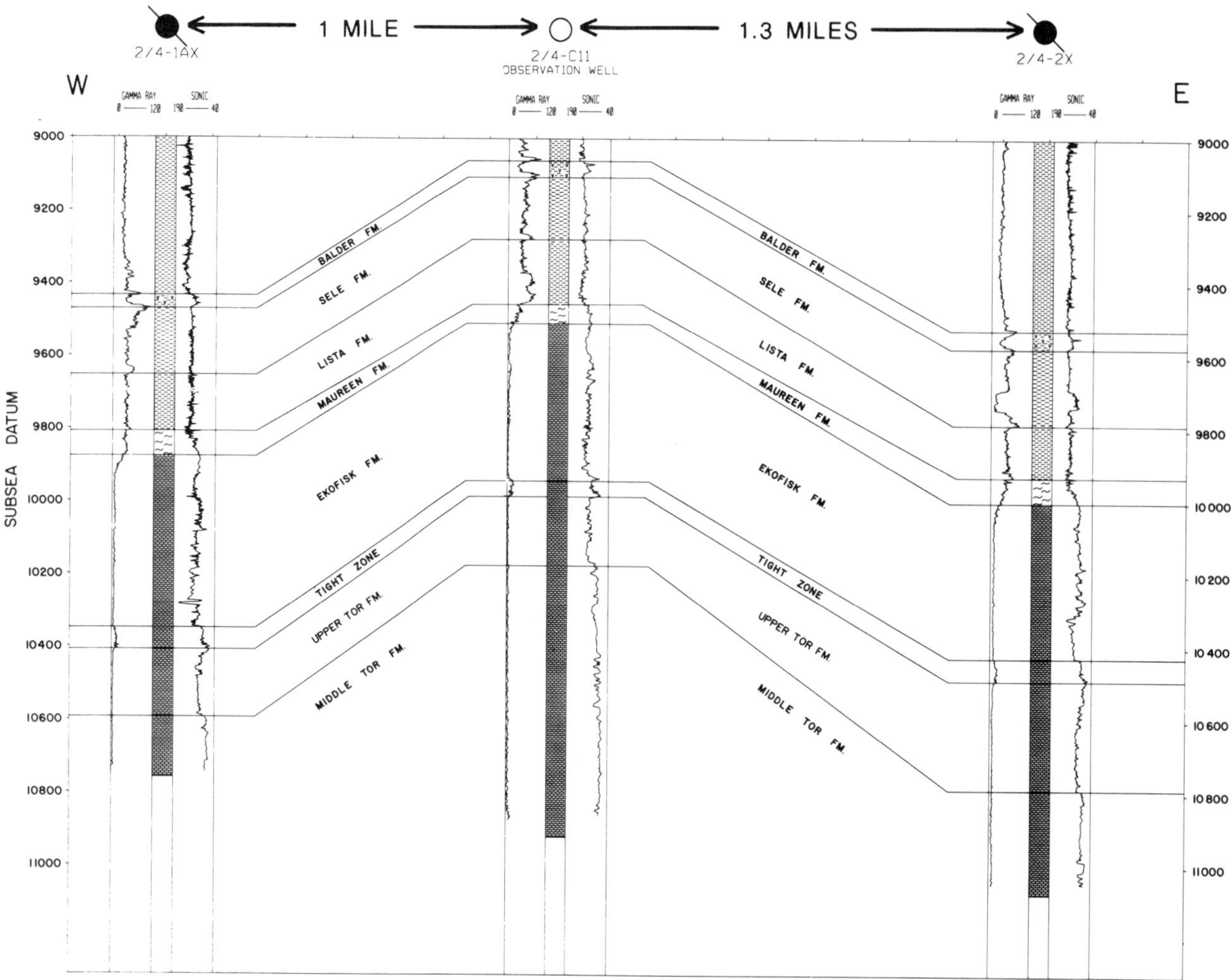

Figure 6. East-west cross section across Ekofisk field. Line of cross section is shown in Figure 3.

The Tor Formation (Maastrichtian) and Ekofisk Formation (Danian) form the reservoir. These intervals have been informally subdivided into members correlatable throughout the Ekofisk area on the basis of porosity and gamma ray logs (Figure 5) . The Tor Formation is divided into three units, of which the uppermost (TA) member is hydrocarbon-bearing. The Ekofisk Formation is subdivided into five members, each hydrocarbon-bearing, but with varying reservoir properties and production potential. The stacked debris flows of the reworked Maastrichtian layer (ED) form the single most important production interval in the Ekofisk Formation. This member is separated from the underlying Tor Formation by the Tight Zone, a nearly impermeable layer of pelagic chalk relatively rich in silica and clay, which acts as an important restriction to pressure continuity between the main producing intervals over much of the field. The Tight Zone is, however, absent in some areas along faults (for example, no Tight Zone was penetrated by the 2/4 K-13B well).

Depositional Setting

Paleoenvironmental indicators in the chalk include the bioclastic grains that constitute its matrix and various ichnofossils representing the tracks, trails, and burrows of an active infauna. The bulk of the chalk consists of two types of skeletal remains, both made of calcium carbonate. The dominant skeletal contributors are coccolithophorid (golden-brown) algae. These live in the photic zone of open ocean waters, encapsulated in a coccosphere (Figure 8A), composed of individual platelets known as coccoliths (Figure 8B). When the single-celled algae die, or are eaten by larger organisms, the coccospheres and coccoliths settle to the seafloor to form a coccolith ooze. Holocene coccolith oozes accumulate in the deep ocean between the depths of a few hundred and 4000–6000 m. Shallower, pelagic sediments are diluted by influx of shallow-water sediments from continental shelves and insular platforms; deeper, they are dissolved by acidic ocean waters (below the calcite compensation depth [CCD], the rate of dissolution equals the rate of supply, and few carbonate grains

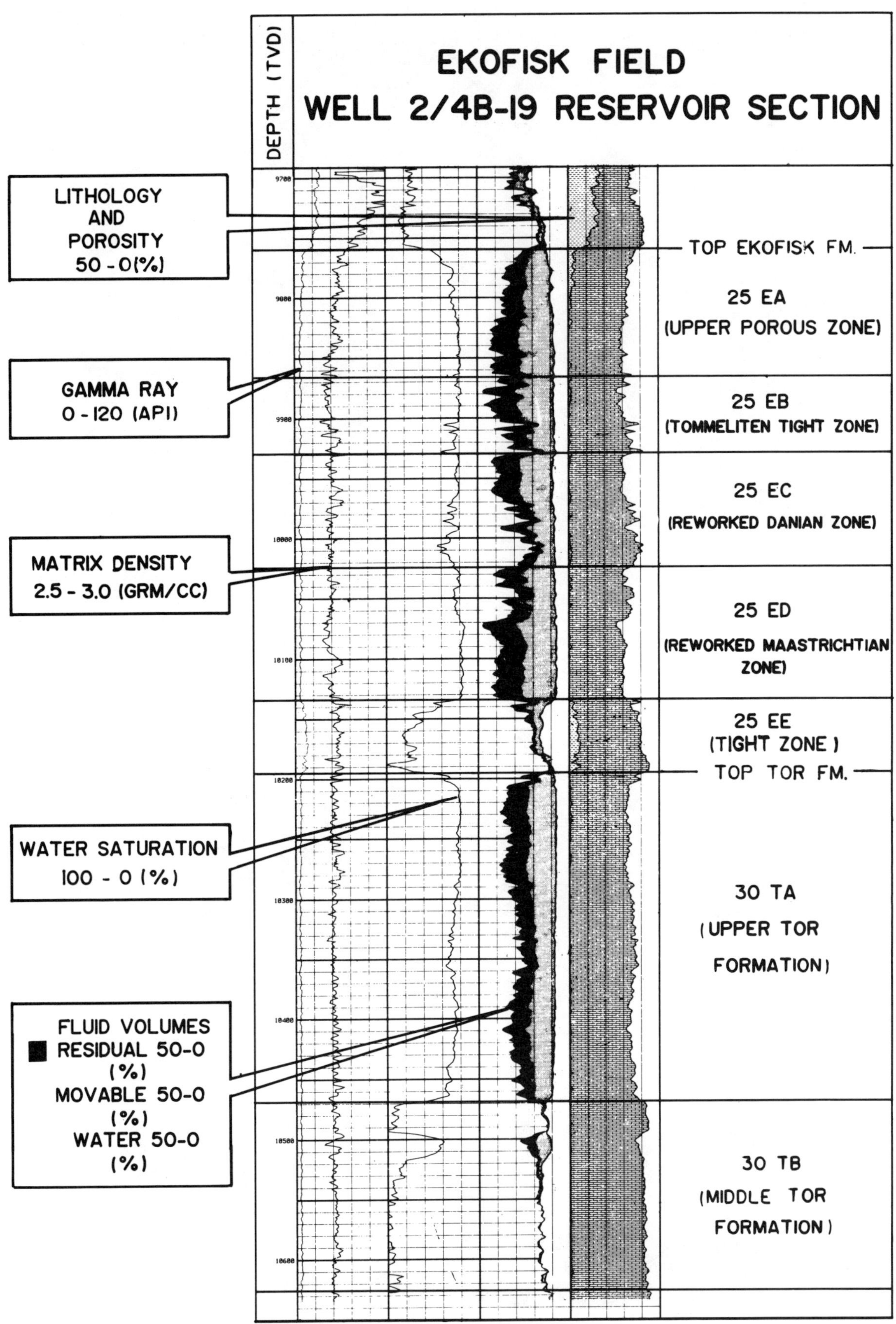

Figure 7. Detailed reservoir stratigraphy of the Ekofisk 2/4 B-19 well (Phillips Group zonation).

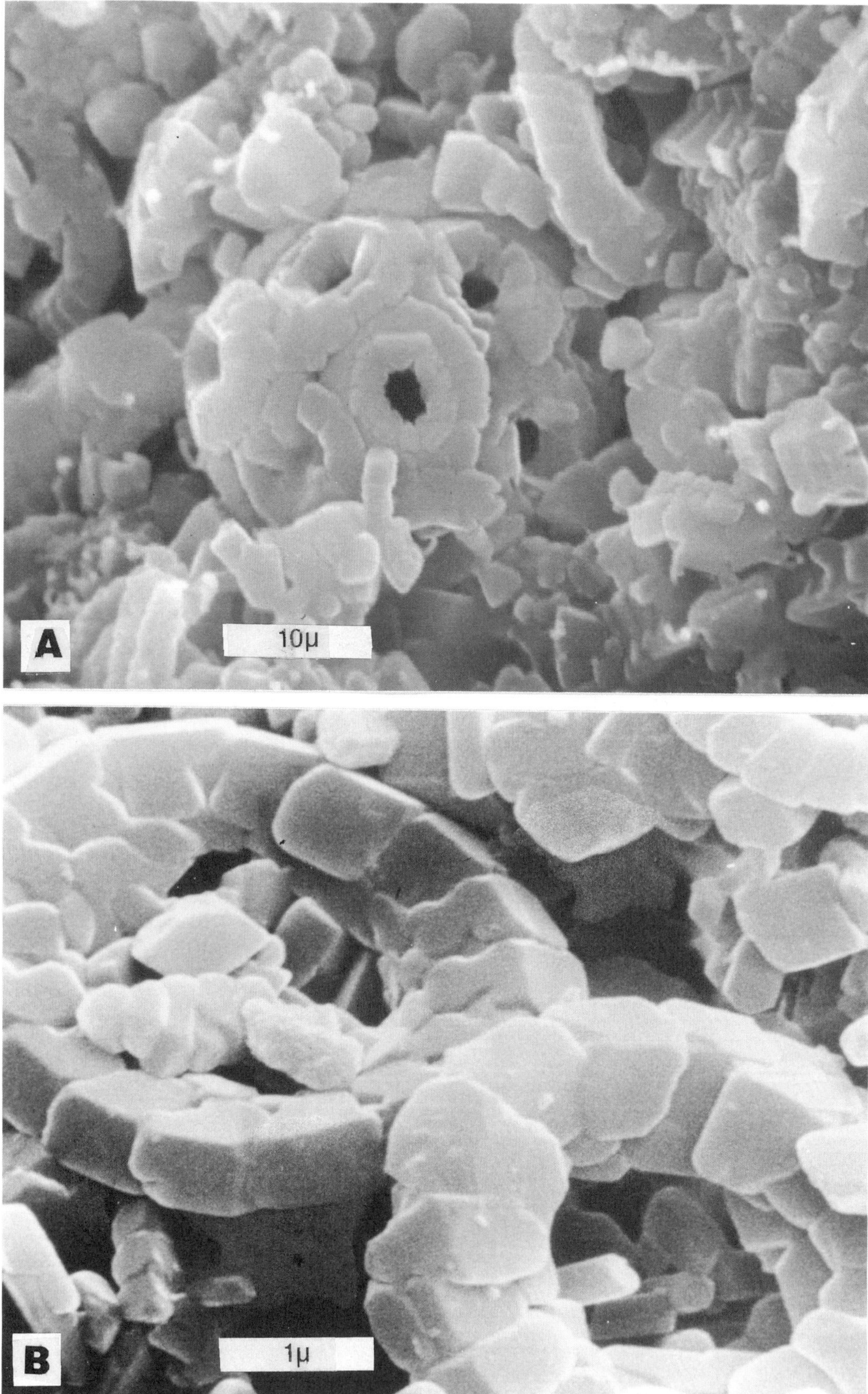

Figure 8. Typical SEM appearance of chalk constituents. (A) Intact coccosphere. (B) Disaggregated coccoliths.

are preserved). In modern coccolith oozes, and in the North Sea chalk, the other significant skeletal contributors are single-celled foraminifera. These live throughout the water column, some co-existing with the algae in near-surface waters (pelagic foraminifera), others living on or in the bottom sediments (benthic foraminifera). Their multichambered tests grow to a diameter of several millimeters (Figure 9).

Ichnofossils in the chalk indicate that the bottom conditions varied from soupy to firm, and from oxic to anoxic. Common ichnogenera in chalk cores from Ekofisk area wells include *Thalassinoides*, *Zoophycos*, *Planolites*, and *Chondrites* (Figure 10). According to Ekdale and Bromley (1984), each of these, when dominating the assemblage, defines a generic ichnofacies. Together, these four ichnofacies characterize a deep-water softground deposit, at least periodically deficient in oxygen (Ekdale, 1985).

It is widely accepted that chalks of the Central graben were deposited in open, unrestricted water 300–700 m deep (Hancock and Scholle, 1975; Hancock, 1976). Initial deposition was pelagic, resulting from a steady rain of disaggregated coccolithophorid skeletons settling individually or in fecal pellets. A coccolith ooze formed on the seabed, and either (1) became partially lithified in situ during periods of quiescence, or (2) was remobilized by seismic activity or slope instability into a variety of gravity deposits moved as slumps, slides, or suspensions (Figure 11). Kennedy (1985) recognized two dominant sedimentary facies:

1. Autochthonous pelagic chalks, which are typically well-cemented, laminated, and intensely burrowed by an active infauna.
2. Allochthonous reworked chalks, consisting of rapidly deposited sediments with complex internal fabrics that reflect both depositional processes and subsequent bioturbation. Reworked chalks display the best reservoir quality.

Bioturbation was responsible for destroying much of the internal fabric of the chalk, hampering the identification of individual sedimentary units both in cores and on logs. Nevertheless, coarse sequences or stacks of redeposited chalks are recognizable.

Further indicators of a basinal or toe-of-slope depositional environment include a variety of slump and resedimentation features, such as calcarenite layers, chalk-pebble conglomerates, shear zones, soft-sediment folds, deformed ichnofossils, and early fractures filled with injected sediment (Watts et al., 1980; Schatzinger et al., 1985; Brewster et al., 1986; Bromley and Ekdale, 1987).

Diagenesis

As described by Feazel et al. (1985), Scholle (1977), Neugebauer (1974), and Schlanger and Douglas (1974), the diagenetic history of a chalk that is never exposed to meteoric waters is a one-way trip from highly porous carbonate ooze to cemented limestone. Initial porosity in modern coccolith ooze may be as high as 75–80%. After dewatering to a grain-supported matrix, the sediment retains 40–50% porosity (Feazel et al., 1985; Scholle et al., 1983). Some intervals in the North Sea chalk retain porosity exceeding 40%: Apparently these horizons have escaped significant diagenetic alteration since early dewatering.

Diagenesis that occludes chalk pores results from two types of compaction: mechanical and chemical. Mechanical compaction precedes lithification and is accomplished at shallow burial depth by dewatering in response to overburden pressures. Chemical compaction is accomplished by solution-transfer (Durney, 1972): dissolution of calcium carbonate at points of greatest stress—grain contacts—and reprecipitation in lower stress regimes in adjacent pores. Chemical compaction has produced a variety of macroscopic features in the chalk, from wispy solution seams ("horsetails") to interpenetrating stylolites.

Chalk buried to the 10,000 ft (3 km) depth of the Ekofisk reservoir has usually lost all significant porosity as the processes mentioned above have collapsed or cemented the intergranular pores (Scholle, 1977). Preservation of porosity exceeding 40% in the Ekofisk reservoir was linked by Scholle (1977) to overpressure, and its reduction of the effective stress at grain contacts, thus reducing solution-transfer of calcite. Feazel and Schatzinger (1985) acknowledged the role of overpressure, but also noted that within approximately the same pressure regime, chalk from below the oil-water contact was heavily overgrown with secondary calcite and had lost almost all porosity, whereas chalk within the oil column remained porous. Both overpressure and hydrocarbon displacement of connate pore waters are responsible for porosity preservation in the chalk.

Porosity and Permeability

The pores in the chalk and especially pore throats (Figure 12) are exceedingly small (often less than one micrometer across); consequently, permeabilities are low (sometimes below detection limits). Pores may be interparticle (primarily those between coccolith fragments) or intraparticle (primarily those within foraminiferal tests). The effective pores, or those that are interconnected, are mostly interparticle, and range in throat diameter from 0.1 to 1.0 micrometers, as a consequence of the 1–20 micrometer size of the algal remains (Price et al., 1976).

Porosity of pelagic chalk varies from less than 5% off-structure to 25% at the crest. Allochthonous chalk contains porosity ranging from 15 to more than 45%, depending on structural position. Porosity distribution in the equivalent reservoir of the West Ekofisk field is described in detail by D'Heur (in "West Ekofisk Field" of the *Atlas of Petroleum Geology*). Permeability is less than 0. 1 md in much of the chalk matrix but ranges as high as 100 md (total permeability) (Figure 13). An average permeability value for the matrix is 1.0 md (Feazel et al., 1985). Permeability is enhanced by a pervasive natural

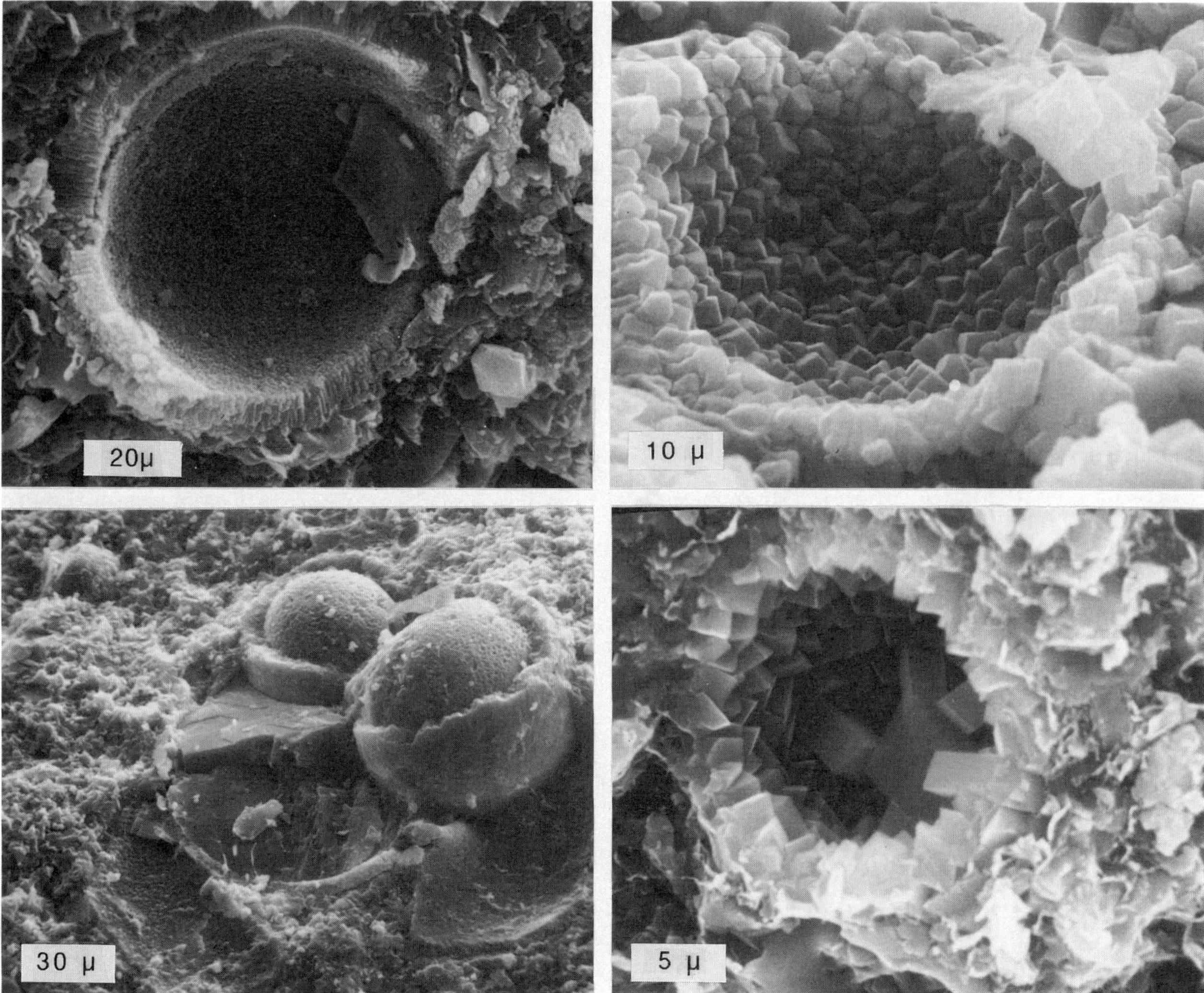

Figure 9. Photomicrograph of pelagic foraminiferal tests, showing occlusion of intraparticle porosity, progressing clockwise from upper left: nearly empty chamber; initial overgrowth of calcite crystals in test walls; growth of individual calcite crystals into void; complete occlusion (micro-steinkerns).

fracture network, particularly well-developed near the crest of the structure.

Fractures

Cores from Ekofisk field clearly demonstrate that the chalk reservoirs are intensely fractured. This fracturing is regarded as essential to field productivity, owing to the low matrix permeabilities typical of chalks.

Four types of fractures are documented (Feazel and Farrell, 1988). Tectonic fractures are ubiquitous throughout the Tor and Ekofisk formations, commonly occurring as cross-cutting conjugate sets dipping 60–75°. These fractures are typically open and enhance both horizontal and vertical permeability. Stylolite-associated fractures are also important to reservoir productivity. They are typically subvertical open fractures up to 20 cm in height, commonly extending from the apices of stylolite teeth. Unlike tectonic fractures, stylolite-associated fractures do not cross-cut stylolite boundaries, and therefore preferentially enhance horizontal permeability. Healed fractures are less common, but reflect early fracturing in semilithified sediments and rapid filling with comminuted chalk matrix. The tectonic and stylolite-associated fractures are commonly filled with sparry calcite or redeposited silica. These fractures act as permeability barriers. Irregular fractures are those which do not conform to the above groupings. Their origins and distribution are poorly understood, and although typically present as open fractures, they are least important as permeability conduits.

Tectonic fracture intensity is clearly related to sedimentology and depositional processes and is greatest in the higher porosity, allochthonous members. Distribution of these fractures is assumed to be controlled by a complex interaction of regional

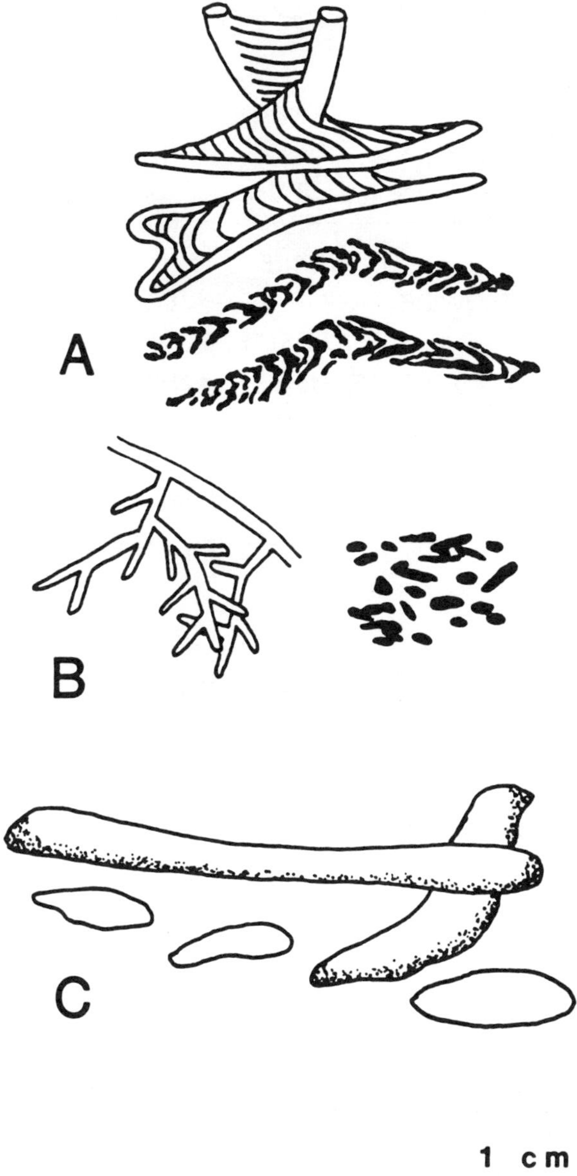

Figure 10. Three common ichnofossils seen in chalk cores (isometric reconstructions and typical cross sections): (A) *Zoophycos*. (B) *Chondrites*. (C) *Planolites*. (From Schatzinger et al., 1985.)

stresses and location on the structure. Stylolite-associated fractures are also lithology-dependent, being almost exclusive to the Tor Formation.

Understanding of fracture geometry remains incomplete and is currently being intensely examined. Studies designed to model fracture orientation and field anisotropy are underway in order to optimize field waterflooding.

Pressure

The Ekofisk reservoir is overpressured. Initial pressure was 7035 psi at -10,400 ft. The initial pressure gradient within the reservoir was 0.366 psi/ft. By mid-1987, after 16 years of production, pore pressure had declined to approximately 3800 psi throughout most of the reservoir. However, pressure in the southern part of the Tor Formation can be as much as 1000 psi higher.

Natural gas has been injected into the Ekofisk Formation in varying amounts since 1975 and has helped maintain pressure. A waterflood of the Tor Formation was begun in 1987 and will maintain reservoir pressure in the Tor at approximately 4000 psi. Expansion of the waterflood into the Ekofisk Formation (layer ED) is anticipated and will maintain pressure in that part of the reservoir at 4000 psi. Other pressure maintenance programs, including injection of nitrogen, are also under consideration.

In addition to injection programs, other means of pressure support are active. RFT pressure data collected from beneath the pay and from flank locations show partially reduced pressures, suggesting that aquifer influx accounts for some pressure support. Also, formation compressibility is an unusually large consideration at Ekofisk. Compressibilities range from 1×10^{-6} psi^{-1} (elastic) to 2.5×10^{-5} psi^{-1} (pore collapse), and estimates of maximum compaction in the reservoir to date vary from 15 to 25 ft. While this compaction has caused the much-publicized seabed subsidence at Ekofisk, it has also provided the reservoir with a significant degree of pressure support.

Pay Zone Thickness

The 2/4 B-19 well (Figure 7) shows typical pay zones encountered at Ekofisk: 384 ft of net pay in the Ekofisk Formation and 284 ft in the Tor Formation, separated by 50 ft of the Ekofisk Tight Zone. In the center of the field, net thickness exceeds 900 ft. Further toward the flanks, pay thickness decreases gradually. There is no well-defined oil-water contact; rather, average porosities decline and water saturations increase gradually throughout the section from higher to lower structural position. Each of the units described within the Ekofisk and Tor formations has its own porosity and water saturation distribution, and each deteriorates beneath the net pay cutoffs independently, generally beginning with the lowest layers and ascending in stratigraphic order.

Reservoir Fluids

Ekofisk field initially produced oil of 36.3° API gravity, with a solution gas-oil ratio of 1547 scf/bbl. Sulfur content of the oil is low (0.21 wt%), as are the amounts of nickel (5.04 ppm) and vanadium (1.95 ppm). Pour point of the oil is 68°F (20°C). Almost no water is produced (BS&W = 1 vol%).

Source

Hughes et al. (1985) described the geochemistry of Ekofisk area oils and suggested that compositional variations among oils from different fields reflect systematic differences in thermal maturity (of source and/or reservoir). Other, minor, compositional

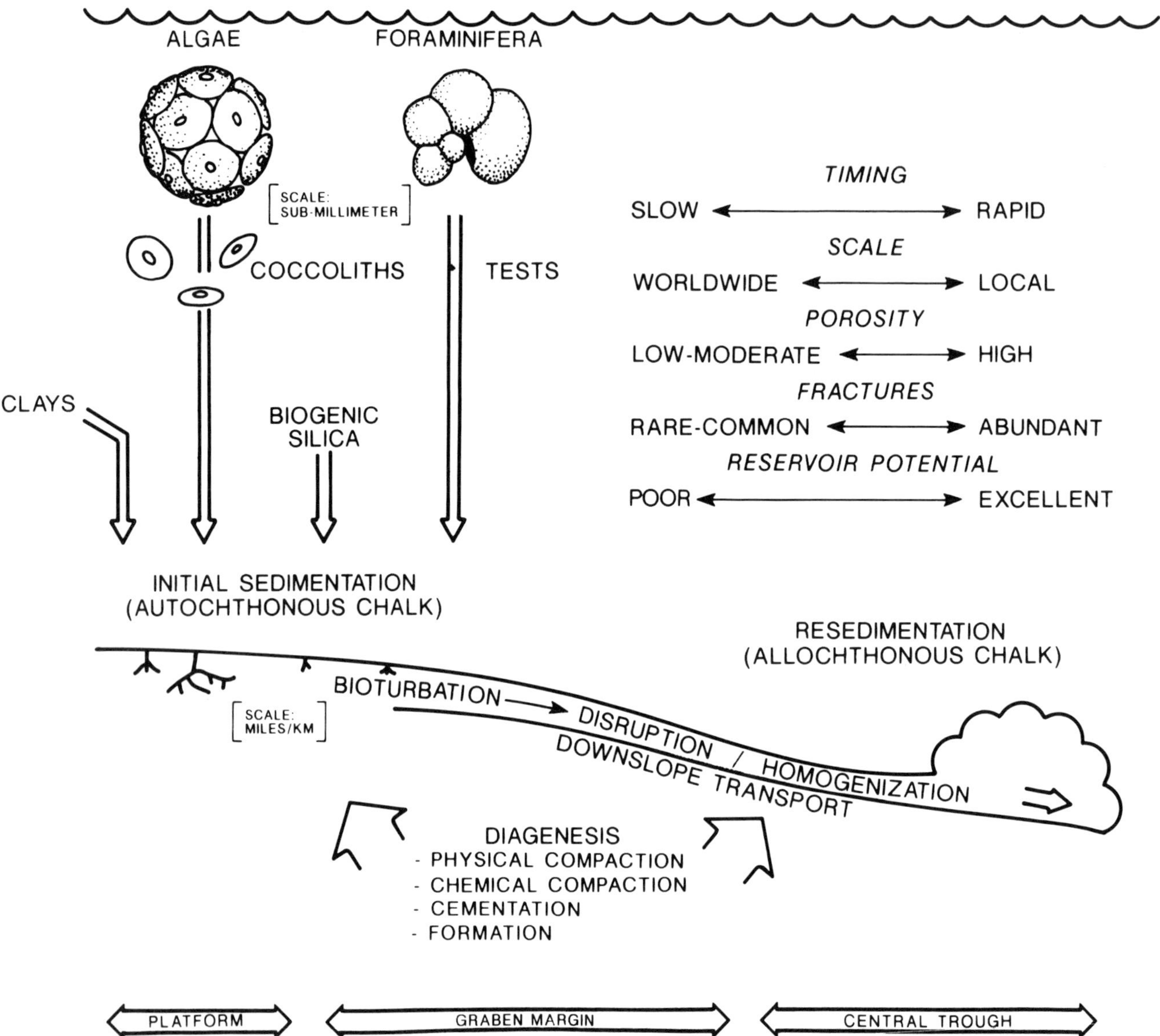

Figure 11. Schematic illustration of sedimentation and resedimentation of chalk within the Central trough. Resedimented intervals form the best petroleum reservoirs, because they are the least cemented and are therefore highly porous and intensely fractured.

variations were ascribed to differing precursor compounds in the source rocks.

The similarity of two families of biomarker compounds ("chemical fossils") in Ekofisk crude oil and in solvent extracts from an inferred source interval in the Mandal Formation of Late Jurassic age is shown in Figure 14. Vitrinite reflectance values from this interval suggest peak oil generation (R_o = 0.93–1.16%) in the present thermal regime. Hughes et al. (1985) suggested that the maturity of the Ekofisk oil, based on molecular ratios, should correspond to source rock vitrinite reflectance values of approximately 0.55–1.00%. Burial history analysis and thermal modeling of the Greater Ekofisk area (Figure 15), based on a crestal well, suggests that most of the oil trapped at Ekofisk field was generated during the past 15 million years. The role of hydrocarbons in preserving porosity would therefore have become effective fairly late in the chalk's diagenetic history. Earlier migration is possible if laterally adjacent, but deeper, drainage areas are included, as these source areas attained thermal maturity somewhat earlier than the crestal area modeled in Figure 15.

Fluid Flow Characteristics

Fluid flow within Ekofisk is dominated by three key factors: the natural fracture system, hydraulic stimulation of the wells, and the presence of the Ekofisk Tight Zone as an imperfect barrier between the field's two reservoirs.

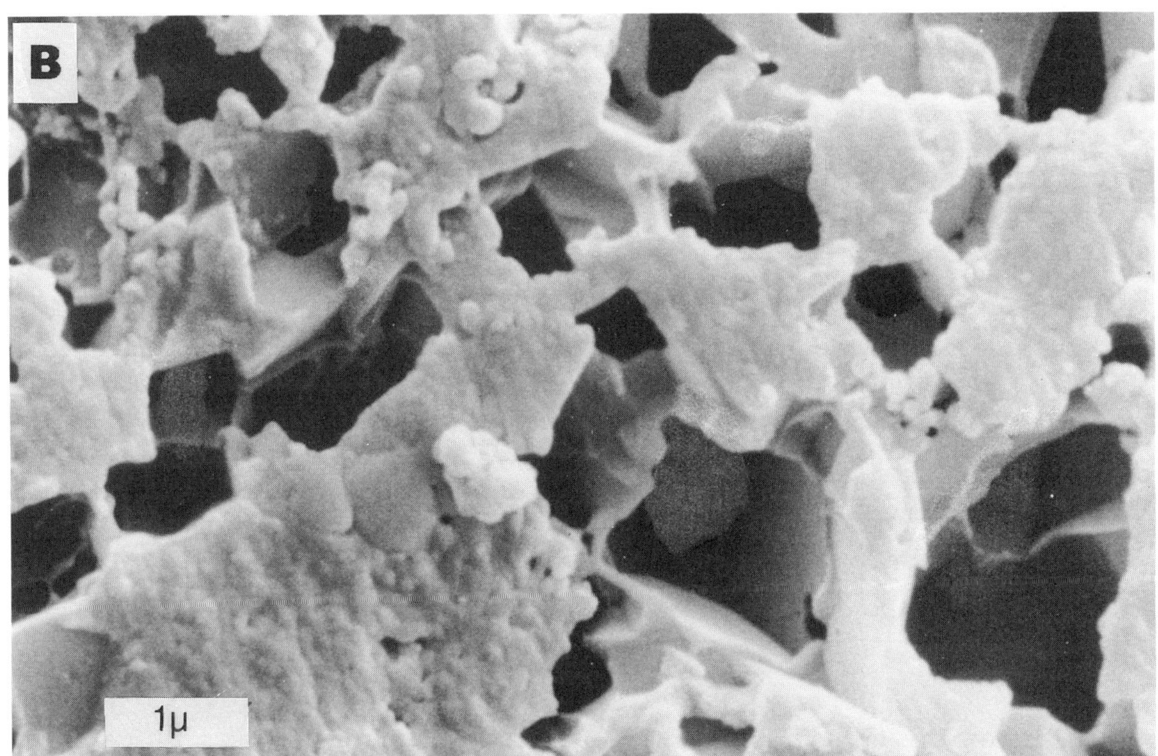

Figure 12. SEM photomicrograph of resin pore casts (chalk removed by dissolution). Note interconnected pore network (A) surrounding skeletal grains, and (B) between disaggregated crystal elements of coccoliths.

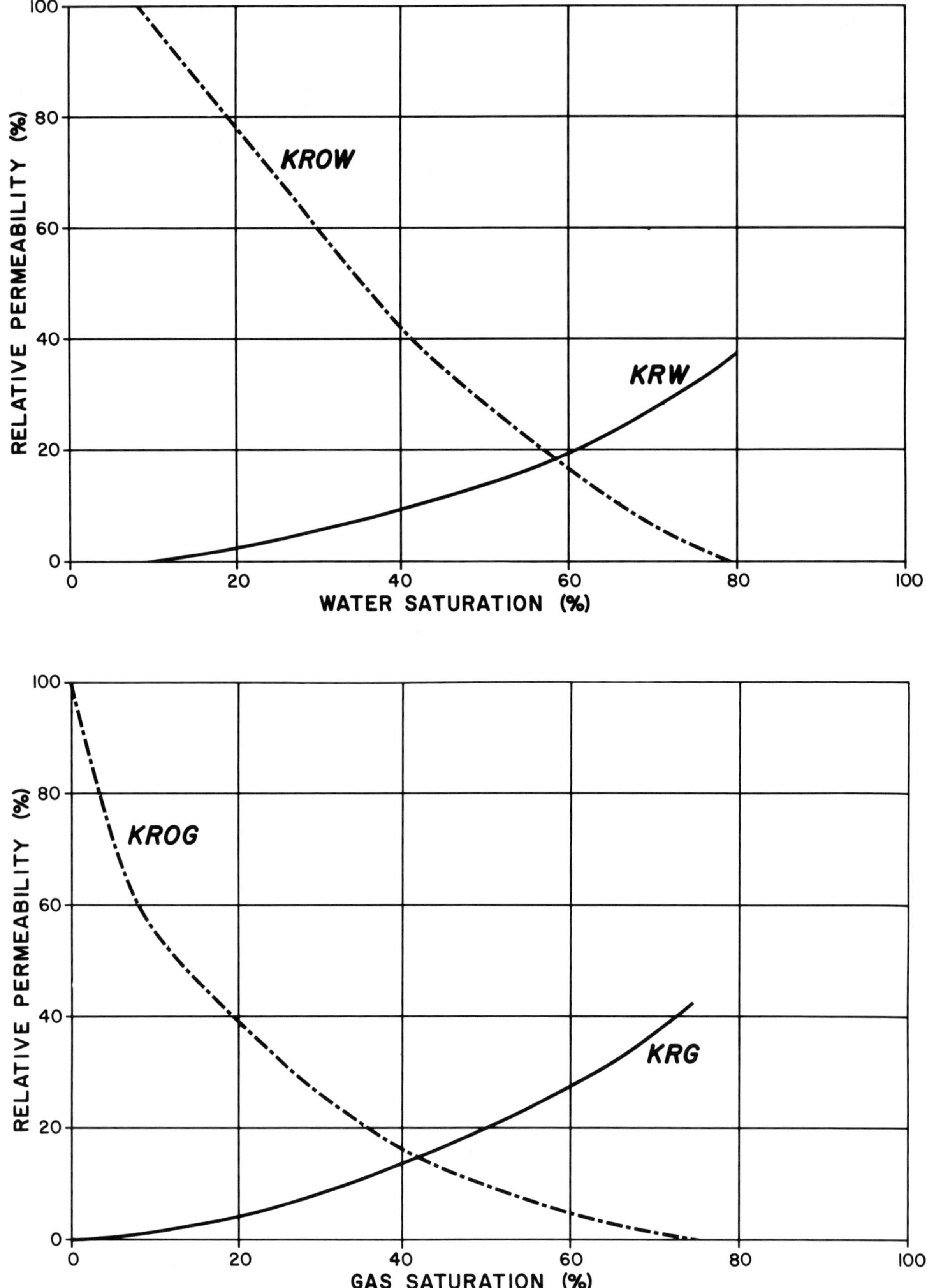

Figure 13. Relative permeability of Ekofisk chalk. (A) KRW, relative permeability to water; KROW, relative permeability to oil in the presence of water. (B) KRG, relative permeability to gas; KROG, relative permeability to oil in the presence of gas (all with irreducible water present).

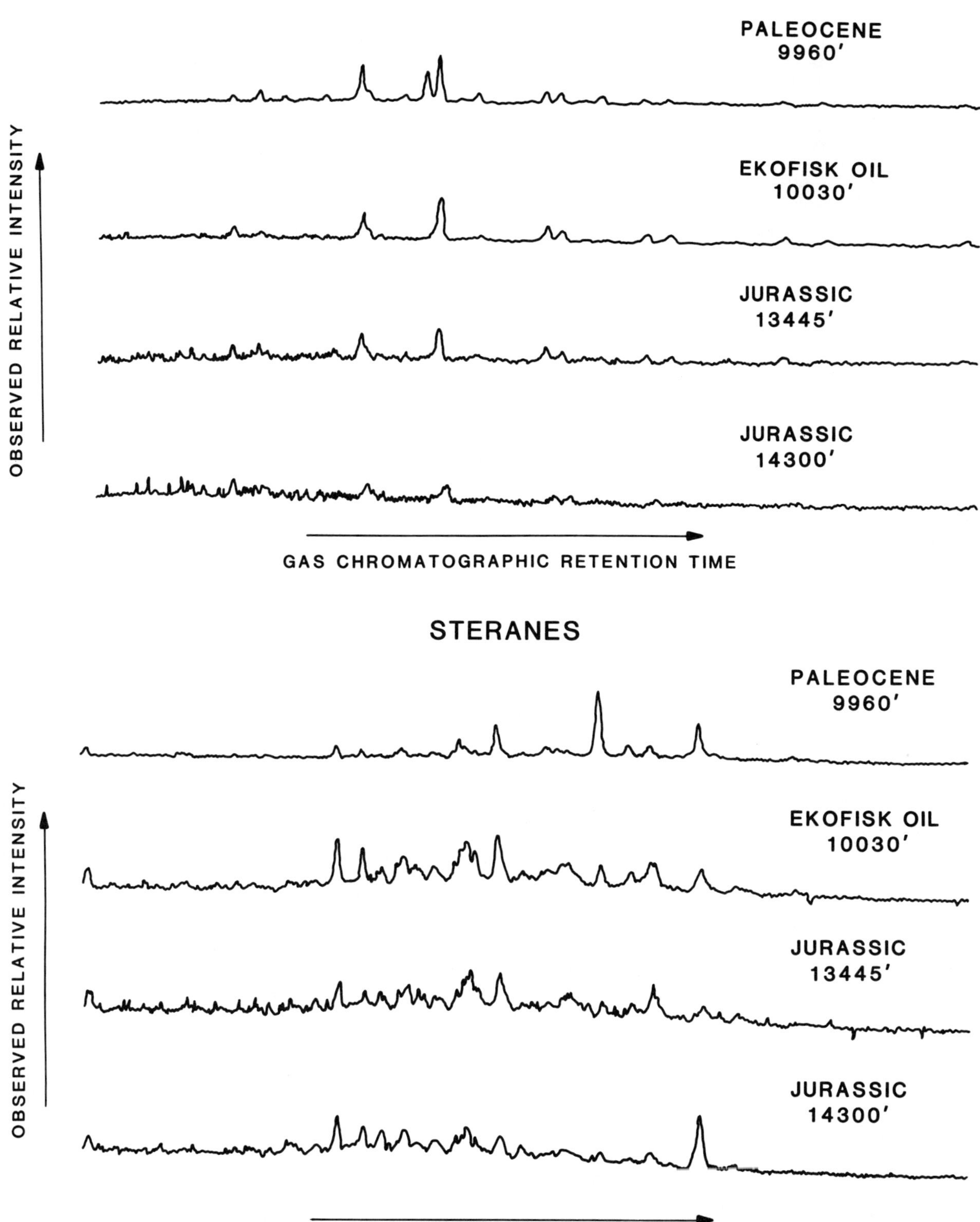

Figure 14. Biomarker "fingerprints" (chemical fossils): gas chromatograph-mass spectrometer separations of two families of compounds (steranes and hopanes) that demonstrate correlation between oil produced from Ekofisk field and extracts from shales of the Mandal Formation (Upper Jurassic), penetrated approximately 4000 ft below the reservoir. Note similarity of peak patterns between the oil and the topmost Jurassic sample, and the dissimilarity between the oil and the Paleocene shales overlying the reservoir.

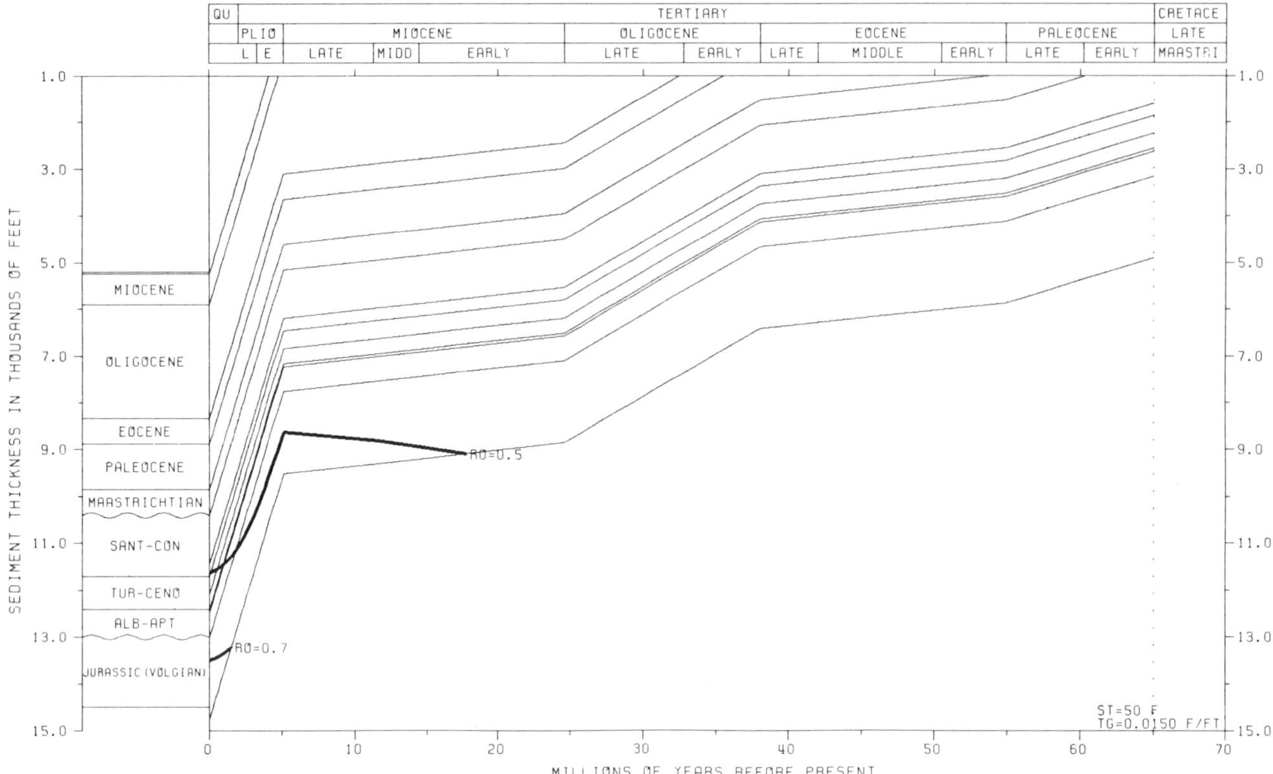

Figure 15. Burial and thermal history of the Ekofisk 2/4 B-19 well, showing the Upper Jurassic section entering the oil window ($R_o = 0.5$) in early- to middle-Miocene time, and reaching peak oil generation ($R_o = 0.7$) as recently as middle-Pliocene time. This well is near the structural crest; a more basinal well would probably show more pronounced thermal effects.

The natural fracture system of Ekofisk has been described and is recognized as critical to field performance. The chalk matrix contains effectively all of the pore volume. The fracture system provides most of the conductivity. Due to fracturing, permeability (kh) results calculated from well test analysis can be as much as 50 times higher than kh estimates based on core analysis of plugs cut from the matrix. Although variable by layer and areally, fracture intensity generally increases with higher structural position. Thomas et al. (1984) and Dangerfield and Brown (1985) discussed matrix and fracture behavior.

To maximize field performance, production wells are stimulated with hydraulic acid fracturing techniques. Typically, several short intervals are perforated and the stimulation is performed in repeated stages, using ball sealers to ensure diversion and fracturing of all intervals. Skin values achieved range from -4.5 to -5.0. Earlier development wells could not be treated as effectively by the techniques and equipment then available: A skin of -4.0 was more common. Nevertheless, production rates from these early, high-pressure wells ranged from 10,000 to 15,000 BOPD at solution GOR. Today's typical flow rates are 3000 BOPD at a GOR of 7500 SCF/STB.

Until the early 1980s, the Ekofisk Tight Zone was considered a continuous, complete barrier to flow between the Ekofisk and Tor formations. Depletion of the two reservoirs was simultaneous but independent. During 1983-1984, drilling and vertical seismic profiling demonstrated the existence of the northwest-trending crestal graben shown in Figure 3. Subsequent information from geophysics, RFT logging, and stimulation studies has shown other faults or fractured areas within the Tight Zone that were not previously detectable. These allow local communication between the Ekofisk and Tor formations and are included in reservoir simulation as open "windows" through which fluids pass.

Development Considerations

With an area of 12,000 net acres and 39 producing wells, average well spacing is approximately 300 ac. This is, however, a limitation imposed by the current production facilities. The three drilling and production platforms have a combined total of 54 drilling slots, some of which are used as gas injectors. Others are mechanically disabled or occupied by dry holes. This drainage density results in a producing life of 50 to 60 years for a producing Ekofisk field platform (Figure 16). Therefore, plans are underway as part of the waterflood program to increase the number of producing wells by inserting additional slots between existing ones. This will reduce well spacing to approximately 215 ac.

The current well pattern has the gas injection wells in the top of the structure, surrounded by most of the producers, which occupy intermediate structural locations. Ekofisk field does not have a gas cap. The gas produced is from high-GOR wells (>8000 SCF/STB) in both the Ekofisk and Tor formations. Of this, approximately 25% is reinjected. Only a few wells are drilled on the flanks, as effective permeability deteriorates rapidly downdip and well productivity is low. This injection pattern is evolving into a staggered-line-drive waterflood pattern, as mentioned in the *Waterflood Summary* section, below. Additionally, since numerous small faults have now been identified by geophysical work, potential new locations are examined to identify local geophysical anomalies, avoid faults, and optimize well placement.

Recovery factors for Ekofisk and other surrounding chalk fields are relatively low. Primary depletion is expected to recover approximately 18% of the original oil in place. However, gas injection began in 1975 and water injection in 1987. The combined effect of these programs has raised the estimated recovery factor to 23%. Additional enhanced recovery projects are being considered, including water injection into the lower interval of the Ekofisk Formation and injection of nitrogen into the top of the reservoir to further improve recovery (Pekot and Gersib, 1987). The recovery factors mentioned above are conservative. They are premised on the end of the field's life occurring in the year 2011, when the existing production license expires. However, Ekofisk can continue to produce at economically attractive rates for many years beyond 2011, and the field's ultimate recovery factor will be significantly higher.

EKOFISK FIELD SUBSIDENCE

Subsidence of the seabed at Ekofisk field was identified in 1984. This prompted an extensive program of geological and engineering studies designed to record the magnitude of field subsidence, predict future subsidence behavior, and recommend corrective action.

Seabed subsidence has occurred in response to hydrocarbon production and reservoir pressure depletion. Compaction of the high-porosity reservoir chalks has been transmitted upward through 3 km of Tertiary overburden to create a bowl-shaped depression with a present maximum of 4 m of subsidence at the seabed. Numerical modeling has predicted ultimate seabed subsidence of 6 m.

As a result of seabed subsidence, the air gap between the cellar decks of the platforms and sea level was reduced. This prompted concern about potential damage from storm waves. To counter this, the five platforms supported by steel gravity jackets were simultaneously lifted by hydraulic jacks. Leg extensions 18 ft long were inserted between the severed jacket legs and the raised platforms. Eighteen months of planning and preparation culminated in September 1987, when this project was brought to a spectacularly successful conclusion.

While monitoring of the subsidence continues, further work is proceeding to secure the Ekofisk Tank. Built as a concrete gravity structure providing both oil storage and processing facilities, the Tank could not be jacked up. A circular concrete wall was floated out to Ekofisk in two halves and joined to form a protective barrier around the Tank.

WATERFLOOD SUMMARY

In early 1984, the Phillips Group initiated a seawater-injection program for Ekofisk field. The original program called for 20 wells to be drilled from a new platform capable of injecting up to 375,000 BWPD into the Tor Formation. The proposed well plan was arranged to form a staggered-line-drive pattern oriented north-south in the northern half of the field. This well plan was based on pressure-test data suggesting a strong north-south anisotropy of critical reservoir parameters.

The new water injection platform was commissioned in 1987. Currently four wells are completed and operating, injecting a maximum of 29,000 BWPD per well.

EXPLORATION CONCEPTS

With a few exceptions (Austin Chalk, U.S. Gulf Coast; Niobrara Chalk, U.S. interior), prior to the discovery of Ekofisk field, chalk was generally considered a rock with low reservoir potential. Its ultrafine grain size and its typical burial history (rapid porosity loss with increasing depth) made chalk an unattractive target. With the Ekofisk discovery, the North Sea chalk became a major exploration play. In the past decade, a tremendous amount of chalk research has focused on deposition, diagenesis, and reservoir behavior. One key finding that affects exploration strategy is that the most productive chalk intervals have been resedimented. Thus, paleo-lows that served as catch-basins for chalk debris flows and turbidites are the most attractive areas if they are now structurally high.

Significantly, analogs to the Ekofisk reservoir can be sought only in Cretaceous and younger basins,

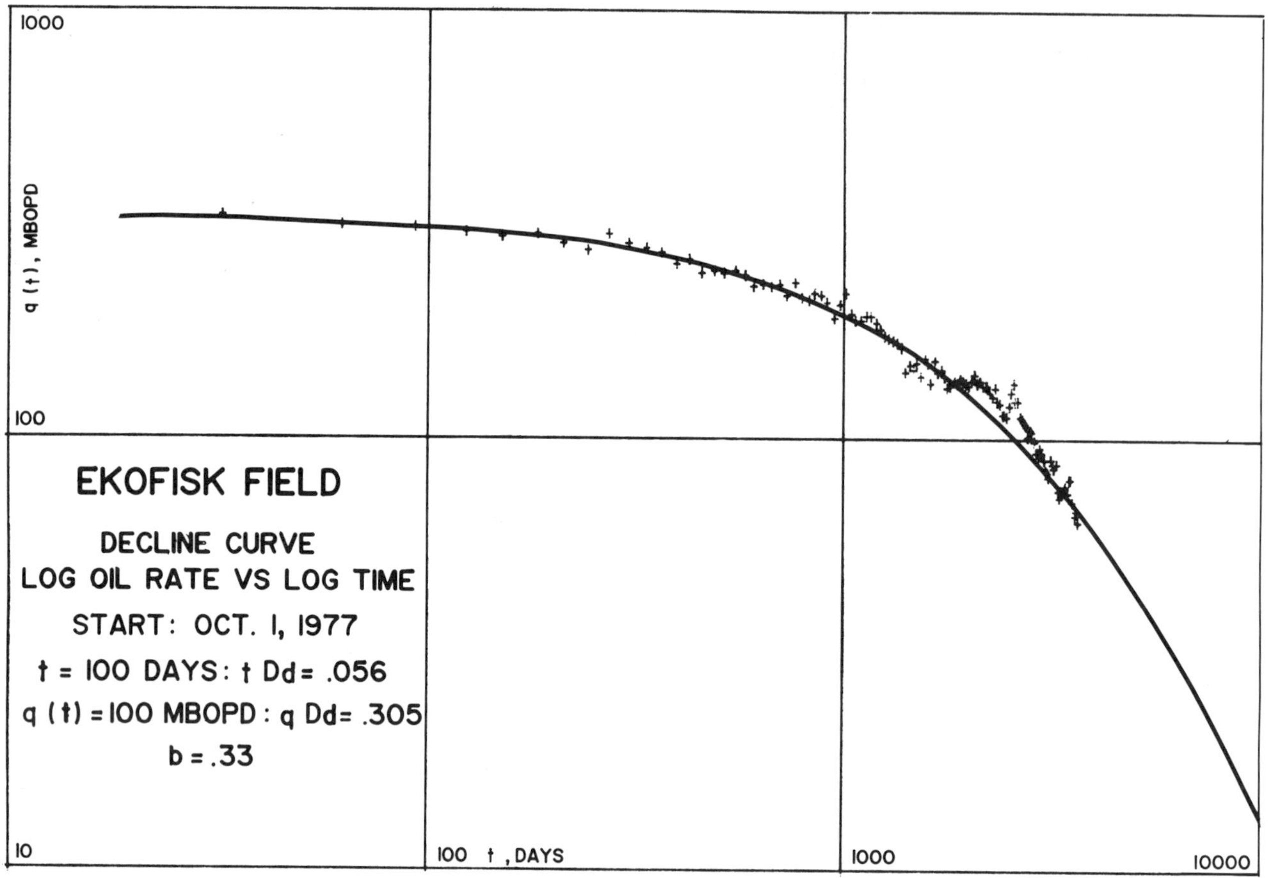

Figure 16. Production decline curve for Ekofisk field.

as true pelagic chalk-forming organisms evolved relatively late in the earth's history. As drilling moves deeper into continental slopes and rises, however, the likelihood of penetrating pelagic sediments increases. It is probable that additional significant chalk reservoirs remain to be discovered.

ACKNOWLEDGMENTS

We are grateful for the assistance of our present and former colleagues at Phillips Petroleum Company, particularly Helen Farrell, Bill Hughes, Rick Schatzinger, and Joe Shveima. The statements in this paper are those of the authors and do not necessarily reflect the opinions of the co-venturers.

REFERENCES CITED

Brewster, J., J. Dangerfield, and H. Farrell, 1986, The geology and geophysics of the Ekofisk Field waterflood: Marine and Petroleum Geology, v. 3, p. 139-169.

Bromley, R. G., and A. A. Ekdale, 1987, Mass transport in European Cretaceous chalk; fabric criteria for its recognition: Sedimentology, v. 34, p. 1079-1092.

Dangerfield, J. A., and D. A. Brown, 1985, The Ekofisk Field: North Sea Oil and Gas Reservoirs Seminar Proceedings: Norwegian Institute of Technology, Trondheim, p. 3-22.

D'Heur, M., 1984, Porosity and hydrocarbon distribution in the North Sea chalk reservoirs: Marine and Petroleum Geology, v. 1, p. 211-238.

Durney, D. W., 1972, Solution-transfer, an important geological deformation mechanism: Nature, v. 235, p. 315-317.

Ekdale, A. A., 1985, Paleoecology of the marine endobenthos: Palaeogeography, Palaeoclimatology, Palaeoecology, v. 50, p. 63-81.

Ekdale, A. A., and R. G. Bromley, 1984, Cretaceous chalk ichnofacies in northern Europe: Geobios, Special Memoir 8, p. 201-204.

Feazel, C. T., and H. E. Farrell, 1988, Chalk from the Ekofisk Area, North Sea: nannofossils + micropores = giant fields, *in* A. J. Lomando and P. M. Harris, eds., Giant oil and gas fields: Society of Economic Paleontologists and Mineralogists Core Workshop 12, p. 155-178.

Feazel, C. T., and R. A. Schatzinger, 1985, Prevention of carbonate cementation in petroleum reservoirs, *in* N. Schneidermann and P. M. Harris, eds., Carbonate cements: Society of Economic Paleontologists and Mineralogists Special Publication 36, p. 97-106.

Feazel, C. T., J. Keany, and R. M. Peterson, 1985, Cretaceous and Tertiary chalk of the Ekofisk Field area, central North Sea, *in* P. O. Roehl and P. W. Choquette, eds., Carbonate petroleum reservoirs: New York, Springer-Verlag, p. 495-507.

Hancock, J. M., 1976, The petrology of the chalk: Proceedings of Geological Association, v. 86, p. 499-535.

Hancock, J. M., and P. A. Scholle, 1975, Chalk of the North Sea, *in* A. W. Woodland, ed., Petroleum and the continental shelf of Europe: New York, John Wiley & Sons, v. 1, p. 413-425.

Hatton, I. R., 1986, Geometry of allochthonous Chalk Group members, Central Trough, North Sea: Marine and Petroleum Geology, v. 3, p. 79-98.

Hughes, W. B., A. G. Holba, D. E. Miller, and J. S. Richardson, 1985, Geochemistry of greater Ekofisk crude oils, *in* Petroleum geochemistry in exploration of the Norwegian Shelf: Norwegian Petroleum Society, Graham and Trotman, p. 75-92.

Kennedy, W. J., 1985, Sedimentology of the Late Cretaceous and Early Paleocene Chalk Group, North Sea Central Graben: North Sea Chalk Symposium, Book I, p. 1-35.

Neugebauer J., 1974, Some aspects of cementation in chalk, *in* K. J. Hsu and H. C. Jenkyns, eds., Pelagic sediments: on land and under the sea: International Association of Sedimentologists Special Publication 1, p. 149-176.

Pekot, L. J., and G. A. Gersib, 1987, Ekofisk, *in* A. M. Spencer et al., eds., Geology of the Norwegian oil and gas fields: London, Graham and Trotman, p. 73-87.

Price, M., M. J. Bird, and S. S. D. Foster, 1976, Chalk pore-size measurements and their significance: Water Services, v. 80, n. 968, p. 596-600.

Schatzinger, R. A., C. T. Feazel, and W. E. Henry, 1985, Evidence of resedimentation in chalk from the Central Graben, North Sea, *in* P. D. Crevello and P. M. Harris, eds., Deep-water carbonates: buildups, turbidites, debris flows and chalks—a core workshop: Society of Economic Paleontologists and Mineralogists Core Workshop 6, p. 342-385.

Schlanger, S. O., and R. G. Douglas, 1974, Pelagic ooze-chalk-limestone transition and its implications for marine stratigraphy, *in* K. J. Hsu and H. C. Jenkyns, eds., Pelagic sediments: on land and under the sea: International Association of Sedimentologists Special Publication 1, p. 117-148.

Scholle, P. A., 1977, Chalk diagenesis and its relation to petroleum exploration: oil from chalks, a modern miracle?: American Association of Petroleum Geologists Bulletin, v. 61, p. 982-1009.

Scholle, P. A., M. A. Arthur, and A. A. Ekdale, 1983, Pelagic environment, *in* P. A. Scholle, D. G. Bebout, and C. H. Moore, carbonate depositional environments: American Association of Petroleum Geologists Memoir 33, p. 620-691.

Thomas, L. K., T. N. Dixon, C. E. Evans, and M. E. Vienot, 1984, Ekofisk waterflood pilot: Society of Petroleum Engineers Publication 13120.

Van den Bark, E., and O. D. Thomas, 1981, Ekofisk: first of the giant oil fields in western Europe: American Association of Petroleum Geologists Bulletin, v. 65, p. 2341-2363.

Watts, N. L., J. F. Lapre, F. S. Van Schijndel-Goester, and A. Ford, 1980, Upper Cretaceous and Lower Tertiary chalks of the Albuskjell area, North Sea: deposition in a slope and base-of-slope environment: Geology, v. 8, p. 217-221.

Appendix 1. Field Description

Field name .... *Ekofisk*

Ultimate recoverable reserves .... *1210 MM bbl (192 × 10^6 Sm^3) gas (NPD estimates)*

Field location:

- **Country** .... *Norway*
- **State** .... *Continental Shelf*
- **Basin/Province** .... *Central graben*

Field discovery:

- **Year first pay discovered** .... *(Paleocene) Danian Ekofisk Formation chalk 1969*
- **Year second pay discovered** .... *(Upper Cretaceous) Maastrichtian Tor Formation chalk 1970*
- **Third pay** .... *NA*

Discovery well name and general location

- **First pay** .... *2/4-1AX; Block 2/4, Norway*
- **Second pay** .... *2/4-2X; Block 2/4, Norway*
- **Third pay** .... *NA*

Discovery well operator .... *Phillips Petroleum*

- **Second pay** .... *Phillips Petroleum*
- **Third pay** .... *NA*

IP in barrels per day and/or cubic feet or cubic meters per day:

- **First pay** .... *1071 BOPD; 5.9 MSCFD*
- **Second pay** .... *2064 BOPD; 2.2 MSCFD*
- **Third pay** .... *NA*

All other zones with shows of oil and gas in the field:

Age	Formation	Type of Show
Miocene	*Utsira Fm. Eq.*	*Oil and gas*
Eocene	*Balder Fm.*	*Oil and gas*
Paleocene	*Lista Fm.*	*Oil and gas*

Geologic concept leading to discovery and method or methods used to delineate prospect, e.g., surface geology, subsurface geology, seeps, magnetic data, gravity data, seismic data, seismic refraction, nontechnical:

The primary objective was to test Paleocene age sandstones encountered in wells to the northwest (Cod field). The secondary objective was to test an anticlinal structure seen on reflection seismic data at the top of the chalk reflector level.

Structure:

Province/basin type (see St. John, Bally, and Klemme, 1984) .... *Intracratonic rift basin*

Tectonic history

The Central graben formed in response to an extensional rifting episode initiated during the Permian-Triassic and continuous throughout the Mesozoic. Contemporaneous strike-slip, fault inversion, and late Mesozoic-Miocene halokinesis influenced regional sedimentary patterns. Extensive basinal subsidence and tectonic quiescence proceeded thoughout the Tertiary, interrupted only by an Oligocene compressional event causing regional arching.

Regional structure

Centrally located within the Central graben, a northwest-southeast-striking complex of en echelon normal faults. Ekofisk field is one of many anticlinal structures resulting from late Mesozoic halokinesis.

Local structure

The Ekofisk field is a north-south elongated domal structure at the top of the chalk level, with maximum structural dips of 8–10°. The field exhibits four-way structural closure with a vertical relief of 876 ft.

Trap

Trap type(s) .... *Ekofisk field exhibits one anticlinal trap enclosing two pay zones*

Basin stratigraphy (major stratigraphic intervals from surface to deepest penetration in field):

Chronostratigraphy	Formation	Depth to Top in ft subsea
Recent-middle Miocene	*Nordland Group*	*Surface to*
Middle Miocene-early Eocene	*Hordaland Group*	*-5341*
Early Eocene-Late Danian	*Rogaland Group*	*-9060*
Danian-Cenomanian	*Chalk Group*	*-9479*
Albian-Portlandian	*Cromer Knoll Group*	*-12,638*
Ryazanian-Callovian	*Tyne Group*	*-13,230*

Location of well in field .... *NA*

Reservoir characteristics:

Number of reservoirs *2*

Formations .... *Ekofisk Formation; Tor Formation*

Ages .... *Paleocene (Danian); Upper Cretaceous (Maastrichtian)*

Depths to tops of reservoirs .... *Ekofisk, -9479 ft (2889 m); Tor, -9921 ft (3024 m)*

Gross thickness (top to bottom of producing interval) .... 950 ft (290 m)

Net thickness—total thickness of producing zones

Average .... *575 ft (175 m)*

Maximum .... *900 ft (275 m)*

Average

Maximum

Lithology

Biomicritic limestone (chalk) composed of low magnesium calcite from disaggregated, microscopic fragments of coccolithophorid algae. With minor, yet significant foraminifera, skeletal debris, clay, chert, and disseminated silica.

Porosity type .... *Reservoir porosity is primarily interparticle porosity with minor associated fracture porosity*

Average porosity .... *Ekofisk, 32%; Tor, 30%*

Average permeability .... *1.0 md; effective permeability up to 150 md dependent on fracture intensity*

Seals:

Upper

Formation, fault, or other feature .... *Maureen Formation equivalent*

Lithology .... *Calcarous claystone*

Lateral

Formation, fault, or other feature

Lithology

Source:

Formation and age .... *Mandal Formation, Late Jurassic*

Lithology .... *Shale*

Average total organic carbon (TOC) *2.08%*

Maximum TOC *2.91%*

Kerogen type (I, II, or III) .... *II*

Vitrinite reflectance (maturation) .... $R_o = 0.65–0.88\%$

Time of hydrocarbon expulsion .... *Miocene*

Present depth to top of source .... 13,360 ft (4072 m)
Thickness .... 46 ft (14 m)
Potential yield .... 45–82 BOE/ac-ft (cumulative)

Appendix 2. Production Data

Field name .... *Ekofisk*

Field size:

- **Proved acres** .... *12,000 (4850 ha)*
- **Number of wells all years** .... *59 (including for sidetracks)*
- **Current number of wells** *42*
- **Well spacing** .... *286 ac*
- **Ultimate recoverable** .... *1210 million bbl; 6332 bscf*
- **Cumulative production** .... *802 million bbl; 3140 bscf*
- **Annual production** .... *26 million bbl; 255 bscf*
- **Present decline rate** *10%*
  - **Initial decline rate** *19%*
  - **Overall decline rate** *16%*
- **Annual water production** .... *1.3 MM bbl*
- **In place, total reserves** .... *Stock tank 7100 million bbl*
- **In place, per acre-foot** .... *2140 reservoir bbl/ac-ft*
- **Primary recovery** .... *1094 million bbl; 6239 bscf*
- **Secondary recovery** .... *1210 million bbl; 6332 bscf*
- **Enhanced recovery** .... *NA*
- **Cumulative water production** .... *23 MM bbl*

Drilling and casing practices:

Amount of surface casing set .... *5000 ft*

Casing program

30-in. drive pipe to 500 ft TVD; 20-in. conductor to 1600 ft TVD; 13⅜-in. surface csg. to 5000 ft TVD; 9⅝-in. intermediate csg. to 10,000 ft TVD; 7-in. production csg. or liner to TD at 11,200 ft TVD

Drilling mud

To 13⅜-in. casing point, sea water and bentonite (native mud); to 9⅝-in. casing point, inhibited sea-water base to prevent shale swelling (Shaletrol/Drispac typical); to TD, sea water gelled with cellulosic polymer (Drispac typical)

Bit program

Long-tooth tri-cone to 8000 ft TVD, near top Paleocene (Hughes XIG typical); short-tooth tri-cone to 10,000 ft TVD, top chalk (Hughes XV typical); tungsten carbide insert to TD (Hughes J22, J33 typical)

High pressure zones .... *Early Miocene–early Paleocene, 3500 ft TVD–9500 ft TVD*

Completion practices:

Interval(s) perforated

Lower Ekofisk Formation, middle to lower part of Tor Formation pay to maintain low GOR. Upper intervals of Ekofisk and Tor formations may be perforated in low structure wells.

Well treatment .... *Multistage acid fracture stimulation*

Formation evaluation:

Logging suites

Uphole section: DIL/LSS/GR/CAL
Reservoir Section: DLL/MSFL/GR
LDL/CNL/GR
LSS/GR (optional)
SHDT/GR (optional)
RFT/GR

Testing practices

Complete in Ekofisk Formation, perform drawdown/buildup test. Add Tor Formation completion and retest commingled well. Production log. Thermal neutron decay time (TDT) log. Test program repeated approximately every 3 years. Static pressure surveys recorded after long shutdowns.

Mud logging techniques

Measurement while drilling gamma ray log in final 1000 ft of 12-in. hole for precise determination of 9⅝-in. casing point

Oil characteristics:

Type .... *Paraffinic-naphthenic*
API gravity .... *36°*
Base .... *Intermediate*
Initial GOR .... *1550 SCF/STB*
Sulfur, wt% *0.18*
Viscosity, SUS .... *61.6 at 70°F, 47.2 at 100°F*
Pour point .... *25°F*
Gas-oil distillate .... *25% vol*

Field characteristics:

Average elevation .... *−10,000 ft subsea*
Initial pressure .... *7035 psi*
Present pressure .... *3500 psi*
Pressure gradient .... *0.685 psi/ft*
Temperature .... *268°F*
Geothermal gradient .... *0.0268°F/ft*
Drive .... *Solution gas, gas injection, waterflood*
Oil column thickness .... *1300 ft*
Oil-water contact .... *−10,785 ft, variable*
Connate water *8%*
Water salinity, TDS .... *61,000 ppm*
Resistivity of water .... *0.042 ohm-m at 245°F*
Bulk volume water (%) *2.5*

Transportation method and market for oil and gas:

34-in. oil pipeline 220 mi to Teeside, England; 36-in. gas pipeline 274 mi to Emden, West Germany.

Eldfisk Field—Norway
Central Graben, North Sea

MICHEL D'HEUR
Petrofina S.A.
Bruxelles, Belgium

FIELD CLASSIFICATION

BASIN: North Sea
BASIN TYPE: Rift
RESERVOIR ROCK TYPE: Chalk (Fractured)
RESERVOIR ENVIRONMENT OF DEPOSITION: Chalk Resedimented by Debris Flow
RESERVOIR AGE: Cretaceous-Paleocene
PETROLEUM TYPE: Oil and Gas
TRAP TYPE: Elongate Anticline

LOCATION AND RECOVERY

The Eldfisk field (rank 225) is located in the southernmost part of the Norwegian sector of the North Sea, 320 km southwest of the Norwegian coast. The field is also situated 15 km south of the giant Ekofisk field (rank 98) and 15 km northwest of the Valhall field (Figure 1). The field and the seven other fields that produce from chalks of early Paleocene and Late Cretaceous age occur in an area commonly referred to as the "Greater Ekofisk area."

Eldfisk lies entirely in Block 2/7. Water depth in the area is 230 ft (70 m).

The ultimate recovery of the field will be 400–450 $\times$ 10^6 STB oil plus 1.6–1.8 $\times$ 10^{12} cf gas given the current production scenario of simple pressure depletion and solution gas drive. A variety of secondary recovery mechanisms are under study.

HISTORY

Pre-Discovery

In 1965, during the first round of licensing in Norway, Phillips, Petrofina, and Agip were assigned licence PL018, which covers Blocks 1/5, 2/4, 2/7, and 7/11 on the continental shelf.

The three companies were joined by the Petronord Group in 1967, and the present participation interests are: Phillips Petroleum Co. of Norway (36.96%); Norske Fina (30%); Norsk Agip (13.04%); Elf Aquitaine Norge (8.1%); Norsk Hydro (6.7%); Total Marine Norsk (4.05%); Eurafrep Norge (0.45%); Coparex Norge (0.4%); and Cofranord (0.3%). Phillips is the operator.

A large structure was identified in Block 2/7 by the first phase of seismic acquisition (Figures 2 and 3).

Exploration drilling in Norway started in 1965. The first well drilled by the group on the Norwegian Shelf was Well 16/11-1; it was a dry hole. Three years later, in 1968, the group discovered gas in Paleocene sandstone at Cod in Block 7/11 (small map in Figure 1). Another discovery in the same reservoir had taken place in the U.K. sector, 80 km south of Cod. Since Blocks 2/4 and 2/7 were located between these two discoveries, the upper Paleocene sandstone became the primary objective in the area.

In July 1969, the Amoco group drilled Well 2/11-1 southeast of the Phillips group Blocks 2/4 and 2/7. Well 2/11-1 tested 900 BOPD from a 50 ft (15 m) thick net pay interval from the top section of the Upper Cretaceous chalk.

In October 1969, the Phillips group drilled its first well on Block 2/4 on the flank of the large Ekofisk dome. The primary objective, the upper Paleocene sandstone, was not present, but from 10,000 to 10,700 ft (3050 to 3250 m), the well encountered a very porous chalk reservoir with very high oil saturation. The net pay was thicker than 600 ft (180 m) with porosities ranging from 30 to 42%. The discovery of Ekofisk gave rise to a reactivation of the exploration in the central North Sea.

Discovery

Ten months after the Ekofisk discovery, the Phillips group started exploration drilling in the adjacent Block 2/7, to test the large Eldfisk structure. The exploration wells are shown in Figure 1.

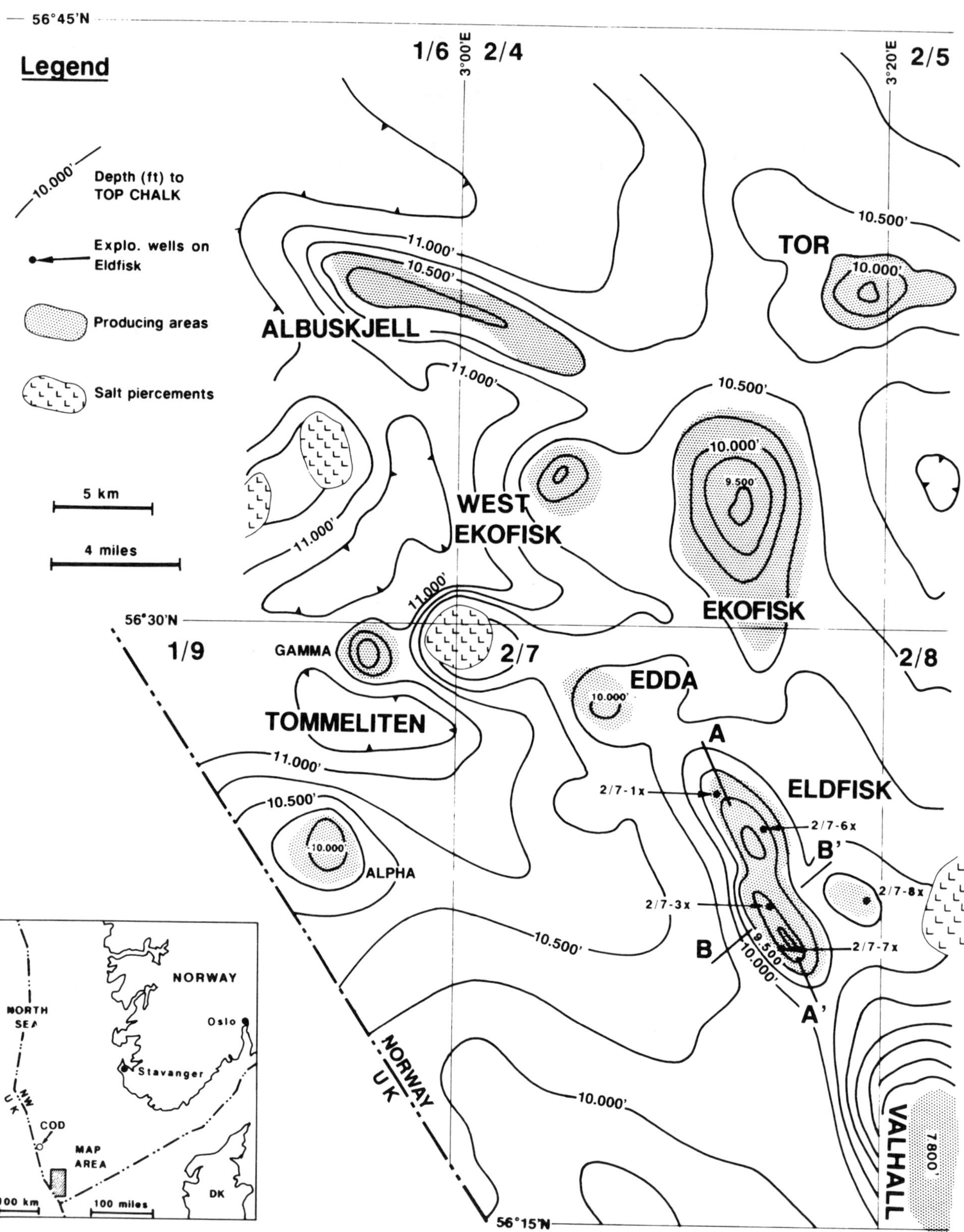

Figure 1. Location map of the "Greater Ekofisk" area. Contours indicate the depth of the top of the Chalk Group.

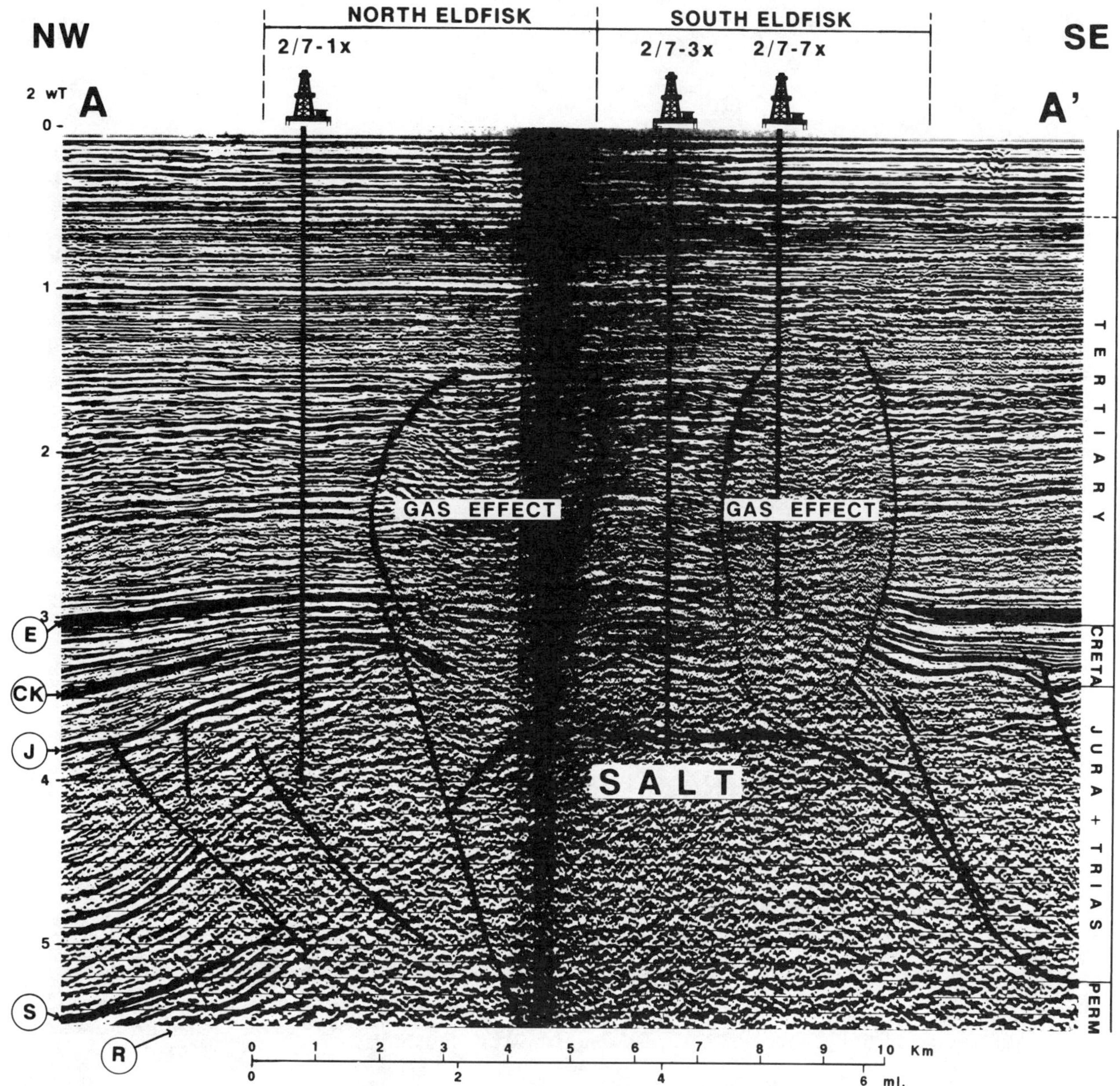

Figure 2. Axial northwest-southeast seismic section (location A-A′ on Figure 1). E, CK, J, S, and R are reflectors near or at the top of the Ekofisk Formation, Cromer Knoll Group, Jurassic, salt, and Rotliegendes, respectively.

The first well, 2/7-1, was drilled in 1970 on the northwestern edge of a disturbed area of seismic response that we now know was due to gas charging of the Tertiary shales. The original interpretation of this gas effect was a complex fault system. The well encountered less than 100 ft of net pay chalk in the Ekofisk Formation* (Danian) and Tor Formation (Maastrichtian). In addition to being thin, the reservoir was of poor quality and the well was not successfully tested. Below the reservoir interval the well penetrated 1200 ft (365 m) of tight Upper Cretaceous chalk, 920 ft (280 m) of Lower Cretaceous marls and shales, and bottomed in the Upper Jurassic shales at a depth of 15,000 ft (4570 m) (Figure 4). The second well on Eldfisk, 2/7-3, is considered the discovery well of the field. It was drilled along the axis of the southern half of the structure, in the center of a disturbed seismic area in 1972. At a depth of 9100 ft (2770 m) subsea, the well penetrated a 300 ft (90 m) thick chalk reservoir (Danian + Maastrichtian age) with porosities ranging from 30% to 40% and high oil saturation. In addition, a secondary pay

*The lithostratigraphy conforms to Deegan and Scull (1977) except where otherwise stated.

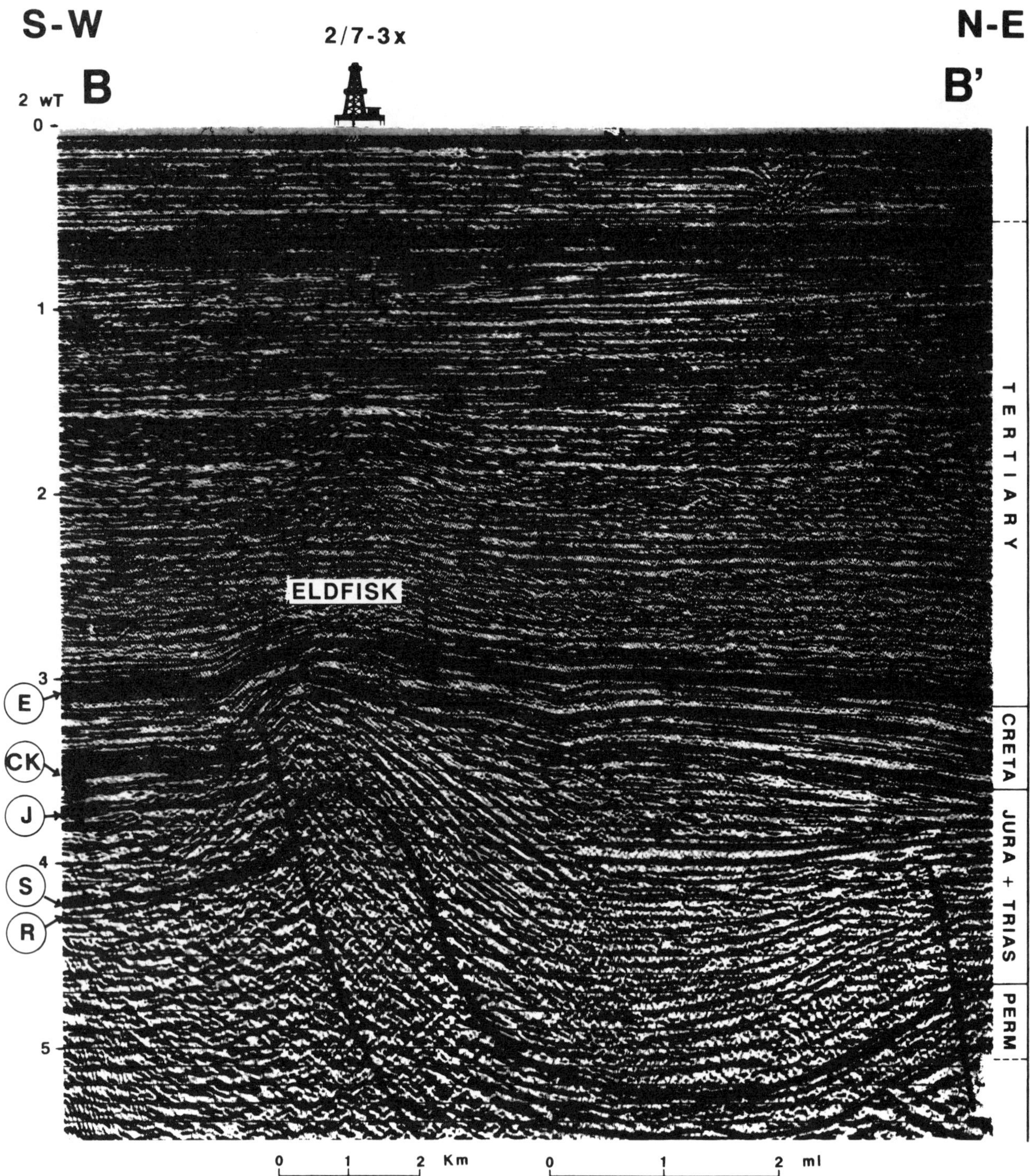

Figure 3. Transverse southwest-northeast seismic section (location B-B′ on Figure 1). Horizons E to R as in Figure 2.

zone, 50 ft (15 m) thick, was found in Campanian chalk, but the rock had only fair reservoir quality. Three intervals of the main pay were tested with individual flow rates of 4200, 4400, and 5000 BOPD. The Campanian pay (Hod Formation) flowed oil at a rate of 400 BOPD. Below the chalk reservoirs, the well encountered the Cretaceous sequence recognized in Well 2/7-1 underlain by 2880 ft (865 m) of Upper Jurassic shales and sandstones resting on Permian evaporites. Most of the Upper Jurassic sandstones were oil bearing but tight with the exception of a 50 ft (15 m) thick interval that produced at an

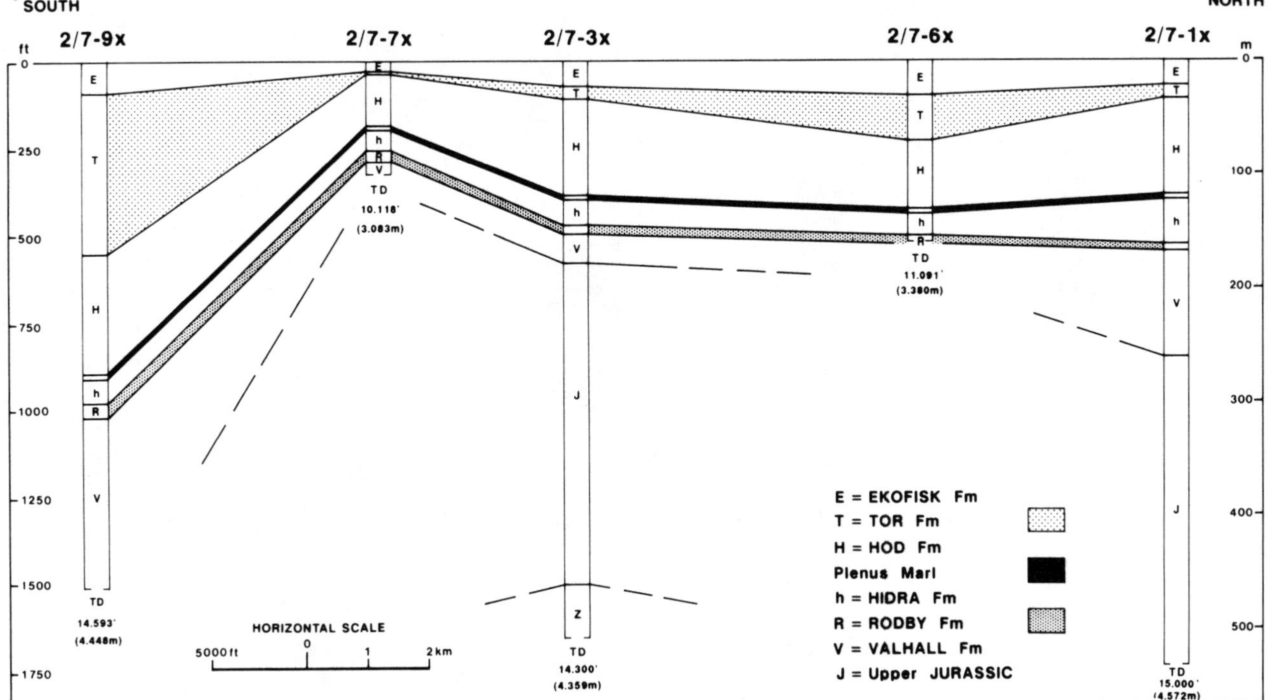

Figure 4. Mesozoic and lower Tertiary sequences penetrated in the Eldfisk exploration and appraisal wells. The section is flattened at the top of the Chalk Group to illustrate the thickness variation of the underlying formations.

unstabilized rate of 920 BOPD. A thicker sandstone interval, later named the Eldfisk Formation (Vollset and Doré, 1984), was unfortunately water bearing.

The two appraisal wells drilled in 1973, 2/7-6 and 2/7-7, also gave good results but demonstrated very large variations in reservoir thickness, stratigraphy, and quality. In Well 2/7-6, the Tor Formation was more than 380 ft (116 m) thick while in Wells 2/7-3 and 2/7-7, it was 32 ft (10 m) and 20 ft (6 m) thick, respectively. The significance of these thickness variations was not understood until the early eighties.

Well 2/7-8 was drilled in 1973 on the lower relief East Eldfisk structure, east of the major Eldfisk accumulation. The reservoir was only 200 ft (60 m) thick, and porosity and oil saturation were much lower than in the Eldfisk wells.

Porosity and oil saturation logs of the major exploration and appraisal wells are shown in Figure 5.

Post-Discovery

Forty-nine development wells were initially planned. Three permanent steel platforms were installed: Drilling/production platform 2/7B was set on the northern part of the field, and drilling platform 2/7A and production process platform 2/7FTP were installed over the southern part of Eldfisk (Figure 6).

The platform and well locations were based on the "total net pay thickness" interpretation of 1975 shown in Figure 7. Development drilling started in 1978. Several of the first development wells gave disappointing results because the complex geology of Eldfisk was not understood. For example, the first development well drilled on the eastern flank (2/7-A12) was a dry hole, and the vertical Well 2/7-A6 encountered only 200 ft (60 m) of net pay instead of the predicted 600 ft (180 m) (Figure 7).

Detailed studies in 1980–1981 led to a better geological model and consequently to better results. A total of 40 producing wells have now been drilled. They are completed either in the Ekofisk Formation, the Tor Formation, the Hod Formation, or a combination of these three.

Since the discovery of Eldfisk, several small culminations to the west and southwest have been tested without real success. The Eldfisk production is pumped to the "Ekofisk Center" through a 24-in. oil line and a 30-in. gas line.

Cumulative production between 1979 and 1988 has been approximately 200 million bbl of oil and 450 bcf of gas.

DISCOVERY METHODS

Then

The presence of a thick Tertiary basin in the central North Sea was identified before the first

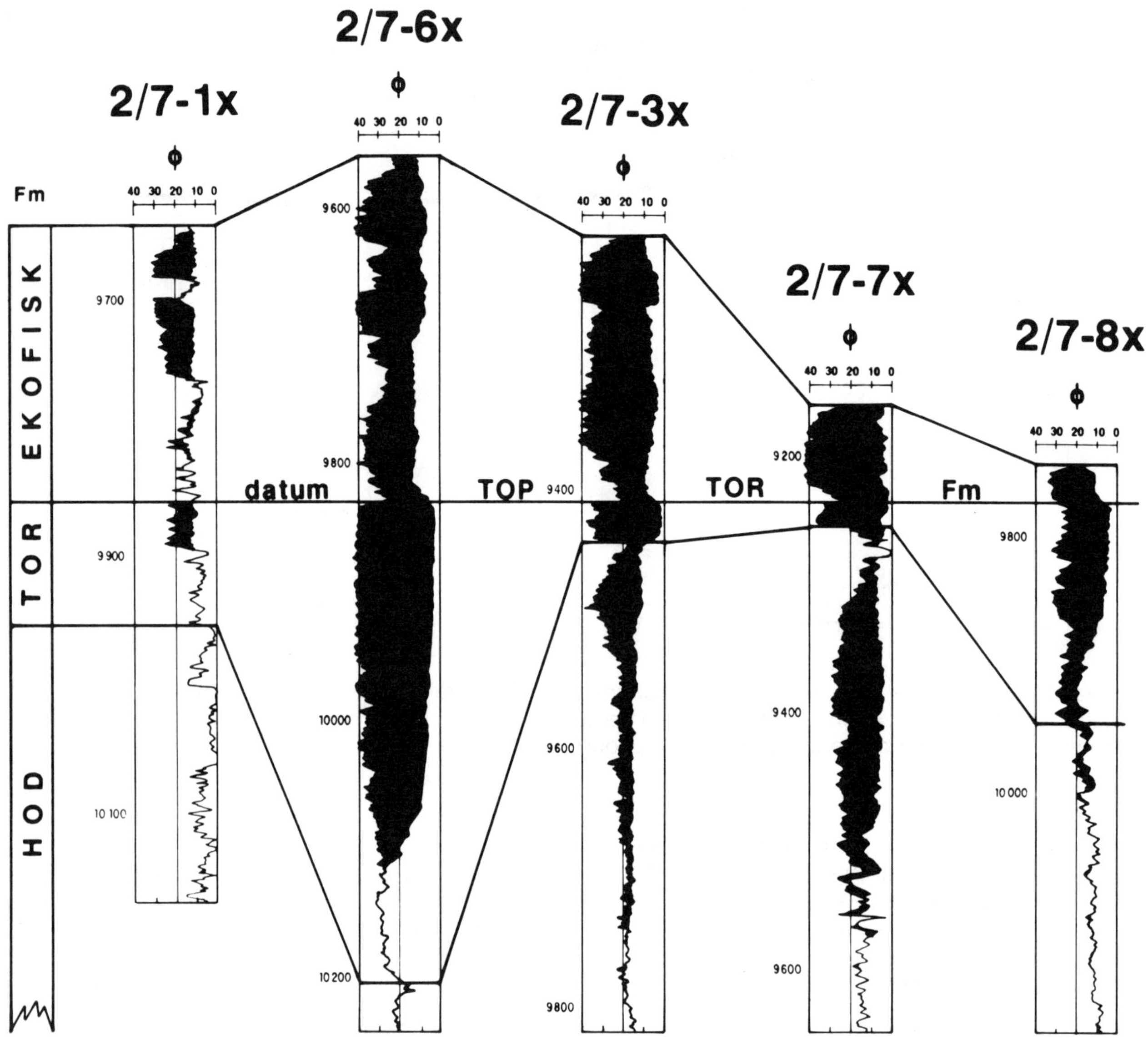

Figure 5. Porosity and hydrocarbon saturation logs of the reservoir in the Eldfisk exploration and appraisal wells (location in Figure 1).

round of Norwegian licensing (1965), but its shape was not well defined until the late sixties.

At that time, exploration concentrated on the southern North Sea where gas had been discovered in sandstones of Early Permian and Triassic ages. The source rock for the gas was the Carboniferous coal measures.

In the central North Sea, the coal was not present, and the Permian sandstones were too deep to be an attractive objective. Salt diapirs had been identified by geophysical means and, provided that Mesozoic or Tertiary reservoir and source rocks existed, large hydrocarbon accumulations were possible. The reasons why exploration started earlier in the southern North Sea rather than in the central area were both geological and technical. To the south, water depth is shallower, the sea is calmer, and the prospects are close to the shoreline. In addition, oil and gas had been found onshore in the United Kingdom and in the Netherlands. Interpolation of geological and geophysical data was simpler in the southern area than extrapolation toward the central North Sea where very little was known.

In 1965, the Eldfisk structure was the only one identified in Block 2/7, based on a 5×5 km seismic grid.

A 1.7×1.7 km grid was shot in the late sixties and showed a number of small satellite culminations of low structural relief. Eldfisk was undoubtedly the largest prospect in the block.

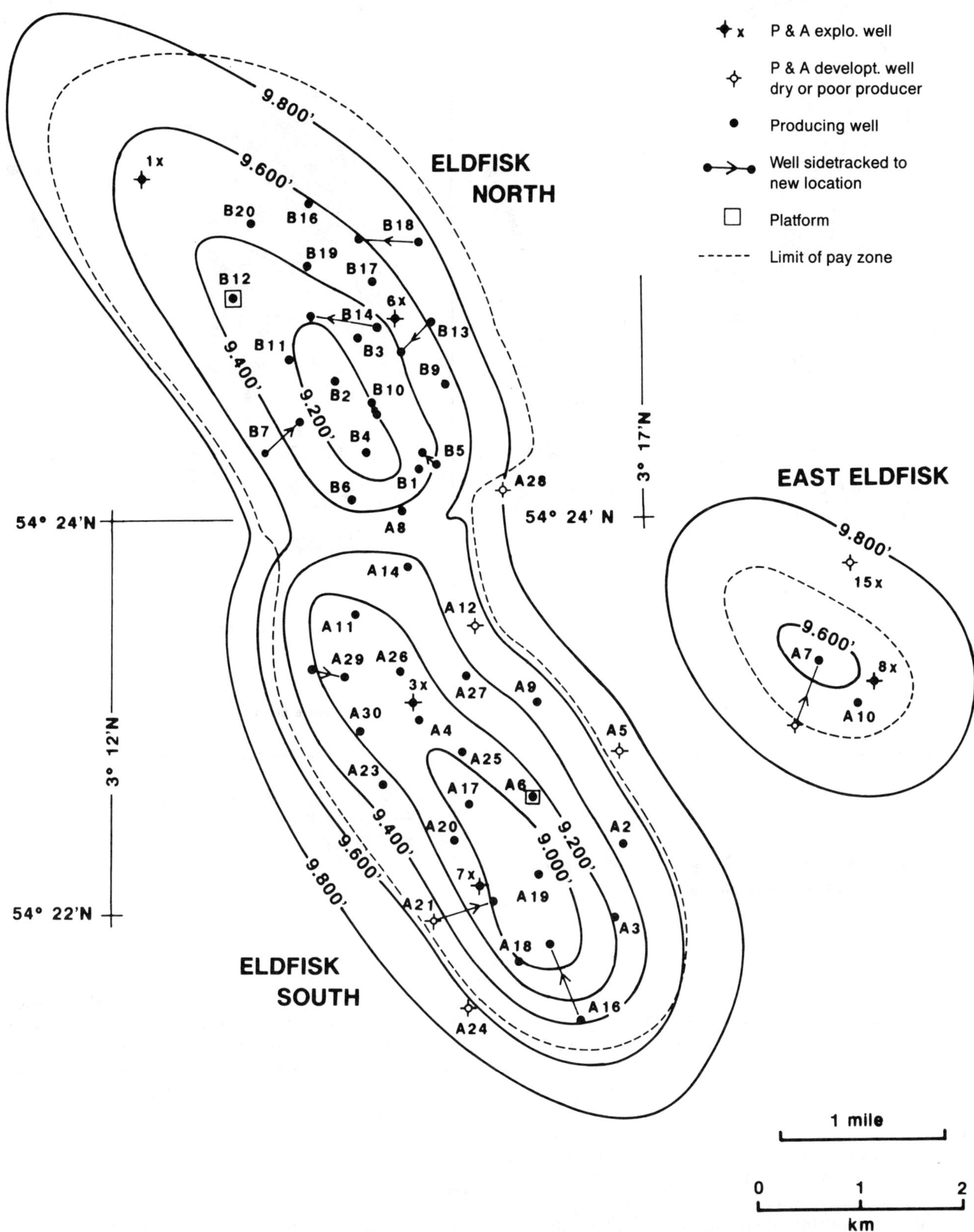

Figure 6. Top Ekofisk Formation reservoir. Current development map of the Eldfisk field. East Eldfisk is drained by two wells drilled from platform A located on Eldfisk South. Contour Interval, 200 ft. Contours below -9800 ft omitted. Depths are in feet subsea.

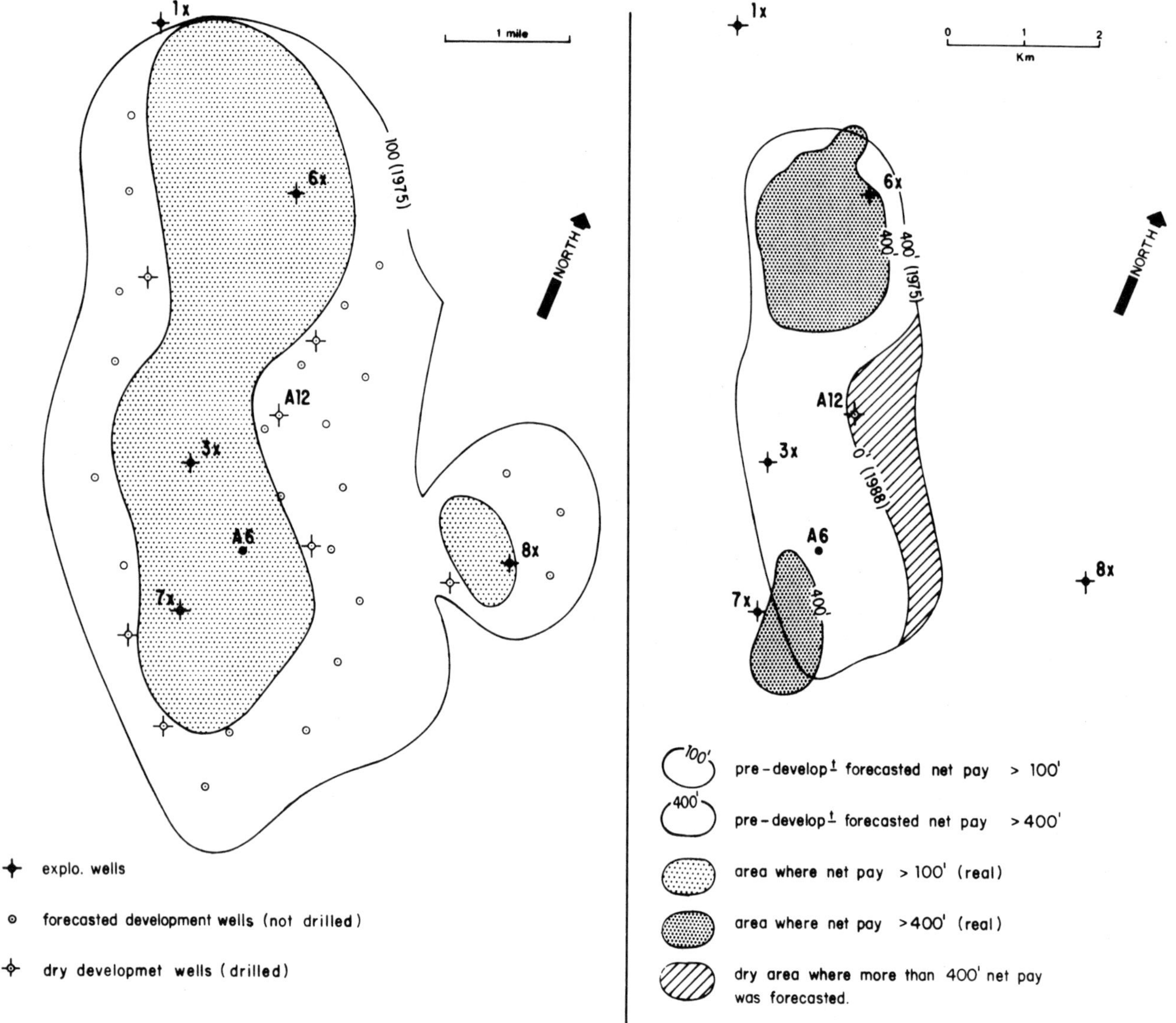

Figure 7. Comparison of net pay estimates made in 1975 (five wells) and 1988 (54 wells).

Unfortunately, the seismic signal was distorted or destroyed in the crestal areas by a strong gas effect in the Tertiary section. This disruption of surface seismic data was, and remains, a considerable impediment to geological interpretation of the field.

In 1967, the objective in Block 2/4 and 2/7 was the upper Paleocene sandstone following the gas discovery at Cod. In 1969, drilling at Ekofisk showed that the sandstone did not extend south of Cod. However, the first Ekofisk well discovered a thick oil column in porous chalk of early Paleocene and Late Cretaceous ages, and these became the main objective of Eldfisk and the satellite structures. Since then, ten significant discoveries have been made in the chalk, but until the late seventies, the peculiarities of the chalk reservoir of the Ekofisk area were poorly understood. The porosity was abnormally high (up to 50% instead of the 5–10% expected for such a depth) and the chalk was stratified. The lack of understanding of the rapid porosity decrease flankward and of the drastic variations of hydrocarbon saturation with rock type led to poor results in the first phase of development drilling.

Now

The variation in porosity and hydrocarbon saturation within chalk reservoirs was better defined and understood in the early eighties. The identification of sedimentary, tectonic, and diagenetic factors related to reservoir quality now allows optimization of development drilling, which is still in progress on several fields.

The unique characteristics of the chalk and their interpretation are discussed in the reservoir section of this article. Improvements in the geophysical interpretation and evaluations were less significant; reshooting and reprocessing were not very helpful. In 1983, the first vertical seismic profile (VSP) was conducted in an Eldfisk well. Vertical seismic profiles helped to better define the structure at a local scale and indicated faults that may have an effect on production (Brewster and Dangerfield, 1984).

Surface Manifestations

Gas seeps on the sea floor have been observed in the northwestern corner of Block 2/7, 15 km northwest of Eldfisk (Hovland and Sommerville, 1985).

However, at this location a Zechstein salt plug has upwelled through the chalk reservoir to a depth of 2600 ft (790 m), where it has drastically disturbed the Tertiary strata that generally create the upper seal of the chalk fields.

It is unlikely that such surface manifestations occur above the Eldfisk field, but the presence of a gas chimney in the Tertiary between 4500 ft (1370 m) and 7500 ft (2280 m) subsea gives a clear seismic indication that gas has escaped from a nearby accumulation. Such important gas chimneys also exist above the chalk reservoirs of the Ekofisk, Albuskjell, Valhall, and Tommeliten fields.

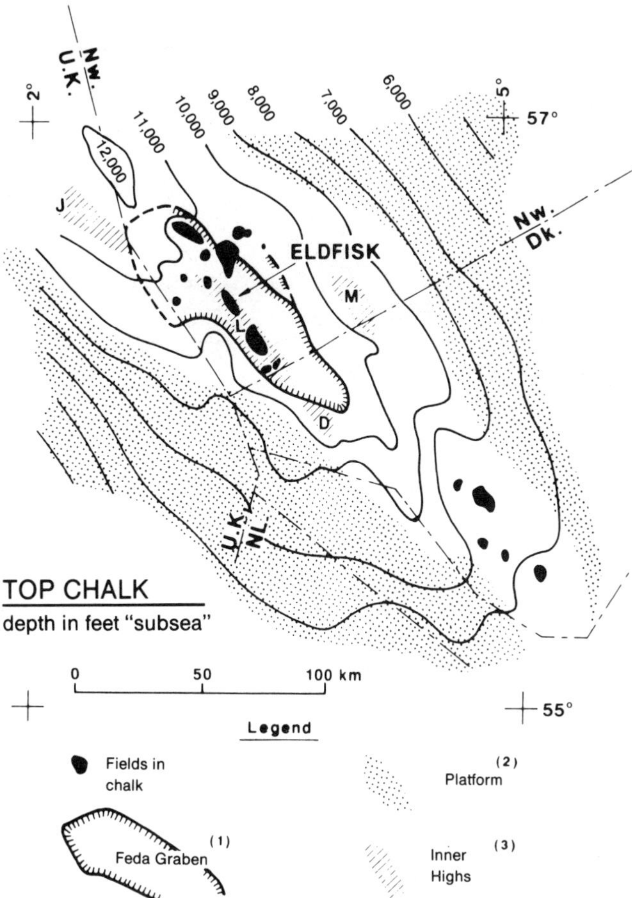

Figure 8. Structure on the top of the chalk in the southern part of the Central trough (surrounded by the shaded areas representing the platforms to the northeast and southwest). (1) is the main part of the Feda graben, a pre-Middle Cretaceous feature described by Gowers and Saebøe (1985), (2) and (3) are post-Middle Cretaceous highs in the Central trough.

STRUCTURE

Tectonic History

The Eldfisk field is located in the southern part of an area loosely referred to as the Central graben (Figure 8). As far as the field is concerned, the area is better defined as the Central trough (Ronnevik et al., 1975). Details on the structural evolution have been published by Gowers and Saebøe in 1985.

The Central trough is an elongated area that has been subsiding since Cretaceous time. It overlies a complicated graben system initiated during the first of a series of rifting phases that occurred through Mesozoic time. Eldfisk, along with several other Norwegian chalk fields, is located above the Feda graben (Gowers and Saebøe, 1985). During the Permian, it was only a small embayment where halite was deposited. This salt became mobile during Late Triassic; its halokinesis is partly responsible for most of the structures in the area. The Upper Permian evaporites were in turn covered by continental redbeds and some shales during Triassic time.

The intracratonic phase ceased in the Triassic when the rifting phase started.

Lower Jurassic marine shale may have been present over the whole area, but doming during Middle Jurassic time resulted in severe erosion that removed the Lower Jurassic sequence and part of the Triassic. Rifting and a major transgression occurred during Late Jurassic time.

The sea covered the whole area, with land emerging only far to the south, west, and east. These land masses gave erosion products that have been recognized in the Eldfisk appraisal Well 2/7-3, where the Eldfisk Formation, of Kimmeridgian to early Volgian age, contains 200 ft (60 m) of very sandy beds that are probably deep-water turbidites. The lithostratigraphy is given in Figure 9. The Upper Jurassic terminated with the deposition of anaerobic clay of the Mandal Formation (previously called the Kimmeridge Clay Formation), which is the prolific source rock of the central and northern North Sea. The thickness of the Upper Jurassic section below Eldfisk locally exceeds 3000 ft (900 m). The area continued to subside during Cretaceous and Tertiary times.

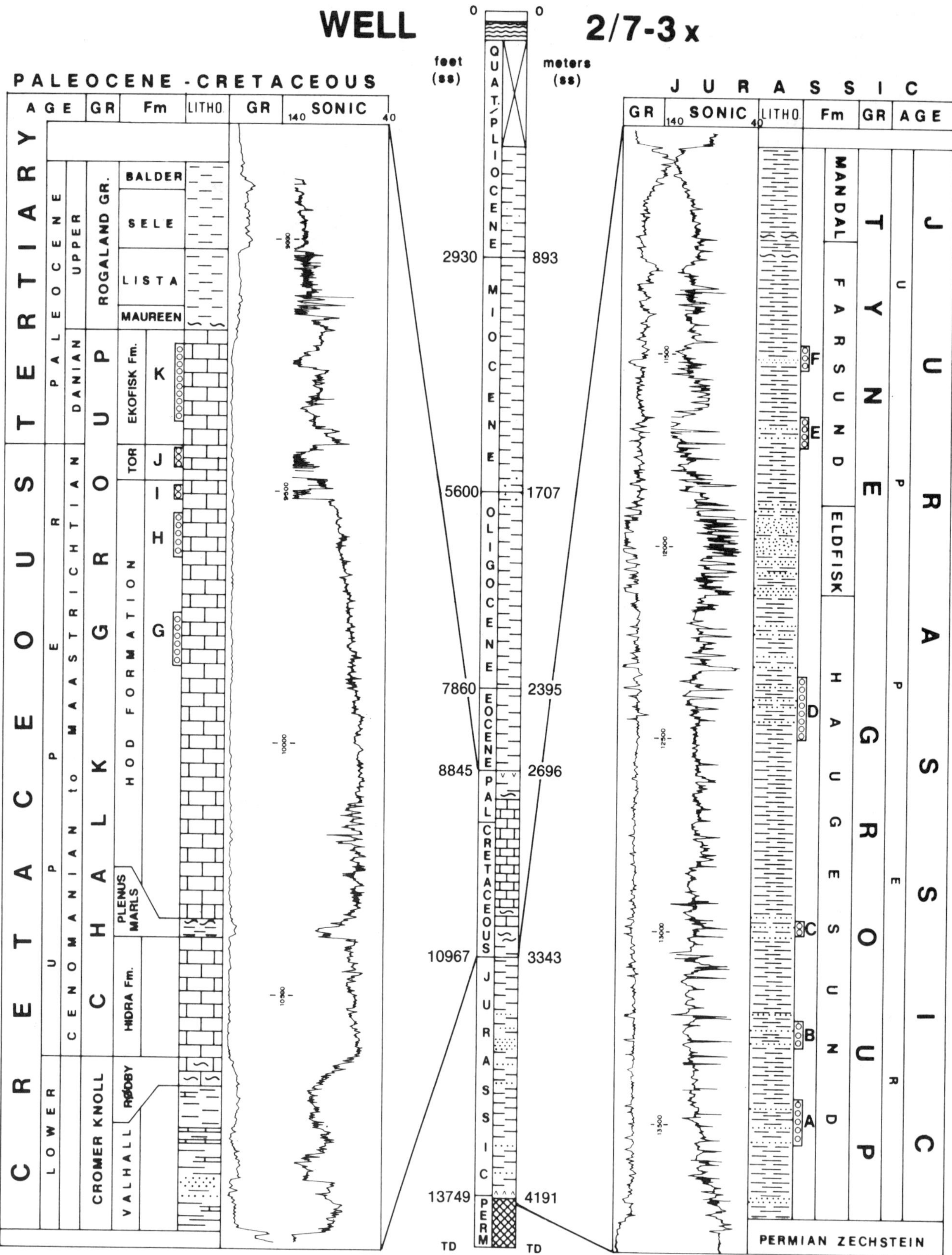

Figure 9. Lithostratigraphy of the exploration Well 2/7-3X. Perforations shown as open arches with lettered units. See Figure 1 for location.

Laramide movements began in the Senonian, climaxed in the late Maastrichtian, and ended in Paleocene time. Together with halokinesis of the Zechstein (Permian), they are responsible for the seafloor instability that resulted in the mobilization of previously deposited chalk. This resuspended material was then brought into the deepest parts of the trough by gravity currents and was deposited as allochthonous chalk. These reworked sediments are the best reservoir intervals in the area. The extension of the trough ceased in Paleocene time and the ensuing subsidence resulted in abundant clay and silt sedimentation.

Regional Structure

Local inversions occurred in the southern part of the Feda graben in Late Jurassic/Early Cretaceous time. Of major importance for Eldfisk is the reactivation during Campanian time of inversion along the major fault bounding the Feda graben to the southwest (Figure 8). This created a 50 km long tilted fault block, trending north-northwest, called the Lindesnes ridge (Ofstad, 1983). The pronounced seabed topography of the Lindesnes ridge strongly influenced the deposition of the allochthonous chalk during Maastrichtian times. South of Eldfisk, the ridge was still active during the Paleocene. The Lindesnes ridge is responsible for controlling reservoir facies and distribution in the Hod, Valhall, Eldfisk, Edda, and Tommeliten fields. Unfortunately, this was not known when the first development wells were drilled.

Local Structure

Halokinesis in Eldfisk occurred in several phases but not during the deposition of the reservoir layers. At the time of deposition, the paleorelief of the tilted fault block was the main tectonic feature. However, in the northeastern part of the structure, a very deep collapsed zone of unclear origin existed during late Campanian times over a 2.5 km^2 area (discussed further in the section related to reservoir depositional setting).

The present-day structure (Figure 6) is a 12 km long anticline trending northwest-southeast. It is cut by a strike-slip fault creating the North Eldfisk and South Eldfisk substructures. South Eldfisk is flanked by a small low-relief eastern dome (East Eldfisk). The structural relief of South Eldfisk exceeds 1200 ft (850 m). The structural axis is nearly parallel to the crest of the paleoridge, at a distance of about 1500 ft (450 m) to the east. It is worth noting that Edda, North Eldfisk, South Eldfisk, and Valhall are "en echelon" along the Lindesnes ridge with respective culminations at 10,000 ft (3040 m), 9200 ft (2800 m), 8900 ft (2700 m), and 7800 ft (2370 m) (Figure 1).

STRATIGRAPHY

The lithostratigraphy of Well 2/7-3x is given in Figure 9.

Permian

Fine- to medium-grained sandstones of the Rotliegendes Group resting on a reddish to greenish basalt have been penetrated in Well 2/7-2, 10 km southwest of Eldfisk.

The Zechstein Group, of unknown thickness, is composed of Upper Permian evaporites. Well 2/7-3 bottomed in halite. This salt is responsible for intense diapirism in the area.

Triassic

Redbeds of nonmarine origin overlay the Zechstein evaporites in the vicinity of Eldfisk, but they are not present below the field, due to erosion.

Jurassic

The oldest Jurassic rocks identified in the Eldfisk structure are of early Kimmeridgian age; the Lower and Middle Jurassic deposits may have been removed by erosion at the end of Middle Jurassic time. In Eldfisk, the Upper Jurassic strata belong to the Tyne Group (equivalent to the Humber Group and Forth Group). It is 2800 ft (850 m) thick and consists of four formations mainly of Kimmeridgian/Volgian age. Well 2/7-3 is a reference well or a type well for three of these four formations, which have fair to excellent source rock characteristics and contain sandstones of poor to fair reservoir quality.

The Haugesund Formation (1600 ft, 480 m) consists of light-gray to dark-brown marine shale, often carbonaceous and calcareous with thin sandstone interbeds. The sequence becomes siltier and sandier upwards.

The Eldfisk Formation (Vollset and Doré, 1984), 230 ft (70 m) thick, consists of marine sandstone with interbeds of gray shales and calcareous streaks. The sandstones are dark yellow to brown, fine to coarse grained, angular and poorly sorted, probably turbiditic. They are restricted to the Eldfisk area.

The Farsund Formation (Vollset and Doré, 1984) 700 ft (210 m) thick, is a marine sequence of dark gray shales with calcareous streaks and numerous sandstone stringers in the lower part.

The Mandal Formation (Hamar et al., 1982), previously referred to as the Kimmeridge Clay Formation, 30 to 300 ft (9 to 90 m) thick, consists of dark to black shales and carbonaceous claystone deposited in anaerobic marine conditions. They are of Volgian to Ryazanian age and have a very high organic content that makes them the most prolific source rock of the central and northern North Sea.

Cretaceous

The Cretaceous is composed of the Cromer Knoll Group and of the Chalk Group, respectively of Early and Late Cretaceous age. The Chalk Group also is comprised of chalks of the lower Paleocene (Danian).

The Cromer Knoll Group consists of gray marls and gray to black shales of the Valhall Formation, 300 to 800 ft (90 to 240 m) thick overlain by reddish marls of the Rödby Formation, 50 to 100 ft (15 to 30 m) thick. In Well 2/7-1, the Rödby Formation also is comprised of some sandstones and siltstones. In Well 2/7-9, southwest of Eldfisk, sandstones of Cretaceous to Jurassic age have been penetrated below the chalk; these are clear, angular, fine- to medium-grained, and locally cemented by calcite.

The Chalk Group is comprised of five formations:

1. The lower one, the Hidra Formation, 190 to 380 ft (55 to 115 m) thick, consists of argillaceous chalk grading upward into cleaner light gray chalk.
2. The Plenus Marl Formation, 20 to 70 ft (6 to 20 m) thick, consists of dark gray to black shales with thin chalk beds.
3. The Hod, Tor, and Ekofisk formations, which contain the Eldfisk hydrocarbon accumulations, consist of chalk varying from argillaceous to pure. All typical North Sea chalk facies have been identified in these formations, from autochthonous to distal allochthonous (Kennedy, 1980, 1987). The Hod Formation has an average thickness of 900 ft (270 m) and is of Turonian to Campanian age.
4. The Tor Formation, up to 500 ft (150 m) thick, has been deposited during Late Campanian and Maastrichtian times.
5. The Ekofisk Formation comprises approximately 250 ft (75 m) of early Paleocene (Danian) chalk.

Detailed lithostratigraphy of the Hod, Tor, and Ekofisk formations is given in the *Reservoir* section.

Upsection chalks of the Ekofisk Formation pass upward into marls of the Maureen Formation and represent a transition between a 60 million year period of carbonate deposition and an additional 60 million year period of clastic sedimentation during the remainder of the Cenozoic.

Tertiary

Above the Chalk Group, which includes carbonates deposited in the earliest Tertiary times, the monotonous 8800 ft (2682 m) thick sequence is divided into three groups.

The Rogaland Group of Paleocene age is 300 ft (90 m) thick and consists of claystone and gray to greenish shale, underlain by the lower Paleocene marls of the Maureen Formation and overlain by tuffaceous shales (ash beds) of the Balder Formation. Locally, the Rogaland Group contains rare siltstone and sandstone streaks that are lateral equivalents of the Forties Formation sandstone, which contains major hydrocarbon accumulations to the North.

The Hordaland Group (Eocene to early Miocene) is 3500 ft (1050 m) thick and consists of gray and brown claystone, locally silty with frequent limestone stringers in the lower half. The claystones are generally highly overpressured and contain gas and oil that has presumably escaped from the underlying reservoirs. A thick interval of Aquitanian age is more silty than the other intervals and apparently contains higher gas concentration, disturbing the seismic signal over parts of the fields.

The Nordland Group (middle Miocene to Recent) is 5000 ft (1520 m) thick and consists of gray clay with silt and sand intervals in the upper part. This group contains much fewer limestone beds than the two lower groups.

TRAP

Trap Type

The Eldfisk field consists of an elongate anticlinal trap cut by a strike-slip fault that creates a permeability barrier between the northern and southern parts of the structure. It is flanked by a small eastern dome separated from the main feature by a pronounced saddle. As such, the field can be subdivided into three distinct accumulations: North, South, and East Eldfisk, which have vertical closures of 900 ft (270 m), 800 ft (240 m), and 200 ft (60 m), respectively (Figure 6).

The total productive area is about 4450 ac (18 km^2). There is a marked unconformity between the Hod and Tor formations and also between the Tor and Ekofisk formations. The unconformities have produced discontinuous pressure barriers over the depletion period. Original pressure profiles showed a continuous gradient between the three formations. Pinchout and truncations exist within the Hod, Tor, and Ekofisk formations, but the traps are essentially structural.

The three areas are characterized by distinct oil-water levels. In fact, as in most of the Central trough chalk fields, the oil-water contacts are not clearly defined since the chalk that has not been invaded by oil has undergone late diagenesis that often results in a total loss of effective porosity (D'Heur, 1984). It is therefore more appropriate to call them paleo— or "frozen"—oil-water contacts and aquifers.

The upper seal is formed by a thick sequence of Tertiary claystones. However, gas has escaped from the reservoir through fractures and has been trapped into lower Miocene silty claystone and silts that probably have since lost any significant permeability.

Reservoir

Stratigraphy and Lithology

The chalk reservoir is a very fine grained rock essentially made up of coccoliths that are skeletal debris of unicellular algae (Figures 10 and 11).

The size of the coccoliths generally ranges between 0.5 and 2 microns, but entire coccospheres as well as chambers of foraminifera and other bioclasts provide much larger pores dispersed in the clay-sized calcite matrix.

The Eldfisk reservoirs are comprised of the Hod, Tor, and Ekofisk formations, which are estimated to contain approximately 7%, 45%, and 48% of the reserves, respectively.

The Hod Formation can be divided into three members (Ofstad, 1981). The lower Hod consists of clean chalk, often bioturbated and including some turbidites. It has been tested in Well 2/7-7 where a 200 ft (60 m) perforated interval produced 6000 BOPD. The middle Hod consists of alternating layers of clay-poor (pale) and clay-rich (dark) chalks, which are often bioturbated. These chalks are less permeable than the lower Hod chalks. The upper Hod member is rather similar in lithology to the underlying units, but the chalk is generally cleaner and contains debris flow beds. The upper Hod is locally eroded on the paleoridge (see geological model, Figure 12). Detailed correlations in the Hod Formation have been established by Brewster and Dangerfield (1984).

The Tor Formation of late Campanian to Maastrichtian age is 10 to 500 ft (3 to 152 m) thick and is also composed of three members recognized by differences in lithology and porosity resulting in a typical log response (D'Heur, 1980). The lower Tor member consists of alternating low-porosity pelagic chalk and high-porosity debris flow chalk. Conglomeratic chalk a few feet thick is locally present at the base of the member. The middle Tor member mainly consists of clean and porous debris flow chalk, interbedded with more distal allochthonous facies. The upper Tor contains by far the best reservoirs. It is essentially made of clean, debris flow chalks that show evidence of slumping and plastic deformation.

The Ekofisk Formation of Danian (lower Paleocene) age has an average thickness of 200 ft (60 m) and also is comprised of three subdivisions. The lower member, more or less equivalent to the Ekofisk "Tight Zone" (sensu lato), consists of fair to low-porosity chalk, locally argillaceous and/or siliceous. Upward the member becomes cleaner and more porous since the content of reworked Daniän and Maastrichtian chalk increases. The base of the lower member acts as a permeability restriction between the high-quality Tor formation reservoir and the less productive reservoir of the middle Ekofisk member.

The middle Ekofisk contains abundant redeposited material of Danian age and less clay and pelagic chalk than the underlying member except at its top where porosity decreases as clay content increases.

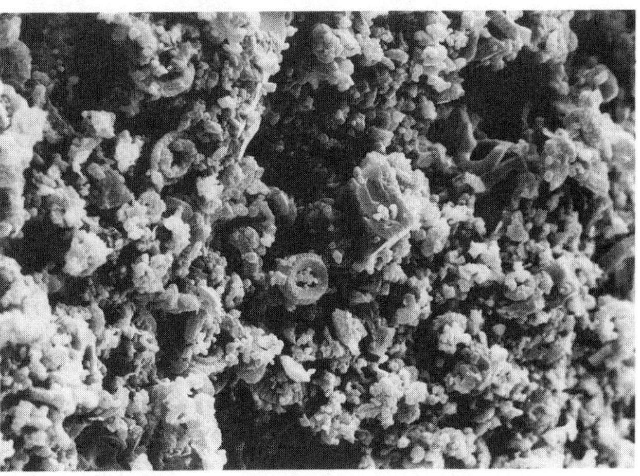

Figure 10. SEM photograph of a Danian chalk (Well 2/7-B11, middle Ekofisk member). The porosity is approximately 33%, magnification 2000×.

Figure 11. Same sample, magnification 10,000×.

The upper Ekofisk member is a very consistent layer of clean and high-porosity chalk in the entire Greater Ekofisk area.

Depositional Setting

The depositional setting is of prime importance as far as reservoir characteristics of the chalk are concerned. Instability of the platform around the Central trough and tectonic pulses in the trough itself resulted in the resuspension of previously deposited sediments. The resuspended coccolith ooze was then brought into the deep areas of the trough by gravity currents and redeposited as allochthonous chalk of various types: debris flows, turbidites, mud cloud sediments, etc. Tectonic activity also produced slumps and slide sheets of chalk. During the periods of quiescence between episodes of sea-floor instability, normal pelagic sedimentation of chalk took place.

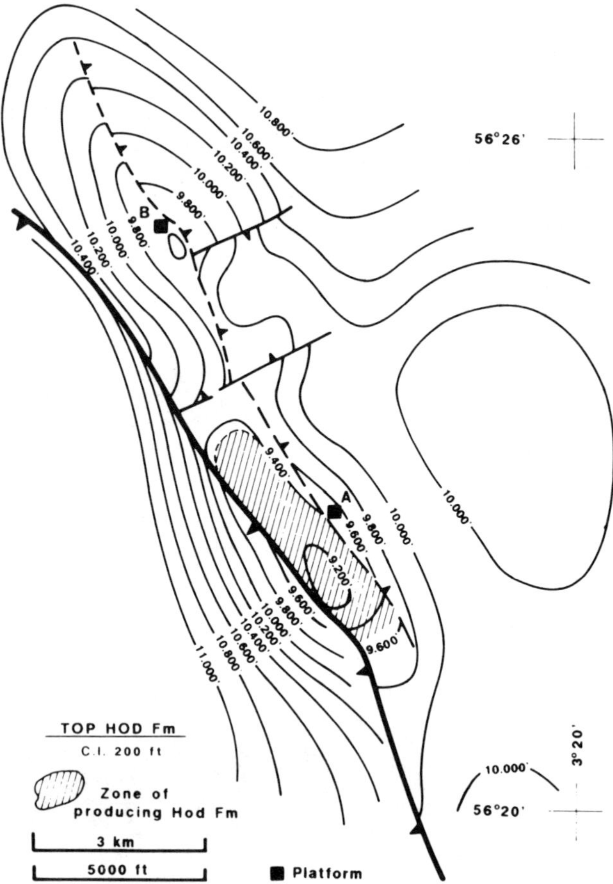

Figure 12. Depth map of the top of the Hod Formation showing faults and the productive area of the Hod Formation. The eastern north-northwest-trending fault (dotted line) is from Brewster and Dangerfield (1984). Structure is in feet subsea.

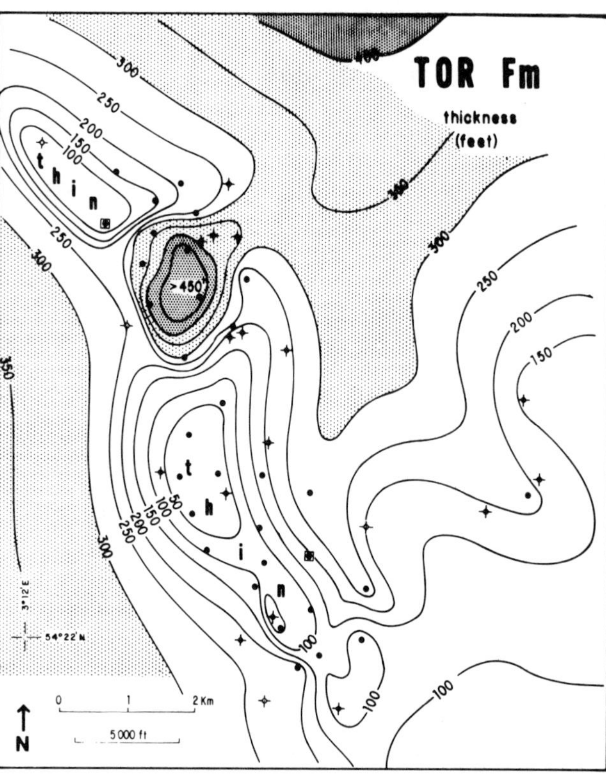

Figure 13. Thickness map of the Tor Formation, showing the thinning over the paleostructure and toward the southeast (Valhall). The thickest section has been encountered in the "collapsed area" of North Eldfisk, which was a pronounced low during Maastrichtian time. Contour interval, 50 ft. Light dot pattern indicates thickness between 300 ft and 400 ft. Dark dot pattern indicates thickness greater than 400 ft.

As a result, the Eldfisk field chalk reservoir consists of alternating rock types:

- Slowly deposited autochthonous pelagic chalk that today corresponds to low porosity (5% to 25%) beds.
- Rapidly deposited allochthonous chalk that generally corresponds to the high porosity (25% to 45%) layers.

In this context, the Eldfisk field is characterized by its location west and northwest of the source areas of the allochthonous sediments, and the existence of a paleostructure of the tilt fault block type (the Lindesnes ridge) during deposition of the main reservoir layers.

As a result, the Ekofisk and Tor formation chalks thin toward the southeast and westward onto the gently dipping eastern flank of the ridge and are the thinnest over the paleocrest as shown on the thickness maps of Figures 13 and 14.

In addition, the coarser chalk material (coccolith aggregates of the debris flows) is more abundant on the eastern slope than on the western flank of the paleorelief resulting in better reservoir characteristics east of the crest.

A strong local thickness (and porosity) anomaly is present in the northeastern part of the field. This is due to the existence of a pre-Maastrichtian collapse in this area, probably of halokinetic origin. The deep local trough was filled by reworked chalk during Maastrichtian time. As much as 550 ft (165 m) of chalk of the Tor Formation has been penetrated in Well 2/7-B10 compared to 70 ft (20 m) at the 2/7-B12 location, less than 2 km to the northwest.

A slight anomaly is also recognized on the Ekofisk Formation thickness map. It is probably due to differential compaction of the Tor Formation deposits before the end of Danian time.

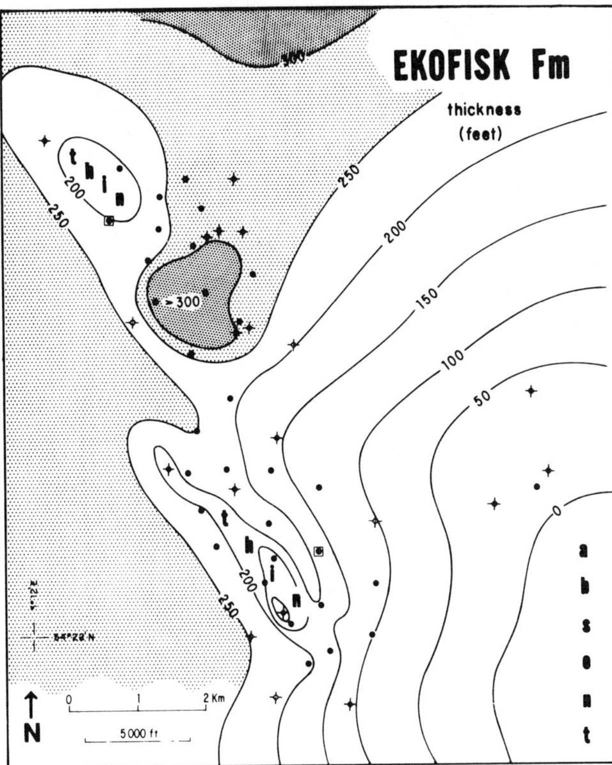

Figure 14. Thickness map of the Ekofisk formation showing similar but less pronounced trends than those of the Tor Formation.

Porosity Type and Diagenesis

Under normal circumstances, chalk buried to 9840 ft (3000 m) would have porosity reduced to 12% (± 6%) by predominantly "chemical compaction" (pressure dissolution followed by calcite precipitation in adjacent pore space), the escape of fluids, and mechanical compaction.

Preservation of porosity as high as 40% is due to the combined effect of several favorable factors (Scholle, 1977):

- The Central Trough chalk is mainly low magnesium calcite that is chemically stable.
- The clay content is often low, which favors spot-welding between calcite grains (Mapstone, 1975). Spot-welding results in an increased resistance to compaction.
- The pore fluid is overpressured because of rapid subsidence of the trough and rapid sedimentation. As a result, the effective stress on the chalk matrix is reduced.
- Entry of oil into the reservoir inhibited the process of chemical compaction.

The present-day porosity distribution is the result of two main distinct factors that played a role at different times in the geological history of the field.

1. Synsedimentary variations occurred in the distribution of the various (favorable and unfavorable) chalk facies, as discussed above.
2. Differential porosity preservation occurred when oil migrated into the reservoir and after.

The structurally high parts of the trap were invaded by oil prior to the lower parts and chemical compaction was inhibited upstructure whereas it was still reducing porosity along the flank. As a result, there is a progressive decrease of porosity for each reservoir layer, from the crest to the flank of the field (Figures 15 and 16). However, there is a strong porosity anomaly in the collapsed zone in the northeastern part of the field. Higher than expected porosity values in this area and in the northeastern flank are due to the thick deposition of favorable rock type in the local trough and east of the Lindesnes ridge. Porosity cross-sections are shown in Figures 17, 18, and 19.

Fractures

The Eldfisk field reservoirs are naturally fractured. Well tests indicate permeabilities up to 300 md while the matrix permeability of the best reservoir is generally less than 10 md. The high permeability fracture system is critical to the field productivity and acts as the collecting network through which fluids travel from the low permeability matrix to the wellbore. To further improve productivity, the wells are stimulated with a hydraulic fracture treatment and acid.

Analyses by Farrell (Phillips Internal Report) of cores from chalk fields in the area have shown that stylolite-associated fractures produced by extension during diapirism are important for productivity. Other tectonic fractures, mainly fault related, create main conduits for flow.

In Eldfisk, core analyses on a limited number of wells suggest the following conclusions:

- The variation in the fracture network may be associated with limestone hardness.
- The predominant dip is subvertical.
- A well-developed fracture network, which includes 90% of the observed fractures, is restricted to the soft friable chalk of the Ekofisk and Tor formations.
- A more restricted fracture network was observed in the Hod Formation.
- Eighty percent of the observed fractures are infilled with sparite or clay; they were probably formed before the diagenesis of the chalk was completed.
- There is evidence of several phases of fracturing and faulting.

Unfortunately, there is not yet enough core material to describe in detail the fracture network and the general fracture characteristics at Eldfisk.

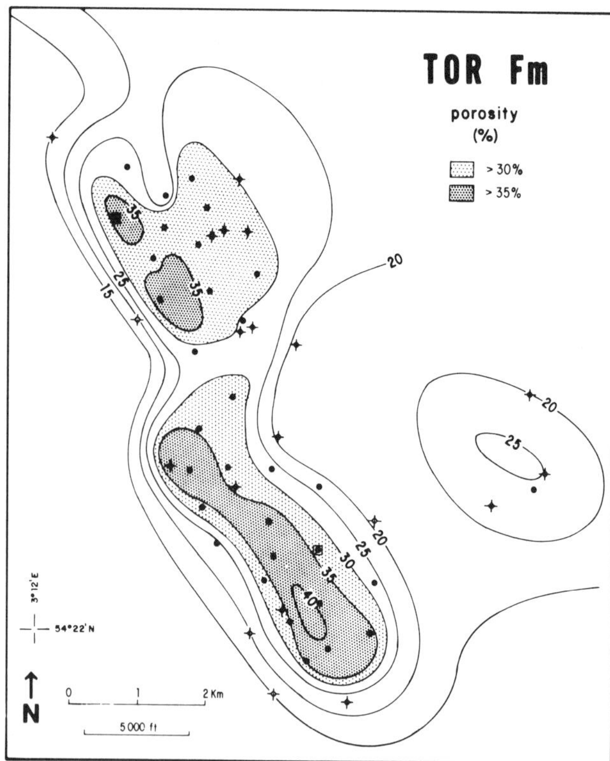

Figure 15. Average porosity map of the Tor Formation. The highest porosity values are present at the crest of the structures as a result of differential porosity preservation. A porosity high also characterizes the "collapsed area," due to high pore pressure in the rapidly sedimented favorable chalk facies. Contour interval, 5%.

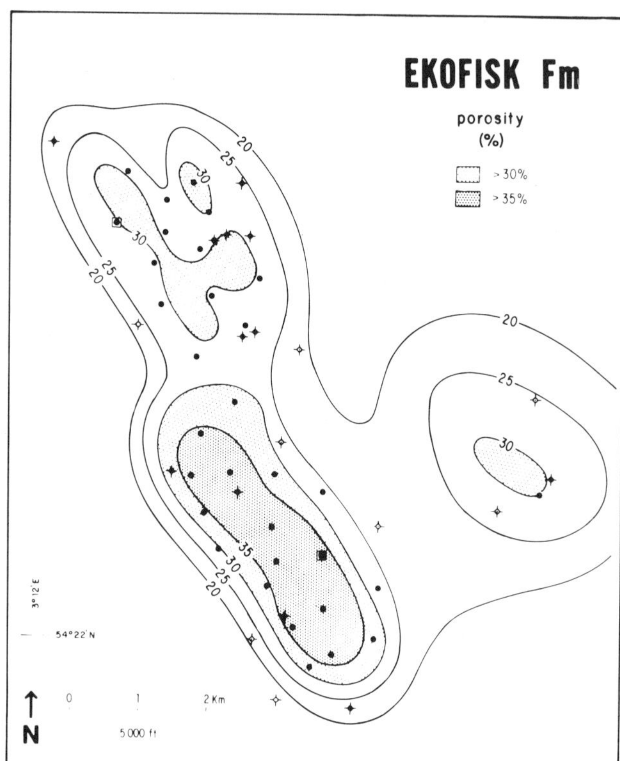

Figure 16. Average porosity map of the Ekofisk Formation showing the same trends as those of the Tor Formation. Contour interval, 5%.

A much better definition has been obtained for the Ekofisk field (Feazel and Farrell, 1988).

Porosity/Permeability

Matrix permeability of the chalk measured on cores varies from .01 md to 15 md. The average permeability is higher in the Tor Formation than in the Ekofisk Formation. Since the porosity decreases toward the flanks of the structure, matrix permeability is also expected to do so and the trends should be reinforced by a decrease of pore size (even at the same porosity value) that is assumed to occur on the flank. In fact, permeability decreases toward the flank in both North Eldfisk and South Eldfisk, as expected.

Observation of cores of the central area of Eldfisk has led to the following conclusions (Figure 20):

- For a porosity value of 40%, the permeability of the Tor Formation is approximately 5 to 10 md whereas it is only 1.2 to 1.5 md for the Ekofisk Formation.
- A permeability of 1 md is obtained for porosities of 23% to 28% in the Tor Formation, whereas it is obtained for porosities of 37% to 39% in the Ekofisk Formation.
- For porosity values lower than 35%, the vertical permeability is often lower than the horizontal permeability except in the massive debris flows.
- For any given Danian reservoir layer, the permeability is slightly better in North Eldfisk than in South Eldfisk. The permeability of the Tor Formation seems to be less sensitive to geographical location.

These observations have been made on four crestal or near crestal wells. It is doubtful that flank wells follow the same porosity-permeability relationships.

Pay Zone Thickness

A peculiarity of the chalk fields is that the oil-water transition zone is commonly much thicker along the flank of the structure than along the central area. In part, this is due to the general porosity reduction toward the flanks. However, at the same height above the free water level, rocks with similar porosity values exhibit higher water saturation in the flank than in the central area of the field. This is attributed to changes in pore morphology and pore throat size that could be explained by differences in stress regime as the salt rose through the enclosing sediments (D'Heur, 1980).

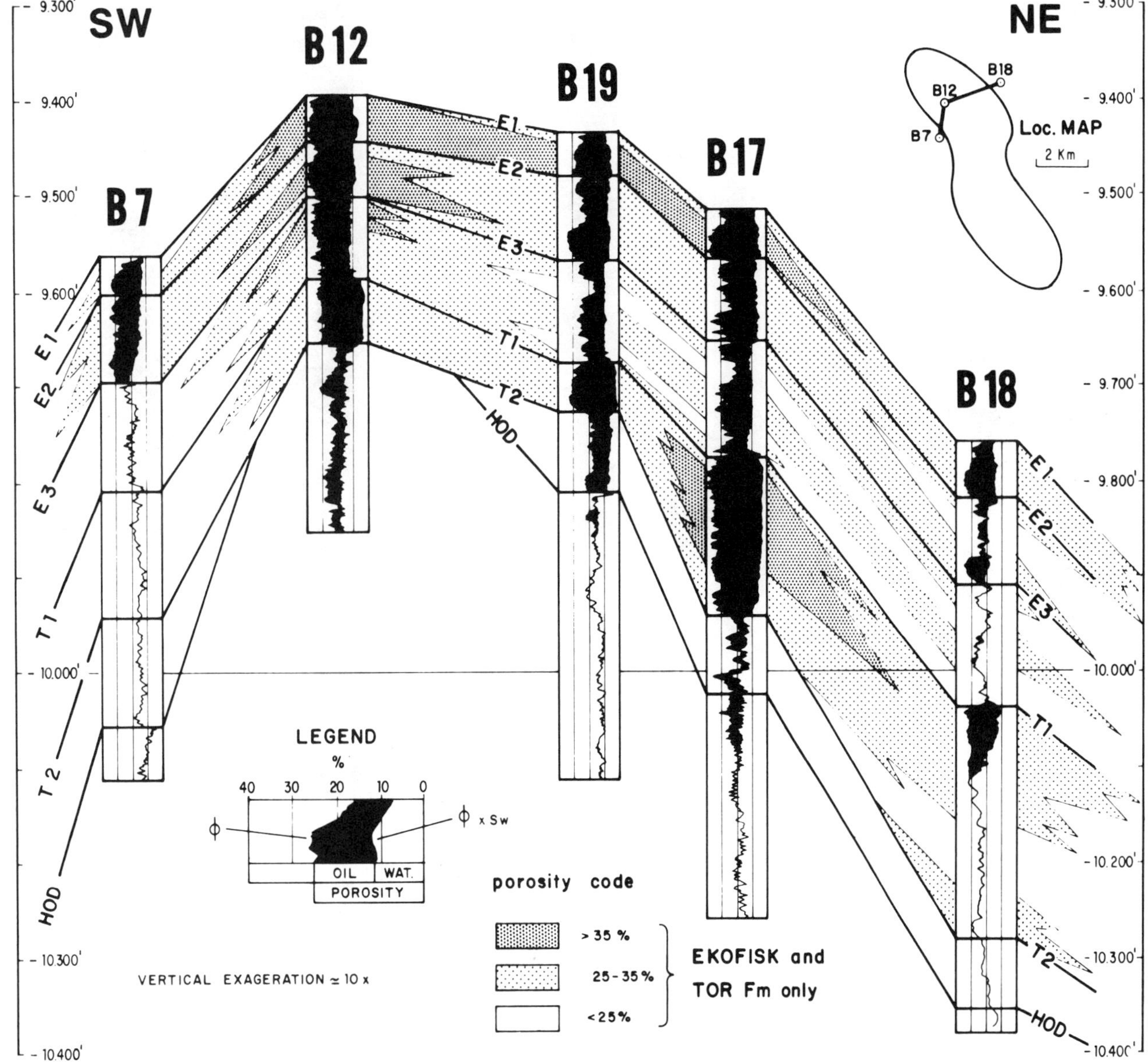

Figure 17. Porosity section across Eldfisk North. Ekofisk "E" and Tor "T" zones are shown.

Both factors, porosity decrease toward the flanks and change in capillary pressure behavior, result in a curved base of the pay zone. Since the three producing formations of Eldfisk are of distinct rock types, depth and slope of this curved base of the pay changes as a function of the distribution of the three formations. This is illustrated in Figure 21 (right side) and Figure 22. In fact, subtle rock type variations occur within the three producing formations, and each of them may be subdivided into at least three units having distinct depth, porosity, and saturation relationships that make the situation even more complicated than is shown schematically in Figure 22.

In addition to vertical variation of rock type from layer to layer, lateral variation also occurs within each layer. As a result, depth and shape of the curved base of the pay zone for a reservoir layer also vary from place to place as a function of the distribution of facies within each layer. For example, since the source areas for the best reservoir sedimentary facies—the debris flow deposits—were located to the east, northeast, and southeast of Eldfisk, the eastern flank of the paleostructure locally contains better reservoir facies than the western flank. This often results in reduced thickness of the oil transition zone and subsequent thickening of the pay zone—in the eastern flank compared to the western flank as

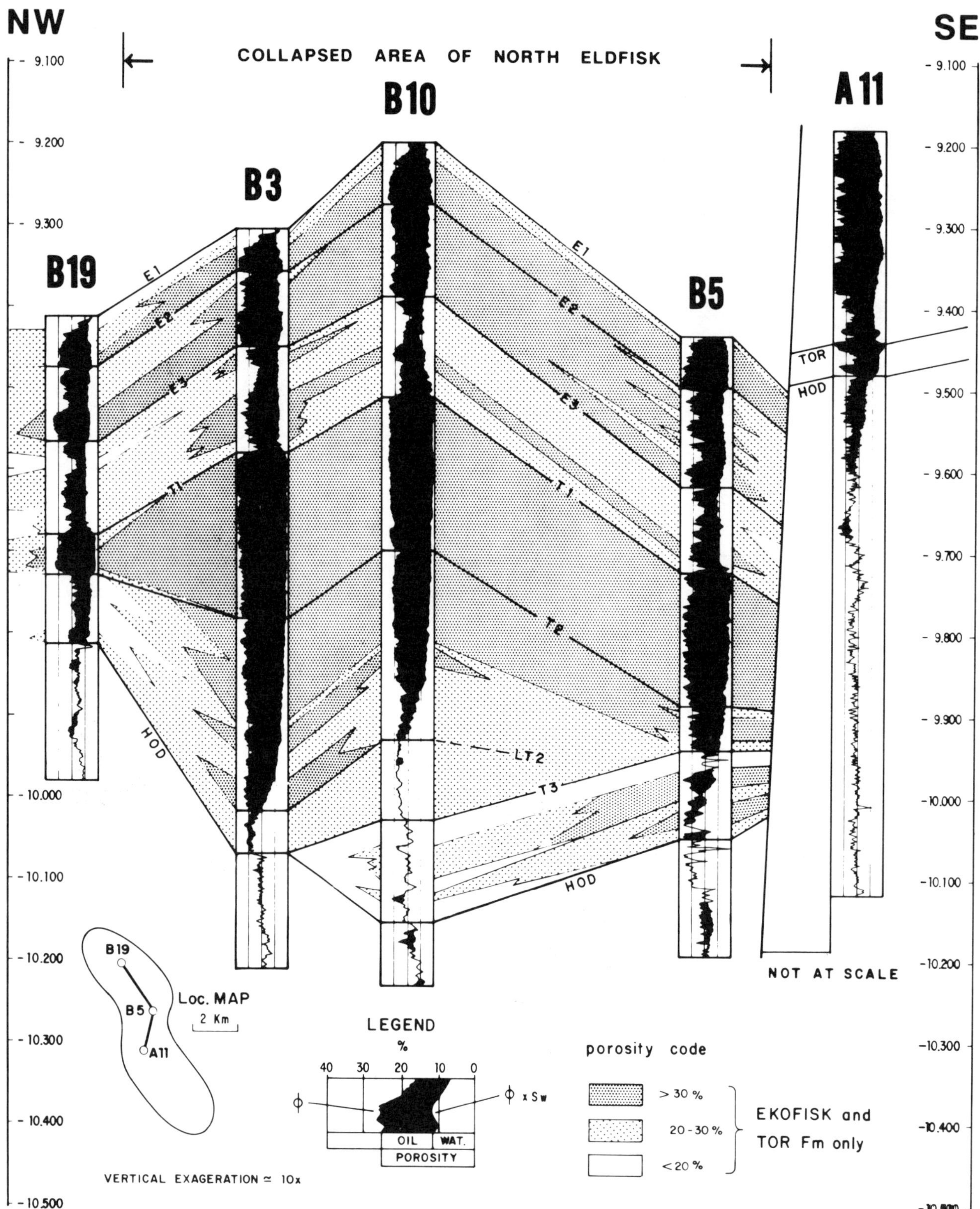

Figure 18. Porosity section across the "collapsed area."

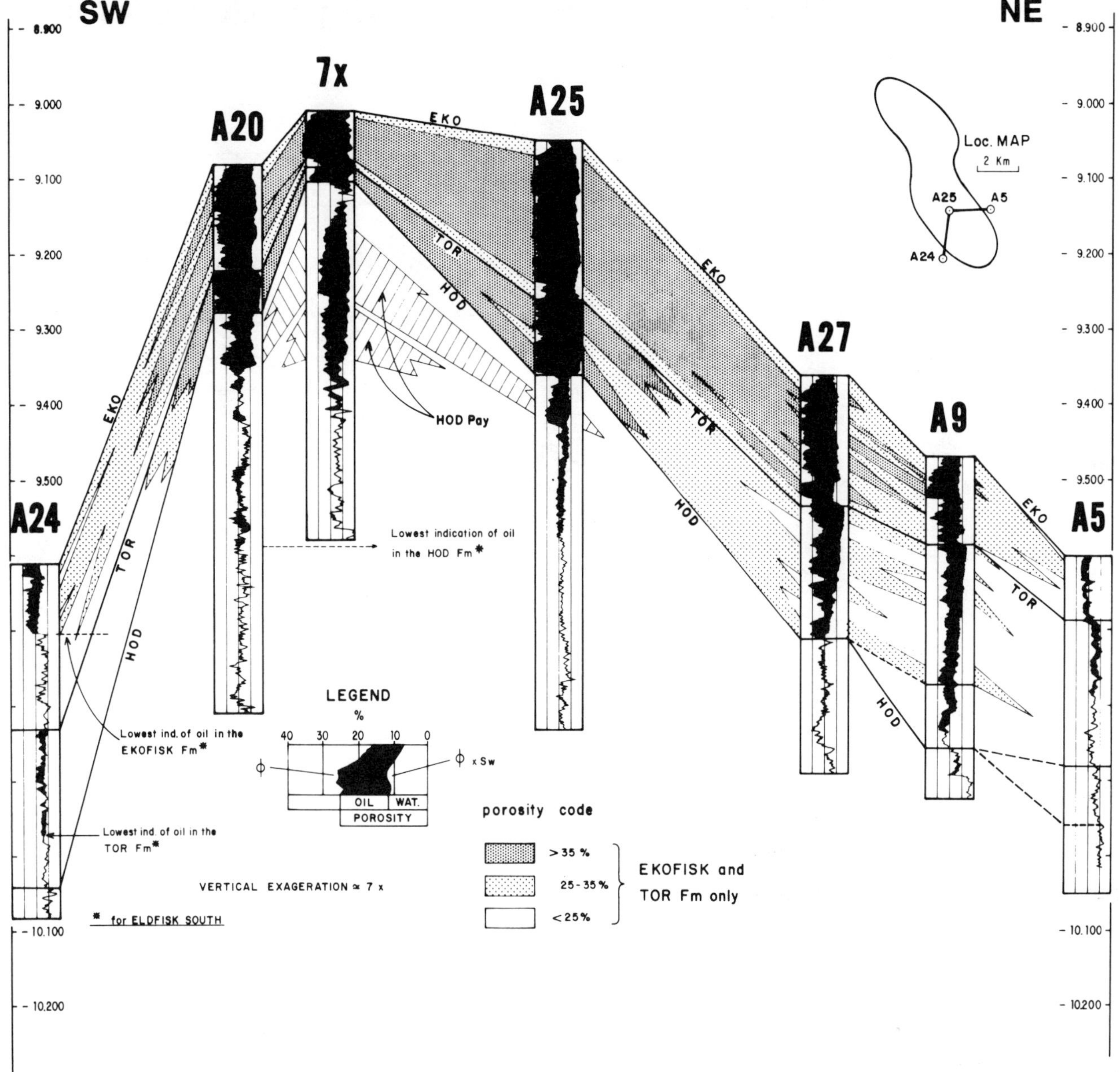

Figure 19. Porosity section across Eldfisk South.

illustrated in Figure 21 (compare Wells B7 and B13). Lateral permeability barriers in the saddle area or along faults are responsible for differences in elevation of the water level between North, South, and East Eldfisk. North Eldfisk has the deepest free water level (10,150 ft, 3090 m) and South Eldfisk has the shallowest (9950 ft, 3033 m). If a single water saturation cut-off of 50% is considered as pay discriminator, the maximum gross and net pay thicknesses have been encountered in the central area of North Eldfisk with 650 ft (200 m) and 600 ft (180 m), respectively. In South Eldfisk they are 460 ft (140 m) and 440 ft (130 m), respectively (Figure 23).

The gross pay thickness of East Eldfisk does not exceed 200 ft (60 m).

Hydrocarbons

The Eldfisk oil is a paraffinic undersaturated volatile oil. The reservoir temperature is 119°C, the initial pressure was 6815 psi at 9200 ft (2800 m), and the initial gas/oil ratio was 1804 scf/STB.

The crude oil gravity is 36° API. The sulfur and nitrogen contents in weight are 0.22% and 0.14%, respectively.

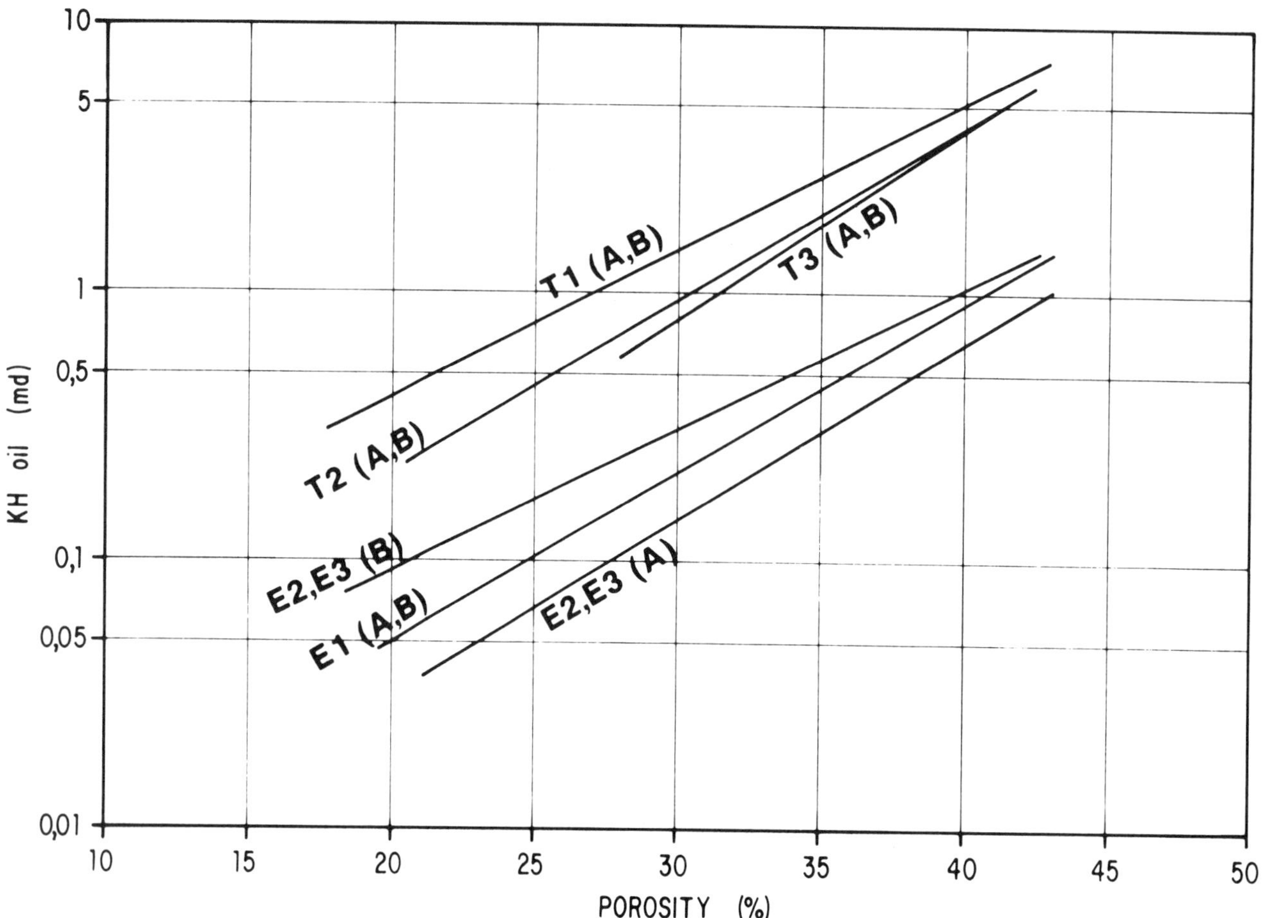

Figure 20. Porosity/permeability relationship by reservoir layers for Eldfisk crestal wells. Tor "T" and Ekofisk "E" formations are shown. KH is horizontal permeability to liquid in md. (A and B) refers to areas as seen in Figures 17 and 18.

Fluid Flow Characteristics

Together with differences in elevation of the water tables and with the geochemical analyses, hydrodynamic data have indicated that at least three separate reservoir areas exist in the field.

Brewster and Dangerfield (1984) already indicated that within North and South Eldfisk, partial permeability barriers locally separate the Ekofisk, Tor, and Hod formations. Significant lateral variations in production characteristics also occur within the reservoir, especially in the Ekofisk Formation of North Eldfisk.

The Tor Formation, which contains approximately 50% of the reserves, is the most productive reservoir; it is followed by the lower and middle members of the Ekofisk Formation. In areas where fracture intensity is low, initial production rates of 5000 to 10,000 BOPD have been achieved. In areas of high fracture intensity, some wells completed in both the Tor and Ekofisk reservoirs have produced at initial rates exceeding 20,000 BOPD.

The Hod Formation, which is productive only in the crestal area of South Eldfisk, is of lesser productivity. Flow rates from the middle member of the Hod Formation are generally of a few hundred barrels per day.

For the lower Hod member, a flow rate of 6000 BOPD has been achieved in an area of high fracture intensity. Such high production rates for the Hod Formation are, however, somewhat questionable for various technical reasons relating to possible flow contribution from the overlying Tor Formation.

Development Considerations

The need for natural fracturing of the chalk together with thick and porous net pay has led to drilling most of the development wells in the crestal

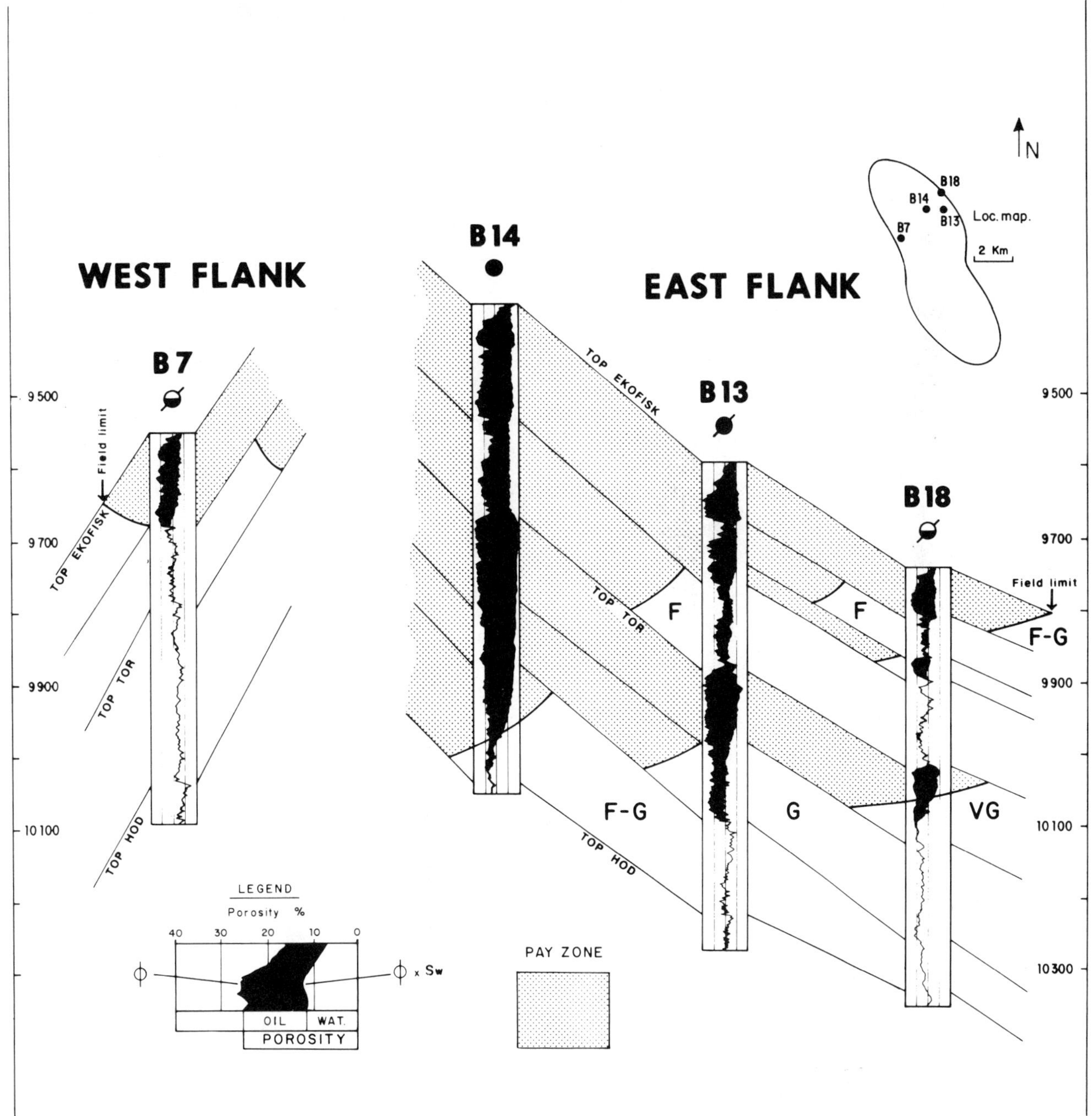

Figure 21. Hydrocarbon distribution in Eldfisk North at different structural elevations and for different rock types. Note that although it is at the same structural elevation as Well B13, Well B7 contains much less oil. This may be due to subtle rock type variation within the Ekofisk and Tor formations or to isolation of the B7 area. Reservoir quality: P, poor; F, fair; F-G, fair to good; G, good; VG, very good.

or near-crestal areas of the field. The low permeability of the chalk, however, necessitates some wells toward the flank of the structure, especially in areas where good porosities are locally preserved. Since porosity trends were not clearly defined in the early phase of development, several flank wells gave negative results; they later have been abandoned or side-tracked. With 39 producing wells, 18 on North Eldfisk, 19 on South Eldfisk, and 2 on East Eldfisk, it is expected that a recovery factor of 18–22% may be achieved by natural depletion (Figure 24).

The average well spacing is about 1600 ft. The best productive areas are the "collapsed" area of North Eldfisk and the crestal area of South Eldfisk.

Production data and logs suggest that movable water is present in the east flank of North Eldfisk in the Tor Formation. Up to now, the Ekofisk Formation has not given such indications. The

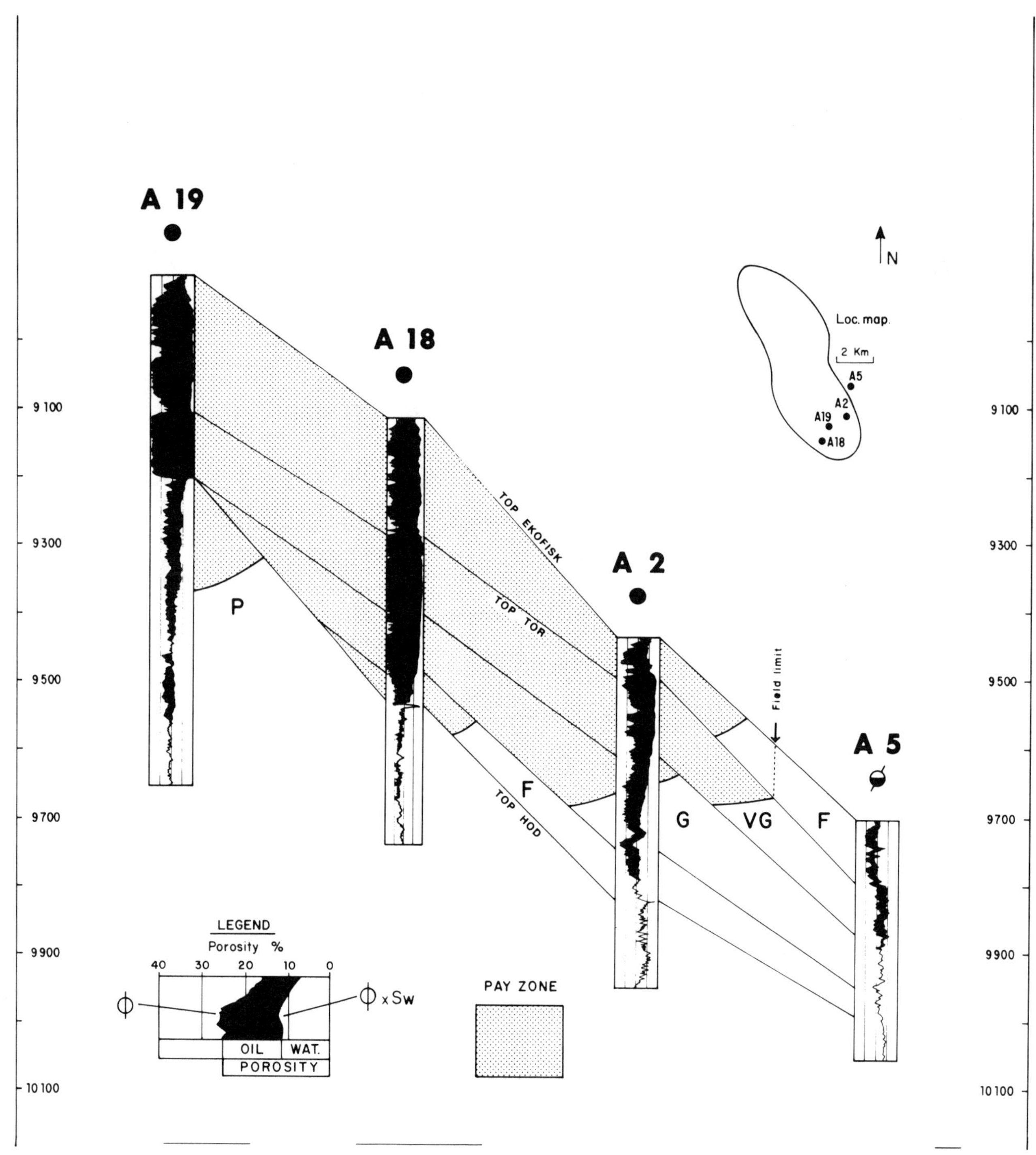

Figure 22. Eldfisk South: variation of oil saturation with rock type and structural elevation. Reservoir quality as per Figure 21.

present plan involves draining the reservoir using solution gas drive only, but gas injection is being studied.

Unfavorable wettability and imbibition characteristics may preclude water injection.

Subsidence of the sea floor resulting from compaction of the chalk reservoir is taking place at Eldfisk. However, below and near the platforms, the reservoir is rather thin, and no modification will be needed for maintaining production safety as was the case for Ekofisk.

Faults

The fault pattern of Eldfisk is not yet very well defined, mainly due to the gas effect in the Tertiary

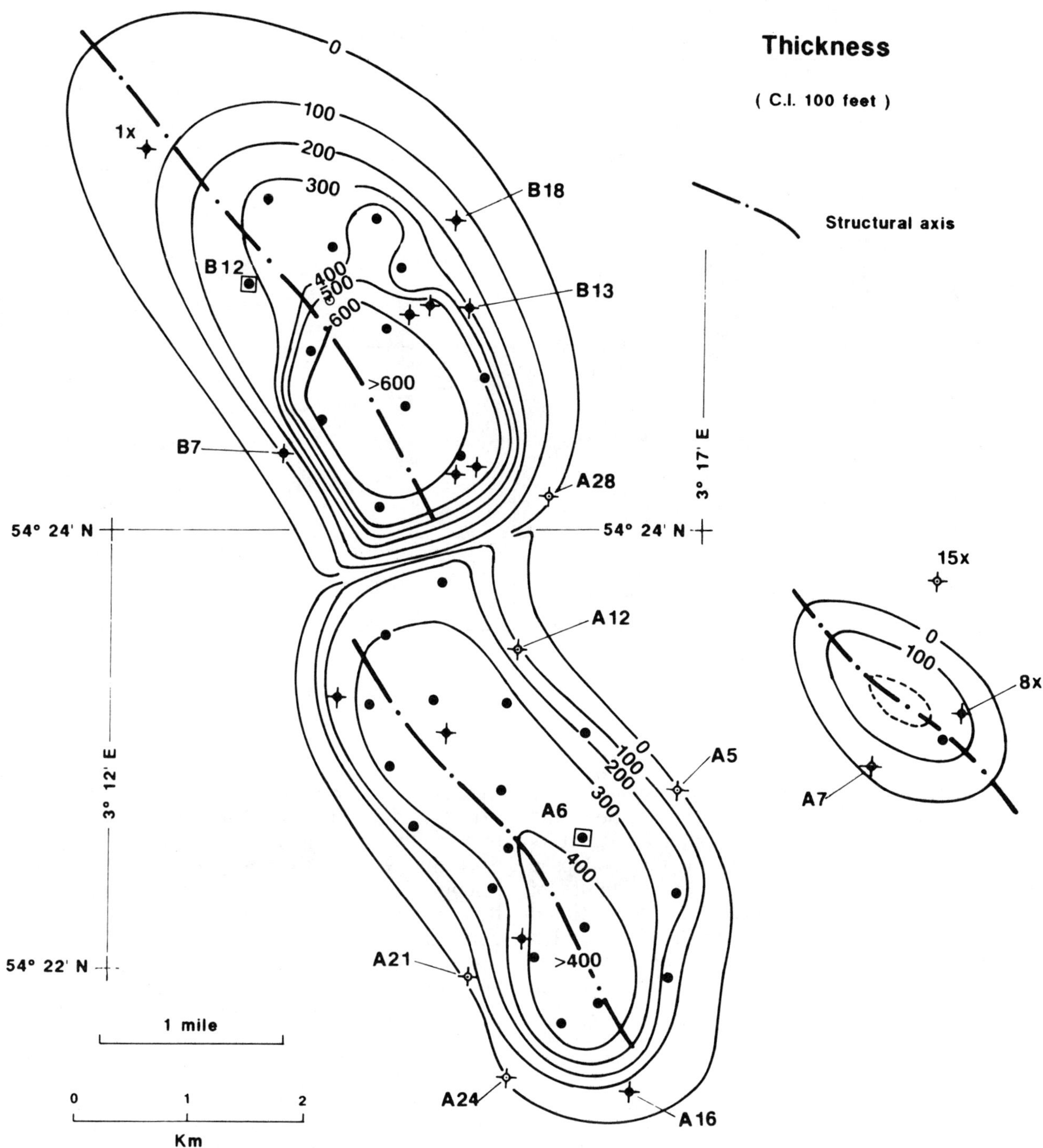

Figure 23. Total net pay thickness map.

that precludes seismic mapping. The fact that the drastic thickness variations from well to well are due to sedimentary factors does not necessarily imply the presence of faults except at some specific locations.

Nevertheless, faults of various tectonic origins are certainly present within the producing area. A major down-to-the-west fault bounding the Lindesnes ridge on the west side of the field affects the pre-Maastrichtian strata. Associated antithetic normal faults and crestal grabens may be present locally (Brewster and Dangerfield, 1984). A strike-slip fault is responsible for the nonalignment of the north-south Eldfisk structural axis. The latter fault or an

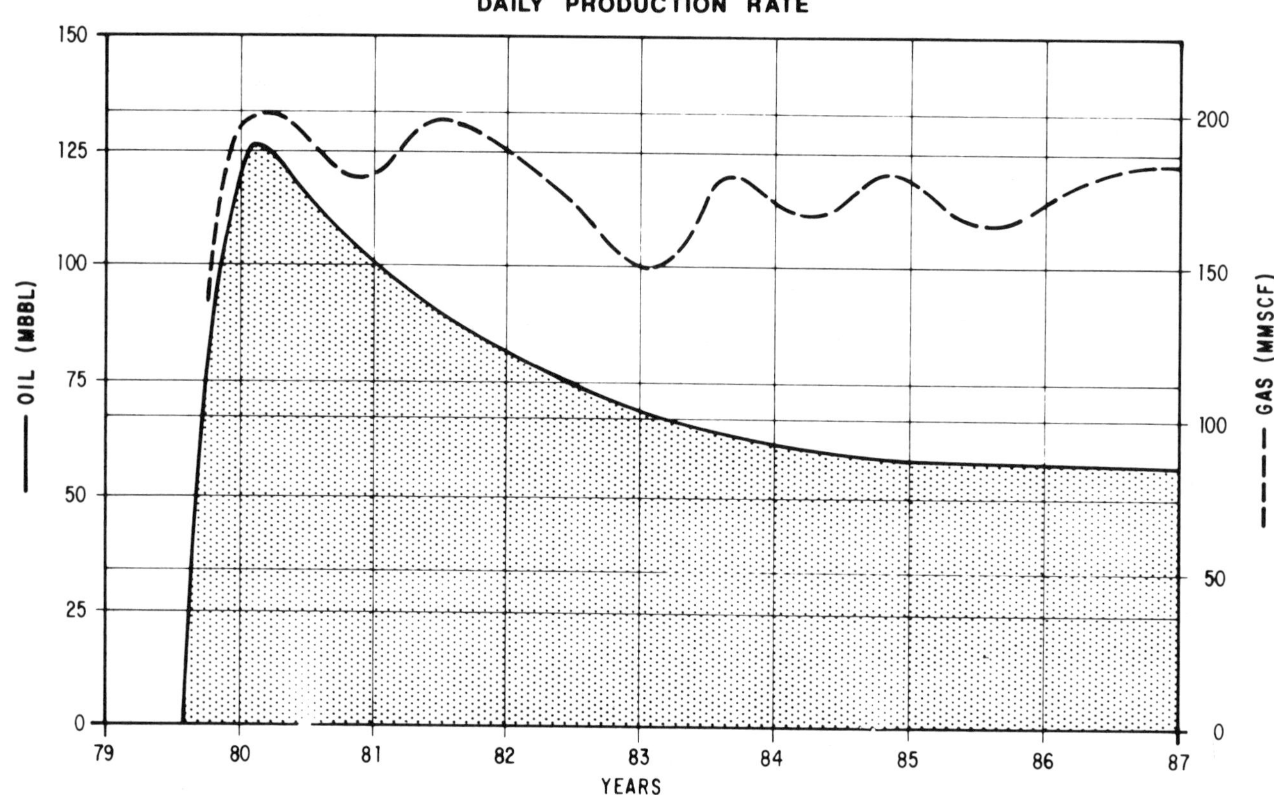

Figure 24. Production history.

associated down-to-the-north fault is responsible for the permeability barrier between the northern and southern areas. More recent extensional faults related to halokinesis may be present. The "collapsed area" in North Eldfisk is bounded by faults of Campanian age, and associated syndepositional faults certainly affect the Tor Formation. Above such a paleorelief, it would be surprising that fault movements would not have been reactivated during Danian time. However, faults affecting the top of the Ekofisk Formation are not currently mapped (Figure 7). Brewster and Dangerfield (1984) have shown three structural interpretations made by the operator in June, September, and December 1979, based on the same data. The large differences demonstrate the difficulty of mapping the faults. Vertical seismic profiles can locally improve the definition. A summary of fault interpretations of the Ekofisk Formation is given in Figure 25. None of these faults has been retained in the current interpretation. Faults affecting the Hod Formation are shown in Figure 12.

Source

The source of the oil and gas of the Central trough chalk fields is the "Hot Shale" of Late Jurassic age. Conclusive correlations between Jurassic shale extract and the Ekofisk crude oil have been published by Van den Bark and Thomas (1981).

In the Greater Ekofisk area, the "Hot Shale" belongs to the Mandal Formation of the Tyne Group. It is also often referred to as the "Kimmeridge Clay" Formation or equivalent of the Humber Group.

It is the best source in terms of richness, thickness, and type. Total organic carbon (TOC) is frequently in the range of 5-12%, and the organic type is predominantly oil prone.

Other formations of the Tyne Group also contain source rocks, but these are of fair to poor quality. The thickness of the Tyne Group in the vicinity of Eldfisk exceeds 1000 m, and the top of the Mandal Formation is locally deeper than 5000 m. A regional study of the Late Jurassic source rocks of the Norwegian continental shelf has been published by Thomas et al. in 1985. They noted that the fields appear to be restricted to locations within or above the current hydrocarbon-generating Jurassic area. Considering that the hydrocarbon accumulations of the greater Ekofisk area lie mainly in the upper Cretaceous and Paleocene reservoirs, a strong element of vertical migration is implied with little apparent lateral migration. However, the location of the field close to the major regional faults suggests that eastward migration from deep Jurassic areas in

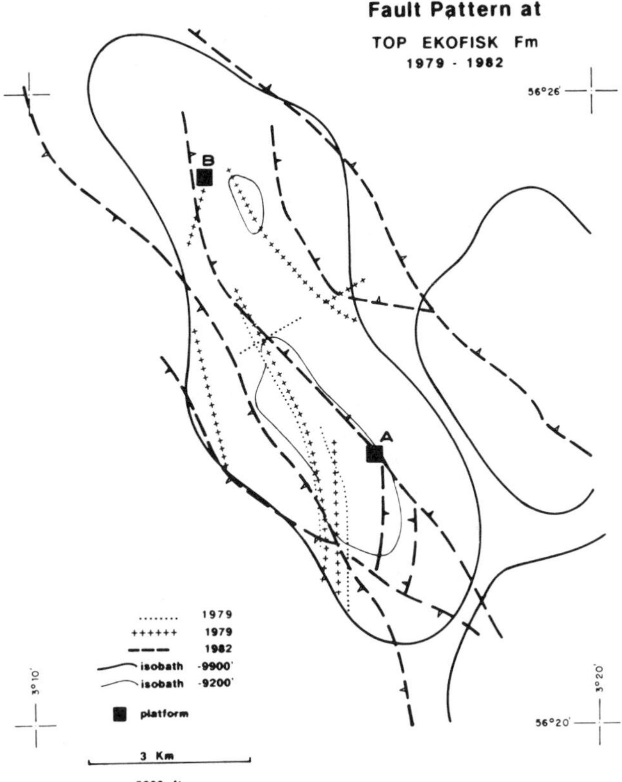

Figure 25. Various fault interpretations at the top of the Ekofisk Formation level. None of these faults have been maintained in the current interpretation.

Block 2/7 may also have taken place. In these deep western areas, significant amounts of oil have been generated during the Paleocene whereas the peak of oil generation just below Eldfisk was reached during Miocene–Pliocene times. The timing of oil entry into the reservoir is of importance in evaluating the relative roles of hydrocarbons and overpressure in preserving high porosity in the chalk.

The blanket distribution of the high-porosity debris flows over large areas together with the en echelon geometry of Tommeliten, Edda, Eldfisk, and Valhall along the Lindesnes ridge favors early oil entry in the crestal part of Eldfisk. Oil generated in the deep northwestern area could have spilled over from Blocks 1/9 and 1/6 into Eldfisk before expulsion began below the field.

Significant differences exist in the geochemistry of Jurassic source rocks and chalk oils in Eldfisk and subtle differences occur between oils in the different chalk formations as outlined by Thomas et al. (1985). These present-day differences can be attributed to differences in the timing of expulsion and in the variations of organic content areally within the Jurassic shales. Extracts of oil from the Tertiary cap rock show characteristics similar to the oil of the chalk reservoirs, proving that the hydrocarbons in the Tertiary have leaked from the reservoirs.

EXPLORATION CONCEPTS

Regional Play

Porous chalks containing significant amounts of producable hydrocarbons have been found only in a restricted area of the North Sea Central trough. The fields are concentrated in the southern part of the trough where six conditions are fulfilled:

1. Proximity of the margins of the trough or structural highs within the trough. Instability of the sea floor of the margins, the surrounding platform, or the inner highs have resulted in resuspension of pelagic chalk and redeposition as favorable reservoir facies.
2. Relatively deep area where the allochthonous chalk may accumulate and build up thick reservoir layers.
3. Absence of good and/or thick reservoir below or above the chalk. If they had been present, the hydrocarbons would not have had to force entry into the relatively low matrix permeability chalk.
4. Tertiary diapirism that is necessary to sufficiently fracture the chalk resulting in reservoirs for migrating hydrocarbons.
5. The organic-rich upper Jurassic strata are deeply buried and have started to expel oil since Eocene–Oligocene time.
6. Thick Tertiary clay to provide the vertical seal.

In any other part of the Central trough, at least one of these conditions is not fulfilled. However, Hatton (1986) showed that along the margins of the trough or close to inner highs, allochthonous chalk may be present. Trapping conditions, mainly stratigraphic, may occur. The chalk may then be hydrocarbon bearing, especially if there are no thick and/or good reservoir rocks in the Jurassic, Lower Cretaceous, or Paleocene series.

General Application of Geologic Parameters

Eldfisk is much more complicated than the layer-cake type of reservoirs of the Ekofisk and West Ekofisk fields due to the existence of an accentuated relief during sedimentation.

Another peculiarity of Eldfisk is its proximity to the higher Valhall structure that had been rising during late Maastrichtian and Danian time, providing reworked chalk material to the Eldfisk area.

With the exception of these peculiarities, the geological parameters of the Eldfisk reservoir follows the usual trends established in the greater Ekofisk area: porosity decreases from crest to flank, favorable reservoir thickness and facies exist to the east as in Edda and Ekofisk, sufficient fractures within reservoirs, deeply buried organic-rich source, and a good vertical seal.

Lessons

A chalk field requires a prudent approach for choosing the location of development wells.

Starting from a provisional development plan, well locations must be continually reevaluated during the phase of development drilling based on trend analyses.

The complexity of the geological history of Eldfisk was initially underestimated, which resulted in several dry holes being drilled in both the eastern and western flank of the structure in the early stage of development. Another consequence is the drastic reduction in the original in-place reserve estimate that took place between 1975 and 1980.

ACKNOWLEDGMENTS

The author thanks Petrofina and the other members of the Phillips Norway Group for their permission to publish this field study. I am grateful to John Brewster and Lawrence Pekot for comments on the manuscript and to Dominique Moens and Joas de Lucena for their help after on the manuscript. The opinions expressed are not necessarily those of the partners.

REFERENCES CITED

Brewster, J., and J. Dangerfield, 1984, Chalk fields along the Lindesnes Ridge, Eldfisk: Marine Petroleum Geology, v. 1, p. 239-278.

Deegan, C. E., and B. J. Scull, 1977, A standard lithostratigraphic nomenclature for the Central and Northern North Sea: Rep Inst. Geol. Sci. UK 77/25, 33 p.

D'Heur, M., 1980, Chalk reservoir of the West Ekofisk Field: the sedimentation of the North Sea reservoir rocks: NPF Norwegian Petroleum Society, 20 p.

D'Heur, M., 1984, Porosity and hydrocarbon distribution in the North Sea chalk reservoirs: Marine Petroleum Geology, v. 1, p. 211-238.

Farrell, H., 1981-1989, Phillips Group internal reports, confidential.

Feazel, C. T., and H. Farrell, in press, Chalk from the Ekofisk Area, North Sea: nannofossils + micropores = giant fields: SEPM Core Workshop Notes.

Gowers, M. B., and A. Saebøe, 1985, On the structural evolution of the Central Trough in the Norwegian and Danish Sectors of the North Sea: Marine Petroleum Geology, v. 2, p. 298-318.

Hamar, G. P., T. Fjaeran and A. Hesjedal, 1982, Jurassic stratigraphy and tectonics of the south-southeastern Norwegian offshore: Petroleum geology of the southeastern North Sea and adjacent onshore areas: Geologisches Mijnbow, v. 62, p. 103-114.

Hatton, I. R., 1986, Geometry of allochthonous chalk group members, Central Trough, North Sea: Marine Petroleum Geology, v. 3, p. 79-98.

Hovland, M., and J. H. Sommerville, 1985, Characteristics of two natural gas seepages in the North Sea: Marine Petroleum Geology, v. 2, p. 319-326.

Kennedy, W. J., 1980, Aspect of chalk sedimentation in the Southern Norwegian offshore, *in* The sedimentation of the North Sea reservoir rocks: Norwegian Petroleum Society.

Kennedy, W. J., 1987, Late Cretaceous and Early Paleocene Chalk Group sedimentation in the Greater Ekofisk Area, North Sea Central Graben: Bull. Centres Rech. Explor. Prod. Elf Aquitaine, v. 11, p. 91-126.

Mapstone, N. B., 1975, Diagenetic history of a North Sea chalk: Sedimentology, v. 22, p. 601-613.

Ofstad, K., 1981, The Eldfisk Area: NPD paper N. 30, Norwegian Petroleum Directorate.

Ofstad, K., 1983, The southernmost part of the Norwegian section of the Central Trough: NPD paper N. 32, Norwegian Petroleum Directorate.

Rønnevik, H. C., W. Van den Bosch, and E. H. Bandlien, 1975, A proposed nomenclature for the main structural features in the Norwegian North Sea: Jurassic Northern North Sea Symposium, NPF (Norwegian Petroleum Society).

SUGGESTED READINGS

D'Heur, M., 1986, The Norwegian chalk fields: habitat of hydrocarbons on the Norwegian continental shelf, (A. M. Spencer et al., eds.), Norwegian Petroleum Society, p. 77-89. An overview of the geology of the ten Norwegian chalk fields.

Hancock, J. M., 1976, The petrology of the chalk: Proceedings of the Geological Association, v. 86, p. 499-535. General description of the chalk components.

Hardman, R. F. P., 1982, Chalk reservoirs of the North Sea: Bulletin of the Geological Society of Denmark, v. 30, p. 119-137. An overview of the reservoir geology of the chalk of the Central trough.

Nygaard, E., K. Lieberkind, and P. Frykman, 1983, Sedimentology and reservoir parameters of the chalk group in the Danish Central Graben: Geologie Mijnbow, v. 62, p. 177-190. Geology of the Danish Chalk reservoirs.

Skovbro, B., 1983, Depositional conditions during Chalk sedimentation in the Ekofisk Area, Norwegian North Sea: Geologie Mijnbow, v. 62, p. 169-175. Geological setting of the Chalk Group in the greater Ekofisk area.

Appendix 1. Field Description

Field name .... *Eldfisk*

Ultimate recoverable reserves .... *350 million bbl + 12 bcf*

Field location:

- **Country** .... *Norway (offshore)*
- **State** .... *North Sea*
- **Basin/Province** .... *Central trough*

Field discovery:

- **Year first pay discovered** .... *(Paleocene) Danian Ekofisk Formation chalk 1970*
- **Year second pay discovered** .... *(Upper Cretaceous) Maastrichtian Tor Formation chalk 1972*
- **Third pay** .... *(Upper Cretaceous) Campanian Hod Formation chalk 1972*

Discovery well name and general location

- **First pay** .... *2/7-1, 330 km southwest of Stavanger (Norway) and 8 km south of giant Ekofisk field*
- **Second pay** .... *2/7-3, 5 km southeast of Well 2/7-1*
- **Third pay** .... *NA*

Discovery well operator .... *Phillips Petroleum*

- **Second pay** .... *Phillips Petroleum*
- **Third pay** .... *NA*

IP in barrels per day and/or cubic feet or cubic meters per day:

- **First pay** .... *15,000 BOPD (Northern Crest; best well); 25,000 BOPD (Southern Crest; best well)*
- **Second pay** .... *6000 BOPD (best well)*
- **Third pay** .... *NA*

All other zones with shows of oil and gas in the field:

Age	Formation	Type of Show
Late Jurassic	*Eldfisk*	*Oil tested*

Geologic concept leading to discovery and method or methods used to delineate prospect, e.g., surface geology, subsurface geology, seeps, magnetic data, gravity data, seismic data, seismic refraction, nontechnical:

Diapiric structure identified by seismic in the mid-sixties. Paleocene sandstone found gas bearing 80 km northwest (COD) was initially the objective. Drilling at Ekofisk in 1969 showed the sandstone did not extend southward but discovered oil in high porosity chalk (lower Paleocene and Upper Cretaceous). This became the objective at Eldfisk.

Structure:

Province/basin type (see St. John, Bally, and Klemme, 1984)

Cratonic basin located on earlier rift graben system. Bally 1211; Klemme IIB.

Tectonic history

Central trough formed in Middle Cretaceous time over an early Mesozoic rift feature named Central graben. Instability within the trough and along its margins is responsible for the abundance of favorable (reworked) chalk facies (reservoir).

Regional structure

Field located over a regional feature of the tilted fault block type that strongly influenced the distribution of the reservoir rock. Halokinesis of the Permian salt responsible for generation of trap and fracturation (productivity).

Local structure

Large northwest-trending anticline cut by dextral strike-slip fault and flanked by a small low-relief eastern dome.

Trap

Trap type(s)

Three anticlines with multiple pays locally combine with truncation, sedimentary facies, and diagenetic variation resulting in porosity/permeability destruction.

Basin stratigraphy (major stratigraphic intervals from surface to deepest penetration in field):

Chronostratigraphy	Formation	Depth to Top* in ft ss
Middle Miocene-Recent	*Nordland Group*	*Sea floor*
Eocene-early Miocene	*Hordaland Group*	*5000*
Early Paleocene	*Ekofisk Formation*	*9000*
Late Cretaceous	*Tor Formation*	*9200*
Late Jurassic	*Mandal Formation*	*10,800*
Late Permian	*Zechstein Group*	*13,650*

**In the central area of the field.*

ELDFISK

Location of well in field .... *Middle of southern area*

Reservoir characteristics:

Number of reservoirs .... *2 or 3 depending on location in the field*

Formations .... *Ekofisk, Tor, Hod*

Ages .... *Danian (early Paleocene), Maastrichtian and Campanian (Late Cretaceous)*

Depths to tops of reservoirs .... *Ekofisk, 9000 ft; Tor, 9200 ft; Hod, 9600 ft*

Gross thickness (top to bottom of producing interval) .... *Max., 650 ft*

Net thickness—total thickness of producing zones

Average .... *400 ft*

Maximum .... *630 ft*

Average ....

Maximum

Lithology .... *Resedimented (allochthonous) chalk, mainly of debris-flow type*

Porosity type .... *Mainly interparticle*

Average porosity .... *Ekofisk and Tor, 30%; Hod, 18%*

Average permeability .... *1-2 md (matrix)*

Seals:

Upper

Formation, fault, or other feature .... *Upper Paleocene*

Lithology .... *Claystone/shale*

Lateral

Formation, fault, or other feature .... *Cementation*

Lithology .... *Chalk*

Source:

Formation and age .... *Mandal Formation (Late Jurassic)*

Lithology .... *Shale*

Average total organic carbon (TOC) *6%*

Maximum TOC *20%*

Kerogen type (I, II, or III) .... *II*

Vitrinite reflectance (maturation) .... $R_o = 0.9\text{-}1.1$

Time of hydrocarbon expulsion .... *NA*

Present depth to top of source .... *10,800 ft (central part) to 15,000 ft (neighboring syncline)*

Thickness .... *100 ft (excellent) + 800 ft (fair to good)*

Potential yield .... *300 BAF extractable organic matter*

Appendix 2. Production Data

Field name .... *Eldfisk*

Field size:

- **Proved acres** *4500*
- **Number of wells all years** *50*
- **Current number of wells** *40*
- **Well spacing** .... *2300 ft*
- **Ultimate recoverable** .... *400–450 million bbl + 1600–1800 bcf*
- **Cumulative production (to 1/87)** .... *190 million bbl + 400 bcf*
- **Annual production (1986)** .... *20 million bbl + 60 bcf*
- **Present decline rate** *13.7%*
  - **Initial decline rate** *14.5%*
  - **Overall decline rate** .... *NA*
- **Annual water production** .... *NA*
- **In place, total reserves** .... *3600 million bbl (1800 STB)*
- **Net rock volume** .... *2000 ac-ft*
- **Primary recovery** .... *400–450 million bbl + 1600–1800 bcf*
- **Secondary recovery** .... *NA*
- **Enhanced recovery** .... *NA*
- **Cumulative water production** .... *NA*

Drilling and casing practices:

- **Amount of surface casing set** .... *30-in. csg shoe at 600 ft; 20-in. csg shoe at 1200 ft*
- **Casing program** .... *(Vertical depth to shoe) 13⅜-in. at 4800 ft; 9⅝-in. at 9000 ft; 7-in. at 9200 ft; 5-in. at 10,200 ft*
- **Drilling mud** .... *Sea water + attapulg. except paraffin based in 12-in. hole*
- **Bit program** .... *12-in. h, long milled tooth bit; 5⅞-in. h, navidrill + stratapax bit*
- **High pressure zones** .... *4500–9000 ft*

Completion practices:

- **Interval(s) perforated** .... *Ekofisk, Tor, Hod, or combination of these 3 (up to 300 ft thick)*
- **Well treatment** .... *Acid stimulation*

Formation evaluation:

- **Logging suites** .... *Overburden: DIL/LSS/GR/CAL, LDT/CNL/GR; reservoir: DLL/MSFL/GR, LDT/CNL/GR, LSS/GR, RFT, SHDT*
- **Testing practices** .... *Selective cased hole DST + total pay deliverability test after stimulation + pressure B.U.*
- **Mud logging techniques** .... *Total concept mud logging on all critical wells*

Oil characteristics:

- **Type** .... *NA*
- **API gravity** .... *36.4°*
- **Base** .... *Paraffinic*
- **Initial GOR** *1804*
- **Sulfur, wt%** *0.22*
- **Viscosity, SUS** *40.91*
- **Pour point** .... *25°F at 100°F*
- **Gas-oil distillate** *31.4*

Field characteristics:

- **Average elevation** .... *9200 ft*

Initial pressure .... *6815 psi*
Present pressure .... *4429 psi*
Pressure gradient .... *7.8 psi/ft*
Temperature .... *119°C*
Geothermal gradient .... *1.95°F/100 ft*
Drive .... *Solution gas drive*
Oil column thickness .... *1200 ft*
Oil-water contact .... *10,100 ft (lowest)*
Connate water *3–30%*
Water salinity, TDS .... *52,000 ppm NaCl equivalent*
Resistivity of water .... *0.42 ohm-m*
Bulk volume water (%) *5%*

Transportation method and market for oil and gas:

18 km oil line 24-in. + 18 km gas line 30-in. to the Ekofisk Center. From Ekofisk, a 36-in., 274 mi gas line to Germany and a 34-in., 220 mi oil line to England.

West Ekofisk Field—Norway
Central Graben, North Sea

MICHEL D'HEUR
Petrofina/Fina Exploration Norway
Brussels, Belgium

FIELD CLASSIFICATION

BASIN: North Sea
BASIN TYPE: Rift
RESERVOIR ROCK TYPE: Limestone (Chalk)
RESERVOIR ENVIRONMENT OF DEPOSITION: Resedimented Chalk
RESERVOIR AGE: Paleocene
PETROLEUM TYPE: Gas and Condensate
TRAP TYPE: Dome Overlying Salt Diapir

LOCATION

The West Ekofisk field (rank 313) is located in the Central trough of the Permian basin in the southern part of the Norwegian sector of the North Sea, 310 km southwest of the Norwegian coast. The field is also situated 7 km west of the giant Ekofisk field (rank 98, Carmalt et al., 1986). The field and the seven other fields that produce from chalk of early Paleocene and Late Cretaceous age occur in an area commonly referred to as the "Greater Ekofisk area" (Figure 1). West Ekofisk lies entirely in Block 2/4. Water depth in the area is 230 ft (70 m). The ultimate recovery of the field will be approximately 65-75 million bbl of condensate and 1 tcf gas.

HISTORY

Pre-Discovery

In 1965, during the first round of licencing in Norway, Licence PL 018 covering Block 1/5, 2/4, 2/7, and 7/11 of the Norwegian Continental Shelf was granted to Phillips, Petrofina, and Agip.

These companies were joined by the Petronord Group in 1967, and the present participation interests are: Phillips Petroleum Company Norway (36.96%); Norske Fina A/S (30%); Norsk Agip A/S (13.04%); Elf Aquitaine Norge A/S (8.1%); Norsk Hydro A/S (6.7%); Total Marine Norsk (4.05%); and others (1.15%). Phillips is the operator.

Exploration drilling in Norway started in 1965. Three years later, the Group made the first discovery at Cod (Block 7/11) where the Paleocene sandstone contained gas. Another discovery in the same reservoir had taken place in the U.K. sector, 80 km south of Cod. Since Block 2/4 is located between these two finds, the upper Paleocene sandstones became the primary objective in the area.

In 1969, the Group drilled its first well on the flank of the largest structure identified by seismic surveys in Block 2/4, Ekofisk (Figure 2). The primary objective, the Paleocene sandstone, was not present but from 10,000 ft to 10,700 ft (3050-3261m) the well encountered a very porous chalk reservoir with very high hydrocarbon saturation. The net pay was thicker than 600 ft (183 m), with porosities ranging from 30 to 42%.

Discovery

Ten months after the Ekofisk discovery, the Group started exploration drilling on the adjacent structure to the west, the West Ekofisk. Well 2/4-5X (also referred to as 2/4-6) was drilled near the crest of a small circular dome (Figure 2) and reached the chalk reservoir in October 1970 at a depth of 10,120 ft (3085 m) subsea. The well penetrated 650 ft (198 m) of very porous, hydrocarbon-bearing chalk of Danian (early Paleocene) and Maastrichtian (Late Cretaceous) age. The reservoir interval correlates very well with that of the Ekofisk field and has similar porosity but a thinner oil column (Figure 3). The well was drilled to 11,100 ft (3383 m) subsea TD in the Maastrichtian and was successfully tested over five intervals that were perforated selectively.

The well flowed (3900 BOPD and 12 MMcfg/d through a 1-in. choke) after acidizing a 70 ft (21 m) interval in the Danian chalk. Porosity and hydrocarbon saturation logs of well 2/4-5X are shown on Figure 3.

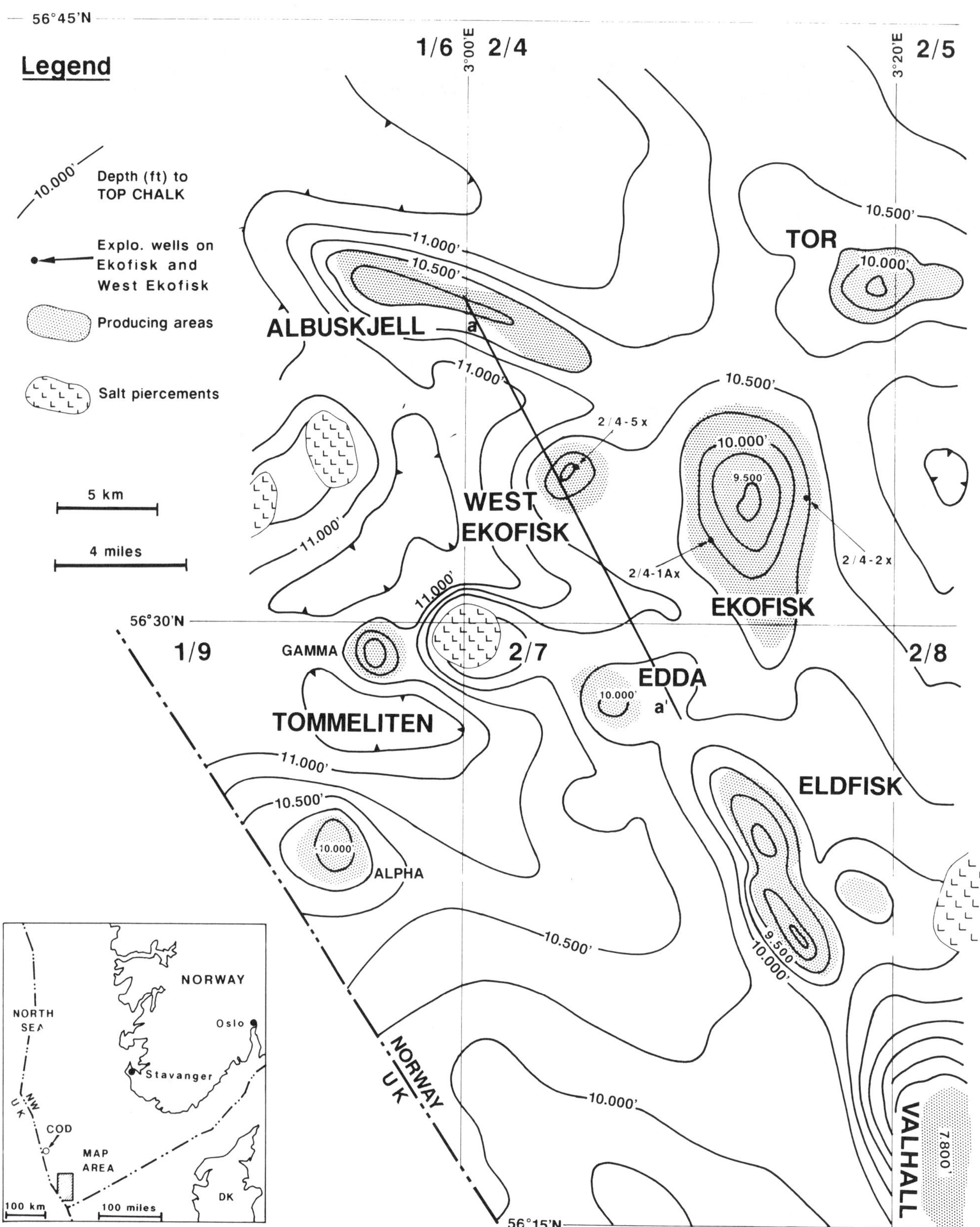

Figure 1. Location map of the Greater Ekofisk area. Contours indicate the depth of the top of the Chalk group in feet. Line aa′ shows the location of the seismic section of the Figure 6A, B, C.

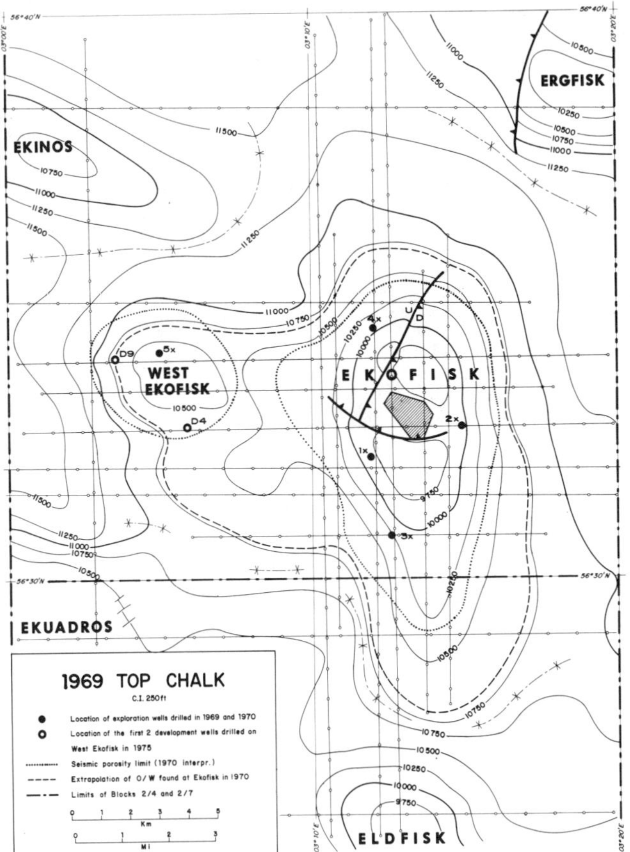

Figure 2. A 1969 structure map at top reservoir level showing the Ekofisk discovery and appraisal wells and the West Ekofisk discovery well 5X. (Depths are subsea.) The locations of the first two development wells of West Ekofisk, D4 and D9, drilled in 1975 are also indicated. The dotted lines around Ekofisk and West Ekofisk are the limits of the porosity areas based on seismic characteristics. The broken line near the -10,750 ft contour is the extrapolation of the oil-water contact found in the second Ekofisk well 2X. Both lines later proved to be inaccurate as limits of the West Ekofisk field.

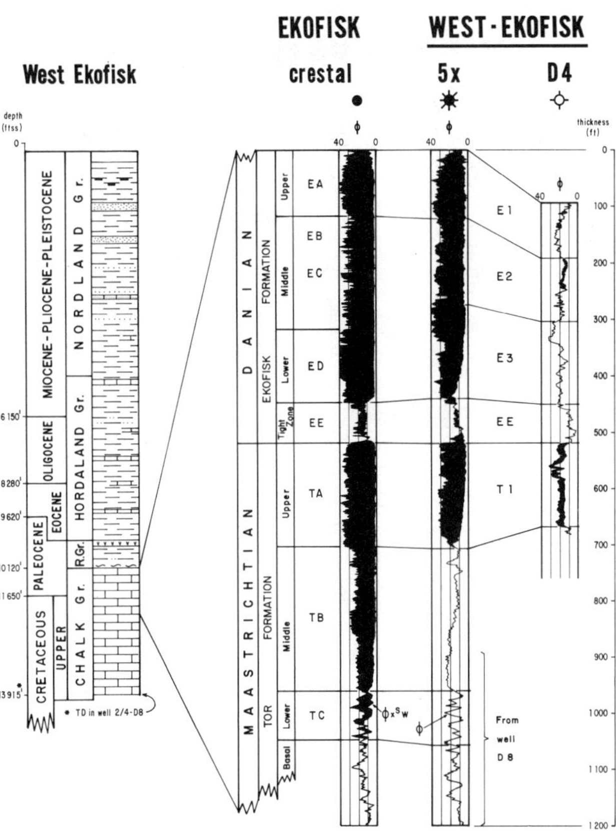

Figure 3. General lithostratigraphy of the West Ekofisk field based on the exploration well 5X and on the vertical development well D8. Reservoir zones, porosity, and hydrocarbon saturation are also shown. A comparison with the crestal Ekofisk well illustrates the similarity between the two wells whereas the reservoir in the development well D4 located in between is thinner, less porous, and contains only minor amounts of hydrocarbon.

Owing to its proximity to the Ekofisk field, West Ekofisk was declared to be commercial without any appraisal well being drilled, which may be unique for a North Sea field. The lateral extent of the field was an assumption based on the so-called limit of seismic porosity (Figure 2), a boundary that later proved to be very imprecise. The original recoverable reserves were then estimated at 180 million STB of oil and 2.1 tcf of gas, which is more than twice the present estimate of original reserves.

Post-Discovery

A single 15-slot drilling production platform was installed above the middle of the structure, and development drilling started in October 1974. The first two development wells were drilled on the flanks of the structure in 1974–1975 with negative results (Figure 4). Well D4 reached the top of the reservoir more than 160 ft (49 m) above the base of the hydrocarbon zone identified in the upper reservoir by the exploration well; however, it failed to find a producible interval and the hydrocarbon saturation was lower than 40% through the reservoir. Well D9, drilled to the west, reached the chalk 130 ft (40 m) higher than well D4, but most of the reservoir was in the transition zone and the well has never been turned over to production. The reasons for the drilling failures were that the porosity of the chalk reservoir rapidly decreases flankward and that the thickness of the oil-water transition zone increases accordingly. These facts were not known at the time. The results of these two development wells led to the drastic reduction in the reserve estimate, following reinterpretation and reduction in interpreted reservoir size.

The ten remaining wells drilled from 1975 to 1977 were located so as to penetrate the reservoir at higher structural elevations (Figure 5) and thus they were

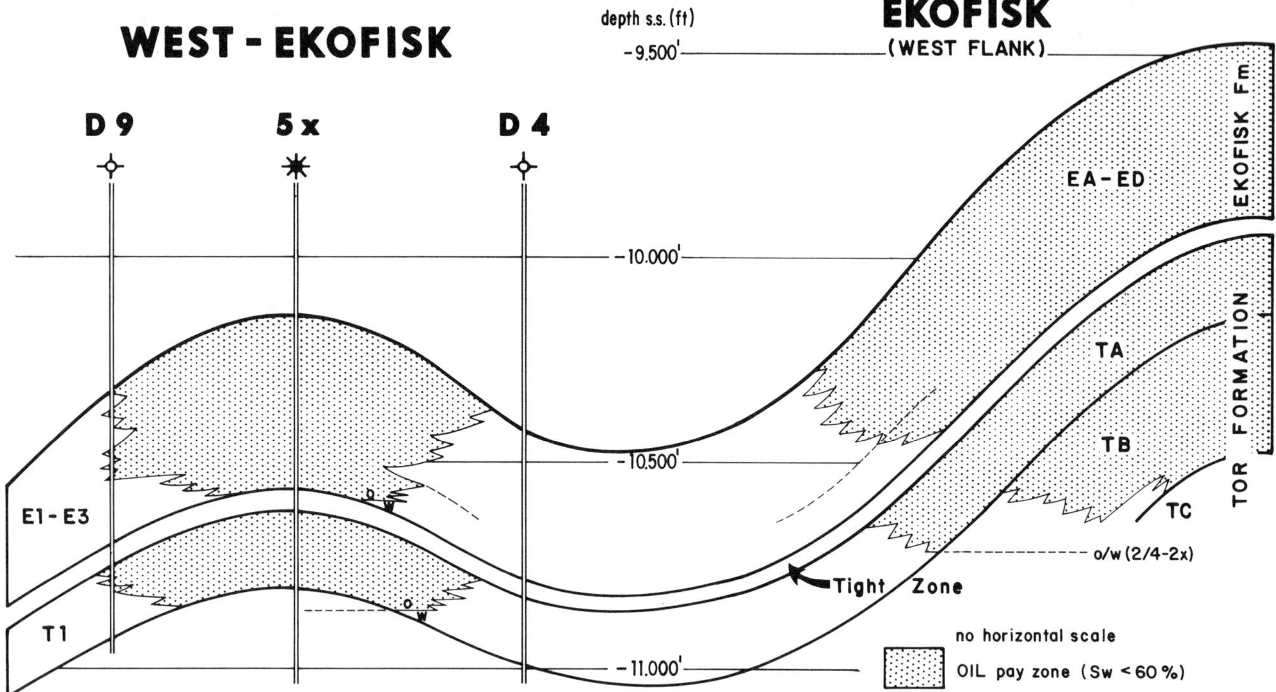

Figure 4. Structural sketch showing the elevations and pay zones of the first two development wells D4 and D9 compared to the discovery well 5X. The pay zones of Ekofisk and West Ekofisk are represented by shading.

successful. Production began in 1977 and reached a peak of 85,000 BOPD and 533 MMcfg/d in 1978. Additional infill and sidetrack wells were drilled between 1978 and 1987. At the end of 1987, 56 million bbl oil and 875 bcf of gas have been produced from the West Ekofisk field.

DISCOVERY METHODS

Then (1965-1975)

In the mid-sixties, exploration was concentrated in the southern North Sea where gas had been discovered in sandstones of Early Permian and Triassic ages. The source rocks for the gas were the Carboniferous coal measures. In the central North Sea, the coal was not present and the Permian sandstones were too deep to be an attractive objective. However, salt diapirs had been identified by geophysical means and, provided that Mesozoic or Tertiary reservoir and source rocks existed, large hydrocarbon accumulations were possible. Several structures had been identified in Block 2/4 based on a 5 × 5 km seismic grid. A 1.7 × 1.7 km grid shot in the late sixties helped to better define a large structure named Ekofisk and a small western dome called West Ekofisk.

The first well, drilled in 1969 on Block 2/4, was designed to test the large Ekofisk structure. As discussed in the previous section, the primary objective, the Paleocene sandstone, was not encountered, but the well penetrated a thick oil column in porous chalks of early Paleocene and Late Cretaceous age. The chalk reservoir discovered at Ekofisk became the only objective at West Ekofisk, which was discovered in November 1970.

The peculiarities of the chalk reservoir of the Ekofisk area, however, remained poorly understood until the mid-seventies and the initial estimate of West Ekofisk was made based on the extrapolation of the apparent oil-water contact found in the Ekofisk exploration wells and on the so-called limit of seismic porosity (Figure 2). In fact, the high porosity of the chalk creates a marked increase in seismic amplitude (Figure 6), and tentative maps were made of the limits of the high-porosity chalk. Unfortunately, this method was not precise enough, and, although located inside the seismic porosity zone, the first two development wells did not encounter good reservoirs.

A positive aspect of the seismic surveys at West Ekofisk is that the reflection quality is generally good and that the extremely low seismic velocity (due to gas effect) characterizing the mid-Tertiary section above most of the other chalk fields of the area is absent or much less pronounced above West Ekofisk.

The lack of understanding that porosity decreases abruptly flankward and that hydrocarbon saturation

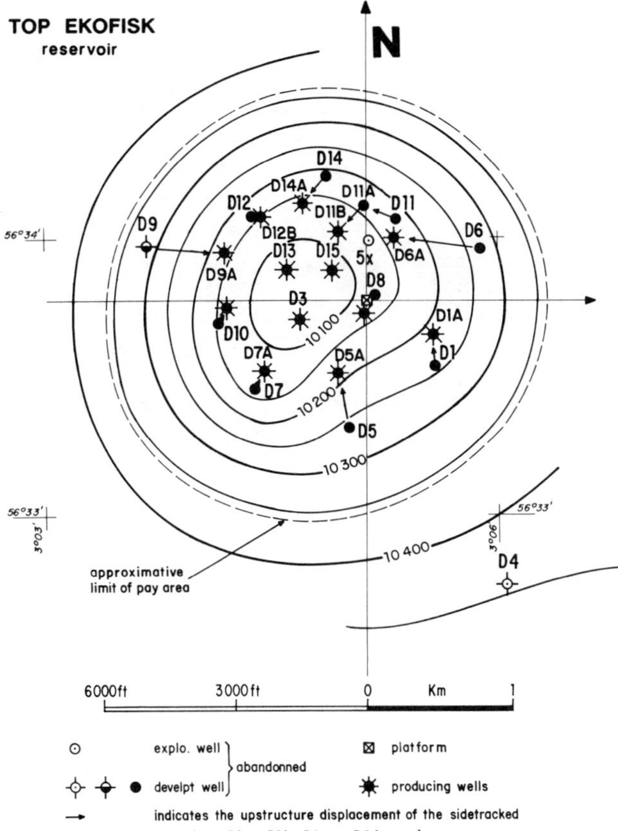

Figure 5. The 1988 development map. Contours indicate the subsea depth of the top of the reservoir. All well locations are shown at the top of the reservoir level. The arrows indicate the updip displacement of several wells that have been sidetracked. Contour interval is 50 ft.

varies drastically with subtle changes of rock type led to the poor results of the first development wells.

Now

West Ekofisk became the first chalk field to be fully developed. By 1977, 13 wells had been drilled on this small circular structure of less than 15 km^2. The chalk formations appeared to correlate perfectly all over the field, and thickness, porosity, matrix, permeability, and hydrocarbon saturation decrease progressively from the crestal area toward the flank in all directions (D'Heur, 1978, 1980). At the same time, studies of chalk diagenesis (Scholle, 1977) and sedimentology (Kennedy, 1980) gave the basis for a valid interpretation. The density of the data, the simplicity of the structural geometry, and the absence of regional variation allowed the elaboration of a clear geological model for West Ekofisk that has been proposed as a simple geological archetype for the chalk fields of the area (D'Heur, 1984). This model was very helpful for the optimization of development drilling of the more than 100 wells drilled since then in other chalk fields. The unique characteristics of the chalk reservoir and their interpretation are discussed later in the *Reservoir* section and in the *Exploration Concepts, Lessons.*

Surface Manifestations

Gas seeps on the seafloor have been observed 9 km southwest of West Ekofisk (Hovland and Sommerville, 1985). At this location a salt plug has upwelled through the chalk to a depth of 2600 ft where it has drastically disturbed the Tertiary strata that generally create the upper seal of the chalk fields. It is unlikely that such surface manifestation occurs elsewhere in the area, but the presence of a gas chimney in the Tertiary above several other chalk fields is still a clear indication that gas has locally escaped from the reservoir. The absence of a gas chimney at West Ekofisk probably precludes any surface manifestation above West Ekofisk.

STRUCTURE

Tectonic History

West Ekofisk, as are the other chalk fields of Norway, Denmark, and England, is located in the Central trough of the North Sea (Figure 7). Gowers and Saeboe (1985) have divided the Central trough into 17 zones with distinct responses of the basement to the various tectonic episodes. West Ekofisk, together with Ekofisk and Albuskjell, is located just north of the Feda graben in an area identified as a complex faulted block in which the basement dips toward its northern bounding fault. This being so, the sedimentological and structural history of these three fields differs significantly from that of the Tommeliten, Eldfisk, and Valhall chalk fields to the south.

The oldest strata penetrated at West Ekofisk are of Cenomanian age. However, Permian salt has been found at depths of 15,650 ft and 5400 ft (4770 and 1646 m), respectively, in Albuskjell, 6 mi (10 km) to the northwest and in well 2/7-13, 6 mi (10 km) to the southwest. The salt is generally overlain by Upper Jurassic sediments and a nearly complete sequence of the Cretaceous and Tertiary aged sediments, 12,000 to 15,000 ft (3658 to 4572 m) thick. The Zechstein salt was deposited in latest Paleozoic times in the intracratonic Northern Permian basin of the North Sea. The salt was subject to intense halokinesis from Jurassic times until late Tertiary time; it is responsible for doming the chalk layer in most of the fields in the area.

The intracratonic Paleozoic sag phase was followed by major rifting that started in Triassic times with the formation of the Central graben. Doming of the area in mid-Jurassic times has resulted in erosion

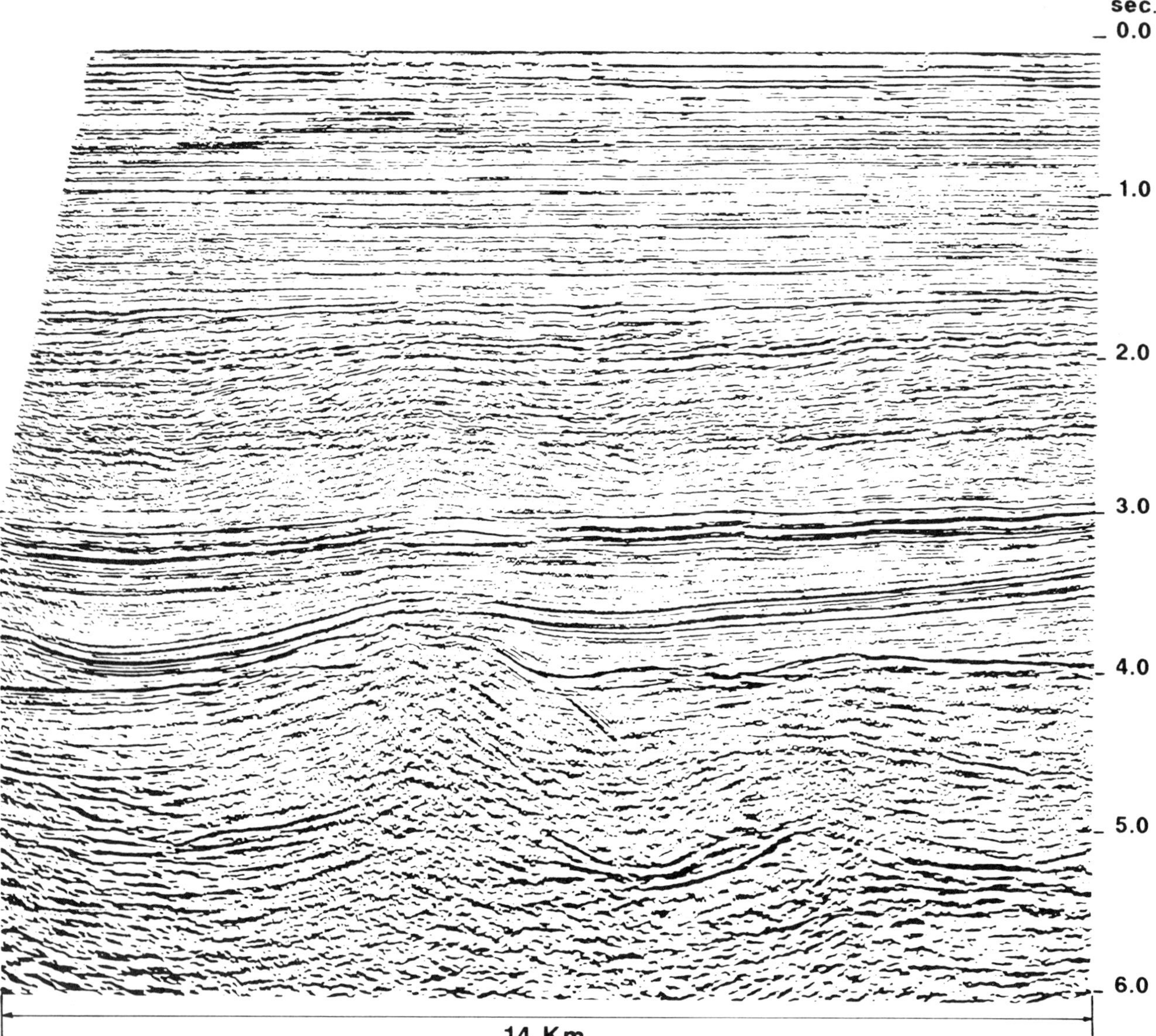

Figure 6A. Uninterpreted regional seismic line northwest-southeast through the West Ekofisk area. The location of this seismic section is shown by line aa′ on Figure 1.

of the Triassic and Lower Jurassic sediments. Crustal extension and movement of the various tectonic elements of the Central graben continued during Late Jurassic and Early Cretaceous times. By mid-Cretaceous times, the area was subject to a much more widespread and regular subsidence and developed as the Central trough. Subsidence continued during Late Cretaceous and Tertiary times.

Of real interest for the chalk field province is its location in the southern part of the trough in a narrow but relatively deep area bounded on the east and west by shallower platforms. During Maastrichtian and Danian times, these salt platforms and some structural elements inside the trough were subject to seafloor instability caused by tectonic pulses. The recently deposited chalk ooze was then resuspended or eroded and transported to the deepest areas in the trough where they formed thick and sometimes continuous chalk layers. As discussed later, these allochthonous (redeposited) chalks contain reservoirs of good quality whereas the "normal" pelagic (autochthonous) chalk beds are generally tight.

Regional Structure

West Ekofisk and Ekofisk are located just west of the base of the eastern margin slope of the Central trough (Figure 7), and during Late Cretaceous and early Tertiary, the area received a tremendous quantity of the remobilized chalk coming from the

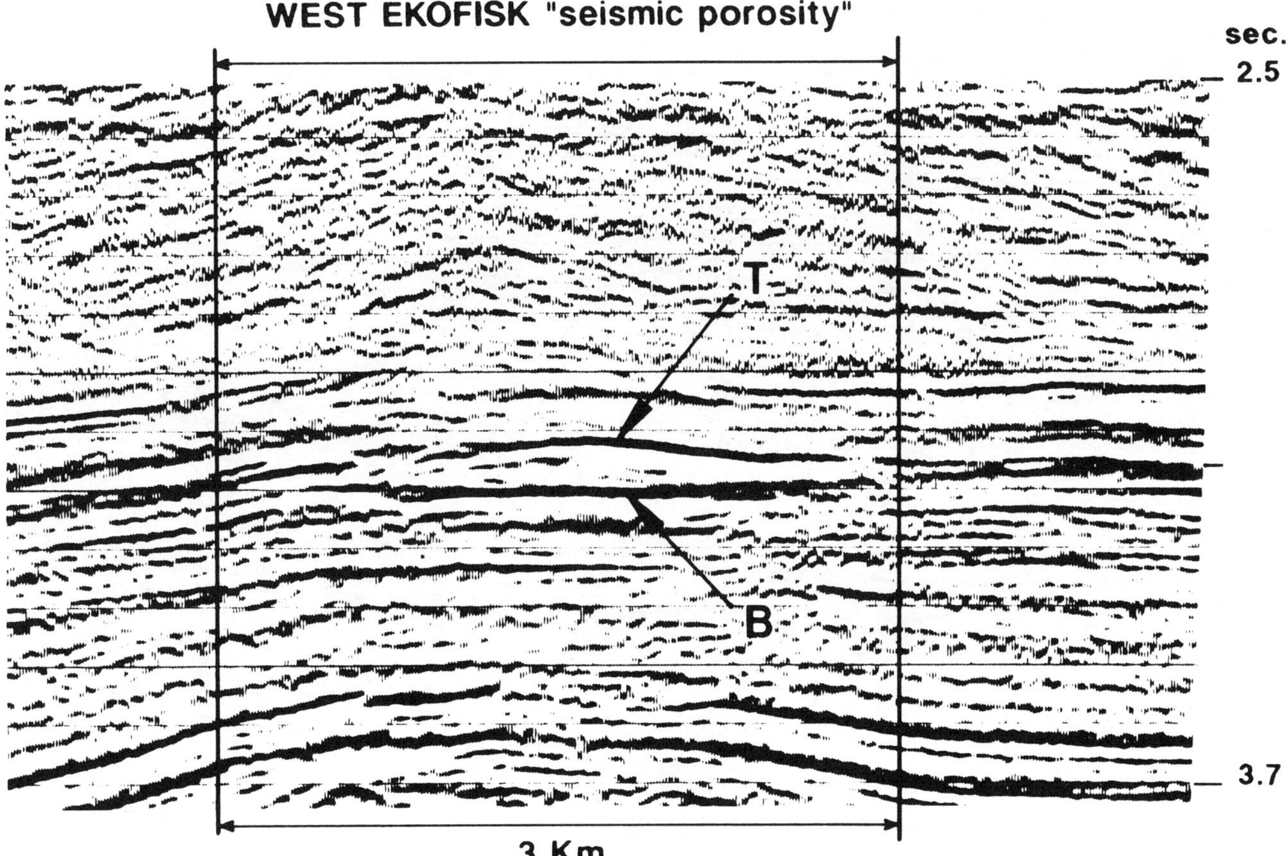

Figure 6B. Enlargement from 2.5 sec to 3.7 sec across the West Ekofisk structure. T and B are reflections at the top and the base of the hydrocarbon-bearing chalk reservoir.

eastern platform redeposited as debris flows, turbidite, and mud-flow sediments. Because of the existence of a northwest-trending paleo-high barrier southwest of West Ekofisk (Lindesnes ridge), the West Ekofisk–Ekofisk area became a depocenter for the allochthonous chalk.

It has been demonstrated that during deposition of the reservoir layers, the seafloor was rather flat at West Ekofisk (D'Heur, 1980). The halokinesis responsible for the trap took place during Tertiary times.

Local Structure

The West Ekofisk structure is a nearly circular dome of halokinetic origin with a diameter of about 3 mi (5 km). The vertical closure at the top of the chalk level is 400 ft (122 m). At the base of the reservoir, the vertical closure is at approximately 260 ft (79 m), the latter value representing the part of the structural relief that results from the halokinesis of the Zechstein salt. The 140 ft (43 m) difference in structural closure at the top as compared to the base of the reservoir is due only to differential compaction of the reservoir layers (discussed later in *Reservoir Thickness*).

The West Ekofisk and Ekofisk structures are separated by a pronounced saddle (Figure 4). In the saddle area, most of the chalk has a low porosity that creates a permeability barrier between the two fields.

STRATIGRAPHY

The lithostratigraphy of well 2/4-D8, a near crestal development well, is given in Figure 3. Rocks older than Late Cretaceous have not been penetrated in West Ekofisk, but they have been drilled in the vicinity of the field. The lithostratigraphic nomenclature used follows Deegan and Scull (1977). Fine- to medium-grained sandstone of the Rotliegendes Group (Lower Permian) resting on a reddish to greenish basalt have been drilled by well 2/7-2, 33 km to the south. The Zechstein salt (Late Permian), responsible for intense diapirism in the area, has been reached at Albuskjell and Tommeliten, 10 km to the northwest and southwest, respectively, and at Tor and Eldfisk 18 km to the northeast and southeast (Figure 1). Triassic redbeds of continental origin may overlie the Zechstein evaporites beneath the field.

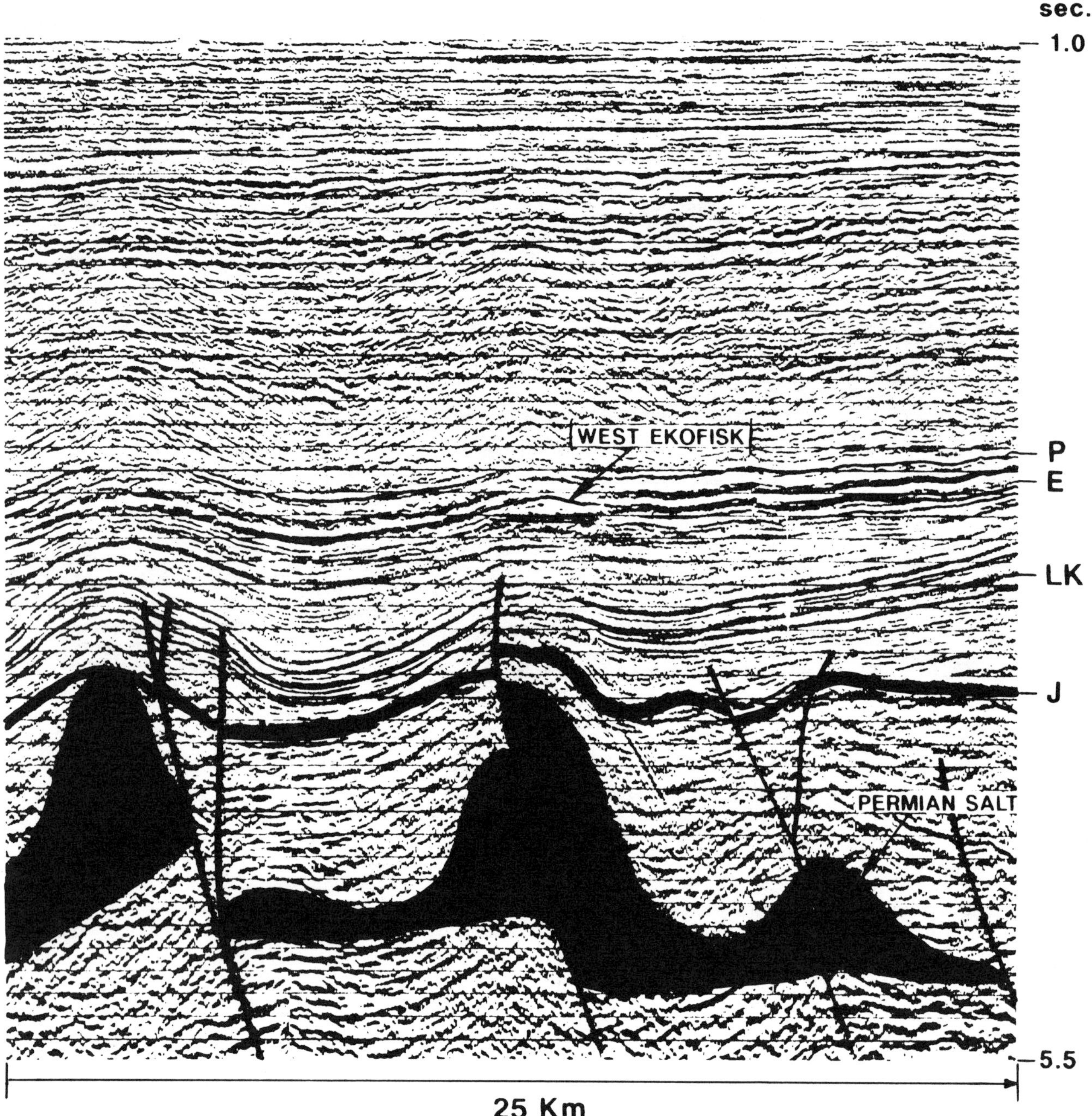

Figure 6C. Regional seismic line across the West Ekofisk area. Interpreted horizons P, E, LK, J, and S correspond to top Paleocene, top Ekofisk Formation, base Chalk Group, top Jurassic, and Salt, respectively.

The Jurassic rocks identified in the Greater Ekofisk area are mainly of early Kimmeridgian to Volgian age. These are marine claystone and shales that have fair to excellent source rock characteristics. The Cretaceous section is composed of the Cromer Knoll Group and of the Chalk Group of Early and Late Cretaceous ages, respectively. The Chalk Group is also comprised of chalk of lower Paleocene Danian age. The Cromer Knoll, which has not been reached at West Ekofisk, probably consists of gray marls and shales of the Valhall Formation overlain by reddish marls of the Rodby Formation.

The Chalk Group, which has been almost entirely penetrated, is comprised of five formations:

1. The lowest one, the Hidra Formation, more than 400 ft (122 m) thick, consists of argillaceous chalk grading upward into cleaner light gray chalk of Cenomanian age.

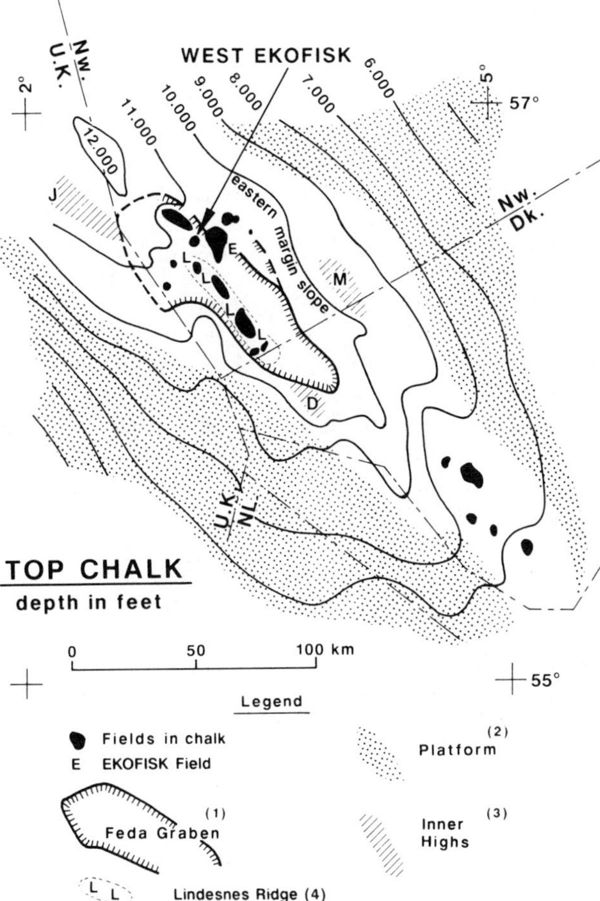

Figure 7. Geological setting of the southern part of the Central trough (surrounded by shaded areas representing platforms to the northeast and southwest). (1) is the main part of the Feda graben, a pre- mid-Cretaceous feature (Gowers and Saeboe, 1985); (2) and (3) are post- mid-Cretaceous features: D, J, and M are respectively the Dogger, Josephine, and Mandal highs; (4) the Lindesnes ridge has been formed by inversion during Campanian time.

2. The Plenus Marl Formation, 30 ft (9 m) thick, consists of dark gray shales with thin chalk beds of late Cenomanian to Turonian age.
3. The Hod Formation, 1500 ft (457 m) thick, consists of more or less argillaceous chalk of Turonian to Campanian age.
4. The Tor Formation, 1300 ft (396 m) thick, is made up of clean chalk deposited during late Campanian and Maastrichtian times.
5. The Ekofisk Formation, of Danian (early Paleocene) age, is comprised of 500 ft (152 m) of chalk with up to 50% silica content.

Detailed lithostratigraphy of the reservoir formations is given in the *Reservoir* section.

The Chalk Group is overlain by a monotonous 9800 ft (2987 m) thick Cainozoic sequence that has been divided into three groups:

1. The Rogaland Group, 500 ft (152 m) thick, is comprised of the lower Paleocene marls of the Maureen Formation overlain by gray claystone and shales of the Lista and Sele formations and by the "ash beds" (shales with tuffaceous material and diatom limestone stringers) of the Balder Formation.
2. The Hordaland Group (Eocene to early Miocene), 4400 ft (1341 m) thick, consists of gray and brown claystone, locally silty with frequent limestone stringers and rare silty shales.
3. The Nordland Group (middle Miocene–Recent), 5000 ft (1524 m) thick, consists mainly of gray clay with siltstone and sandstone intervals in the upper part.

TRAP

Trap Type

The West Ekofisk field is a domal structural trap created by upwelling of the Permian Zechstein salt during Eocene to early Miocene time. The dome has a 3 mi (5 km) diameter at the top reservoir level, but the circular productive area is limited to a 2 mi (3.2 km) diameter. Tertiary shales seal the top of the reservoir. Laterally, the reservoir is sealed by chalk of equivalent age that is tight as a result of late diagenesis in the aquifer (Figure 4). Downward the Maastrichtian chalk has also lost its permeability by pressure solution and recrystallization in the aquifer (discussed later).

The Ekofisk "Tight Zone" (Figure 3, EE, and Figure 4), which is a 50 ft (15 m) thick interval of argillaceous and siliceous chalk with thin stringers of interbedded shales at the base of the Danian, separates the Ekofisk Formation reservoir from the Tor Formation reservoir beneath. These two distinct zones have different oil-water contacts. The upper reservoir contains hydrocarbons down to -10,550 ft (-3216 m), which is 130 ft (40 m) deeper than the closure at the top of the chalk level. The lower reservoir, of Maastrichtian age, is hydrocarbon-bearing down to -10,900 ft (-3322 m).

Reservoir

Stratigraphy, Lithology

The West Ekofisk field reservoir consists of chalk of the Danian Ekofisk Formation (early Paleocene) and Maastrichtian Tor Formation (Late Cretaceous).

The chalk is a calcilutite made up primarily of coccoliths, which are debris of tests of unicellular algae (coccolithophoridae). The size of the coccoliths generally ranges between 0.5 and 2 microns, but entire coccospheres as well as chambers of foraminifera and other bioclasts provide much larger pores dispersed in the clay-sized matrix. The relatively coarse carbonate debris include foraminiferal and molluscan remains as well as sponge, radiolarian, and bryozoan fragments.

The noncarbonate fraction is generally less than 5% in the Tor Formation but locally exceeds 50% in the Ekofisk Formation. The noncarbonate components are mainly clay, which reduces the permeability, and quartz, which increases rock strength. The petrology of the North Sea chalk has been described by Hancock (1976) and others.

The chalks of the Ekofisk and Tor formations have been subdivided into eight members that are generally characterized by subtle or significant variations in lithology (Figure 3). They are mainly related to different sedimentary facies with corresponding, distinctive initial porosities. The difference in initial porosity from bed to bed is still apparent in the hydrocarbon zone while it may be masked by recrystallization in the aquifer reservoir. In other words, the West Ekofisk reservoir is of layer-cake type in that it consists of more porous and less porous beds, depending on variations of sedimentary conditions. Most of the following information on the lithology of the Tor Formation is based on observations made in the other fields of the surrounding North Sea, as only 20 ft (6 m) of the Tor Formation have been cored in West Ekofisk.

(The nature of varying porosities and porosity/permeability relationships in the reservoirs is very important. They are discussed in this stratigraphic, lithologic subsection only in general and relative terms; they are more fully discussed, quantified, and illustrated by accompanying figures in the *Reservoir* section.)

The basal member of the Tor Formation is mainly composed of pelagic (autochthonous) chalk, slowly deposited, probably as a rain of fecal pellets. This rock has no reservoir quality except for rare thin debris flow (allochthonous) intercalations. No part of this member contains hydrocarbon in the Ekofisk and West Ekofisk fields.

The lower member of the Tor Formation has a very typical log signature since it is composed of several porous debris flows 6 to 12 ft (2-4 m) thick, intercalated with tight pelagic chalk. The porous beds contain oil in the Ekofisk field but not in West Ekofisk.

The 220 ft (67 m) thick middle member of the Tor Formation is probably composed of diluted and slumped debris flows and mud flows mixed with pelagic chalk and some turbidites. This member of fair to poor porosity produces oil in the central area of the Ekofisk but is water-bearing in West Ekofisk.

The upper Tor member, 150 ft to 200 ft (46-61 m) thick, is the only Maastrichtian hydrocarbon reservoir in the West Ekofisk field. It is thought to consist of rather homogeneous chalk interpreted as a complex sequence of massive slumped debris flows with interbeds of turbidites and mud flows. Its uppermost 20 ft (6 m) have been cored and are described in well D8 as brownish-white, soft chalk with stylolites. As interpreted from logs, this cored interval is probably not representative of the underlying 180 ft (55 m) of highly porous, clean (96% $CaCo_3$) chalk, which contains the best reservoir.

The Tor Formation is overlain by a "Tight Zone," some 60 ft (18 m) thick, of which less than 10 ft (3 m) have been cored. This interval, which correlates peak to peak in log curves in the Albuskjell/West Ekofisk/Ekofisk/Edda area, is well known and consists mainly of pelagic chalk, locally silicified and containing variable quantities of clay and a trace of dolomite and feldspar. The chalk is generally laminated (varved) with black shale partings. It also contains thin turbidites, distal debris flows and, occasionally, conglomerates, and hard-ground. This layer creates a permeability barrier between the Maastrichtian and Danian reservoirs; it is usually referred to as the "Tight Zone" (sensu stricto) or layer EE in the Phillips Group nomenclature.

The lower member of the Ekofisk Formation (layer ED), approximately 150 ft thick, has been cored between -10,514 ft and -10,564 ft (-3205 and -3220 m) in exploration well 2/4-5X. The sequence consists of buff slumped debris-flows and originally burrowed chalk showing a wide range of plastic deformation features, shear lamination, and slump brecciation. This interval is characterized by very high porosity values and low clay and silica content. It contains predominantly Maastrichtian taxa that have been redeposited during early Danian times. Although most of the Ekofisk and Tor formations chalks are redeposited allochthonous sediments, the lower Ekofisk member is the only one to be usually referred to as the "Reworked Zone," which is somewhat misleading. This layer thins drastically in all directions away from the Ekofisk/West Ekofisk area.

The middle Ekofisk member, 100 to 150 ft (30-46 m) thick, has been cored over a 90 ft (27 m) interval in well 5X. It consists of buff chalk, slumped throughout. Lithologies include burrowed chalks, large bioclasts, poor micrites, and debris flows. The slumped interval is overlain by buff, bioturbated chalk with poorly defined rhythms, recording a phase of pelagic deposition. The middle Ekofisk member is characterized by high clay and silica content (up to 60% quartz) and by a lower average porosity compared to the overlying and underlying members.

The upper member of the Ekofisk Formation is usually a homogeneous chalk with little internal structure. It is approximately 100 ft (30 m) thick, the lower half of which has good to very good porosities and is interpreted as debris flows. The upper half contains laminated mudflows and varved pelagic chalk with burrows and thin calcareous turbidites. The uppermost section shows a progressive increase in clay content and passes into marls and shale of the Maureen Formation.

Depositional Setting

The depositional setting of the chalk reservoir of the Central trough was discussed and clarified in May 1980 at the NPF (Norwegian Petroleum Society)

Seminar on North Sea Reservoir Rocks. D'Heur (1980) pointed out the "stratified" porosity of the chalk over wide areas and suggested variation in sedimentation rate to explain the alternation of high porosity (rapidly deposited) and low porosity (slowly deposited) chalks. Kennedy (1980) reported he had identified distinct porous debris flows interbedded with slowly deposited tight pelagic chalk in the Tor field. Further discussions led to the concept that the porous chalks were made of redeposited material whereas the pelagic chalk was generally tight compared to the pelagic chalk. The higher original porosity of the redeposited chalk and its better resistance to porosity destruction with depth result from at least five factors:

1. The prelithified structure of the original sediment (Kennedy, 1987).
2. The winnowing from its associated clay, since clay reduces the efficiency of "spot welding" between calcite grains (Mapstone, 1975).
3. Lower silica and aragonite content, a significant source for early cement (Hardman, 1982).
4. Very rapid accumulation in a narrow subsiding area that favored overpressuring of pore fluid.
5. Reduced activity of burrowing organisms that otherwise would have favored input of early cement by open burrows.

Some of these factors will be further discussed in the *Reservoirs* section.

Kennedy (1987) based refinements of his depositional model on more observations while D'Heur (1984) detailed the correlations over the whole Greater Ekofisk area and Hatton (1986) extended these into the British sector of the Central trough.

An overall description of the depositional model is given in Kennedy (1987). It may be briefly summarized as follows for the Maastrichtian and Danian times: While coccolith aggregates were deposited everywhere to form the normal pelagic chalk, local and regional gravitational seafloor instabilities provoked resuspension of the coccolith ooze. Because of the topography of the seafloor, mass movements were induced by basinward slopes, fault displacements, local inversions, and halokinesis of the Permian salt. As a result, debris flows, turbidites, mud-flow, and mud-cloud deposits were distributed in the Central trough according to topography and to the distance from the source of the allochthonous material. None of the pelagic chalks constitute a good reservoir rock in the area.

Conversely, the allochthonous deposits may be characterized by excellent porosity (20 to 50%) and satisfactory matrix permeability (>1 md). Mass slides, slump folds, and other internal structures are common. In most cases, the slumps were not surface flows of material but semi-dewatered plastic deformations by movement beneath an overburden thickness of meters to tens of meters.

The reservoir layers of West Ekofisk were deposited over a flat seafloor as proven by the constant depositional thickness of each unit (D'Heur, 1980). Most of the Maastrichtian debris flows probably had their source on the platform to the east, whereas the Danian flows came from several distinct sources: the eastern platform, the Lindesnes ridge to the south, and local active diapirs in the surrounding area. Long episodes of quiescence existed during early Maastrichtian and early Danian times. Shorter periods of quiescence alternating with episodes of seafloor instabilities occurred throughout the Danian.

The basal Danian Tight Zone (EE) was deposited during a long period of quiescence in the central part of the basin while the Danian sea regressed after the widespread Maastrichtian transgression. The so-called "Reworked Zone" (ED) overlying the "Tight Zone" is the witness of a major local seafloor instability that occurred east and/or south of the West Ekofisk field. The West Ekofisk/Ekofisk area was the depocenter of this typical reworked layer during early Danian time. An influx of cold water from the north resulted from the acceleration of the opening of the Norwegian Sea in late Danian time and led to the end of a 35 million year period of chalk deposition in the North Sea.

Porosity, Diagenesis

Chalk overlain by 9000 ft (2743 m) of sediments normally has a porosity reduced to 5% to 15% by dewatering, compaction, pressure solution, and recrystallization in pore space. The most fortunate anomaly of the Central trough chalk is its very high degree of porosity preservation, locally up to 50%. Initial high percentage porosity of the sediment was due to the large proportion of rapidly sedimented allochthonous material that passed quickly through the water sediment interface and escaped bioturbation and significant early cementation.

Most of the factors preserving the chalk porosity in the Central trough have been explained by Scholle (1977) based on a limited number of data on the North Sea chalk fields. Additional causes have been found by Mapstone (1975), Hardman (1982), and others. The critical factors for the high porosity preservation are:

- The low magnesium calcite composition that makes the sediments chemically stable.
- The favorable nature of the pore fluids.
- "Spot-welding" at grain contacts, which created a rigid framework resisting compaction during a period of time.
- The low-clay content that favored the spot-welding.
- The development of overpressure in the chalk layer that reduced the effective stress.
- Entry of oil into the reservoir, which inhibited the process of porosity reduction by pressure solution.

The present-day porosity distribution in the West Ekofisk field is the result of the combined effect of (1) the distribution of the various sedimentary facies and (2) the differential porosity preservation.

The various members of the Ekofisk and Tor formations had distinctive average initial porosities because they were deposited with distinct ratios of allochthonous to autochthonous material; this explains the porosity layering. Since it has been established that no seafloor paleorelief existed at West Ekofisk during Maastrichtian and Danian times and, as the field is of small areal size, no lateral initial porosity variation existed within individual chalk beds. The present-day lateral porosity variations are therefore related to differential porosity preservation only.

Porosity decrease from crest to flank of the structure may also be explained by the combined effects of two factors:

1. The progressive invasion of the trap by oil, from top to bottom, with a corresponding progressive inhibition of pressure solution–recrystallization at deeper and deeper levels and lower and lower porosity values while the porosity was still decreasing in the paleoaquifer.
2. Development of differential overpressure within the trap; it is therefore not surprising that the porosity maps of each reservoir unit (Figure 8) fit remarkably well the structural shape of the reservoir, as this shape has probably remained stable in a similar configuration since the invasion of the reservoir by hydrocarbons.

The various reservoir units follow the same trend of porosity decrease with depth: for example, 14 porosity points from crest to flank for the upper Ekofisk member and 10 points for the upper Tor member. However, as the structural relief diminishes with depth (as a result of porosity preservation), the amount of porosity reduction with depth is similar for all members, i.e., approximately 5 porosity points per 100 ft of structural elevation. A north-south and an east-west porosity section (Figures 9 and 10) illustrate the porosity distribution in the field.

Reservoir Thickness

The reduction of porosity resulting from differential porosity preservation necessarily corresponds to a reduction of rock thickness. Therefore, variations of thickness of a chalk layer are also due to two main factors: (1) sedimentary thickness variation (depositional), and (2) differential porosity preservation (diagenetic).

As a first approximation, the sedimentary thickness variation may be evaluated after having removed the effect of differential porosity preservation from the present-day thickness variation.

Assuming that calcite transfer during diagenesis occurs over very short distances only, the relationship between the initial (depositional) thickness D and the present-day thickness P is given by:

$$D = P\left[\frac{1 - \phi_P}{1 - \phi_D}\right]$$

where ϕ_D and ϕ_P are, respectively, the average initial porosity and the average present-day porosity of the chalk layer. Because there has been no lateral variation of facies, i.e., of initial porosity over the small West Ekofisk area, it may be considered that the average initial porosity of the coccolith ooze was constant (e.g., 70%) over the field area. A value of 70% may be considered as a first approximation. Then:

$$D = P\left[\frac{1 - \phi_P}{0.3}\right]$$

In the crestal well D8 and in the lowest flank well D4, where the present-day thickness of the Ekofisk Formation (excluding the "Tight Zone") varies from 452 ft to 366 ft (138 to 111 m), respectively, the depositional thickness was:

in D8: $D = 452\left[\frac{1 - 0.33}{0.3}\right]$

and in D4: $D = 366\left[\frac{1 - 0.19}{0.3}\right]$

which in both cases gives an approximate value of 1000 ft (305 m). As very similar values are obtained for intermediate well locations, this demonstrates that the depositional thickness may be considered as constant in the field area and that the present-day thickness variations are due to differential compaction only. It is confirmed by the relationship between thickness and porosity shown in Figure 11. Therefore, the thickness maps of the reservoirs (Figure 12) show the same trends as the porosity maps which themselves reflect the structural shape.

The "gross interval" thickness reduction toward the periphery of the dome results in a thinning of the "net reservoir." This "net reservoir" thinning is accentuated by the reduction of porosity and is evident in the net/gross ratio from crest to flank.

Another consequence of the thinning of the reservoir layers flankward owing to differential porosity preservation is that the structural relief is less pronounced at the base of the reservoir than at the top. In West Ekofisk, the base of the reservoir shows a structural relief of 250 ft (76 m) whereas it reaches 400 ft at the top of the Ekofisk Formation. This means that at the top of the chalk level, more than a third of the structural relief is due to differential porosity preservation.

Fractures

The existence of a high-permeability fracture system is critical to the productivity of a chalk reservoir since it acts as the collecting network through which fluids travel from the low-permeability matrix to the well bore. Cores have been

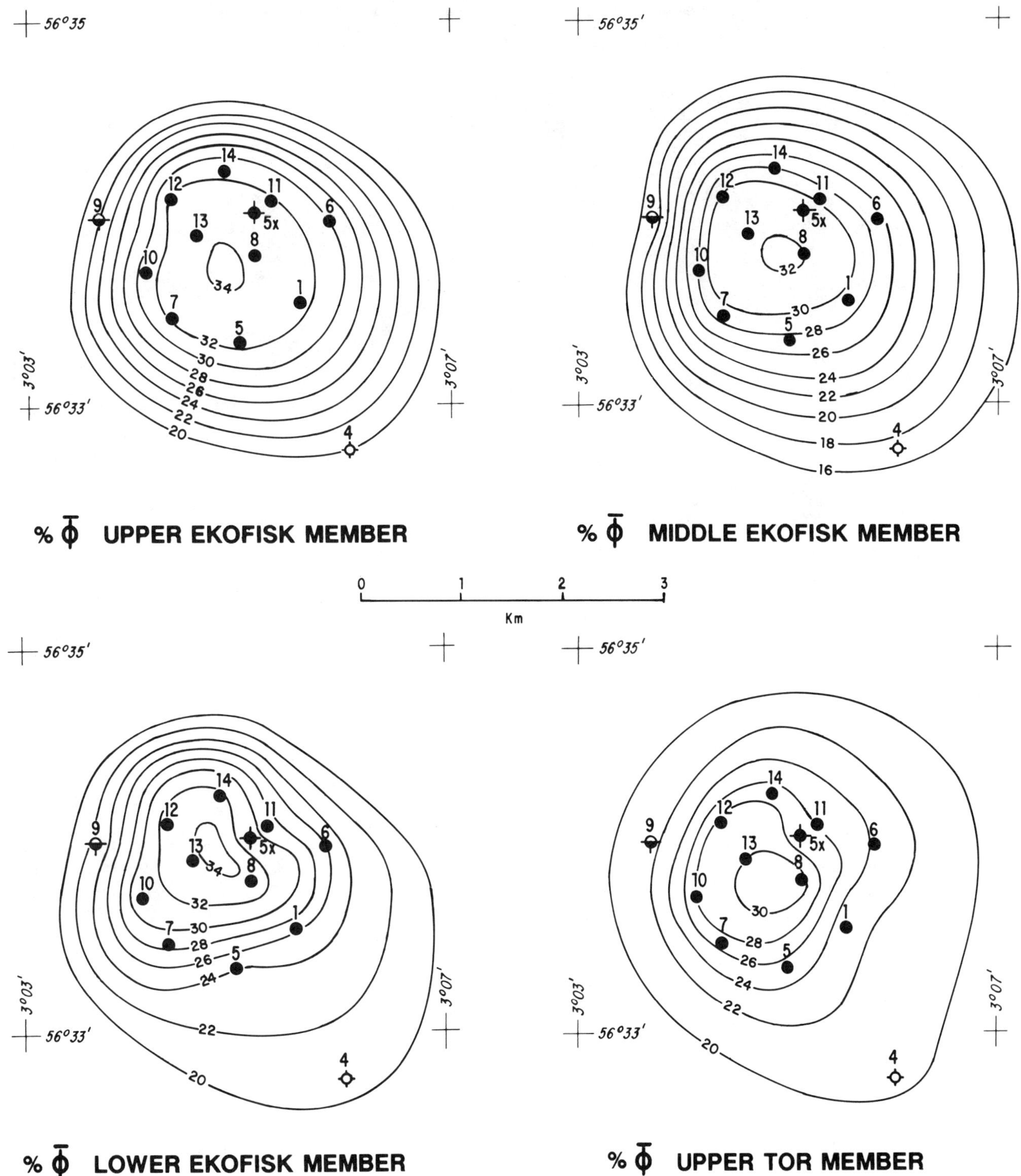

Figure 8. Average gross porosity maps of the four reservoir units. Contour intervals are 2%.

cut in only two wells in West Ekofisk, the exploration well 5X and the vertical development well D8. These are two crestal wells, and no information has been collected in flank or peripheral wells. In addition, the cored interval is thin compared to the total reservoir thickness and no representative information has been gained in the Tor Formation. Fracture study of the West Ekofisk field therefore gives only a very local knowledge of the Ekofisk Formation; much more is known from cores taken in the Ekofisk field, which is also described in this *Atlas of Oil and Gas Fields* by C.T. Feazel et al.

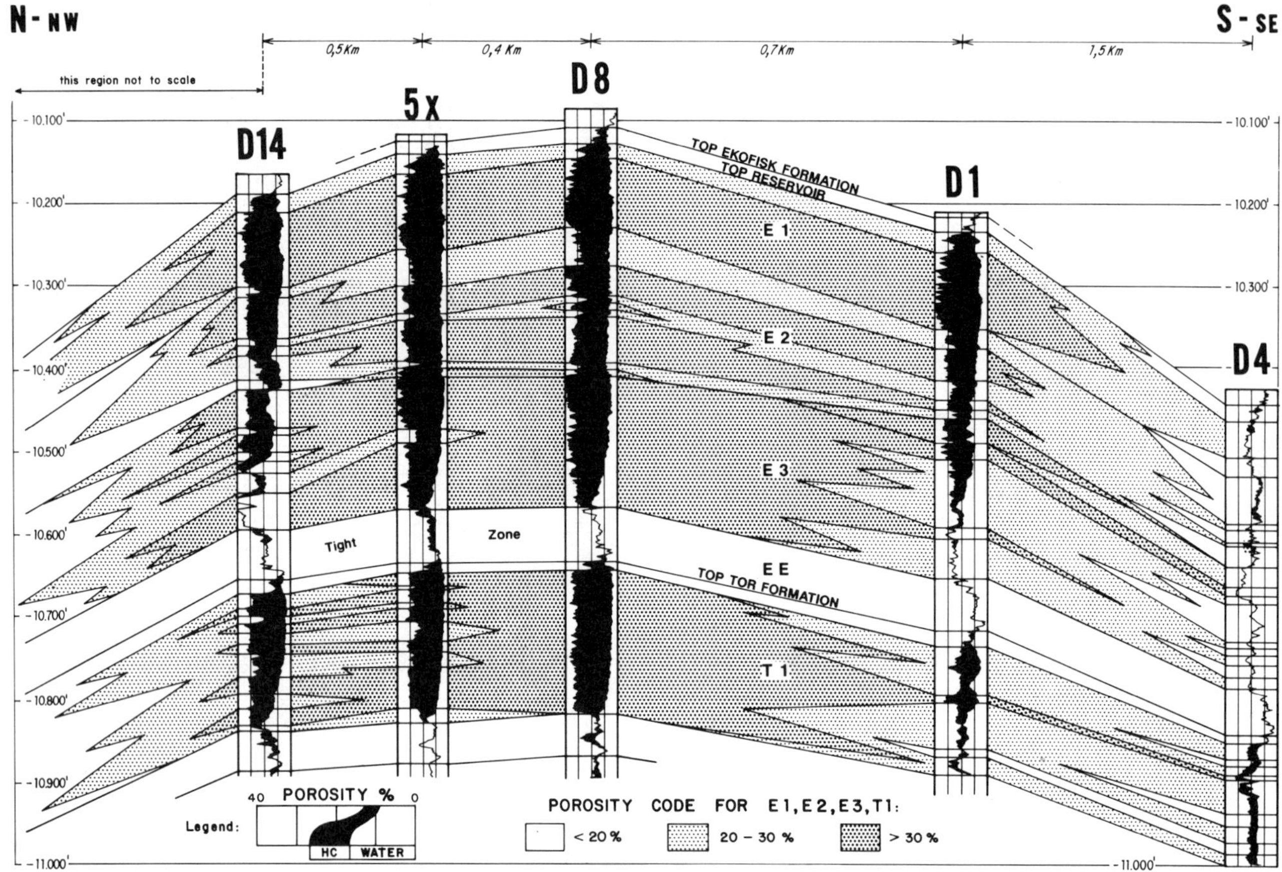

Figure 9. North-northwest to south-southwest porosity section across the field. (Well locations can be seen on Figure 5.)

In West Ekofisk, the cores taken in the crestal area gave the following indications:

- Fracture dip is generally 80° ± 10°.
- Most of the fractures observed in the upper and middle Ekofisk members are tectonic and irregular fractures, whereas stylolite associated fractures are frequent in the lower (reworked) Ekofisk member.
- 50% of the fractures are in part mineralized and 30% are totally filled by calcite.
- The fracture spacing is between 1 and 10 cm for 90% of the fractures (closed and open).
- Their length is frequently between 2 and 10 cm.
- Fracture and stylolite densities generally increase as rock hardness increases.
- The average stylolite frequency is 1.7 per foot and 90% of the stylolites have a frequency of less than 4 per foot.

These observations are probably not representative of the entire field area.

Porosity/Permeability

The relationship between porosity and permeability is available for the crestal area only. It is shown on Figure 13 for the Ekofisk Formation in the exploration well 2/4-5X. In this area, the matrix permeability ranges from 0.02 to 10 md. A permeability of 1 md is reached for a porosity of 32%, which is the average porosity of the Ekofisk Formation reservoir. A porosity of 20% corresponds to a permeability of 0.05 md. Rocks of similar porosity located at similar elevation above the oil-water contact are, however, characterized by a higher water content on the flank than in the central area of the dome (Figures 14 and 15). This indicates that the capillary pressure system varies in the same direction. Because there is no reason to assume that the calcite grains and pores are smaller on the flank than in the central area with the same porosity value, the change in capillary pressure system may be rather a consequence of variation in pore morphology and pore throat size. This modification may be due to differences of stress regimes: In the central area, the upper part of the chalk layer is under extension and the extension decreases in the flank to a neutral surface; further toward the syncline, the reservoir may undergo compaction and the porous network may be squeezed and flattened.

The lower Ekofisk and upper Tor members have the best productivity. Despite its high porosity, the

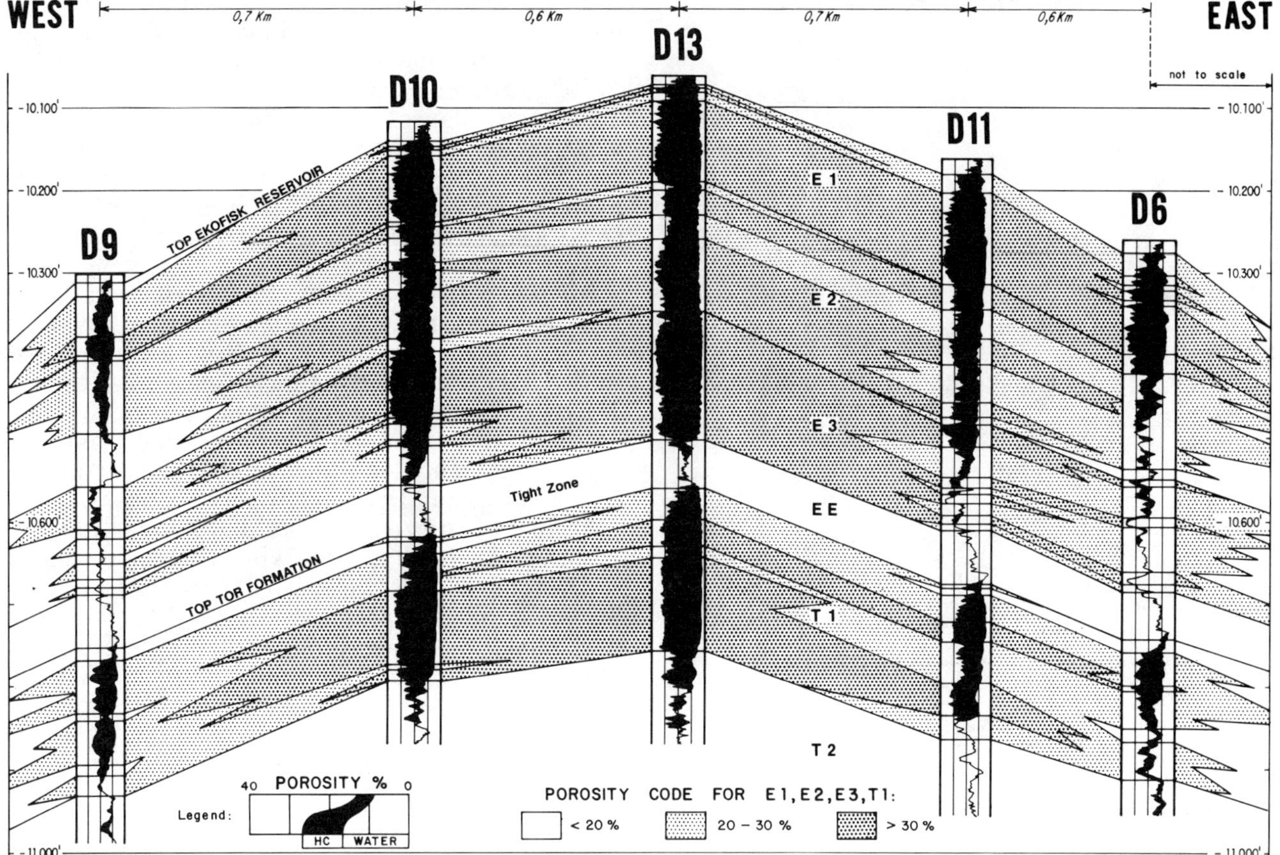

Figure 10. West-east porosity section across the field showing the progressive porosity decrease in the flank of the structure. (Well locations shown on Figure 5.)

upper member of the Ekofisk Formation has a lower permeability resulting from the abundance of laminated rock types.

Pay Zone Thickness

The Danian and Maastrichtian reservoirs are separated by the "Tight Zone," which, due to its high content of laminated chalk, turbidites, hard-ground, clay, and shaly partings, creates a permeability barrier.

In the central area of the field, the different water tables of the Ekofisk and Tor formations reservoirs are located at -10,550 ft and -10,900 ft (-3216 and -3322 m), respectively. As a result of the progressive change of the capillary pressure system, the transition zone thickens from some tens of feet in the central area to more than 200 ft in the periphery of the field. This outward thickening corresponds to a progressive rise of the base of the pay zone, as shown in Figure 16. The curved shapes of the bottom of the 90%, 80%, and 40% in hydrocarbon saturation envelopes are shown in Figures 14 and 15. The depths of these envelopes also depend on and correspond to the rock types that characterize the various reservoir layers. Logically, the thickest pay zone for each reservoir is observed in the central area of the field where the net to gross ratio is close to 1. The maximum net pay thicknesses are 430 ft and 190 ft (131 and 58 m) for the Ekofisk and Tor formations, respectively (Figure 17). With an average porosity of 32% and an average water saturation of 15%, these pay thicknesses combined correspond to a maximum hydrocarbon saturated pore thickness of 170 ft (52 m). The net-pay thickness maps are shown in Figure 17.

Hydrocarbons

West Ekofisk is a rich-gas condensate field. The initial GOR was 4100 SCF/STB. It increased to 18,000 SCF/STB in 1979 and exceeds 45,000 SCF/STB in 1988. The initial oil and gas specific gravities were 0.8 and 0.7, respectively. In fact, the West Ekofisk fluid was a near-critical system, initially unsaturated. Bubble point and dew point pressures were 6900 and 6440 psi, respectively.

Fluid Flow Characteristics

As discussed before, the "Tight Zone" (EE) at the base of the Danian interval acts as a vertical permeability barrier between the Ekofisk and Tor formations reservoirs. Most of the chalk below the oil-water contact consists of dead end pore volume (connected to the bulk of the pore volume in one

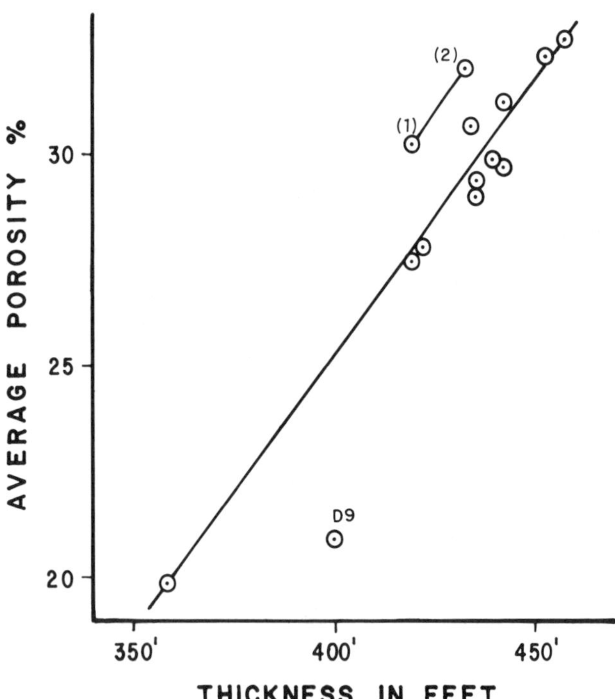

Figure 11. Relationship between thickness and porosity for the Ekofisk Formation (excluding the "Tight Zone") based on data before 1980. Points 1 and 2 are located in the northwestern area where regional depositional thickness of the chalk slightly increased.

direction only) and it provides only a limited aquifer support. This resulted in a drastic pressure drop between the years 1977 (7200 psi) and 1988 (2000 psi) at a reference depth of -10,400 ft (-3170 m). The best production well flowed 15,000 BOPD and 120 MMcfg/day from a 220 ft (67 m) perforated interval, through a 1-in. choke. Selective tests in the discovery well on 100 ft (30 m) of upper Tor interval, 70 ft (21 m) of lower Ekofisk, and 90 ft (27 m) of upper Ekofisk gave respectively 3600 BOPD, 3900 BOPD, and 3000 BOPD through a 1-in. choke after acidification.

Development Considerations

Owing to its proximity to the giant Ekofisk field, West Ekofisk was declared commercial without drilling any appraisal wells. When the decision was made to develop the field based on the results of one exploration well only, none of the peculiarities of the chalk reservoirs were known and the reserves were later shown to have been overestimated by 100%.

A drilling/production platform (2/4-Delta) was installed above the center of the structure in 1974, and eight development wells were drilled between 1974 and 1976. After the negative results of the first two development wells, the average spacing was revised to 2500 ft (762 m). Four additional infill wells were drilled in 1976 and 1977 as part of a development phase referred to as the "modified development plan" or MDP. Full development of the field was reached in 1977 with ten producing wells.

Casing collapses became frequent after 1980, mainly as a consequence of reservoir compaction and overburden reaction (see the following section). Seven wells where casing collapse occurred have been sidetracked to reach the reservoir at higher structural elevations (indicated by arrows on Figure 5).

In 1986 and 1988, two new crestal wells were drilled to satisfy an increased need for additional gas. Peak production was reached in 1978 with rates of 533 MMcfg/day and 85,000 bbl of condensate/day (Figure 18). By 1988, the gas and condensate production rates declined to 145 MMcfg/day and 3200 bbl/day, respectively. The cumulative production by 1988 was 880 bcf of gas and 57 million bbl of condensate. By year 2011 it is expected that 1 tcf of gas and 65–75 million bbl of condensate will have been produced.

The production is transported to the Ekofisk center by a 24-in. oil and gas line. From Ekofisk, oil is piped to Teeside, England, through a 34-in. line, 220 mi (354 km) long. Gas is transported to Emden, Germany, through a 36-in. line, 274 mi (441 km) long.

Reservoir Compaction and Seafloor Subsidence

Seabed subsidence was recognized at the Ekofisk field in 1984. It is a consequence of reservoir compaction caused by a drastic drop of fluid pressure after ten years of production. The reservoir was initially overpressured and fluid production resulted in a progressive increase of stress on the chalk matrix. The reservoir compaction becomes severe because the reservoir is very thick and very porous, at least at the center of the fields. In addition, the chalk reservoir, having lost its porosity and permeability where oil has not been present, is sealed downward and laterally in most places and has only limited aquifer support. Reservoir compaction is transmitted to the surface through the thick Tertiary rocks, which are mainly soft and underconsolidated. West Ekofisk, owing to its smaller size, has a lesser degree of undercompaction of the overburden and because of the presence of hard limestone stringers through the Tertiary sequence, an arching effect develops above the reservoir so that only a part of reservoir compaction is transmitted to the seabed. In 1988, after 12 years of production from West Ekofisk, compaction and subsidence at the center of the field was approximately 20 ft and 3 ft (6 and 1 m), respectively.

Drilling of new (sidetracked) wells allows comparisons with the original (preproduction) holes. Because of small but significant original thickness variations over short distances, thickness comparison between an old and a new well are not relevant. However, initial porosities, reflecting depositional facies and diagenesis, are not considered to vary over short distances, and comparison between preproduction

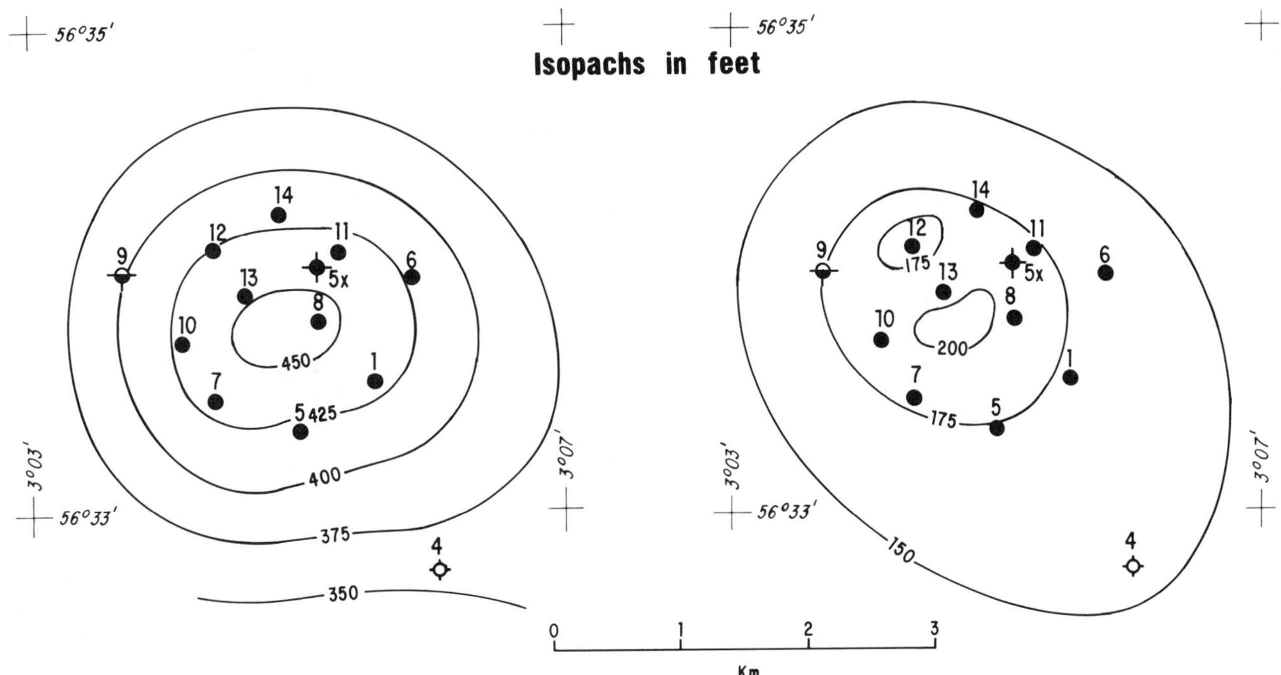

Figure 12. Gross thickness maps of the upper member of the Tor Formation and the Ekofisk Formation (excluding the "Tight Zone").

and present-day porosities of neighboring wells may indicate porosity loss caused by compaction, and by calculation, an absolute value of reservoir compaction. An example in West Ekofisk is given by wells D12 and D12B, drilled in 1977 and 1986, respectively (Figure 19). The compaction of the Ekofisk Formation at that location is evaluated at 12 ft (4 m) after 9 years of production. Compaction and the resulting decrease of porosity, thickness, and structural relief is such that new data cannot directly be used to refine the original mapping. For this reason, all the maps presented in this field description are based on precompaction data only.

Faults

It is quite possible that faults affect the West Ekofisk reservoir; however, none have been identified yet, neither from seismic nor from well data. Growth faults, contemporaneous with chalk deposition during Danian and Maastrichtian times, are less probable, as the variation of reservoir thickness seems to be the result of differential porosity preservation only. An important but much older fault affecting the pre-Cretaceous section may be present to the north of West Ekofisk, trending from Ekofisk to Albuskjell; its consequences on the West Ekofisk reservoir, if any, are not known. For the time being, the reservoir can be modeled without taking any faults into account.

Source

The source for the oil and gas of the Central graben is the shale of the Mandal Formation, of Late Jurassic age. Van den Bark and Thomas (1981) have shown a convincing correlation between a Jurassic-age shale extract and an Ekofisk crude oil, proving the Ekofisk crude oil was sourced by these shales and not by the Paleogene shales, as had been suggested before (Byrd, 1975). Only one well, Ekofisk B19, has been drilled to the Jurassic in the immediate vicinity of West Ekofisk. Pekot and Gersib (1987) report that geochemical analyses from this well indicate that the shales between 13,400 and 14,300 ft (4084 m and 4359 m) are thermally mature and have fair to very good remaining potential to generate oil and gas.

Molecular maturity indicators show a progressive decrease of maturity by field from West Ekofisk southward to Eldfisk. Pekot and Gersib (1987) report that this variation is probably due either to differences in thermal maturity and/or to thermal alteration in the reservoir. They also mention that the West Ekofisk and Ekofisk crudes may have a greater terrestrial input and occurrence of oxidation than crudes from Edda and Eldfisk.

The timing of emplacement of hydrocarbon into the reservoir is of importance for establishing the role of hydrocarbon in preserving the very high porosity of the chalk. Most of the information

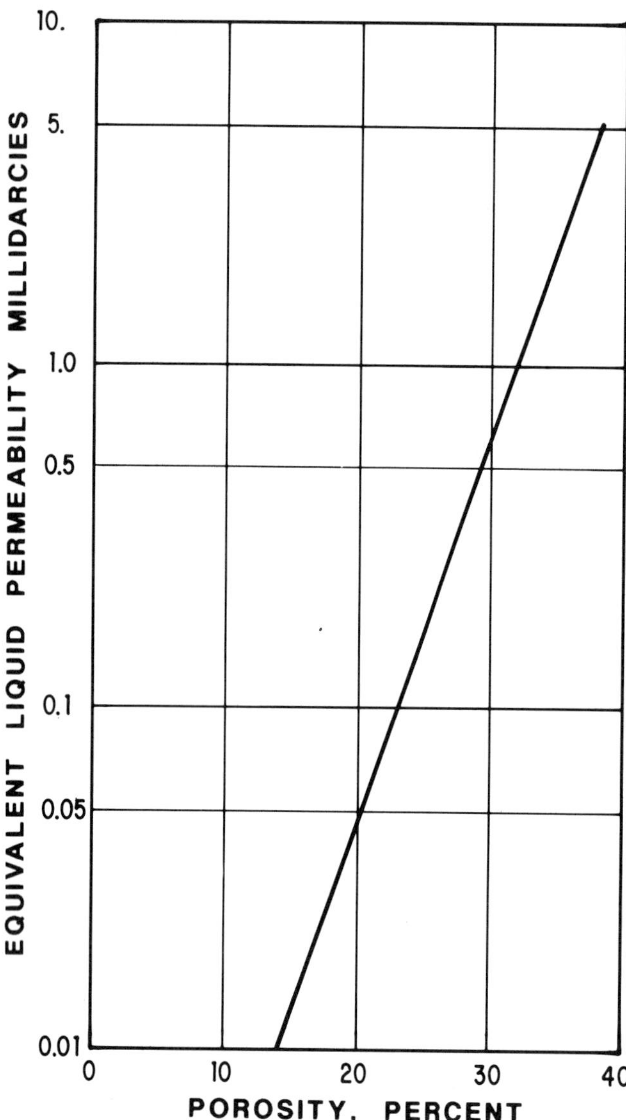

Figure 13. Porosity/permeability relationship in the Ekofisk Formation of the exploration well 5X.

available for the source rock is based on data from wells drilled on structures of high relief. It is therefore obvious that the maturity of the source rock is higher in the synclines and in deeper areas of the trough, west of West Ekofisk. For most if not all the chalk fields, the vertical component of migration has probably been predominant in the areas of intense halokinesis and in those where major faults affect the Cretaceous strata. However, vertical migration from just below the fields cannot explain entry of oil into the reservoir early enough to fully justify the high level of porosity preservation. Fortunately, it has been established that overpressure and spot-welding have delayed pressure solution.

In fact, oil may have been present in the crestal part of the trap earlier than is generally believed; that would imply only a secondary subhorizontal component of migration in the upper part of the chalk layer from the western area where the source rock expelled oil much earlier than the source rock below the fields. Because of the geometry of the Central trough during early Tertiary times, it is very probable that such a horizontal component is responsible for early emplacement of oil in the higher part of the traps. West of West Ekofisk, the Upper Jurassic shales already generated significant amounts of oil during Paleocene and Eocene times, while below the structures, expulsion is thought to have started in Oligocene-Miocene times. As is the case for Albuskjell, which is also a gas-condensate field, West Ekofisk lies just east of the deepest area of the Central graben where the Upper Jurassic sediments are now at depth greater than 17,000 ft.

Between West Ekofisk and Ekofisk, there is a saddle; before the chalk lost its permeability in the saddle area as a result of late diagenesis, oil may have been flushed by gas from West Ekofisk to Ekofisk. Another point is that, in West Ekofisk, the Tertiary seal has been perfectly effective whereas in Ekofisk a significant amount of gas has migrated into the overburden, forming a gas chimney. A full, refined geochemical model is necessary to clarify this matter in the entire Greater Ekofisk area.

EXPLORATION CONCEPTS

Regional Play

Porous chalk containing significant amounts of producible hydrocarbons have been found only in a restricted area of the Central trough. The fields are concentrated in the southern part of the trough where six conditions are fulfilled:

1. Proximity of the margins of the trough or structural highs within the trough; instability of the seafloor of the margins, the surrounding platform, or the inner highs have resulted in resuspension of pelagic chalk and redeposition as favorable reservoir facies.
2. Relatively deep area where the allochthonous chalk may accumulate and build up thick reservoir layers.
3. Absence of good and/or thick reservoir below or above the chalk, so that hydrocarbons would not have migrated preferentially into the other reservoirs instead of into the relatively low-matrix permeability chalk.
4. Tertiary diapirism necessary to sufficiently fracture the chalk and to give the rock a satisfactory productivity.
5. Organic-rich Upper Jurassic strata deeply buried and having begun to expel oil since Eocene-Oligocene time.
6. Thick Tertiary clay providing the vertical seal.

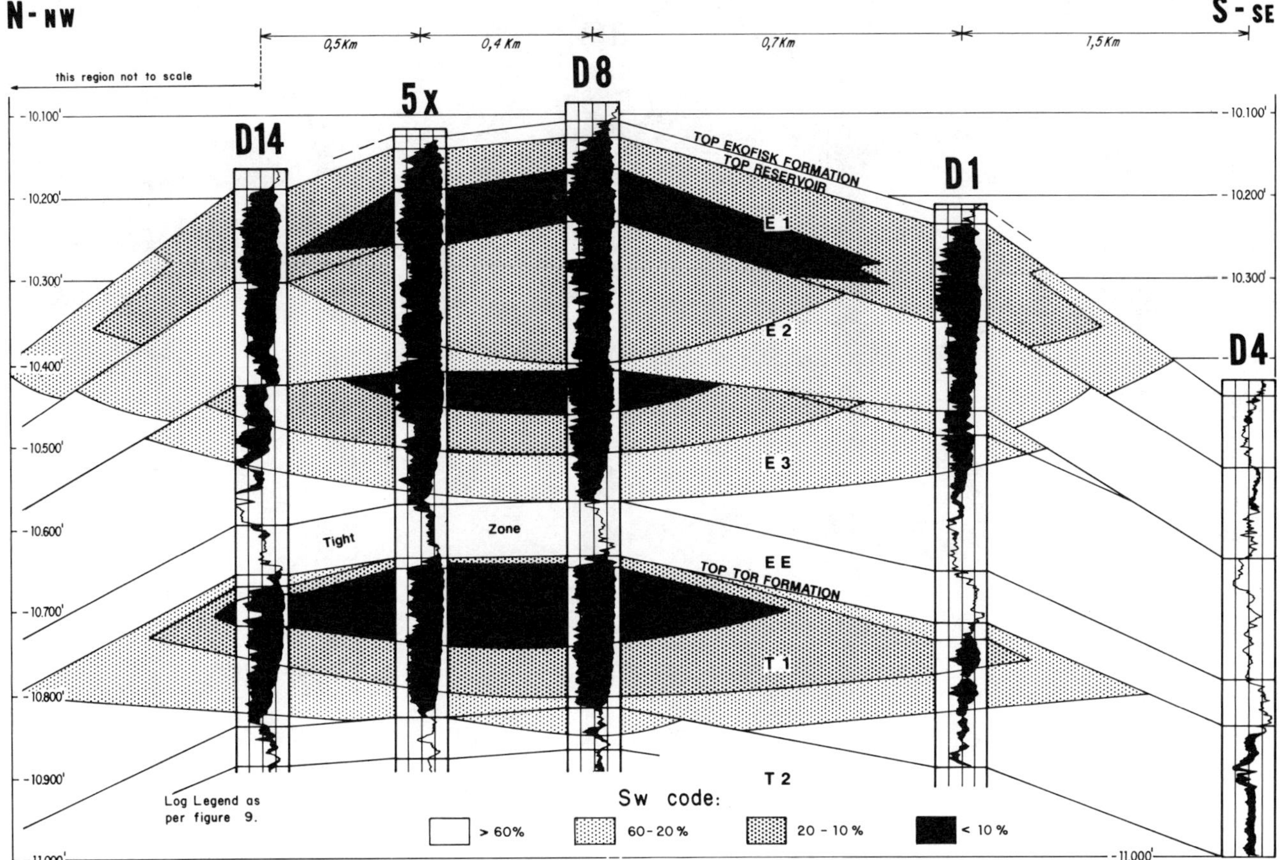

Figure 14. North-northwest to south-southeast section showing the hydrocarbon saturation envelopes for the different reservoir zones (expressed inversely as water saturations). (Well locations shown on Figure 5.)

In any other part of the Central trough, at least one of these conditions is not fulfilled. However, trapping conditions, mainly stratigraphic, may occur along the margins of the trough or close to inner highs, and the chalk may be hydrocarbon-bearing, especially if there are no thick and/or good reservoir rocks in the Jurassic, Lower Cretaceous, or upper Paleocene series.

General Application of Geologic Parameters

The West Ekofisk reservoir is of a layer-cake type. Each individual layer has its own initial petrophysical characteristics, depending on its depositional mode. Chalk reservoirs strongly differ from the other rock types; this fact has needed several years to be fully understood. Porosity decrease with depth for a given facies depends both on mechanical and "chemical" compaction. A peculiarity of the chalk reservoir is a rapid decrease of porosity, gross thickness, net to gross ratio, permeability, and hydrocarbon saturation from the central area toward the periphery.

West Ekofisk became the first chalk field to be fully developed in the North Sea. Because of the numerous wells drilled on this small circular structure of rather simple geological history, West Ekofisk gave the basis for the development of a detailed geological model for chalk reservoirs (D'Heur, 1980).

Lessons

Most of the exploration and appraisal wells drilled on the chalk structures were located in the crestal area. The first two development wells of West Ekofisk, drilled on the flank of the structure, gave negative results but provided data for understanding the lateral variation of the reservoir characteristics. The simple "base case" geological model established for West Ekofisk has since then been helpful in the optimization of development drilling in the other fields of the Central trough, where more than 100 wells still were to be drilled. This model cannot be applied without modification because each field has its specific characteristics resulting from its structural history, its distance to the source of

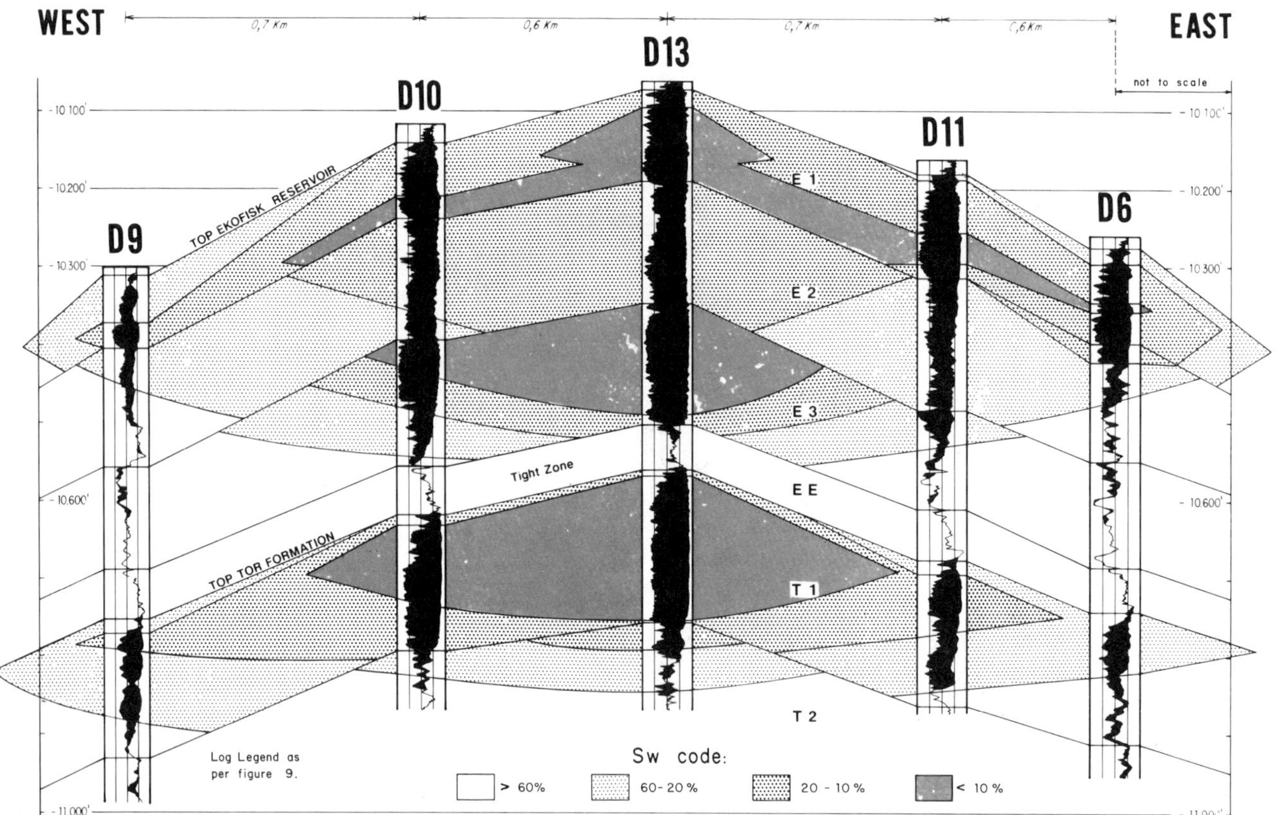

Figure 15. West-east section with hydrocarbon saturation logs and envelopes (expressed as water saturation). (Well locations shown on Figure 5.)

allochthonous chalk, and from geological factors affecting the field before, during, and after deposition of the reservoir layers. However, the general trends and the methodology derived from the model apply in most cases.

In general, the crestal or near-crestal areas contain the best reservoirs, whereas the flank wells are often poorly productive or even dry. Therefore, starting from a provisional development plan, well locations must be continually reevaluated during the development drilling phase, based on trend analysis. Only drilling step by step away from the central area toward the flank will minimize the risk of drilling dry holes. The misunderstanding of the chalk peculiarities also resulted in wrong reserve estimates in the seventies. For example, Edda and West Ekofisk reserves were first overestimated by 100%. New discoveries can now be better evaluated.

ACKNOWLEDGMENTS

The author thanks Petrofina and the other members of the Phillips Norway Group for their permission to publish this paper. He is also grateful to Walter Ziegler (Petrofina) for comments on the manuscript and to Joao de Lucena and Dominique Moens for their help.

REFERENCES

Byrd, W. D., 1975, Geology of the Ekofisk Field, *in* A. W. Woodland, ed., Petroleum and continental shelf of northwest Europe, Vol. 1, Geology: Barking, Applied Science, p. 439-446.

Carmalt, S. W., and B. St. John, 1986, Giant oil and gas fields, *in* Future petroleum provinces in the world: American Association of Petroleum Geologists Memoir 40.

Deegan, C. E., and B. J. Scull, 1977, A standard lithostratigraphic nomenclature for the central and northern North Sea: Rep. Inst. Geol. Sci. UK 77/25.

D'Heur, M., 1978, Etude geologique de West Ekofisk: Petrofina internal report 5441/78 MD/PV.

D'Heur, M., 1980, Chalk reservoir of the West Ekofisk Field: the sedimentation of the North Sea reservoir rocks: (Norsk Petroleumsforening), article X, 20 p.

D'Heur, M., 1984, Porosity and hydrocarbon distribution in the North Sea chalk reservoirs: Marine Petroleum Geology, v. 1, p. 211-238.

Feazel, C. T., and H. E. Farrell, 1988, Chalk from the Ekofisk area, North Sea: nannofossils + micropores = giant fields, *in* A. J. Lomando and P. M. Harris, eds., SEPM Core Workshop Notes, p. 155-178.

Gowers, M. B., and A. Saeboe, 1985, On the structural evolution of the Central Trough: in the Norwegian and Danish sectors

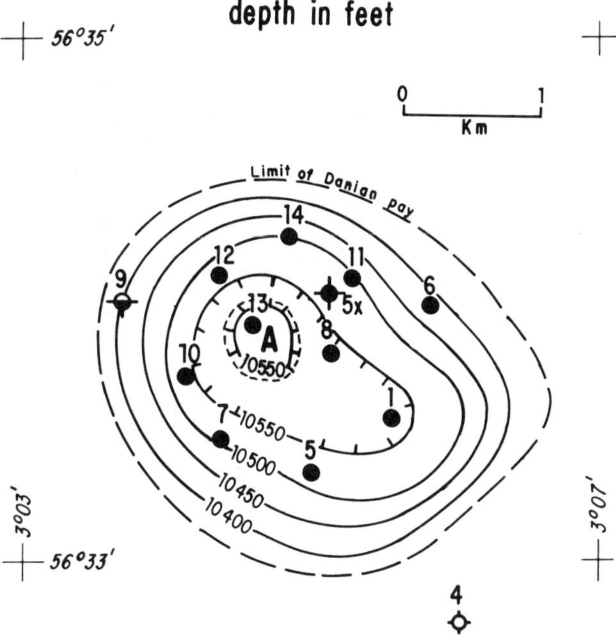

Figure 16. Subsea depth map of the base of the Ekofisk Formation pay zone showing a progressive rise from the central area of the field to the periphery. In the restricted area A in the center, the base of the pay zone corresponds to the base of the reservoir. (Depths in feet subsea with 50 ft contour interval.)

of the North Sea: Marine Petroleum Geology, v. 2, p. 298–318.

Hancock, J. M., 1976, The petrology of the chalk: Proceedings of the Geological Association, v. 86, p. 499–535.

Hardman, R. F. P., 1982, Chalk reservoirs of the North Sea: Bulletin of the Geological Society of Denmark, v. 30, p. 119–137.

Hatton, I. R., 1986, Geometry of allochthonous chalk group members, Central Trough, North Sea: Marine Petroleum Geology, v. 3, p. 79–98.

Hovland, M., and J. H. Sommerville, 1985, Characteristics of two natural gas seepages in the North Sea: Marine Petroleum Geology, v. 2, p. 319–326.

Kennedy, W. J., 1980, Aspects of chalk sedimentation in the southern Norwegian offshore: the sedimentation of the North Sea reservoir rocks: Norsk Petroleumsforening, article IX, 29 p.

Kennedy, W. J., 1987, Sedimentology of Late Cretaceous-Paleocene chalk reservoirs, North Sea Central Graben, *in* J. Brooks and K. W. Glennie, eds., Petroleum geology of northwest Europe: London, Graham and Trotman, p. 469–483.

Mapstone, N. B., 1975, Diagenetic history of a North Sea chalk: Sedimentology, v. 22, p. 601–613.

Pekot, L. J., and G. A. Gersib, 1987, Ekofisk, *in* A. M. Spencer et al., eds., Geology of the Norwegian oil and gas fields: London, Graham and Trotman, p. 73–87.

Scholle, P. A., 1977, Chalk diagenesis and its relation to petroleum exploration: oil from chalks, a modern miracle? American Association of Petroleum Geologists Bulletin, v. 61, p. 982–1009.

Van Den Bark, E., and O. D. Thomas, 1981, Ekofisk: first of the giant oil fields in western Europe: American Association of Petroleum Geologists Bulletin, v. 65, p. 169–175.

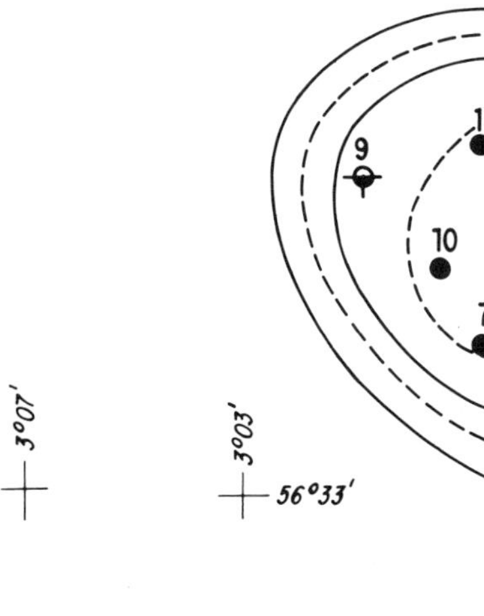

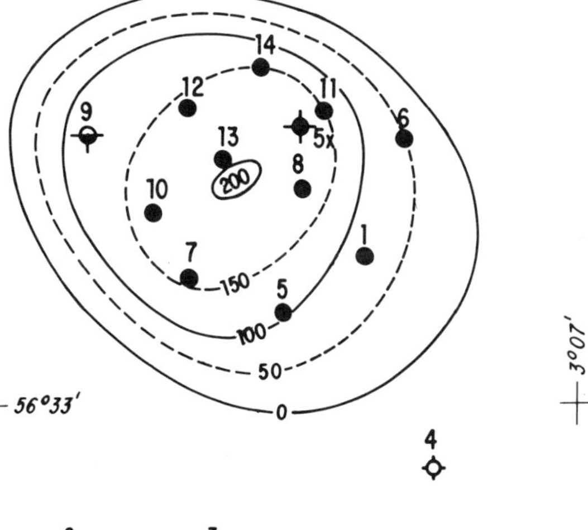

Figure 17. Net pay thickness of the Ekofisk Formation and Tor Formation reservoirs.

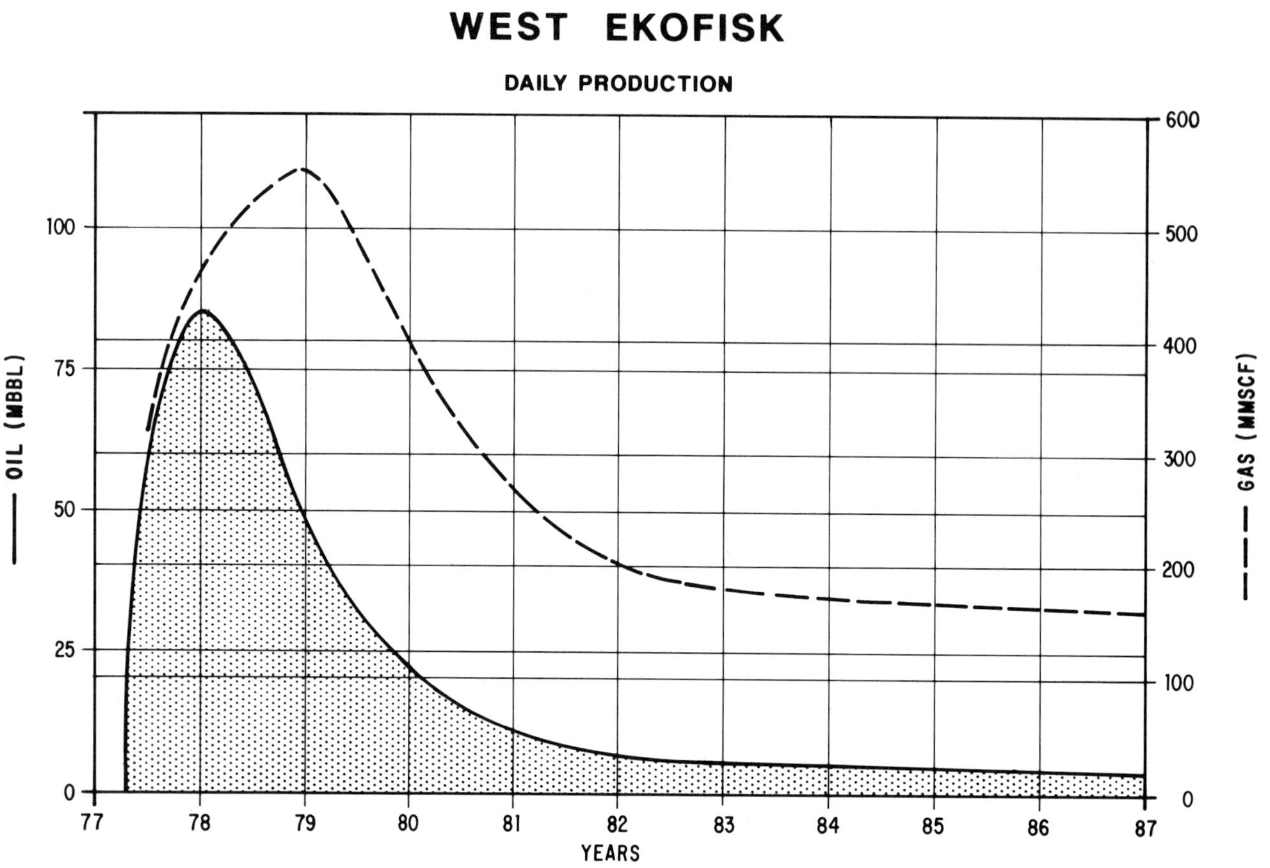

WEST EKOFISK

DAILY PRODUCTION

Figure 18. Production history.

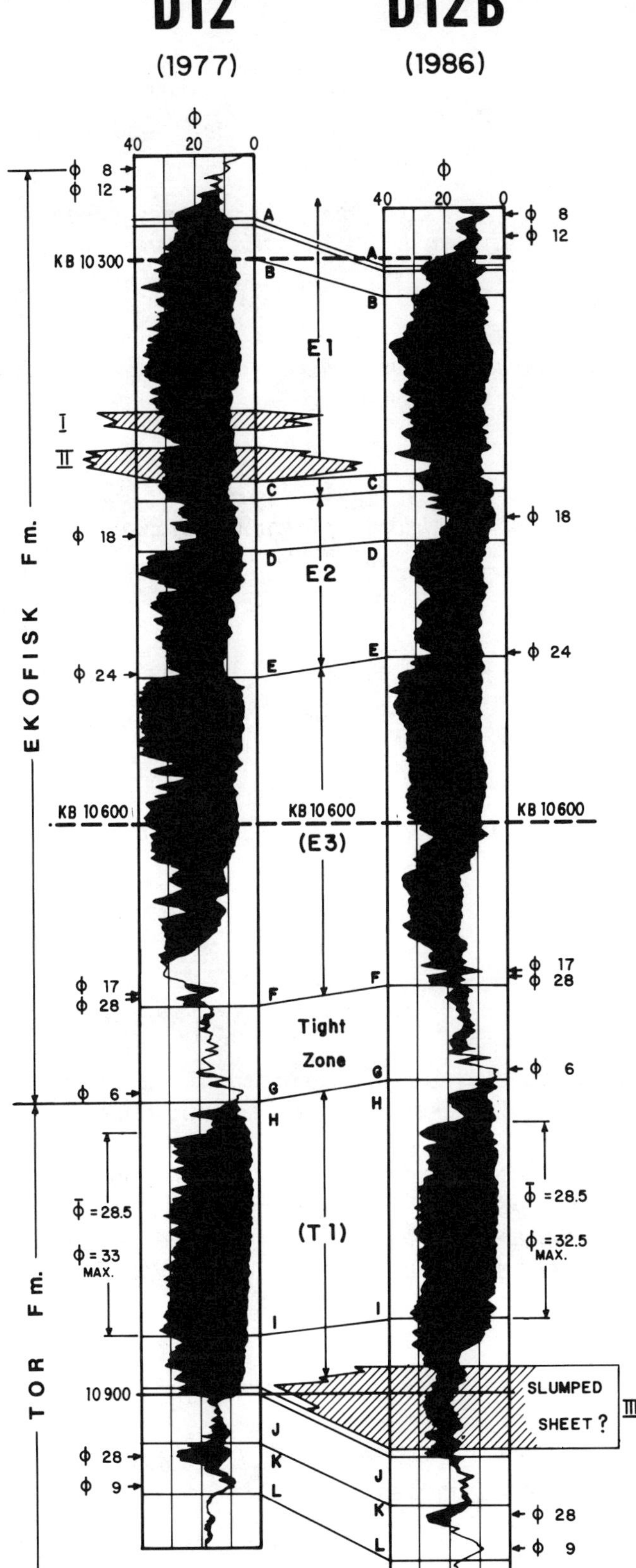

Figure 19. Comparison of porosity logs of wells D12 drilled in 1977 and D12B drilled in 1986 in the immediate vicinity of D12. The two logs are suitable for comparison as the porosity of correlatable beds of poor to fair porosity (= noncompactible) are identical in both wells. Detailed correlations (A to L) help the comparison. Compaction estimate by thickness comparison is not possible because of the presence of additional chalk bodies (-I and II) in the Ekofisk Formation of D12 and (III) in the Tor Formation of D12B. For the Ekofisk Formation, the porosity of the most porous beds of D12 has decreased as a result of reservoir compaction as shown in D12B. The decrease of porosity corresponds to a compaction of 12 ft between 1977 and 1986. There is no evidence of reservoir compaction for the Tor Formation.

WEST EKOFISK

Appendix 1. Field Description

Field name .... *West Ekofisk*

Ultimate recoverable reserves .... *65 million bbl condensate + 1050 bcf gas*

Field location:

Country .... *Norway (offshore)*

State

Basin/Province .... *Central trough*

Field discovery:

Year first pay discovered .... *Paleocene Danian Ekofisk Formation chalk 1970*

Year second pay discovered .... *Upper Cretaceous Maastrichtian Tor Formation chalk 1970*

Third pay

Discovery well name and general location:

First pay .... *2/4-5X, 315 km southwest of Stavanger (Norway), 7 km west of Ekofisk field; lat. 56°34′00″N, long. 3°05′08″E*

Second pay

Third pay

Discovery well operator .... *Phillips Petroleum*

Second pay

Third pay

IP in barrels per day and/or cubic feet or cubic meters per day:

First pay .... *14,000 BOPD (discovery well); 120 × 10⁶ cfd gas*

Second pay

Third pay

All other zones with shows of oil and gas in the field:

Age	Formation	Type of Show
No significant shows below the pay zone		

Geologic concept leading to discovery and method or methods used to delineate prospect, e.g., surface geology, subsurface geology, seeps, magnetic data, gravity data, seismic data, seismic refraction, nontechnical:

Structure identified by seismic in the mid-sixties. Upper Paleocene sandstones found gas-bearing 80 km to the northwest (Cod) was the objective until 1969. Then, the exploration well drilled on the nearby Ekofisk structure showed that these sands were not present in the area but that a thick chalk sequence was oil-bearing. This lower Tertiary-Upper Cretaceous chalk became the only objective in the West Ekofisk structure.

Structure:

Province/basin type ...... *Cratonic basin located over earlier rift graben system (Bally 1211, Klemme IIB)*

Tectonic history

Central trough formed in mid-Cretaceous time over an early Mesozoic rift system. Reactivation of old faults and halokinesis of Permian salt within and around the trough was responsible for the favorable (reservoir) chalk facies deposited in the West Ekofisk area.

Regional structure

West Ekofisk is located in the area of depot-center of the favorable allochthonous chalk facies that comprise the best reservoirs. Seafloor at the time of deposition was rather flat in this area.

Local structure

West Ekofisk is a small circular dome formed by halokinesis of the Permian salt during the Tertiary, after deposition of the reservoir layers.

Trap:

Trap type(s)

Structural trap. A tight layer, 60 ft thick, separates the upper and lower reservoirs. Lateral porosity reduction in the flank area and beyond the pay zone results in the field being sealed in all directions.

Basin stratigraphy (major stratigraphic intervals from surface to deepest penetration in field):

Chronostratigraphy	Formation	Depth to Top in ft
M. Miocene-Recent	*Nordland Group*	*Seafloor*
Eocene-early Miocene	*Hordaland Group*	*5600*
Early Paleocene	*Ekofisk Formation*	*10,100*
Late Cretaceous	*Tor Formation*	*10,600*
Late Cretaceous	*Hod Formation*	*11,850*
Late Cretaceous	*Hidra Formation*	*13,420*

Reservoir characteristics:

Number of reservoirs *2*
Formations .... *Ekofisk Formation; Tor Formation (upper member)*
Ages .... *Danian (early Paleocene) and Maastrichtian (Late Cretaceous)*
Depths to tops of reservoirs .... *10,100 and 10,600 ft (in the central area)*
Gross thickness (top to bottom of producing interval) .... *700 ft (in the central area)*
Net thickness—total thickness of producing zones
Average .... *360 ft*
Maximum .... *640 ft (in the central area)*
Average
Maximum
Lithology .... *Resedimented (allochthonous) chalk mainly of debris flow type*
Porosity type .... *Mainly interparticles*
Average porosity *30%*
Average permeability .... *1-2 md (matrix)*

Seals:

Upper
Formation, fault, or other feature .... *Upper Paleocene formation*
Lithology .... *Claystone/shale*
Lateral
Formation, fault, or other feature .... *Cementation*
Lithology .... *Chalk*

Source:

Formation and age .... *Mandal Formation (Late Jurassic)*
Lithology .... *Anaerobic shale*
Average total organic carbon (TOC) *2.5%*
Maximum TOC *20%*
Kerogen type (I, II, or III) .... *II*
Vitrinite reflectance (maturation) .... $R_o = 0.7\text{–}0.9$
Time of hydrocarbon expulsion .... *Eocene*
Present depth to top of source .... *14,000 ft*
Thickness .... *100 ft (excellent) + 800 ft (fair to good)*
Potential yield .... *100-300 BAF*

WEST EKOFISK

Appendix 2. Production Data

Field name .... *West Ekofisk*

Field size:

- **Proved acres** .... *3000 (1215 ha)*
- **Number of wells all years** *22*
- **Current number of wells** *12*
- **Well spacing** .... *1200 ft*
- **Ultimate recoverable** .... *65 million bbl + 1050 bcf*
- ***Cumulative production (1/1987)*** .... *55 million bbl + 825 bcf*
- **Annual production (1986/1987)** .... *1.4 million bbl + 53 bcf*
- **Present decline rate** *14%*
  - **Initial decline rate** *36%*
  - **Overall decline rate** *96%*
- **Annual water production** .... *None*
- **In place, total reserves** .... *950 million bbl (reservoir conditions)*
- **In place, per acre-foot** *122,450*
- **Primary recovery** .... *65 million bbl + 1050 bcf*
- **Secondary recovery**
- **Enhanced recovery**
- **Cumulative water production**

WEST EKOFISK

Drilling and casing practices:

- **Amount of surface casing set** .... *30-in. csg shoe at 500 ft*
- **Casing program** .... *(Vertical depth to shoe, from KB) 20-in. csg at 1600 ft, 13⅜-in. csg at 5000 ft, 9⅝-in. csg at 10,100 ft, 7-in. csg at 11,000 ft*
- **Drilling mud** .... *Overburden: seawater + attapugite; reservoir: paraffin based*
- **Bit program** .... *Overburden: long-milled tooth bit; reservoir: navidrill + stratop*
- **High pressure zones** .... *4800–9200 ft*

Completion practices:

- **Interval(s) perforated** .... *100–250 ft in Ekofisk and Tor formations*
- **Well treatment** .... *Acid stimulation*

Formation evaluation:

- **Logging suites** .... *Overburden: DIL/LSS/GR/CAL; reservoir: DLL/MSFL/GR, LDT/CNL/GR*
- **Testing practices** .... *Selective production test on Tor Fm.; build-up; production test on Tor and Ekofisk; build-up*
- **Mud logging techniques** .... *On critical wells only*

Oil characteristics:

- **Type** .... *Corrosive condensate associated with gas*
- **API gravity** *48*
- **Base** .... *Paraffinic*
- **Initial GOR** .... *4100 SCF/bbl (730 m³/m³)*
- **Sulfur, wt%** *0.09*
- **Viscosity, SUS** .... *36.6 at 100°F*
- **Pour point** .... *13°F*
- **Gas-oil distillate** .... *23.2 vol% (23.8 wt%)*

Field characteristics:

- **Average elevation** .... *10,400 ft ss (3170 m)*
- **Initial pressure** .... *7200 psi (496 bars)*

Present pressure .... *2000 psi*
Pressure gradient .... *0.09 psi/ft*
Temperature .... *266°F (130°C)*
Geothermal gradient .... *0.026°F/ft*
Drive .... *Gas*
Oil column thickness .... *900 ft (from top of pay in crestal area to base of pay in flank)*
Oil-water contact .... *(Lowest) 10,925 ft ss*
Connate water .... *10–20% (central area)*
Water salinity, TDS .... *60,000 mg/L*
Resistivity of water .... *0.45 ohm-m at 118°C*
Bulk volume water (%) *22%*

Transportation method and market for oil and gas:
Oil and gas transported to the Ekofisk Center by a 24-in. line 8 km long. From there, gas is transported to Emden (Germany) and oil to Teeside (England) by pipelines 440 km and 350 km long (respectively).

Tommeliten Gamma Field—Norway
Central Graben, North Sea

ERIK B. NIELSEN
Statoil
Stavanger, Norway

GUY FRASELLE
Fina Exploration Norway A/S
Stavanger, Norway

LAURA VIANELLO
Norsk Agip A/S
Forus, Norway

FIELD CLASSIFICATION

BASIN: North Sea
BASIN TYPE: Rift
RESERVOIR ROCK TYPE: Limestone
RESERVOIR ENVIRONMENT OF DEPOSITION: Resedimented Chalk
RESERVOIR AGE: Paleocene and Cretaceous
PETROLEUM TYPE: Oil and Gas
TRAP TYPE: Dome Overlying Salt Diapir

LOCATION

The Tommeliten Gamma field lies in Block 1/9 in the southernmost part of the Norwegian North Sea, 300 km (~190 mi) from the southwest coast of Norway. The Tommeliten Gamma field is part of a chalk fields complex, the Greater Ekofisk Complex (Albuskjell, Tor, S.E. Tor, W. Ekofisk, Ekofisk, Edda, Tommeliten Alfa, Eldfisk, Valhall, and Hod) situated within the Central Trough structural province of the North Sea (Figure 1). Water depth over the field is 75 m (245 ft).

The Tommeliten Gamma field is a moderate-sized hydrocarbon accumulation with an estimated ultimate recovery of 11 bScm (390 bScf) gas and 5 MMScm (32 MMbbl) oil.

HISTORY

Pre-Discovery

The first North Sea well to establish the presence of oil in chalk was the A-1 well drilled in 1966 by Dansk Undergrunds Consortium (DUC) in the Danish chalk field complex (Damtoft et al., 1987) (Figure 1). It encountered oil and gas shows in Paleocene chalk of the Ekofisk Formation, but the discovery, named Kraka, was not declared commercial until 1985.

Ekofisk, the largest chalk field in the North Sea and the first discovery in the Norwegian sector, was discovered in 1969 by the Phillips Group. By 1975, seven fields had been discovered in the area, in Upper Cretaceous and Paleocene chalk reservoirs of the Hod, Tor, and Ekofisk formations, located within structures created by salt movements or uplift related to inversion (see *Tectonic History*).

The first detailed exploration of Block 1/9 started in 1974 with the shooting of a seismic survey, consisting of a 1 × 1 km grid oriented northwest-southeast. This survey revealed the presence of four structures (Alfa, Beta, Delta, and Gamma) at the top of the Chalk Group (Figure 2).

Seismic anomalies caused by gas (extremely low velocity) were seen in the crestal areas and in the Tertiary section above the Alfa, Delta, and Gamma structures (Figures 3 and 4). Salt diapirs were interpreted below these three structures. Similar seismic anomalies were present over five of the already discovered fields (Albuskjell, Ekofisk, Eldfisk, Valhall, and Hod), but not over the Tor or Edda fields.

Block 1/9 was awarded in 1976 as Production Licence 044 to a group comprising Den norske stats

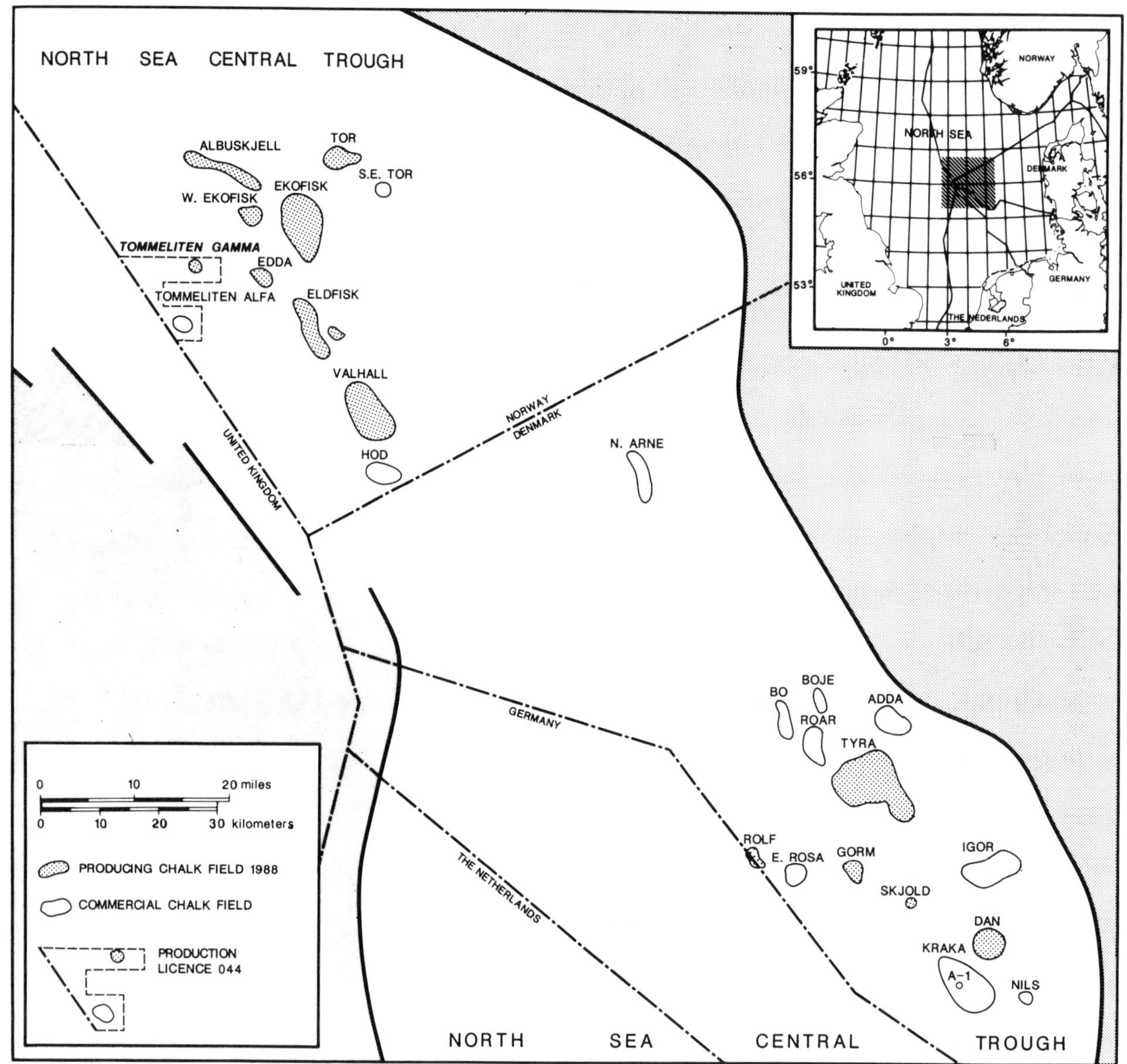

Figure 1. Location map of the northwest European structural province of the North Sea Central trough, highlighting producing and commercial chalk fields in the North Sea. Compiled from Damtoft et al. (1987) and D'Heur (1987a).

oljeselskap a.s—Statoil (50%) as Operator, Phillips Petroleum Norsk A/S (25.87%), Norske Fina A/S (15%), and Norsk Agip A/S (9.13%). A part of the block was relinquished in 1982, with Production Licence 044 continuing for the remaining area (Figure 2). The 1/9-1 well drilled in 1976 discovered the Tommeliten Alfa field, which was gas condensate-bearing in the Tor and Ekofisk formations.

Before the Gamma structure was drilled, detailed studies of unmigrated and migrated seismic lines and gravity data were carried out, and corrections for the effect of gas on velocity were evaluated, using nearby wells for velocity information. Special attention was paid to the top of the salt, the existence of a cap rock, and the thickness of the salt.

Discovery

The 1/9-4 wildcat, drilled in 1977, was planned as a high flank well (Figure 5) and had the Tor and Ekofisk formations of Maastrichtian and Danian age, respectively (Figure 6), as the primary objectives. The well reached the top of the Ekofisk Formation at 3090 mss (10,138 ft), 110 m (360 ft) below prognosis, and encountered gas condensate in the 95 m (313 ft) thick Ekofisk Formation and in the 100 m (328 ft) thick

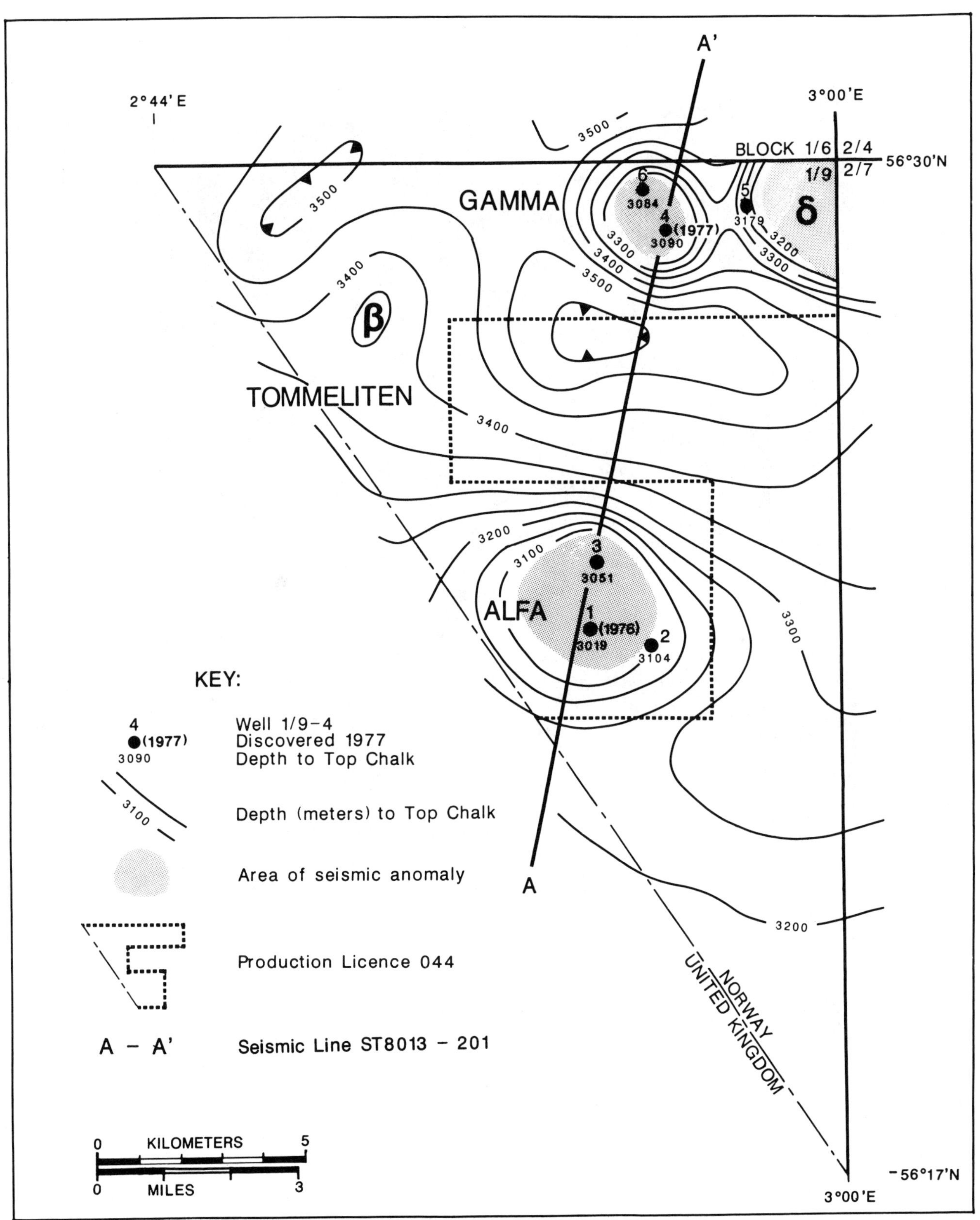

Figure 2. Structure map of Top Chalk Group in Block 1/9 (1983 interpretation). Note areas of seismic anomaly caused by gas effect. A-A' seismic line in Figures 3 and 4.

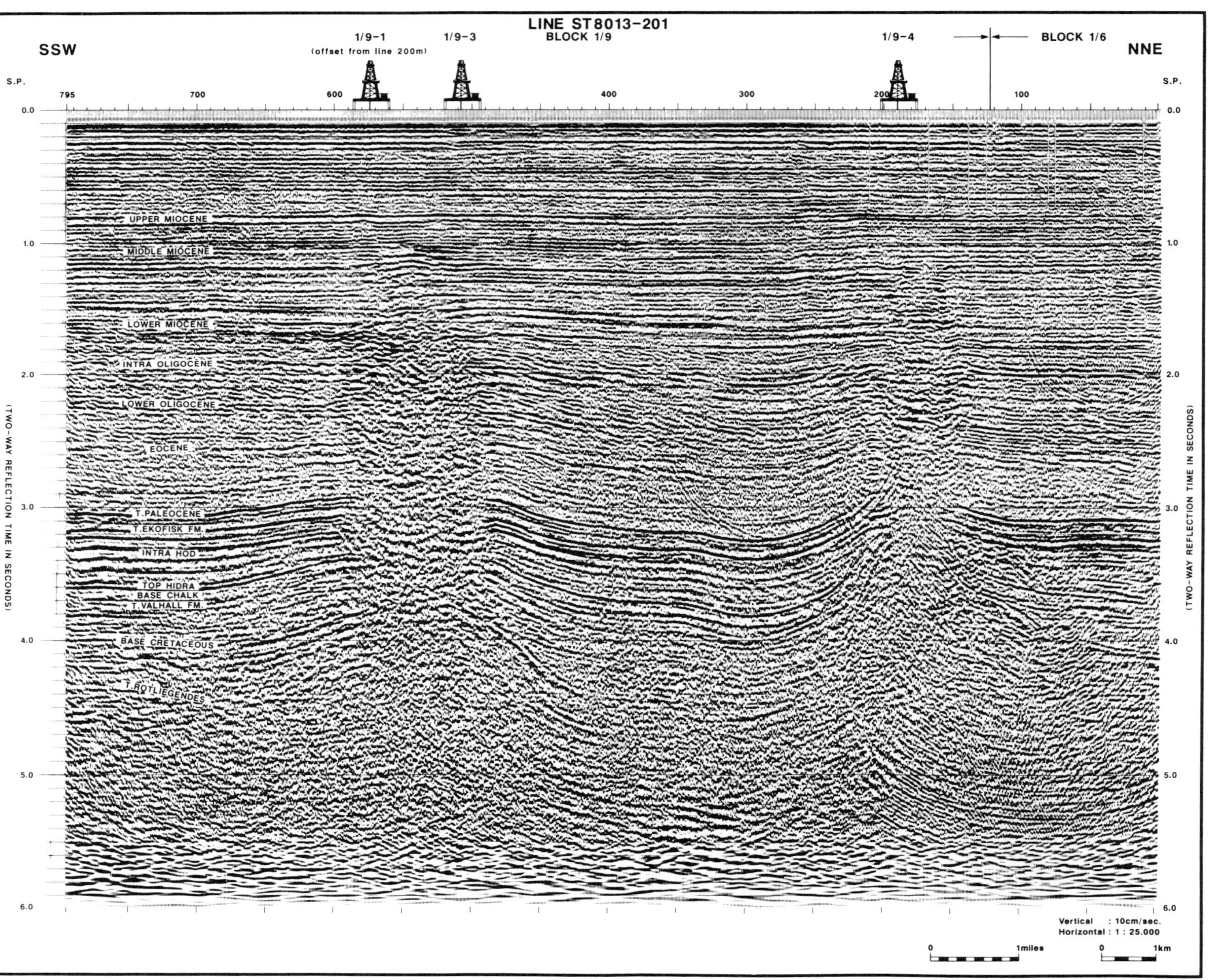

Figure 3. Seismic cross section through Alfa and Gamma structures in Block 1/9 (location on Figure 2).

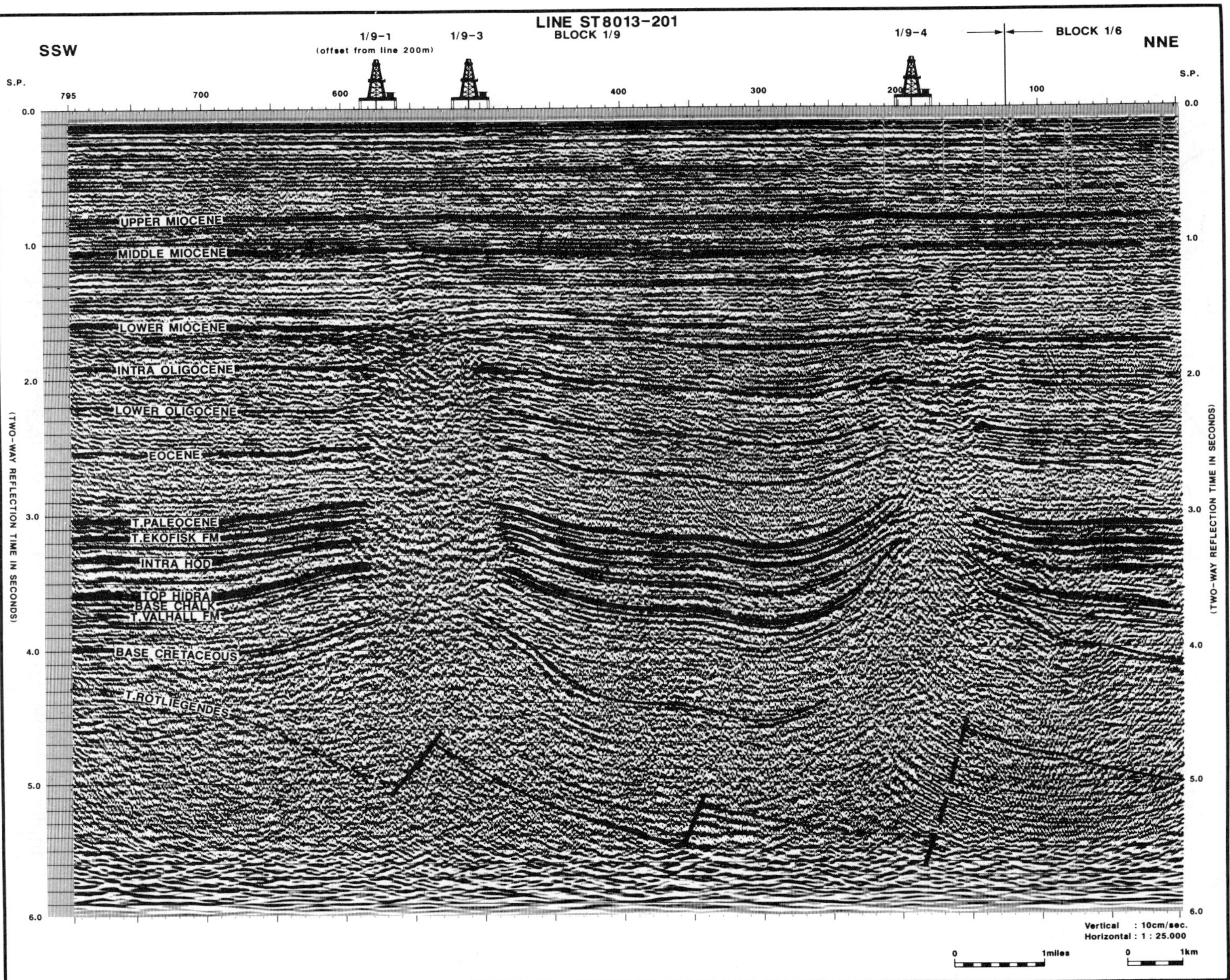

Figure 4. Seismic cross section through Alfa and Gamma structures in Block 1/9 (location in Figure 2).

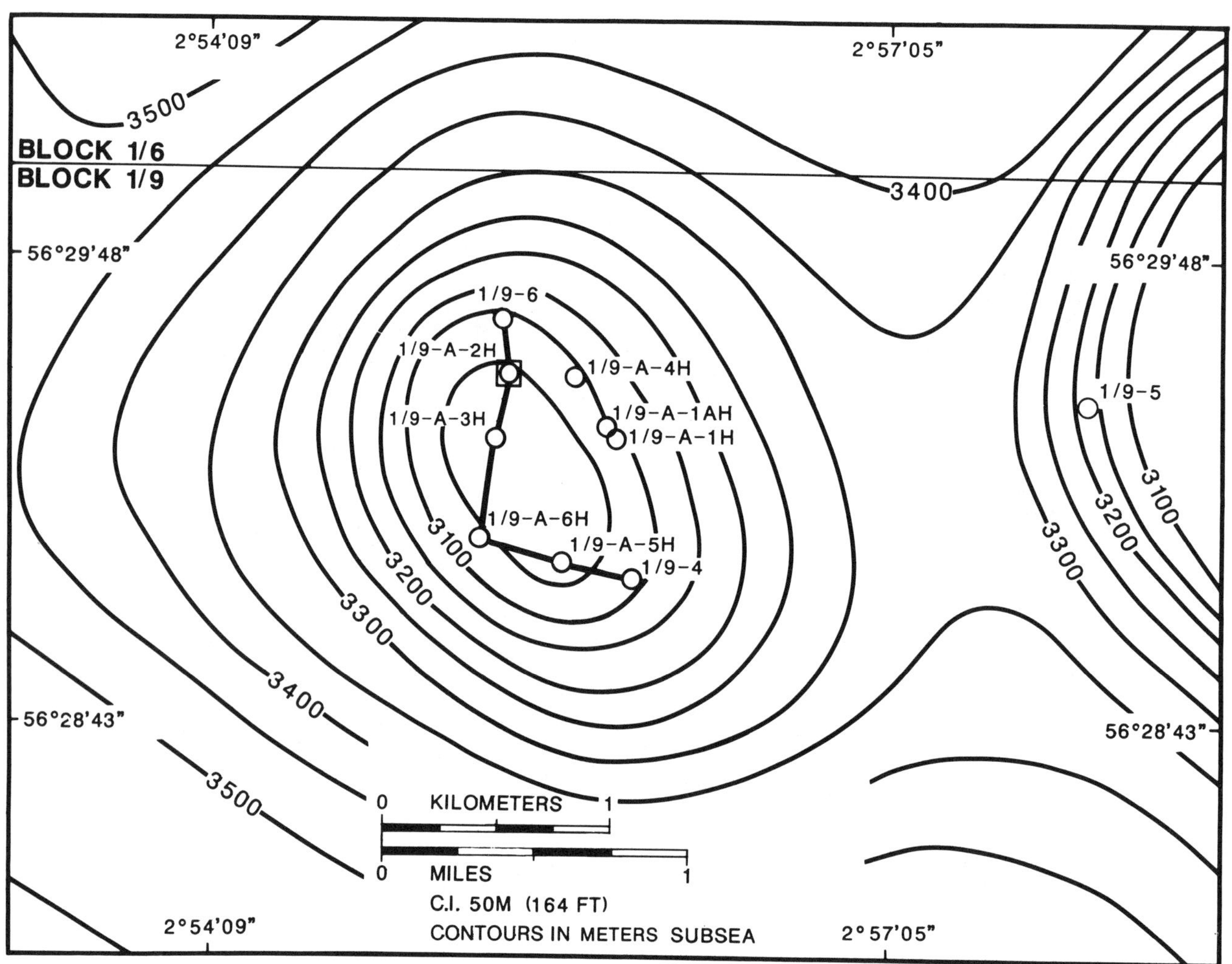

Figure 5. Structure map (depth) of the Top Ekofisk Formation (Top Chalk Group), Tommeliten Gamma field (1988 interpretation). Open box, location of the subsea A-template. The line joining the wells shows the location of the structural cross section in Figures 7 and 8.

Tor Formation (see structure cross sections in Figures 7 and 8).

Below the hydrocarbon-bearing section, the well penetrated tight chalk in the Hod and Hidra formations (Figure 6). The base of the Upper Cretaceous was at 3621 mss (11,880 ft), base Lower Cretaceous at 3675 mss (12,057 ft), and total depth at 3685 mss (12,090 ft) in the Upper Permian Zechstein salt. Jurassic and Triassic deposits were not present.

Four drill-stem tests were performed. The best Ekofisk Formation test (DST No. 4 in Figures 7 and 8) flowed 500 MScm/d (17.7 MMScf/d) of gas and 215 Scm/d (1350 bbl/d) of condensate from a 17 m (56 ft) perforated interval, while the best Tor Formation test (DST No. 2 in Figure 8) flowed 700 MScm/d (24.7 MMScf/d) of gas and 655 Scm/d (4120 bbl/d) of condensate from a 20 m (65 ft) perforated interval. The well was suspended in January 1978. The Tommeliten Gamma field was in 1977 the fifteenth North Sea chalk field to be discovered.

Post-Discovery

In 1978, appraisal well 1/9-5 was drilled on the west flank of the Delta structure (Figure 2). Although good porosities were preserved in the Ekofisk Formation, the Tor Formation was tighter and the entire section was water-bearing.

Before drilling the 1/9-6 appraisal well on the Tommeliten Gamma structure, a radial pattern seismic survey was shot in 1981 to improve the structural interpretation, and to aid in mapping the extent of hydrocarbons and the porosity changes on the flanks.

The 1/9-6 well, drilled in 1982 as a deviated well from a location near the 1/9-4 discovery well, was planned to pass through the gas-affected area above the reservoir and to test the northwest flank of the accumulation.

The Top Ekofisk Formation was reached at 3084 mss (10,118 ft) and the well then penetrated a 71 m (233 ft) thick (true vertical thickness,TVT)

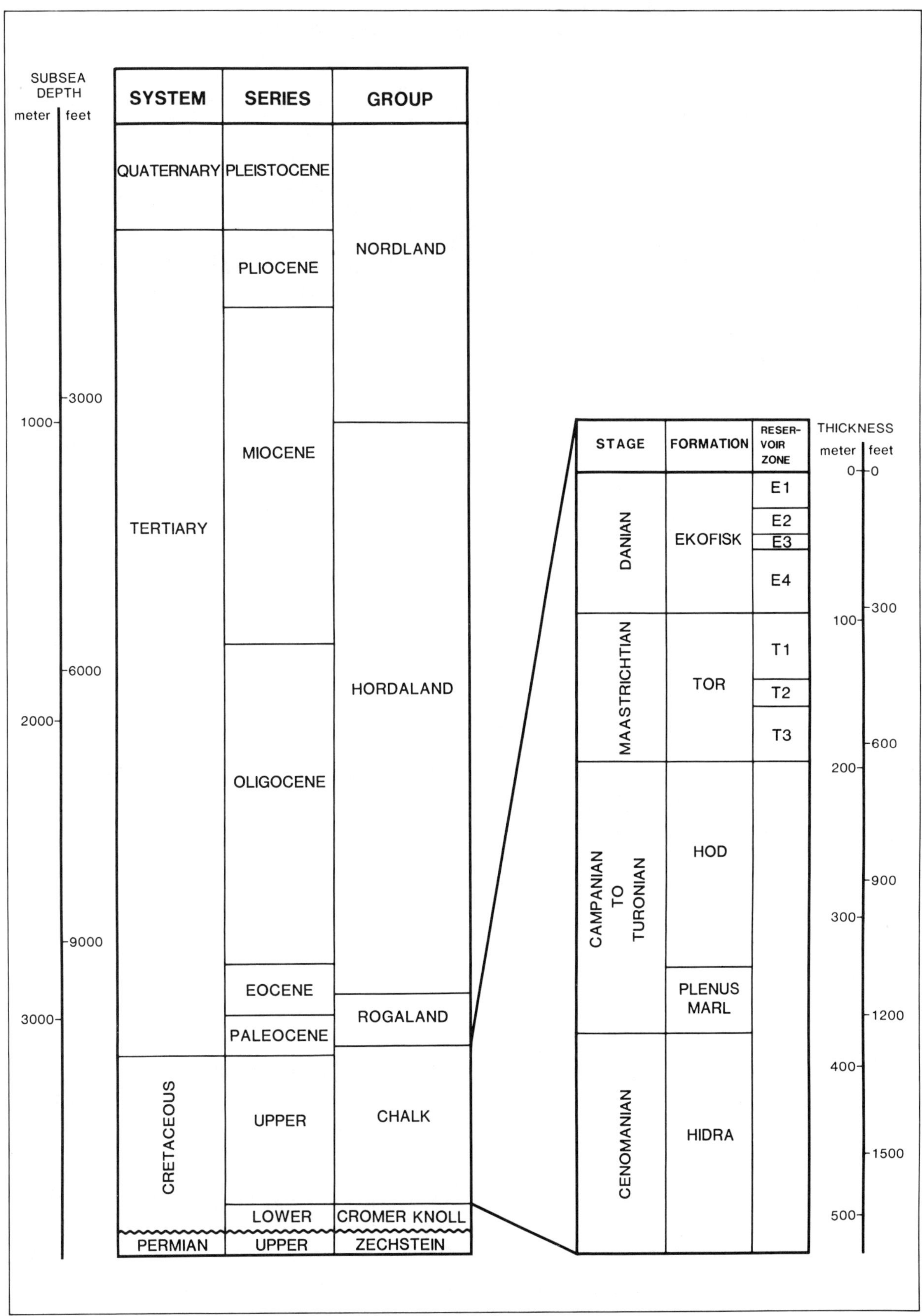

Figure 6. Stratigraphic column of discovery well 1/9-4.

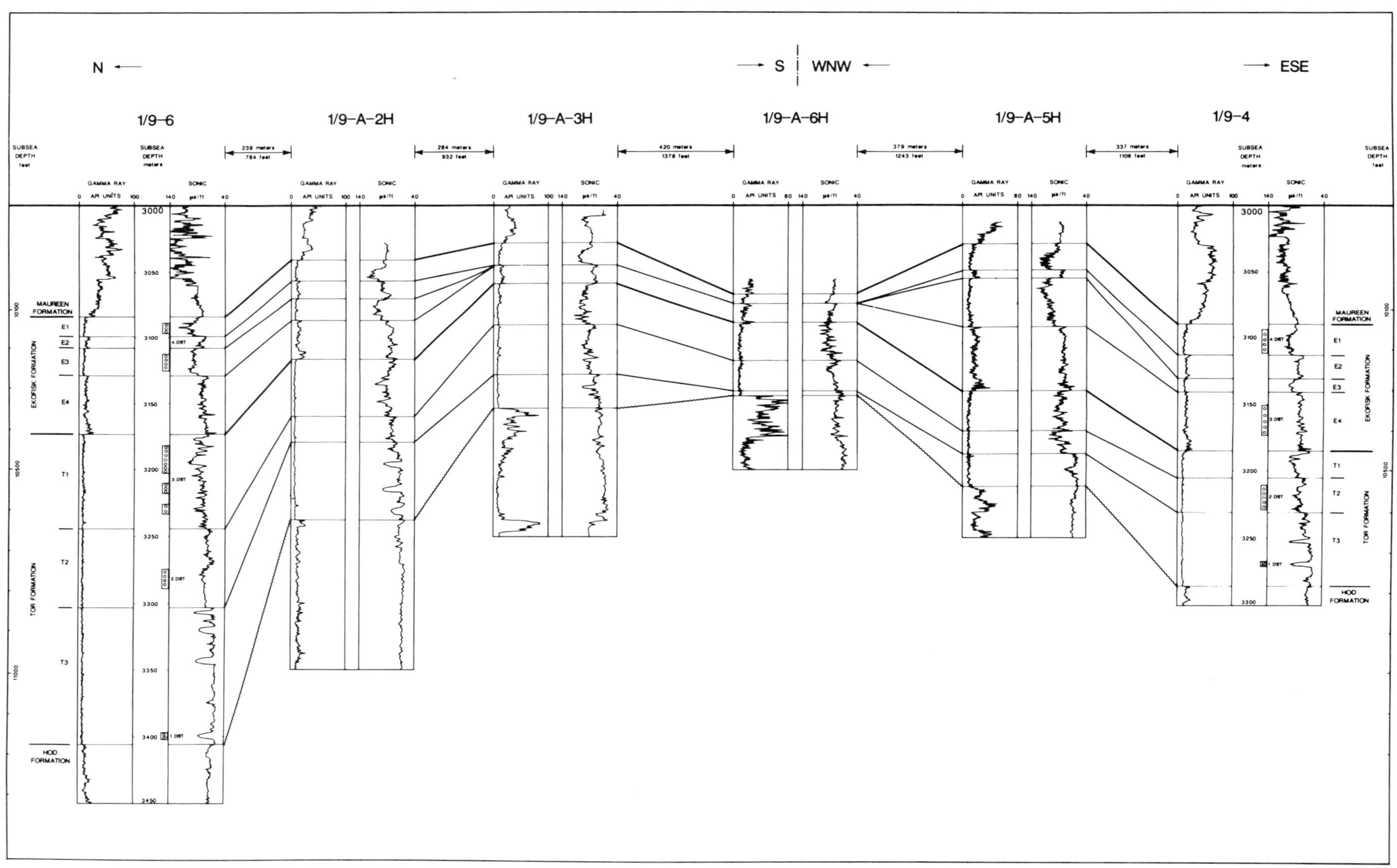

Figure 7. Structural cross section showing log correlation, Tommeliten Gamma field. Location of the cross section is shown in Figure 5. The correlations of the Ekofisk and Tor formations and of the reservoir zones are illustrated by the Gamma Ray and Sonic logs. The drill-stem test intervals (DST No. 1-4) in well 1/9-4 and 1/9-6 are indicated.

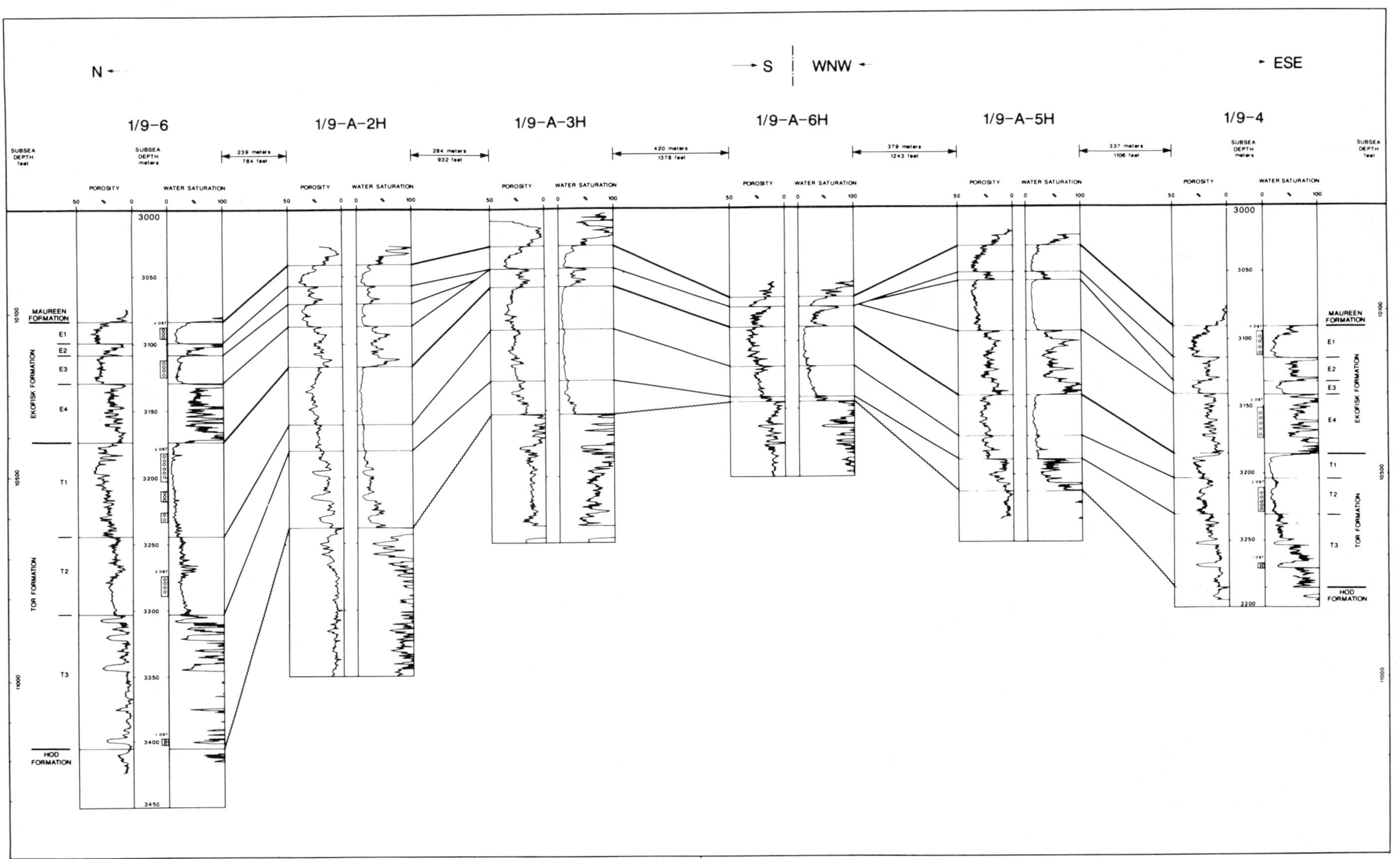

Figure 8. Structural cross section showing porosity and water saturation, Tommeliten Gamma field. Location of the cross section is shown in Figure 5. The distribution of porosity and water saturation in the Ekofisk and Tor formations are illustrated by the log derived porosity and water saturation logs. The drill-stem test intervals (DST No. 1-4) in well 1/9-4 and 1/9-6 are indicated.

hydrocarbon-bearing Ekofisk Formation, and a 186 m (610 ft) (TVT) hydrocarbon-bearing Tor Formation. Total depth was reached at 3502 mss (11,490 ft) in the Hod Formation. The top of the Ekofisk Formation was 106 m (348 ft) above prognosis and had a structural dip to the north. This clearly showed a greater than expected uncertainty of the structural interpretation within the gas-affected area. A vertical seismic profile (VSP) survey was shot, but it was unsuccessful. Four drill-stem tests were performed. The best Ekofisk Formation test (DST No. 4 in Figure 8) flowed 850 MScm/d (30 MMScf/d) of gas and 605 Scm/d (3800 bbl/d) of condensate from a 24 m (80 ft) perforated interval, and the best Tor Formation test (DST No. 3 in Figure 8) flowed 850 MScm/d (30 MMScf/d) of gas and 700 Scm/d (4400 bbl/d) of condensate from a 46 m (150 ft) perforated interval. The well was suspended in December 1982.

The Tommeliten Gamma field was declared commercial in 1986, and development drilling started the same year. The participation interests in the development of the Tommeliten Gamma field with Statoil as Operator are Statoil (70.64%), Norske Fina A/S (20.23%), and Norsk Agip A/S (9.13%).

Detailed sedimentological and fracture studies during 1986 revealed that salt movements had probably been active during the deposition of the Tor and Ekofisk formations. Based on this, a study of the growth history of the Tommeliten Gamma structure was initiated.

It was also decided to obtain detailed information on structural dip, fractures, stratigraphy, and facies, as well as perform reservoir modeling and simulations during the development drilling.

No drill-stem tests were run during development drilling, but the log analyses show that both the Ekofisk and the Tor formations are hydrocarbon-bearing and have low water saturation (Figure 8). By April 1988, six production wells had been drilled from a subsea template at the well 1/9-A-2H location (Figure 5), and completion and testing of all the wells was underway.

The field is early in the line of a new generation of chalk fields being developed and will in October 1988 be the thirteenth such North Sea field in production.

The chalk fields in the North Sea (Figure 1) produce from intervals in the Hod Formation, Tor Formation, and Ekofisk Formation. The reservoirs are at depths from 2000 mss (~6600 ft) in the Danish sector down to 3500 mss (~11,500 ft) in the Norwegian sector. The fields contain recoverable reserves of about 321 MMScm (2020 MMStb) of oil and 270 bScm (9.5 TScf) of gas (D'Heur, 1986).

DISCOVERY METHOD

Then

In the mid-1970s the chalk exploration had reached a mature stage with more than ten discoveries in the North Sea. As a result it was the chalk of the Gamma structure that was the primary exploration objective. Good porosity was expected in the upper part of the chalk sequence; excellent source potential had been confirmed in the area and leaching of gas was seen above the structure. Furthermore, a central collapse area above the diapir was not anticipated.

In the mid-1980s the development of chalk fields had become more complicated than first thought. Although there were some successes, there was also a series of disappointments caused by major variations in thickness, porosity, hydrocarbon saturation, fracturing, and formation strength within the productive zones.

To reduce the chance of upset with the economically sensitive Tommeliten Gamma field, it was necessary to develop a geological model with far more detail than normal at so early a stage in a chalk field development.

It was decided to prolong this evaluation into the first development phase by:

- Reevaluating the biostratigraphy, facies, and dipmeter logs in the two older wells (1/9-4 and 6).
- Providing a detailed understanding through time of the growth of the structure.
- Collecting relevant, consistent data in the first four development wells to identify anticipated, major variations in the reservoir.
- Performing modeling of reservoir geometry and quality during the drilling of these wells.

Now

The successful discovery of the Tommeliten Gamma field benefited from the experience gained in exploration on other chalk fields. Now, the knowledge acquired from the Tommeliten Gamma field geological development studies is such that it can make a significant contribution to the development of the other chalk fields.

The first phase of the geological modeling and reservoir simulation was completed during a break after the drilling of the first four development wells in 1987. The estimate of the net hydrocarbon pore volume was lower than previous estimates and more than two-thirds of the volume was found to be in the northern part of the field. These results influenced the second development phase. This was originally planned to include four additional development wells drilled a few years later from a new subsea template on the southern part of the field.

Instead, it was decided to complete the development of the field by drilling only two more wells in the southern part of the field. These were drilled from the already installed subsea template to drain the reservoir where the reservoir thickness and quality were now anticipated to be marginal.

The final geological reservoir modeling, presented in this paper, has revealed important conclusions concerning the reservoir geometry and quality. These are important for the optimalization of development drilling and production in new chalk fields and the chalk fields still being developed. In particular, far greater attention has to be paid to reservoir geological modeling in the early appraisal and development phase.

STRUCTURE

Tectonic History

The Tommeliten Gamma field is located on the western side of the North Sea Central trough, referred to as the Central graben. Gowers and Sæbøe in 1985 have published a regional structural interpretation of the area.

The graben system was probably initiated in Late Permian time and established the main structural framework for the area (Figure 9).

Zechstein salt was thick and provided the source for later diapirism in areas north of the Mid-North Sea high/Ringkøbing-Fyn high arch and in the Søgne basin and Tail End graben. Over the Mid-North Sea high/Ringkøbing-Fyn high arch, Zechstein evaporites were relatively thin, and here there is no evidence for later halokinesis. On the structural highs (Vigeland ridge, Grensen nose, and Dogger high), the Zechstein either consisted of marginal evaporitic facies with thin carbonate-dominated sequences or was occasionally absent.

The Triassic and Early Jurassic structural pattern was similar to that of Late Permian time with continued greater subsidence of the basinal areas compared to the established highs. Halokinetic movements are partially responsible for local structuring.

According to Gowers and Sæbøe (1985), the major tectonic movement during the early Middle Jurassic was the middle Kimmerian uplift that corresponds to an important east-west crestal doming centered on the middle North Sea high.

During the Late Jurassic, there was a major rifting phase and the sea transgressed over most of the area. The existing tectonic pattern was complicated by a change of the fault pattern as a west-northwest trend appeared and local subgrabens, transfer zones, and block tilting developed. The upper Jurassic sediments locally exceeded 3000 m (~9900 ft) in structural lows.

Gowers and Sæbøe (1985) recognized considerable tectonic activity during the Early Cretaceous, with widespread faulting and halokinesis resulting in significant local thickness variations.

During the Late Cretaceous and early Paleocene, sediment thickness variations were caused by major subsidence patterns, structural inversion, salt diapir growth, differential compaction, and intrachalk erosion. The main subsidence followed a northwest-southeast trend with a clearly defined depocenter north-northeast of Block 1/9. The areas of local Late Cretaceous inversion resulted from a reversal of existing normal faults. These inversions have been interpreted as being due to a simple reversal of wrench movements such that former extensional basins in a right-lateral regime became compressional under a left-lateral regime (Gowers and Sæbøe, 1985).

The Central trough continued to subside during the Tertiary along the same trends as during the Late Cretaceous (Hardman, 1982). Inversion trends, and in particular the Lindesnes ridge, continued to be pronounced during the early Tertiary, providing a source of reworked sediments for the surrounding areas. Halokinetic activity persisted until the middle Tertiary.

Regional Structure

The Tommeliten Gamma field is located in the transfer block intermediate between the U.K. Central graben and the Feda graben and north of the Grensen nose (Figure 9).

Seismic lines provide the basis for the regional structural understanding. The seismic horizons can be most reliably interpreted down to the post-Middle Jurassic section (Figures 3 and 4).

The Late Jurassic–earliest Cretaceous extensional phase strongly affected the Block 1/9 area. The variation of sediment thickness shows that the transfer block was tilted toward the northeast. The eastern limit of this block is a NNW-SSE normal fault downthrown to the west that is parallel to the main trend into the area. East-west transfer faults limit the block to the north and to the south. The Tommeliten Gamma and Delta structures overlie the east-west fault downthrown to the south, which limits the block to the north, while the Tommeliten Alfa structure lies in the central part of the block (Olsen et al., in preparation).

Owing to the block tilting, salt migrated up-dip, which resulted in the formation of the Tommeliten Alfa structure, localized above a basement fault. Tilting and differential sediment loading also caused salt to flow northward against the footwall of the northern fault, contributing to the start of the Gamma and Delta domes (Olsen et al., in preparation).

During the Late Jurassic and Early Cretaceous, salt movements were responsible for thinning of sediment onto crestal areas. In Block 1/9, the Lower Cretaceous sequence thins from the north to south. The sediments thickening to the Tommeliten Gamma structure periphery could be due to a local structural high and differential compaction of the underlying sediments. Lower Cretaceous sediments are missing in the central area of the Tommeliten Gamma structure, probably as a result of pulses of halokinetic movements that occurred during Cretaceous times.

From the Middle Cretaceous to early Paleocene, chalk deposition in this area was strongly influenced by topography created by salt movement and inversion. Chalk layers generally thin toward the crestal areas of moving salt structures such as the Tommeliten Gamma and Alfa fields. It is probable that additional small faults formed in response to

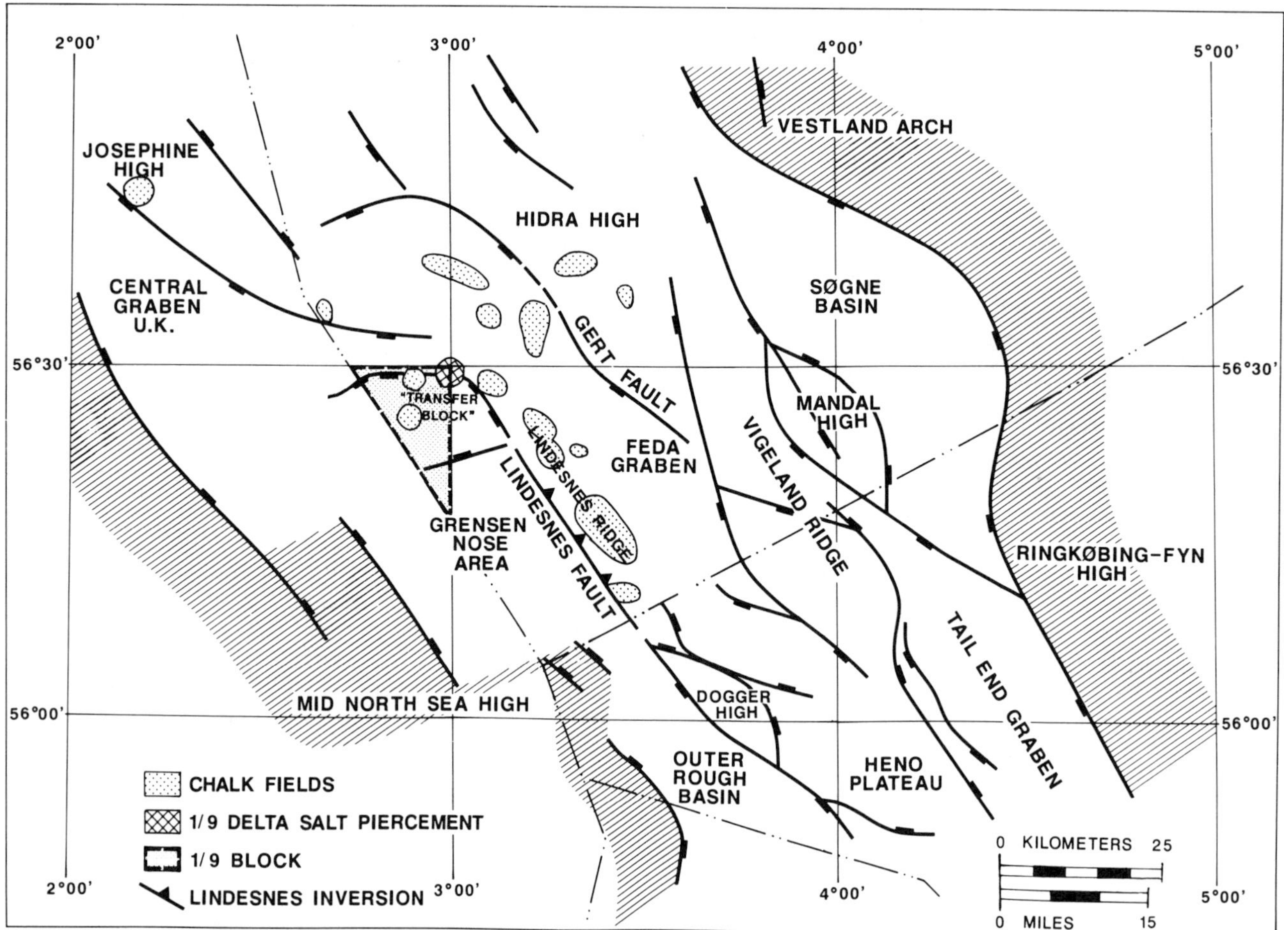

Figure 9. Central trough, structural sketch at Middle-Late Cretaceous time. Modified and compiled from Gowers and Sæbøe, Damtoft et al., and Olsen et al. (in preparation).

the large-scale events. These faults, however, cannot be defined on the seismic owing to the effect of gas in the Tertiary claystones.

Some slight halokinetic movements took place in the period from the Paleocene to the end of the Oligocene in the area, but they are considered to be post-diapiric adjustments. Salt withdrawal into the Delta salt piercement probably took place during the late Paleocene-Eocene.

From the Miocene to Present, no particular movements have been detected as the Tommeliten Gamma structural relief was progressively buried.

Local Structure

The Tommeliten Gamma structure is an anticline, slightly elongated in a west-east direction. At the top of the reservoir, the structural dips on the northern, northeastern, and eastern flanks are around 15°.

Structural dips in the southwestern flank are around 25°. Due to synsedimentary infill and differential compaction during the growth of the structure, the dips are slightly higher for the older horizons.

The structure is the result of the combination of faulting and halokinesis with faulting starting during Jurassic times. The field is located above the east-west-trending transfer fault that has a dip-slip component toward the south and a sinistral strike-slip component.

Though the structure is mainly due to diapirism, the presence of the transfer fault beneath the field is reflected in the slight east-west elongation of the structure.

The tilting of the downthrown block combined with fault movement caused a large amount of salt to migrate northward into the Tommeliten Gamma field structure. As the diapir developed with time, it caused significant local variations in the thickness of the individual formations.

STRATIGRAPHY

The stratigraphic section in the Tommeliten Gamma field consists of Permian to Recent units with

Jurassic and Triassic deposits absent in wells owing to salt piercement. The missing section therefore includes the main source rocks for hydrocarbons, which are present off-flank in the Late Jurassic, dark shale of the Mandal Formation.

The description and thicknesses of the lithostratigraphic units, given below, relate mainly to well 1/9-4, which shows the most complete sequence for the Tommeliten Gamma field (Figure 6). The stratigraphic nomenclature used is based on that given by Deegan and Scull (1977) and by Vollset and Doré (1984).

The part of the Zechstein Group that has been penetrated (in well 1/9-4 and 1/9-A-3H) is characterized by massive halite. Age: Late Permian.

The Cromer Knoll Group is 54 m (177 ft) thick in well 1/9-4, compared to 470 m (1542 ft) on the Tommeliten Alfa structure. It consists of calcareous shale and marl belonging to the Rodby Formation and a dark gray shale of the underlying Valhall Formation. Age: Portlandian to Albian.

The Chalk Group, a calcilutite sequence reaching a thickness of 500–2000 m (1640–6560 ft) in the Central trough and 531 m (1742 ft) in well 1/9-4, comprises the Hidra Formation, Plenus Marl Formation, Hod Formation, Tor Formation, and Ekofisk Formation. Age: Cenomanian to Danian.

In the shelf regions surrounding the Central trough, the sediment was pelagic chalk as seen onshore in northwestern Europe (Hancock, 1975). The accumulations in the Central trough are a mixture of pelagic and resedimented chalk. Skovbro (1983) summarized the chalk sedimentation in the Norwegian North Sea. Carbonate sedimentation ceased at the end of the early Paleocene.

The chalk fields in the North Sea (Figure 1) produce from intervals in the upper part of the Chalk Group, e.g., in the Hod Formation, Tor Formation, and Ekofisk Formation.

The Rogaland Group, the Hordaland Group, and the Nordland Group having a total thickness of 3015 m (9892 ft) in well 1/9-4 are predominantly composed of clay, claystone, and shale, which are silty in part. Age: Paleocene to Recent.

These units are more than 3000 m (~9900 ft) thick in the central part of the elongate central North Sea basin, where subsidence dominated throughout the Tertiary and Quaternary.

TRAP

Trap Type

The Tommeliten Gamma field is a structural trap. At the reservoir level, the trapping geometry is that of an almost symmetrical anticlinal dome overlying a salt diapir. The crest is at about 3010 mss (9875 ft), with the lowest closing contour for the top of the Ekofisk Formation at 3350 mss (10,990 ft). The top of the Zechstein salt in the crestal area is encountered at 3250 mss (10,660 ft). The field has an area of about 6.5 km^2 (1600 ac) and a vertical extent of nearly 400 m (1300 ft).

The vertical seal is provided by the 3000 m (~9900 ft) thick sequence of over-pressured Tertiary shales and claystones. The sealing capacity is probably enhanced by a marked pore pressure regression starting 250 m (820 ft) above the reservoir.

A downdip sealing effect caused by cementation and compaction is seen in all chalk layers with the rate of reservoir deterioration dependent on the depositional mode of the chalk. The lower, argillaceous part of Ekofisk Formation deteriorates most quickly and acts as a vertical seal for the underlying Tor Formation in downflank areas.

None of the productive zones in the reservoir are expected to have pay down to their structural spill-point. Net pay cut-offs are taken at 10% porosity in the Tor Formation and at 15% porosity in the Ekofisk Formation or at 50% water saturation.

The Tor Formation sits unconformably on the low porosity Hod Formation (5–10%). This unconformity marks the lower stratigraphic limit of the productive hydrocarbon column in the field.

Halokinesis started in the Early Cretaceous with a diapiric phase during the Late Cretaceous to Paleocene synchronous with the deposition of the Tor and Ekofisk formations. Smaller scale salt movements, which continued until the early Miocene, are attributed largely to post-diapiric adjustments. These latest movements have not given rise to a significant redistribution of already accumulated hydrocarbons, as seen in the Ekofisk field (D'Heur, 1984) and in the Albuskjell field (D'Heur, 1987b).

The hydrocarbons probably migrated into the reservoir via high porosity/high permeability chalk layers with large areal extent and via fracture zones. These layers have the highest porosity and the lowest downflank rate of decrease in porosity. In some of the major fracture and fault zones, the porosity is markedly higher than the fieldwide average.

Reservoir

Stratigraphy

The Tor and Ekofisk formations of Maastrichtian and Danian ages, between the depths of 3000 and 3400 mss (~9900–11,200 ft), constitute the Tommeliten Gamma field reservoir.

The Tor Formation consists of interbedded chalk and limestone (soft and hard calcilutite) with chalk dominating the upper part of the formation. Stylolites and thin solution seams are frequently found in the interval, while joints become more and more frequent toward the top. The correlation of the reservoir zones is based on biostratigraphy, facies analysis, mineralogy, and log response (Figure 7). In the Tor Formation, only a few biostratigraphic events are correlatable. However, log response tied to pelagic and single-event gravity flow units have allowed correlation of three reservoir zones (T3, T2, and T1 upward).

The lower part of the Ekofisk Formation is an argillaceous, firm to hard limestone that is overlain by interbedded chalk and limestone. In the Ekofisk Formation, the biostratigraphic zonation has a higher resolution and can be used in the correlation of the two high porosity reservoir zones (E3 and E1).

Depositional Setting and Facies

The model for the depositional setting and facies in the Tommeliten area has to a large degree followed the development in understanding of the deposition of the chalks in the Central trough. Between 1979 and 1983 there were a series of important breakthroughs as recorded in the following papers:

- Perch-Nielsen et al. (1979) showed that the main reservoir interval in the Ekofisk Formation of the Ekofisk field was a sequence of reworked Maastrichtian material.
- Watts et al. (1980) correlated stacked slumps in the Ekofisk field with a mass slide in the Albuskjell field, resedimentation taking place in a slope and base of slope setting.
- Kennedy (1980) published a study of the sedimentology and distribution of pelagic and resedimented facies.
- Hardman (1982) presented a discussion on the control of deposition and composition on chalk reservoir quality.
- Nygaard et al. (1983) compared resedimentation processes in clastic and chalk deposits.

By the time 1/9-4 and 1/9-6 had been drilled, the Tommeliten Gamma area was considered by Brewster and Dangerfield (1984) and by D'Heur (1984) to be on the northwestern termination of the Lindesnes ridge inversion feature. D'Heur and Pekot (1987) concluded that the Tommeliten Gamma area was located on a half-horst type paleostructure (the northwestern end of the Lindesnes ridge) that existed during the deposition of the Tor and Ekofisk formations. A similar interpretation of the structural setting had already been presented for the nearby Edda field (D'Heur et al., 1985).

Kennedy (1987) presented a regional depositional model for the Greater Ekofisk area. Here the Tommeliten Gamma field was shown to be southwest of a slope-centered basin in the Ekofisk–W. Ekofisk area, and just within the distal edge of significant turbidite development sourced from the northeast.

As a result of intensive in-field studies, the model for the depositional setting of the Tommeliten Gamma has now been established in much greater detail. The facies succession in the field reference well 1/9-6 and the TVT isopach maps of the reservoir zones are shown in Figures 10 and 11.

The facies succession in the Tor Formation and the seismic interpretation in general (Olsen et al., in preparation) suggests that the structure had developed as a local, salt-induced paleohigh within the basin at this time. In the lower part of the Tor Formation (T3), an onlapping sequence of resedimented facies including the basinal facies (pebbly chalks and graded or laminated calcarenites) is recognized. This is followed by a sequence of laminated, homogeneous and pelagic chalks that were deposited on the high and not mixed with basinal sediments. Since this succession is found in the cored intervals of three flank wells, the high must have been fairly broad at this time.

During deposition of the Ekofisk Formation there are indications that the crestal area was affected by local faulting probably caused by salt withdrawal and doming. As a result, a basinal setting was developed in the near crestal areas as shown by the facies succession and by the thicknesses of the reservoir zones (Figures 10 and 11).

The Tommeliten Gamma field is an excellent example of the significant effect very localized halokinetic movements during deposition of the reservoir sequence can have on geometry of the chalk reservoir units and on the distribution of facies.

When reviewing the facies, it is important to recognize that the resedimentation of chalk in the North Sea Central trough involved extremely large volumes, several linear- or point-source areas (Kennedy, 1987), many major sediment gravity flow events (Hatton, 1986), a predominantly very fine grain size (0.1–30.0 microns), and mineralogically stable, nonflocculating source material.

Pelagic and resedimented chalk (autochthonous and allochthonous of some authors) are well represented in the Tommeliten Gamma field, as shown in Figures 10 and 12. Based on facies analysis, it is estimated that the Tor Formation consists predominantly of resedimented chalk (75%) and the Ekofisk Formation consists predominantly of pelagic chalk (70%). The resedimented material in both the Tor and Ekofisk formations is primarily intraformational.

The pelagic facies is distinguished by a range of preservational styles of the biogenic structures (Ekdale et al., 1984), together with unsorted constituent particles in the matrix. The pelagic facies includes periodite chalk and bioturbated chalk. The periodite chalk is also bioturbated but has rhythmic alternation of darker and lighter units. The low porosity periodite chalk is only found in the Ekofisk Formation (E2 and E4), while the bioturbated chalk is found throughout the reservoir section. In the Tor Formation, the pelagic chalk has low porosity. An exception to the rule that all pelagic facies have low porosity (D'Heur, 1984) is seen in the upper part of the Ekofisk Formation (zone E1). It is likely that the high porosity pelagic chalk in this zone was originally a very pure coccolith ooze, since the intercoccolith porosity is still very well preserved. The fieldwide variation in average porosity with depth in the reservoir zones consisting predominantly of pelagic facies (E1, E2, and E4) are 4–5 porosity units per 100 m (328 ft).

The resedimented facies constitute a spectrum of chalk types, distinguished by primary structures and textures, and include homogeneous, laminated, and

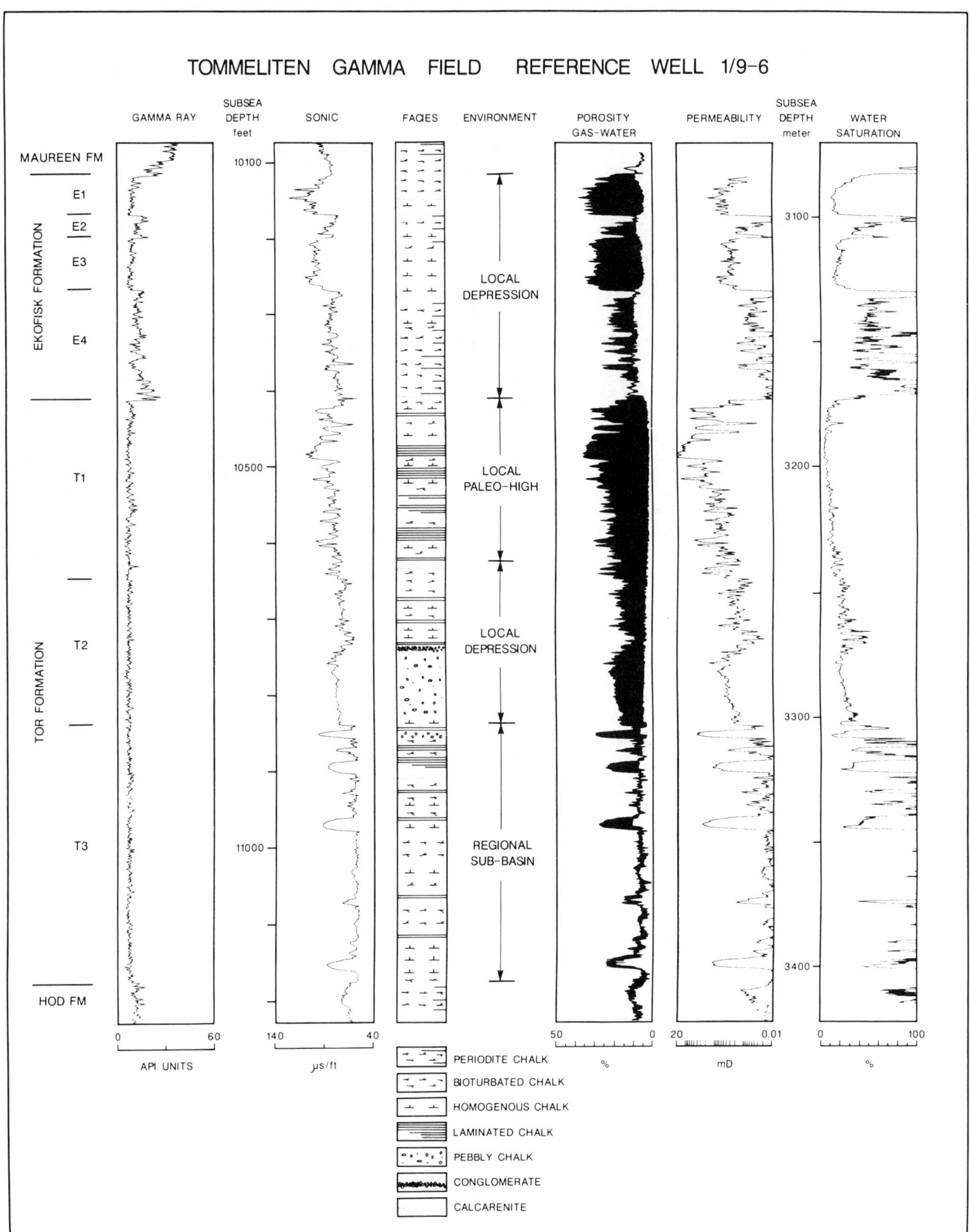

Figure 10. Type log, Tommeliten Gamma field reference well 1/9-6 showing environmental interpretation, and distribution of porosity, hydrocarbons, permeability, and water saturation in the reservoir interval of the Ekofisk and Tor formations. Facies distribution based on core description (compiled from wells 1/9-6 and 1/9-A-1H) and dipmeter log (SHDT) analyses. See also the spectrum of chalk facies in Figure 12.

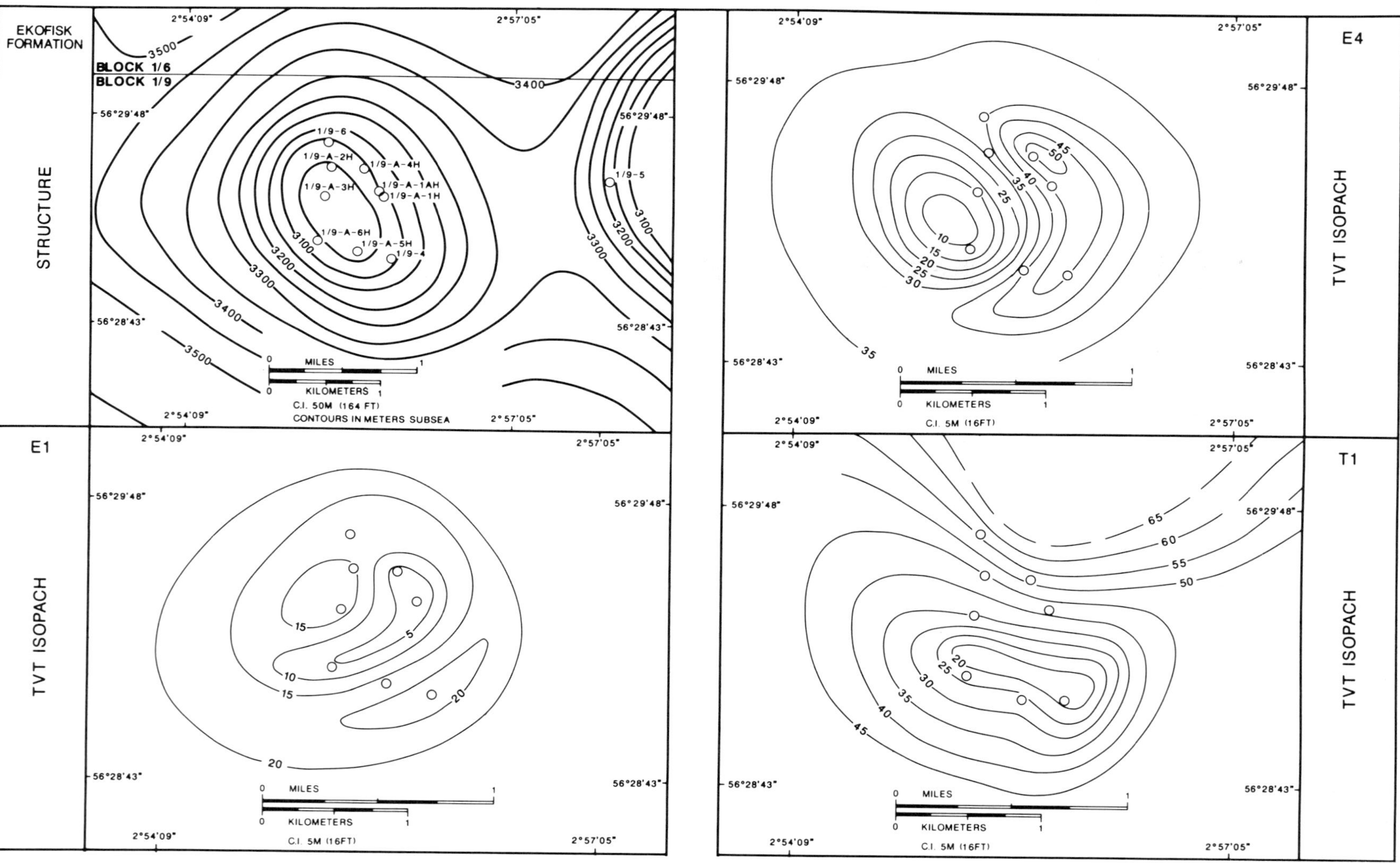

Figure 11. Structure map (depth) of the Top Ekofisk Formation (upper left) and TVT isopach maps of the reservoir zones (E1, E2, E3, E4, T1, T2, and T3 downward) in the Ekofisk and Tor formations. Contour interval in the TVT isopach maps is 5 m (16 ft) except in T3, where it is 10 m (33 ft).

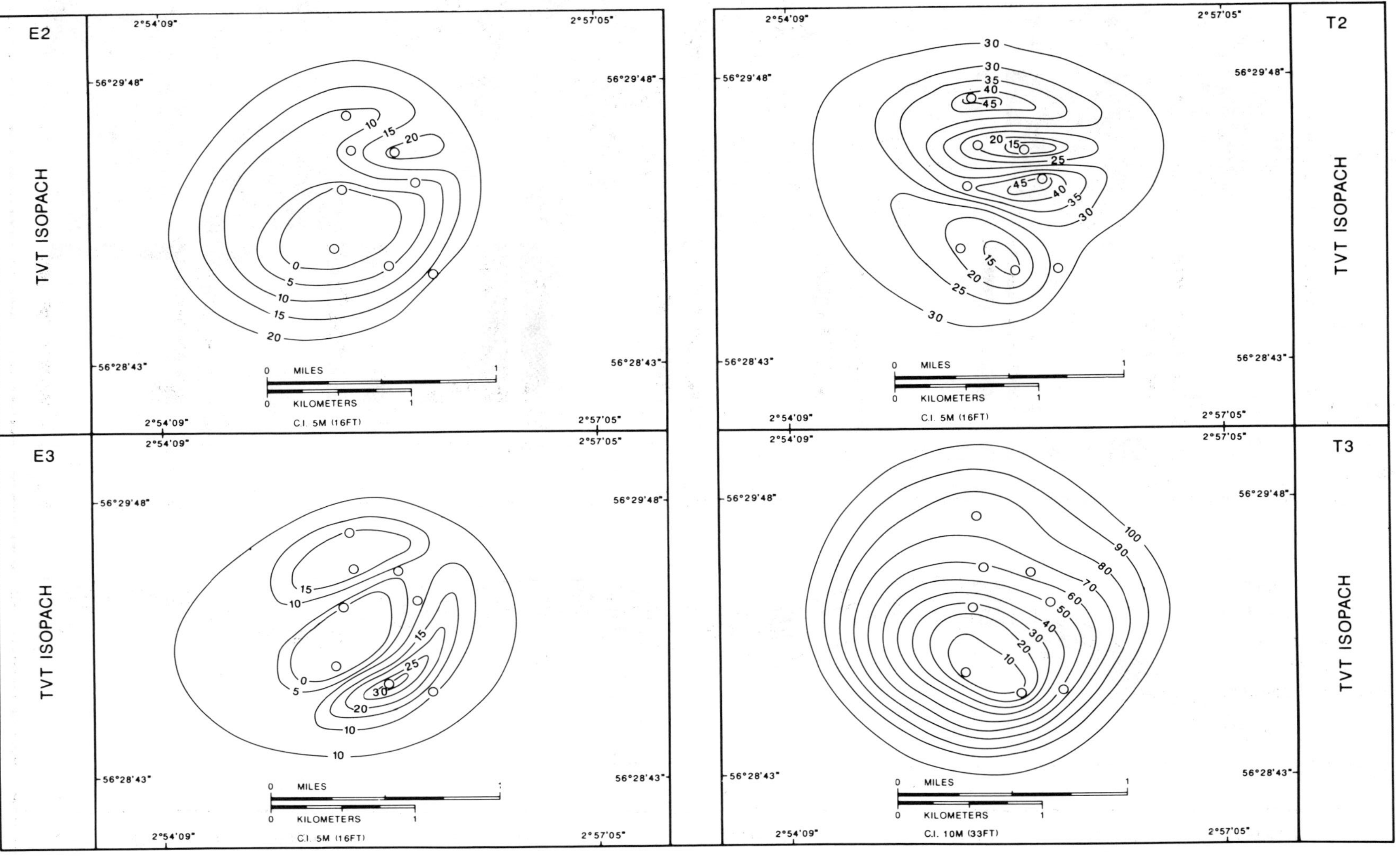

Figure 11. (Continued)

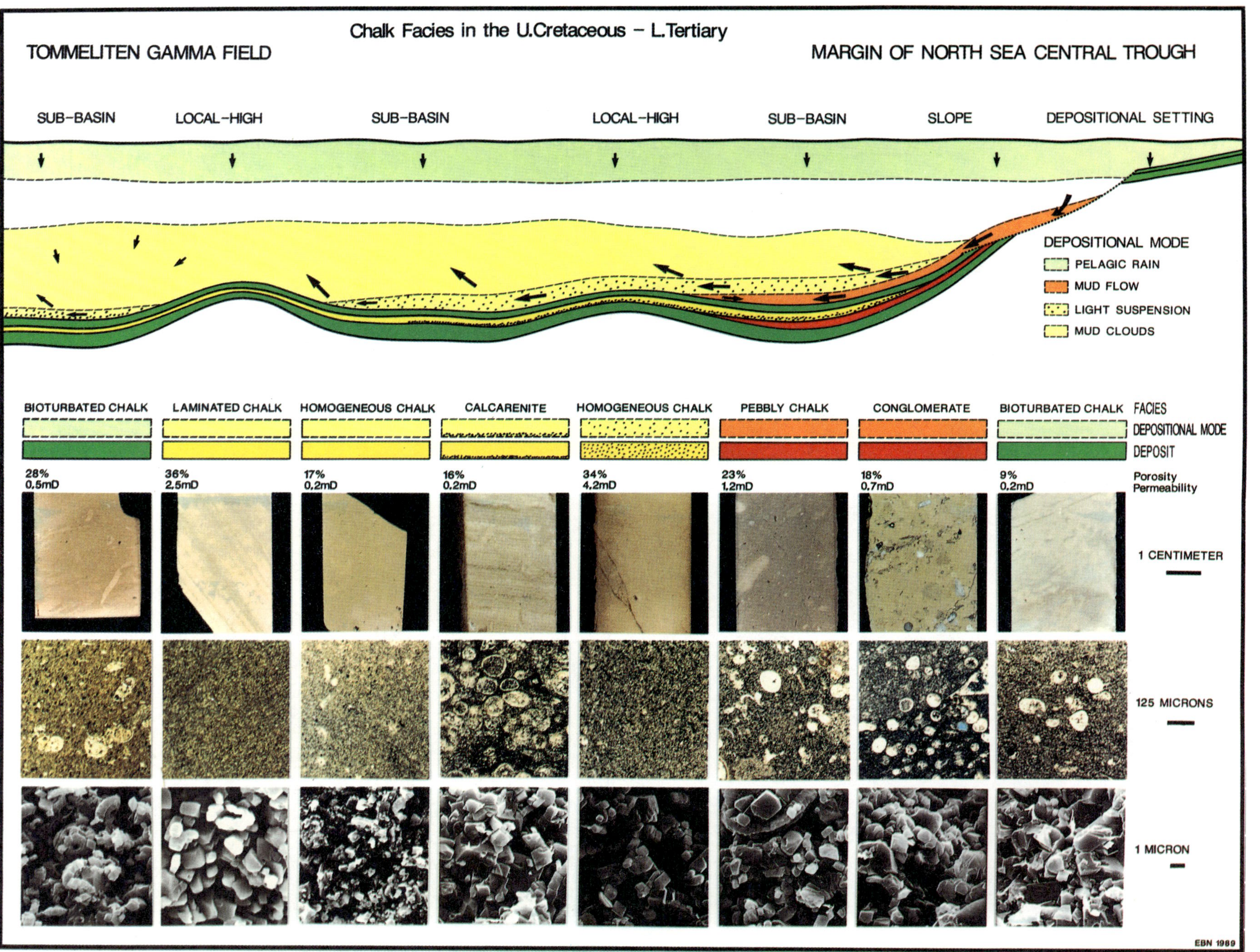

Chalk Facies in the U.Cretaceous – L.Tertiary
TOMMELITEN GAMMA FIELD
MARGIN OF NORTH SEA CENTRAL TROUGH
SUB-BASIN
LOCAL-HIGH
SUB-BASIN
LOCAL-HIGH
SUB-BASIN
SLOPE
DEPOSITIONAL SETTING
DEPOSITIONAL MODE
PELAGIC RAIN
MUD FLOW
LIGHT SUSPENSION
MUD CLOUDS
BIOTURBATED CHALK
LAMINATED CHALK
HOMOGENEOUS CHALK
CALCARENITE
HOMOGENEOUS CHALK
PEBBLY CHALK
CONGLOMERATE
BIOTURBATED CHALK
FACIES
DEPOSITIONAL MODE
DEPOSIT
28% 0,5mD
36% 2,5mD
17% 0,2mD
16% 0,2mD
34% 4,2mD
23% 1,2mD
18% 0,7mD
9% 0,2mD
Porosity
Permeability
1 CENTIMETER
125 MICRONS
1 MICRON
EBN 1989

pebbly chalk; conglomerates with chalk clasts; and graded or laminated calcarenites. The distribution of the resedimented facies in well 1/9-6 is shown in Figure 10. Here the variation in porosity in the resedimented facies is considerable, but most marked in T3, compared to T1 and T2.

In the reservoir zones with prevailing resedimented chalk facies the gradients are 12, 8, and 6 porosity units per 100 m (328 ft) in E3, T1, and T2, respectively, and in the single-event gravity flow layers in the lower part of the Tor Formation the gradient is 2 porosity units per 100 m (328 ft).

The spectrum of chalk facies from Tommeliten Gamma and elsewhere, shown in Figure 12, include the main types of resedimented chalk in the field, together with a tentative model for their depositional mode and setting. It is tentative because studies of facies distribution and of primary and secondary structures and textures in resedimented chalk are in a very early stage, compared to recent studies of carbonate turbidites, pelagic turbidites, carbonate suspensions, and associated facies (Straaten, 1971; Heath and Mullins, 1984; Stow et al., 1984; Buchbinder et al., 1988).

The main conclusion is that resedimented chalk from mudflows and suspensions (that differentiate into denser and lighter parts) can be successively trapped in subbasins and depressions in front of the area where the precursor of the suspensions, a slide or a slump, originated. The mud clouds created at the same time will spread over larger areas. In a single gravity-flow event, the successive catchment of parts of the flow and the differentiation during the travel of the suspensions gives a lateral facies variation in the basin settings from unsorted ooze with all size fractions, to ooze with lesser and lesser amounts of the finest fractions, and finally to a sorted calcarenite. This lateral basin facies succession will not be developed when the mass flow is completely entrapped in a nearby subbasin or depression. The deposition of these basin facies is followed by long-term settling of mud consisting of finer and finer fractions (5–0.1 microns) both over the basins and the highs. The lateral and vertical variations in initial permeability and porosity will vary considerably within the deposits of a single major event of resedimentation (see Figure 12).

Composition, Texture, and Porosity Types

Because the porosity varies considerably (Figures 8 and 10), the porosity types will be dealt with together with the composition and matrix texture of the reservoir rocks.

The chalk can be classified after Folk (1962) as fossiliferous micrites and biomicrites, with the micrite matrix built up mainly of debris from planktonic coccolithophorids. In the Tor Formation, the debris consist of fragments and platelets of coccoliths with an interparticle porosity. In the Ekofisk Formation, the debris range from well-preserved coccospheres and coccoliths to platelets that give rise to both an inter- and intracoccolith porosity. The coarse biogenic fraction occurs in variable amounts (1-20%) and characteristically includes in the Tor Formation planktonic foraminifera and calcispheres, and in the Ekofisk Formation planktonic foraminifera and radiolarians.

Mineralogically, the chalk contains predominantly low magnesian calcite. The noncarbonate fraction in the chalk is disseminated either primarily throughout the matrix as terrigenous clay, or diagenetically as biogenic silica. The Tor Formation chalk has 1-6% free silica and 1-3% clay, and the Ekofisk Formation chalk has 4–48% free silica and 3–13% clay.

Most productive chalks are relatively pure deposits of low magnesian calcite (Hardman, 1982), and this stable mineralogy accounts for the relatively good preservation of the porosity. Just as important for preservation of the microporosity is that the stable mineralogy may also account for some preservation of primary particle size and sorting.

The size of the primary particles in the matrix cannot always be determined with certainty, but examples of well-sorted particles, consisting of crystal-aggregates or single crystals, are frequently observed in the Tor Formation, and they do also occur in the Ekofisk Formation. The preserved sedimentary textures indicate that some resedimented chalk facies may have had initial interparticle porosity around 40%, and therefore have experienced only minor bulk changes in porosity during burial (Figure 10).

What can be determined with certainty is the average crystal size in the matrix. In this field, it is 1.7 microns in the Tor Formation and 0.9 microns in the Ekofisk Formation. The crystals are generally smaller and rounded in the high porosity zones especially in the Ekofisk Formation, where confining pressure solution (see Mimran, 1977) can be very pronounced.

A general comparison of the two formations gives a markedly lower matrix permeability in the Ekofisk Formation than in the Tor Formation for the same porosity, which reflects the crystal size, the size

Figure 12. Above, tentative model illustrating the different depositional modes developed during a single major gravity flow event in a schematic west-east-oriented frame through the North Sea Central trough. The arrows indicate the movement of particles in the mud flow, the light suspension, the mud cloud and in the pelagic rain before settling. Below, a spectrum of the chalk facies seen in the Tommeliten Gamma field, together with an interpretation of their depositional mode and their porosity and permeability. The chalk spectrum is illustrated with core photographs, thin-section photomicrographs, and scanning electron photomicrographs.

distribution, and type of particles. The average crystal size in the Tor Formation is probably smaller in this field, where a greater part of the Tor Formation is believed to have been deposited from light suspensions and mud clouds than in the Ekofisk and W. Ekofisk fields, where the Tor Formation is deposited in a basin setting.

The bulk changes in matrix porosity in the North Sea reservoir chalks during deep burial, varying pressure conditions, and hydrocarbon emplacement are not very well documented. Reduction in porosity owing to mechanical compaction and pressure-solution precipitation, and preservation of porosity are discussed in D'Heur (1984). Mimran (1977) interpreted calcite migration in chalk as a result of variations in pore pressures and tectonic stresses. Calcite dissolved in areas of high pore pressures is reprecipitated in areas with low tectonic stresses. In the overpressured Tommeliten Gamma field, this calcite migration might explain the high crestal porosities and the downflank porosity decrease.

Porosity changes caused by subtle variations in pore pressures and tectonic stresses within the reservoir chalks of other fields have not been reported, in the context either of productive and nonproductive zones or structural elevation.

Fractures and Faults

The deformation of the reservoir by fracturing is the result of post-diapiric movements, faulting, and differential compaction. Owing to the lack of reliable seismic data, the fault pattern and the associated deformation zones are not mappable. However data on faults, deformation zones, joints, and fractures associated with stylolites have been determined from logs or cores in all the wells. Faults with a vertical displacement of about 20 m (66 ft) have been identified in the northern part of the field. An open fracture system was seen only in connection with a fault at the top of the Hod Formation, and here this fracture system was hydrocarbon-bearing in the Hod Formation.

A positive correlation is seen in cores between the proportion of resedimented intervals in the reservoir zones and the frequency of joints. The main intervals with joints seen in cores are also detectable on logs.

Only future production data will show the magnitude of enhancement in permeability from the joints in the reservoir. It is expected that the fracture density will not be as high as that seen in the Ekofisk field (Brewster et al., 1986) but it will be higher than observed in the Albuskjell field, where only stylolite-associated fractures are expected to contribute to permeability enhancement (Watts, 1983).

Porosity and Permeability

The matrix porosity decreases downflank in the Tommeliten Gamma field reservoirs. The porosity and the water saturation have a very high negative correlation, as in all chalk fields (Figure 8). The porosity and permeability show a positive correlation, but for the same porosities, the permeability is markedly lower in the Ekofisk Formation than in the Tor Formation (Figure 10).

The fieldwide variation in average porosity with depth in the productive zones of the Ekofisk Formation (E1 and E3) are 4 and 12 porosity units per 100 m (328 ft), and the crestal porosities are 33% (E1) and 36% (E3). For the Tor Formation (T1, T2, and T3) the values are 8, 6, and 3 porosity units per 100 m (328 ft), and here the crestal porosities are 31% (T1), 27% (T2), and 22% (T3).

It is not clear why there is such a difference between the different reservoir zones, but the resedimented chalk facies do have a potential for greater diagenetic alteration than the pelagic chalk facies in this reservoir.

Pay-Zone Thicknesses

In the crestal area, the gross thicknesses of all the reservoir zones in the Ekofisk and Tor formations are the same as the net pay thicknesses. Downflank the net pay thicknesses will initially increase, because the gross thicknesses increase markedly from 60 m (197 ft) to more than 260 m (853 ft), and then decrease gradually owing to reservoir deterioration. The average net pay thicknesses are 69 m (226 ft) and 125 m (410 ft) for the Ekofisk and Tor formations, respectively.

Hydrocarbon Characteristics

Since at the time of writing production has not started, the characteristics of the hydrocarbons summarized in Appendix 2 are those derived from test-produced fluids and gas.

Source Rocks

In common with the rest of the Central trough, the Mandal Formation is considered the only significant hydrocarbon source rock in the Block 1/9 area. This formation is equivalent to the upper part of the Kimmeridge Clay Formation and is Volgian to Ryazanian in age. It consists of dark shale interbedded with brown, very argillaceous limestone and thin sandy layers. It is characterized by a high level of radioactivity (high gamma ray readings), which is typical for shales deposited under anoxic conditions and having high organic content. The formation was deposited in a marine environment with high organic productivity and restricted bottom circulation. Terrestrial organic material is also observed.

The thickness of the Mandal Formation in this area ranges from 10 to 70 m (33-230 ft), with the variations being governed by structural patterns (Figure 13). It is thicker in the deep part of the transfer block where this formation is buried to a minimum depth of 4000 mss (~13,000 ft).

It is estimated that close to the area, the total organic carbon (TOC) of the Mandal Formation is

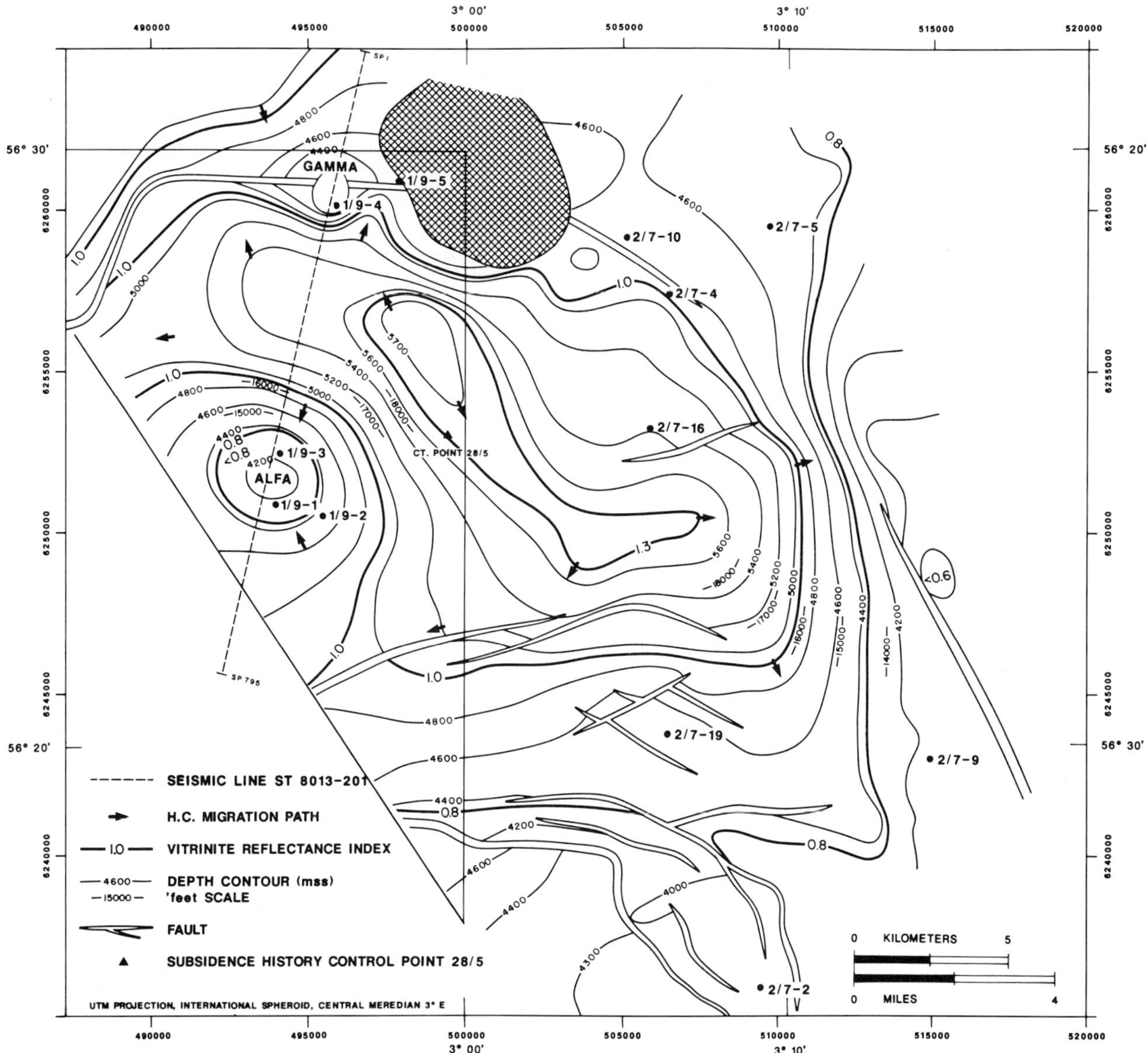

Figure 13. Maturity map at top Jurassic (base Cretaceous).

around 5 to 6%. The TOC reported from the 1/9-3 well (Tommeliten Alfa structure) is, however, between 1.3 and 2.9%, probably due to the marginal western position of the transfer block related to the center of the Central trough. The higher sand content in the formation might be indicating a local facies variation again reflecting the marginal position of the block.

The most abundant kerogen type in the area is a mixture of type II oil-prone degraded liptinite plus gas condensate-prone exinite and type III oil-prone vitrinite. Regional data indicate that the content of gas-prone type III increases to the north and west in the Central trough. The hydrogen index of the source rock from the 1/9-3 well is 243, which matches well with type II/III kerogen (see Figure 14).

The source rock maturity varies in the Norwegian Central trough area. Oil fields are more abundant in its southern part while gas fields are more present in the northern part of the trough. This variation is mainly due to the subsidence history of the Jurassic source rock.

Thermal maturity profiles have been generated for several control locations (30 points) in the Block 1/9 area along seismic lines. One of them is shown in Figure 15.

This figure shows that the peak of oil generation ($R_o = 0.8$) was probably reached when the top Mandal

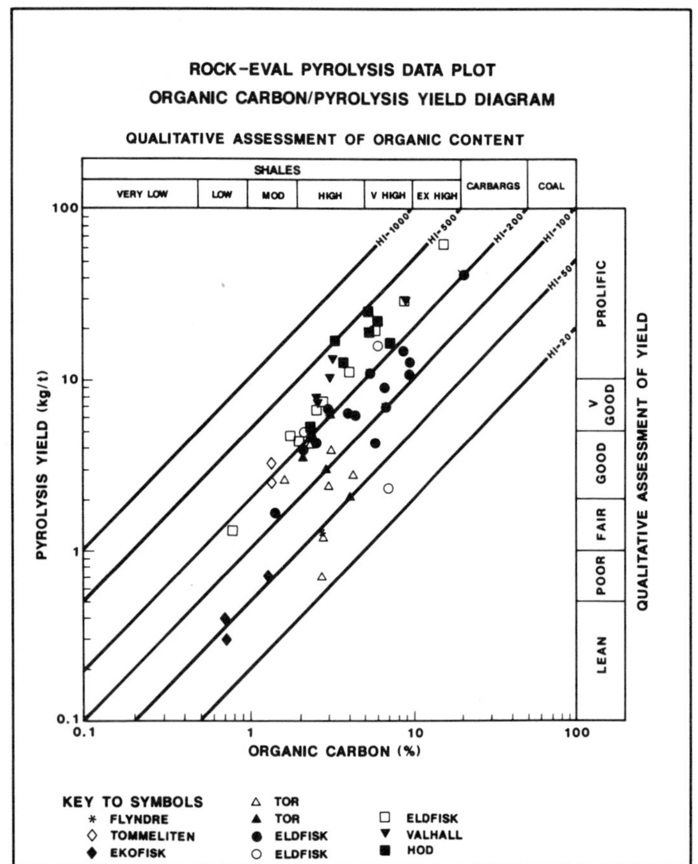

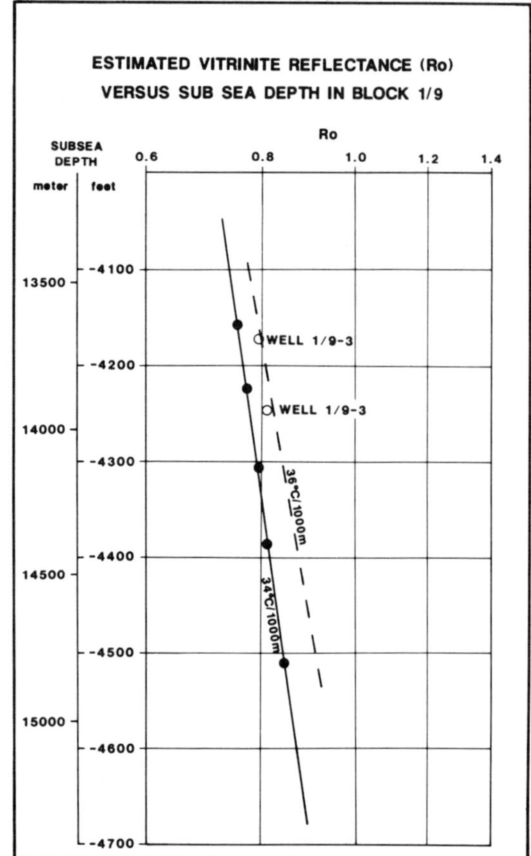

Figure 14. Geochemistry Rock-Eval pyrolysis data plot, estimated vitrinite reflectance versus depth.

Formation was at ~4420 mss (14,500 ft), which corresponds to the late Oligocene. The probable onset of gas generation ($R_0 = 1.0$) was reached with further burial to ~5010 mss (16,437 ft), which corresponds to the mid-Miocene.

The temperature gradient used here is 34°C/km, which was based on a regional understanding and supported by well data. Vitrinite reflectance (R_0) data from well 1/9-3 are limited, but these suggest a slightly higher temperature gradient of ~35.5°C/km. The maturity distribution map in Figure 13 presents the most conservative view of the areal extent of mature Jurassic source rocks.

It is assumed that the generated hydrocarbons migrated up into the overlying chalk via faults and fractures related to the salt diapir.

EXPLORATION CONCEPTS

Regional Play

The regional chalk play, which includes the Tommeliten Gamma field, requires the presence of source potential in the basin, or subbasins, below the reservoir formations. Periodic synsedimentary tectonic movements occurred during the deposition of the reservoir formations in a pelagic realm. These movements gave rise to a differentiation between high and low porosity calcilutites within subbasins and over the intervening paleohighs. Trap-forming salt movements or tilting of fault blocks occurred during, and in a period after, the deposition of the calcilutites. A good, overpressured, shale seal occurred above the reservoir and tight chalk facies within the pay section.

General Application of Geologic Parameters

Comparison with the other chalk fields in the area shows that there is a whole spectrum of pelagic to resedimented reservoir chalk facies and a highly variable, reservoir deterioration with increasing structural depth in the different reservoir zones. The presence of gas in the crestal areas and in the Tertiary section of Tommeliten Gamma field is also seen in the Albuskjell, Ekofisk, Eldfisk, Valhall, Hod, and Tommeliten Alfa fields. The condensing of the Tor and Ekofisk formations indicating structural growth during the deposition of these formations is observed in the Tor, Eldfisk, Valhall, Hod, Dan, Gorm, and

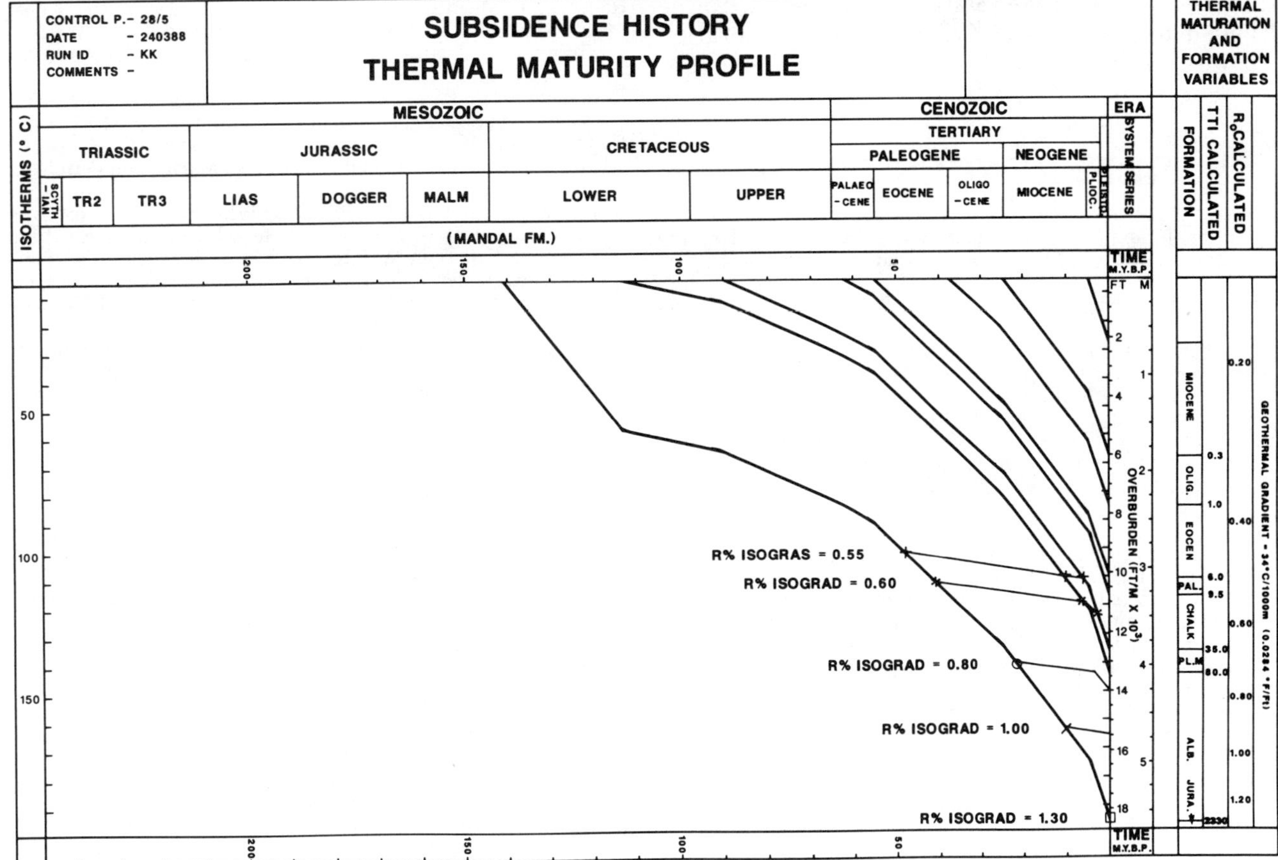

Figure 15. Subsidence history—thermal maturity profile (point location in Figure 14).

Kraka fields. Synsedimentary tectonic movements on the paleohighs giving rise to local highs and depressions have been described at formation level in the Eldfisk and Valhall fields.

Lessons

This field is unique because of the detailed, fieldwide data of facies distribution, geometry, and reservoir quality within seven correlatable chalk reservoir zones. It has, for example, been shown that resedimented chalk facies deposited from light suspensions and mud clouds on a paleohigh actually have fair reservoir quality and that the resedimented material in the reservoir formations is predominantly intraformational.

The concepts or lessons that can be developed from the Tommeliten Gamma field and applied elsewhere are:

1. The importance of having a strong emphasis on the reservoir geological modeling during appraisal and early development drilling. This allows identification of any major asymmetry in the hydrocarbon distribution due to the facies distribution not paralleling the shape of the structural trap.
2. The identification of potential for "upgrading" the chalk reservoir quality on paleohighs. Here synsedimentary tectonic movements can create very local depressions, where lower porosity deposits are resedimented as fair to good porosity reservoir chalk. A similar "upgrading" is also likely due to local synsedimentary movements in a subbasin.
3. The possibility of recognizing the influence of diapiric pulses and salt readjustments on structural relief and formation geometries in the area just above a salt diapir. This was achieved by a detailed interpretation of the depositional setting of the chalk facies in conjunction with structural data.

ACKNOWLEDGMENTS

We would like to thank Statoil, Fina Exploration Norway A/S, and Norsk Agip A/S for their kind permission to publish this paper, although the opinions expressed herein are solely our responsibility and are not necessarily those of any of the above mentioned companies. We are also grateful to colleagues for their assistance, and we particularly thank Charles Jourdan (Statoil) for his help with the English text and Knut Kvingan (Fina Exploration Norway) for his advice in the geochemistry analysis.

REFERENCES CITED

Brewster, J., and J. A. Dangerfield, 1984, Chalk fields along the Lindesnes Ridge, Eldfisk: Marine and Petroleum Geology, v. 1, p. 239-278.

Brewster, J., J. Dangerfield, and H. Farrell, 1986, The geology and geophysics of the Ekofisk Field waterflood: Marine and Petroleum Geology, v. 3, p. 139-170.

Bromley, R. G., and A. A. Ekdale, 1987, Mass transport in European Cretaceous chalk; fabric criteria for its recognition: Sedimentology, v. 34, p. 1079-1092.

Buchbinder, B., C. Benjamini, Y. Mimran, and G. Gvirtzman, 1988, Mass transport in Eocene pelagic chalk on the northwestern edge of the Arabian platform, Shefela area, Israel: Sedimentology, v. 35, p. 257-274.

Damtoft, K., C. Andersen, and E. Thomsen, 1987, Prospectivity and hydrocarbon plays of the Danish Central Trough, *in* J. Brooks, and K. W. Glennie, eds., Petroleum geology of north west Europe: London, Graham & Trotman, p. 403-417.

Deegan, C. E., and B. J. Scull, 1977, A standard lithostratigraphic nomenclature for the Central and Northern North Sea: Report 77/25 Institute of Geological Sciences - UK, and Bulletin 1 Oljedirektoratet - N, 36 pp.

D'Heur, M., 1984, Porosity and hydrocarbon distribution in the North Sea chalk reservoirs: Marine and Petroleum Geology, v. 1, p. 211-238.

D'Heur, M., 1986, The Norwegian chalk fields, *in* A. M. Spencer et al., eds., Habitat of hydrocarbons on the Norwegian continental shelf: London, Graham & Trotman, p. 77-89.

D'Heur, M., 1987a, West Ekofisk, *in* A. M. Spencer et al., eds., Geology of the Norwegian oil and gas fields: London, Graham & Trotman, p. 165-175.

D'Heur, M., 1987b, Albuskjell, *in* A. M. Spencer et al., eds., Geology of the Norwegian oil and gas fields: London, Graham & Trotman, p. 39-50.

D'Heur, M., and L. J. Pekot, 1987, Tommeliten, *in* A. M. Spencer et al., eds., Geology of the Norwegian oil and gas fields: London, Graham & Trotman, p. 117-128.

D'Heur, M., L. de Walque, and F. Michaud, 1985, Geology of the Edda Field Reservoir: Marine and Petroleum Geology, v. 2, p. 327-340.

Ekdale, A. A., R. G. Bromley, and S. G. Pemberton, 1984, Ichnology: trace fossils in sedimentology and stratigraphy: Short Course Notes, Society of Economic Paleontologists and Mineralogists, 15, 317 pp.

Folk, R. L., 1962, Spectral subdivision of limestone types, *in* W. E. Ham, ed., Classification of carbonate rocks: American Association of Petroleum Geologists Memoir 1, p. 62-84.

Gowers, M. B., and A. Sæbøe, 1985, On the structural evolution of the Central Trough in the Norwegian and Danish sectors of the North Sea: Marine and Petroleum Geology, v. 2, p. 298-318.

Hancock, J. M., 1975, The sequence of facies in the Upper Cretaceous of northern Europe compared with that in Western Interior: Geological Association of Canada Special Paper No. 13, p. 83-118.

Hardman, R. F. P., 1982, Chalk reservoirs of the North Sea: Geological Society of Denmark Bulletin, v. 30, p. 119-137.

Hatton, I. R., 1986, Geometry of allochthonous Chalk Group members, Central Trough, North Sea: Marine and Petroleum Geology, v. 3, p. 79-98.

Heath, K. C., and H. T. Mullins, 1984, Open-ocean, off-bank transport of fine-grained carbonate sediment in the Northern Bahamas, *in* D. A. V. Stow and D. J. W. Piper, eds., Fine-grained sediments: deep-water processes and facies: Geological Society Special Publication No. 15, Oxford, Blackwell Scientific Publications, p. 199-208.

Kennedy, W. J., 1980, Aspects of chalk sedimentation in the southern Norwegian offshore, *in* The sedimentation of the North Sea reservoir rocks, Geilo, 11-14 May, 1980, Norwegian Petroleum Society, Oslo, 29 pp.

Kennedy, W. J., 1987, Sedimentology of Late Cretaceous-Palaeocene Chalk reservoirs, North Sea Central Graben, *in* J. Brooks and K. W. Glennie, eds., Petroleum geology of north west Europe: London, Graham & Trotman, p. 469-481.

Mimran, Y., 1977, Chalk deformations and large-scale migration of calcium carbonates: Sedimentology, v. 24, p. 333-360.

Nygaard, E., K. Lieberkind, and P. Frykman, 1983, Sedimentology and reservoir parameters of the Chalk Group in the Danish Central Graben: Geologie en Mijnbouw, v. 62, p.177-190.

Olsen, T. S., A. Roberts, E. B. Nielsen, J. Høyen, and T. A. Knai, in preparation, The detailed structural growth of the Tommeliten Fields, Norwegian Central Graben.

Perch-Nielsen, K., K. Ulleberg, and J. E. Evensen, 1979, Comments on "The terminal Cretaceous event: a geologic problem with an oceanographic solution" (Gartner and Keany, 1978): Cretaceous-Tertiary Boundary Events Symposium, Copenhagen 1979, II. Proceedings, p. 106-111.

Skovbro, B., 1983, Depositional conditions during Chalk sedimentation in the Ekofisk area Norwegian North Sea: Geologie en Mijnbouw, v. 62, p. 169-175.

Stow, D. A. V., F. C. Wezel, D. Savelli, S. C. R. Rainey, and G. Angell, 1984, Depositional model for calcilutites: Scaglia Rossa limestones, Umbro-Marchean Apennines, *in* D. A. V. Stow and D. J. W. Piper, eds., Fine-grained sediments: deep-water processes and facies: Geological Society Special Publication No. 15, Oxford, Blackwell Scientific Publications, p. 223-241.

Straaten, L. M. J. U. van, 1971, Origin of Solnhofen limestone: Geologie en Mijnbouw, v. 50, p. 3-8.

Vollset, J., and A. G. Doré, eds., 1984, A revised Triassic and Jurassic lithostratigraphic nomenclature for the Norwegian North Sea: Norwegian Petroleum Directorate Bulletin 3, Stavanger, 52 pp.

Watts, N. L., 1983, Microfractures in chalks of Albuskjell field, Norwegian Sector, North Sea: possible origin and distribution: American Association of Petroleum Geologists Bulletin, v. 67, p. 201-234.

Watts, N. L., J. F. Lapre, F. S. van Schijndel-Goester, and A. Ford, 1980, Upper Cretaceous and lower Tertiary chalks of the Albuskjell area, North Sea: deposition in a slope and a base-of-slope environment: Geology, v. 8, p. 217-221.

Appendix 1. Field Description

Field name .... *Tommeliten Gamma*

Ultimate recoverable reserves .... *390 bcf (11 bSm^3) and 32 MM bbl (5 MM Sm^3)*

Field location:

Country .... *Norway offshore*

State

Basin/Province .... *North Sea/Central trough (Central graben)*

Field discovery:

Year first pay discovered .... *(Paleocene) Danian Ekofisk Formation chalk 1977*

Year second pay discovered .... *(Upper Cretaceous) Maastrichtian Tor Formation chalk 1977*

Third pay .... *NA*

Discovery well name and general location

First pay .... *1/9-4(fourth well drilled in Block 1/9) 300 km SW of Stavanger*

Second pay *1/9-4*

Third pay .... *NA*

Discovery well operator .... *Statoil*

Second pay .... *Statoil*

Third pay .... *NA*

IP in barrels per day and/or cubic feet or cubic meters per day:

First pay .... *17.7 mmcf (500,000 Sm^3) and 1350 bbl (215 Sm^3)*

Second pay .... *24.7 mmcf (700,000 Sm^3) and 4120 bbl (655 Sm^3)*

Third pay .... *NA*

All other zones with shows of oil and gas in the field:

Age	Formation	Type of Show
None		

Geologic concept leading to discovery and method or methods used to delineate prospect, e.g., surface geology, subsurface geology, seeps, magnetic data, gravity data, seismic data, seismic refraction, nontechnical:

The structural interpretation gave top structure and the lowest closing contour at Top Chalk Group at −9515 ft (−2900 m ss) and −10,991 ft (−3350 m ss), respectively. The productive reservoirs in the region were known to be the chalk of the Ekofisk Formation and the Tor Formation, e.g., the upper part of the Chalk Group.

Structure:

Province/basin type (see St. John, Bally, and Klemme, 1984)

Bally 1211, Klemme IIB/IIIA; cratonic basin in the context of an earlier rift graben complex.

Tectonic history

The rift complex, previously named Central graben, was initiated in Later Permian–early Mesozoic times. Rejuvenation took place during Late Jurassic to mid-Cretaceous times and complexed the Central trough into several subunits. These tectonic episodes played a major role in determining the distribution and quality of the chalk facies and reservoirs.

Regional structure

Reverse tectonic movements following normal movements created a regional NNW-SSE half-horst feature that, combined with halokinesis of Permian salt, are responsible for facies, reservoir rock quality distribution, and early field structuration.

Local structure

Slightly NNW-SSE-elongated anticlinal dome—limited to the east by a saddle, connecting the field structure with a salt piercement to the east.

Trap

Trap type(s)

Anticlinal trap with two pays: erosional/nondepositional unconformities occur in the center of the field due to synsedimentary movements. Porosity and permeability distribution are related to facies distributions and diagenetic effects.

Basin stratigraphy (major stratigraphic intervals from surface to deepest penetration in field):

Chronostratigraphy	Formation	Depth to Top in ft (m) ss
Middle Miocene-Recent	*Nordland Group*	*-246 (-75)*
Eocene-early Miocene	*Hordaland Group*	*-3281 (-1000)*
Paleocene-early Eocene	*Rogaland Group*	*-9567 (-2916)*
Cenomanian-Danian	*Chalk Group*	*-10,138 (-3090)*
Portlandian-Albanian	*Cromer Knoll Group*	*-11,880 (-3621)*
Upper Permian	*Zechstein Group*	*-12,057 (-3675)*

Location of well in field .... *NA*

Reservoir characteristics:

Number of reservoirs *2*

Formations .... *Ekofisk Formation and Tor Formation*

Ages .... *Ekofisk, Danian; Tor, Maastrichtian*

Depths to tops of reservoirs .... *Ekofisk, -9875 ft (-3010 m) ss; Tor, -9941 ft (-3030 m) ss*

Gross thickness (top to bottom of producing interval)

Average .... *Ekofisk, 226 ft (69 m); Tor, 410 ft (125 m)*

Maximum .... *Ekofisk, 315 ft (96 m); Tor, 610 ft (186 m)*

Net thickness—total thickness of producing zones

Average .... *Ekofisk, 190 ft (58 m); Tor 360 ft (110 m)*

Maximum .... *Ekofisk, 243 ft (74 m); Tor 423 ft (129 m)*

Lithology .... *Pelagic and resedimented chalks, sparse biomicrite, micritic fraction sorted/unsorted*

Porosity type .... *Ekofisk, inter- and intracoccolith porosity; Tor, interparticle porosity*

Average porosity .... *20%; Ekofisk, 23%; Tor, 19%*

Average permeability .... *Ekofisk, 0.16 md; Tor, 0.49 md*

Seals:

Upper

Formation, fault, or other feature .... *Maureen Formation equivalent (Paleocene)*

Lithology .... *Claystone and shale*

Lateral

Formation, fault, or other feature .... *Decrease of porosity toward the flanks (compaction and cementation)*

Lithology .... *Chalk*

Source:

Formation and age .... *Mandal Formation, Late Jurassic*

Lithology .... *Shale*

Average total organic carbon (TOC) *4-5%*

Maximum TOC *10%*

Kerogen type (I, II, or III) .... *III/II*

Vitrinite reflectance (maturation) .... $R_o = 0.9\text{–}1.1$

Time of hydrocarbon expulsion .... *NA*

Present depth to top of source .... *-12,140 ft (-3700 m) ss*

Thickness .... *165 ft (50 m); 328 ft (100 m) in central basin*

Potential yield .... *8.9 kg/T*

Appendix 2. Production Data

Field name .... *Tommeliten Gamma*

Field size:

- **Proved acres** .... *1600 (650 ha)*
- **Number of wells all years** *8*
- **Current number of wells (as of year)** *6*
- **Well spacing** .... *30 ac (350 m between wells)*
- **Ultimate recoverable** .... *390 bcf and 32 MM bbl*
- **Cumulative production (to 1/1/87)** .... *Shut in*
- **Annual production (1988)** .... *NA*
- **Present decline rate** .... *NA*
  - **Initial decline rate** .... *NA*
  - **Overall decline rate** .... *NA*
- **Annual water production** .... *NA*
- **In place, total reserves** .... *740 bcf*
- **In place, per acre-foot** .... *NA*
- **Primary recovery** .... *390 bcf and 32 MM bbl*
- **Secondary recovery** .... *NA*
- **Enhanced recovery** .... *NA*
- **Cumulative water production** .... *NA*

Drilling and casing practices:

- **Amount of surface casing set** .... *722 ft (220 m)*
- **Casing program**

  30-in. at 722 ft (220 m); 20-in. at 3281 ft (1000 m); 13⅜-in. at 6332 ft (1930 m); 9⅝-in. at 10,007 ft (3050 m); 7-in. liner at 11,221 ft (3420 m)
- **Drilling mud** .... *Water-based*
- **Bit program** .... *36-in., 26-in., 17½-in., 12¼-in., 8½-in.*
- **High pressure zones** .... *−7710 ft (−2350 m) ss; mud weight, 15.5 lb/gal (1.86 SG)*

Completion practices:

- **Interval(s) perforated** .... *NA*
- **Well treatment** .... *Mechanical fractured and acidized*

Formation evaluation:

- **Logging suites** .... *DIFL/BHC-GR-SP-CAL; DLL/MLL-GR; CDL/CNL-GR; HR-DIP; Circumferential Acoustic log; Coregun*
- **Testing practices** .... *DSTs in the exploration wells 1/9-4 sidetrack and 1/9-6*
- **Mud logging techniques** .... *Formation evaluation log, pressure evaluation log; drilling data pressure log; temperature data log; computer logs*

Oil characteristics:

- **Type** .... *Paraffinic*
- **API gravity** .... *50-35°*
- **Base** .... *Paraffins*
- **Initial GOR** .... *5880 scf/bbl*
- **Sulfur, wt%** *0*
- **Viscosity, SUS** .... *0.0573 cp*
- **Pour point** .... *NA*
- **Gas-oil distillate** .... *NA*

Field characteristics:

- **Average elevation** .... *−10,500 ft (−3200 m) ss*
- **Initial pressure** .... *7050 psi (48.6 MPa)*

Present pressure .... *NA*
Pressure gradient .... *0.17 psi/ft (3.85 KPa/m)*
Temperature .... *264°F at –10,170 ft (129°C at –3100 m) ss*
Geothermal gradient .... *0.02°F/ft*
Drive .... *Gas expansion*
Oil column thickness .... *1100 ft (335 m)*
Oil-water contact .... *–10,975 ft (–3345 m) ss*
Connate water .... *NA*
Water salinity, TDS .... *95,000 ppm NaCl equivalents*
Resistivity of water .... *Ekofisk, 0.025 ohm-m; Tor, 0.023 ohm-m at 264°F*
Bulk volume water (%) *25%*

Transportation method and market for oil and gas:
Subsea pipelines via the Edda field installation.

Fulmar Field—U. K.
South Central Graben, North Sea

M. MEHENNI
W. Y. ROODENBURG
Shell U. K. Exploration and Production
Altens, Aberdeen, United Kingdom

FIELD CLASSIFICATION

BASIN: North Sea
BASIN TYPE: Rift
RESERVOIR ROCK TYPE: Sandstone
RESERVOIR ENVIRONMENT OF DEPOSITION: Shallow Marine and Deep Water Turbidites
RESERVOIR AGE: Jurassic
PETROLEUM TYPE: Oil
TRAP TYPE: Faulted Anticline

LOCATION

The Fulmar field is located in the central North Sea, U.K. Sector, in water depths of around 270 ft (82 m). The field lies 170 mi (270 km) southeast of Aberdeen and 7 mi (11 km) northeast of the Auk field and is situated on a fault-bounded terrace of the western margin of the southwest Central graben, adjacent to the Auk horst (Figure 1). Fulmar field occurs mainly within block 30/16 (Shell/Esso) and extends northward into block 30/11b (Amoco/Mobil/Texas Eastern/Amerada Hess/Fulmar Oil Co.). The field is unitized with the Shell/Esso share of the equity presently determined at approximately 94%. The Fulmar field is operated by Shell U.K. Exploration and Production* on behalf of the Shell/Esso North Sea Venture and the Fulmar Unit.

The field is of relatively small areal extent (4.4 mi^2; 11.3 km^2), yet has a hydrocarbon column of up to 900 ft (274 m). The ultimate recovery of hydrocarbons from Fulmar is estimated at 427 million bbl oil (68 million m^3) and 262 bcf (7.4×10^9 m^3) of associated wet gas.

HISTORY

Pre-Discovery

The Auk horst was the first structure to be tested in concession block 30/16, early in 1971. It lies at the margin of the Central graben, is considerably larger than the Fulmar structure, and constituted an obvious target for exploration. The reservoir objective in Auk was the Permian eolian sandstones of the Rotliegend Group, which were already established as the major reservoir horizon for southern North Sea gas fields. Commercially producible oil was discovered in Permian Zechstein carbonates and to a lesser extent in the underlying Rotliegend sandstones (Buchanan and Hoogteyling, 1979).

In the early seventies, little was known of the Jurassic sequence in the Central graben area; where it had been penetrated, it consisted dominantly of shale. However, by 1974 the possible existence of an upper Jurassic sandstone play in the area was becoming apparent:

- Gulf/Phillips drilled well 30/17a-1 (1973) to test a closure at Upper Permian level. It encountered 300 ft (90 m) of Upper Jurassic sandstones (with a porosity of 19% but no oil shows).
- In early 1974, Shell/Esso drilled well 29/20-1, west of the Auk field, with an Upper Permian objective. The well penetrated 290 ft (88 m) of Jurassic sandstone, but there was no closure at Jurassic level and no shows were observed.
- During appraisal drilling of the Hamilton Brothers' Argyll field (southeast of Auk), well 30/24-6 found Upper Jurassic sands unconformably truncated below the Lower Cretaceous. A 24 ft (7 m) interval of these sands flowed 1200 bbl (191 m^3) of oil/day (Pennington, 1975).

The presence of good-quality sands and the occurrence of nearby mature oil source rocks made any closure at Upper Jurassic level a potential hydrocarbon trap. Consequently, the Fulmar prospect, in relatively shallow water and close to the Auk

*Shell Expro operates on behalf of Shell U.K. Limited and Esso Exploration Production U.K. Limited.

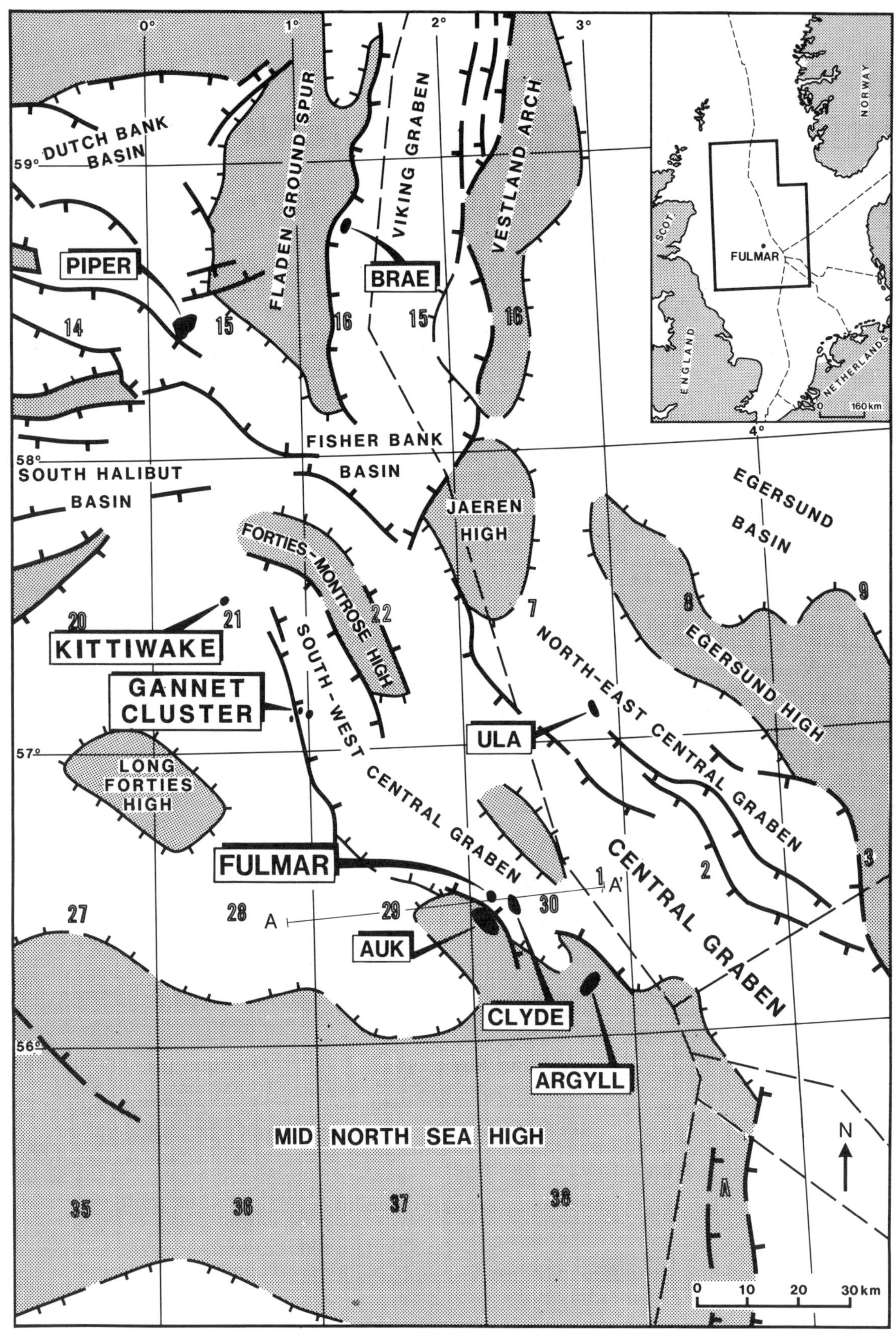

Figure 1. Megatectonic framework of the Central North Sea and the location of the Fulmar field. A–A′ is the location of the cross section shown by Figure 4. After Johnson et al., 1986. By permission of the publishers, Butterworth & Co. (Publishers) Ltd. ©.

production facilities, became an attractive drilling target for Shell/Esso. A further stimulus was that block 30/16 formed part of a license from which a 50% relinquishment would have to be made in June 1976. The known existence of extensive diapirism of Permian Zechstein salt throughout the Central graben area and the domal shape of the Fulmar structure led to the belief that it was salt-induced. If salt diapirism had not affected the Upper Jurassic too drastically, it was predicted on the basis of surrounding wells that up to 300 ft (90 m) of sandstone might be encountered.

Discovery

The Fulmar discovery well, 30/16-6 (Figure 2), was spudded during August 1975 by the Shell semi-submersible Staflo. Jurassic sandstones were encountered at 10,162 ft subsea (3097 m subsea), the upper 89 ft (27 m) of which were thinly bedded and intercalated with shales. The well then penetrated a massive 901 ft (275 m) fine-grained, well-sorted, clean sandstone, thus giving a gross sand interval of 990 ft (302 m), approximately 500 ft (150 m) of which was cored.

A sharp oil-water contact (OWC) was seen at 10,830 ft subsea (3301 m subsea). Logs showed the oil-bearing sandstones to have an average porosity of 24% and an oil saturation of 88%. Core measurements confirmed these figures and showed permeabilities of 1–2 darcys to be common. The total hydrocarbon column in the well was 668 ft (204 m) with 607 ft (185 m) net oil-bearing sand.

It was initially planned to suspend the well and test it later, but hole problems precluded testing and the well was abandoned.

Post-Discovery

The first appraisal well, 30/16-7, was drilled in 1977 approximately 600 m (2000 ft) southwest of the discovery well (Figure 2). The well encountered an oil-bearing "Upper Reservoir" 139 ft (42 m) thick, separated from a water-bearing "Main Reservoir" by 94 ft (29 m) of Kimmeridge shale (Johnson et al., 1986). A water-up-to (WUT) at 10,860 ft subsea (3310 m subsea) was considered consistent with the OWC at 10,830 ft subsea (3301 m subsea) as seen in the discovery well. Cores from the discovery and appraisal well caused considerable debate as to the shallow- or deep-water origin of the predominantly massive Upper Jurassic reservoir sandstones. Simultaneously, a 3D production seismic grid consisting of 94 lines at 75 m spacing was shot over the structure to facilitate later field development should the prospect prove to be commercially viable.

Commercial potential of the field was confirmed when well 30/16-7 flowed 8500 bbl (1350 m^3)/day of 40° API oil during a production test. With originally estimated oil-in-place figures of between 800 and 1000 million stb (125–160 million m^3), the field could now be recognized as a North Sea giant. Development plans for the field were submitted to the U.K. Department of Energy for approval in November 1977.

The first stage of development consisted of the predrilling of four oil producers during 1978–1979 from a six-slot, subsea template installed on the seabed in July 1978. Well T4, on the northern flank of the field (Figure 2), was surprising in that it encountered a shallow OWC of 10,562 ft subsea (3219 m subsea) in poor-quality sandstones. Data from these wells, three of which were cored, formed the basis of a geological model of the reservoir that was used in subsequent development drilling and for unitization. The main massive reservoir sandstones were interpreted as a stacked series of shallow-marine deposits, overlain in the west by deeper-water mass-flow sands encased in Kimmeridge shale (Johnson et al., 1986).

It is worth noting that during 1978, BNOC* (as operator for Britoil/Shell/Esso) discovered the analogous Clyde field in the neighboring block 30/17b.

By 1980, a 36-slot platform had been installed on Fulmar from which 18 wells were drilled between 1982 and 1984. These included one gas injector needed to minimize gas flaring prior to the completion of a gas-export pipeline; ten peripheral water injectors to maintain reservoir pressure as early reservoir performance had indicated only limited natural water drive; and seven oil producers with initial completion intervals low enough to avoid gas breakthrough from the gas cap created by injection and high enough to delay water influx as long as possible.

In 1984, drilling was suspended after completing 23 development wells, since combined oil production exceeded the capacity of the surface facilities. In 1986, the gas-export line became operational. The 1987 gas export averaged over 80 MMscf/day. In early 1987, daily oil production started to fall below the processing capacity for the first time as water cut and average GOR increased. In April 1987, a second phase of development drilling and workovers was initiated.

Results to date are encouraging with nine new wells performing better than expected and allowing production to average around 160,000 bbl (25,400 m^3) oil/day.

DISCOVERY METHOD

The Fulmar field was the first commercial discovery of oil in Jurassic reservoirs in the Central graben of the North Sea. The discovery resulted from careful geological review of an area that had already passed through its initial phase of exploration for

*BNOC: British National Oil Corporation; became Britoil in August 1982.

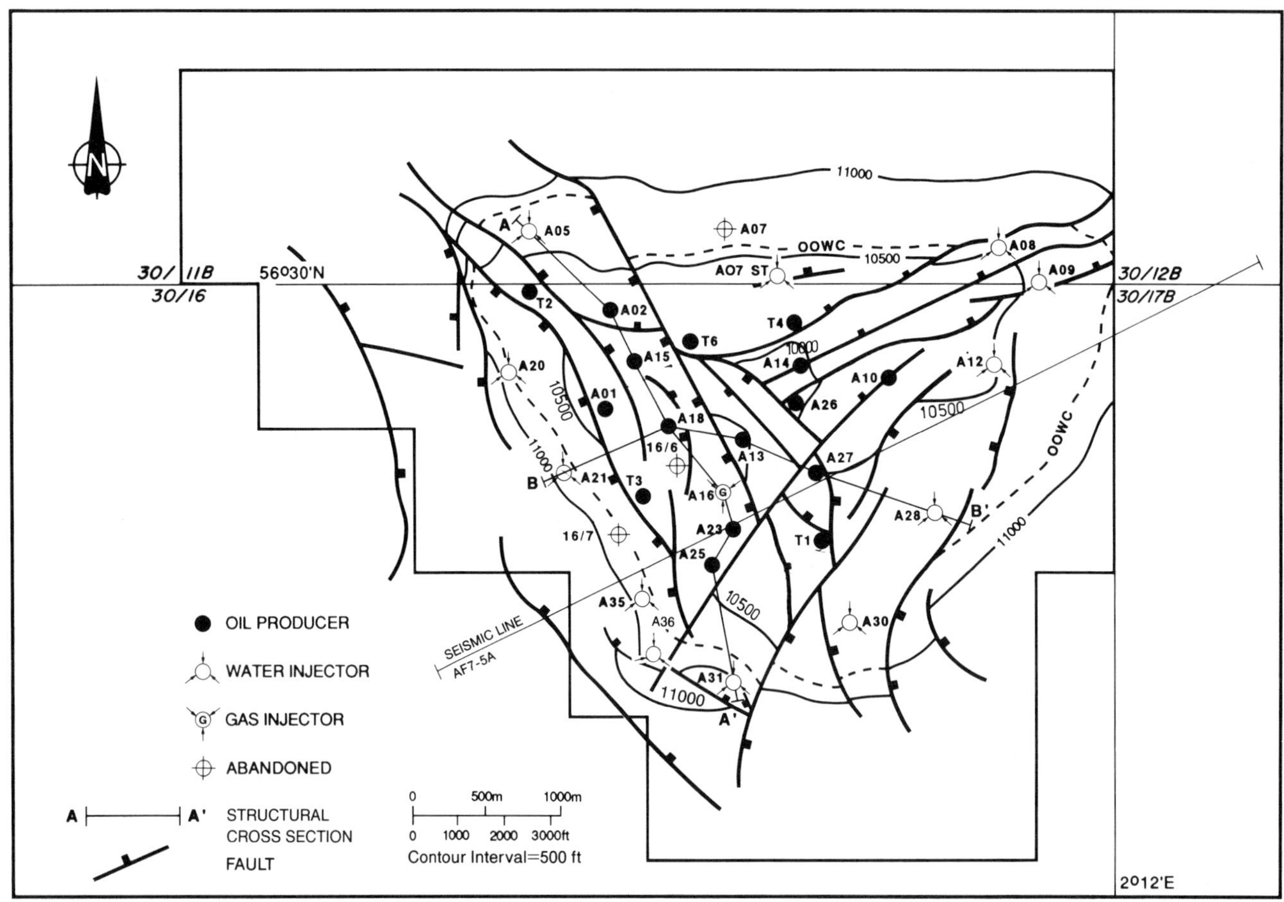

Figure 2. Top Fulmar Formation map showing the field outline, the original oil-water contact (OOWC), the well locations, and position of cross sections. The discovery well (16/6) is shown abandoned near the center of the field. A-A′ and B-B′ are the locations of cross sections shown by Figure 17. Contour values are subsea level.

Permian targets and that had a lower exploration priority in comparison with the Brent province of the northern North Sea. Geological review revealed the presence of potential Upper Jurassic sandstone reservoirs in proximity to mature Upper Jurassic source rocks and structures probably resulting from salt movement. From contemporary (1970) 2D seismic, it was possible to map the domal Fulmar closure at the base of the Cretaceous level and to suggest that its pod-like shape was salt-induced.

STRUCTURE

The pre-Permian geological history of the North Sea is summarized by Glennie (1984). Permian deposition occurred in two major basins, the Northern and Southern Permian basins, separated by the mid North Sea high. A system of grabens, of which the most important were the Viking and the Central grabens, cut across the two basins and were to dominate the later structural history of the area. Their complexities and evolution can best be seen in the atlas maps of Ziegler (1982).

The Fulmar area was located toward the southern margin of the Northern Permian basin where the Lower Permian Rotliegend eolian sands developed in a postorogenic intracratonic setting. The setting can be classified as 1211 under the modified scheme of Bally and Snelson (1980) (i.e., basin located on earlier rifted graben of a rigid, pre-Mesozoic continental lithosphere) and as 3A, using the scheme of Klemme (1971) (i.e., Continental rifted basin; Craton and accreted zone rift).

A rapid transgression of the Zechstein sea with ensuing evaporitic conditions resulted in the deposition of a thick sequence of evaporites. Later movement and removal of Zechstein salt was one of the major factors affecting the facies and thicknesses of the overlying Triassic and Jurassic sediments (Figure 3).

The Zechstein evaporites were followed by nonmarine Triassic sediments that accumulated in rim synclines adjacent to salt swells and diapirs. During the Middle Jurassic, the area was gradually

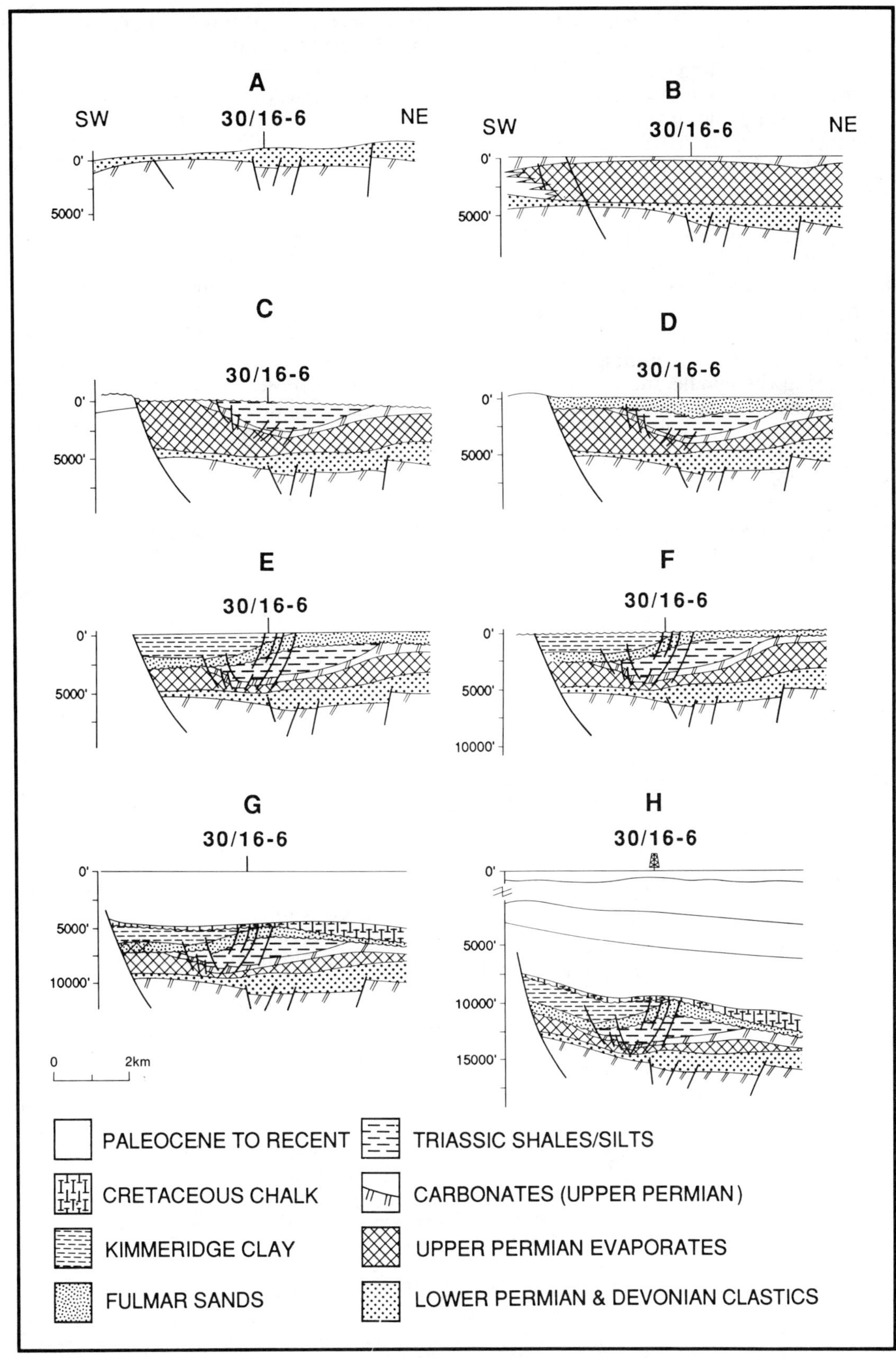

Figure 3. Conceptual model of the geological evolution of the Fulmar field. (A) Lower Permian Rotliegend. (B) Upper Permian Zechstein evaporites and carbonates. (C) Triassic, salt withdrawal producing salt ridges and synclines; early Cimmerian Unconformity. (D) Upper Jurassic Fulmar sands; continuation of salt withdrawal. (E) Upper Jurassic Kimmeridge Clay Formation. (F) Late Cimmerian Unconformity. (G) Cretaceous Chalk Group. (H) Present.

uplifted. Fluvio-lacustrine conditions led to salt solution and Triassic erosion; continual salt withdrawal resulted in "grounding" of the Triassic rim synclines either on a Zechstein carbonate base or on thin immobile salt, and inversion of the rim synclines commenced (Figure 3C). Such rim synclines became topographic highs, and eroded siliciclastics were deposited in erosional and/or salt solution hollows.

The Upper Jurassic marine transgression led to the infill of further rim synclines with reworked continental sands. Locally, the delicate balance between continuing subsidence and the rate of sedimentation resulted in amplified shallow-marine sand sequences such as the Fulmar Formation (Figure 3D). During the middle and upper Kimmeridgian (Late Jurassic), epicontinental marine conditions were again established and the sea transgressed the remaining highs blanketing the area with organic-rich muds (Figure 3E).

Whereas Early Cretaceous sediments accumulated in a tectonically active basin, regional subsidence over the area was the dominant feature throughout the Late Cretaceous. Finally, isostatic adjustments resulting from an increasing load of sediments and water were the cause of further subsidence during the Cenozoic.

Regional Structure

The regional structure is characterized by a series of major eastward-dipping faults that form a series of terraces between the Auk platform and the Central graben (Figure 4). These faults are complemented by a series of antithetic faults of Permian to Early Cretaceous age. They also affect subsequent sedimentation.

Late Permian topography strongly influenced the Zechstein depositional pattern with thick halite deposits in the centers of the basins and shelf carbonates along the graben margins. Major Triassic depocenters developed in the Fulmar, Clyde, and block 30/17a areas where withdrawal of Zechstein salt occurred. Reactivation of faults in the Jurassic and Early Cretaceous (Cimmerian tectonic phase) resulted in fault block uplift and subsidence within an overall setting of basin subsidence. Within the basins, local structure influenced by halokinesis resulted in successive salt withdrawal synclines that governed the pattern of Triassic and Jurassic sediment input and ultimately led to the creation of the Fulmar, Clyde, and block 30/17a structures.

Local Structure

The Fulmar structure (Figures 2 and 5) is a dome with pronounced flanks dipping to the southwest (18–25°), south (12°), and north (less than 10°). The structure is conformably overlain by the Kimmeridge Clay Formation on the western flank, but in the north and northeast, the Fulmar sequence has been significantly eroded with progressive truncation of the stratigraphic units.

The field is cut by a series of northwest–southeast-trending, eastward-dipping normal faults (range of throws 30–300 ft, 10–90 m). A number of similarly trending antithetic faults help define the west flank. A set of northeast–southwest-trending faults with a strike-slip component is present in the northeast and crestal areas of the field. The left lateral movement along these faults is estimated to be some 450 m.

The structural closure of the Fulmar field is interpreted to be an inverted, salt-withdrawal syncline (i.e., a turtle-back structure) that formed during the Mesozoic and lower Tertiary. The following observations suggest this: (1) the closure is detached; folding does not involve the Paleozoic sequence, (2) folding by inversion of a Triassic-Jurassic salt-withdrawal basin explains the superposition of the crest of the structure on the axis of the Triassic sedimentary pod, and (3) a well in the Clyde field, which is interpreted to have a similar structural genesis, has penetrated salt.

STRATIGRAPHY

From the seabed down to approximately 9000 ft (2750 m), the drilled sequence is dominated by clays and claystones of Quaternary and Tertiary age (Nordland and Hordaland Groups). They are generally slightly silty, calcareous, and glauconitic, with occasional bioclasts. These overlie an average 550 ft (170 m) thick section of green to gray-brown calcareous claystones of the Rogaland and Montrose groups (Figure 6).

The Cretaceous Chalk Group comprises 200 to 250 ft (60–75 m) of slightly argillaceous white to pink-gray mudstones and packstones with abundant shell fragments that overlie the older Upper Jurassic Humber Group unconformably.

The Kimmeridge Clay Formation of the Humber Group shows a wide variation of thicknesses, ranging from 0 to 750 ft (0–230 m) over the field area. It is a high gamma-ray, olive-black to gray silty claystone, carbonaceous and glauconitic. The Kimmeridge Clay Formation is both the source and cap rock for the Fulmar accumulation. The claystone sequence contains a wedge-shaped sequence of mass-flow sandstones up to 180 ft (55 m) thick. This reservoir represents a turbiditic wedge built out along the edge of the Auk horst and extending over the southwestern part of the field. The main reservoir, the Fulmar Formation of Kimmeridgian age, is a fine- to medium-grained arkosic sandstone locally cemented by dolomite and chalcedony.

Triassic Smith Bank shales underlie the Upper Jurassic. They are at least 1500 ft (450 m) thick and mainly consist of silty, red-brown claystones, gray sandstone interbeds, and limestone streaks. Dolo-

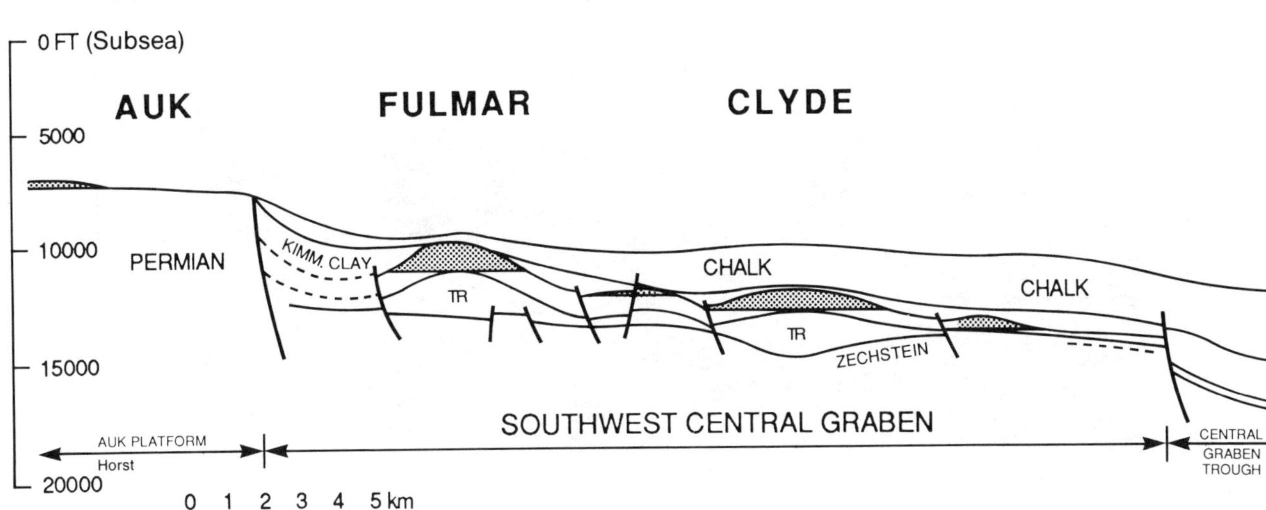

Figure 4. West-east geological cross section through the Auk-Fulmar-Clyde area illustrating the cascade nature of the accumulations. The location of the cross section is shown on Figure 1. Cross-hatch areas are hydrocarbon accumulations in Fulmar sandstone reservoir. After Johnson et al., 1986. By permission of the publishers, Butterworth & Co. (Publishers) Ltd. ©.

mites of the Zechstein Group of Upper Permian age are the oldest rocks that have been penetrated in the field.

Reservoirs elsewhere in the central North Sea range in age and facies from Permian eolian sandstones and shallow-marine carbonates, through Triassic fluvial sandstones, Middle Jurassic coastal and deltaic deposits, Upper Cretaceous allochthonous chalks, to early Tertiary submarine fan deposits.

The dominant source rock for the hydrocarbons present in the Central graben area and indeed for the central and northern North Sea is the Upper Jurassic Kimmeridge Clay Formation. It has developed as an organic-rich black shale and is informally called the "hot shale" because of its high radioactive log response. Oxfordian shales have also been found to be rich gas-prone source rocks.

Carboniferous Coal Measures, known to be the major source rocks for the gas fields of the southern North Sea, are presently not known to have contributed to hydrocarbon generation in the Central graben. Higher in the section, Triassic sediments have no source potential. The Lower Cretaceous, not present in the Fulmar area, is believed to be gas-prone. Finally, the Tertiary is unlikely to be an important source rock because of its low level of maturity.

TRAP

The Fulmar field is essentially an anticlinal closure with an unconformity trap element in the north and northeast. The lateral and upper seals are formed by the Kimmeridge Clay to the west and by the Cretaceous Chalk to the north and northeast.

The main part of the field has an oil-water contact (OWC) at 10,840 ft subsea (3304 m subsea), but a shallower contact at 10,562 ft subsea (3219 m subsea) occurs in the northern block and is possibly caused by entrapment of formation water during oil migration (Johnson et al., 1986). In the wedge-shaped sandstone reservoir contained in the Kimmeridge Clay Formation, the OWC is taken at 10,861 ft subsea (3310 m subsea), although shallower oil-water contacts have been encountered locally.

RESERVOIR

Lithostratigraphy

The Fulmar sands comprise shallow-marine sandstones of the Fulmar Formation and sandstone turbidites contained within the overlying Kimmeridge Clay Formation. Major changes in facies and thickness occur across a northwest-southeast-trending "hinge line" that effectively divides the field into two equal-sized areas. The Fulmar Formation in the northeast is an easily correlatable layered sandstone complex. In contrast, the southwestern side of the field has the appearance of a thick pile of undifferentiated sandstones (Figure 7). The sandstone turbidites of the Kimmeridge Clay Formation are limited to the southwest of the field.

A major review of the reservoir geology was initiated in 1987 to refine the existing reservoir model (Johnson, et al., 1986). This study was based on all

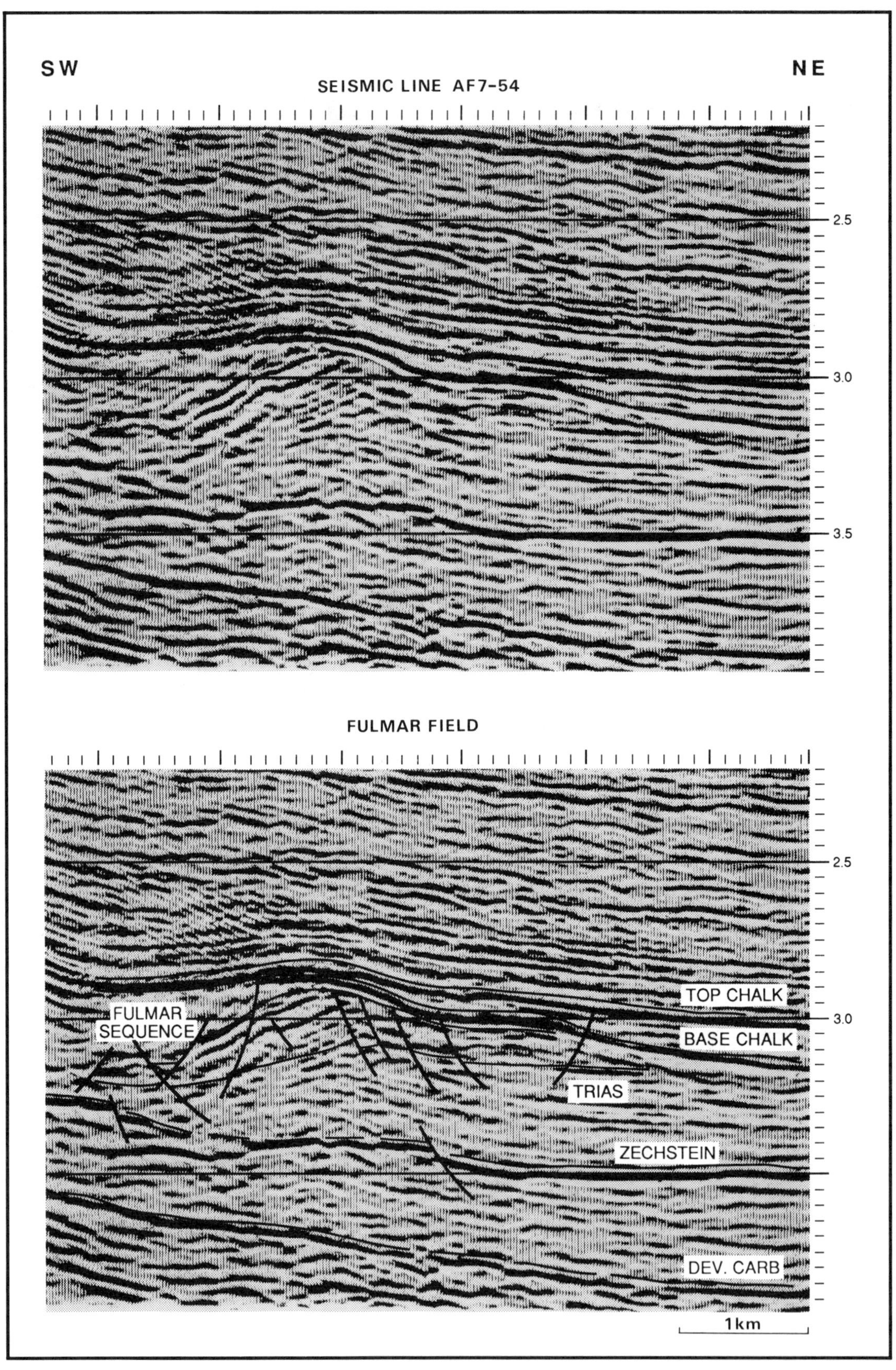

Figure 5. Southwest-northeast seismic line (AF7-54) across the Fulmar structure. The location of the line is shown on Figure 2.

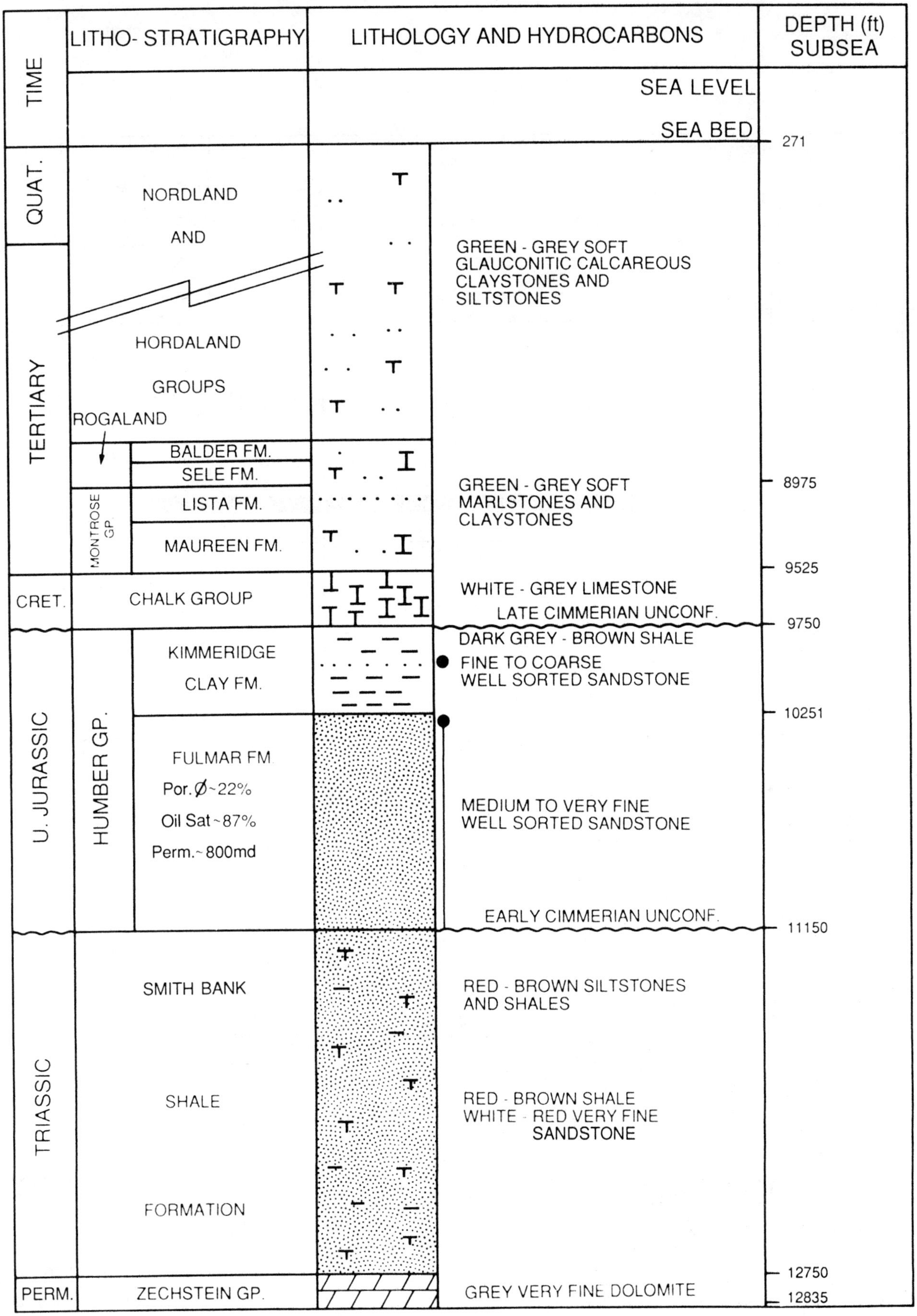

TIME	LITHO- STRATIGRAPHY		LITHOLOGY AND HYDROCARBONS	DEPTH (ft) SUBSEA
			SEA LEVEL	
			SEA BED	271
QUAT. / TERTIARY	NORDLAND AND HORDALAND GROUPS		GREEN - GREY SOFT GLAUCONITIC CALCAREOUS CLAYSTONES AND SILTSTONES	
TERTIARY	ROGALAND	BALDER FM.	GREEN - GREY SOFT MARLSTONES AND CLAYSTONES	
		SELE FM.		8975
	MONTROSE GP.	LISTA FM.		
		MAUREEN FM.		9525
CRET.	CHALK GROUP		WHITE - GREY LIMESTONE	
			LATE CIMMERIAN UNCONF.	9750
U. JURASSIC	HUMBER GP.	KIMMERIDGE CLAY FM.	DARK GREY - BROWN SHALE FINE TO COARSE WELL SORTED SANDSTONE	10251
		FULMAR FM. Por. Ø ~ 22% Oil Sat ~ 87% Perm. ~ 800md	MEDIUM TO VERY FINE WELL SORTED SANDSTONE	
			EARLY CIMMERIAN UNCONF.	11150
TRIASSIC	SMITH BANK SHALE FORMATION		RED - BROWN SILTSTONES AND SHALES RED - BROWN SHALE WHITE - RED VERY FINE SANDSTONE	12750
PERM.	ZECHSTEIN GP.		GREY VERY FINE DOLOMITE	12835

Figure 6. Fulmar field generalized stratigraphy.

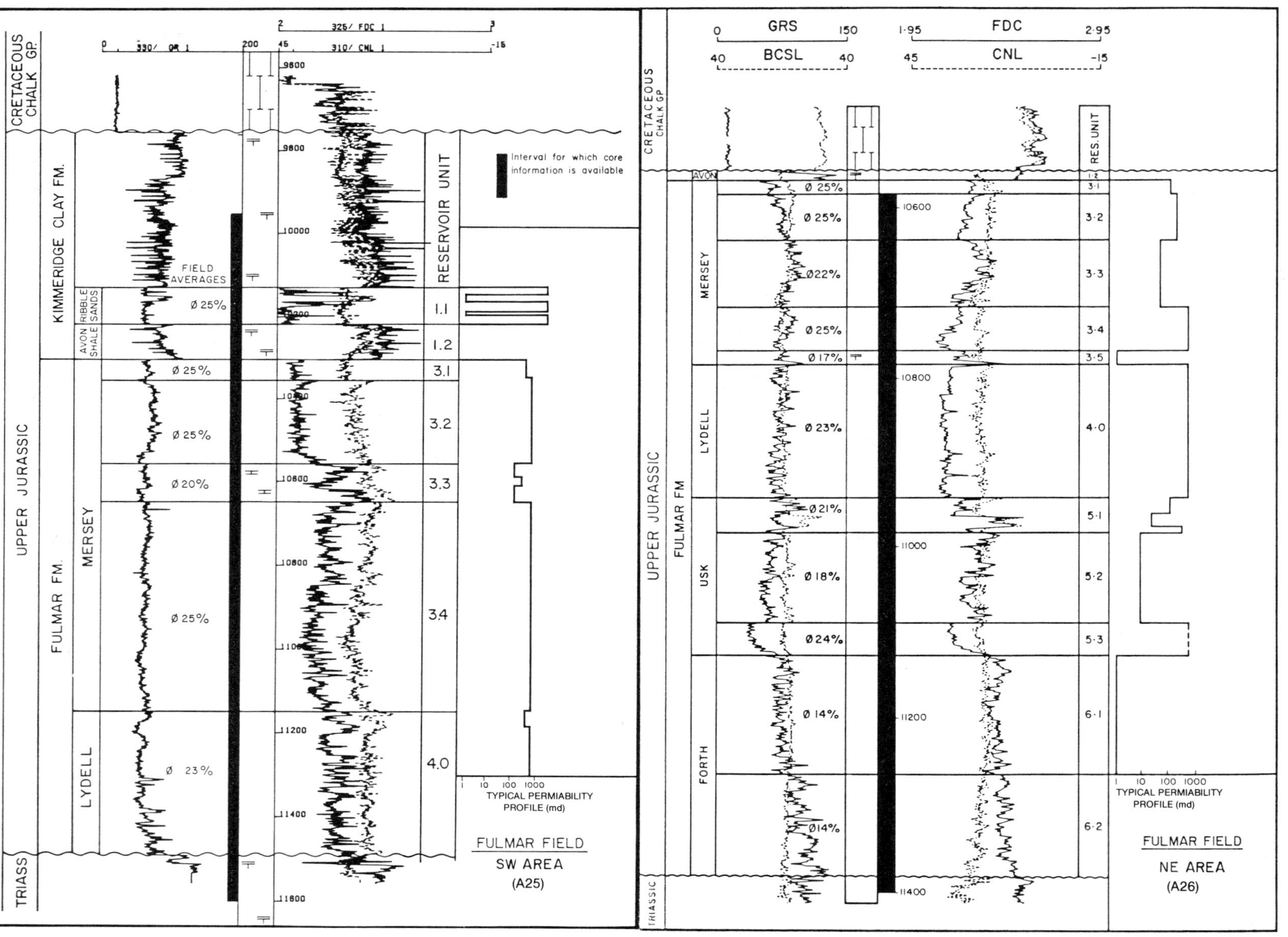

Figure 7. Reservoir type logs for the Fulmar field illustrating the contrasting sequences between the southwest and northeast flanks.

available core (6425 ft), log, and production data. Fourteen reservoir units were distinguished based on lithofacies and rock quality (Figures 7 and 8). The units can be identified from a combination of log traces and have well-defined porosity-permeability relationships (Figure 9). A palynological study provided valuable support in establishing a stratigraphic framework for correlation purposes (Figure 10).

The Fulmar sands of the Fulmar field contain three genetically distinctive facies associations (Figure 11): shallow-marine sandstones consisting of large-scale coarsening-upward sequences (reservoir units 3, 4, 5, and 6), argillaceous and cemented sandstones deposited in a lower-energy, shallow-marine environment (Clyde facies or reservoir unit 2), and turbiditic sandstones deposited in an anoxic, deeper-marine environment (Ribble or reservoir unit 1.1) (Figure 7).

Units 6.2 and 6.1 comprise very fine grained argillaceous and glauconitic sandstones, which are strongly bioturbated (*Chondrites* and rare *Zoophycus*) and only rarely show ripple cross-lamination. These argillaceous sandstones pass upward into a fine-grained, clean, very well sorted, moderately bioturbated sandstone (Unit 5.3). This regressive cycle unconformably onlaps the Triassic and is separated from an overlying major regressive cycle by an interbedded sandstone/shale interval (Unit 5.2) that forms a localized restriction to fluid flow.

The major regressive cycle (Units 5.1 and 4.0) consists of fine- to medium-grained, generally massive and mottled sandstones. Individual burrow types are difficult to distinguish. The trace fossils identified most closely resemble Cruziana and Skolithos burrow assemblages. Physical sedimentary structures are relatively uncommon but include parallel stratification to low-angle cross-stratification.

In the northwest of the field this main phase of basin shallowing was terminated by a thin fine-grained, argillaceous sandstone (Unit 3.5), but in the southwest sandstone accumulation continued. This fine-grained unit again forms a further restriction to fluid flow.

The upper part of the Fulmar Formation (Units 3.4, 3.3, 3.2, and 3.1) shows a stacked accumulation of coarsening- and occasionally fining-upward sequences. Grain sizes range from very fine to medium. The sandstones are bioturbated and massive.

Unit 2.0, the northeastern lateral equivalent of units 3.4 to 3.1, is a poorly sorted, fine-grained, extensively bioturbated, argillaceous sandstone with early diagenetic calcite and silica cementation derived from disseminated bivalve shells and siliceous sponge remains.

Unit 1.1 of the Kimmeridge Clay Formation is separated from the Fulmar Formation by a thin interval of finely laminated non-bioturbated shales (Unit 1.2). The turbiditic sandstones of Unit 1.1 are fine to medium grained and are interbedded with thin laminated shales and siltstones. The sandstones are poorly to well sorted and can contain abundant clay clasts, but within the sandstones there is little clay of either detrital or authigenic origin. The majority of sandstones consist of massive and laminated beds and seem not to be bioturbated.

Environment of Deposition

The Upper Jurassic sandstones of the Fulmar field were most likely derived from the reworking of Triassic sediments that had previously been deposited on the Auk platform. The sandstones accumulated in a relatively shallow sea.

The rapidly subsiding Fulmar basin is located on the western margin of the southwest Central graben. Relatively slow sedimentation rates allowed time for biogenic reworking of the Fulmar sands, resulting in destruction of primary sedimentary structures.

The Fulmar Formation can be interpreted as a major regressive sequence developed in a moderate-energy, offshore to nearshore shallow-marine environment (Figure 12A). A Fulmar-related shoreline has, however, not been positively identified. The large-scale (50–500 ft, 15–150 m) coarsening-upward sequences represent a gradual change in depositional conditions from low to high energy, caused by the progradation of shallow-marine ridges or sheet sands. The principal depositional mechanism of these sediments is thought to be periodic storm-induced currents rather than persistent longshore or tidal currents. The shallow-marine sandstones interfinger with heterogeneous, argillaceous, and early cemented sandstones deposited in a low-energy environment (Unit 2.0).

Basin deepening and anoxic conditions allowed the deposition of the organic-rich Kimmeridge Clay Formation. The sands of Unit 1.1 were introduced into this subsiding basin by sediment gravity-flow processes (Figure 12B). These flows may have been initiated seismically as a result of movements along the Auk fault. Rapid salt removal in the southwest of the field, consequent downwarping of the Triassic surface, and a continuous supply of sand has led to a thick, more or less homogeneous sand accumulation in the southwest and a thinner, distinctly layered sequence in the northwest of the field. Isopach maps indicate a westward shift in the sedimentary depocenter toward the Auk platform (Figures 13 and 14). It appears that the salt removal, rate of sediment supply, and faulting are all interlinked in a complex synsedimentary manner.

Porosity and Permeability

The major controls on porosity and permeability are the texture of the original sediments (grain size and clay content) and the presence of authigenic cements, particularly dolomite (Figures 15 and 16). The majority of the diagenetic changes that the sediments have undergone occurred during very early

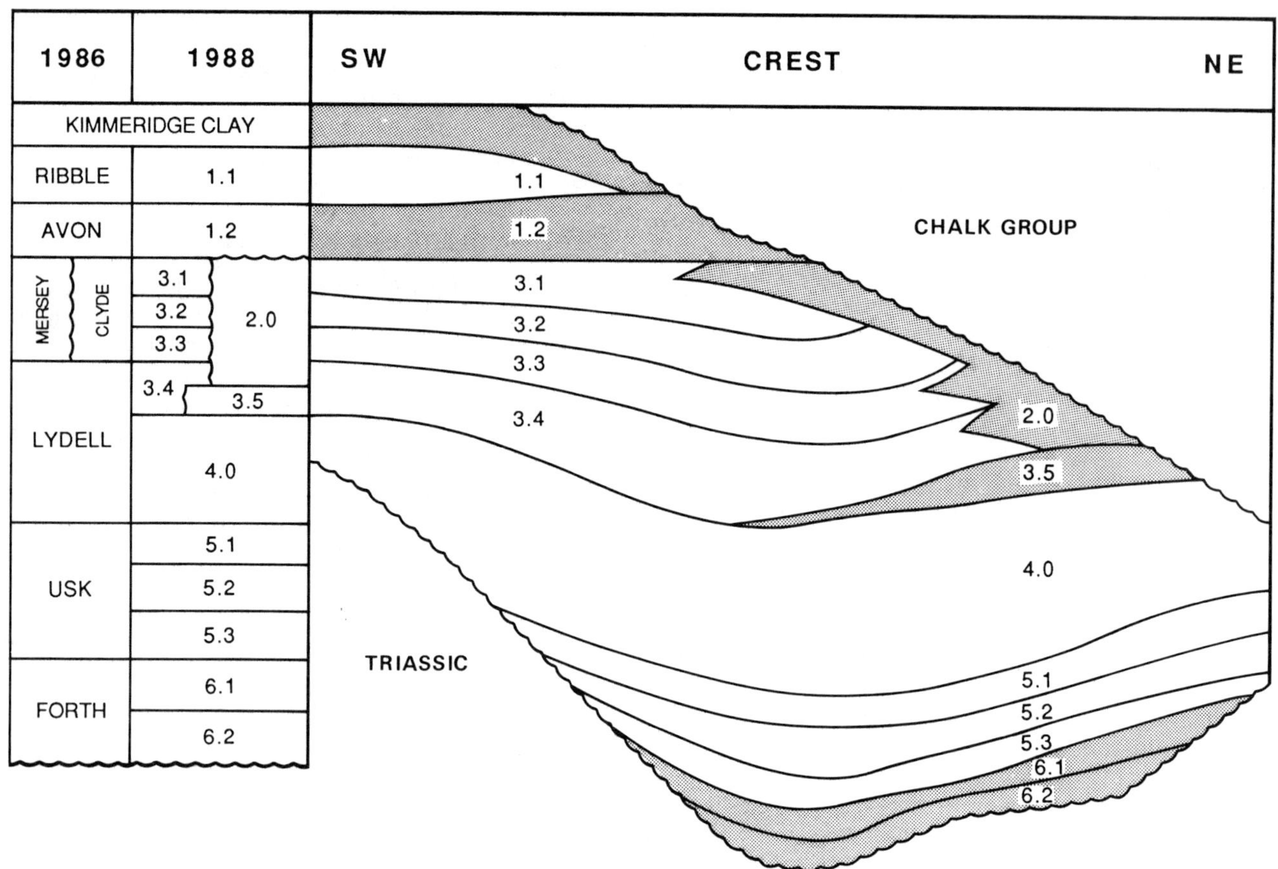

Figure 8. Reservoir unit nomenclature for the Fulmar field (1986 nomenclature after Johnson et al.). Cross hatch areas are poor- to non-reservoir rock. By permission of the publishers, Butterworth & Co. (Publishers) Ltd. ©.

burial (Johnson et al., 1986; Stewart, 1986). This early diagenesis is the result of rapid burial in a closed system and has given rise to dispersed or locally concentrated authigenic minerals. The low degree of grain-to-grain pressure solution and stylolite formation and the lack of pervasive illite suggests that a late diagenetic stage has not yet been reached. The Fulmar Formation in the Fulmar field is overpressured (1100 psi above hydrostatic). It is possible that the early development and maintenance of overpressures have prevented severe compaction by supporting some of the overburden load. As a consequence, the reservoir sandstones display excellent properties with porosities ranging between 15 and 30% (average 22%) and permeabilities between 10 and 3000 md (average 800 md).

Faults and Fractures

The Fulmar field has been subjected to several phases of faulting that have combined to produce a complex and dense fault pattern. Seismic has been used quite extensively to map the fault system. However, this was difficult within the Fulmar sands since most of the faulting pre-dates the Cimmerian unconformity and cannot be traced into the Cretaceous. More recently, in-house reprocessing and interactive interpretation have improved both the quality of the data and confidence in locating and correlating faults (Figure 17).

The normal fault system observed in the field is not considered to be sealing as indicated by the general field pressure history. The presence of a shallow OWC in the northern part of the field, however, appears to be partly due to the sealing capacity of the northeast-southwest-trending set of faults that exhibit a wrench component and a significant compressional seal effect (fault smearing of shalier units and grain diminution along the faults).

The majority of fractures observed in cores are largely attributed to syn- and early post-depositional dewatering (Johnson et al., 1986; Stewart, 1986). Shear zones and narrow, steeply inclined fractures are observed, often in association with fluidization pipes and massive, dewatered intervals. Additionally,

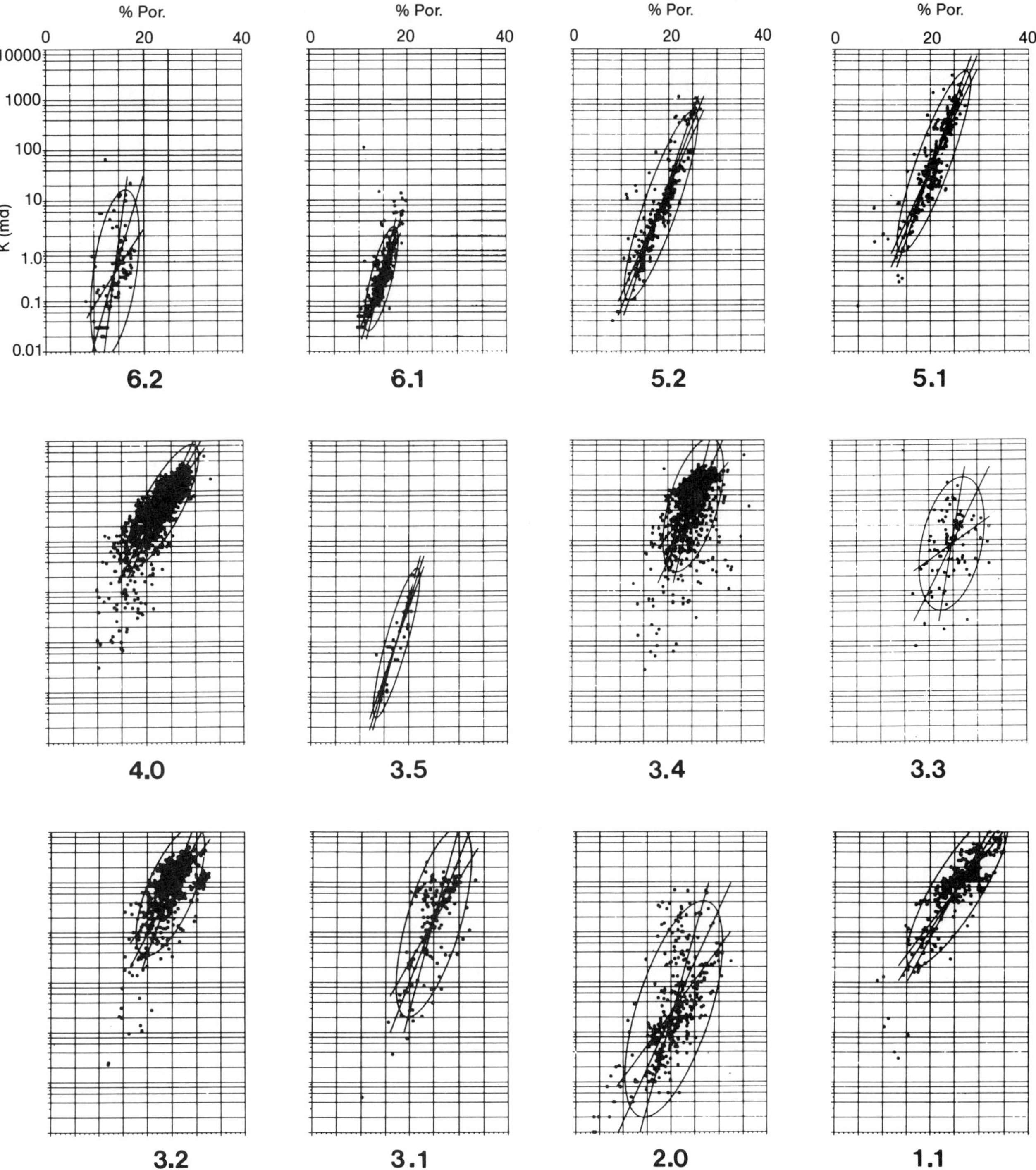

Figure 9. Porosity-permeability cross-plots of Fulmar reservoir units. The number of each plot refers to the reservoir unit nomenclature shown by Figure 8. Porosity and permeability axes and values along all axes of all plots are the same as for reservoir Unit 6.2.

some of the fractures may have formed in response to growth faulting. This would imply that the fractures and soft deformation structures are concentrated in belts directly on the downthrown side of and parallel to the large normal faults in the western part of the field. Fracturing of the sediments has led to reductions in permeability across the features, especially where concentration of fine-grained material and dolomitization within the fracture has occurred. Because of their limited lateral extent, they will not be significant in affecting fieldwide flow.

Hydrocarbon Characteristics

The oil found in Fulmar is highly undersaturated with a bubble-point pressure at reservoir temperature of 1800 psia (12.4 MPa), substantially below the initial reservoir pressure of 5700 psia (39.3 MPa) at a datum of 10,500 ft subsea and a gas-oil-ratio (GOR) of 614 scf/stb. The 40° API light oil has a viscosity of 0.44 cp (mPa.s). It contains 1.5 mg/L of H_2S and is saturated with asphaltenes at reservoir pressure. A gas chromatogram of the Fulmar oil is displayed as Figure 18.

Fluid Flow Characteristics

Reservoir pressures corrected to a common datum tend to cluster within 50 psia (345 kPa) of the general field trend, suggesting that faults are generally not sealing; more direct evidence for this has recently emerged from pulse tests between a number of wells.

A number of pressure measurements have indicated that Units 5.1 and 5.2 lag behind the main pressure trend. Unit 1.1, which has only just been brought on to production, has shown a similar pressure decline to the sandstones of the Fulmar Formation with a time lag of approximately three months.

The numerous dolomitic and siliceous streaks, slip planes, and fractures observed in cores probably hamper fluid flow locally. Recent thermal neutron decay time (TDT) logging has shown that oil in Unit 5.3 is sometimes by-passed; also, parts of Units 2.0 and 3.3 were by-passed by gas as the cap expanded down into the better 4.0 sands.

Productivity indices of 50–100 bbl/day/psi (1.2–2.3 m^3/day/kPa) are routinely achievable and wells can flow at 30,000 bbl/day (4800 m^3/d). For a number of wells, production has now been reduced to 7000–10,000 bbl/day (1100–1600 m^3/day) in order to limit water cuts and gas-oil ratios, thereby meeting facility constraints.

Development Considerations

The initial development plan was based on flank water injection to maintain reservoir pressure and to sweep what was considered to be a homogeneous sandstone sequence. Drainage areas and possible geological inhomogeneities were ignored. Primary recovery by depletion/compaction and aquifer drive is thought unlikely to have exceeded 10%. Secondary recovery (water/gas injection) is expected to reach nearly 60%. At the end of 1987, energy contributions were estimated as follows: 67% from water injection, 18% from gas injection, 10% from natural water influx, and 5% from oil/rock expansion. Owing to the lack of aquifer support, water injection is necessary to attain high oil recovery efficiencies from Fulmar. With injection, oil recovery from the field will reach 55 to 60% of the oil-in-place.

SOURCE BED

The source rock for the Fulmar oil is the black shales of the Kimmeridge Clay Formation. Oil-migration from the deeper parts of the Central graben to the east and northeast of the field and into the trap was completed by early Tertiary time (Figure 19).

The average total organic carbon (TOC) of the Kimmeridge Clay Formation is 5% with peaks of 10%; the rock has a type II kerogen, an average yield of 50 bbl/acre-ft (6.4×10^6 m^3 oil/km^3) per 1% TOC, and a vitrinite reflectance of 0.75.

EXPLORATION AND DEVELOPMENT CONCEPTS

The Fulmar discovery established a new play in the Central graben area. The play has continued to prove successful in the Clyde, Kittiwake, and Gannet West oil fields and in a number of less well defined discoveries.

Originally, samples analyzed for palynology were largely barren. However, improvements in preparation techniques and fresh palynological studies of the Fulmar sandstones have now yielded stratigraphic and facies information valuable for reservoir correlation and regional geological understanding.

The development of the field has justified the need to try to continuously improve the structural and sedimentological description of the reservoir in order to predict and rectify occurrences of gas and water breakthrough and oil by-pass.

Today's technical knowledge would have given us more precise tools (interactive seismic interpretation among others) to better position and optimally perforate wells. But in the case of Fulmar, exploration and appraisal philosophies were soundly based with most of the predictions made early in the life of the field still valid.

ACKNOWLEDGMENTS

The authors wish to thank the managements of Shell U.K. Exploration and Production, Esso Exploration and Production U.K., Amoco (U.K.) Exploration Co., Fulmar Oil Company, Amerada Hess (U.K.) Limited, and Texas Eastern North Sea Inc. for permission to publish this paper.

The authors wish to stress that the views expressed within this publication are those of the

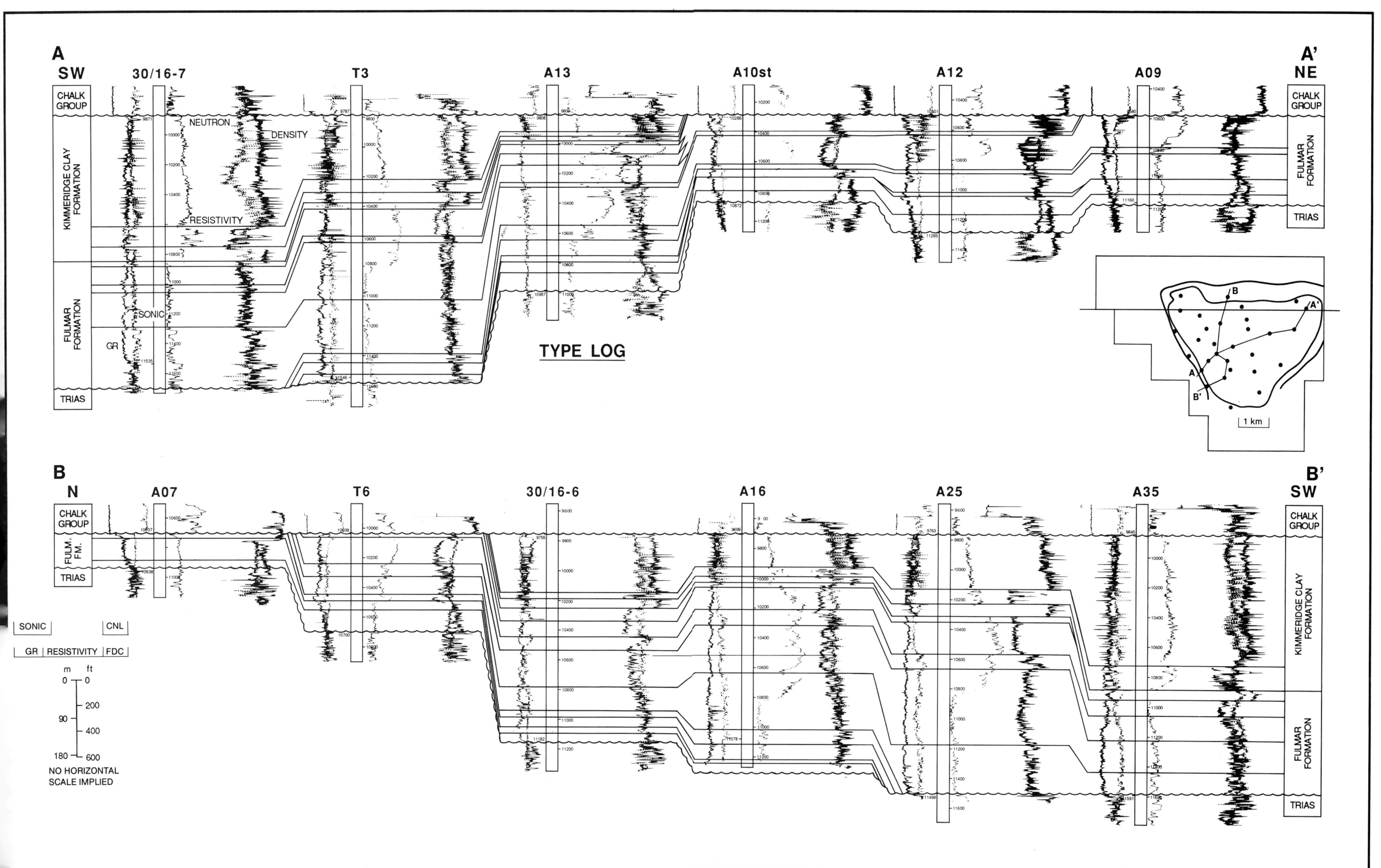

Figure 10. Log correlation panels through the Fulmar field.

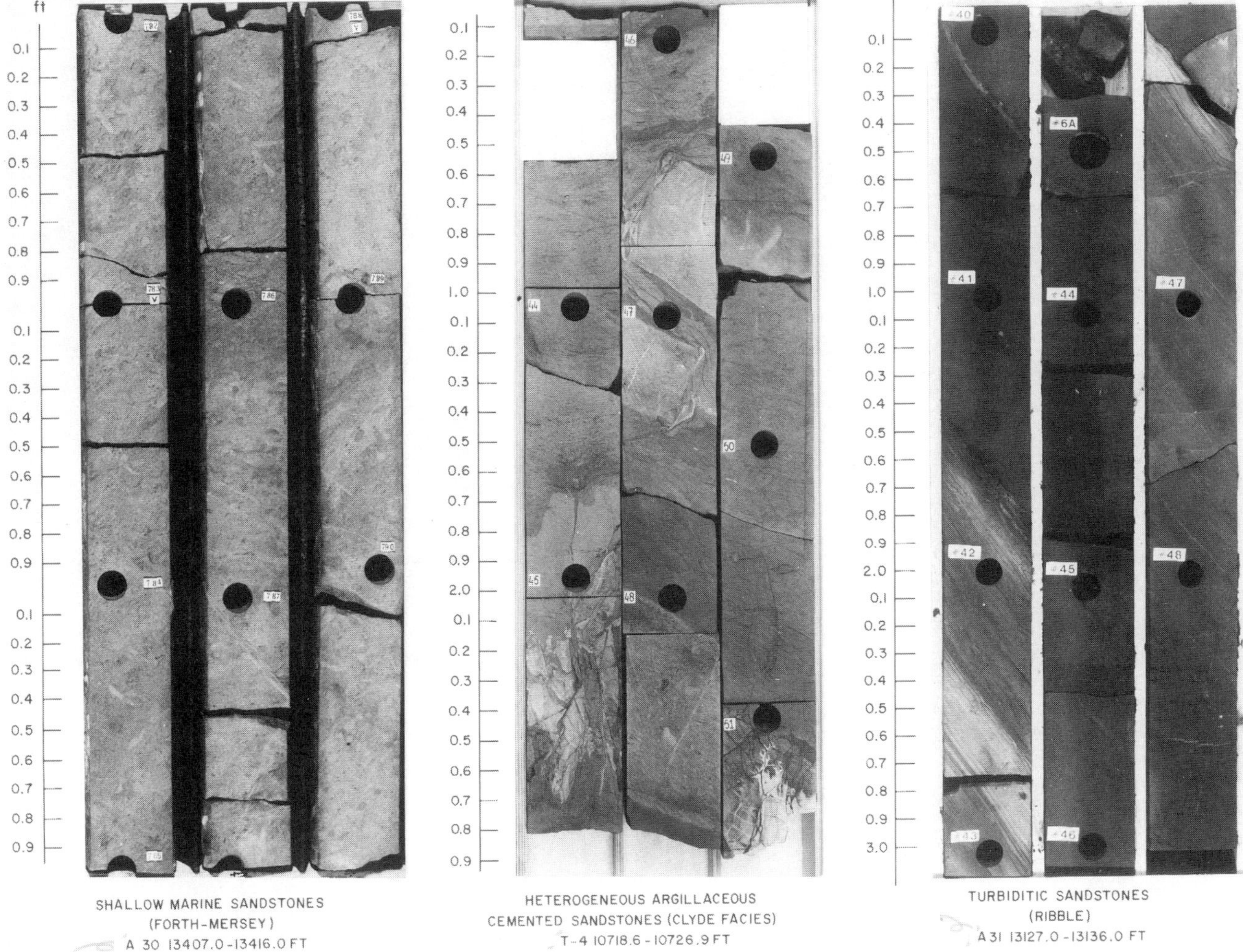

Figure 11. Core photographs of three facies associations recognized in the Fulmar field.

operator and may not represent the views of the other Fulmar participants.

REFERENCES

Bally, A. W., and S. Snelson, 1980, Realms of subsidence, *in* A. D. Miall, ed., Facts and principles of world petroleum occurrence: Canadian Society of Petroleum Geologists Memoir 6, p. 9–94.

Buchanan, R., and L. Hoogteyling, 1979, Auk field development: a case history illustrating the need for a flexible plan: Journal of Petroleum Technology, October 1979, p. 1305–1312.

Glennie, K. W., ed., 1984, Introduction to the petroleum geology of the North Sea: Oxford, Blackwell Scientific Publications, 236 p.

Johnson, H. D., T. A. Mackay, and D. J. Stewart, 1986, The Fulmar oil field (Central North Sea): geological aspects of its discovery, appraisal and development: Marine and Petroleum Geology, v. 3, p. 99–125.

Klemme, H. D., 1971, What giants and their basins have in common: Oil and Gas Journal, v. 69, n. 9, 10, 11; pt. 1, p. 85–90; pt. 2, p. 103-110; pt. 3, p. 96–100.

Pennington, J. J., 1975, The geology of the Argyll field, *in* A. W. Woodland, ed., Petroleum and the continental shelf of northwest Europe: London, Institute of Petroleum, p. 285–294.

Stewart, D. J., 1986, Diagenesis of the shallow marine Fulmar Formation in the Central North Sea: Clay Minerals, v. 21, p. 1-28.

Ziegler, P. A., 1982, Geological atlas of western and central Europe: Amsterdam, Elsevier, 130 p.

SUGGESTED READINGS

Brooks, J., and K. W. Glennie, eds., 1987, Petroleum geology of north west Europe: Proceedings of the 3rd Conference on Petroleum Geology of North West Europe, 1986, Graham and Trotman, London. The latest in a series of conference volumes regarding North Sea geology.

Johnson, H. D., and D. J. Stewart, 1985, The role of clastic sedimentology in the exploration and production of oil and gas in the North Sea, *in* P. J. Brenchley and B. P. J. Williams, eds., Sedimentology: recent developments and applied aspects:

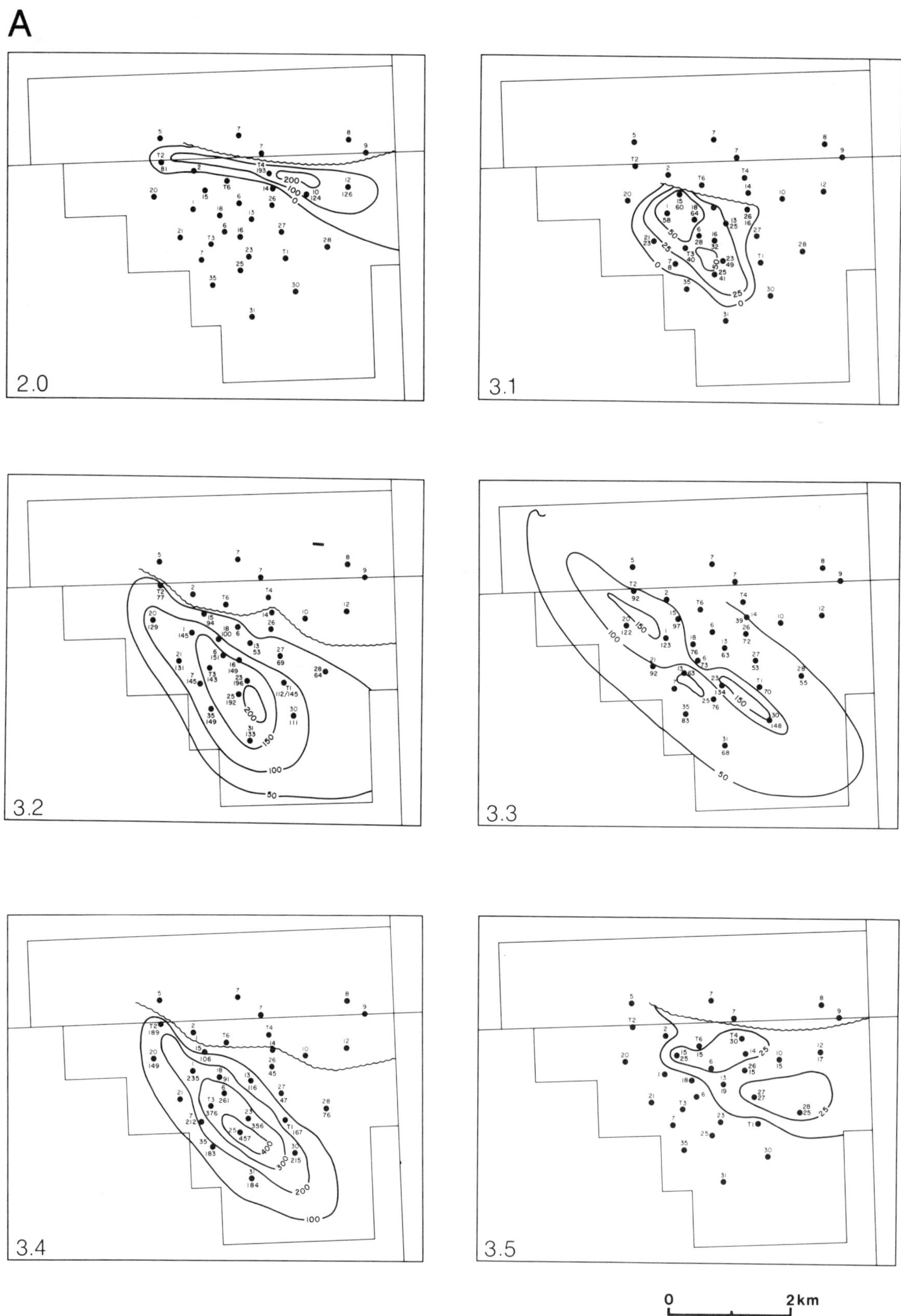

Figure 12. (A) and (B) Isopach maps of Fulmar reservoir units illustrating the lensoid shape of the sand pods and a general westward shift of the sedimentary depocenter. The decimal number on each map identifies the individual sand pods corresponding to the reservoir unit nomenclature shown by Figure 8. Contour intervals: 2.0, 100 ft; 3.1, 25 ft; 3.2, 50 ft; 3.3, 50 ft; 3.4, 100 ft; 3.5, 25 ft; 4.0, 100 ft; 5.1, 50 ft; 5.2, 50 ft; 5.8, 25 ft; 6.1, 50 ft; 6.2, 50 ft.

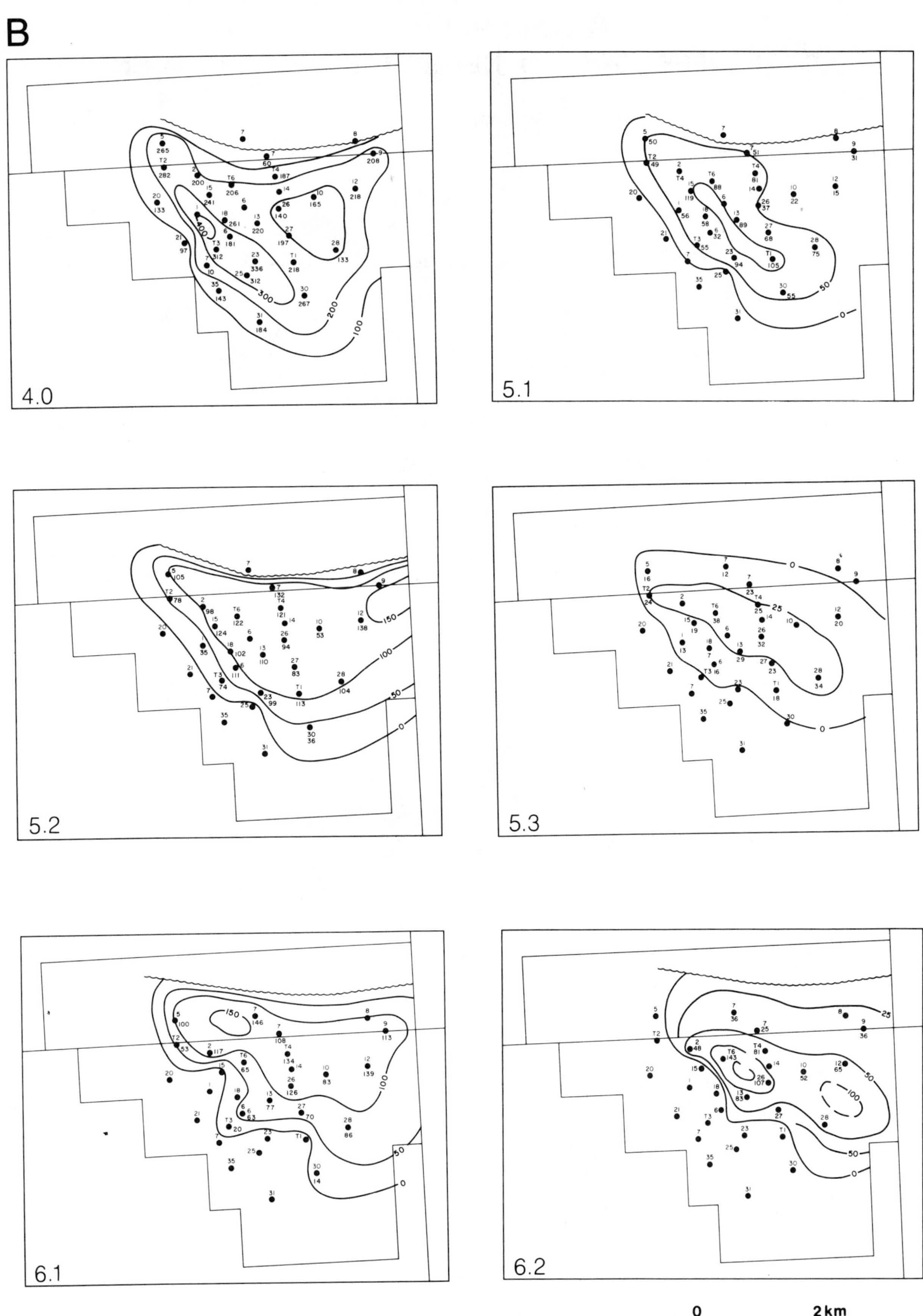

Figure 12. (Continued)

A. MAIN SANDS : DISTAL OFFSHORE TO NEARSHORE ENVIRONMENT

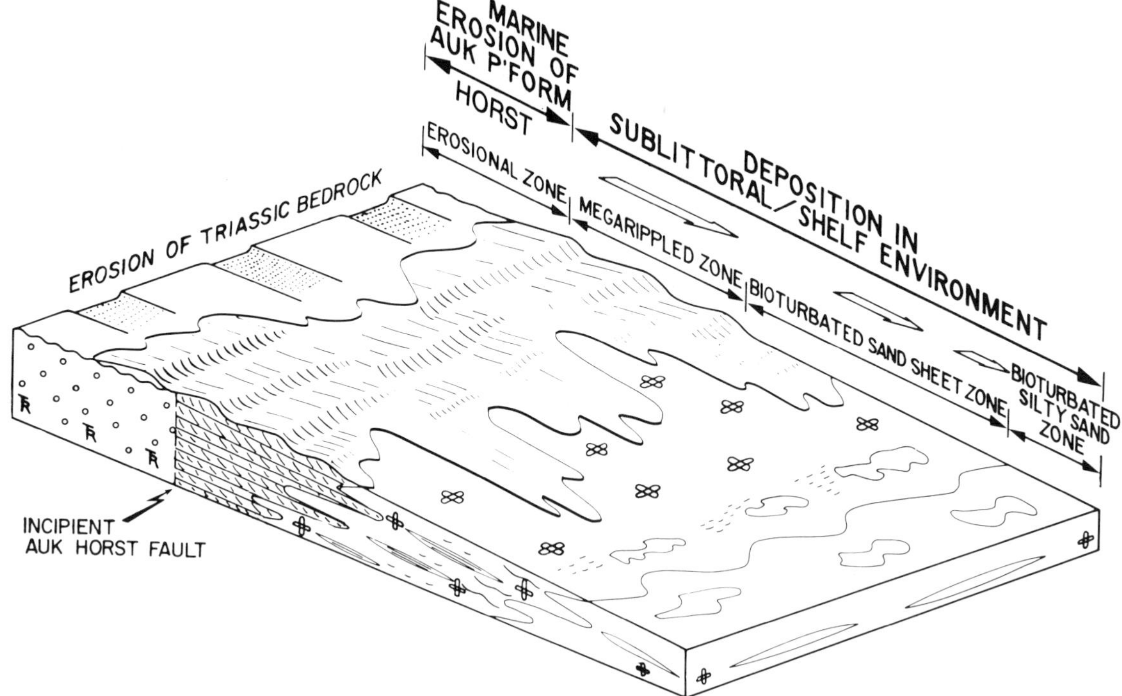

B. UNIT 1.1: MASS-EMPLACED SANDS DEEPER MARINE ENVIRONMENT

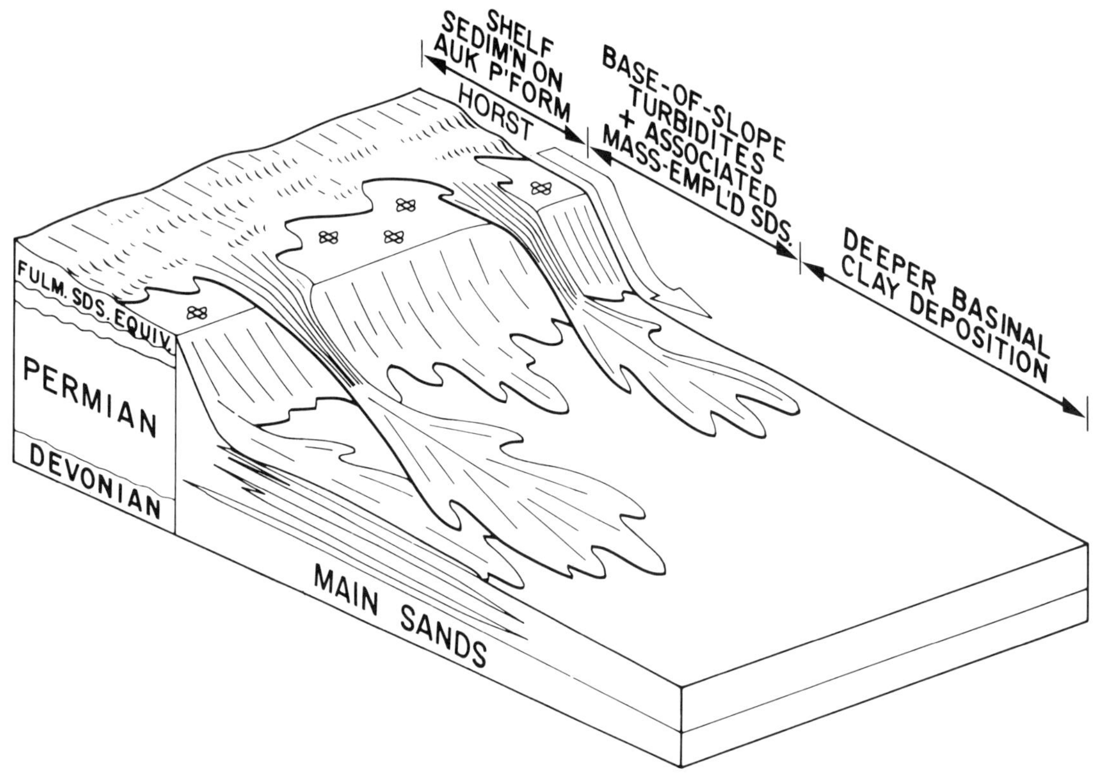

Figure 13. (A) Schematic paleogeographic reconstruction off the massive, bioturbated, occasionally cross-bedded Fulmar sands. (B) The massive, parallel laminated, mass-emplaced Kimmeridge Clay Formation sands.

SW NE

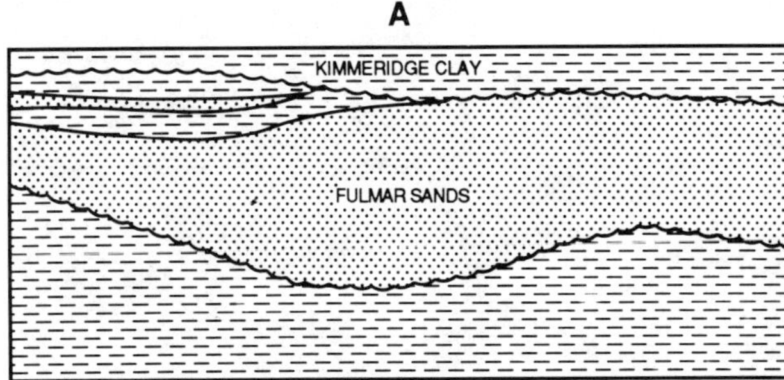

1977 : ONE MAIN AMPLIFIED SEQUENCE AND ONE SUBORDINATE SAND.

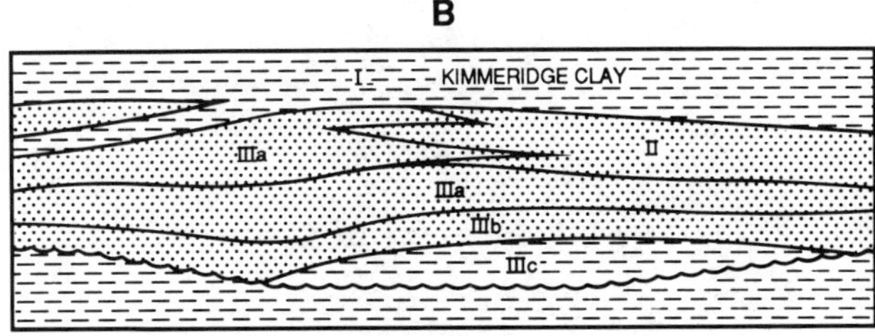

1986 : THREE GENETICALLY DEFINED UNITS.

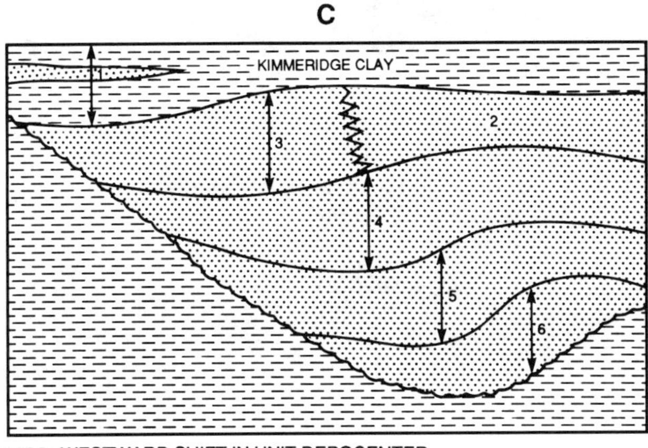

1988 : WESTWARD SHIFT IN UNIT DEPOCENTER.

0 1 km

Figure 14. A simplified cross section of the Fulmar sand "pod" illustrating the evolution in reservoir correlation subdivision for Fulmar and the general westward shift of the sedimentary depocenter. The numbered units shown in the 1988 interpretation are for correlation purposes and correspond to those shown in Figure 8.

Oxford, Blackwell Scientific Publications, p. 249-310. A review of depositional models of major clastic reservoirs in the U.K. sector of the North Sea.

Johnson, H. D., T. A. Mackay, and D. J. Stewart, 1986, The Fulmar oilfield (Central North Sea): geological aspects of its discovery, appraisal and development: Marine and Petroleum Geology, v. 3, p. 99–125. An excellent description of the appraisal history of the Fulmar field.

Kleppe, J., et al., eds., 1987, North Sea Oil and Gas Reservoirs: Proceedings of Norwegian Institute of Technology (NTH) conference (1985), Graham and Trotman, London, 352 p. Presents a variety of petroleum geoscience and engineering aspects of North Sea reservoirs, particularly from fields in the Norwegian Sector.

Figure 15. (A) Photomicrograph and SEM illustrating a typical well-sorted clay-free texture in a medium-grained arkosic sandstone from Unit 1.1. Porosity, 28.1%; permeability 4300 md. (B) Photomicrograph and SEM illustrating a typical moderate- to well-sorted texture in a medium-grained, arkosic sandstone with a minor amount of illite and mica from Unit 4.0. Porosity, 25.8%; permeability, 1700 md.

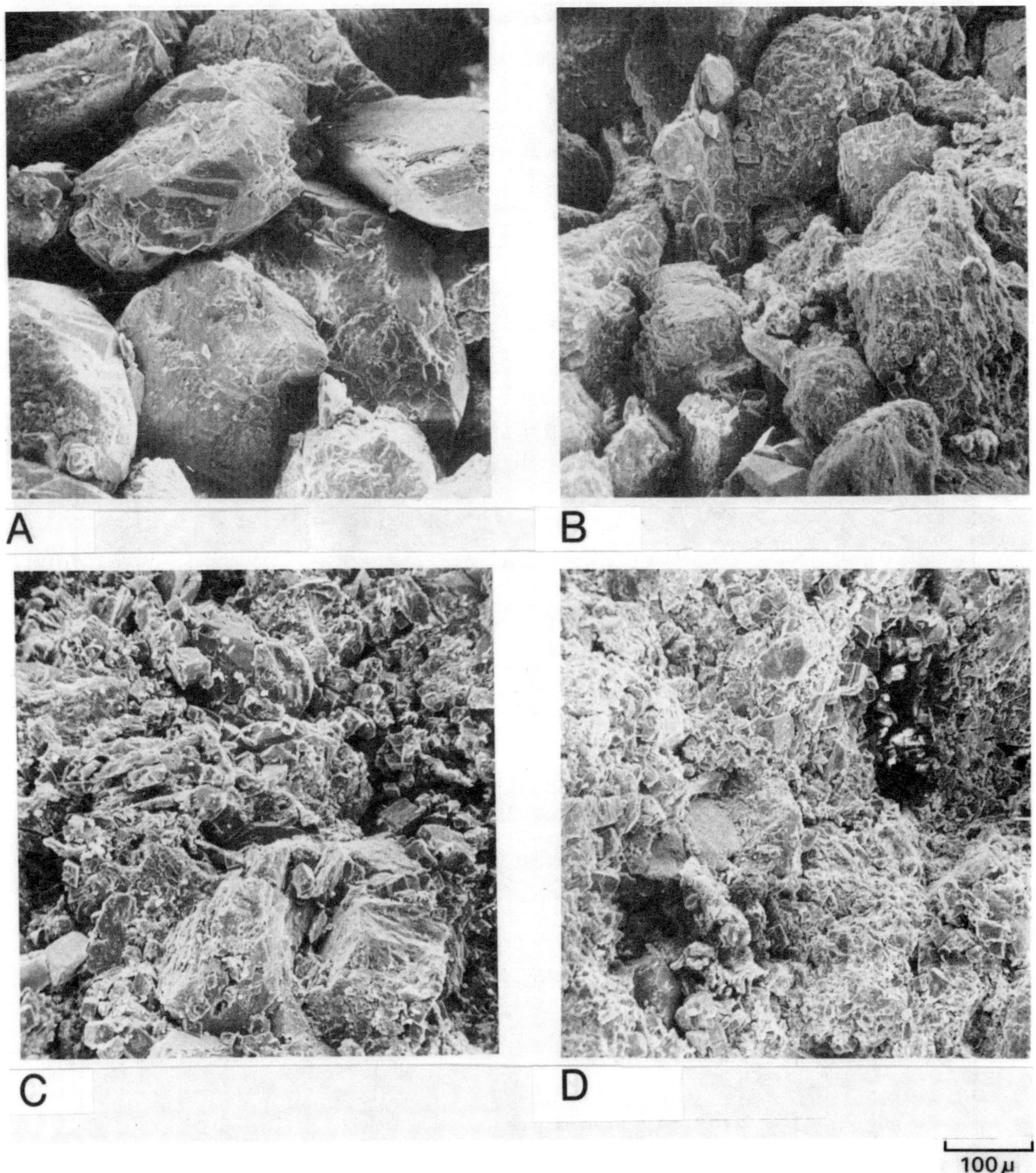

Figure 16. Series of SEM photographs illustrating progressive dolomite authigenesis. (A) No dolomite: porosity, 28%, permeability, 4000 md. (B) Dolomite 4% bv.; porosity, 28%; permeability, 1100 md. (C) Dolomite 21% bv.; porosity, 24%; permeability 320 md. (D) Dolomite 32% bv.; porosity, 17.9%; permeability, 6.6 md.

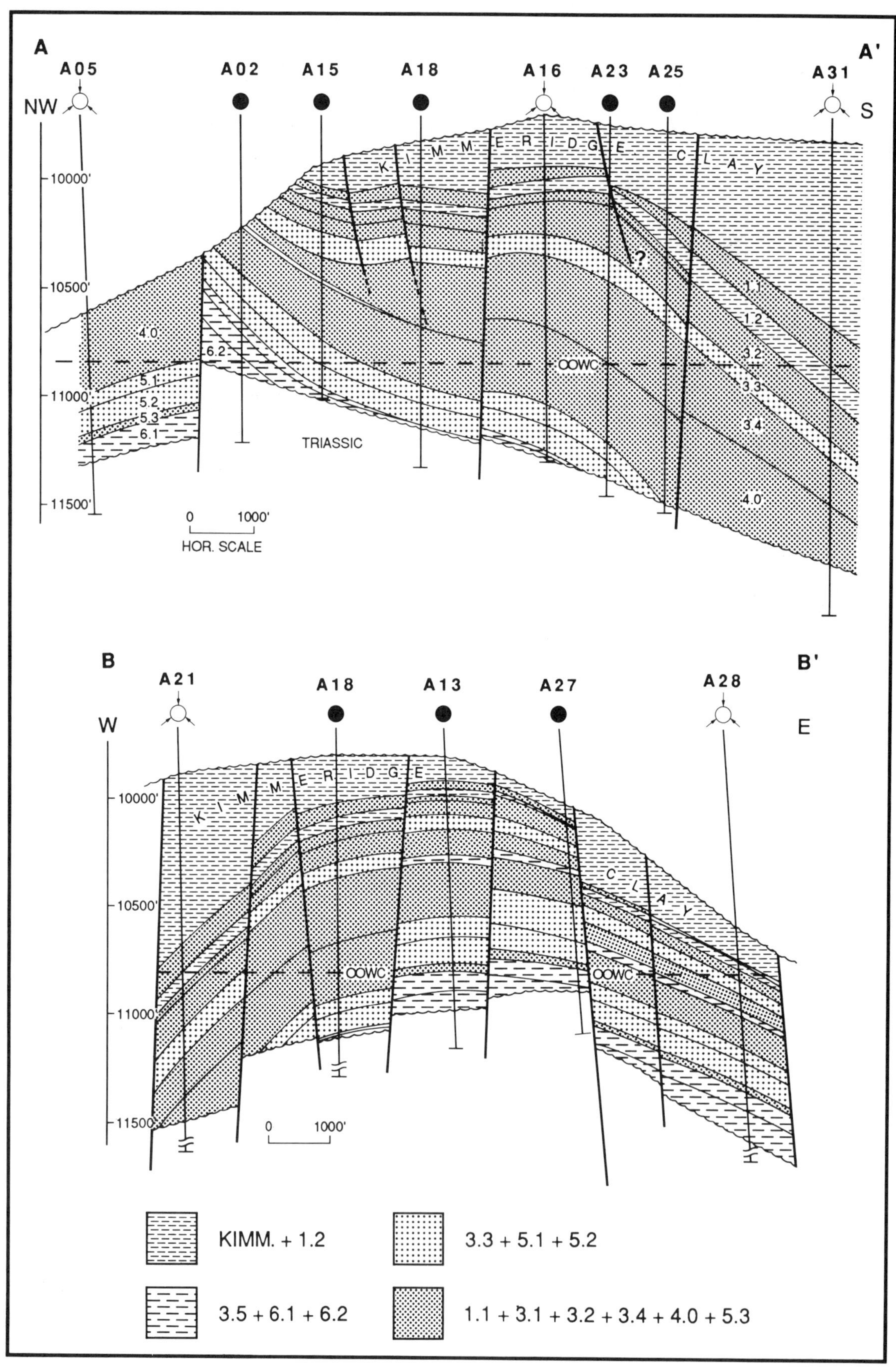

Figure 17. Structural cross sections through the Fulmar field. Locations of the lines are shown on Figure 2.

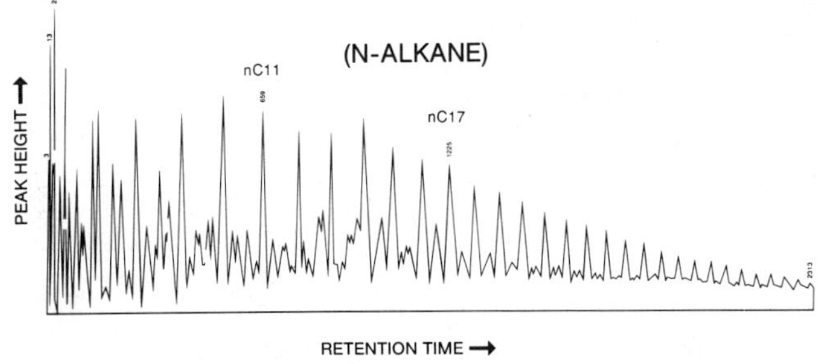

Figure 18. Representative gas chromatograph of oil sample from the Fulmar reservoir.

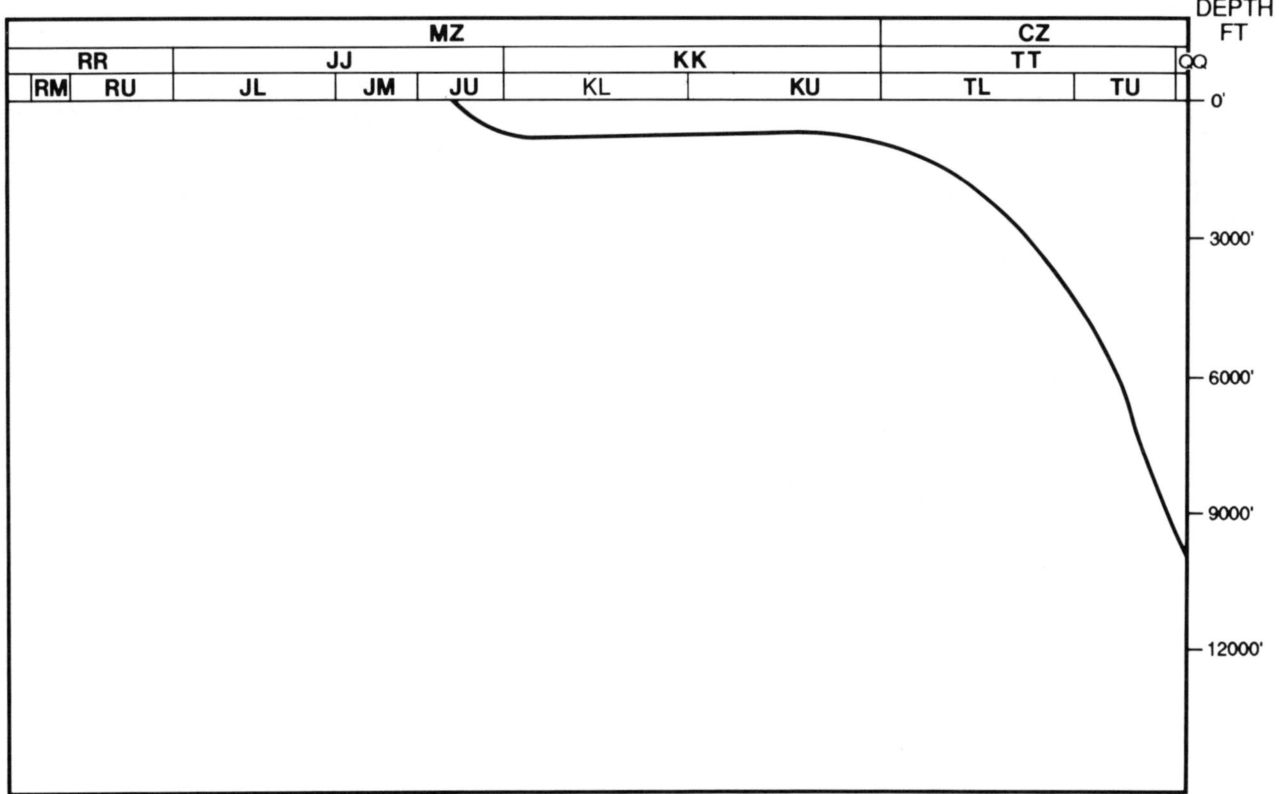

Figure 19. Burial history curve for the Fulmar reservoir. MZ, Mesozoic; CZ, Cenozoic; RR, Triassic; JJ, Jurassic; KK, Cretaceous; TT, Tertiary; QQ, Quaternary; L, Lower; M, Middle; U, Upper.

Appendix 1. Field Description

Field name .... *Fulmar field*

Ultimate recoverable reserves .... *427 MMstb (68 × 10^6 m^3)*

Field location:

- **Country** .... *United Kingdom*
- **State** .... *Offshore block 30/16, 30/11b*
- **Basin/Province** .... *North Sea Central graben*

Field discovery:

- **Year first pay discovered** .... *Upper Jurassic Fulmar Formation sandstones 1975*
- **Year second pay discovered** .... *Upper Jurassic Kimmeridge Clay sandstone turbidites 1977*
- **Third pay** .... *None*

IP in barrels per day and/or cubic feet or cubic meters per day:

- **First pay** .... *160,000 bbl/day*
- **Second pay** .... *8500 bbl/day*
- **Third pay**

All other zones with shows of oil and gas in the field:

Age	Formation	Type of Show
None		

Geologic concept leading to discovery and method or methods used to delineate prospect, e.g., surface geology, subsurface geology, seeps, magnetic data, gravity data, seismic data, seismic refraction, nontechnical:

Presence of Jurassic shallow-marine sand deposition downdip of the Auk platform inferred from wells 30/17a-1, 29/20-1, and from block 30/24 (Argyll field) in addition to a structure seen on seismic.

Structure:

Province/basin type .... *Central North Sea/Graben Margin*

Tectonic history

The main events from late Paleozoic to Upper Cretaceous: the Permian Rotliegendes sandstones were deposited in post-orogenic intracontinental basins. A rapid Zechstein transgression resulted in the widespread deposition of evaporites (Ziegler, 1978). Triassic continental deposits accumulated in rim synclines next to salt diapirs and swells. After the mid-Jurassic uplift and erosional events, Upper Jurassic sands derived from Triassic continental deposits accumulated in Upper Jurassic shallow seas. Finally, this was covered by organic-rich muds of Late Jurassic age followed by Cretaceous muds and chalks.

Regional structure

The regional structure is strongly influenced by half-graben tectonics and halokinesis. The field is on a terrace which forms the western margin of the southwest Central graben, a series of east-shading step faults. Salt movements led to expulsion of salt and deposition of a thick section of sediments in the central area of the block. Further expulsion of salt from the periphery(?) caused the rims to sag and form a rim syncline.

Local structure

The Fulmar structure is a domal anticline, transected by numerous faults; the field is triangular in shape, with a steep southwest flank (up to 25° dip); the reservoir formations are substantially eroded to the north and east.

Trap:

Trap type(s) .... *Single anticlinal trap with truncation element*

Basin stratigraphy (major stratigraphic intervals from surface to deepest penetration in field):
(Well 30/16-6)

Chronostratigraphy	Formation	Depth to Top in ft ss
Quaternary/Tertiary	*Nordland and Hordaland Groups*	
Tertiary	*Montrose Group*	*8975*
Upper Cretaceous	*Chalk Group*	*9525*
Upper Jurassic	*Humber Group*	*9750*
Triassic	*Smith Bank Formation*	*11,150*
Permian	*Zechstein Group*	*12,750*

Reservoir characteristics:

Number of reservoirs .... *1 (principally Fulmar Formation sandstone with 14 identified sand units; but also a sand of the overlying Kimmeridge Clay Formation)*

Formations .... *Fulmar, Kimmeridge Clay*

Ages .... *Upper Jurassic*

Depths to tops of reservoirs .... *9950 ft ss (crest)*

Gross thickness (top to bottom of producing interval) .... *500–1200 ft*

Net thickness—total thickness of producing zones

Average .... *450 ft*

Maximum .... *950 ft*

Average

Maximum

Lithology .... *Fine- to medium-grained, well-sorted sandstones*

Porosity type .... *Intergranular with minor moldic porosity*

Average porosity *22%*

Average permeability .... *800 md*

Seals:

Upper

Formation, fault, or other feature .... *Kimmeridge Clay or Chalk*

Lithology .... *Shale, chalk*

Lateral

Formation, fault, or other feature .... *As above*

Lithology .... *Shale, chalk*

Source:

Formation and age .... *Kimmeridge Clay, Kimmeridgian*

Lithology .... *Clay*

Average total organic carbon (TOC) *5%*

Maximum TOC *10%*

Kerogen type (I, II, or III) .... *II*

Vitrinite reflectance (maturation) .... $R_o = 0.75$

Time of hydrocarbon expulsion .... *Early Tertiary*

Present depth to top of source .... *9800 ft*

Thickness .... *0–700 ft*

Potential yield .... *250 bbl/ac-ft*

Appendix 2. Production Data

Field name .... *Fulmar field*

Field size:

- **Proved acres** .... *2000 (2800 including Ribble) (810 ha or 1134 ha including Ribble)*
- **Number of wells all years** ..... *30 (as of 1/1/88 excluding exploration wells and drilling wells in progress)*
- **Current number of wells (1/1/89)** .... *32 (typically 2000 ft between producers)*
- **Well spacing** .... *80 ac/well*
- **Ultimate recoverable** .... *427 million bbl oil; 262 bcf gas*
- **Cumulative production (1/1/89)** .... *319 million bbl*
- **Annual production** .... *57 million bbl oil; 35 bcf gas*
- **Present decline rate**
  - **Initial decline rate** .... *Field on plateau production (zero decline)*
  - **Overall decline rate**
- **Annual water production** .... *4 million bbl*
- **In place, total reserves** .... *800 million bbl*
- **In place, per acre-foot**
- **Primary recovery** *5%*
- **Secondary recovery** *48%*
- **Enhanced recovery**
- **Cumulative water production (1/1/89)** .... *20 million bbl*
- **Cumulative water injection (1/1/89)** .... *360 million bbl*
- **Cumulative gas injection (1/1/89)** .... *80 bcf*

Drilling and casing practices:

- **Amount of surface casing set** .... *Approx. 500 ft (30 in.)*
- **Casing program**
  20-in. at approx. 2000 BDF; 13⅜-in. at 6000 ft (to case off swelling clays); 9⅝-in. at 9500 ft (in chalk, above reservoir); 7-in. to TD
- **Drilling mud** .... *Oil-base mud*
- **Bit program** .... *17½-in., 12¼-in., 8½-in., rotary type*
- **High pressure zones** ... *Main sands 1100 psia above hydrostatic Triassic contains sands at initial pressure*

Completion practices: .... *Single; side pocket mandrel added in the case of injectors; 5-in. liner for through tubing workover*

- **Interval(s) perforated** .... *Main sands intervals; later recompletion in the Ribble*
- **Well treatment** .... *Acid stimulation tried in poor injector wells; scale inhibitor treatment in oil producers when required*

Formation evaluation:

- **Logging suites** .... *From 9⅝-in. shoe: MSFL/DLL (DIL)/LDT/CNL/GR/SP/BHC/RFT/CBL; routine TDT monitoring; production logging for problem wells*
- **Testing practices** .... *Twice monthly well gauging; pressure build-up surveys, surface fall-off surveys, static bottom hole pressure surveys*
- **Mud logging techniques** .... *None in development wells*

Oil characteristics:

- **API gravity** .... *40°*
- **Base** .... *Naphthenic-paraffinic*
- **Initial GOR** .... *614 scf/stb*
- **Sulfur, wt%** .... *1.5 mg/L in fluid, 10–15 ppm H_2S in export gas*
- **Viscosity, SUS** .... *0.44 cp oil, 103 gas, 130 water*
- **Pour point** .... *ASTM max: 3°C; ASTM min: 36°C; cloud point: 19.6°C*

FULMAR

Gas-oil distillate

Field characteristics:

Average elevation .... *9950 TVD ft ss*
Initial pressure .... *5700 psi*
Present pressure .... *5000 psi*
Pressure gradient .... *0.57 psi/ft*
Temperature .... *260°F at 10,500 ft subsea*
Geothermal gradient .... *1.6°F/100 ft*
Drive .... *Water injection (70%); gas injection (20%); aquifer (10%)*
Oil column thickness .... *900 ft*
Oil-water contact .... *10,560, 10,840 ft*
Connate water *15%*
Water salinity, TDS .... *138,000 ppm NaCl eq.*
Resistivity of water .... *0.02 ohm (0.05 at 25°C)*
Bulk volume water (%) .... *3.75% (SW $\times$ θ = 0.15 $\times$ 0.25)*

Transportation method and market for oil and gas:

Oil (+ Clyde + Auk) to floating storage unit via a single anchor leg mooring; gas (+ Clyde) export to St. Fergus (290 km) via 20-in. gasline (22-7-86)

Midgard Field—Norway
Southwest Central Graben, North Sea

O. F. EKERN
Saga Petroleum A.S.
Høvik, Norway

FIELD CLASSIFICATION

BASIN: Mid-Norway Continental Margin
BASIN TYPE: Passive Margin
RESERVOIR ROCK TYPE: Sandstone
RESERVOIR ENVIRONMENT OF DEPOSITION: Marginal Marine/Estuarine and Wave-Dominated Shoreline
RESERVOIR AGE: Jurassic
PETROLEUM TYPE: Gas and Condensate
TRAP TYPE: Fault Traps

LOCATION

The Midgard field is located offshore mid-Norway approximately 230 km northwest of Trondheim (Figure 1). The field is situated on the Halten terrace, which is located between the Trøndelag platform to the east and the Vøring basin to the west. The most significant additional discoveries in this province are the Draugen and Njord oil fields, the oil- and gas-bearing Heidrun field, and the condensate-dominated Smørbukk field. The major part of the Midgard field is located in blocks 6507/11 and 6407/2, both operated by Saga Petroleum (Figure 2). The total areal extent of the field is in excess of 47 km^2. The estimated recoverable gas reserves are 109×10^9 Sm^3. In addition, a condensate recovery of 15×10^6 Sm^3 is estimated.

HISTORY

Pre-Discovery

The first well on the Norwegian continental shelf was drilled in 1966. The exploration drilling was restricted for the following 15 years to the area south of the 62nd parallel. However, it was evident by the early 1970s that a significant sedimentary basin existed beneath the mid-Norway continental margin. The thickness and distribution of these sediments were first assessed using gravimetric and magnetic data (Grønlie and Ramberg, 1970; Åm, 1970). Seismic reflection data soon became the major source of information. Initially, beginning in 1970, seismic data were mainly acquired by the Norwegian Petroleum Directorate, but later seismic data were also acquired by the oil industry. Interpretations made from regional seismic data were published by Rønnevik et al. (1975), Rønnevik and Navrestad (1977), and Jørgensen and Navrestad (1979, 1981). Another important element in the database was established through a regional geological mapping program of the shallow bedrock carried out by the Continental Shelf Institute (IKU). The program identified the age of sedimentary units subcropping below the Quaternary cover by the use of sparker seismic, sampling, and coring. This work is summarized by Bugge et al. (1984).

The work carried out up to 1979 clearly demonstrated that the conditions for hydrocarbon discoveries on the mid-Norway continental margin were present. Hence, high expectations existed prior to the opening of the area north of the 62nd parallel through the 5th Concession Round. The first licence in the Haltenbank area, block 6507/12, was awarded to a group with Saga Petroleum as operator in January 1980. The first well was completed the same year but was disappointingly dry.

Discovery

The dry status of the first well drilled in the area was explained by a lower than predicted maturation level of the Jurassic source rocks. Hence, interest was now focused on blocks with structures associated with drainage areas where source rock intervals were buried at greater depths.

The second part of the 5th Concession Round included block 6507/11, awarded in March 1981. Well 6507/11-1 was drilled by Saga Petroleum the same year. The well tested gas with a flow rate of 0.76×10^6 Sm^3 from the northern segment of a horst block complex (Figures 2 and 3). The discovery was later named Midgard field; Midgard is the name of the human world in Norse mythology. This was the first hydrocarbon discovery offshore mid-Norway.

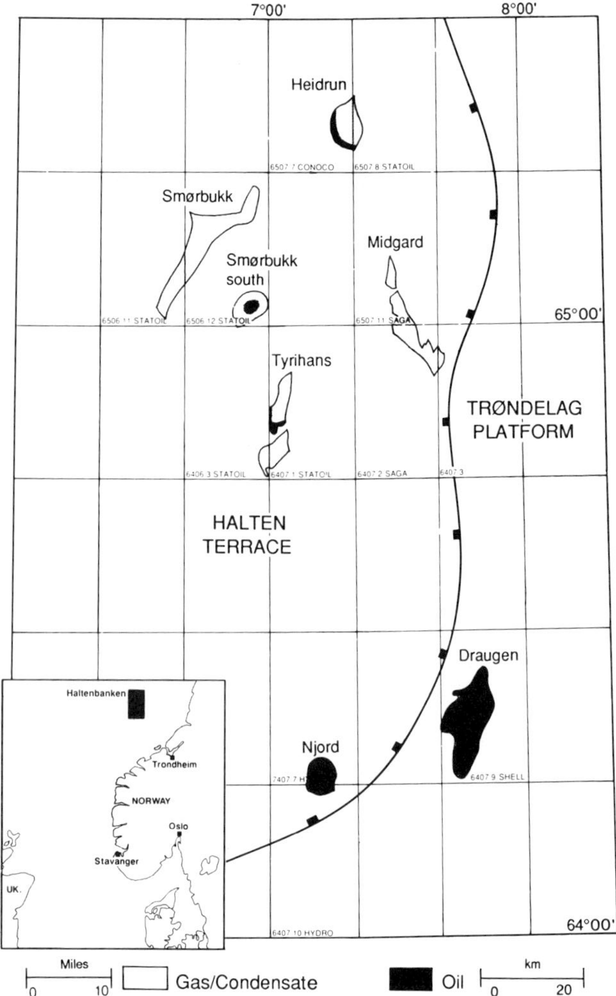

Figure 1. Location map showing the position of Midgard field and other finds in the Haltenbank area with main structural feature.

The gas was encountered in a sandstone sequence of Jurassic age. Both the Middle Jurassic Fangst Group and the Lower Jurassic Tilje Formation were hydrocarbon-bearing (Figure 4). In the well, the two units covered the intervals 2320.5-2378 m and 2465-2563 m, respectively. The two formations were in pressure communication with a gas-water contact at 2493 m.

The lithostratigraphic nomenclature used in the present description of the Midgard field and the Haltenbank area is in accordance with a recent revision that has been formally introduced by the Norwegian Petroleum Directorate (Dalland et al., 1988). Earlier studies of the Midgard field (Ekern, 1987) have been carried out using previous informal names (Figure 4).

Post-Discovery

The interpretation of the seismic reflection data clearly indicated that the field was likely to continue into the neighboring block to the south. This block was awarded following the third part of the 5th Concession Round in 1982, with Saga Petroleum as operator.

Three additional wells have been drilled. Well 6407/2-2 was drilled in 1983, proving gas-condensate in the Gamma segment (Figure 2). In 1985, well 6507/11-3 was drilled, proving gas-condensate with a basal oil-leg in the Beta segment. The continuation of the field into the southern Delta segment was demonstrated in well 6407/2-3 in 1987.

Several additional discoveries have been made within the same play. The most important are Smørbukk field, which contains gas-condensate and was identified in 1984, and Heidrun oil and gas field in 1985.

Another important discovery, the Draugen field, was made on the western margin of the Trøndelag platform in 1984 with oil accumulation in the Upper Jurassic Rogn Formation. Both Heidrun and Draugen field have been declared commercial by their respective owner groups, and their development applications will probably be decided by the Norwegian Parliament during 1988. Development of the Midgard field will depend strongly on the market situation for gas. Saga Petroleum has established various development alternatives for the Midgard field. These are designed to fit in with the regional development schemes and the various gas offtake rates from the area. Earliest possible production start-up is late 1994.

DISCOVERY METHOD

In the mid-Norway region, onshore geology consists mainly of crystalline basement. Apart from local finds of ice-transported blocks of Middle Jurassic rocks recorded onshore in the mid-Norway region (Kjerulf, 1870), there are no significant exposures of sedimentary sequences. The assessment of the thickness and distribution of offshore sedimentary basins was first made using gravimetric and magnetic data, as described above. Increased understanding of the plate tectonics led to the realization that the mid-Norwegian shelf was a part of a larger passive continental margin. The geological concept for the area was based on structural styles expected within such a basin category.

Prospective structures, however, were first recognized using reflection seismic data acquired since 1970.

STRUCTURE

The Midgard field is located on Halten terrace constituting a part of a tectonically complex zone called the Kristiansund-Bodø fault complex

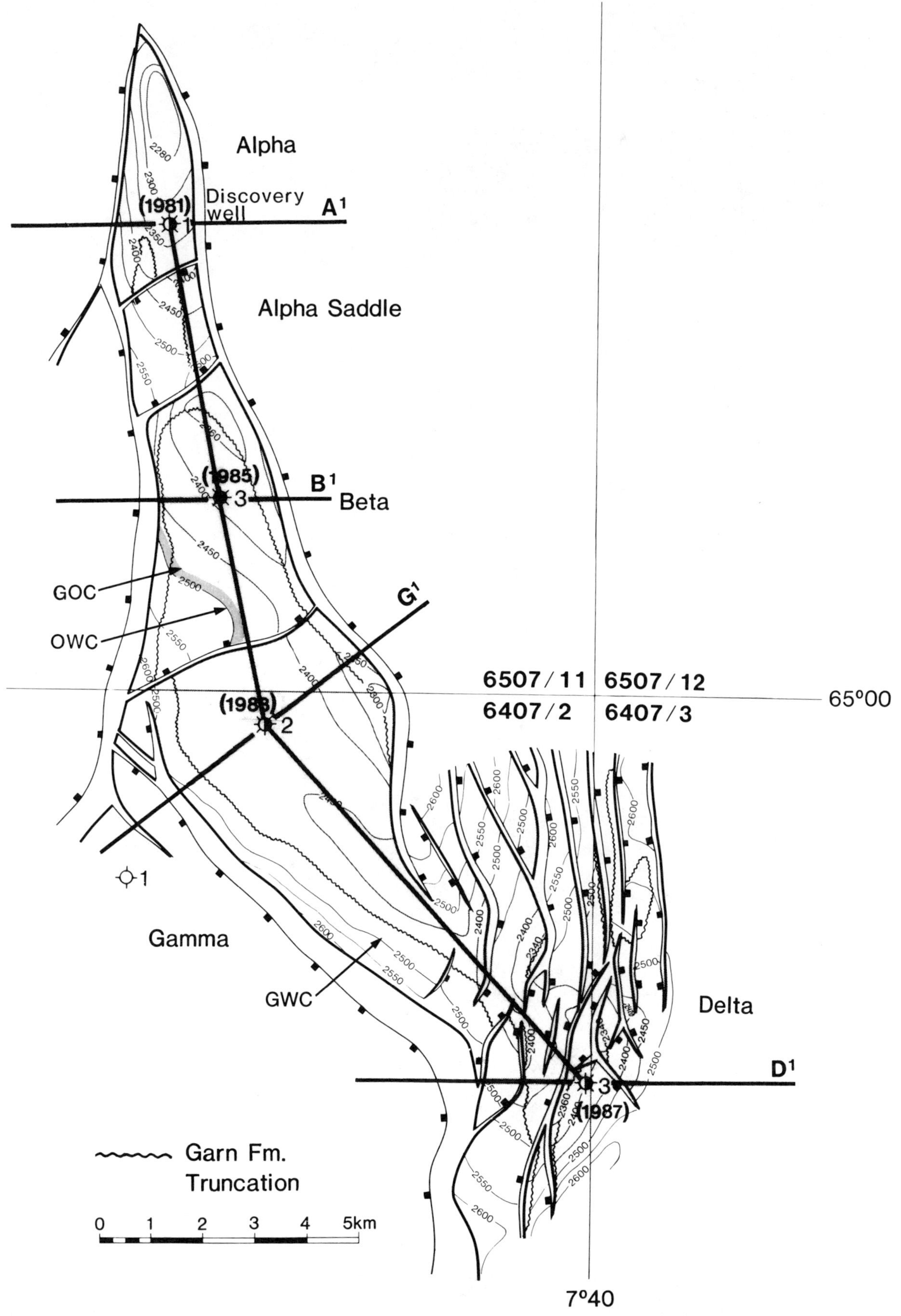

Figure 2. Structure map of Midgard field with subsea depths contoured in meters on top of the reservoir. Structural compartment names of the field are indicated. The gas-water contacts in each different segment are marked. The basal oil-leg in the Beta segment is shown. Lines of the cross sections illustrated in Figures 3 and 10 are marked.

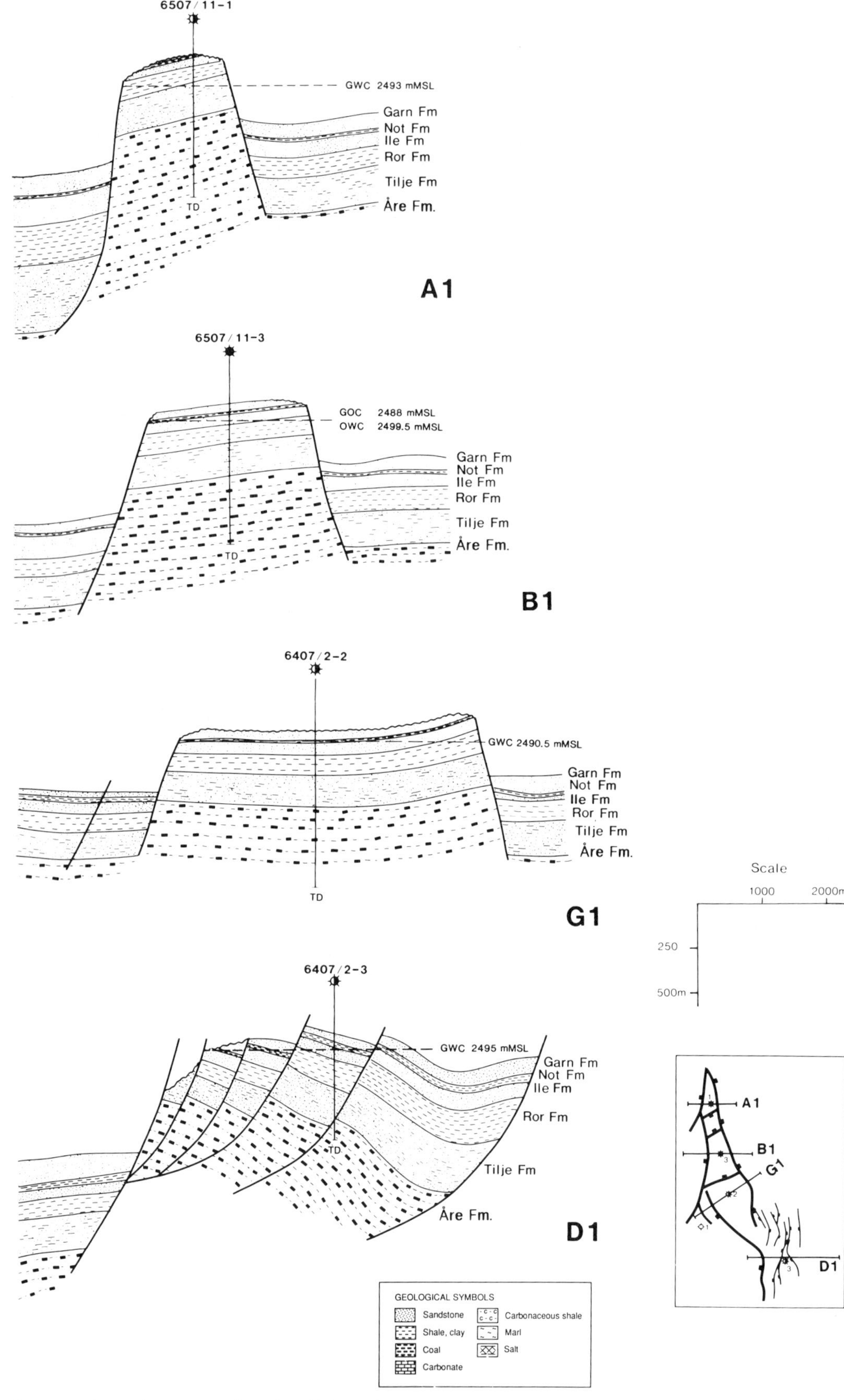

Figure 3. Structural sections crossing the Alpha, Beta, Gamma, and Delta segments of Midgard field. Locations of cross sections are shown on Figure 2.

SYSTEM	STAGE	PAY ZONES	LITOSTRATIGRAPHIC UNITS: REVISED GP.	REVISED FM.	INFORMAL. GP.	INFORMAL. FM.	
CR.							
JURASSIC	KIM.-RYA		VIKING	SPEKK	GRIP	NESNA	
	BATH.-OXF.			MELKE		ENGEL-VÆR	
	AALENIAN - BAJOCIAN	☼	FANGST	GARN	HALTEN	TOMMA	UNIT I
		☼		NOT			U.II
		☼		ILE			UNIT III
		☼					
	TOARCIAN		BÅT	ROR		LEKA	
	HETANGIAN - PLIENSBACHIAN	☼		TILJE		ALDRA	
		☼		ÅRE			
						HITRA	

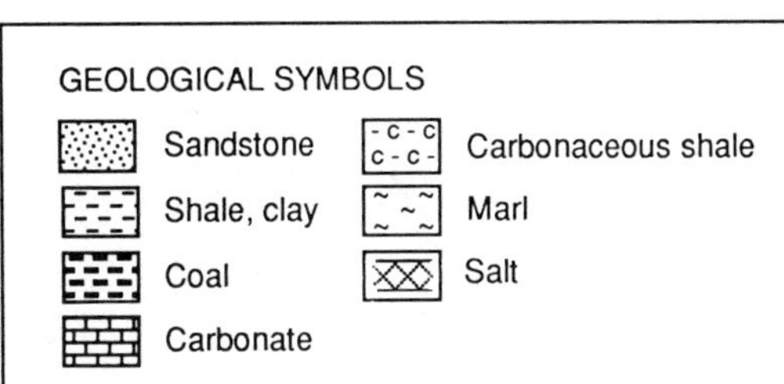

Figure 4. Jurassic stratigraphy of the Midgard field. Pay zones in the Middle and Lower Jurassic intervals are indicated.

(Gabrielsen et al., 1984). This faulted area separates the Trøndelag platform to the east from the Vøring basin to the west (Figure 5). The Vøring basin is limited to the west by Tertiary oceanic crust.

The basement in the area is believed to consist of rocks deformed during the Caledonian orogenesis that ended in the Late Silurian-Early Devonian.

The evolution of the mid-Norway continental margin was dominated by wrenching in the Devonian. The uplifted Caledonian mountain chain was subject to extensive erosion, and coarse clastics were deposited in intramontane basins. These fault-bounded basins have been identified on the western coast of Norway.

The period of wrench tectonics was followed by crustal extension and rifting. Bukovics et al. (1984) have suggested that the extensional phase had already begun by the late Carboniferous. By analogy with eastern Greenland, large Permian-Carboniferous half-grabens are anticipated. Seismic lines from

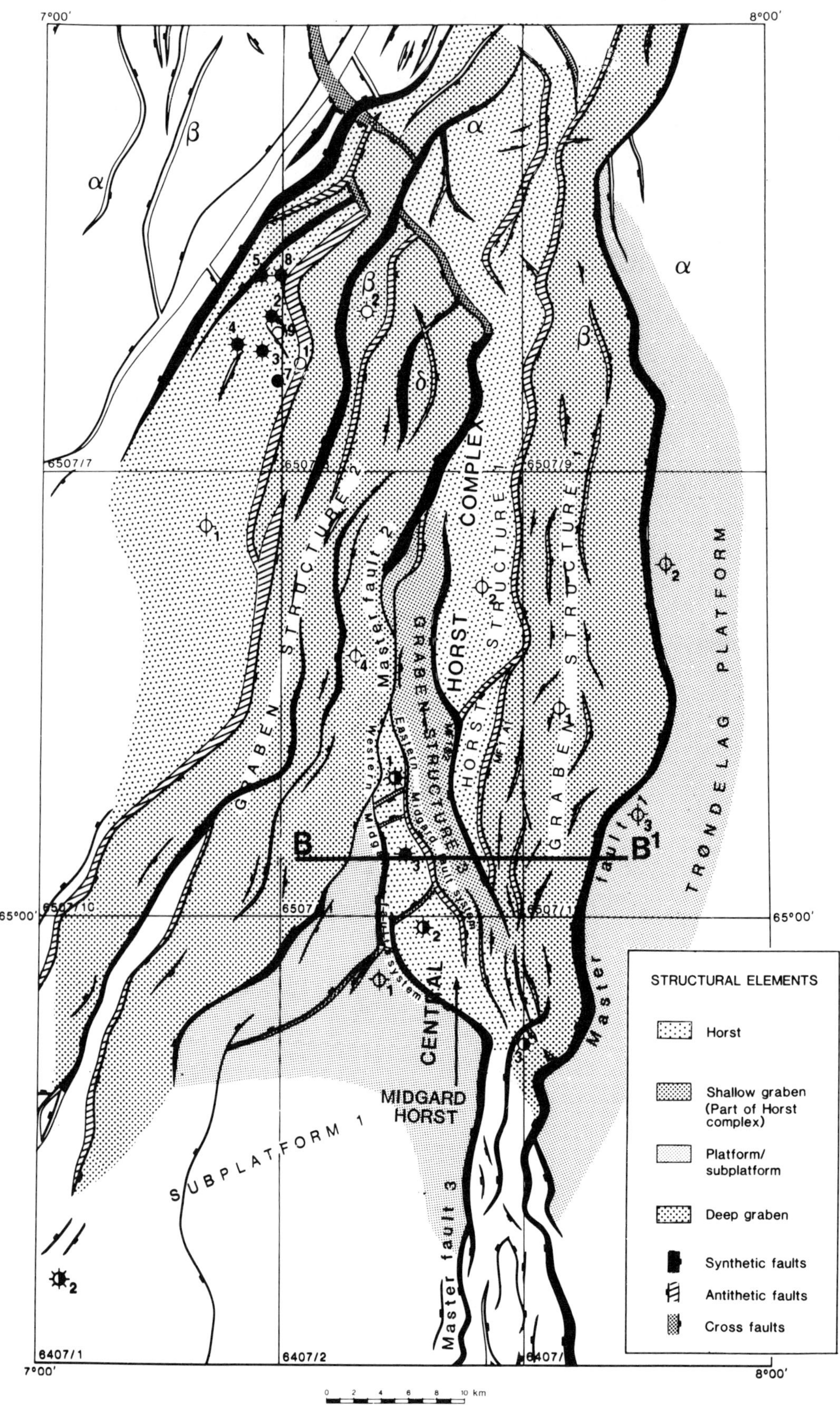

Figure 5. The main regional structural elements of the mid-Norway continental shelf. The figure illustrates horst and graben features and the different fault categories. The position of the cross section on Figure 6 is shown.

the Trøndelag platform east and south of the Midgard field support the existence of half-grabens of at least Permian–Triassic age.

The main phase of extension was initiated in the Late Triassic. Horst and graben development is documented by seismic and well data on the Halten terrace. Extension accelerated from Middle to Late Jurassic, expressed by an angular unconformity at the top Middle Jurassic level. Rift movement is manifested by syndepositional listric faulting on a few master faults (Gabrielsen and Robinson, 1984).

Compensatory antithetic faults also developed. Some of the structures might have been formed or modified by later movement of the Triassic salt that is developed in the sedimentary sequence as a result of a Middle Triassic transgression (Hollander, 1984).

Renewed rifting took place in Early Cretaceous further to the west, and the major subsidence in this phase affected the Vøring basin. This pattern was succeded by a more regional subsidence also affecting the Halten terrace and Trøndelag platform during Late Cretaceous. Rifting activity increased related to the break-up between Norway and Greenland during the latest Cretaceous and Paleocene. Crustal separation and drifting probably began by early Eocene. An important element of the structural evolution is the considerable subsidence during the Pliocene on the Halten terrace. This subsidence, probably of thermal origin, has been on the order of 1.5 km during the last 3.5 million years, and it had a major impact on the source rock maturation pattern.

The Midgard field is composed of a series of fault blocks along a NNW-SEE trend, separated by cross faults. Faulting was initiated along three master faults (Figures 5 and 6). Two of these faults constitute the western limit of the field and have a listric geometry. The third master fault represents the boundary between the Halten terrace and the Trøndelag platform.

The master faults are believed to have a common detachment zone, probably in the Triassic salt. Continued movement on these faults was followed by activation of accommodation structures. The associated rollover anticlines that developed in the hanging-walls are cut by second-order antithetic and third-order synthetic faults. Minor strike-slip movement along tear-faults in the area is also indicated.

By the end of Late Jurassic, movement had ceased along several of the faults, and only the master faults or their later equivalents remained active, first as faults and later as axes of rotation. The Midgard field is divided into five structural elements (Alpha, Alpha Saddle, Beta, Gamma, and Delta). While the central and northern part of the field is developed as horst blocks with clearly defined eastern and western limits (Figure 7), the Delta structure in the southeast is more complex. This structure is cut by numerous synthetic and antithetic faults.

STRATIGRAPHY

All the producible hydrocarbons on the Halten terrace have been discovered in formations of Jurassic age. Most of the reserves have been found in the Middle Jurassic Garn and Ile Formations. Reserves are also associated with the Lower Jurassic Tilje and Åre formations (Figure 8). The hydrocarbons in the Midgard field are found in four lithostratigraphic units. The wells have demonstrated that the units can be looked upon as one reservoir as the same pressure regime and hydrocarbon contacts have been documented. In the western part of the Halten terrace, additional reserves are found in the Tofte Formation (Aasheim et al., 1986). This unit is a sandy equivalent of the shaly Ror Formation found in the Midgard area. In the Draugen field, the hydrocarbons are predominantly in the Upper Jurassic sandstones (Ellenor and Mozetic, 1986).

The deepest well in the Midgard field has been drilled 75 m into the Triassic gray beds. In the surrounding area the deepest penetration has been 2.5 km into the Triassic terminating in the redbeds. A geohistory plot of a Midgard well illustrates that the reservoir is presently at its maximum burial (Figure 9).

The subsidence rate was high during deposition of the Upper Triassic–Lower Jurassic Åre Formation and continued at a high rate up to the Middle Jurassic (Ror Formation and Fangst Group). The rate of burial was abruptly reduced in the Upper Jurassic (Viking Group). This situation continued through the deposition of the Lower Cretaceous Cromer Knoll Group. The sedimentation rate increased in the Upper Cretaceous (Shetland Group) and further in the Tertiary up to Pliocene with deposition of the Rogaland and Hordaland groups. In the Pliocene the subsidence increased significantly, resulting in the deposition of 1100 m of the Naust Formation in the Midgard wells.

There are two potential source rocks in the Haltenbank area, the Upper Jurassic Spekk Formation and the Lower Jurassic Åre Formation.

The Spekk Formation is a marine organic-rich hot-shale equivalent to the Draupne and Kimmeridge Clay Formations found in the North Sea. The kerogen type is classified as type II-III, and the high hydrogen indices indicate oil generation capacity. The Åre Formation contains abundant beds of coal and carbonaceous shales. The hydrogen index values and the kerogen type, which is dominated by humic organic matter classified as type III, indicate a source rock with potential mainly for gas generation.

TRAP

The Midgard field constitutes a structural trap divided into individual segments (Figure 2). The four

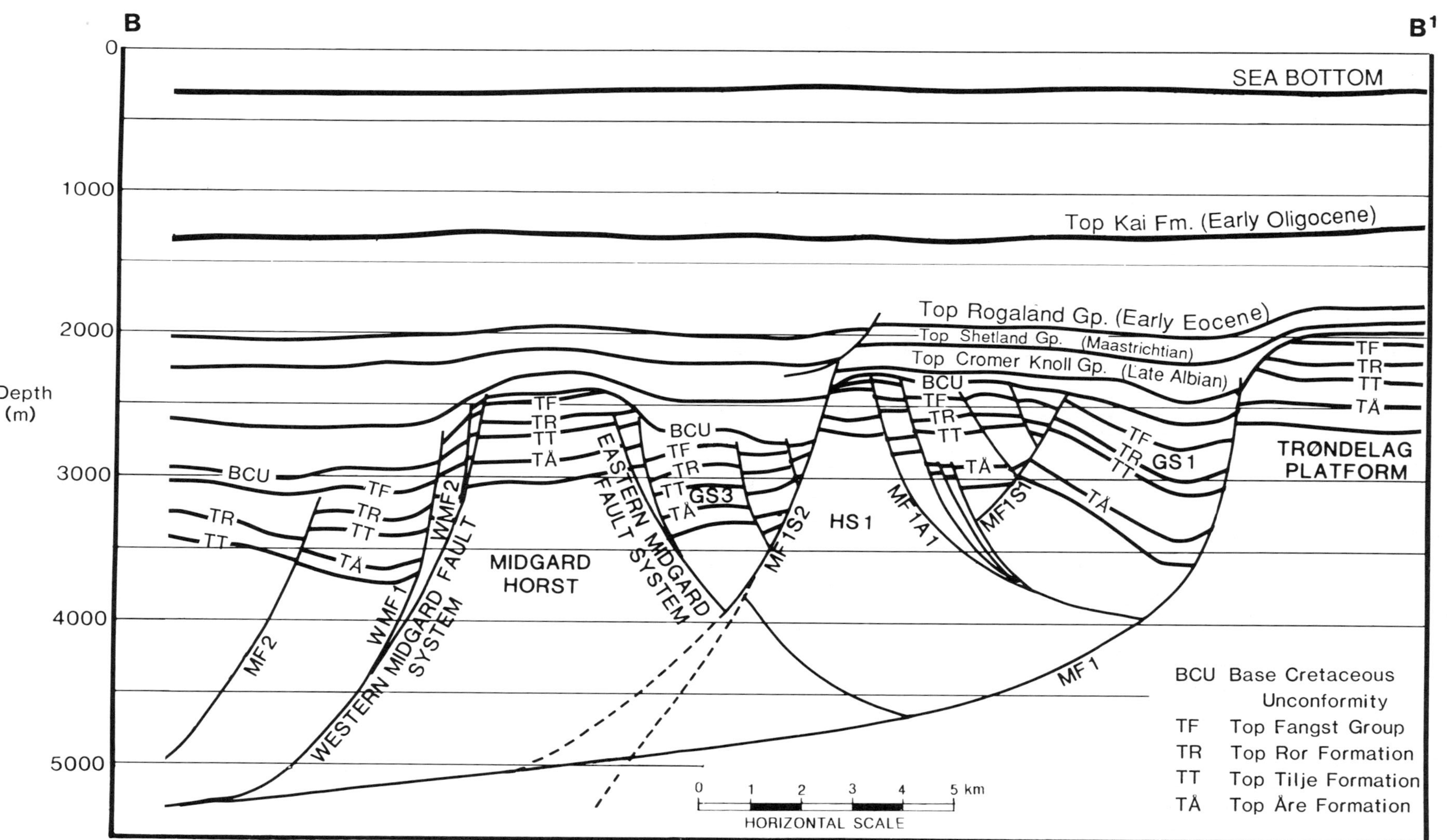

Figure 6. Structural cross section showing the Midgard horst and other neighboring structural elements. The main controlling faults are marked. The position of the section is shown on Figure 5.

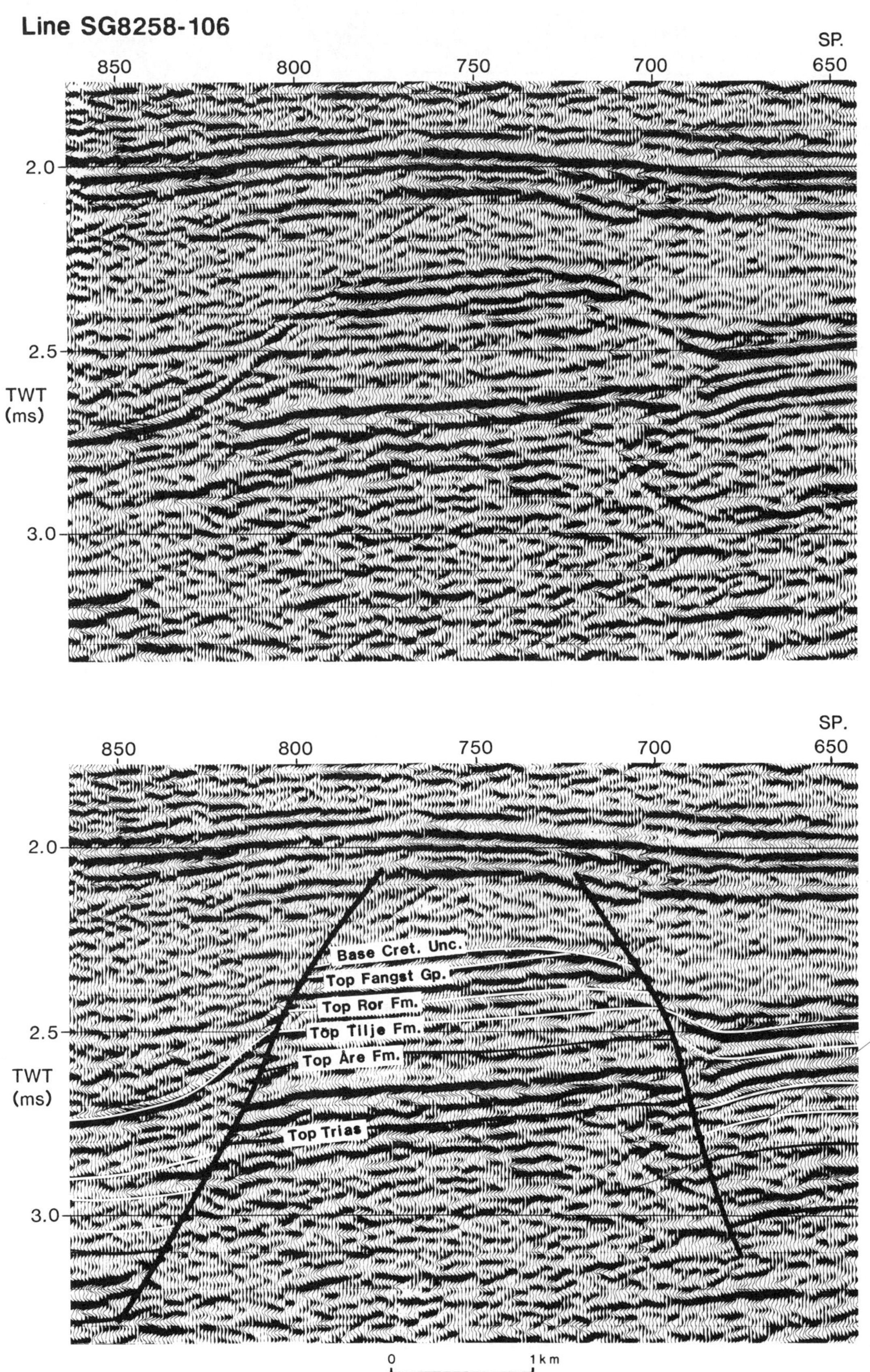

Figure 7. Seismic line across the Beta segment of the Midgard field. The section illustrates the well-defined eastern and western limits of the horst block. The internal part of the horst block is apparently unfaulted. The line is equivalent to the B1 cross section shown on Figure 3.

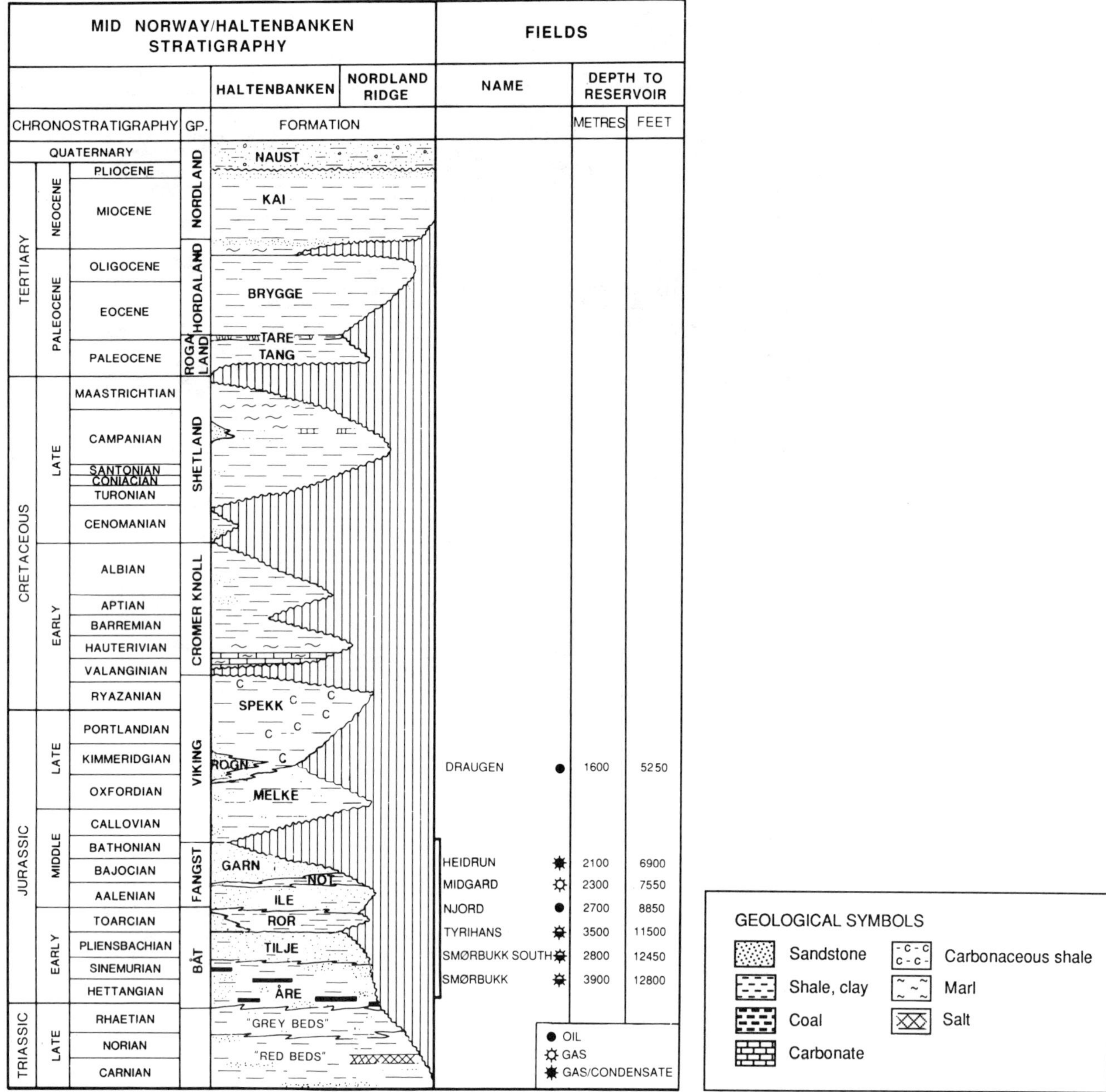

Figure 8. Generalized stratigraphy for the Haltenbanken area. The nomenclature is based on the revised names approved by the Norwegian Petroleum Directorate in 1988.

northern compartments (Alpha, Alpha Saddle, Beta, and Gamma) are developed as horst blocks with major eastern and western bounding faults. The Delta segment to the south is anticlinal and is composed of fault segments separated by listric faults. The reservoirs in the fault blocks are juxtaposed and sealed laterally to the east and west by the Lower Cretaceous shales of the Cromer Knoll Group. Top seal is the Upper Jurassic Viking Group. This shale unit is thin and is bounded by unconformities at its base and top. The unit may be absent over the highest part of the Alpha segment. In this area, the top seal will be the Lower Cretaceous shales.

The hydrocarbon-water contacts show slight variation across the field. A gas-water contact was identified in the Tilje Formation at 2493 m in the Alpha segment. This corresponds to a possible spill-point in the northeastern part of the segment as indicated from the fault plane analysis. In the Beta segment an 11.5 m oil-leg was penetrated below the gas-oil contact at 2488 m. The present mapping does not give evidence for any communication in the hydrocarbon phase between the Alpha and Beta segments. In the Gamma segment no hydrocarbon-contacts have been identified because the potential contact is located in the shaly Not Formation of the

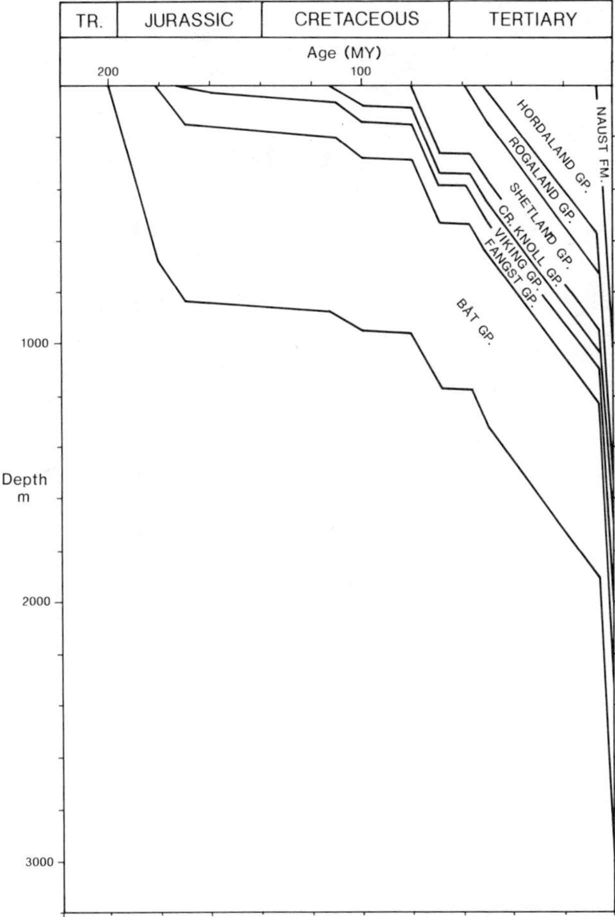

Figure 9. Geohistory plot for the Midgard field.

Fangst Group in the well testing this segment. A possible gas contact at 2490.5 m is interpreted from the pressure data. In the Delta segment, a gas-water contact was identified at 2495 m, which corresponds to the lowest structural closure of the Delta segment. The Beta, Gamma, and Delta segments are probably in communication despite the different observed contacts. These minor differences are probably caused by each segment being associated with individual drainage areas and the fact that hydrocarbon migration has been a late process that is still going on in the area.

RESERVOIRS

The reservoirs in the Midgard field are mainly found within three units: the Lower Jurassic Tilje and the Middle Jurassic Ile and Garn Formations (Figure 10). Minor reserves are also found in an updip structural position in the Lower Jurassic Åre Formation in the Alpha segment. The Middle Jurassic units are the most important, holding 96% of the reserves. These formations can be correlated across large areas of the Haltenbanken.

The deposition of the Tilje Formation took place during a transgression from the northwest that gradually inundated the existing low-relief delta/coastal plain and resulted in a low-energy marginal marine/estuarine environment. The formation increases in thickness toward the southeast, averaging 125 m in the Midgard field (Figure 11). The lower part of the formation is composed of fine-grained sandstones with a high siltstone content. The upper part of the formation consists of several upward-coarsening sequences grading up to coarse sands. This part of the formation is interpreted to represent interbedded shelf, shoreface, and delta front deposits formed by a prograding coastline with a high clastic input (Figure 12). The reservoir quality is very variable. In each sequence, the quality generally improves from poor to good upward (Figure 13). The Tilje Formation constitutes a reservoir only in the northern Alpha segment. In this segment, the Late Jurassic uplift and subsequent erosion has removed the major part of the Fangst Group.

The Fangst Group is separated from the Tilje Formation by the Ror Formation. The Ror Formation comprises shelf muds with occasional storm influenced fine-grained sandstone beds. This unit does not have any reservoir potential in the Midgard field. In the western part of the Haltenbank area, however, an equivalent sandy facies is developed.

The Fangst Group contains three formations: the Ile Formation, the Not Formation, and the Garn Formation. The Fangst Group generally thins toward the northeast and is on average approximately 140 m thick in the Midgard field (Figure 14).

The Ile Formation consists of good reservoir quality sandstones that exhibit fairly uniform properties across the Midgard field. The basal part of the Ile Formation represents a progradational tidally influenced delta complex (Figure 15), while the middle part consists of more wave-influenced delta front sediments. The upper part is dominated by middle to upper shoreface deposits, and exhibits the best and most uniform reservoir properties (Figures 16 and 17).

The Not Formation is generally a nonreservoir unit. The uppermost part does, however, contain some continuous siltstone and sandstone layers and may potentially contribute slightly to the reserves.

The formation was deposited in an inner shelf or lagoonal environment and consists predominantly of pyritic mudstones. The uppermost part consists of delta front sediments that reflect the onset of renewed delta progradation. The thickness of the formation varies across the field with a maximum thickness of 28 m on the southern Delta segment.

The Garn Formation contains sandstones of good to excellent reservoir quality. They are interpreted to represent a wave-dominated upper shoreface environment that was formed either by a delta front or a coastal barrier bar complex (Figure 18). The thickness across the field is uniform, ranging from 45.5 to 54 m in the wells. The formation is not

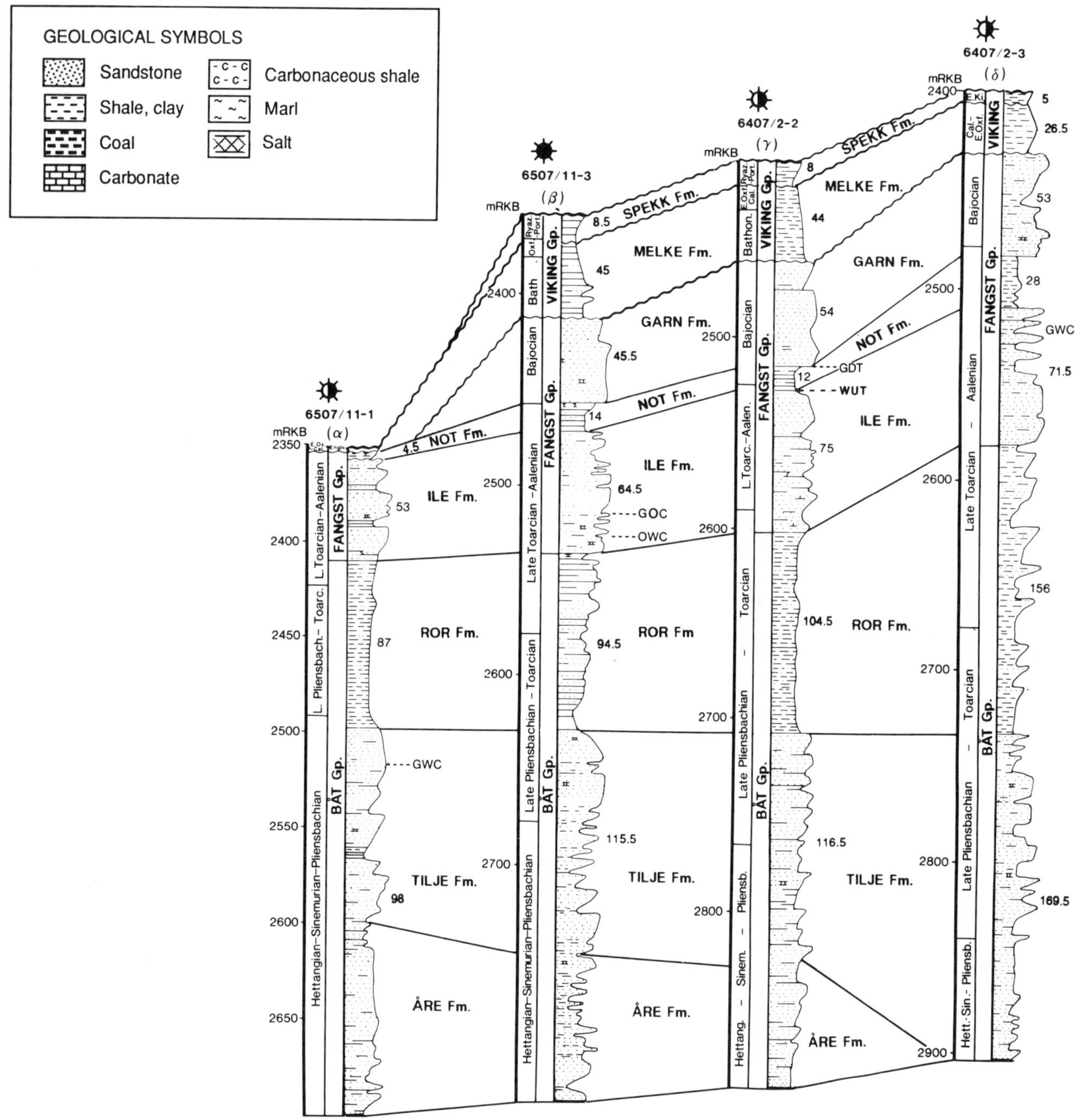

Figure 10. Geological correlation of the Midgard field wells. The position of the cross section is shown in Figure 2.

represented in well 6507/11-1 on the Alpha segment due to erosion.

The lithofacies and the sedimentary facies are the most important factors controlling variations in reservoir properties within the Midgard field (Figure 19). Diagenesis has influenced the reservoir quality, but compaction and mineral growth have reduced the reservoir properties only by a moderate degree. This can be explained by the short period of deeper burial of the reservoir. Dissolution of feldspars and carbonate cements have improved the reservoir properties with development of secondary porosity.

Primary intergranular pores, partly modified by dissolution of early precipitated calcite cement, constitute the predominant porosity type in the entire reservoir interval. Secondary porosity, formed by dissolution of unstable detrital grains (mainly K-feldspar) is found throughout, but to a variable degree in the different lithologies. Intercrystalline microporosity associated with kaolinite platelets and expanded

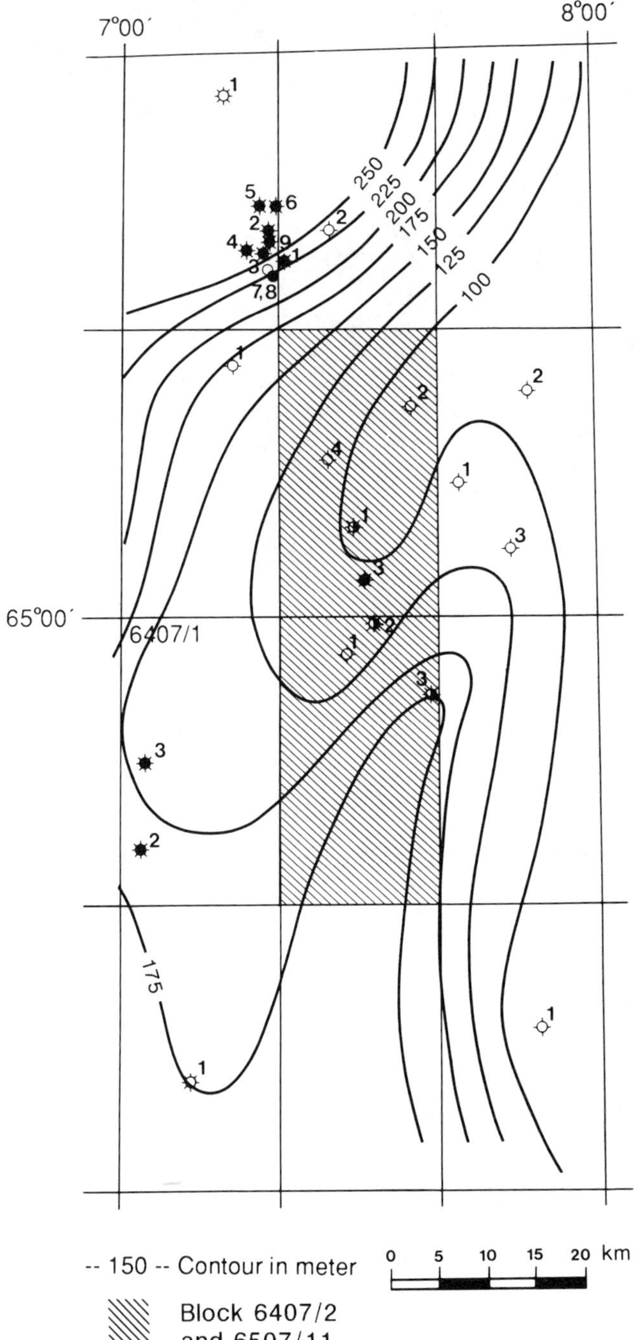

Figure 11. Gross isochore map of the Tilje Formation. The Midgard field is located within the two shaded blocks of the figure.

mica flakes is occasionally found in most of the sequences. The most important contributor to microporosity seems to be detrital clays within the sandstones. Dispersed microporosity is especially found in the sandstones of the Ile and the Tilje formations.

The gas composition in the Midgard field is dominated by methane constituting approximately 82%. Ethane also contributes a significant portion (9%). Gas wetness is estimated to be 15% in the gas-condensate. The condensate has a density of 0.81 g/cc (API 44°). A gas-condensate ratio of 5235 Sm^3/Sm^3 has been calculated from a simulation of the processing planned for the field development. The oil-leg identified in the Beta segment is moderately heavy with a density of 0.87 g/cc (API 31°). A gas-oil ratio of 161 Sm^3/Sm^3 has been measured. The gas associated with the oil has a wetness of 28%.

There is good correlation between the different hydrocarbon phases as well as between the individual segments, indicating a common source for the hydrocarbons. The gas in the individual reservoir zones is at dew-point, except for the gas in the Garn Formation in the Delta segment. This may indicate a restriction in the communication into this part of the structure.

The thickest gross gas column is found in the northern part of the Alpha segment where more than 220 m is present. As this part of the field has the poorest reservoir properties, the largest net gas column is interpreted to occur centrally on the Delta segment where a hydrocarbon pore volume value of more than 35 m may be present (Figure 20).

The reservoir units are anticipated to be continuous high porosity and permeability zones. The Not and the Ror formations are the only major continuous impermeable units. The resulting geometry indicates that the field will mainly be supported by an edge water drive. Only the Tilje Formation in the Alpha segment and the Garn and Ile formations in the Delta segment will be supported by any sizeable aquifer. However, the exact aquifer strength for the different portions of the field is difficult to assess because the pressure support across the different faults is unknown. The production tests have confirmed excellent production characteristics. For the production planning, a general maximum well rate of 3×10^6 Sm^3/day has been used.

The uncertainties related to the aquifer mean that the recovery factors cannot be determined exactly. The field is planned to be produced through pressure depletion, and a recovery factor of 0.73 has been estimated for a full field development. Due to its limited thickness, the oil zone is presently not regarded as an economical production target.

FAULTS

The main faults defining the eastern and western margin of the three northern horst blocks have significant throws and are easily recognized and correlated using seismic data (Figure 7). The horst block segments are separated by northeast-southwest-trending cross-faults on which the throws are less pronounced. The seismic data give no evidence for any significant faults within each horst block segment.

The cores from the wells do not indicate the presence of fractures or faults serving as reservoir barriers. The

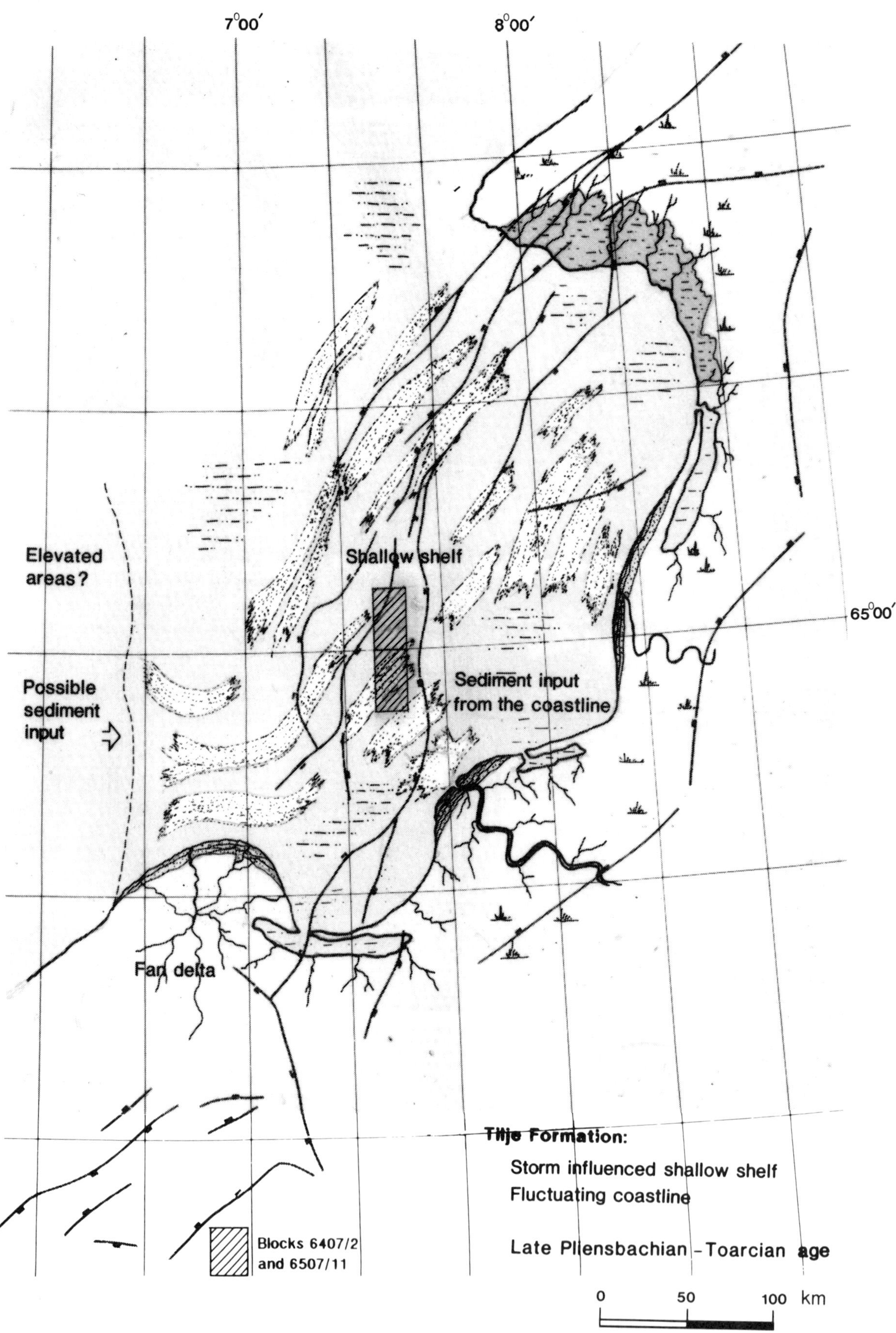

Figure 12. Paleogeographic reconstruction illustrating environments of deposition of the Tilje Formation in late Pliensbachian-Toarcian time.

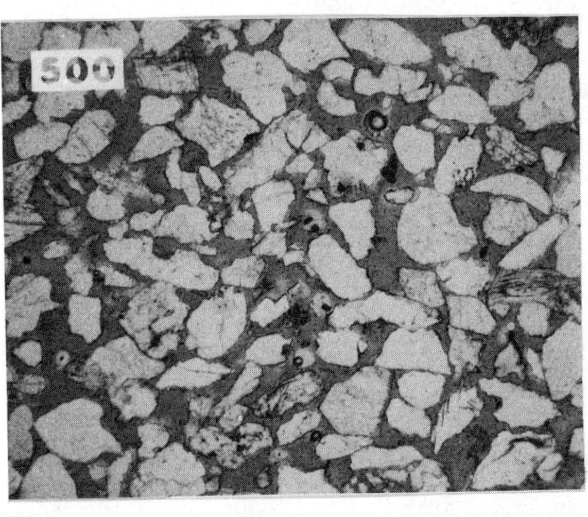

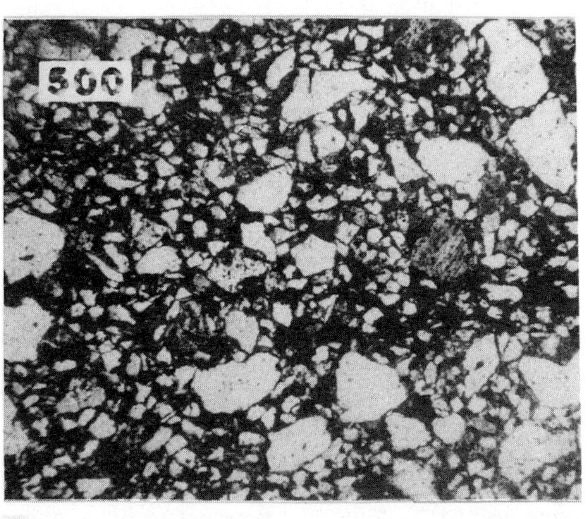

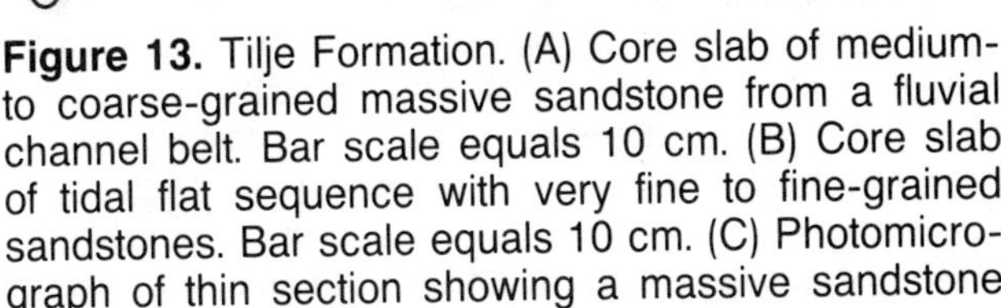

Figure 13. Tilje Formation. (A) Core slab of medium- to coarse-grained massive sandstone from a fluvial channel belt. Bar scale equals 10 cm. (B) Core slab of tidal flat sequence with very fine to fine-grained sandstones. Bar scale equals 10 cm. (C) Photomicrograph of thin section showing a massive sandstone that is medium grained and moderately sorted (porosity: 32.0%; permeability: 8082 md). Bar scale equals 500 microns. (D) Photomicrograph of thin section showing a very poorly sorted subfeldspathic wacke composed of very coarse sand to silt grains (porosity: 13.3%; permeability: 2.90 md). Bar scale equals 500 microns.

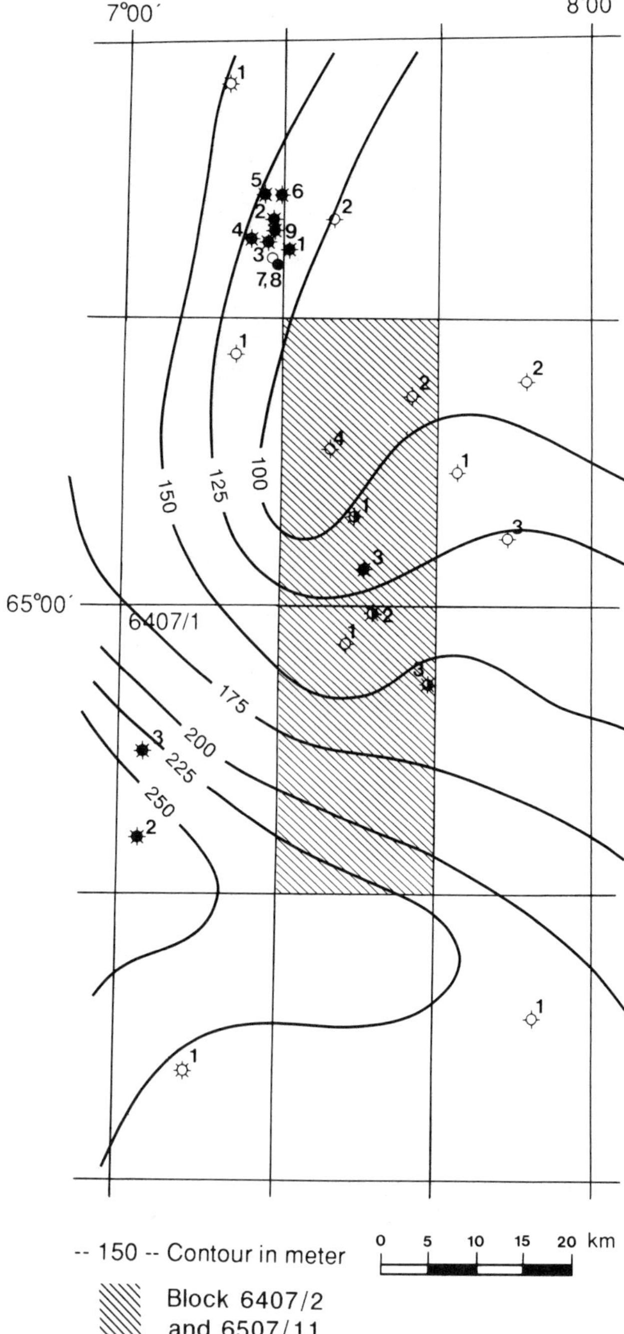

Figure 14. Gross isochore map of the Fangst Group. The unit is eroded in well 6507/11-1 and only a reduced sequence of 59 m is preserved. The Midgard field is located within the two shaded blocks of the figure.

Delta segment to the southeast is structurally far more complex than the horst blocks. A series of north–south-trending listric faults divide this part of the field into several fault-bounded segments. The correlation of the faults is difficult and was first solved following the interpretation of the 3D survey acquired over this part of the field in 1986. The potential sealing capacity of the faults is of importance both to the hydrocarbon distribution and flow, as well as the aquifer strength. Generally, the faults are believed to be nonsealing where a sand-sand contact exists due to high reservoir quality of the sands. Particular attention has been focused on the cross-fault between the Beta and Gamma segments due to the possible extension of the oil-leg observed in the Beta segment. The well in the Gamma segment encountered a gas-filled Garn Formation. The potential hydrocarbon contacts would be within the shaly Not Formation, and consequently the well gave no evidence as to the existence of any oil layer. There are minor areas with sand-sand contact across the fault within the Fangst Group above the hydrocarbon-water contacts (Figure 21).

The separate drainage area for the Gamma structure is too limited to account for the observed hydrocarbons in this structure (Figure 22). Hence, the general migration into this compartment probably took place through the Beta segment and/or the Delta segment. The fault between the Beta and Gamma segments is probably nonsealing although it might act as a preferential seal that prevents the migration of oil to the Gamma segment.

The different dew-point observed for the gas tested in the Garn Formation in the Delta segment does indicate that faults in this area might have sealing properties. Most likely, these characteristics are associated with a north–south-trending continuous fault zone running centrally across the segment west of well 6407/2-3.

SOURCES

There are two potential source rocks within the drainage area of the Midgard field: the Upper Jurassic Spekk Formation and the Lower Jurassic Åre Formation. The Spekk Formation has a typical hot shale character on the gamma logs. The total organic carbon (TOC) is on average 8% with a pyrolysis yield of 60 mg/g rock in the drainage area. Kerogen type II-III and hydrogen indices above 500 indicate high oil generation capacity (Figure 23). The source rocks of the Åre Formation are defined as the intervals having densities of less than 2 g/cc. The TOC varies significantly between the shales and the coals, with an average of 8%. Pyrolysis yield has been measured in the wells with average between 25 and 116 mg/g rock. The interpreted hydrogen index of around 250 and a type III kerogen indicate gas generation potential.

The Åre Formation has been mature for oil generation since Early Cretaceous time in the deeper parts of the drainage area. However, temperature and generation rate were both too low to allow any significant migration of hydrocarbons. Rapid subsidence started approximately 3.5 million years ago and continued as the Naust Formation was deposited. This resulted in a pronounced increase in temperature and brought the Åre Formation into an

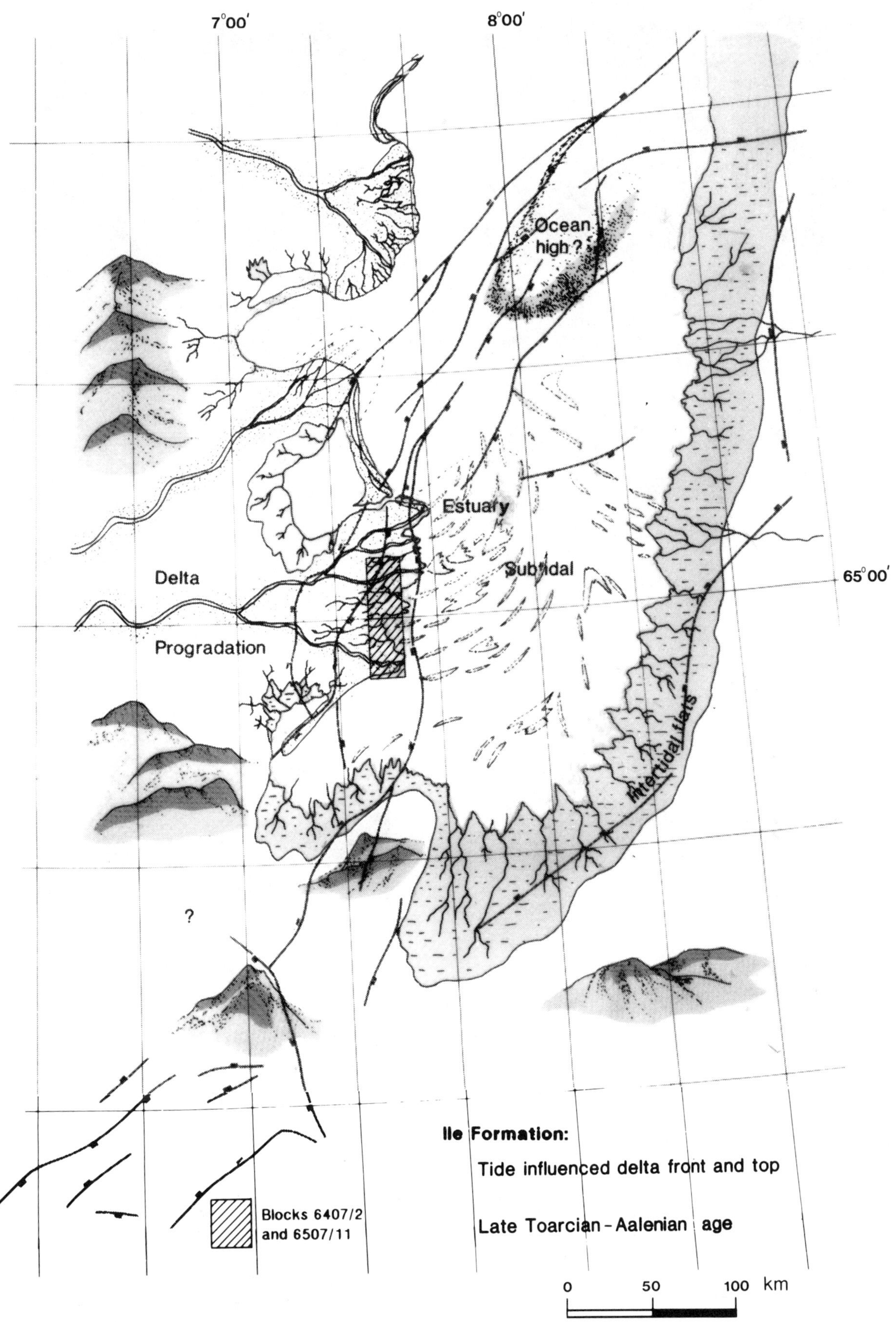

Figure 15. Paleogeographic reconstruction illustrating the environments of deposition of the Ile Formation in late Toarcian-Aalenian time. The Midgard field is located within the two shaded blocks of the figure.

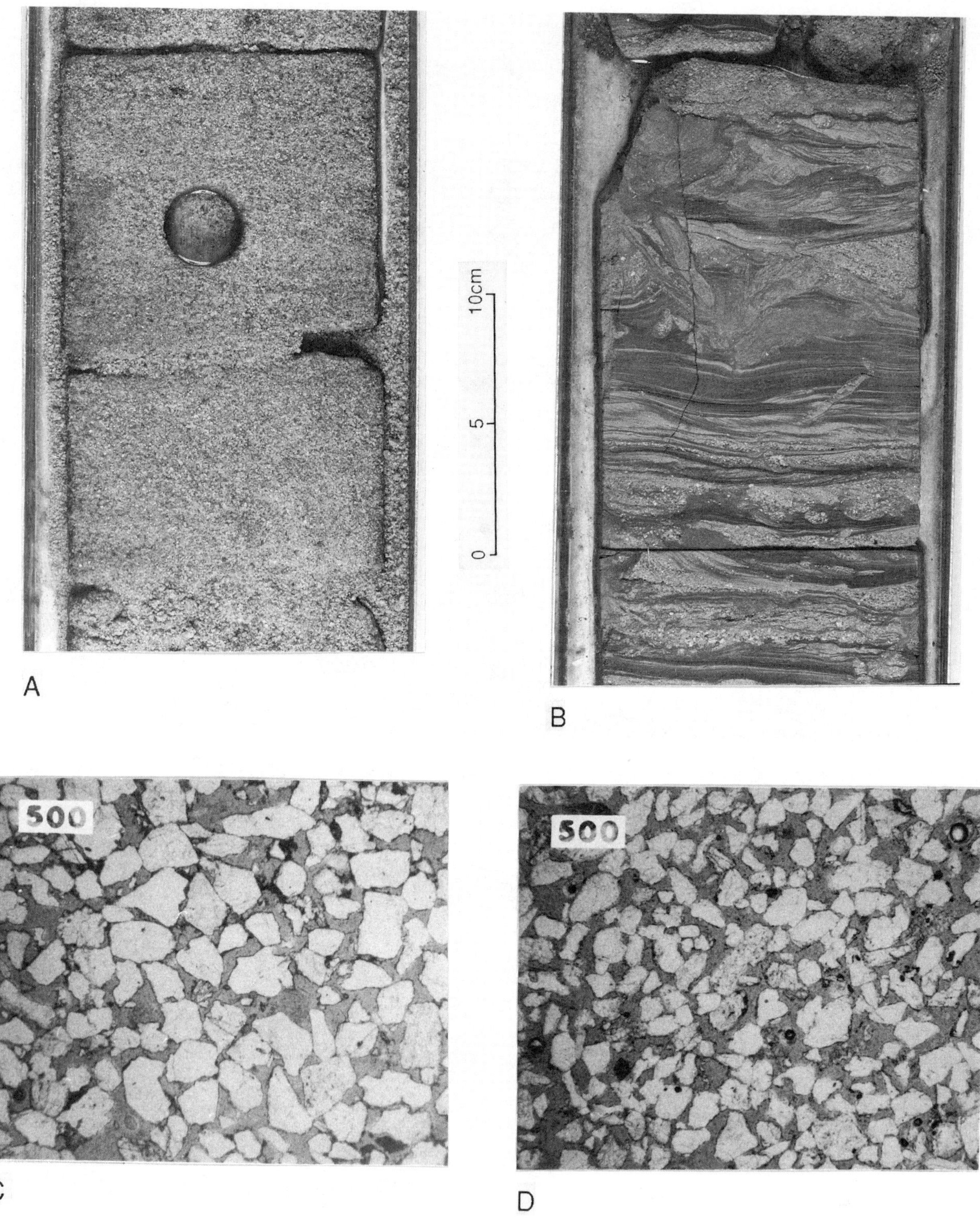

Figure 16. Ile Formation. (A) Core slab of tidally influenced delta front sandstone. The sands are very coarse and were sorted. Bar scale equals 10 cm. (B) Core slab of storm lags in bioturbated fine-grained sandstone and argillaceous siltstone layers. Bar scale equals 10 cm. (C) Photomicrograph of thin section showing moderately sorted, medium-grained sand in which secondary grain dissolution pores (center) form an effective supplement to the primary interparticle pore network (porosity: 33.4%; permeability: 1082 md). Bar scale equals 500 microns. (D) Photomicrograph of thin section showing well-sorted, medium-grained sand. The porosity is 28.9% and comprises clean, well-connected interparticle pores. Permeability is 4760 md. Bar scale equals 500 microns.

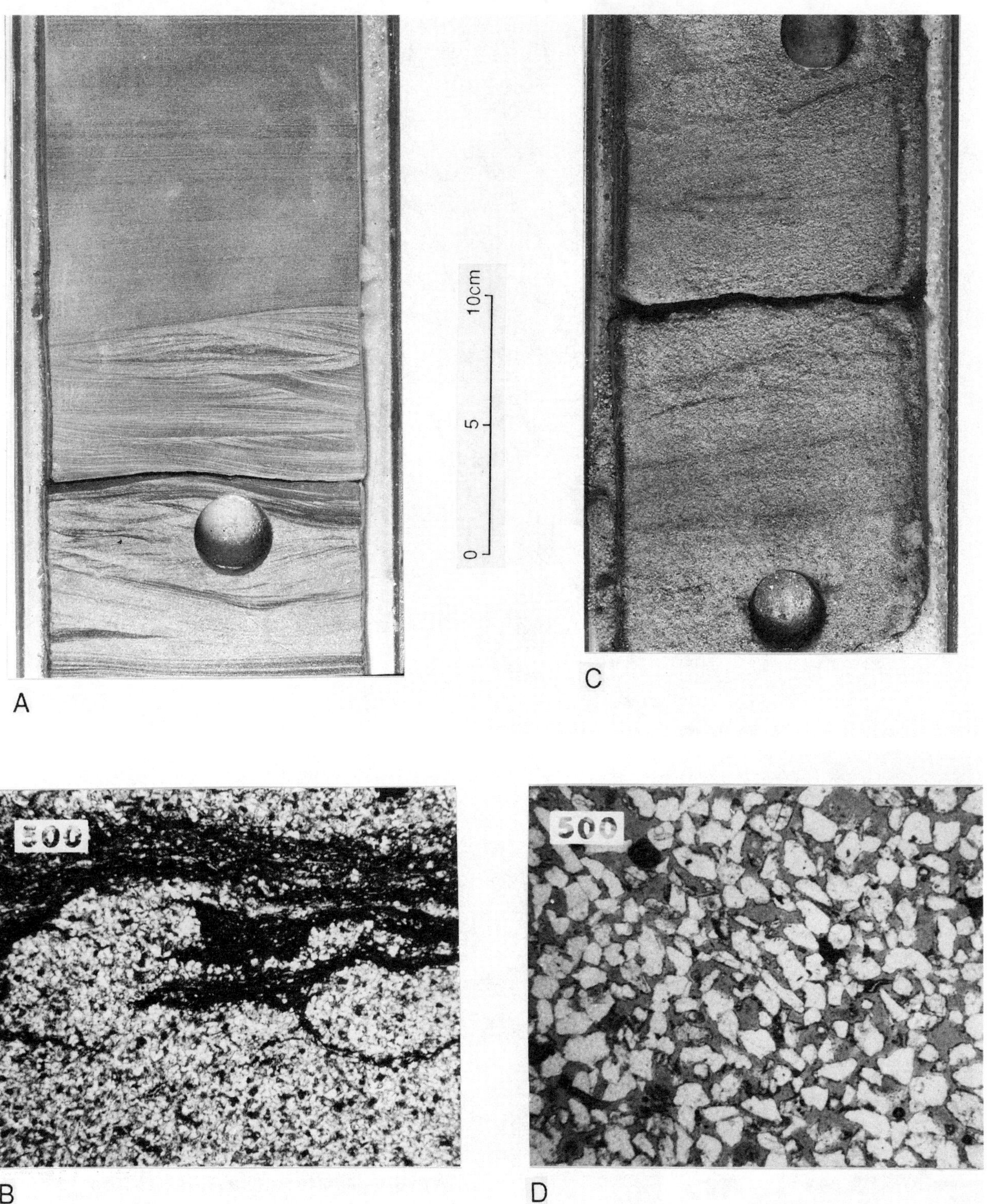

Figure 17. (A) Ile Formation. Core slab of wave-rippled, fine-grained silty sandstone overlain by fine-grained parallel laminated sandstone. Bar scale equals 10 cm. (B) Ile Formation. Photomicrograph of thin section showing a well-sorted, argillaceous siltstone that displays a typical bioturbation fabric (porosity: 16.3%; permeability: 1.08 md). Bar scale equals 500 microns. (C) Garn Formation. Core slab of a medium-grained well-sorted sandstone deposited in middle to upper shoreface. Bar scale equals 10 cm. (D) Garn Formation. Photomicrograph of thin section showing a well-sorted, fine- to medium-grained sandstone that has a high, remnant primary interparticle porosity, supplemented by grain-dissolution voids (porosity: 33.4%; permeability: 7071 md). Bar scale equals 500 microns.

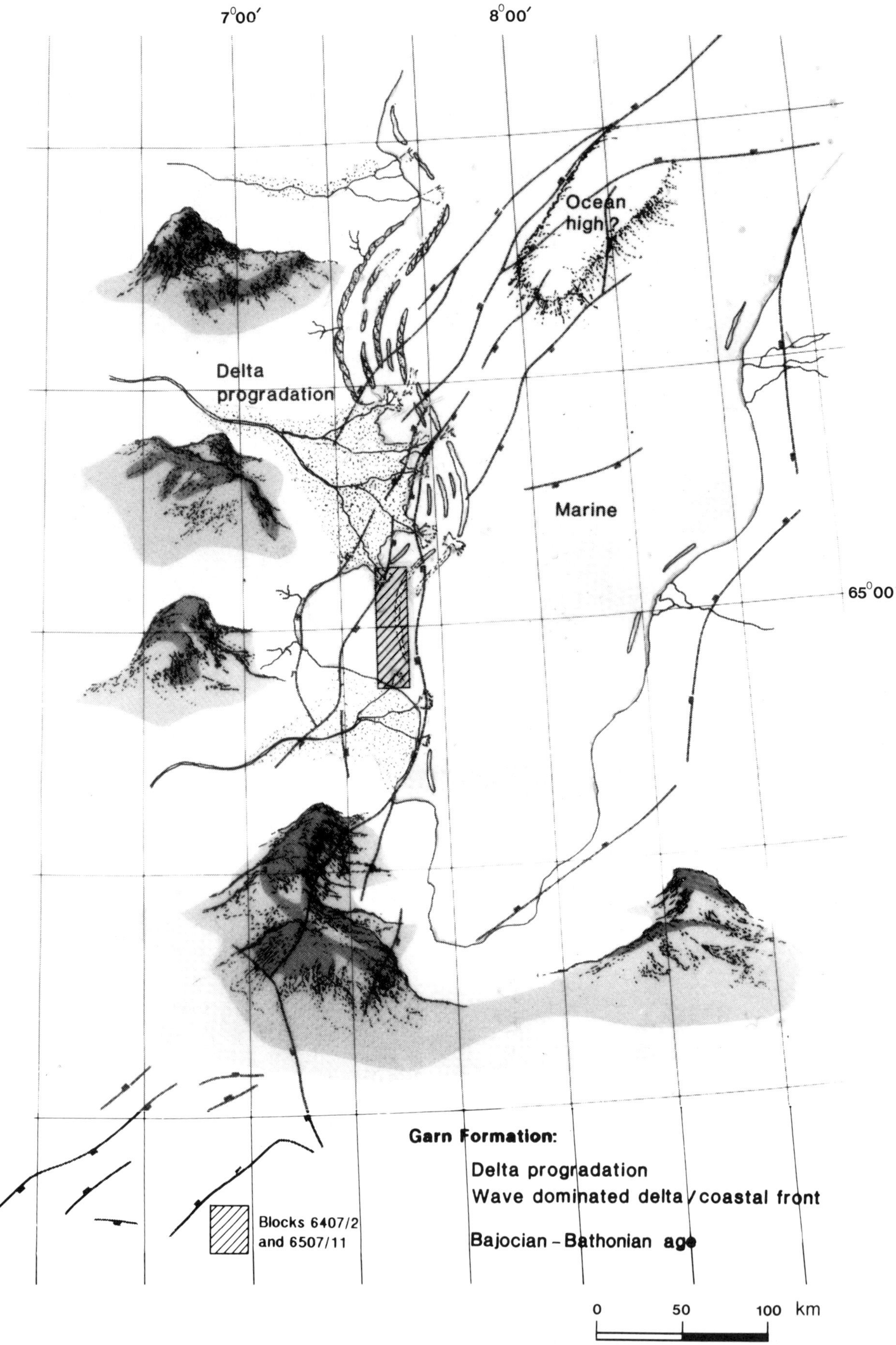

Figure 18. Paleogeographic reconstruction illustrating the environments of deposition of the Garn Formation in Bajocian-Bathonian time.

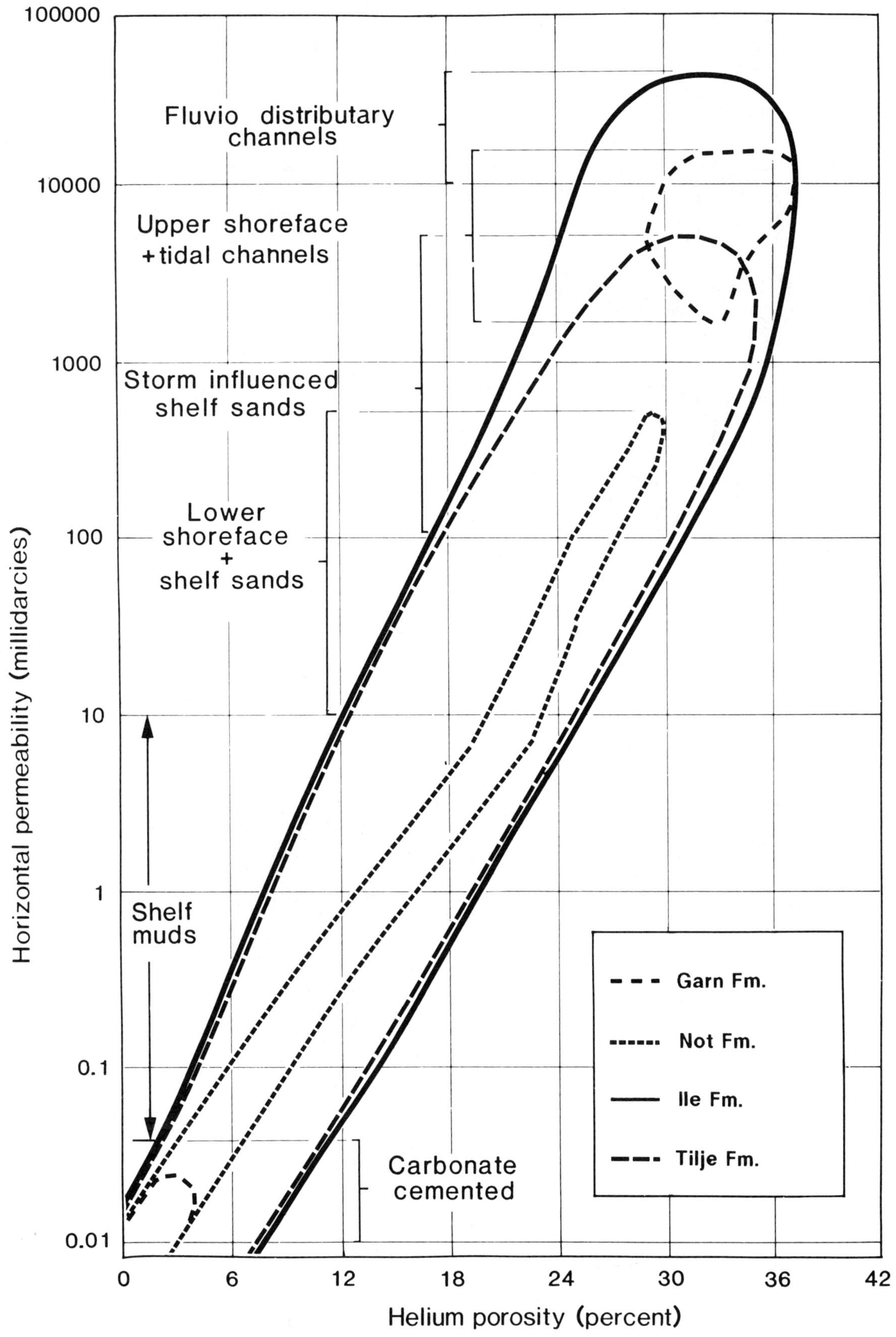

Figure 19. Porosity vs. permeability for different lithologies and sedimentary facies within the Midgard field.

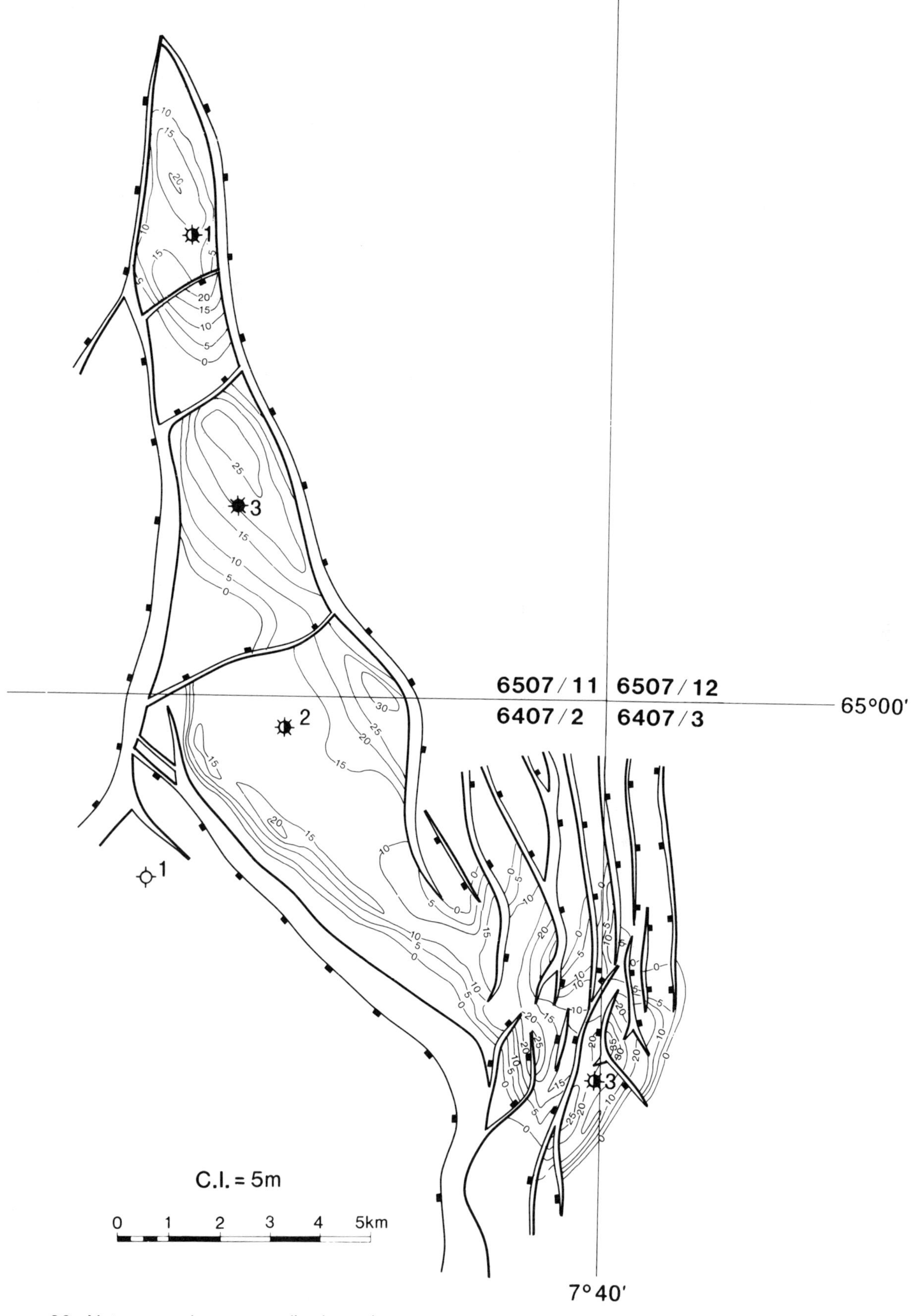

Figure 20. Net gas column map (hydrocarbon pore volume) of the Midgard field. The greatest thickness is located in the Delta segment to the southeast.

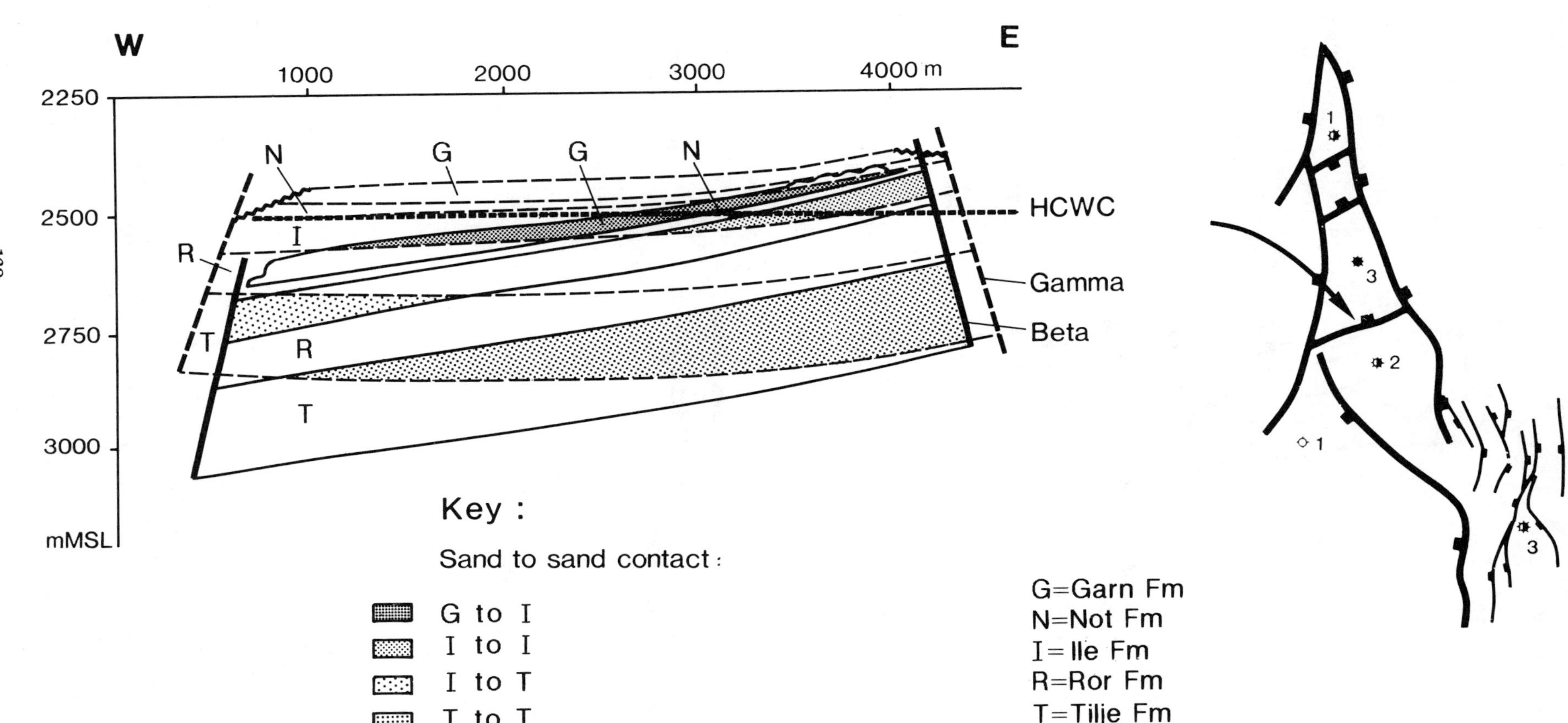

Figure 21. Fault plane section along the fault separating the Beta and Gamma segments.

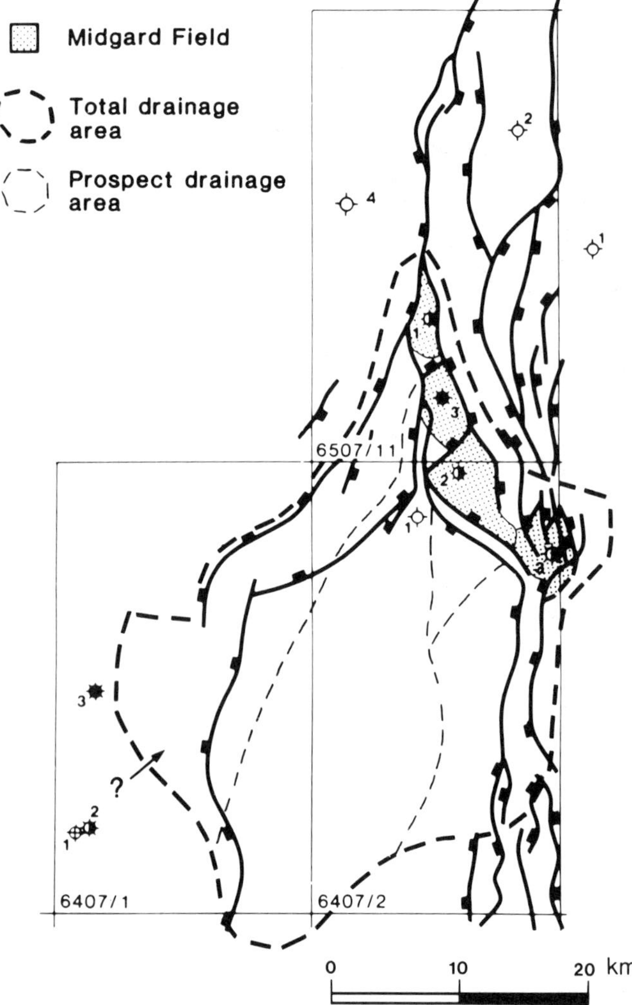

Figure 22. Drainage area for the Midgard field and the individual structural segments.

active gas- and condensate-generating stage. A similar development took place for the Spekk Formation, which entered the oil-generating window during Pliocene subsidence. The formation has been in a position for condensate and gas generation in the deeper parts of the drainage area during the last 2 million years.

The general maturation trend for the Haltenbanken area based on vitrinite reflectance values (R_o) from coals in the Åre Formation indicates that early oil generation ($R_o = 0.62$) occurs from 3550 m (Figure 24). Maximum oil generation takes place in the interval 4000–4300 m ($0.86 < R_o < 1.0$).

The rapid subsidence that induced efficient generation of hydrocarbons from both source rocks is also thought to have initiated the migration process. At this stage, the Midgard structure was fully developed and ready for entrapment. The individual drainage areas for each of the Midgard compartments have different extent and source rocks maturation levels. Hence, the Gamma segment apparently has an insignificant drainage area and is probably dependent on being filled either through the Beta or Delta segments or both (Figure 22). The Beta structure has by far the largest and the deepest part of the total Midgard field drainage area. The existence of the oil-leg in this segment can be readily explained by the different geometries of the drainage areas.

The Åre Formation in particular has potential for gas-condensate generation, and this indicates that it is the major source for the hydrocarbons encountered in the Midgard field. This is confirmed by correlation studies between the reservoir fluids and source rock extracts. Both the pristane to phytane ratios as well as carbon isotopes of C15+ hydrocarbons support this conclusion. The gas-condensate has pristane to phytane ratios between 1.8 and 2.4, which give good correlation to the Åre Formation. A ratio of 1.6 in the oil suggests a slightly higher contribution from the marine Spekk Formation (Figure 25).

EXPLORATION CONCEPTS

The main reservoirs of the Halten terrace are represented by sandstone sequences of Early and Middle Jurassic age. This is equivalent to the stratigraphy known from the North Sea area indicating deposition under similar major geological constraints. The traps of the Halten terrace are mainly fault-bounded. A significant throw can exist across these faults at Jurassic levels. There is also a general increase in Jurassic sediment depth toward the west. The development and preservation of porosity and permeability are clearly depth-dependent. Typical porosities of 30% exist at 2500 m in the Midgard field. A porosity depth gradient of 0.8%/100 m has been established for the area (Bjørlykke et al., 1986) indicating that the economic basement (10% porosity) is at approximately 4500 m for these reservoirs.

It is now well established that both the Upper Jurassic Spekk Formation and the Lower Jurassic Åre Formation have contributed significantly to hydrocarbon accumulation in the Haltenbanken area. Both units have generated both oil and gas depending on the maturation development within each drainage area. The early exploration of the Haltenbanken area was mainly based on North Sea experience. Hence, the Upper Jurassic shales were believed to have the greatest generating potential, and a maturity/depth pattern equivalent to that seen in the North Sea was expected. As a consequence, the block east of the Midgard field was prioritized in the first concession round. The pronounced effect of the major Pliocene burial leading to an increased depth to the hydrocarbon-generating zones was not recognized prior to the first wildcat. The results of the first well moved the attention westward to structures surrounded by potential source rocks that had been buried at greater depths. The drilling results

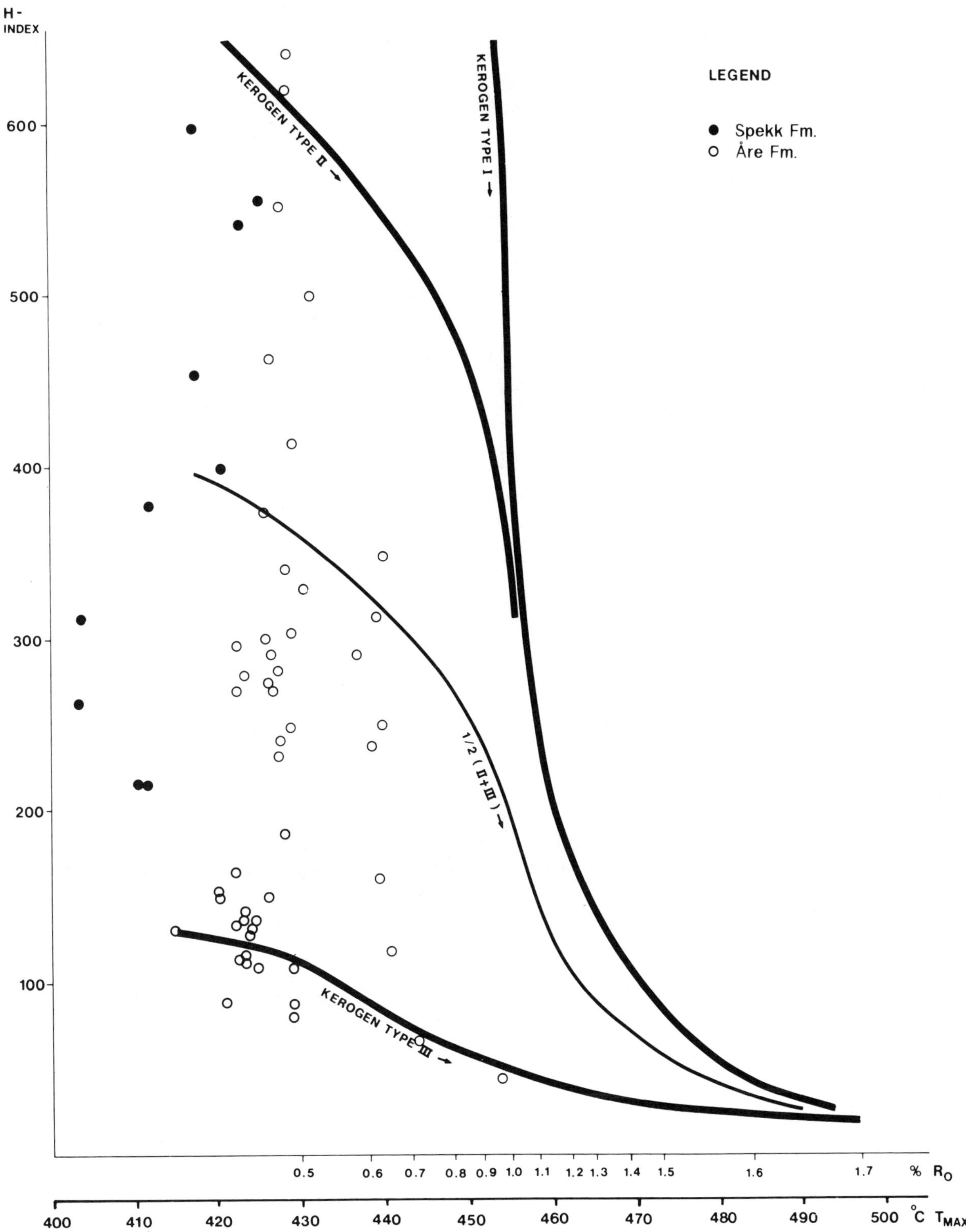

Figure 23. Hydrogen index vs. T_{max} for the Åre Formation and the Spekk Formation.

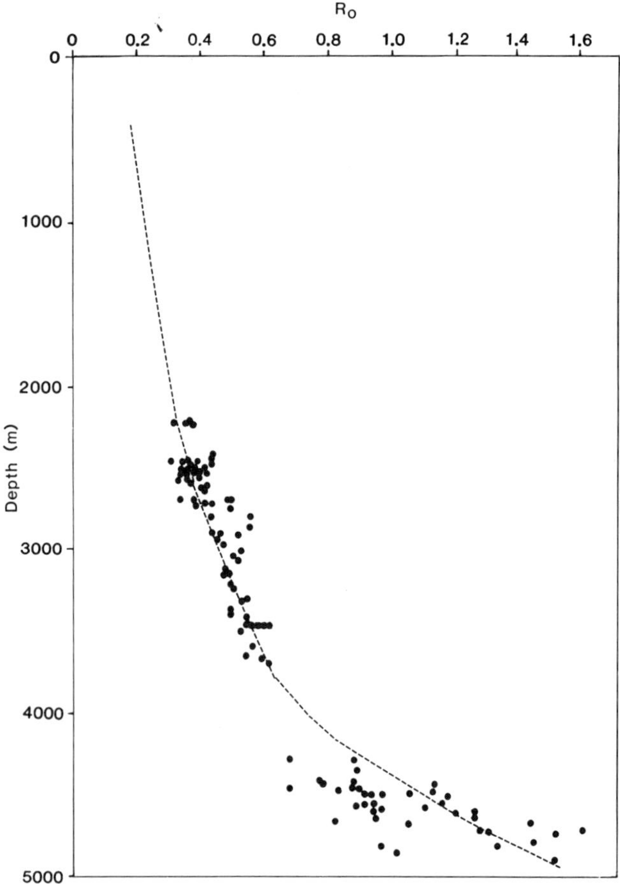

Figure 24. Maturation pattern for the Haltenbanken area based on vitrinite reflectivity values from coals in the Åre Formation.

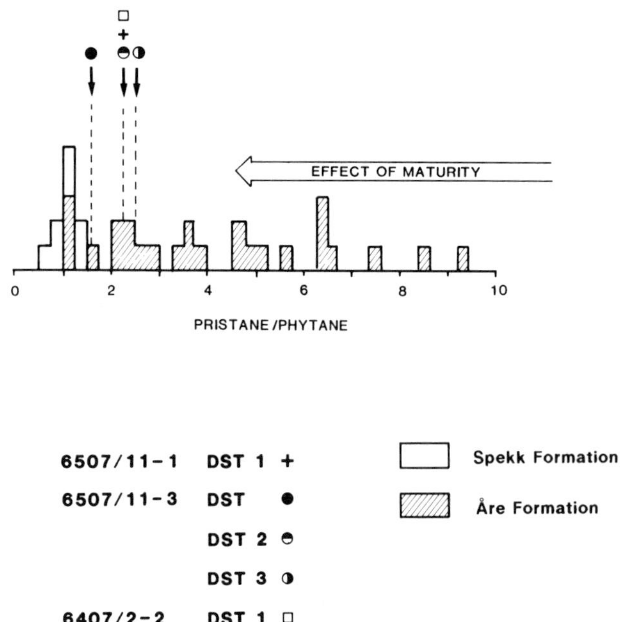

Figure 25. Pristane/phytane ratio in the reservoir fluids and source rocks in the Midgard field.

documented a significant source rock potential in the Lower Jurassic interval with facies development different from the North Sea. The discovery of the Midgard field documented the generating potential of the Lower Jurassic. However, this discovery together with the Tyrihans field led to a general opinion that the Haltenbanken was gas-prone. This understanding had some effect on exploration strategy and prioritization of individual companies.

The discovery of Draugen oil field in 1984 dramatically changed this gas-label and refocused attention on the province. More detailed basin modeling prior to the first exploration drilling would probably have led to a more optimized location for the first wells, possibly leading up to a direct discovery of the Midgard field. Using the full suite of data from the first exploration well, it should also have been possible to develop a more sophisticated analysis of each drainage area using PVT-modeling (Heum et al., 1986). Using this approach, it should be possible to predict more accurately hydrocarbon type and quantity in the different prospects.

ACKNOWLEDGMENTS

The author would like to thank Saga Petroleum as operator for the Midgard licences as well as the participating companies Statoil, Shell, Arco, Deminex, Agip, and Norsk Hydro for permission to publish this paper. The study is based on work carried out by several past and present Saga colleagues. Their efforts are highly appreciated.

REFERENCES CITED

Aasheim, S. M., A. Dalland, A. Netland, and A. Thon, 1986, The Smørbukk gas/condensate discovery, Haltenbanken, *in* A. M. Spencer, et al., eds., Habitat of hydrocarbons on the Norwegian Continental Shelf: Norwegian Petroleum Society, Graham and Trotman, p. 299-305.

Åm, K., 1970, Aeromagnetic investigation on the continental shelf of Norway Stad-Lofoten (62-60°): Nor. Geol. Unders., v. 266, p. 49-61.

Bjørlykke, K., P. Aagard, H. Dypvik, D.S. Hastings, and A. Harper, 1986, Diagenesis and reservoir properties of Jurassic sandstones from the Haltenbanken area, offshore mid Norway, *in* A. M. Spencer, et al., eds, Habitat of hydrocarbons on the Norwegian Continental Shelf: Norwegian Petroleum Society, Graham and Trotman, p. 275-286.

Bugge, T., R. Knarud, and A. Mørk, 1984, Bedrock geology on the mid-Norwegian Continental Shelf, *in* A. M. Spencer, et al., eds., Petroleum geology of the North European Margin: Norwegian Petroleum Society, Graham and Trotman, p. 253-270.

Bukovics, C., N. D. Shaw, E. G. Cartier, and P. A. Ziegler, 1984, Structure and development of the mid-Norway Continental

Margin, *in* A. M. Spencer, et al., eds., Petroleum geology of the North European Margin: Norwegian Petroleum Society, Graham and Trotman, p. 407-423.

Dalland, A., D. Worsley, and K. Ofstad, 1988, A lithostratigraphic scheme for the Mesozoic and Cenozoic sucession offshore mid- and northern Norway: NPD-bulletin No. 4.

Ekern, O. F., 1987, Midgard, *in* A. M. Spencer, et al., eds., Geology of the Norwegian oil and gas fields: Norwegian Petroleum Society, Graham and Trotman, p. 403-410.

Ellenor, D. W., and A. Mozetic, 1986, The Draugen oil discovery, *in* A. M. Spencer, et al., Habitat of hydrocarbons on the Norwegian Continental Shelf: Norwegian Petroleum Society, Graham and Trotman, p. 313-316.

Gabrielsen, R. H., and C. Robinson, 1984, Tectonic inhomogeneities of the Kristiansund-Bodø Fault Complex, offshore mid-Norway, *in* A. M. Spencer, et al., eds., Petroleum geology of the North European Margin: Norwegian Petroleum Society, Graham and Trotman, p. 397-406.

Gabrielsen, R. H., R. Færseth, G. Hamar, and H. C. Rønnevik, 1984, Nomenclature of the main structural features on the Norwegian Continental Shelf north of the 62nd parallel, *in* A. M. Spencer, et al., eds., Petroleum geology of the North European Margin: Norwegian Petroleum Society, Graham and Trotman, p. 41-60.

Grønlie, G., and I. B. Ramberg, 1970, Gravity indications of deep sedimentary basins below the Norwegian Continental Shelf and the Vøring Plateau: Nor. Geol. Tidsskr., v. 50, p. 375-391.

Heum, O. R., A. Dalland, and K. K. Meisingset, 1986, Habitat of hydrocarbons at Haltenbanken (PVT modelling as a predictive tool in hydrocarbon exploration), *in* A. M. Spencer, et al., eds., Habitat of hydrocarbons on the Norwegian Continental Shelf: Norwegian Petroleum Society, Graham and Trotman, p. 259-274.

Hollander, N. B., 1984, Geohistory and hydrocarbon evaluation of the Haltenbanken area, *in* A. M. Spencer, et al., eds., Petroleum geology of the North European Margin: Norwegian Petroleum Society, Graham and Trotman, p. 383-388.

Jørgensen, F., and T. Navrestad, 1979, Main structural elements and sedimentary succession of the shelf outside Nordland, *in* Norwegian Sea Symposium: Norwegian Petroleum Society, Article 11.

Jørgensen, F., and T. Navrestad, 1981, The geology of the Norwegian Shelf between 62°N and the Lofoten Islands, *in* L. V. Illing and G. D. Hobson, eds., Petroleum geology of the Continental Shelf of North-West Europe: Heyden, p. 407-413.

Kjerulf, T., 1870, Undersøkelse av nogle Kulslags og Torv. (Study of some samples of coal and peat.) Vidensk. Selsk. Forh. Christiania 1870, p. 404-413.

Rønnevik, H. C., and T. Navrestad, 1977, Geology of the Norwegian Shelf between 62-69°N: Geojournal, v.1, p. 33-46.

Rønnevik, H. C., E. I. Bergsager, A. Moe, O. Vrebø, and T. Navrestad, 1975, The geology of the Norwegian Continental Shelf, *in* A. W. Woodland, ed., Petroleum and the continental shelf of north-west Europe, Vol. 1 Geology: Applied Science Publishers, p. 117-129.

Appendix 1. Field Description

Field name .... *Midgard*

Ultimate recoverable reserves .... *109 × 10^9 Sm^3 gas and 15 × 10^6 Sm^3 condensate*

Field location:

- **Country** .... *Norway*
- **State** ....
- **Basin/Province** .... *Halten terrace*

Field discovery:

- **Year first pay discovered** .... *L. and M. Jurassic Garn, Ile, Åre, and Tilje formations 1981*
- **Year second pay discovered** .... *NA*
- **Third pay** .... *NA*

Discovery well name and general location

- **First pay** .... *6507/11-1, 230 km northwest of Trondheim, Norway*
- **Second pay** .... *NA*
- **Third pay** .... *NA*

Discovery well operator .... *Saga Petroleum*

- **Second pay** .... *NA*
- **Third pay** .... *NA*

IP in barrels per day and/or cubic feet or cubic meters per day:

- **First pay** .... *2.0 × 10^6 Sm^3/d (estimated max. initial gas rate for individual wells)*
- **Second pay** .... *NA*
- **Third pay** .... *NA*

All other zones with shows of oil and gas in the field:

Age	Formation	Type of Show
None		

Geologic concept leading to discovery and method or methods used to delineate prospect, e.g., surface geology, subsurface geology, seeps, magnetic data, gravity data, seismic data, seismic refraction, nontechnical:

Primary assessment of the basin was made based on magnetic and gravity data. The definition of the structure was made through seismic data.

MIDGARD

Structure:

Province/basin type (see St. John, Bally, and Klemme, 1984)

Bally 1141; Klemme IIIc

Tectonic history

The main phase of extension started in Late Triassic. The main block faulting occurred during the middle Late Jurassic. The stretching phase terminated during Cretaceous. Subsidence continued through Cenozoic time and culminated in rapid regional subsidence in late Pliocene to Pleistocene time.

Regional structure

East on Halten terrace, 10 km west of the Trøndelag platform.

Local structure

Northwest-trending horst feature separated in segments by northeast crossing faults.

Trap

- **Trap type(s)** .... *4 fault traps (3 horst segment traps, 1 rotated fault segment complex trap)*

Basin stratigraphy (major stratigraphic intervals from surface to deepest penetration in field):

Chronostratigraphy	Formation	Depth to Top in m
Quaternary	*Naust Formation*	*298*
Tertiary	*Kai Formation*	*1374*
Tertiary	*Hordaland Group*	*1582*
Tertiary	*Rogaland Group*	*1978*
Cretaceous	*Shetland Group*	*2087*
Cretaceous	*Cromer Knoll Group*	*2302*
Jurassic	*Viking Group*	*2384*
Jurassic	*Fangst Group*	*2435*
Jurassic	*Båt Group*	*2576*
Triassic	*Gray Beds/Redbeds*	*3249*

Location of well in field .... *NA*

Reservoir characteristics:

Number of reservoirs *1*

Formations

Garn Formation, Ile formation, Tilje Formation, Åre Formation; the formations are treated as one reservoir based on identical pressure regime and hydrocarbon contacts.

Ages .... *Early-Middle Jurassic*

Depths to tops of reservoirs .... *2280 m*

Gross thickness (top to bottom of producing interval) .... *220 m (max column)*

Net thickness—total thickness of producing zones

Average .... *Garn Formation, 54 m; Ile Formation, 62 m*

Maximum .... *Garn Fm., 64 m; Ile Fm., 68 m; Tilje Fm., 17 m (in 6507/11-1 only)*

Average ....

Maximum ....

Lithology .... *Fine to medium occ. coarse-grained, moderate sorted sandstones*

Porosity type .... *Intergranular*

Average porosity .... *Garn Fm., 29%; Ile Fm., 26%; Tilje Fm., 25%*

Average permeability .... *Garn Fm., 6650 md; Ile Fm., 4600 md; Tilje Fm., 2200 md*

Seals:

Upper

Formation, fault, or other feature .... *Formation*

Lithology .... *Shale*

Lateral

Formation, fault, or other feature .... *Fault*

Lithology .... *Shale*

Source:

Formation and age .... *Åre Fm., Early Jurassic; Spekk Fm., Late Jurassic*

Lithology .... *Shale*

Average total organic carbon (TOC) .... *Åre Fm., 8%; Spekk Fm., 8%*

Maximum TOC .... *Åre Fm., 70%; Spekk Fm., 11%*

Kerogen type (I, II, or III) .... *Åre Fm., III; Spekk Fm., II*

Vitrinite reflectance (maturation) .... *Åre Fm., R_o = 0.55; Spekk Fm., immature on structure*

Time of hydrocarbon expulsion .... *Initiated 3.5 million years ago*

Present depth to top of source .... *Åre Fm., 2800 m; Spekk Fm., 2300 m on structure*

Thickness .... *Åre Fm., 350 m; Spekk Fm., 10 m*

Potential yield .... *NA*

Appendix 2. Production Data

Field name .... *Midgard*

Field size:

- **Proved acres** .... *47 km²*
- **Number of wells all years** *17*
- **Current number of wells** .... *NA*
- **Well spacing** .... *Cluster production*
- **Ultimate recoverable** .... *109×10^9 Sm^3 gas; 15×10^6 Sm^3 condensate*
- **Cumulative production** .... *NA*
- **Annual production** .... *10×10^9 Sm^3*
- **Present decline rate** .... *NA*
  - **Initial decline rate** .... *Const. rate*
  - **Overall decline rate** .... *NA*
- **Annual water production** .... *NA*
- **In place, total reserves** .... *152×10^9 gas; 29×10^6 condensate; 8×10^6 oil*
- **In place, per acre-foot** .... *NA*
- **Primary recovery** .... *109×10^9 Sm^3 gas; 15×10^6 Sm^3 condensate*
- **Secondary recovery** .... *NA*
- **Enhanced recovery** .... *NA*
- **Cumulative water production** .... *NA*

Drilling and casing practices:

- **Amount of surface casing set** .... *4800 m*
- **Casing program**

  30-in. to 400 m; 20-in. to 900 m; 13⅜-in. to 1950 m; 9⅝-in. to 2300 m; 7-in. to 2600 m
- **Drilling mud** .... *Gyp/polymer mud*
- **Bit program** .... *36-in. H.O., 17½-in. pilot hole, open to 26-in.; 12¼-in.; 8½-in.*
- **High pressure zones** .... *Max. 1.5 g/cc in Upper Cretaceous*

Completion practices:

- **Interval(s) perforated** .... *Full phase*
- **Well treatment** .... *NA*

Formation evaluation:

- **Logging suites**

  36-in.–12¼-in.: MWD/DIFL/LSBHC/GR/(CDL-CNL)
  8½-in.: + DLL/MLL/Z/DENS/SPECTRA/FMT/VSP/COREGUN
- **Testing practices** .... *3–10 m to confirm production profiles*
- **Mud logging techniques**

  Computerized unit, continuous ditch gas and chromatograph gas analysis, preparation of dried, wet, and canned samples.

Oil characteristics:

- **Type** .... *NA*

  (Tissot and Welte Classification in "Petroleum Formation and Occurrence," 1984, Springer-Verlag, p. 419)
- **API gravity** .... *0.76 specific gravity*
- **Base** .... *NA*
- **Initial GOR** .... *5235 Sm^3/Sm^3 (gas-condensate ratio)*
- **Sulfur, wt%** .... *NA*
- **Viscosity, SUS** .... *0.024 cp*
- **Pour point** .... *NA*
- **Gas-oil distillate** .... *NA*

Field characteristics:

- **Average elevation** .... *2490 m*
- **Initial pressure** .... *252 bar*
- **Present pressure** .... *252 bar*
- **Pressure gradient** .... *0.101 bar/m*
- **Temperature** .... *91°C*
- **Geothermal gradient** .... *0.0365°C/m*
- **Drive** .... *Volumetric expansion and moderate water drive*
- **Oil column thickness** .... *Max. gas 220 m*
- **Oil-water contact** .... *2490–2500 m*
- **Connate water** .... *Garn Fm., 7%; Ile Fm., 16%; Tilje Fm., 20%*
- **Water salinity, TDS** .... *NA*
- **Resistivity of water** .... *NA*
- **Bulk volume water (%)** .... *NA*

Transportation method and market for oil and gas:

Pipeline to Norwegian mainland; further to U.K., Sweden, or domestic use for electricity.

MIDGARD

Sleipner Øst Field—Norway
Viking Graben, North Sea

O. J. ØSTVEDT
Esso Norge A.S.
Forus, Norway

S. EVENSEN
H. K. A. RANAWEERA
Statoil
Stavanger, Norway

FIELD CLASSIFICATION

BASIN: North Sea
BASIN TYPE: Rift
RESERVOIR ROCK TYPE: Sandstone
RESERVOIR ENVIRONMENT OF DEPOSITION: Submarine Fan and Marine Shoreline
RESERVOIR AGE: Paleocene and Jurassic
PETROLEUM TYPE: Gas/Condensate
TRAP TYPE: Anticline with Pinchout and Faulted Anticline

LOCATION

The Sleipner Øst gas-condensate field is located in Block 15/9 in the Norwegian sector of the North Sea (Figure 1). It is part of Production License (PL) 046 and was previously referred to as the 15/9 Gamma field. The field is situated on the Ling high, which forms the eastern margin of the South Viking graben (Figure 2). The Viking graben is a major structural element in the northern North Sea along which many large petroleum fields are located.

Several gas-bearing structures have been discovered in the Sleipner area. Sleipner Vest and Sleipner Øst (Øst means east) are the largest accumulations, whereas the others are relatively small and only economically attractive as potential satellite developments to the larger fields.

The hydrocarbons in the Sleipner Øst field are trapped within a primary Paleocene reservoir and a secondary Jurassic reservoir. Estimated ultimate recovery for the field is 2.1 tcf (60×10^9 Sm^3) of wet gas, which after processing is equivalent to 1.8 tcf (51×10^9 Sm^3) of dry gas and 211 million bbl (23×10^6 metric tons) of condensate. The planned production start-up date is 1993.

HISTORY

Pre-Discovery

Production Licence PLO46, comprising Block 15/9 and the neighboring Block 15/8, was awarded in 1976 to the following group:

Statoil (Operator)	50%
Esso Norge a.s	40%
Norsk Hydro Produksjon a.s	10%

In early 1988, Statoil sold a portion of its interest to Elf Aquitaine Norge a.s and Total Marine Norsk a.s, which together with the implementation of a sliding scale (an option given by the Norwegian authorities to increase Statoil's share in commercial fields) resulted in the current share of interest as given below:

Statoil (Operator)	49.6%
Esso Norge a.s	30.4%
Norsk Hydro Produksjon a.s	10.0%
Elf Aquitaine Norge a.s	9.0%
Total Marine Norsk a.s	1.0%

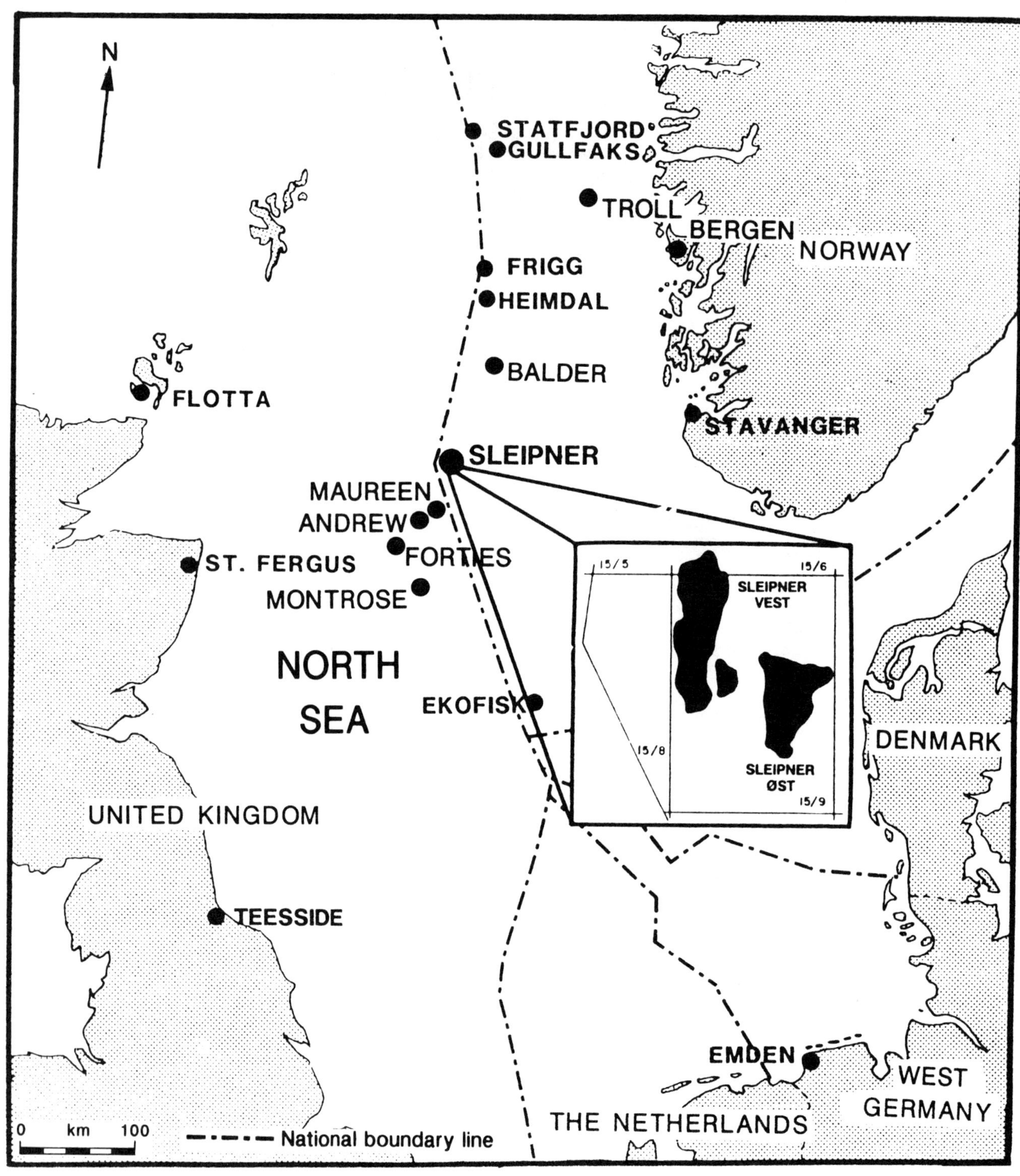

Figure 1. General location map showing main fields.

The Sleipner Vest field was discovered by Well 15/6-3 drilled by Esso in 1974. Seismic interpretations indicated that the field extended southward into Block 15/9, which at that time was open acreage. During 1975, a detailed seismic survey was shot in order to map the discovery and other prospects in the open acreage. Several prospects were mapped, and Esso and Statoil among other companies decided to apply for the two open Blocks 15/9 and 15/8.

In December 1976, the two blocks were awarded to the initial three companies (see details above). With this an active phase of exploration started in the licence area. To date, a total of 18 wells have been drilled in Block 15/9, one in Block 15/8, and three

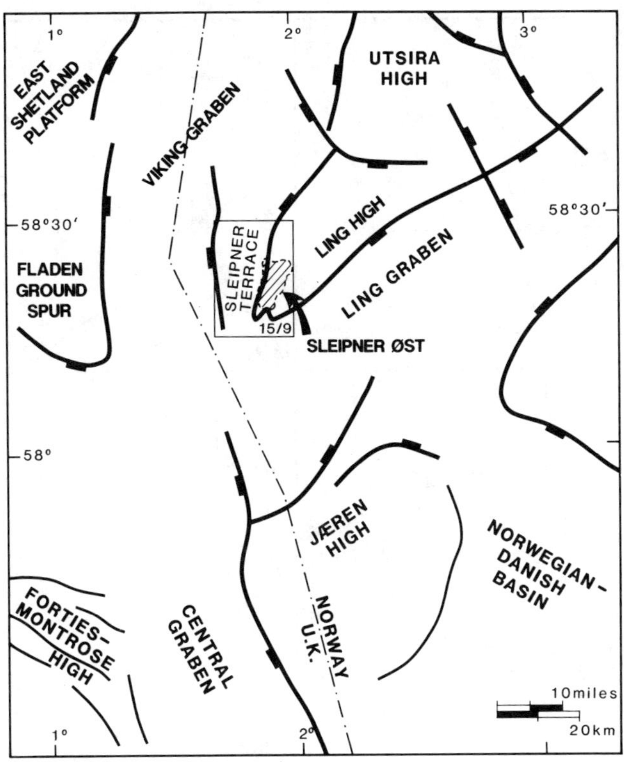

Figure 2. Regional structural elements. Block 15/9 and Sleipner Øst field are outlined. Normal faults are shown with downthrown side marked by ticks.

in the portion of Sleipner Vest that lies in Block 15/6 (Figure 3). Successive exploration and appraisal wells in Block 15/9 have proven that the major reserves lie to the south and east of the original discovery in Block 15/6. Thus, the areal distribution of proven reserves has changed considerably through time.

DISCOVERY AND EXPLORATION METHODS

The initial seismic mapping of the Ling high indicated a structural closure at the Jurassic/Triassic level, but it was not until 1981 that the prospect was investigated by drilling wells. Geological models based on data from wells on the Sleipner terrace postulated a Jurassic and possibly a Triassic sandstone reservoir. These sandstones were the target for the first well, 15/9-9, drilled on the Ling high during the summer of 1981. However, the well penetrated 100 m of gas-bearing Paleocene sandstones of the Heimdal Formation. This was unexpected because Paleocene structures on the Sleipner terrace had all been dry. No Jurassic sandstones were encountered in the well, and the Triassic sandstones were water-bearing. As a result of the well, the existing seismic database was infilled to create a 0.5 × 0.5 km grid. Three subsequent wells, 15/9-11, -13, and -16, drilled in the Sleipner Øst area during 1981 and 1982 confirmed the presence of a thick Paleocene gas-bearing sandstone with good reservoir properties. Two of the wells, 15/9-11 and 15/9-13, also proved the existence of gas-bearing Middle Jurassic sandstones, with a gross thickness of about 35 m. In order to map the field accurately it was decided in 1982 to aquire a 3D seismic survey that covered the whole structure. The 3D data form the basis for the current geophysical mapping of the field. The following reflectors have been mapped: top Paleocene, top Heimdal sandstones, top Chalk, top Lower Cretaceous, base Cretaceous, and top Jurassic/Triassic sandstones. A direct hydrocarbon indicator (flat spot) is recognizable in parts of the Paleocene sandstones. The pinchout of the Heimdal sandstones is recognized as a down-lapping event on the top of the Chalk reflector (Figure 4).

STRUCTURE

The present structural configuration in the Sleipner Øst area is largely related to extensional rifting in the Viking graben area. The first stage of rifting took place in Early Permian time when intracratonic basins were formed (Ziegler, 1981), but the major rifting events took place during the Jurassic and Early Cretaceous.

A major thermal event occurring in the Middle Jurassic was followed by rifting and subsidence of the North Sea graben system. In response to the rifting, normal faults trending north-northeast-south-southwest were formed along the margin of the Viking graben. The fault blocks are rotated and step down from the Ling high to the Sleipner terrace and the Viking graben (Figures 2 and 5). Another set of faults trending west-northwest-east-southeast developed parallel to the direction of rifting in the Norwegian-Danish basin (Figure 2) and are present in the eastern part of the block.

During this tectonic event, referred to as the "mid-Kimmerian phase," sediments were eroded from the uplifted Utsira high and redeposited in the subsiding Viking graben. Dating of the sediments on the Ling high (Figure 2) is ambiguous but indicates that no deposition took place during the Early and Middle Jurassic until a major transgression occurred in the late Middle Jurassic and shallow-marine sands of the Hugin Formation were deposited. These sands, which form the secondary reservoir in the field, are overlain by marine shales deposited during a continued sea-level rise in the Late Jurassic.

The last major rifting event, the "late Kimmerian phase," took place during the Late Jurassic to Early Cretaceous. In addition to reactivation of the existing fault pattern, left-lateral strike-slip movements occurred along the WNW-ESE-trending faults due to different local rates of extension (Pegrum and

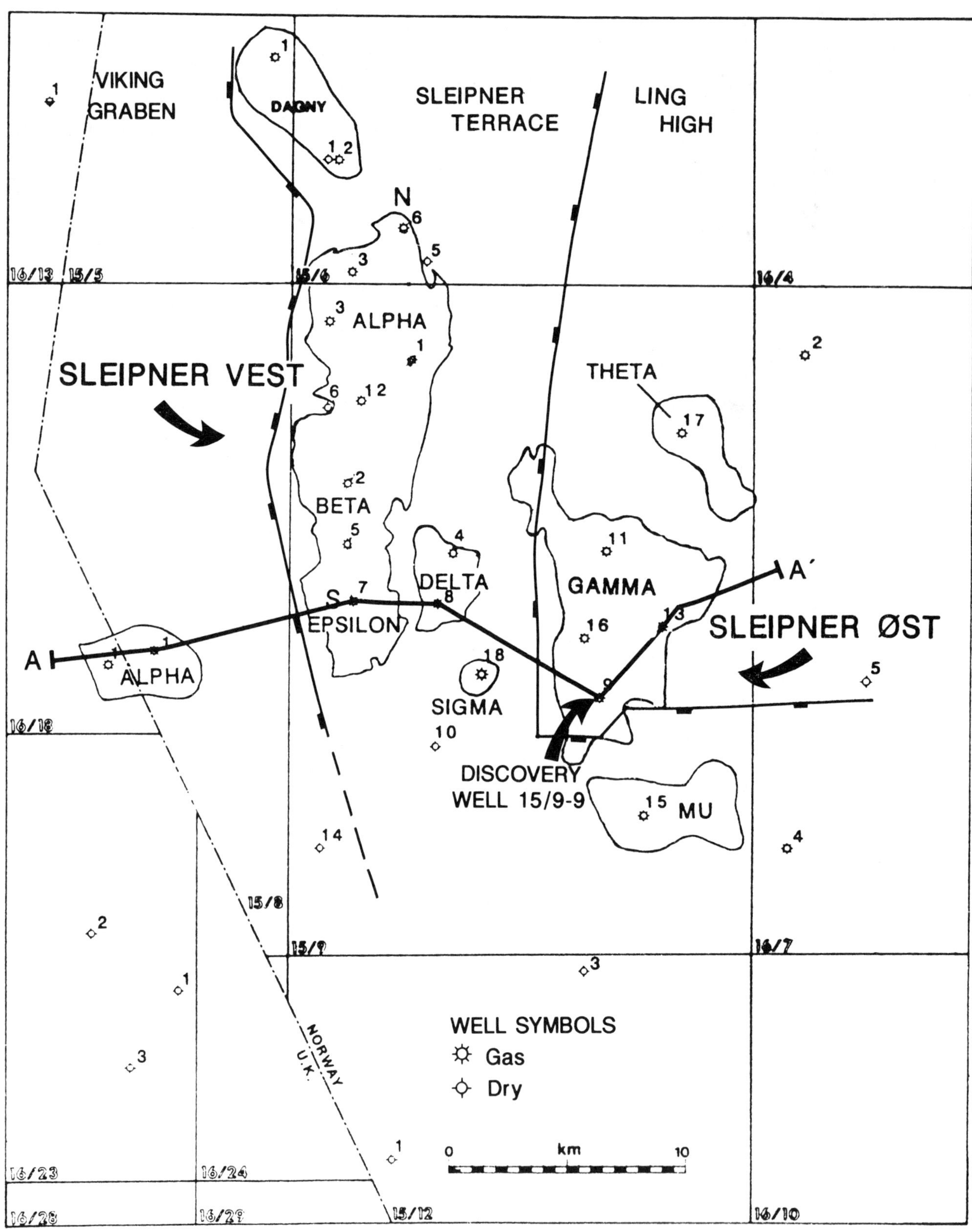

Figure 3. Outline of main fields in the Sleipner area including major structural elements. Line A-A′ shows location of cross-section, Figure 5.

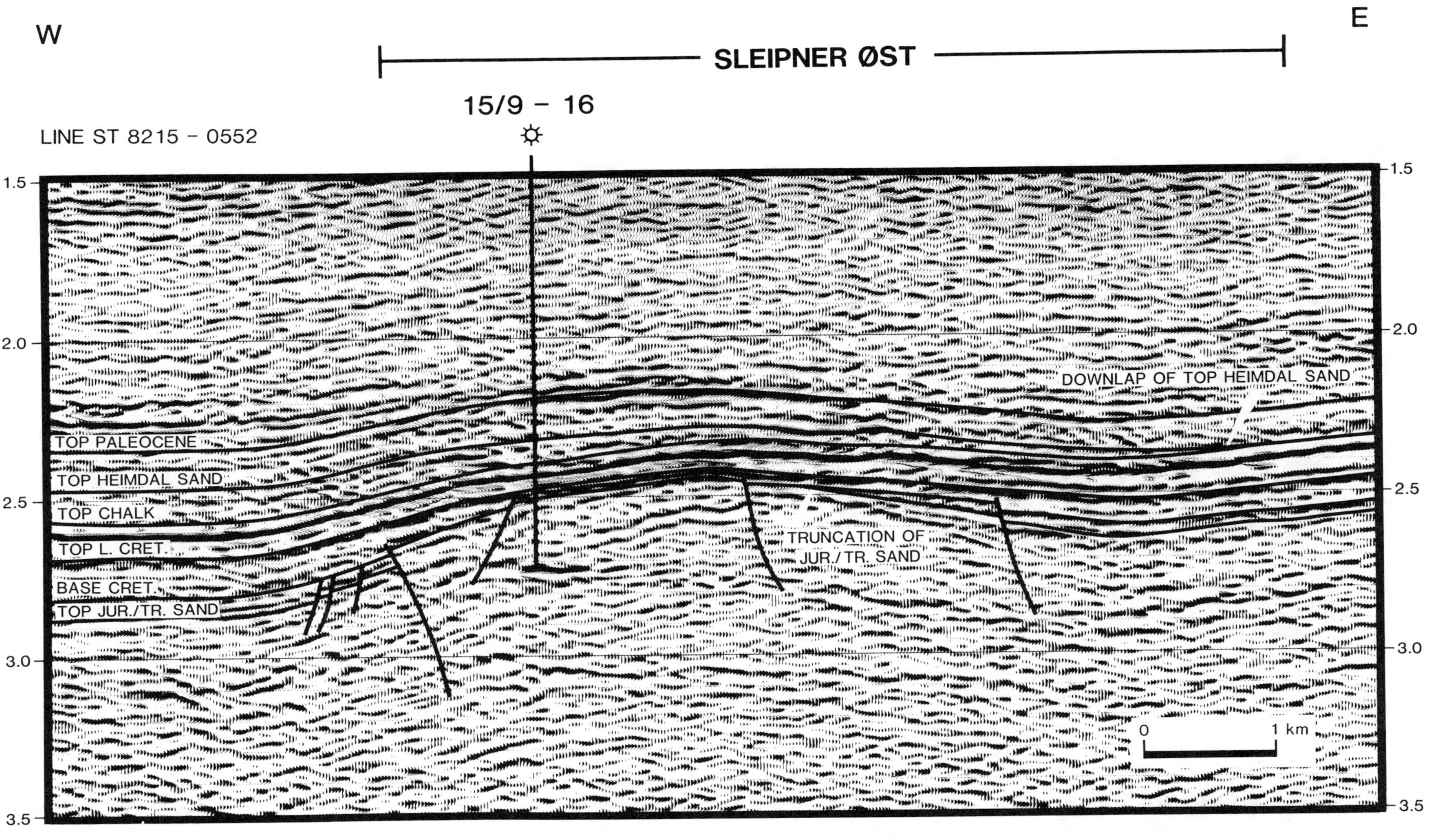

Figure 4A. Interpreted east-west seismic line. A flat event occurs at 2.35 seconds.

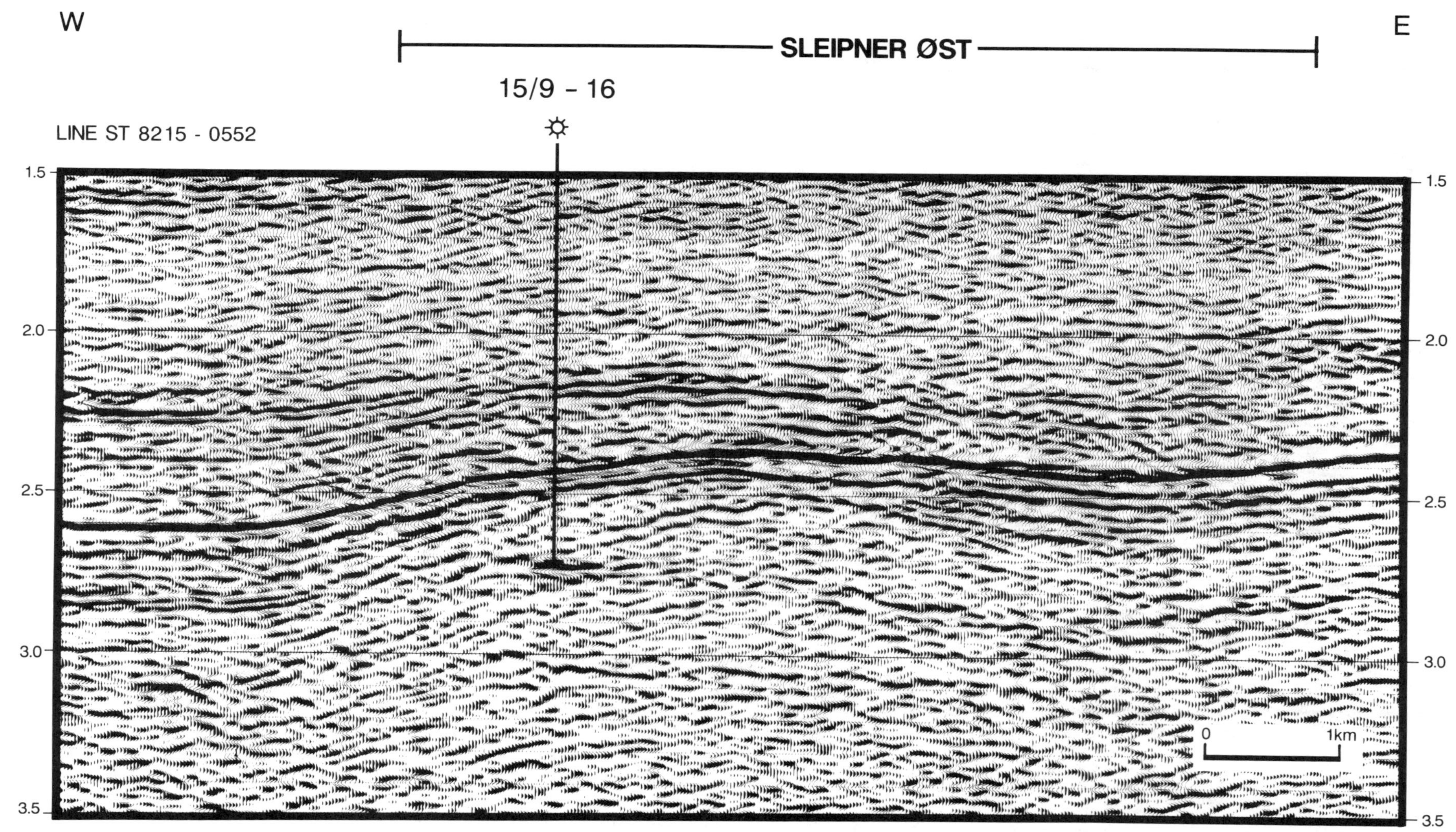

Figure 4B. Uninterpreted seismic line. Location of line is shown in Figures 7 and 8.

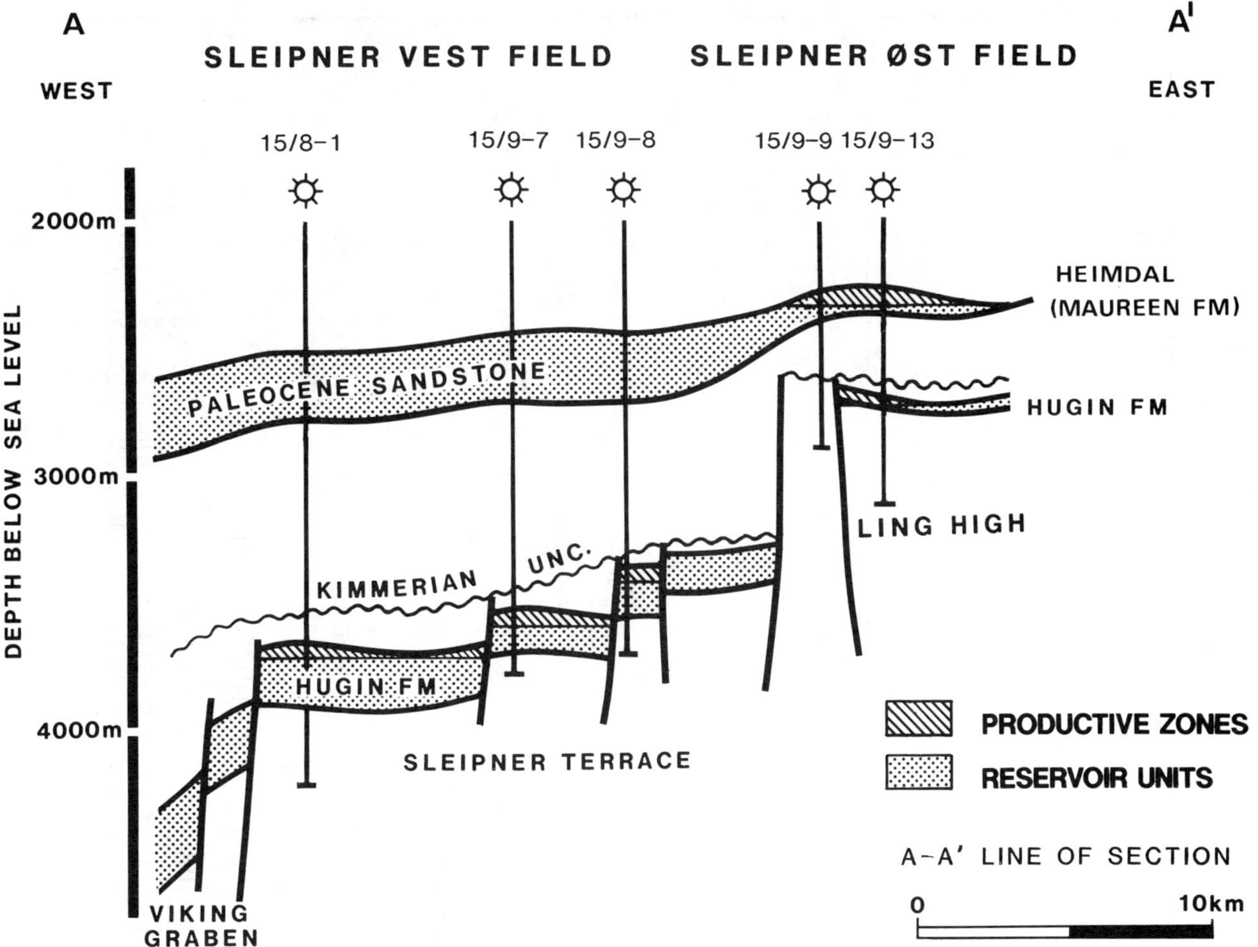

Figure 5. Generalized structural cross section. Line of section shown on Figure 3.

Ljones, 1984). Several episodes of sea level drops accompanied the rifting. The resultant "late Kimmerian" unconformity is both locally and regionally significant and recognizable on seismic over most of the North Sea. On the crestal part of the Sleipner Øst structure, the Jurassic section is completely removed by truncation (Figure 5).

At the end of the rifting phase, the Ling high was an easterly tilted structural high with the Sleipner terrace downfaulted to the west and the Ling graben downfaulted to the south (Figure 2).

Steepening of preexisting faults is thought to have occurred in the mid-Cretaceous as a result of the early Alpine orogeny and the concomitant reversal of the stress field (Pegrum and Ljones, 1984). During the Late Cretaceous, regional subsidence took place in the Viking graben and adjacent areas, including the Utsira high. The subsidence was probably related to the onset of rifting in the North Atlantic (Ziegler, 1981).

In the Late Cretaceous-early Tertiary, the Laramide orogenic event and drops in eustatic sea level led to uplift and erosion of the East Shetland platform and Fladen ground spur (Figure 2). Sandy sediments were shed eastward as submarine fan complexes, some of which extended into Block 15/9. These sands constitute the main reservoir in the field.

The Paleocene trap was formed in mid-Tertiary time when transtensional stress resulted in gentle folding of the sediments into domal structures. These movements most likely resulted from the late Eocene Alpine phase (Pegrum and Ljones, 1984). This was the last episode of significant tectonic activity in the area.

STRATIGRAPHY

The stratigraphic section present in the field is shown in Figure 6. The lithostratigraphic nomenclature follows that of Deegan and Scull (1977) as revised by Vollset and Doré (1984).

The oldest formation penetrated in the field is the Lower Permian Rotliegendes Group, consisting of continental sandstones and conglomerates. The overlying marginal marine carbonates and evaporites

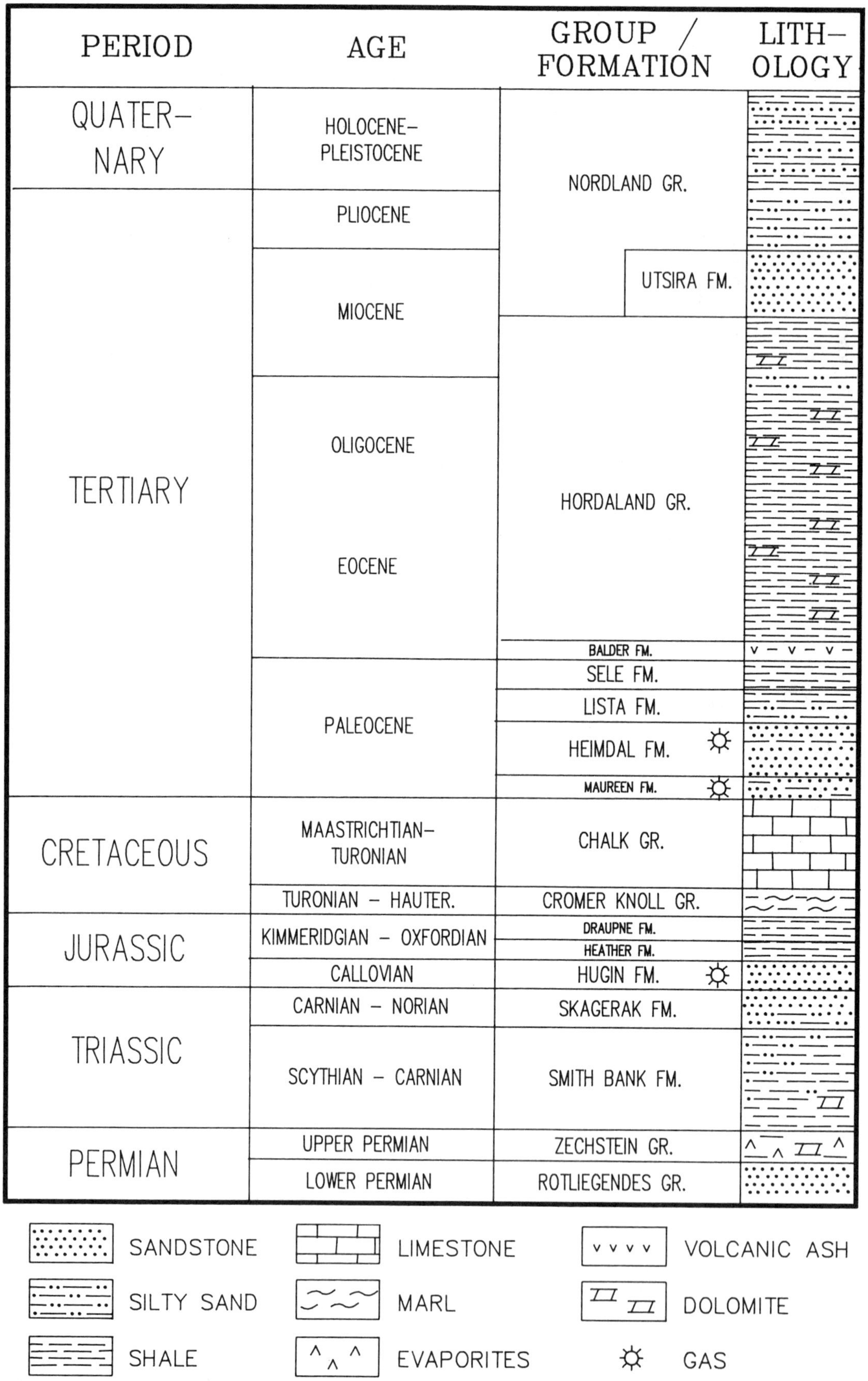

PERIOD	AGE	GROUP / FORMATION	LITHOLOGY
QUATERNARY	HOLOCENE–PLEISTOCENE	NORDLAND GR.	
TERTIARY	PLIOCENE	NORDLAND GR.	
	MIOCENE	NORDLAND GR. / UTSIRA FM.	
	OLIGOCENE	HORDALAND GR.	
	EOCENE	HORDALAND GR.	
		BALDER FM.	
	PALEOCENE	SELE FM.	
		LISTA FM.	
		HEIMDAL FM. ☼	
		MAUREEN FM. ☼	
CRETACEOUS	MAASTRICHTIAN–TURONIAN	CHALK GR.	
	TURONIAN – HAUTER.	CROMER KNOLL GR.	
JURASSIC	KIMMERIDGIAN – OXFORDIAN	DRAUPNE FM.	
		HEATHER FM.	
	CALLOVIAN	HUGIN FM. ☼	
TRIASSIC	CARNIAN – NORIAN	SKAGERAK FM.	
	SCYTHIAN – CARNIAN	SMITH BANK FM.	
PERMIAN	UPPER PERMIAN	ZECHSTEIN GR.	
	LOWER PERMIAN	ROTLIEGENDES GR.	

SANDSTONE LIMESTONE VOLCANIC ASH
SILTY SAND MARL DOLOMITE
SHALE EVAPORITES ☼ GAS

Figure 6. Generalized stratigraphy of Sleipner Øst.

of the Upper Permian Zechstein Group reach a thickness of about 50 m. The Lower to Middle Triassic Smith Bank Formation is a 300–500 m thick sequence of red argillaceous sediments deposited in a lacustrine or muddy flood-plain environment. Silty sands of braided streams of flood-plain origin assigned to the Upper Triassic Skagerak Formation are only present on the flanks of the Ling high.

The Lower Jurassic and most of the Middle Jurassic sections are absent on the Ling high where the upper Middle Jurassic Hugin Formation of Callovian age rests directly on Triassic rocks. The Hugin Formation, which makes up the secondary reservoir in the field, is 35 m thick in Well 15/9-11, but thins slightly toward the south and is absent on the crestal part of the structure.

On the Sleipner terrace, the Hugin Formation is 50–200 m thick and constitutes the reservoir in the Sleipner Vest field. The underlying Middle Jurassic Sleipner Formation is coal rich and is considered a possible source for the gas in the Sleipner area. Middle Jurassic sandstones are in general a very important play in the Viking graben area.

Upper Jurassic shales are only a few tens of meters thick but are important cap and source rocks. The Draupne Formation is the major source rock in the central and northern North Sea. It has probably contributed to the gas in the Sleipner fields.

Cretaceous marls and chalk have a total thickness of 200–300 m and are overlain by lower Tertiary (Paleocene) submarine fan sandstones of the Maureen and Heimdal formations. Both formations thin rapidly and pinch out to the southeast.

Paleocene sandstones are important hydrocarbon reservoirs in the North Sea. The major fields are Heimdal and Balder north of Sleipner Øst and Maureen, Andrew, Forties, and Montrose fields in the U.K. sector south of Sleipner Øst. The sandstones are capped by upper Paleocene shales.

Volcanic ash (tuff) characterizes the lower Eocene Balder Formation, which is an excellent marker bed and produces a strong regional seismic reflector. The overlying stratigraphic section consists dominantly of marine shales and marls with the 250 m thick upper Miocene Utsira Formation as the only major sandstone interval. The Pliocene and Pleistocene sequence consists of alternating shales and sandstones.

TRAP

The Paleocene reservoir is a combined structural and stratigraphic trap with an asymmetric domal shape (Figure 7). Structural dips are steepest to the west and quite gentle in the other directions. The southeastern extent of the field is partly determined by the depositional limit of the sand that forms the stratigraphic element of the trap. The trap is filled to the structural spillpoint at 2417 m (subsea), located in the northern part. The crest of the reservoir occurs at 2260 m (subsea) and the total closure height is, hence, 157 m. Paleocene shales of the Lista and Sele formations overlie the Heimdal sandstones and form a good seal. The structure was formed in the mid-Tertiary at about the time of the onset of hydrocarbon generation in the Viking graben.

The Jurassic reservoir sandstones are faulted and folded into an overall asymmetric domal structure with the steepest dips to the west (Figure 8). Upper Jurassic shales seal the reservoir. Reservoir sandstones are absent in the crestal part of the structure, due to erosional truncation and/or sandstone pinchout. The structural spillpoint is at about 2800 m (subsea) and the crest of the reservoir at 2645 m (subsea).

RESERVOIRS

Sandstones of the Paleocene Heimdal Formation represent the primary reservoir in the Sleipner Øst field. The reservoir, which has a maximum gross thickness of 120 m, thins rapidly toward the southeast. A shaly unit in the upper part of the formation can be correlated between the wells and is used to separate the upper and lower Heimdal Formation (Figure 9). This shale unit has not represented a permeability barrier during geologic time but may act as a barrier to vertical flow during production.

Middle to Upper Jurassic sandstones of the Hugin Formation form the secondary reservoir, which has an average gross thickness of 35 m. A wide range of reservoir properties are predicted due to facies variations. Although good permeabilities were measured in the reservoir, faults and sandstone discontinuities may act as permeability barriers.

Paleocene Reservoir

Lithofacies

The sandstones of the Paleocene Heimdal Formation were deposited as part of a sand-rich submarine fan complex, derived from the Fladen ground spur and the East Shetland platform. The Sleipner Øst structure is located in the distal part of the fan. Despite its distance from the source area (approximately 50 km), the Heimdal Formation is extremely sand rich (Figure 10) with very good reservoir properties. The sands were probably transported to the area from the northwest by high-density turbidity currents and deposited in front of the main fan complex. The sandstones are abutting the northwesterly dipping paleoslope of the chalk surface. Three main lithofacies are recognized in the Heimdal Formation from core data: massive sandstones, shales and claystones, and interbedded siltstones and shales.

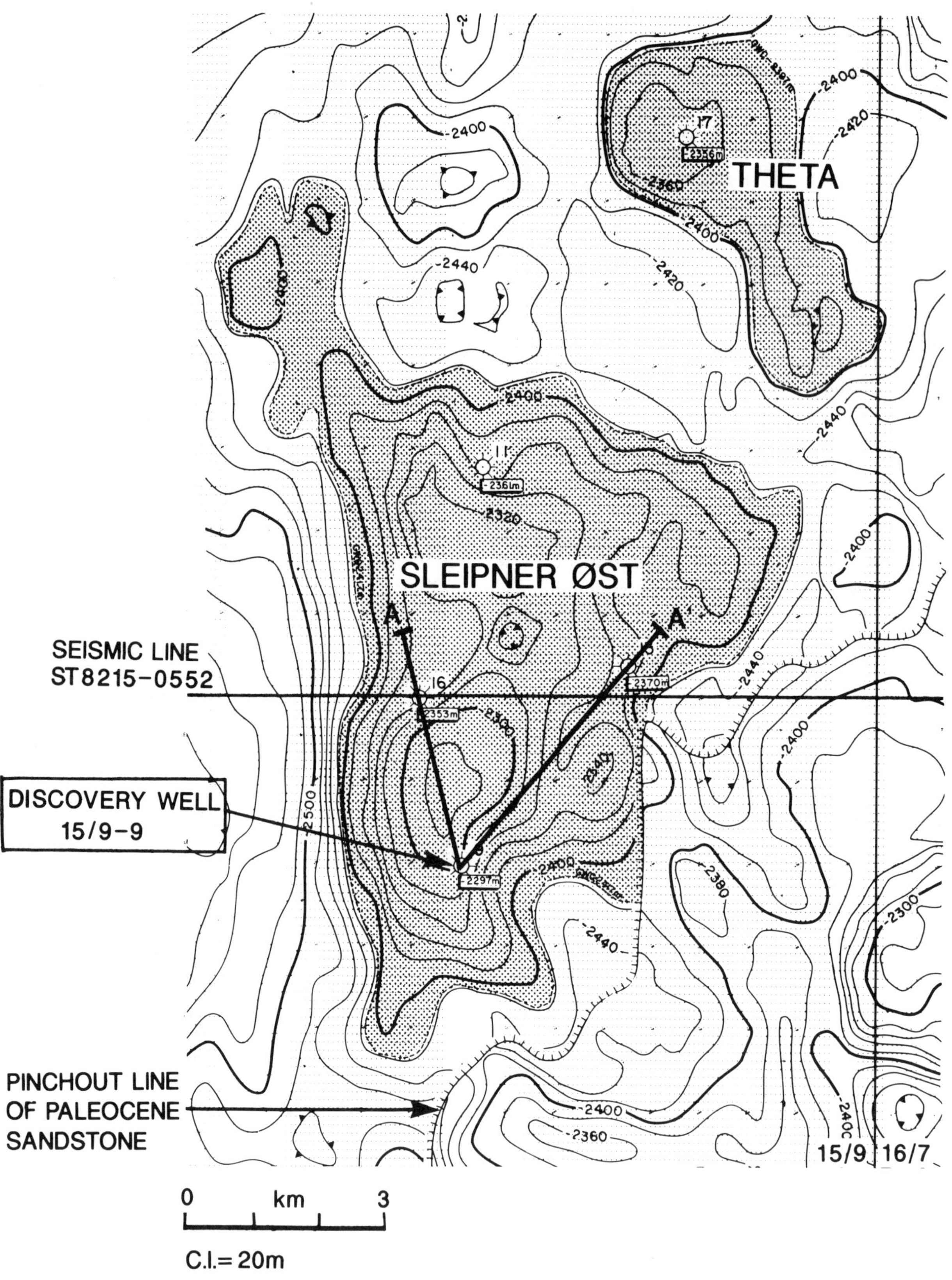

Figure 7. Structural map of the top of the Heimdal Formation. East of the Heimdal Formation pinchout line, the map represents the top of the Chalk Group. Contours in meters subsea. The stippled area shows the outline of the Sleipner Øst field and the Theta structure. Theta is a separate structure, not included in Sleipner Øst. Line A-A′ shows location of cross section, Figure 9. Seismic line is shown in Figure 4. This map represents Statoil's interpretation.

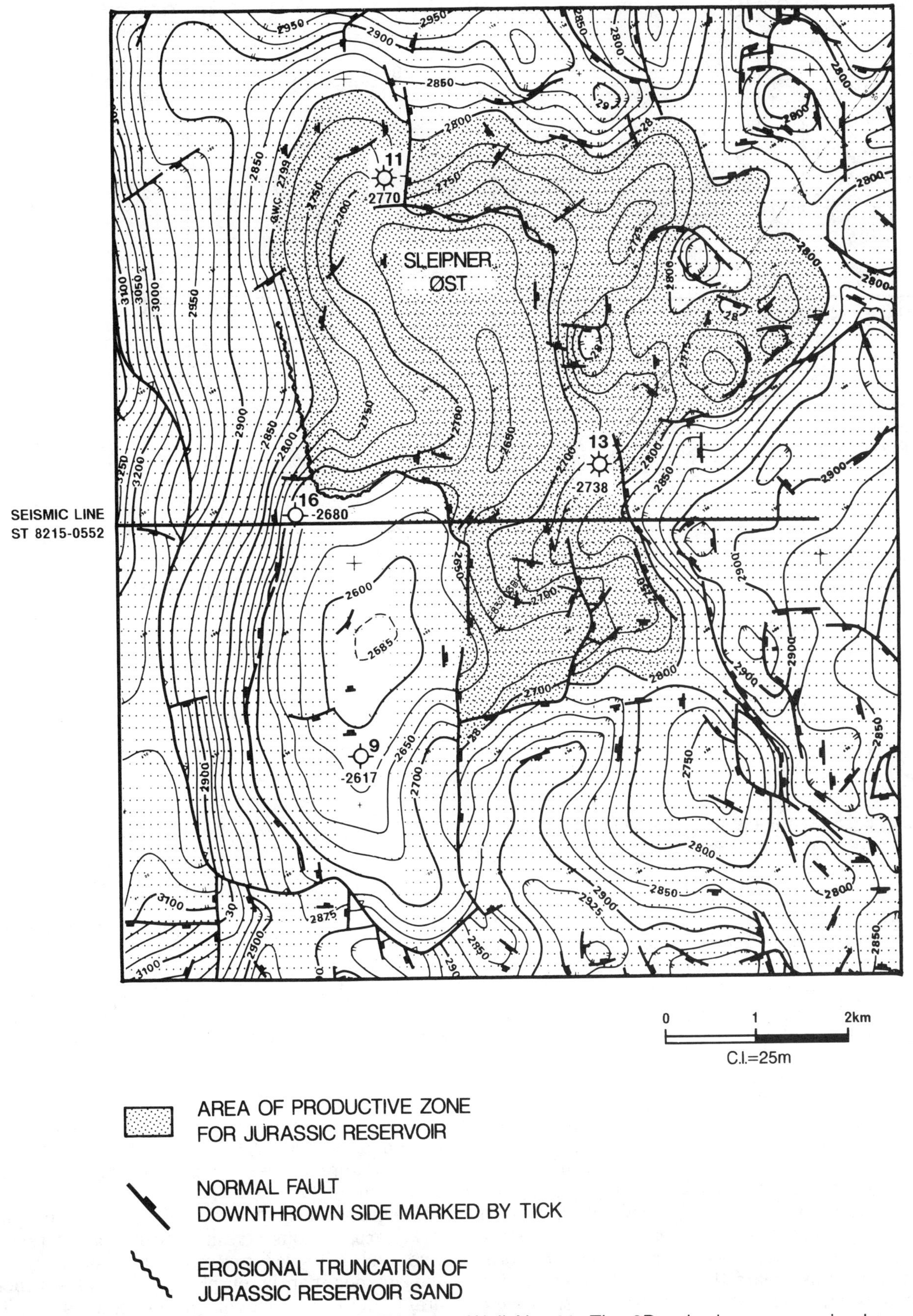

Figure 8. Structural depth map of top Jurassic/Triassic sandstone. Contours in meters subsea. Seismic line shown in Figure 4. Discovery well for this reservoir is Well No. 11. The 3D seismic coverage is shown by larger spaced dot pattern.

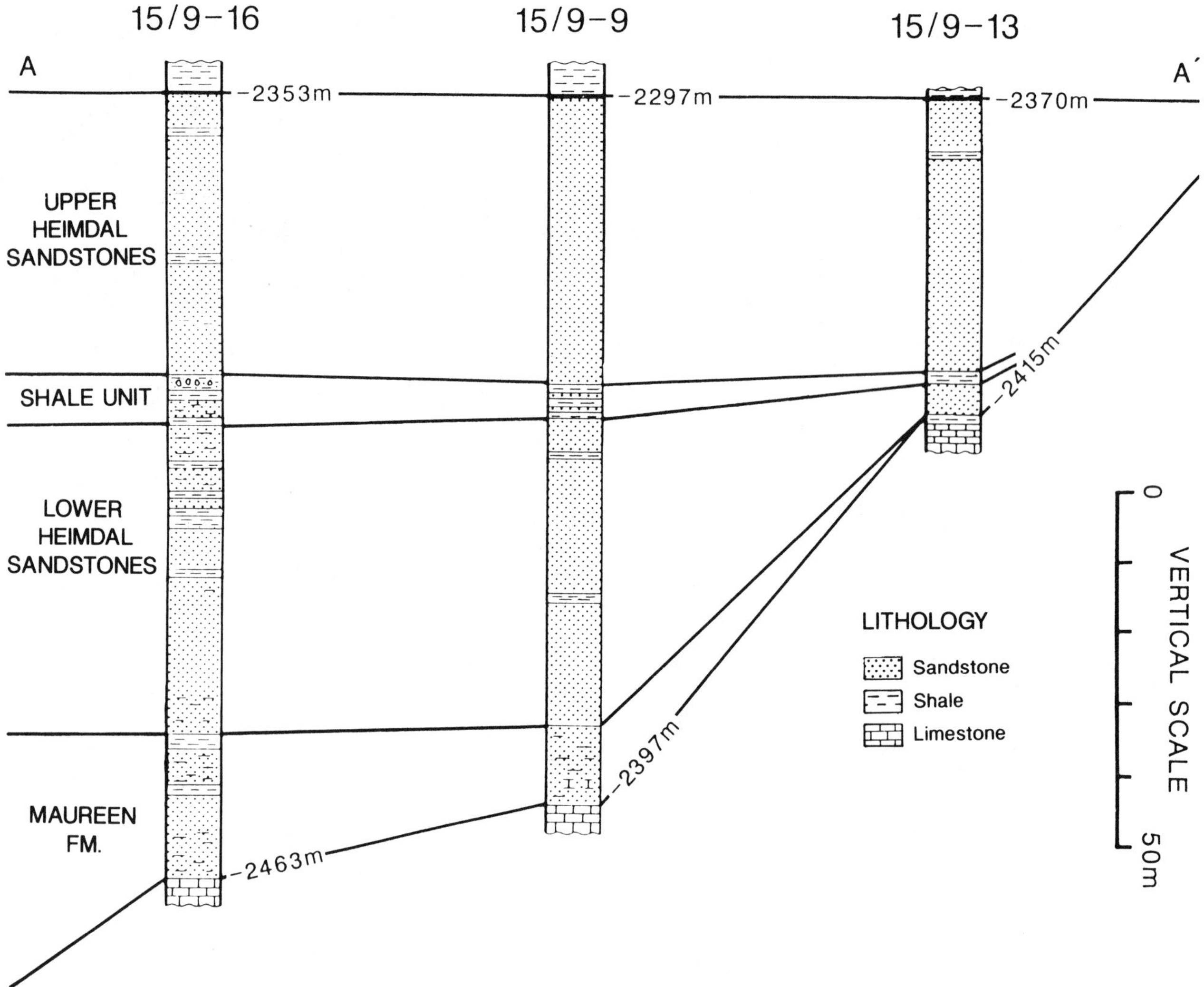

Figure 9. Stratigraphic cross-section showing the Paleocene sandstone units. The location is shown in Figure 7.

Massive Sandstones—Sandstones of this facies are fine to fine-medium grained, moderately to poorly sorted, and make up the dominant part of the Heimdal Formation. A wide variety of water escape structures are abundant including dish structures, consolidation laminae and clay-enriched primary laminae, fluidization channels and layers. Dish structures appear as 1-2 mm thick diffuse dark gray to greenish gray laminae. Individual dishes occur as broad, continuous, concave-up gently curving laminae (Figure 11) and are often associated with consolidation laminae. Consolidation laminae appear as dark-colored, horizontal to inclined, subparallel to slightly wavy or irregular laminae. They are the most abundant fluid escape structures within this facies. The fluidization channels distinguished in the cores are described as pillars that are vertical or steeply dipping (Figure 11). Rip up clasts of shale and laminated siltstone occur in units up to 0.5 m thick within this lithofacies. The presence of various fluid escape structures suggests that these beds were deposited very rapidly by high-density turbidity currents (Lowe, 1975).

Slurry sandstones occurring within the facies consist of layers 0.1-3 m thick of poorly sorted, muddy, fine-grained sandstones. The slurry sandstones are believed to have been deposited from muddy, high-density turbidity currents.

Shales and Claystones—This facies is comprised of medium-dark gray shales and light medium green shales (Figure 10, best example between 2425 m to 2430 m on the log).

The gray shales appear to be waxy with irregular dark, carbonaceous streaks and minor fractions of sand and silt-sized material. The sand fraction consists dominantly of planktonic and benthonic foraminifers in various stages of preservation. The silt fraction consists mostly of terrigenous and biogenic material. *Chondrites* commonly occur throughout. Pyritized wood fragments occur rarely.

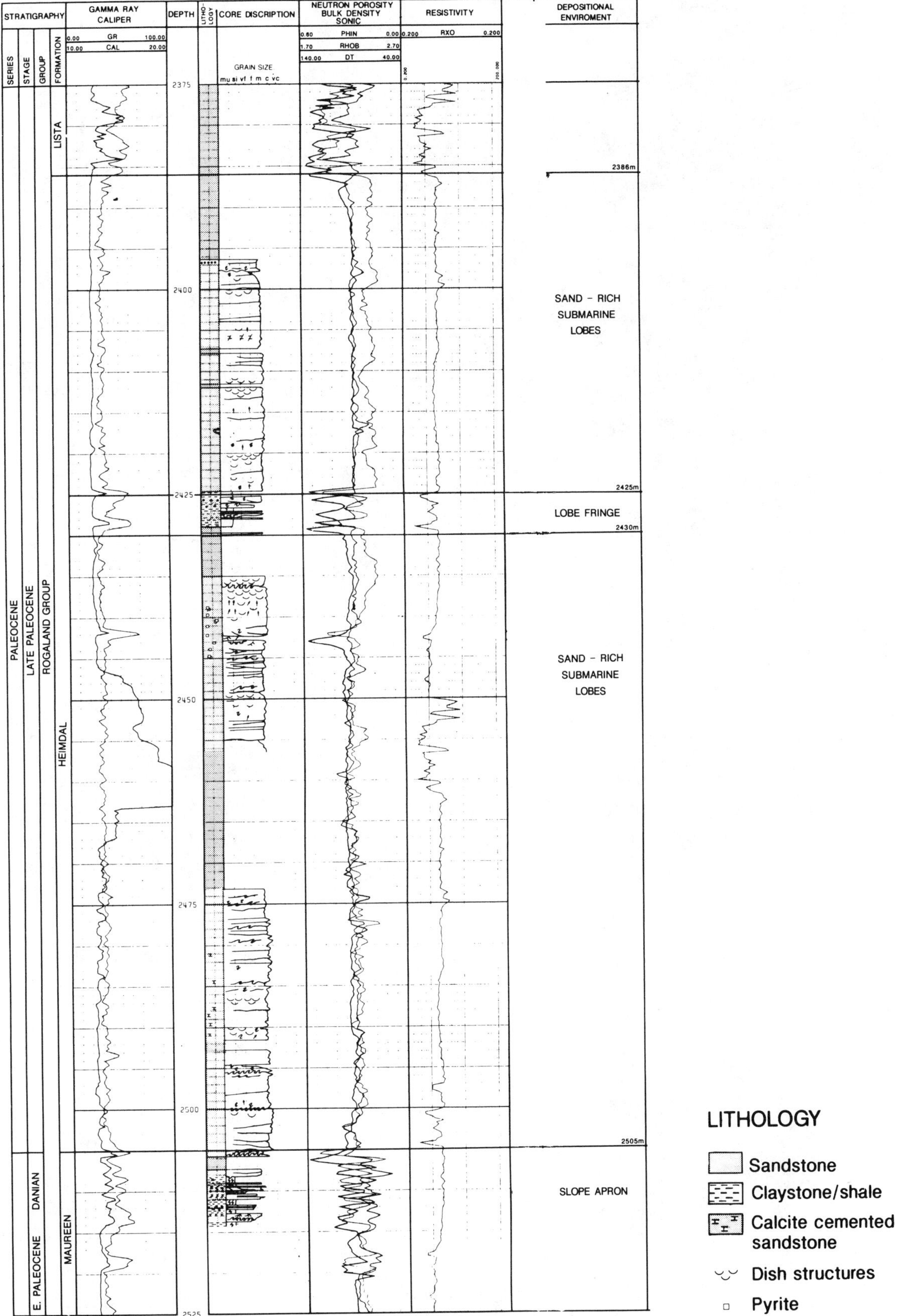

Figure 10. Data summary chart of Paleocene section, Well 15/9-11.

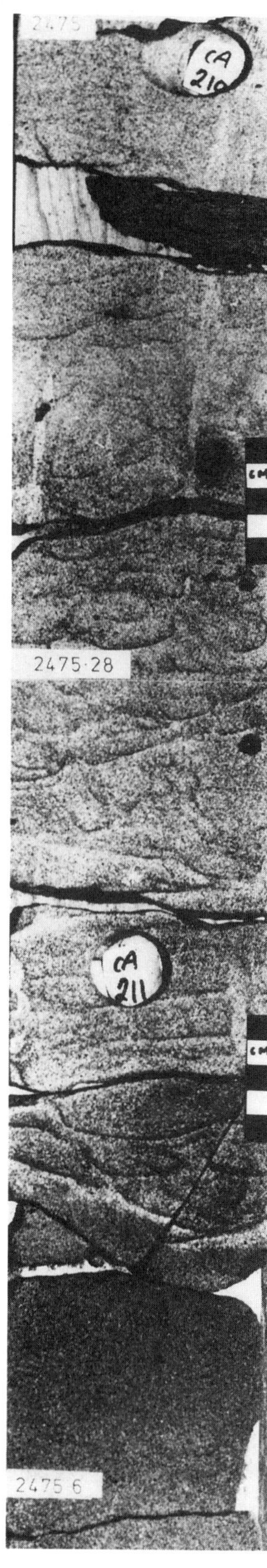

Figure 11. Core photograph of the Massive Sandstone Facies. Note the dish structures showing strongly curved concave-up laminae, and pillars occurring between the upward-curving margins of the dishes. Photo by Statoil.

The green shales are waxy in appearance with irregular dark gray carbonaceous streaks and minor sand and silty fractions as in the gray shales. They are further characterized by a diverse and abundant trace fossil association of *Chondrites*, *Zoophycos*, and *Skolithos*. Rare, thin medium-grained sandstone layers are interbedded with the shale. Calcite is common as fracture fillings. Bed thickness ranges from 0.2 to 3.5 m. Both gray and green shales are believed to be of hemipelagic origin.

Interbedded Siltstones and Shales—Silt-laminated shale occurs in 0.5 m to 10 m thick units. Moderately sorted coarse silt to very fine sandstones occur as parallel laminae in calcareous shale (Figure 12). Microfossils, chiefly foraminifera, are present in the shale intervals. This facies is interpreted as the deposits of low-density turbidity currents.

Mineralogy and Diagenesis—The composition of the sandstone is similar in all wells. Quartz is the dominant detrital mineral and represents 75%–85% of the total. Feldspar in the form of orthoclase, microcline, and plagioclase is the second most abundant detrital mineral and forms between 7% and 12% of the total. Some of the feldspar grains have been altered (Figure 13B) either by sericitization or by dissolution. A range of rock fragment types is present (trace to 3%) that includes granite, metamorphic rock fragments (Figure 13A), chert, and small argillaceous fragments. Fine-grained extrusive volcanic fragments are also common. Mica (muscovite), glauconite, and heavy minerals such as garnet, zircon, and tourmaline are present as traces.

Detrital clay is present in many of the sandstones varying from a trace to 16%. When present, it is fairly evenly distributed, giving a visible microporosity in thin section. Some of the sandstones contain a trace to 2% kaolinite.

Chlorite authigenesis is widespread (Figure 13C), and as shown by petrographic data it seems to be concentrated in dish-structure horizons.

Feldspar is also found in a dissolved state (Figure 13B) producing minor secondary porosity and is usually coated by authigenic chlorite. Authigenic kaolinite is only abundant in the uppermost sandstones of Well 15/9-11 and occurs as isolated vermicules or booklets. Quartz cementation is rarely well developed and appears to be the last-formed cement.

The reservoir sandstones from Wells 15/9-11, -13, and -16 have average porosities of 23%, 26%, and 25%, respectively. The permeability ranges from 200 md to 1500 md. From the available data, it is evident that the controls on porosity and permeability appear to be primary, such as grain size, sorting, and detrital clay. Diagenesis in the form of authigenic clay formation and mineral cementation is minor and would have only a small effect on petrophysical properties.

Figure 12. Core photograph of interbedded siltstones and shales facies, with irregularly bedded sandstones. Photo by Statoil.

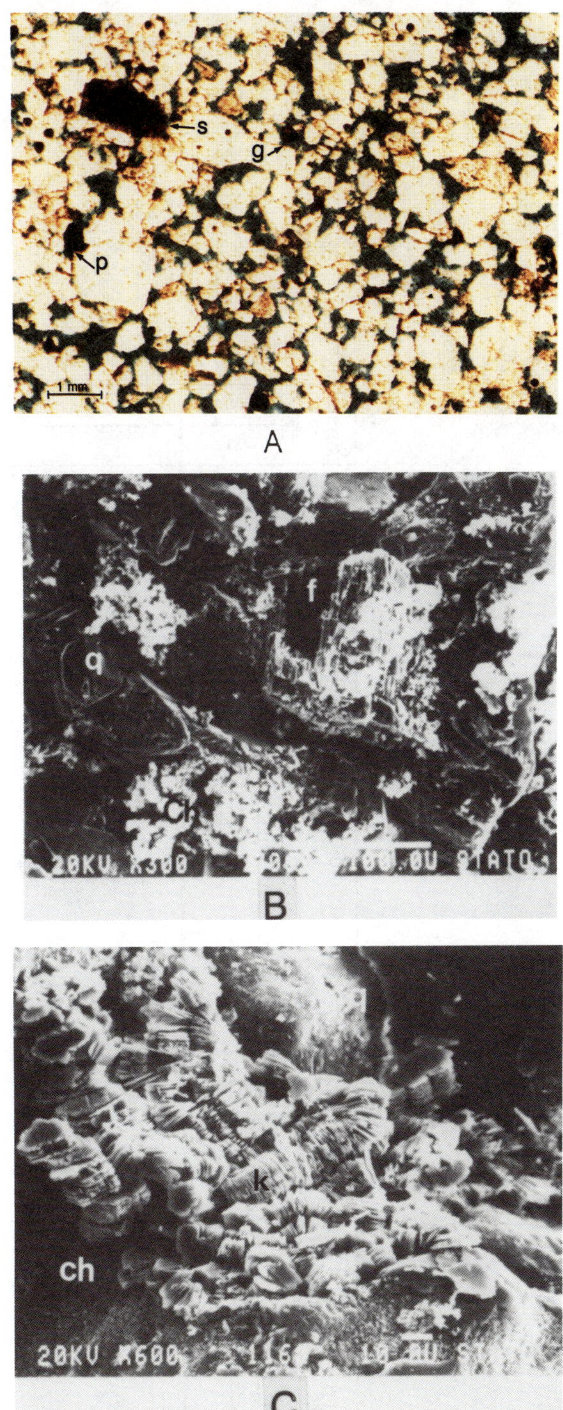

Figure 13. (A) Photomicrograph thin section of a fine-grained sandstone with very little detrital clay, exhibiting excellent porosity. Note the presence of small detrital garnet grains (g), argillaceous rock fragment (s) partially replaced by pyrite (p). (B) SEM photomicrograph of A showing altered ("rotten") feldspar (f), quartz overgrowth (q), and chlorite (ch). (C) SEM photomicrograph of a fine-grained sandstone showing pore-filling kaolinite (k) postdating grain-coating chlorite (ch). Quartz overgrowth (q) postdates chlorite. Bar scale equals 100 microns. Photos by Statoil.

Jurassic Reservoir

Lithofacies

The Hugin Formation sandstone reservoir has been penetrated by only two wells in the Sleipner Øst field. This formation is interpreted to represent shallow-marine shoreface sands and coastal plain sediments (Figure 14) that were deposited as a result of successive relative rises in the sea level during the Callovian-Oxfordian age. Three major phases of sediment accumulation are recognized in the Sleipner

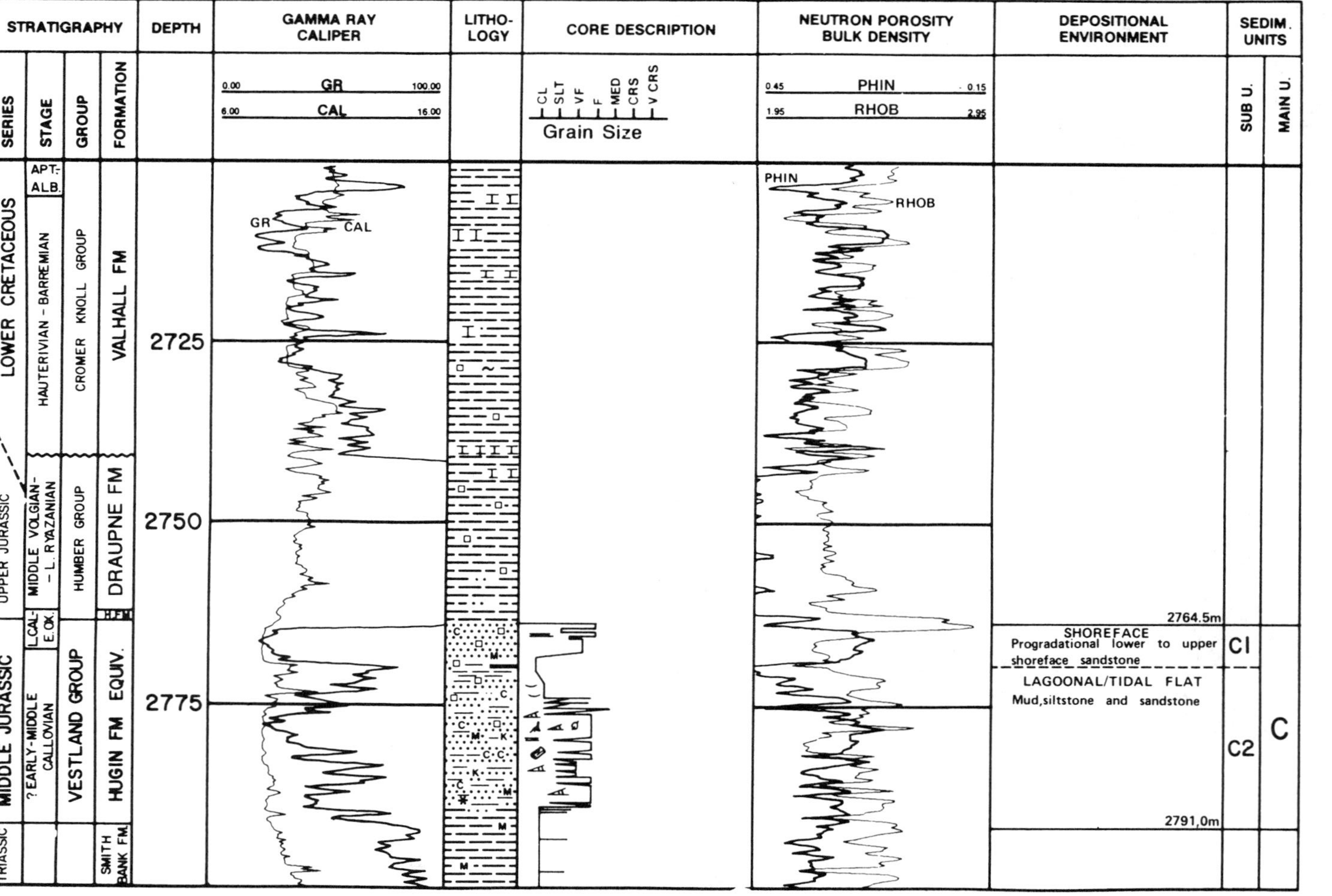

Figure 14. Data summary chart of Mesozoic section, Well 15/9-13.

area. The Hugin Formation sandstone in the Sleipner Øst field was deposited during the last phase.

The shoreface sandstones generally consist of very fine to fine, argillaceous and micaceous sandstones and medium to pebbly sandstones. Bioturbation is common and has obliterated most of the sedimentary structures in the fine-grained sandstones. Both the fine- and medium-grained sandstones can show parallel to low-angle cross-bedding and occasional graded bedding. The coastal plain deposits are comprised of lagoonal and tidal flat facies. These consist of interbedded fine- to very fine grained argillaceous sandstones and shales. The sandstones are thinly bedded and sometimes have scoured bases and rip-up mud clasts. Sedimentary structures such as ripple lamination, flaser bedding, and convolute lamination are present. Bioturbation is weak while organic debris and disseminated coal are present throughout. Interbedded shales are dark in color, pyritic, and contain siderite nodules. These shales occur as 2-4 m thick units and are moderately to highly bioturbated.

Mineralogy and Diagenesis

The composition of the sandstones is the same in both wells of Sleipner Øst and is similar to the Hugin Formation in the Sleipner Vest field. The fine-grained shore face sandstones are well sorted with a very low abundance of authigenic clay matrix, trace amounts of carbonate, and high porosity. The coastal plain sandstones are poorly sorted and contain abundant mica, chloritic mud clasts, brown pore-lining clays, illite and chlorite, and have a moderate to low porosity.

Well data indicate lateral and vertical variations in reservoir properties that makes it difficult to predict field-wide reservoir properties. Estimated average thickness of effective (net) reservoir vs. total (gross) thickness of the reservoir unit is 78% and the average porosity is 22%. Permeability measurements indicate a range from 100 to 400 md in the reservoir, although faults and sandstone discontinuities may represent permeability barriers.

SOURCE

Geochemical studies suggest that the Sleipner Øst gases were derived from a mixed source of Jurassic shales and coals in the Viking graben and deeply buried parts of the Sleipner terrace.

The richest source rock in the area is the Upper Jurassic Draupne Formation (Figures 6 and 15) that has up to 10% TOC and a dominantly type II kerogen. The underlying Heather Formation has about 5% TOC and mostly a type II/III kerogen. On the Sleipner terrace, coals are present in the Middle Jurassic Sleipner Formation and in the overlying Callovian-Hugin equivalent. Average gross thickness of the source rocks is about 500 m and the average potential yield is estimated to be 28.5 m^3 gas/km^3 rock.

Hydrocarbons have been expelled from source rocks in the Viking graben during the last 40 m.y. (Figure 16). The gases probably migrated through permeable sandstones and up fault zones and fractures to successively shallower reservoirs.

GAS CHARACTERISTICS

The gas has a GOR of 3018 Sm^3/m^3 and an average methane content of 78%. It is further characterized by a high N_2 content (1-2 mol%), low aromaticity, and high average pristane/phytane ratio (Figure 15).

The hydrocarbons in the Jurassic and the Paleocene reservoir are very similar in composition except for a lower CO_2 content in the Paleocene reservoir. This is probably due to a loss of CO_2 by dissolution during migration through the Cretaceous chalk interval. Pressure in the Paleocene reservoir is equal to the hydrostatic pressure. In the Jurassic reservoir, the pressure is 20 bar (294 psi) higher than the hydrostatic pressure.

EXPLORATION CONCEPTS

Middle Jurassic sandstone reservoirs in rotated fault blocks were the original exploration target in the Sleipner Vest area. This was also the case for the first well on Sleipner Øst. The Paleocene trap was recognized from the seismic data, but previous drilling in the Sleipner area had not resulted in any hydrocarbon discoveries in Paleocene structures. Hydrocarbons are trapped in Paleocene sandstones both north and south of Sleipner, but these were not considered important for this area because the Cretaceous chalk was expected to act as a barrier for hydrocarbon migration. In Sleipner Øst, a flat event was recognized on some of the seismic lines but was not interpreted as a hydrocarbon indicator.

The first well drilled on Sleipner Øst was a surprise in two ways. A thick gas column was discovered in the Paleocene, and the Jurassic reservoir was absent on the crestal part of the structure. Calibrating the seismic with the well data showed that the flat spot in the Paleocene corresponded to the gas-water contact. Migration of hydrocarbons to the Paleocene reservoir is believed to have taken place along the fault zone separating the Ling high from the Sleipner terrace (Figures 3 and 5).

In hindsight, it can be argued that more attention should have been paid to the seismic flat spot and the mapped Paleocene closure, and less to assumed migration problems.

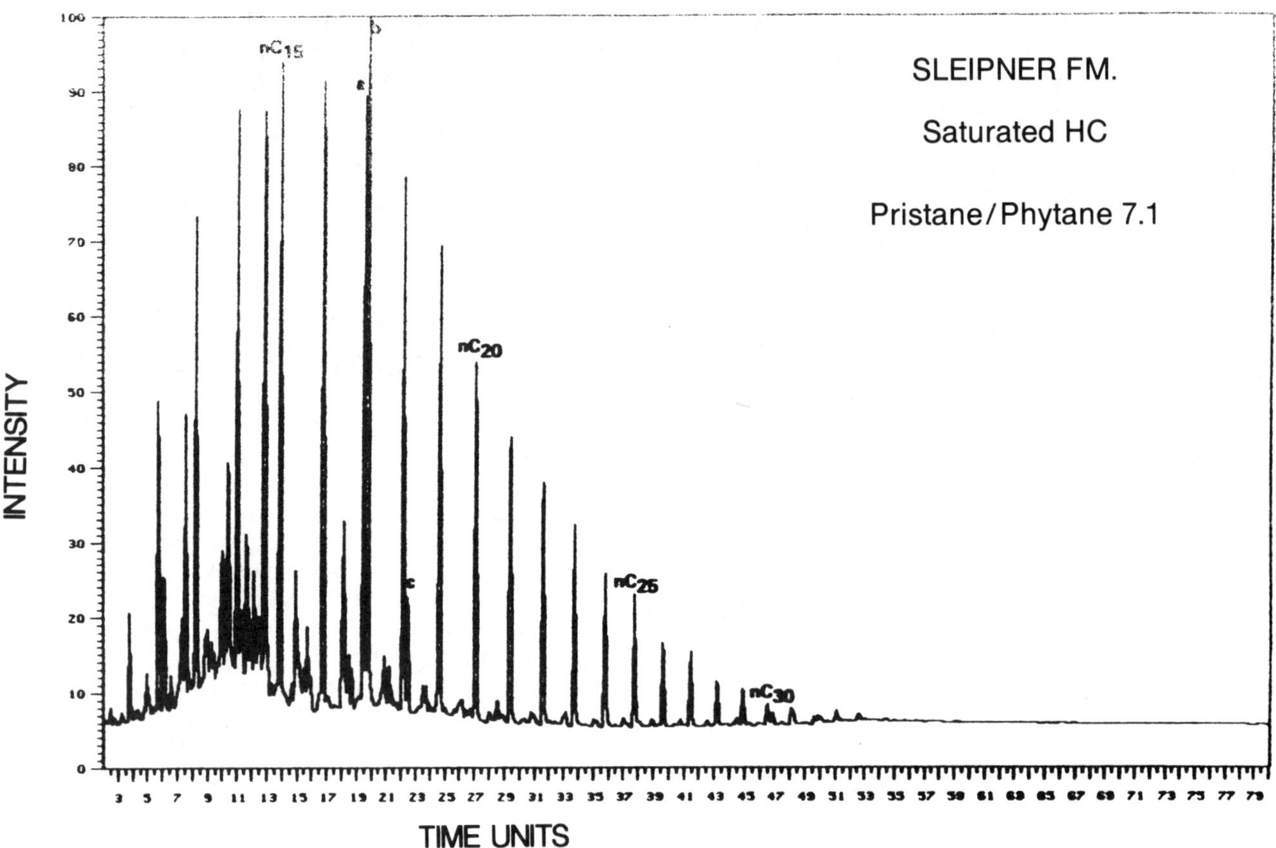

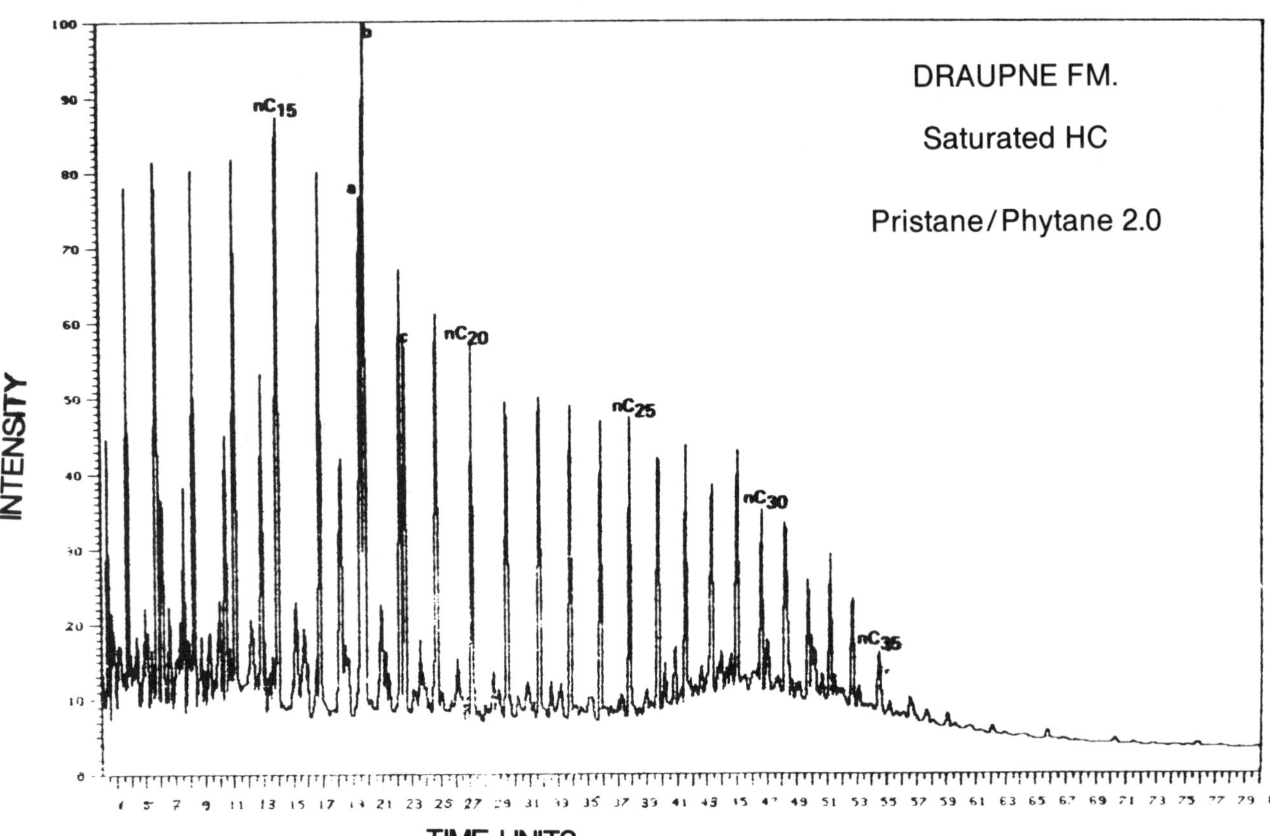

Figure 15. Gas chromatograph for condensate and gas from the Sleipner Formation (top) and the Draupne Formation (bottom), Well 15/8-1 (Figure 3).

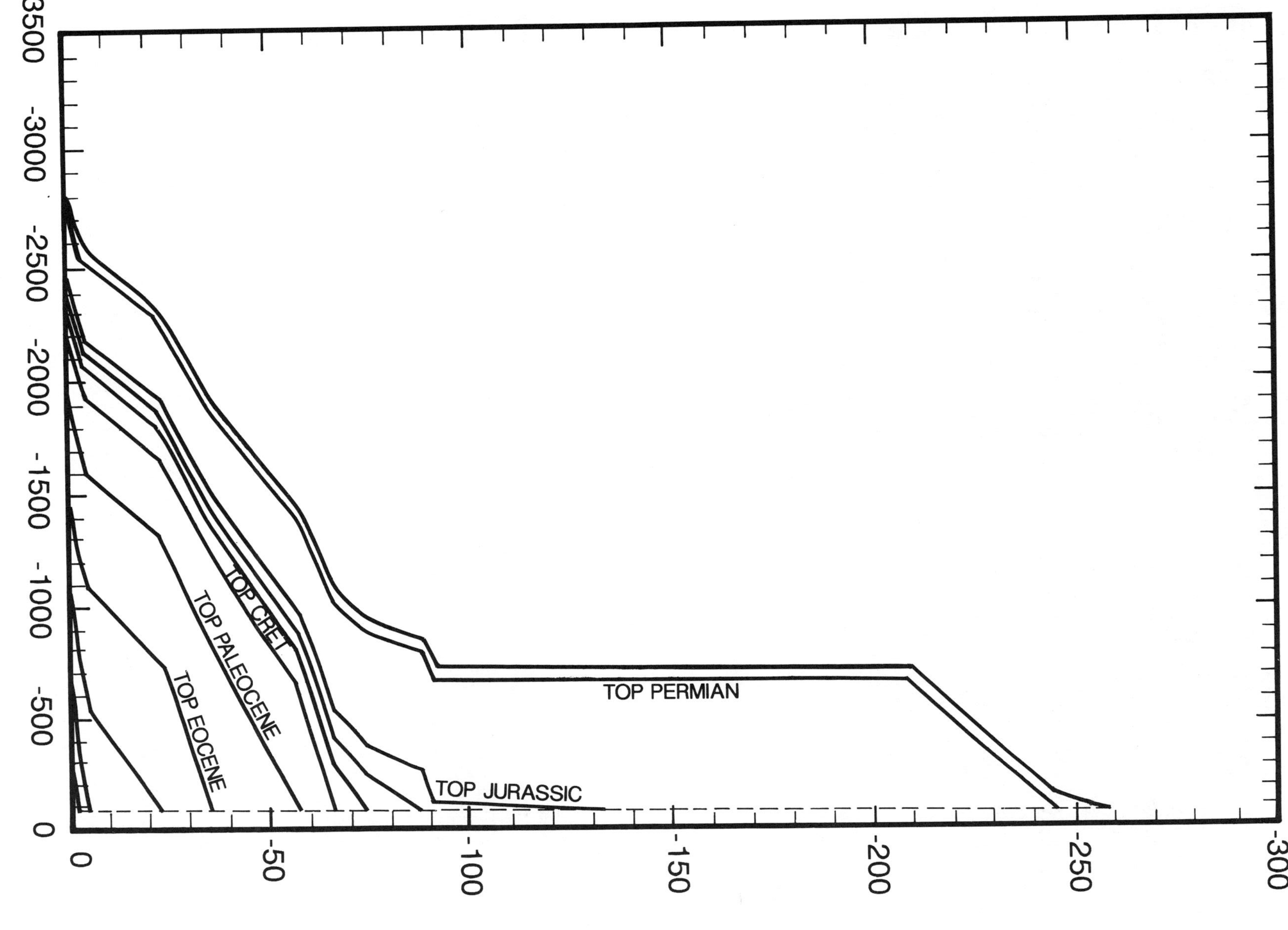

Figure 16. Geohistory diagram for Well 15/9-16.

ACKNOWLEDGMENTS

This field description is based on studies and interpretations performed by geologists and geophysicists of Esso Norge a.s and Den Norske Stats Oljeselskap A/S (Statoil). The authors wish to thank the PL 046 partners; Statoil, Esso Norge a.s, Norsk Hydro, Elf Aquitaine Norge a.s, and Total Marine Norsk for permission to publish this field study.

REFERENCES CITED

Deegan, C. E., and B. J. Scull, 1977, A standard lithostratigraphic nomenclature for the Central and Northern North Sea: Institute of Geological Sciences, Report 77/25. The Jurassic and part of the Triassic is revised by Vollset and Doré (1984).

Lowe, D. R., 1975, Water escape structures in coarse-grained sediments: Sedimentology, v. 22, p. 157-204.

Pegrum, R. M., and T. E. Ljones, 1984, 15/9 Gamma gas field offshore Norway; a new trap type for North Sea basin with regional implications: American Association of Petroleum Geologists Bulletin, v. 68, n. 7, p. 874-902. A detailed description of the geology in Block 15/9, with emphasis on the structural development.

Vollset, J., and A. G. Doré, 1984, A revised Triassic and Jurassic lithostratigraphic nomenclature for the Norwegian North Sea: Norwegian Petroleum Directorate Bulletin No. 3, Sept. 1984.

Ziegler, P. A., 1981, Evolution of sediment basins in north-west Europe, *in* L. V. Illiang and G. D. Hobson, eds., Petroleum geology of the continental shelf of north-west Europe: The Institute of Petroleum, London: Heyden & Son Ltd., pp. 3-39.

SUGGESTED READINGS

Ranaweera, H. K. A., 1987, Sleipner Vest, *in* A. M. Spencer, et al., eds., Geology of the Norwegian oil and gas fields: Norwegian Petroleum Society. London, Graham and Trotman, p. 253-263. Description of the Sleipner Vest field.

Østvedt, O. J., 1987, Sleipner Øst, *in* A. M . Spencer, et al., eds., Geology of the Norwegian oil and gas fields: Norwegian Petroleum Society: London, Graham and Trotman, p. 234-252. Description of the Sleipner Øst field.

Appendix 1. Field Description

Field name .... *Sleipner Øst*

Ultimate recoverable reserves .... *2.1 tcf/60 × 10^9 Sm^3 wet gas*

Field location:

- **Country** .... *Norway*
- **State**
- **Basin/Province** .... *South Viking graben, North Sea*

Field discovery:

- **Year first pay discovered** .... *Paleocene Heimdal Formation 1981*
- **Year second pay discovered** .... *Middle-Upper Jurassic Hugin Formation 1981*
- **Third pay** .... *NA*

Discovery well name and general location

- **First pay** .... *Well 15/9-9, 58°20′47.4″N, 01°53′24.6″E*
- **Second pay** .... *Well 15/9-11, 58°24′02.5″N, 01°53′41.8″E*
- **Third pay** .... *NA*

Discovery well operator .... *Den norske stats oljeselskap a.s (Statoil)*

- **Second pay** .... *Den norske stats oljeselskap a.s (Statoil)*
- **Third pay** .... *NA*

IP in barrels per day and/or cubic feet or cubic meters per day:

- **First pay** .... *No production data; the field will be on production from 1993*

Test data from well 15/9-13, choke size = 25.4 mm (1 in.)

Paleocene pay: *Gas—760,000 Sm^3/d (26.8 mmcf/d)*
Condensate—400 Sm^3/d (14,125 SCF/d)

Jurassic pay: *Gas—804,000 Sm^3/d (28.4 mmcf/d)*
Condensate—388 Sm^3/d (13,700 SCF/d)

- **Second pay** .... *NA*
- **Third pay** .... *NA*

All other zones with shows of oil and gas in the field:

Age	Formation	Type of Show
None		

Geologic concept leading to discovery and method or methods used to delineate prospect, e.g., surface geology, subsurface geology, seeps, magnetic data, gravity data, seismic data, seismic refraction, nontechnical:

Seismic data indicate closure on Paleocene, Jurassic, and pre-Jurassic rocks. The prospect is located on a structural high, separated by a large fault from previous Jurassic gas discoveries to the west. The concept was that gas had migrated up along the bounding fault into Permian and Jurassic sandstones. Migration along fault planes was confirmed by other wells in the area. As no hydrocarbons previously had been encountered in Paleocene sandstones in this area, it was not regarded as a potential pay zone.

Structure:

Province/basin type (see St. John, Bally, and Klemme, 1984)

South Viking graben, North Sea/intracratonic rift basin

Tectonic history

The Permian and Triassic were tectonically relatively quiet periods. The Utsira high remained a structural high relative to the area to the west. During Jurassic times extensional rifting in the Viking graben caused north-south-trending rotated fault blocks to form between the graben axis and the Utsira high. The rifting ceased in the Late Jurassic and little tectonism occurred until middle Tertiary times, when a compressional phase caused domal structures to form.

Regional structure

The structure is located on a southwestern extension of the Utsira high, separated from the South Viking graben by a series of rotated downward-stepping fault blocks.

Local structure

Paleocene pay: Asymmetric anticline with steepest dip to the west. Jurassic pay: Faulted anticline.

Trap

Trap type(s) .... *1 combined structural and stratigraphic trap (Paleocene pay); 1 structural with a possible stratigraphic component (Jurassic pay)*

Basin stratigraphy (major stratigraphic intervals from surface to deepest penetration in field):

Chronostratigraphy	Formation	Depth to Top in m subsea
Miocene	*Utsira Formation*	*820*
Paleocene	*Heimdal Formation*	*2260*
L. Cretaceous	*Chalk Group*	*2440*
M.-L. Jurassic	*Hugin Formation*	*2650*
E.-L. Triassic	*Smith Bank Formation*	*2700*
L. Permian	*Zechstein Group*	*2900*

Location of well in field .... *NA*

Reservoir characteristics:

Number of reservoirs *2*

Formations .... *Heimdal; Hugin*

Ages .... *Heimdal, Tertiary (Paleocene); Hugin, Middle-Upper Jurassic (Callovian-Oxfordian)*

Depths to tops of reservoirs .... *Heimdal, -2260 m ss; Hugin, -2650 m ss*

Gross thickness (top to bottom of producing interval) .... *100 m (Heimdal) + 35 m (Hugin)*

Net thickness—total thickness of producing zones

Average .... *58.7-67 m*

Maximum .... *149-175 m*

Average

Maximum

Lithology

Heimdal, fine- to medium-grained well-sorted quartzitic sandstone; Hugin, very fine to fine-grained, argillaceous and micaceous sandstone and medium-grained to pebbly sandstone

Porosity type .... *Both pays, intergranular porosity*

Average porosity .... *Heimdal, 26%; Hugin, 22%*

Average permeability .... *Heimdal, 400 md; Hugin, 100-400 md*

Seals:

Upper

Formation, fault, or other feature .... *Heimdal, formation seal (Lista Fm., Paleocene); Hugin, Formation seal (Draupne Fm., Kimmeridgian, U. Jurassic)*

Lithology .... *Shale*

Lateral

Formation, fault, or other feature .... *Heimdal, stratigraphic seal (Lista Fm.); Hugin, sealing faults and possible stratigraphic seal (Draupne Fm.)*

Lithology .... *Shale*

Source:

- **Formation and age** .... *Draupne, U. Jurassic; Heather, U. Jurassic; Hugin equivalent, M.–U. Jurassic; Sleipner, M. Jurassic*
- **Lithology** .... *Draupne and Heather, shale; Hugin and Sleipner, shale and coal*
- **Average total organic carbon (TOC)** *5%*
- **Maximum TOC** *10.2%*
- **Kerogen type (I, II, or III)** .... *III and II/III*
- **Vitrinite reflectance (maturation)** .... $R_o = 1.2–1.5$
- **Time of hydrocarbon expulsion** .... *Approx. 40 m.y.b.p. to present*
- **Present depth to top of source** .... *4200–5500 m*
- **Thickness** .... *500 m average gross thickness*
- **Potential yield** .... *28.5 m^3 gas/km^3 rock*

Appendix 2. Production Data

Field name .... *Sleipner Øst*

Field size:

- **Proved acres** .... *14,579 (59 km^2)*
- **Number of wells all years** .... *4 (planning to drill 19 production wells)*
- **Current number of wells** .... *4 exploration/appraisal wells*
- **Well spacing** .... *3.5 km*
- **Ultimate recoverable** .... *2120 bcf (60×10^9 Sm^3) of wet gas*
- **Cumulative production** .... *NA*
- **Annual production** .... *230 bcf (6.5×10^9 Sm^3) (planned)*
- **Present decline rate** .... *NA*
  - **Initial decline rate** .... *NA*
  - **Overall decline rate** .... *NA*
- **Annual water production** .... *NA*
- **In place, total reserves** .... *3438 bcf (97×10^9 Sm^3)*
- **In place, per acre-foot** .... *NA*
- **Primary recovery** .... *2120 bcf (60×10^9 Sm^3)*
- **Secondary recovery** .... *NA*
- **Enhanced recovery** .... *NA*
- **Cumulative water production** .... *NA*

Drilling and casing practices for the exploration and appraisal wells:

- **Amount of surface casing set** .... *60 m subsea-floor*
- **Casing program**

  20-in. shoe, 400 m ssfl (base of Nordland Gp., Pliocene); 13⅜-in., 1110 m ssfl (Hordaland Gp. shale, Miocene–Oligocene); 9⅝-in., 2500 m ssfl (base of Chalk Gp., U. Cretaceous); if testing, 7-in. liner to T.D.
- **Drilling mud** .... *Saltwater base*
- **Bit program** .... *Steel tooth bits, mostly journal bearing*
- **High pressure zones** .... *None*

Completion practices:

- **Interval(s) perforated** .... *NA*
- **Well treatment** .... *NA*

Formation evaluation:

- **Logging suites** .... *From top to T.D., ISF/sonic-GR, FDC/CNL, VSP; in Paleocene, Cretaceous, and Jurassic section, DLL/MSFL, dipmeter; RFT in reservoirs and zones of interest*

Testing practices .... *Production tests were carried out for each of the reservoirs in 3 of 4 wells in the field*

Mud logging techniques .... *Recording of drilling rate, lithology, cuttings, gas chromatography, H_2S detection and shale density*

Oil characteristics:

Type .... *NA*
(Tissot and Welte Classification in "Petroleum Formation and Occurrence," 1984, Springer-Verlag, p. 419)

API gravity .... *NA*

Base .... *NA*

Initial GOR .... *NA*

Sulfur, wt% .... *NA*

Viscosity, SUS .... *NA*

Pour point .... *NA*

Gas-oil distillate .... *NA*

Field characteristics:

Average elevation .... *Paleocene pay, -2338 m ss; Jurassic pay, -2722 m ss*

Initial pressure at GWC .... *Paleocene pay, 3554 psi; Jurassic pay, 4352 psi, 290 psi above hydrostatic pressure*

Present pressure .... *As above*

Pressure gradient .... *0.45 psi/ft*

Temperature .... *Paleocene pay, 93°C/100°F at GWC; Jurassic pay, 104°C/219°F at GWC*

Geothermal gradient .... *0.036°C/m*

Drive .... *Water drive (Paleocene pay); volumetric expansion (Paleocene and Jurassic pays)*

Gas column thickness .... *Paleocene pay, 157 m; Jurassic pay, 155 m*

Gas-water contact .... *Paleocene pay, -2417 m ss; Jurassic pay, ≈-2805 m ss*

Connate water .... *Paleocene pay, 33%; Jurassic pay, 22%*

Water salinity, TDS *6.21%*

Resistivity of water .... *0.0473 ohm-m at 93°C/200°F*

Bulk volume water (%) .... *Paleocene pay, 33%; Jurassic pay, 22%*

Transportation method and market for oil and gas:

The gas will be transported in pipelines (Zeepipe, Statpipe, and Norpipe systems) to Emden in Holland and Zeebrugge in Belgium. A pipeline will be built to transport condensate from Sleipner Øst to Kårstø in Norway where it will be processed.

Sleipner Vest Field—Norway
Southern Viking Graben, North Sea

H. K. A. RANAWEERA
Statoil
Stavanger, Norway

FIELD CLASSIFICATION

BASIN: North Sea
BASIN TYPE: Rift
RESERVOIR ROCK TYPE: Sandstone
RESERVOIR ENVIRONMENT OF DEPOSITION: Nearshore Marine
RESERVOIR AGE: Jurassic
PETROLEUM TYPE: Gas and Condensate
TRAP TYPE: Faulted Anticline

LOCATION

The Jurassic Sleipner Vest gas-condensate field lies in the Norwegian Sector of the North Sea, approximately 240 km (150 mi) west-southwest of Stavanger (Figure 1) at lat. 58°22'N and long. 01°45'E. The field as currently defined is situated mainly in block 15/9 with a small portion extending northward into block 15/6. Water depths in the field area are approximately 110 m (360 ft). Marginal marine sandstones of Callovian age form the main reservoir in the Sleipner Vest field. Estimated recoverable reserves are 126 $\times 10^9$ Sm^3 (approximately 4.5 tcf) gas and 29 $\times 10^6$ tons (approximately 200 million bbl) condensate. The field is named after the eight-legged horse of the god Odin in Norse mythology.

HISTORY

Pre-Discovery

Licence PL046, comprising block 15/9 and the neighboring block 15/8, was first awarded in 1976 to the following group:

Statoil (Operator)	50%
Esso Norge a.s	40%
Norsk Hydro Produksjon a.s.	10%

In early 1988, the group was extended to give the following division of interests:

Statoil (Operator)	49.6%
Esso Norge a.s	30.4%
Norsk Hydro Produksjon a.s	10.0%
Elf Aquitaine Norge a.s	9.0%
Total Marine Norsk a.s	1.0%

Exploration drilling in 1974 in the neighboring block to the north, 15/6 (PL029 - Esso, 100%), resulted in Jurassic sandstone gas discoveries in wells 15/6-2 and 15/6-3. Well 15/6-3 encountered a 74 m (240 ft) thick gas-condensate column, and subsequent seismic mapping indicated that the structure extended southward into block 15/9 (Figure 2). The discovery was informally called the Sleipner field.

Early seismic interpretation in block 15/9 defined a major north–south-trending fault system that separated the Sleipner terrace in the west from the Utsira high fault block in the east (Figure 3). In addition, Jurassic strata were thought to be thin or absent on the Utsira high and in the southern half of block 15/9. For these reasons, early exploration was directed toward the known gas-condensate bearing sandstones of the Sleipner terrace.

Discovery

After the licence award, the first well, 15/9-1, was spudded in 1977 on the Sleipner Alpha prospect (Figure 2). It encountered a gas-condensate column of about 100 m (300 ft) underlain by thin oil-bearing sandstones. The internal reserve estimates for the Alpha-structure had to be reduced, however, in late 1977 after well 15/6-5 proved to be dry. In 1978, well 15/9-2 proved the presence of gas-condensate in the Beta structure (Figure 2). Reserves on the Alpha structure were again decreased after the drilling of well 15/9-3. Consequently, further exploration was directed toward the other Jurassic structural culminations south of the Alpha structure. The next two wells, 15/9-4 and 15/9-5, drilled in 1979, proved the presence of reserves in the Delta and Beta structures, respectively.

Well 15/9-6, drilled during 1980 between the Alpha and Beta structures, was located in a down-faulted area that proved to be dry. The results of the initial

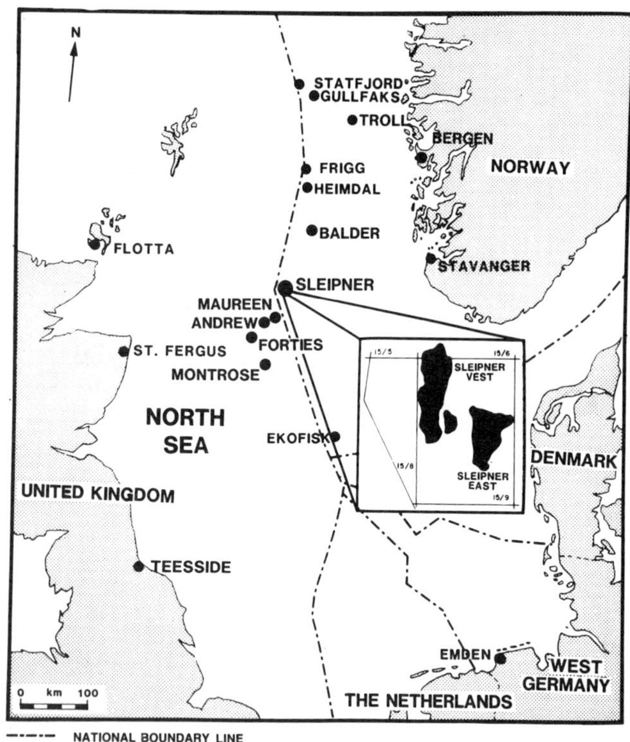

Figure 1. General location map.

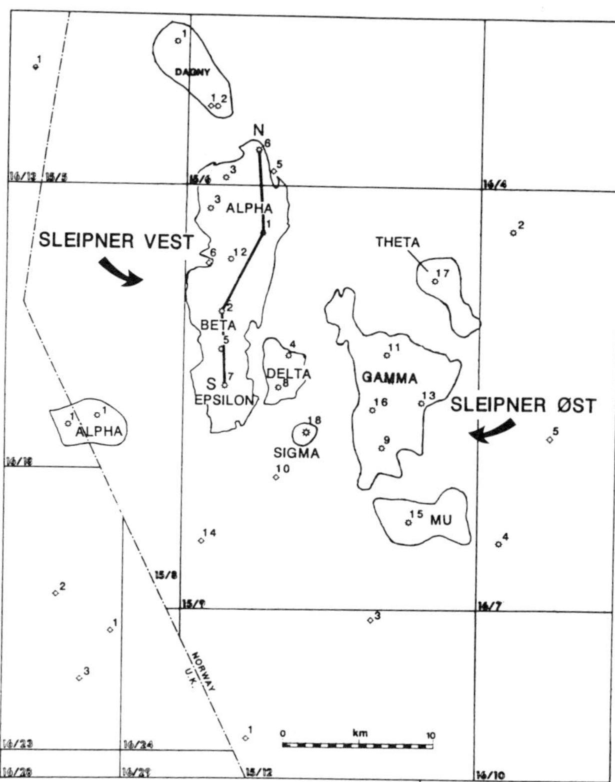

Figure 2. Location map of production Licence 046.

phase of exploration have been described by Larsen and Jaarvik (1981).

At this time, it was apparent that the 1975 seismic data were of too poor a quality for detailed reservoir mapping. Consequently, 2250 km (1400 mi) of 3D seismic data on a 75 m (250 ft) grid were acquired on the Alpha structure during 1980. A further 3050 km (1900 mi) were shot during 1981 to cover the remaining area of Sleipner Vest. Interpretation of these seismic data provides the basis for the present reservoir mapping. In the same year, well 15/9-9 was drilled on the Gamma structure (now designated Sleipner Øst field), a structural culmination on the upthrown Utsira block, to test for hydrocarbons in possible Permian–Triassic and Jurassic intervals. Although no hydrocarbons were discovered in these intervals, well 15/9-9 did penetrate a significant gas-bearing interval within porous Paleocene sandstones with a gross pay thickness of 103 m (338 ft) (Pegrum and Ljones, 1984). Following this gas discovery, the eastern half of block 15/9 was remapped and an additional seismic program shot in September 1981. The extent of the Sleipner Øst field was proven by well 15/9-11 in 1981 and wells 15/9-13 and 15/9-16 in 1982. Wells 15/9-13 and 15/9-11 proved the presence of additional hydrocarbons in Middle Jurassic sandstones.

During this period of exploration of Sleipner Øst, activity continued on Sleipner Vest. Early in 1982, well 15/9-12 confirmed the new seismic interpretation that no structural closure existed between the Alpha and the Beta structures. Well 15/6-6 confirmed the presence of gas-condensate in the northern area of the Alpha structure, although the gas-water contact proved to be higher than expected.

During this period, drilling activity also extended to previously unexplored areas. Wells 15/9-10 and 15/9-14 south of Sleipner Vest proved to be dry, while hydrocarbons were located in well 15/8-1. Hydrocarbons were also proved in wells 15/9-15 and 15/9-17, drilled on the Mu and Theta structures, respectively (Figure 2). The latest well, 15/9-18, located southeast of Sleipner Vest, only found a minor accumulation.

In 1982, an additional 3D seismic survey was acquired which covered the remaining area of block 15/9.

DISCOVERY METHOD

The discovery by well 15/6-3 of the Sleipner Vest field in 1974, which apparently extended into the neighboring block 15/9, led to the acquisition of a detailed seismic survey over the area in 1975. A 1 km by 1 km (3280 ft by 3280 ft) seismic grid covering the field area was acquired in 1975 and 1976. These seismic data provided the basic data that were used until 1980 when it became apparent that the quality of the data was too poor for detailed reservoir mapping. This led to a 3D seismic survey in 1980 that covered the Alpha structure.

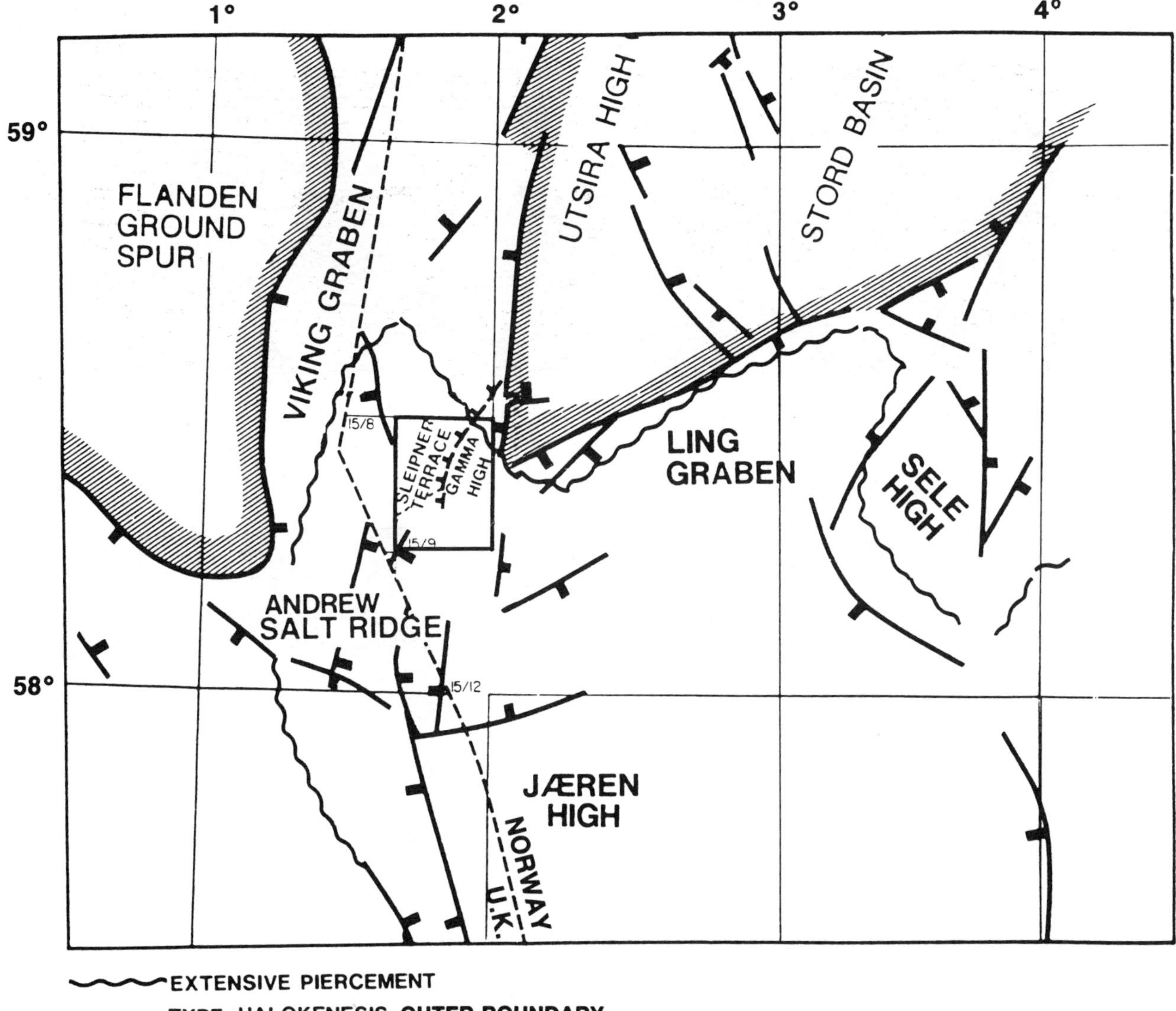

Figure 3. Structural framework of the Sleipner Vest field. The East Shetland platform lies to the west of the area shown.

The first 3D seismic survey of the Sleipner area was acquired and processed by GECO. The survey consisted of 170 lines, totaling 2250 km (1400 mi), with an east-west shot line orientation. The shot point interval was 25 m (82 ft) and the line spacing 75 m (245 ft). Six horizons were interpreted: top Ekofisk Formation, top Sola Formation, top Jurassic, top Heather Formation, top Sleipner Formation, and top Triassic. The mapping showed that the Alpha structure is broadly domal with two main fault trends, north-south and east-west, present in the block. The post-Jurassic and top Jurassic reflectors are broken by a few small faults. The top Heather and top Sleipner Formation horizons are much more faulted. A second 3D survey, in 1981, covered the Beta, Delta, and Epsilon structures (Figure 2). The data were acquired and processed by GSI. The survey consisted of 179 lines with a total length of 3046 km (1890 mi). The direction and the spacing used was the same as for the previous survey. The same six horizons were again "picked" (Figure 4). The interpretation confirmed the extensively faulted domal structure of the Alpha culmination (Figure 5). It also showed a north-south elongated dome-shaped structure with a fairly flat top (Beta), a steeply dipping half dome-shaped structure (Epsilon), and a horst structure (Delta).

A third 3D survey was acquired in 1982 and covered the Sleipner Øst field and the southern part of block 15/9. It consists of 5979 km (3710 mi) of fully 3D migrated rows.

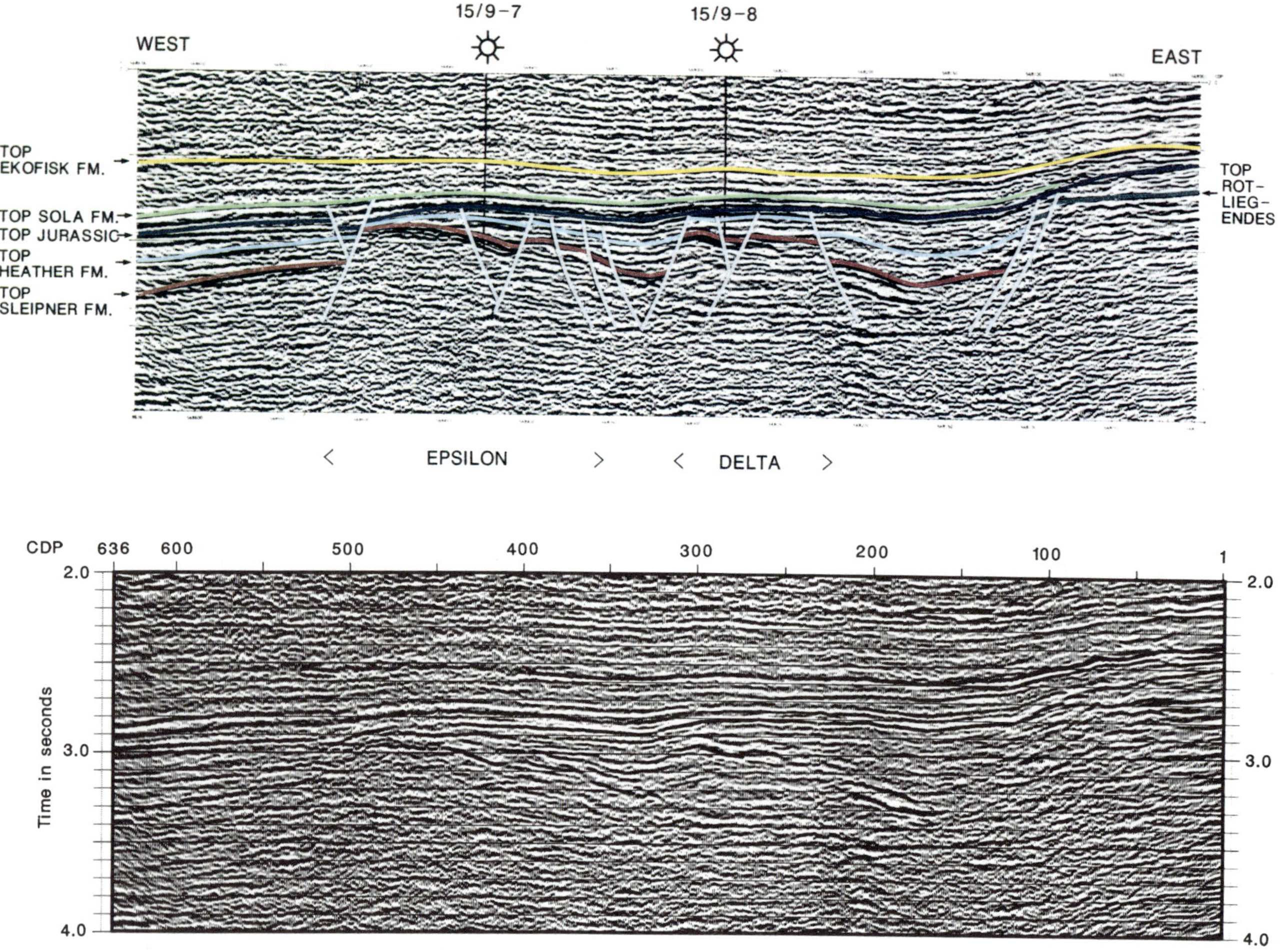

Figure 4. 3D seismic section through Epsilon (well 15/9-7) and Delta (well 15/9-8) structures (see Figure 2 for well locations). (A) Interpreted line. (B) Uninterpreted line.

STRUCTURE

Sleipner Vest is situated near the southern end of the Viking graben where several major structural elements converge. These include the Utsira high and Ling graben to the east, the East Shetland platform and the Flanden ground spur to the west, and the Andrew salt ridge and Jaeren high to the south (Figure 3). The structural pattern within the Sleipner area is dominated by two major basement controlled fault blocks; namely, the Gamma high to the east and the Sleipner terrace to the west. The westerly downthrown Sleipner terrace has a preserved cover of Jurassic rocks of variable thickness.

The structural configuration of the Sleipner terrace is further complicated by the presence of a thick sequence of Permian Zechstein evaporites. Rifting affected the block 15/9 area periodically from the Early Permian until Late Jurassic. Maximum extension was reached in the Late Jurassic. Differential rates of rifting led to localized compression and strike-slip fault movements and, perhaps, rejuvenation of older basement faults. This extensional event in the Late Jurassic probably triggered movement in the underlying Zechstein evaporites, deforming the Triassic and Jurassic cover by diapirism in the Sleipner area.

Reversal of motion on formerly normal faults, the result of compressional movements in the early Middle Cretaceous, resulted in further structuring at Jurassic–Triassic levels. A period of relative tectonic quiescence followed during the Late Cretaceous, reflecting a major change in the regional stress field, owing perhaps to the initial opening of the North Atlantic Ocean.

Rift movements within the North Sea had essentially ceased by the Early Cretaceous. At the end of the Late Cretaceous, the Laramide orogenic event resulted in an uplift and stripping of the East Shetland platform. This event led to coarse clastic input into the Sleipner area in the form of Paleocene submarine fan deposits. A further rejuvenation of the fault system occurred in block 15/9 during the mid-Tertiary.

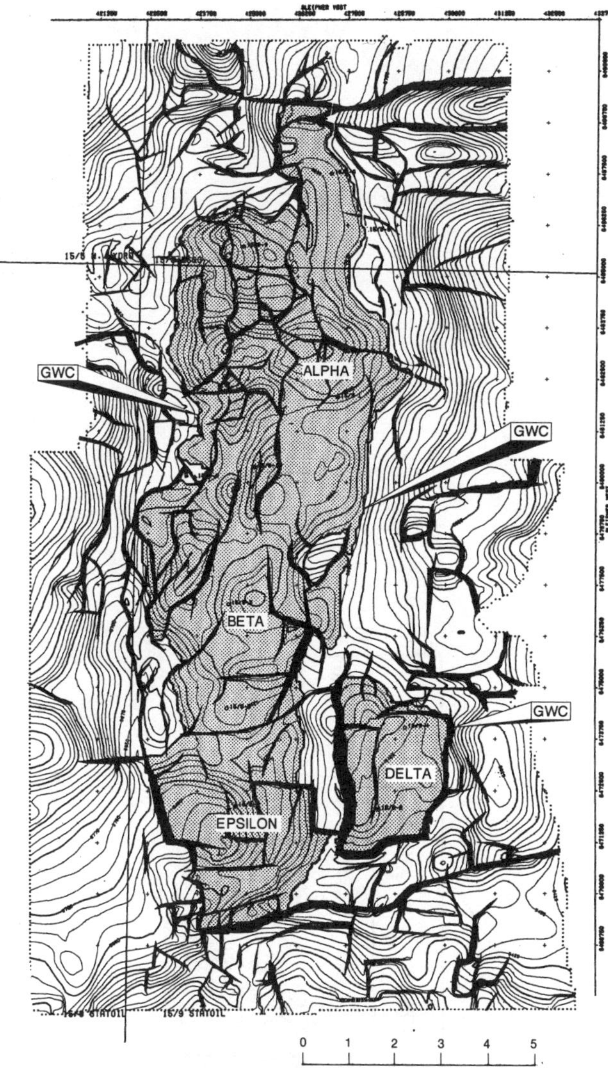

Figure 5. Top reservoir depth (structure) map of Sleipner Vest field. A gas-water contact is shown along the margin of the field.

STRATIGRAPHY

A generalized stratigraphic column for the field area is given in Figure 6. The lithostratigraphic nomenclature used is that proposed by Deegan and Scull (1977) and revised by Vollset and Doré (1984).

Pre-Triassic rocks have not yet been penetrated on the Sleipner terrace in block 15/9. Sleipner Øst well 15/9-9 penetrated the Lower Permian Rotliegendes, which is comprised of interbedded red sandstones and breccias deposited in an interpreted terrestrial to fluvial/lacustrine environment. The Zechstein sequence in this well was composed of interbedded dolomites and anhydrites with a basal, thin, dark-colored, radioactive shale, ascribed to the Kupferschiefer Formation.

Triassic strata are present throughout block 15/9, but their stratigraphy is poorly understood. Most of the wells on the Sleipner terrace drilled a few tens of meters of Triassic redbed shales, siltstones, and sandstones. Well 15/9-9 penetrated palaeontologically barren sandstones of the Skagerak Formation, which is underlain by thick red shale beds of the Smith Bank Formation.

Jurassic strata are well developed in Sleipner Vest. The oldest Jurassic strata encountered belong to the Middle Jurassic Bathonian–Bajocian Sleipner Formation (Figure 6), comprising shales, siltstones, and thick coal beds with minor sandstones deposited in delta top/coastal plain environments. The Callovian Hugin Formation, which forms the main reservoir, comprises shallow-marine shelf sandstones and shales, barrier-bar sandstones, back-barrier sandstones, siltstones, and shales. These were deposited as a result of three transgressive sedimentary cycles. The Upper Jurassic strata encountered in Sleipner Vest are the Oxfordian Heather Formation shales and the Kimmeridgian and Volgian Draupne Formation clays.

Lower Cretaceous strata in the 15/9 area are comprised of marine shales, marls, and siltstones. The Upper Cretaceous consists of a thick development of the Cenomanian to Maastrichtian Chalk group.

A complete lower Paleocene sequence (Figure 6) is encountered in the Sleipner Vest field and is represented by the Maureen, Heimdal, Lista, and Sele formations, with the Heimdal Formation forming the reservoir in the Sleipner Øst field. The Balder Formation straddles the Paleocene/Eocene boundary and is overlain by an Eocene and Oligocene sequence consisting of monotonous sections of clays and silts with stringers of sand and limestones. The Miocene sequence in block 15/9 comprises a lower group of marine shales and an upper group of shelf sands. The latter corresponds to the Utsira Formation.

TRAP

The Sleipner Vest field contains several dome-shaped culminations formed by pre-Cretaceous movements (Figure 5). The rifting and localized strike/slip movements in the Late Jurassic are believed to have triggered the diapiric movement in the underlying Zechstein evaporites, deforming the Triassic and Jurassic cover in the Sleipner area (Figure 7). The Zechstein salt acted as a zone of structural décollement on the Sleipner terrace, with the post-Zechstein strata being deformed independently of the Pre-Zechstein "basement." Finally, late Eocene compressive movements are thought to have rejuvenated the Mesozoic strike-slip faults, inducing salt movement and forming a series of en echelon domal structures (Pegrum and Ljones, 1984). The Upper Jurassic shales overlying the reservoir rocks act as a seal.

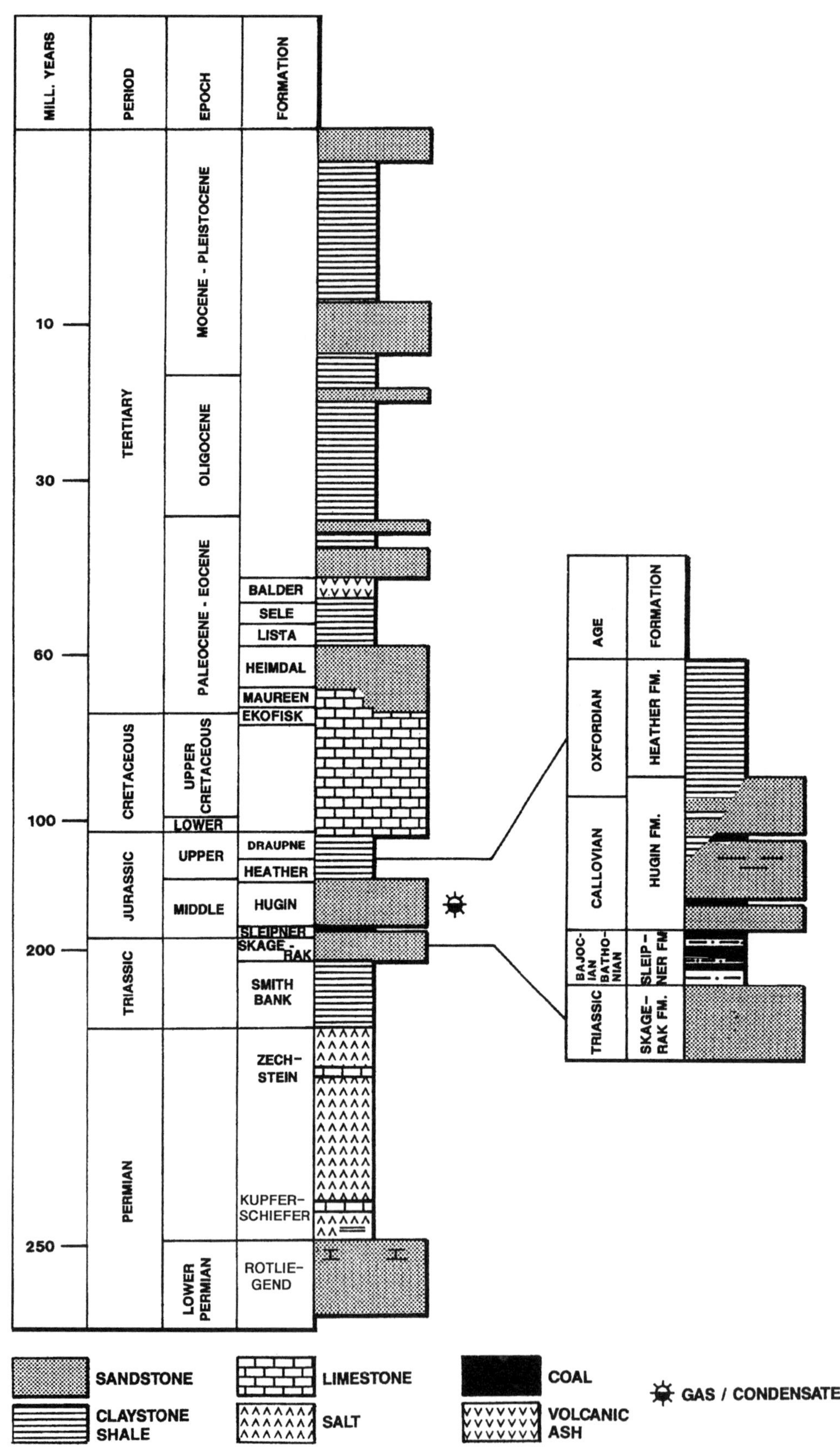

Figure 6. Generalized stratigraphic column for the Southern Viking graben. The main reservoir of the Sleipner Vest field is the Hugin Formation.

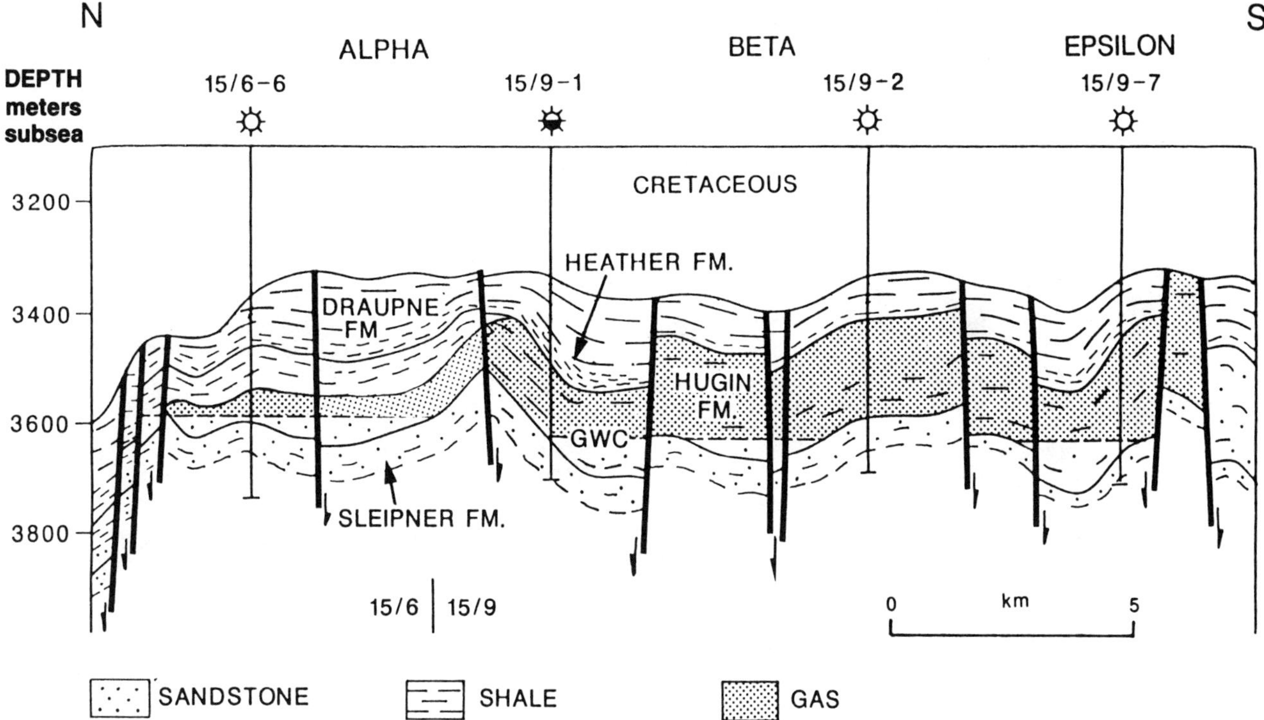

Figure 7. North-south stratigraphic and structural cross section through Sleipner Vest field (see Figure 2 for well locations and line of section).

Reservoir

The seismic interpretation and well information show that the depth to the top of the reservoir is variable (Figure 5), which reflects the structural complexity of the field. In the Alpha structure, the top of the pay zone is at 3450 m subsea (mss) (11,320 ft subsea). Further south in the Epsilon structure it is at 3377 mss (11,080 ft subsea). The Delta structure at the top of the reservoir is at 3425 mss (11,240 ft subsea). The total thickness of the Hugin Formation (Figure 8) varies from 54 to 203 m (175–665 ft).

Stratigraphy and Facies

The main reservoir in the Sleipner Vest field, the Hugin Formation, was deposited during a general transgressive period in the Callovian–early Oxfordian. It is dominated by coastal-nearshore marine sandstones, with the depositional environments ranging from coastal marsh and back-barrier through barrier and offshore bars to offshore sand and mud sheets (Figure 9). The formation is divided into three main units, A, B, and C (Figure 10), corresponding to three major phases of sediment accumulation. These three phases were the result of three episodes of marine transgression separated by three regressive progradational pulses.

Unit A—In the early Callovian, the sea transgressed the Sleipner area from the north, reaching as far south as wells 15/9-10 and 15/9-14 (Figure 9). A progradation of the shoreline toward the north followed, depositing unit A, which is dominated by shoreface sandstones (Figure 11) and is associated with some back-barrier lagoonal (Figure 12) and coastal marsh deposits. Little deposition took place north of block 15/9 during this phase. Unit A comprises units of storm origin, which are overlain by very fine grained, argillaceous, burrowed shoreface sandstones. These pass up into fine- and medium-grained, locally pebbly sandstones, characterized by planar cross-bedded, fining-upward, graded units (Figure 11). The whole sequence is capped by a thick coal unit. South of well 15/9-10, unit A is composed of back-barrier lagoonal facies. This facies comprises an upward-fining graded sandstone unit generally poorly sorted, with bioturbated and siderite cemented intervals (Figure 12). These stacked washover sands are intercalated with siltstone, shale, and thin coal.

Unit B—A second transgression followed after the deposition of unit A and covered approximately the same area as the first. A coastal barrier bar complex was developed, trending east-northeast in block 15/9 and centered in the area of well 15/9-2 (Figure 13). Thick, clean sands of shoreface/foreshore and tidal inlet (Figure 14) origin were deposited. South of the barrier, a first phase of estuarine deposition was followed by the development of lagoonal, tidal delta, and coastal marsh environments. In the north, in block 15/6, shelf sands were deposited in an offshore

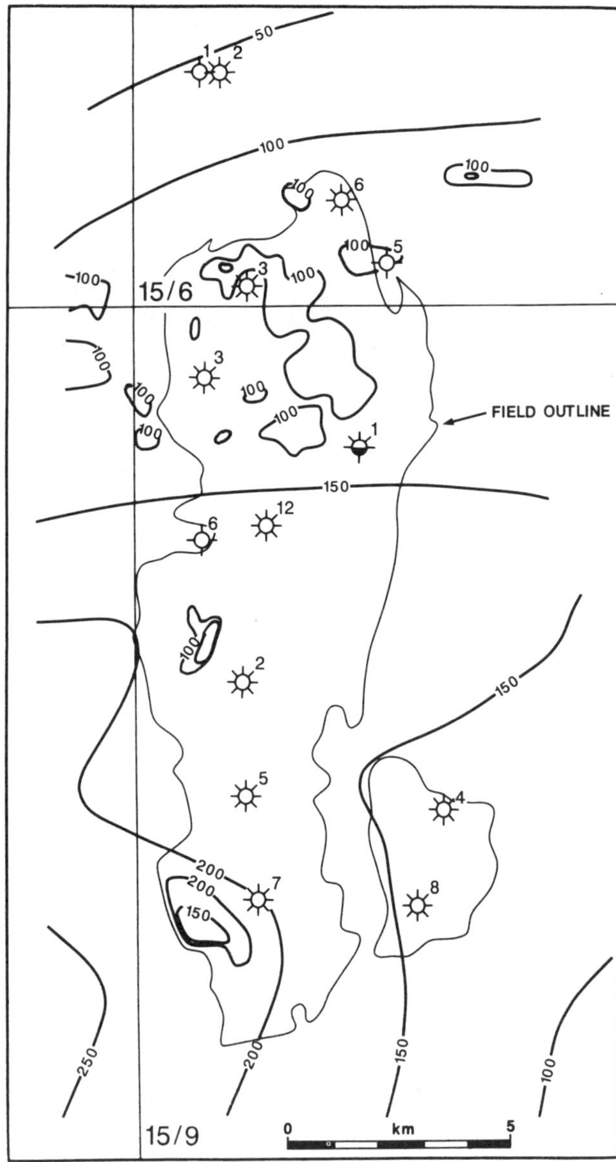

Figure 8. Isopach map showing the total thickness of the Hugin Formation in Sleipner Vest field.

environment. Core data and depositional models presented for unit B suggest the major portion is dominated by fine- to medium-grained, occasionally pebbly, graded fining-up units with planar to trough cross-bedding separated by mud drapes (Figure 15). This is overlain by a coarsening-upward, argillaceous, carbonaceous, burrowed sandstone. Above this is an interbedded thoroughly burrowed and bioturbated organic-rich siltstone and mudstone that becomes progressively siltier and bioturbated upwards.

Unit C—During the latter part of the Callovian, there was a third transgressive phase that eventually flooded the whole Sleipner area. The general transgression was followed by three pulses of regression causing substantial sediment accumulation. The sediments of unit C are composed of lower to upper shoreface sandstones that have a thick development in the southern part of the Sleipner field. No sands were deposited north of well 15/9-12. Unit C comprises fine to very fine argillaceous and micaceous sandstone and medium to coarse sandstone. Bioturbation is common and has obliterated bedding and primary sedimentary structures. The vertical arrangements of this unit begin with a basal lag overlain by burrowed fine-grained sandstone that passes upward into cross-bedded medium sandstones.

Mineralogy

Quartz is the most abundant detrital mineral in the Hugin Formation sandstones, usually followed by feldspar. Normal unstrained quartz is the most abundant type while strained and polycrystalline quartz are less common. Both orthoclase and microcline feldspars occur (Figure 11B), with some plagioclase and perthites. Minor, more scattered detrital grains include mica, rock fragments, ooliths, altered glauconite, and heavy minerals including tourmaline and zircons. The detrital heavy mineral composition, combined with the presence of granitic and extrusive volcanic rock fragments, suggests that the source rocks for the sandstones were largely igneous.

Evidence of diagenesis is widespread. The main diagenetic phases are quartz overgrowth development, siderite, pyrite, vitrinite gel, kaolinite, dolomite, illite, leucoxene, anatase, and feldspar dissolution (Figure 11B). Some of these exhibit more than one phase of development, depending upon variations in depositional and diagenetic history. In general, quartz overgrowths and later phases of pyrite and siderite leucoxene and feldspar dissolutions are common in tidal inlet (Figure 14) and shoreface facies (Figure 11). Early pyrite and siderite are restricted mainly to lagoonal facies (Figure 12). Illite results mainly from alteration of small (3–4 micron) kaolinite platelets, and the incidence of illite is greater in the offshore marine sheet sandstones. Vitrinite gel occurs in sandstones within a few meters of coal or swamp facies from which it is considered to be derived.

Reservoir Characteristics

The reservoir quality is generally good throughout the Sleipner Vest field even though there are slight variations in the rock properties in different wells. The main controls on porosity are predominantly primary in origin. The effective porosity varies between 16 and 19%. Water saturation is 15%. Permeability within the reservoir is generally good (50–400 md). The good quality reservoir lithologies are sometimes separated by both persistent and discontinuous shale or shalier horizons with low permeability. Faulting in the region can also control permeability.

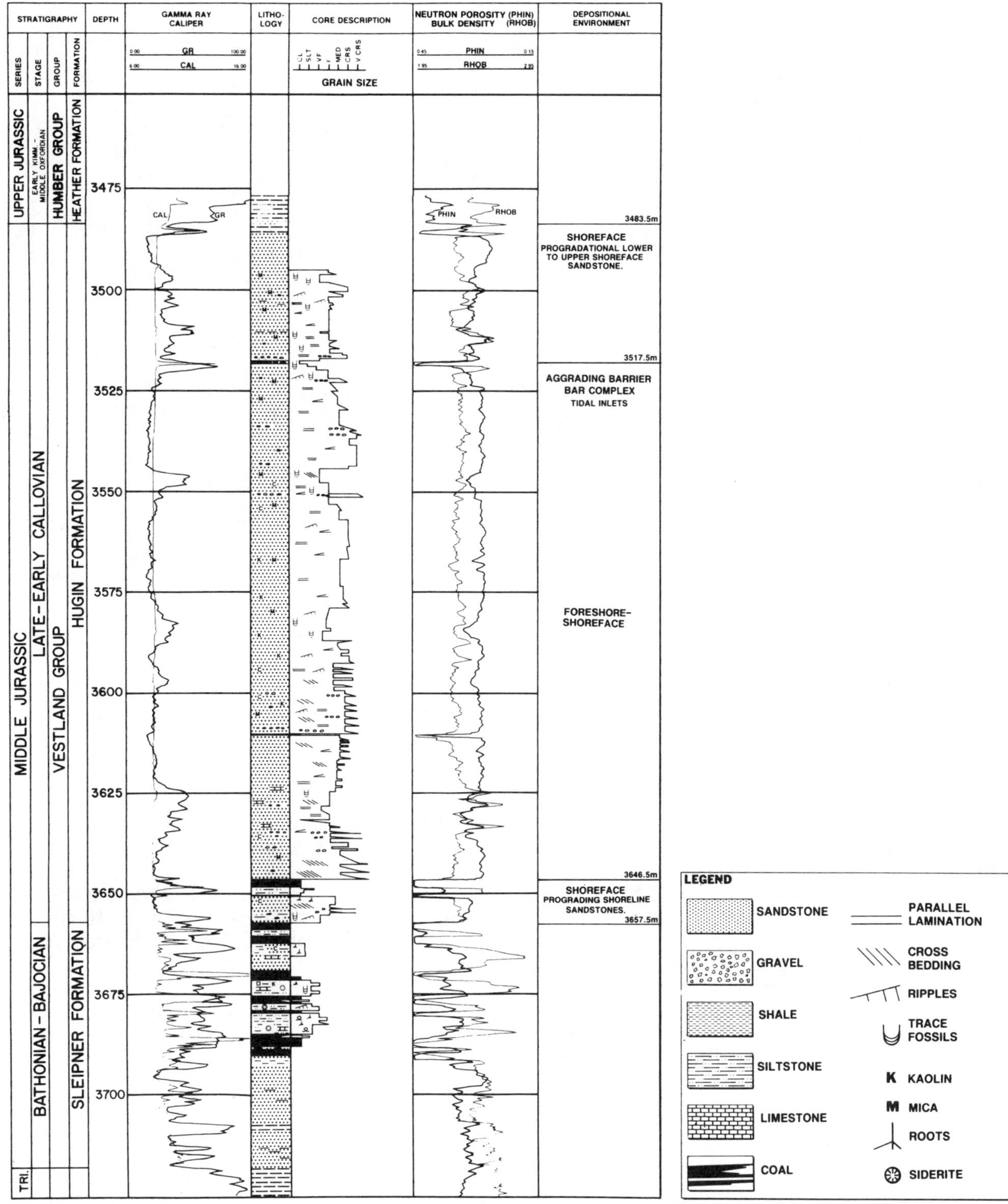

Figure 9. Data summary chart from well 15/9-2 showing variation of the gamma-ray and neutron-density logs and their components with regard to lithofacies.

The pressure in the Sleipner Vest reservoir is 70 bar (1015 pound-force/in^2) higher than the hydrostatic pressure (Figure 16). Pressure distribution in the reservoir is complex, as indicated by varying pressure measurements from the different wells. Some of the pressure variations can be attributed to imprecision of measurement but real differences are thought to exist, giving compartments different pressure regimes.

For reserve calculation purposes, the pressure model shown in Figure 16 has been used. The model indicates the existence of two separate pressure

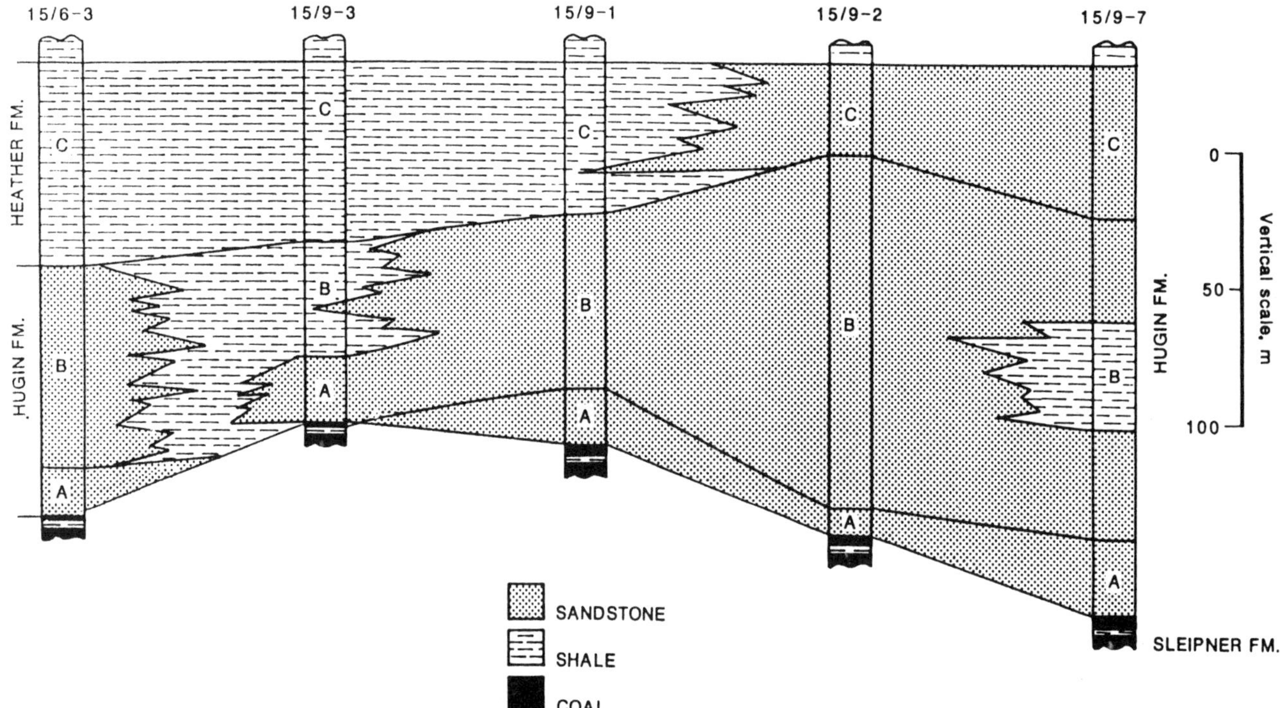

Figure 10. Stratigraphic cross section showing major reservoir units. Line of cross section can be seen in Figure 2.

systems in the Alpha structure. Pressure measurements from well 15/6-6 support the definition of a separate pressure regime in the northern part of the Alpha structure. The Delta structure has its own pressure regime.

Gas and Condensate Characteristics

The Sleipner gas condensate is paraffinic and has a GOR of 3484 Sm^3/m^3 (19,560 cfg/bbl oil) and average dry gas value of 82%. It also has a low N_2 content (0.7 mol%), high aromaticity in C_2–C_8 hydrocarbon fraction, and low average pristane/phytane ratio in C_{15+} fraction (Figure 17).

Source

The condensates in Sleipner Vest were derived from both the Draupne and Hugin/Sleipner formations and were generated during the late Tertiary (Figure 18) from a type II/III kerogen. The contribution from the Draupne Formation increases significantly to the north into block 15/6. In block 15/9 the accumulated hydrocarbons were generated from high maturity source rocks, with late gas from the deepest Hugin/Sleipner formations. Early and mid-mature hydrocarbons from the Draupne Formation were generated in the downfaulted areas between Sleipner Vest and the Utsira high in block 15/9 and migrated into the 15/9-1 and 15/9-3 wells. This explains the oil observed in 15/9-1. Early and mid-mature condensate generated in the Hugin and Sleipner formations migrated into the adjacent Alpha structure in block 15/8. Late condensates and gas generated from the Hugin/Sleipner formations in the graben bypassed the block 15/8 Alpha structure and migrated into the Sleipner Vest area.

Reserves

Recoverable reserves in Sleipner Vest are estimated to be 126×10^9 Sm^3 (approximately 4.5 tcf) gas and 29×10^6 tons (approximately 200 million bbl) condensate. Owing to the complexity of the reservoir, the production wells will be positioned so as to cover most of the field; between 35 and 45 wells will be needed to drain the reservoir. The field is assumed to have a very small aquifer and, therefore, will be produced by gas depletion. The Sleipner Vest field was declared commercial in 1982. The final development plan for the field has not yet been determined and will depend on the market possibilities for the sale of the gas.

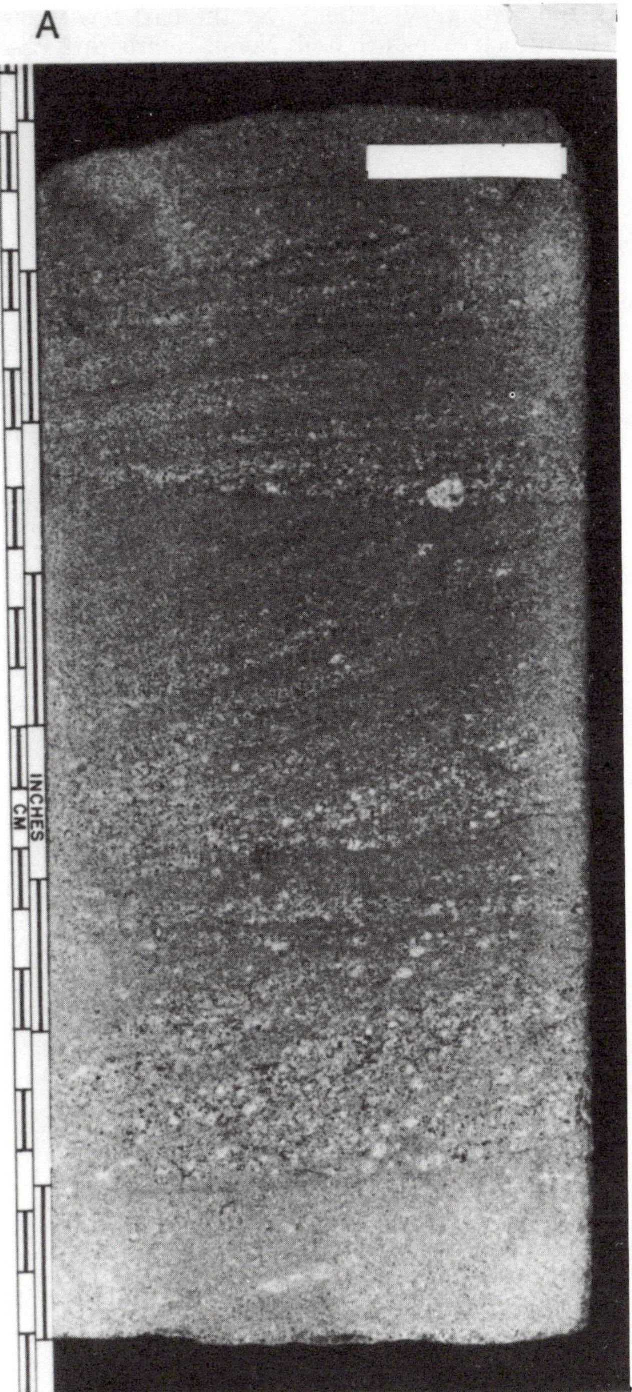

Figure 11. Prograding shoreface facies. (A) Core slab of clean medium- and coarse-grained sandstone showing well-preserved primary sedimentary structures such as planar cross-bedding and fining-upward graded units. (B) Photomicrograph of 11A showing partial dissolution of microcline feldspar and the resulting porosity. (C) SEM photo showing good connecting intergranular porosity, initial quartz overgrowths, and small patches of kaolinite.

EXPLORATION CONCEPTS

Early exploration around the Sleipner area was based on the 2D seismic interpretation that defined four different structures: Alpha, Beta, Epsilon, and Delta. Because of the negative result of the well 15/9-6, drilled in a down-faulted area between the Alpha and Beta structures, it was evident that the seismic data were of poor quality. The 3D seismic survey that followed this led not only to the discovery of the adjacent Sleipner Øst field, but also to other structures such as Mu, Sigma, and Beta in block 15/9 and Alpha in block 15/8. It also confirmed that there was no structural closure between Alpha, Beta, and Epsilon structures. There was a reduction in calculated reserves on the flanks of the Alpha and Delta structures, but at the same time, there was an increase in the central part of the field.

ACKNOWLEDGMENTS

This article is based on the results of contributions made by many people from Statoil who have worked

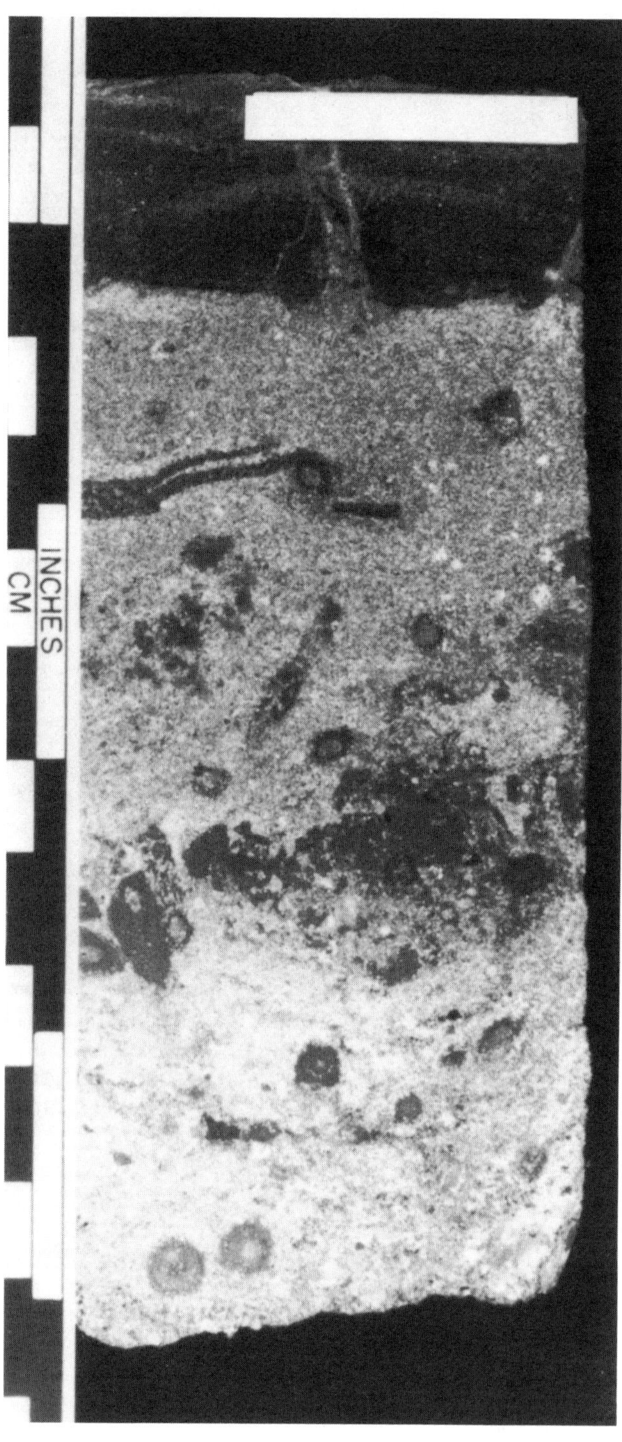

Figure 12. Back-barrier lagoonal facies. Core slab of an upward-fining graded sandstone unit generally poorly sorted and considered to be stacked washovers. Siderite cementation causes the sandstone to have an orange-brown color. The dark spherical structures are horizontal burrows.

on the Sleipner Vest field over the past few years. The author thanks Statoil, Norsk Hydro, and Esso Norge a.s, Elf Aquitaine Norge a.s, and Total Marine Norsk a.s for giving permission to publish this paper.

REFERENCES

Deegan, C. E., and B. J. Scull, 1977, A standard lithostratigraphic nomenclature for the Central and Northern North Sea: Rep. Inst. Geol. Sci., v. 77/25.

Larsen, R. M., and L. J. Jaarvik, 1981, The geology of the Sleipner Field Complex, *in* Norwegian Symposium on Exploration (Norsk Petroleumsforening), Article 15, 31 p.

Pegrum, R. M., and T. E. Ljones, 1984, 15/9 Gamma gasfield offshore Norway, new trap type for North Sea basin with regional implications: American Association of Petroleum Geologists Bulletin, v. 68, n. 7, p. 874-902.

Ranaweera, H. K. A., 1987, Sleipner Vest, *in* Geology of the Norwegian oil and gas field (Norsk Petroleumsforening), Article 22, p. 253-264.

Vollset, J., and A. G. Doré, 1984, A revised Triassic and Jurassic lithostratigraphic nomenclature for the Norwegian North Sea: Norwegian Petroleum Directorate, Bulletin 3.

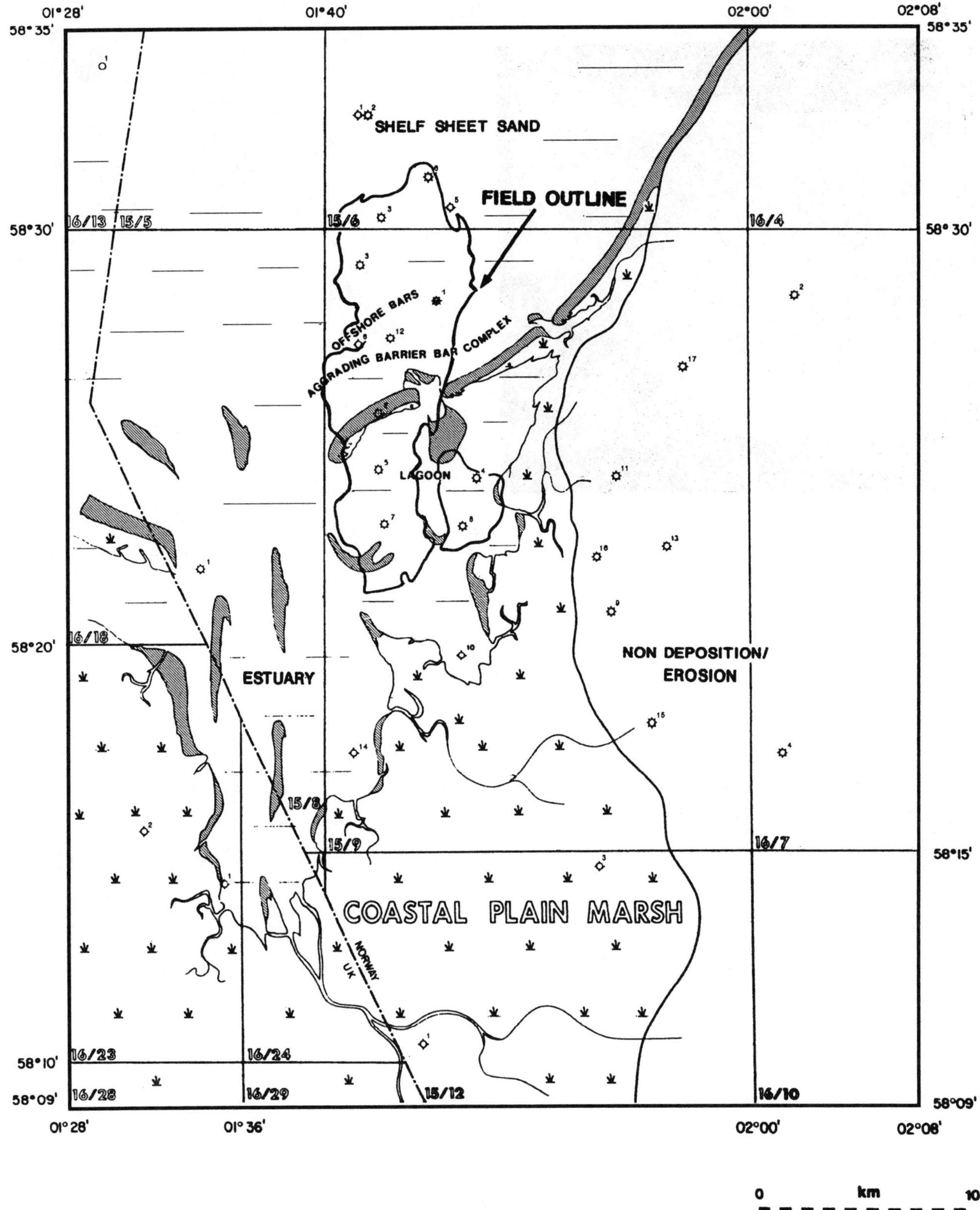

Figure 13. Paleogeographic reconstruction of Hugin Formation.

14 A

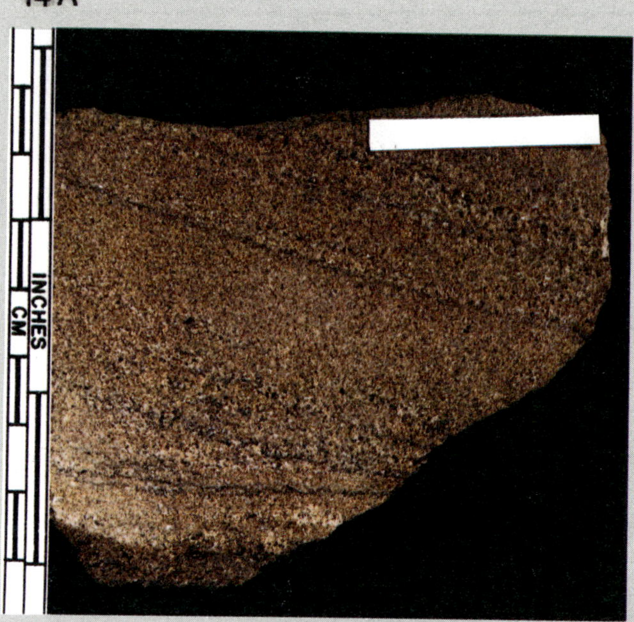

14B

Figure 14. Tidal inlet fill facies. (A) Core slab of a medium to coarse-grained sandstone with cross-bedding and double mud drapes. (B) SEM of 14A showing minor quartz overgrowth development, authigenic clay, and good interconnecting pores. Bar scale = 10 microns.

Appendix 1. Field Description

Field name .... *Sleipner Vest field*

Ultimate recoverable reserves .... *126 × 10⁹Sm³ (4.5 tcf) gas, 29 × 10⁶ tons (200 million bbl) condensate*

Field location:

- **Country** .... *Norway*
- **State**
- **Basin/Province** .... *Southern Viking graben, North Sea*

Field discovery:

- **Year first pay discovered** .... *Middle-Upper Jurassic Hugin Formation 1974*
- **Year second pay discovered** .... *None*
- **Third pay** .... *None*

Discovery well name and general location:

- **First pay** .... *Well 15/6-3: 58°30′15.9″N, 01°42′39.5″E*
- **Second pay**
- **Third pay**

Discovery well operator ..... *Esso Norge a.s (the Sleipner field extends into the 15/9 block, and the operator for the whole field is Den norske stats oljeselskap a.s [Statoil])*

- **Second pay**
- **Third pay**

IP in barrels per day and/or cubic feet or cubic meters per day:

- **First pay** .... *No production data; the field is planned to begin production in early 1990s*
- **Second pay**
- **Third pay**

Figure 15. Marine sheet sandstone facies. Core slab of a fine-grained, micaceous sandstone showing how the original fabric has been altered by a *Diplocraterion* burrow.

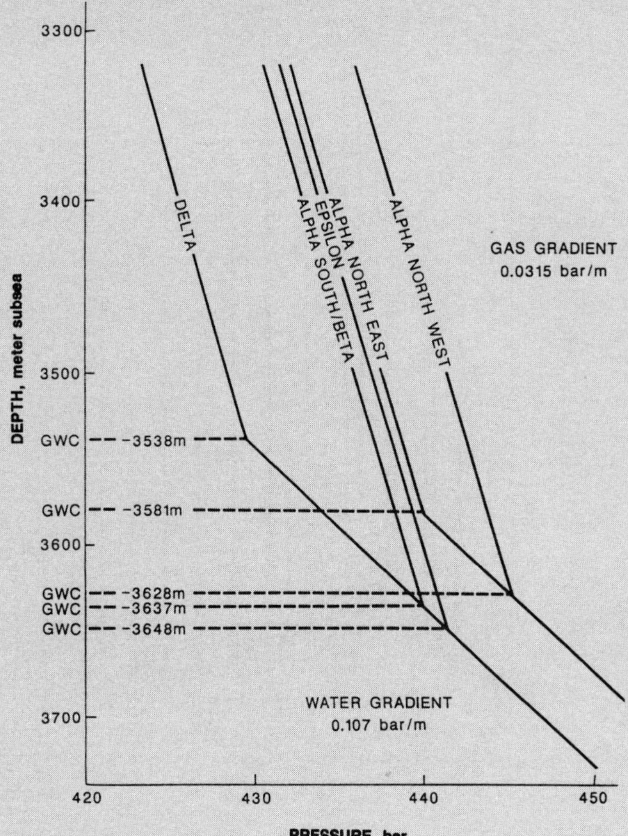

Figure 16. Pressure model for Sleipner Vest field.

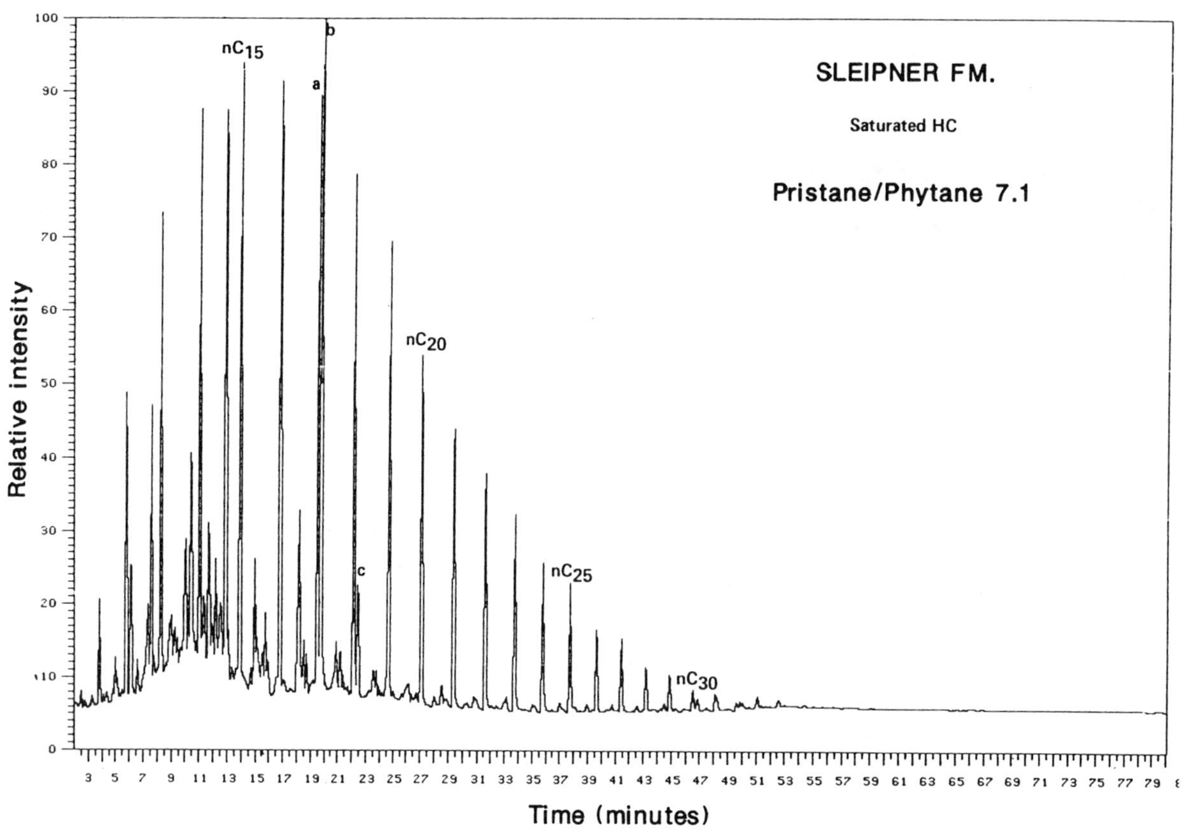

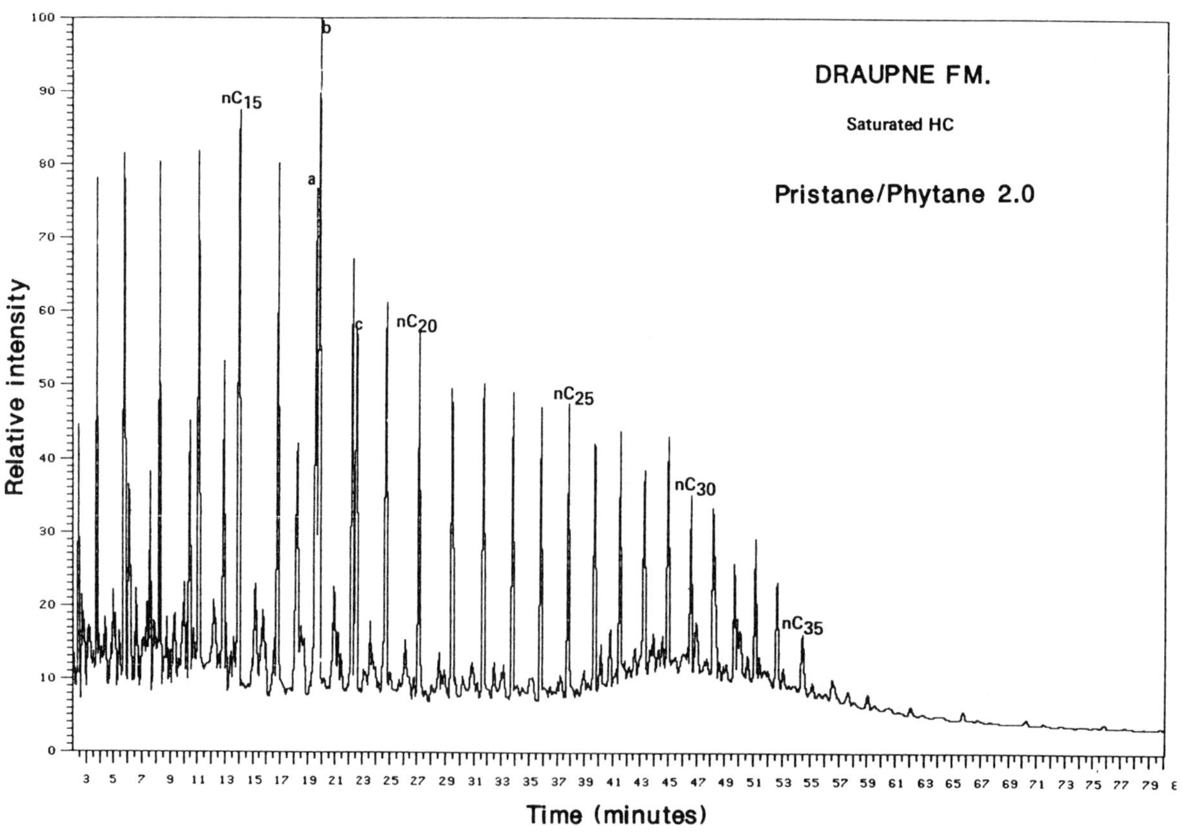

Figure 17. Gas chromatogram for condensate and gas from Sleipner and Draupne formations.

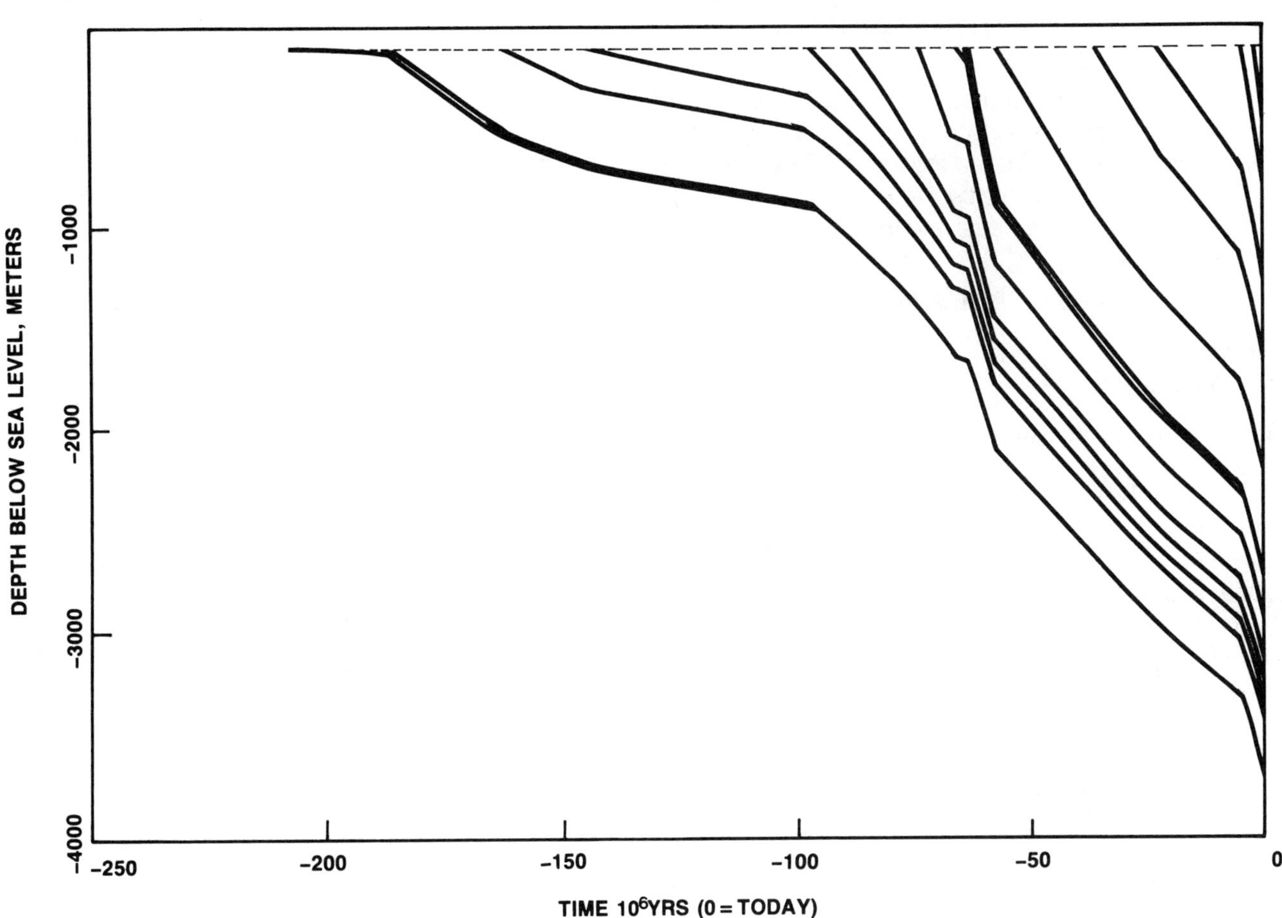

Figure 18. Geohistory for well 15/9-2 in Sleipner Vest field.

All other zones with shows of oil and gas in the field:

Age	Formation	Type of Show
None		

Geologic concept leading to discovery and method or methods used to delineate prospect, e.g., surface geology, subsurface geology, seeps, magnetic data, gravity data, seismic data, seismic refraction, nontechnical:

Early seismic interpretation indicated a closure on Jurassic and pre-Jurassic rocks in blocks 15/6 and 15/9. It also defined a major north-south-trending fault system that separated the Sleipner terrace in the west from the Utsira high fault block in the east. The seismic control together with regional well data indicated that Jurassic sediments were thin or absent on the Utsira high block. The structural and stratigraphic studies also indicated that the Callovian sandstones would be thin or absent in the southern half of block 15/9. For these reasons, early exploration was directed toward the known gas/condensate-bearing sands of the Sleipner terrace.

Structure:

Province/basin type .... *South Viking graben, North Sea/intracratonic*

Tectonic history

Extensional rifting affected the area periodically from Early Permian until Late Jurassic times and reached a maximum in Late Jurassic, with differential rates of rifting leading to localized compression and strike-

slip movements. Reversal of normal faults due to compressional movements in early mid-Cretaceous times resulted in structuring at Jurassic-Triassic levels.

Regional structure

Sleipner Vest is situated near the southern end of the Viking graben at the convergence of several major structural elements such as the Utsira high and Ling graben to the east, the East Shetland platform and Flanden ground spar to the west, and the Andrew salt ridge and Jaeren high to the south. The structural pattern within the Sleipner area is dominated by two major basement controlled fault blocks, namely the Gamma high to the east and the Sleipner terrace to the west.

Local structure

Jurassic pay: domal structures

Trap:

Trap type(s) .... *Combined structural and stratigraphic trap*

Basin stratigraphy (major stratigraphic intervals from surface to deepest penetration in field):

Chronostratigraphy	Formation	Depth to Top in ft (m) ss
Miocene	*Utsira Formation*	*2620 (799)*
Paleocene	*Heimdal Formation*	*7760 (2366)*
Upper Cretaceous	*Chalk Group*	*9050 (2758)*
Upper Jurassic	*Draupne Formation*	*11,000 (3352)*
Middle Jurassic	*Hugin Formation*	*11,500 (3505)*
Triassic	*Skagerraak Formation*	*12,060 (3676)*

Location of well in field

Reservoir characteristics:

Number of reservoirs *1*

Formations .... *Hugin*

Ages .... *Middle and Upper Jurassic (Callovian and Oxfordian)*

Depths to tops of reservoirs .... *3450 m ss (11,320 ft ss)*

Gross thickness (top to bottom of producing interval) .... *203 m (665 ft)*

Net thickness—total thickness of producing zones

Average .... *70–158 m (230–375 ft)*

Maximum .... *88–160 m (290–525 ft)*

Average

Maximum

Lithology .... *Very fine to fine-grained, argillaceous, micaceous sandstone and medium-grained to pebbly subarkosic sandstone*

Porosity type .... *Intergranular porosity*

Average porosity *22%*

Average permeability .... *10–430 md*

Seals:

Upper

Formation, fault, or other feature .... *Formation seal (Draupne Fm., Kimmeridgian, U. Jurassic)*

Lithology .... *Shale*

Lateral

Formation, fault, or other feature .... *Sealing faults and formation seal (Draupne Fm.)*

Lithology .... *Shale*

Source:

Formation and age .... *Draupne and Heather, Upper Jurassic; Hugin and Sleipner, Middle Jurassic*

Lithology .... *Draupne and Heather, shale and radioactive shale; Hugin and Sleipner, shale and coal*

Average total organic carbon (TOC) *5%*

Maximum TOC *10.2*
Kerogen type (I, II, or III) .... *III and II/III*
Vitrinite reflectance (maturation) .... $R_o = 1.2\text{–}1.5$
Time of hydrocarbon expulsion .... *40 MYBP to present*
Present depth to top of source .... *12,800–15,240 ft (4200–5000 m)*
Thickness .... *1500 ft (500 m) average gross thickness*
Potential yield .... *28.5 m^3 gas/km^3 (4.2 mcf/mi^3) rock*

Appendix 2. Production Data

Field name .... *Sleipner Vest field*

Field size:

Proved acres .... *19,768 (80 km^2)*
Number of wells all years *14*
Current number of wells .... *14 exploration/appraisal wells*
Well spacing .... *3 km*
Ultimate recoverable .... *126 $\times$ 10^9 Sm^3 (4.5 tcf) gas and 29 $\times$ 10^6 tons (200 million bbl) condensate*
Planned annual production .... *325 bcf (9.2 $\times$ 10^9 Sm^3)*
Present decline rate
Initial decline rate
Overall decline rate
Annual water production
In place, total reserves .... *6925 bcf (196 $\times$ 10^9 Sm^3)*
In place, per acre-foot
Primary recovery .... *4450 bcf (126 $\times$ 10^9 Sm^3)*
Secondary recovery
Enhanced recovery
Cumulative water production

Drilling and casing practices:

Amount of surface casing set .... *200 ft (60 m) subsea-floor (ssfl)*

Casing program

20-in. shoe: 450 m ssfl (lower Nordland Group, Pliocene); 13⅜-in.: 1170 m (3840 ft) ssfl (Hordaland Group shale, Miocene-Oligocene); 9⅝-in.: 2800 m (9190 ft) ssfl (top Chalk Group, Upper Cretaceous); if testing: 7-in. liner to TD

Drilling mud .... *Salt water base*
Bit program .... *Steel tooth bits, mostly journal bearing*
High pressure zones .... *Middle to Upper Jurassic sandstone reservoir*

Completion practices:

Interval(s) perforated
Well treatment

Formation evaluation:

Logging suites .... *From top to TD: ISF/Sonic-GR, FDC/CNL, VSP; in Paleocene, Cretaceous, and Jurassic intervals: DLL/MSFL, HDT; RFT in reservoirs and zones of interest*
Testing practices .... *Production tests (DST) were carried out for the reservoir in 7 of the 14 wells in the field*

Mud logging techniques

Normal mud logging techniques such as recording drilling parameters, logging the well for hydrocarbon shows, *collecting samples, preparing sample logs, and conducting other services throughout drilling operations; additional requirements are shale density shale factor measurement, H_2S detection, chromatography, and calcimetry*

Oil characteristics:

Type
API gravity
Base
Initial GOR
Sulfur, wt%
Viscosity, SUS
Pour point
Gas-oil distillate

Field characteristics:

Average elevation .... *3400 m (11,155 ft) ss*
Initial pressure at gas-water contact .... *6484 psi (447 bar)*
Present pressure .... *Initial pressure*
Pressure gradient .... *0.45 psi/ft (0.107 bar/m)*
Temperature .... *123°C (253°F)*
Geothermal gradient .... *0.036°C/m (0.0198°F/ft)*
Drive .... *Water drive and gas depletion*
Gas column thickness .... *200 m (655 ft)*
Gas-water contact .... *3581 m (11,750 ft) ss in the north; 3637 m (11,935 ft) ss in the south*
Connate water *22%*
Water salinity, TDS *7%*
Resistivity of water .... *0.031 ohm at 123°C*
Bulk volume water (%) *22%*

Transportation method and market for oil and gas:

The gas is planned to be transported in pipelines (Zeepipe, Statpipe, and Norpipe systems) to Emden in Holland and Zeebrugge in Belgium, the same as from Sleipner Øst. The same pipeline that will be built from Sleipner Øst to the Ula pipeline will be used to transport condensate via Ekofisk to Teeside, England.

Jackson Field—Australia
Cooper-Eromanga Basins, Central Australia

J. W. HUNT, D. A. GUTHRIE
Esso Australia Limited
Sydney, Australia

A. P. DODMAN
Santos Limited
Adelaide, Australia

FIELD CLASSIFICATION

BASIN: Eromanga
BASIN TYPE: Cratonic Sag
RESERVOIR ROCK TYPE: Sandstone
RESERVOIR ENVIRONMENT OF DEPOSITION: Fluvial (Braided)
RESERVOIR AGE: Jurassic
PETROLEUM TYPE: Oil
TRAP TYPE: Anticline with Minor Stratigraphic Component

INTRODUCTION

The Jackson oil field is one of Australia's largest onshore oil fields. It contains multiple reservoirs in the lower part of the oil-prone, Mesozoic-aged epicratonic Eromanga basin. The Eromanga basin overlies the more restricted intracratonic Permian-Triassic Cooper basin, one of several major Australian coal basins of the Gondwana super-continent. The Cooper basin, in contrast, is gas prone and is Australia's most prolific onshore hydrocarbon basin. It hosts the Tirrawarra oil field, also discussed in this Atlas.

Most likely recoverable reserves in the Eromanga basin where it overlies the Cooper basin are about 100 bcfg (2.8×10^9 m^3) and 150 million barrels of black oil. Most likely recoverable reserves in the Cooper basin are about 9 tcfg (2.52×10^{11} m^3) and 300 million barrels of gas-liquids and volatile oil. The Jackson oil field contains about 110 million barrels of oil in place with an estimated ultimate recovery of about 40 million barrels.

LOCATION

The Jackson oil field is located in southeastern central Australia (Figure 1) within the Naccowlah Block of Authority to Prospect (ATP) 259P in southwest Queensland (Figure 2).

It lies near the southeastern margin of the Permian-Triassic Cooper basin on the eastern end of a major west-northwest-trending structural high, the Pepita-Naccowlah-Jackson (PNJ) trend (Figure 2). This trend, and associated northeast-southwest Gidgealpa-Merrimelia-Innamincka (GMI) trend, are the most productive and prospective trends in the Cooper and Eromanga basins for both oil and gas.

Other major fields within the Cooper and Eromanga basins are the Tirrawarra oil field and the Gidgealpa, Merrimelia, Daralingie, Moomba, Big Lake, Toolachee, Challum, and Wackett oil and gas, or gas fields (Figure 2).

EXPLORATION HISTORY

Pre-Discovery

The license area ATP 259P is part of a larger permit originally granted in 1958 to a joint venture of SANTOS Ltd. and Delhi Petroleum Pty. Ltd. Preliminary exploration consisted of gravity and magnetic surveys in combination with surface geology mapping. This was followed by the acquisition of single fold, analog, seismic data and the drilling of a number of exploration wells.

The first exploration success was in 1963 with the discovery in South Australia of gas from the Permian section in the Gidgealpa 2 well. Further gas discoveries in Permian age sediments in the Moomba field led to the installation of a treatment facility and the construction of a gas pipeline from Moomba to Adelaide and, later, to Sydney (Figure 1). Permian oil discoveries were also made, notably the Tirrawarra field. At that time, however, the oil could not be extracted

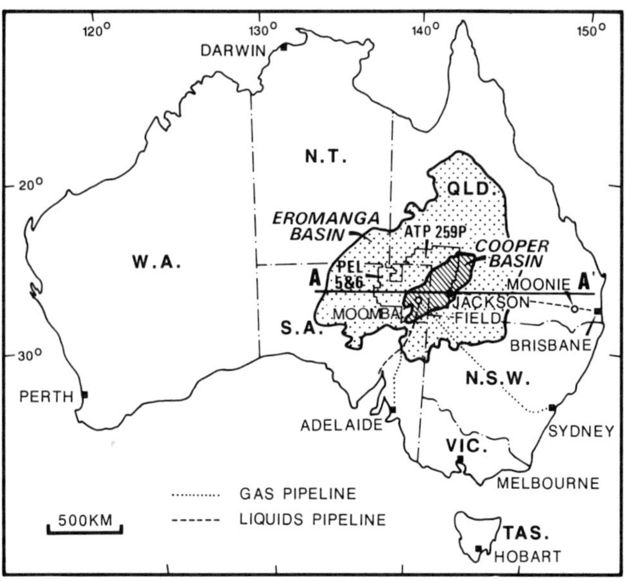

Figure 1. Location of the Jackson oil field, ATP 259P and the Cooper and Eromanga basins in central Australia. Australian states: QLD, Queensland; NSW, New South Wales; VIC, Victoria; TAS, Tasmania; SA, South Australia; WA, Western Australia; and NT, Northern Territory.

economically from the tight Permian reservoirs and so was of less significance than the gas.

Reservoir rocks of the Eromanga basin form the major aquifers of the Great Artesian Basin, the largest artesian basin in the world. Despite numerous shows during drilling, the Eromanga basin was considered by most early explorers to be devoid of economic hydrocarbon accumulations because of suspected flushing by artesian water. The first Eromanga basin discovery was of minor gas (about 15 bcf in place) in the Jurassic Namur Sandstone of the Namur field in 1976. The first oil discovery followed in 1978 from the Jurassic Hutton Sandstone (and subsequently the overlying Namur Sandstone) in the Strzelecki field (South Australia), which also contains Permian hydrocarbons. In the same year in Queensland, and closer to the Jackson area, Wackett 1 flowed gas from the Jurassic Birkhead Formation, with live oil shows from a thin sandstone band in the same sequence. Thus, by 1979, most exploration wells in PELs 5 & 6 in South Australia and ATP 259P in Queensland were drilled with dual Permian and Mesozoic objectives.

Discovery

In 1979, a portion of the Queensland lease area (originally granted as ATP 66P and ATP 67P) was renewed as ATP 259P. A 20% interest in the southeastern portion of the area, known as the Naccowlah Block (Figure 2), was farmed out in return for a work program of additional seismic acquisition and exploratory drilling.

Jackson 1, the second well in this farmout agreement, was drilled crestally on the northern culmination of the Jackson anticline in December 1981 as a dual test for Eromanga basin oil and gas and for Cooper basin gas and condensate. The well resulted in the discovery of three separate oil pools, in the Cretaceous Murta Member, the Jurassic Westbourne Formation, and the Jurassic Hutton Sandstone. The Mesozoic stratigraphy of the discovery well is shown in Figure 3, and drill-stem test results and oil characteristics are listed in Table 1. Although shows were observed in the Permian section, it produced water on test.

Delhi Petroleum Pty. Ltd. (now a subsidiary of Esso Australia Ltd.) was exploration operator for the discovery, and the field is now operated by SANTOS Ltd., on behalf of a consortium of the above, Ampol Exploration Ltd., Claremont Petroleum N.L., and Oil Company of Australia N.L.

Post-Discovery

Jackson 1 was completed from the Hutton Sandstone and the Westbourne Formation as a single completion with a down-hole sliding sleeve assembly to enable production from either formation. A post-drill depth map on the "C" horizon (Figure 4B) was the basis for setting subsequent appraisal and exploratory drilling on the structure. Initial drilling in the Jackson field concentrated on appraisal and development of the Hutton Sandstone. The Westbourne and Murta reservoirs were secondary objectives because of their lower reserves and productivity. Initial well spacing was about 80 ac (32.5 ha), though recently, infill drilling has been undertaken to improve drainage of the Hutton and Westbourne reservoirs.

Many wells were completed with a single 7-in. production casing and single tubing strings. Surface casing (13 in.) was run to about 600 ft (185 m). Where wells had more than one productive zone, the lower zone was often flowed up the tubing, and the upper zone flowed up the tubing/production casing annulus. As water cuts rose, pumping equipment became necessary to maintain production rates, and dual tubing strings were installed in a number of wells. Beam pumps are the most common pumping equipment, though electric submersible pumps have been used in several high productivity wells. The wells are perforated underbalanced using KCl brine as a completion fluid in most instances. Perforation density is 4 shots/ft (spf) with either 2⅛-in. through tubing guns or 4-in. casing guns. A few wells were perforated with tubing conveyed guns.

The water cut and pressure history of the Hutton indicates an active bottom and edge water drive reservoir. High permeability streaks and shale barriers influence the water cut behavior and well

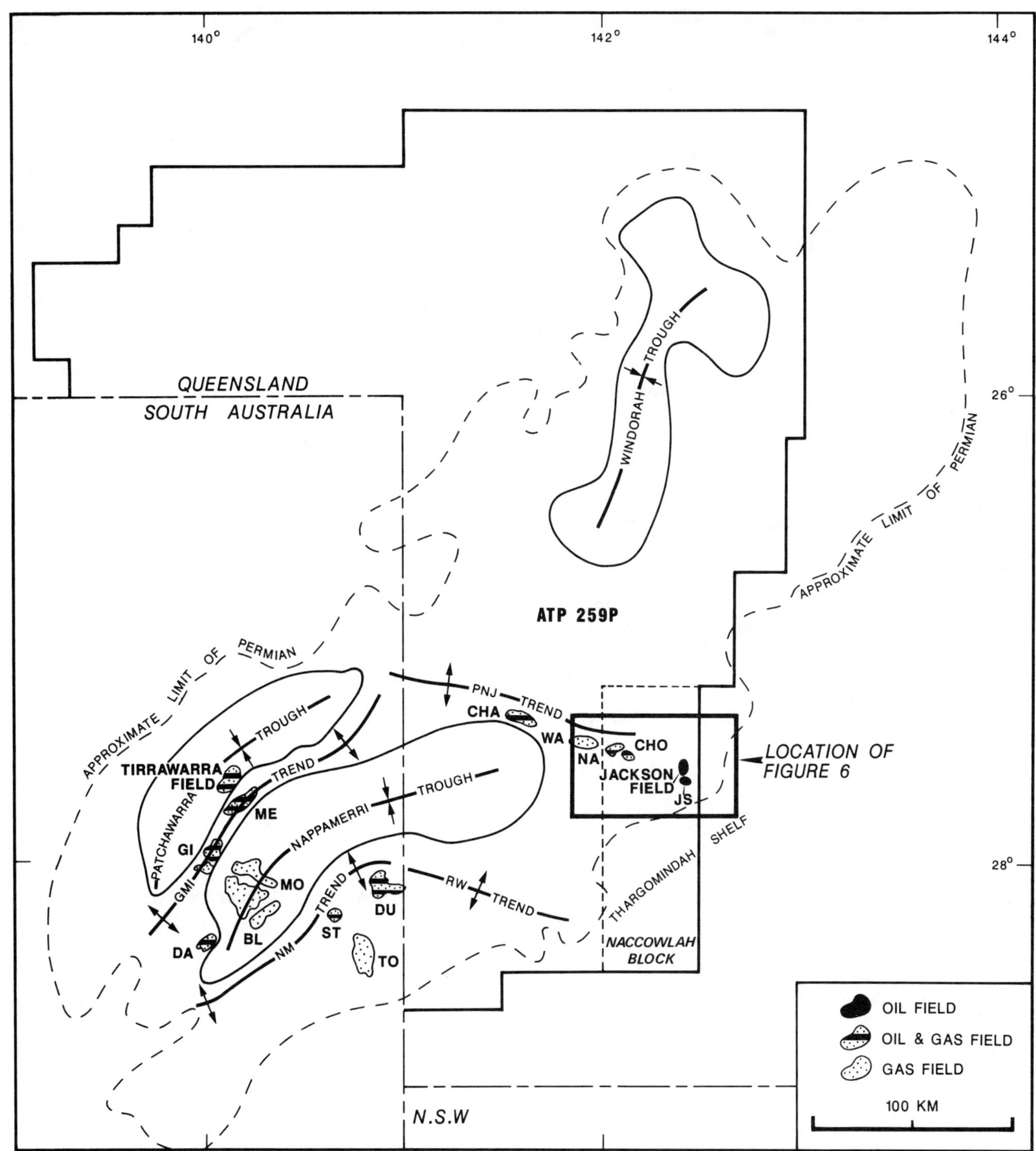

Figure 2. Regional geological setting and major oil and gas fields of the Cooper and Eromanga basins. PNJ, Pepita-Naccowlah-Jackson Trend; GMI, Gidgealpa-Merrimelia-Innamincka Trend; NM, Nappacoongee-Murteree Trend; RW, Roseneath-Wolgolla trend; DA, Daralingie field; BL, Big Lake field; MO, Moomba field; GI, Gidgealpa field; ME, Merrimelia field; ST, Strzelecki field; DU, Dullingari field; TO, Toolachee field; CHA, Challum field; WA, Wackett field; NA, Naccowlah field; CHO, Chookoo field; JS, Jackson South field.

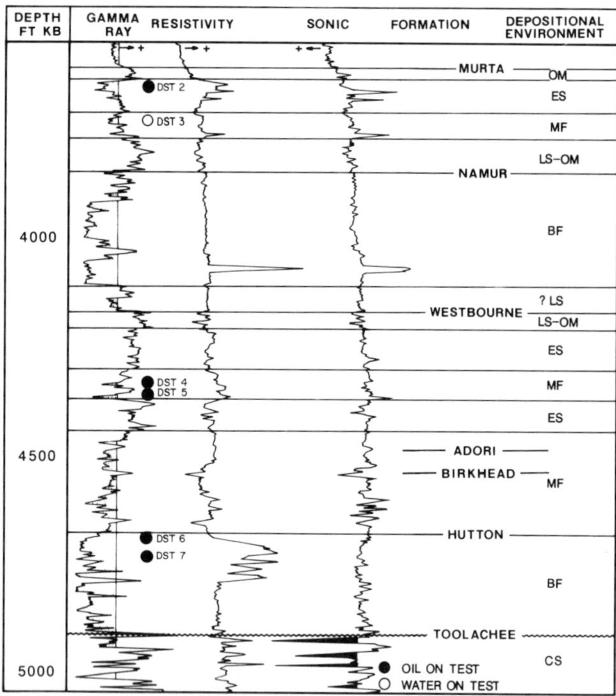

Figure 3. Stratigraphic column, discoveries, and depositional environments for Jackson 1. Depositional environments from G. Powis (pers. comm.): MF, meandering fluvial; BF, braided fluvial, ES, estuarine; LS, lower shoreface; OM, offshore marine; RM, restricted marine; CS, coal swamp. DST, drill-stem test.

productivity. Shale barriers have been utilized in selecting perforation intervals to delay water coning.

DISCOVERY METHOD

The Jackson anticline was first recognized on the basis of surface geology mapping. It was, however, one of a number of large surface features and lay over 120 mi (192 km) from the early exploration successes at Gidgealpa and Moomba. It was not until 1979 that regional seismic data were acquired in the area and a subsurface feature confirmed. In early 1981, a 2×2 km (1.2×1.2 mi) grid of 12-fold vibroseis* data was recorded that defined a northerly trending anticline with two major culminations, Jackson and Jackson South. A pre-drill time map on the "C" horizon (top of Cretaceous Cadna-owie Formation, Figure 8) is shown in Figure 4A.

Owing to the uniform nature of sediments above the Cadna-owie Formation, variation in average velocity to the "C" horizon is minor. The pre-drill time structure map of the "C" horizon (Figure 4A) was therefore a reliable guide to the form of the "C" horizon depth structure map (Figure 4B). Because of the conformable nature of the Eromanga basin sediments below the Cadna-owie Formation, the "C" horizon time map was also used as a guide to the form of structure for the main reservoirs. Subsequent refinement of the Jackson field structure maps has been due largely to increased seismic coverage and more detailed mapping at reservoir levels.

* Trademark of Continental Oil Company.

Table 1. Results of drill stem tests, Jackson 1. The Transition Beds are now called the Cadna-owie Formation, and the Mooga is called the Hooray Sandstone.

Text No.	Interval (ft) Drillers Depths	Formation	Top Choke	Bottom Choke	Rev. Circ.	Results
1	3367-3438	Transition Beds	0.5″	0.75″	No	50′MW*, 370′W, 400′ VSGCW, 45′ VSGCM
2	3625-3753	Mooga: Murta Member	0.5″	0.75″	Yes	GTS 42 mins RTSTM OTS 89 mins 338BOPD 51° API @ 60° F
3	3712-3742	Mooga: Murta Member	0.5″	0.75″	No	WTS 42 mins 1600 BWPD
4	4312-4367	Westbourne	0.5″	0.75″	Yes	GTS 146 mins RTSTM OTS 160 mins 188BOPD 41.1° API @ 60° F
5	4368-4429	Westbourne	0.5″	0.75″	Yes	GTS 31 mins RTSTM OTS 32 mins 1165BOPD 41.1° API @ 60° F
6	4688-4713	Hutton Sandstone	0.5″	0.75″	Yes	GTS 29 mins RTSTM OTS 30 mins 1325BOPD 39.9° API @ 60° F
7	4687-4773	Hutton Sandstone	0.5″	0.75″	Yes	OTS 12 mins 155PSI 2357 BOPD 40° API @ 60° F

*MW, muddy water; VSGCW, very slightly gas cut water; RTSTM, rate too small to measure; BOPD, bbl oil per day; BWPD, bbl water per day; GTS, gas to surface; OTS, oil to surface; WTS, water to surface.

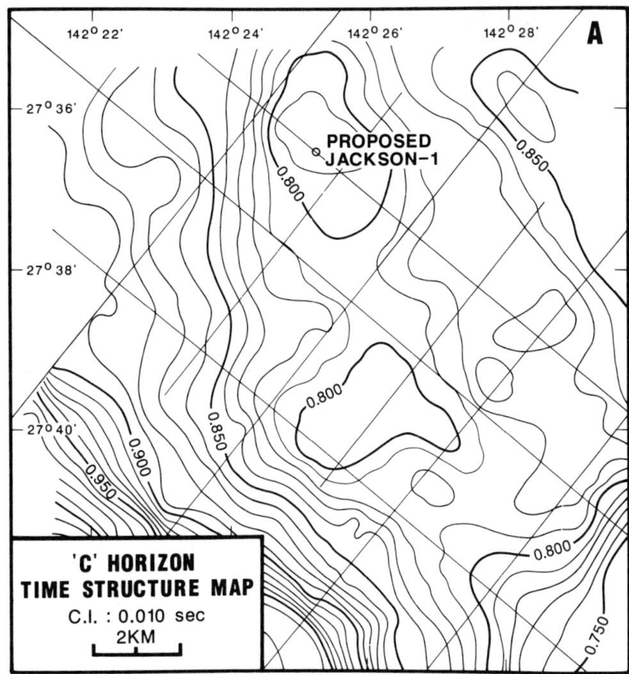

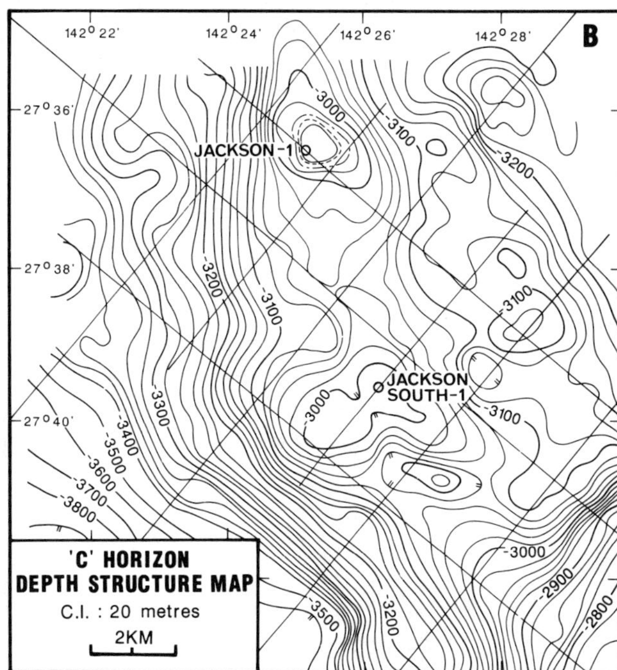

Figure 4. Jackson and Jackson South fields structure maps. (A) Pre-drill time structure map after M. Micenko in Moore (1981). (B) Post-drill depth structure map after M. Micenko and R. Halyburton in Robinson (1982).

STRUCTURE

Although the oil in Jackson field is reservoired in the Eromanga basin, it is most likely derived from the underlying Cooper basin. The following discussion therefore includes both basins. The location of the structural and stratigraphic cross sections to be discussed is shown in Figure 5, with a summary of discoveries and subcrop edges in the area.

Cooper Basin

In the middle to late Carboniferous, the Kanimblan orogeny affected much of eastern Australia. The event produced extensional block faulting along major structural lineaments in the Cooper basin basement, widespread subsidence, and the initiation of sedimentation in the Cooper basin. Extensional block faulting or rifting continued throughout the Early Permian culminating with the development of the Daralingie unconformity in the Kungurian to Ufimian. Deposition during the Early Permian pre- and synrift phases consisted mainly of fluvial with subordinate deltaic and lacustrine sediments. In the Jackson area Early Permian sedimentation was initially restricted to the Nappamerri trough in the south and the Karwin trough, a deep but narrow depocenter immediately southwest of Jackson (Figure 6). From these troughs the Early Permian sediments thin by onlap and interval thinning toward the stable cratonic platform in the east called the Thargominda shelf (Figure 6), and over basement horst blocks of the PNJ trend to the north.

By the late Early Permian, rifting had largely ceased in the Cooper basin. The Daralingie unconformity represents the final phase of tectonism and a brief hiatus during which erosion occurred on some uplifted blocks. At Jackson this is represented by the erosion of at most 300 ft (92 m) of section. In the Late Permian the Cooper basin entered a postrift sag phase of tectonism. Deposition of terrestrial sediments recommenced by sectional onlap of structural highs surrounding the slowly subsiding troughs, and by onlap at the basin margins. Late Permian sediments are therefore more widespread than those of the Early Permian. Early Triassic sediments conformably overlie the Late Permian sediments.

In the Middle Triassic, a northerly tilt of the Cooper basin took place that shifted the depocenter from the Nappamerri trough area in the southern Cooper basin to the Windorah trough of the northern Cooper basin (Figure 2). Terrestrial deposition continued following a short lacunae but was halted in the Middle to Late Triassic by renewed tectonism. The Cooper basin experienced major regional tilting to the northeast with extensive uplift and erosion along the southern basin margins and along the major structural trends. This led to breaching of the regional Triassic seal to the Cooper basin sediments (Figure 5), and, in some areas, to angular truncation of the underlying Permian sediments.

The presence of normal faults that intersect the Triassic sediments in some areas indicates the uplift was associated with extensional tectonics. Although wrenching has previously been suggested to explain the Late Triassic tectonism (Nelson, 1985), the

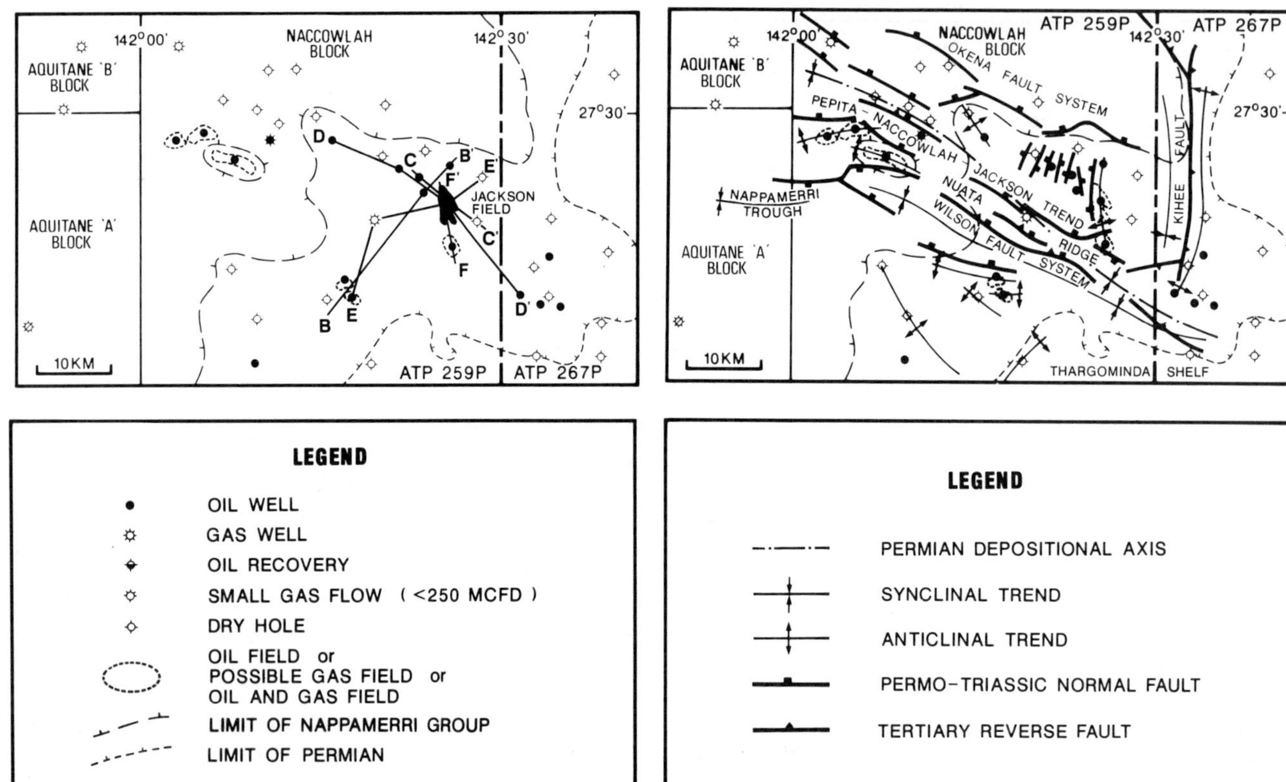

Figure 5. Location of wells, fields, and subcrop edges in the vicinity of Jackson field, and cross sections shown in Figures 7, 11, and 12.

Figure 6. Structural elements of the Jackson field and environs. The inverted Karwin trough is centered along the Nuata Ridge.

multidirectional fault pattern of the Jackson area (Figure 6) is not compatible with significant strike-slip (cf. Harding and Lowell, 1979; Harding, 1984) and a dominantly extensional interpretation is preferred.

Eromanga Basin

Sedimentation of the Eromanga basin sequence commenced in the Early Jurassic following epierogenic downwarping of the Australian craton. Deposition of this continental sequence occurred in a quiescent tectonic regime with only minor intermittent structural adjustments along major structural trends. Subtle syndepositional drape over paleohighs also occurred at this time as a result of sediment loading on and compaction of the thick Permian coal-bearing sequences.

Deposition of the Eromanga basin sequence continued through the Cretaceous until the onset of the Kosciuskan orogeny in eastern Australia. During this time up to three phases of tectonic activity have been suggested by Moore and Pitt (1984). Intermittent uplift imposed a westerly regional tilt to the Eromanga basin. Early tectonic activity is thought to have been extensional, which is supported by the continued development of normal faults above the existing Permian horst blocks, by the development of detached and keystone faults throughout the Cretaceous section, and by the preservation of a thick Tertiary section (up to 1000 ft or 305 m) only over the Cooper and Eromanga depocenters. From the Miocene onward the increasing intensity of the Kosciuskan orogeny caused crustal shortening and imposed compressional tectonics on the Eromanga basin. This resulted in the development of reverse faults that characteristically propagate to the surface indicating this to be the most recent tectonic influence. Late compression also caused structural inversion of the Karwin trough, the Permian depocenter to the southwest of Jackson, the only established case of structural inversion in the Cooper basin. Studies of earthquake epicenters in southeastern Australia also support the contention that the area is currently under the influence of a compressional tectonic regime.

Jackson Field

Following the middle Carboniferous Kanimblan orogeny, the Jackson structure was formed as one of several minor basement horst blocks between the major fault systems of the PNJ trend. Lower Permian sediments onlap and cover the Jackson horst block indicating its positive relief. Mild uplift in the late Early Permian enhanced structural closure on

Jackson, which has remained a positive feature since then. During subsequent tectonism the greatest deformation occurred along major bounding faults of the PNJ trend, namely the Okena fault to the north, the Wilson fault to the southwest, and the Kihee fault to the east (Figures 6 and 7). Deformation of the Jackson feature was slight, with minor faulting and folding over the basement horst (Figure 7).

During the Late Triassic tectonic event, the PNJ trend was uplifted. Triassic sediments were totally removed from the Jackson area as well as 100 ft or more of the underlying Late Permian rocks. Truncation of the Triassic Nappamerri Group around Jackson (Figure 5) and associated and subsequent reactivation of the Early Permian faults is postulated to have later facilitated migration of Permian sourced hydrocarbons to the overlying Eromanga basin reservoirs.

The area was structurally quiescent until the Tertiary when an early phase of the Kosciuskan orogeny reactivated many of the older extensional faults. This enhanced structural closure over many basement horst blocks, including Jackson. A Late Tertiary compressional phase of tectonism reactivated normal faults along the Wilson fault system as reverse faults leading to structural inversion of the Karwin trough southwest of Jackson (Figure 7A).

The Jackson structure at the eastern end of the inverted depocenter was simultaneously elevated to the highest point on the PNJ trend (Figure 7). Inversion of the trough, which contained a large volume of mature Permian sediments, biased all migration pathways to its northern margin. Any hydrocarbons spilled from existing traps in the "Karwin trough" by the process of structural inversion may therefore have migrated toward Jackson. Importantly, the bounding faults on Jackson itself were not reactivated during this later phase of Tertiary tectonism, thus leaving the thin Eromanga basin seals at Jackson structurally intact.

STRATIGRAPHY

The regional geology of the Cooper and Eromanga basins has previously been discussed by Thornton (1979) and Moore and Pitt (1984). Figure 8 shows a generalized stratigraphic column for the Cooper and Eromanga basins, showing the principal discovery and source horizons, lithology, and depositional environments. The distribution of discoveries by basin and their relationship to important subcrop edges previously discussed are shown in Figure 9. A schematic regional cross section over both basins is shown in Figure 10, and detailed local stratigraphic cross sections are shown in Figure 11.

Cooper Basin

The Cooper basin is a northeast-trending basin that contains over 6000 ft (2000 m) of fluvial and lacustrine sediments of late Carboniferous to Middle Triassic age. The basal units of the Cooper basin are the glaciofluvial Merrimelia Formation and Tirrawarra Sandstone (Williams and Wild, 1984). These basal units, which are absent in the Jackson area (Figure 11), are overlain by a succession of coaly, fluvial to deltaic units, including the important Patchawarra and Toolachee formations that host major gas and gas-liquids and minor oil reserves.

The thin but widespread lacustrine Murteree and Roseneath Shales that are contained in the Early Permian succession act as regional seals and have poor source potential. The Late Permian Toolachee Formation is conformably overlain by Late Permian to Early Triassic rocks of the Nappamerri Group, consisting of interbedded sandstones, siltstones, and shales with poor hydrocarbon source potential. Cooper basin deposition was terminated in the Late Triassic. Subcrop of the Permian sediments and the Nappamerri Group around the Jackson field and the location of Permian discoveries is shown in Figure 9.

Thickness changes in the Permian were controlled in large part by syndepositional faulting, which accounts for the horst and graben style of structuring apparent in the Cooper basin in Figure 11. This geometry has resulted in significant variation in source rock thickness and maturity and in reservoir thickness off structure. For instance, although the Patchawarra Formation at Jackson (Figure 11B) is fairly thin and marginally mature, a much thicker mature to overmature section, possibly a major source area for Jackson field, is present in the Karwin well to the southwest. Additional reservoir is commonly developed within closure on the flanks of the major structures, particularly in the Patchawarra Formation, owing to onlap.

Eromanga Basin

The Early Jurassic to Early Cretaceous Eromanga basin covers about 415,000 m^2 (1,100,000 km^2) of central Australia (Figures 1 and 2) and contains a succession of terrestrial and marine sediments up to 7500 ft (2500 m) thick. The Eromanga basin succession is comprised of a sand-prone lower part (Figure 10) with subordinate shaly units, overlain by a thick section of Cretaceous marine shales. The lower sand-prone part forms the major aquifers of the larger Great Artesian Basin, which includes the Eromanga, Surat, and Carpentaria basins. These aquifers are recharged by meteoric water in the east and flow in a general southwesterly direction (Habermehl, 1986) at an average rate of about 1 m per year. The finer-grained fluviatile and lacustrine sediments in the lower part provide the seals and source rocks for some of the oil accumulations in the Eromanga basin. The finer-grained sediments are generally silty rather than clayey and have limited sealing capacity as a result.

Deposition was initiated with meandering fluvial sandstones and shales of the Poolawanna Formation

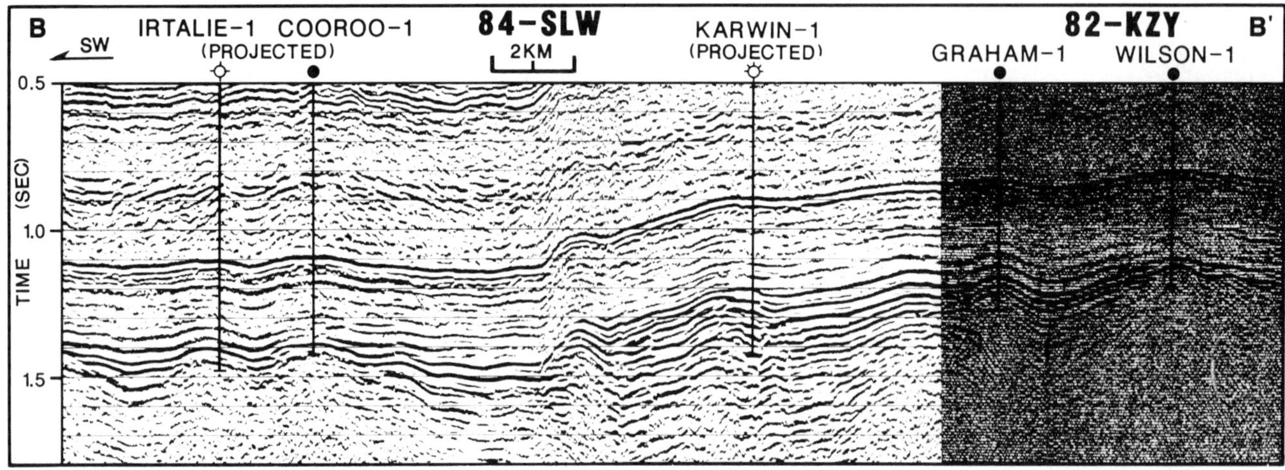

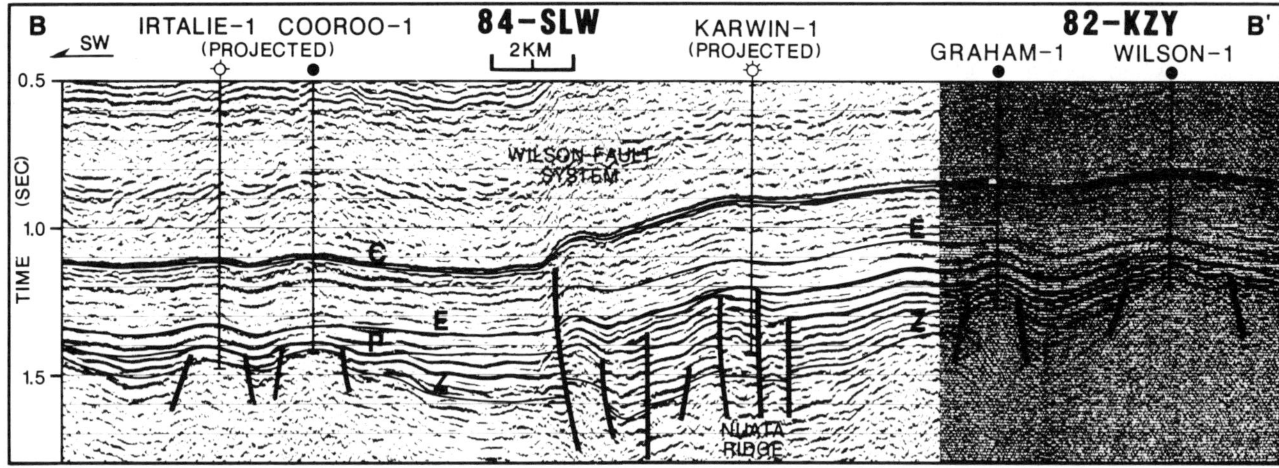

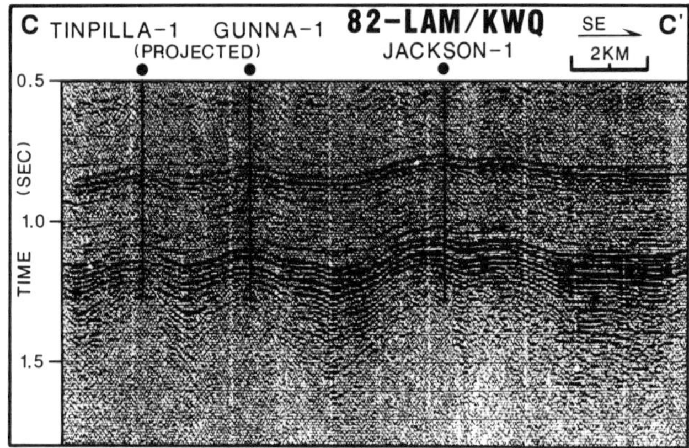

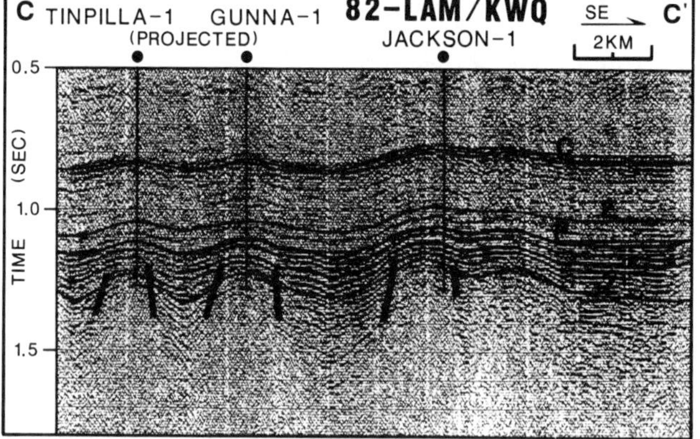

Figure 7. Seismic sections, uninterpreted and interpreted, across Jackson field. (A) B–B′ Irtalie to Wilson, showing structural inversion of the Wilson fault system and inversion of the Karwin trough. (B) C–C′ Tinpilla to Jackson. Locations are shown in Figure 5 and seismic codes are shown in Figure 8.

(or "basal Jurassic" unit), which hosts several small oil pools in the vicinity of Jackson. Three sandstone to shale sequences were then deposited: the Hutton Sandstone to Birkhead Formation, the Adori Sandstone to Westbourne Formation, and the Hooray Sandstone (previously the Namur Sandstone Member of the Mooga Formation) to Murta Member. The Hutton Sandstone contains major oil reserves in the Jackson area, and smaller reserves are present in the Westbourne Formation and Murta Member.

The Hutton, Adori, and Hooray sandstones were deposited in a very extensive braided fluvial environment of great lateral extent, while the finer-grained Birkhead Formation, Westbourne Formation, and Murta Member were deposited in dominantly meandering fluvial to shallow-marine environments.

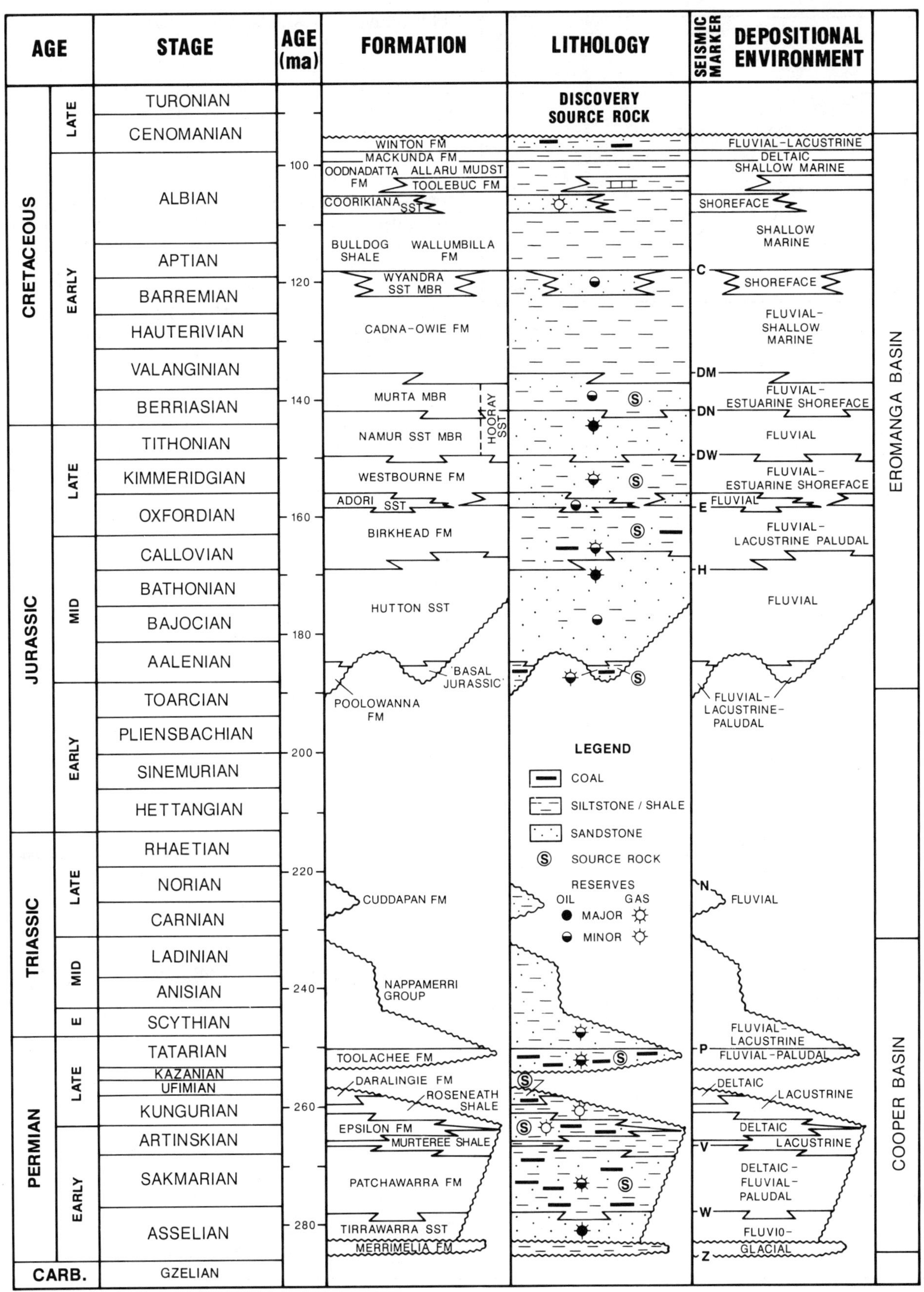

Figure 8. Generalized stratigraphic column showing discoveries, source rocks, lithologies, depositional environments, and seismic codes for the Cooper and Eromanga basins. FM, formation; SST, sandstone; MBR, member.

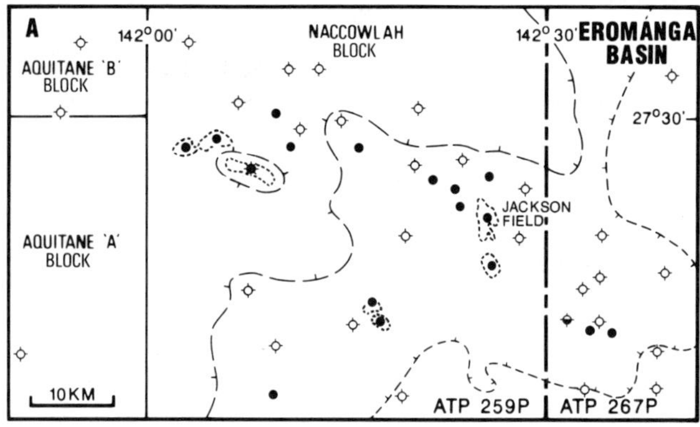

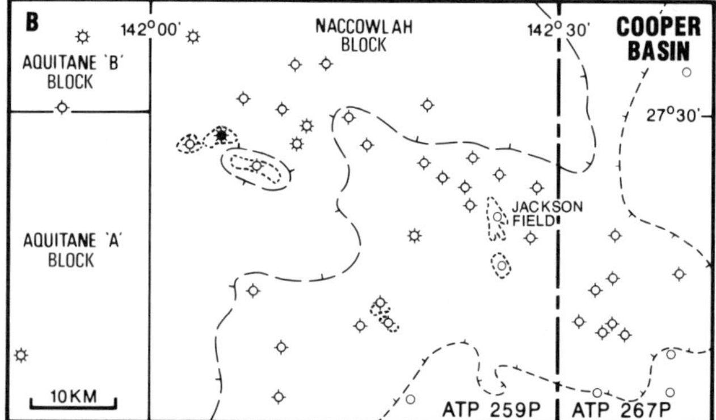

Figure 9. Discovery maps for the Eromanga and Cooper basins in the vicinity of Jackson field. Open circle shows absence or nonpenetration of formation.

The above lithostratigraphic variations are attributed to eustatic sea level variations (Exxon, 1976). The first evidence of marine sedimentation is found in the Westbourne Formation and increases progressively up the section. The overlying Cretaceous rocks were deposited rapidly in a predominantly restricted marine environment. Regression during the Late Cretaceous resulted in deposition of the dominantly terrestrial Winton Formation. Only minor Tertiary deposition took place over the Eromanga basin, which was largely an erosional domain at that time.

Reservoirs of the lower Eromanga basin succession are mineralogically mature, containing 70 to 80% quartz on average, with subordinate lithic and feldspar grains. Intervening finer-grained rocks in the Jurassic section and the marine Cretaceous rocks in general are composed of volcaniclastic material derived from the orogenic province of eastern Australia. The Eromanga basin hydrocarbon traps are generally only partly filled, mostly containing 20 to 40 ft (6 to 12 m) oil columns, irrespective of aerial trap size. Recovery is mainly by water drive.

Thickness and associated stratigraphic variation in the Eromanga basin is subtle (Figure 11) and is usually taken to reflect drape and compaction over the structures of the underlying Cooper basin.

TRAP

The Jackson field is primarily a structural trap with a subordinate, stratigraphic component, shown schematically in cross section in Figure 12. The pool

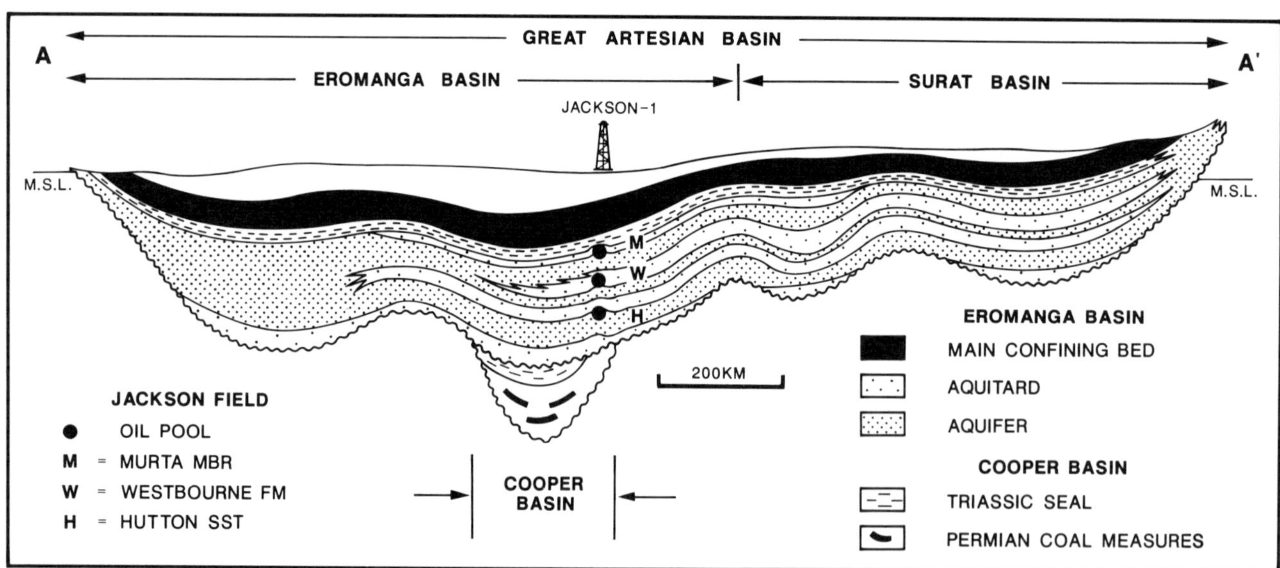

Figure 10. Generalized structural cross section showing the main Eromanga (Great Artesian) basin aquifer system and its relationship to the underlying Cooper basin, after Bowering (1982). Location of cross section A-A′ is shown in Figure 1. Published with the permission of the Australian Petroleum Exploration Association Limited.

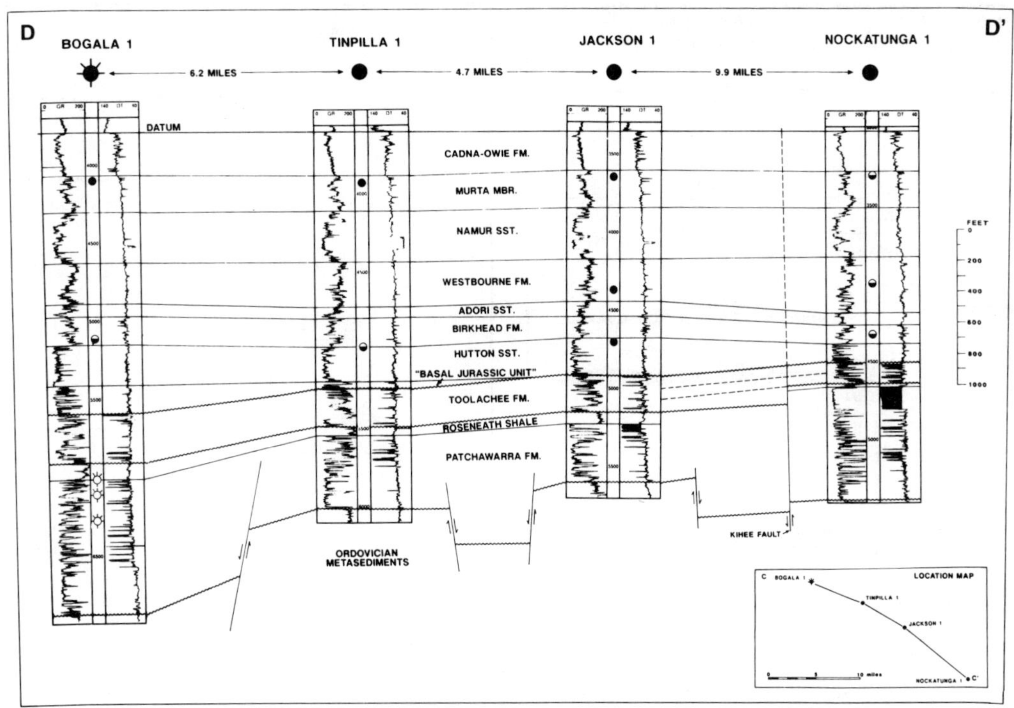

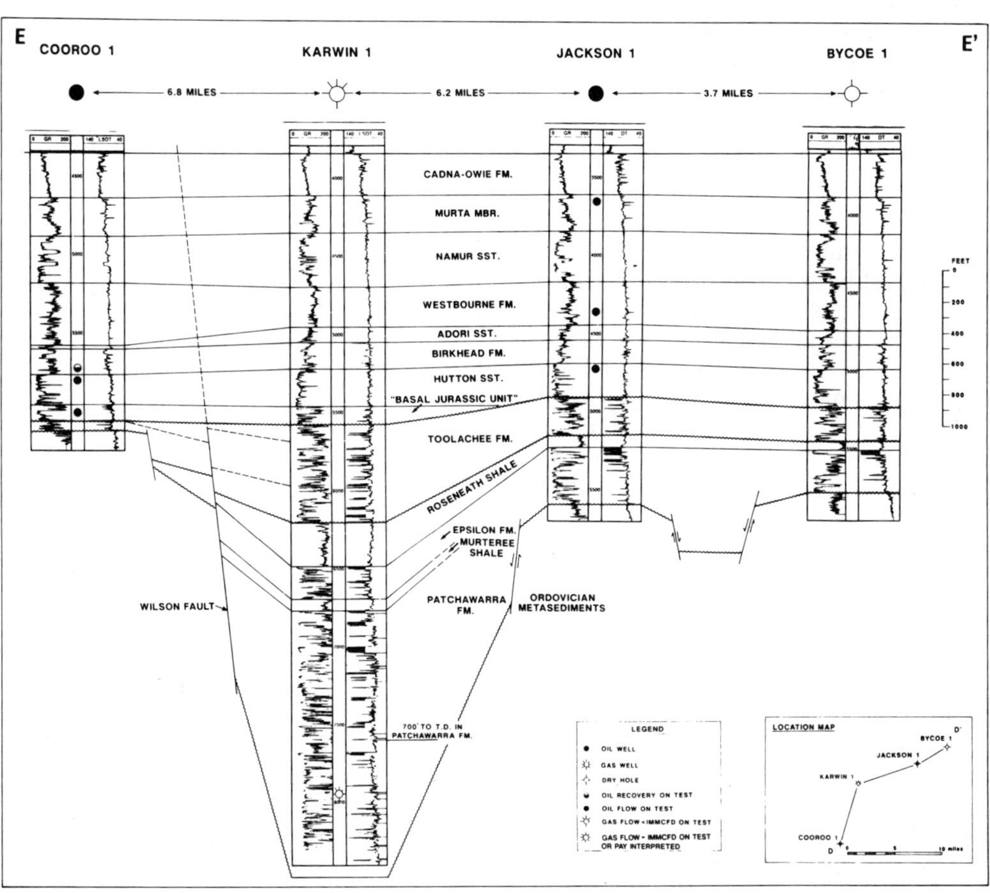

Figure 11. Stratigraphic geological sections over the Jackson field and environs. (A) D-D′ Bogala to Nockatunga. The Kihee Fault on D-D′ was initiated in the Triassic as a reverse fault, and reactivated in the Tertiary; and E-E′ Cooroo to Bycoe. (B) The Wilson fault on E-E′ was initiated in the Permian as a normal fault and reactivated in the Tertiary as a reverse fault. Locations of the cross sections D-D′ and E-E′ are shown in Figure 5.

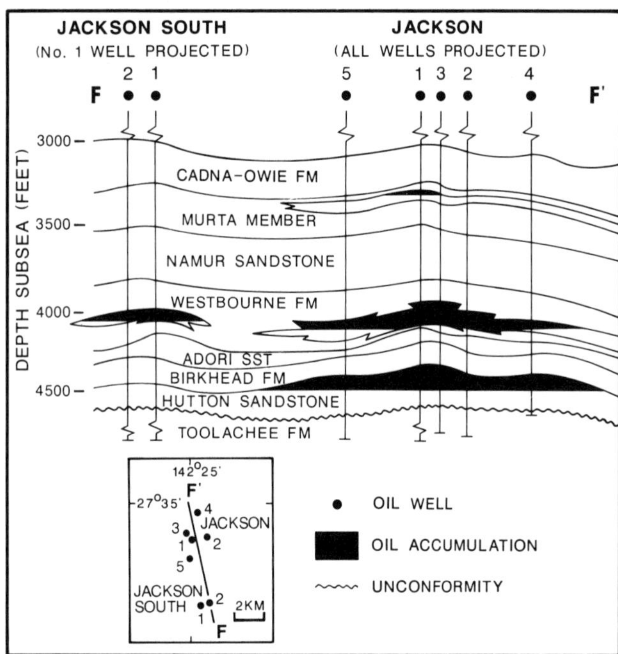

Figure 12. Generalized structural cross section F-F′ over the Jackson and Jackson South fields after Halyburton and Robertson (1984). Location of cross section F-F′ is shown in Figure 5. Published with the permission of the Australian Petroleum Exploration Association Limited.

configuration of the Hutton Sandstone is largely anticlinal with the Birkhead Formation providing top seal. However, localized sand development in the basal portion of the Birkhead Formation significantly enhances the structural closure in some areas. Figure 13 shows a depth map on the top of Hutton porosity; the oil pool is some 4 mi long by 2 mi wide (6.5 km by 3 km) with an areal closure of about 2500 ac (4 mi^2 or 1012 ha) and a maximum vertical relief of approximately 135 ft (40 m). Structural dips on its flanks range from 1.5° to 4.5°. A significant stratigraphic component of trapping is evident in the overlying Westbourne and Murta reservoirs, which contain several stratigraphically isolated pay sandstones.

Although local top seal is provided by silty to shaly lithologies within the Birkhead, Westbourne, and Murta, the same formations all appear to be allowing vertical migration of hydrocarbons. The evidence for this was at first circumstantially based on the presence of multiple pools over mature Permian subcrop. The geochemical similarity of some of the stacked pools, for example the Hutton and Westbourne oils in the Jackson field, which will both be discussed later, also supports the concept. Because the same formations are involved as seal and migration pathway, lateral continuity of the predominantly fluvial facies is probably poor.

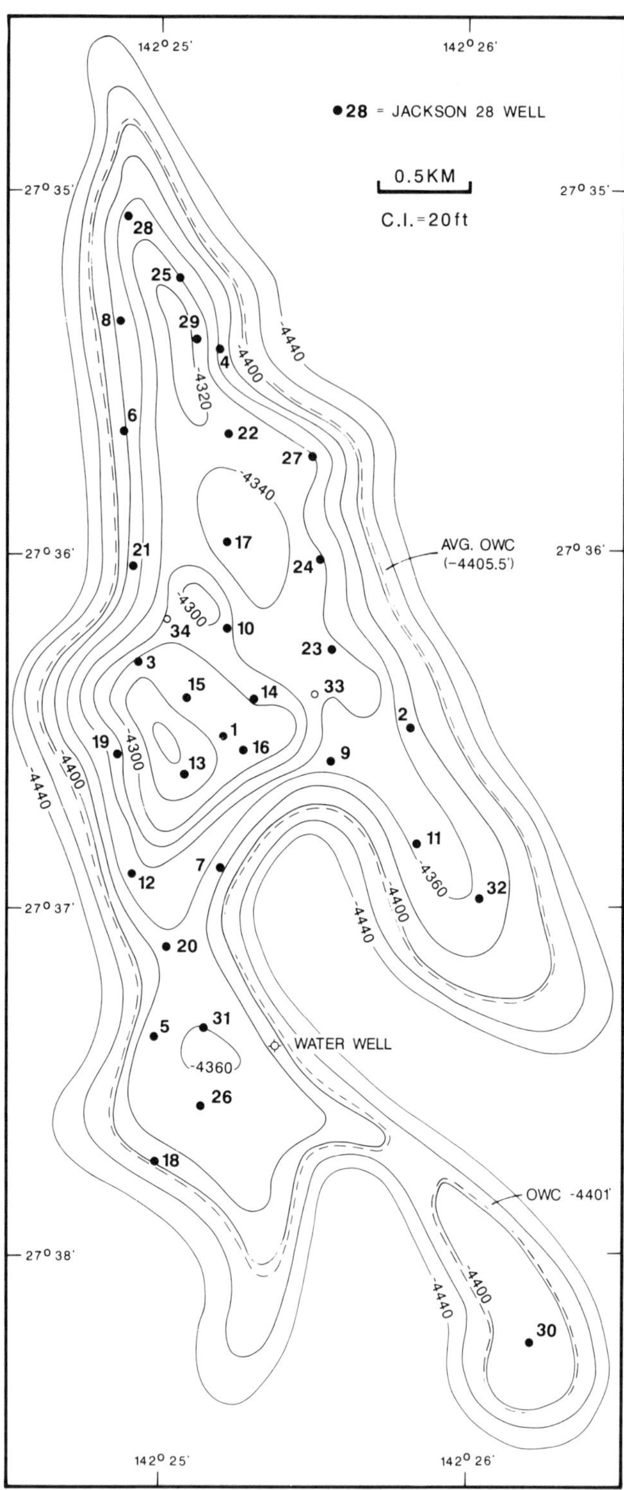

Figure 13. Depth structure map on top of Hutton porosity for the Jackson field. Contour interval, 20 ft. After R. Donaleshen and C. Grasso in Holton (1988).

RESERVOIR

The reservoirs of Jackson field have previously been discussed by Halyburton and Robertson (1984), including modal petrographic analyses for the Murta, Westbourne, and Hutton reservoirs and clay mineralogy for selected samples. Log response over the Hutton, Westbourne, and Murta reservoirs in Jackson 1 is shown in Figure 14.

The reservoirs consist mainly of moderately sorted, very fine- to medium-grained quartzose sandstones, with minor detrital feldspar, mica, and lithics. Clay minerals are common, particularly in the finer-grained rocks. Authigenic kaolinite, illite-smectite, and smectite are the main clays present. Minor amounts of dolomite and siderite occur as cement and grain replacements.

Loss of primary porosity in the reservoirs is mainly controlled by depth of burial, and quartz overgrowths are the commonest cause of porosity reduction. Loss of porosity is greatest in the finer-grained sandstones, and some secondary porosity has been created in the coarser sandstones by dissolution of feldspars and carbonate cement. The regional pattern of porosity variation in the top of the Hutton sandstone in the vicinity of Jackson field is shown in Figure 15. Decreasing porosity to the north, west, and south coincides generally with increasing burial depth.

Hutton Sandstone

The Hutton Sandstone consists of fine- to coarse-grained light brown quartzose sandstones, with minor lithic grains and feldspar. The matrix is mainly kaolinite. The sandstone was deposited in a braided fluvial environment, shows a blocky gamma signature (Figure 14), and has excellent reservoir characteristics. The upper portion of the Hutton Sandstone is transitional with the basal section meandering fluvial facies of the Birkhead Formation, resulting in some stratigraphic variation in the top of porosity. This is illustrated in the stratigraphic cross section in Figure 16. The pay zone reaches a maximum thickness of about 110 ft (34 m). Reserves in the reservoir have been estimated from hydrocarbon pore volume maps (pay thickness × porosity × oil saturation) instead of the more commonly used isopachs of pay, because of lateral variation of porosity and water saturation.

The Hutton Sandstone is very clean except for the shaly section in the top as shown by the gamma ray response in Figure 14. Details of drilling and log response and reservoir facies are shown in Figure 17. Rate of penetration decreases and quality deteriorates markedly above 4755 ft (1450 m) (Figure 17A and 17B) owing to the presence of silty laminae and siderite cement (Figure 17C). Below 4765 ft (1453 m) rate of penetration increases and reservoir quality improves in the cleaner braided fluvial facies (Figure 17D). Sonic derived porosities range from 17 to 24% and calculated water saturations from 10 to 42% (Halyburton and Robertson, 1984). Core derived porosities generally range from 18 to 22%, permeabilities from 100 to 3000 md (Figure 18A), and water saturations from 31 to 54%. Average relative permeability and water saturation are shown in Figure 17B, and residual oil saturation and permeability in Figure 17C.

Prior to 1988 (pre-Jackson 31), reserves in the Hutton reservoir were calculated using a cementation exponent, "m," of 1.89 and a saturation exponent, "n," of 2.21 (at overburden pressure). These values were based on detailed log analysis and special core analysis, assuming that salinity of connate water in the reservoir was the same as that in the underlying aquifer—about 2000 ppm NaCl equivalent—giving an Rw of 2.6 @ 75°F for the Hutton. However, capillary pressure analysis of cores from the field (Figure 18D) confirmed that connate water salinities were approximately 7000 to 10,000 ppm NaCl equivalent (Neumann, 1987) and that "m" and "n" values for log interpretation of the clean sandstones should be 2.0 and 2.0, respectively (cf. Gravestock and Alexander, 1988). This resulted in a significant upward revision of reserves for the field. The reason for the salinity variation is that the oil was emplaced prior to the influx of fresh artesian water.

Cutoffs used to calculate effective net oil pay in the Hutton are porosity greater than or equal to 12%, permeability greater than or equal to 10.0 md, shale volume less than or equal to 45%, and water saturation less than or equal to 70%. The very low mobility of water, at approximately 0.8 relative to oil at reservoir conditions, creates a favorable environment for oil displacement (Figure 18B).The residual oil saturation after water flooding is about 30% (Figure 18C). The quality of the reservoir is also reflected in the capillary pressure curve in Figure 18D. Oil in place is about 94 million barrels of which 38 million barrels is estimated to be recoverable.

Westbourne Formation

The Westbourne Formation reservoirs are comprised of very fine to fine-grained sandstones in the upper portion of the formation. These are generally well cemented and exhibit tight to poor porosity. The middle to lower part of the formation contains poor to moderately well sorted coarse-grained sandstone with weak siliceous cement and little argillaceous matrix. Porosity in this zone is generally fair but rarely good. The sandstones range in thickness from 1 to 20 ft (0.3 to 6.1 m) and most appear to have limited lateral continuity across the field. Compositionally these are quartzose with occasional carbonaceous and highly argillaceous partings.

Log response over the Westbourne reservoir is shown in Figure 14. Because of the thin-bedded nature of some of the sandstones and their feldspar and mica content, gamma log response is often anomalous. Since the Eromanga basin connate waters are extremely fresh (about 1250 ppm chloride, 2400 ppm bicarbonate), detection of hydrocarbons by conventional methods is difficult.

The Westbourne Formation contains both low-energy, meandering fluvial to deltaic and offshore

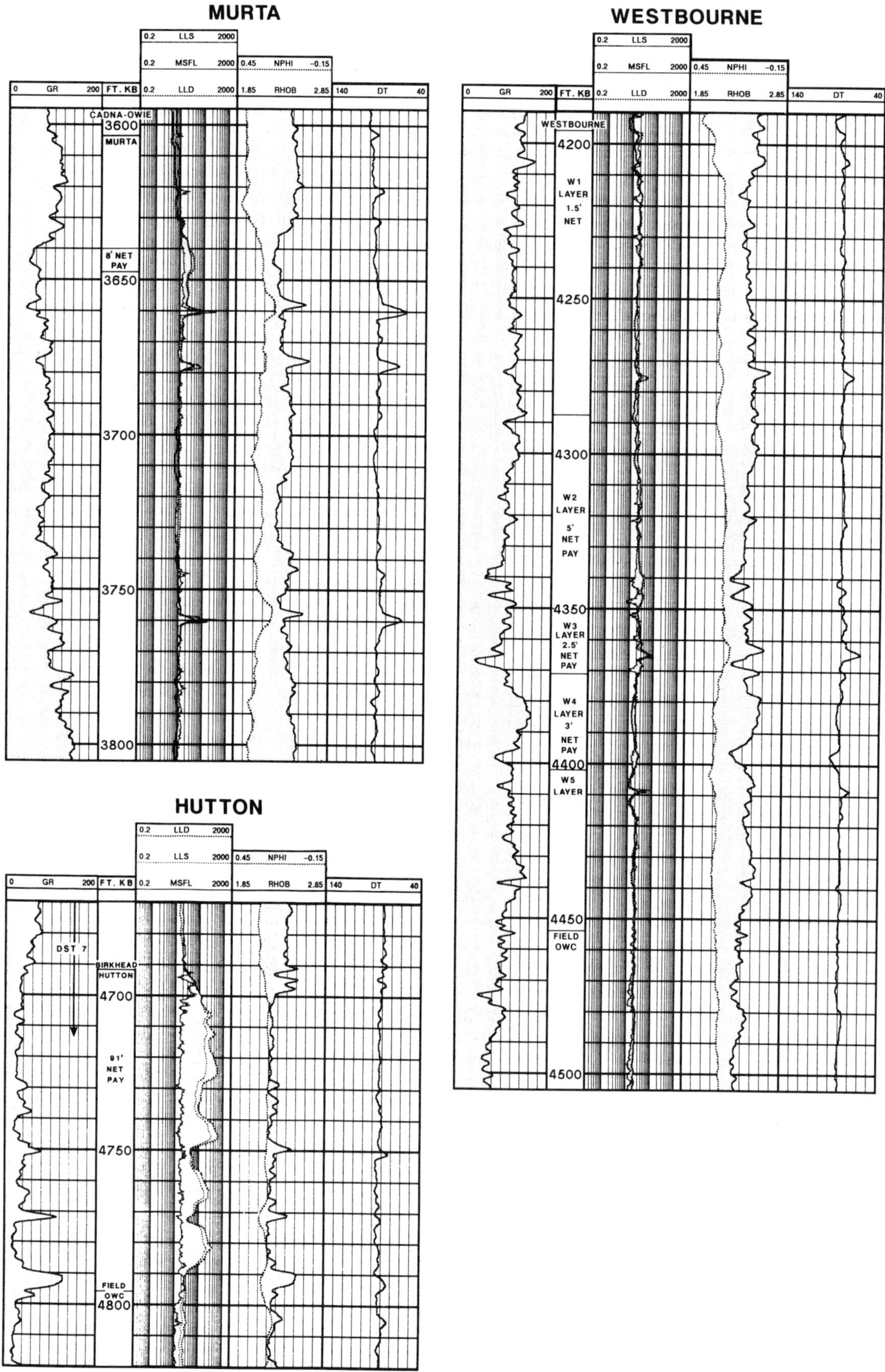

Figure 14. Log response over the Murta, Westbourne, and Hutton sands in Jackson 1.

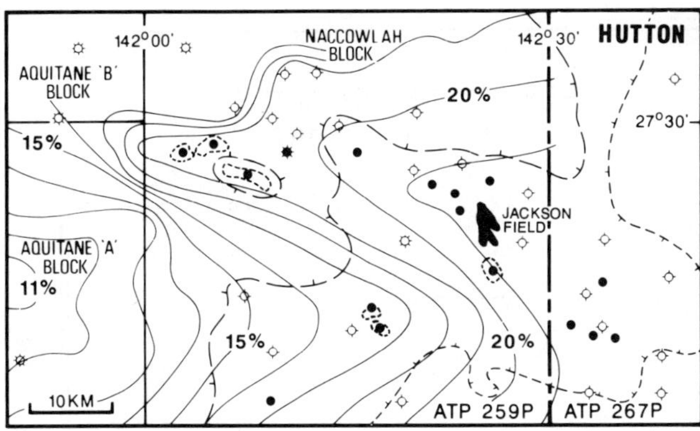

Figure 15. Isoporosity map for the top of Hutton Sandstone in Jackson field.

marine sandstones. The point bar sandstones are confined mainly to the lower part of the formation. The Westbourne Formation reservoirs have been mapped using a resistivity anomaly in the upper part, for structural control. The formation is divided into five stratigraphically defined layers (Figure 19) for the purpose of mapping and reserves calculation (Donaleshen, 1987), and several independent oil-water contacts are recognized. Most reserves in the upper part occur in the upper channel sand (layer 2), which is probably continuous in its area of development. Most reserves in the lower part are found in the lower channel sand (layer 4), although many other thin interbedded pay sandstones are present in the section.

Because of the thinly bedded nature of many of the Westbourne Formation pay sandstones, it has been difficult to establish their petrophysical and production characteristics. Most of the thinner interbedded sandstones are beyond the resolution of conventional porosity tools. Pay is determined from a combination of DST information and production data in conjunction with the spherically focused micro log (MSFL) tool to determine water saturation of the thin interbedded sandstones. In Jackson 31, drilling and gamma response (Figure 20A) only partly correspond. The vertical distribution of reservoir

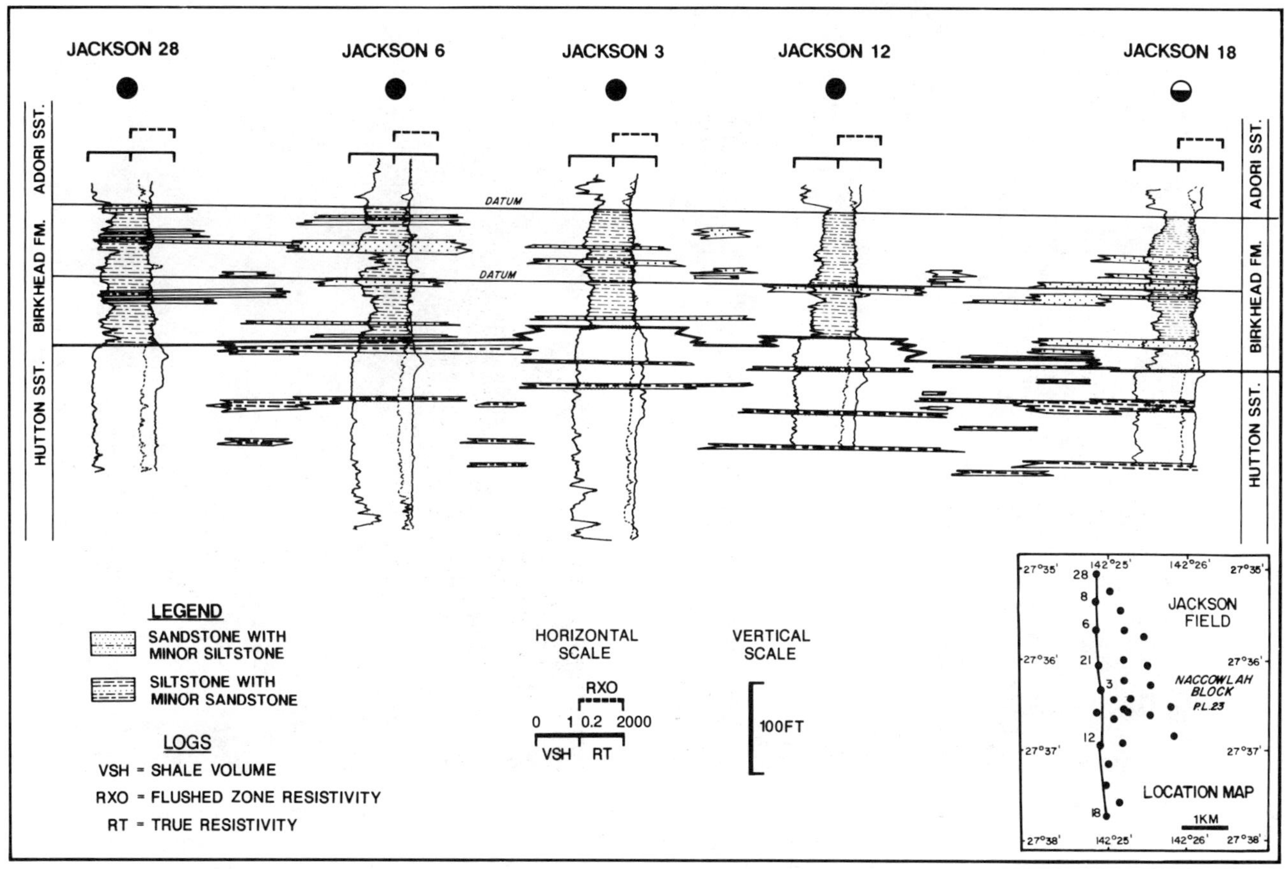

Figure 16. Stratigraphic cross section over Jackson field showing the Hutton to Birkhead transition.

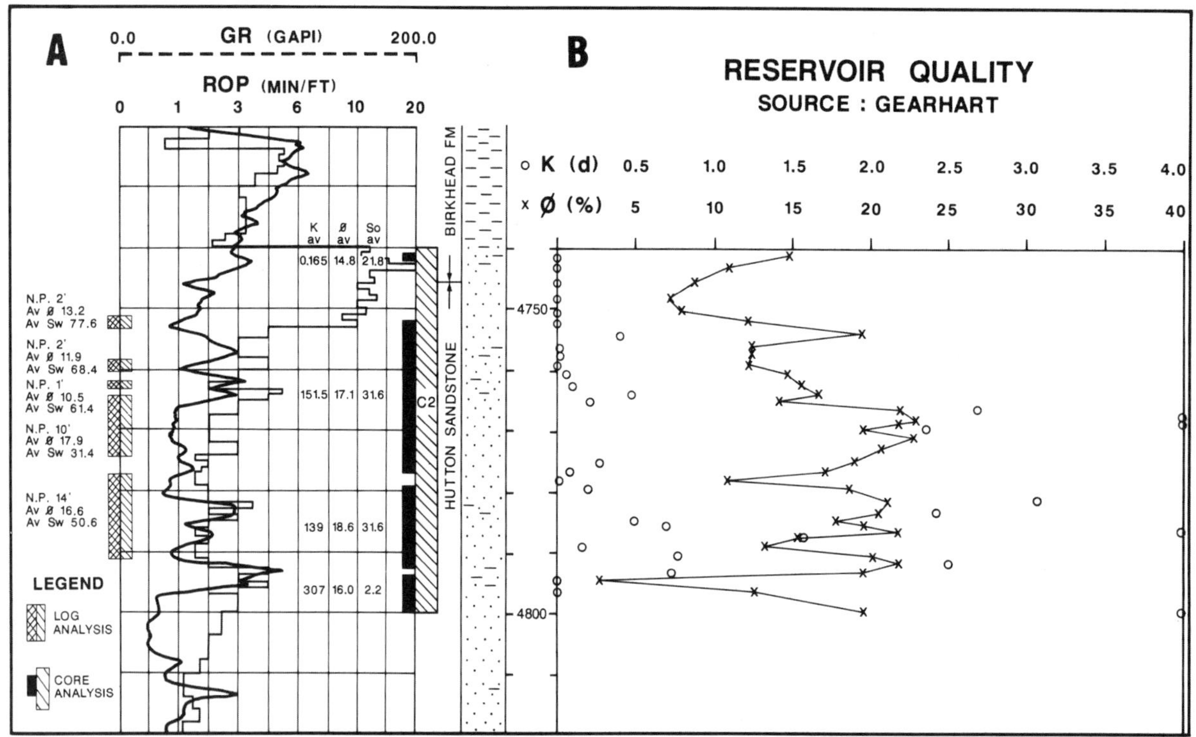

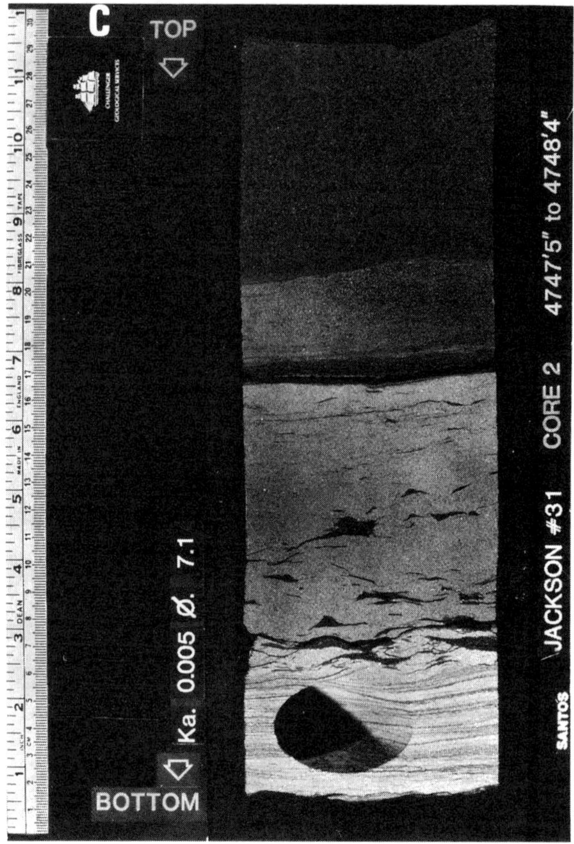

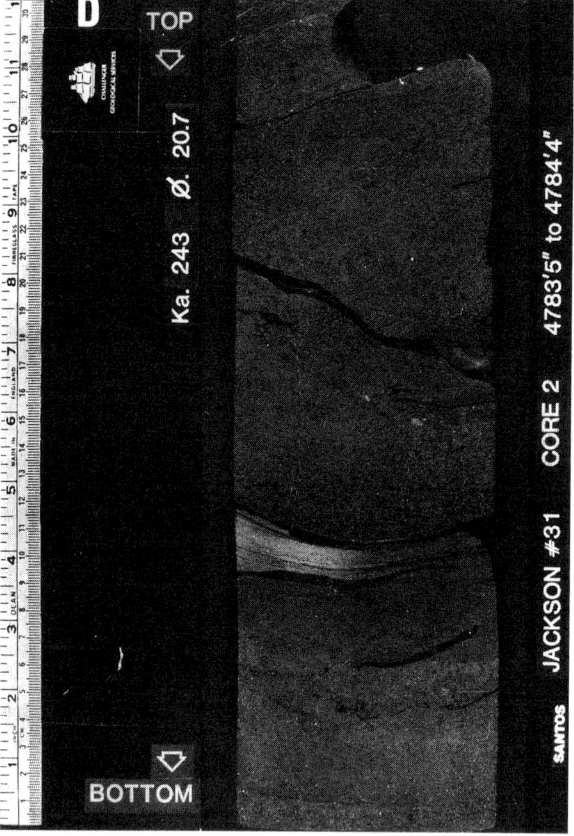

Figure 17. Reservoir properties for the Hutton Sandstone in Jackson 31. (A) Log and drilling response (ROP, rate of penetration). (B) Porosity and permeability profile based on core samples. (C) Meandering fluvial facies comprising tight laminated sandstone (lower quarter), overlain by productive sandstone with rip-up clasts, and tight siderite cemented sandstone in the upper half. (D) Braided fluvial facies, comprising productive medium- to coarse-grained sandstone with isolated clay drapes and ripple laminae.

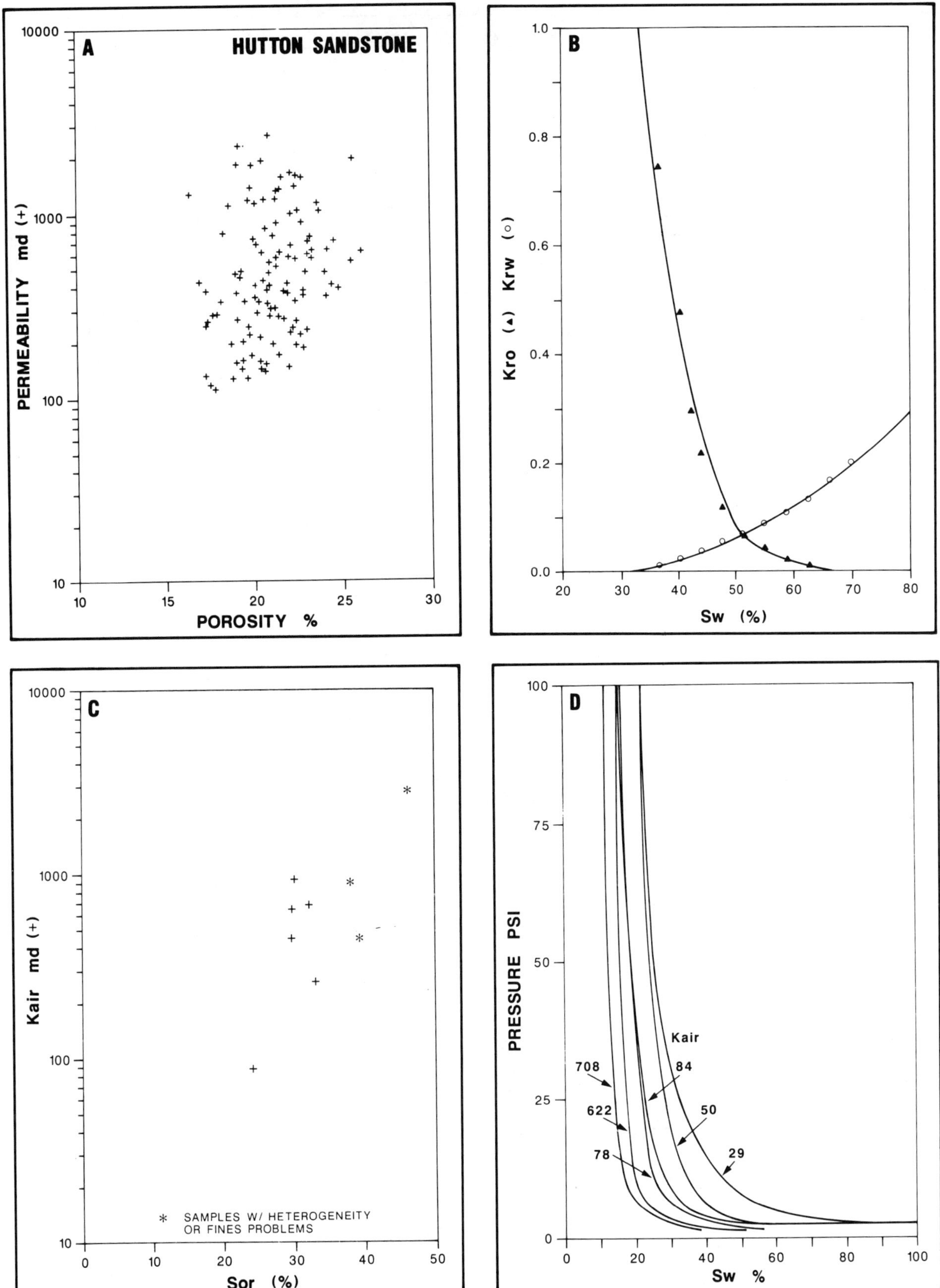

Figure 18. Reservoir quality parameters for the Hutton Sandstone in the Jackson field. (A) Porosity permeability relationship based on core data. (B) Average relative permeability and water saturation. (C) Residual oil saturation and permeability. (D) Capillary pressure curves. B, C, and D are based on samples from Jackson 2, 19, and 28.

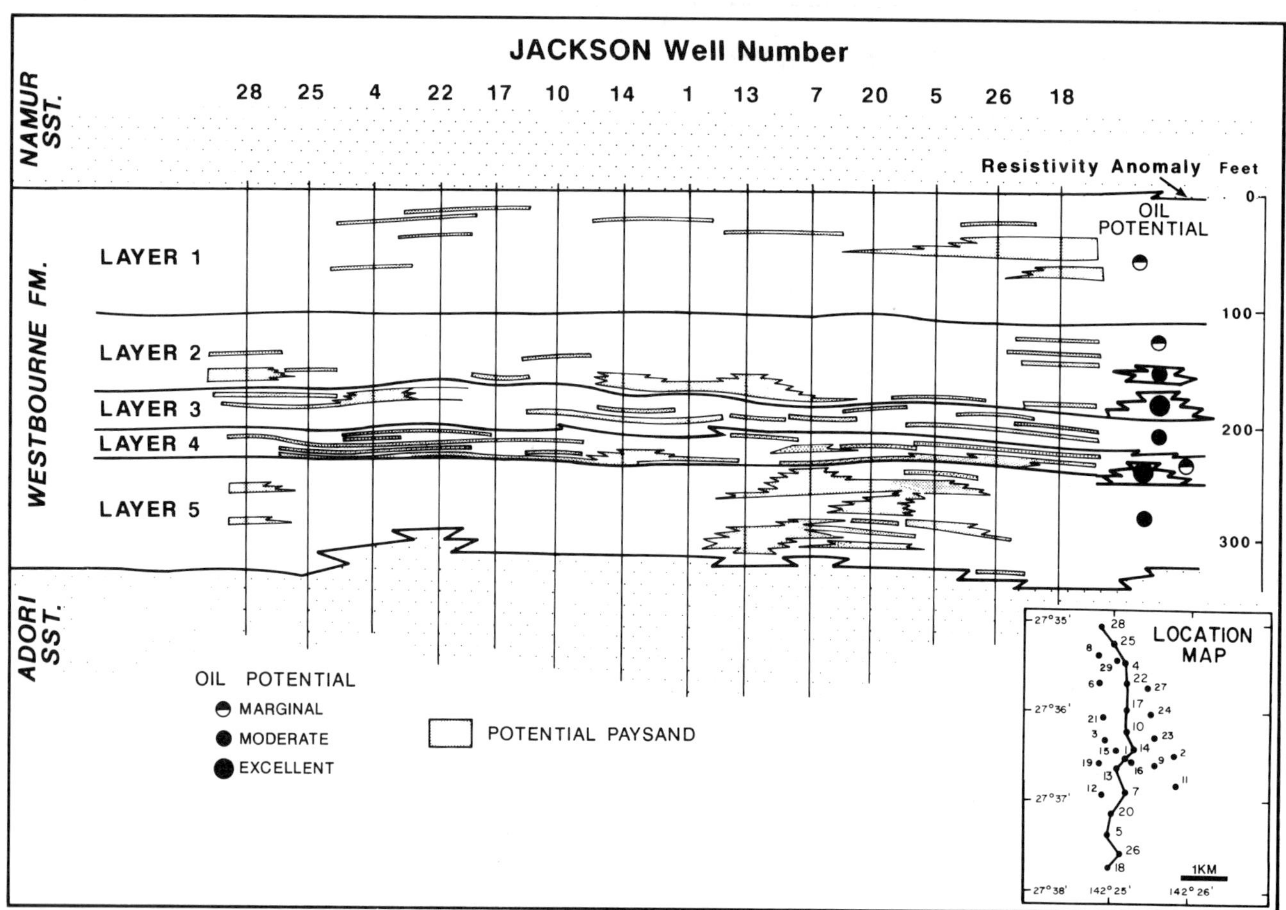

Figure 19. Stratigraphic cross section over Jackson field showing reservoir development in the Westbourne Formation. Approximate ultimate reserves are: layer 1, very little; layer 2, 250,000 bbl primary, 1,250,000 bbl with water flood; layer 3, 750,000 bbl primary; layer 4, 1,000,000 bbl primary; and layer 5, <100,000 bbl primary.

porosity and permeability is highly variable (Figure 20B) owing to the thinly bedded character of the coarser producing sandstones (Figures 20C and 20D).

Thus reservoir sandstone development varies considerably with resultant widespread variation in reservoir quality. Pay sands, where present, show log-derived porosities of 12 to 15% (Halyburton and Robertson, 1984) and log-derived water saturations of 50 to 80%. General criteria used to define pay are porosity greater than or equal to 12%, permeability greater than or equal to 2–5 md, and water saturation less than or equal to 68%. However, coring is necessary to accurately determine reservoir properties in each well. Oil in place in the Westbourne Formation is about 13 million barrels contained mainly in the upper and lower channel sands. Estimated ultimate recoverable oil is about 3 million barrels.

Murta Member

The Murta Member consists of interbedded sandstone and siltstone with minor claystone. The sandstone is quartzose, fine- to very fine grained, and poorly to moderately well sorted. Potash feldspar and minor mica are present as framework grains, and authigenic clay matrix is an important constituent of the finer-grained sandstones. The Murta reservoir is also difficult to evaluate on logs owing to the presence of radiometrically "hot" sandstones and the thinly interbedded nature of the sandstones (Figure 14).

The Murta Member contains sandstone deposited in restricted marine, shoreface, deltaic, and fluvial environments, with a resultant large variation in reservoir quality and continuity. Because the Murta contains only small reserves of recoverable oil, it is not discussed in detail. Oil in place is about 2 million barrels, of which about 0.7 million barrels are estimated to be recoverable.

PRODUCTION

The Jackson wells were drilled with KCl-polymer mud and cased, usually with 7-in. casing. The Hutton

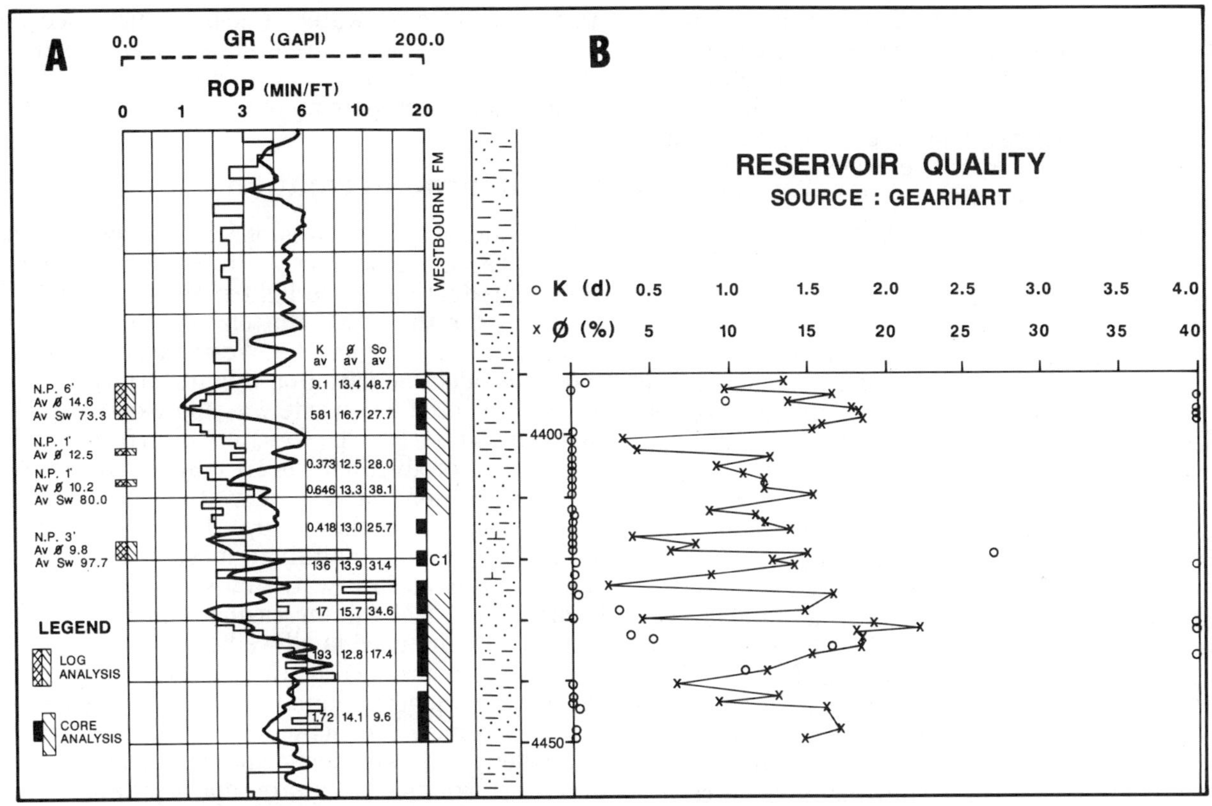

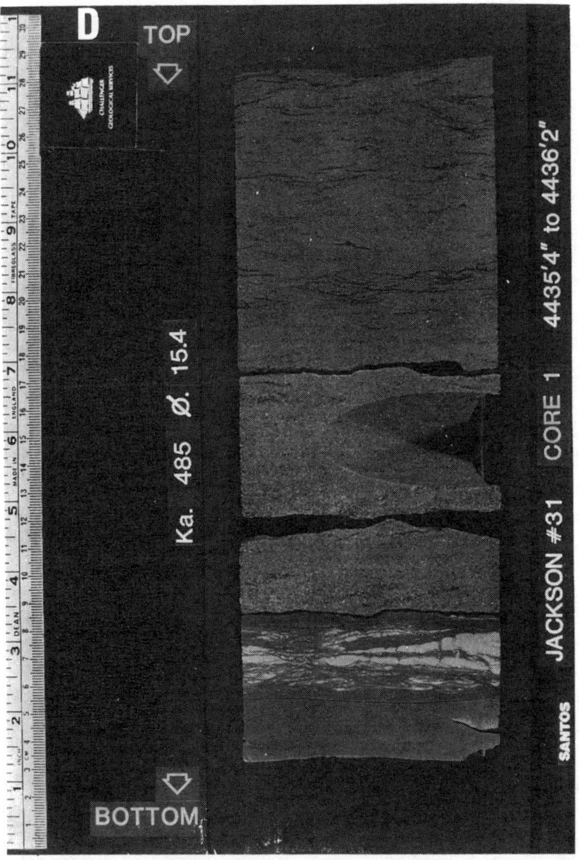

Figure 20. Reservoir properties of the Westbourne Formation in Jackson 31. (A) Log and drilling response (ROP, rate of penetration). (B) Porosity and permeability profile based on core samples. (C) Thinly bedded oil-saturated sandstone (upper middle) from layer 3. (D) Base of oil saturated sandstone in layer 2.

and Murta wells are perforated underbalanced, using KCl brine as a completion fluid, with a perforation density of 4 spf with either 2½-in. through tubing guns or 4-in. casing guns. Twelve spf have been shown to be successful for the thin sands of the Westbourne Formation. The wells are flowed to recover all completion fluid, then tested and placed on production.

Hutton Sandstone

The initial pressure of the Hutton reservoir, estimated from the number 1 and 2 wells, was about 2215 psig (152.8 bar) at 4350 ft (676 m) subsea (4730 ft [1443 m] KB average). The fluid offtake rate (i.e., rate of fluid production) per well was initially based on a maximum drawdown of 250 psi (17.25 bar) or a maximum oil rate of 1500 bbl oil per day. These limits were set to provide the desired oil rate with the available wells, given production facility constraints at the time, and to provide a stable basis for evaluation of reservoir and well performance, thereby assisting preparation of further field development plans. With ongoing production and increasing water cut owing to reservoir pressure drawdown, choke sizes were increased to maintain the target oil rate. When the flowing tubing pressure of a well reached the system pressure of 45 psi (3.1 bar) and the well was no longer meeting its assigned offtake rate, a beam pumping unit was installed. To 1 November 1985, production from the Hutton reservoir was 7.25 million barrels of oil and 1.9 million barrels of water with negligible gas.

The reservoir has strong bottom and edge water drive. Water cut from the Hutton reservoir is due to water coning from the underlying aquifer and channeling from high permeability streaks. Shale barriers within the reservoir are of limited extent, and these delay the water coning effect. Computer simulation studies indicated that the coning is not rate sensitive and will not have a detrimental effect on ultimate recovery. Consequently, production rates were increased from their initial low level. At the end of 1987, 18.5 million barrels of oil and 17.8 million barrels of water had been produced from the reservoir. The average production rate was 12,800 BOPD and 37,200 BWPD (50,000 bbl fluid/day) from 27 completions (Figure 21). The local peaks in production (Figure 21) correspond to times when new wells or pumps were brought online.

Average field reservoir pressure initially declined with production, then stabilized at approximately 1800 psi (124.2 bar) at 50,000 bbl fluid/day. The Hutton aquifer is extensive and continuous throughout the Eromanga basin and provides an essentially infinite pressure source. The aquifer is approximately 250 psi (17.2 bar) overpressured relative to a sea level datum, because the recharge zone for the aquifer is above sea level.

Electric submersible and beam pumps have been installed progressively to maintain desired fluid production rate as water cuts have increased and as field reservoir pressure declined. Well productivities range from 0.1 to 10.0 BOPD/PSI drawdown; the average is approximately 2 BOPD/PSI.

Westbourne Formation

Development of the Westbourne has been secondary to the more productive Hutton Reservoir and also to production of the lighter Murta oil (used as a diluent for the waxy Hutton oil). Consequently, production rates from the Westbourne reservoirs show considerable variation (Figure 21). At the end of 1987, the production rate was approximately 1000 BOPD plus 1700 BWPD (water cut 63%) from 8 wells. Cumulative production was 1.1 million barrels of oil and 0.6 million barrels of water. The original completions were flowed up the Hutton tubing production casing annulus; most declined very quickly even at low water cuts, owing to water loading. To offset the increasing water cuts, beam pumps were installed in all wells.

The Westbourne Formation hosts a complex set of generally noncommunicating multilayered reservoir sandstones. Production from the sandstones is comingled in the well bores, and individual reservoir unit performance data are not available. Most sandstones have strong edge aquifer pressure support; a few produce by oil expansion drive. Recoveries are expected to vary from 5% to 25% of the initial oil in place in the different sandstone bodies.

Murta Member

Production from the Murta reservoir was accelerated early in the field life because the lighter Murta oil is a good diluent for the waxy, high pour point Hutton oil. The initial oil offtake rate was approximately 500 BOPD from two wells, and this has been maintained as a total fluid rate to date. At the end of 1987 the rate was 160 BOPD and 380 BWPD (water cut 71%) (Figure 21), for a cumulative production of 0.4 million barrels of oil and 0.3 million barrels of water.

Little information is available on the drive mechanism of the reservoir. Most wells exhibit good pressure support that is thought to be provided by edge water drive. Production rates have been maintained by the addition of beam pumping equipment.

SOURCE ROCKS

Since the discovery of oil in the Eromanga basin, there has been disagreement over the likely source of the oil, some workers proposing the thin intraformational shales of the Eromanga basin, and others the abundant shales of the underlying Cooper basin. The liquids potential of the ubiquitous Permian coals

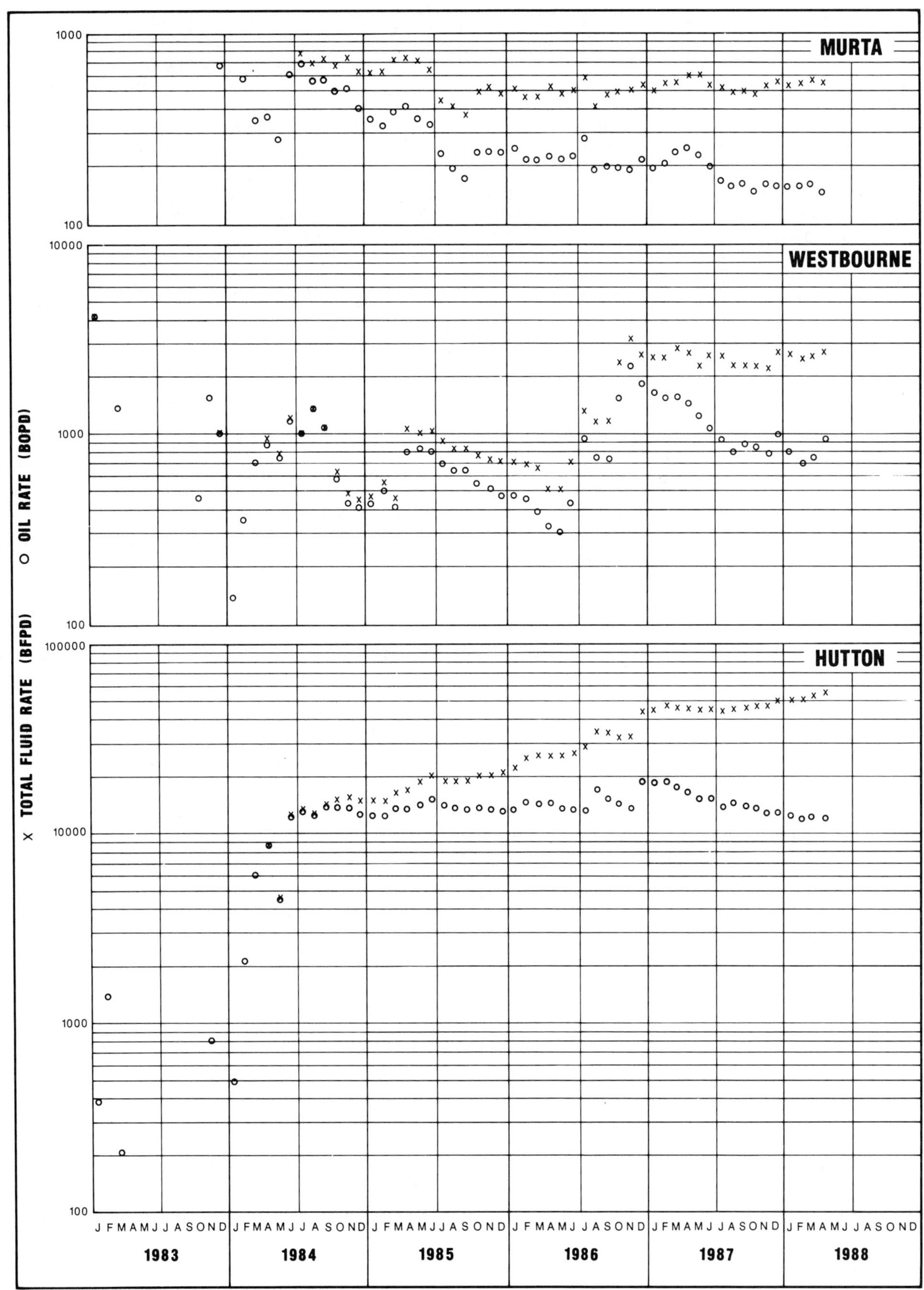

Figure 21. Production history for reservoirs in the Jackson field.

is usually assumed to be minor, although this view has recently been challenged (e.g. Smyth, 1983; Cook and Struckmeyer, 1986; Taylor et al., 1988).

The cause of the debate has been that although Eromanga basin source rocks are oil prone and Cooper source rocks gas prone, general geological consideration (migration pathways and play geometry) suggested a Permian source for many fields. The characteristics of Cooper and Eromanga basin source rocks have been previously discussed by Kantsler et al. (1983), Smyth et al. (1984), Vincent et al. (1985), Taylor et al. (1988), Powell et al. (1989), and Michaelsen and McKirdy (1989). The source rocks have been characterized by their total organic carbon (TOC) content, Rock-Eval pyrolysis, and dispersed organic matter (DOM) descriptions, obtained using reflected and transmitted light techniques, mainly on cuttings samples. More recently Alexander et al. (1988), Powell et al. (1989), and Jenkins (1989) have discussed aspects of the aromatic and naphthenic biomarker geochemistry of the source rocks.

The above studies have shown that the Cooper and Eromanga basin source rocks in general contain sufficient TOC to qualify as good potential source rocks. The composition of DOM in each formation is generally distinctive, albeit with much in common between the units, because the DOM is generally dominated by higher plant material.

Cooper Basin

Cooper basin hydrocarbons are assumed to originate from abundant organic matter (3 to 6% TOC by weight) in gas-prone fluvial shales and coals of the Toolachee and Patchawarra formations. The organic matter is dominated by vitrinite and inertinite (type III kerogen of Tissot and Welte, 1978), with minor amounts of sporinite and cutinite (<10% by volume) derived from a terrestrial flora. Local concentrations of exinite and bacterially biodegraded organic matter with significant oil generation potential have been proposed as the source of oil found in Permian reservoirs (Kantsler et al., 1983; Taylor et al., 1988).

The chemical composition of selected Permian source rocks based on core samples is shown in Figure 22. The majority of samples (47) consist of mixed type II/III material. The eight samples enclosed by the dashed line in the upper part of the diagram are coals, and their high HIs are an artifact of high TOC contents (McKirdy et al., 1985). The six samples enclosed by the dashed line in the lower part of the diagram are from the widespread but thin Permian lacustrine shales (Murteree and Roseneath shales) and show little potential to produce hydrocarbons.

Eromanga Basin

Organic matter in the Eromanga source rocks is also derived principally from a terrestrial flora. However, evolution of the plants from Permian to Jurassic resulted in a generally oil-prone source type in the Eromanga basin. Exinite, particularly resinite, is commonly a major component of DOM in the Jurassic rocks, inertinite is only minor, and vitrinite is often resinous (suberinite). Sporinite is the dominant exinite in the Murta Member. This results in an overall hydrogen-prone character and better apparent liquid hydrocarbon potential than for the Permian.

Potential source rocks in the Eromanga basin succession are the fluvial and lacustrine shales (and minor coals) of the Poolawanna, Birkhead, and Westbourne formations and the Murta Member. The source potential of the Cretaceous marine shales is poor. These shales are immature and contain less than 1.5% TOC on average, comprising mainly oxidized higher plant material (inertinite—type III).

The best source potential is found in the Jurassic Birkhead Formation and the Cretaceous Murta Member. The Westbourne Formation has a low TOC content, which also consists mainly of oxidized type III material, while the Poolawanna is generally thin and discontinuous (although the most mature).

The Birkhead Formation shales and coals were deposited in low-energy meandering fluvial to shallow lacustrine environment. The Birkhead Formation contains an average of about 1.5% TOC dispersed in the shales, reaching a maximum of about 4.0%. The organic matter comprises a mixture of type III (vitrinite) and type II organic matter, the type II material consisting of sporinite and resinite, with rare alginite (Smyth et al., 1984; Cook and Struckmeyer, 1986).

The Murta Member consists of finely laminated shale, siltstone, and sandstone deposited in proximal to distal lacustrine environments (Ambrose et al., 1986; Mount, 1982). It contains about 0.6% TOC on average reaching a maximum of up to 4% in some thin shaly laminae. The organic matter comprises a mixture of type III (inertinite) and II (mainly sporinite) material, with a reasonable potential to produce liquid hydrocarbons.

A summary of the chemical composition of Eromanga basin source rocks based on Rock-Eval pyrolysis is shown in Figure 22. The 28 samples below the type III line are from the gas-prone marine Cretaceous shales. The remaining samples show a higher average HI than for the Permian (Figure 22). Several samples, mainly from the Murta Member, with DOM comprised mainly of sporinite, show the highest HIs.

MATURITY

Source maturity in the Cooper and Eromanga basins has been defined principally by vitrinite reflectance (R_v) (Kantsler et al., 1983; Vincent et al., 1985). Vitrinite reflectance data are available for more than 100 wells in the basins, and profiles have been established from 15 to 30 samples (usually cuttings)

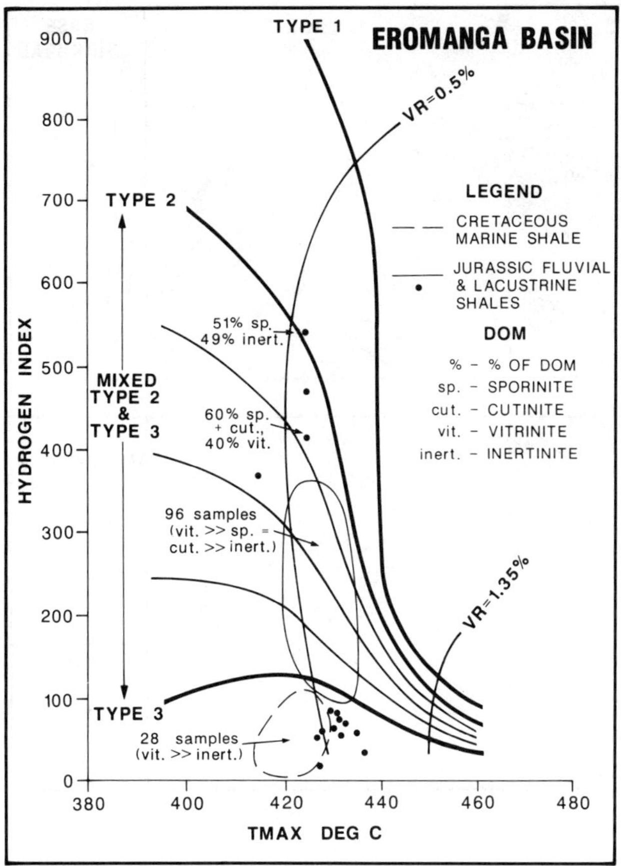

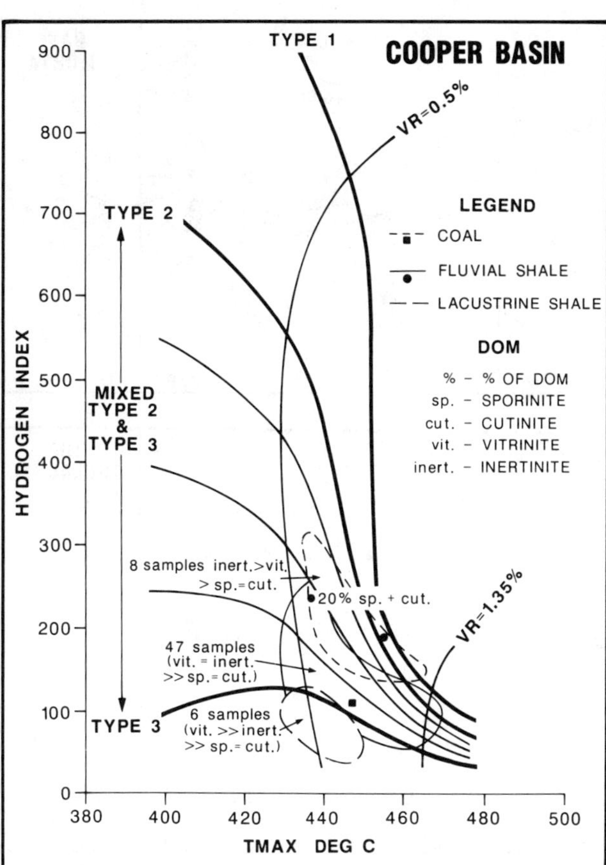

Figure 22. Kerogen type and maturity of Cooper and Eromanga basin source rocks based on hydrogen index (type) and T_{max} (maturity) data from Rock-Eval analysis. >, greater than; and >> much greater than.

from each well. The presence of coals throughout the main intervals of interest—the lower part of the Eromanga basin and the Cooper basin—has resulted in the collection of a reliable data set. Source maturity has been mapped for the principal source units (Figure 23), and the distribution and composition of Permian hydrocarbons is closely related to maturation (Hunt et al., 1989).

The maturity profile for Jackson 1 is shown in Figure 24. An oil window for terrestrial source rocks has traditionally been defined by the maturation levels R_v 0.6% to 1.2%. If hydrocarbon accumulations, however, are correlated with local source maturity in the Cooper basin, accumulations are absent at maturities of <0.8% R_v, and significant volumes appear at about 0.95% R_v (allowing for a migration distance of about 10 km or 6.2 mi). Thus, it appears that the Permian section at Jackson is only marginally mature for hydrocarbon generation. Maturity in the Nappamerri trough to the southwest, however, is much higher, and both the Toolachee and particularly the Patchawarra Formation are at generative levels of maturation near to Jackson (Figure 23).

The organic matter in Eromanga basin source rocks is generally more exinite rich (oil prone) than for that in the Cooper basin, and several studies quoted in Vincent et al. (1985) have proposed that initial generation of oil from labile resinite which is common in the Birkhead Formation can take place at maturation levels as low as 0.4%. Powell et al. (1989) proposed significant generation in the Murta Member, which contains bacterially enriched lipids, at 0.5 to 0.6% R_v. Thus the Birkhead Formation and the Murta Member may be oil mature in the vicinity of Jackson (Figure 23). However, although the Eromanga basin source rocks are more oil prone than those of the Cooper basin, they are not particularly rich. Given their low level of maturity, it is unlikely that they have contributed large volumes to the oil reservoired in the Eromanga basin. Their contribution is discussed further in the oil-source correlation section.

OIL CHARACTERISTICS

Oils found in Jurassic and Cretaceous reservoirs are mostly light (39° to 56° API), mature, paraffinic to naphthenic, condensate-like crudes. Some such as the Jackson Hutton and Westbourne oils have high wax contents (Figure 25) that account for their higher pour points (23° and 10° C, respectively) and API

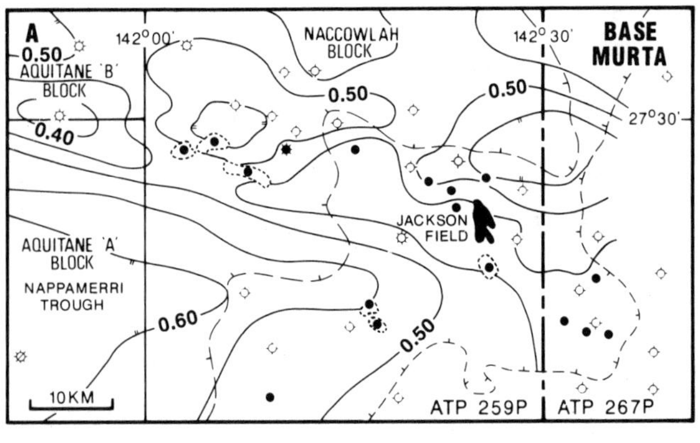

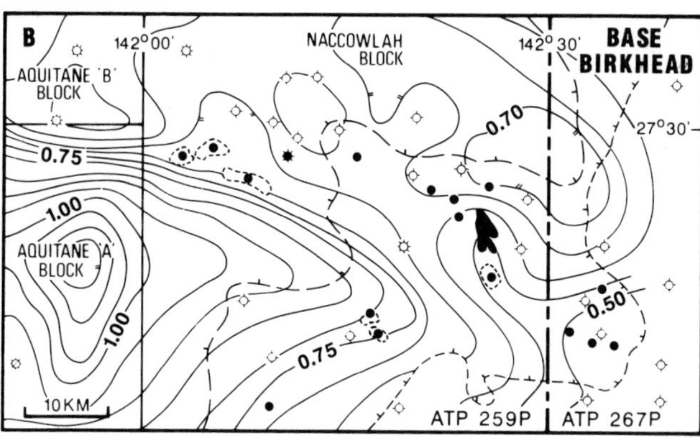

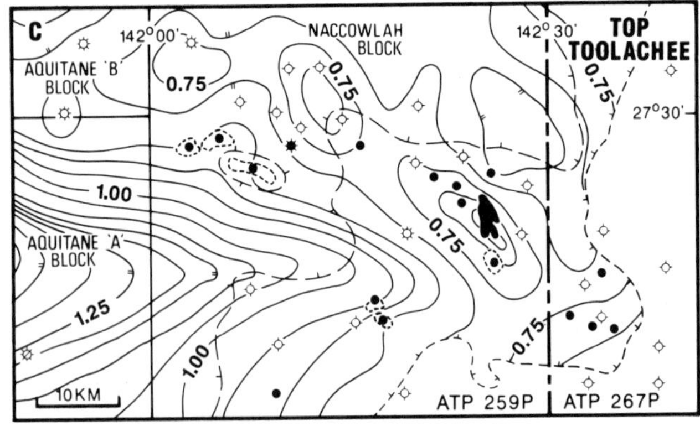

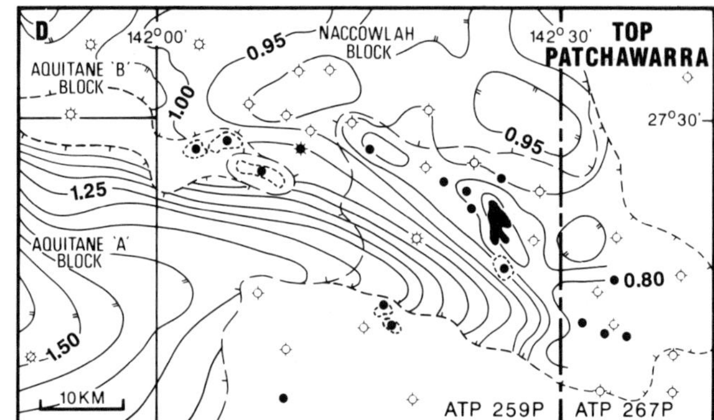

Figure 23. Maturity maps showing vitrinite isoreflectance contours for selected source horizons. Contour interval, 0.05%. Well symbols are defined in Figure 9.

gravities (40° and 48°, respectively), compared with the more naphthenic Jackson Murta oil (0° C and 50° API). The oils all have low sulfur contents (< 1%).

The lack of gasoline-range fraction in the waxy oils and lack of benzene and toluene in the gasoline-range fraction (Figure 26) are interpreted as evidence of water washing. This characteristic is present even in oils trapped in stratigraphically isolated thin sands of the Westbourne and Murta Member. Water washing is a result of the presence of the low salinity of the artesian water and may account for the extremely low GORs of the Jurassic and Cretaceous oils (10 SCF/STB in the Hutton to 70 SCF/STB in the Murta reservoir). The oil zone connate water is, however, thought to be saline (Neumann, 1987), which suggests oil migration prior to onset of the aquifer and water washing in situ.

Consistent with the low gas content, solution gas drive in the reservoirs is minimal. Formation volume factors for the reservoirs are also low, being approximately 1.1 RB/STB. Viscosities are relatively low, ranging from 0.7 cp for the Murta to 1.5 cp for the Hutton at reservoir conditions (Figure 27).

OIL-SOURCE CORRELATIONS

Kantsler et al. (1983), Vincent et al. (1984), Alexander et al. (1988), Powell et al. (1989), and Jenkins (1989) have discussed the detailed geochemistry of Cooper and Eromanga basin hydrocarbons including the Jackson oils. The Jurassic and Cretaceous oils show moderate to low pristane/n-C17 ratios, low phytane/n-C18 ratios, and intermediate pristane/phytane ratios (Kantsler et al., 1983), consistent with a bacterially enriched land plant origin (type II/type III kerogen). Kantsler et al. (1983) concluded that the Hutton oils were probably in large part sourced from deeper, more mature source rocks, but that the Murta oils may be locally sourced from bacterially enriched, early mature source rocks.

Vincent et al. (1983) showed that the biomarker distributions in the Cooper and Eromanga basin oils were similar and thus of little use in highlighting differences in source. The biomarkers are consistent with a terrestrial origin, as opposed to marine, with a significant contribution from lipid-enriched bacterial biomass.

More recently, Alexander et al. (1988) have correlated oils of the Cooper and Eromanga basins using aromatic biomarkers. They found that the Jackson oils lacked aromatic biomarkers characteristic of Jurassic source rocks, with the exception of the Murta oil which had a minor Jurassic biomarker contribution. Jenkins (1989) has also recognized age-specific isopimarane and trisnorhopane biomarkers present in Eromanga basin source rocks. The isopimarane biomarker is an aromatic compound derived probably from resin. It is characteristic of

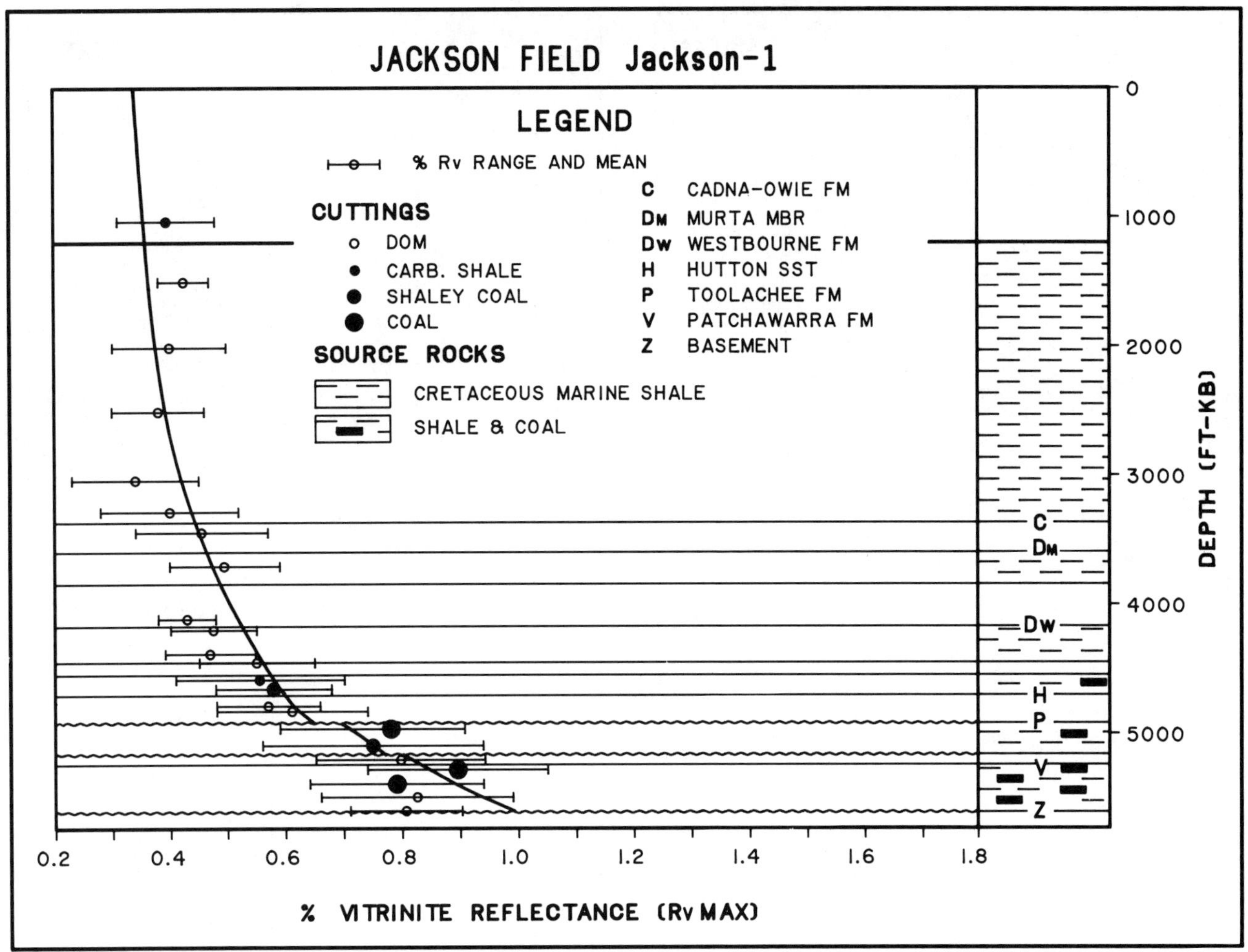

Figure 24. Maturity profile of Jackson 1 based on vitrinite reflectance of 30 ft composite cutting samples.

the Mesozoic source rocks and marks the evolution and dominance of resinous coniferous Araucacian flora by the Jurassic. Importantly, this biomarker is absent in the Jackson Hutton and Westbourne oils, which is interpreted as evidence of their Permian origin (Jenkins, 1989). The naphthenic trisnorhopane biomarker is probably related to bacterial remains (Powell et al., 1989) and is restricted to the Murta Member. It is present to some degree in many Murta oils, including the Jackson Murta oil. Powell et al. (1989) argue for in situ sourcing of Murta oils at low levels of maturity ($R_v = 0.5$ to 0.6%) from bacterial biomass, based on chemical maturity, elemental pyrolysis-GC, and GCMS data.

Given the possibility of vertical migration and mixing of Jurassic and Permian oils and the absence of a definitive Permian biomarker, comprehensive oil-source correlations for the Cooper and Eromanga basins have yet to be achieved. The balance of evidence to date, however, suggests a major contribution of Permian sourced hydrocarbons is present in the Eromanga basin, and the Jackson field in particular.

GEOTHERMAL GRADIENTS

Geothermal gradients in the Cooper and Eromanga basins range from 1.5 to 3.0°F/100 ft (3.0 to 6.0°C/100 m), based on corrected bottom hole temperatures (Kantsler et al., 1983; Pitt, 1986). The geothermal gradient at Jackson is about 2.9°F/100 ft (5°C/100 m). These gradients are high by world standards and, in the Moomba area of the Nappamerri trough, have been attributed to radiogenic decay in high level intrusions (Middleton, 1979).

The Cooper and Eromanga basins have undergone a complex thermal history (Kantsler et al., 1983; Middleton, 1979), and most maturation modeling studies for the basins invoke a recent heating event—0 to 5 Ma (Pitt, 1986)—to match present-day maturity.

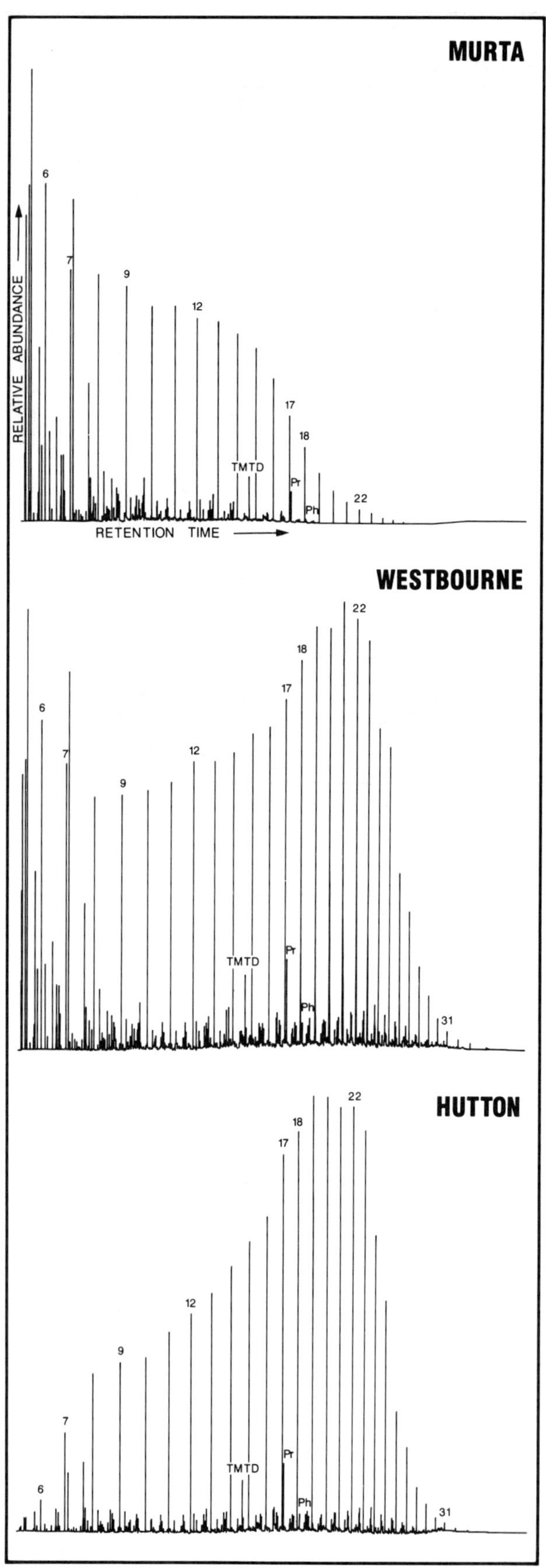

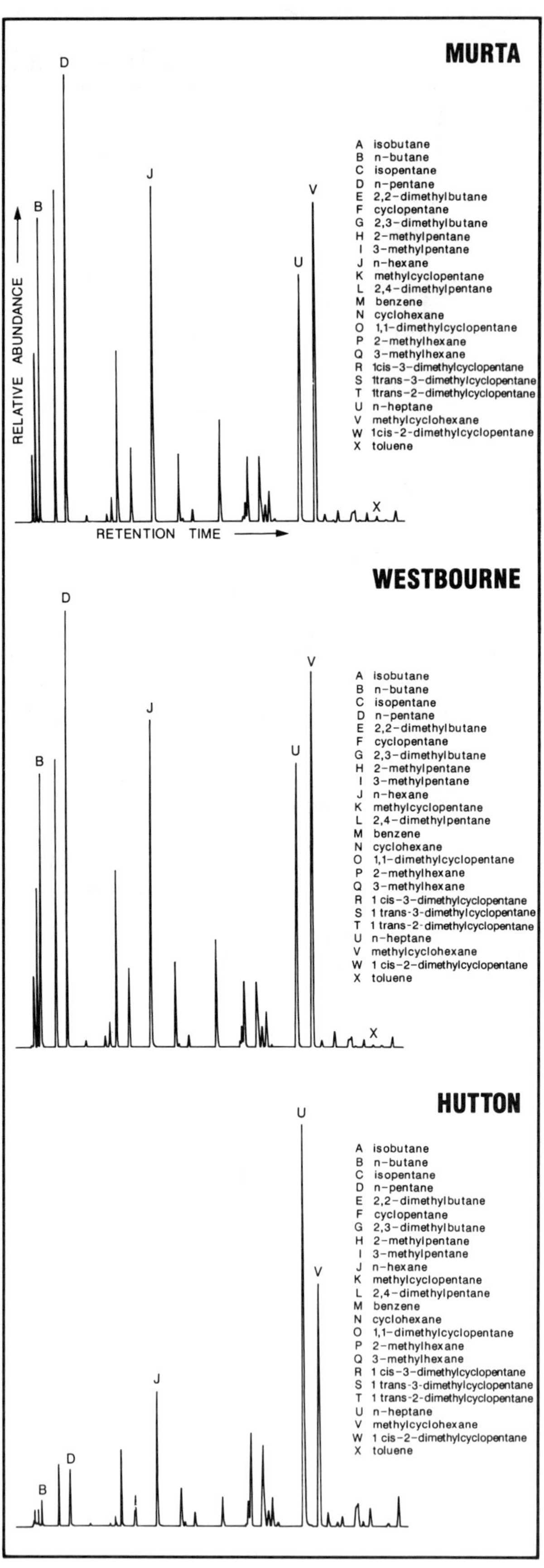

Figure 25. Whole-oil gas chromatograms of Jackson field oils.

Figure 26. Gasoline-range gas-liquid chromatograms of Jackson field oils.

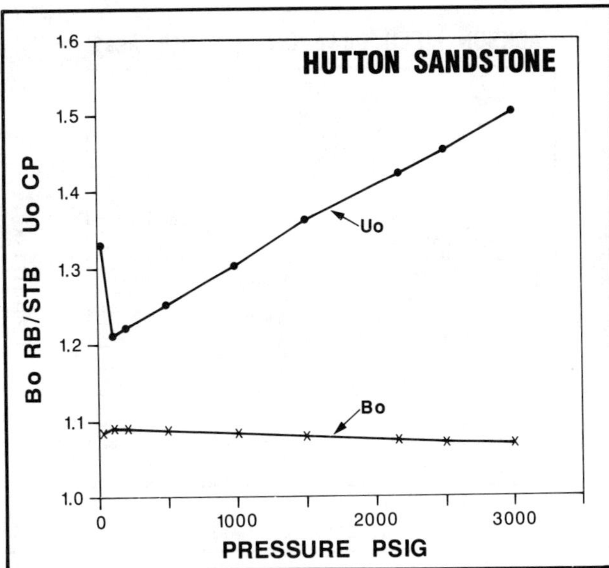

Figure 27. Viscosity and barrels of oil for Hutton Sandstone oil, Jackson field.

This is supported by a recent assessment of thermal history in the southwestern Cooper and Eromanga basin based on apatite fission track analysis (Duddy, in press).

TIMING OF GENERATION

Based on the previous discussion of source maturity, both the Cooper and Eromanga basins sections are traditionally considered to be early mature to mature for oil generation in the vicinity of Jackson. The timing of generation is considered here using the maturation modeling technique of Lopatin. More detailed treatments of maturation modeling in the Cooper and Eromanga basins can be found in Pitt (1986), Kantsler et al. (1983), and Vincent et al. (1985).

Burial history and maturation models for Jackson field and for a nearby (10 km) down-dip "Karwin trough" to the southwest are shown in Figure 28. It is possible to estimate section loss at the Early Permian and Late Triassic unconformities from the vitrinite reflectance profile in Figure 24. Maturation modeling, however, is insensitive to the Permian burial and thermal history because the Permian section was only shallowly buried until the Cretaceous. The timing of generation is most sensitive to the recent temperature and uplift history (0 to 100 Ma) and these are both poorly known.

The magnitude of Tertiary uplift at Jackson has been estimated in Figure 28 from a consideration of interval thickness ratios in the Eromanga basin, although the precise timing of the uplift and erosion is speculative. The temperature history for the model is simply the proportion of the present well temperature profile (0.85 in Jackson 1, and 0.90 in Karwin 1) required to match the observed and modeled vitrinite reflectance profiles.

The timing of generation for Cooper basin source rocks, estimated from Figure 28, assuming onset of significant generation at 0.8 R_v, is about 10 Ma at both Jackson and Karwin for the top Toolachee Formation, and 50 and 100 Ma, respectively, for the Patchawarra Formation. The Permian section at Jackson is interpreted to be water saturated, however, which suggests that primary migration has yet to take place and that the Jackson field may have been sourced from the deeper areas of Permian to the south and west.

The timing of generation in Eromanga basin source rocks at Jackson is speculative, depending on the level of maturation assigned to peak generation. Using a maturity level of 0.6% R_v, based on a correlation between Jurassic oil accumulations and local source maturity, the Murta is immature at both Jackson and Karwin, but the Birkhead passed through the oil window between about 60 and 85 Ma, respectively. This correlation ignores the possibility that some or most of the oil may have migrated from the underlying Permian section.

MIGRATION AND ENTRAPMENT

The timing of generation is critical to an understanding of migration and entrapment. The timing of generation in the Cooper and Eromanga basin source rocks, ranging from about 90 Ma to the present, covers a critical but poorly understood interval in the structural history of the basins, namely the period of Tertiary structuring.

The problem of migration pathways to the Jackson field is usually addressed using present-day structure maps as a guide, which assumes recent generation. Figure 29 shows postulated migration pathways based on the form of the top Cadna-owie and top Patchawarra structure maps, which both show the same overall pattern. The steepest gradients are out of the Nappamerri trough where migrating hydrocarbons from the Eromanga or Cooper basins would have encountered conjugate fault blocks of the Wilson Fault system that define the Nuata ridge (which contains the "Karwin trough"). This ridge provides a focus for migrating hydrocarbons, and a tight syncline to the north of and parallel to the Nuata ridge provides a barrier to the northward migration of hydrocarbons.

Thus hydrocarbons migrating within the Patchawarra Formation and encountering the Nuata ridge may have either leaked up fault planes (to the Late Permian or the Eromanga basin section) or accumulated downdip, trapped by structural and fault seals. Hydrocarbons migrating within the Toolachee Formation would have access to the overlying

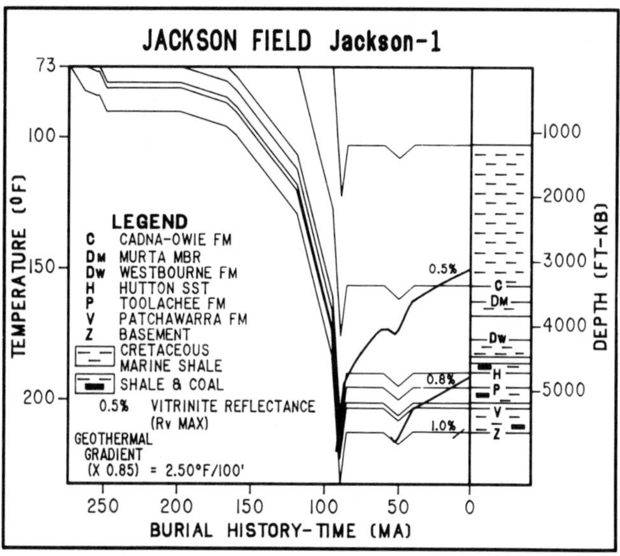

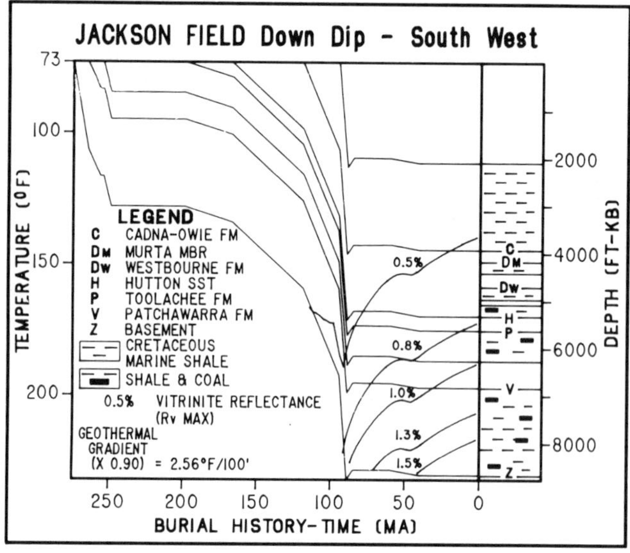

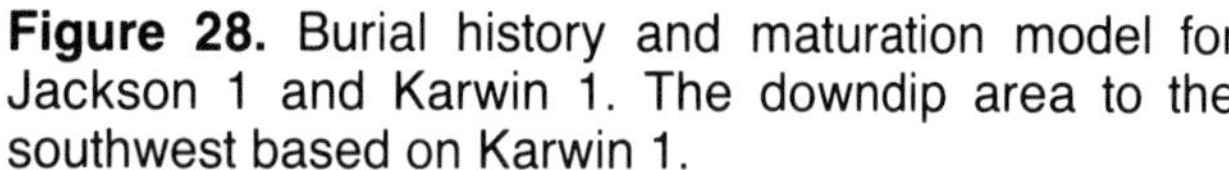
Figure 28. Burial history and maturation model for Jackson 1 and Karwin 1. The downdip area to the southwest based on Karwin 1.

Eromanga basin sequence via erosionally truncated beds along the Nuata ridge. Once in the Eromanga basin sequence the present-day migration pathway leads directly toward the larger Jackson structure. Owing to the structural geometry of the Wilson fault and its conjugate faults, it is probable that hydrocarbons generated within the Permian sediments of the Nappamerri trough cannot access Jackson directly through the Permian section.

A more difficult problem is assessing the paleostructure because Tertiary uplift is estimated at up to 2500 ft (760 m) in the Eromanga basin overlying the eastern part of the Cooper basin. Up to three Tertiary uplift events have been postulated (Moore and Pitt, 1984), but the relative magnitude and timing of each event are poorly known. A conformable interval that makes up the lower third of the Eromanga basin succession, the top Cadna-owie Formation to top Hutton Sandstone interval, may be used as a guide to Late Cretaceous migration pathways (Figure 29). This interval is generally accepted to be representative of the remainder of the succession, because structuring of the Eromanga basin was minor prior to the Tertiary.

The identification of migration pathways to Jackson remains problematical, given the patchy nature of discoveries along proposed routes, probably a result of the poorly understood Tertiary structural history. The low fill factor for the larger Jackson–Jackson South structure (Figures 4 and 13) may also reflect Tertiary post-generation enhancement of existing structures, although an alternative explanation of insufficient generation is equally plausible.

EXPLORATION CONCEPTS

Regional Play

With the exception of being the largest, Jackson field has many elements in common with other Eromanga basin fields. The play type that characterizes Jackson may be designated "Cooper source, Eromanga trap." The essential elements of this play are access to liquids-mature Permian source rocks; lateral and vertical migration from the Cooper to the Eromanga basin, either via subcrop in the absence of the Nappamerri Group or via faults; accumulation in anticlinal traps within the Eromanga; and, in many places, vertical migration within the Eromanga to give vertically stacked pools.

At Jackson, although the Nappamerri Group is absent, the subcropping Toolachee Formation is immature, and a more complex fault-controlled migration pathway from the Early Permian Patchawarra Formation is postulated in this paper.

Application of Geologic Parameters

The structure of Jackson field, as of most Eromanga basin fields, is of a simple anticline formed by drape, compaction, and growth of older Permian and pre-Permian structures. A minor stratigraphic component of trapping is present owing to facies variation within the reservoir/seal units.

In common with many other Eromanga structures, including both fields and dry holes, Jackson shows

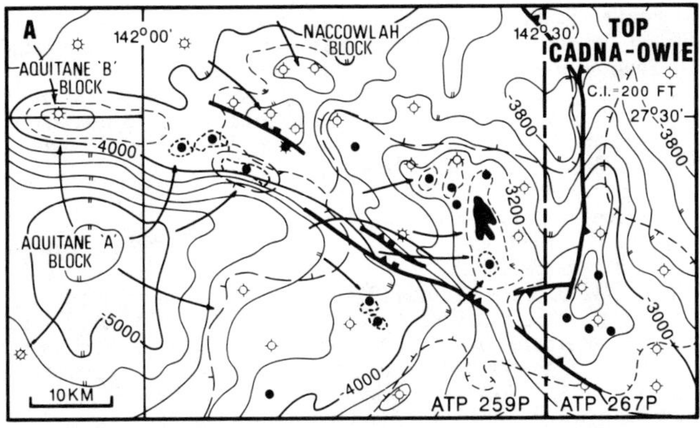

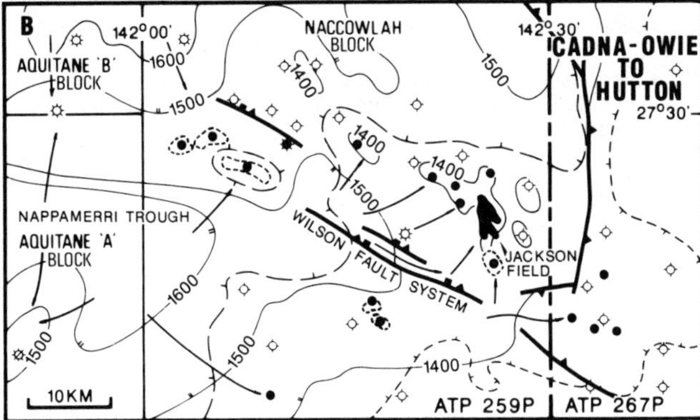

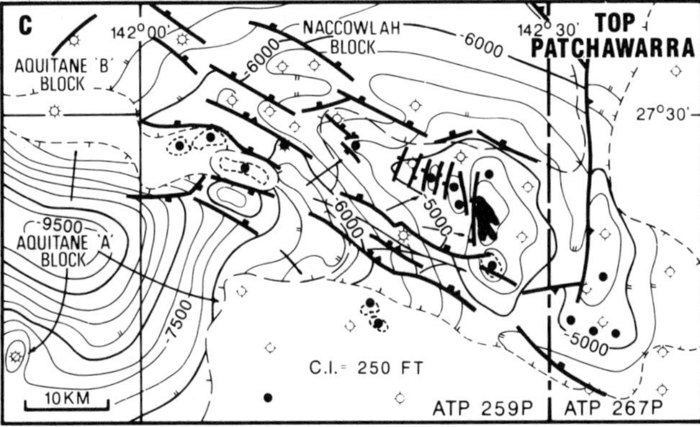

Figure 29. Depth and interval maps for (A) top Cadna-owie, contour interval 200 ft; (B) the Cadna-owie to Hutton interval, contour interval 100 ft; and (C) the top Patchawarra, contour interval 250 ft. Arrows indicate migration directions based on the form of the map.

significant Tertiary enhancement of the older structure. The structural feature unique to Jackson is the presence within its drainage area of a Permian depocenter, inverted during the Tertiary. This Permian trough contains a thick mature source section.

Jackson field also shows the presence of vertically stacked pools, a feature of many Eromanga fields. This pattern is interpreted as evidence of vertical migration within the Eromanga basin and implies that the finer-grained fluvial top seal facies have limited lateral continuity.

The Jackson, Hutton, and Westbourne reservoirs contain waxy oils characteristic of terrestrial source rocks, generated at higher levels of maturity ($>0.8\%$ R_v). The lighter Murta oil is thought to be derived from local bacterially enriched terrestrial organic matter at lower levels of maturity ($<0.6\%$ R_v) in common with many other Murta oils in the basin.

All Eromanga oils are water washed, consistent with the presence of a strong aquifer flow in the Eromanga basin. However, connate water in the oil zone at Jackson is thought to be saline, and the water washing (but not flushing) must have taken place in situ. This also highlights the probable poor lateral continuity of the fluvial seal facies, since the oils are water washed in spite of their apparent stratigraphic isolation.

The occurrence of water washing and vertical migration cautions care in the interpretation of geochemical data in the basin. Both factors may selectively affect source/maturity parameters, and the mixing of Permian liquids with early mature Eromanga liquids cannot be discounted.

Lessons

Although not the first Eromanga oil field discovered, it is worth remembering that conventional wisdom only a few years before the Jackson discovery held that oil would not be found in the Eromanga basin because of the absence of marine/source rocks and because any traps would be flushed by the aquifer. Thus, several of the initial wells in the basins were drilled without geological evaluation of the Eromanga basin sequence, bypassing Eromanga oil pools.

In general terms, Jackson is one of several major productive culminations on the major intrabasinal structural trends that cut the Cooper and Eromanga basins. In this respect it is located on a highly prospective fairway of large early structures, which contain shallowly buried high quality reservoirs and are juxtaposed with deeply buried mature terrestrial source rocks in both basins.

The reason that Jackson field contains the largest hydrocarbon column in the Eromanga basin, while still being only partly filled in common with all other Eromanga pools, is, however, still speculative. At the time of its discovery it was the largest of several equally prospective closures in the area, differentiated only by the fact that it was further marginwards. It was several years later that drilling showed the presence of a previously unrecognized, inverted Permian depocenter within the drainage area of Jackson.

In this paper we have speculated that the reason for the size of the Jackson field is related to the Tertiary inversion of the mature Permian trough and associated structural enhancement of Jackson nearby. However, while the occurrence of both factors together appears to be unique in the basins, it is not clear that they could be successfully predicted even today.

ACKNOWLEDGMENTS

The Jackson field study has been compiled with the permission and assistance of Esso Australia Limited and its subsidiary Delhi Petroleum Pty. Ltd., and the approval of Santos Limited. Many explorationists in the above companies and Claremont Petroleum N.L. contributed to the study via proprietary reports and discussion; the advice and comments of B. J. Burns, N. Milne, I. R. Mortimore, D. Phelps, R. Page, G. Powis, H. R. B. Wecker, A. P. Whittle, and C. C. Yew are acknowledged in particular. The drafting was undertaken by the Esso and Delhi exploration and drafting departments.

REFERENCES

Alexander, R., A. V. Larcher, I. Kagi., and P. L. Price, 1988, The use of plant-derived biomarkers for correlation of oils with source rocks in Cooper-Eromanga Basin system, Australia: APEA Journal, v. 28, p. 310-324.

Ambrose, G., R. Suttill, and I. Lavering, 1986, A review of the Early Cretaceous Murta Member in the Southern Eromanga Basin, *in* D. I. Gravestock, P. S. Moore, and G. M. Pitt, eds., Contributions to the geology and hydrocarbon potential of the Eromanga Basin: Geological Society of Australia Special Publication No. 12, p. 71-84.

Bowering, O. J. W., 1982, Hydrodynamics and hydrocarbon migration—a model for the Eromanga Basin: APEA Journal, v. 22(1), p. 227-236.

Cook, A. C., 1982, The nature and significance of organic facies in the Eromanga Basin, *in* D. I. Gravestock, P. S. Moore, and G. M. Pitt, eds., Contributions to the geology and hydrocarbon potential of the Eromanga Basin: Geological Society of Australia Special Publication No. 12, p. 203-220.

Cook, A. C., and H. Struckmeyer, 1986, The role of coals as a source rock for oil, *in* R. C. Glenie, ed., Technical papers presented at PESA Symposium, 14-15 November, 1985, Melbourne, 469 p.

Delhi Petroleum Pty. Ltd., 1987, Naccowlah Block Summary Report: Delhi Petroleum Pty. Ltd., unpublished report.

Donaleshen, R. R., 1987, Jackson-Westbourne field geological status report: Delhi Petroleum Pty. Ltd., unpublished report, 40 p.

Exxon, N. F., 1976, Geology of the Surat Basin in Queensland: Australian Bureau of Mineral Resources, Geology and Geophysics, Bulletin, 166 p.

Fisher, R. W., & Associates, 1985, A technical review of the Naccowlah Block, ATP 259P, Queensland: Delhi Petroleum Pty. Ltd., unpublished report.

Flook, M., 1986, Porosity study in ATP 259P Queensland: Delhi Petroleum Pty. Ltd., unpublished report, 15 p.

Gravestock, D. I., and E. M. Alexander, 1988, Eromanga Basin, South Australia—core and well log study: National Energy Research, Development and Demonstration Programme Project No. 820, unpublished report, 108 p.

Habermehl, M. A., 1986, Regional ground water movement, hydrochemistry and hydrocarbon migration in the Eromanga Basin, *in* D. I. Gravestock, P. S. Moore, and G. M. Pitt, eds, Contributions to the geology and hydrocarbon potential of the Eromanga Basin: Geological Society of Australia Special Publication No. 12, p. 353-376.

Halyburton, R. V., and A. L. Robertson, 1984, Geology of the Jackson Oil field: APEA Journal, v. 24(1), p. 259-265.

Harding, T. P., 1984, Graben hydrocarbon occurrences and structural style: American Association of Petroleum Geologists Bulletin, v. 68, p. 333-362.

Harding, T. P., and J. D. Lowell, 1979, Structural styles, their plate tectonic habitats, and hydrocarbon traps in petroleum provinces: American Association of Petroleum Geologists Bulletin, v. 63, p. 1016-1058.

Hollingsworth, R. J. S., 1982, The Jackson Oil Discovery: Delhi Petroleum Pty. Ltd., unpublished report.

Holton, G., 1988, Jackson 31 well completion report: Santos Limited, unpublished report, 10 p.

Hunt, J. W., R. S. Heath, and P. McKenzie, 1989, Geological controls on the distribution and composition of Cooper Basin hydrocarbons, *in* B. J. O'Neil, ed., The Cooper and Eromanga basins, Australia: Proceedings of Petroleum Exploration Society of Australia, Society of Petroleum Engineers, Australian Society of Exploration Geophysicists (SA Branches), Adelaide, p. 509-523.

Jenkins, C. C., 1989, Geochemical correlation of source rocks and crude oils from the Cooper and Eromanga Basins, *in* B. J. O'Neil, ed., The Cooper and Eromanga basins, Australia: Proceedings of Petroleum Exploration Society of Australia, Society of Petroleum Engineers, Australian Society of Exploration Geophysicists (SA Branches), Adelaide, p. 525-540.

Kantsler, A. J., T. J. C. Prudence, A. C. Cook, and M. Zwigulis, 1983, Hydrocarbon habitat of the Cooper/Eromanga Basin: APEA Journal, v. 23(1), p. 75-92.

Martin, K. R., 1983, Petrology of the Murta Member, Westbourne Formation and Hutton Sandstone in Jackson No. 1 and Jackson South No. 1, Eromanga Basin, southwest Queensland: Delhi Petroleum Pty. Ltd., unpublished report.

McKirdy, D. M., 1982, Aspects of the source rock and petroleum geochemistry of the Eromanga Basin, *in* P. J. Moore and T. J. Mount, compilers, Eromanga Basin Symposium, Summary Papers: Geological Society of Australia and Petroleum Exploration Society of Australia, p. 258-259.

McKirdy, D. M., J. K. Emmett, B. A. Mooney, R. E. Cox, and B. L. Watson, 1984, Organic geochemical facies of the Cretaceous Bulldog shale, western Eromanga Basin, South Australia, *in* D. I. Gravestock, P. S. Moore, and G. M. Pitt, eds., Contributions to the geology and hydrocarbon potential of the Eromanga Basin: Geological Society of Australia Special Publication No. 12, p. 287-304.

Michaelsen, B. H., and D. M. McKirdy, 1989, Organic facies and petroleum geochemistry of the lacustrine Murta Member (Mooga Formation) in the Eromanga Basin, Australia, *in* B. J. O'Neil, ed., The Cooper and Eromanga basins, Australia: Proceedings of Petroleum Exploration Society of Australia, Society of Petroleum Engineers, Australian Society of Exploration Geophysicists (SA Branches), Adelaide, p. 541-558.

Middleton, M. F., 1979. Heat flow in the Moomba, Big Lake, and Toolachee gas fields of the Cooper Basin and complications for hydrocarbon maturation: Australian Society of Exploration Geophysicists, Bulletin 10, p. 149-155.

Moore, P. S., 1981, Jackson 1 drilling proposal: Delhi Petroleum Pty. Ltd., unpublished report, 6 p.

Moore, P. S., and G. M. Pitt, 1984, Cretaceous of the Eromanga Basin—implications for hydrocarbon exploration: APEA Journal, v. 24(1), p. 358-376.

Mount, T. J., 1982, Geology of the Dullingari Murta oilfield, *in* P. J. Moore and T. J. Mount, compilers, Eromanga Basin Symposium, Summary Papers: Geological Society of Australia and Petroleum Exploration Society of Australia, p. 356-374.

Nelson, A. W., 1985, Tectonics of the Jackson-Naccowlah Area, southwest Queensland and their implications for hydrocarbon accumulations: APEA Journal, v. 25(1), p. 85-94.

Neumann, R. G., 1987, A review of log analysis techniques in the Eromanga Basin: Esso Australia Limited, unpublished report, 25 p.

Pitt, G. M., 1986, Geothermal gradients, geothermal histories and the timing of thermal maturation in the Eromanga-Cooper Basins, *in* D. I. Gravestock, P. S. Moore, and G. M. Pitt, eds, Contributions to the geology and hydrocarbon potential of the Eromanga Basin: Geological Society of Australia Special Publication No. 12, p. 323-351.

Powell, T. G., C. J. Boreham, D. M. McKirdy, B. H. Michaelsen, and R. E. Summons, 1989, Petroleum geochemistry of the Murta Member, Mooga Formation and associated oils, Eromanga Basin, Australia: APEA Journal, v. 29(1), p. 114-129.

Robinson, S., 1982, Well completion report, Jackson No. 1: Delhi Petroleum Pty. Ltd., unpublished report, 20 p.
Smyth, M., 1983, Nature of source material for hydrocarbons in Cooper Basin, Australia: American Association of Petroleum Geologists Bulletin, v. 67, p. 1422–1428.
Smyth, M., A. C. Cook, and R. P. Philp, 1984, Birkhead revisited: petrological and geochemical studies of the Birkhead Formation, Eromanga Basin: APEA Journal, v. 24(1), p. 230–242.
Taylor, G. H., S. Y. Liu, and M. Smyth, 1988, New light on the origin of Cooper Basin oil: APEA Journal, v. 28(1), p. 202–209.
Thornton, R. N. C., 1979, Regional stratigraphic analysis of the Gidgealpa Group, southern Cooper basin, Australia: South Australian Geological Survey Bulletin No. 9, 140 p.
Vincent, P. S., I. R. Mortimore, and D. M. McKirdy, 1985, Hydrocarbon generation, migration and entrapment in the Jackson-Naccowlah area, ATP 259P, Southwestern Queensland: APEA Journal, v. 25(1), 62–84.
Williams, B. P. J., and E. K. Wild, 1984, The Terrawarra Sandstone and Merrimelia Formation of the southern Cooper Basin, South Australia—the sedimentation and evolution of a glaciofluvial system: APEA Journal, v. 24, n. 1, p. 377-392.

SUGGESTED READINGS

Halyburton, R. V., and A. L. Robertson, 1984, Geology of the Jackson Oil field. Discussion of Jackson field reservoir geology soon after discovery.
Kantsler, A. J., T. J. C. Prudence, A. C. Cook, and M. Zwigulis, 1983, Hydrocarbon habitat of the Cooper/Eromanga Basin. Summary of general controls of hydrocarbon distribution in the Cooper and Eromanga Basins.
Vincent, P. S., I. R. Mortimore, and D. M. McKirdy, 1985. Hydrocarbon generation, migration and entrapment in the Jackson-Naccowlah area, ATP 259P, Southwestern Queensland. Comprehensive discussion of hydrocarbon occurrences in the Jackson area.

JACKSON

Appendix 1. Field Description

Field name .... *Jackson*

Ultimate recoverable reserves .... *40,000,000 BO*

Field location:

- **Country** .... *Australia*
- **State** .... *Queensland*
- **Basin/Province** .... *Eromanga basin*

Field discovery:

- **Year first pay discovered** .... *Cretaceous Murta Member 1981*
- **Year second pay discovered** .... *Jurassic Westbourne Formation 1981*
- **Third pay** .... *Jurassic Hutton Sandstone 1981*

Discovery well name and general location

- **First pay** .... *Jackson No. 1, Murta Mbr., 230 km east of Moomba, 1150 km west of Brisbane*
- **Second pay** .... *Jackson No. 1, Westbourne Formation*
- **Third pay** .... *Jackson No. 1, Hutton Sandstone*

Discovery well operator .... *Delhi Petroleum Pty. Ltd.*

- **Second pay** .... *Delhi Petroleum Pty. Ltd.*
- **Third pay** .... *Delhi Petroleum Pty. Ltd.*

IP in barrels per day and/or cubic feet or cubic meters per day:

- **First pay** .... *679 BPD (December 1983)*
- **Second pay** .... *4180 BPD (January 1983)*
- **Third pay** .... *290 BPD (December 1982)*

All other zones with shows of oil and gas in the field:

Age	Formation	Type of Show
Early Permian	*Patchawarra Fm.*	*Trace fluorescence, gas*

Geologic concept leading to discovery and method or methods used to delineate prospect, e.g., surface geology, subsurface geology, seeps, magnetic data, gravity data, seismic data, seismic refraction, nontechnical:

Jackson No. 1 was located on the time crest of a seismically defined 4-way dip-closed anticline, which also had surface expression owing to Tertiary uplift. Following small oil discoveries in the Eromanga basin and gas and oil in the Cooper basin, the well was drilled as an Eromanga oil test and Cooper gas test.

Structure:

Province/basin type (see St. John, Bally, and Klemme, 1984)

Klemme 3A/2B; Bally Cooper basin (529) 1211, Eromanga basin (Surat) (674) 121

Tectonic history

Late Carboniferous tectonics created a local basement horst at Jackson amid an extensional, subsiding basin. Structural episodes during the late Early Permian and Late Triassic reactivated and reinforced the feature. During the Tertiary, compressional tectonics inverted the local depocenter to the south of Jackson and elevated Jackson to the highest point on the regional Pepita-Naccowlah-Jackson trend.

Regional structure

Eastern end of the Pepita-Naccowlah-Jackson trend.

Local structure

Anticlinal closure on basement horst block, called Jackson–Jackson South structure.

Trap

Trap type(s)

The trap is a 4-way dip-closed anticline with multiple pays. The top of porosity in the Hutton Sandstone is stratigraphically controlled, as is the distribution of pay sands in the Westbourne Formation and Murta Member.

Basin stratigraphy (major stratigraphic intervals from surface to deepest penetration in field):

Chronostratigraphy	Formation	Depth to Top in ft
Early Cretaceous	*Cadna-owie Fm.*	*3354*
Early Cretaceous	*Murta Member*	*3604*
Late Jurassic	*Westbourne Fm.*	*4169*
Middle Jurassic	*Hutton Sandstone*	*4692*
Late Permian	*Toolachee Fm.*	*4912*
Early Permian	*Patchawarra Fm.*	*5230*
Pre-Permian	*Basement*	*5598*

Location of well in field

Reservoir characteristics:

Number of reservoirs *3*

Formation .... *Murta Member*

Age .... *Early Cretaceous*

Depths to tops of reservoirs .... *3625 ft (1104.9 m)*

Gross thickness (top to bottom of producing interval) .... *28 ft (8.5 m)*

Net thickness—total thickness of producing zones

Average .... *8 ft (2.4 m)*

Maximum

Average

Maximum

Lithology .... *Quartzose sandstone, fine- to very fine grained, poorly to moderately well sorted*

Porosity type .... *Primary intergranular, minor secondary dissolution*

Average porosity *19%*

Average permeability .... *0.2–120 md*

Formation .... *Westbourne Formation*

Age .... *Late Jurassic*

Depths to tops of reservoirs .... *4403 ft (1342 m)*

Gross thickness (top to bottom of producing interval) .... *82 ft (24.9 m)*

Net thickness—total thickness of producing zones

Average .... *13 ft (4 m)*

Maximum

Average

Maximum

Lithology .... *Quartzose sandstone, fine- to very fine grained, poorly to moderately well sorted*

Porosity type .... *Primary intergranular, minor secondary dissolution*

Average porosity *13.4%*

Average permeability .... *0.5–1500 md*

Formation .... *Hutton Sandstone*

Age .... *Mid-Jurassic*

Depths to tops of reservoirs .... *4688 ft (1428.9 m)*

Gross thickness (top to bottom of producing interval) .... *89 ft (27.1 m)*

Net thickness—total thickness of producing zones

Average .... *89 ft (27.1 m)*

Maximum
Average
Maximum
Lithology .... *Quartzose sandstone, fine- to coarse-grained, moderately well sorted*
Porosity type .... *Primary intergranular, minor secondary dissolution*
Average porosity *20.8%*
Average permeability .... *4–1780 md*

Seals:

Upper
Formation, fault, or other feature .... *Murta Member*
Lithology .... *Siltstone, shale*

Lateral
Formation, fault, or other feature .... *Murta Member*
Lithology .... *Siltstone, shale*

Upper
Formation, fault, or other feature .... *Westbourne Formation*
Lithology .... *Siltstone, shale*

Lateral
Formation, fault, or other feature .... *Westbourne Formation*
Lithology .... *Siltstone, shale*

Upper
Formation, fault, or other feature .... *Birkhead Formation*
Lithology .... *Siltstone, shale*

Lateral
Formation, fault, or other feature .... *Birkhead Formation*
Lithology .... *Siltstone, shale*

Source:

Formation and age .... *Murta Member, Early Cretaceous*
Lithology .... *100 ft (30.5 m) silt and shale, 140 ft (43 m) sand*
Average total organic carbon (TOC) *1%*
Maximum TOC *4%*
Kerogen type (I, II, or III) .... *III = II*
Vitrinite reflectance (maturation) .... *R_o = 0.50 to 0.65% down-dip*
Time of hydrocarbon expulsion .... *40 to 65 Ma down-dip at 0.50%*
Present depth to top of source .... *3604 to 4487 ft (1099.7 to 1368.5 m) down-dip*
Thickness .... *100 ft (30.5 m) shale*
Potential yield .... *18 bbl oil/ac-ft at peak liquids yield*

Formation and age .... *Westbourne Formation, Late Jurassic*
Lithology .... *shale = silt > sand, 170 to 220 ft (52 to 67 m) down-dip*
Average total organic carbon (TOC) *1%*
Maximum TOC *4%*
Kerogen type (I, II, or III) .... *III >> II*
Vitrinite reflectance (maturation) .... *R_o = 0.55 to 0.80% down-dip*
Time of hydrocarbon expulsion .... *70–80 Ma down-dip at 0.5% R_o*
Present depth to top of source .... *4169 to 4987 ft (1271.5 to 1521.0 m) down-dip*
Thickness .... *150 to 200 ft (46 to 61 m) shale/silt down-dip*
Potential yield .... *66 bbl oil/ac-ft at peak liquids yield*

Formation and age .... *Birkhead Formation, Middle Jurassic*
Lithology .... *30 ft (9.1 m) shale, 6 ft (1.8 m) coal, 120 ft (36.6 m) sand*

Average total organic carbon (TOC) 4%
Maximum TOC .... Coaly
Kerogen type (I, II, or III) .... III > II
Vitrinite reflectance (maturation) .... R_o = 0.60 to 0.90% down-dip
Time of hydrocarbon expulsion .... 80 to 85 Ma down-dip at 0.5% R_o
Present depth to top of source .... 4539 to 5457 ft (1384.4 to 1664.4 m) down-dip
Thickness .... 30 ft (9.1 m) to 80 ft (24.4 m) shale down-dip
Potential yield .... 89 bbl oil/ac-ft at peak liquids yield

Formation and age .... Toolachee Formation, Late Permian
Lithology .... 130 ft (39.6 m) shale, 40 ft (14.0 m) coal, 170 ft (51.8 m) sand
Average total organic carbon (TOC) .... 8% in shale
Maximum TOC .... Coaly
Kerogen type (I, II, or III) .... III >> II
Vitrinite reflectance (maturation) .... R_o = 0.7 to 1.1% down-dip
Time of hydrocarbon expulsion .... 15 to 65 Ma down-dip at 0.8% R_o
Present depth to top of source .... 4912 to 6277 ft down-dip
Thickness .... 130 to 100 ft (39.6 to 30.5 m) shale; 40 (14.0 m) to 50 ft coal down-dip
Potential yield .... 80 bbl oil/ac-ft at peak liquids yield

Formation and age .... Patchawarra Formation, Early Permian
Lithology .... 80 ft (24.4 m) coal, 80 ft (24.4 m) shale, 210 ft (64.0 m) sand
Average total organic carbon (TOC) 8%
Maximum TOC .... Coaly
Kerogen type (I, II, or III) .... III >> II
Vitrinite reflectance (maturation) .... R_o = 0.85 to 1.35% down-dip
Time of hydrocarbon expulsion .... 40 to 85 Ma down-dip
Present depth to top of source .... 5230 to 6967 ft (1595.2 to 2124.9 m) down-dip
Thickness .... 80–540 ft (24.4–164.7 m) shale; 80–280 ft (24.4–85.4 m) coal down-dip
Potential yield .... 80 bbl oil/ac-ft at peak liquids yield

Appendix 2. Production Data

Field name .... Jackson

Field size:

Proved acres .... ≈1993 ac
Number of wells all years 34
Current number of wells 34
Well spacing .... 60–80 ac
Ultimate recoverable .... 41 million bbl
Cumulative production .... 24 million bbl
Annual production .... 5.4 million bbl
Present decline rate .... 25% p.a.
Initial decline rate
Overall decline rate .... 12% p.a.
Annual water production .... 5 million bbl
In place, total reserves .... 110 million bbl
In place, per acre-foot .... ≈1040 bbl/ac-ft

Primary recovery .... *41 million bbl*
Secondary recovery .... *Undefined*
Enhanced recovery .... *Undefined*
Cumulative water production .... *31 million bbl*

Drilling and casing practices:

Amount of surface casing set .... *600 ft*
Casing program .... *7-in. to T.D.; 13-in. surface casing*
Drilling mud .... *KCl—polymer*
Bit program (1989) .... *Top hole: 12.25-in. I 1.1.4; to TD: 8.5-in. I 4.3.7; core: 8.5-in. ACC EHSTARR*
High pressure zones .... *Nil*

Completion practices:

Interval(s) perforated .... *Variable*
Well treatment .... *Nil*

Formation evaluation:

Logging suites .... *Gamma ray, LDL-CNL, sonic*
Testing practices .... *DST zones selected as required to confirm productivity; production test on completion*

Mud logging techniques

Standard mudlogging unit equipped with total gas detector and gas chromatograph; drilling parameters and ditch cuttings monitored from surface to TD

Oil characteristics:

Type .... *Murta: paraffinic-naphthenic*
(Tissot and Welte Classification in "Petroleum Formation and Occurrence," 1984, Springer-Verlag, p. 419)
API gravity .... *47°*
Base
Initial GOR .... *70 (scf/stb)*
Sulfur, wt% .... *<0.5*
Viscosity, SUS .... *0.5 cp*
Pour point .... *-8°C (18°F)*
Gas-oil distillate

Type .... *Westbourne: paraffinic*
(Tissot and Welte Classification in "Petroleum Formation and Occurrence," 1984, Springer-Verlag, p. 419)
API gravity .... *41°*
Base
Initial GOR .... *3 (scf/stb)*
Sulfur, wt% *0.5*
Viscosity, SUS
Pour point .... *25°C (77°F)*
Gas-oil distillate

Type .... *Hutton: paraffinic*
(Tissot and Welte Classification in "Petroleum Formation and Occurrence," 1984, Springer-Verlag, p. 419)
API gravity .... *39°*
Base
Initial GOR .... *6 (scf/stb)*
Sulfur, wt% .... *<0.5*
Viscosity, SUS .... *1.5 cp*
Pour point .... *25°C (77°F)*
Gas-oil distillate

Field characteristics:

Average elevation .... *250 ft*
Initial pressure .... *Murta, 1730 psi; Westbourne, 2070 psi*
Present pressure .... *Murta, 1730 psi; Westbourne variable*
Pressure gradient .... *0.33 psi/ft*
Temperature .... *230°F*
Geothermal gradient .... *2.93°F/100 ft*
Drive .... *Water*
Oil column thickness .... *Murta, 8 ft; Westbourne, 17 ft; Hutton, 20–80 ft*
Oil-water contact .... *Murta, 3630 ft; Westbourne, 4420 ft; Hutton, 4680 ft*
Connate water *10–30%*
Water salinity, TDS *1500–2500*
Resistivity of water
Bulk volume water (%)

Transportation method and market for oil and gas:
Pipeline.

Palm Valley Gas Field—Australia
Amadeus Basin, Northern Territory

R. F. DO ROZARIO
Magellan Petroleum Australia Pty. Ltd.
Brisbane, Queensland, Australia

FIELD CLASSIFICATION

BASIN: Amadeus
BASIN TYPE: Foredeep
RESERVOIR ROCK TYPE: Sandstone
RESERVOIR ENVIRONMENT OF DEPOSITION: Intertidal to Nearshore Marine
RESERVOIR AGE: Ordovician
PETROLEUM TYPE: Gas
TRAP TYPE: Anticline with Stratigraphic Component

LOCATION

The Palm Valley gas field is situated in the Amadeus basin, Northern Territory, near the center of Australia (Figure 1). The arcuate-shaped anticline (Figures 2 and 3) has strong surface expression and is one of a series of en echelon, generally westerly to northwesterly trending structures exposed in the basin's central province (Figure 4). The field is connected to the nearest town, Alice Springs, 75 mi (120 km) to the northeast by an 8-in. natural gas pipeline. A second pipeline, a 14-in. trunk line, runs 808 mi (1300 km) north from the field to the city of Darwin on Australia's northern coast. The other major field discovered in the Amadeus basin to date, the Mereenie field, is located 68 mi (110 km) to the west and produces both oil and gas from the same Ordovician formations that are productive at Palm Valley.

The Palm Valley field is unique in Australia, stemming from its predominant dependence on natural fracture permeability. Well bores that have intersected major fractures have yielded large gas flows, ranging up to 137 million standard cubic feet of gas per day (MMscfg/day) in the Palm Valley No. 6 well. Production test analysis and reservoir simulation studies (H. K. van Poollen and Associates, 1985) indicate the proved recoverable gas reserves for the field to be approximately 325 billion standard cubic feet (bscf), essentially comprising the gas contained within the fracture system and the more permeable zones of the matrix rock. The van Poollen study estimated the possible gas in place in both the fracture system and the total matrix rock at 1.8 trillion standard cubic feet (tscf). It is believed that a substantial portion of the in-place gas ultimately will be converted to recoverable reserves.

HISTORY

Pre-Discovery

Petroleum exploration in the Amadeus basin began in the late 1950s when Frome-Broken Hill Pty. Ltd. undertook limited geological and geophysical reconnaissance surveys over a two-year period. In 1960, Magellan Petroleum Corporation, on recommendation from U.S. petroleum consultant Dr. Duncan McNaughton, acquired petroleum prospecting rights over the north-central portion of the basin, in which the Palm Valley and Mereenie fields later were discovered (Figure 5). Exploration commenced with detailed surface mapping, photogeologic interpretation, and stratigraphic and structural studies, supplemented by seismic and gravity traverses. These studies led to the definition of a number of prospects, many of which were surface structures. The Palm Valley anticline was one of these latter structures, with the outcropping Devonian Hermannsburg Sandstone having a maximum vertical relief of approximately 1100 ft (335 m) above the adjacent Missionary Plain to the north. From the early surface mapping and photo geology, the anticline was identified as an arcuate generally westerly trending flexure having poorly defined axial plunge, particularly to the west (Figure 6). Its axis, nevertheless, could be traced for over 25 mi (40 km).

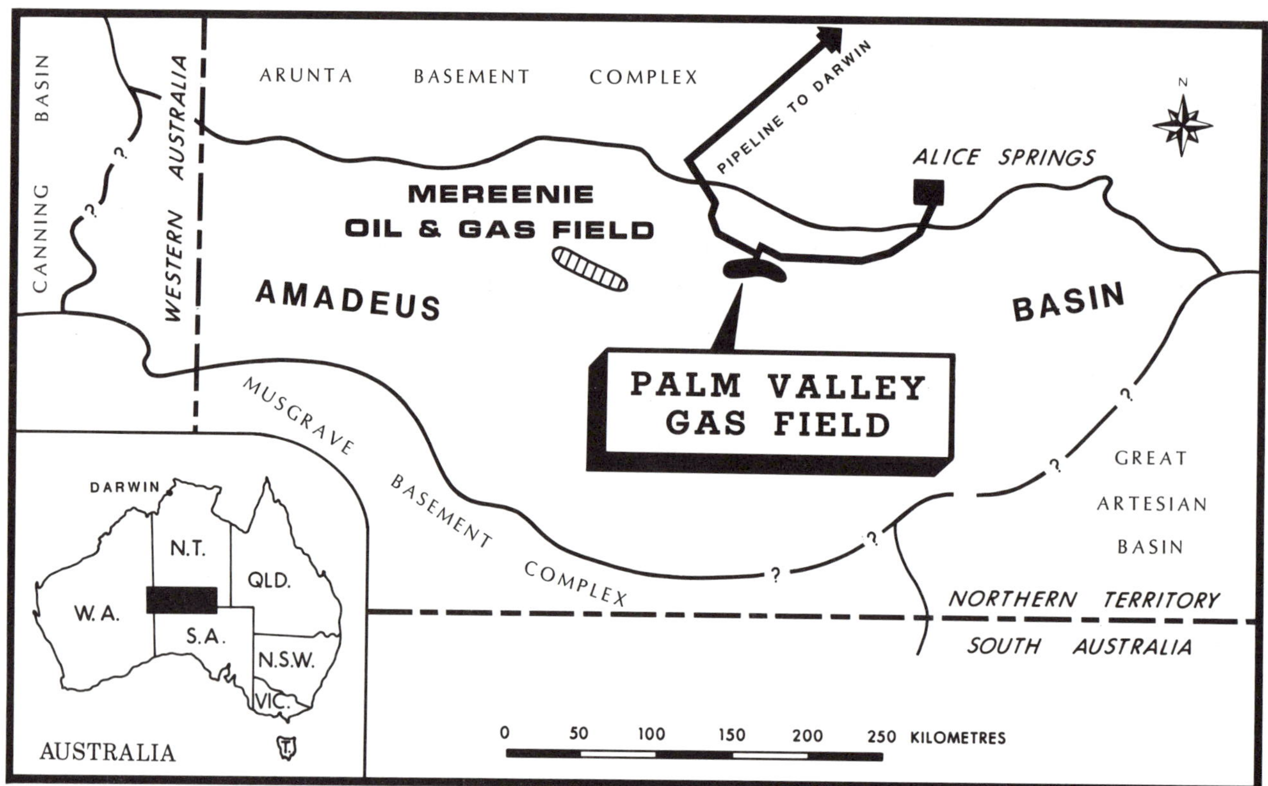

Figure 1. Location map of the Palm Valley gas field in the northern Amadeus basin, Central Australia.

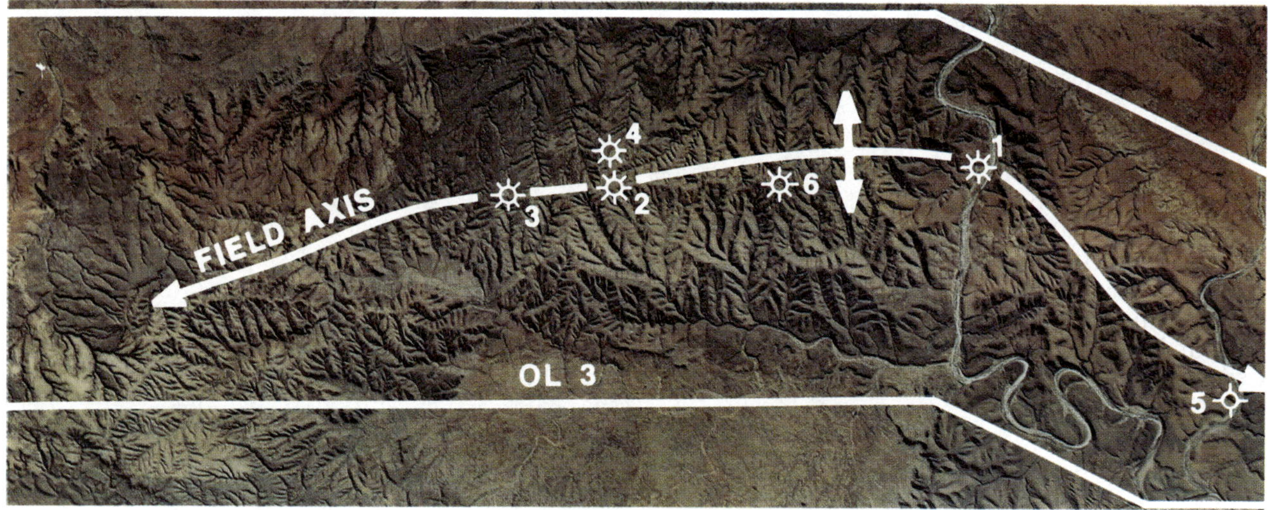

Figure 2. Air photograph of the Palm Valley field within Oil License 3 (OL). Note the arcuate shape of the anticline and its steeply dissected topography. Top is north. Area shown is about 15 × 40 km.

Drilling in the basin commenced in 1963 with the Ooraminna No. 1 well, which tested a small flow of gas from the Proterozoic Areyonga Formation. In the next year, the first discovery in the basin was made at Mereenie when the Mereenie No. 1 well, located at a crestal position on a surface anticline, flowed up to 11 MMcfg/day from Ordovician sandstones.

Discovery/Discovery Method

The Palm Valley field was discovered in 1965, with the drilling of Palm Valley No. 1 under the direction of the Magellan Petroleum Chief Geologist at the time, Mr. Roy Hopkins. The discovery well was located proximal to the surface anticlinal axis, down the eastern plunge from the apex (Figure 6). Prior

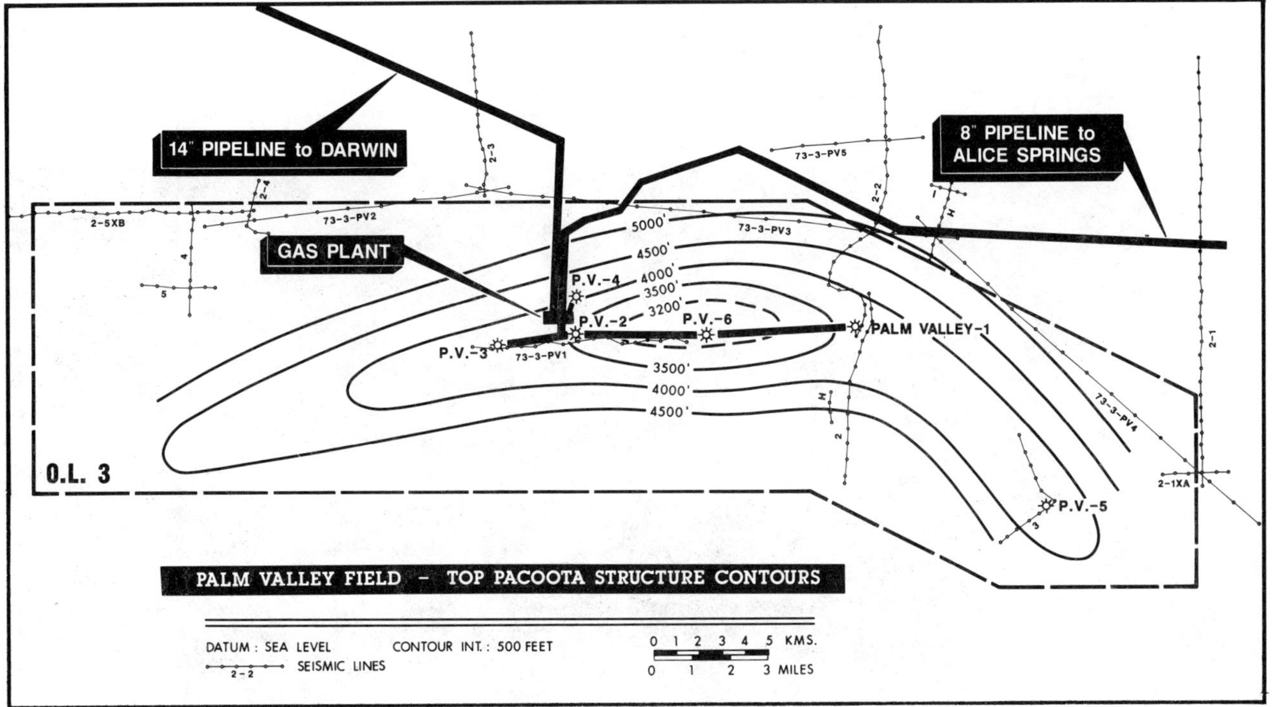

Figure 3. Structure map of the Palm Valley field at the Ordovician top Pacoota Sandstone horizon. Note sparseness of seismic data owing to rugged surface topography.

to drilling, the structure had been defined primarily by surface and photogeologic studies. Seismic control was limited to three short lines, Lines H, H, and I (Figure 3), shot by the Australian Government Bureau of Mineral Resources. The combined data showed strong dip on the respective north and south anticlinal flanks, but axial plunge, particularly to the west, was poorly defined. This led to uncertainty in determining the magnitude of structural closure, which was further complicated at depth because of unconformities in the Devonian sequence overlying the objective Ordovician section (Figure 7).

The discovery well, Palm Valley-1, was completed as a shut-in gas well after reaching a total depth of 6658 ft (2029 m) in the Ordovician Pacoota Sandstone. Gas was encountered in the Ordovician Lower Stairway Sandstone, the basal Horn Valley Siltstone, and the Pacoota Sandstone (Figure 8). The maximum flow recorded was 11.7 MMcfg/day from an open-hole test while drilling with air, near the base of the Horn Valley Siltstone. Subsequent production tests conducted over two intervals in the Lower Stairway Sandstone and upper part of the Pacoota Sandstone, after the well had been mudded up and killed, yielded maximum flows of 3.6 and 3.7 MMcfg/day, respectively, indicating extensive formation damage to the reservoir.

Studies of cores and wireline logs from all three productive zones showed generally very low porosities and permeabilities. Consequently, productivity was attributed mainly to fractures that were identified, both macroscopically and microscopically, from cores (Figure 9). This was confirmed by later reservoir testing.

Post-Discovery

Owing to the lack of commercially viable markets, appraisal drilling proceeded slowly over a number of years with the next two wells, Palm Valley Nos. 2 and 3, being drilled in 1970 and 1973, respectively. These wells also were located close to the surface structural axis and again encountered significant gas flows from Ordovician reservoir rocks. The No. 2 well, in fact, recorded an Australian onshore record flow at the time of 69.7 MMcfg/day, after intersecting large open fracturing in the uppermost Pacoota Sandstone.

Additional seismic was also acquired during the period 1966 to 1973 to further define the anticline's subsurface structure. Owing to the ruggedness of the terrain, however, it was limited to accessible routes such as along part of the field's surface axis or outer flanks and down river beds (Figure 3). These confirmed roll-over in a north-south direction (lines 2, 2-2, and 3, Figures 10 and 11), but there were insufficient data to map axial plunge.

Following the signing in 1981 of a contract to supply gas to the town of Alice Springs, small-scale gas production commenced in mid-1983, with volumes currently averaging 3 to 4 MMcfg/day. Prospects of more substantial gas sales to the northern coastal city of Darwin and towns in

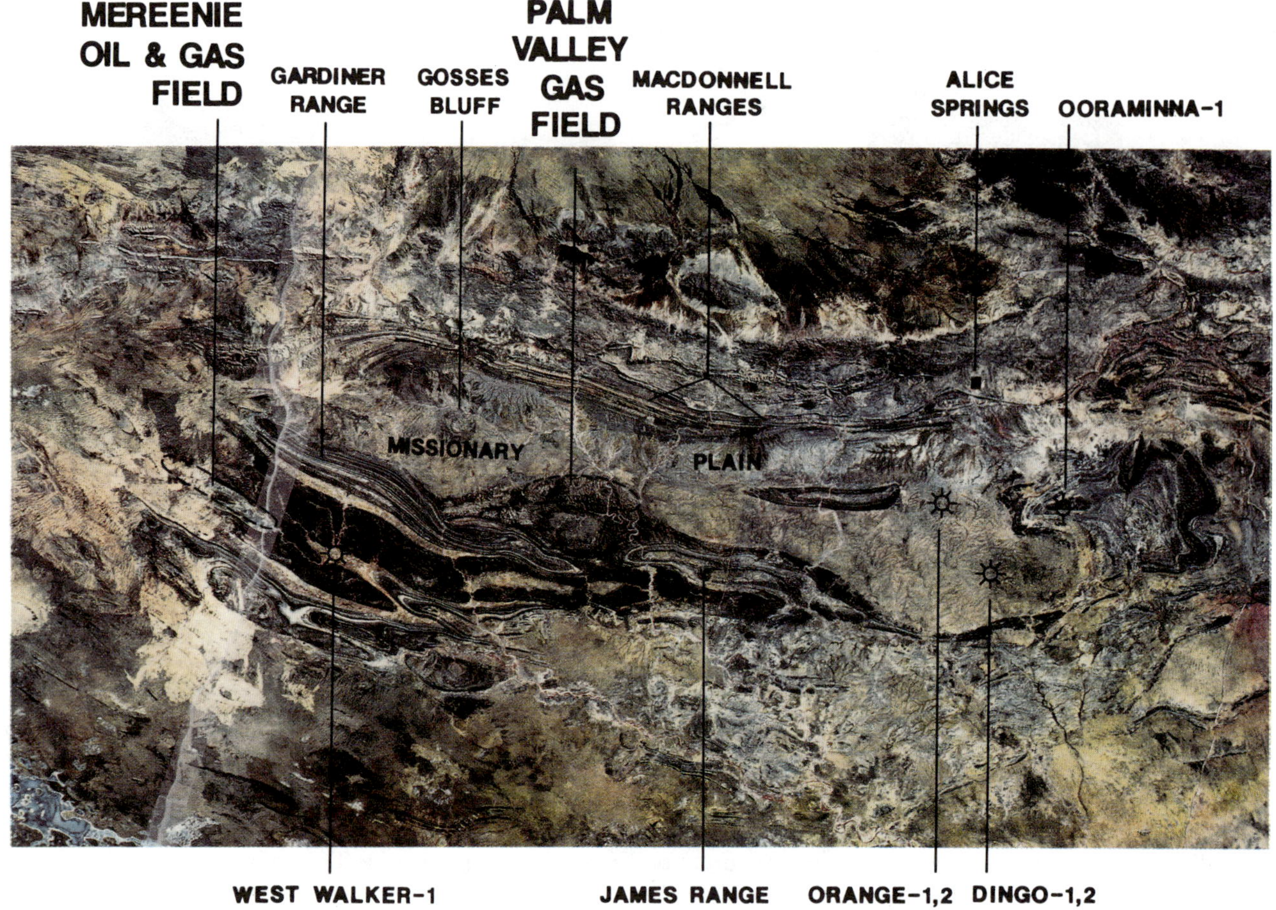

Figure 4. LANDSAT image of the central and eastern Amadeus basin with significant hydrocarbon discoveries marked. Note the excellent surface outcrop and sinuous configuration of structures in the central part of the basin. Area shown is about 350 × 650 km.

between, as an alternative fuel to petroleum liquids for electricity generation, resulted in three additional appraisal/development wells being drilled between late 1984 and early 1986—Palm Valley Nos. 4, 5, and 6—in order to confirm reserve estimates and increase field deliverability. Palm Valley No. 4 was located on the field's northern flank, whereas Nos. 5 and 6 again were drilled close to the surface structural axis (Figure 2). All wells penetrated a thick gas-bearing Ordovician section; however, only Palm Valley No. 5, situated on the southeastern nose of the field, failed to encounter commercial gas flows owing to a lack of open fracturing. The structural and stratigraphic distribution of gas flows from all field wells is illustrated in Figure 12, and test results are summarized in Table 1. The success in significantly increasing deliverability from the field enabled the Darwin gas supply contract to be signed in 1985. Subsequently, production from the field has increased to an average of 20 MMscfg/day, following completion of the connecting trunk pipeline in the latter part of 1986.

REGIONAL SETTING

The regional geology of the Amadeus basin has been described previously by the Bureau of Mineral Resources (Wells et al., 1970; Forman and Shaw, 1973), and Magellan Petroleum (McNaughton et al., 1968; Froelich and Kreig, 1969; Pearson and Benbow, 1976). More recent work on hydrocarbon generation and source potential within the basin has been published by Pancontinental Petroleum Limited (Schroder and Gorter, 1984; Gorter, 1984) and Jackson et al. (1984).

The basin is an east-west-trending, intracratonic depression in central Australia, containing approximately 38,000 ft (11,500 m) of upper Proterozoic to

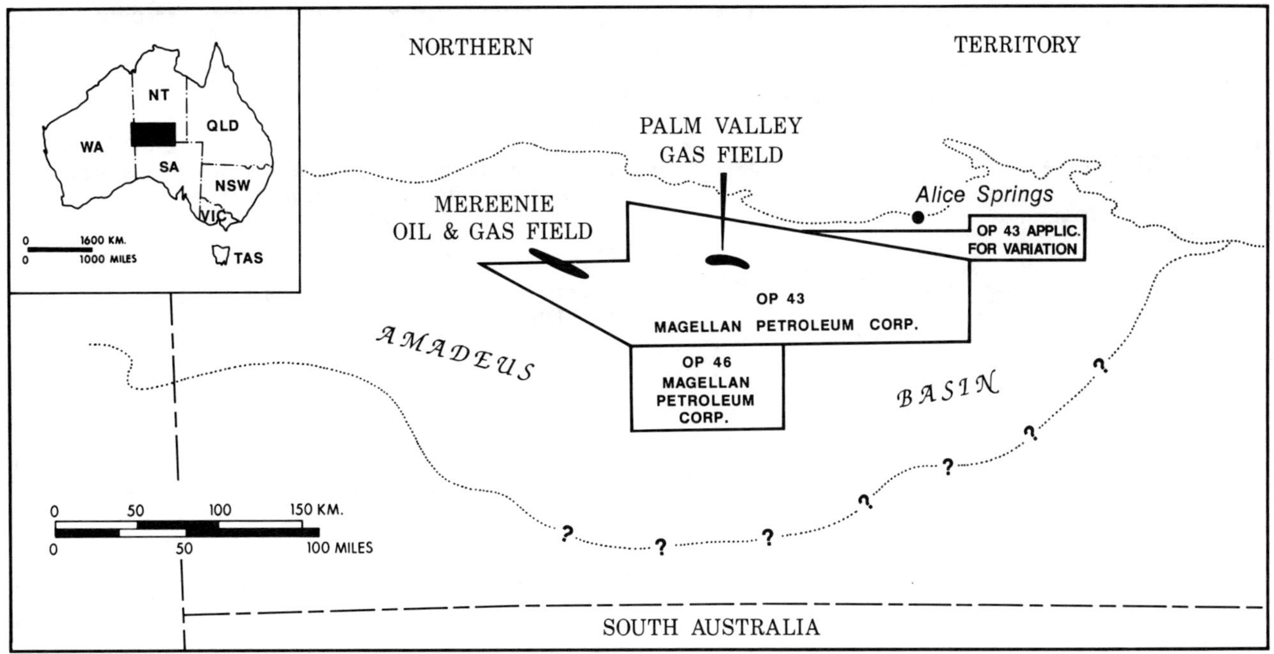

Figure 5. Location of the original petroleum exploration permits (OP 43 and 46) in the Amadeus basin acquired by Magelllan Petroleum Corporation in 1960. The position of the later discovered Mereenie and Palm Valley fields has been indicated.

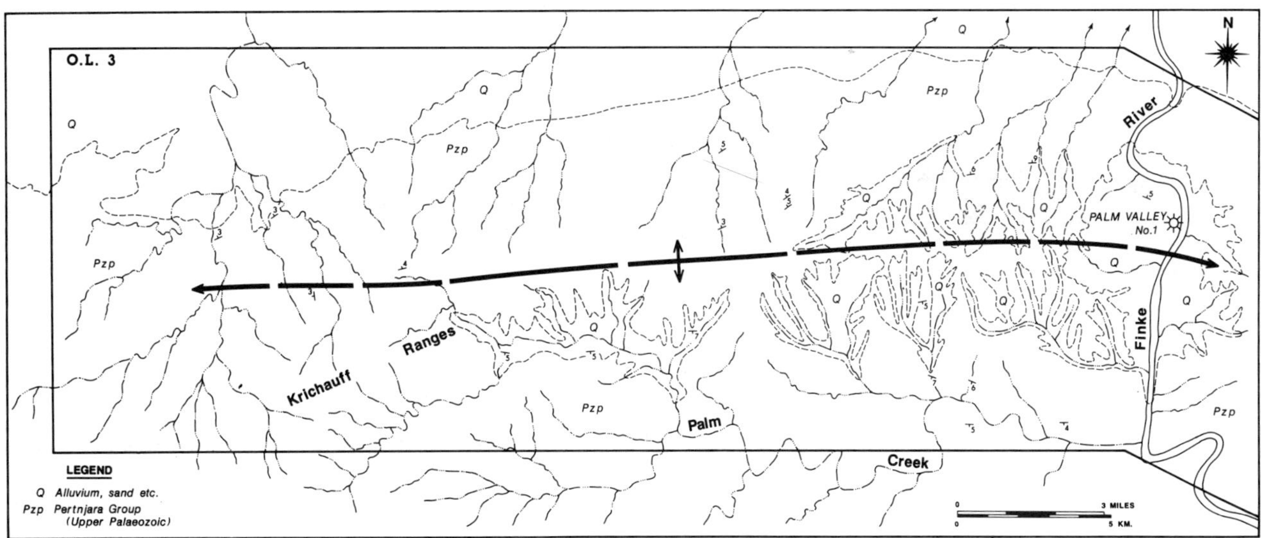

Figure 6. Surface geological map of the Palm Valley structure with the location of the discovery well, Palm Valley No. 1. (From R. M. Hopkins, 1961.) The western plunge of the structure was poorly defined from outcrop data.

Upper Devonian shallow-marine to continental sediments. The basin is a remnant of a once much larger depositional province that covered a vast area in central Australia. It is bound by the Precambrian igneous and metamorphic rocks of the Arunta Complex to the north and the Musgrave Complex to the south (Figure 1). To the east, sediments are truncated and overlapped by Mesozoic sediments of the Great Artesian basin (Pearson and Benbow, 1976), whereas westwards, the sedimentary section thins over basement and merges with the predominantly Paleozoic Canning basin. The distribution of upper Proterozoic sediments was controlled by an east-west-trending hinge line across the north-central part

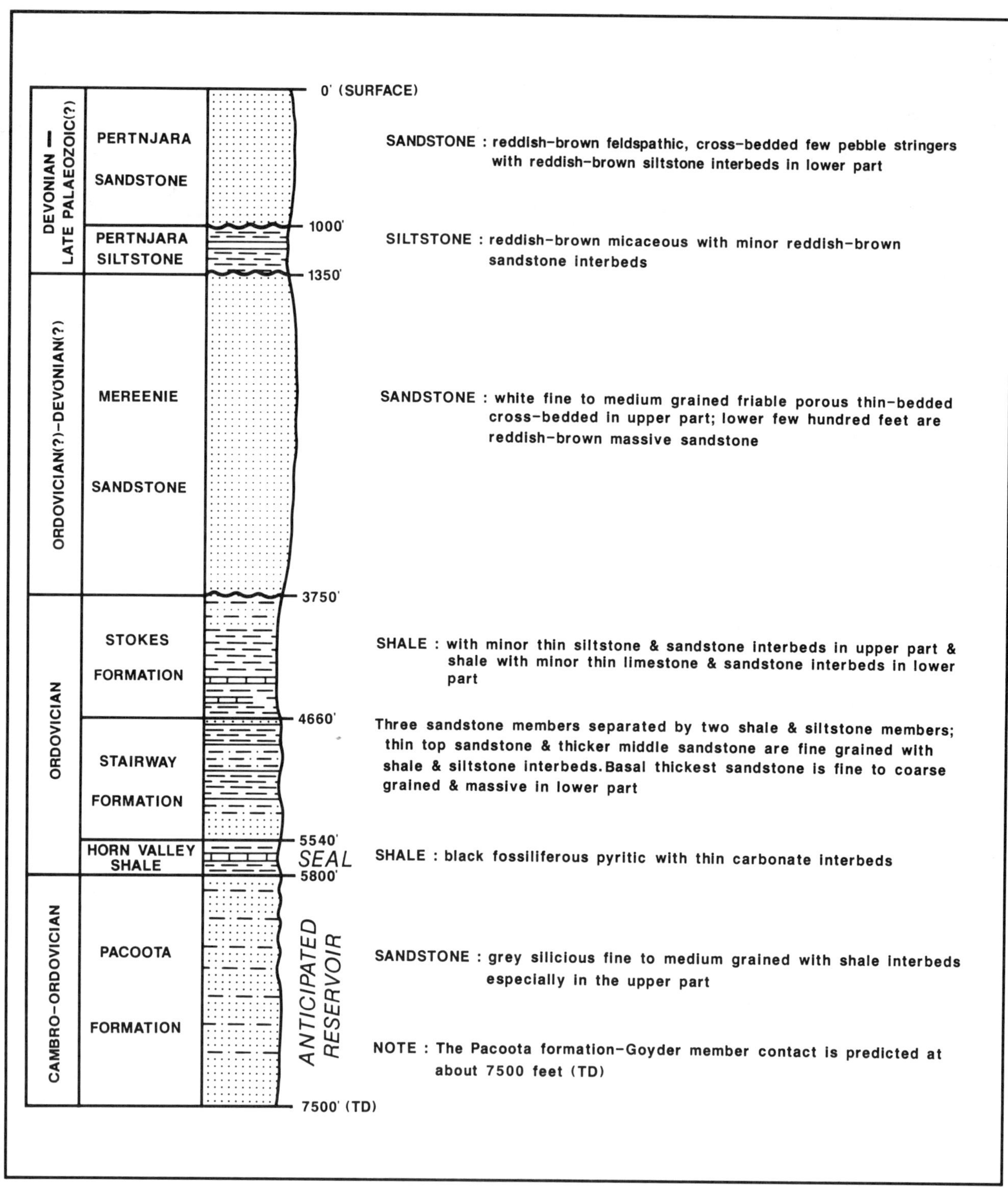

Figure 7. Predicted stratigraphic section, Palm Valley No. 1. (From Application for a Stratigraphic Drilling Subsidy, Palm Valley No. 1, 1964.) Note expected unconformities in the Devonian sediments overlying the anticipated Cambrian-Ordovician reservoir rocks.

of the basin from which the thickness of section increases both northward and southward (Jackson et al., 1984). The Paleozoic section is thickest along the present northern margin of the basin, the region in which the Palm Valley field is located (Figure 13). This is an area of generally sinuous en echelon style structuring bordering the Missionary Plains syncline, immediately south of the MacDonnell Ranges. The basin's stratigraphic sequence, the stratigraphic distribution of hydrocarbon occurrences, and the major tectonic episodes are summarized in Figure 14.

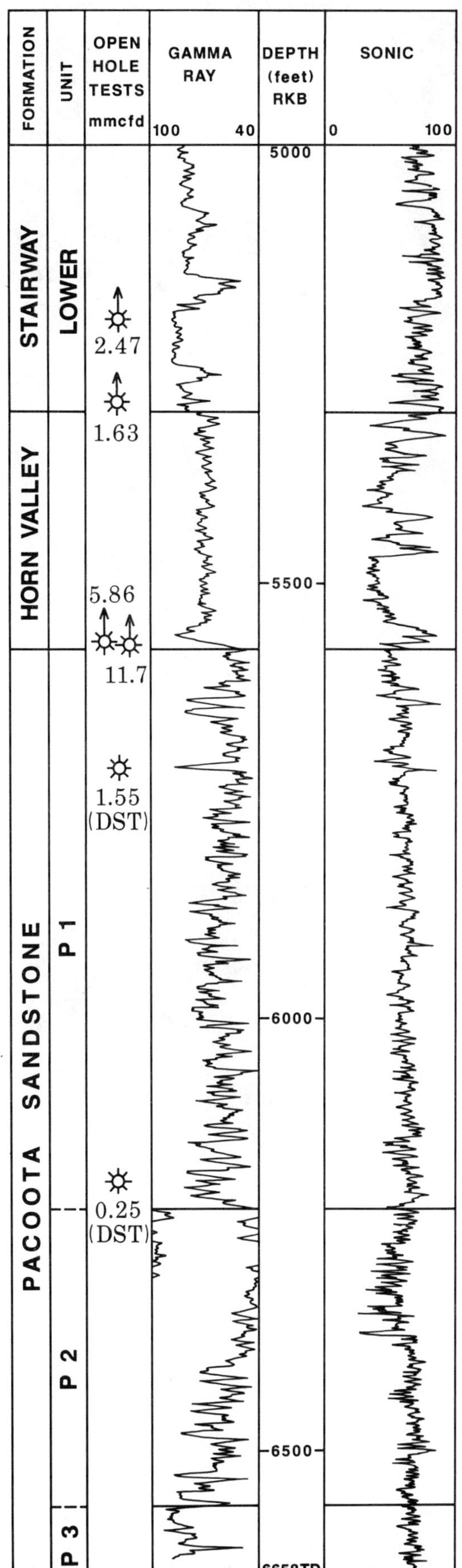

Figure 8. Wireline log section over the reservoir zone, Palm Valley No. 1. Gas flows (cumulative) recorded while air drilling are noted, as well as flows on drill-stem test after the well was mudded up and killed.

STRUCTURE

Regional

The structural expression of the Amadeus basin is strikingly manifested in the surface outcrop of the north-bounding MacDonnell Ranges and the prominent system of extensive folds that parallel or subparallel the basin margins. This is clearly illustrated in the composite LANDSAT photo of the area (Figure 4) on which the location of the Palm Valley field and other hydrocarbon discoveries has been noted.

The tectonic evolution of the basin is complex, owing to the influence of at least six orogenic episodes ranging from the late Proterozoic Areyonga Movement to the Late Devonian to Early Carboniferous Alice Springs orogeny. Of these, the Petermann Ranges and Alice Springs orogenies are considered to have been the two major diastrophic events that had the greatest effect on structural development within the basin.

The late Proterozoic Petermann Ranges orogeny resulted in the termination of deposition in the southwest owing to associated mountain building, recumbent folding, and northward thrusting. Salt within the Bitter Springs Formation is thought to have served as a decollement surface for this thrusting. Gorter (1984) suggests, however, that deep crustal movements were responsible for much of the deformation in the Amadeus basin, except in the northeastern region. Structural features related to the Petermann Ranges orogeny are evident predominantly in the southwest of the basin. These are generally east-west-striking anticlines showing thinning of upper Proterozoic sediments over the crest.

Structures initiated during the late Proterozoic were subsequently modified and new ones created during the Late Devonian to Early Carboniferous Alice Springs orogeny. In the eastern part of the basin, recumbent nappes and overthrusts formed in the Arunta Complex, while folds, strike-slip faults, and thrust sheets deformed the overlying sedimentary section. In the central, northern, and western parts of the basin, numerous large-scale folds were generated, often well-exposed at the surface. Fold axes trend westerly in the center of the basin; however, further west they are aligned in a more northwesterly direction. In some cases, structuring has been complicated by halokinesis, resulting in salt-cored anticlines.

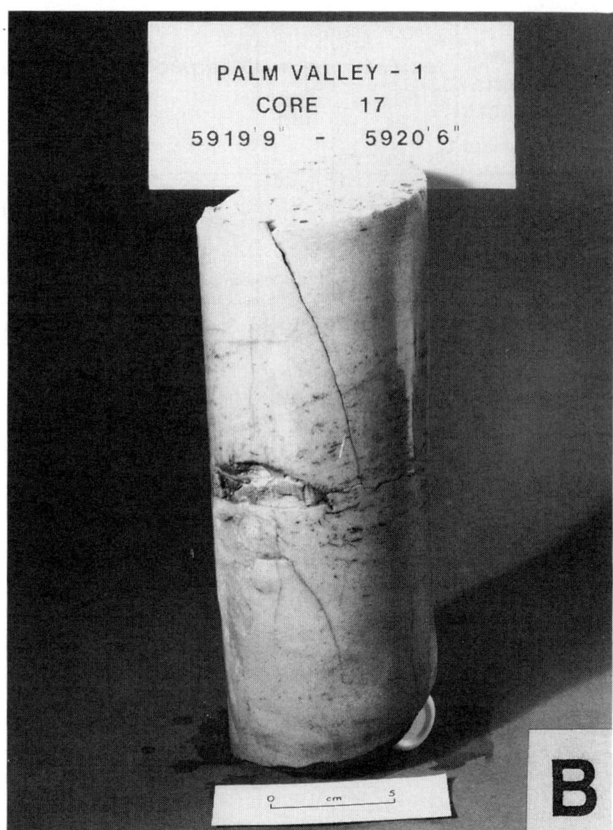

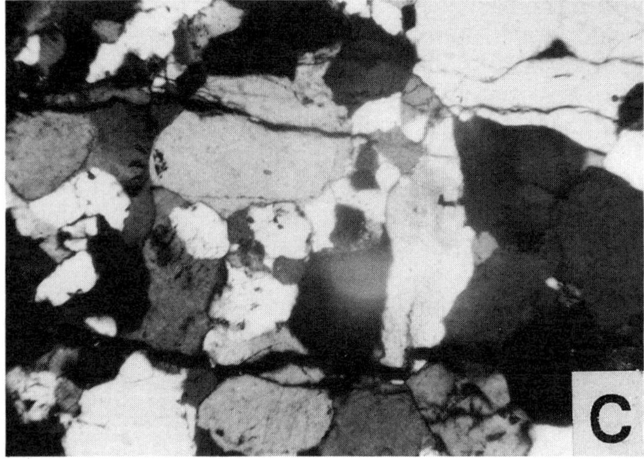

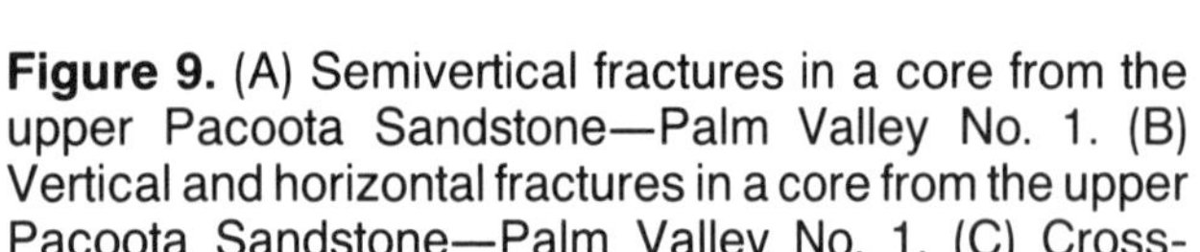

Figure 9. (A) Semivertical fractures in a core from the upper Pacoota Sandstone—Palm Valley No. 1. (B) Vertical and horizontal fractures in a core from the upper Pacoota Sandstone—Palm Valley No. 1. (C) Cross-polarized light photomicrograph of quartzite from Pacoota Sandstone showing hairline open fracture and welded grain contacts. Width of field is 2 mm.

Local

The system of folds in the Palm Valley region (Figure 4) generally was formed by compression from the northeast during the initial stages of the Alice Springs Orogeny (Bradshaw and Evans, 1988). Subsequent dextral strike-slip movement resulting from southeasterly directed compressive forces in the latter stages of the Alice Springs orogeny initiated en echelon style folding in the Gardiner and James Ranges.

It commonly has been considered that the Palm Valley anticline was a contemporaneous part of this en echelon fold system. Recently, however, a Magellan worker has proposed that the structure may be a younger flexure that formed from halokinetic activity commencing during late or post Alice Springs

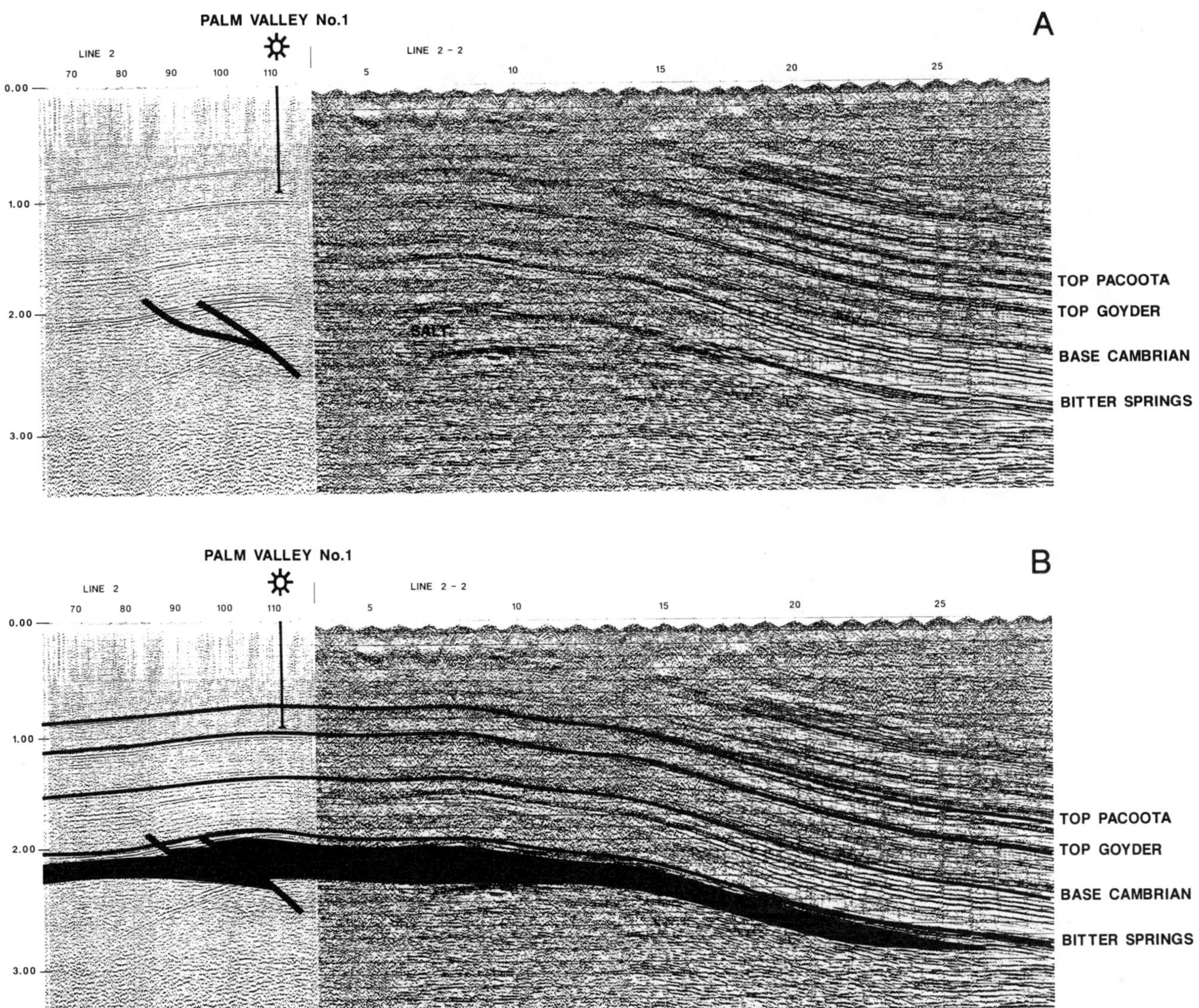

Figure 10. Seismic section of lines 2 and 2-2 across the axis of the Palm Valley anticline (see Figure 3 for location). (A) is uninterpreted. (B) is interpreted. The flattening of the structure along the crest is only apparent because of the curvature of the seismic lines that follow the river beds. Note presence of salt pillowing interpreted (L. E. Roe, 1988) in the Proterozoic Bitter Springs Formation.

orogeny time. It is suggested that such an origin could explain the relative lack of breaching of the anticline by erosion, as compared to all other adjacent uplifted features, and also explain why the major axial fracture system is open and conductive to fluids, even though the trend of these fractures essentially is perpendicular to the regional principal horizontal compressive stress field now active in the rocks.

STRATIGRAPHY

The stratigraphic sequence of the Amadeus basin comprises Proterozoic to upper Paleozoic shallow-marine to continental sediments (Figure 14). In the Palm Valley field, only the Devonian Pertnjara and Mereenie units and the Ordovician Larapinta Group have been penetrated by wells drilled to date. The paleogeography of the hydrocarbon-bearing Larapinta Group is illustrated in Figure 15. Within this section, gas has been encountered in the lower part of the Stairway Sandstone, the basal Horn Valley Siltstone, and the Pacoota Sandstone. These formations are described briefly in the following paragraphs.

The *Stairway Sandstone* ranges in thickness between 969 ft and 1018 ft (295–310 m) in wells drilled in the Palm Valley field to date and is a shallow, intertidal, marine sequence, of sandstones, siltstones, and shales. A basal quartzitic unit, approximately 130 ft (40 m) thick, is occasionally pyritized and fractured and has associated good gas production. Sandstones are generally clear, white to pale yellow

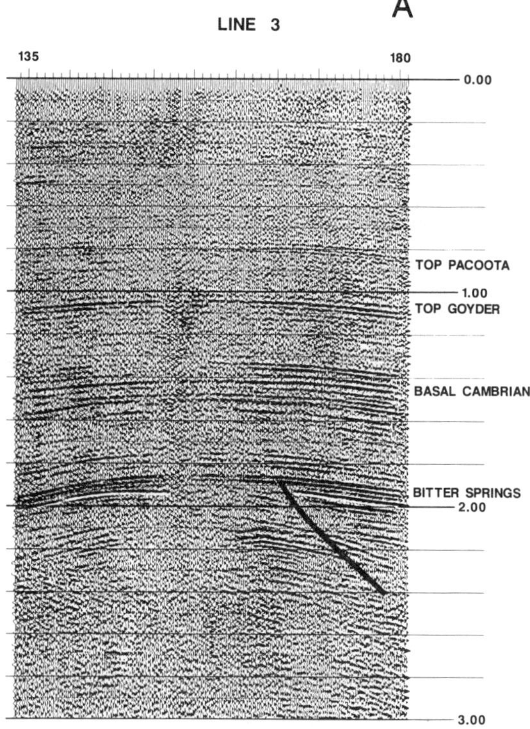

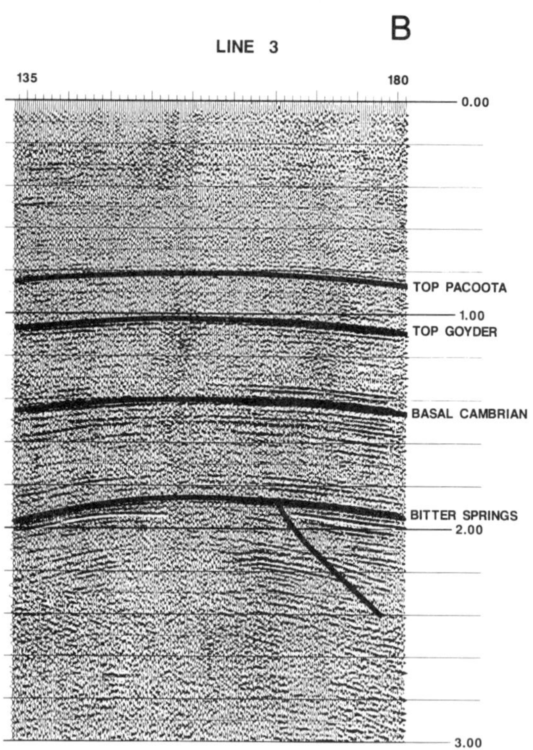

Figure 11. Seismic section of line 3 across the eastern nose of the Palm Valley anticline (see Figure 3 for location). (A) is uninterpreted. (B) is interpreted. (Interpretation by L. E. Roe, 1988.) Note relatively uniform flexuring and probable faulting in the Proterozoic Bitter Springs Formation.

and orange, very fine to medium- and occasionally coarse-grained, moderately to well sorted, with subangular to subrounded dominantly quartz grains. They are moderately to well cemented with silica resulting in fair to poor porosities. Interbedded with the sandstones are thin to occasionally thick siltstone units. These are medium to dark gray and gray-brown, hard, siliceous, and micaceous, becoming slightly calcareous with depth. Shales are medium to dark gray and brown, hard, and brittle. The contact with the underlying Horn Valley Siltstone is gradational.

The *Horn Valley Siltstone* consists of euxinic siltstone and shale with thin interbeds of limestone and dolomite. The formation was deposited on a shallow-marine shelf and ranges in thickness between 330 and 373 ft (100–114 m) at Palm Valley. A limestone and dolomite layer 10 to 20 ft (3-6 m) thick at the base of the formation is gas productive in Palm Valley No. 1.

The siltstones are medium to dark gray and black and are locally carbonaceous and bituminous. They are hard, blocky, micromicaceous, and calcareous throughout, as well as containing pyrite both in disseminated and nodular form. The shales have all the above characteristics but are fissile. The black color of the siltstones and shales, together with the high percentage of pyrite, suggest a strongly reducing environment of deposition.

The limestones encountered were white, buff, and gray in color, translucent, hard, micritic to cryptocrystalline, and appear throughout the lower half of the formation as discrete thin beds. The basal carbonate unit marks the boundary and is conformable with the underlying Pacoota Sandstone.

The *Pacoota Sandstone*, the main gas-bearing reservoir, consists of nearshore, marine, interbedded quartzose sandstones, siltstones, and shales. The formation is subdivided into distinct units on the basis of lithology and log character (Figure 8). No wells drilled in the Palm Valley field have penetrated the entire Pacoota section, which is estimated to be approximately 1700 ft (518 m) thick from seismic and regional data. Gas flows have so far been encountered in the upper three Pacoota units.

The sandstones are predominantly clear, yellow, and pale gray, becoming pink to medium red-brown where ferruginous. They are generally very fine to fine, occasionally medium to coarse grained, subangular to subrounded, moderately well sorted, and generally well cemented with silica, displaying common quartz overgrowths. Glauconite is abundant in the middle Pacoota Sandstone P2 unit.

Siltstones are dark gray, black, and occasionally red-brown, hard, blocky to subfissile and generally siliceous. They are also pyritic and micromicaceous and grade in part to and are interbedded with very fine sandstones. Shales are pale to dark gray and black, dark green and gray-green, hard, brittle, fissile, and micromicaceous, and are generally thinly interbedded with the siltstones.

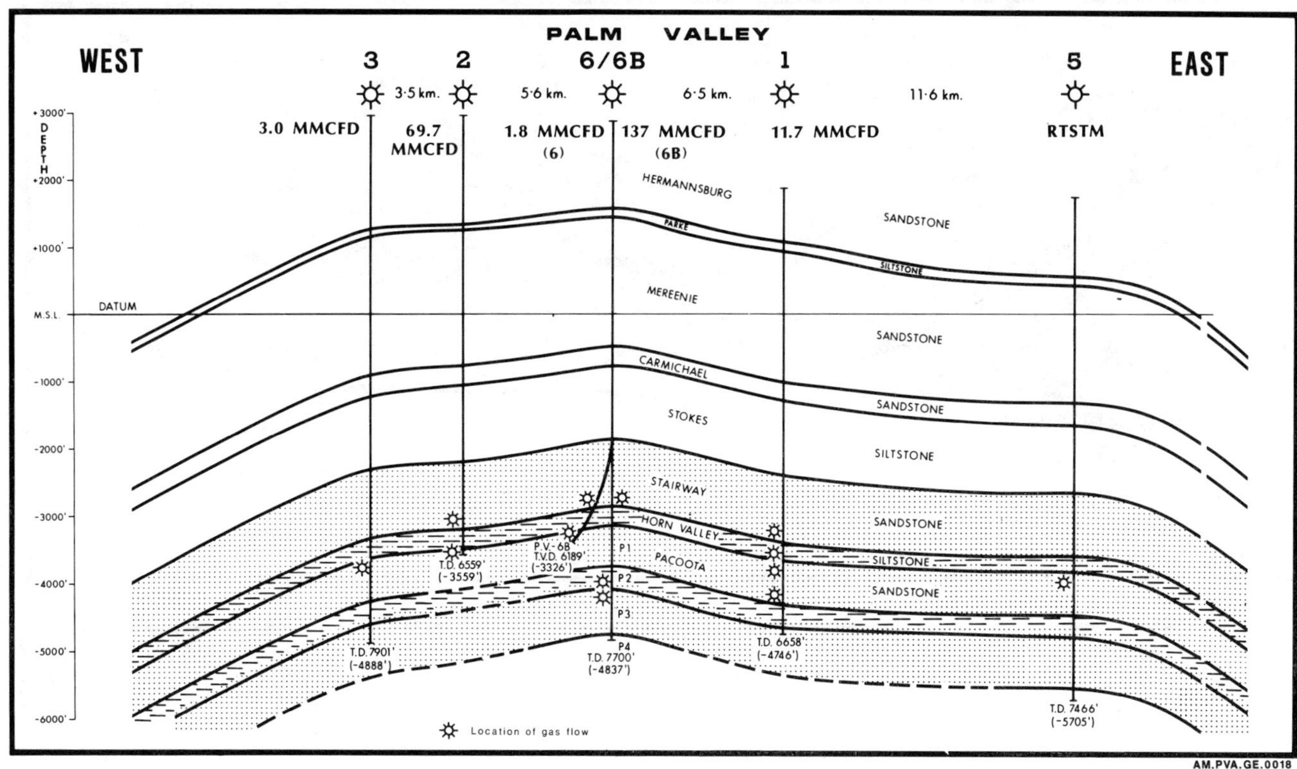

Figure 12. Diagrammatic west-east cross section of the Palm Valley field showing stratigraphic distribution of gas flows in axial wells.

Table 1. Flow tests results summary—Palm Valley field.

WELL	YEAR COMPLETED	MAXIMUM FLOW RATE (MMSCF) FROM OPEN HOLE TEST						AOF MMSCF (July 1986)
		L. STAIRWAY	HORN VALLEY	PACOOTA P1 UNIT	P2 UNIT	P3 UNIT	P4 UNIT	
PALM VALLEY 1	1965	2.47	11.7	(1.55)	-	PP	NP	37
PALM VALLEY 2	1970	3.86	-	69.7	NP	NP	NP	47.8
PALM VALLEY 3	1973	-	-	3.0	-	PP	NP	8.7
PALM VALLEY 4	1985	-	-	1.5	-	PP	NP	3.0
PALM VALLEY 5	1985	-	-	RTSTM	-	-	PP	-
PALM VALLEY 6	1986	0.069	-	-	1.3	1.82	PP	Plugged
PALM VALLEY 6B (Deviated Well)	1986	12.1	-	137	NP	NP	NP	8.2

PP : Partially Penetrated
NP : Not Penetrated

RTSTM : Rate Too Small To Measure
( ) : From Drill Stem Test
AOF : Absolute Open Flow

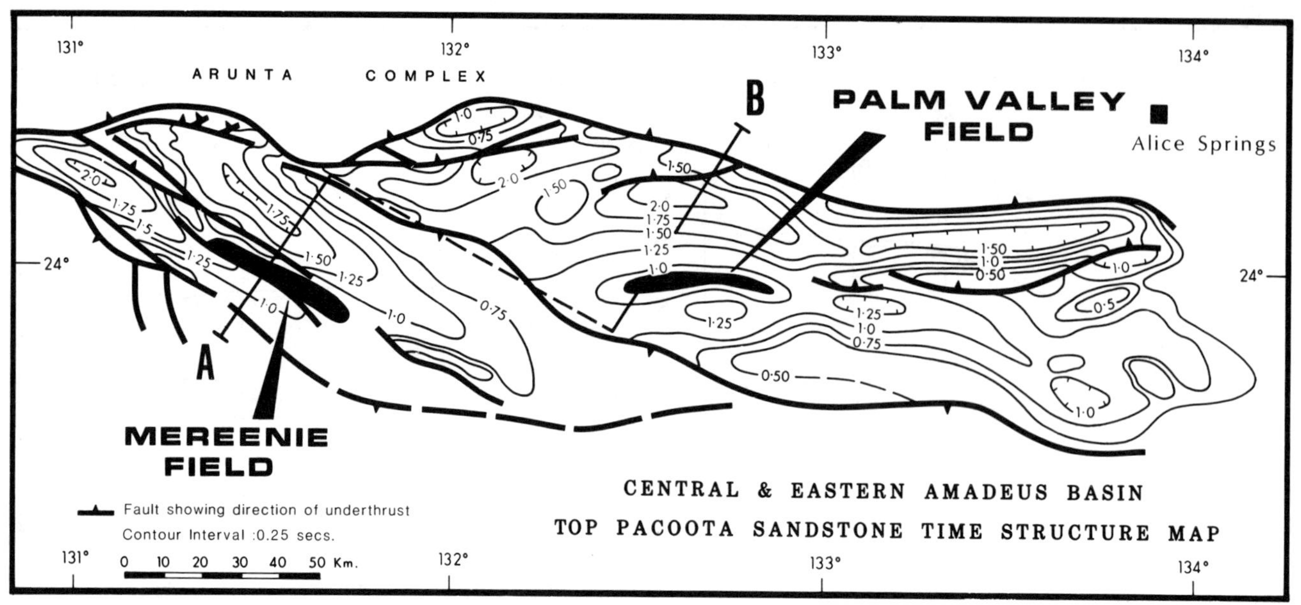

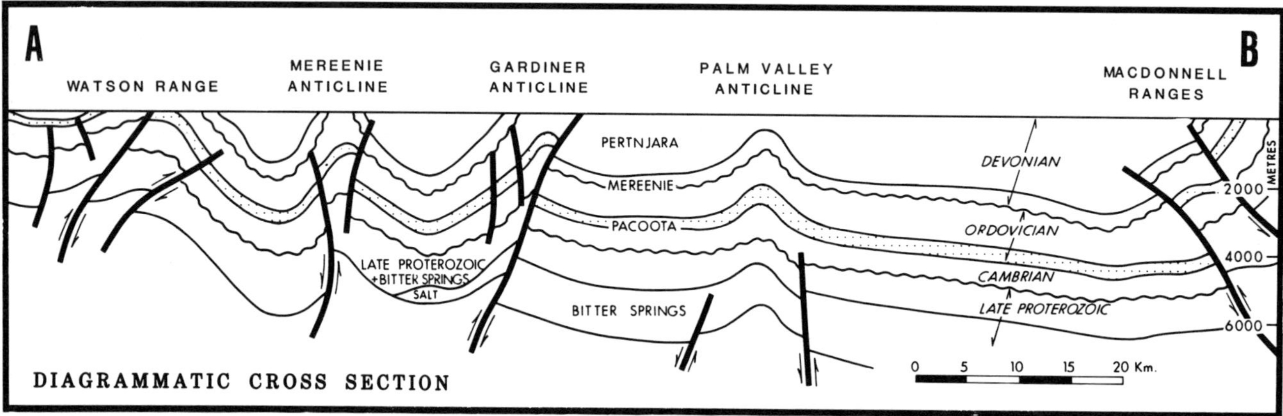

Figure 13. Central and eastern Amadeus basin top Pacoota Sandstone (main reservoir formation) time structure map and diagrammatic cross section. (After Schroder and Gorter, 1984.) Note increase in sediment thickness toward the basin's northern margin and presence of extensive thrust faulting.

TRAP

The Palm Valley field is a combination structural/stratigraphic trap with gas being reservoired in an anticlinal closure and the distribution of productive zones controlled by fracture porosity and permeability. North and south surface flank dips mapped from outcrop are confirmed in the subsurface at the Stairway/Pacoota Sandstone reservoir level by the small amount of seismic existing over the field (Figure 3) and control from the one existing flank well—Palm Valley No. 4. There is insufficient seismic control to map plunge in an east-west strike direction; however, data from the five axial wells indicate its existence at the top Pacoota Sandstone horizon over a distance of at least 17 mi (27 km).

The reservoir is sealed by siltstones and tight sandstones of the Horn Valley and Middle Stairway formations, overlain by shales of the Stokes formation, with gas flows confined to the Lower Stairway and underlying Pacoota Sandstone (Figure 12). In one well, Palm Valley No. 1, a gas flow also was recorded from the dolomite marker bed at the base of the Horn Valley Siltstone, owing to the presence of fractures in that brittle carbonate unit.

The maximum gross hydrocarbon column encountered in the Palm Valley field to date is in the No. 6 crestal well and is approximately 1600 ft (487 m), extending from the Lower Stairway to the upper Pacoota P3 unit. Although wells have penetrated below this level, there has been no measurable increase in gas production. In addition, no gas-water contact has been established so far, as water has been recovered from only one well, Palm Valley No. 1, in a drill-stem test of the Pacoota P2 unit. This water is believed to be stratigraphically isolated, based on pressure data and the lack of any other water influx in the field.

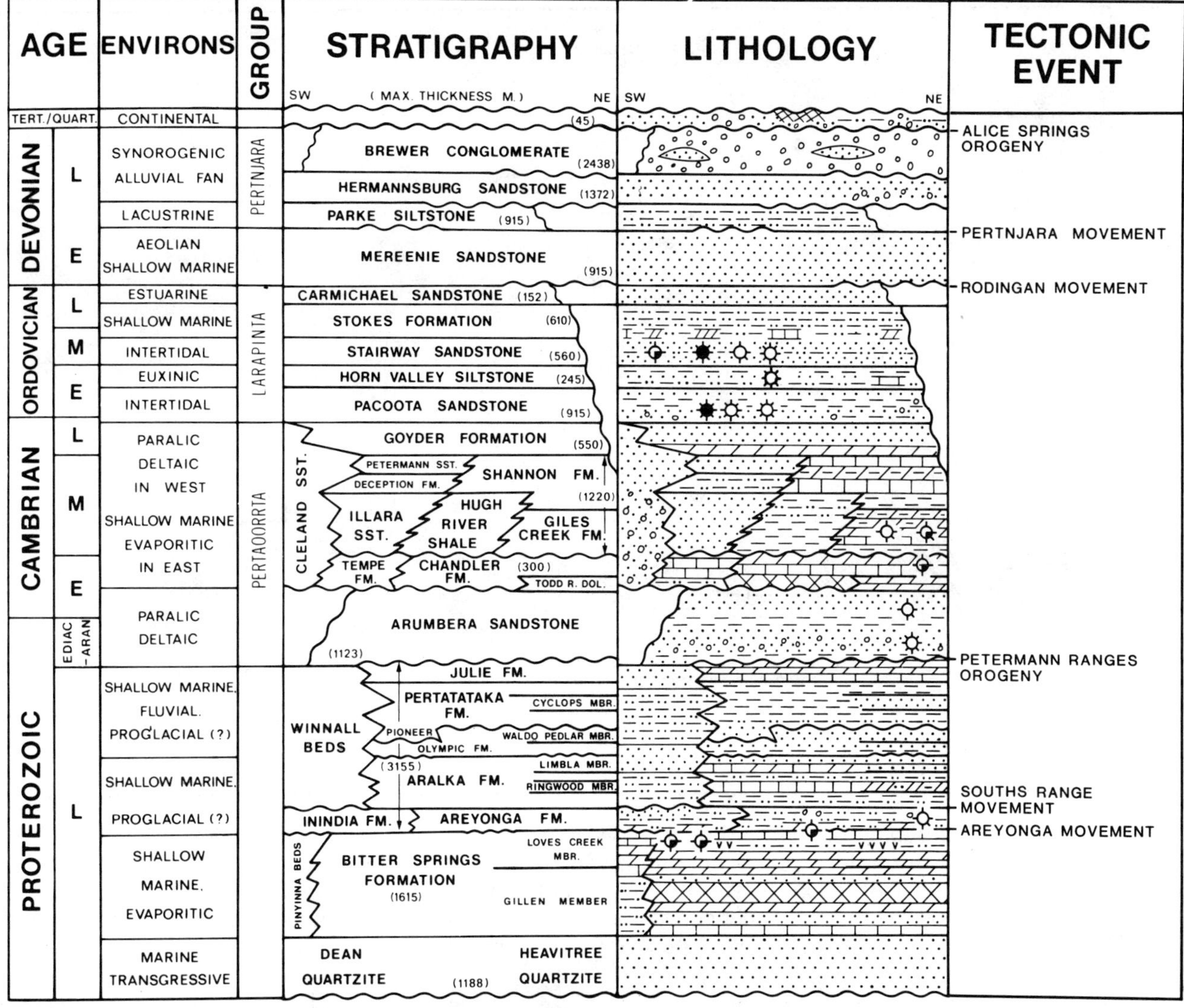

Figure 14. Stratigraphic table, Amadeus basin, including the distribution of oil and gas occurrences. (After Jackson et al., 1984.) Six major tectonic episodes have been recognized, with the most recent, the Alice Springs orogeny, considered chiefly responsible for thrusting, large-scale folding, and wrench movement in the basin's central and northern areas.

As discussed earlier, trap formation may have occurred later than the main and most active stage of the Alice Springs orogeny, which was responsible for the bulk of the deformation in the central and northern parts of the basin. Tensional forces resulting from vertical movement possibly created the fracture system that plays such an important role in the Palm Valley reservoir and its associated high gas deliverability. The nature of the fracturing and its ramifications with respect to field development and, indeed, to exploration in other areas is the subject of on-going research by Magellan.

RESERVOIR

The reservoir for the Palm Valley field is both multilayered and heterogeneous, with gas being contained in fractures and low porosity Ordovician quartzose sandstones and carbonates. The reservoir fits into the restricted category of fields worldwide characterized by predominantly fracture dependent permeability.

The reservoir sandstones usually contain some detrital clay matrix and are well cemented, with generally very little original pore space remaining (Figure 9). Original porosities were probably in the range of 20 to 30%. Subsequent burial and diagenesis, however, have resulted in infilling of the pore spaces by carbonate and silica (Figure 16), the latter deposited as crystal overgrowths caused by pressure solution and suturing. As a consequence, present-day porosities are very low, averaging 2 to 4%, with associated permeabilities in the order of 0.1 md. This is illustrated in Table 2, a summary of the core analysis results from Palm Valley No. 1. Table 3 presents a representative mega- and microscopic

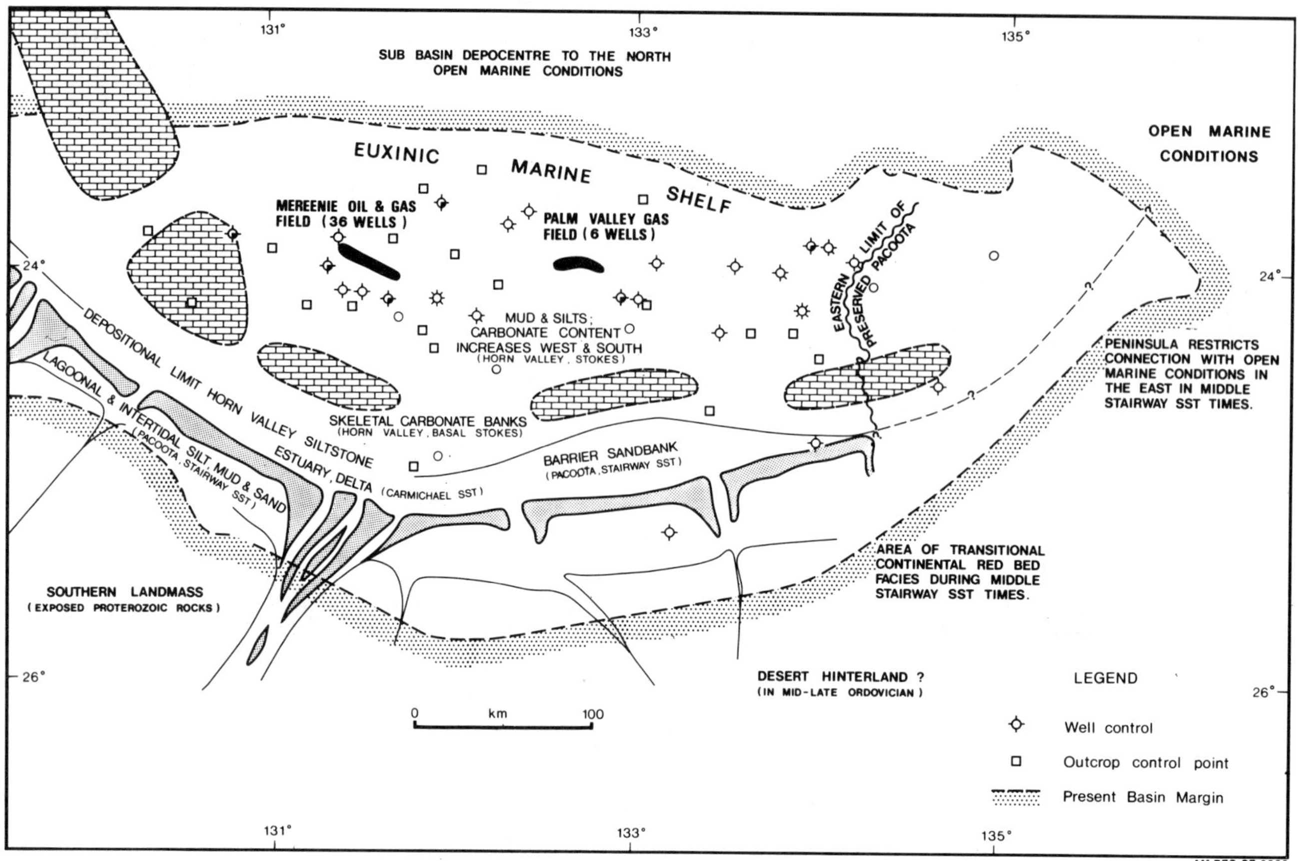

Figure 15. Paleogeography of the hydrocarbon-bearing Ordovician Larapinta Group, maximum transgressive phase. (After Jackson et al., 1984.) A marine transgression from the north commenced in the Early Ordovician resulting in the deposition of euxinic marls, muds, and silts (Horn Valley Siltstone) on an open shelf beneath well-oxygenated upper waters that supported a rich benthonic fauna. Intertidal flats and barrier bars (Pacoota Sandstone) formed landward of this open shelf.

description of a core sample taken from the Upper Pacoota P1 unit in the same well. Note the presence of feldspar as an accessory mineral that may constitute up to 10% of the finer grained, more argillaceous sandstones.

The low observed porosities and permeabilities cannot account for the significant flows of gas encountered while drilling. It was concluded from examination of cores and thin sections that fractures were chiefly responsible for productivity in the reservoir. Figure 9 illustrates fracture occurrence in two cores cut in the Pacoota Sandstone in Palm Valley No. 1. Fractures are predominantly vertical to subvertical, although some subhorizontal fracturing is present. The existence of fracturing at the microscopic level was confirmed by thin-section analyses of cores, in which fine hairline fractures can be seen cutting across the welded quartz grains (Figure 9). Fractures also were detected from the comprehensive suite of wireline logs run in more recent wells. The results of consequent subsurface fracture analysis studies have been documented in a paper by Do Rozario and Baird (1987), some conclusions from which are summarized below.

Fractures

There are two dominant fracture trends at reservoir depths in the Palm Valley field. These are north-northwest and northeast to east-northwest, both of which generally parallel major trends mapped from surface outcrop (Figure 17). Fracture occurrence varies from zones of shearing that appear to parallel bedding planes and may be correlative from one well to another, to vertical or semivertical fractures that intersect the borehole diagonally. The latter two fracture types are the most common in the Palm Valley field. Well performance, however, depends not only on the intersection of these fractures but also on their openness; that is, their relative permeabilities and concentration over the zone penetrated. Consequently, flow rates would be expected to vary significantly across the field, as demonstrated in the six wells drilled to date, with gas rates ranging from too small to measure up to 137 MMcf/day (Table 1).

Since few wells have been drilled in the field so far, it is difficult to draw definite conclusions as to the relationship between well location and produc-

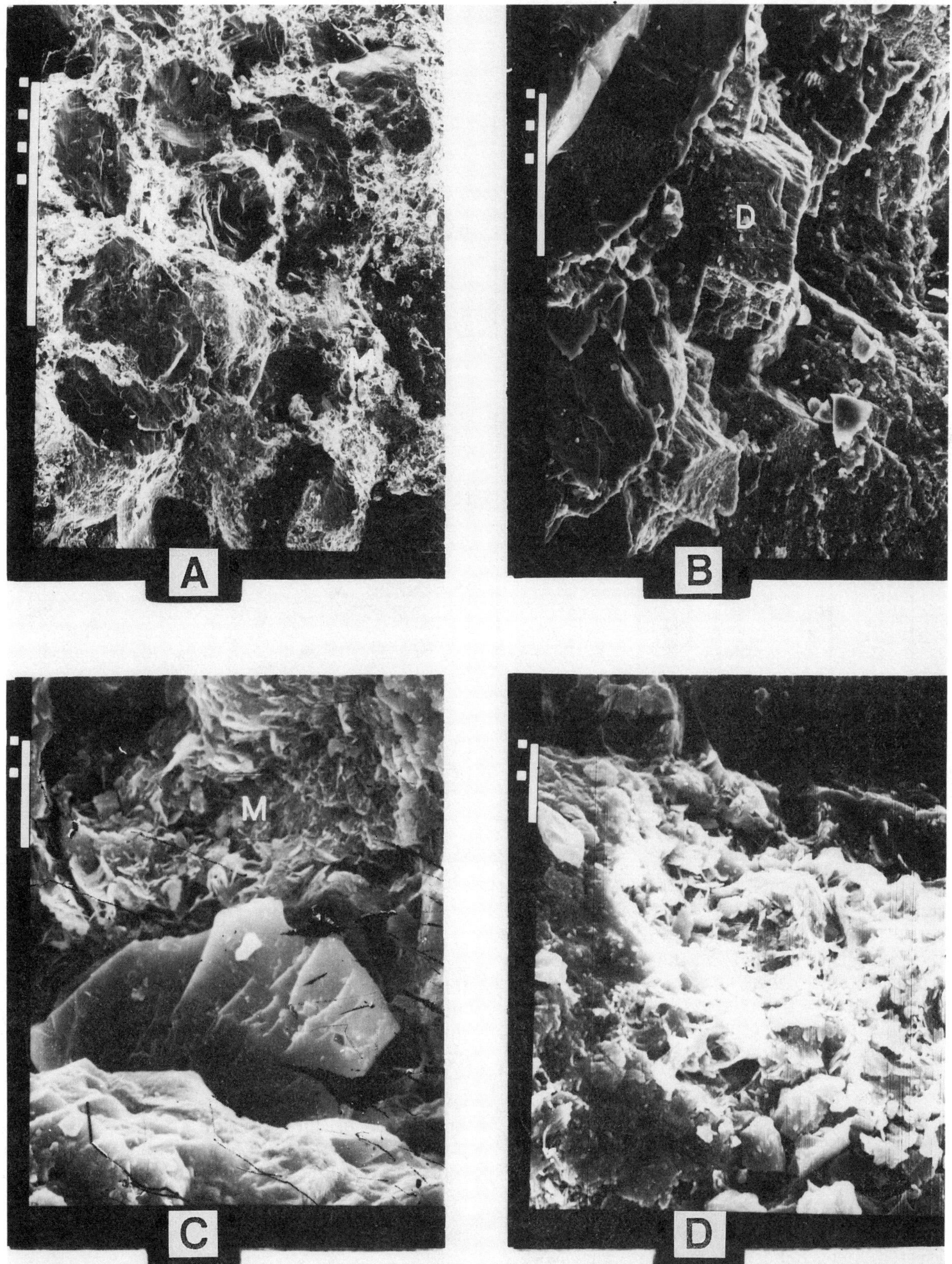

Figure 16. Scanning electron micrograph photographs from the Pacoota Sandstone, Palm Valley No. 1. (From Davies, Almon and Associates, 1980.) (A) Primary sandstone intergranular porosity largely filled by detrital clay matrix, "M," during burial. Magnification 45×. (B) Example of primary pore filled by dolomite cement, "D." Magnification 300×. (C) and (D) Examples of primary porosity largely filled by illite clay matrix. Magnification 2000× and 1500×, respectively.

Table 2. Core analysis data—Palm Valley No. 1. Porosities are generally very low to low in the reservoir, averaging 2 to 4%. Permeabilities are very low, averaging 0.1 md.

SAMPLE No.	DEPTH ft (KB)	PERMEABILITY MD	POROSITY %	FORMATION
41	5626.4	0.1	1.8	Pacoota P1
42	5627.3	0.1	2.0	"
43	5636.5	-	5.0	"
44	5637.5	0.1	2.1	"
1	5728	0.1	6.5	"
2	5744	0.1	4.5	"
3	5748	0.1	3.1	"
4	5756	0.1	5.2	"
5	5758	0.1	1.7	"
6	5770	0.1	2.4	"
7	5779	0.1	2.3	"
8	5780	0.1	2.3	"
9	5895	0.1	6.0	"
10	5898	0.1	6.8	"
11	5903	0.1	4.4	"
12	5906	0.1	2.9	"
13	5908	0.1	5.4	"
14	5910	0.1	1.1	"
15	5912	0.1	4.2	"
16	5914	0.1	6.7	"
17	5918	0.1	1.7	"
18	5920	0.1	2.5	"
19	5922	0.1	3.5	"
20	6167	0.1	1.7	Pacoota P2
21	6304	0.1	2.2	Pacoota P3
22	6347	0.1	4.4	"
23	6353	0.1	4.8	"
24	6355	0.1	4.9	"
25	6357	0.1	4.1	"
26	6361	0.1	4.4	"
27	6365	0.1	3.5	"
28	6367	0.1	3.0	"
29	6369	0.1	4.7	"
30	6370	0.1	4.6	"
31	6373	0.1	3.8	"
32	6375	0.1	6.2	"
33	6378	0.1	3.9	"
34	6383	0.1	4.0	"
35	6385	0.1	3.1	"
36	6388	0.1	3.3	"
37	6391	0.1	5.4	"
38	6471	0.1	1.7	"
39	6472	0.1	1.7	"
40	6475	0.1	1.5	"

tivity. Nevertheless, the data show that the wells located along the surface structural axis (Palm Valley Nos. 1, 2, 3, and 6B) encountered strong gas flows, whereas those away from an axial position (Palm Valley Nos. 4, 5, and 6) did not (Figure 17A). Palm Valley Nos. 2 and 3 also were located in a zone of major east-northeast surface fracturing which, if related to fracture intensity in the subsurface, may have contributed to their success. Palm Valley No. 5, although close to the axis on the field's southeastern nose, is in an area of sparse surface fracturing which, if reflected at depth, may explain its very small flow rate (RTSTM). The Palm Valley No. 6 well, situated a short distance south of the surface crest, recorded a relatively small flow (maximum 1.82 MMcfg/day), but encountered substantial increases in productivity from all reservoir zones in a hole (No. 6B) deviated toward the field's axis (see Figure 18), confirming that an interval of greater fracture concentration and/or more "open" fractures had been intersected. An earlier deviation, Palm Valley No. 6A, with a smaller stepout, resulted in no improvement in flow rate to the original vertical hole, thus negating the possible theory that any deviation alone would proportionately increase productivity as a result of the intersection of a greater number of high-angle fractures.

Table 3. Representative mega- and microscopic description of core sample from the upper Pacoota Sandstone Pl unit, Palm Valley No. 1 (Froelich and Bryan, 1965). Note the presence of feldspar as an accessory mineral that may constitute up to 10% of the finer grained, more argillaceous sandstones.

CORE NO. 15 : DEPTH: 5637.5 ft. (1718.3m).

FORMATION : UPPER PACOOTA SANDSTONE.

Megascopic Description:

Sandstone, white to light grey, well-sorted medium-grained vitreous orthoquartzite. Originally very clean sand, presently fairly tight because of cementation by welding.

Microscopic Description:

The predominant mineral is fairly well-sorted, medium-grained quartz. Coarse, well-rounded quartz grains are common, and fine sub-angular grains are sparsely scattered in the interstices. Original grain size and rounded shapes are well preserved, but secondary quartz overgrowths in optical continuity with the well-rounded grains have produced a mosaic appearance; moderate cementation by welding has produced straight crystal contacts, incipient sutures, rare flattened grains and pressure solution concavities. There are minor amounts of twinned medium-grained plagioclase feldspar with minor sericitic alteration. Quartz grains have well-rounded borders where finely cyrstalline carbonate forms matrix in pod 1.6mm in diameter; there is some replacement of the quartz and feldspar (?) grains by carbonate cement.

Framework:

Quartz grains (90%); 0.32mm, medium (65%) 0.72mm, coarse (20%); 0.16mm fine (3%) Feldspar (2%) medium to coarse.

Matrix or cement:

Silica cement (9%) as quartz overgrowths and welding the sand grains and pore filling. Carbonate cement (1%) as groundmass surrounding rounded quartz grains.

Accessories:

Feldspar with sericite, anhydrite and barite (?), pyrite, collophane, zircon, rutile, titanite, tourmaline and magnetite.

Porosity:

Nil to trace of intergranular which may be caused by plucking during preparation of thin section. Core analysis indicates a porosity of 2.1% and a permeability of less than 0.1 md. Rock is essentially tight.

Classification:

Welded orthoquartzite: Very clean, fairly well-sorted, medium to coarse-grained welded bimodal orthoquartzite with scattered rounded coarse quartz grains.

The tentative conclusion may thus be drawn that a narrow zone of open fracturing exists along the axis of the anticline—most probably because of tensional forces associated with anticlinal flexure—which results in high well productivity when intersected. Although fractures are still present outside this axial zone, they are appreciably less permeable, and productivity would be significantly lower. There may exist other areas lower down on the flanks of the anticline that are also highly productive owing to greater subsurface fracture concentration and permeability; however, this is yet to be determined.

SOURCE

Formations with significant source rock potential are present in the Ordovician Larapinta Group at Palm Valley and elsewhere in the Amadeus basin. Of particular importance are the euxinic black shales

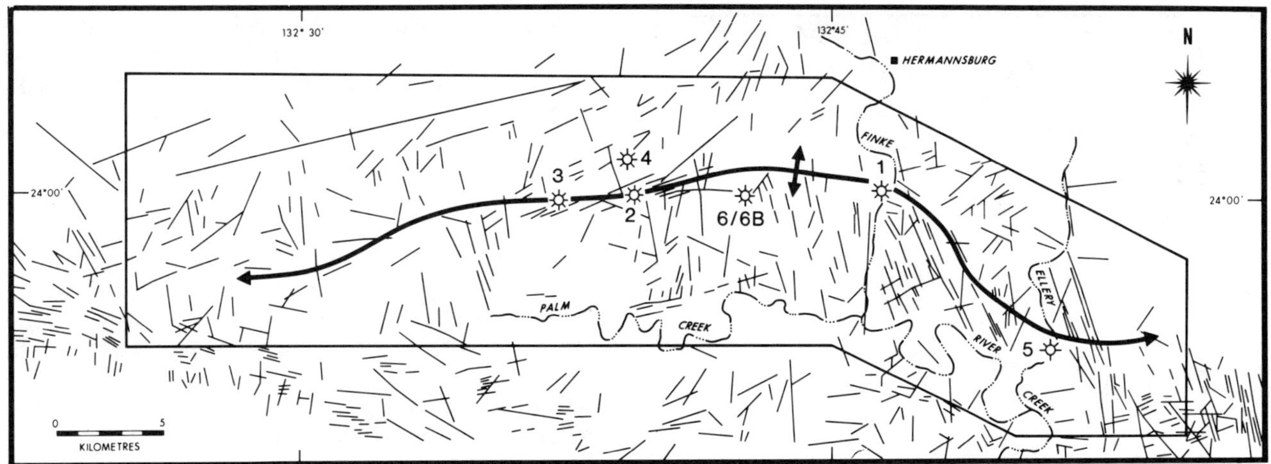

Figure 17A. Major surface fractures identified from photogeology, Palm Valley anticline. (From Geophoto Services, 1973.) Note strong east-northeasterly fracture development along axis in vicinity of Palm Valley Nos. 2 and 3.

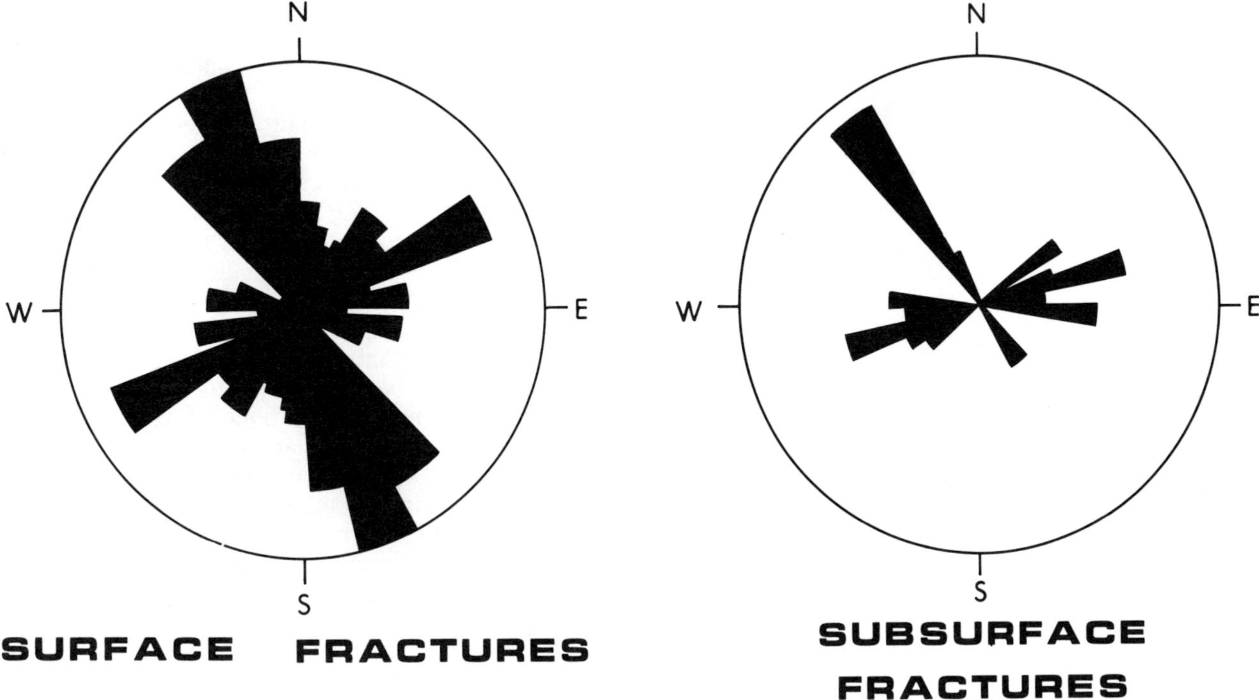

Figure 17B. Orientation comparison of surface and subsurface fracture trends taken from azimuth frequency rose diagrams (Do Rozario and Baird, 1987). Note good correlation of the two principal fracture directions.

and siltstones of the Horn Valley Formation. Lesser quality source rocks occur within the underlying Cambrian and upper Proterozoic sequences. Only formations of the Larapinta Group have been intersected in wells drilled on the field to date, and results of geochemical analysis of selected cores from two of these are presented in Table 4. Geochemical data from Cambrian and Proterozoic rocks in other wells in the basin are well documented in the reference paper by Jackson et al. (1984).

From the Palm Valley data, it can be concluded that the Ordovician source rocks are all overmature, with the Horn Valley Formation having the highest total organic carbon content (TOC). Kerogen type is predominantly type III, i.e., gas prone; however, farther west the Horn Valley Formation is more oil prone, containing type II kerogen with "amorphous algal sapropel" a major component. Although not penetrated at Palm Valley, Cambrian and Proterozoic source rocks occur at depths greater than approx-

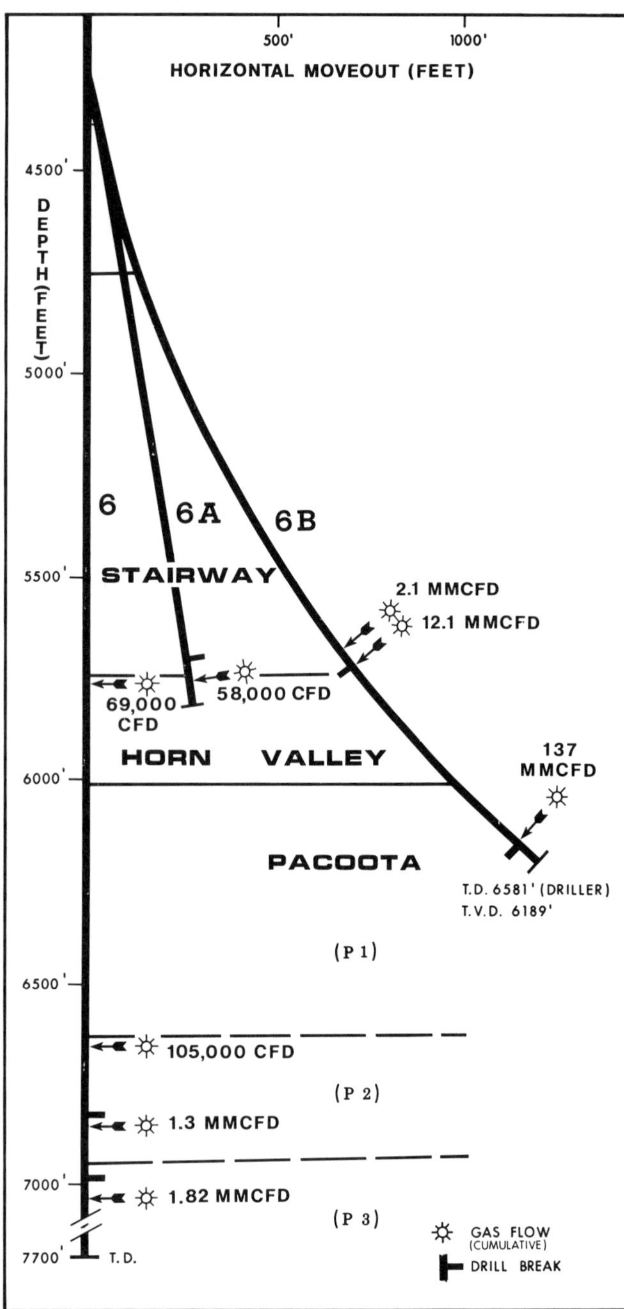

Figure 18. Well deviation plot—Palm Valley Nos. 6, 6A, 6B—showing location of drill breaks and cumulative gas flows (Do Rozario and Baird, 1987). Note major increases in gas flows in No. 6B, compared to original vertical well and lesser deviated hole.

imately 10,000 ft (3000 m) and are undoubtedly late overmature. From available geochemical data, they are of generally lower organic richness compared to the Ordovician source rocks (Figure 19).

Burial-history reconstruction (Figure 20) shows that Proterozoic rocks, in this case the Bitter Springs Formation, would have been at a late stage of gas generation during the main phase of the Alice Springs orogeny (350 million years before present), whereas the Ordovician Horn Valley Siltstone would have been in the early gas generative stage. Liquid hydrocarbons trapped at this time would have been subject to continued thermal alteration and gasification in the reservoir until the present time, hence explaining the very dry, predominantly methane gas characteristic of the field (Figure 21). Hydrocarbons generated from Ordovician source rocks, consequently, are most likely to occupy the majority of the Palm Valley reservoir, considering the short migration path to Ordovician sandstones of the Pacoota and Stairway formations. Although small, there may have been some remaining late stage contribution from the underlying Proterozoic and Cambrian source rocks, if vertical migration has occurred.

RESERVES

Historically, reserve estimations for the Palm Valley field have varied significantly, owing to the problems in deriving definitive parameters for the fractured reservoir and the relatively few wells drilled on the structure. Studies undertaken between 1970 and 1984 have resulted in reserve figures ranging from 82 to 10,000 bscf based on volumetrics and 20 to 100 bscf using material balance (Table 5). This wide variation is typical of fractured reservoirs and reflects the difficulty in modeling a heterogeneous, multilayered reservoir.

Volumetric estimates were based on parameters derived from log analyses, well test results, and a reservoir area of between 50,000 and 77,500 acres (202 and 314 km^2). Calculations of reservoir area relied, however, on the earlier assumption that a gas-water contact was present at -4280 ft, based on water recovered from drill-stem tests in Palm Valley No. 1. Material balance estimates were determined from static pressure extrapolation of pressure buildups.

In 1983, Magellan's reservoir engineer, Mr. Larry Franks, recognized that these earlier methods of reserve estimation were inappropriate for the fractured nature of the Palm Valley reservoir, and, instead, through the use of reservoir simulation, it was concluded that recoverable reserves in the structure were approximately 300 bscf, significantly higher than the results of earlier studies that relied on material balance calculations for proven reserves. This was confirmed by a reserves study undertaken by petroleum engineering consultants, H. K. van Poollen and Associates, in August 1984.

Results and methodology for this latter study were presented in a paper by Sabet and Franks, given at the Australian Petroleum Exploration Association Conference in 1985. As well as proving recoverable reserves of between 262 and 356 bscf of gas in the fracture system and higher permeability matrix rock, the study indicated the possible gas-in-place to range between 726 and 1106 bscf. No gas-water contact was indicated. Moreover, production projections from the

Table 4. Geochemical source rock data for selected Ordovician cores from the Palm Valley Nos. 1 and 3 wells. Data from Jackson et al. (1984). Mean maximum reflectance (in oil) determined from graptolite zooclasts — (G). Note source rocks are all overmature and gas prone.

FORMATION	WELL	DEPTH (m)	TOC %	KEROGEN H/C atomic	KEROGEN $\delta^{13}C_{PDB}$ ‰	KEROGEN R_0 max %	KEROGEN TYPE	HYDROCARBON YIELD mg/g TOC	HYDROCARBON YIELD %EOM	SOURCE RATING maturity/quality
STAIRWAY SANDSTONE	**Palm Valley-1**	**1407.3**	**0.26**					**14**	**68.6**	**overmature/ dry gas**
		1507.2	**0.31**	**0.41**			**III**	**4**	**44.8**	
		1584.4	**0.23**							
HORN VALLEY SILTSTONE	**Palm Valley-1**	**1697.1**	**0.07**			**2.37(G)**				**overmature/ dry gas**
		1703.5	**0.29**					**5**	**56.0**	
		1716.9	**0.67**	**0.65**	**−29.1**			**1**	**48.5**	
PACOOTA SANDSTONE	**Palm Valley-1**	**1746.5**	**0.47**					**<1**	**63.4**	**overmature/ dry gas**
		1756.9	**0.21**							
		1879.7	**0.36**					**2**	**25**	
		1939.4	**0.26**					**6**	**63.7**	
		1972.7	**0.29**					**14**	**54.2**	
	Palm Valley-3	**2070.2**	**0.20**	**0.60**			**III**	**2**	**31.8**	
		2077.2	**0.42**	**0.63**	**−28.8**	**2.05(G)**	**III**	**2**	**25.8**	
		2082.1	**0.27**	**0.65**		**2.00(G)**	**III**	**3**	**54.6**	

simulation work showed that the reservoir was capable of sustaining a daily production rate of 32 MMscf for more than 20 years, thus finally enabling firm markets to be sought.

EXPLORATION CONCEPTS

The Palm Valley field is a fractured reservoir that was discovered in the early stages of exploration in the Amadeus basin when conventional traps were being sought utilizing conventional methods, principally surface mapping and seismic surveying.

Prior to drilling, there were no indications that well productivity would be heavily dependent on fracture permeability, and it was only after completion of the first well, as described in earlier chapters, that the importance of fractures was recognized. Subsequent wells in the field have highlighted the uncertainties in developing fractured reservoirs; namely, the large variation in productivity from closely spaced wells (e.g., Palm Valley No. 2 and No. 4) or those in a similar structural position (e.g., Palm Valley No. 5 and Palm Valley No. 1). Insufficient wells have been drilled in the field to date to provide reliable data on reservoir productivity/variability over the entire structure; however, as mentioned previously, the present wells do appear to confirm the existence of a zone of open fracturing along the anticline's axis, which would be a target for future drilling. Recent drilling results also indicate the production benefits to be achieved by deviating wells into known highly fractured zones or across suspected open fractures, a technique that will probably be utilized more frequently in the future.

Development of the Palm Valley field has shown the importance of understanding the nature and behavior of fractures in a fractured reservoir, both for economic well planning, good well performance prediction, and reliable reserves estimation. As more knowledge of the field is acquired through further drilling, testing, and long-term production, a more accurate model will be constructed enabling a better understanding of the complexities and ultimate performance of the heterogeneous reservoir. This knowledge also may be applied to other exploration prospects suspected of having fracture porosity and permeability.

ACKNOWLEDGMENTS

The author wishes to thank the Palm Valley Joint Venture for permission to publish this paper and, in particular, the staff of Magellan Petroleum Australia Limited for their contribution to and assistance in preparation of the manuscript.

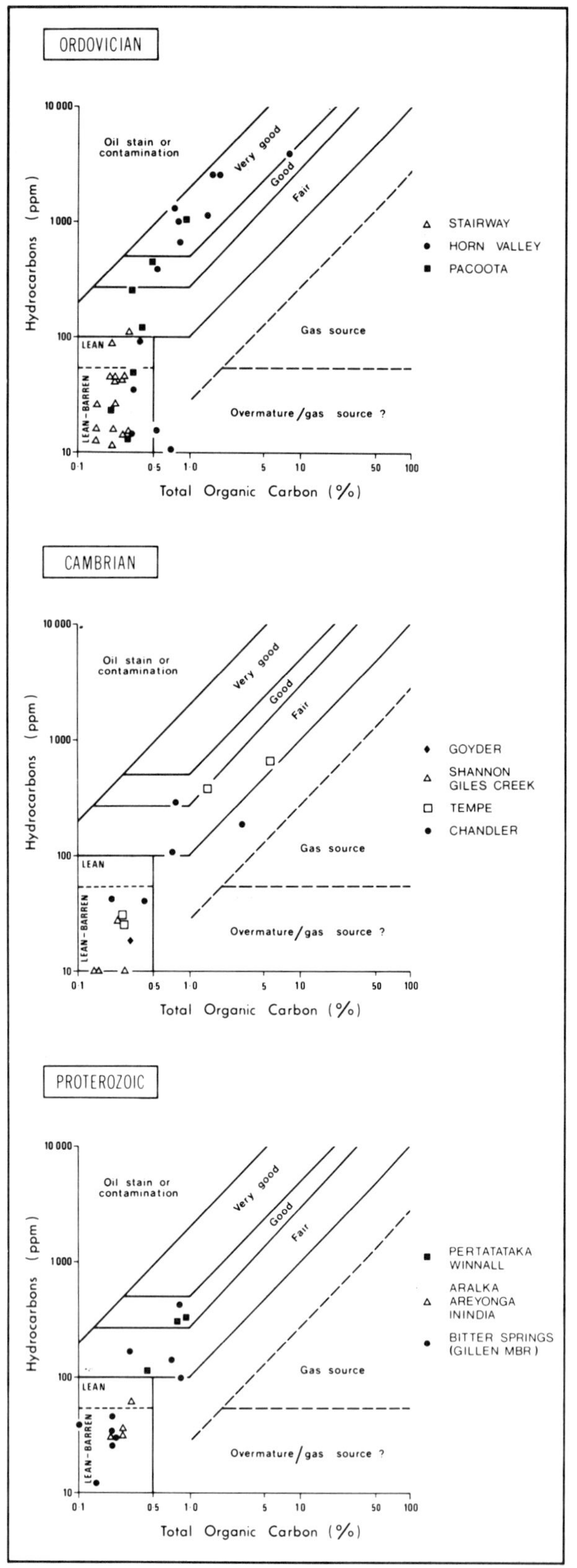

Figure 19. Comparison of organic richness potential of source rocks in the Amadeus basin. (After Jackson et al., 1984.) Ordovician source rocks, in particular the Horn Valley Siltstone, have good to very good source richness.

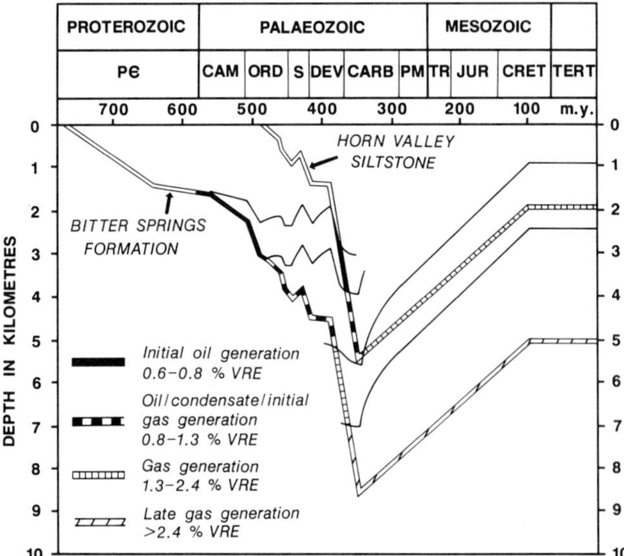

Figure 20. Burial and maturation history plot for the Palm Valley field (based on Palm Valley No. 2). Assumed geothermal gradient of 2.9° C/100 m. (After Jackson et al, 1984.)

REFERENCES

Bradshaw, J. D., and P. R. Evans, 1988, Palaeozoic tectonics, Amadeus Basin, Central Australia: Australian Petroleum Exploration Association Journal, v. 28(1), p. 267-282.

Davies, Almon & Associates, 1980, Pacoota Sandstone study East Mereenie and Palm Valley fields, Amadeus Basin, Northern Territory, Australia: Unpublished report prepared for Pancontinental Petroleum, 52 p.

Do Rozario, R. F., and B. W. Baird, 1987, The detection and significance of fractures in the Palm Valley gas field: Australian Petroleum Exploration Association Journal, v. 27(1), p. 264-280.

Forman, D. J., and R. D. Shaw, 1973, Deformation of the crust and mantle: Australian Bureau of Mineral Resources Bulletin, v. 144, 18 p.

Froelich, A. J., and W. B. Bryan, 1965, Petrographic study of thin sections Stairway, Horn Valley and Pacoota Formations, Palm Valley No. 1: Unpublished report for Magellan Petroleum N.T. Pty. Ltd., 87 p.

Froelich, A. J., and E. A. Kreig, 1969, Geophysical-geologic study of northern Amadeus Trough, Australia: American Association of Petroleum Geologists Bulletin, v. 53, p. 1978-2004.

Geophoto Services, 1973, Fracture Analysis Palm Valley—Gardiner—James Range Anticlines, Northern Territory, Australia: Unpublished report prepared for Duncan McNaughton, Petroleum Consultant, 8 p.

Gorter, J. D., 1984, Source potential of the Horn Valley Siltstone, Amadeus Basin: Australian Petroleum Exploration Association Journal, v. 24(1), p. 66-90.

Jackson, K. S., D. M. McKirdy, and J. A. Deckelman, 1984, Hydrocarbon generation in the Amadeus Basin, Central Australia: Australian Petroleum Exploration Association Journal, v. 24(1), p. 42-65.

McNaughton, D. A., 1962, Petroleum Prospects Oil Permits 43 and 46, Northern Territory Australia: Unpublished report for Magellan Petroleum Corporation, 53 p.

McNaughton, D. A., R. M. Quinlan, R. M. Hopkins, and A. T. Wells, 1968, Evolution of salt anticlines and salt domes in the Amadeus Basin, Central Australia: Geological Society of America Special Paper, v. 88, p. 229-247.

GAS	MOL.%
Oxygen plus Argon	<0.01
Nitrogen	2.01
Carbon Dioxide	0.18
Methane	87.20
Ethane	8.90
Propane	1.23
I-Butane	0.12
N-Butane	0.24
I-Pentane	0.04
N-Pentane	0.03
Hexane	0.03
Heptanes	0.01
Octanes & Higher	0.00

Figure 21. Representative gas analysis, Palm Valley field. Gas is very dry, predominantly methane.

Narr, W., and J. B. Currie, 1982, Origin of porosity—example from Altamont Field, Utah: American Association of Petroleum Geologists Bulletin, v. 66/9, p. 1231-1247.

Overby, W. K., and R. L. Rough, 1971, Prediction of oil and gas bearing rock fractures from surface structural features: U.S. Bureau of Minerology Report on Investigation 7500.

Pearson, T. R., and D. D. Benbow, 1976, Amadeus Basin, *in* R. B. Leslie, H. J. Evans, and C. L. Knight, eds., Economic geology of Australia and Papua and New Guinea, 3. Petroleum: Australasian Institute of Mining and Metallurgy, Monograph Series 7, p. 216-225.

Ranneft, T., 1964, Application for a stratigraphic drilling subsidy, Palm Valley No. 1—an exploratory test for stratigraphic information in Oil Permit 43, Amadeus Basin, Northern Territory: Unpublished report for Magellan Petroleum (N.T.) Pty. Ltd., 10 p.

Rough, R. L., 1974, Stress measurements at Palm Valley, Gosses Bluff and Ooraminna structures in Amadeus Basin, Northern Territory, Australia: Unpublished report for Magellan Petroleum Australia Limited, 30 p.

Schroder, R. J., and J. D. Gorter, 1984, A review of the recent exploration and hydrocarbon potential of the Amadeus Basin, Northern Territory: Australian Petroleum Exploration Association Journal, v. 24(1), p. 19-41.

Turpie, A., and F. J. Moss, 1963, Palm Valley—Hermannsburg Seismic Survey, Northern Territory 1961: Australian Bureau of Mineral Resources Record No. 1963/5, 4 p.

Table 5. History of recoverable reserves estimates for the Palm Valley field. Reserves are in billion standard cubic feet of gas in place.

YEAR	VOLUMETRICS	MATERIAL BALANCE	MODELING
1970	1000-9000	—	—
1971	8000	—	—
1974	5000-10,000	100	—
1974	82	96	—
1975	370-750 in matrix and 76-117 in fractures	20 to 60	—
1976	117 (minimum)	50	—
1976	—	—	46
1981	360	47	—
1984	261-367	54	—

Van Poollen & Associates, Inc., 1975, Engineering and geological analysis. Palm Valley Field, Amadeus Basin, Australia: Unpublished report for Magellan Petroleum Australia Limited, 36 p.

Van Poollen & Associates, Inc., 1985, Palm Valley reserves update: Unpublished report for Magellan Petroleum Australia Limited, 7 p.

Wells, A. T., D. J. Forman, L. C. Ranford, and P. J. Cook, 1970, Geology of the Amadeus Basin, Central Australia: Australian Bureau of Mineral Resources Bulletin, v. 100, 207 p.

SUGGESTED READINGS

Bueno, E., and J. Avila, 1987, An integrated study of a naturally fractured reservoir La Paz Field, Western Venezuela: American Association of Petroleum Geologists Research Conference on Naturally Fractured Reservoirs, 20 p. Good case history of a fractured field.

Kostura, J. R., and J. H. Ravenscroft, 1977, Fracture controlled production: American Association of Petroleum Geologists Reprint Series No. 21, 221 p. Compilation of significant papers on various aspects of fractures and fractured reservoirs.

Plumb, R. A., and S. H. Hickman, 1985, The correspondence between stress-induced borehole elongation determined by the four-arm dipmeter and the borehole televiewer in the Auburn geothermal well: Journal of Geophysical Research, v. 90, p. 5513-5521. Good documentation of relationship between borehole ovality and stress direction.

Sibbit, A. M., and O. Faivre, 1985, The dual laterolog response in fractured rocks: SPWLA 26th Logging Symposium, 33 p. Presents some solutions to log analysis problems in fractured reservoirs.

PALM VALLEY

Appendix 1. Field Description

Field name .... *Palm Valley field*

Ultimate recoverable reserves .... *325 billion ft³ gas*

Field location:

Country .... *Australia*
State .... *Northern Territory*
Basin/Province .... *Amadeus*

Field discovery:

Year first pay discovered .... *Ordovician lower Stairway Sandstone 1965*
Year second pay discovered .... *Ordovician basal Horn Valley Siltstone 1965*
Third pay .... *Ordovician Pacoota Sandstone 1965*

Discovery well name and general location:

First pay .... *Palm Valley No. 1, 75 mi west of Alice Springs, lat. 24°00′01″S; long. 132°46′11.4″E*
Second pay
Third pay .... *Three pay zones encountered in discovery well: Ordovician Lower Stairway Sandstone, basal Horn Valley Siltstone, Upper Pacoota Sandstone*

Discovery well operator .... *Magellan Petroleum (N.T.) Pty. Ltd.*
Second pay
Third pay

IP in barrels per day and/or cubic feet or cubic meters per day:

First pay .... *11.7 MMcfg/day (cumulative flow from open-hole test)*
Second pay
Third pay

All other zones with shows of oil and gas in the field:

Age	Formation	Type of Show
None		

Geologic concept leading to discovery and method or methods used to delineate prospect, e.g., surface geology, subsurface geology, seeps, magnetic data, gravity data, seismic data, seismic refraction, nontechnical:

Surface expressed anticline recognized from air photography. Initially mapped using surface geology and limited seismic data owing to rugged topography.

Structure:

Province/basin type .... *Bally 41; Klemme IIa*

Tectonic history

The Late Devonian to Early Carboniferous Alice Springs orogeny resulted in the formation of large-scale folds in the central, northern, and western parts of the basin. Thrusting and wrench-style displacements were also initiated with some halokinesis owing to presence of salt in Cambrian/Proterozoic sediments.

Regional structure

Field is on the southern flank of the Missionary Plain and may have formed later than the en echelon Gardiner/James Range anticlinal trend to the south.

Local structure

Arcuate low-relief surface anticline with general east-west trend and poorly defined western nose. Surface dips average 3–7°.

Trap:

Trap type(s)

Trap is structural/stratigraphic with gas reservoired in an anticlinal closure and distribution of productive zones controlled by fracture porosity and permeability.

Basin stratigraphy (major stratigraphic intervals from surface to deepest penetration in field): (Palm Valley No. 1)

Chronostratigraphy	Formation	Depth to Top in ft
Devonian	*Hermannsburg*	*Surface*
Devonian	*Parke*	*858*
Devonian	*Mereenie*	*1006*
Ordovician	*Carmichael*	*2901*
Ordovician	*Stokes*	*3214*
Ordovician	*Stairway*	*4211*
Ordovician	*Horn Valley*	*5288*
Ordovician	*Pacoota*	*5561*

Location of well in field

Reservoir characteristics:

Number of reservoirs .... *1 (fractured)*
Formations .... *Stairway, Horn Valley, Pacoota*
Ages .... *Ordovician*
Depths to tops of reservoirs .... *See Basin stratigraphy, above*
Gross thickness (top to bottom of producing interval) .... *1600 ft (Palm Valley No. 6)*
Net thickness—total thickness of producing zones
Average .... *Fracture dependent*
Maximum
Average
Maximum

Lithology

Very fine to medium-grained quartzose sandstones, well cemented with carbonate and silica (Stairway and Pacoota formations); the Horn Valley basal productive unit is a dolomitized shelfal carbonate

Porosity type .... *Fracture porosity dominant; generally low intergranular porosity also present*
Average porosity .... *3% intergranular; 0.002% fracture*
Average permeability .... *0.1 md intergranular; 48 md fracture*

Seals:

Upper
Formation, fault, or other feature .... *Middle Stairway/Horn Valley*
Lithology .... *Siltstones*

Lateral
Formation, fault, or other feature .... *Fracture permeability dependent*
Lithology

Source:

Formation and age .... *Horn Valley (Ordovician)*
Lithology .... *Siltstone, minor carbonates*
Average total organic carbon (TOC) *0.3%*
Maximum TOC *0.7%*
Kerogen type (I, II, or III) .... *III*
Vitrinite reflectance (maturation) .... $R_o = 2.2$
Time of hydrocarbon expulsion .... *350–100 m.y.b.p.*
Present depth to top of source .... *5288 ft (Palm Valley No. 1)*

Thickness .... *350 ft (avg.)*
Potential yield

Appendix 2. Production Data

Field name .... *Palm Valley field*

Field size:

- **Proved acres** *50,000*
- **Number of wells all years** *6*
- **Current number of producing wells** *5*
- **Well spacing** .... *Min. 0.9 mi; max. 7.3 mi*
- **Ultimate recoverable** .... *325 bcf*
- **Cumulative production** .... *13.2 bcf*
- **Annual production (1989)** .... *8.0 bcf*
- **Present decline rate** *2.8%*
  - **Initial decline rate**
  - **Overall decline rate**
- **Annual water production (1989)** .... *2549 bbl*
- **In place, total reserves** .... *1800 bcf (possible)*
- **In place, per acre-foot**
- **Primary recovery**
- **Secondary recovery**
- **Enhanced recovery**
- **Cumulative water production**

Drilling and casing practices:

- **Amount of surface casing set** .... *500 ft*
- **Casing program** .... *13⅜-in. to 500 ft; 9⅝-in. to 4100 ft; 7-in. to TD; 2⅞-in. tubing*
- **Drilling mud** .... *Air drilling/KCL—polymer baracarb*
- **Bit program** .... *Sealed Journal bearing AIDC 837*
- **High pressure zones** .... *None*

Completion practices:

- **Interval(s) perforated** .... *Major fracture locations; variable between wells*
- **Well treatment** .... *Acid*

Formation evaluation:

- **Logging suites** .... *Older wells: GR, sonic, sidewall neutron, induction, microresistivity; Modern wells: GR, sonic, density, neutron, dual laterolog-msfl, dipmeter, fracture identification log*
- **Testing practices** .... *Open-hole testing while air drilling; drill-stem testing otherwise; wells are normally production tested*
- **Mud logging techniques** .... *Standard mud logging unit used; samples caught off blooie line while air drilling*

Typical gas analysis:

Gas	Mol%
Oxygen plus argon	*<0.01*
Nitrogen	*2.01*
Carbon dioxide	*0.18*
Methane	*87.20*
Ethane	*8.90*

Propane	*1.23*
I-Butane	*0.12*
N-Butane	*0.24*
I-Pentane	*0.04*
N-Pentane	*0.03*
Hexane	*0.03*
Heptanes	*0.01*
Octanes and higher	*0.00*

Initial GOR
Sulfur, wt%
Viscosity, SUS
Pour point
Gas-oil distillate

Field characteristics:

Average elevation .... *2540 ft*
Initial pressure .... *2875 psi*
Present pressure (1987) .... *2788 psi*
Pressure gradient .... *0.468 psi/ft*
Temperature .... *158°F*
Geothermal gradient .... *0.026°F/ft*
Drive .... *Depletion drive*
Gas column thickness .... *1600 ft*
Gas-water contact .... *None proven*
Connate water
Water salinity, TDS
Resistivity of water
Bulk volume water (%)

Transportation method and market for oil and gas:

74.5 mi 8-in. pipeline to town of Alice Springs; 807.5 mi 14-in. pipeline to city of Darwin

Uzen Field—U.S.S.R.
Middle Caspian Basin, South Mangyshlak Region

GREGORY F. ULMISHEK
U. S. Geological Survey
Denver, Colorado

FIELD CLASSIFICATION

BASIN: Middle Caspian
BASIN TYPE: Foredeep
RESERVOIR ROCK TYPE: Sandstone
RESERVOIR ENVIRONMENT OF DEPOSITION: Alluvial, Fluvial, and Lacustrine
RESERVOIR AGE: Triassic to Cretaceous
PETROLEUM TYPE: Oil and Gas
TRAP TYPE: Anticline

LOCATION

Uzen field is located east of the central Caspian Sea in the Shevchenko administrative region, Kazakh S.S.R.; it is the largest oil field in the Middle Caspian basin (Figure 1). The field is situated in the South Mangyshlak region (subbasin) on a structural terrace between strongly folded Triassic rocks exposed north of the terrace and the Mangyshlak trough, a deep Mesozoic-Cenozoic trough to the south (Figure 2). Another giant field located about 48 km (30 mi) west on the same structure is the Zhetybay oil-gas field. Besides these two giants, several oil, gas, and gas-condensate fields of much smaller size have been discovered in the South Mangyshlak region.

Total areal extent of the Uzen field is 252 km^2 (97 mi^2). The field, along with all other fields of the region, is operated by a local management unit (Ob'yedinenie) located in the Shevchenko city on the shore of the Caspian Sea.

Oil reserves are considered a state secret in the Soviet Union; thus, no data on reserves of the field have been published in the Soviet literature. Reserves in place have been assessed by Ulmishek and Harrison (1981a, 1981b) at 7.5 billion bbl, with an expected recovery factor of 25 to 26%. Carmalt and St. John (1986) ranked the Uzen field as the 129th largest field in the world, with recoverable reserves of 1.875 billion bbl of oil.

HISTORY

Pre-Discovery

The first oil flow in the Middle Caspian basin was obtained in 1893 from middle Miocene rocks in the Starogroznenskoye field (Figure 1). Shallow oil pools in folds of the Caucasus foredeep remained the main exploration target until the late 1940s. A number of such pools were found in the Terek-Sunzha and South Dagestan regions (Figure 1). Deeper drilling began soon after World War II and resulted in the discovery of major oil reserves in the Mesozoic section of the Terek-Sunzha region, and large gas reserves in the lower Tertiary strata of the Stavropol region. At the same time, multiple oil and gas-condensate discoveries in the Arzgir-Prikumsk region demonstrated that the productivity is not confined exclusively to the foredeep, as many geologists had believed, but extends into the basin foreland. These discoveries paved the way for initiating exploration in remote semidesert areas east of the Caspian Sea.

Discovery

The Uzen anticlinal structure was inferred from surface geologic mapping of exposed Upper Cretaceous and Tertiary rocks by S. N. Alekseichik in 1941.

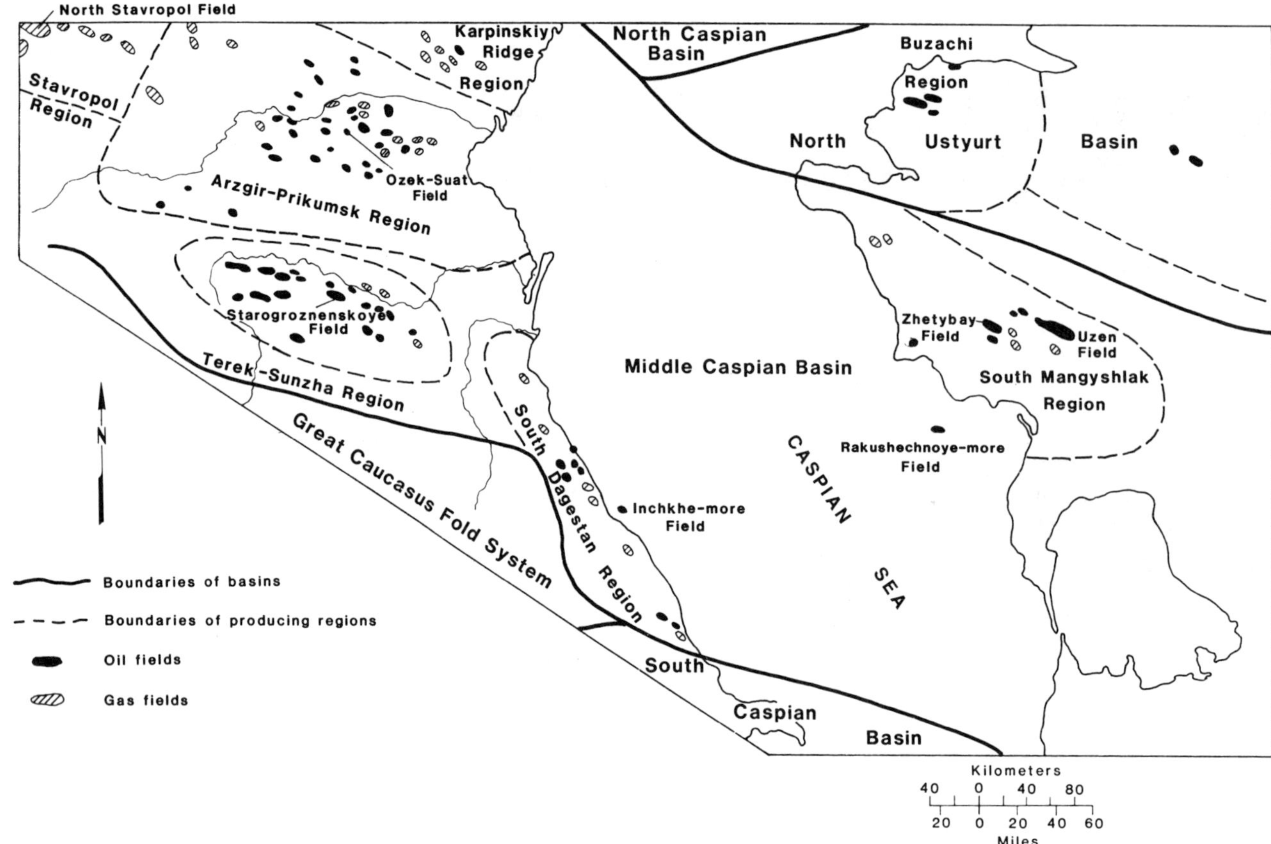

Figure 1. Middle Caspian basin and its producing regions.

Geologic studies of the region were interrupted by World War II and were resumed in 1950. Detailed mapping of the structure, with shallow (18 to 91 m or 60 to 300 ft) core drilling, was undertaken from 1950 to 1954. Core drilling to depths of several hundred meters on the Zhetybay and Uzen structures (Figure 2) began in 1957, contemporaneously with areal geophysical investigations. Some 395 km (247 mi) of reflection seismic data were run, with the distances between transverse profiles ranging from 2.5 to 4 km (1.6 to 2.5 mi). Longitudinal profiles were run along the axis and flanks of the structure. Forty-six core wells, with a total depth of 12,000 m (39,000 ft) and a maximum depth of 985 m (3230 ft), were drilled. Core samples were taken from marker beds. These activities detailed the structure on Cretaceous horizons and in 1960 revealed gas pools in Cretaceous rocks of the Uzen field.

Judging from gas flows in core wells, the Uzen field was originally considered to be mainly a gas prospect. Deep drilling in the field began in 1961. The first wildcat (well no. 1), located near the crest of the closure, was drilled to a depth of 1772 m (5814 ft). The well penetrated a thick clastic section of Middle Jurassic age overlain by Upper Jurassic carbonates and shales. The section contained several tens of interbedded oil-saturated potential reservoirs. The well was perforated from 1248 to 1261 m (4094 to 4137 ft) and tested 2295 bbl/day of oil in stratum XVI (of Bathonian age). The next three wells (nos. 2, 5, and 22) were tested and confirmed the presence of oil in other pays, i.e., in strata XIII, XIV, XV, and XVII. It became apparent that the new discovery was a giant field. The discovery of the Uzen field occurred at a time when the addition of oil reserves in the Volga-Ural province (then the main producing region of the U.S.S.R.) had sharply decreased. The Uzen discovery stimulated rapid development of exploratory activities in this new region.

Post-Discovery

Exploratory drilling in the Uzen field began in 1962, on the basis of data obtained from the first four wildcats. The first exploration stage involved drilling of 26 wells to depths of 1450 to 2500 m (4760 to 8200 ft). Most of the wells were located in rows across the highest part of the field, called the Main cupola (Russian geologists use the term *cupola* to refer to local highs). The distance between rows varied from 3 to 5 km (1.9 to 3.1 mi). By early 1964, this stage of exploration had delineated the part of the field in the highest area of structural closure.

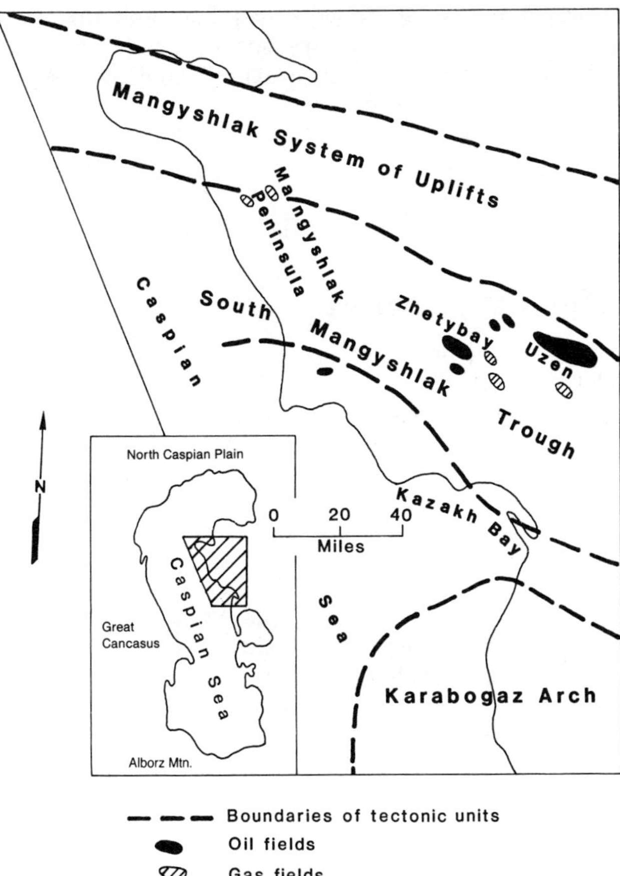

Figure 2. Index map of the South Mangyshlak subbasin.

During the next exploration stage (1964–1965), 33 more wells were drilled between previously completed rows of wells and on the flanks and plunges of the structure. This drilling phase identified new pools in strata XX and XXI, and discovered an oil field (Karamandybas field) just west of Uzen along the same structural trend. Gas pools in the Cretaceous section were explored by 14 wells. The drilling indicated that these pools are confined only to the Main cupola and gas reserves do not exceed 275 bcf (Chakabayev et al., 1977). In 1965, the Uzen field was turned over to the Ministry of Oil Industry for development and production.

Uzen oil has a very high paraffin content (up to 28%). The paraffin tends to precipitate and plug the permeability in less permeable beds, as a result of either a pressure decrease or cooling of the reservoir. Therefore, a hot waterflood was planned from the outset of production. The waterflood, however, was delayed for several years because of water supply problems, and full-scale heating of injection water was implemented in 1983.

About 800 producing wells and 260 injection wells were drilled during original development of the field. Injection wells were drilled in rows across the axis of the structure; they sectioned the field into nine productive bands 4 km (2.5 mi) wide. Spacing on the first phase of development varied from 84 acres (34 ha)/well for strata XIII and XIV to 69 acres (28 ha)/well for strata XV and XVI and to 113 acres (46 ha)/well for strata XVII and XVIII. Many wells were perforated in two adjoining strata for commingled production.

Casing practices include setting surface casing to 180 m (590 ft) with 273 mm casing, and casing to about 1375 m (4500 ft) with 168 mm casing. Several types of completions were tried. Hydraulic fracturing using hydrofluoric and hydrochloric acid additives proved to be most effective.

The second phase of field development began in 1972 after the inefficiency of the existing production system had become obvious. Additional rows of injection wells sectioned the field further into bands 2 km (1.25 mi) wide. Infill drilling of more production wells began at about the same time and continues to the present. Many additional injection wells have also been drilled.

The strong inhomogeneity of reservoir rocks (permeability varies from a few md to 1000 to 1200 md) resulted in low effectiveness of waterflooding. Injected water channeled into most permeable beds. The zone of flooding greatly expanded, with concomitant ineffectiveness of sweep. Water breakthrough to producing wells took place along very narrow intervals. The total watered-out thickness of pays was extremely low and varied from 2 to 5 m (Ilyaev et al., 1975). Most of the produced oil came from reservoir beds with permeabilities greater than 300 md, while less permeable reservoirs barely responded. An extensive program of drilling production and injection wells with completions in low permeability intervals has been undertaken during recent years (Aitkulov et al., 1982; Batyurbayev, 1982). The maximum rate of production was achieved in 1975 when the Uzen field produced 120 million bbl of oil. But by 1976, a sharp decline of production had already begun. By 1980, production had decreased by 37% to 75 million bbl, and to 66 million bbl by 1984. More recent data are not available, but production has continued to decrease. The rate of decline may be somewhat less, partly because of increased production from small pools in the lower part of the Jurassic that were found during development drilling. The cumulative output from the Uzen field is probably close to 1400–1450 million bbl of oil.

REGIONAL GEOLOGY

Tectonic History

The Middle Caspian basin formed over a Hercynian accreted terrane that is known as the Scythian (west of the Caspian Sea) and Turanian (east of the sea) plates. The South Mangyshlak subbasin is located in the western part of the Turanian plate. The basement of the South Mangyshlak subbasin has been drilled in only a few locations on the margins of the trough. The basement comprises strongly deformed and metamorphosed Paleozoic clastics and carbonates cut by granite intrusions. The deformation and metamorphism probably occurred in the late Paleozoic (certainly before the Late Permian). The exact nature of the basement in the central part of the South Mangyshlak trough remains unknown. It may be formed by the same folded and metamorphosed rocks (Letavin, 1980), or, according to other interpretations (Akramkhodzhayev and Yuldashev, 1984), it may be composed of undeformed Paleozoic rocks covering a median massif (microcontinent).

The second, taphrogenic stage of development of the South Mangyshlak subbasin (as well as all the Middle Caspian basin) took place during Late Permian and Triassic time. A large graben-rift formed in the region in an area presently occupied by the Mangyshlak system of uplifts (Figure 3). The Zhetybay step, where the Uzen field is situated, was located on the southern shoulder of the rift. The rift is filled by continental clastics of Late Permian–Early Triassic (Induan) age, and late Early Triassic (Olenekian) through Late Triassic marine clastics, carbonates, and, in places, volcanics. Total thickness of this sequence in the central part of the rift exceeds 9 km (5.6 mi). Outside the rift, the thickness decreases abruptly. At the end of Triassic time, the rift was inverted, sectioned into relatively narrow horsts and grabens, strongly folded, and eroded.

The third stage of geologic history began in Early Jurassic time and continues to the present. The area of the inverted Triassic rift remained highly uplifted, with little or no Jurassic–Tertiary sediments deposited over it. The deep South Mangyshlak trough formed between this uplift and the Karabogaz arch to the south. The structural Zhetybay step developed between the inverted rift and the trough. The step is separated by deep faults from both of these structures. The trough and the step are covered by thick, gently deformed sedimentary rocks of Early Jurassic through Tertiary age. Two unconformities, pre-Cretaceous and pre-middle Miocene, were important for formation of structural traps.

The Middle Caspian basin as a whole is classified as 221 under the modified scheme of Bally and Snelson (1980) (i.e., perisutural basins on rigid lithosphere associated with formation of compressional megasuture; foredeep and underlying platform sediments, or moat on continental crust adjacent to a-subduction margin; ramp with buried grabens, but with little or no blockfaulting). The basin is classified as IICa using the scheme of Klemme (1971) (i.e., continental multicycle basins; crustal collision zone-convergent plate margin; closed). The South Mangyshlak subbasin (which also can be considered a separate basin) does not fit well into either classification. The Alpine foredeep of the western Middle Caspian basin does not extend into the South Mangyshlak subbasin. Formation of the subbasin is clearly connected with Late Permian–Triassic rifting, but the rift itself was folded and uplifted. In the present-day structure, the rift lies outside the subbasin. Basins of this type do not belong in existing basin classification schemes.

STRUCTURE

Regional Structure

In the present-day structure, the South Mangyshlak trough consists of two large depressions separated by a structural saddle (Figures 3 and 4). The depth to the top of the Triassic in the Zhazgurly depression exceeds 5 km (3.1 mi). The onshore part of the Segendyk depression is slightly shallower; its offshore continuation is poorly known. The Karabogaz arch to the south of the trough is a highly uplifted structure on which Cretaceous rocks overlie the pre-Triassic basement at a depth of a few hundred meters. The Peschanomys uplift is separated from the Karabogaz arch by the deep and poorly studied Kazakh depression, which is offshore. On the top of the uplift, Jurassic rocks overlie a relatively thin Triassic section at a depth of slightly more than 3 km (1.9 mi). The uplift was considered a prime exploration target for a number of years, but only one small discovery in the Triassic section has been made onshore. Offshore exploration appears to have been unsuccessful and failed to establish commercial fields. The Mangyshlak system of uplifts consists of two chains of swells that are separated by the narrow and shallow Chakyrgan trough. Strongly deformed Upper Permian–Triassic rocks are exposed on the north chain of swells and occur at a very shallow depth on the Bekebashkuduk swell.

The Zhetybay step is a structural terrace that controls major oil and gas fields of the region. On the north and south, the terrace is bounded by steep flexures dipping southward (Figure 5). Structural surfaces in the Jurassic–Cretaceous section are gently inclined to the south. The step contains a number of local structures that are predominantly assembled into two zones parallel to the strike of the step. The largest structures, the Uzen and Zhetybay anticlines, are located in the northern and southern zones, respectively. All other structures are smaller. Local structures of the Zhetybay step contain about a dozen discovered oil and gas fields, but the main reserves are controlled by the Zhetybay and Uzen anticlines. No hydrocarbons have been found in the South Mangyshlak trough. Pools of heavy, biodegraded oil are known north of the Zhetybay step (Tuybedzhik structure).

Local Structure

The Uzen field is associated with a large anticlinal fold that is about 45 km (28 mi) long and 9 km (5.6 mi) wide (Figure 6). The fold is significantly asymmet-

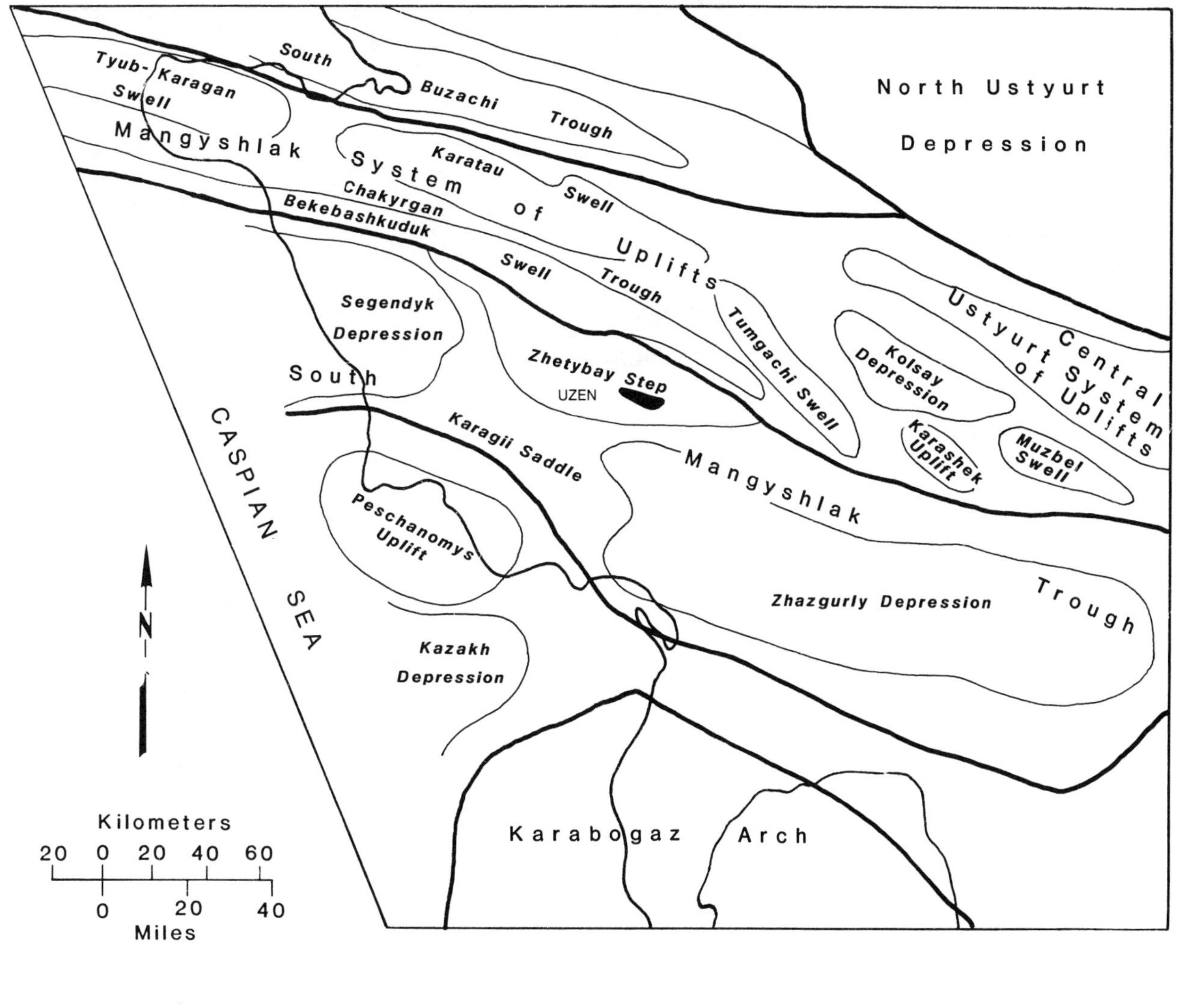

Figure 3. Main structural units of the South Mangyshlak region. See index map, Figure 2, for location. (After Krylov, 1971.)

rical, with the crest (Main cupola) located in the eastern part of the structure. The northern flank dips gently at 1.5 to 2°; whereas, the southern flank is steeper, with dip angles reaching 6 to 8°. The western plunge of the fold is complicated by two subsidiary closures, the Khumuryn and Parsymuryn cupolas. The Karamandybas structure is separated from the Uzen anticline by a fault. Hydrocarbon pools of the former have no hydrodynamic connection with the Uzen pools, and therefore are considered a separate field. This general structural configuration is conformable through all the Jurassic-Cretaceous section. Underlying Triassic rocks have a different structural configuration that is poorly known but is believed to have northern to northwestern trend (Klychnikov, 1982).

Faulting is probably rather intense in significantly deformed Triassic rocks. Dip angles in Triassic cores commonly reach 50 to 70°. The presence of faults with small displacements is also interpreted to occur in the Jurassic section, especially in its lower part. This interpretation is based primarily on well log correlation and production data. Yuferov et al. (1974) concluded that the distribution of small oil pools in the lower part of the Jurassic section is controlled by small faults (Figures 6 and 8). Most, if not all, of these faults die out in the upper part of the Jurassic section.

The Uzen fold began to form in Jurassic (Gribkov and Lazarev, 1968) or Early Cretaceous (Dmitriyev, 1985) time and continued through the Tertiary. However, the two principal phases of deformations were connected with the main unconformities in pre-Cretaceous and pre-middle Miocene times. These two phases account for the major part of the structural closure.

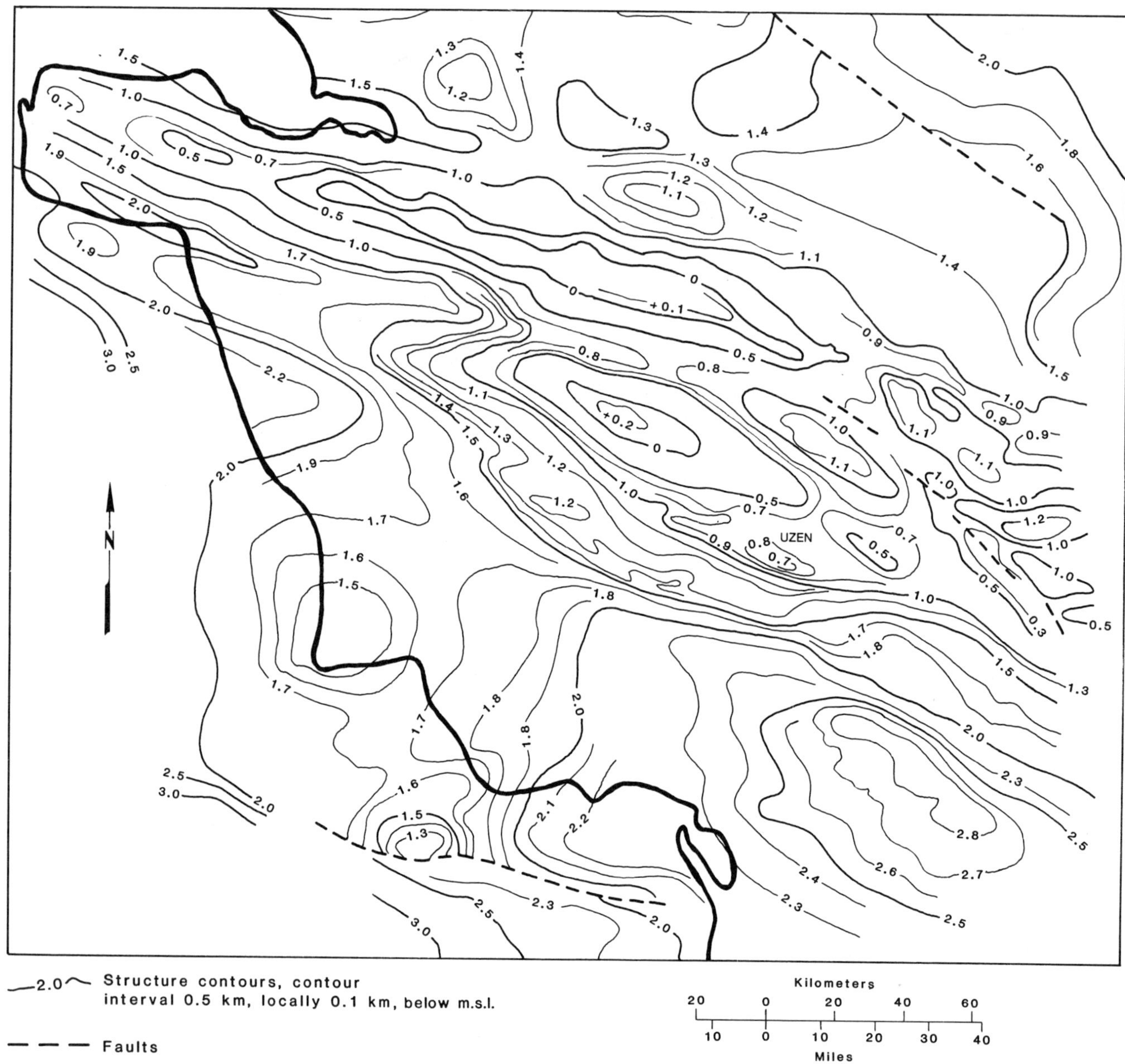

Figure 4. Contour map of the South Mangyshlak region. Contours on the base of the Hauterivian (reflector III-g). (After Sorotskaya, 1968.)

STRATIGRAPHY

Oil and gas pools on the Zhetybay step have been discovered over a large stratigraphic interval from the Triassic to the Turonian. Well over 90% of discovered hydrocarbons, however, are concentrated in the Jurassic section beneath the upper Callovian–Kimmeridgian regional seal (Dikenstein et al., 1983). All known pools in the Jurassic and Cretaceous sections are controlled by structural traps. Exploratory drilling into the Triassic section has also concentrated on local structures. It seems, however, that hydrocarbon pools in this section are chiefly controlled by zones of fracturing (Timurziyev, 1984).

The presence of Upper Permian–Induan continental rocks known in exposures of the Mangyshlak uplifts has not been proved on the Zhetybay step (Orudzheva et al., 1985). Deformed, dense, gray and variegated shales and siltstones of Early Triassic age occur at the bottom of the drilled section. Their penetrated thickness exceeds 1500 m (5000 ft) in the Uzen field. This unit does not contain reservoir rocks and only noncommercial flows of hydrocarbons have been tested from it. Overlying rocks are carbonates and shales of the upper Olenekian–Middle Triassic. They are found in the southern part of the step, and pinch out under the pre-Jurassic unconformity in its northern part. Fractured and cavernous carbonate

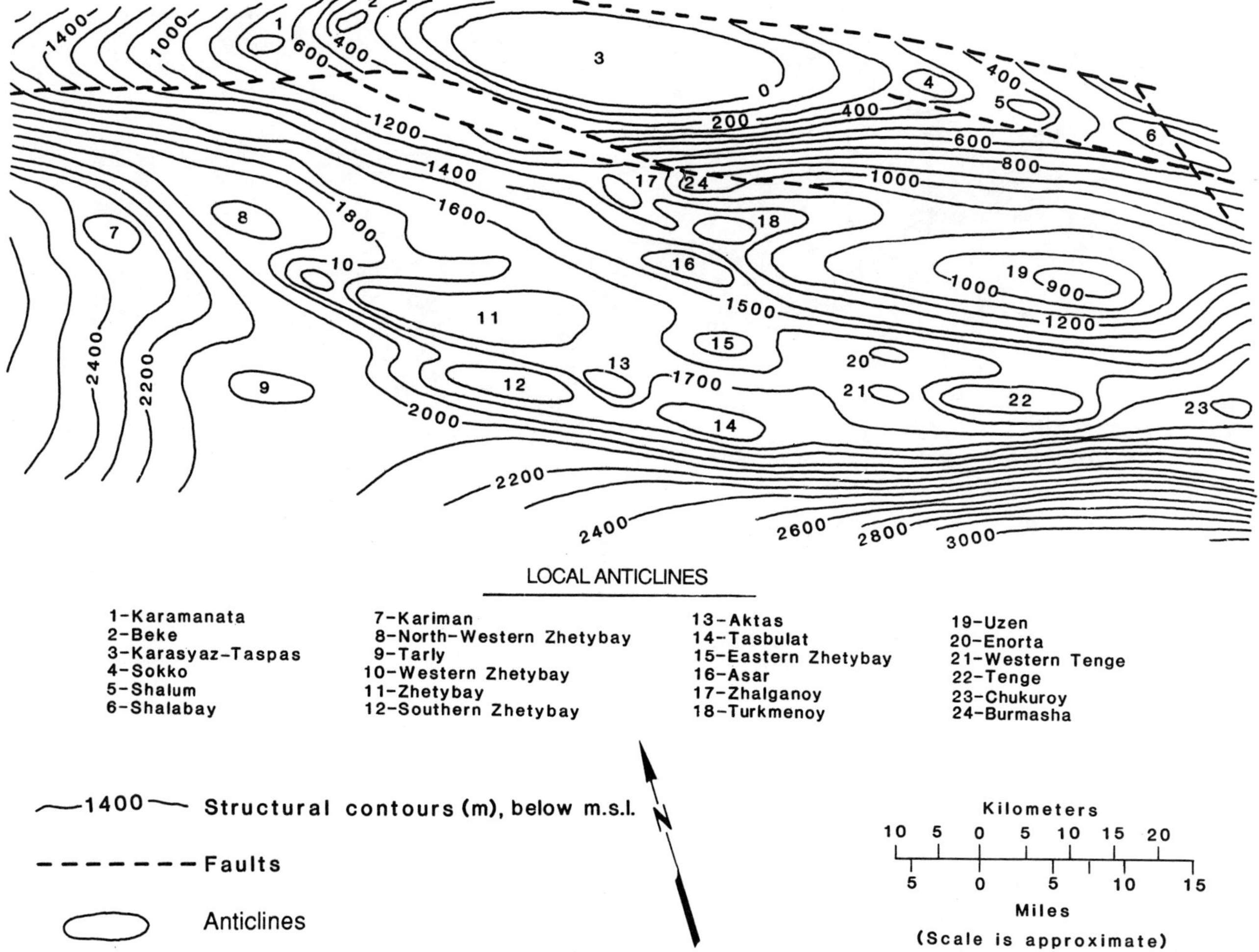

Figure 5. Contour map of the Zhetybay step. See Figure 3 for location. Contours on the Oxfordian limestone (reflector III). (After Yuferov et al., 1977.)

beds in this unit contain relatively small hydrocarbon pools in a few fields. Upper Triassic clastics and volcanics are present south of the step, in the South Mangyshlak trough.

The main productive Lower Jurassic through lower Callovian section of the Zhetybay step is up to 1000 m (3300 ft) thick. The section consists of an irregular alternation of sandstones, siltstones, and shales. Rocks of terrestrial origin (mainly alluvial facies) predominate in the section, but marine interbeds occur in its upper part. The section includes 13 sandstone packages that contain oil and gas pools in different fields (Figure 7).

The upper Callovian through Kimmeridgian interval consists chiefly of shales with beds of marls and limestones. These rocks form the main regional seal for hydrocarbon fields in underlying clastics. The thickness varies from over 300 m (1000 ft) on the southern part of the step to less than 100 m (330 ft) in the Uzen field. Probably because of inadequate thickness, the Upper Jurassic seal in the Uzen field leaks gas into Cretaceous reservoirs.

The overlying Cretaceous section is predominantly marine clastic rocks (terrestrial in the Barremian), with carbonate intervals in the Valanginian-Hauterivian and the upper part of the Upper Cretaceous. A dozen reservoir strata are identified in this section. Most of them contain gas pools in the Uzen field, but are essentially nonproductive in other fields of the step.

TRAP

The Uzen field is controlled by a typical structural trap. The closure of the trap on the top of stratum XIII is 290 m (951 ft); it increases downward in the section. Upper Jurassic shales and carbonates form the seal. Oil in strata XIII-XVIII fill the trap to the spillpoint.

The main oil reserves of the Uzen field are found in strata XIII-XVIII occurring in the Middle Jurassic-lower Callovian section. Pools in these strata have

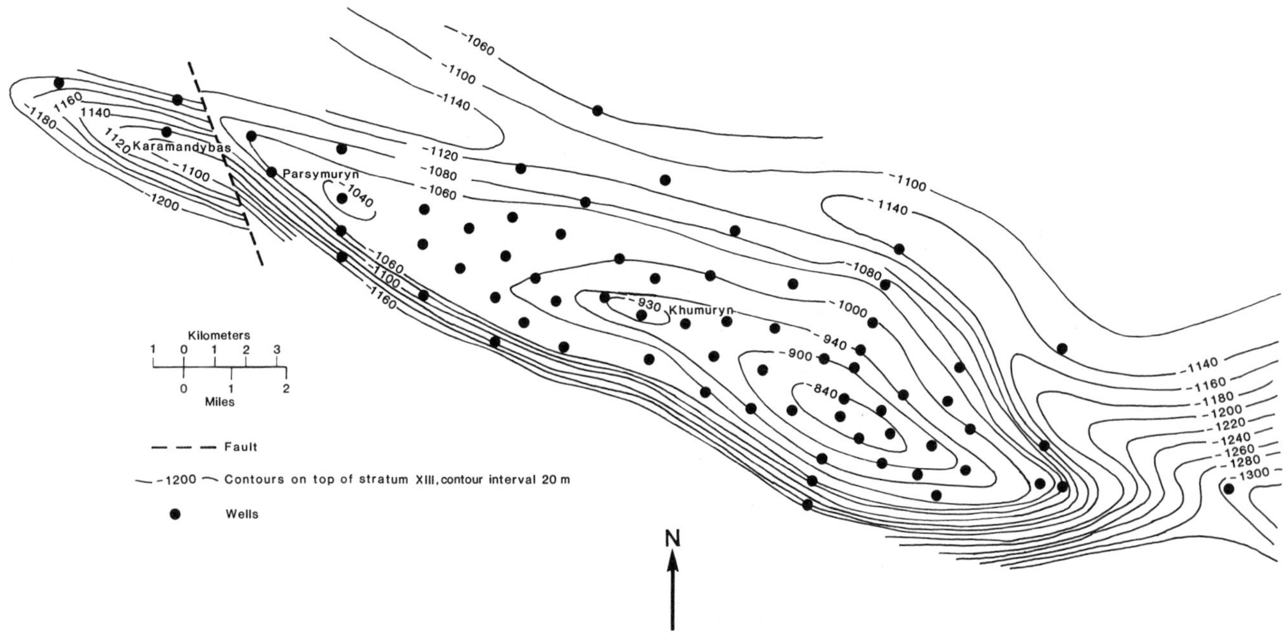

Figure 6. Contour map of the Uzen field. Contours on top of stratum XIII. (After Makhambetov, 1968.)

a common oil-water contact at a depth varying in different parts of the field from 1124 to 1150 m (3688 to 3773 ft) below sea level (Figures 8 and 9). Accordingly, the areas of the pools decrease in successively deeper reservoirs. Collectively, all of the pools actually form a single pool with an oil column 311 m (1020 ft) high. The common oil-water contact indicates a hydrodynamic connection between the strata. The pools in strata XVI and XVII contain small gas caps, but the oil in these pools is undersaturated by gas. This implies the absence of gas/oil equilibrium in the field and suggests the recent redistribution of oil and gas.

Initial reservoir pressure in strata XIII–XVIII ranged from 9610 to 12,061 kPa (1394 to 1749 psi). Original saturation pressure is estimated at 7355 to 10,885 kPa (1067–1579 psi). Initial solution gas-oil ratio was from 291 to 347 standard cubic feet/stock tank barrel. The oil formation volume factor is estimated at about 1.2 reservoir barrel/stock tank barrel.

Oil and gas pools in the underlying part of the Jurassic section are controlled by local highs (cupolas) within the larger anticline, and presumably by small faults (Figure 8). The pools do not contain significant oil reserves. Oils of these pools are lighter (40.5° API) and contain smaller amounts of resins and paraffin. The oil flow rates from Triassic rocks are considered noncommercial. Gas pools in the Jurassic section are small and could scarcely be of economic interest in this remote region.

Main gas pools in the Cretaceous section are found in strata VIII, X, XI, and XII on the Main cupola of the field (Figure 9). These pools probably result from leakage of gas from Jurassic rocks. The leakage was related to the significant thinning of the Upper Jurassic seal over the Uzen fold, compared with other structures of the Zhetybay step. Gas in these strata consists of 70 to 98% methane, 1.8 to 7.6% nitrogen, 0.12 to 2.0% carbon dioxide, and 0.024 to 0.016% helium.

RESERVOIRS

Twenty-five reservoir strata are identified in the Uzen field, and 18 of them contain hydrocarbon pools. The upper 12 reservoir strata occur in the Cretaceous section; they contain relatively small gas pools only on the Main cupola of the field (Figure 9). The other 13 strata occur in the Jurassic part of the section and are mainly oil productive. The reservoir rocks are primarily polymictic sandstones and siltstones with a high content of clayey material. Thicker shale beds separate reservoir strata from each other; thinner shales are present within the pays.

Strata I–XII (Figure 7) occur at depths ranging from 190 to 900 m (620 to 2950 ft). They consist of friable sandstones and siltstones with porosities of 26 to 34% and permeabilities of 200 to 600 md. The individual pays range in thickness from 10 to 50 m (33 to 160 ft). Each pay includes from one (stratum V) to four or five (strata III and VIII) sandstone beds.

Stratum XIII in the lower Callovian section and stratum XIV in the upper Bathonian contain more than 60% of the oil reserves of the field. The average thickness of stratum XIII is 35 m (115 ft), and the

Stratigraphic Column				Reservoir Strata: Regional Nomenclature	Reservoir Strata: Uzen Nomenclature
Cretaceous	Upper		Danian		
			Maestrichtian		
			Campanian		
			Santonian		
			Coniacian		
			Turonian	M-I	I
			Cenomanian	M-II	II
	Lower		Albian	M-III	III
				M-IV	IV
				M-V	V
				M-VI	VI
				M-VII	VII
				M-VIII	VIII
				M-IX	IX
				M-X	X
			Aptian		
		Neocomian	Barremian	M-XI	XI
			Hauterivian		
			Valanginian	M-XII	XII
Jurassic	Upper (Malm)		Kimmeridgian		
			Oxfordian		
			Callovian	Yu-I	XIII
				Yu-II	XIV
	Middle (Dogger)		Bathonian	Yu-III	XV
				Yu-IV	XVI
			Bajocian	Yu-V	XVII
				Yu-VI	XVIII
				Yu-VII	XIX
				Yu-VIII	XX
				Yu-IX	XXI
				Yu-X	XXII
			Aalenian	Yu-XI	XXIII
				Yu-XII	XXIV
	Lower (Liassic)			Yu-XIII	XXV
Triassic				Not Identified	

Figure 7. Stratigraphic column and reservoir strata of the South Mangyshlak subbasin. (The Callovian is regarded to be Upper Jurassic in the U.S.S.R.)

average net pay is 11.6 m (38 ft). Stratum XIII is characterized by rapid variability in reservoir properties (Figure 10). From one to twelve sandstone layers constitute the overall pay zone in different wells. Many individual sandstones rapidly pinch out laterally. The thickest sandstones have the best permeability; they form elongated bodies (Figure 11) that are interpreted to be river channel deposits (Kalugin et al., 1975; Yuferov et al., 1977).

Strata XIV-XVIII have characteristics in common with stratum XIII. All sandstones are very nonpersistent laterally. Channel sandstones have been positively identified in stratum XIV, but probably are also present in the other pays. Thicknesses of the strata vary from 40 to 66 m (130 to 215 ft) and net pays range from 17.8 to 31.5 m (58 to 103 ft).

All of the reservoir rocks of strata XIII-XVIII are poorly to moderately sorted; they are fine- to medium-grained sandstones and siltstones with clayey and calcareous cement. The clay content may reach 40% and more. Reservoir rocks are characterized by sharply varying permeabilities that range from a few millidarcys to 1000 to 1200 md. Permeability strongly depends on the amount of clayey cement, but does not correlate with porosity. The latter is essentially more uniform for different sandstones; porosity can vary from 18 to 23%, but commonly, it is 21 to 22%. Porosity is intergranular; fracturing is minor or absent. Deposition of all Lower to Middle Jurassic rocks occurred in various facies zones of alluvial plains, such as in braided rivers, shallow lakes, and swamps, under generally humid climatic conditions (Aktanova, 1968).

The underlying strata (XIX-XXV) contain relatively small reserves. Very few data on their reservoir properties have been published. Generally, they are rather similar to rocks of the main pays, but reportedly are even more variable in reservoir properties (Aleksin et al.,1969).

SOURCE

Source rocks for hydrocarbons in the South Mangyshlak subbasin have not been positively identified by geochemical methods. The most probable source seems to be Lower to Middle Jurassic terrestrial clastics with dominant type III and subordinate type II kerogens. The content of organic carbon in these rocks generally varies from 1 to 1.5%. Most of the section is located within the oil window on the Zhetybay step, but the lower part of the source-rock interval is supermature in the South Mangyshlak trough (Geodekyan et al., 1978). The high content of paraffin in all Mangyshlak oils suggests a terrestrial source.

An opposing point of view is that Triassic marine shales were the source for both Triassic and Jurassic hydrocarbons (Timurziyev, 1986). On the basis of geochemical data, Triassic and Jurassic oils have common characteristics as well as differences. These differences can be attributed either to different sources for the oils (Kordus et al., 1973) or to the higher maturity of Triassic rocks (Timurziyev, 1986). Modern oil-source rock correlation techniques are needed to solve this problem.

Oil produced from the Uzen field is of the paraffinic type (Tissot and Welte, 1984). Gravity of degassed oil is 36.5 to 33° API; under reservoir conditions, it is 54 to 50.5° API. The oil has high resin (9.7 to 21.1%) and paraffin (up to 28%) contents and a low sulfur content (0.1 to 0.24%). Viscosity of the oil in the reservoir is 3.4 to 4.2 cp. Degassed oil congeals at a temperature of 25 to 30°C (77 to 86°F). Saturation of connate water in productive reservoir rocks varies from 30 to 38% of pore space. The average geothermal gradient in the Uzen field is 2.1°F/100 ft (38°C/km). In the top pay (stratum XIII), the temperature ranges from 53.3 to 68°C (128 to 154°F).

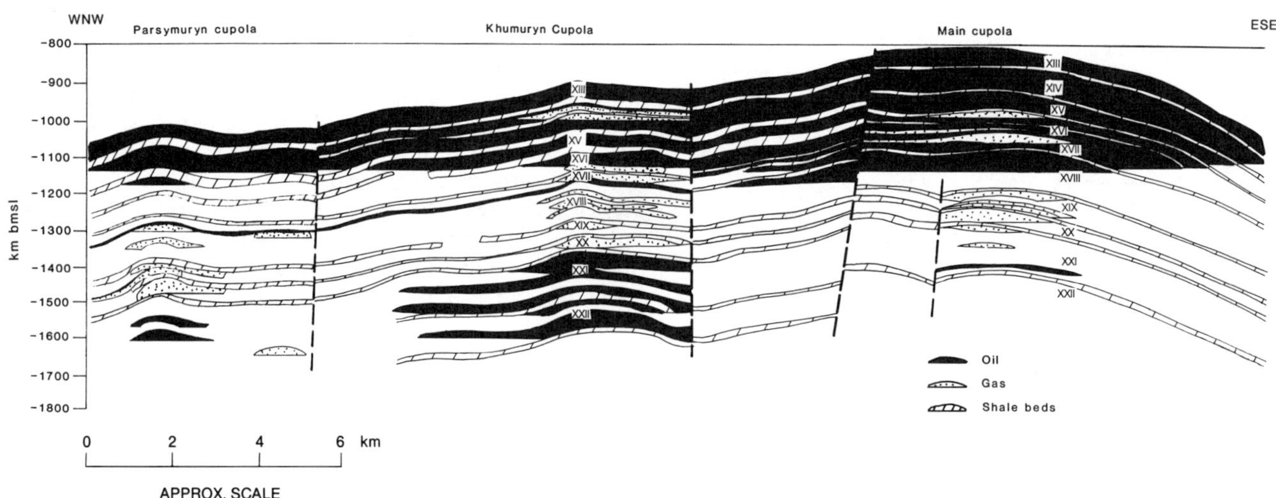

Figure 8. Cross section along the long axis of the Uzen field. (After Yuferov et al., 1974.)

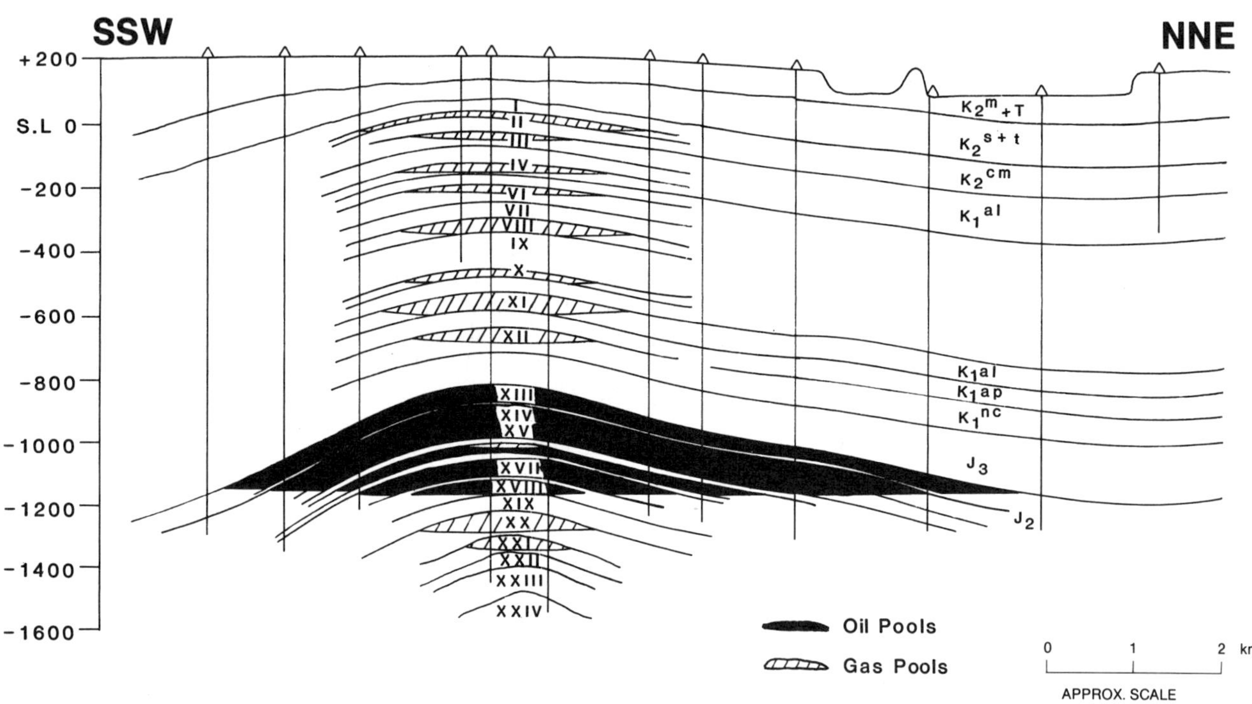

Figure 9. Tranverse cross section through the Main cupola of the Uzen field. (After Makhambetov, 1968.)

EXPLORATION AND DEVELOPMENT CONCEPTS

Discovery of the Uzen and Zhetybay fields demonstrates that the largest fields in a frontier region that are connected with simple structural traps are found very early in the exploration process. These two giant fields contain the bulk of oil and gas reserves of the South Mangyshlak subbasin. Probably, the geologic conditions that resulted in formation of these fields have not been well understood. Almost all presently discovered oil and gas fields occur on the Zhetybay structural step. Numerous attempts to explore for oil and gas in the South Mangyshlak trough and its southern flank have failed to result in a single discovery, despite

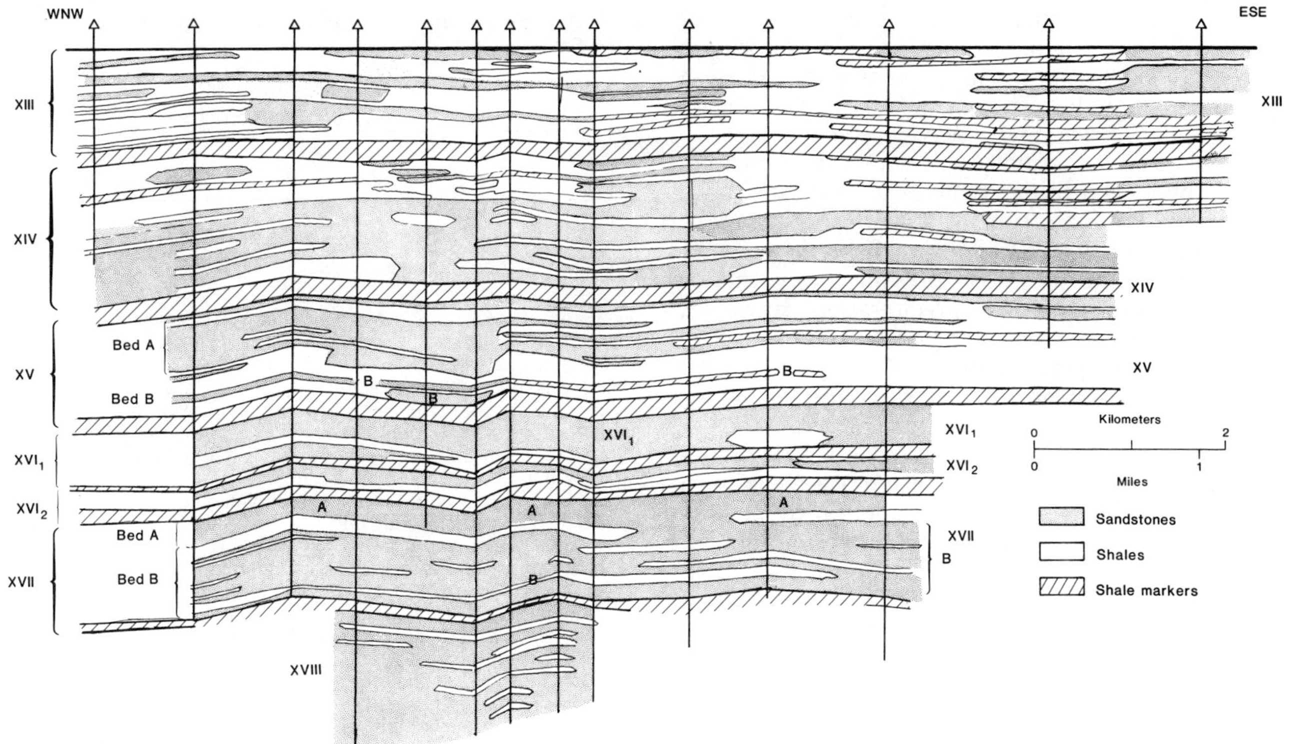

Figure 10. Stratigraphic cross section showing the reservoir structure of the Uzen field. (After Bykov et al., 1968.) The location is approximately the same as the section shown on Figure 8.

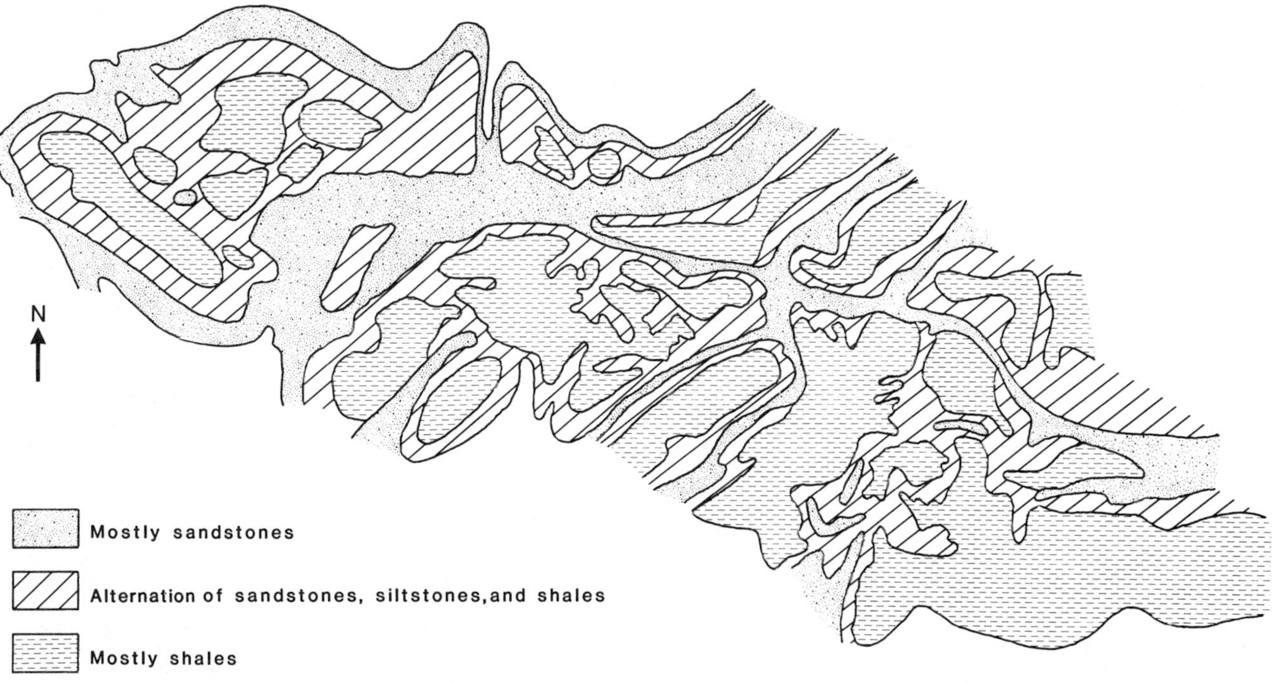

Figure 11. Channel sandstones in stratum XIII of the Uzen field. (After Kalugin et al., 1975.)

the similar stratigraphy and the presence of structural traps. Reliable identification of source rocks and modeling of the maturation and migration processes can explain this phenomenon and provide explorationists with new play concepts.

The major production problem in the Uzen field is the pronounced inhomogeneity of reservoir rocks in the main pay zones coupled with the high content of paraffin. Hot waterflooding in the early stage of production was designed for pressure maintenance and prevention of paraffin precipitation. However, the effectiveness of the waterflood was precluded by sharply varying permeabilities of reservoir rocks, which led to rapid breakthrough of injected water into production wells through very narrow zones of high permeability. As a result, the volumetric sweep efficiency appeared to be very low and reservoirs with permeability of less than 50 md were almost nonproductive. Based on data of Surguchev et al. (1978), reservoir rocks of low permeability in the field contain over 20% of oil in place (Table 1). Most of the produced oil has been recovered from reservoir rocks with permeabilities of more than 150 md. A low recovery factor from the field can be expected.

Another major deficiency of the development plan was the location of the injection wells in rows that cut the field into linear blocks—a pattern that did not take into account the channel-like distribution of best reservoir rocks. This resulted in an ineffective sweeping and required large additional investments in the redesign of the injection pattern.

Table 1. Distribution of oil in place in reservoir rocks of the Uzen field.

	Intervals of permeability, in md				
	10-20	20-50	50-150	150-400	>400
Oil in place, in %	8.7	12.8	30.8	27.0	20.7

REFERENCES

Aitkulov, A. U., Yu. P. Kislyakov, and Yu. P. Kovalskiy, 1982, Implementation of waterflooding in the Uzen field: Neftepromyslovoye Delo, n. 3, p. 2-4.

Akramkhodzhayev, A. M., and Zh. Yu. Yuldashev, 1984, Geologic framework and hydrocarbon potential of the Turanian plate from the standpoint of new global tectonics, *in* M. V. Muratov, A. L. Yanshin, and R. G. Garetskiy, eds., Tektonika molodykh platform (Tectonics of young platforms): Moscow, Nauka, p. 64-68.

Aktanova, S. A., 1968, Lithology of Jurassic rocks of the Uzen field: Izvestiya Akademii Nauk Kazakhskoy SSR, ser. geol., n. 2, p. 51-56.

Aleksin, A. G., G. T. Voronova, and S. I. Chechetkin, 1969, Experience with exploration for the multi-strata oil-gas fields: the Uzen-Karamandybas field of South Mangyshlak: Geologiya Nefti i Gaza, n. 8, p. 19-23.

Bally, A. W., and S. Snelson, 1980, Realms of subsidence, *in* A. D. Miall, ed., Facts and principles of world petroleum occurrence: Canadian Society of Petroleum Geologists Memoir 6, p. 9-94.

Batyurbayev, M. D., 1982, Increase of the production effectiveness of the Uzen and Karamandybas fields: Neftyanoye Khoziaystvo, n. 9, p. 9-12.

Bykov, N. Ye., T. P. Borovleva, R. V. Polikarpova, and G. V. Vorontsova, 1968, Differentiation of the producing sequence of the Uzen field: Neftegazovaya Geologiya i Geofizika, n. 3, p. 24-28.

Carmalt, S. W., and B. St. John, 1986, Giant oil and gas fields, *in* M. T. Halbouty, ed., Future petroleum provinces of the world: American Association of Petroleum Geologists Memoir 40, p. 11-54.

Chakabayev, S. Ye., T. N. Dzhumagaliyev, and E. S. Votsalevskiy, 1977, Oil of Kazakhstan: Sovetskaya Geologiya, n. 8, p. 3-8.

Dikenstein, G. Kh., S. P. Maximov, and V. V. Semenovich, eds., 1983, Neftegazonosnye provintsii SSSR (Petroleum provinces of the USSR): Moscow, Nedra, 271 p.

Dmitriyev, L. P., 1985, Obosnovanie ratsionalnogo kompleksa i metodiki poiskov zalezhey nefti i gaza v nizkopronitsaemykh kollektorakh (Rational methods of exploration for oil and gas in low permeable reservoir rocks): VNIIOENG, Vyp. 5(78), Moscow, 58 p.

Geodekyan, A. A., Yu. M. Berlin, V. L. Pilyak, and G. F. Ulmishek, 1978, Generation of oil and gas in geologic history of the Caspian region, *in* A. A. Geodekyan, ed., Protsessy neftegazoobrazovaniya v akvatorii Kaspiyskogo Morya (Processes of oil and gas generation in the Caspian Sea): Institute of Oceanology, Moscow, p. 77-116.

Gribkov, V. V., and V. S. Lazarev, 1968, Formation of some local structures in Mangyshlak, *in* V. S. Muromtsev, ed., Osobennosti geologicheskogo stroeniya i otsenka neftegazonosnosti Mangyshlaka (Geologic framework and petroleum potential of Mangyshlak): Leningrad, Nedra, p. 140-146.

Ilyaev, V. I., Ye. K. Ogay, and G. G. Parshina, 1975, Features of watering-out of reservoirs of the Uzen field: Neftepromyslovoye Delo, n. 5, p. 3-6.

Kalugin, A. K., V. L. Kuzmin, and Yu. K. Yuferov, 1975, Methods and directions of exploration in South Mangyshlak: Geologiya Nefti i Gaza, n. 11, p. 1-7.

Klemme, H. D., 1971, What giants and their basins have in common: Oil and Gas Journal, v. 69, n. 9, 10, 11; pt. 1, p. 85-90; pt. 2, p. 103-110; pt. 3, p. 96-100.

Klychnikov, A. V., 1982, Some structural and lithologic peculiarities of Triassic rocks of the Zhetybay-Uzen tectonic zone: Neftegazovaya Geologiya i Geofizika, n. 9, p. 11-13.

Kordus, V. I., V. V. Gribkov, A. I. Bogomolov, and E. H. Chikhacheva, 1973, Geochemical characteristics of oils and organic matter of Triassic and Jurassic rocks of the Uzen field in connection with formation of pools: Geologiya Nefti i Gaza, n. 2, p. 34-38.

Krylov, N. A., 1971, Obshchiye osobennosti tektoniki i neftegazonosnosti molodykh platform (General features of tectonics and the oil and gas distribution on the young platforms): Moscow, Nauka, 156 p.

Letavin, A. I., 1980, Fundament molodoy platformy yuga SSSR (Basement of the young platform of the southern USSR): Moscow, Nauka, 151 p.

Makhambetov, Kh. M., 1968, Some methodological problems of exploration of the Uzen field: Nauchno-tekhnicheskiy sbornik po dobyche nefti, n. 32, Moscow, Nedra, p. 12-18.

Orudzheva, D. S., V. I. Popkov, and A. A. Rabinovich, 1985, New data on the geology and hydrocarbon potential of pre-Jurassic rocks of South Mangyshlak: Geologiya Nefti i Gaza, n. 7, p. 17-22.

Sorotskaya, A. V., 1968, Tectonics of subsided areas of the Jurassic-lower Miocene structural complex from seismic data, *in* V. S. Muromtsev, ed., Osobennosti geologicheskogo stroyeniya i otsenka neftegazonosnosti Mangyshlaka (Geologic framework and petroleum potential of Mangyshlak): Leningrad, Nedra, p. 151-155.

Surguchev, M. L., A. V. Chernitskiy, and H. K. Sizova, 1978, Distribution of oil reserves and recoverability of oil from reservoirs of the Uzen field: Geologiya Nefti i Gaza, n. 8, p. 1-5.

Timurziyev, A. I., 1984, Reservoirs and hydrocarbon pools in low permeable rocks and the improvement of methods of their forecasting: Geologiya Nefti i Gaza, n. 11, p. 49-54.

Timurziyev, A. I., 1986, Mechanism of formation of oil and gas fields of South Mangyshlak: Geologiya Nefti i Gaza, n. 10, p. 25-31.

Tissot, B. P., and D. H. Welte, 1984, Petroleum formation and occurrence, second edition: Berlin-Heidelberg-New York-Tokyo, Springer-Verlag, 699 p.

Ulmishek, G., and W. Harrison, 1981a, Petroleum geology and resource assessment of the Middle Caspian basin, USSR, with special emphasis on the Uzen field: Argonne National Laboratory Report ANL/ES-116, 145 p.

Ulmishek, G., and W. Harrison, 1981b, Uzen development gives new insight into projecting future Soviet oil output: Oil and Gas Journal, August 24, p. 148, 151–152, 154.

Yuferov, Yu. K., L. P. Dmitriyev, and A. A. Rabinovich, 1974, Role of faults in the formation and distribution of oil and gas pools in South Mangyshlak: Geologiya Nefti i Gaza, n. 4, p. 18–24.

Yuferov, Yu. K., K. Kh. Boranbayev, O. P. Korchin, and A. Ye. Dmitriyev, 1977, Channel oil and gas pools in South Mangyshlak: Neftegazovaya Geologiya i Geofizika, n. 9, p. 17–21.

Appendix 1. Field Description

Field name .... *Uzen (Uzenskoye)*

Ultimate recoverable reserves .... *Approximately 2,000 million bbl of oil and 8 billion m³ of nonassociated gas*

Field location:

- **Country** .... *U.S.S.R.*
- **State** .... *Shevchenko Administrative Region, Kazakh S.S.R.*
- **Basin/Province** .... *Middle Caspian basin, Mangyshlak subbasin*

Field discovery:

- **Year first pay discovered** .... *First group, Cretaceous zones (mainly gas) 1938*
- **Year second pay discovered** .... *Second group, M. Jurassic clastics (mainly oil) 1961*
- **Third pay** .... *Third group (mainly oil) 1962*

Discovery well name and general location:

- **First pay** .... *(first group of pays) A core well*
- **Second pay** .... *(second group of pays) Uzen No. 1, on crest of the structure*
- **Third pay** .... *(third group of pays) A development well*

Discovery well operator .... *Ministry of Geology of the U.S.S.R.*

- **Second pay** .... *Ministry of Geology of the U.S.S.R.*
- **Third pay** .... *Ministry of Oil Industry of the U.S.S.R.*

IP in barrels per day and/or cubic feet or cubic meters per day:

- **First pay** .... *(first group of pays) 10,000 to 35,000 m³/day*
- **Second pay** .... *(second group of pays) 750 to 1100 bbl/day*
- **Third pay** .... *(third group of pays) 50 to 1100 bbl/day*

All other zones with shows of oil and gas in the field:

Age	Formation	Type of Show
Lower Triassic		*Oil*

Geologic concept leading to discovery and method or methods used to delineate prospect, e.g., surface geology, subsurface geology, seeps, magnetic data, gravity data, seismic data, seismic refraction, nontechnical:

Surface mapping performed during 1937–1941 indicated the presence of a large anticlinal structure.

Structure:

Province/basin type (see St. John, Bally, and Klemme, 1984)

The Mangyshlak subbasin differs substantially from the rest of the Middle Caspian basin. The subbasin does not fit well into either classification scheme.

Tectonic history

(1) Hercynian accreted zone (pre-Late Permian); (2) Late Permian–Triassic rifting and formation of deep grabens; (3) pre-Jurassic inversion and folding of the rifts, partial truncation of the Upper Permian–Triassic section; (4) Jurassic–early Miocene subsidence south of the inverted rifts and formation of the subbasin; (5) pre-middle Miocene deformations; (6) deposition of thin post-lower Miocene sediments; (7) recent uplift.

Regional structure

Structural step between the Central Mangyshlak uplift (inverted and folded Late Permian–Triassic rift) and the South Mangyshlak trough.

Local structure

WNW-trending anticlinal fold 45 km long and 9 km wide. Northern limb dips 1.5 to 2°, southern limb dips 6 to 8°. Faulting in the post-Triassic section is minor.

Trap

Trap type(s) .... *Anticlinal trap with multiple pays; fault control for minor pools in the third group of pays is inferred*

Basin stratigraphy (major stratigraphic intervals from surface to deepest penetration in field).

Chronostratigraphy	Formation	Depth to Top in m
Tertiary		*0*
Upper Cretaceous		*0-120*
Aptian-Albian		*200-380*
Neocomian		*800-900*
Upper Jurassic		*900-1050*
Lower-Middle Jurassic		*1100-1350*
Triassic		*2150-2250*

Location of well in field .... *NA*

Reservoir characteristics:

Number of reservoirs .... *20 to 21*

Formations .... *NA*

Ages .... *Upper Cretaceous (Cenomanian-Middle Jurassic [Bajocian])*

Depths to tops of reservoirs

First group of pays .... *172 m*

Second group of pays .... *1050 m*

Third group of pays .... *1520 m*

Gross thickness (top to bottom of producing interval)

First group of pays .... *730 m*

Second group of pays .... *310 m*

Third group of pays .... *320 m*

Net thickness—total thickness of producing zones

Average .... *(first group of pays) NA*

Maximum .... *(first group of pays) 270 m*

Average .... *(second group of pays) NA*

Maximum .... *(second group of pays) 270 m*

Lithology .... *Fine- to medium-grained, well- to moderately sorted feldspathic graywacke and graywacke*

Porosity type .... *Intergranular*

Average porosity *21.5%*

Average permeability .... *235 md*

Seals:

Upper

Formation, fault or other feature .... *Main regional seal is the upper Callovian-Kimmeridgian*

Lithology .... *Shales, marls, limestones; other shales seal individual pools*

Lateral

Formation, fault or other feature .... *Small faults supposedly seal laterally some pools in the third group of pays*

Lithology .... *NA*

Source:

Formation and age .... *Lower-Middle Jurassic, Triassic (?)*

Lithology .... *Dark continental to shallow-marine shales (Jurassic); black marine shales and marls (Triassic)*

Average total organic carbon (TOC) .... *1.0 to 1.3% (Jurassic); 1.4 to 7.5% (Triassic)*

Maximum TOC .... *NA*

Kerogen type (I, II or III) .... *II and III (Jurassic); II (Triassic)*

Vitrinite reflectance (maturation) .... *R_o = 0.6 to 0.8 (Jurassic)*

Time of hydrocarbon expulsion .... *Mainly Cretaceous (for Jurassic source)*
Present depth to top of source .... *1100 to 1400 m*
Thickness .... *1000 to 1300 m*
Potential yield .... *NA*

Appendix 2. Production Data

Field name .... *Uzen (Uzenskoye)*

Field size:

- **Proved acres** .... *25,170 ha*
- **Number of wells all years** *2000*
- **Current number of wells (as of)** *1000*
- **Well spacing** .... *Originally 28-42 ha/well (1968-1969), then density increased*
- **Ultimate recoverable** .... *approximately 2000 million bbl*
- **Cumulative production** .... *1350 to 1400 million bbl (to 1/1/1985)*
- **Annual production** .... *66 million bbl (1984)*
- **Present decline rate** .... *NA*
  - **Initial decline rate** .... *7.4% (1976)*
  - **Overall decline rate** .... *45 to 50%*
- **Annual water production** .... *190 million bbl (1979)*
- **In place, total reserves** .... *about 7500 million bbl*
- **In place, per acre-foot** .... *NA*
- **Primary recovery** .... *NA*
- **Secondary recovery** .... *about 2000 million bbl*
- **Enhanced recovery** .... *NA*
- **Cumulative water production** .... *NA*

Drilling and casing practices: *(second group of pays)*

- **Amount of surface casing set** .... *180 m*
- **Casing program** .... *Surface casing 273 mm—180 m, casing 168 mm—1300 to 1450 m*
- **Drilling mud** .... *Drilling mud with specific weight 1.25 g/cm³, water with addition of caustic and lignite*
- **Bit program** .... *Turbodrill*
- **High pressure zones** .... *None*

Completion practices:

- **Interval(s) perforated** .... *NA*
- **Well treatment** .... *Multifrac by HCl + HF solution with sand; and other techniques*

Formation evaluation:

- **Logging suites** .... *NA*
- **Testing practices** .... *NA*
- **Mud logging techniques** .... *NA*

Oil characteristics:

- **Type** (Tissot and Welte classification *in* "Petroleum Formation and Occurrence," 1984, Springer-Verlag, p. 419) .... *(second group of pays) Paraffinic*
- **API gravity** .... *(second group of pays) 33 to 36.5°*
- **Base** .... *(second group of pays) NA*
- **Initial GOR** .... *(second group of pays) 72 to 85 m^3/m^3*
- **Sulfur, wt %** .... *(second group of pays) 0.1 to 0.24*
- **Viscosity, SUS** .... *(second group of pays) 3.4 to 4.2 cp (in reservoir conditions)*

Pour point .... *(second group of pays) 25 to 30°C*
Gas-oil distillate .... *(second group of pays) NA*

Field characteristics:

Average elevation .... *(second group of pays) 200 to 240 m*
Initial pressure .... *(second group of pays) 98 to 123 kg/cm²*
Present pressure .... *(second group of pays) NA*
Pressure gradient .... *(second group of pays) NA*
Temperature .... *(second group of pays) 53.5 to 72°C*
Geothermal gradient .... *(second group of pays) 3.8°C/100 m*
Drive .... *(second group of pays) Water flooding from the beginning of production*
Oil column thickness .... *(second group of pays) 100.5 m*
Oil-water contact .... *(second group of pays) 1140 to 1150 m bsl*
Connate water .... *(second group of pays) 30 to 38%*
Water salinity, TDS .... *(second group of pays) NA*
Resistivity of water .... *(second group of pays) NA*
Bulk volume water (%) .... *(second group of pays) NA*

Transportation method and market for oil and gas:

Oil is transported through heated pipeline to refineries in Kuybyshev City.

Appleton Field—U.S.A.
Gulf of Mexico Basin, Alabama

T. J. PETTA
Union Texas Petroleum
Houston, Texas

S. D. RAPP
Western Atlas Core Laboratories
Irving, Texas

FIELD CLASSIFICATION

BASIN: Gulf of Mexico
BASIN TYPE: Extensional
RESERVOIR ROCK TYPE: Dolomite
RESERVOIR ENVIRONMENT OF DEPOSITION: Shoal, Beach
RESERVOIR AGE: Jurassic
PETROLEUM TYPE: Oil
TRAP TYPE: Combination

LOCATION

Appleton field is located in Escambia County, Alabama, approximately 70 mi (112 km) northeast of Mobile and 20 mi (32 km) north of Jay field (Figure 1). We estimate the four wells will produce 3 million bbl of oil from the Smackover Formation. Folded Appalachian igneous and metamorphic rocks underlie this portion of Alabama (Adams et al., 1926). Partial erosion of the northeast-southwest-trending crystalline basement complex created structural anomalies. High-energy carbonate shoals accumulated over the positive relief structures during the Middle Jurassic. These shoals are composed of peloids and ooid grainstones, algal debris, and algal boundstones. Abrupt lateral change to peloid mudstones is common. Lower Haynesville Formation anhydrites (Figure 2) provide the top and lateral seal for the high-energy carbonate reservoirs of the Smackover. Appleton is one of many Smackover oil fields in Alabama where reservoir occurrence and hydrocarbon accumulation are closely related.

HISTORY

Pre-Discovery

The first Smackover oil produced in Alabama was from fields in Choctaw County (Moore, 1971). Discoveries such as Choctaw Ridge field (1967) (Figure 1) resulted from wells drilled to extend the Smackover salt ridge and anticline play eastward from Mississippi. Typical exploration methodology was to first interpret the position of salt anticlines using gravity data. Next, CDP seismic data were acquired to confirm the structure. The subsequent drill site was located on the crest of the structure. Until the discovery of Jay field in 1970, all Alabama Smackover production was in Choctaw County in reservoirs between 11,000 and 12,000 ft (3350 and 3650 m) in depth.

Discovery of Jay field moved the productive limits of the Smackover play 100 mi (160 km) eastward (Figure 1). Productive limits also moved downward because some of the reservoirs at Jay produce from 15,500 ft (4725 m). Fields within the Jay complex contain over 500 million barrels of oil equivalent (MMBOE) of recoverable reserves. This discovery caused a significant increase in exploratory activity in the Alabama-Florida area.

As oil companies continued to acquire CDP data, they recognized numerous structural anomalies updip from the Jay complex. Smackover reservoirs occur over an eroded crystalline basement surface rather than salt anticlines in this area called the Conecuh Ridge (Figure 1). The sharply defined structural anomalies associated with the eroded basement are small compared to the salt anticlines. Most oil companies preferred to drill the salt anticlines because of their larger size.

Although Getty discovered Vocation field (estimated ultimate recovery 4 MMBOE) in 1971, no other significant accumulations were found on the Conecuh Ridge until 1983. Getty used regional facies analysis to define where the Smackover was apt to be dolomitized. Dolomite reservoirs were preferred over limestone reservoirs. Seismic data acquired from 1970 to 1980 is generally of poor quality. Although numerous wells were drilled during the decade after

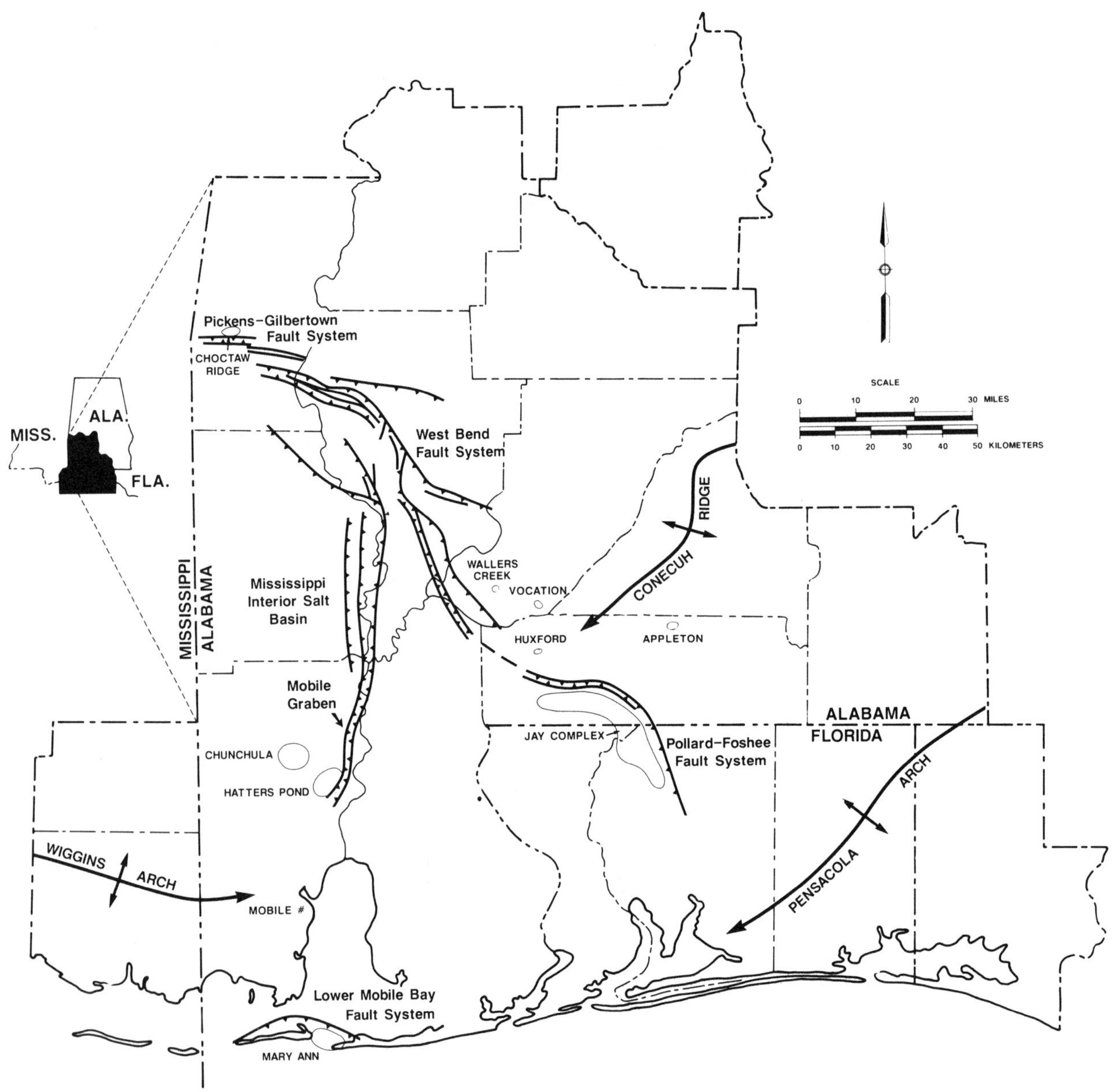

Figure 1. Index map of Mississippi, Alabama, and Florida showing the relationship of Appleton field to the Conecuh Ridge, the Jay complex, and major structural features.

the Vocation discovery, no other significant oil accumulations were found.

Because of the improved quality of seismic data gathered from 1980 to 1982, Texaco, Inc. embarked on an extensive drilling program in the Conecuh Ridge area. This program resulted in the discovery of Huxford and Appleton fields in 1983 (Figure 1). Similar exploration strategies resulted in Strago's discovery of Wallers Creek in 1985. Southern Union Exploration, Moore-McCormack, Zinn Petroleum, and Coastal Petroleum successfully completed new field wildcats during 1987 and 1988. Seismic data quality continues to improve, and more subtle subsurface anomalies are recognized.

Discovery

Discovery of Appleton field was made in 1983 by the Texaco, 1-D.W. McMillan 2-14 in Section 2-T3N-R9E, Escambia County, Alabama. The well was drilled to 13,080 ft (3987 m) and bottomed in Precambrian igneous rock that is unconformably overlain by the Smackover (Figure 3). Perforations were through 7-in. casing and production was made through 2⅞-in. tubing run to a depth of 12,850 ft (3917 m). Initial production rate was 661 BOPD and 1060 MCFGD through an 18/64-in. choke. The porosity zone at Appleton (Figure 4) coincides with thin Smackover over the crest of a positive basement

ERATHEM	SYSTEM	SERIES	GEOLOGIC UNIT	
CENOZOIC	QUATERNARY	HOLOCENE PLEISTOCENE	UNDIFFERENTIATED "CITRONELLE"	
	TERTIARY	PLIOCENE		
		MIOCENE	PENSACOLA CLAY	"Meyer sand"
				"Escambia sand"
				"Luce sand"
				"Amos sand"
		OLIGOCENE	UNDIFFERENTIATED	
		EOCENE	JACKSON GROUP	
			CLAIBORNE GROUP	
		PALEOCENE	WILCOX GROUP	
			MIDWAY GROUP	
MESOZOIC	CRETACEOUS	UPPER	SELMA GROUP	
			EUTAW FORMATION	
			TUSCALOOSA GROUP	Upper
				Marine
				Lower
		LOWER	LOWER CRETACEOUS UNDIFFERENTIATED	
			COTTON VALLEY GROUP	
	JURASSIC	UPPER	HAYNESVILLE FORMATION	
			BUCKNER ANHYDRITE	
			SMACKOVER FORMATION	
			NORPHLET FORMATION	
			"PINE HILL ANHYDRITE"	
			LOUANN SALT	
	TRIASSIC	MIDDLE	WERNER FORMATION	
			EAGLE MILLS FORMATION	
			BASEMENT COMPLEX	

Figure 2. Columnar section of southwest Alabama.

feature. The Norphlet pinches out along the flanks of the structure. All of the best production in this portion of Alabama is located on the crest of positive features having no Norphlet sandstone. Where the Norphlet is absent, the Smackover usually contains a grainstone facies.

Post-Discovery

Development of Appleton field proceeded slowly. Texaco realized the carbonate reservoirs probably covered less than 640 ac (260 ha.). Having drilled four producers and one dry hole between 1983 and 1986, Texaco farmed out several locations. This strategy caused the drilling of several nonproductive wells. The exception was the Smackover discovery of Burnt Corn Creek field (Figure 3). This marginal field is apparently on a structural feature separate from Appleton.

DISCOVERY METHOD

Interpretation of regional lithofacies maps, gravity data, and analog fields led Texaco to acquire a proprietary seismic data network in Alabama from 1980 to 1982. Using seismic analysis, numerous structures including Huxford and Appleton were defined. Typically, a decrease in seismic interval time between the Smackover and Haynesville events equates to the presence of Smackover reservoir.

STRUCTURE

Tectonic History

Appleton field is in the northeastern Gulf of Mexico basin. Here, the Gulf Coastal Plain overlies the southwestward-plunging Appalachian Piedmont province. The field lies north of Pollard–Foshee fault system (Figure 1). Structures north of the fault system have basement rocks at their core rather than salt. Folded Archean and Algonkian gneiss and schist of the Appalachians comprise the basement (Adams et al., 1926).

Regional Structure

Partial erosion of the basement occurred prior to and during Norphlet deposition. Norphlet sandstones and conglomerates occur on the flanks of the basement feature like Appleton (Figure 4). Smackover grainstones are concentrated around and over these erosional remnants. The basement features were, for the most part, covered by Haynesville evaporites and clastics (Figure 2). Because of differential compaction, the positive relief of these structures is often

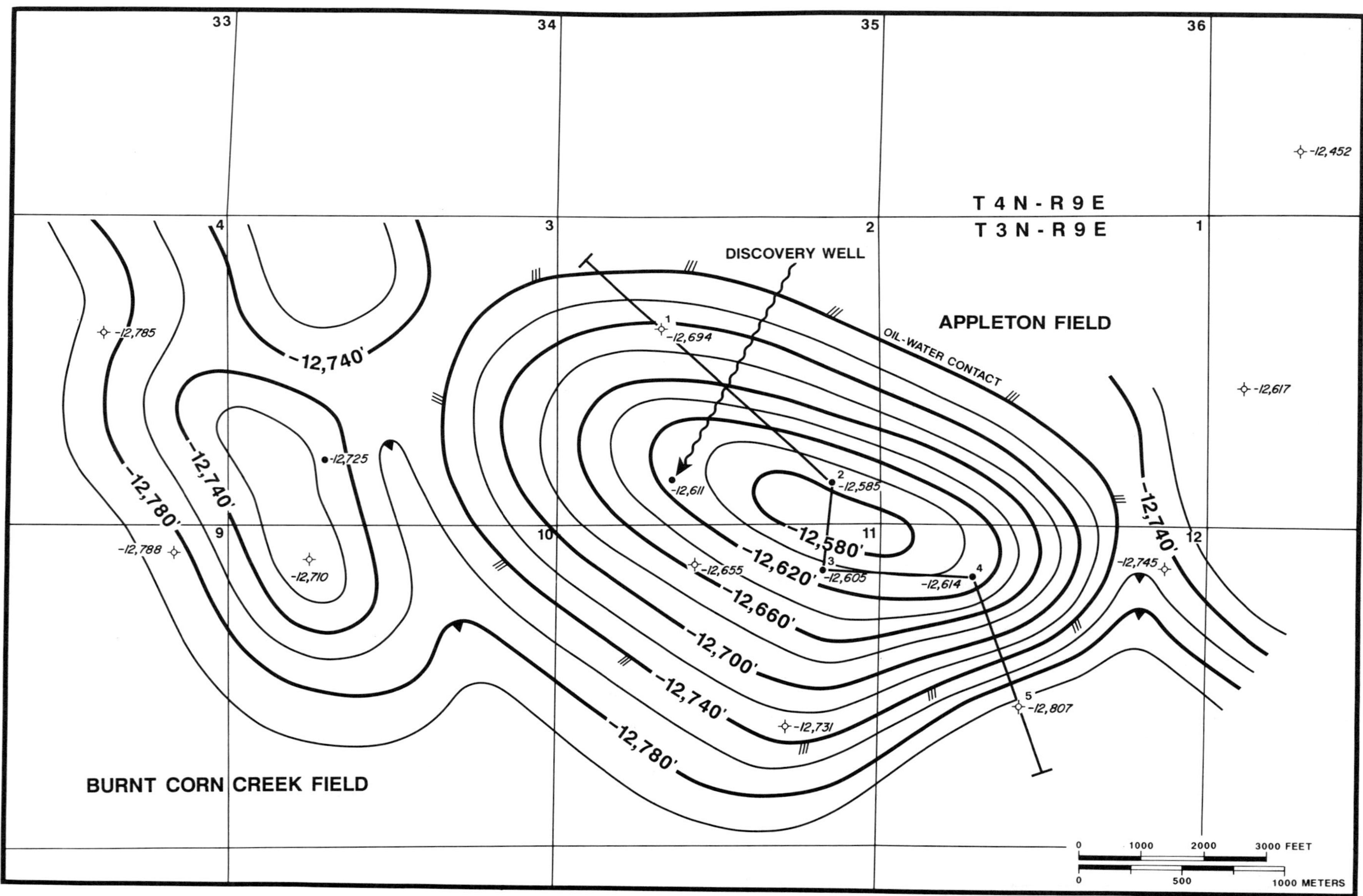

Figure 3. Structure map: top Smackover, Appleton field. The Texaco 1-D.W. McMillan 2-14 was the discovery well for this field. Contours (20 ft [6 m] interval) indicate 160 ft (49 m) of closure. The oil-water contact is at -12,740 ft (-3908 m). The oil-water contact outlines the Appleton field.

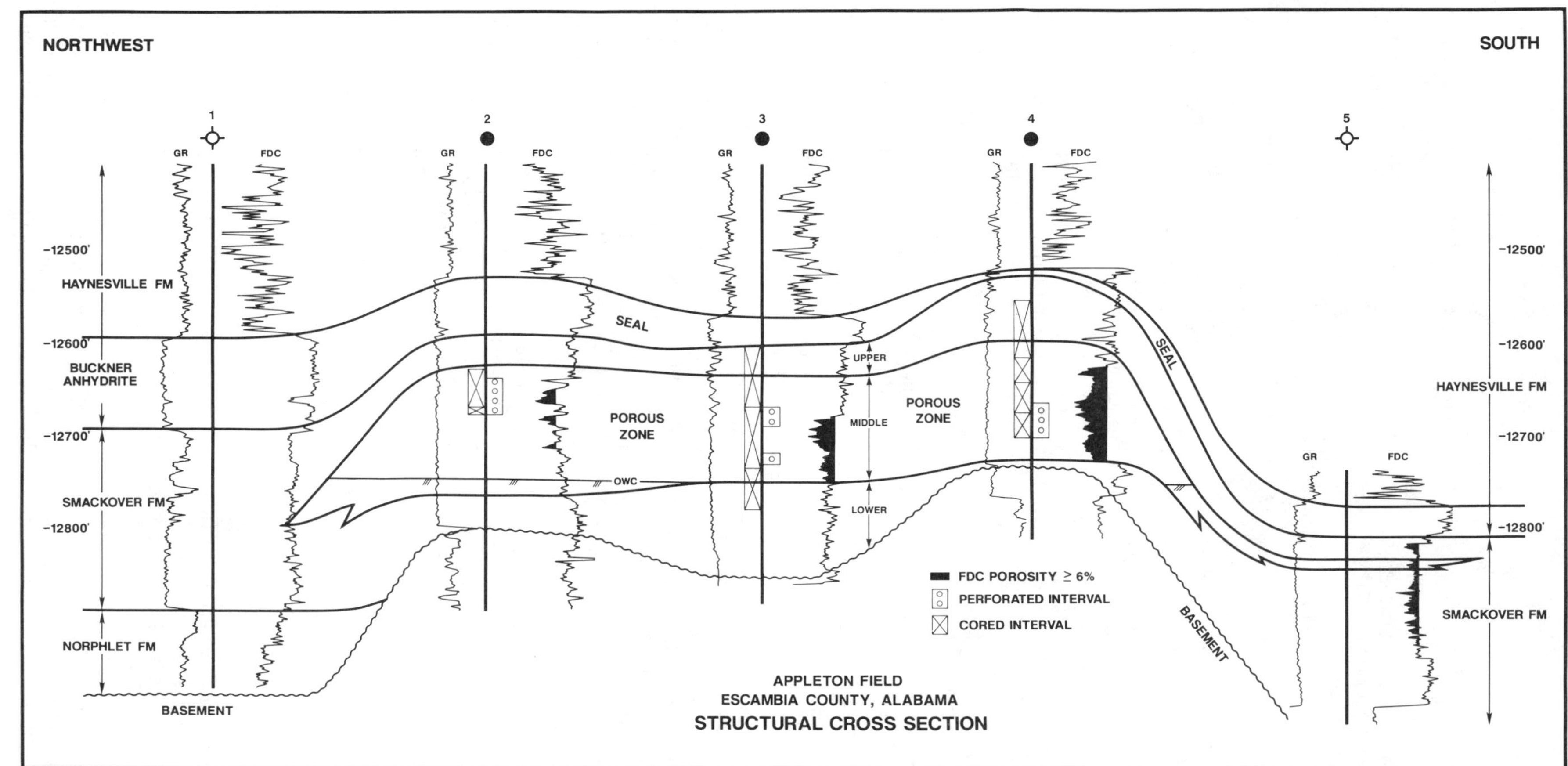

Figure 4. Structural cross section oriented along regional strike showing the lateral extent of the Porous Smackover facies at Appleton field. The cross section location is shown by Figure 3 and all subsequent maps. Well numbers on the cross section correspond to those on the maps. Reservoir quality porosity at Appleton is mostly confined to the middle Smackover interval. Dolomite algal boundstone, peloid packstone, and peloid oolite grainstone facies occur within the middle Smackover. Generally, muddier textures are predominant in both the lower and upper Smackover intervals.

preserved. Therefore, isopach and/or isochron thins that can be mapped at several horizons are coincident with the positive structures.

Local Structure

Several maps can be used to demonstrate the passive drape structural style characteristic of Appleton and all other Smackover fields in the region. The top of the Smackover and the top of the reservoir have 160 and 120 ft (39 and 47 m) of counter-regional closure, respectively (Figures 3 and 5).

The lowermost anhydrite unit of the overlying Haynesville Formation thins by 70 ft (21 m) over the structure (Figure 6). The 125 ft (38 m) of closure apparent in the Haynesville to Smackover interval (Figure 7) increases to nearly 150 ft (46 m) within the lower Tuscaloosa to Smackover interval (Figure 8). Passive drape and differential compaction were responsible for preservation of this closure during southwestward tilting of the Gulf of Mexico basin.

STRATIGRAPHY

The columnar section (Figure 2) illustrates the rock sequence of southwest Alabama. The basement complex that forms the Conecuh Ridge was deeply eroded during the Upper Jurassic. Erosion supplied sediment deposited as continental clastics called the Norphlet Formation.

Clastic deposition was interrupted by a major marine transgression during which carbonate muds of the Smackover Formation were deposited. When the rate of transgression slowed, the high-energy carbonate facies of the Smackover accreted upward and spread laterally (Moore, 1984).

Smackover fields in southwest Alabama typically produce from high-energy grainstone facies that accumulated on a carbonate ramp. However, the distribution of grainstones is not within a belt of high energy, oriented parallel to paleoshoreline. Reservoirs are instead associated with paleotopography that was created by either basement erosion or salt anticlines.

Within Appleton field, the Smackover Formation is divided into three facies. The lower Smackover is composed of slightly to pervasively dolomitized peloid wackestones and packstones (Saller and Moore, 1986; this study). Packstones frequently contain layers of oncolites that are easily recognized in limestones. When the packstones have been dolomitized, "ghosts" of the oncolites are identified. Many of the oncolites and "ghosts" have been flattened because of compaction. The numerous peloids and oncolites suggest lower Smackover deposition took place in a low- to moderate-energy, restricted subtidal environment. A broad lagoon is the modern analog for this type of facies. Deposition of the lower Smackover occurred over a broad, low-relief ramp. Regional circulation barriers such as basement (Wiggins Arch) and shoreline (Conecuh Ridge) features and salt anticline grainstone barriers (grainstones of the Jay trend) modified local facies distribution.

The middle Smackover is mostly dolomite that was deposited as peloid packstones, peloid grainstones, and algal boundstones. Abundant blue-green algal structures and lack of normal marine fossils suggest that hypersaline conditions prevailed. This portion of the Smackover, which is thickest on the flanks of the Appleton basement feature, was deposited in a shallow subtidal environment. Most porosity in Appleton field occurs within the middle Smackover interval. Reservoir dolomites are concentrated over and adjacent to the underlying basement feature. The reservoir facies thins rapidly on the fields' flanks, and peloid wackestones and packstones of the lower Smackover become the dominant facies.

The uppermost Smackover consists of a wide variety of facies including ooid and peloid packstones and grainstones, and algal boundstones, overlain by peloid wackestones and limestones. Much of the upper Smackover has wavy and planar bedding, mudcracks, and fenestral vugs (Green, 1985). Marine vadose pisolites (Figure 9A) occur in the #2 Texaco McMillan Trust 11-1 well. These pisolites probably formed in an upper intertidal to lower supratidal setting. This facies is overlain by a sequence of peloid grainstones and packstones and by algal boundstones. The algal boundstones are overlain by dolomitized peloid wackestones and mudstones.

This sequence of facies suggests the crest of Appleton field was subaerially exposed. Exposure was due to either lateral progradation of high energy grainstones or a minor regression. The overlying dolomites, deposited as wackestones and mudstones, are probably tidal flat deposits. The shallow subtidal to intertidal and supratidal deposits are thickest on the crest of the Appleton structure and appear to grade laterally into high-energy packstones and grainstones off the structures' flanks.

The lower Smackover at Appleton field does not contain any organic-rich limestones characteristic of Smackover source beds throughout the Gulf of Mexico basin. Hydrocarbons probably matured within downdip, organic-rich Smackover limestones and migrated through stylolites and fractures. Vertical fractures are common in the lower Smackover at Appleton field. These fractures could have provided the conduit for hydrocarbons to migrate into the reservoir facies at Appleton.

TRAP

One hundred and sixty feet (49 m) of structural closure is mapped at the top of the Smackover (Figure 3). A top and lateral seal is provided by the Buckner Anhydrite. This basal anhydrite member

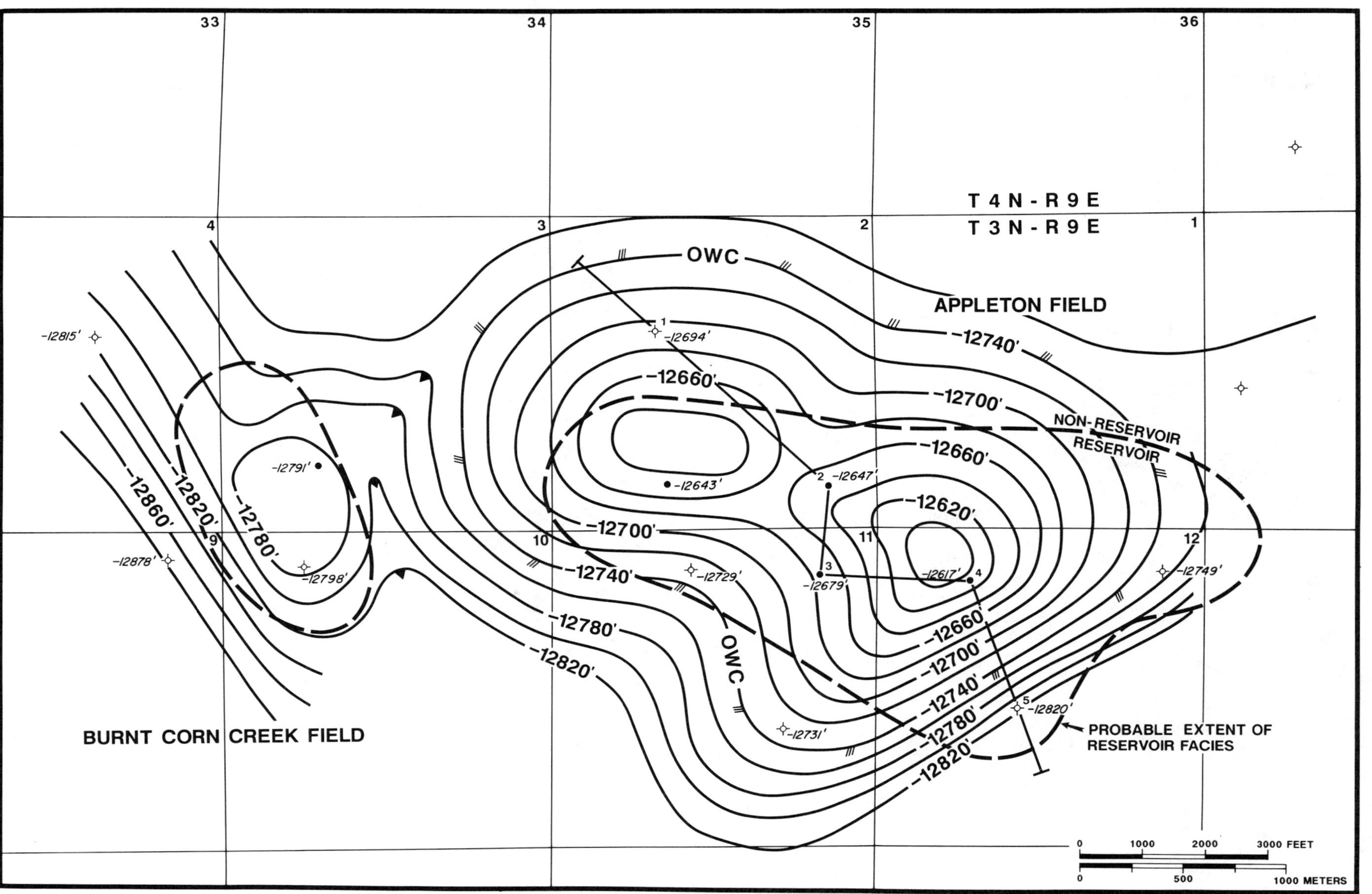

Figure 5. Structure map: top of the Smackover porous facies. Slightly over 120 ft (37 m) of closure is present. Contour interval is 20 ft (6 m).

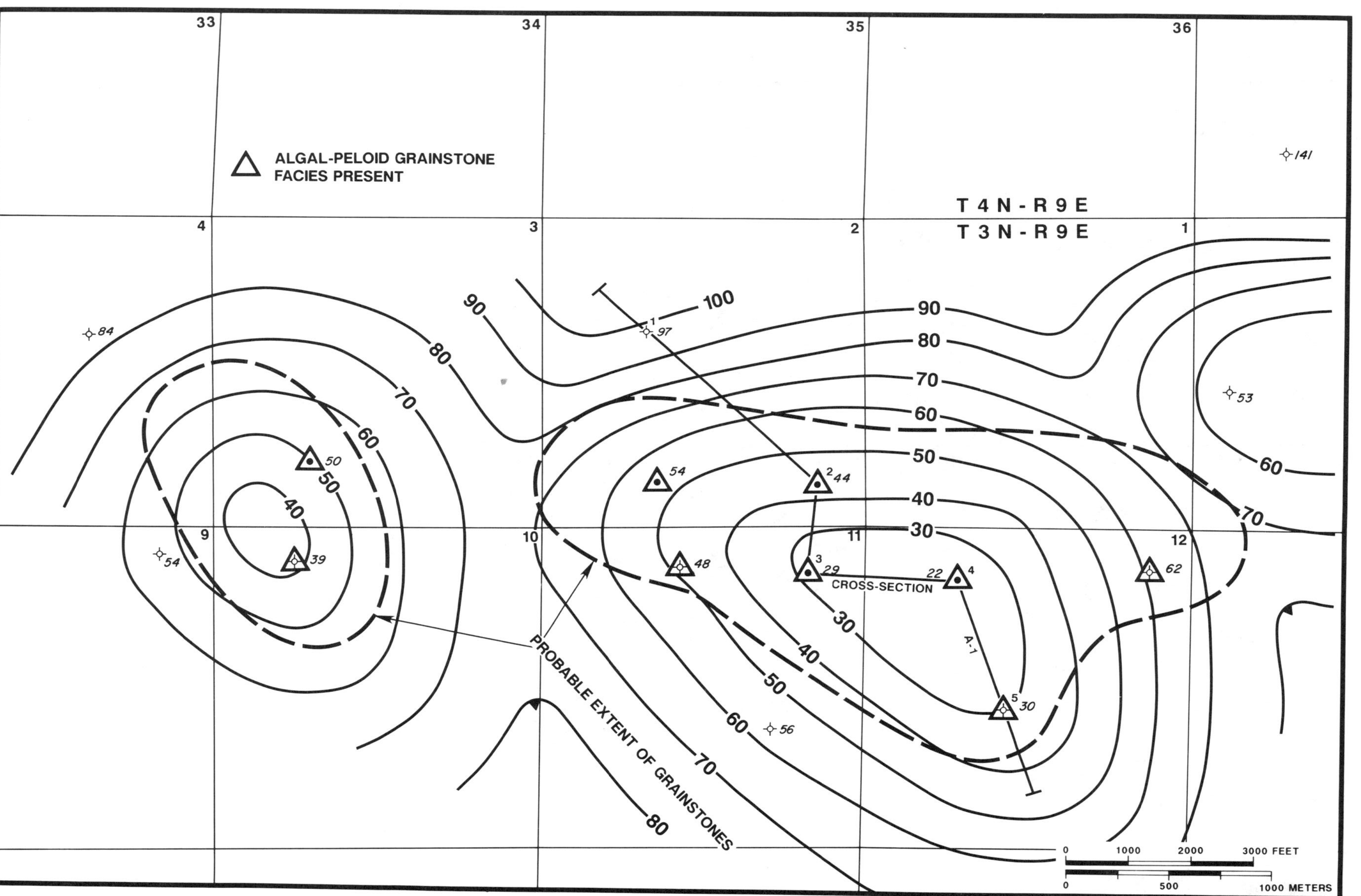

Figure 6. Isopach map: Buckner–Smackover. At least 70 ft (21 m) of early (depositional) relief is shown. Contour interval is 10 ft (3 m).

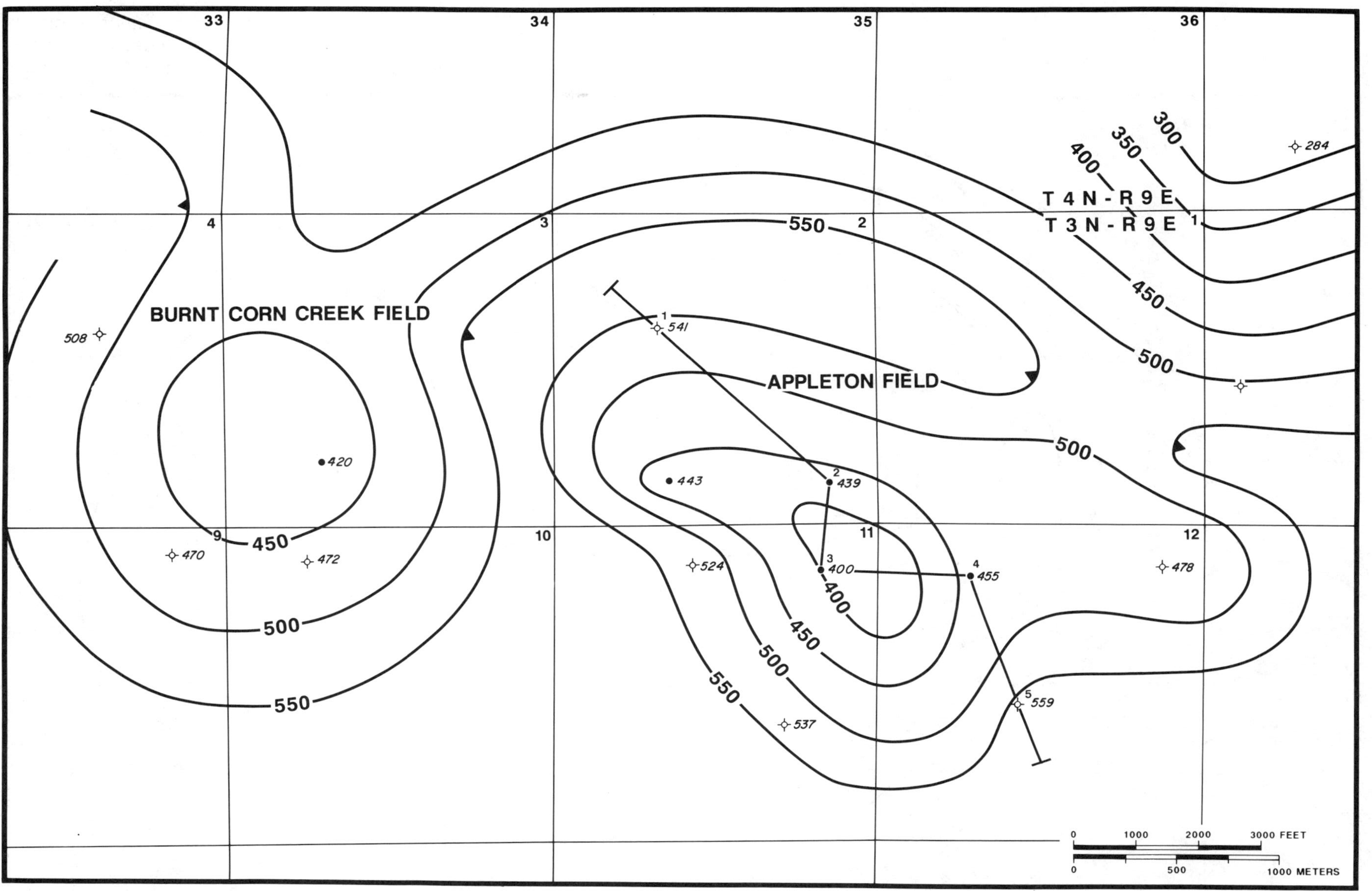

Figure 7. Isopach map: The Haynesville-Smackover interval is thin over the structure. The amount of thinning is 140 ft (43 m), nearly double that shown by the Buckner-Smackover map. Most of the 160 ft (49 m) of Smackover closure present today was, therefore, created by differential compaction by the time Haynesville deposition had ceased. Contour interval is 50 ft (16 m).

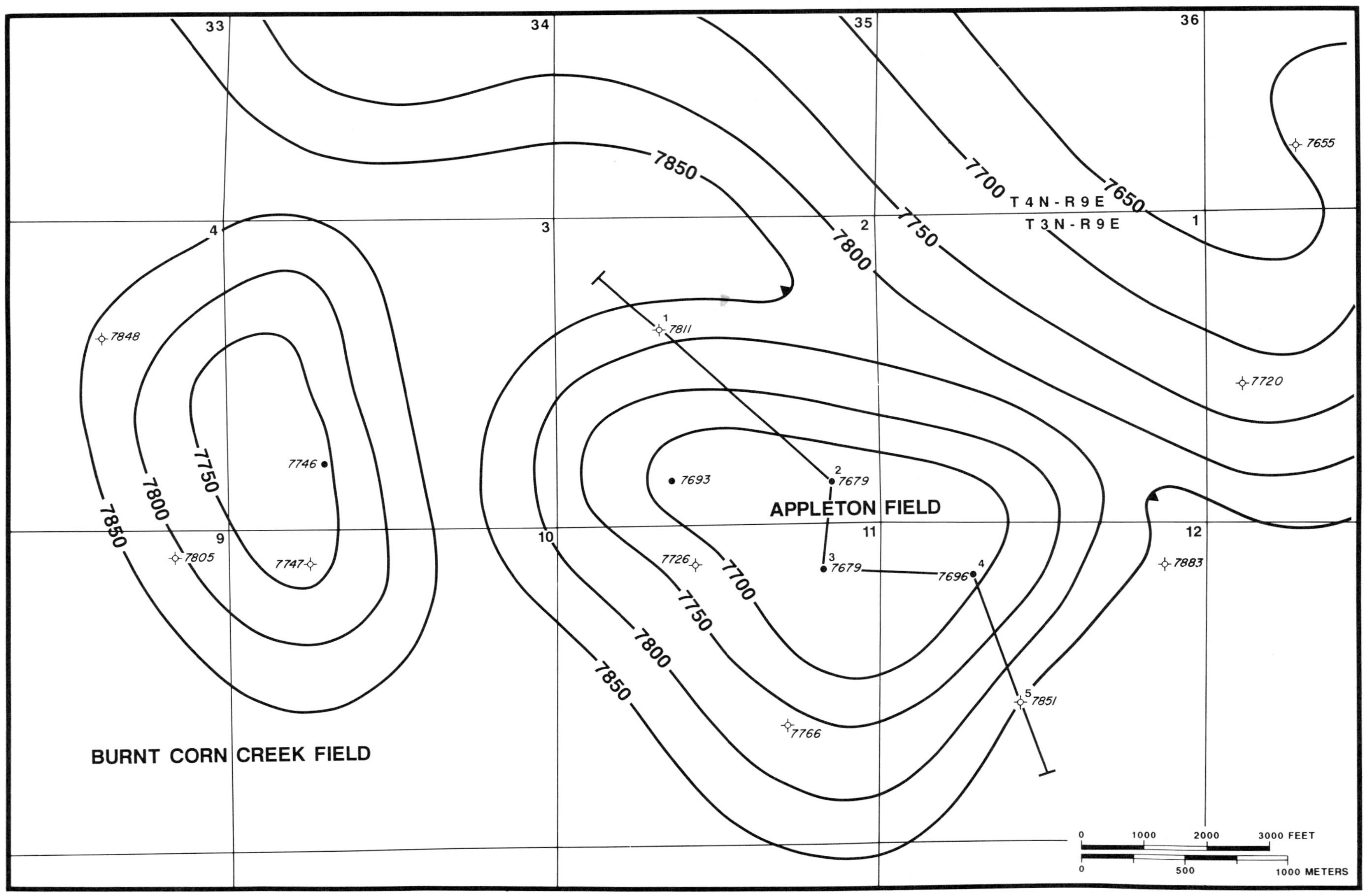

Figure 8. Isopach map: Lower Tuscaloosa-Smackover. This interval thins by 150 ft (46 m) over the structure. Contour interval is 50 ft (16 m).

of the Haynesville Formation varies from 30 ft (9 m) thick directly atop the structure to over 100 ft (30 m) thick to the north. The oil-water contact of -12,740 ft (-3883 m) is coincident with the small structural sag north and east of the field. The relationship of the thicker anhydrite with a structural low demonstrates the dependency of the trap on early structure.

If this structure had not occurred close to the paleoshoreline, the depression that contains thicker anhydrite might have been the depositional site for carbonate muds. Carbonate mudstones, both limy and dolomitic, in southwest Alabama do not usually act as good lateral seals. Fractures and stylolites that could have transmitted oil migrating into traps could also have provided avenues of escape for oils to lead from accumulations.

RESERVOIR

A dolomite reservoir, deposited as algal boundstone and peloid packstone/grainstone, coincides with a basement high at Appleton field (Figure 5). The reservoir pinches out away from the crest (Figure 2). A much smaller reservoir is present at Burnt Corn Creek field 1.5 mi (2.4 km) west of Appleton field (Figure 2).

Dolomite intercrystalline porosity is most common at Appleton field. Intercrystalline porosity (pores preserved among dolomite rhombs) is most commonly associated with algal boundstone (Figure 9B) and peloid packstones/grainstone facies of the middle Smackover. Vuggy and, possibly, moldic porosity (Figure 9C) is often associated with intercrystalline porosity. These combinations of porosity enhance the overall reservoir quality.

Measured porosity values within porous dolomites of the middle Smackover are generally greater than 10% and may be as great as 25%. Permeability values as high as 4000 md have been measured in vuggy dolomites of the middle Smackover.

Nondolomitized grainstones, which occur in the lower portions of the upper Smackover on the crest and in the uppermost Smackover on the fields' flanks, can contain interparticle pores and moldic pores (Figure 9D). Interparticle pore volume is often reduced by circumgranular equant calcite cements. Moldic pores are often present in peloid grainstone in which original interparticle porosity has been partially to completely occluded by circumgranular, calcite cements. These moldic limestones have high porosity and low permeability. Porosity values as high as 20% with permeability values less than 1.0 md are common.

The variability of pore types and porosity and permeability values is a result of diagenesis. Dolomitization has been the most important diagenetic process in the development and preservation of reservoir rocks. Dolomite occurs throughout the Smackover Formation and is not limited to any particular facies. Porous dolomites are most prevalent in crestal wells and in those located along Appleton's southern flank. Saller and Moore (1986) concluded from petrologic and geochemical evidence that dolomitization was a result of hypersaline waters moving through the carbonate strata shortly after burial. The mechanism invoked, "evaporative pumping," was caused by intense evaporation from supratidal environments. Exposed surfaces on the Appleton structure and the sabkha immediately north of the field were areas where this mechanism might have operated.

Pisolites occur in the lower part of the upper Smackover in the Texaco #2 McMillan Trust 11-1 well on the crest of the Appleton structure. These pisolites probably developed in an upper intertidal to lower supratidal setting and indicate a period of subaerial exposure. This exposed surface may have created the area necessary to draw hypersaline fluids upward through sediments by capillary pressure (evaporative pumping). Intensive evaporation in an upper intertidal and supratidal environment would also allow the accumulation and downward movement of relatively dense hypersaline waters into the underlying sediments.

The presence of moldic and vuggy porosity suggests that these dolomitizing waters were probably undersaturated with respect to calcite and aragonite. In addition, the hydraulic gradient, whether created by evaporative pumping or brine reflux, was greater in sediments on the south side of the field because of the more prevalent dolomitization observed in the middle Smackover on that flank. The dolomitized interval containing pisolites is overlain by slightly dolomitic peloid packstones and grainstones as well as algal boundstones. This sequence implies that a minor transgression followed the period of subaerial exposure. This interval grades upward into nonporous dolomite, which appears to have been peloid wackestones and mudstones deposited in a tidal flat environment. Therefore, dolomitization at Appleton may have been associated with two phases of exposed tidal flat deposits, both of which could have created concentrations of dolomitizing, hypersaline waters. Dolomitization of the middle Smackover peloid packstone, grainstones, and algal boundstones created a high-quality reservoir. In contrast, dolomitization of mudstones and wackestones has created dolomite having low porosity and permeability.

Moldic porosity preserved in upper Smackover ooid/peloid grainstones on the flanks of Appleton field resulted from the leaching of peloids and cementation with circumgranular calcite cements. Grains are generally undercompacted and often completely supported by cements. Leaching and cementation occurred soon after deposition and certainly before significant burial. Pore waters undersaturated with respect to aragonite and/or calcite percolated downward through these grainstones. An "inverted"

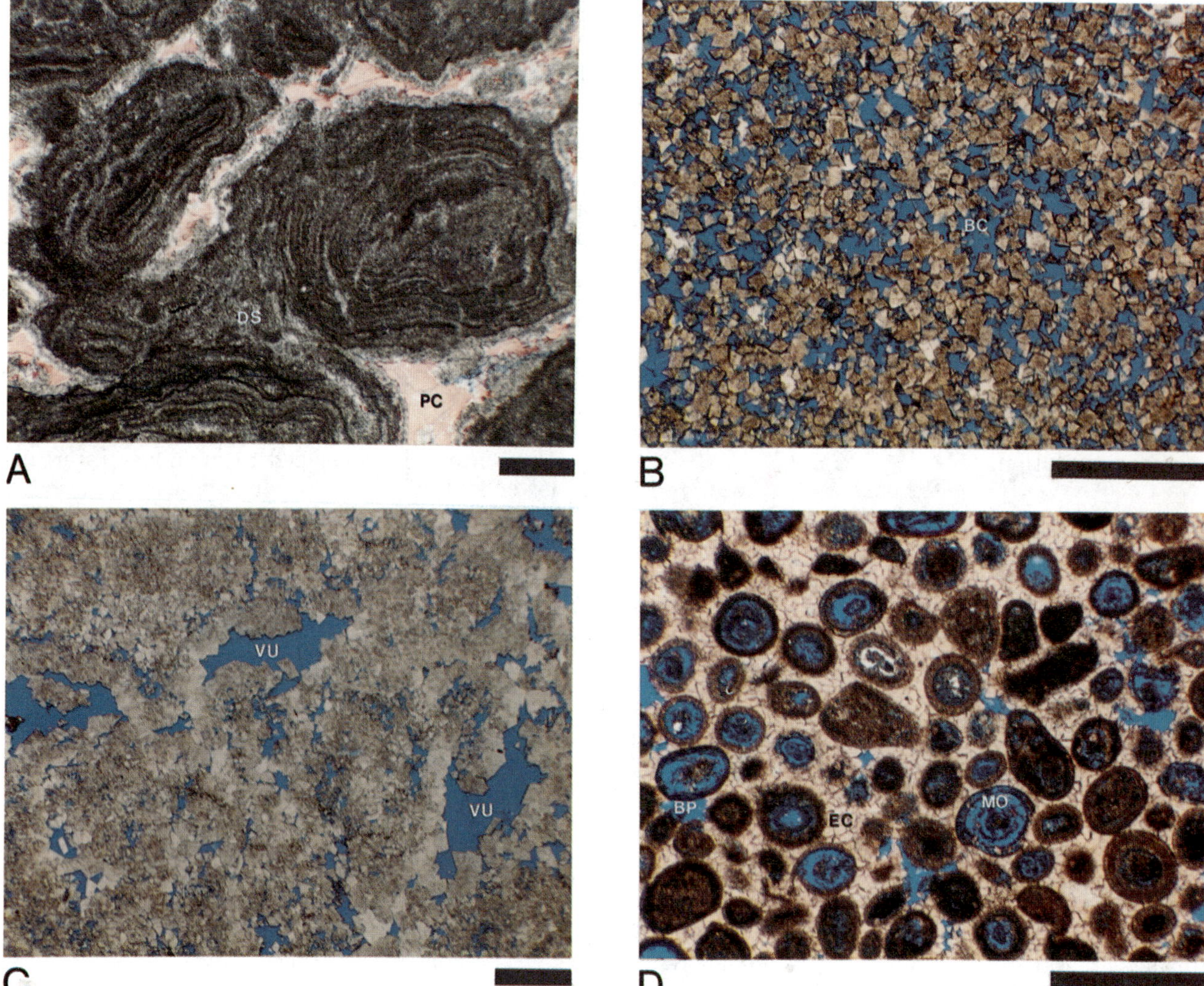

Figure 9. Thin-section photomicrographs displaying microfacies and porosity types. (A) Pisoids interpreted to have formed in the marine vadose environment. Note the irregular shapes and perched diagenetic sediment (DS). Fenestral porosity has been completely filled with poikilotopic calcite spar (PC). Texaco #2 McMillan Trust 11-1, 12,944 ft (3946 m). (B) Pervasively dolomitized peloid packstone with significant intercrystalline porosity (BC). Larger pores were probably created by dissolution of allochems and/or micrite. Texaco #4 McMillan Trust 12-11, 13,144 ft (4007 m). (C) Pervasively dolomitized algal boundstone with vuggy pores (VU) and intercrystalline pores. Texaco #2 McMillan Trust 11-1, 13,027 ft (3972 m). (D) Pellet ooid grainstone. Partial or complete leaching of many ooids has created oomoldic pores (MO). Most of the original interparticle porosity has been filled with fine to medium crystalline, equant calcite cement (EC), leaving few interparticle pores (BP). Texaco #4 McMillan Trust 11-12, 13,061.5 ft (3982 m). The bar scale is 1.0 mm on each of these photomicrographs.

texture was created where pores are present where grains originally occurred. Cements fill original pore spaces. Finely crystalline equant calcite cement in the original interparticle pore spaces suggests that these cements were precipitated from meteoric phreatic waters. This texture has high porosity but low permeability.

SOURCE

Petroleum source beds occur in the lower portion of the Smackover throughout the Gulf of Mexico basin (Oehler, 1984). These beds are generally dark brown to black, mud-dominated, laminated limestones. These source rocks were deposited in an anoxic and perhaps hypersaline subtidal environment that preserved algal kerogen (Sassen and Moore, 1988).

Sassen (1987) demonstrated that stylolitic zones within the lower Smackover Formation can contain over 10% total organic carbon (TOC), although the mean TOC value is only 0.5%. Sassen suggests that pressure solution of Smackover limestones concentrated insoluble rock components including kerogen along stylolites. Stylolites have been identified as avenues for hydrocarbon migration in other carbonate sequences (Grabowski, 1984; Palacas et al., 1984).

Hydrocarbons entrapped at Appleton field were generated in stylolitic limestones of the lower Smackover Formation that occur downdip (Claypool and Mancini, 1989). Stylolites may have afforded lateral permeability to hydrocarbons, whereas fractures allow vertical migration into the reservoir beds.

A burial-history curve for Appleton field (Figure 10) shows that primary expulsion of oil should have occurred during the earliest Tertiary. The lower Tuscaloosa to Smackover isopach (Figure 8) shows that the Appleton feature was in the path of hydrocarbon migration.

PRODUCTION

The pay zone in Appleton field varies in thickness from zero, off the structure, to a maximum of 122 ft (37 m) over the basement feature. Average thickness is 93 ft (28 m). Oil is 49° API and the initial GOR was 1594. Sulfur, present within H_2S, comprises 1.75% by weight. All of the fields within the Conecuh Ridge area are water driven, and Appleton is no exception. The field is presently producing nearly 36,000 bbl of oil and 67 mmcf of gas monthly from four wells (Estes, 1989). The four wells are flowing through choke sizes adjusted to conform with the 400 BOPD allowable rate set by the state and by the rate at which the gas plant can be operated. Dolomite reservoirs in Appleton and nearby fields have good lateral continuity. Spacing for oil wells is 160 ac (65 ha.). Recovery of 30-40% of the oil in place can be expected so that the entire field should produce 3 MMBOE. Over 1.7 MMBOE had been produced through March 1989 (Figure 11).

EXPLORATION CONCEPTS

The largest accumulations in the Conecuh Ridge area of southwest Alabama occur in Wallers Creek, Vocation, Huxford, and Appleton fields (Figure 1). At least 15 smaller fields have been discovered in the area but are not shown. As structures related to eroded basement features are mapped using seismic and subsurface data, they are drilled. Exploration companies can expect to find numerous fields in the area as more data become available.

Because of the small size (320-640 ac [130-260 ha.]) of these fields and the relatively high cost of seismic acquisition ($6,000-$7,000 per mi) many companies have relied on group shoot data and other industry seismic data to build a regional framework of seismic coverage. Once anomalies are mapped and subsurface facies and porosity maps examined, companies usually acquire proprietary seismic data and take leases. Sometimes leasing precedes seismic data acquisition.

The preferred areas or "fairways" in which to focus exploratory efforts are the southwestward-plunging erosional noses developed atop the larger Conecuh Ridge feature. The noses can be mapped using gravity, magnetic, seismic, and subsurface data. All 19 fields discovered so far in the Conecuh Ridge area are related to smaller, erosional positive features on the noses. Although the fields are of small areal extent and a closely spaced network of seismic data is needed to define each anomaly, exploration can be very profitable.

Drilling costs are low because no intermediate casing is needed and no unusual drilling problems exist in this area. Bottom water drive efficiently sweeps these reservoirs, and recovery factors above 30% are common. Field life is usually 13 years, and an average field will pay out in 2-3 years.

Therefore, exploration for this type of trap should continue in this area for at least the next decade. High success rates of exploratory drilling will be the reward for those companies willing to commit the funds necessary to properly define anomalies. The geologic processes and settings responsible for the fields on the Conecuh Ridge in Alabama may be repeated on the lower relief Pensacola arch in Florida (Figure 1). This feature has been sparsely tested with negative results.

ACKNOWLEDGMENTS

Union Texas Petroleum and Western Atlas Core Laboratories permitted publication of this paper. Texaco personnel including Freddie W. Pellegrin and Clifford Crowe in New Orleans provided some of the field production data. Thoughtful reviews were provided by Susan Longacre and Ted Beaumont. A special thanks to them. Barbara Sloan of Union Texas Petroleum provided drafting assistance. Special thanks to Virginia Curtis of Union Texas for typing the manuscript.

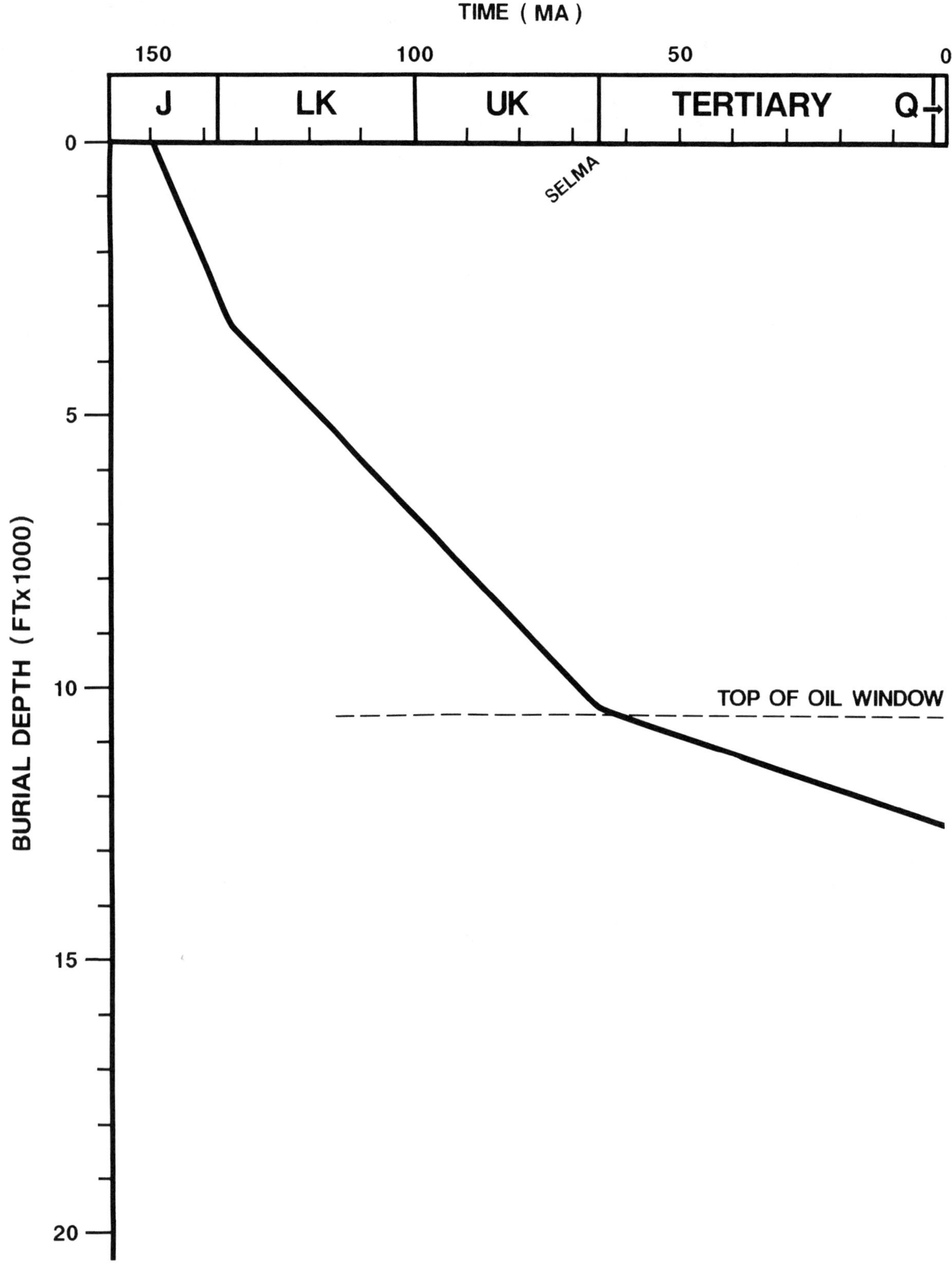

Figure 10. Burial-history curve for Appleton field. Main oil generation from the lower Smackover began during the Paleocene.

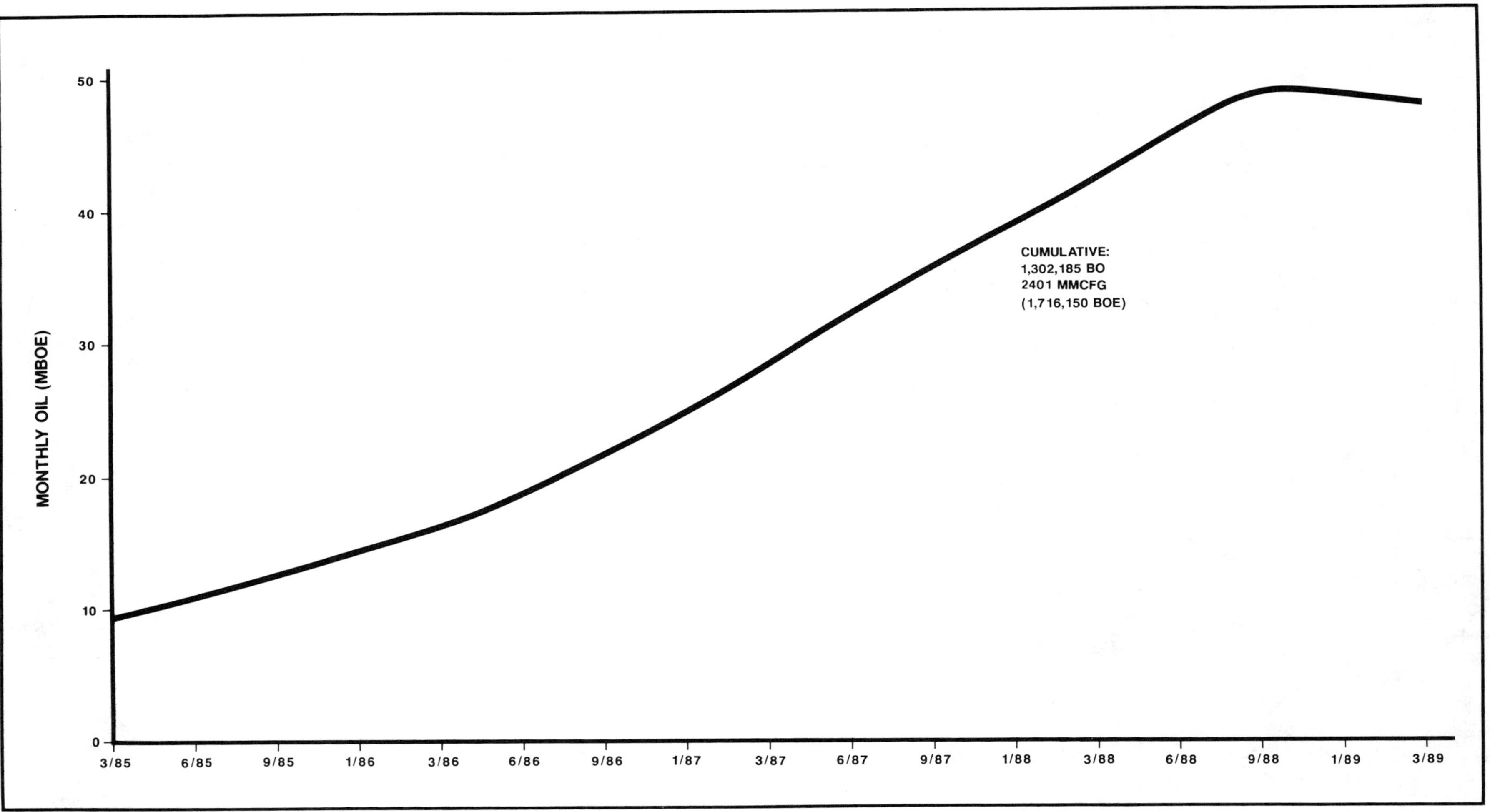

Figure 11. Monthly production curve for Appleton field.

REFERENCES CITED

Adams, G. I., C. Butts, L. W. Stephenson, and W. Cooke, 1926, Geology of Alabama: Geological Survey of Alabama Special Report 14, 312 p.

Baria, L. R., D. L. Stoudt, P. M. Harris, and P. D. Crevello, 1982, Upper Jurassic reefs of the Smackover Formation, United States Gulf Coast: American Association of Petroleum Geologists Bulletin, v. 66, p. 1449–1482.

Barret, M. L., 1986, Replacement geometry and fabrics of the Smackover (Jurassic) dolomite, southern Alabama: GCAGS Transactions, v. 36, p. 9–18.

Claypool, G. E., and E. A. Mancini, 1989, Geochemical relationship of petroleum in Mesozoic reservoirs to carbonate source rocks of Jurassic Smackover formation, southwestern Alabama: America Association of Petroleum Geologists Bulletin, v. 73, p. 904–924.

Estes, R., 1989, Monthly production report: State Oil and Gas Board of Alabama, May edition, 82 p.

Grabowski, G. J., 1984, Generation and migration of hydrocarbons in Upper Cretaceous Austin Chalk, south-central Texas, *in* J. G. Palacas, ed., Petroleum geochemistry and source rock potential of carbonate rocks: American Association of Petroleum Geologists Studies in Geology 18, p. 97–115.

Green, A. L., 1985, Influence of basement islands and ridges on Smackover deposition, southwest Alabama: American Association of Petroleum Geologists Bulletin, v. 69, p. 259.

Moore, C. H., 1984, The Upper Smackover of the Gulf rim; depositional systems, diagenesis, porosity evolution, and hydrocarbon production, *in* P. S. Ventress, D. G. Bebout, B. F. Perkins, and C. H. Moore, Research conference proceedings, p. 283–308.

Moore, D. B., 1971, Subsurface geology of Southwest Alabama: Geological Survey of Alabama Bulletin 99, 39 p.

Oehler, J. H., 1984, Carbonate source rocks in the Jurassic Smackover trend of Mississippi, Alabama, and Florida, *in* J. G. Palacas, ed., Petroleum geochemistry and source rock potential of carbonate rocks: American Association of Petroleum Geologists Studies in Geology 18, p. 63-69.

Palacas, J. G., D. E. Anders, and J. D. King, 1984, South Florida Basin—a prime example of carbonate source rocks for petroleum, *in* J. G. Palacas, ed., Petroleum geochemistry and source rock potential of carbonate rocks: American Association of Petroleum Geologists Studies in Geology 18, p. 71–96.

Saller, A. H., and B. R. Moore, 1986, Dolomitization in the Smackover formation, Escambia County, Alabama: GCAGS Transactions, v. 36, p. 275-282.

Texaco Inc., 1987, Alabama Oil and Gas Board, Exhibit G.2 Docket #7-22-8732.

Appendix 1. Field Description

Field name .... *Appleton*

Ultimate recoverable reserves .... *NA*

Field location:

- **Country** .... *United States*
- **State** .... *Alabama*
- **Basin/Province** .... *Gulf of Mexico*

Field discovery:

- **Year first pay discovered** .... *Upper Jurassic Smackover dolomite 1983*
- **Year second pay discovered** .... *NA*
- **Third pay** .... *NA*

Discovery well name and general location

- **First pay** .... *Texaco, 1 D. W. McMillan 2-14, SESW Sec. 2-T3N-R3E*
- **Second pay** .... *NA*
- **Third pay** .... *NA*

Discovery well operator .... *Texaco Inc.*

- **Second pay** .... *NA*
- **Third pay** .... *NA*

IP in barrels per day and/or cubic feet or cubic meters per day:

- **First pay** .... *661 BOPD and 1060 MCFGD*
- **Second pay** .... *NA*
- **Third pay** .... *NA*

All other zones with shows of oil and gas in the field:

Age	Formation	Type of Show
None		

APPLETON

Geologic concept leading to discovery and method or methods used to delineate prospect, e.g., surface geology, subsurface geology, seeps, magnetic data, gravity data, seismic data, seismic refraction, nontechnical:

Gravity data and subsurface mapping preceded a regional seismic survey. Lines acquired during the survey defined the Appleton structure as well as others in the area.

Structure:

Province/basin type (see St. John, Bally, and Klemme, 1984)

Gulf Coast Province/Type II a

Tectonic history

Triassic rifting of the southern end of the folded Appalachian Piedmont caused topographic high to form. These positive areas became sources of clastic sediments that underlie the Smackover. High-energy carbonates were concentrated over these high areas.

Regional structure

Appleton field is located within a positive salient of the Conecuh Ridge that plunges southwestward beneath the Gulf of Mexico coastal plain.

Local structure

Appleton is a small anticlinal closure on a southwest-trending basement fold.

Trap

Trap type(s) .... *1 combination trap*

Basin stratigraphy (major stratigraphic intervals from surface to deepest penetration in field):

Chronostratigraphy	Formation	Depth to Top in ft (m)
Pliocene-Pleistocene	*Citronelle*	*Surface*
Upper Cretaceous	*Selma*	*-2600 (-792)*
Upper Cretaceous	*Lower Tuscaloosa*	*-4918 (-1499)*
Upper Jurassic-Lower Cretaceous	*Cotton Valley*	*-9700 (-2956)*
Upper Jurassic	*Haynesville*	*-12,168 (-3708)*
Upper Jurassic	*Buckner*	*-12,557 (-3827)*
Upper Jurassic	*Smackover*	*-12,611 (-3843)*
Upper Jurassic	*Norphlet*	*-12,887 (-3927)*
Algonkian	*Basement Complex*	*-12,989 (-3959)*

Location of well in field .... *West flank of the anticline*

Reservoir characteristics:

Number of reservoirs .... *1 in the southeast-southwest section 2, T3N-R9E*

Formations .... *Smackover*

Ages .... *Jurassic*

Depths to tops of reservoirs .... *-12,617 ft (-3845 m)*

Gross thickness (top to bottom of producing interval) .... *184 ft (56 m)*

Net thickness—total thickness of producing zones

Average .... *93 ft (28 m)*

Maximum .... *122 ft (37 m)*

Average ....

Maximum ....

Lithology .... *Dolomitized algal peloid ooid grainstone*

Porosity type .... *Pay zone composed of vuggy and dolomite intercrystalline porosity*

Average porosity *16%*

Average permeability .... *245 md*

Seals:

Upper

Formation, fault, or other feature .... *Buckner*

Lithology .... *Anhydrite*

Lateral

Formation, fault, or other feature .... *Buckner and upper Smackover*

Lithology .... *Anhydrite and limy dolomite*

Source:

Formation and age .... *Lower Smackover*

Lithology .... *Laminated carbonate mudstones*

Average total organic carbon (TOC) .... *0.51% (10.3% in stylolites)*

Maximum TOC .... *2.52% (63.3% in stylolites)*

Kerogen type (I, II, or III) .... *Type I*

Vitrinite reflectance (maturation) .... *R_o = 0.70 (est.)*

Time of hydrocarbon expulsion .... *Early Tertiary*

Present depth to top of source .... *-13,000 ft (-3962 m)*

Thickness .... *80-100 ft (24-30 m)*

Potential yield .... *NA*

Appendix 2. Production Data

Field name .... *Appleton*

Field size:

- **Proved acres** .... *640 ac (259 ha.)*
- **Number of wells all years** .... *NA*
- **Current number of wells (as of year)** *4*
- **Well spacing** .... *160 ac (65 ha.)*
- **Ultimate recoverable** .... *3.0 million bbl*
- **Cumulative production (to 3/89)** .... *1.7 million bbl*
- **Annual production (1987)** .... *0.533 million bbl*
- **Present decline rate** .... *
  - **Initial decline rate** .... *
  - **Overall decline rate** .... *

**A true decline rate is unavailable because of the capacity of the gas plant (sulfur removal). Although the plant's capacity is 1500 BOPD mechanical problems have prevented production at those rates.*

- **Annual water production** .... *None*
- **In place, total reserves** .... *5.316 million bbl*
- **In place, per acre-foot** *448*
- **Primary recovery** .... *2.6 million bbl*
- **Secondary recovery** .... *NA*
- **Enhanced recovery** .... *NA*
- **Cumulative water production** .... *148,000 bbl*

Drilling and casing practices:

- **Amount of surface casing set** .... *16-in. at 117 ft (36 m)*
- **Casing program** .... *10¾-in. to 2893 ft (882 m); 7-in. to TD; 2⅞-in. tubing to 12,850 ft (3917 m)*
- **Drilling mud** .... *Ligno sulfonate*
- **Bit program** .... *Variable (no unusual problems)*
- **High pressure zones** .... *None*

Completion practices:

- **Interval(s) perforated** .... *Middle Smackover porosity zone (12,902–12,938 ft; 3914–3925 m)*
- **Well treatment** .... *NA*

Formation evaluation:

- **Logging suites** .... *DILL-Sonic; CNL-FDC or CNL-LDT; dipmeter*
- **Testing practices** .... *Tested on various chokes after perforating through production tubing*
- **Mud logging techniques** .... *Standard*

Oil characteristics:

- **Type** .... *NA*
  (Tissot and Welte Classification in "Petroleum Formation and Occurrence," 1984, Springer-Verlag, p. 419)
- **API gravity** .... *49°*
- **Base** .... *NA*
- **Initial GOR** *1594*
- **Sulfur, wt%** *1.75*
- **Viscosity, SUS** .... *NA*
- **Pour point** .... *NA*
- **Gas-oil distillate** .... *NA*

Field characteristics:

- **Average elevation** .... *200 ft (61 m)*
- **Initial pressure** .... *6264 psi (43,290 kPa)*
- **Present pressure (10/87)** .... *5871 psi (40,580 kPa)*
- **Pressure gradient** .... *0.46 psi/ft*
- **Temperature** .... *221°F*
- **Geothermal gradient** .... *0.0121°F/ft*
- **Drive** .... *Bottom water drive*
- **Oil column thickness** .... *122 ft (37 m)*
- **Oil-water contact** .... *−12,739 ft (−3883 m)*
- **Connate water** .... *NA*
- **Water salinity, TDS** .... *NA*
- **Resistivity of water** .... *NA*
- **Bulk volume water (%)** .... *NA*

Transportation method and market for oil and gas:

Oil is transported by truck, gas by pipeline.

Helez-Brur-Kokhav Field—Israel
Southern Coastal Plain

Y. GILBOA
Lapidoth
Israel Oil Prospectors Corp. Ltd.

H. FLIGELMAN
Oil Exploration (Investments) Ltd.

B. DERIN
Israel National Oil Company

FIELD CLASSIFICATION

BASIN: Eastern Mediterranean
BASIN TYPE: Passive Margin
RESERVOIR ROCK TYPE: Sandstone, Some Carbonates
RESERVOIR ENVIRONMENT OF DEPOSITION: Platform
RESERVOIR AGE: Lower Cretaceous
PETROLEUM TYPE: Oil
TRAP TYPE: Stratigraphic-Structural

LOCATION

The Helez field, a complex of three fields, Helez, Brur, and Kokhav, is located some 55 km (34 mi) south of Tel Aviv and 12 km (7.5 mi) east of the Mediterranean coast line. It was the first oil field discovered in the Eastern Mediterranean (1955) and is Israel's most significant oil-producing field.

The main producing formations are Neocomian sand beds and fringing dolomitic reef, with Oxfordian barrier reefs and Middle Jurassic calcarenite of secondary importance. The producing beds are overlain by Cretaceous and Tertiary sediments, the latter becoming very thick to the west. The field is 11 km (7 mi) long and 1 to 1.8 km (0.6 to 1.1 mi) wide with a producing area of 12.5 km^2 (30.9 mi^2) (Figure 1).

The field is operated by Lapidoth-Israel Prospectors Corp. Ltd. Recently, Lapidoth granted limited farmouts to other operators (Naphtha-Israel Petroleum Crop. Ltd. and Delek Oil Exploration Ltd.) for exploration and development activities. The ultimate recovery is estimated to be about 19 million bbl.

HISTORY

Pre-Discovery

During the British Mandate government, the Iraq Petroleum Co. (IPC) carried out gravity surveys in the coastal area of Israel. The surveys revealed a 50 km (31 mi) long trend of positive anomalies in its southern part, between Ashqelon and the Sinai border. Seismic surveys across the gravity maximum of the Huleiqat feature (now Helez) led to the spudding of the Huleiqat #1 well on 25 September 1947 (Tschopp, 1956). The first target, the Cenomanian dolomites, which are the country's major aquifer, were encountered containing brackish water, and the underlying Albian-Aptian beds were found to be predominantly carbonates with no hydrocarbon shows. At a depth of 1055 m (3641 ft), 11¾-in. casing was set. Because of political unrest, drilling was suspended without any positive results in February 1948.

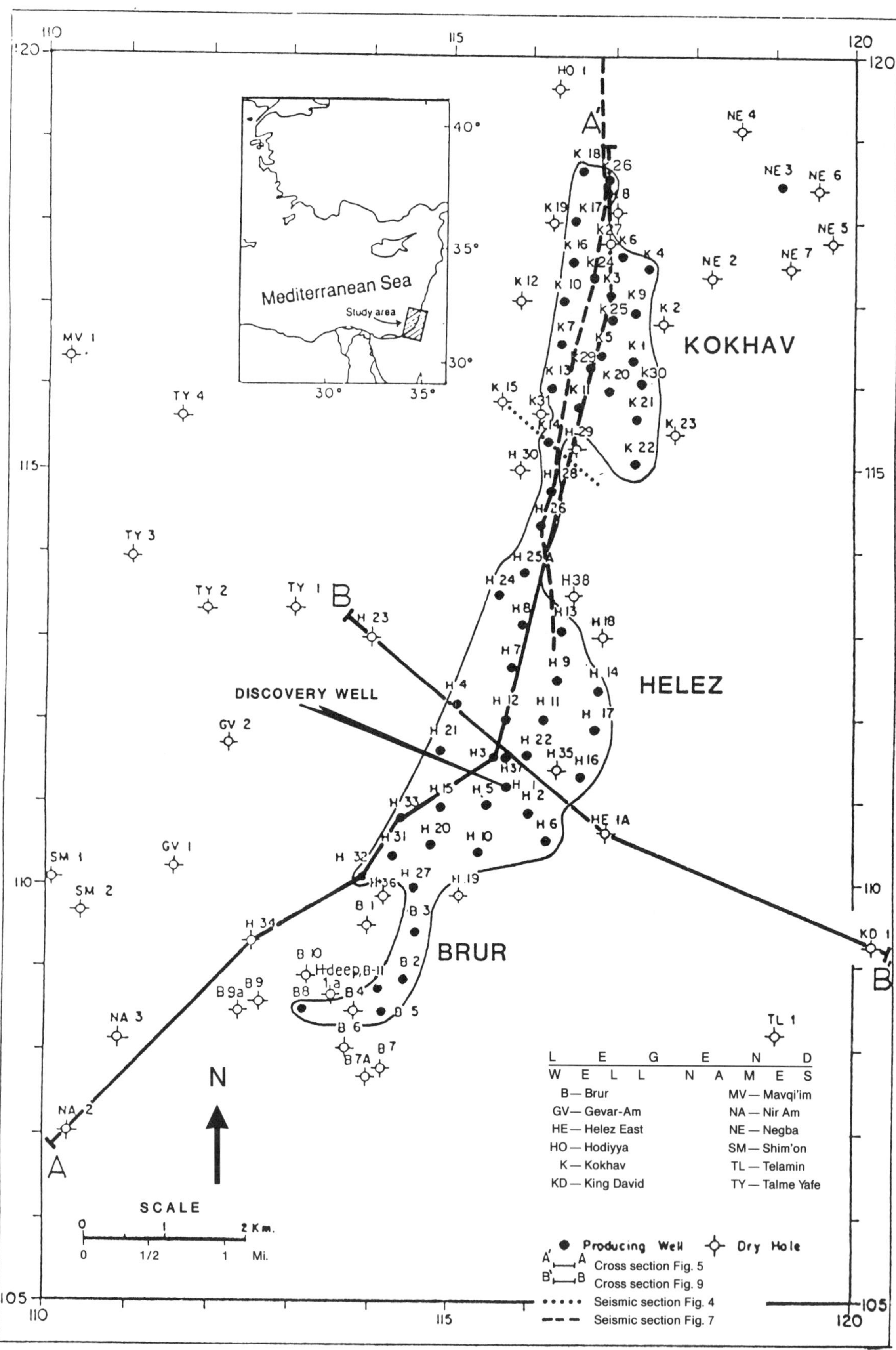

Figure 1. Map of the Helez oil field showing well locations, geologic cross sections, and seismic sections.

After the establishment of the State of Israel in 1948, the Weizmann Institute of Sciences carried out seismic surveys in the area. Based on these findings, the Beeri-Helez-Negba area was qualified as a good prospect (Ball and Ball, 1953). The gravity maximum prospect in that area was granted to the joint enterprise of Lapidoth and Israel Oil Prospectors (I.O.P.).

The coincidence of seismic structures with a gravity maximum, plus gas seepages detected during structural hole drilling, led Lapidoth-I.O.P. in 1954 to drill a test well at Beeri, 24 km (15 mi) south of the Huleiqat well. The Beeri 1 well was abandoned at a depth of 3646 m (11,962 ft) in Lower Jurassic beds without encountering any particular indications of hydrocarbons.

In spite of the disappointing results, Lapidoth's chief geologist, H. J. Tschopp, insisted on the oil possibilities of this major structural trend. Upon concluding that the Huleiqat well did not penetrate the sand-bearing section of Neocomian age, he recommended deepening it.

Discovery

With the data from five structural holes, the drilling of Huleiqat 1 well was resumed on 26 August 1955 and the well, renamed Helez 1, was deepened to 1515 m (4971 ft) (Figure 1) and completed as an oil producer on 12 October 1955. The initial production of the discovery well was approximately 400 bbl/d of 29° API oil from the porous Middle sandstone of the Helez Formation (Valanginian–Barremian age) at the depth of 1480 m (4856 ft) (Figure 2).

Post-Discovery

Field expansion and subsequent development started in 1955 and proceeded in several stages. Initial development of the field was concentrated in the central part, the Helez field, and in the southern part, the Brur field (Figure 1).

1. 1955–August 1962: Drilling of 29 development and stepout wells in the Helez area and six wells in the Brur area. Production was limited to the "middle sand member" of the Helez Formation (the "K," "W," "A," and "Z" sands, Figure 2). The northernmost stepout well (Helez 29 in Block E, Figure 3) was dry and considered the boundary of the field in that direction. One well, Helez 22 (TD 4477 m, 14,690 ft) in Block B tested deeper formations of Early Jurassic age without encountering new pay zones.
2. September 1962–1968: The drilling of 25 additional wells, following the oil discovery of Kokhav 1, located 1.2 km (0.75 mi) to the northeast of Helez 29. This well was recommended by W. Randall, Chief Geologist of Lapidoth, as another stepout trial to the north of the Helez field. As a result, 24 wells were drilled and production found mainly in two new horizons: the lower sandstone member, "B" sandstone, and the Kokhav dolomite (Figure 2). In addition there was minor production from the top Jurassic limestone. Only one well (Kokhav 7) was deepened to the Middle Jurassic and found no significant shows. Simultaneously, 14 wells were drilled in the Helez and Brur fields, some of them producing from the Kokhav dolomite and Jurassic limestones (Brur Calcarenite). Weeks (1962) was the first to define the Helez field as a hinge zone, and thus established a new exploration approach for the area. The hinge zone reflects the unstable conditions of the shelf margins.
3. 1969–end 1983: Except for one well in the Kokhav field (Kokhav 25) and a deep test to the basement in the southern portion (Helez Deep 1A), which was drilled by O.E.I.L., no drilling took place. The Helez Deep 1A was dry and abandoned at total depth of 6093 m (19,991 ft) in basement rocks.
4. End 1983–today: Drilling was resumed after granting farmouts to other operators. To date seven wells have been drilled (five in Kokhav, one in Helez, and one in Brur) with some success. Lapidoth itself opened new pay zones and deepened existing wells. This last phase of drilling is mainly responsible for the present field production rate of 450 bbl/d (October 1988). Older wells, each producing less than 8 bbl/d, are noncommercial.

A total of 82 wells were drilled in the Helez field, 55 of them producers.

Twenty-one wildcats surrounding the Helez field were drilled and abandoned within the Helez lease, finding no production.

All of the wells were drilled with rotary equipment using water-based mud. A typical completion consists of 300 m (985 ft) of surface casing (13⅜-in. or 9⅝-in.), a 1000 to 1200 m (3280 to 3940 ft) intermediate casing (9⅝-in. or 7-in.) to seal off the Upper Cretaceous loss of circulation zone, and a 7-in. or 5½-in. producing string to total depth (1500 to 1700 m, 4920 to 5580 ft). The most common tubing size is 2⅞-in. Some of the wells were dual producers, and in a few, several reservoirs were commingled.

Only electric resistivity logs and micrologs, and occasionally sonic and GR/neutron logs were run in the older wells (to 1970). A fuller suite of logs (induction, gamma ray, compensated neutron, sonic, density, and dipmeter) is now commonly run.

STRUCTURE

The Helez field is located on a faulted anticline, tilted with a gentle dip to the east and downfaulted to the west (Figure 3). The folding is related to deep-seated compressional faults (Figure 4). Truncated Upper Jurassic beds reflect the high position of the structure during the Late Jurassic and probably

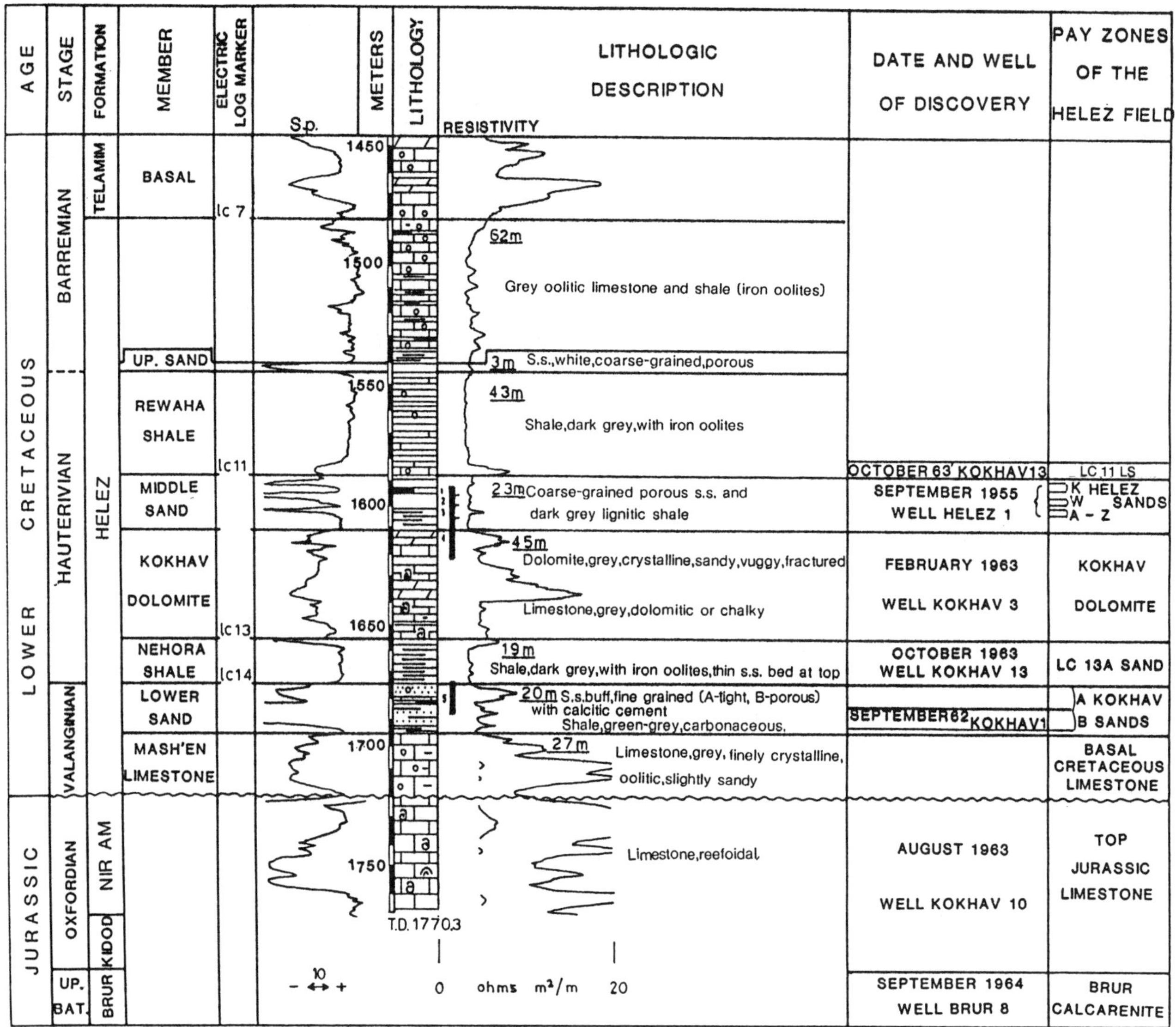

Figure 2. Producing zone of the Helez field, presented on the log of Kokhav-2 well (see Figure 1). In this well the Gevar-Am Formation is missing. Kidod shale and Brur not penetrated. (Modified from Shenav, 1971.)

earlier (Rosenberg, 1979). Pre-Cretaceous erosion carved a canyon that crosses the Helez oil field in a northwesterly direction (Figures 5 and 6). Within the field this canyon formed a channel 16 km (9.9 mi) long and 7 km (4.3 mi) wide with wall slopes of about 40°. The canyon is filled with up to 1000 m (3280 ft) of shales belonging to the Lower Cretaceous Gevar-Am Formation.

The Helez structure is characterized by two axial trends (Figure 3), a north-south Jurassic trend, predominant in the northern part of Helez and Kokhav, and a northeast-southwest Late Cretaceous-early Tertiary structural trend, dominant in the southern blocks of Helez and Brur.

The known faults that affected the area are related to the following events:

1. Uplifting and tilting at the end of the Jurassic.
2. The Alpine folding phase of Late Cretaceous and early Tertiary times.
3. Neogene tensional movements (Rosenberg, 1979).

During the Neogene events, transverse faults divided the Helez field into several blocks (Figures 3 and 7).

STRATIGRAPHY

Israel is located on the northern margin of the Arabo-Nubian massif that stabilized during the Precambrian, some 600 m.y.a. The geological history of the country is closely related to the interplay of this huge, rigid cratonic mass and the sea lying to the north and northwest of it. The vast, shallow epeiric seas that characterized the area during long

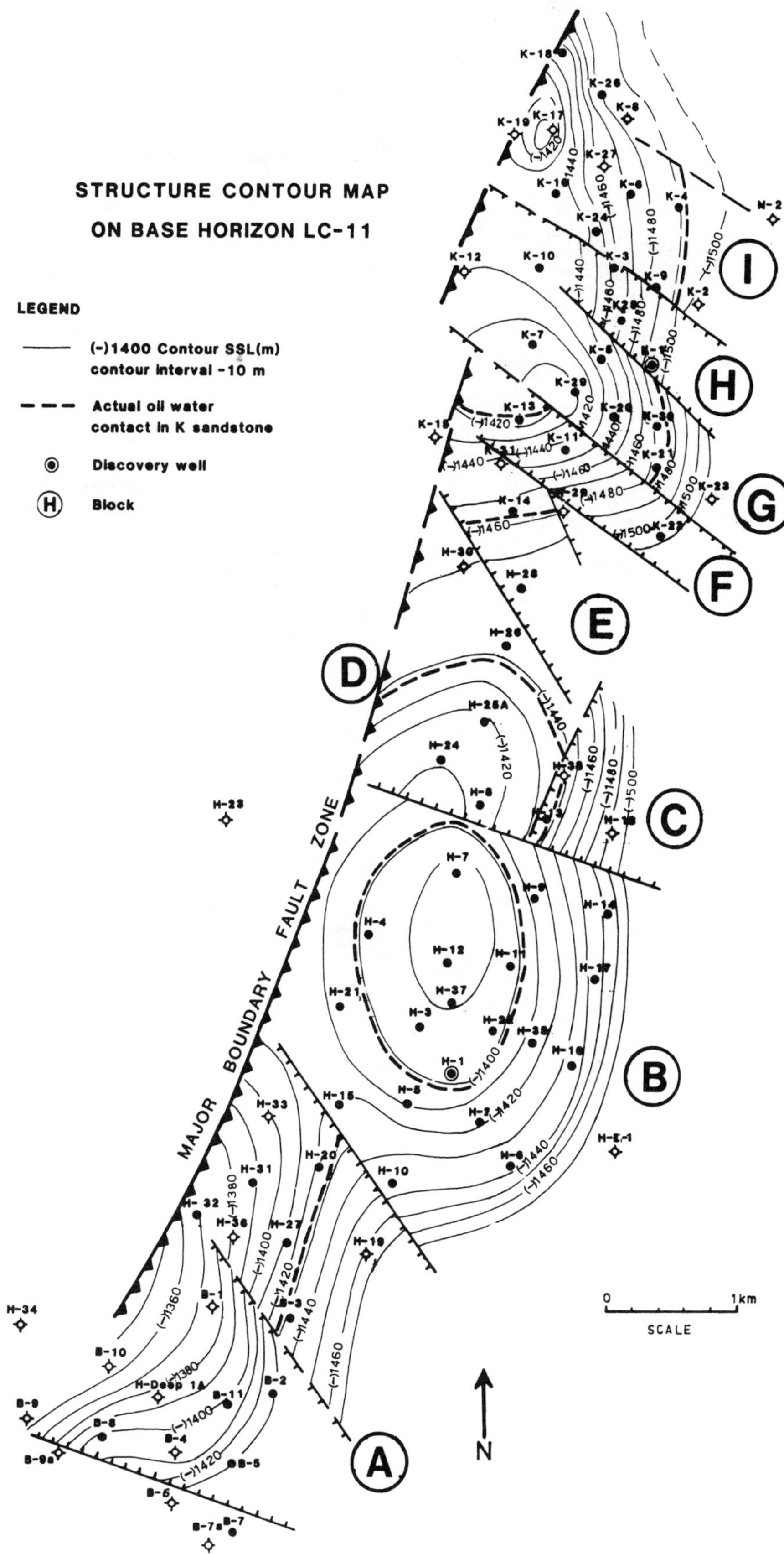

Figure 3. Helez-Brur-Kokhav field structure contour map on base horizon LC-11. C.I. = 10 m.

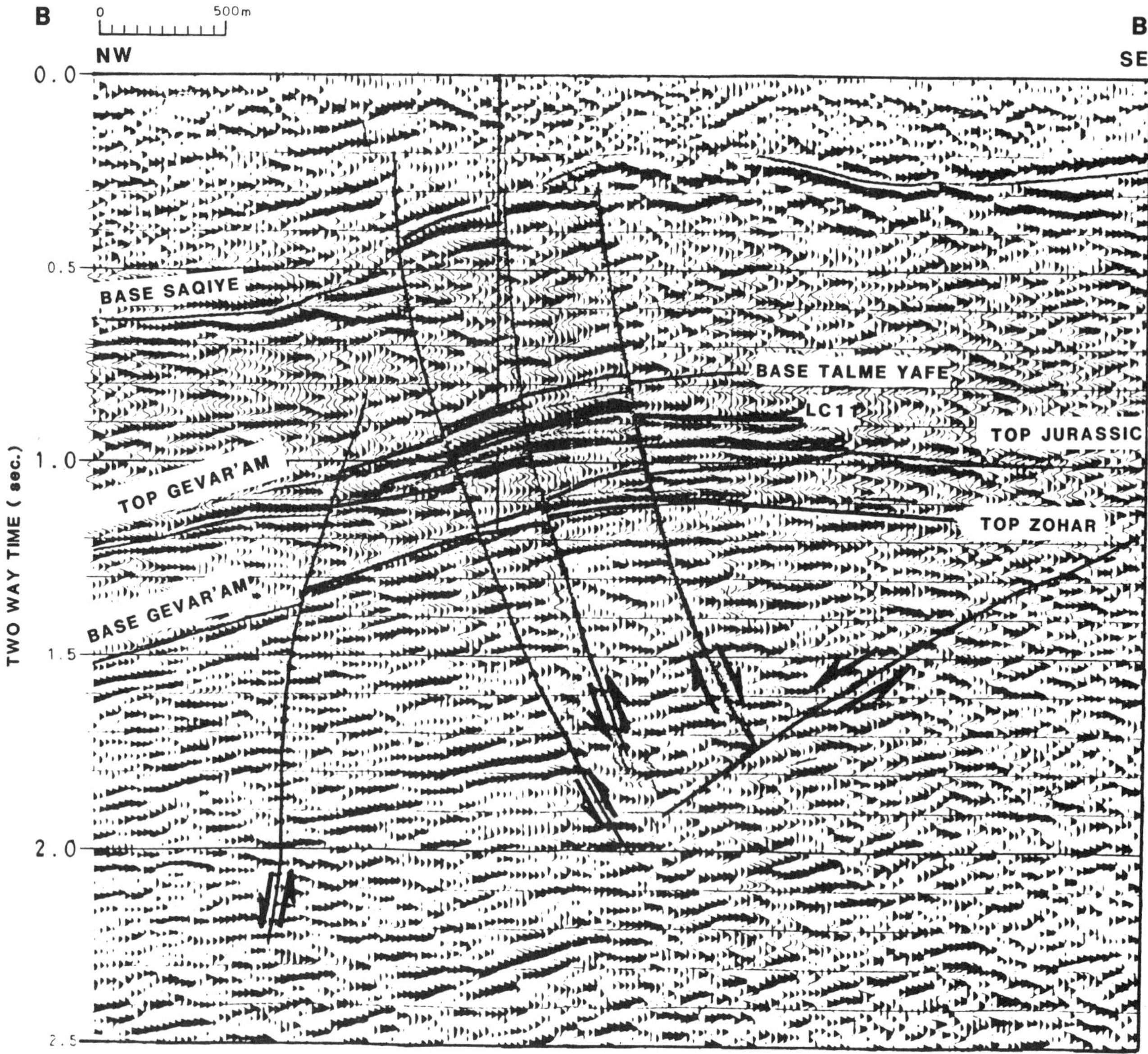

Figure 4. Kokhav field. (A) Interpreted transverse seismic cross section. (B) Uninterpreted seismic cross section. (For location, see Figure 1, B–B'.)

geological periods are a direct result of the shield's proximity. Most of the sediments were deposited on a platform under varying continental and epicontinental environments. The stratigraphic column in the platform area is considered to be not more than 6.5 km (21,000 ft) thick, mostly of Mesozoic age. In the early Tertiary, on the other hand, the topographic relief was high and the sea began to retreat. West of this platform, beyond the hinge line, a thicker sedimentary column was deposited on the Cretaceous and Neogene continental slope and outer shelf underlying the present-day coastal plain and shelf (Bein and Gvirtzman, 1977).

The Arabo-Nubian massif contributed enormous quantities of clastic sediments of Mesozoic age to the shallow, adjacent basin. In the Tertiary the Nile River also distributed the massif's clastics on the ancient shores of Israel. The clastics intermingled with shallow sea carbonates of biogenic origin, and their very fine detrital derivates also spilled over the shelf edge, accumulating at the base of the continental slope (Bein and Weiler, 1976).

The Helez field is located on the edge of this platform. Here, only one well (Helez Deep 1A) penetrated the entire sedimentary column, commencing in Pleistocene sediments and drilling through the Middle Triassic (Paleozoic and early Triassic sediments are missing). The stratigraphy of the area (Table 1) is summarized below.

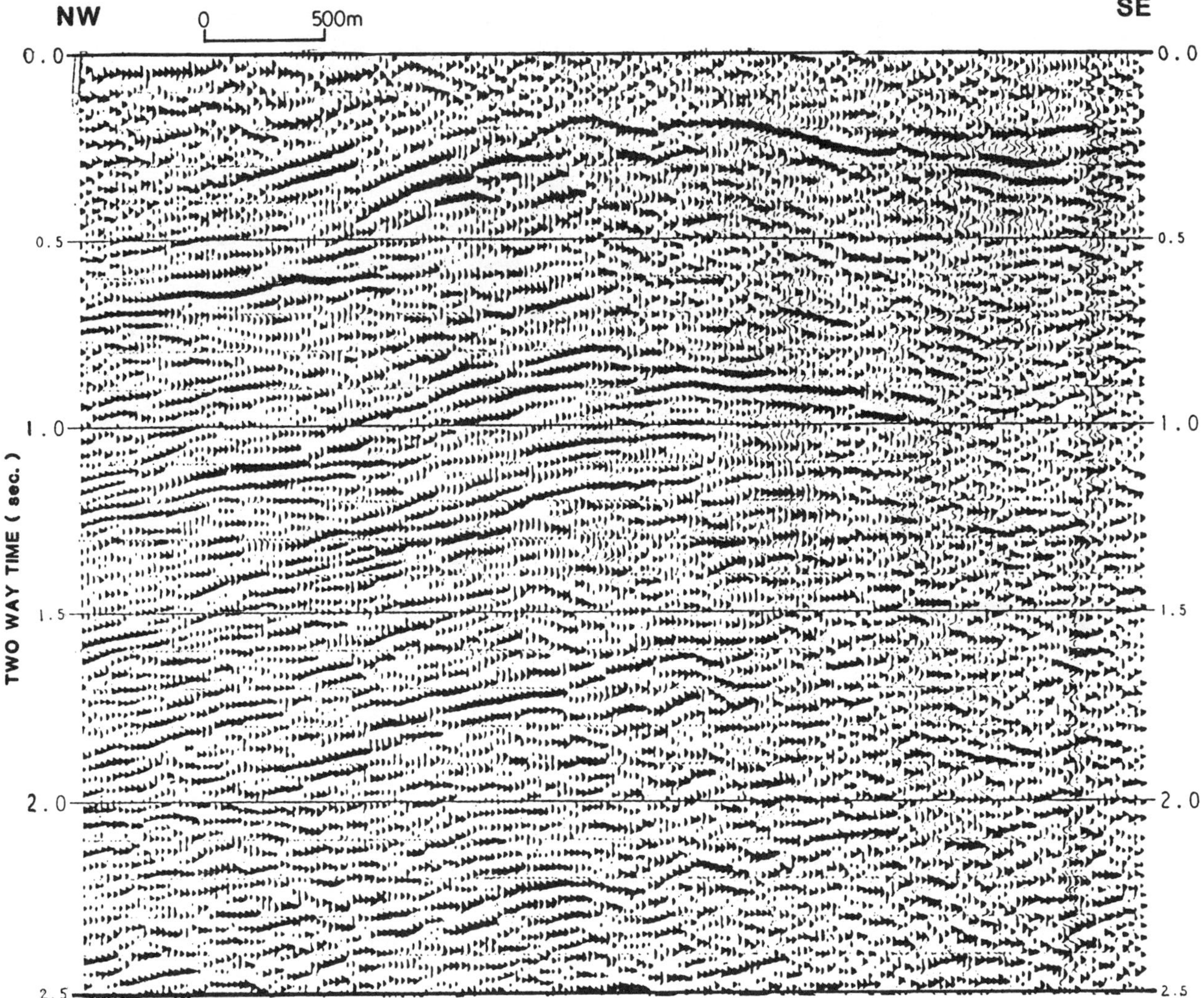

Figure 4B.

Paleozoic

Acid volcanics (Erez Porphyr), probably of Paleozoic age, overlie older, green Doroth schists at a depth of 6000 m (19,690 ft) in Helez Deep 1A.

Triassic

The Erez volcanics are overlain by a 315 m (1033 ft) thick section of Middle Triassic Or Haner Conglomerate (Garfunkel and Derin, 1983). The missing Paleozoic section was eroded during intra-Triassic movements. A sequence of Upper Triassic shallow-marine dolomites and limestones overlies the conglomerate.

Jurassic

The Jurassic sequence in the Helez area is extremely thick, being almost 3000 m (9840 ft). It is subdivided into three distinct units (Derin and Gerry, 1972; Derin, 1974; Rahamim, 1973):

1. The Lower Jurassic unit penetrated completely in the Helez Deep 1A well consists of 1400 m (4590 ft) of shallow-marine carbonates.
2. The Middle Jurassic unit is characterized by intercalation of open-marine spiculitic limestone (Barnea Limestone) and shale, as well as high energy, oolitic shoal and shallow shelf sediments (Derin, 1974).
3. In the Late Jurassic, the Helez area was located on a high energy barrier reef trend (Derin, 1974). This threshold zone separates a deep, open-marine area and probably a "starved basin" to the west from a shallow, inner shale basin and shelf carbonates to the east (Figure 8).

The uplift and tilting at the end of the Jurassic exposed Jurassic rocks in parts of the Helez field area

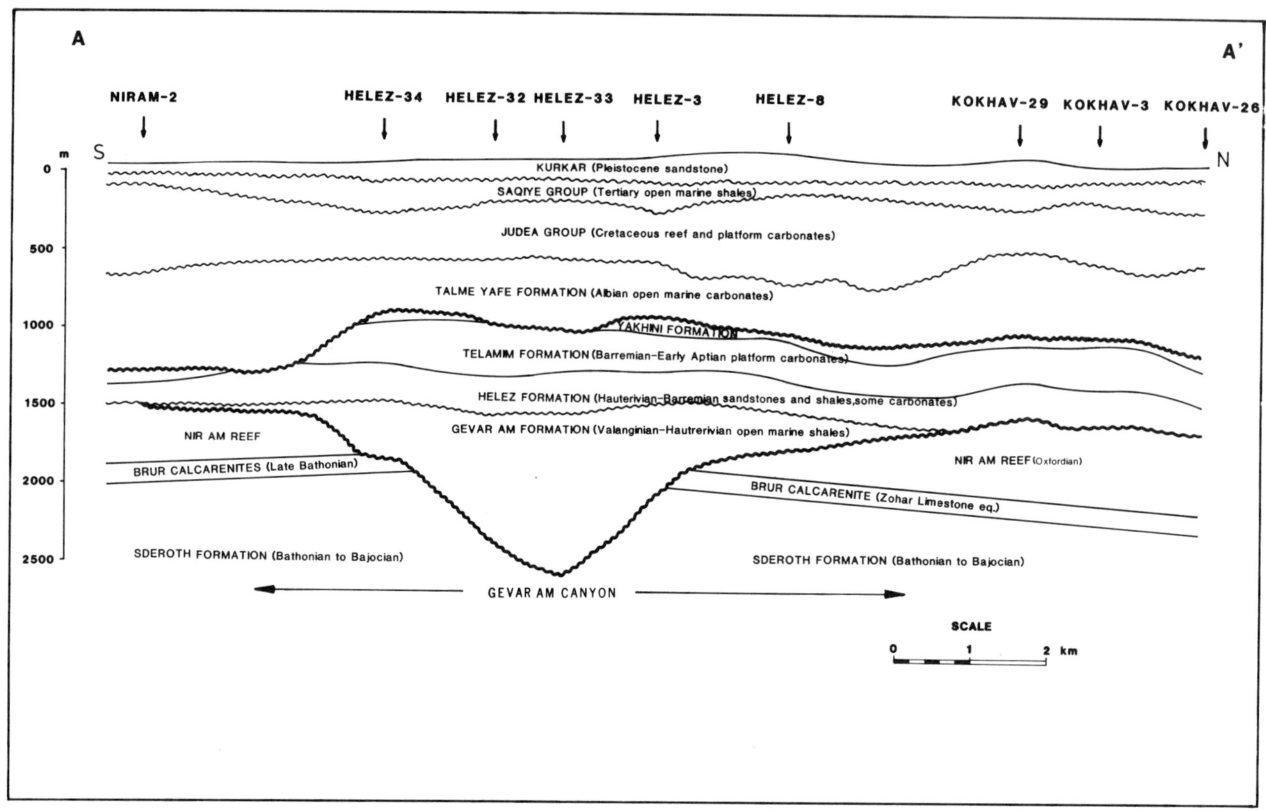

Figure 5. Geologic cross section in the Helez oil field showing the position and shape of the Gevar-Am canyon. (Location is shown on Figure 1.)

to subaerial erosion, forming the latest Jurassic-earliest Cretaceous unconformity phase. This erosional phase probably coincides with the initial development of a submarine canyon (Gevar-Am channel) and the karstic phenomena in the Jurassic Nir-Am Barrier Reef.

Lower Cretaceous

During Early Cretaceous time, as in the Late Jurassic, the Helez field area was located on a "shelf break" or hinge zone (Garfunkel and Derin, 1983). Here Lower Cretaceous platform sediments were deposited over a gently inclined Jurassic erosional surface, and the shaly facies equivalent was formed on the basinward side. During the Neocomian-Lower Aptian, the interplay between the rate of subsidence and the rate of supply of sedimentary material as well as temporary periods of minor uplift continuously shifted the position of the hinge line and along with it, the related environments (Grader and Reiss, 1958; Z. Cohen, 1964, 1969, 1971, 1976; Gvirtzman and Klang, 1972; Bein, 1974; Bein and Weiler, 1976; Bein and Gvirtzman, 1977; A. Cohen, 1983).

The Lower Cretaceous is divided into five formations, easily distinguished by their lithological properties (Cohen, 1971 and Table 1):

1. Gevar-Am Formation: Berriasian to Barremian; 0-940 m (3084 ft). A series of gray silty shales bounding the field to the west and filling the old erosional Gevar-Am channel that traverses the Helez field. Consequent dark shales with a few pyritic sand lenses filled the deep basin west of the continental slope and the submarine channels that cut it, followed by the sedimentation of the transgressive Helez Formation that extended far to the east (Shenav, 1971; Cohen, 1969, 1976). Simultaneously the shaly Gevar-Am Formation continued to be deposited in the deeper basin (Figures 6 and 9).
2. Helez Formation: Valanginian-Barremian; 160-370 m (525-1215 ft). This formation is comprised of a series of alternating beds of shale, sandstone, and a subordinate amount of limestone and dolomite. Overall, the sediments tend to be thinner continental sediments in the east, and in the west, thicker, marine deposits (Figure 2).
3. Telamim Formation: Upper Barremian-Lower Aptian; 160-290 m (525-950 ft). This formation

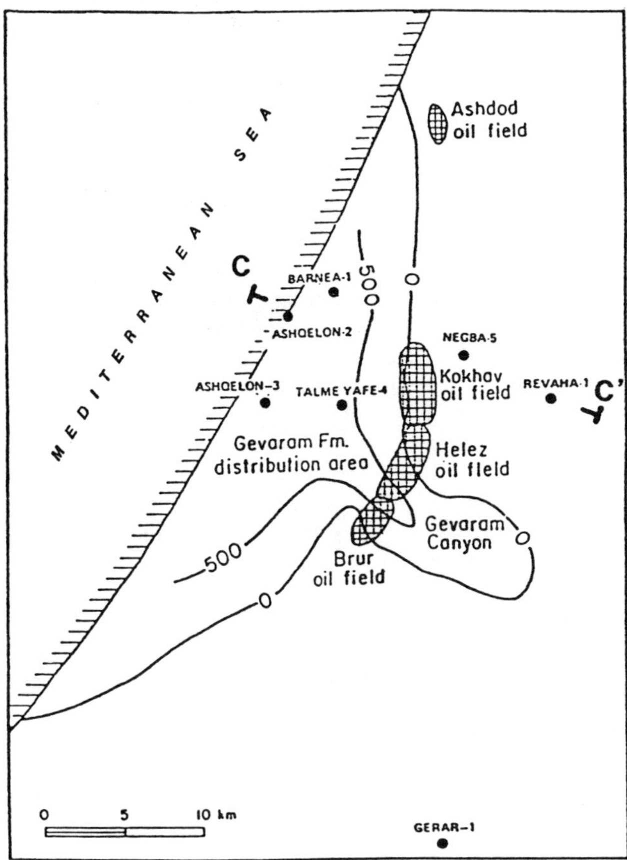

Figure 6. Thickness of the Gevar-Am Formation in the Helez area (contour interval, 500 m). C-C′ indicates location of section in Figure 11. (From Bein and Sofer, 1987.)

is comprised of a section containing dolomitic, oolitic, or sandy limestone with several sandstone beds at the base and a reef bank at the top.

4. Yakhini Formation: Upper Aptian; 430–530 m (1410–1740 ft). This formation is comprised of a sequence of limestone, chalky limestone, dolomitic limestone, dolomite, and marl.
5. Talme Yafe Formation: Late Aptian-Albian; 0–850 m (0–2790 ft). This formation is comprised of a series of light gray marlstones with streaks of gray to buff pellitic limestone, thickening westward in the form of a wedge.

Generally, two environments of deposition can be distinguished in the Helez area: (1) platform or shallow marine-littoral deposits of shale, sandstone, limestone, dolomite, and dolomitic reefoidal limestone of the Helez, Telamim, and Yakhini formations (Bein, 1974), and (2) deep basin sediments deposited west of the continental slope consisting of dark shales of Berriasian-Aptian age (Gevar-Am Formation) and clastic marls with conglomerate and carbonate detritus of Albian age (Talme Yafe Formation). The two basinal formations overlie unconformably older strata without being congruent in their extension or direction of development (Figures 5 and 9).

Upper Cretaceous

The Upper Cretaceous sequence consists of Cenomanian-Turonian-aged (Judea Group) shelf carbonates (dolomites and limestones).

Senonian beds were not encountered in the field area.

Tertiary

Paleocene and Eocene formations are missing in the Helez area and a thick section of shale and marl of late Eocene to Pleistocene age (Saqiye Group) overlie unconformably the Cretaceous beds. The Tertiary sequence is clearly divided into two main parts separated by an unconformity: (1) the late Eocene-Miocene and (2) Pliocene-Pleistocene. Late Miocene-aged evaporites are developed generally west of the Helez field as well as in the whole Mediterranean basin (the Messinian crisis).

Pleistocene-Holocene

The youngest sediments in the area consist of calcareous sandstones (Kurkar Group) covered by soils.

TRAP

The Helez field is a combination stratigraphic-structural trap located on a northeast-southwest-trending faulted anticline that is tilted gently to the east and downfaulted to the west (Figures 3, 4, and 7).

The structural configuration of the field area was probably developed during the Late Cretaceous to the Eocene folding phase, coinciding with the above-mentioned depositional hinge belt. Prominent Upper Jurassic reef fronts acted as a barrier to the Lower Cretaceous sands carried from the land by the rivers (Rosenberg, 1979). The importance of the Jurassic barrier reefs and the Lower Cretaceous patch reef (Kokhav Dolomite) does not lie only in their reservoir characteristics but in their effect on the sedimentation on the lagoonal (eastern) side of the oil-bearing sand beds. Under these environmental conditions the sands were deposited westward toward the crests of the underlying reefs. Since thinning and shaling out of these sands occurred in the same direction (updip), stratigraphic traps were later formed. The Helez sandstone reservoir rocks themselves are sealed by overlying and interbedded shales (Shenav, 1971, and Figure 2).

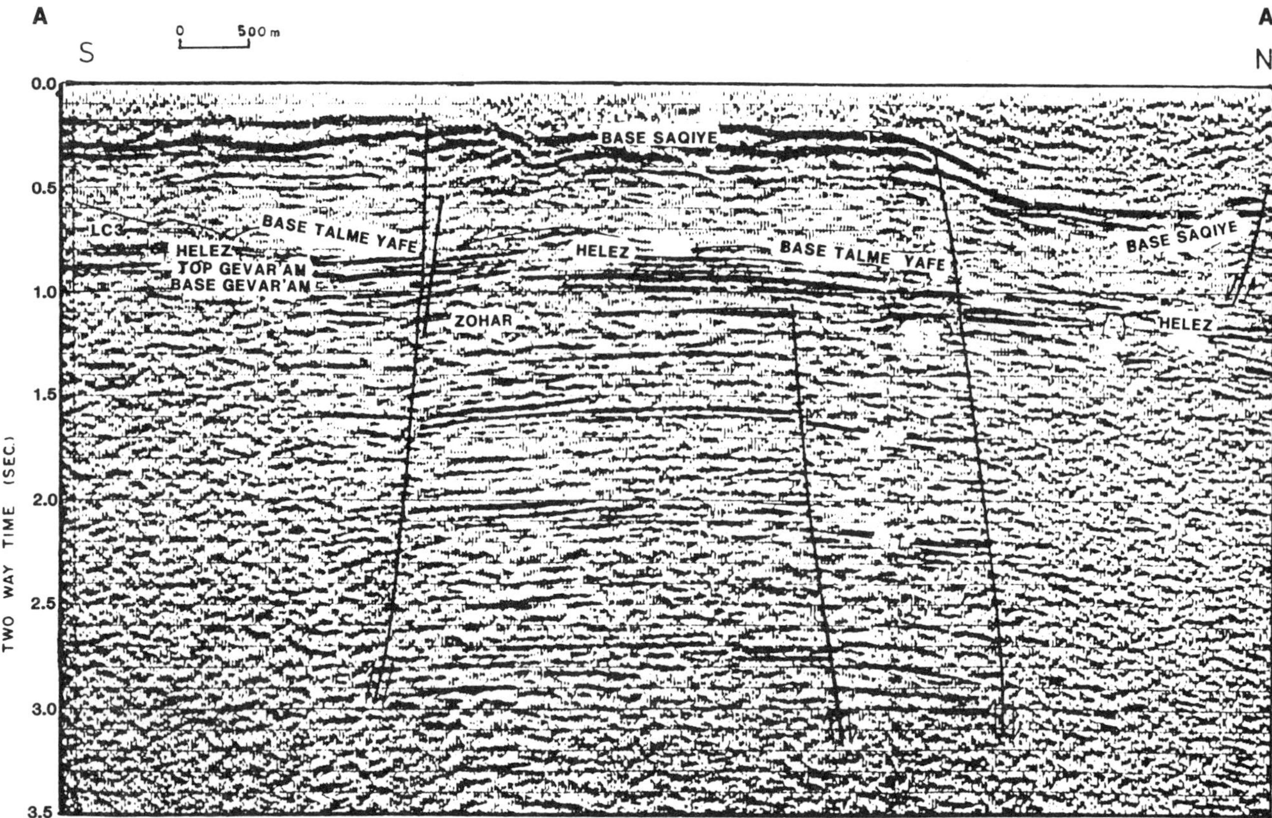

Figure 7. Helez-Kokhav field. (A) Interpreted longitudinal seismic cross section. (B) Uninterpreted seismic cross section. (For location see Figure 1, A–A′.)

SOURCE

The Helez oil source rock and the oil migration path remain a matter for discussion. As described above, part of the Helez field is crossed by the deep Gevar-Am canyon that cut approximately 1000 m (3280 ft) into Jurassic limestones during Early Cretaceous time, which was subsequently filled by shales of the Gevar-Am Formation (Cohen, 1976). The dark shale basinal environment of Gevar-Am deposition is limited to the east by the hinge belt of Late Jurassic age. In places these shales interfinger with the Helez Formation in which most of the oil has been found, and they also lie in contact with Jurassic limestones containing oil produced in the Helez-Kokhav trend as well as in the Ashdod field wells farther north (Figure 6). Therefore, it was logical to conclude that the Gevar-Am shales are the source rocks of the Helez oil (Cohen, 1971, 1976). Later, Amit (1978) noted that the Gevar-Am shales in the Helez area are immature for oil generation, while in the west, near Ashqelon where they are buried much deeper, they have a higher degree of maturation. These facts led Amit (1978) to consider the more deeply buried Gevar-Am shale in the west as the potential source rock.

In a recent study, Bein and Sofer (1987) have made significant progress in understanding the mechanism of the Helez oil migration. They analyzed extracts of the Middle Jurassic Barnea Formation and found a geochemical similarity between them and the oil accumulations and shows found in the Helez-Kokhav, Ashdod, and Ashqelon wells (Figures 6 and 10). Average total organic carbon (TOC) of the Barnea limestone is 0.5%, reaching a maximum of 2.6%, and its kerogen type is II. No similarity with the hydrocarbons was found in the extracts of the Gevar-Am Formation (Figure 10).

OIL MIGRATION

These findings led Bein and Sofer (1987) to suggest that the oil generated in the west at the depth of 4500–5000 m (14,760–16,400 ft) by a common source rock, the Jurassic Barnea Limestone, was later expelled, migrated updip eastwards below the blanket of the Gevar-Am shale, and then accumulated in the Helez sands and Upper Jurassic limestones. The migration was possibly aided by faulting and fracturing (Figure 11).

However, it is not yet clear whether the bitumen extracted from the Barnea Limestone is indigenous or a product of migration from other sources. Therefore, in addition to the Barnea Limestone and

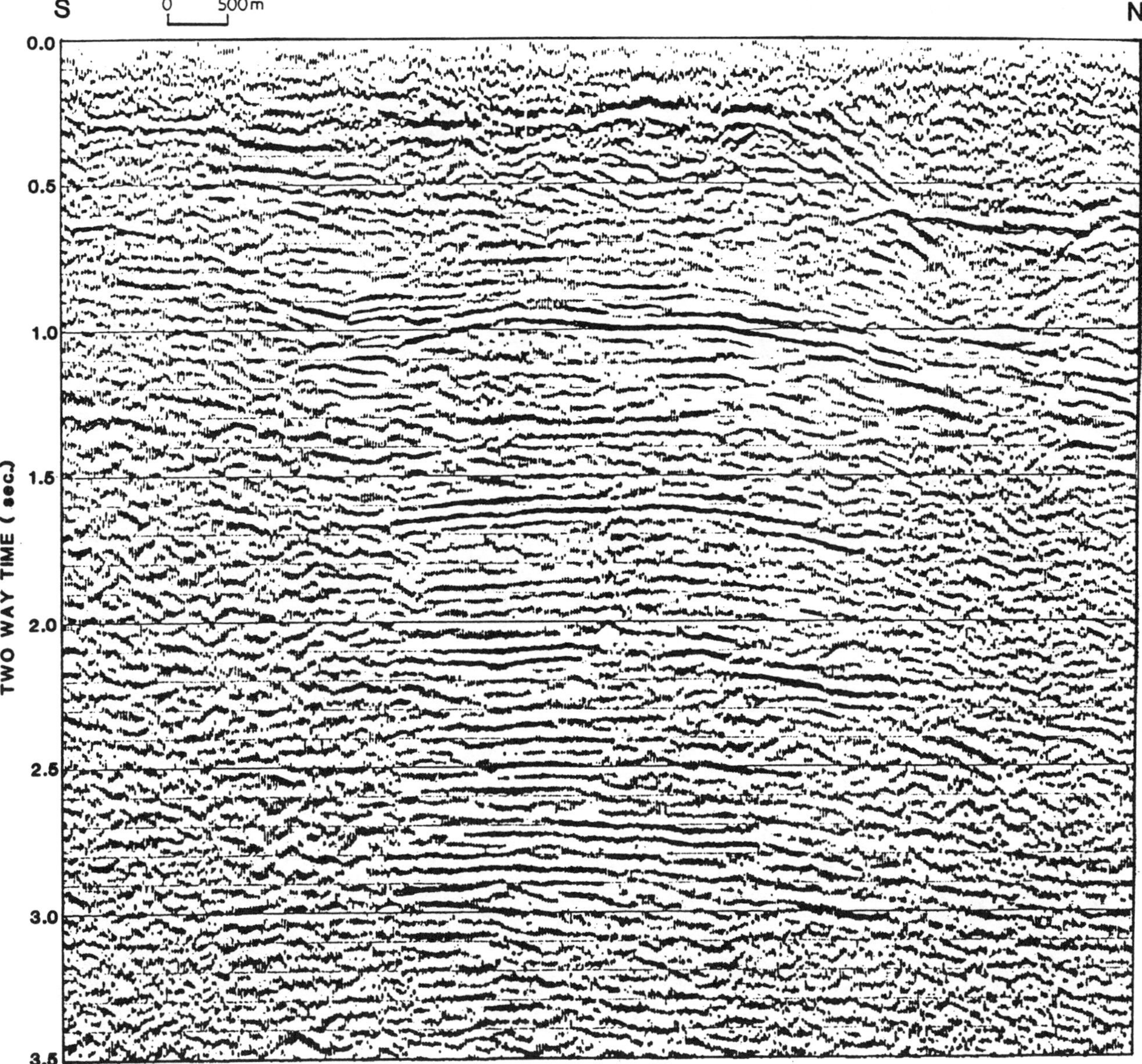

Figure 7B.

Gevar-Am shales, deeper Triassic beds are considered a potential source. The existence of such Triassic source rocks best explains the oil shows in the Lower Jurassic in the Ramallah wildcat (north of Jerusalem) and in the Triassic in Ga'ash 2 well located in the central Coastal Plain (May, 1981, 1984).

Because the exact source rock of the Helez oil is not clearly identified, the timing of migration and the trapping can only be speculative. It seems that the oil migrated and was trapped in the Helez field reservoirs during the Neogene, after the major tectonic movements. Intensive Oligocene-early Miocene erosion enabled the flushing of the formation's earlier fluids by the intrusion of the late Neogene sea that penetrated the exposed older (probably Cretaceous) beds. It is most likely that the oil trapping process occurred through the post-folding tensional transverse faults that cross the Helez field, ascending from deep layers in the western basin.

RESERVOIRS

General Description

The Helez field produced oil from the Lower Cretaceous Helez Formation (several sandstones and two carbonate zones) and to a lesser extent from Upper and Middle Jurassic porous limestones (Figure 2). The Helez sandstones are the main oil reservoirs. The

Table 1. Generalized stratigraphic column for the Helez area.

Age			Group or Formation: Basin	Group or Formation: Platform
Pleistocene-Holocene				Kurkar
Neogene			Saqiye	
Upper Cretaceous		Cenomanian-Turonian		Judea
Lower Cretaceous		Albian	Talme Yafe	Yakhini
		Upper Aptian		
		Lower Aptian	Gevar-Am	Telamim
	Neocomian	Barremian		
		Hauterivian		Helez
		Valanginian		
		Berriasian		
Jurassic	Late	Oxfordian		Nir-Am/Beer Sheva Kidod
	Middle	Bathonian		Karmon
		Bajocian	Barnea	Sderoth
		Aalinian		Upper Nirim
	Upper	Lias		Nirim Upper/Lower Mish'hor
Triassic	Late			Shefayim (Mohilla equiv.)
	Middle			Or Haner Cgl.
Paleozoic ?			Erez Porphyr	
Precambrian			Doroth Schist	

upper sandstone reservoirs ("K," "W," and "A-Z") are related to the middle sandstone member and are found at an average depth of 1500-1550 m (4920-5090 ft) (1390-1490 m or 4560-4890 ft SSL). These sandstone beds, ranging in thickness from 1 to 12 m (3 to 40 ft), are separated by shale layers (Figure 2). The "A-Z" sandstone that appears as one sandstone body off structure to the east is separated updip into two sandstones, "A" and "Z" (see Figure 12). The "B" sandstone (the lower sand member) is the most important producing sandstone in the Kokhav field and is found at an average depth of 1650 m (5400 ft) (1525-1600 m or 5000-5250 ft SSL).

A permeability barrier, the Gevar-Am shale, is located at the updip western limit of the sandstone. Commonly, production becomes poorer toward the west in spite of a relatively higher structural position owing to increasing shaliness or pinchout of the producing sandstones. Conversely, on the eastern and downdip flank of the structure, the sandstones have better porosity.

Transverse adjustment faults that divide the Helez structure into blocks separate the Lower Cretaceous sandstone pay zones into different reservoirs. These relatively small faults were located on the basis of production anomalies, seismic interpretation, structural mis-ties, and reservoir pressure differences that could not be explained otherwise (Figure 3).

The LC 11 Limestone is found immediately above the middle sand member and produced from only one well (Kokhav 13).

The reefoidal Kokhav Dolomite developed on the margin of the elevated platform and extends along the Helez field's anticlinal structure. Its productive limits are affected by the same faulting system as the producing Helez sands as well as by lithologic changes into limestone or shale.

The basal Lower Cretaceous oolitic limestone, the Mashen Member of the Helez Formation, and the Jurassic Nir-Am Reef are considered as one thick continuous reservoir underlain by one common aquifer, since no apparent separation between the two formations exists. Even faulting does not necessarily isolate the oil trapped in different fault blocks owing to the considerable thickness of the Jurassic beds and the apparently common fracture systems of both formations. In most areas, the fractured limestones of the two formations have a very low matrix porosity. Any significant porosity, therefore, is mainly secondary and developed in fractures and solution vugs. The oil-producing limestones were found at the depth of 1600-1700 m (5250-5580 ft).

The Middle Jurassic Brur Calcarenite producing area is limited to the southern extreme of the Helez field.

Petrography of Sandstone Reservoirs

The major reservoirs of the Helez field are the middle and lower sand members of the Helez Formation (Shenav, 1971, and Figure 2). The sandstones of the Helez Formation constitute about 10% of the total formation and are partly mixed with tuff. Quartz, which is the dominant mineral, was

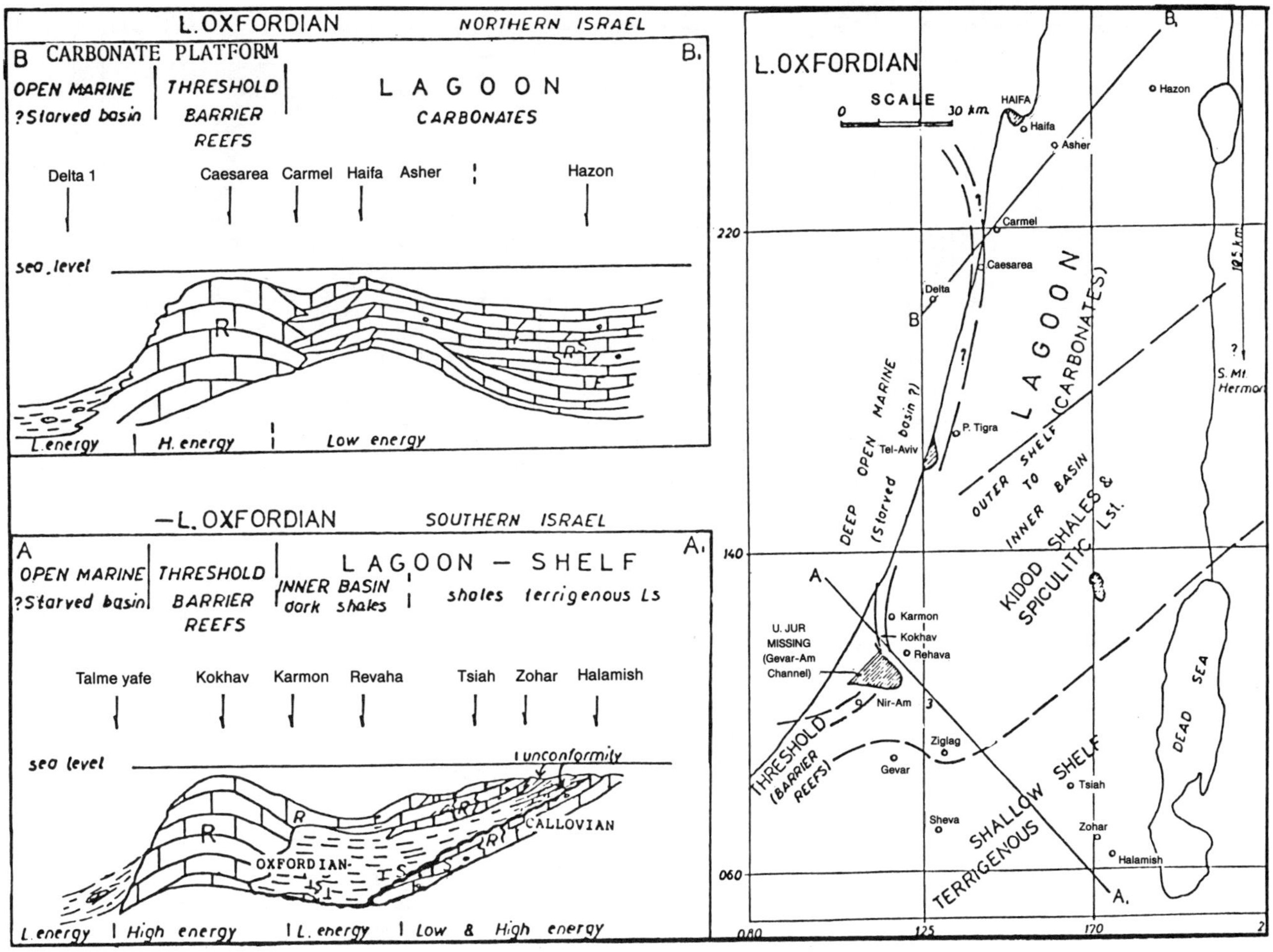

Figure 8. Upper Jurassic environments of deposition. R, reef. (From Derin, 1974.)

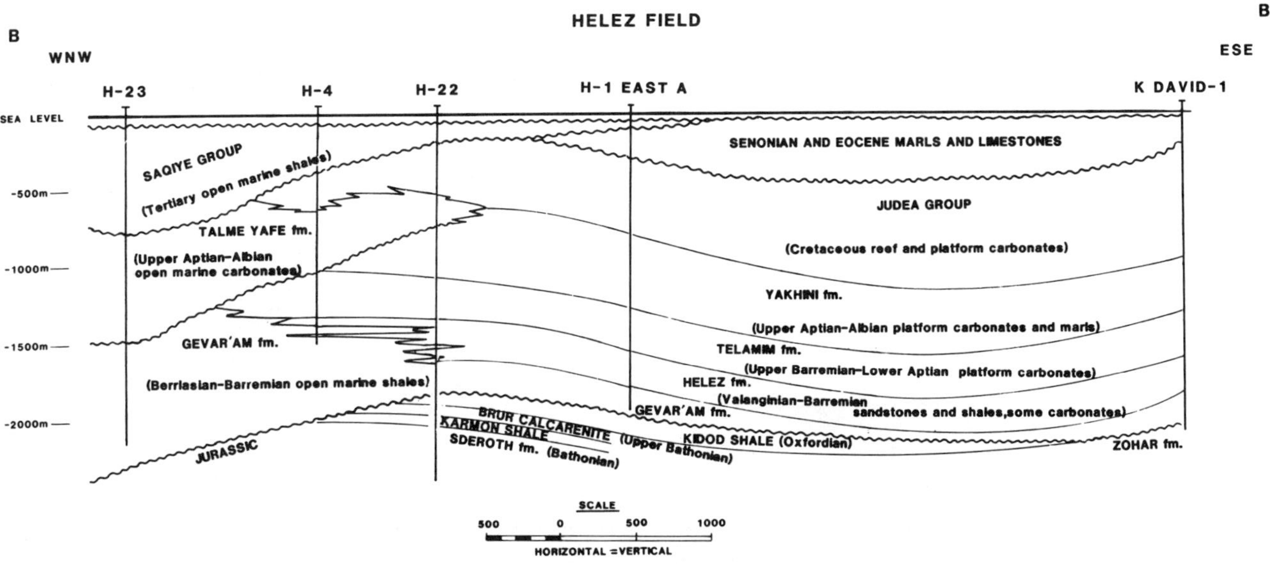

Figure 9. Geologic cross section through the Gevar-Am canyon, showing erosional contact with underlying Jurassic rocks. (Location is shown in Figure 1.) (Modified from Cohen, 1976.)

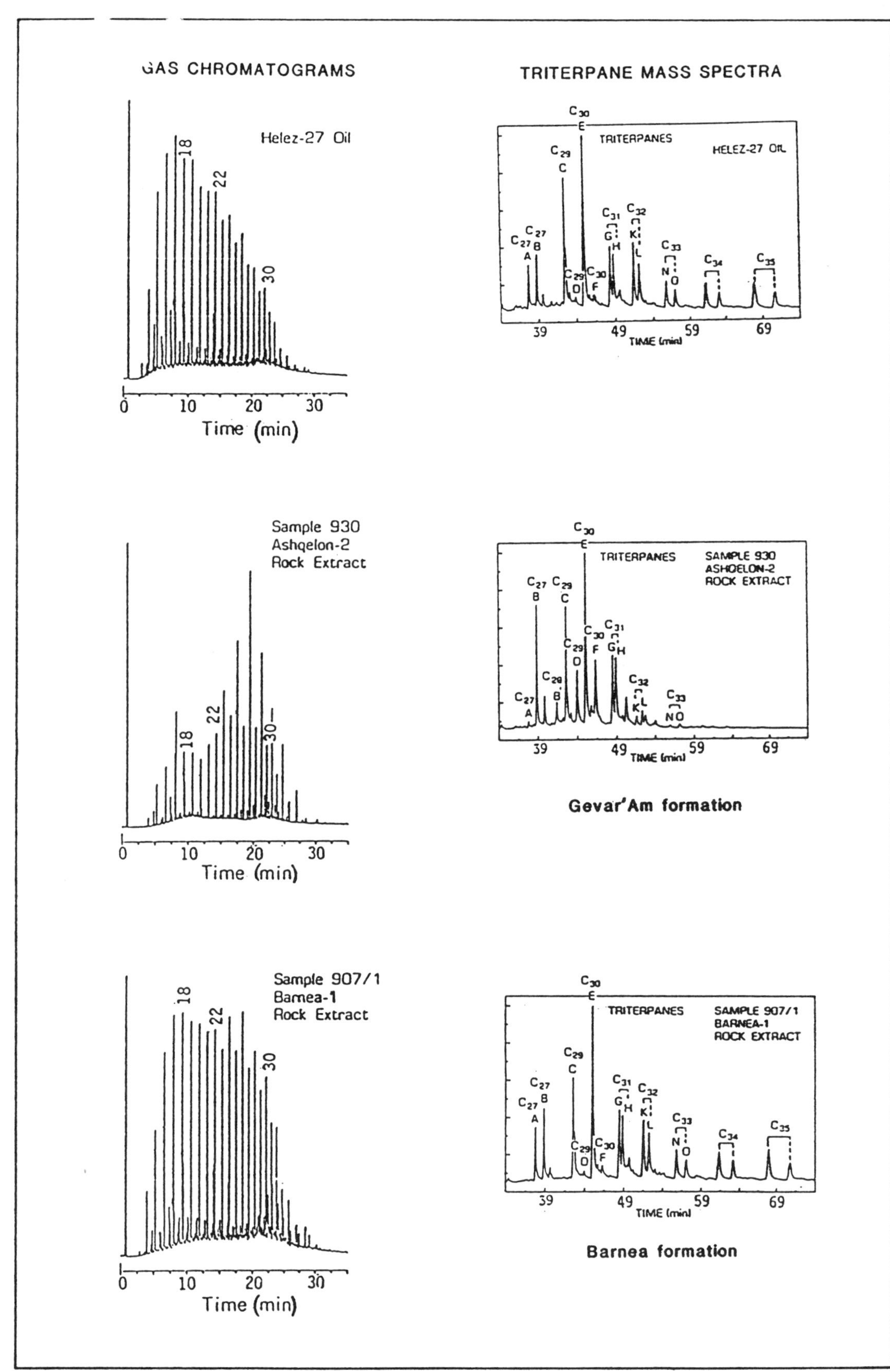

Figure 10. Analysis of oil from the Helez 27 well and rock extracts from the Gevar-Am and Barnea formations. (From Bein and Sofer, 1987.)

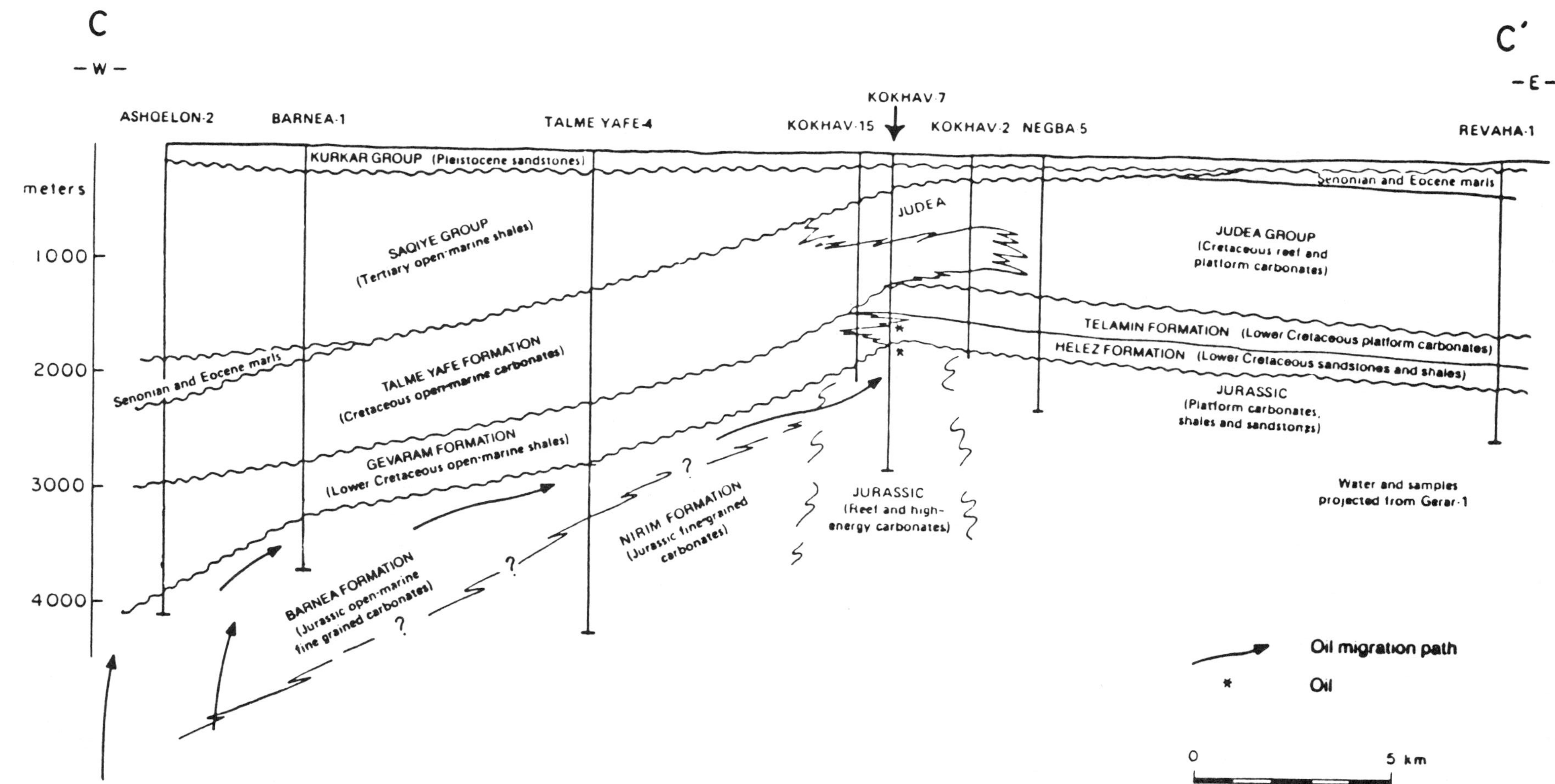

Figure 11. Oil migration paths in the Helez area. (Location is shown on Figure 6.) (After Bein and Sofer, 1987.)

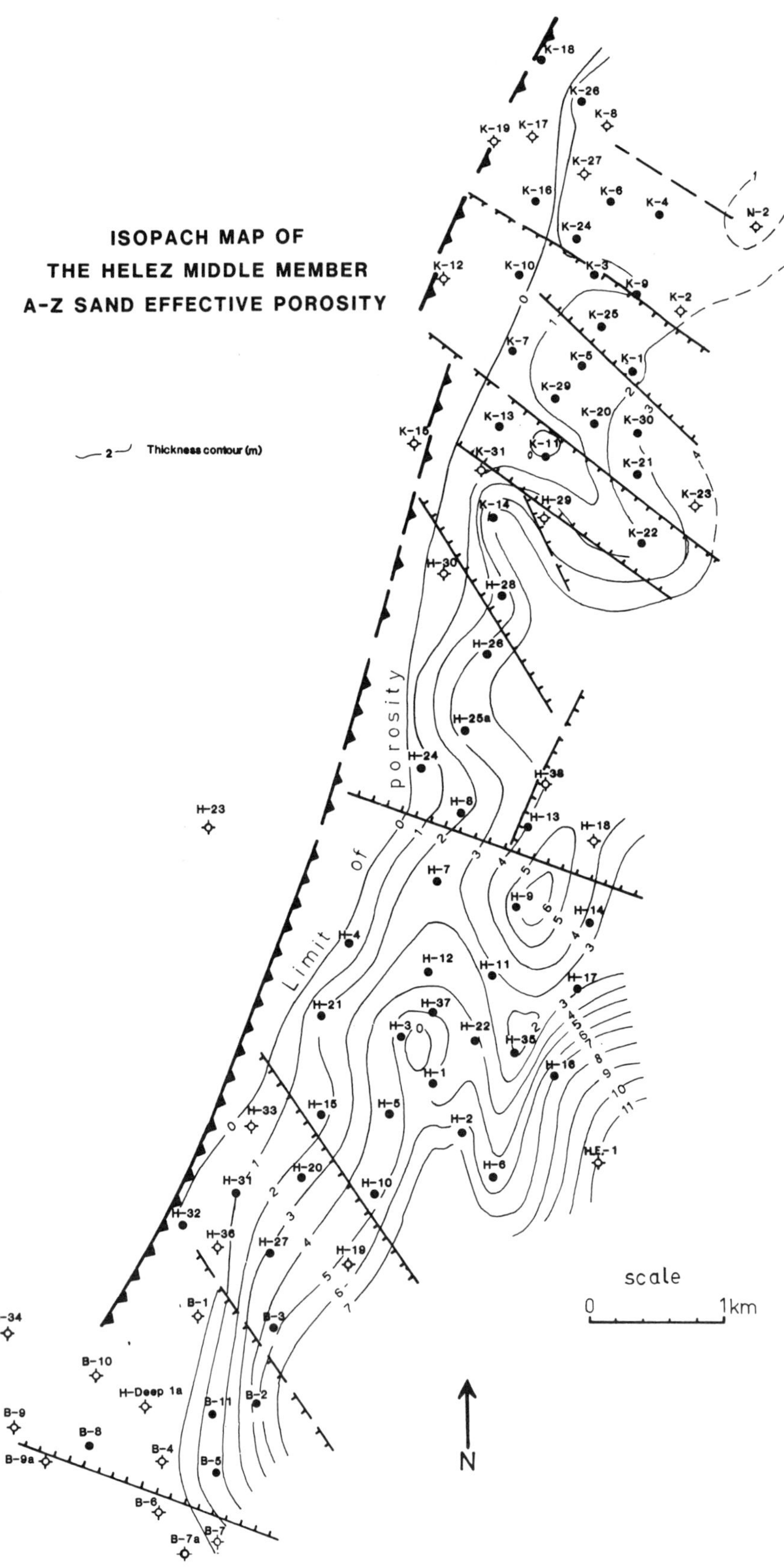

Figure 12. Helez-Brur-Kokhav field. Isopach map of the Helez middle member "A-Z" sand effective porosity. C.I. = 1 m.

derived by weathering and redeposition of Nubian sandstones exposed to the east and southeast. Other grains found in the sandstone include sanidine, volcanic rock fragments, and various carbonate, phosphate, and iron (oolite) allochems. Calcite cement tends to be found in marine sandstones, whereas dolomite cement is found in sandstones that were deposited in the coastal area (Shenav, 1971).

Four porosity types were defined, of which intergranular and intercrystalline porosities are the principal types.

The reservoir sandstones of the Helez Formation were divided by Shenav (1971) into three types (Figure 13):

Type I — Sandstone includes part of the lower sands and part of the middle sand "K." Deposited in an offshore marine environment, it contains marine fossils and is cemented generally by sparry calcite. Average grain size is medium to fine. The grains are moderately to poorly sorted. Porosity is intercrystalline and reaches 16%. Permeability is less than 30 md.

Type II — Sandstone includes middle member sands, "A-Z," "W," and part of the "K" sand, deposited in a tidal channel and/or lagoonal environment. It is composed of sands of medium grain size, with moderate or poor sorting and few marine fossils. Cement is calcitic and/or dolomitic. Porosity values may reach 30% (average 16%), and permeability may be up to 2000 md (average 50 md). The lower values are related to moderate sorting and to high clay content.

Type III — Sandstone includes part of the middle member sands "A-Z," "W," and the lower sand ("B" or Kokhav sand) and is probably of eolian origin, deposited in a coastal area. These sands, with average mean grain size ranging from very fine to medium, do not contain marine fossils. The grains are mostly well sorted, loosely packed, and with low cement content. Porosity can reach 32% (average 24%), and permeability may exceed 2000 md (average 200 md).

Most of the oil (about 38 million bbl of oil in place) is found in the more porous sandstones (types III and II) that were deposited in coastal environment, either eolian, tidal channels, or lagoons. Marine sandstones (type I), where the porosity values are generally low, contain almost no oil.

Rock and Fluid Characteristics

Based on core analysis data in conjunction with log calculations, typical values of porosity and original water saturation for each zone have been obtained (Table 2). Permeability values are significantly variable in the sandstone reservoirs, as illustrated by their distribution in the Helez main producer, "A-Z" sand (Table 3). Saturation dependent data of oil/water relative permeability is not available. A PVT analysis has never been performed on the Helez oil. Physical reservoir conditions including temperature, pressure, and fluid properties measured in the field are as follows:

Bottomhole temperature	150°F
Initial pressure	2000 psig @ 1402 m SSL
Oil gravity	27.5-31.0 API
Solution gas-oil ratio	200-275 SCF/STB
Gas gravity	0.75 (air = 1)

Based on the above data, the bubble point pressure is estimated at 1000-1500 psia, the original formation volume factor at 1.05-1.15, and the original oil viscosity at about 2.0 cp. A typical composition of the Helez oil is presented in Table 4.

Two types of formation water are encountered in the Helez field: Helez Formation water related to Group I, and the Jurassic formation water related to Group III (Fleisher, 1987). Analysis of most of the Helez Formation water (Group I) shows total dissolved solids values of around 55 g/L and characteristic ionic ratios of Na/Cl = 0.86-0.97, (Ca+Mg)/Na = 0.12-0.19, and Cl/Br = around 300 (Table 5). The formation brines of Group III are characterized by the ratio Na/Cl = 0.75, a high Ca/Mg ratio, (Ca+Mg/Na) > 0.27 $\delta^{18}O$, and δD values suggesting the presence of meteoric water components. Waters of a similar composition to Group III are present in the Paleozoic, Triassic, and Jurassic strata of Israel.

Production and Reserves

The Helez field was put on production in October 1955 and to the end of July 1988 it has produced approximately 16.5 million barrels of oil. About 97% of the oil comes from Lower Cretaceous beds (Helez Formation, Figure 2), while the remaining 3% comes from Upper and Middle Jurassic limestones.

The original oil in place of the four principal producing zones of the Helez field, the "A-Z," "B," "K," and "W" sands, was estimated volumetrically to be about 40.3 million barrels (Table 2). Each zone is divided into multiple reservoirs separated by the various faults. For the most part, the fault system leaves the reservoirs open downdip to a large aquifer to the east, thus allowing an active water influx. The various reservoirs are characterized by the different water/oil levels and exhibit different pressure-production decline behavior. Original pressure in the main "A-Z" sand was approximately 2000 psig at 1402 m (4600 ft) SSL. According to its pressure history it appears that the water influx arrested the pressure decline at 900 to 1000 psi and pressure increases were later observed as offtake rates declined. The lowest reservoir pressures

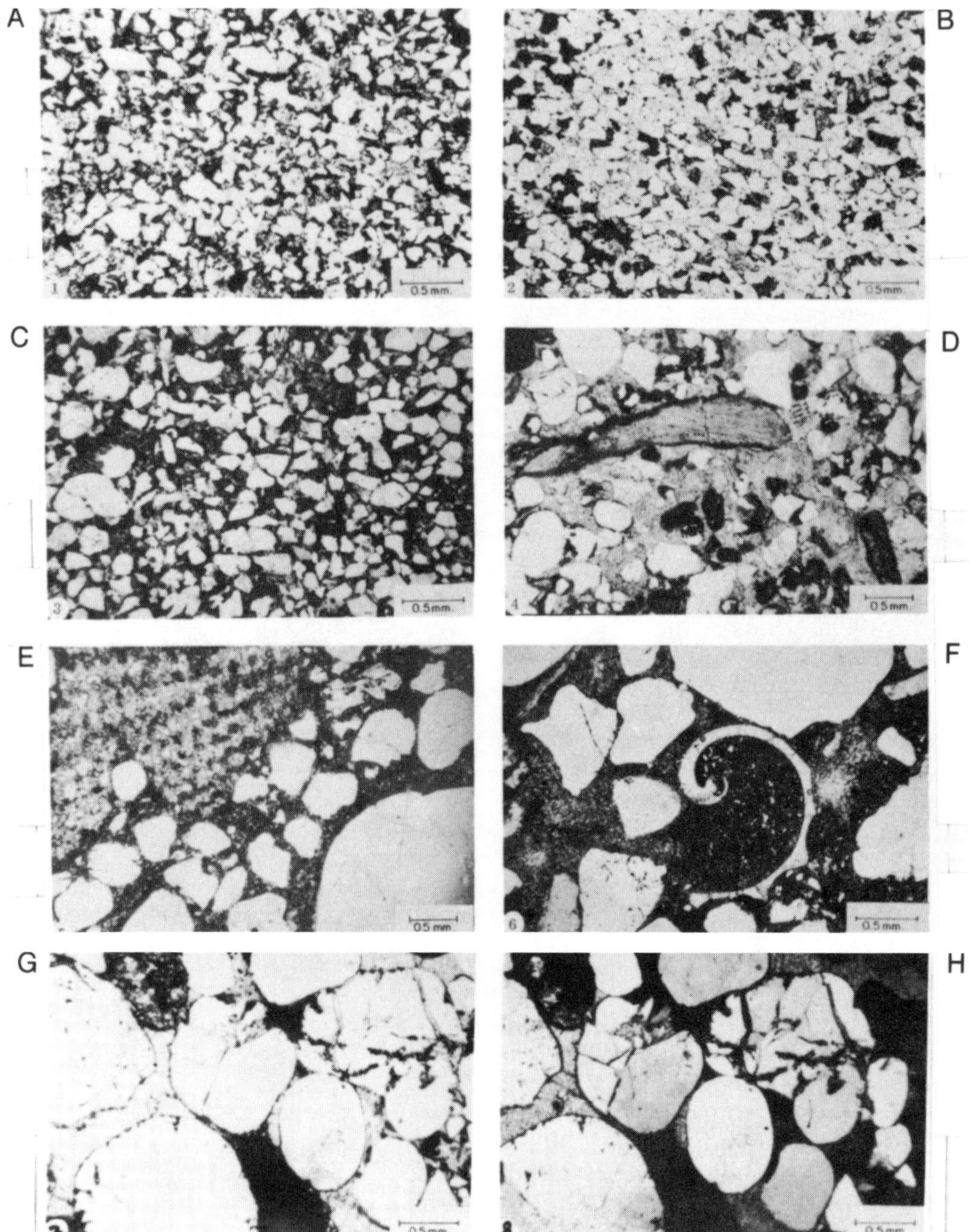

Figure 13. (A) Type III sandstone (eolian coastal sands). The cement crystals are dolomite (plug 29, Kokhav 6, 1651 m. The lower sand member—"Kokhav B"). (B) Type III sandstone (eolian coastal sands). The cement consists of dolomite crystals and some pyrite (plug 1, Kokhav 9, 1661 m. The lower sand member—"Kokhav B"). (C) Type III sandstone (eolian coastal sands); general view. The cement is dolomite and some pyrite (plug 6, Kokhav 6, 1669 m. The lower sand member—"Kokhav B"). (D) Type I sandstone (offshore marine sand). This rock is transitional between sandstone and limestone. The components are quartz grains, faunal fragments, and iron oolites. The cement consists of dense calcite crystals (sample HS-54, Kokhav 2, 1677 m. The lower sand member—"Kokhav A"). (E) Type I sandstone (offshore marine sand). The grains are quartz (poorly sorted, in part well rounded) and a big coral fragment that has undergone dolomitization. The cement consists of dolomite crystals (sample HS-466, Kokhav 7, 1517 m. The middle sand member—"Helez A"). (F) Type I sandstone (offshore marine sand). The grains are quartz and faunal fragments. Micritic calcite that forms the matrix has undergone recrystallization (plug 1, Kokhav 6, 1558 m. The middle sand member—"Helez W"). (G) Type II sandstone (tidal channel or lagoonal sand). The grains are quartz and volcanic rock fragments rich in pyrite. The cement is a single (poikilotopic) calcite crystal. The quartz grains are well rounded, have a high sphericity, and are partly shattered (sample HS-458, Helez 8, 1568 m. The middle sand member—"Helez A"). (H) As in G crossed nicols (sample HS-458, Helez 8, 1568 m. The middle sand member—"Helez A"). (From Shenav, 1971.)

Table 2. Helez oil field tabulation of reservoir parameters and production (end of July 1988).

Pay Zone and Fault Segment	Acre Feet	Porosity, Fraction	Water Saturation	Original Barrels per Acre Feet	Original Oil in Place MSTB	Estimated Cumulative Production MSTB	Remaining Oil in Place MSTB	Percent Produced	Remaining Oil Saturation Percent
"K" Sand									
Segment A	454	0.220*	0.370*	1,014.4	461	366.3	94.7	79	13
Segment B	3,818	0.220*	0.370*	1,014.4	3,873	1.070.1	2,807.9	28	45
Segment C+D	235	0.133*	0.420*	564.6	133	5.0	128	28	57
Segment E	397	0.133*	0.420*	564.6	224	310.0	-86	138**	-22
Segment G	332	0.133*	0.420*	564.6	187	90.0	97	48	30
Segment H	146	0.133*	0.420*	564.6	82	0.0	82	0	58
Segment I	316	0.133*	0.420*	564.6	178	17.0	161.0	10	53
Total	5,698				5,138	1,858.4	3,279.6	36	33
"W" Sand									
Segment B	2,553	0.240	0.360	1,089.0	2,780	1.116.0	1,664	40	37
Segment C+D	705*	0.124*	0.430	517.3	365	29.8	335.2	8	52
Segment E+F	1,078*	0.124*	0.430	517.3	558	768.6	-210.6	138**	-22
Segment G	446	0.268	0.343	1,288.7	575	284.6	290.4	50	33
Total	4,782				4,278	2,189.0	2,079.0	52	25
"A-Z" Sands									
Segment A	3,520	0.220*	0.380*	998.3	3,514	1,282.7	2,231.3	37	39
Segment B	16,026	0.220*	0.380*	998.3	15,999	7,195.8	8,803.2	45	34
Segment C	1,872	0.160	0.430	667.5	1,250	135.0	1,115.0	11	51
Segment D	1,965	0.160	0.430	667.5	1,312	835.0	476.1	64	21
Segment E	1,056	0.126	0.430	525.6	1,555	198.2	356.8	36	37
Segment F	664	0.126	0.430	525.6	1.349	—	349.0	0	57
Segment G	1,699	0.160*	0.390*	714.3	1,214	182.3	1,031.7	15	52
Segment H	237	0.165*	0.410*	712.5	169	13.2	155.8	8	54
Segment I	190	0.170*	0.430*	709.2	135	40.5	94.5	30	40
Total					24,497	9,883.6	14,613.4	40	43
"B" Sand									
Segment F	1,100	0.268	0.441*	1,096.5	1,206	277.6	928.4	23	43
Segment G	1,794	0.272*	0.438*	1,118.8	2,007	149.0	1,858.0	7	52
Segment H	1,079	0.276*	0.434*	1,143.3	1,234	516.8	717.2	42	24
Segment I	1,947	0.220	0.390	982.2	1,912	371.3	1,540.7	20	49
Total					6,359	1,314.7	5,044.3	21	42

(continued)

Table 2. (Continued)

Pay Zone and Fault Segment[1]	Acre Feet	Porosity, Fraction	Water Saturation	Original Barrels per Acre Feet	Original Oil in Place MSTB	Estimated Cumulative Production MSTB	Remaining Oil in Place MSTB	Percent Produced	Remaining Oil Saturation Percent
Kokhav Dolomite									
Segment A	877	0.1	0.45	460.5	360	104.1	255.9	29	39
Segment B	4,370	0.1	0.45	408.5	1,785	5.9	1,779.1	0.3	54.8
Segment D	800	0.1	0.45	405.0	324	7.2	316.8	2.2	53.8
Segment F	3,250	0.1	0.45	502.5	1,308	—	1,308.0	0.0	55
Segment G	7,750	0.1	0.45	398.7	3,090	383.3	2,706.7	12.4	48.2
Segment H	3,920	0.1	0.45	400.5	1,570	131.0	1,439.0	8.3	50.4
Segment I	4,900	0.1	0.45	400.2	1,961	6.9	1,954.1	0.4	54.8
Total					10,398	638.4	9,759.6	6.1	

Oil volume factor used in computations is 1.06
[1]The reservoirs consist of pay zones ("K" sand etc.) divided by faults into segments on Blocks (A,B, etc.; see Figure 3).
*Value is estimated.
**Cumulative production obviously includes oil originating from other sources.

Table 3. Helez oil field permeability distribution in Helez "A" sand

Upper Limit of X Class	Median of X Class	Sampling Frequency	Smoothed Frequency
.0709	.0533	.0180	.0385
.1256	.0943	.0721	.0451
.2224	.1671	.0541	.0517
.3939	.2960	.0360	.0581
.6976	.5242	.0721	.0641
1.2356	.9284	.0721	.0693
2.1885	1.6444	.0721	.0736
3.8762	2.9126	.0991	.0766
6.4655	5.1587	.0541	.0781
12.1600	9.1370	.0090	.0782
21.5376	16.1833	.0811	.0768
38.1470	28.6635	.0360	.0740
67.5651	50.7681	.0270	.0699
119.0698	89.9194	.0360	.0648
211.9567	159.2634	.1081	.0589
375.4131	282.0839	.0631	.0525
664.9237	499.6209	.0450	.0459
1177.6985	884.9178	.0360	.0394
2085.9143	1567.3475	.0000	.0331
3694.5266	2776.0522	.0090	.0273

The sample mean is 7.1225
The sample median is 4.0654
The standard deviation of sample data is 64.8705
Ratio of mean to median is 1.3998
The coefficient of variation is 147.9826
The 95% confidence limits on estimate of true mean are 4.1220 and 12.3071
The 99% confidence limits on estimate of true mean are 3.4543 and 14.6860

Source: Courtesy of Lapidoth.

Table 4. Helez oil field separator oil analysis
Podbielniak Analysis

Component	Mol%	Wt%	Liq Vol%
Ethane	0.21	0.03	0.06
Propane	0.56	0.10	0.18
Iso-butane	0.85	0.21	0.32
N-butane	0.99	0.24	0.36
Iso-pentane	1.41	0.43	0.59
N-pentane	1.98	0.60	0.83
Hexanes	5.65	2.03	2.68
Heptanes plus	88.35	96.36	94.98
	100.00	100.00	100.00

Heptanes Plus (calculated)	
Specific gravity at 60°F (water = 1)	0.8860
API gravity at 60°F	28.207
Molecular weight	261.000
Ft^3 vap/gal at 14.65 psia and 60°F	10.77
Total Sample	
Liq. specific gravity at 60°F (water = 1)	0.8733
API gravity at 60°F	30.535
Ft^3 vap/gal at 14.65 psia	11.583
Specific gravity as a vapor	10.752
Molecular weight	239.288

Table 5. Chemical composition of formation waters from oil wells in the Helez field.

No.	B.H.	Stratlgr.	Depth m	TDS	Ca	Mg	Na	K	Cl	SO_4	HCO_3	Cl/Br	Na/Cl	Ca/Na	Ca+Mg/Na	^{18}O	D	Group
1	Brur 2	L.Cr(Tl)	1142-1154	51.5	61	34	796	—	886	6	3	—	0.89	0.08	0.12	+1.92	—	I
2	Brur 2	L.Cr(Hz)	1500-1502	53.3	66	38	804	7.9	913	0	3	—	0.88	0.08	0.13	+1.42	—	I
3	Helez 1A	L.Cr(Hz)	1669-1679	55.6	93	41	807	13.1	938	12	2	207	0.86	0.12	0.17	+1.19		I
4	Helez 2	L.Cr(Hz)	1521-1522	64.1	72	48	957	10.0	1108	1	3	275	0.86	0.08	0.13	+1.64	-6.01	I
5	Helez 6	L.Cr(Hz)	1536-1548	52.3	71	42	862	—	971	9	4	—	0.89	0.08	0.13	+1.68	—	I
6	Helez 10	L.Cr(Hz)	1540-1545	64.1	91	46	970	9.4	1095	—	3	—	0.89	0.09	0.14	+2.55	—	I
7	Helez 20	L.Cr(Hz)	1507-1530	56.6	80	44	855	—	976	2	1	—	0.88	0.09	0.15	+5.33	+21.3	I
8	Helez 21	L.Cr(Hz)	1520-1524	61.7	93	54	975	—	1121	—	1	—	0.87	0.10	0.15	+4.31	+16.3	I
9	Helez 22	L.Cr(Hz)	—	61.4	67	45	922	10.3	1058	0	4	265	0.87	0.07	0.12	+1.50	+15.35	I
10	Helez 22	J(Zr)	2000-2014	45.5	82	43	658	6.5	783	3	3	—	0.84	0.12	0.19	—	—	I
11	Helez 22	J(Nir)	3377-3408	64.2	515	60	507	18.6	1106	9	2	—	0.46	1.01	1.13	—	—	III
12	Helez 25A	L.Cr(Hz)	1571-1575	71.1	111	50	1007	6.5	1261	2	1	—	0.80	0.11	0.16	+2.58	—	I
13	Helez 25A	L.Cr(Hz)	1576-1578	74.5	112	51	1134	—	1294	1	1	—	0.88	0.10	0.14	+2.65	—	I
14	Helez 27	L.Cr(Hz)	1505	57.6	69	42	870	8.7	974	11	2	—	0.89	0.08	0.13	+1.55	—	I
15	Helez 29	L.Cr(Hz)	1587-1603	47.9	55	37	731	9.2	800	10	3	—	0.91	0.08	0.13	+0.98	+1.5	I
16	Kokhav 1	L.Cr(Hz)	1593-1600	37.2	43	30	544	12.2	640	1	3	—	0.85	0.08	0.13	+0.02	+2.4	I
17	Kokhav 3	L.Cr(Hz)	1560-1590	29.5	35	24	426	6.8	477	27	3	—	0.89	0.08	0.14	- 2.19	—	I
18	Kokhav 7	L.Cr(Hz)	1493-1510	82.8	143	77	1200	12.2	1424	2	4	282	0.84	0.12	0.18	+1.02	—	I
19	Kokhav 7	J(Zr)	2485-2495	110.0	578	126	1195	23.6	1902	10	3	137	0.63	0.48	0.59	—	—	III
20	Kokhav 11	L.Cr(Hz)	1547-1549	47.0	54	34	713	8.7	812	10	4	260	0.88	0.08	0.12	+1.00	-2.68	I

Abbreviations used in Table 5:
Time Units:
L.Cr, Lower Cretaceous
J, Jurassic
Rock Units:
Tl, Telamim Fm.
Hz, Helez Fm.
Zr, Zohar Fm.
Nir, Nirim Fm.
Source: From Fleisher, 1987.

measured were those of the "K" sand, being less than 500 psig at 1400 m (4595 ft) SSL in some fault blocks. Because no increase in gas/oil ratio has been reported, it is believed that the Helez oil was undersaturated during its history in most of the reservoir.

The Kokhav Dolomite volume was estimated in a similar manner to be about 10.0 million barrels of oil in place whereas the volume of the oil in the Jurassic limestone is still unknown.

At their current status of development the primary recovery from the reservoirs is calculated to vary from 5 to more than 50% of the original oil in place. The higher values occur in the reservoirs that are adequately developed, where the number and the location of completions allow high areal sweep efficiency for the encroaching water. In addition, in some fault segments that are evidently in partial contact with an aquifer, the producing mechanism is a combination of pressure depletion and water drive, and the primary recovery is low. The main producing formations mentioned above are nearing depletion. Most of the swept area should be at or near residual oil saturation although isolated sand lenses or low permeability streaks might currently have very high oil saturation.

Assuming an ultimate recovery of 45% for the sand zones, and 10% for the Kokhav Dolomite, about 19 million stock tank barrels of oil are expected to be produced.

EXPLORATION AND DEVELOPMENT CONCEPTS

The Helez field is nearing the end of its primary production phase (16.5 MMBO cumulative production at the end of 1988 vs. 19 MMBO of estimated recoverable reserves). Nevertheless, because of the field structure, which is composed of numerous, separate small reservoirs, some of the fault segments in certain reservoirs have not been adequately drained by the existing wells. The "attic" oil in these segments, which could amount to more than 1 million barrels, is planned to be produced by: (1) drilling new wells, (2) opening new pay zones in existing wells, and (3) deepening or recompleting existing wells.

A second possibility for increasing current production is the application of secondary recovery by water flooding in fault blocks where the natural water drive is weak. The thin Helez "K" sandstone seems to be a good candidate for such an operation.

Even after primary recovery has been made more efficient by adding new completions, 40 to 60% of the original oil in place will remain in the ground. Because these reservoirs will have been swept by encroaching water, secondary recovery was discarded and only a tertiary recovery scheme should be considered. Miscible displacement with carbon dioxide seems to be the most efficient known method for increasing production in the Helez field. The additional oil recovery might be from 5 to 10 million barrels. A more definite estimate can only be made after further data concerning saturation distribution are obtained. Perhaps most important is the feasibility of tertiary recovery, which is closely related to oil prices and availability; under current conditions, the application of tertiary procedures is not economic.

ACKNOWLEDGMENTS

Mr. J. Fisher, Managing Director of the Israel National Oil Company Ltd. and Mr. E. Roih, Managing Director of Lapidoth Israel Oil Prospectors Corporation Ltd., permitted the publication of this paper.

Unpublished studies and reports of Lapidoth Israel Oil Prospectors Corporation Ltd. are the basis of the Helez field presentation.

REFERENCES

Amit, O., 1978, Organochemical evaluation of Gevar-Am shales (Lower Cretaceous), Israel, as possible oil source rock: American Association of Petroleum Geologists Bulletin, v. 62, p. 827-836.

Ball, M. W., and D. Ball, 1953, Oil prospects of Israel: American Association of Petroleum Geologists Bulletin, v. 37, n. 1, p. 1-113.

Bein, A., 1974, Reef development in the Judea Group from the Carmel and the Coastal Plain, Israel: Ph.D. thesis, the Hebrew University, Jerusalem (in Hebrew).

Bein, A., and G. Gvirtzman, 1977, A Mesozoic fossil edge of the Arabian plate along the Levant coastline and its bearing on the evolution of the eastern Mediterranean, *in* B. Biju-Duval and L. Montadert, eds., Structural history of the Mediterranean basins: Paris, Edition Technip, p. 95-110.

Bein, A., and Z. Sofer, 1987, Origin of oils in the Helez Region, Israel—implication for exploration in the Eastern Mediterranean: American Association of Petroleum Geologists Bulletin, v. 71, p. 65-75.

Bein, A., and Y. Weiler, 1976, The Cretaceous Talme Yafe Formation: a contour current shape sedimentary prism of calcareous detritus at the continental margin of the Arabian Craton: Sedimentology, v. 23, n. 4, p. 511-532.

Cohen, A., 1983, Lithofacies and environments of deposition of the Lower Cretaceous Helez & Telamim Formations, Israel: GSI Report S/5/83.

Cohen, Z., 1964, The geology of Heletz-Brur-Kokhav field: Isr. J. Earth-Sci, v. 13, p. 179-183.

Cohen, Z., 1969, The geology of the Lower Cretaceous in the Coastal Plain: Israel Inst. Petroleum Research & Geophysics (unpublished).

Cohen, Z., 1971, The geology of the Lower Cretaceous of the Heletz Field, Israel: Ph.D. thesis, the Hebrew University, Jerusalem (in Hebrew).

Cohen, Z., 1976, Early Cretaceous buried canyon: influence on accumulation of hydrocarbons in Helez oil field, Israel. American Association of Petroleum Geologists Bulletin, v. 60, p. 108-114.

Derin, B., 1974, The Jurassic of central and northern Israel: Ph.D. thesis, the Hebrew University, Jerusalem (in Hebrew, with English abstract), 152 p.

Derin, B., and E. Gerry, 1972, Jurassic biostratigraphy and environments of deposition in Israel: Israel Inst. Petroleum Micropaleontological Lab, Report 2/72, 5 p.

Fleisher, E., 1987, The geochemistry of formation water in Israel: OEIL Report 87/20.

Garfunkel, Ẑ., and B. Derin, 1983, Permian-early Mesozoic tectonism and continental margin formation in Israel and its implication for the history of the Eastern Mediterranean, *in* J. E. Dixon and R. H. F. Robertson, eds., The geological evolution of the Eastern Mediterranean: Geological Society (London) Spec. Publication No. 12, p. 187-202.

Grader, P., and Z. Reiss, 1958, The Lower Cretaceous of the Heletz area: Geological Survey of Israel Bulletin No. 16, 14 p.

Gruy H. J. & Associates Inc., 1975, A preliminary study of enhanced recovery from the Heletz-Brur-Kokhav fields, Israel (submitted to Lapidoth).

Gvirtzman, G., and A. Klang, 1972, A structural and depositional hinge line along the Coastal Plain of Israel, evidenced by magneto-tellurics: Geological Survey of Israel Bulletin No. 55, 18 p.

Lapidoth-Israel Oil Prospectors Corporation Ltd., 1969, Various reports and maps (unpublished).

May, P. R., 1981, Ashdod-Gan Yavne area: review, evaluation and recommendations: Oil Exploration (Investments) Ltd. Report 81/31, 56 p.

May, P. R., 1984, Hydrocarbon migration and primary trapping in central Coastal Plain and Hashefela region, Israel: Oil Exploration (Investments) Ltd. Report 84/29, 11 p.

Rahamim, S., 1973, Geological report on Talme Yafe 4—deepened well: Lapidoth.

Rosenberg, E., 1979, The remaining oil potential of the Helez field—geology and exploration. Report to Lapidoth.

Shenav, H., 1971, Lower Cretaceous sandstone reservoirs, Israel: petrography, porosity, permeability: American Association of Petroleum Geologists Bulletin, v. 55, p. 2194-2224.

Starinsky, A., 1974, Relationship between Ca-chloride brines and sedimentary rocks in Israel: Ph.D. thesis, the Hebrew University, Jerusalem (in Hebrew, with English abstract), 84 p.

Tschopp, H. J., 1956, The oilfield of Heletz, Israel: Bull. Swiss Assoc. Pet. Geol., v. 22, p. 41-54.

Weeks, L. G., 1962, Israel's petroleum problem: Report to Ministry of Development.

Appendix 1. Field Description

Field name .... *Helez-Brur-Kokhav*

Ultimate recoverable reserves .... *19 million bbl*

Field location:

- **Country** .... *Israel*
- **State**
- **Basin/Province** .... *Eastern Mediterranean*

Field discovery:

- **Year first pay discovered** .... *(L. Cret.) Valanginian/Barremian Helez Formation 1955*
- **Year second pay discovered** .... *Helez Formation zones and Jur. Brur Calcarenite 1962*
- **Third pay** .... *Additional zones of L. Cret. and Jur. February 1963*
- **Fourth pay** .... *August 1963*
- **Fifth pay** .... *October 1963*

Discovery well name and general location:

- **First pay** .... *Helez 1, 10 km southeast of Ashqelon*
- **Second pay** .... *Kokhav 1*
- **Third pay** .... *Kokhav 5*
- **Fourth pay** .... *Kokhav 10*
- **Fifth pay** .... *Kokhav 13*

Discovery well operator .... *Lapidoth (for all pays)*

- **Second pay**
- **Third pay**

IP in barrels per day and/or cubic feet or cubic meters per day:

- **First pay** .... *400 BOPD (A-Z sand)*
- **Second pay** .... *190 BOPD (B sand)*
- **Third pay** .... *200 BOPD (K dolomite)*
- **Fourth pay** .... *500 BOPD in Kokhav 16 (Jurassic ls.)*
- **Fifth pay** .... *140 BOPD (LC 11 ls.)*

All other zones with shows of oil and gas in the field:

Age	Formation	Type of Show
Upper Barremian-lower Aptian	*Telamim*	*Oil*
Valanginian-Barremian	*Helez Lc 13A*	*Oil*
Jurassic	*Brur calcarenite*	*Oil*

Geologic concept leading to discovery and method or methods used to delineate prospect, e.g., surface geology, subsurface geology, seeps, magnetic data, gravity data, seismic data, seismic refraction, nontechnical:

Structure inferred and mapped using surface, gravity data complemented by seismics during trend development.

HELEZ-BRUR-KOKHAV

Structure:

Province/basin type .... *Bally 1141; Klemme IIC a/b*

Tectonic history

Elevated structure since Jurassic. Upper Jurassic high was truncated and deep canyon was formed (Gevar-Am channel). Faulting is related to uplifting and tilting at the end of the Jurassic, to the folding of the Cretaceous-lower Tertiary and to Neogene movements.

Regional structure

Field lies on a hinge line, defining boundary between platform and basinal environments.

Local structure

Northeast-southwest- and north-south-trending asymmetrical anticline with a gently dipping eastern limb and a downfaulted western limb.

Trap:

Trap type(s)

On east flank of anticlinal uplift multiple traps (1) middle and lower sand members of the Helez Formation pinch out to west (strato-structural), (2) carbonate facies of Helez Formation, and (3) Jurassic reef.

Basin stratigraphy (major stratigraphic intervals from surface to deepest penetration in field):

Chronostratigraphy	Formation	Depth to Top in m
Pleistocene	*Kurkar*	*0*
Neogene	*Saqlye*	*100-120*
Cenomanian-Turonian	*Judea Group*	*250-300*
Upper Aptian-Albian	*Talme Yafe*	*500-700*
Upper Aptian-Albian	*Yakhini*	*600-1250*
Upper Barremian-lower Aptian	*Telamim*	*950-1400*
Valanginian-Barremian	*Helez*	*1380-1500*
Berriasian-lower Aptian	*Gevar-Am*	*1430-1700*
Oxfordian	*Nir Am-Beer Sheva*	*1600-1900*
Oxfordian	*Kidod*	*1800-2030*
Upper Bathonian	*Zohar or Brur*	*1830-2450*
Bathonian	*Karmon*	*1950*
Bathonian-Bajocian	*Sderoth*	*2000*
Aalenian-Lias.	*Nirim*	*3300*
Lias.	*Mish'hor*	*4770*
Late Triassic	*Shefayim (Mohilla equiv.)*	*4800*
Middle Triassic	*Or Haner Cgl.*	*5440*
Paleozoic	*Erez Porphyr*	*5766*
Precambrian	*Dorath Schist*	*5798*

Reservoir characteristics:

Number of reservoirs .... *Four sands, one dolomite, one limestone, two limestones*

Formations .... *Helez, Nir-Am, Brur*

Ages .... *Lower Cretaceous, Upper and Middle Jurassic*

Depths to tops of reservoirs .... *Helez Formation, 1480-1600 m; Upper Jurassic, 1600 m*

Gross thickness (top to bottom of producing interval) .... *100 m within Helez Formation; 140 m including top Jurassic*

Net thickness—total thickness of producing zones

Average .... *Sand pay zone, 15 m; Kokhav dolomite, 4 m*

Maximum .... *Sand pay zone, 20 m; Kokhav dolomite, 15 m*

Average

Maximum

Lithology

Middle (Helez K, W, A-Z pay zones) and lower (Kokhav B) sand members; orthoquartzitic, some subarkosic with volcanic fragments, some carbonate cement, porosity around 20%. The carbonate pay zone—reef complexes (Kokhav dolomite and Upper Jurassic Nir-Am Formation) and Brur calcarenite.

Porosity type .... *Intercrystalline and intergranular porosity*

Average porosity .... *16-32% (sand reservoirs)*

Average permeability .... *30-2000 md (sand reservoirs)*

Seals:

Upper

Formation, fault, or other feature .... *Helez Formation and faults*

Lithology .... *Shales*

Lateral

Formation, fault, or other feature .... *Permeability loss due to facies changes and pinchout*

Lithology .... *Shales*

Source:

Formation and age .... *Unconfirmed; thus far can only suggest a possible source or a conduit (Middle Jurassic Barnea Formation)*

Lithology .... *Limestone*

Average total organic carbon (TOC) *0.5%*

Maximum TOC *2.6%*

Kerogen type (I, II, or III) .... *II*

Vitrinite reflectance (maturation) .... $R_o = 0.52–0.75$ *(immature in sampled location)*

Time of hydrocarbon expulsion

Present depth to top of source .... *4000–5000 m*

Thickness

Potential yield

Appendix 2. Production Data

Field name .... *Helez-Brur-Kokhav*

Field size:

Proved acres .... *1250 ha*

Number of wells all years .... *82, of these 55 producers*

Current number of wells .... *9 producers*

Well spacing .... *Most on 50 ac, some 90 ac*

Ultimate recoverable .... *19 million bbl*

Cumulative production (1988) .... *16.5 million bbl*

Annual production (1988) .... *0.15 million bbl*

Present decline rate *15%*

Initial decline rate .... *10% (est.)*

Overall decline rate .... *15% (est.)*

Annual water production .... *0.115 million bbl*

In place, total reserves .... *50 million bbl*

In place, per acre-foot .... *712 bbl*

Primary recovery .... *19 million bbl*

Secondary recovery

Enhanced recovery .... *5–10 million bbl*

Cumulative water production .... *30 million bbl*

Drilling and casing practices:

Amount of surface casing set .... *250–300 m*

Casing program .... *13⅜-in. or 9⅝-in. to 300 m; 9⅝-in. or 7-in. to 1000–1250 m; 7-in. or 5½-in. to TD; 2⅞-in. tubing*

Drilling mud .... *Freshwater based*

Bit program .... *Varies*

High pressure zones .... *None*

Total loss circulation zone .... *Upper Cretaceous section 250-800 m drilling blind*

Completion practices:

Interval(s) perforated .... *Each pay zone separately*

Well treatment ..... *Acidized in dolomite and limestone, also in calcareous sandstone; scarcely fractured*

Formation evaluation:

Logging suites .... *Older wells: GR, sonic, neutron, resistivity, microlog; modern wells: GR, CNL, density, sonic, induction, dipmeter*

Testing practices .... *Typically production tested (mainly swabbing); frequently DST while drilling*

Mud logging techniques

Not typically done in uppermost 250-300 m and while drilling blind between 300 and 1000 m; deeper—continuous with gas detector and chromatograph

Oil characteristics:

Type .... *Normal hydrocarbon*

API gravity *27.5-31.0*

Base

Initial GOR .... *200-275 SCF/STB*

Sulfur, wt%

Viscosity, SUS .... *2.0 cp*

Pour point

Gas-oil distillate

Field characteristics:

Average elevation .... *1200-1300 m*

Initial pressure .... *2000 psi*

Present pressure .... *900-1100 psi*

Pressure gradient

Temperature .... *150°F*

Geothermal gradient

Drive .... *Water*

Oil column thickness .... *Multiple thin pay zones*

Oil-water contact .... *Multiple*

Connate water

Water salinity, TDS .... *55,000 ppm but variable*

Resistivity of water .... *0.065 ohm, 150°F*

Bulk volume water (%)

Transportation method and market for oil and gas:

Pipeline

Lindsborg Field—U.S.A.
Salina Basin, Kansas

K. DAVID NEWELL
Kansas Geological Survey
Lawrence, Kansas

FIELD CLASSIFICATION

BASIN: Salina
BASIN TYPE: Cratonic Sag
RESERVOIR ROCK TYPE: Dolomite, Limestone, and Sandstone
RESERVOIR ENVIRONMENT OF DEPOSITION: Subtidal, Open Marine, and Shoreface
RESERVOIR AGE: Ordovician
PETROLEUM TYPE: Oil
TRAP TYPE: Anticline with Stratigraphic Controls

LOCATION

The Lindsborg field (15.19 million bbl cumulative production; Beene, 1989) is a narrow north–south-trending field that straddles the McPherson and Saline County line in central Kansas, U.S.A. (Figure 1). It is the largest of a north-northeast–south-southwest-trending group of oil fields that is located on an ill-defined structural saddle between the prolific Sedgwick basin to the south and the relatively barren Salina basin to the north. Other major fields in this cluster include the Salina field (4.57 million bbl cumulative production; Beene, 1989), which is on-trend and to the north of the Lindsborg field, and the Smolan field (5.61 million bbl cumulative production; Beene, 1989). The Smolan field is on a geologic structure immediately west of the anticline on which the Lindsborg field is partly located (Figure 2).

The Lindsborg field has produced 15.19 million bbl of oil since its discovery in 1938 (Beene, 1989) and may have an ultimate recovery of approximately 16 million bbl of oil. Total areal extent of the field is approximately 10,080 ac. It presently is being produced by several small independent oil companies. Cumulatively, 318 wells have been drilled to produce oil from this field since its discovery in 1938. In 1988, 74 wells were still active, averaging 2.67 BOPD per well.

HISTORY

Pre-Discovery

The anticlinal trend on which the Lindsborg field is partly located was first detected with a magnetometer survey by Dixie Oil Company in 1928 (Figure 3A) (Lee H. Cornell, Wichita, KS, personal communication). Dixie (later to become part of Standard Oil Company of Indiana) then followed with a shallow-core-drill survey in 1929 (Figure 3B) in order to confirm the possible presence of a geologic structure indicated by the magnetometer map. At this time, core drilling was a very common exploration technique used to find subtle near-surface expressions of structures in this region. Core drilling was particularly successful in finding structural traps farther south in central and southern McPherson County. Koester (1934) reports that over 50 core drills were in operation in McPherson County alone during 1930, and most all the oil fields in the county were defined by core drilling prior to development.

Core drilling was not as immediately successful in Saline County as it was in McPherson County. Based on their core-drill survey, Dixie Oil Company in 1929 drilled the #1 Muir, a wildcat well in SW SE SW sec 6-T.15S.-R.2W. to first test the Lindsborg anticline (Figure 3). This well was a dry hole but had a show of oil in Ordovician Maquoketa–Viola carbonate rocks at 3290 ft (1003 m). This well location was eventually encompassed in the productive area of the Salina field that was discovered 14 years later in 1943. Dixie drilled a second test in 1929 6.5 mi (10.5 km) farther south-southwest (Figure 3). This well, the #1 Olsson in NW NW NE sec 10-T.16S.-R.3W., encountered oil at 3305 ft (1008 m) in Upper Ordovician Maquoketa dolomite and is credited as being the discovery well for the 1.47 million bbl Olsson field (Figures 1 and 2), but it pumped only about 5 bbl a day in initial tests and was abandoned in late 1930 after producing only 1405 bbl of oil (Folger and Hall, 1933). The Olsson field, which produces oil from both Viola Limestone and Maquoketa dolomite, subsequently did not undergo major development until the late 1940s and early 1950s. Two additional tests, the Dixie Oil Company #1 Miller in NE SE SE sec 9-T.16S.-R.3W. and the #1 Martin

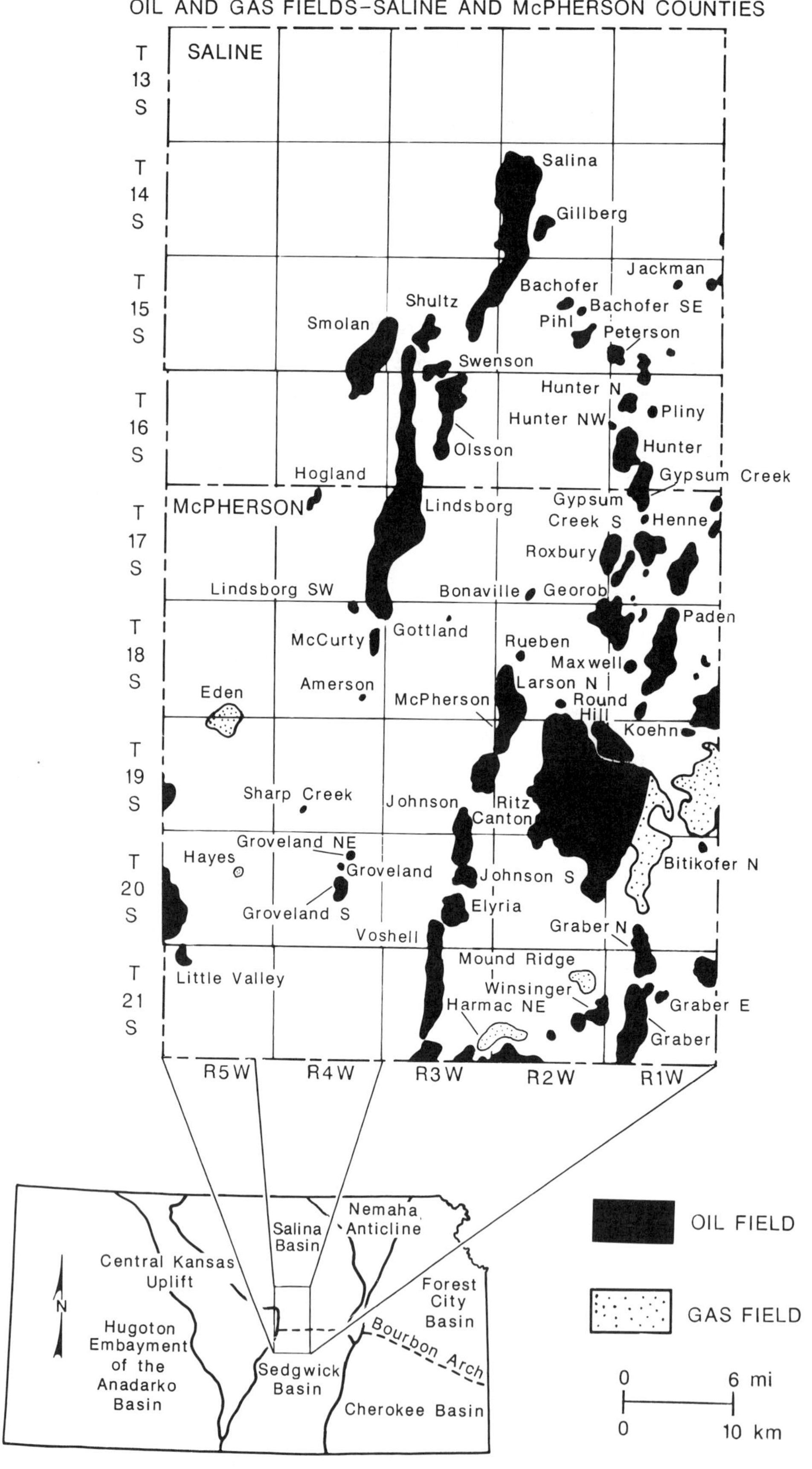

Figure 1. Location of study area in context with major Late Mississippian and Early Pennsylvanian tectonic features in Kansas. Locations of oil and gas fields are from Paul et al. (1982).

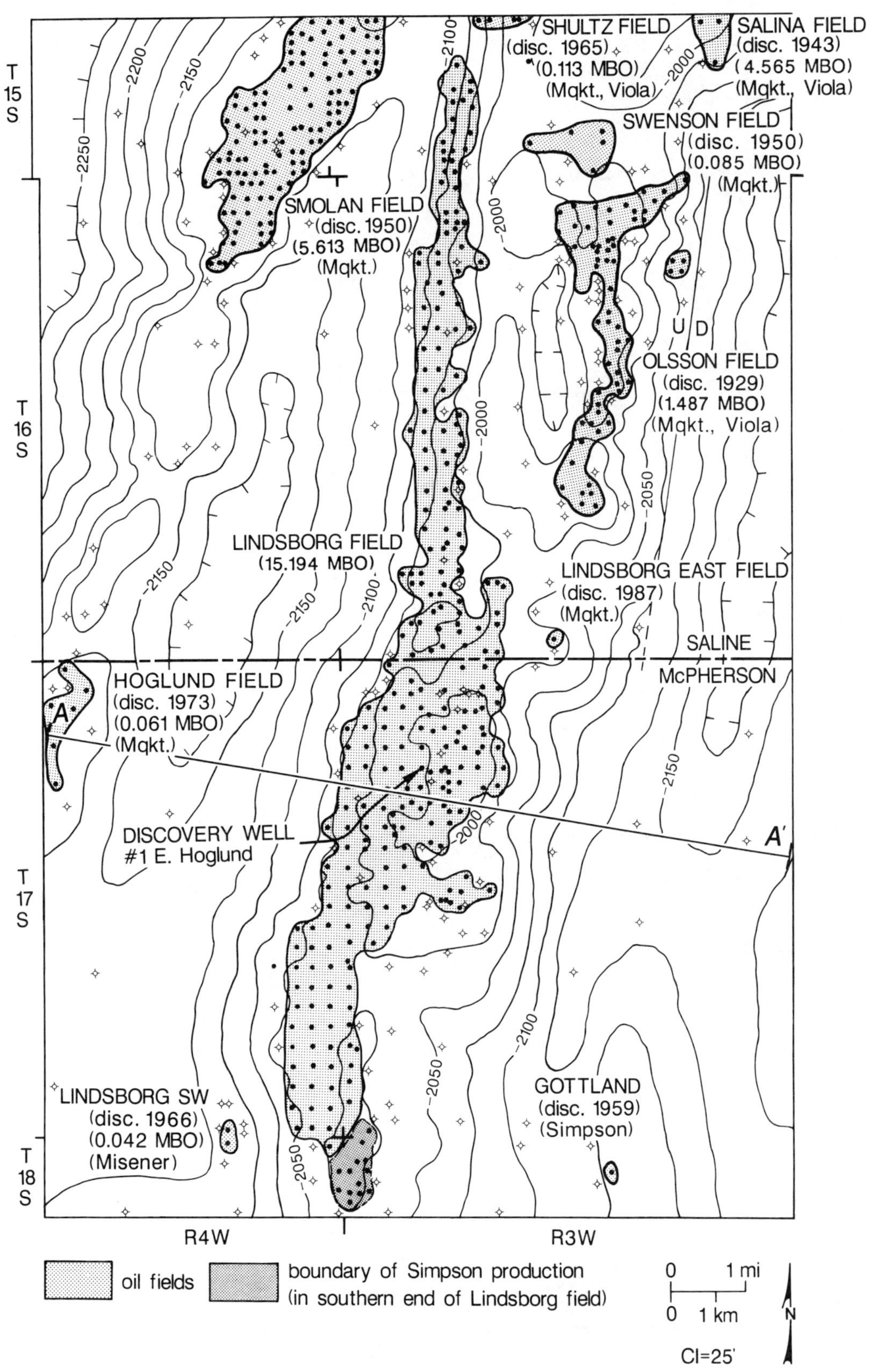

Figure 2. Structure map of Maquoketa dolomite in vicinity of Lindsborg field (modified from Thatcher, 1961) with outlines of production for Lindsborg and nearby oil fields. See Figure 13 for limits of production for each pay zone in the central part of the Lindsborg field. Cumulative production of nearby oil fields from Beene (1989).

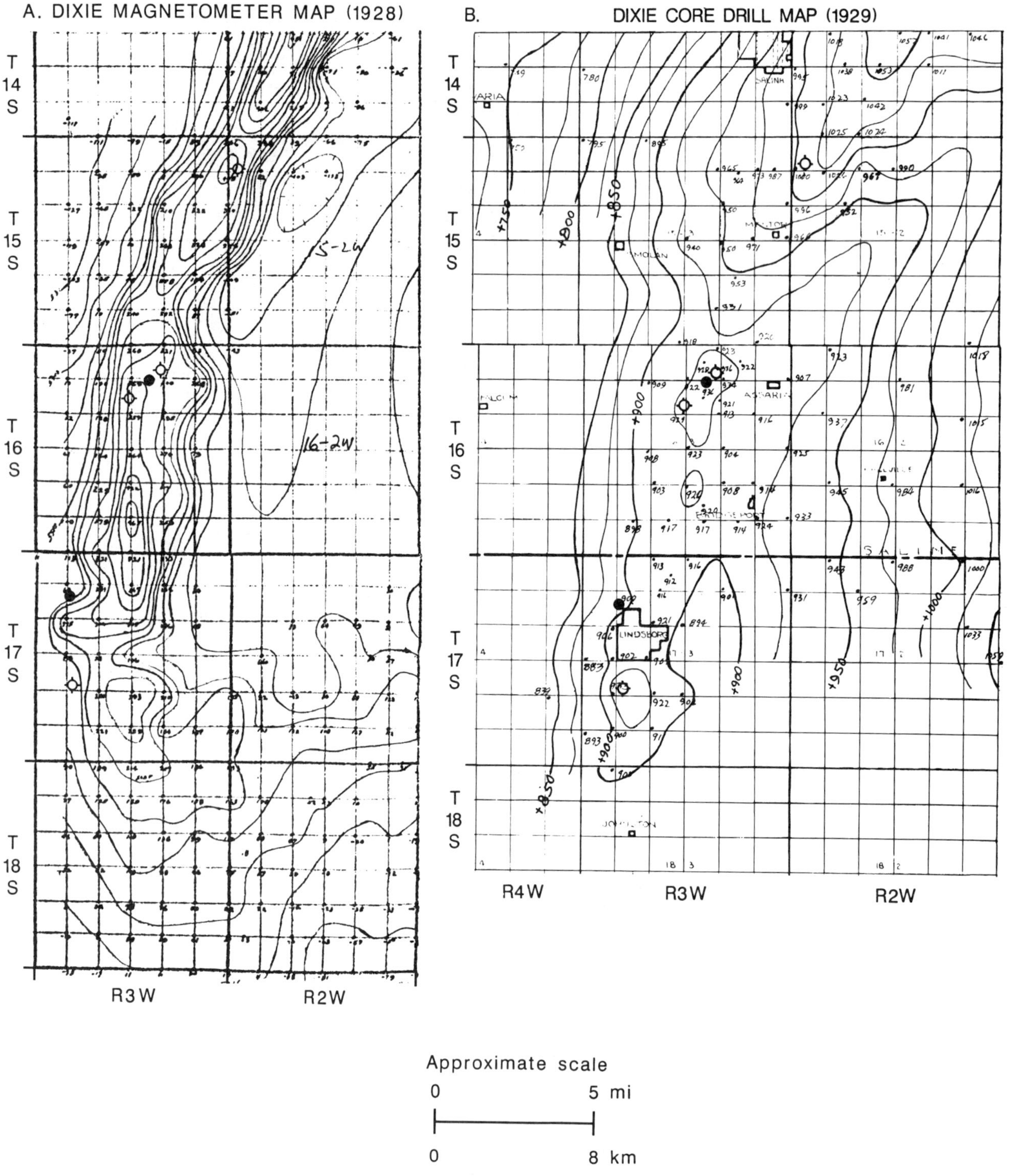

A. DIXIE MAGNETOMETER MAP (1928)

T 14 S

T 15 S

T 16 S

T 17 S

T 18 S

R3W R2W

B. DIXIE CORE DRILL MAP (1929)

T 14 S

T 15 S

T 16 S

T 17 S

T 18 S

R4W R3W R2W

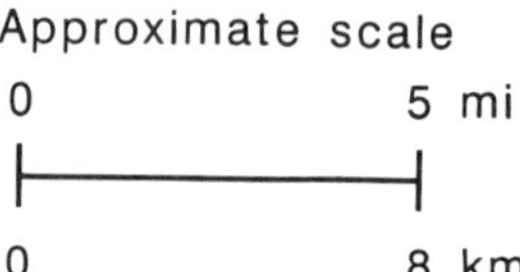

Figure 3. (A) Segment of a magnetic intensity map made in 1928 from a Dixie Oil Company survey. Measurements were made with a surface magnetometer and readings were taken along roads at section intersections that are spaced about 1 mi (1.6 km) apart. Map courtesy of its author, Lee H. Cornell (Wichita, KS). (B) Segment of 1929 core drill map from Dixie Oil Company. Contour interval is 25 ft. Surface mapped is the top of the Hollenberg Limestone Member of the Permian Sumner Group, which is at a depth of 200-600 ft (60-180 m) in the vicinity of Lindsborg, KS. Locations of the five early wells drilled by Dixie Oil Company are shown on each map. Both maps are displayed at approximately the same scale.

in NE SE SE sec 3-T.16S.-R.3W., were attempted in 1930 and 1931, respectively, to extend the production found at the #1 Olsson well (Figure 3). Results of these wells and another drilled by Dixie in 1930 in SW SW sec 20-T.17S.-R.3W. (Figure 3) were not encouraging, so exploration efforts in the area languished for most of the 1930s.

In mid-1937, a regional correlation-point reflection seismographic survey of the Salina basin was conducted by Carter Oil Company (later to become part of Exxon Company, U.S.A.). At that time, the Salina basin had only been recognized for a few years (it was first publicly reported in Barwick, 1928) and its petroleum potential was still very much unassessed. The seismic survey used dynamite for a source and utilized wells penetrating the Viola or deeper formations for stratigraphic control (Brewer, 1959). Interpretation of geologic structure was done by mapping on good reflectors such as the one that originates at the base of the Upper Ordovician Maquoketa shale and at the top of the subjacent carbonate section composed of Maquoketa dolomite and Middle Ordovician Viola Limestone.

Discovery

A seismic closure was defined on the Lindsborg anticline just north of the town of Lindsborg on the basis of the correlation-point seismic data (Figure 4) and was subsequently tested by the Carter Oil Company #1 E. Hoglund in W2 SW NW sec 8-T.17S.-R.3W. On 2 February 1938, this well became the discovery well for the Lindsborg field when it tested 345 bbl of 36° gravity oil and approximately 50 bbl of water a day from the Maquoketa dolomite at 3347–3354 ft (1022–1023 m).

Initial development of the field was rather slow with only three offset wells subsequently drilled—one in 1938, one in 1940, and one in 1941 (Thatcher, 1961). At this time, Carter Oil Company did not feel that the rate of pay-out from the field would justify its investment, so it farmed out part of its interest to M&L Oil Company, an independent (Lee H. Cornell, personal communication). Ver Wiebe (1943) reports that the field had a significant revival in 1942 when offset wells drilled by M&L Oil Company came in as "good producers." Scout cards indicate that Maquoketa-producing wells drilled that year by M&L had initial potentials of up to 883 BOPD.

In July 1942, the Auto Ordnance #1 Johnson well in W2 SW NE sec 8-T.17S.-R.3W. opened a new pay zone at 3447–3460 ft (1051–1055 m) from sandstone in the Middle Ordovician Simpson Group. Initial potential from the Simpson pay zone was reported to be 2247 BOPD. Initial production potentials from this new pay zone were quite high. The first seven Simpson producers completed in 1942 reported initial potentials averaging 1922 BOPD.

In 1944(?), the Viola Limestone was added as a third producing zone in the field. The discovery well for this zone is unclear because early wells usually identify the overlying Maquoketa dolomite as Viola. The earliest well that unequivocally produced oil from the Viola Limestone appears to be the #1 Farmers Bank in W2 NW SW sec 17-T.17S.-R.3W., which was drilled to the Maquoketa in 1942 and subsequently deepened to the Simpson in 1944 by Bay Petroleum (later to be part of Tenneco). This well reported 663 bbl a day initial potential production from 3391.5–3392.5 ft (1034.4–1034.7 m).

Post-Discovery

As field development progressed and limits of production became established, it became evident that the Viola and Simpson oil pools are structurally trapped in a culmination on the Lindsborg anticline. However, most of the Maquoketa production is present in off-structure positions on the west flank of the anticline, hence Maquoketa oil is concentrated in a structural-stratigraphic trap (Figure 2).

A second phase of exploration occurred in the early 1950s (Figure 5) with the discovery of several small fields due north of the original Lindsborg field. (These small fields were subsequently found to be contiguous with and were assimilated into the Lindsborg field.) In July 1951, Musgrove Petroleum Company brought in the #1 Holm as the discovery well for the Holm field in NW NW NW sec 32-T.16S.-R.3W. This well was drilled on acreage farmed out from Bay Petroleum (Al Abercrombie, Wichita, KS, personal communication) and initially produced 207 BOPD from Maquoketa dolomite at 3406–3412 ft (1039–1041 m). This was quickly followed by the discovery of the Holm Southeast field in January 1952 by the Bay Petroleum Company #1 Holt in NW SE sec 32-T.16S.-R.3W., which produced 17 BOPD from Maquoketa dolomite at 3388–3398 ft (1033–1036 m). National Associated Petroleum Company followed in May 1952 with the discovery of the Holm North field by the #1 Nelson in NE SW sec 20-T.16S.-R.3W., which initially produced 44 BOPD from Maquoketa dolomite at 3427–3437 ft (1045–1048 m). A fourth discovery, dubbed the Salemsborg field, was made at SW SW SW sec 5-T.16S.-R.3W. in November 1952 by the Phillips and Sanderson #1 Johnson. This well initially produced 120 BOPD from Maquoketa dolomite at 3436–3442 ft (1048–1050 m). The reason why these exploration wells were drilled is not readily apparent, but they were probably "acreage plays" in which oil companies farmed out a portion of their leased acreage to other companies (usually small independents) agreeing to drill a test well strategically located on that acreage (Lee H. Cornell, personal communication; Al Abercrombie, personal communication). Geological and geophysical reasoning may play only a minor part in placing wells in acreage plays. In particular, the Phillips and Sanderson #1 Johnson well, the discovery well for the Salemsborg field, was promoted as an offset to the then recently discovered Smolan field about 2 mi (3.2 km) to the northwest (Dale Reese, Salina, KS, personal communication).

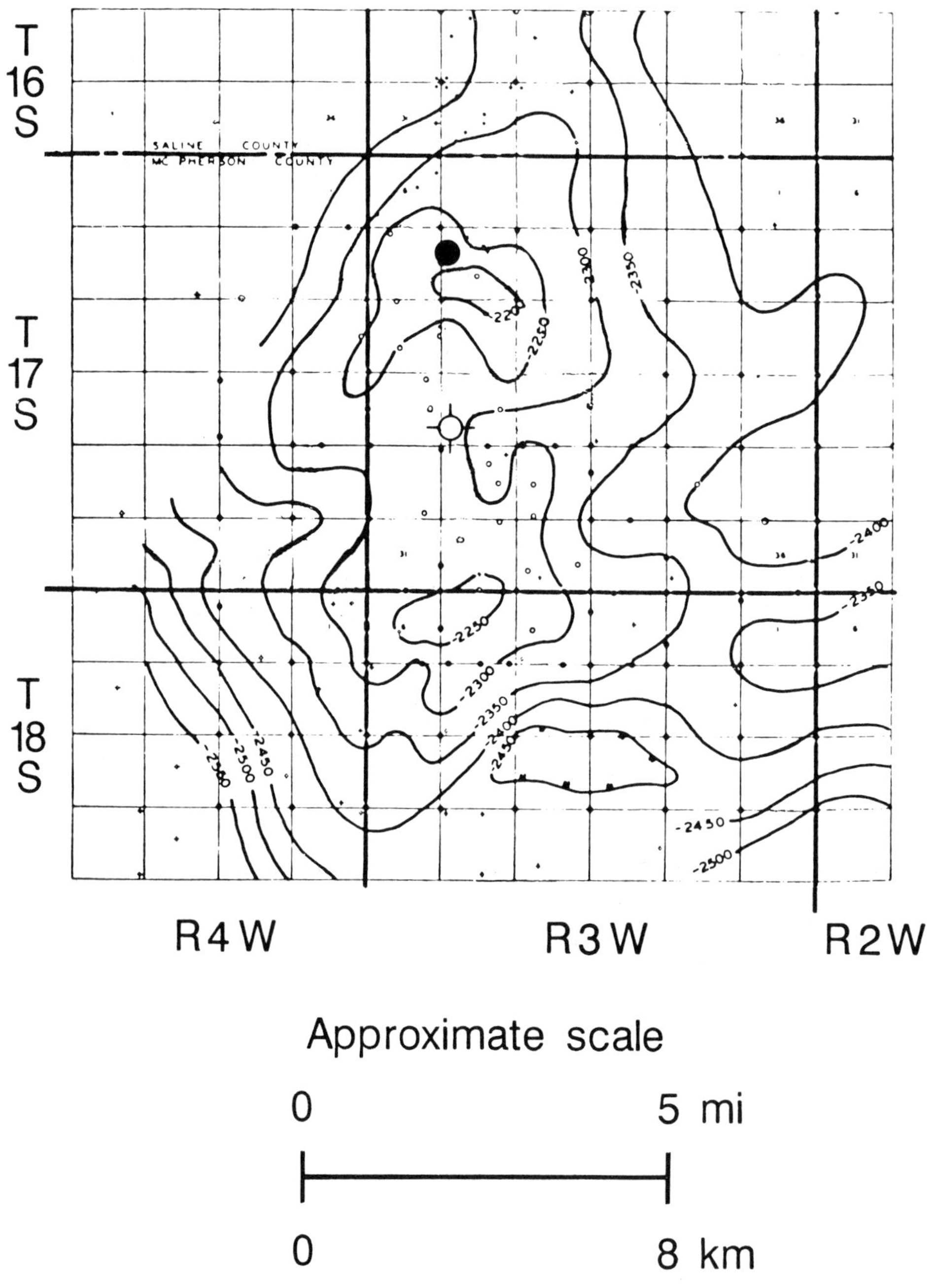

Figure 4. Seismic contour map made by Carter Oil Company in 1937 on top of Maquoketa-Viola carbonate rocks showing closure associated with Lindsborg anticline (from Brewer, 1959). Contour interval is 50 ft (15 m). Locations of Carter Oil Company discovery well and early Dixie Oil Company test wells are shown. (Also see Figure 3.)

With subsequent development drilling along this north-south-trending group of discoveries, it quickly became apparent that a common pay zone in a single oil accumulation was being exploited, so in 1953 these fields and other nearby subsidiary discoveries were combined and subsequently identified as the Salemsborg field (Ver Wiebe et al., 1954). With additional outpost drilling, the Salemsborg field was combined with the Lindsborg field in 1957 (Goebel et al., 1958).

One other field has since been amalgamated with the Lindsborg field. In January 1955, the Rupp-Ferguson Oil Company #1 Landgren in SW SW SW sec 6-T.18S.-R.3W. discovered oil in Simpson sandstone at a depth of 3525–3527 ft (1075–1076 m). This well had an initial pump rate of 25 bbl a day

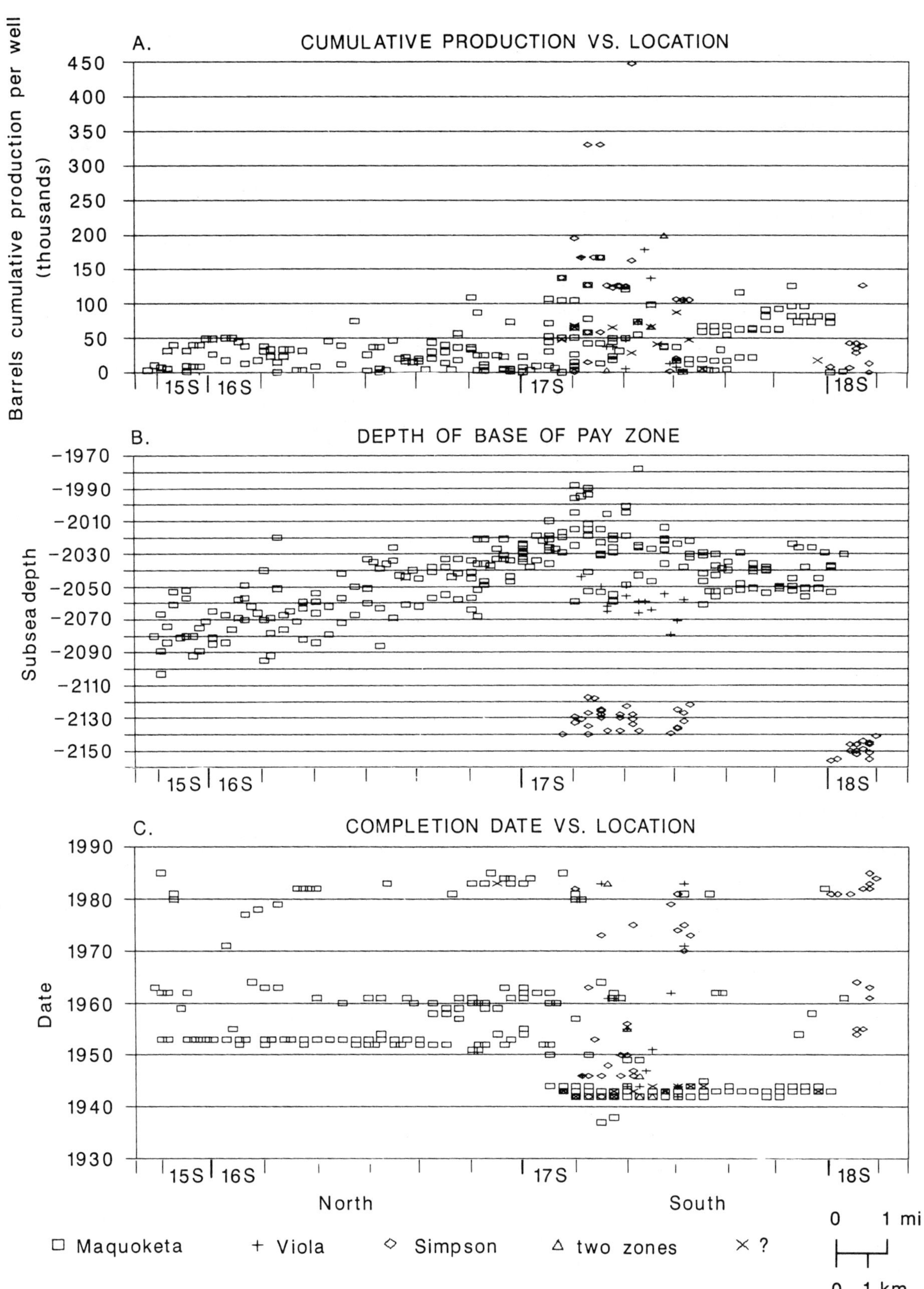

Figure 5. Summaries of various attributes of Lindsborg field on a north-south-trending projection. Each data point represents a single well. The pay zone in which each well is completed is differentiated according to the type of symbol. This information is projected onto a north-south-trending plane regardless of the position of the well along the relatively narrow width of the field. Figure shows (A) cumulative production of wells versus location; (B) subsea depth of base of produced interval versus location; and (C) completion date of wells versus location.

and was the discovery well for the Lindsborg South field. Subsequent wells in the area proved that Maquoketa production of the Lindsborg field geographically overlapped Simpson production of the Lindsborg South field, hence in 1967 the Lindsborg South field was combined with the Lindsborg field (see Figure 2).

Most early wells in the field were completed open-hole with acidization being a common practice to augment production. After 1952, fracturing procedures sometimes used with acidization became common completion practices, particularly in the Maquoketa and Viola carbonates. Simpson sandstone is usually not treated, but a small number of Simpson wells have been acidized. Wells drilled after 1961 were commonly completed by perforating through casing. Generally, approximately 150 ft (45 m) of surface casing (13⅜ in.) is set to the base of Holocene river alluvium and underlying unconsolidated gravels of the Pliocene-Pleistocene "Equus beds." A second string of casing (8⅝ in.) is then set to approximately 350 ft (110 m) to the base of the Hutchinson Salt Member of the Permian Wellington Formation. 5½-in. casing is then set to total depth. Most completions are single-zone completions. Fewer than 5% of the wells in the field are completed in more than one zone (Figure 6).

Older logging suites usually include a resistivity survey with spontaneous potential log, and neutron with gamma ray. Compensated neutron and density logs are now commonly run on newer wells in the field in addition to a resistivity survey. Logs are not run on many production wells. Abbreviated log surveys from the basal part of the Pennsylvanian System or Chattanooga Shale on down to Total Depth also are quite common. Sonic logs and the relatively new compensated neutron and density logs with photoelectric cross-section index (Pe) have only rarely been used in this area of Kansas.

Pressure records from drill-stem tests indicate the Maquoketa pay zone had an initial bottom-hole pressure of approximately 1300 psi (9064.5 kPa) while the Simpson recorded approximately 1525 psi (10,615.8 kPa). Both zones have recorded bottom-hole pressure in drill-stem tests of approximately 350 psi (2514.5 kPa).

Little to no gas is produced with the oil from the three pay zones in the field, but water is produced with oil from all the zones. Production from perimeter wells in particular has generally been poor and water-producing (Ver Wiebe, 1945). The reported initial water-cut in production wells has increased with time. Some wells presently produce greater than 98% water (Dale Reese, personal communication). Water produced in the field has been reinjected into the pay zones, Permian units, Pennsylvanian Lansing-Kansas City groups, and Cambrian-Ordovician Arbuckle Group that underlie the Simpson Group.

Although there is no set spacing pattern in the Lindsborg field, wells are set approximately every 40 ac (16 ha.) (Figure 2). A 10 ac (4 ha.) spacing, similar to that of the nearby Smolan field (Talbott, 1954), is present in the extreme northern end of the field. The first secondary recovery operations started in 1947 and have since been initiated in several localities throughout the field. Most of these operations have been dump floods that reinject formation water recovered during production. Water from near-surface Pliocene-Pleistocene "Equus beds" has also been utilized for injection.

The greatest annual production of the Lindsborg field, 943,680 bbl, occurred in 1944 (Figure 7). Latest production figures in 1986 (Beene, 1989) indicate an annual production of 72,052 bbl. Over its 50 year history, the field has averaged a 7% decline in production per year. Early production declines in the mid-1940s after the year of peak production were as high as 30%. Although most production in the field has been attributed to the Maquoketa dolomite, individual wells producing oil from the Simpson have on the average been much more prolific and have accounted for an increasing percentage of production with time (Figures 5 and 7). Average cumulative production for all Maquoketa wells in the field is 43,800 bbl per well; Viola wells average 48,300 bbl per well, whereas Simpson wells average 98,300 bbl per well.

Cumulative and initial production per well from Maquoketa dolomite in the Salemsborg sector of the field (principally the northern half of the field in T.15S. and T.16S.) has generally been less than that of the same pay zone in the Lindsborg sector (wells in T.17S. and T.18S.) (Figure 5). The Maquoketa pay zone in the Salemsborg sector is farther down structure than in the Lindsborg sector (Figures 2 and 5), hence wells in the Salemsborg sector experience faster water encroachment than structurally higher wells in the Lindsborg sector. Wells in the Lindsborg sector also had greater initial production rates than wells in the Salemsborg sector of the field. This may indicate that Maquoketa dolomite has inherently better permeability and porosity nearer the crest of the Lindsborg anticline. Talbott (1954) also reports the Smolan field, a Maquoketa oil accumulation discovered in 1950 on the crest of a subparallel-trending anticline immediately west of the Lindsborg anticline (Figure 2), had several wells with draw-down potentials of over 1000 bbl a day—considerably greater than nearby structurally higher (but off-structure) Maquoketa wells in the Salemsborg sector of the Lindsborg field.

DISCOVERY METHOD

The geologic concept involved in this discovery was the search for anticlinal closures in a topographically relatively flat area having no surface expression of deeper geologic structures. The magnetometer survey was an efficient and easy method for surveying large areas for more detailed follow-up geologic studies. Core drilling was an effective method for exploring for structural trends in central

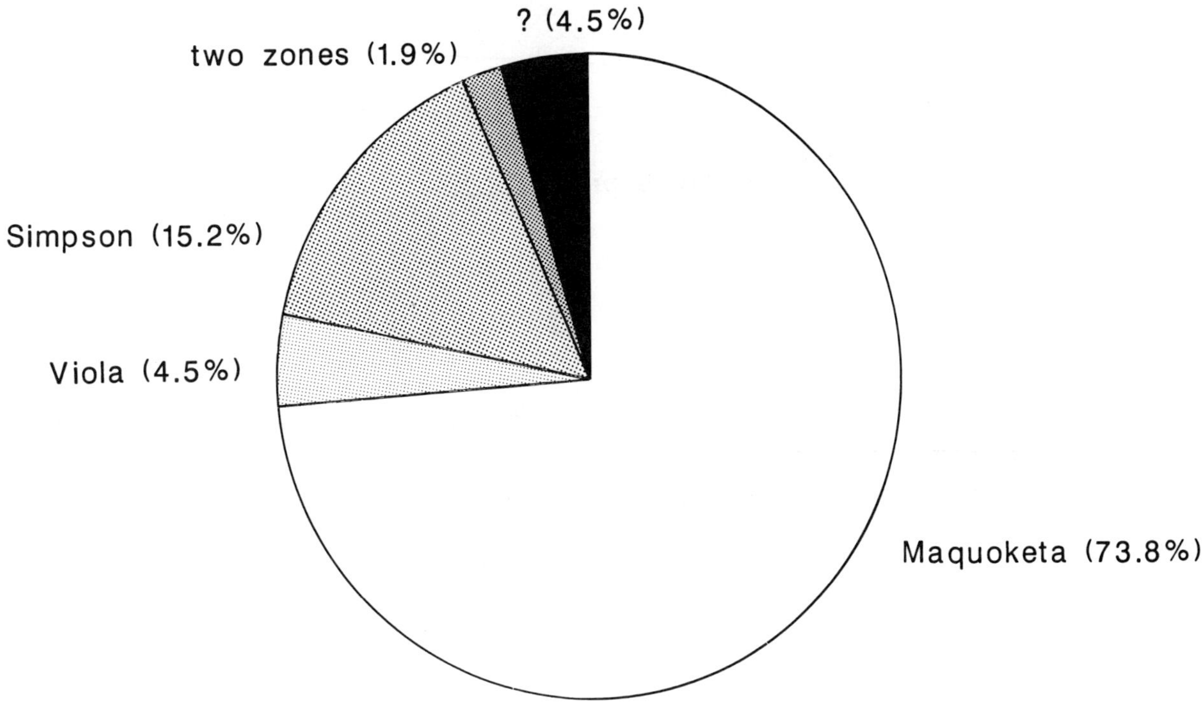

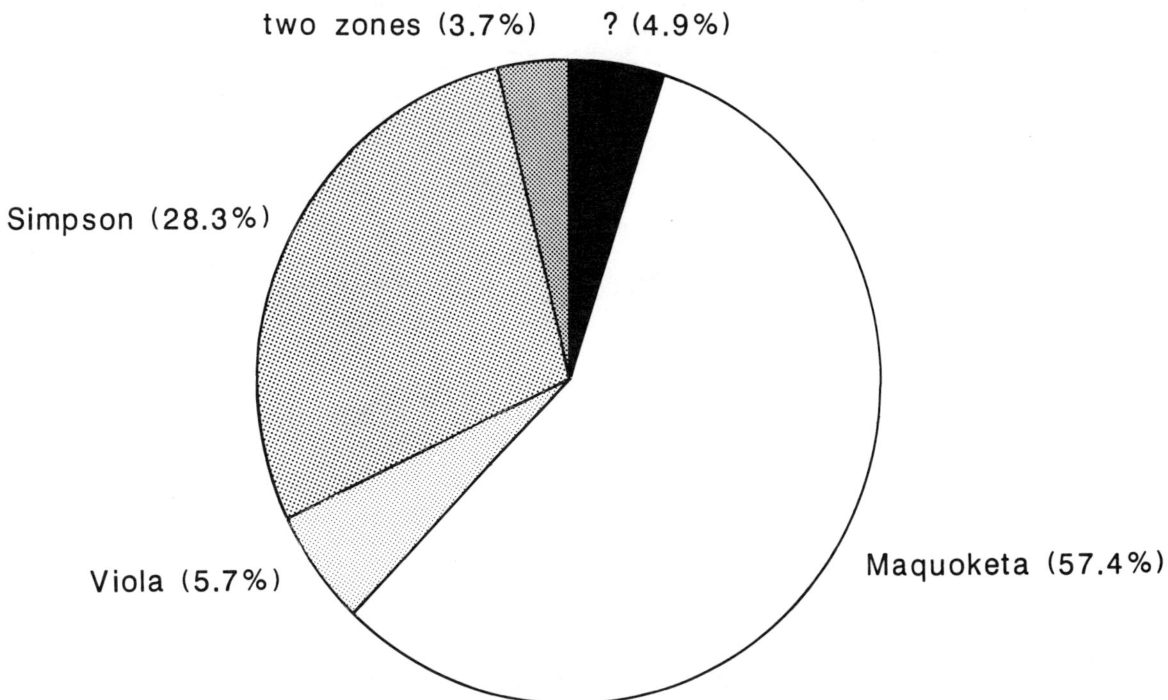

Figure 6. Pie diagrams illustrating percentages of wells completed in each pay zone and percentages of production of entire field from each pay zone.

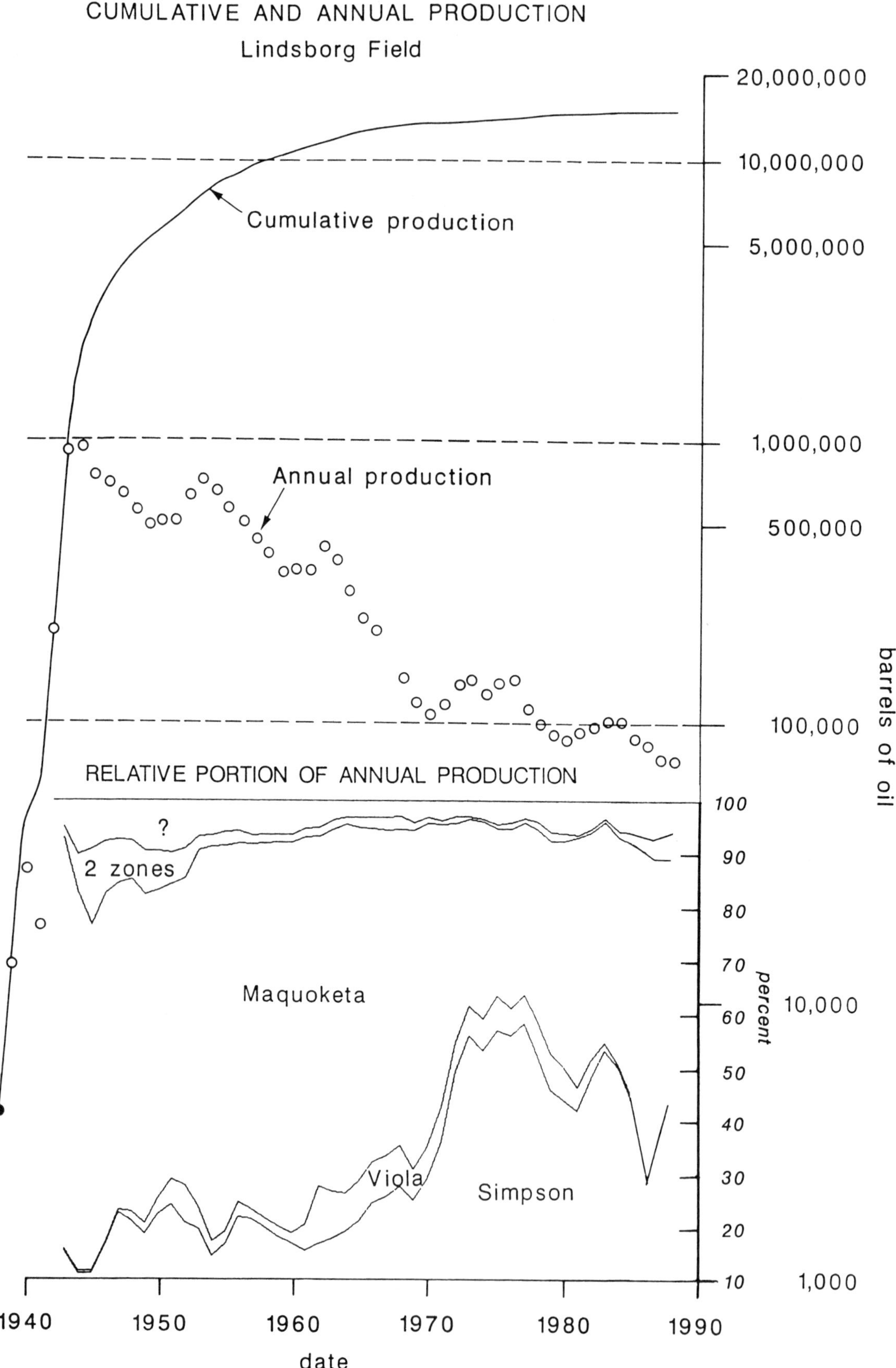

Figure 7. Cumulative and annual production of Lindsborg field (including component fields before they were unified into the Lindsborg field) and relative importance of each pay zone with time.

McPherson County to the south of the Lindsborg field. In this area, limestone beds in the Permian Sumner Group were correlated and mapped by shallow core-drill surveys (Hiestand, 1933). Subtle low-amplitude anticlines or a flattening in the gentle westerly dip of near-surface beds were taken as possible indications of deep structures. In the vicinity of the Lindsborg field, the alluvium-filled valley of the Smoky Hill River and an older buried Pleistocene channel called the McPherson Valley (Lohman, 1942) reduced the effectiveness of core drilling for defining the precise location of culminations on the Lindsborg anticline. The reflection seismic method used by Carter Oil Company, although primitive by today's standards, was quite effective for finalizing locations of wildcat wells after the core-drill survey detected the anticline.

Today's exploration methods would probably have similar success in finding the structurally controlled oil accumulations in the Simpson Sandstone and Viola Limestone. However, the stratigraphically entrapped oil in the thin pay zone at the top of the Maquoketa dolomite still would be difficult to detect prior to drilling, even by current relatively advanced seismic methods. Serendipity plays a part in many discoveries, and the Maquoketa production in the Lindsborg field is no exception.

STRUCTURE

The Lindsborg field is in a broad structural saddle between the Salina and Sedgwick basins. The Salina basin, according to the Bally and Snelson (1980) structural classification of petroleum provinces, is a 1211 basin (i.e., cratonic basin located on an earlier rifted graben). It is categorized as an I basin (cratonic interior basin) according to precepts in Klemme (1971).

The Salina basin, Sedgwick basin, and nearby Central Kansas and Nemaha uplifts (Figure 1) largely were defined as tectonic entities in Late Mississippian to Early Pennsylvanian time (Lee, 1956; Merriam, 1963). Structural development of these major features and smaller structural trends such as the Lindsborg anticline, which parallels the Nemaha uplift, are probably manifestations of the Ouachita orogeny that occurred farther south in Oklahoma.

The unconformity at the base of Pennsylvanian System is an angular unconformity that records most of the movement that formed the more pronounced present-day local structural features. Subsequent minor movement can be recorded at least up through Permian time because core drilling can detect near-surface manifestations of the Lindsborg anticline and nearby parallel-trending McPherson anticline in shallow Permian beds (Hiestand, 1933; Brewer, 1959). Post-Permian structural movement principally involves a component of gentle westward tilt of the entire region and is associated with subsidence of the Denver-Julesburg basin to the west in Colorado (Jewett and Merriam, 1959).

Mild pre-Pennsylvanian activity of the south-southwest-plunging Lindsborg anticline is also recorded as early as Late Devonian to Early Mississippian time by subcrop patterns beneath the angular unconformity at the base of the Chattanooga Shale (Newell, 1987). During deposition of the Ordovician System, the structural grain trended northwest-southeast. At that time, the Salina basin and Forest City basin in northeast Kansas were combined as a large shallow basin called the North Kansas basin (Lee, 1956; Merriam, 1963). The North Kansas basin was separated from the ancestral Anadarko basin by a broad northwest-southeast-trending arch called the Central Kansas arch. This feature extended from the present-day area of the Central Kansas uplift to southeastern Kansas (Merriam, 1963). The Lindsborg area was located on the northeastern flank of the Central Kansas arch, and the northwest-southeast trends that dominated this period of geologic history are manifested on an isopach map of the Maquoketa dolomite (Figure 8).

Precambrian structural features may have played a key role in controlling the location of younger Paleozoic structural features in this part of the mid-continent. Computer-processed magnetic data by Yarger (1983) indicate a strong north-northeast-south-southwest-trending magnetic anomaly spatially coincident with the Lindsborg anticline (Figure 9). This anomaly is the same one detected by the 1928 Dixie Oil Company magnetometer survey (Figure 3A) and is probably due to faulting or intrusive igneous activity occurring about 1100 m.y.b.p. in association with the formation of the Central North American Rift System (cf. Ocola and Meyer, 1973). In the vicinity of the Lindsborg field, the Precambrian surface is approximately 450 ft (140 m) below the base of the Simpson Group. The structural discontinuity associated with this magnetic anomaly could have been intermittently reactivated to accommodate new geologic stresses, particularly those associated with the Ouachita orogeny.

Regional subsurface geologic mapping on the top of the Simpson Group (Figure 10) indicates culminations on the south-southwest-plunging Lindsborg anticline. The nearby McPherson anticline and the Lindsborg anticline generally are north-south-trending structural features. Conversely, northeast-southwest-trending segments on these anticlines are associated with structural saddles. This system of en echelon anticlinal culminations is compatible with a component of north-northeast-south-southwest dextral strike-slip motion along possible underlying Precambrian fault systems associated with both anticlinal trends. Although vertical slip may be the dominant movement on these faults (particularly with the fault associated with the McPherson anticline that shows consistent down-to-the-west offset), a component of dextral slip would cause

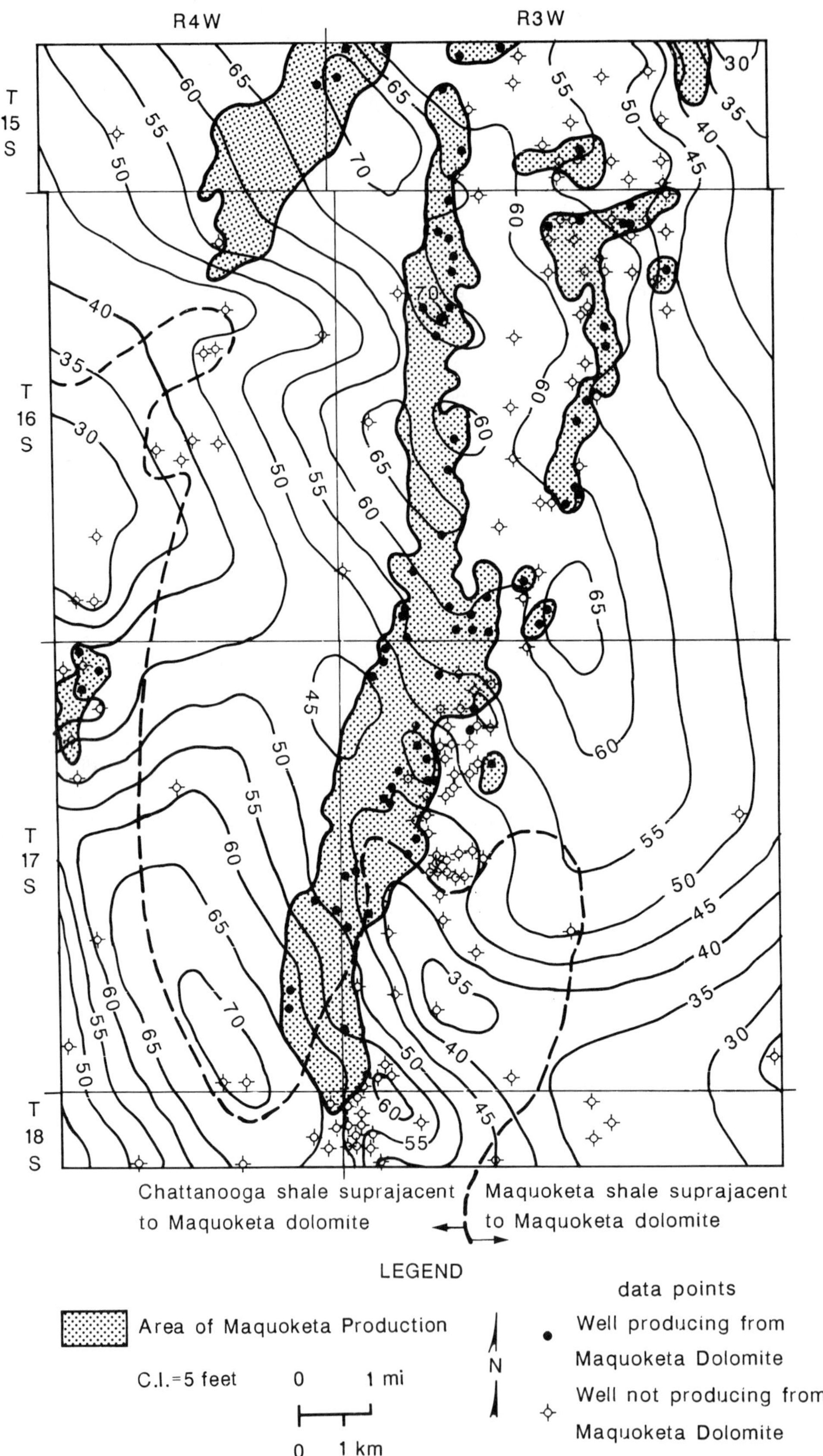

Figure 8. Isopach map of Maquoketa dolomite, identification of strata suprajacent to the Maquoketa dolomite, and present distribution of production in Maquoketa dolomite.

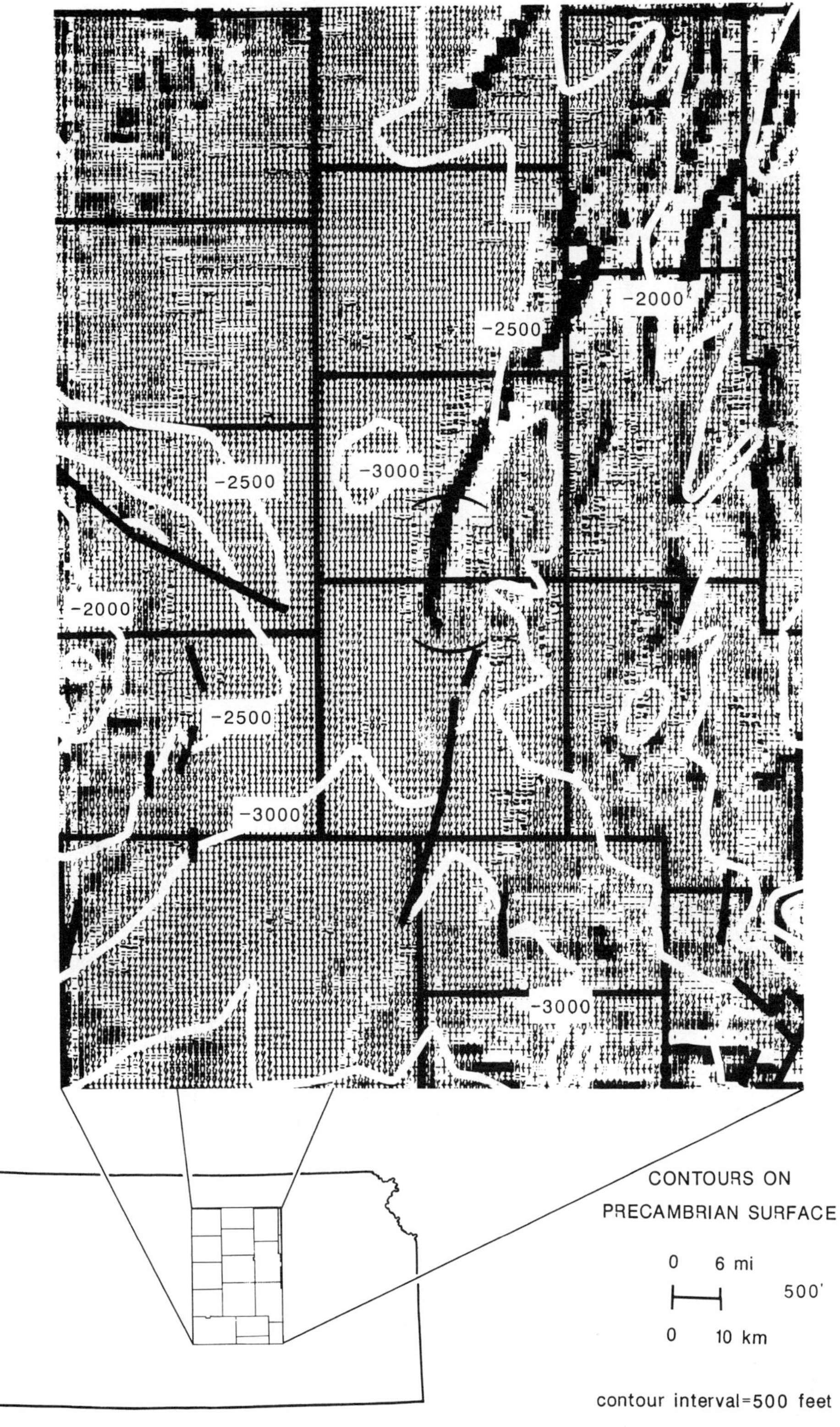

Figure 9. Shaded second vertical-derivative of total magnetic field intensity map of central Kansas (from Yarger, 1983) with contours on Precambrian surface superimposed (from Cole, 1976). This type of data processing accentuates the magnetic signatures of basement features. Note some structural discontinuities have high magnetic intensities and show up as heavily shadowed lines. One of these structural discontinuities underlies the Lindsborg anticline (brackets near center of map note location of Lindsborg field). Area of thick sediments of the Central North American Rift System shows up as a broad magnetically featureless area over several counties.

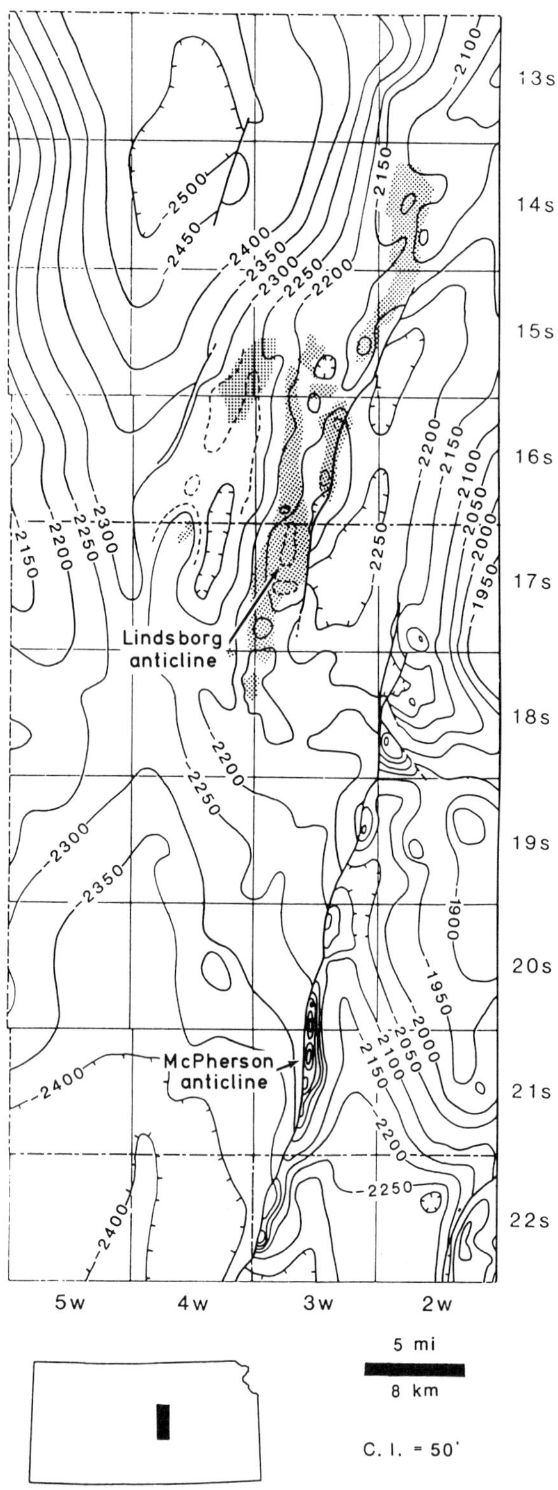

Figure 10. Regional structure map on top of Simpson Group in vicinity of Lindsborg field. North-south-trending fault segments on Lindsborg and nearby McPherson anticlines have closure associated with them. Localization of closures on these north-south-trending segments may be due to a possible component of right-lateral movement. Lindsborg field and nearby fields with production from Middle and Upper Ordovician rocks are shown for perspective. See Figures 1 and 2 for identification of each field.

north-south-trending fault segments to undergo slight compression and consequent uplift, thereby causing a structural closure at that locality. Northeast-southwest-trending segments would experience little or no compression or possibly even slight extension and consequently less uplift. Repeated stratigraphic sections drilled during development of the Voshell field (see Figure 10) on the McPherson anticline are indicative of reverse faulting (Hiestand, 1933). This reverse faulting may be another manifestation of compression along the north-south-trending fault segments.

STRATIGRAPHY

A study of ultimate oil recovery in Kansas (Adler, 1971) indicates it will be from the following rocks: Pennsylvanian rocks, 34%; Mississippian, 14%; Middle and Upper Devonian, 1%; Middle and Upper Ordovician, 11%; and Lower Ordovician, 40%. Roughly 65% of the Pennsylvanian production will be from sandstones in the Cherokee, Marmaton, and Pleasanton groups in the Cherokee basin of eastern Kansas (Figure 11). Carbonate reservoirs in the Arbuckle Group on the Central Kansas uplift will account for about 90% of the oil production from Lower Ordovician rocks. As of 1983, more than 5 billion bbl of oil have been produced in Kansas since production began in the 1860s (Watney and Paul, 1983). Original oil in place as of 1983 is estimated to be 16.6 billion bbl of oil (Watney and Paul, 1983). Most of this production is relatively shallow—from 2000 to 4000 ft (600 to 1200 m) in depth.

Most gas reserves and production in Kansas are concentrated in southwestern and south-central parts of the state. The largest fields are the giant Hugoton and Panoma fields in the Hugoton embayment of the Anadarko basin. The Hugoton field produces from carbonates in the Permian Chase Group, and the underlying Panoma field produces from carbonates in the Permian Council Grove Group. Total Kansas production from these two fields through 1986 respectively totaled 18.39 and 1.19 tcf (Newell et al., 1987a).

The most prolific and geographically widespread oil reservoir in the Sedgwick basin is Mississippian "chat"—a cherty weathered zone developed on pre-Chesterian Mississippian limestones subjacent to the angular unconformity at the base of the Pennsylvanian System (Merriam and Goebel, 1959; Newell et al., 1987a). Either Mississippian or Middle-Upper Ordovician rocks constitute the most important pay zones in the Salina basin, depending on where the east-west-trending boundary is drawn between the Salina and Sedgwick basins. A Mississippian production trend extends northward through eastern McPherson County and into eastern Salina County (Figure 12). A Middle-Upper Ordovician trend primarily attributed to reservoirs developed in Viola limestone and

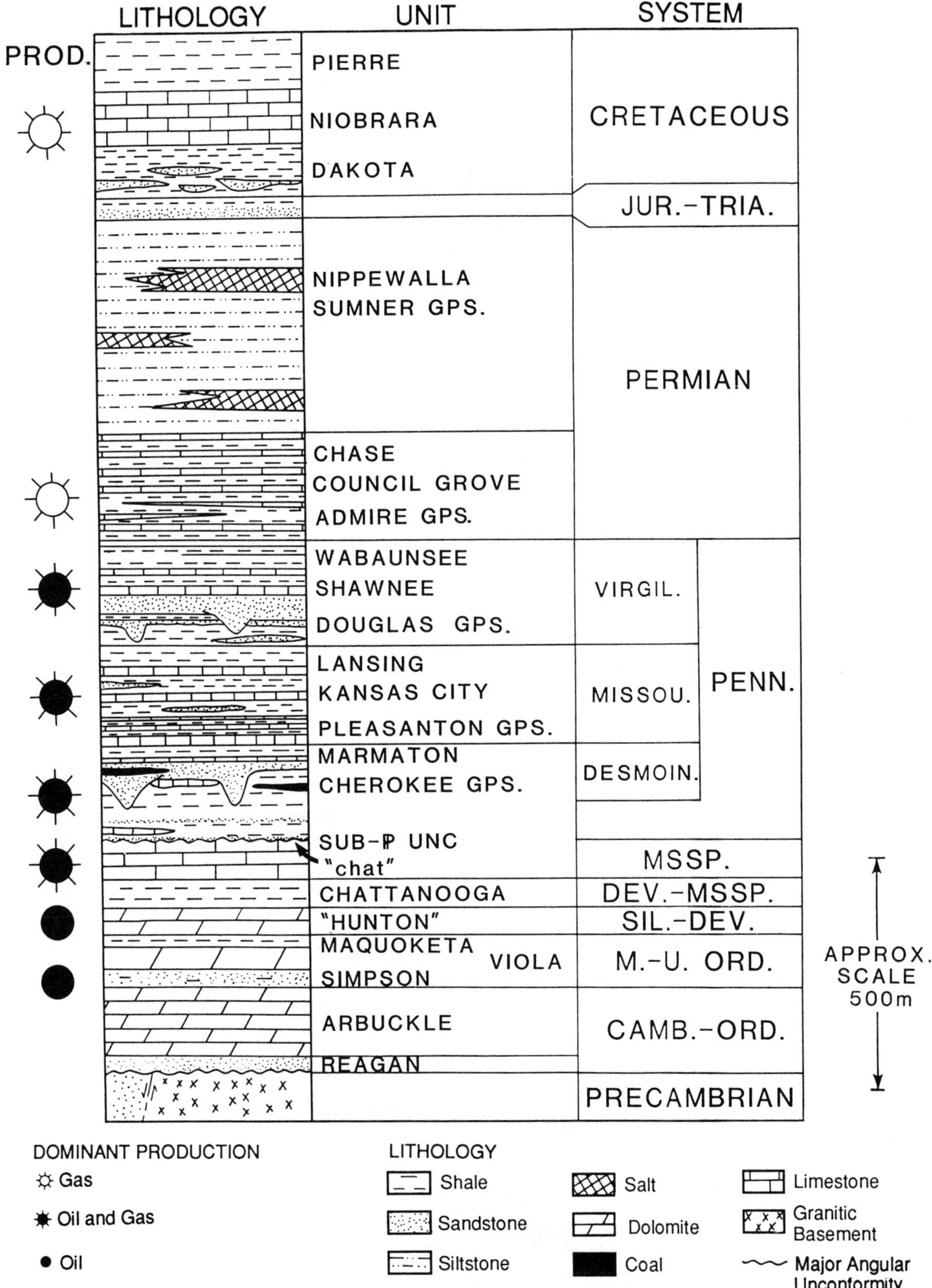

Figure 11. Generalized stratigraphic column for central Kansas. Major type of production from each stratigraphic interval is shown on left.

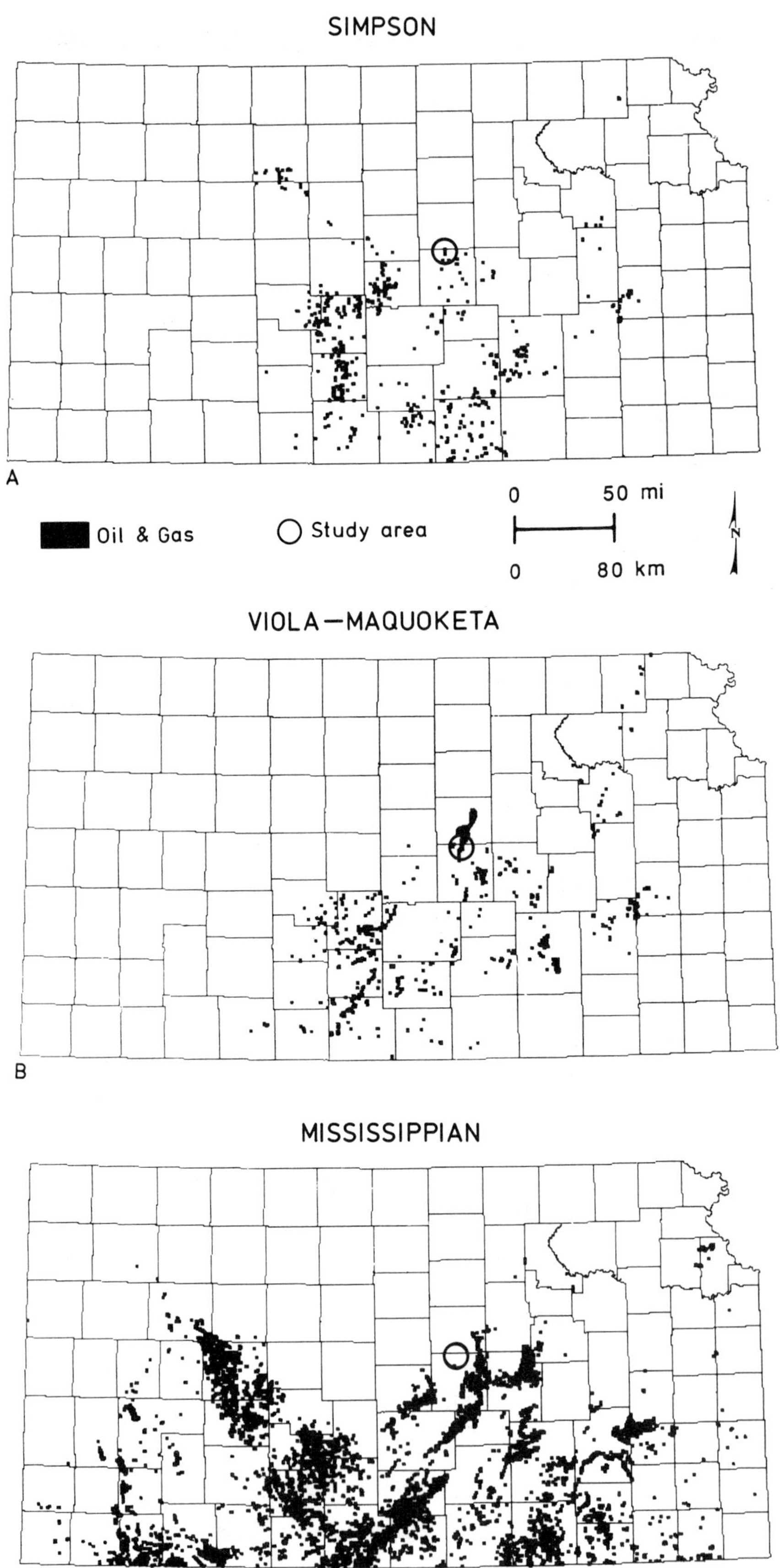

Figure 12. Distribution of production from major stratigraphic intervals in Kansas; (A) Middle Ordovician Simpson; (B) Middle and Upper Ordovician Maquoketa and Viola; (C) Mississippian production.

Maquoketa dolomite is present northwest of this Mississippian trend and extends from central McPherson County through north-central Saline County. The Lindsborg field is the largest field in this Middle–Upper Ordovician trend.

Marine and nonmarine sedimentary rocks of the stable craton underlie all of Kansas. A maximum thickness of 9500 ft (2900 m) of Phanerozoic sedimentary section is found in the southwestern part of the state in the Hugoton embayment of the Anadarko basin (Merriam, 1963). In this area basement rocks are buried to a subsea depth of -6900 ft (-2100 m). Paleozoic rocks in Kansas are economically very important in that they have been a resource for groundwater, quarry stone, lead and zinc, and petroleum. All Paleozoic systems are represented in Kansas; however, many of these systems are incompletely developed due to frequent erosion or nondeposition (Merriam, 1963). Middle and Upper Ordovician rocks in this part of the mid-continent are generally characterized by an alteration of carbonate and siliciclastic strata. The Simpson consists of interbedded shales and clean quartzose sandstones that represent the basal deposit of the Tippecanoe transgression onto the North American craton (Sloss, 1963). As water depths increased, marine lime sediments of the Viola Limestone were deposited. Following a minor erosional event, carbonates and shales of the Maquoketa were respectively deposited. A major hiatus is represented by the unconformity at the base of the overlying Chattanooga Shale. The Chattanooga Shale, which unconformably overlies the Maquoketa in this part of Kansas, is Devonian-Mississippian in age and is part of the Kaskaskia Sequence of the North American craton (Sloss, 1963).

Sandstones in the Simpson Group and porous carbonates of the Maquoketa and Viola are viable reservoirs for oil and gas in the mid-continent (Figure 12; Newell et al., 1987a). Shales within the Simpson have an abundance of oil-prone organic matter and may generate the oil present in adjacent sandstones (Hatch et al., 1987). The Chattanooga Shale is a prolific source of oil and gas in the mid-continent, particularly where it is thermally mature in Oklahoma (Cardott and Lambert, 1985). Marginal maturity of this unit is evident in Kansas (Jenden et al., 1988).

In the vicinity of the Lindsborg field, the Simpson is approximately 90 ft (27 m) thick, but it achieves a maximum thickness of 250 ft (75 m) in south-central Kansas as it thickens southward into the Anadarko basin of Oklahoma (Cole, 1975). The Viola Limestone is approximately 60 ft thick in the study area. A maximum thickness of about 300 ft (90 m) is recorded for the Viola in northeast Kansas (Cole, 1975). Maquoketa dolomite and shale together are approximately 100 ft (30 m) thick in the study area but thin southward due to erosion at the base of the Chattanooga Shale. The Chattanooga Shale is approximately 180 ft (55 m) thick in the vicinity of the Lindsborg field and concomitantly thickens southward as the Maquoketa thins. The Maquoketa Shale reaches a maximum thickness of 170 ft (50 m) in north-central Kansas (Cole, 1975). The Chattanooga Shale thickens to more than 250 ft (75 m) in northeastern Kansas (Lee, 1956).

TRAP

Viola and Simpson oils are structurally trapped in two separate pay zones in a culmination on the Lindsborg anticline (Figures 2 and 13). Another small culmination at the extreme southern end of the field structurally traps a separate pool of Simpson oil (Figure 2). Oil columns in Simpson Sandstone and Viola Limestone in the anticlinal culmination in the central part of the field are 34 ft (10.5 m) and 52 ft (16 m) thick, respectively. The vertical closure of the trap in the central part of the field (i.e., vertical footage from culmination to spillpoint) is estimated to be 60 ft (18 m), as defined by the elevation of the structural saddle (at Simpson level) between this closure and the closure associated with the Olsson field to the north. Original oil-water contacts for the Viola and Simpson oil accumulations are estimated to be at subsea depths of -2079 ft (-634 m) and -2156 ft (-657.5 m), respectively.

The Simpson pool at the southern end of the Lindsborg field (Figure 2) has an oil-column thickness of 19 ft (6 m) in a structural closure of approximately 25 ft (7.5 m). The original oil-water contact of this small Simpson pool is estimated to be at a depth -2156 ft (-657.5 m) subsea.

The seal for vertically migrating hydrocarbons of the Viola pay zone is a 10-20 ft thick (3-6 m) bed of nonporous limestone at the top of the Viola Limestone. A 10-15 ft thick (3-4.5 m) shale at the top of the Simpson Group serves as a vertical seal for migrating oil in the Simpson pay zone (Figure 14).

Maquoketa oil in the Lindsborg field is caught in a structural-stratigraphic trap on the west flank of the Lindsborg anticline (Figures 2 and 13). Total oil column in the Maquoketa is 126 ft (38.5 m) thick. The lowest depth from which oil has been produced in the field is -2103 ft (-641.5 m) subsea. This may not necessarily represent an oil-water contact because the lowest level at which oil is produced along the western flank of the anticline varies with location. The base of production is deepest at the extreme northern end of the field (Figures 2 and 5). It gradually rises 40 ft to a depth of -2060 ft (-628 m) at the south end of the Salemsborg sector of the field (i.e., approximately the boundary line between T.16S. and T.17S.) and then is relatively level at points farther south (Figure 5).

Over most of the field, the upper seal for the oil in the Maquoketa dolomite is the Maquoketa shale—a light gray to maroon dolomitic shale approximately 70 ft (21 m) thick. In the extreme southern end of the field the varicolored to gray Devonian–Mississippian Chattanooga Shale is the upper seal for oil in the Maquoketa dolomite (Figure 8). In this area, the Maquoketa shale is absent due to an erosional

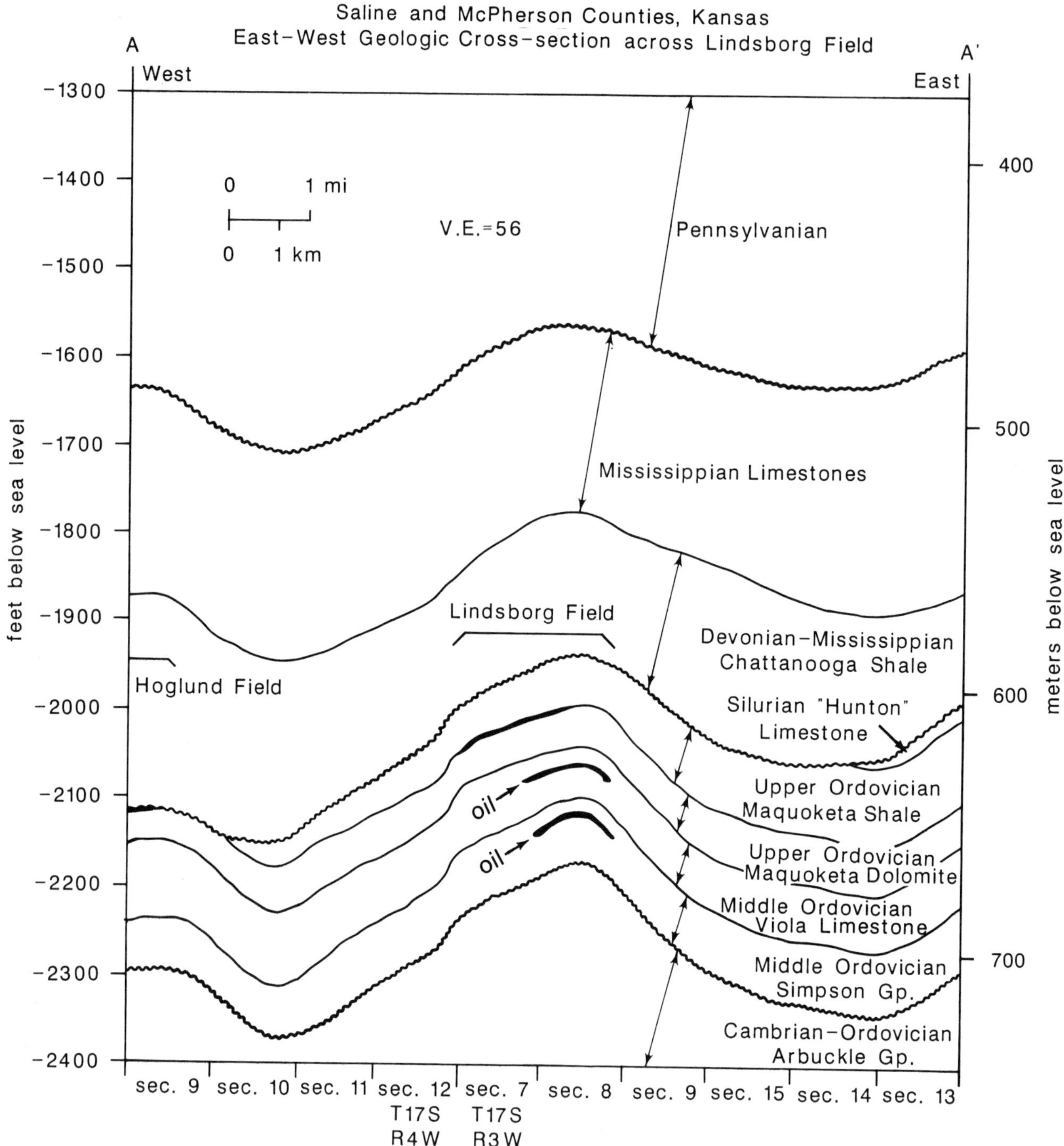

Figure 13. West-northwest-east-southeast geologic cross-section across Lindsborg field showing pay zones. Refer to Figure 2 for location. Vertical exaggeration is 56.

event expressed at the base of the overlying Chattanooga Shale. The erosion and consequent truncation of units at the sub-Chattanooga unconformity is due to a large paleo-valley south and west of the Lindsborg field (Lee, 1956). The Chattanooga thickens into this paleo-valley as underlying units are concomitantly thinned by erosion at its base (see Figure 13).

The updip lateral seal for oil in the Maquoketa dolomite is impermeable and probably nonporous Maquoketa dolomite. The irregular distribution of Maquoketa production in the Lindsborg and nearby fields (Figures 2 and 8) indicates porosity and permeability (and corresponding production) in the Maquoketa dolomite is somewhat erratic and difficult to predict.

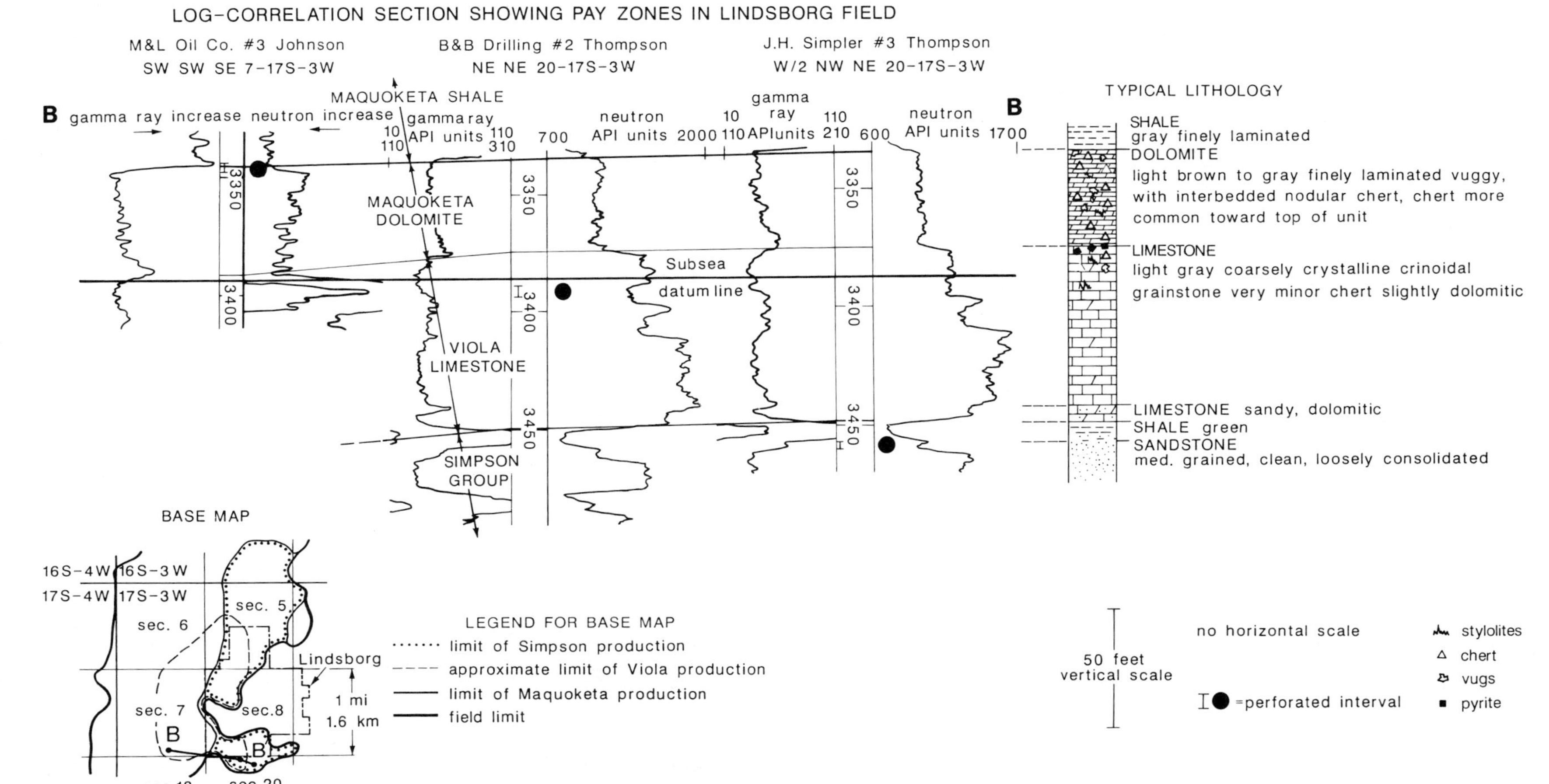

Figure 14. Log-correlation structural cross section showing pay zones in Lindsborg field and interpreted lithologies. Base map shows general spatial distribution of production from each unit at the culmination of the Lindsborg anticline in the central part of the field.

RESERVOIRS

Depths to the top of each reservoir in the Lindsborg field are variable due to combined effects of surface elevation and subsurface structure. Viola production is reached at a depth of 3360 ft (1025 m) at the top of the structural trap in the central part of the field. Simpson oil is found 80 ft (24 m) deeper at 3440 ft (1049 m). The depth to the Simpson pool at the southern end of the field is 3495 ft (1066 m) and constitutes the deepest drilling necessary in the entire field.

Thicknesses of the produced interval for all wells in the field average 6.5 ft (2 m) for the Maquoketa, 10 ft (3 m) for the Viola, and 4.5 ft (1.5 m) for the Simpson. Maximum reported thicknesses of the producing zones in the Maquoketa, Viola, and Simpson are respectively 26 ft (8 m), 18 ft (5.5 m), and 14 ft (4 m).

The typical log signatures for the three reservoir units in the Lindsborg field and the inferred lithology of each of these zones is illustrated in Figure 14. The Maquoketa pay zone is a thin zone at the top of a cherty dolomite that constitutes the Maquoketa dolomite. The Viola pay zone is a porous limestone, dolomite, or dolomitic limestone, 5-20 ft (1.5-6 m) thick, present in the upper half of the Viola Limestone (Figure 14). The Viola can be as thick as 80 ft (24 m) in the field but thins to approximately 50 ft (15 m) at the southern end of the field. The Simpson Group is relatively uniform in thickness in the area of the field (approximately 75 ft; 23 m). The pay zone in this unit is a dolomitic sandstone bed, 15-20 ft (4.5-6 m) thick, that directly underlies a shale at the top of the Simpson.

Log determination of pay-zone lithologies is allowed by a crossplot of apparent matrix photoelectric macrosection (UMAA) and apparent grain density (RHOMAA) from a Schlumberger lithodensity™ log (Figure 15) (cf., Doveton, 1986). Maquoketa dolomite is composed of approximately 30% chert, 60% dolomite, and 10% calcite. Viola is mostly calcite but has some zones with varying portions (approximately 5% to 30%) dolomite. Simpson is mostly composed of quartz, but some zones having variable portions of dolomite, calcite, and clay minerals plot considerably away from the quartz end-member point.

The Maquoketa dolomite is mappable as a distinct lithologic unit over much of the structural saddle that separates the Salina and Sedgwick basins (Lee, 1956). It thins eastward of the Lindsborg anticline and pinches out in eastern Salina County and McPherson County where Maquoketa shale directly overlies Viola Limestone. Immediately west of the Smolan and Lindsborg fields, the Maquoketa dolomite thins due to erosion beneath the angular unconformity at the base of the Chattanooga Shale. An isopach map of the Maquoketa dolomite (Figure 8) indicates this unit is 40-70 ft (12-21 m) thick in the vicinity of the Lindsborg field. Northwest-southeast trends of thinning and thickening also are evident but are not correlative to production trends.

In well cuttings, the Maquoketa dolomite is a gray, slightly silty medium crystalline dolomite with white to gray speckled chert. In places, the unit can be glauconitic and the chert can be fossiliferous. Cores of this unit show that the chert is present as lenticular beds and nodules flattened parallel to bedding (Figure 16). Vugs within these nodules can be quite large (up to 3 inches, 7 cm, in diameter) and may be partly filled with euhedral quartz, dolomite, and minor pyrite crystals. Some vugs, particularly smaller ones, bleed oil immediately after a core has been pulled from a well. Vugs and pinpoint porosity bleeding oil commonly can be observed throughout an entire section of Maquoketa dolomite, but only the very top of this unit has ever been produced in the Lindsborg field. Porosity in the pay zone at the top of the Maquoketa dolomite is estimated from cuttings as being from 5% to 12%. Wireline-log determination of porosity in this reservoir can be ambiguous due to difficulty in estimating the proportions of chert, dolomite, and limestone in the matrix. No core permeability measurements were available from the Maquoketa dolomite reservoir or any other reservoir in the field.

In cores, the Maquoketa dolomite is a fine-grained, finely laminated dolomitic mudstone (Figure 16). It is sparsely fossiliferous but contains minute flakes of organic matter, and some laminations are contorted in what may be burrows. Lack of strong current features and normal marine biota indicate it may be the product of subtidal sedimentation in a restricted euxinic lagoon. Lenticular diagenetic chert nodules are the locus for vugs in this unit. Minor fracture porosity also can be present in the chert nodules. Minor intercrystalline porosity is locally evident in the surrounding fine-grained sucrosic dolomite.

Thatcher (1961) reports that no drastic change occurs in well cuttings or wireline logs of Maquoketa dolomite between productive and nonproductive areas (also see Figure 14). However, he reports that core descriptions from productive wells note the presence of numerous fractures, whereas core descriptions from dry holes seldom mention fractures. The spatial distribution of production on the west flank of the Lindsborg anticline and along the trend of locally steepened dip associated with the Salemsborg sector of the field is compatible with fracture-controlled permeability and porosity. However, the thinness of the pay zone over the entire Lindsborg field, its consistent position at the top of the Maquoketa dolomite, and the presence of on-structure production on the structurally flat crests of anticlines associated with the nearby Smolan and Olsson fields (see Figure 2) argue against fracture porosity and permeability as being the sole control of production in this unit. Diagenesis also is probably very important in determining the distribution of porosity and permeability in the Maquoketa dolomite.

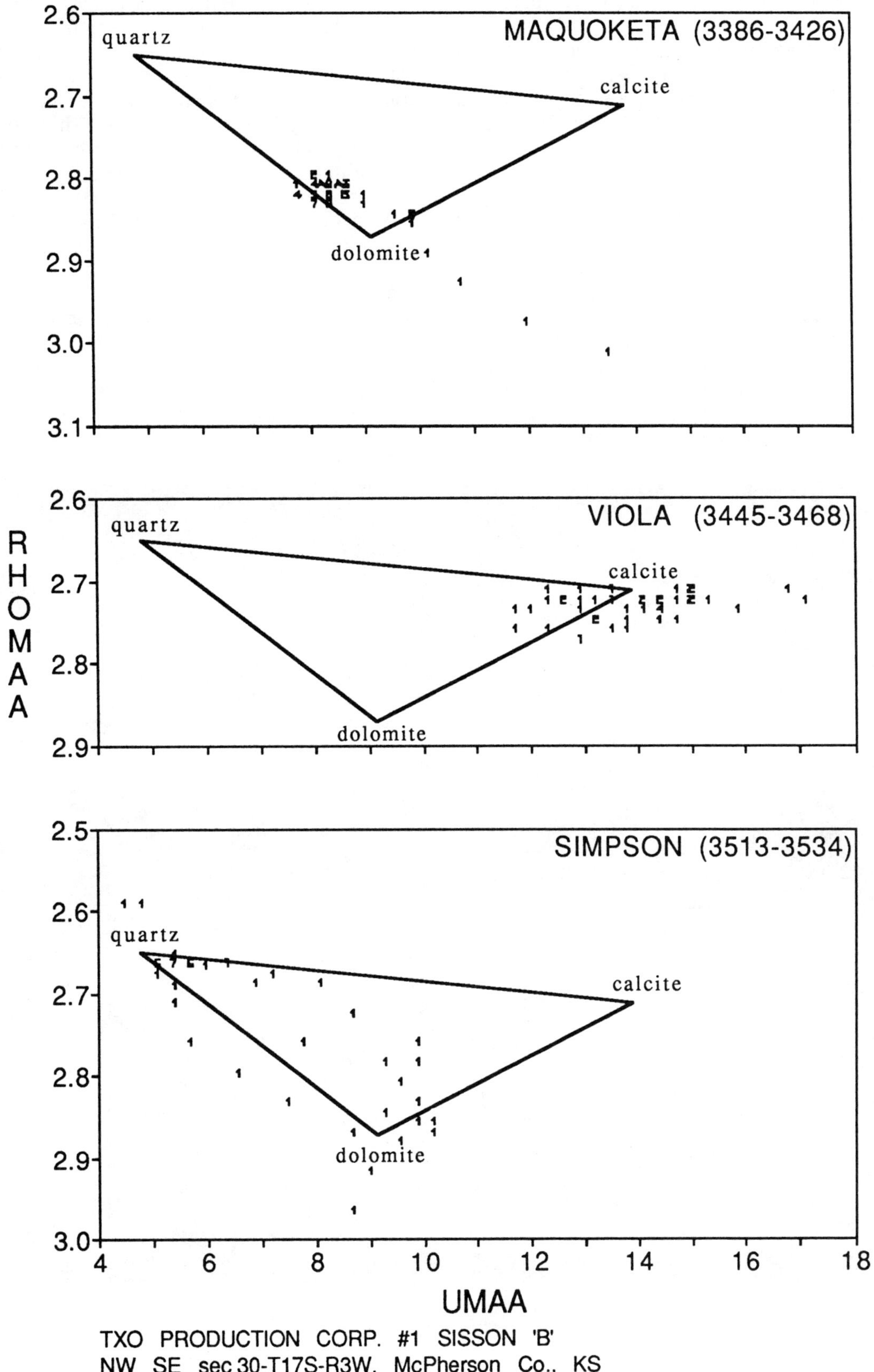

TXO PRODUCTION CORP. #1 SISSON 'B'
NW SE sec 30-T17S-R3W, McPherson Co., KS

Figure 15. RHOMAA-UMAA cross-plot of lithologies in pay zones from lithodensity log (trademark of Schlumberger).

Unfortunately, understanding and consequently predicting porosity in the Maquoketa dolomite is very difficult because no cores are available from any production wells from the Lindsborg field or nearby fields. Cores have been cut in the field but they have been lost or held proprietary. Cores inspected and photographed in this report are taken from two nonproducing wells drilled in the middle of the Maquoketa production fairway, but it is not clear how representative they are with respect to the type and distribution of porosity that would characterize a productive well.

No cores of the pay zone in the Viola Limestone were available for the study, but cuttings descriptions reveal the reservoir in the Viola is a gray to white coarsely crystalline limestone with minor dolomite and chert. Log analyses indicate this reservoir can have as much as 15% porosity. This porosity decreases eastward across the crest of the Lindsborg anticline (see Figure 14) and prohibits production from this unit in the eastern part of the field. The Viola pay zone in the nearby Olsson field is reported to be considerably variable, ranging from fine to coarsely crystalline dolomite or dolomitic limestone (Don Hoy Smith, Wichita, KS, personal communication). The vertical seal for the Viola reservoir in the Lindsborg field, a 10 ft (3 m) thick bed at the top of the Viola Limestone, is a crinoidal grainstone with minor amounts of brachiopod and bryozoan debris. Similar crinoidal grainstones and packstones in Viola Limestone are reported by St. Clair (1985) from other parts of Kansas and are interpreted to represent deposition in moderate- to high-energy open-marine environments.

Cuttings descriptions of the Simpson pay zone in the Lindsborg field indicate it is a well-rounded, medium- to coarse-grained dolomitic quartz sandstone with good porosity (estimated 15% to 18% from log analyses). Cores of Simpson sandstone from wells that have log character similar to the Simpson at

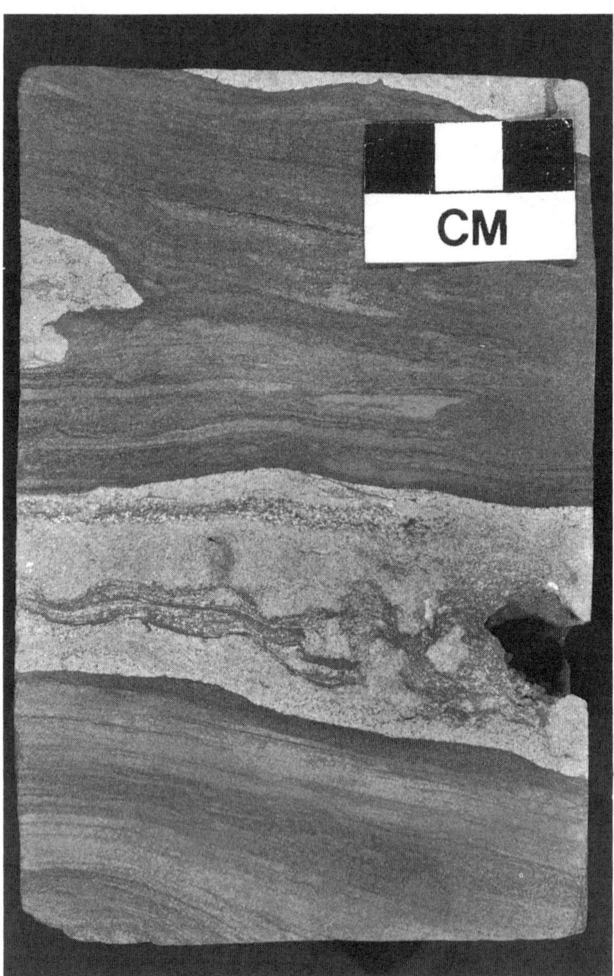

A

B

Figure 16. (A) Core photograph of Maquoketa dolomite (3395 ft; 1035.5 m). Note white chert nodule and associated vug. (B) Crinoidal packstone in Viola Limestone (3444 ft; 1050.5 m). This lithology is interpreted to be the upper seal for the hydrocarbons in the Viola Limestone. Core slabs are from Damac Drilling #1 Sandra Allen well in SW sec 7-T.17S.-R.3W.

Lindsborg field were interpreted by Autio (1985) to represent shoreline, barrier island, or high-energy shelf deposits.

Average decline curves for wells drilled in the 1940s indicate wells completed in the Simpson generally are much more prolific than Maquoketa wells and have lesser rates of production decline (Figure 17A). Although the Simpson decline curve is comparatively erratic due to relatively few wells used in its computation, both reservoirs display a constant rate of production decline with respect to ultimate production per well. Wells completed in the Maquoketa during the 1960s and 1970s also demonstrate similar linear declines in production (Figure 17B). Water drive in all reservoirs in Lindsborg field is indicated by greater initial water cuts in newer production wells (Figure 18). If reservoirs were subject to strong water drive, production decline curves such as those in Figure 17 would ideally be level for a long period of time and then decline sharply (Farina, 1984). A relatively weak water drive is probably indicated for at least the Maquoketa reservoir in the Lindsborg field by its constant rate of production decline.

The average decline curve for Viola wells drilled in the 1940s (Figure 17A) illustrates a productivity intermediate to that of the Simpson and Maquoketa. The sharp increase in annual production evident at 50,000 bbl cumulative production is due to very successful sand-fracturing operations performed on this reservoir in several wells during the early 1960s (Dale Reese, personal communication).

SOURCE

Average API gravity of Maquoketa oil is 36°, but gravities as low as 27° API also have been reported. Gravity of Viola oil ranges from 28° to 34° API. Simpson oil averages 28° API. Limited chemical analyses indicate that sulfur content for all three oils ranges between 0.43 and 0.54 wt%. Pour points for Maquoketa, Viola, and Simpson oils are respectively 10°F (-12°C), 5°F (-15°C), and 5°F (-15°C). Maquoketa oil is *aromatic intermediate* and Simpson oil is *paraffinic* according to crude-oil classification in Tissot and Welte (1984). Viola oil was not sampled. No distillate and very little casing-head gas is produced with the oil from all three pay zones, hence the production gas-oil ratio and any contribution by gas drive to production are virtually negligible.

Gas chromatograms (Figure 19) indicate Simpson oil is different from Maquoketa oil. Simpson oil has the following peculiar characteristics: (1) odd-over-even predominance of normal alkanes in the C11 to C19 range; (2) relatively low amounts of the isoprenoids pristane and phytane, and similar low amounts of branched and cyclic alkanes; and (3) relatively low amounts of saturated hydrocarbons with gas chromatographic retention times greater than that of nC-19. Oils with these attributes are very characteristic of oil generated from Middle Ordovician rocks throughout the mid-continent and eastern United States (Hatch et al., 1987). The most likely source rocks in the hydrocarbon drainage area around the Lindsborg field are shales in the Simpson Group. Content of total organic carbon in shales from the Simpson Group in central Kansas is generally greater than 0.5 wt% (Figure 20) and is therefore sufficiently rich for generation of petroleum. Extracts from Simpson shale also are similar to the oil present in the Simpson reservoir of the Lindsborg field (J.R. Hatch, Denver, CO, personal communication).

Oil in the Maquoketa reservoir is different from that of the Simpson reservoir in that Maquoketa oil has greater quantities of pristane, phytane, and saturated hydrocarbons with molecular weights greater than nC-19 (Figure 19). Like Simpson oil, an odd-over-even preference is exhibited for saturated hydrocarbons with molecular weights less than that of nC-19. This may be due to mixing with Simpson oil or it also may be characteristic of the type of organic matter that generated this oil. The source unit for the Maquoketa oil is more problematical than that of the Simpson oil because available organic analyses from central Kansas (Figure 20) indicate Maquoketa shale may not be sufficiently rich to be a good source rock. Chattanooga Shale displays better organic richness but still may be inadequate to generate significant quantities of hydrocarbons in this region. Shales in both the Chattanooga and Maquoketa, however, are in contact with the reservoir at the top of the Maquoketa dolomite in the hydrocarbon drainage area of the Lindsborg field and other nearby fields with Maquoketa production (Figures 8 and 13), hence a very short and efficient migration path from source rock to reservoir is facilitated for both of these shales. Fine-grained fragments of organic matter compose less than 1% (by visual estimate) of the Maquoketa dolomite itself. No total organic carbon measurements have yet been performed on this unit, so its hydrocarbon source potential is still undetermined.

Pyrolysis measurements by Rock-Eval™ instrumentation indicate Simpson and Maquoketa shales in this area are characterized by Type I (oil-prone) organic matter while Chattanooga shale contains Type II (oil- and gas-prone) organic matter. Maturation data from rocks from the southern end of the Salina basin (Figure 21) indicate both units are marginally mature with respect to hydrocarbon generation and therefore may be capable of generating hydrocarbons present in the Lindsborg field.

A maturation model depicted in Figure 22 corroborates the Rock-Eval data by showing the Simpson could be in very early stages of maturity with respect to hydrocarbon generation. This model assumes the present-day geothermal gradient of approximately 1.4°F/100 ft (25°C/km) (based on bottom-hole temperatures from well logs, drill-stem tests, and regional data from Stavnes and Steeples, 1982) and burial by approximately 3000 ft (900 m) of now-eroded Cretaceous strata. Of course, other models are possible, but if this model reasonably approximates the tectonic and

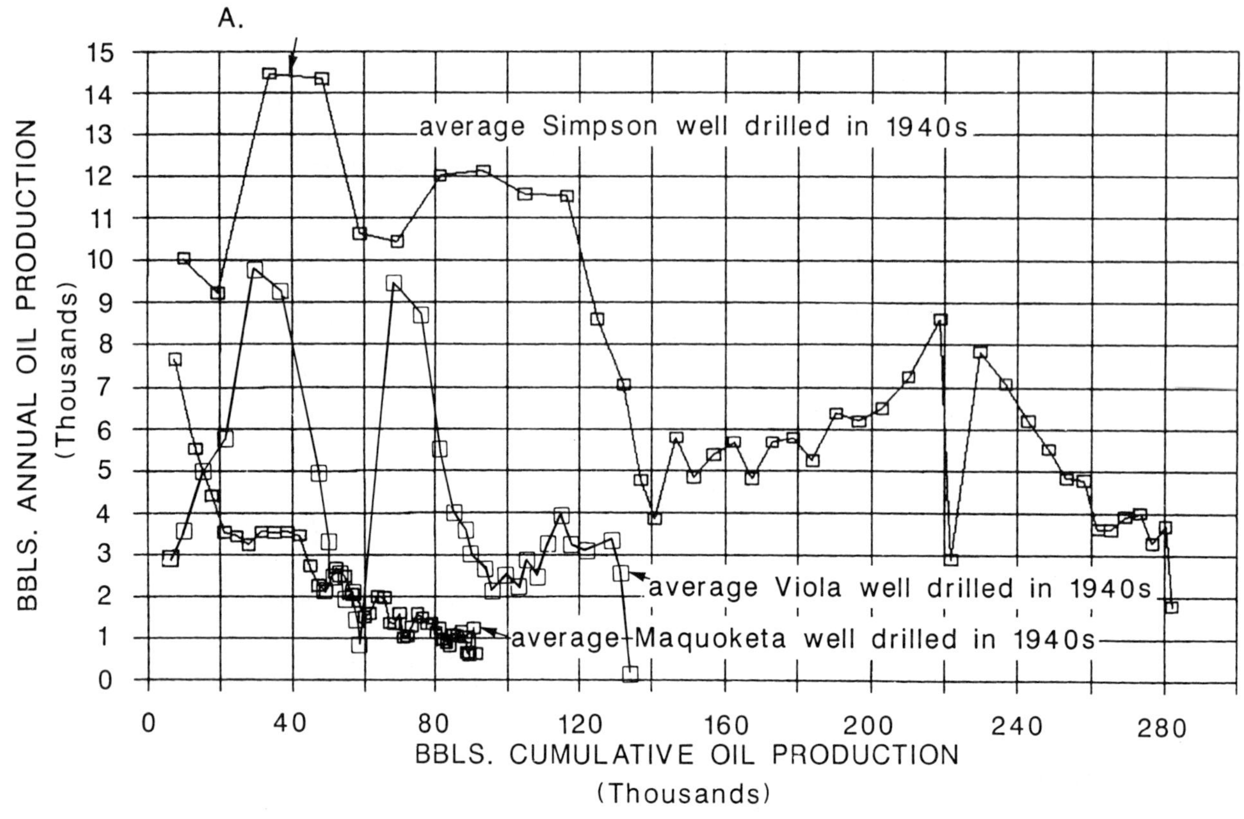

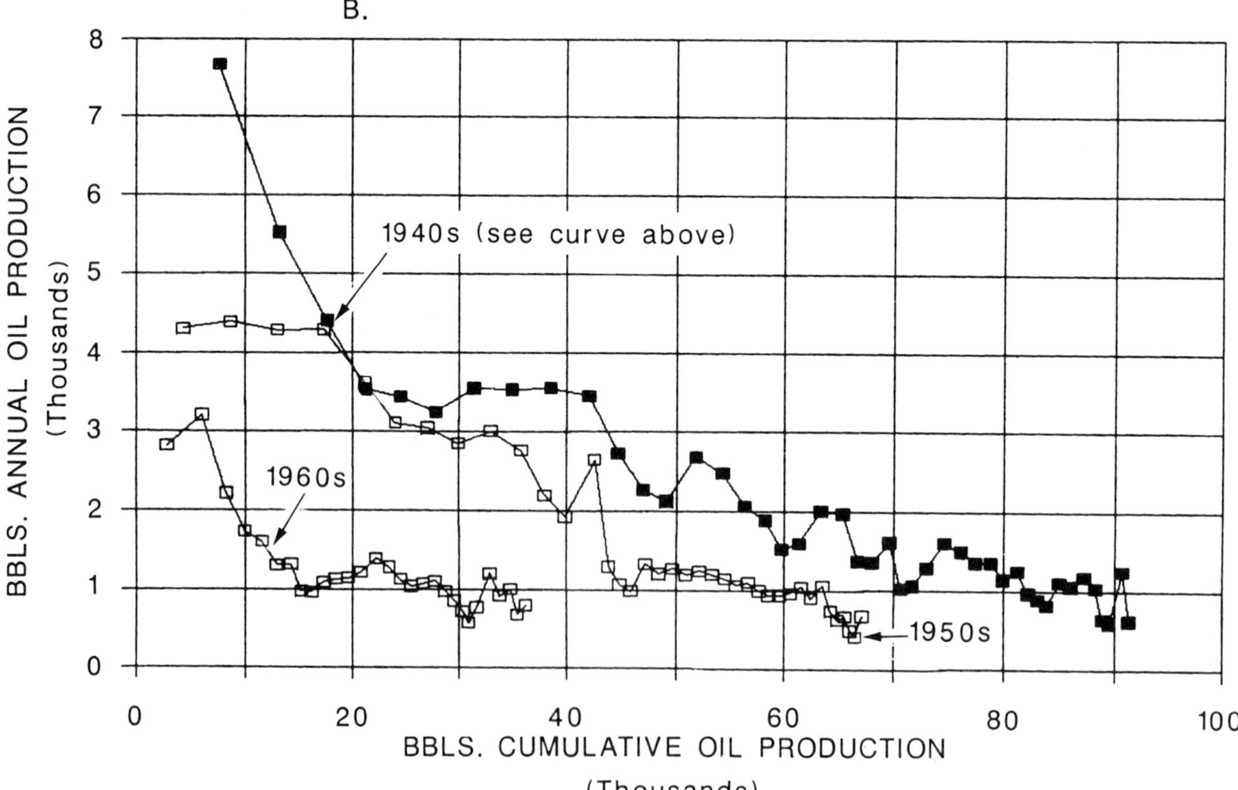

Figure 17. (A) Averaged production decline curves for Maquoketa, Viola, and Simpson wells drilled in Lindsborg field in the 1940s. (B) Averaged production decline curves for Maquoketa wells respectively drilled in Lindsborg field during the 1940s, 1950s, and 1960s. Data taken from all single-well leases with more than 5 years of production history and 5000 bbl cumulative production.

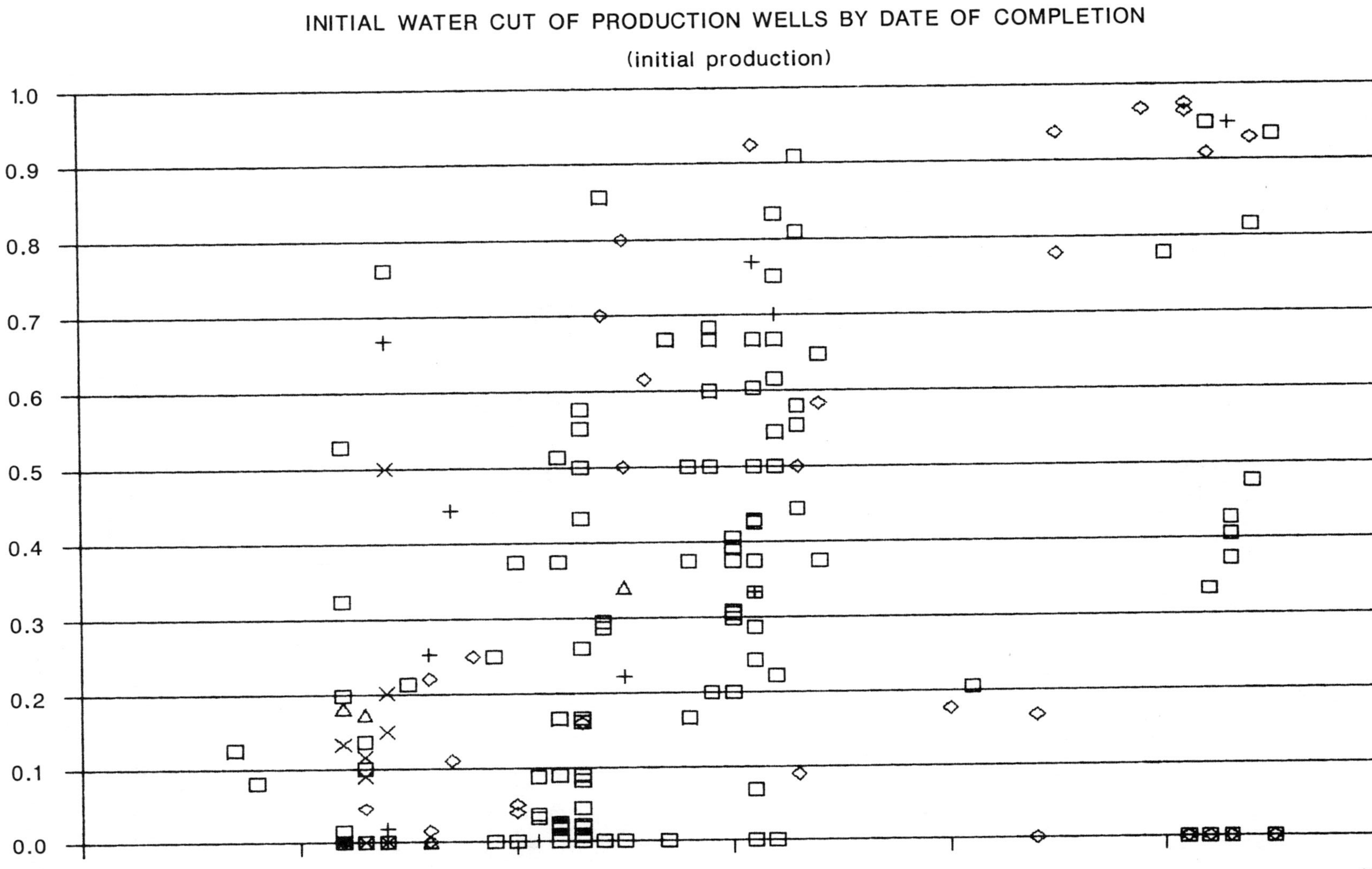

Figure 18. Reported initial water-cut of production wells according to their date of completion. Each data point is differentiated according to pay zone and represents a single well. Many wells showing no co-produced water probably had some associated water, but it was not reported as a percentage of production.

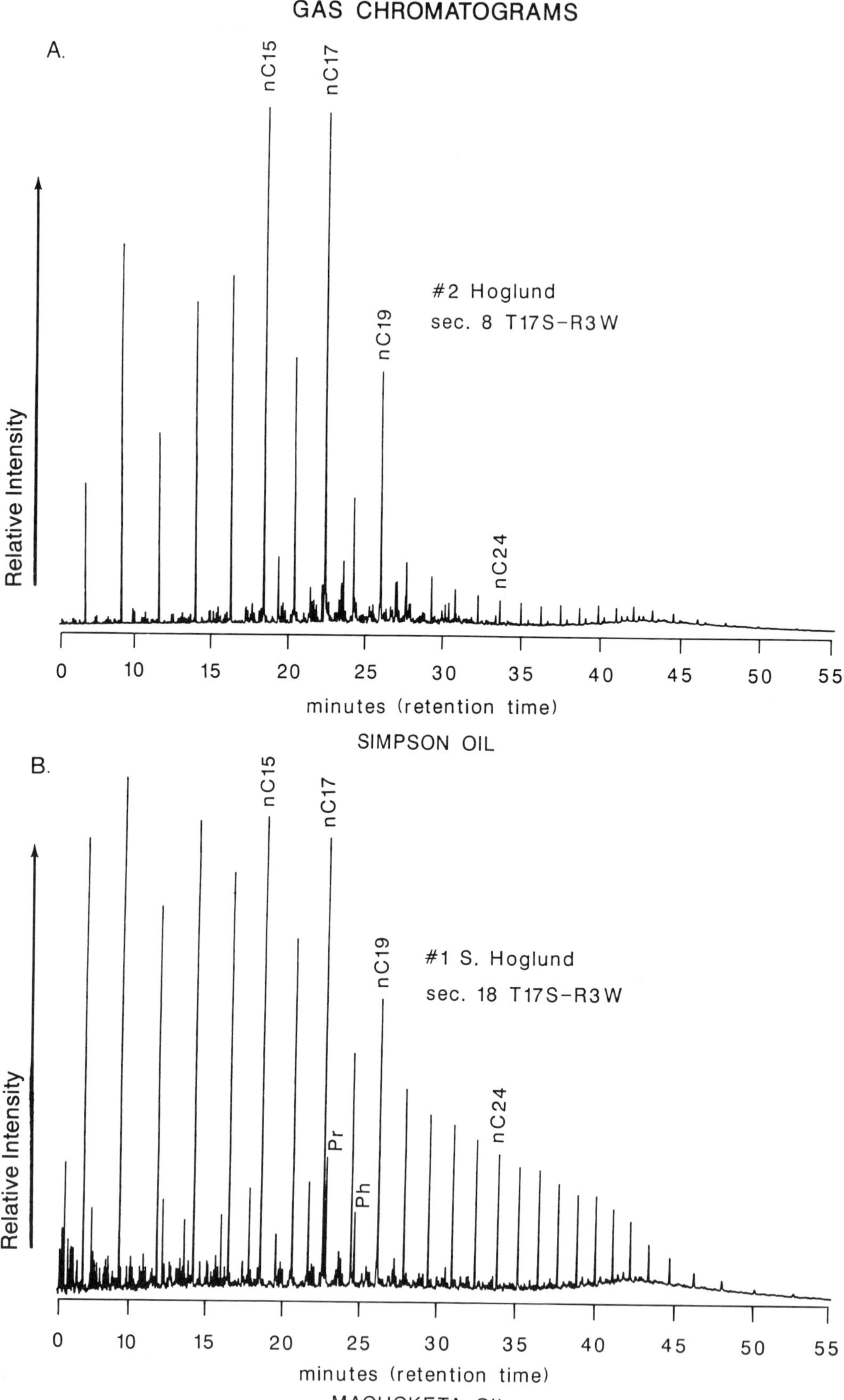

Figure 19. Gas chromatograms of (A) Simpson oil and (B) Maquoketa oil from Lindsborg field (see text for discussion).

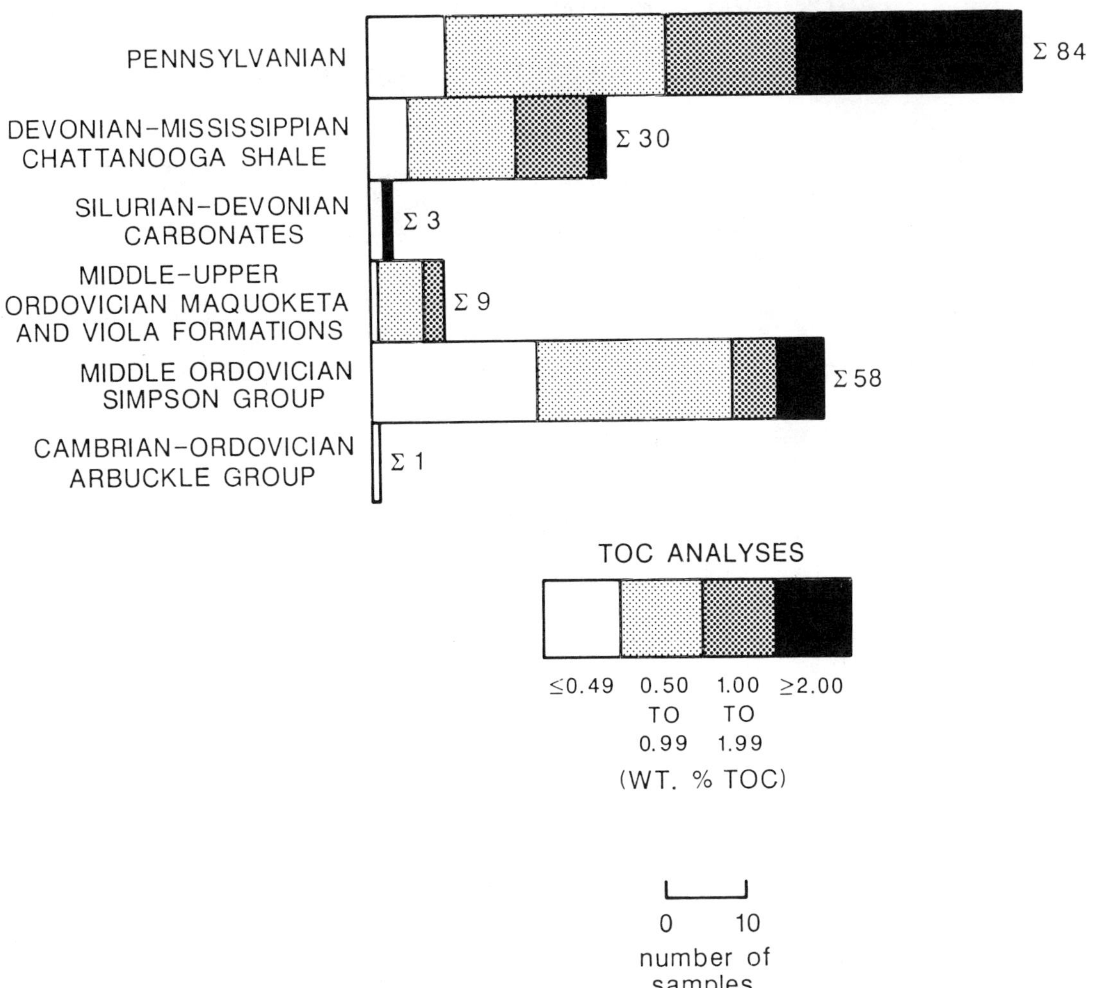

Figure 20. Summary of organic content of samples from major stratigraphic intervals from central Kansas.

thermal history of this region, Ordovician source rocks would presently be entering the hydrocarbon window. Secondary migration of any hydrocarbons expulsed from source rocks would therefore be controlled by present structural configuration of the Lindsborg anticline.

The ultimate source of oil in Kansas is subject to controversy. Some researchers (Rich, 1933; Walters, 1958; Momper, 1978; Price, 1980) support a long-distance migration model in which a large contribution of hydrocarbons in the state came from thermally mature source rocks farther south in the deeper part of the Anadarko basin. This model predicts that the relatively shallow Salina basin will be mostly barren of hydrocarbons because it is a thermally immature basin and is in a migrational "shadow zone" in which northward-migrating hydrocarbons are shunted onto the Central Kansas and Nemaha uplifts. However, the Rock-Eval data and maturation modeling indicate that local generation possibly could be responsible for hydrocarbon accumulation in the Lindsborg field. If local generation from marginally mature source rocks is responsible for the oil in the Lindsborg field, greater maturation could be expected where the Ordovician strata productive in the Lindsborg area is buried more deeply in the relatively unexplored Salina basin to the north.

EXPLORATION CONCEPTS

Stratigraphically controlled traps in Maquoketa dolomite in central Kansas are difficult to predict with present data and exploration techniques. The Lindsborg field, however, indicates that production

ROCK-EVAL MATURATION DATA

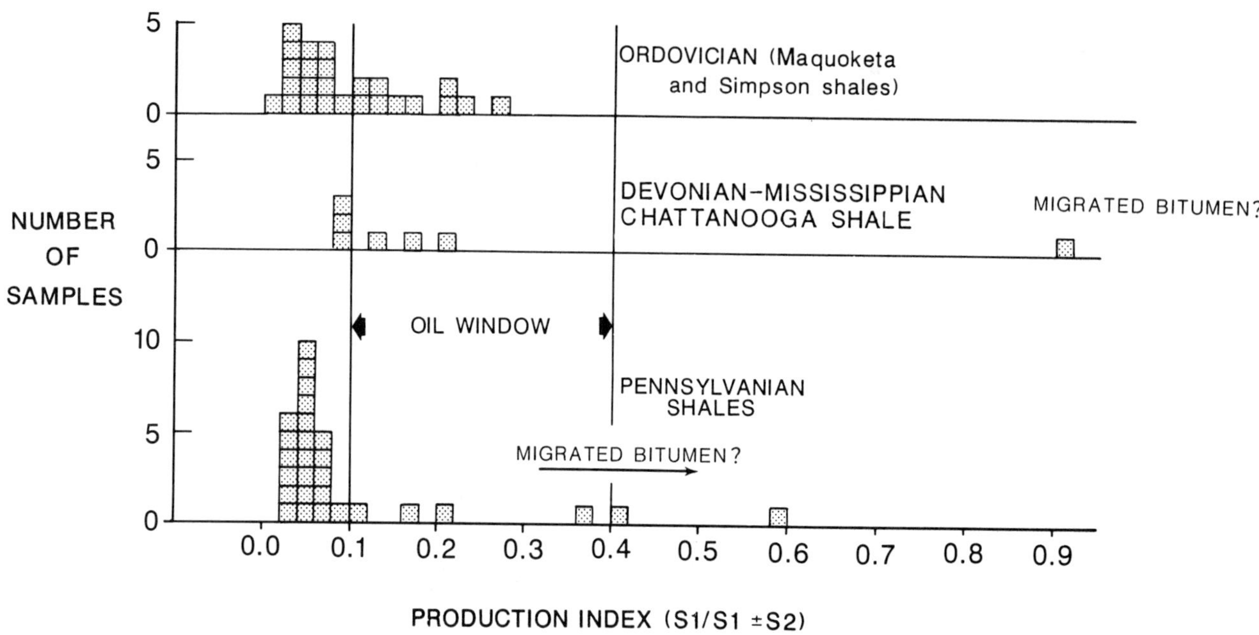

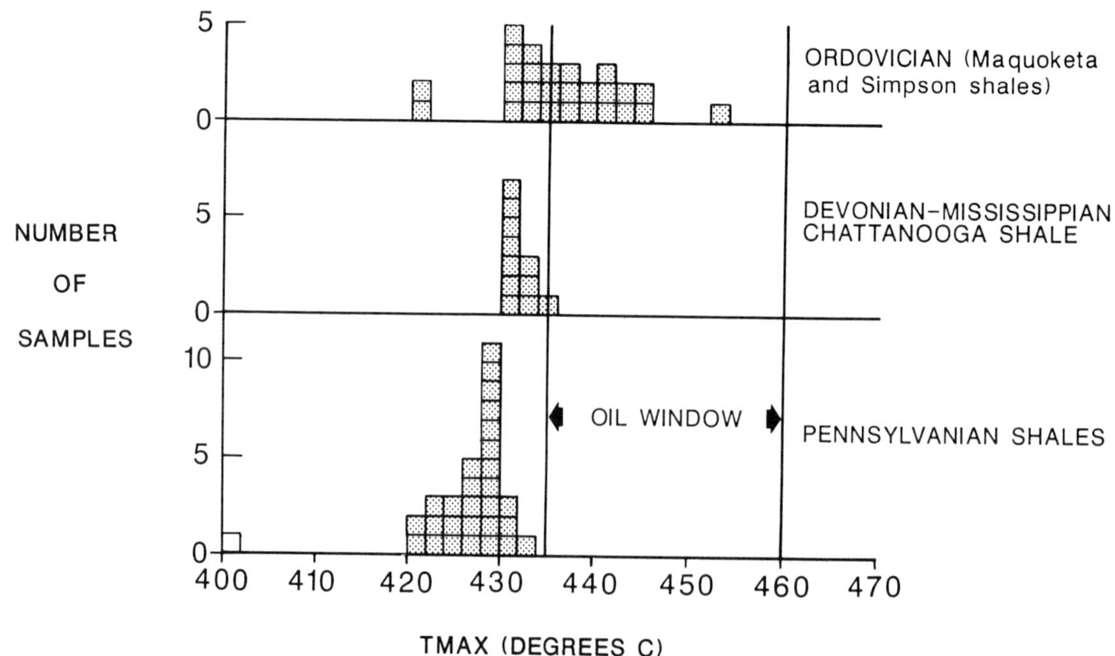

Figure 21. Histograms of Rock-Eval pyrolysis data of major stratigraphic intervals from central Kansas. Tmax and Production Index data and generally accepted oil windows are shown.

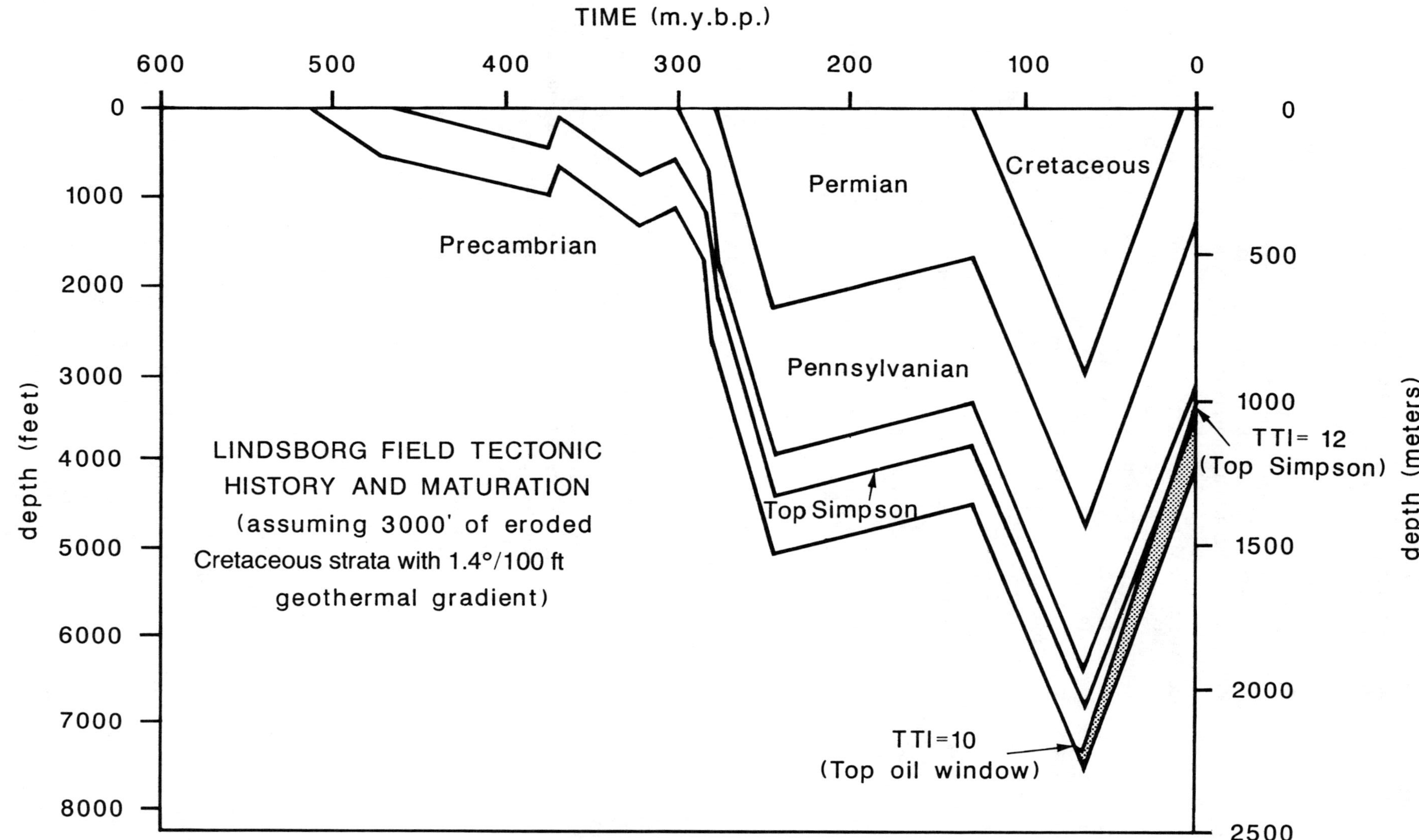

Figure 22. Tectonic subsidence model of Lindsborg field with calculated time-temperature index (TTI) maturation. A TTI of 10 is taken to be the beginning of possible significant oil generation. Note Simpson Group is in very early stages of oil generation. Refer to Waples (1981) for explanation of calculation and correlation of TTIs.

from these traps can be lucrative and long-lived. Modest expansion of Maquoketa production also is possible. A Maquoketa oil discovery in 1987 about 1 mi (1.6 km) east of the Lindsborg field (Figure 2) indicates that production from this unit also may be obtained from parts of the eastern flank of the Lindsborg anticline. The possible association of Maquoketa production with fracture porosity caused by structural flexures or faulting in this region may indicate that other areas both on-structure and on the flanks of structures are prospective for new Maquoketa production. With present knowledge, it may be too daring to place a wildcat well over the relatively steep flank of a major structure on the tenuous hope that a stratigraphic trap in the Maquoketa is there awaiting discovery. However, a reexamination of shows in old dry holes in these structural settings may reveal the presence of previously overlooked reserves. Although present-day reflection seismology may not have the resolution to see a thin porous zone such as the pay zone at the top of the Maquoketa dolomite, it can reveal structural details in previously undrilled areas that may be of importance in developing prospects near dry holes with intriguing shows.

The lack of sufficient data to explain (and consequently predict) porosity distribution in the Maquoketa points to the fact that cores are extremely important in understanding rocks that produce oil. Valuable exploration data can be obtained by coring even in geologically prosaic locations such as infill development wells. New logging tools, such as spectral gamma ray and lithodensity logs, also offer a better understanding of reservoir lithologies and their correlation.

The concept that at least some of the oil produced in the Lindsborg field may be locally generated from shales in the Simpson Group bodes well for the prospectivity of structural trends in the relatively unproductive and under-explored Salina basin. The stratigraphy of the Salina basin is very similar to that of the Forest City basin because both basins were united during early Paleozoic time (Lee, 1956). Ordovician oil similar to that found in the Simpson reservoir at Lindsborg field also is produced in several fields in the Forest City basin (Hatch et al., 1987; Newell et al., 1987b). The marginal maturity that characterizes potential source rocks in the Simpson Group in the structural saddle between the Sedgwick and Salina basins should improve where the Simpson Group is more deeply buried in both basins. Areas of higher geothermal gradient also may improve the maturity of potential source rocks in this region. The high productivity of Simpson wells in the Lindsborg field also should encourage additional exploration for reserves in this unit.

The association of the Lindsborg anticline with a structural trend in underlying Precambrian rocks indicates that the location of other Phanerozoic anticlinal trends may be controlled by structural discontinuities in basement rocks, specifically those associated with the Central North American Rift Systems (CNARS). Computer processing of regional aeromagnetic data has been very successful in discerning Precambrian structural trends in the mid-continent (see Yarger, 1983) and hence may be a cost-effective exploration tool for evaluating vast swaths of territory for subsequent detailed geologic and geophysical analysis. The CNARS itself is a target for petroleum exploration; therefore, any explorationist developing prospects targeted in Phanerozoic strata over the rift should be cognizant of potentially prospective deeper strata. Some wells in the vicinity of the Lindsborg field have been drilled into arkosic sedimentary rocks of the CNARS, but no hydrocarbon shows have been reported in these wells. However, elsewhere in the Salina basin minor hydrocarbon shows have been reported in Precambrian sedimentary rocks of the CNARS (Newell et al., 1988).

The consistent presence of structural closures on north-south-trending segments of anticlines in the region around the Lindsborg field indicates that faulting associated with these anticlines may have a minor component of strike-slip movement. Models of structural development taking into account possible lateral motion on faults may therefore prove to be powerful tools for predicting the location of closures on known and relatively unexplored structural trends in the mid-continent.

ACKNOWLEDGMENTS

The author benefited considerably by conversations with individuals involved with exploration and development of the Lindsborg and nearby fields. These individuals include Al Abercrombie, Jerry Langrehr, Don Melland, Clark Roach, Don Hoy Smith, and Millard Smith. Special thanks go to Jack Wilson, Lee H. Cornell, and Dale Reese who were very generous of their time and information. Helpful and appreciated suggestions on the manuscript were made by W. Lynn Watney, John Doveton, Ronald M. Hedberg, Lee Cornell, and Don Hoy Smith. Crude oil analyses were courtesy of Mobil Oil Corporation and D. N. Bressie, Jr., at National Cooperative Refinery Association. Mark Schoneweis, Renate Hensiek, Michael Lambert, and Bryan Stephens helped with drafting of figures.

REFERENCES CITED

Adler, F. J., 1971, Future petroleum provinces of the mid-continent, region 7, *in* I. H. Cram, ed., Future petroleum provinces of the United States—their geology and potential: American Association of Petroleum Geologists Memoir 15, p. 985-1042.

Autio, W., 1985, Lithofacies and depositional environments of the Simpson Group of central Kansas, *in* M. D. Adkins-Heljeson, ed., Core studies in Kansas—sedimentology and diagenesis of economically important rock strata in Kansas: Kansas Geological Survey Subsurface Geology Series 6, p. 1-7.

Bally, A. W., and S. Snelson, 1980, Realms of subsidence, *in* Facts and principles of world petroleum occurrence: Canadian Society of Petroleum Geologists Memoir 6, p. 9-94.

Barwick, J. S., 1928, The Salina basin of northcentral Kansas: American Association of Petroleum Geologists Bulletin, v. 12, p. 177-199.

Beene, D. L., 1989, 1988 Oil and gas production in Kansas: Kansas Geological Survey Oil and Gas Production Data Set 88, 254 p.

Brewer, R. R., Jr., 1959, A geophysical case history of the Lindsborg field, McPherson County, Kansas, *in* W.W. Hambleton, ed., Symposium on geophysics in Kansas: Kansas Geological Survey Bulletin 137, p. 287-295.

Cardott, B. J., and M. W. Lambert, 1985, Thermal maturation by vitrinite reflectance of Woodford Shale, Anadarko basin, Oklahoma: American Association of Petroleum Geologists Bulletin, v. 69, p. 1982-1998

Cole, V. B., 1975, Subsurface Ordovician-Cambrian rocks in Kansas: Kansas Geological Survey Subsurface Geology Series 2, 18 p.

Cole, V. B., 1976, Configuration of the top of Precambrian rocks in Kansas (1:500,000 map): Kansas Geological Survey Map M-7.

Doveton, J. H., 1986, Log analysis of subsurface geology—concepts and computer methods: John Wiley and Sons, New York, 273 p.

Farina, J. F., 1984, Geological applications of reservoir engineering tools: American Association of Petroleum Geologists Short Course Note Series, 102 p.

Folger, A., and R. H. Hall, 1933, Development of the oil and gas resources of Kansas (in 1928, 1929, and 1930): Kansas Geological Survey Mineral Resources Circular 2, 174 p.

Goebel, E. D., P. L. Hilpman, A. L. Hornbaker, and D. L. Beene, 1958, Oil and gas developments in Kansas during 1957: Kansas Geological Survey Bulletin 133, 264 p.

Hatch, J. R., S. R. Jacobson, B. J. Witzke, J. B. Risatti, D. E. Anders, W. L. Watney, K. D. Newell, and A. K. Vuletich, 1987, Possible Late Middle Ordovician organic carbon isotope excursion; evidence from Ordovician oils and hydrocarbon source rocks, Mid-Continent and east-central United States: American Association of Petroleum Geologists Bulletin, v. 71, p. 1342-1354.

Hiestand, T. C., 1933, Voshell field, McPherson County, Kansas: American Association of Petroleum Geologists Bulletin, v. 17, p. 169-191.

Jendem, P. D., K. D. Newell, I. R. Kaplan, and W. L. Watney, 1988, Composition and stable-isotope geochemistry of natural gases from Kansas, Midcontinent, U.S.A.: Chemical Geology, v. 71, p. 117-147.

Jewett, J. M., and D. F. Merriam, 1959, Geologic framework of Kansas; a review for geophysicists, *in* W.W. Hambleton, ed., Symposium on geophysics in Kansas: Kansas Geological Survey Bulletin 137, p. 9-52.

Klemme, H. D., 1971, What giants and their basins have in common: Oil and Gas Journal, v. 69, n. 9, 10, 11; pt. 1, p. 85-90; pt. 2, p. 103-110; pt. 3, p. 96-100.

Koester, E. A., 1934, Development of the oil and gas resources of Kansas in 1931 and 1932: Kansas Geological Survey Mineral Resources Circular 3, 76 p.

Lee, W., 1956, Stratigraphy and structural development of the Salina basin area: Kansas Geological Survey Bulletin 121, 167 p.

Lohman, S. W., 1942, Ground-water supplies available for national defense industries in south-central Kansas: Kansas Geological Survey Bulletin 41, pt. 1, p. 1-20.

Merriam, D. F., 1963, The geologic history of Kansas: Kansas Geological Survey Bulletin 162, 317 p.

Merriam, D. F., and E. D. Goebel, 1959, Where's the oil in Kansas?: Oil and Gas Journal, v. 57 (March 9), p. 212-218.

Momper, J. A., 1978, Oil migration limitations suggested by geological and geochemical considerations: American Association of Petroleum Geologists Continuing Education Course Note Series 8, p. B8-B60.

Newell, K. D., 1987, Salina basin sub-Chattanooga subcrop map (1:500,000 map): Kansas Geological Survey Open-File Report 87-11.

Newell, K. D., W. L. Watney, S. W. L. Cheng, and R. B. Brownrigg, 1987a, Stratigraphic and spatial distribution of oil and gas production in Kansas: Kansas Geological Survey Subsurface Geology Series 9, 86 p.

Newell, K. D., W. L. Watney, B. P. Stephens, and J. R. Hatch, 1987b, Hydrocarbon potential in Forest City basin: Oil and Gas Journal, v. 88 (October 19), p. 58-62.

Newell, K. D., M. Lambert, and P. Berendsen, 1988, Oil and gas shows in the Salina basin: Kansas Geological Survey Subsurface Geology Series 10, 38 p.

Ocola, L. C., and R. P. Meyer, 1973, Central North American Rift System 1; structure of the axial zone from seismic and gravimetric data: Journal of Geophysical Research, v. 78, p. 5173-5194.

Paul, S. E., L. Chang, and S. Burt, 1982, Oil and gas fields in Kansas (1:500,000 map): Kansas Geological Survey Map M-17.

Price, L., 1980, shelf and shallow basin oil as related to hot-deep origin of petroleum: Journal of Petroleum Geology, v. 2, p. 91-116.

Rich, J. L., 1933, Function of carrier beds in long-distance migration of oil: American Association of Petroleum Geologists Bulletin, v. 15, p. 911-924.

St. Clair, P. N., 1985, Core studies of the Viola limestone in Barber and Pratt counties, south-central Kansas, *in* M. D. Adkins-Heljeson, ed., Core studies in Kansas—sedimentology and diagenesis of economically important rock strata in Kansas: Kansas Geological Survey Subsurface Geology Series 6, p. 8-16.

Sloss, L. L., 1963, Sequences in the cratonic interior of North America: Geological Society of America Bulletin, v. 74, p. 93-111.

Stavnes, S. A., and D. W. Steeples, 1982, Geothermal resources of Kansas (1:500,000 map): National Geophysical Data Center of the National Oceanic and Atmospheric Administration (map can be ordered from Kansas Geological Survey).

Talbott, W. C., 1954, The geology of the Smolan pool area, Saline County, Kansas: Unpublished M.S. thesis, University of Oklahoma, Norman, Oklahoma, 56 p.

Thatcher, R. J., 1961, Correlation of structural history with reservoir fluid distribution in the Lindsborg pool, Kansas: Unpublished M.S. thesis, University of Oklahoma, Norman, Oklahoma, 53 p.

Tissot, B. P., and D. H. Welte, 1984, Petroleum Formation and Occurrence, Second edition: New York, Springer-Verlag, 669 p.

Ver Wiebe, W. A., 1943, Exploration for oil and gas in western Kansas during 1942: Kansas Geological Survey Bulletin 48, 88 p.

Ver Wiebe, W. A., 1945, Exploration for oil and gas in Kansas during 1944: Kansas Geological Survey Bulletin 56, 112 p.

Ver Wiebe, W. A., E. D. Goebel, A. L. Hornbaker, and J. M. Jewett, 1954, Oil and gas developments in Kansas during 1953: Kansas Geological Survey Bulletin 107, 204 p.

Walters, R. F., 1958, Differential entrapment of oil and gas in Arbuckle dolomite of central Kansas: American Association of Petroleum Geologists Bulletin, v. 42, p. 2133-2173.

Waples, D., 1981, Organic geochemistry for exploration geologists: Minneapolis, Burgess Publishing Co., 151 p.

Watney, W. L., and S. E. Paul, 1983, Oil exploration in Kansas—present activity and future potential: Oil and Gas Journal, v. 81 (July 25), p. 193-198.

Yarger, H. L., 1983, Regional interpretation of Kansas aeromagnetic data: Kansas Geological Survey Geophysical Series 1, 35 p.

SUGGESTED READINGS

Brewer, R. R., Jr., 1959, A geophysical case history of the Lindsborg field, McPherson County, Kansas, *in* W. W. Hambleton, ed., Symposium on geophysics in Kansas: Kansas Geological Survey Bulletin 137, p. 287-295. Presents a detailed summary of the geophysical hardware, techniques, and maps used in discovery of the Lindsborg field.

Newell, K. D., W. L. Watney, S. W. L. Cheng, and R. B. Brownrigg, 1987, Stratigraphic and spatial distribution of oil and gas production in Kansas: Kansas Geological Survey Subsurface Geology Series 9, 86 p. Illustrates, with maps and accompanying discussion, the distribution of petroleum production in Kansas according to attributes such as volume, pay zone, depth, type of pay, and date of discovery.

Appendix 1. Field Description

Field name .... *Lindsborg*

Ultimate recoverable reserves .... *~16,000,000 bbl*

Field location:

- **Country** .... *United States*
- **State** .... *Kansas*
- **Basin/Province** .... *Salina basin*

Field discovery:

- **Year first pay discovered** .... *Ordovician Maquoketa dolomite 1938*
- **Year second pay discovered** .... *Ordovician Simpson Group 1942*
- **Third pay** .... *Ordovician Viola* Limestone 1942*

Discovery well name and general location

- **First pay** .... *#1 E. Hoglund (W2 SW SW sec. 8-T17S-R3W) (for Maquoketa)*
- **Second pay** .... *#1 Johnson (W2 SW NE sec. 8-T17S-R3W) (for Simpson)*
- **Third pay** .... *#1 Farmers Bank (W2 NW SW sec. 17-T17S-R3W) (?) (for Viola)**

**Maquoketa dolomite is erroneously called Viola in some early wells; older records are also incomplete or unclear; therefore, exact discovery well for Viola pay is unclear.*

Discovery well operator .... *Carter Oil Company (Maquoketa)*

- **Second pay** .... *Auto Ordnance (Simpson)*
- **Third pay** .... *Bay Petroleum [Tenneco] (Viola)*

IP in barrels per day and/or cubic feet or cubic meters per day:

- **First pay** .... *345 bbl/d (Maquoketa)*
- **Second pay** .... *2247 bbl/d (Simpson)*
- **Third pay** .... *663 bbl/d (Viola)*

All other zones with shows of oil and gas in the field:

Age	Formation	Type of Show
Pennsylvanian (Missourian)	*Lansing-Kansas City Groups*	*Oil stain*
Mississippian-Devonian	*Misener sandstone**	*Oil stain*

**Basal sandstone of Chattanooga Shale*

Geologic concept leading to discovery and method or methods used to delineate prospect, e.g., surface geology, subsurface geology, seeps, magnetic data, gravity data, seismic data, seismic refraction, nontechnical:

Anticline on which the Lindsborg field is partly located was detected with a shallow core-drilling survey in 1929 by the Dixie (Stanolind) Oil Co. In 1937 Carter Oil Company resurveyed this area with a correlation-point reflection seismograph survey using dynamite as a source. A seismic contour map on a reflector from the top of the Viola-Maquoketa carbonate section aided in placing the discovery well. Production was extended off-structure by subsequent development wells.

LINDSBORG

Structure:

Province/basin type (see St. John, Bally, and Klemme, 1984)

1211 basin by Bally and Snelson (1980) classification (i.e., cratonic basin located on earlier rifted graben). I basin by Klemme (1971) classification (i.e., cratonic interior basin).

Tectonic history

Severe NNE-SSW rifting and basaltic igneous activity occurred during Precambrian time (1100 Ma). Pre-Pennsylvanian Phanerozoic time is generally characterized by cratonic carbonate sedimentation and tectonic quiescence. Deformation during Late Mississippian-Early Pennsylvanian time (correlative to Ouachita orogeny) is largely responsible for the present-day structure and follows NNE-SSW Precambrian zones of weakness. Regional tilting generally characterizes later deformation—to SW during Late Pennsylvanian and Permian, W during Mesozoic, NW during Cenozoic.

Regional structure

Lindsborg anticline is located in a broad E-W-trending structural saddle between the Salina basin (to the north) and Sedgwick basin (to the south). It is also approximately equidistant from the Nemaha uplift (to the east) and Central Kansas uplift (to the west). This structural saddle also overlies the extreme southern end of the Precambrian Central North American Rift System (CNARS).

Local structure

River alluvium of the Smoky Hill river valley obscures any local surface structure associated with the field. Outside of the river valley, surface rocks dip homoclinally to the west at less than 5° dip. Lindsborg anticline plunges SSW.

Trap

Trap type(s)

Simpson and Viola oil pools are in a local culmination on the Lindsborg anticline, hence oils in these units are in a structural trap. Maquoketa oil is located off-structure along the zone of maximum flexure on the west flank of the anticline, hence oil in this unit is in a structural-stratigraphic trap. A small structural closure at the southern end of the field produces oil from the Simpson.

Basin stratigraphy (major stratigraphic intervals from surface to deepest penetration in field):

Chronostratigraphy	Formation	Depth to Top
Pennsylvanian (Missourian)	*Lansing-Kansas City Groups*	*2220 ft*
Mississippian (Meramecian)	*Salem(?) Limestone*	*2900 ft*
Devonian-Mississippian	*Chattanooga Shale*	*3110 ft*
Middle Ordovician	*Simpson Group*	*3450 ft*
Cambrian-Ordovician	*Arbuckle Group*	*3520 ft*
Precambrian	*Basement (arkose and "meta-sedimentary rocks" of Rice Formation*	*4020 ft*

Location of well in field .... *NA*

Reservoir characteristics:

Number of reservoirs *3*

Formations .... *Maquoketa Dolomite, Viola Limestone, Simpson Sandstone*

Ages .... *Upper Ordovician, Middle Ordovician, Middle Ordovician, respectively*

Depths to tops of reservoirs

	Avg. Drill Depth	Avg. Subsea Depth
Maquoketa	*3385 ft*	*-2035 ft*
Viola	*3425 ft*	*-2085 ft*
Simpson	*3480 ft*	*-2130 ft*

Gross thickness (top to bottom of producing interval)

Average thickness in feet: Maquoketa Dolomite, 60; Viola Limestone, 50; Simpson Group, 70

Net thickness—total thickness of producing zones

Average .... *Maquoketa, 6.5 ft; Viola, 10 ft; Simpson, 4.5 ft*

Maximum .... *Maquoketa, 26 ft; Viola, 18 ft; Simpson, 14 ft*

Average ....

Maximum ....

Lithology

Maquoketa: gray, cherty, fine to medium crystalline vuggy dolomite
Viola: gray to white, coarse crystalline, slightly dolomitic and cherty limestone
Simpson: white, fine- to medium-grained, well-sorted quartz sandstone

Porosity type

Maquoketa: intercrystalline and vuggy porosity with minor moldic and fracture porosity
Viola: intercrystalline porosity with possibly moldic porosity
Simpson: intergranular porosity

Average porosity .... *Maquoketa, 5-12%; Viola, 12-15%; Simpson, 15-18%*

Average permeability .... *NA*

Seals:

Upper

Formation, fault, or other feature, and lithology

Maquoketa: upper seal is Maquoketa shale over most of the field, Chattanooga Shale in minor part of field (average thickness of either of these units is ~80 ft)
Viola: upper seal is 10-20 ft thick nonporous limestone bed at top of Viola
Simpson: upper seal is 10 ft thick bed of shale at top of Simpson

Lateral

Formation, fault, or other feature, and lithology

Maquoketa reservoir grades laterally to nonporous Maquoketa dolomite; Viola and Simpson have no lateral seals

Source: *Two source units are possible—one for Maquoketa and one for Viola and Simpson*

Formation and age

Maquoketa, possibly a mixture of Simpson (Middle Ordovician) and Chattanooga Shale (Devonian-Mississippian); Simpson and Viola, Simpson Group (Middle Ordovician)

Lithology .... *Shale*
Average total organic carbon (TOC) .... *0.8% for Simpson; 0.6% for Chattanooga**
Maximum TOC .... *15% for Simpson; 3% for Chattanooga**
Kerogen type (I, II, or III) .... *I for Simpson; II for Chattanooga**
Vitrinite reflectance (maturation) .... *~0.6-0.7 (equiv.) for Simpson; ~0.6 for Chattanooga**
**Measurements taken from several wells within 50 mi of Lindsborg field.*
Time of hydrocarbon expulsion .... *Late Cretaceous to present (70 Ma to present)*
Present depth to top of source .... *Simpson, 3550 ft; Chattanooga, 3150 ft*
Thickness .... *Simpson shales, 35-40 ft composite thickness; Chattanooga, 230 ft*
Potential yield .... *NA*

Appendix 2. Production Data

Field name .... *Lindsborg*

Field size:

Proved acres *10,080*
Number of wells all years *318*
Current number of wells (as of 1988) *74*
Well spacing .... *40 ac average; 10 ac spacing in extreme northern part of field*
Ultimate recoverable .... *~16.0 million bbl*
Cumulative production .... *15.19 million bbl through 1988*
Annual production .... *0.072 million bbl in 1988*
Present decline rate ..... *98% (percent present year's production [i.e., 1988] of that of the previous year)*
Initial decline rate .... *~70%*
Overall decline rate *93%*
Annual water production .... *NA*
In place, total reserves .... *NA*
In place, per acre-foot .... *~240 bbl*
Primary recovery .... *NA*
Secondary recovery .... *NA*
Enhanced recovery .... *NA*
Cumulative water production .... *NA*

Drilling and casing practices:

Amount of surface casing set .... *100–175 ft*

Casing program

13⅜-in. casing is usually set to 100–175 ft (to below base of river alluvium of Smoky Hill River and Plio-Pleistocene "Equus beds"); 8⅝-in. casing usually set to 300–375 ft (to below base of bedded salts in Permian section); 5½-in. casing is set to ~3350 ft (either just above top of potential producing zone [followed by open-hole completion], or through the potential producing zone [followed by perforating operations])

Drilling mud .... *Saltwater based chemical or bentonite muds with weight of 9.5–10.5 lb/gal with viscosity between 35 and 45*

Bit program .... *After second string of casing is set to 300–375 ft, one button-bit can usually drill the entire hole*

High pressure zones .... *None*

Completion practices:

Interval(s) perforated

Maquoketa: top 5–6 ft of Maquoketa dolomite

Viola: 8–10 ft thick porous zone located below 10–20 ft thick nonporous bed at top of unit

Simpson: top 5–6 ft of 20–30 ft sandstone, located immediately below a 10 ft shale at top of unit

Well treatment .... *Maquoketa and Viola carbonates are usually sand fractured and acidized; Simpson sandstone is usually untreated but in some cases is acidized*

Formation evaluation:

Logging suites

Older suites usually include lateral and normal resistivity logs with SP, neutron, and GR; more recent suites include compensated neutron/lithodensity with GR and SP, and dual induction suites with SFL

Testing practices

Potential pay zones in Maquoketa and Viola usually are tested either through perforations or open hole; potential pay zones in Simpson are usually tested through perforations but in a few cases (mostly older wells) are tested open-hole

Mud logging techniques

Inspection of drill-cuttings for fluorescence and odor and oil-cut is the primary technique for determining if a well may be worthy of further testing

Oil characteristics:

Type .... *Maquoketa oil is aromatic intermediate; Simpson is paraffinic**

(Tissot and Welte Classification *in* "Petroleum Formation and Occurrence," 1984, Springer-Verlag, p. 419)

**Viola oil is not analyzed, but appears similar to Simpson oil*

API gravity .... *Maquoketa, 27–36°; Viola, 28–34°; Simpson, 28°*

Base .... *NA*

Initial GOR .... *Negligible in all zones*

Sulfur, wt% .... *Maquoketa, 0.54; Viola, 0.50; Simpson, 0.52*

Viscosity, SUS .... *NA*

Pour point .... *Maquoketa, 10°F; Viola, 5°F; Simpson, 5°F*

Gas-oil distillate .... *None in all zones*

Field characteristics:

Average elevation

Average surface elevation of field, 1350 ft
Average subsea depth of pay zones: Maquoketa, -2035 ft; Viola, -2040 ft; Simpson, -2130 ft

Initial pressure .... *Maquoketa, BHP ~1300 psi; Viola, BHP NA: Simpson, BHP ~1525 psi*

Present pressure .... *NA*

Pressure gradient .... *~0.43 psi/ft*

Temperature .... *40-45°C*

Geothermal gradient .... *~25-30°C/km*

Drive .... *Water drive for all three pay zones*

Oil column thickness .... *Maquoketa, 126 ft; Viola, 52 ft; Simpson closure in central part of field, 34 ft; Simpson closure in southern part of field, 19 ft*

Oil-water contact .... *Maquoketa, -2103 ft; Viola, -2079 ft; Simpson closure in central part of field, -2140 ft; Simpson closure in southern part of field, -2156 ft*

Connate water .... *NA*

Water salinity, TDS .... *Maquoketa, 59,860 mg/L; Viola NA; Simpson, 10,940 mg/L*

Resistivity of water .... *Maquoketa, ~0.05; Viola, ~0.08; Simpson, ~0.125*

Bulk volume water (%) .... *NA*

Transportation method and market for oil and gas:

Bulk of oil from field is put in pipeline and sent to an NCRA refinery in McPherson, KS, 10 mi away; minor portion of oil is trucked from stock tanks on individual leases; major products from refinery are gasoline and distillates (including diesel fuel).